Game Theory and International Environmental Cooperation

The International Library of Critical Writings in Economics

Founding Editor: Mark Blaug

Late Professor Emeritus, University of London, UK
Late Professor Emeritus, University of Buckingham, UK

This series is an essential reference source for students, researchers and lecturers in economics. It presents by theme a selection of the most important articles across the entire spectrum of economics. Each volume has been prepared by a leading specialist who has written an authoritative introduction to the literature included.

Wherever possible, the articles in these volumes have been reproduced as originally published using facsimile reproduction, inclusive of footnotes and pagination to facilitate ease of reference.

For a list of all Edward Elgar published titles visit our website at
www.e-elgar.com

Game Theory and International Environmental Cooperation

Essential Readings

Edited by

Michael Finus

Professor, Chair in Environmental Economics, University of Bath, UK

and

Alejandro Caparrós

Associate Research Professor, Consejo Superior de Investigaciones Científicas (CSIC), and Deputy Director, Institute for Public Goods and Policies, Spain

THE INTERNATIONAL LIBRARY OF CRITICAL WRITINGS IN ECONOMICS

An Elgar Research Collection

Cheltenham, UK • Northampton, MA, USA

Published by
Edward Elgar Publishing Limited
The Lypiatts
15 Lansdown Road
Cheltenham
Glos GL50 2JA
UK

Edward Elgar Publishing, Inc.
William Pratt House
9 Dewey Court
Northampton
Massachusetts 01060
USA

A catalogue record for this book is available from the British Library

Library of Congress Control Number: 2014952133

MIX
Paper from responsible sources
FSC www.fsc.org FSC® C013056

ISBN 978 1 78254 509 5

Printed and bound in Great Britain by T.J. International Ltd, Padstow

Contents

Acknowledgements

The editors and publishers wish to thank the authors and the following publishers who have kindly given permission for the use of copyright material.

American Economic Association for articles: Scott Barrett (2006), 'Climate Treaties and "Breakthrough" Technologies', *American Economic Review*, **96** (2), May, 22–5; Michael Kosfeld, Akira Okada, and Arno Riedl (2009), 'Institution Formation in Public Goods Games', *American Economic Review*, **99** (4), September, 1335–55.

Blackwell Publishing Ltd for articles: Francesco Bosello, Barbara Buchner and Carlo Carraro (2003), 'Equity, Development, and Climate Change Control', *Journal of the European Economic Association*, **1** (2–3), April–May, 601–11; Effrosyni Diamantoudi and Eftichios S. Sartzetakis (2006), 'Stable International Environmental Agreements: An Analytical Approach', *Journal of Public Economic Theory*, **8** (2), May, 247–63; Johan Eyckmans and Michael Finus (2006), 'Coalition Formation in a Global Warming Game: How the Design of Protocols Affects the Success of Environmental Treaty-Making', *Natural Resource Modeling*, **19** (3), September, 323–58; Lata Gangadharan and Veronika Nemes (2009), 'Experimental Analysis of Risk and Uncertainty in Provisioning Private and Public Goods', *Economic Inquiry*, **47** (1), January, 146–64; Hans-Peter Weikard (2009), 'Cartel Stability under an Optimal Sharing Rule', *Manchester School*, **77** (5), September, 575–93; Astrid Dannenberg, Andreas Lange and Bodo Sturm (2014), 'Participation and Commitment in Voluntary Coalitions to Provide Public Goods', *Economica*, **81** (322), April, 257–75.

Cambridge University Press for article: Alejandro Caparrós and Jean-Christophe Péreau (2013), 'Forming Coalitions to Negotiate North–South Climate Agreements', *Environment and Development Economics, Special Issue on Strategic Behaviour and Environmental Commons*, **18** (1), February, 69–92.

Duncker and Humblot GmbH for article: Alfred Endres and Cornelia Ohl (2001), 'International Environmental Cooperation in the One Shot Prisoners' Dilemma', *Schmollers Jahrbuch, Zeitschrift für Wirtschafts- und Sozialwissenschaften/Journal of Applied Social Science Studies*, **121** (1), 1–26.

Elsevier Ltd for excerpt and articles: Karl-Göran Mäler (1989), 'The Acid Rain Game', in H. Folmer and E. van Ierland (eds), *Valuation Methods and Policy Making in Environmental Economics*, Chapter 12, 231–52; Carlo Carraro and Domenico Siniscalco (1993), 'Strategies for the International Protection of the Environment', *Journal of Public Economics*, **52** (3), October, 309–28; Rögnvaldur Hannesson (1997), 'Fishing as a Supergame', *Journal of Environmental Economics and Management*, **32** (3), March, 309–22; Stefan Ambec and Yves

Sprumont (2002), 'Sharing a River', *Journal of Economic Theory*, **107** (2), December, 453–62; Michael Finus and Sigve Tjøtta (2003), 'The Oslo Protocol on Sulfur Reduction: The Great Leap Forward?', *Journal of Public Economics*, **87** (9–10), September, 2031–48; Andreas Lange and Carsten Vogt (2003), 'Cooperation in International Environmental Negotiations due to a Preference for Equity', *Journal of Public Economics*, **87** (9–10), September, 2049–67; Marc Germain, Philippe Toint, Henry Tulkens, and Aart de Zeeuw (2003), 'Transfers to Sustain Dynamic Core-Theoretic Cooperation in International Stock Pollutant Control', *Journal of Economic Dynamics and Control*, **28** (1), October, 79–99; Geir B. Asheim, Camilla Bretteville Froyn, Jon Hovi, and Fredric C. Menz (2006), 'Regional versus Global Cooperation for Climate Control', *Journal of Environmental Economics and Management*, **51** (1), January, 93–109; Santiago J. Rubio and Alistair Ulph (2007), 'An Infinite-Horizon Model of Dynamic Membership of International Environmental Agreements', *Journal of Environmental Economics and Management*, **54** (3), November, 296–310; Nori Tarui, Charles F. Mason, Stephen Polasky, and Greg Ellis (2008), 'Cooperation in the Commons with Unobservable Actions', *Journal of Environmental Economics and Management*, **55** (1), January, 37–51; Aart de Zeeuw (2008), 'Dynamic Effects on the Stability of International Environmental Agreements', *Journal of Environmental Economics and Management*, **55** (2), March, 163–74; Anke Gerber and Philipp C. Wichardt (2009), 'Providing Public Goods in the Absence of Strong Institutions', *Journal of Public Economics*, **93** (3–4), April, 429–39; Prajit K. Dutta and Roy Radner (2009), 'A Strategic Analysis of Global Warming: Theory and Some Numbers', *Journal of Economic Behavior and Organization*, **71** (2), August, 187–209; Vincent Boucher and Yann Bramoullé (2010), 'Providing Global Public Goods under Uncertainty', *Journal of Public Economics*, **94** (9–10), October, 591–603; Larry Karp and Leo Simon (2013), 'Participation Games and International Environmental Agreements: A Non-Parametric Model', *Journal of Environmental Economics and Management*, **65** (2), March, 326–44; Michael Finus and Pedro Pintassilgo (2013), 'The Role of Uncertainty and Learning for the Success of International Climate Agreements', *Journal of Public Economics*, **103**, July, 29–43; Scott Barrett (2013), 'Climate Treaties and Approaching Catastrophes', *Journal of Environmental Economics and Management*, **66** (2), September, 235–50.

Mohr Siebeck GmbH and Co. KG for article: Scott Barrett (2002), 'Consensus Treaties', *Journal of Institutional and Theoretical Economics*, **158** (4), December, 529–47.

National Academy of Sciences for article: Alessandro Tavoni, Astrid Dannenberg, Giorgos Kallis, and Andreas Löschel (2011), 'Inequality, Communication, and the Avoidance of Disastrous Climate Change in a Public Goods Game', *Proceedings of the National Academy of Sciences*, **108** (29), July, 11825–9.

Oxford University Press via the Copyright Clearance Center's Rightslink Service for articles: Scott Barrett (1994), 'Self-Enforcing International Environmental Agreements', *Oxford Economic Papers, Special Issue on Environmental Economics*, **46**, October, 878–94; Seong-lin Na and Hyun Song Shin (1998), 'International Environmental Agreements under Uncertainty', *Oxford Economic Papers*, **50** (2), April, 173–85; Santiago J. Rubio and Alistair Ulph (2006), 'Self-Enforcing International Environmental Agreements Revisited', *Oxford Economic Papers*, **58** (2), April, 233–63; Matthew McGinty (2007), 'International

Environmental Agreements among Asymmetric Nations', *Oxford Economic Papers*, **59** (1), January, 45–62; Bård Harstad (2012), 'Climate Contracts: A Game of Emissions, Investments, Negotiations, and Renegotiations', *Review of Economic Studies*, **79** (4), October, 1527–57.

Springer Science and Business Media B.V. for articles: Michael Hoel (1992), 'International Environment Conventions: The Case of Uniform Reductions of Emissions', *Environmental and Resource Economics*, **2** (2), March, 141–59; Henk Folmer, Pierre v. Mouche and Shannon Ragland (1993), 'Interconnected Games and International Environmental Problems', *Environmental and Resource Economics*, **3** (4), August, 313–35; Parkash Chander and Henry Tulkens (1997), 'The Core of an Economy with Multilateral Environmental Externalities', *International Journal of Game Theory*, **26** (3), October, 379–401; Michael Finus and Bianca Rundshagen (1998), 'Toward a Positive Theory of Coalition Formation and Endogenous Instrumental Choice in Global Pollution Control', *Public Choice*, **96** (1–2), July, 145–86; Carsten Helm (2001), 'On the Existence of a Cooperative Solution for a Coalitional Game with Externalities', *International Journal of Game Theory*, **30** (1), September, 141–6; Alejandro Caparrós, Jean-Christophe Péreau and Tarik Tazdaït (2004), 'North–South Climate Change Negotiations: A Sequential Game with Asymmetric Information', *Public Choice*, **121** (3–4), December, 455–80; Carlo Carraro, Johan Eyckmans and Michael Finus (2006), 'Optimal Transfers and Participation Decisions in International Environmental Agreements', *Review of International Organizations*, **1** (4), December, 379–96; Carlo Carraro, Carmen Marchiori and Sonia Oreffice (2009), 'Endogenous Minimum Participation in International Environmental Treaties', *Environmental and Resource Economics*, **42** (3), March, 411–25; David M. McEvoy and John K. Stranlund (2009), 'Self-Enforcing International Environmental Agreements with Costly Monitoring for Compliance', *Environmental and Resource Economics*, **42** (4), April, 491–508; Michèle Breton, Lucia Sbragia and Georges Zaccour (2010), 'A Dynamic Model for International Environmental Agreements', *Environmental and Resource Economics*, **45** (1), January, 25–48; Hans-Peter Weikard, Rob Dellink and Ekko van Ierland (2010), 'Renegotiations in the Greenhouse', *Environmental and Resource Economics*, **45** (4), April, 573–96; Michael Hoel and Aart de Zeeuw (2010), 'Can a Focus on Breakthrough Technologies Improve the Performance of International Environmental Agreements?', *Environmental and Resource Economics*, **47** (3), November, 395–406; Pierre Courtois and Guillaume Haeringer (2012), 'Environmental Cooperation: Ratifying Second-Best Agreements', *Public Choice*, **151** (3–4), June, 565–84; Matthew McGinty, Garrett Milam and Alejandro Gelves (2012), 'Coalition Stability in Public Goods Provision: Testing an Optimal Allocation Rule', *Environmental and Resource Economics*, **52** (3), July, 327–45; Todd L. Cherry and David M. McEvoy (2013), 'Enforcing Compliance with Environmental Agreements in the Absence of Strong Institutions: An Experimental Analysis', *Environmental and Resource Economics*, **54** (1), January, 63–77.

Every effort has been made to trace all the copyright holders but if any have been inadvertently overlooked the publishers will be pleased to make the necessary arrangement at the first opportunity.

In addition the publishers wish to thank the Library of Indiana University at Bloomington, USA, for its assistance in obtaining these articles.

Foreword

In May 2012, Chris Berry from Edward Elgar contacted us to explore the opportunity to edit a volume on Game Theory and the Environment. Having worked in the field for such a long time, it did not take us very long to be convinced that this would be a most exciting project. Upon reflection, we realised that we needed to narrow down the topic and finally ended up with the title 'Game Theory and International Environmental Cooperation: Essential Readings'. We received all the support and help we needed to work out the concept of the volume, to obtain the permission of all publishers to reproduce articles in the volume and received constant encouragement on the way to production. Our work was most enjoyable due to the very professional work of Carrie Talbot (Project Editor) and Chris and their very personal touch in dealing with editors. We, the editors of this volume, would like to express our great appreciation for all the support we received and would like to thank Carrie and Chris for their outstanding work. Finally, we would like to thank Alaa Khourdajie at the University of Bath for research assistance, who supported this project in many ways.

Michael Finus and Alejandro Caparrós
September 2014

Introduction

Michael Finus and Alejandro Caparrós

Preliminaries

This volume presents a collection of articles on 'Game Theory and International Environmental Cooperation'. Environmental problems, like acid rain, global warming, the loss of biodiversity, the overuse of renewable resources like water and fish, are characterised by externalities across agents and nations (Barrett 1990 and Mäler 1990). Pollution and the overuse of resources create negative externalities. Full or partial internalisation of those negative externalities requires coordination and cooperation across agents. Viewed from a different angle, starting from a situation without cooperation, forming formal or informal agreements with the objective to reduce these negative effects creates positive externalities. Whereas in the national context those cooperative arrangements can – at least in theory – be enforced, this is much more difficult in the international context as there is no supranational institution with legal enforcement power. It is for this reason that the term 'self-enforcing agreements' has been coined. Roughly speaking, self-enforcing means that the parties to an agreement – for example, an international environmental treaty, like the Montreal Protocol on the reduction of chlorofluorocarbons (CFCs), the Kyoto Protocol on the reduction of greenhouse gases, the Helsinki or Oslo Protocol on the reduction of sulphur oxides and the UN Fish Stocks Agreement, which deals with the conservation and management of straddling and highly migratory fish stocks – must monitor compliance, decide on measures against violations of treaty obligations and must conduct those measures all by themselves. However, those measures are costly, adding normally to the free-rider problem. In the context of pollution, international treaties have been called international environmental agreements (IEAs), whereas in the context of fishery agreements, the term regional fishery management organisations (RFMOs) has been coined.

Following Finus (2003), international cooperation faces two types of free-rider incentives. The first type is insufficient participation. From a normative point of view, all parties involved in an externality problem should contribute to its solution. For a global externality, like global warming, this implies that all countries should join a climate treaty.[1] Only then can a globally optimal abatement level and a cost-effective allocation of abatement levels be achieved. The Kyoto Protocol and the current struggle to negotiate a follow-up protocol illustrate how far reality is from this normative benchmark. Other examples illustrating this point are abundant. The second type of free-rider incentive concerns insufficient compliance. Frequent violations of the International Convention on the Regulation of Whaling stress that compliance cannot be taken for granted and impairs on the success of international cooperation. As a tendency, the less ambitious (or the less effective) environmental treaties are, the higher is the rate of compliance and participation. Thus from a high compliance and membership record, one cannot infer the success of an environmental treaty – it may just indicate that a treaty codifies what countries would have done under non-cooperative behaviour anyway *(Finus and Tjøtta 2003, Chapter 11)*. Moreover, as Aakre (2013), Aakre and Hovi (2010) and Barrett and Stavins (2003) point out, there may be a trade-off between membership and compliance. Weak enforcement mechanisms may encourage high participation and vice versa.

In the light of the brief description above, it seems 'natural' to use game theory as an analytical tool to study the problems of international cooperation related to environmental and natural resource use issues, as highlighted by the volume edited by Pethig (1992) and the surveys by, for instance, Barrett (2003), Finus (2001, 2003, 2008), Sumaila (1999), Caparrós, Péreau and Tazdaït (2003) and Wagner (2001). The overriding theme is to study the strategic interaction among governments and their representatives in the context of environmental externalities. The central positive research question is: how can we capture and explain the free-riding problems in an international environmental externality game and how are they related to the underlying fundamentals? Answers to this question almost 'naturally' lead to the second normatively motivated research question: how should we design agreements in order to mitigate free-rider problems?

Though it is always difficult to trace the origins in a research field, we start in Part I, Foundations, with papers which appeared around 1990, being well aware that we could have included some earlier papers. We start with the classical paper by *Mäler (1989, Chapter 1)* about the acid rain game, which sets out the fundamental game-theoretic aspects of transboundary pollution, where the degrees of externalities between countries can be quite different and are captured through the transportation matrix. The acid rain game was analysed in many subsequent papers (for example, *Finus and Tjøtta 2003, Chapter 11* and Kaitala, Mäler and Tulkens 1995 to name only a couple) and its importance for the profession cannot be underestimated, as highlighted by Folmer and van Mouche (2002) in their essay in honour of Karl-Göran Mäler. Other classical papers by *Barrett (1994, Chapter 2)*, *Carraro and Siniscalco (1993, Chapter 3)*, *Hoel (1992, Chapter 4)* and *Chander and Tulkens (1997, Chapter 5)* also appear in Part I. The subsequent sections, Part II to Part VIII, look at various extensions. Given the numerous contributions to the field, selecting papers is anything other than a trivial task. We finally decided to apply the following criteria, without claiming that we have given sufficient credit to all authors who deserve it: (a) a paper needs to meet all the criteria of the title of this volume, namely game theory and international environmental cooperation; (b) the analysis of cooperation needs to entail more than two players, rendering the analysis of coalitions interesting; (c) a paper must make a theoretical novel contribution, must capture fundamental features of actual environmental negotiations, must address an interesting policy question and must be a seminal paper in the field; and (d) with one exception (the paper by Karl-Göran Mäler because of its fundamental importance for the field), only peer-reviewed journal articles have been included.

A Methodological Roadmap

Classifying a large number of very diverse papers is not easy. Inspired by game theory textbooks, one could be inclined to group papers into cooperative versus non-cooperative game theory. However, the classical definition of non-cooperative game theory, namely the absence of a third party with enforcement power, would suggest that all papers using concepts of cooperative game theory, like the Shapley value and the core, would have to be judged as inappropriate for the analysis of international environmental cooperation. We believe this would be misleading. Three examples shall illustrate this point. For instance, stability in the sense of the core assumes that after a deviation of a single player or group of players the remaining players play their minimax (α-core) or Nash equilibrium strategy (γ-core), which, in the context of environmental externalities, implies at least an implicit punishment. In contrast, the concept of internal and external stability has often been described as belonging to non-cooperative coalition theory. However, a fundamental assumption of this stability definition, shared by similar concepts and also the core, is that all members honour their obligations, that is, compliance is perfect, and free-riding only concerns leaving an agreement (*McEvoy and Stranlund 2009, Chapter 15*). Moreover, the concept of internal and external stability assumes that once a player leaves the agreement, the remaining players continue to cooperate, which, without doubt, entails a cooperative spirit. Finally, most membership models (explained in more detail below) assume that members to an agreement choose their economic strategies (for example, emissions, R&D investments and fishing efforts) cooperatively and hence efficiently by maximising the aggregate payoff to their coalition. Again, this entails a cooperative spirit. But also most compliance models (explained in more detail below), which analyse whether an agreement can be enforced through various threats of sanctioning non-compliance, typically ignore the process of how membership evolves and start their analysis by considering the grand coalition, a coalition comprising all players, and hence assume full participation.

The brief discussion above suggests at least four conclusions. First, the classification into cooperative versus non-cooperative game theory appears to be not very fruitful for the papers that have been written on the subject. Second, a probably better structure for the time being is suggested by grouping papers into membership models, which focus on the decision of countries as to whether to join an agreement or to remain outside, and with which countries to form an agreement, negotiation models, which model the entire formation process of an agreement as a bargaining process, and compliance models, which analyse whether and how non-compliance can be deterred through threats of punishment. Roughly speaking, with few exceptions (for example, papers capturing elements of different categories such as *Finus and Rundshagen 1998, Chapter 18*, *Kosfeld, Okada and Riedl 2009, Chapter 34* and *McEvoy and Stranlund 2009, Chapter 15*), most papers belong to one of these three categories of models. Third, more work needs to be done to design models that capture membership, negotiations and compliance issues simultaneously, and their strategic interaction.[2] Fourth, on a less critical note, one may argue that membership and compliance models are not that different, which is supported by the fact that a couple of issues have been analysed with both types of models. Already *Barrett's 1994 (Chapter 2)* seminal paper uses both approaches, though most citations refer to his membership model. Both of his models illustrate the free-rider problem and that full cooperation is difficult to obtain. Later papers have looked, for instance, at the effect of second-best designs, in the context of membership models (for example, Altamirano-Cabrera, Finus and Dellink 2008, *Courtois and Haeringer 2012, Chapter 20* and Finus and Maus 2008), but also in the context of compliance models (*Barrett 2002, Chapter 19* and *Finus and Rundshagen 1998, Chapter 18*), and the same is true, for instance, for the issues of multiple coalitions (membership models: for example, *Bosello, Buchner and Carraro 2003, Chapter 35*, Carraro and Marchiori 2003, *Eyckmans and Finus 2006, Chapter 36*, Finus and Rundshagen 2003; compliance model: *Asheim et al. 2006, Chapter 37*) and issue linkage (membership models: Botteon and Carraro 1998, Carraro and Marchiori 2004, Carraro and Siniscalco 1997, Katsoulacos 1997 and Kroeze-Gil and Folmer 1998; compliance models: Cesar and de Zeeuw 1997, Conconi and Perroni 2002, *Folmer, van Mouche and Ragland 1993, Chapter 12*). Conceptually, in membership models the free-rider leaves the agreement and chooses their non-cooperative best reply whereas in compliance models the free-rider remains in the agreement but also chooses their non-cooperative best reply. In compliance models free-riding is followed by punishment and in membership models the non-deviating players adjust their strategies, which, in most models, is nothing other than an implicit punishment. Finally, in some papers, which have not been grouped into negotiation models, at least some aspects of negotiations are captured, as, for instance, in Altamirano-Cabrera, Finus and Dellink (2008), *Courtois and Haeringer (2012, Chapter 20)*, *Finus and Rundshagen (1998, Chapter 18)* and Finus and Maus (2008).

A general characteristic of viewing models is time. In a static model or a one-shot model, the non-cooperative equilibrium is the only possible outcome. Hence, in order to allow for some form of cooperation, some dynamics need to be introduced into a model. Essentially, what is needed is that agents can depart from non-cooperation by agreeing on some form of cooperation, either partial or full cooperation, but can also react if cooperation is not honoured. Cooperation must be conditional; otherwise it would not be rational. Reaction means some form of punishment, even though this is not always explicitly spelled out in some papers. Most of the membership models model time only implicitly. In order to determine whether an agreement is stable, they compare the payoff of members with cooperation with a situation where one or some players have left the agreement and the remaining players have readjusted their economic strategies to the new situation. Similarly, most membership models view the coalition formation process as a two-stage game, where in the first stage players choose their membership and in the second stage their economic strategies, but the passage of time is not explicitly modelled. In contrast, most compliance models capture time explicitly, including discounting. In compliance models, time is an important factor. Some time must elapse between violation of treaty obligations, detection and punishment; otherwise violation could never pay in the first place. Moreover, whether deterrence of violations is successful depends, apart from on the credibility of sanctions, on the weight attached to the short-term gains from non-compliance against the long-term losses from punishment, where relative weights are determined through discounting.

Closely related to the characteristic of time is the structural relation of payoffs (Finus 2003). The easiest assumption is structural independence. For instance, many two-stage membership models assume that payoffs follow from a one-shot decision about emissions levels (for example, the classical papers by *Barrett 1994, Chapter 2*, *Carraro and Siniscalco 1993, Chapter 3* and *Hoel 1992, Chapter 4*, but also later papers by *Carraro, Marchiori and Oreffice 2009, Chapter 24*, *Dimantoudi and Sartzekatis 2006, Chapter 6*, Eichner and Pethig 2013, *Karp and Simon 2013, Chapter 8* and *Rubio and Ulph 2006, Chapter 7*). However, also in repeated games, payoffs at a particular point in time do not depend on decisions taken in previous rounds (for example, *Barrett 1994, Chapter 2*, *2002, Chapter 19*, Endres and Finus 1998, *Finus and Rundshagen 1998, Chapter 18* and *Gerber and Wichardt 2009, Chapter 13*). In the context of emissions, this is closely related to the assumption of flow pollutants. But also in simple fishery models, considering only the steady state and not the transition renders the payoff structure independent (for example, Pintassilgo 2003 and Pintassilgo et al. 2010).

In contrast, in differential (that is, time is continuous) and difference (that is, time is discrete) games, payoffs are structurally dependent (Calvo and Rubio 2012 and Jørgensen, Martín-Herrán and Zaccour 2010). In the context of emissions, this is related to the assumption of stock pollutants and, in a renewable resource model, to the stock of the resource. Some of the sophisticated compliance models assume structural dependence (*Hannesson 1997, Chapter 10* and *Tarui et al. 2008, Chapter 16*), which means that the possibility to enforce an agreement through punishment changes over time. In membership models, structural dependence comes in two versions. The simple version assumes that membership is a one-shot decision based on the discounted payoffs of a differential game (*Carraro, Eyckmans and Finus 2006, Chapter 30*, Eyckmans and Tulkens 2003, Kwon 2006 and Rubio and Casino 2005). Ulph (2004) called this fixed membership. In contrast, the more sophisticated version allows for variable membership (for example, *Germain et al. 2003, Chapter 47*, *Rubio and Ulph 2007, Chapter 45* and *de Zeeuw 2008, Chapter 46*): stability of coalitions has to be tested along the entire time path, allowing for the possibility that countries may enter or leave a coalition over time.

Further characteristics will be discussed below in the context of membership, negotiation and compliance models. At this stage, it suffices to highlight that, in terms of methods, it appears to us that there are two categories: theory and experimental models. Some of the theoretic models are too rich in order to obtain exclusively analytical results, and hence further simulations are necessary. Simulations can be conducted by simply assuming different values of some crucial parameters of models (*Barrett 1994, Chapter 2*, 2001, Lange 2006, *McGinty 2007, Chapter 28*, Rubio and Casino 2005 and *Rubio and Ulph 2007, Chapter 45*), can be more systematic by conducting Monte Carlo Simulations (Dellink and Finus 2012, Dellink, Finus and Olieman 2008 and Kolstad and Ulph 2011), can be based on the calibration of models derived from case studies (Kaitala, Mäler and Tulkens 1995, Kallbekken and J. Hovi 2007, *Mäler 1989, Chapter 1* and Pintassilgo 2003) or from general computable general equilibrium models and integrated assessment models of climate change (*Carraro, Eyckmans and Finus 2006, Chapter 30*, Eyckmans and Tulkens 2003, Finus, Sáiz and Hendrix 2009, *Germain et al. 2003, Chapter 47* and Nagashima et al. 2011). Experimental models typically test the prediction of theory either in a lab or through online surveys (for example, *Cherry and McEvoy 2013, Chapter 14*, Dannenberg 2011, *Dannenberg, Lange and Sturm 2014, Chapter 25*, Dannenberg, Sturm and Vogt 2010, *Kosfeld, Okada and Riedl 2009, Chapter 34*, *Gangadharan and Nemes 2009, Chapter 44*, McEvoy et al. 2010, *McEvoy and Stranlund 2009, Chapter 15*, Milinski et al. 2008 and Milinski, Rohl and Marotzke 2011).

Membership Models

The Characteristic Function

The classical approach of studying the formation of coalitions assumes a transferable utility (TU)-framework and is based on the characteristic function. This function assigns to every coalition a worth, which is the aggregate payoff that a coalition can secure for its members, irrespective of the behaviour of players outside this coalition. The focus of this approach is on the coalition's worth in the grand coalition and how it might be shared among all players. The papers by *Chander and Tulkens (1997, Chapter 5)* and *Germain et al. (2003, Chapter 47)* in this volume, but also papers like Chander and Tulkens (1992, 1995), Eyckmans and Tulkens (2003) and Kaitala and Lindroos (1998), belong to this approach. An allocation of the worth in the grand coalition, which corresponds to the social optimum, is called an imputation. Whether an imputation is stable has been tested in the context of IEAs predominantly with the concept of the core. An imputation lies in the core if no single player or group of players has an incentive to form an alternative agreement, that is, the group of deviators cannot achieve a higher worth for their group. Under the α-core assumption it is assumed that after the deviation the remaining players choose their economic strategies as a form of retaliation (minimax strategies), but also under the γ-core assumption they play their non-cooperative Nash equilibrium strategies. The γ-core can be interpreted as if the remaining players not involved in a deviation break up into singleton players and choose their economic strategies such as to maximise their own payoff.[3] In any case, the core assumes that a deviation triggers punishment. For instance, in a global emission game, emissions of non-deviators will increase after a deviation.

Chander and Tulkens analysed this problem in various papers by considering particular imputations that are derived from a set of transfer rules, which were later termed the Chander–Tulkens transfer rules. Essentially, everybody in the grand coalition receives his or her payoff in the Nash equilibrium, that is, the payoff if all players played as singletons, plus a share of the surplus, which is defined as the difference between the worth in the social optimum coalition and the sum of Nash equilibrium payoffs. For the share, they considered different assumptions, but the most popular seems to be the ratio between the individual marginal damage from emissions and the sum of marginal damages in the social optimum. Essentially, this transfer rule is a Nash bargaining solution for a TU-game with unequal weights. Under mild conditions, *Chander and Tulkens (1997, Chapter 5)* show that their imputation lies in the γ-core, and given that the α-core makes stronger assumptions about the reaction of non-deviating players after a deviation, their imputation also lies in the α-core. *Helm (2001, Chapter 9)* later generalised this result (without assuming a particular imputation) by showing that the γ-core is non-empty, that is, there exists at least one imputation which is immune to deviations by all possible groups of players.

A similar optimistic result is obtained in *Ambec and Sprumont (2002, Chapter 32)*. They analyse the problem of sharing a river that flows through several countries. The authors consider a particular imputation (which they call a 'welfare distribution'), namely one where the marginal contribution vector corresponding to the ordering of the agents along the river satisfies the core lower bounds and the aspiration upper bounds. In their model, the core lower bounds require that no coalition should get less than the welfare it could achieve by using the water it controls, while the aspiration upper bounds demand that no coalition enjoys a higher welfare than what it could achieve in the absence of the remaining agents. This imputation satisfies some axiomatic properties and lies in the core (though it is not the only imputation which lies in the core).

Overall, the core leads to an optimistic view regarding the prospects of cooperation in the light of environmental externalities: the grand coalition, implementing the social optimum, can be achieved if the gains from cooperation are shared in an appropriate manner. Though this result is encouraging from a normative point of view, from a positive point of view it cannot explain the lack of insufficient cooperation from which most transboundary environmental problems suffer.

The Partition Function: A First Approach

An alternative approach is the partition function, which assigns a worth to every coalition and singleton in the game. Different from the characteristic function, the worth depends on the entire coalition structure, in other words, the partition of players. That is, externalities across players and coalitions are explicitly captured. The main focus of this approach is on the prediction of equilibrium coalition structures emerging from some coalition formation process. The early papers by *Barrett (1994, Chapter 2)*, *Carraro and Siniscalco (1993, Chapter 3)* and *Hoel (1992, Chapter 4)*, as well as numerous subsequent papers (for example, Carraro and Machiori 2003, *Carraro, Marchiori and Oreffice 2009, Chapter 24*, *Courtois and Haeringer 2012, Chapter 20*, Finus and Pintassilgo 2012, Finus and Rundshagen 2006b, 2009, Kolstad 2010, *Lange and Vogt 2003, Chapter 33*, McGinty 2011, Pintassilgo and Lindroos 2008 and *Weikard 2009, Chapter 29*), belong to this approach, even though no reference to the 'partition function' is made in the early papers. These papers, but also most subsequent papers, look at a restrictive set of partitions, namely only those with one non-trivial coalition (a coalition of at least two players) and all other players being singletons. They apply the concept of internal and external stability in order to determine stable coalitions. A partition is called internally stable if no member of the coalition has an incentive to leave and externally stable if no outsider has an incentive to join. Incentives are tested based on individual payoffs, called valuations, which are derived from some allocation rule of how the worth of a coalition is split.[4] In contrast to the core, only single (and no group) deviations are considered, with all other players reacting by re-optimising their strategies. All singletons continue to choose their economic strategies by maximising their own payoff, and the coalition maximises the aggregate payoff to their coalition, which comprises now one member less. The implicit punishment is much weaker compared to the core. For instance, in a global emission game, emissions by the non-deviating players increase less after a deviation than under the core assumption. Consequently, deviations become more attractive and therefore, typically, only smaller coalitions and not the grand coalition are internally and externally stable. The effect that only single and no group deviations are considered does not materialise because single deviations are more attractive than group deviations in most environmental externality settings. The main conclusion from this approach is much more pessimistic, which *Barrett (1994, Chapter 2)* called the 'paradox of cooperation'. Whenever cooperation would matter most, that is, the aggregate gain from cooperation would be large, measured as the difference between the aggregate payoff in the social optimum and the Nash equilibrium, stable coalitions achieve only little. This does not mean that stable coalitions are always small; however, if they are large, the aggregate gain from cooperation is small. The later papers by *Diamantoudi and Sartzetakis (2006, Chapter 6)*, *Rubio and Ulph (2006, Chapter 7)* and *Karp and Simon (2013, Chapter 8)*, which generalised *Barrett's 1994 (Chapter 2)* finding in several respects, did not qualify this major conclusion.[5]

The Partition Function: A Second Approach

After the papers by Bloch (1997) and Yi (1997, 2003) on the partition function approach in general, which have been further developed and adapted to the context of IEAs by Carraro and Marchiori (2003), Finus (2003) and Finus and Rundshagen (2003), it has become clear that the partition function approach is conceptually much broader than suggested by the early papers and that the idea of core stability can be integrated in the partition function approach.

Generally speaking, most coalition formation games can be viewed as two-stage games, where in the first stage players choose their membership strategy and in the second stage (which may be divided into further sub-stages) choose their economic strategies. The game is solved by backwards induction. In each stage various assumptions can be made, of which some important ones are displayed in Table 1. The assumptions in the first stage can be modelled through various membership or coalition games, which capture different institutional settings of coalition formation (Finus and Rundshagen 2006a, 2009). For instance, in the single coalition open membership game each player simultaneously announces an address, either zero or one. All players who have announced one form a coalition and all players who have announced zero remain as a singleton. In the multiple coalition open membership game, players can simultaneously announce an address from one to N, and those players who announced the same address form a coalition. In the sequential move unanimity game by Bloch (1996), an initiator proposes a list of potential coalition members, and players are asked sequentially whether they would like to accede to this coalition. If one potential member refuses accession, he or she becomes the new proposer. Once a coalition has formed, coalition formation continues among the remaining players.[6] Typical dichotomous characteristics are listed in Table 1, where those on the right-hand side render models more difficult and those on the left-hand side may be regarded as the standard assumptions.

Given that some coalition has formed in the first stage, the assumptions of the second stage determine the partition function (the worth to every possible partition) and the valuation function (the individual payoffs to every partition through the transfer or sharing rule). Those assumptions, together with the specifics of the underlying economic problem, determine the properties of the partition and valuation function. Those properties are crucial for the incentives to form coalitions in the first stage. It turns out that some economic problems, even though they appear to be quite different, share similar properties. For instance, in positive (negative) externality games, that is, players not involved in the enlargement of coalitions are better (worse) off through such a move, typically, only small (large) coalitions are stable, as players have an incentive to stay outside (join) coalitions. Typical examples of positive externalities include output and price cartels and the provision of public goods. Firms not involved in an output cartel benefit from lower output by the cartel via higher market prices. This is also the driving force in price cartels where the cartel raises prices above non-cooperative levels. Similarly, agents not involved in a public good agreement benefit from higher provision levels of participants. Most environmental externality problems exhibit positive externalities. In contrast, in negative externality games, outsiders have an incentive to join coalitions and therefore most coalition models predict the grand coalition as a stable outcome. Examples include trade agreements, which impose tariffs on imports from outsiders, or R&D collaboration among firms in imperfectly competitive markets, where members gain a comparative advantage over outsiders if the benefits from R&D accrue exclusively to coalition members.

Moving backwards to the first stage, stable coalitions for a particular membership game can be determined by applying various stability concepts, like Nash equilibrium, strong Nash equilibrium, coalition-proof Nash equilibrium, farsighted stable equilibrium or equilibrium binding agreements (Bloch 1997, Finus 2001, *McEvoy and Stranlund 2009, Chapter 15* and Finus and Rundshagen 2003).

Table 1: Assumptions in Membership Models

1. Stage: participation strategies		
(a) sequence	simultaneous	sequential (fixed and flexible membership)
(b) agreements	single	multiple
(c) membership	(i) open	exclusive (majority, unanimity voting; different degree of consensus)
	(ii) isolated	linked
(d) min. participation clause	no	yes
2. Stage: economic strategies		
(a) choice	(i) simultaneous (Nash–Cournot)	sequential (Stackelberg)
	(ii) first-best design (efficient)	second-best designs (e.g. bargaining and modesty)
(b) transfer (sharing) scheme	no	yes (cooperative vs optimal sharing rules)
(c) parameter values	known	unknown (risk-neutrality vs risk-aversion; no vs partial learning)
(d) payoff function	continuous	non-continuous (threshold)
(e) payoff structure	static	dynamic
(f) other payoff components	disutility from free-riding, other regarding preferences, ancillary benefits, …	

Viewed together, coalition formation comprises three ingredients: the coalition or membership game, which defines the membership rules; the property of the partition function (related to the assumption of the second stage); and the equilibrium concept to determine stable coalition structures (Finus and Rundshagen 2006a, 2009). This allows interpreting 'old' concepts in a 'new light'. For instance, the concept of internal and external stability turns out to be a Nash equilibrium in a single coalition open membership game, whereas the γ-core can be reproduced in the context of the partition function as a strong Nash equilibrium in a multiple coalition exclusive membership H-game (for details see Finus and Rundshagen 2009) in which membership is decided with a strong unanimity rule. The possibilities and directions of deviations are captured through the equilibrium concept[7] and the reaction after a deviation follows from the membership rule.[8] This allows for a systematic analysis of how assumptions drive results.

Results Related to the First Stage

For instance, holding the assumptions of the second stage constant (for example, positive externality game), the effect of different membership rules on equilibrium coalition structures can be analysed (see 1. Stage, Table 1). In Finus and Rundshagen (2006a) and Finus, Rundshagen and Eyckmans (2014), it is shown that a sequential coalition formation process opens up strategic options not available in a simultaneous process, which typically lead to outcomes with lower global welfare (Feature 1(a) in Table 1). In *Bosello, Buchner and Carraro (2003, Chapter 35)*, *Eyckmans and Finus (2006, Chapter 36)* and Finus and Rundshagen (2003), it is demonstrated that by giving up the restriction of either joining an agreement or remaining outside, and allowing for the possibility of the coexistence of multiple coalitions, this may lead to equilibrium coalitions structures with multiple coalitions associated with higher global welfare and lower global emissions than single coalitions (Feature 1(b) in Table 1). Those normally excluded from a single agreement may form their own separate agreements, which is individually and globally beneficial. The main conclusion is confirmed in a compliance model by *Asheim et al. (2006, Chapter 37)*, which compares one agreement with two separate ones. In Finus, Altamirano-Cabrera and van Ierland (2005), *Eyckmans and Finus (2006, Chapter 36)*, Finus and Rundshagen (2006a, 2009) and Finus, Sáiz and Hendrix (2009), it is shown that exclusive membership and a high degree of consensus about membership is conducive to the success of agreements (Feature 1(c)(i) in Table 1). Another possibility arises from linking membership in different agreements, sometimes referred to as issue linkage (Botteon and Carraro 1998, Carraro and Marchiori 2004 and Carraro and Siniscalco 1997).[9] The idea is to make membership in a second agreement (for example, trade or R&D agreement) conditional on joining the environmental agreement (Feature 1(c)(ii) in Table 1). If the second agreement possess different properties (which must stem from the properties of the partition function and hence from the assumptions about the second stage of coalition formation), membership in the linked agreement may be higher than in the isolated environmental agreement.[10] Finally, in *Carraro, Marchiori and Oreffice (2009, Chapter 24)* it is shown that the introduction of a minimum participation rule increases the size and the success of stable coalitions in a single coalition open membership game (Feature 1(d) in Table 1). The minimum participation rule is a threshold membership, which needs to be passed before an agreement becomes binding. The threshold increases the loss for a country considering leaving an agreement at the minimum participation level, with an implicit punishment very similar to the γ-core discussed above.

Results Related to the Second Stage

Alternatively, many papers have made the same assumption regarding the first stage, assuming a single coalition open membership game, but considered different assumptions regarding the second stage. See Table 1.

FEATURE 2(A): CHOICE OF ECONOMIC STRATEGIES

Whereas *Carraro and Siniscalco (1993, Chapter 3)* assume a simultaneous choice of emissions (Nash–Cournot assumption), respectively abatement levels, *Barrett (1994, Chapter 2)* assumes a sequential choice (Stackelberg assumption) where the coalition acts as a Stackelberg leader (Feature 2(a)(i) in Table 1). It appears that the Stackelberg assumption provides the coalition with an informal advantage, allowing them to reduce the leakage effect, which leads to larger and more successful coalitions compared to the Nash–Cournot assumption as appears from Table 3.2 in Finus (2003).

Another alternative assumption in the second stage was already suggested by *Hoel (1992, Chapter 4)*, who considers different rules as to how the coalition chooses their emissions, which requires the departure from the assumption of ex-ante symmetric players to render the analysis interesting. Hoel departs from the standard assumption that coalition members choose their economic strategies efficiently by maximising the aggregate payoff to their coalition and considers various second-best designs. The crucial question is whether and which second-best designs may buy higher participation such that overall this may lead to better outcomes. Similar questions have also been analysed, for instance, in Altamirano-Cabrera, Finus and Dellink (2008), *Courtois and Haeringer (2012, Chapter 20)* and Finus and Maus (2008), or, in the context of compliance models, by *Barrett (2002, Chapter 19)* and *Finus and Rundshagen (1998, Chapter 18)*. Roughly speaking, second-best agreements, like agreements on the lowest common denominator (that is, unanimity voting) or median voter proposal (majority voting) in a bargaining process, agreements by consensus or modest emission reductions, may increase membership sufficiently to make a departure from first-best worthwhile. In their experimental models, Dannenberg (2011) and *Dannenberg, Lange and Sturm (2014, Chapter 25)* consider a similar problem, though they do not confirm all positive theoretical predictions.

Another type of second-best agreement is considered in *Barrett (2006, Chapter 26)* and *Hoel and de Zeeuw (2010, Chapter 27)*. Based on the negative conclusion emerging from the standard model, namely that stable coalitions with the objective to reduce emissions are not very successful, they ask the question whether attention should shift to an R&D agreement. In their models, if countries make a sufficiently large financial contribution to an R&D fund, such that a threshold is passed, a breakthrough technology may emerge which is carbon-free. The answer depends on the technology and the externality it generates. Whereas in Barrett's model more positive results are only obtained if there are economies of scale (network externalities), Hoel and de Zeeuw derive more positive conclusions. The fact that an R&D agreement is not a straightforward solution to the free-rider problem does not come totally as a surprise as investment in R&D with spillovers (non-exclusive benefits) basically constitute the provision of a public good (Nagashima et al. 2011).

FEATURE 2(B): TRANSFER AND SHARING SCHEME

Due to its simplicity, the standard and widespread assumption in the context of the partition function is ex-ante symmetric players. Ex-ante symmetry means that all players have the same payoff function, and hence an equal split of the coalitional worth is the natural assumption (for example, *Barrett 1994, Chapter 2, Diamantoudi and Sartzetakis 2006, Chapter 6* and *Rubio and Ulph 2006, Chapter 7*).[11] However, players are different ex-post, depending on whether they are members of a coalition or remain outside. (In the more general case of multiple coalition structures, it depends on the size of the coalition to which a player belongs; see Carraro and Machiori 2003 and Finus and Runsdshagen 2003.) For ex-ante asymmetric players, without any sharing or transfer scheme, coalitions are normally smaller than for symmetry. For a given coalition, this implies an unequal distribution of the gains from cooperation, with those getting a smaller than average share having an incentive to leave the coalition. Through a transfer scheme, such asymmetries can be reduced. For a long time, concepts from cooperative game theory, like Nash bargaining solution, the Shapley value or the Chander–Tulkens transfer rule, have been applied (Altamirano-Cabrera and Finus 2006, Barrett 2001 and Botteon and Carraro 1997, *Eyckmans and Finus 2006, Chapter 36* and Weikard, Finus and Altamirano-Cabrera 2006). The problem is that different transfer rules lead to different coalition structures and it is not clear whether other transfer rules would be even better suited to mitigate the free-rider incentive. Similar to the Chander–Tulkens transfer rule in the context of the core, there was a necessity to develop an 'optimal' transfer rule in the context of internal and external stability. Such a transfer rule or sharing scheme, which Eyckmans and Finus (2004) call an 'almost ideal sharing scheme', was independently developed by Eyckmans and Finus (2004), *McGinty (2007, Chapter 28)* and *Weikard (2009, Chapter 29)* and analysed from various angles, for instance in *Carraro, Eyckmans and Finus (2006, Chapter 30)*, Fuentes-Albero and Rubio (2010), Kolstad (2010) and McGinty (2011), and experimentally tested in *McGinty, Milam and Gelves (2012, Chapter 31)*. Surprisingly, it turns out that the basic structure is very similar to the Chander–Tulkens transfer rule, which, as we have pointed out above, is a Nash bargaining solution in a TU-framework (Eyckmans, Finus and Mallozzi 2012). The only difference is that the threat point or disagreement point is not complete non-cooperation but the payoff if one player leaves the coalition and all other players continue with cooperation. Accordingly, the surplus is defined as the sum of payoffs in a coalition minus the sum of free-rider payoffs. As Eyckmans and Finus (2004) show, weights do not matter for the set of stable coalitions, though equal sharing can be linked to some axiomatic properties, known from solution concepts or sharing rules of cooperative game theory as illustrated in Eyckmans, Finus and Mallozzi (2012). Interestingly, it appears that using such an almost ideal sharing scheme means that asymmetry is no longer an obstacle to cooperation but in fact may become an asset. Asymmetry may now lead to larger stable coalitions than symmetry as the gains from cooperation become larger. See, for instance, *Finus and Pintassilgo (2013, Chapter 39)*, *McGinty (2007, Chapter 28)*, Pintassilgo et al. (2010) and *Weikard (2009, Chapter 29)*.

FEATURE 2(C): PARAMETER VALUES

Another departure from the standard assumption is to consider that some parameters of the payoff function are unknown (Ulph 1998). This can be analysed under risk-neutrality or risk-aversion. Under risk-neutrality, the earliest reference in the context of coalition formation is *Na and Shin (1998, Chapter 38)*, which came to the provocative and counter-intuitive conclusion that – in a strategic context – learning could lead to worse outcomes than if players remain under the veil of ignorance. This conclusion generated quite a number of follow-up papers (Dellink and Finus 2012, Dellink, Finus and Olieman 2008, Kolstad 2007, Kolstad and Ulph 2008, 2011 and Ulph 2004). Viewed together (Finus and Pintassilgo 2012, *2013, Chapter 39*), it appears that four key ingredients drive results. It depends whether there is uncertainty about the benefits or costs of emissions, whether uncertainty concerns the global level or the distribution of benefits and costs, the degree of learning and whether an optimal transfer scheme is available to address asymmetric gains from cooperation under learning. The major conclusion from *Finus and Pintassilgo (2013, Chapter 39)* is much more positive: only under very special conditions does learning lead to worse outcomes than no and partial learning and, if this is the case, the negative effect can be mitigated or even turned into a positive effect through the almost ideal sharing scheme mentioned above.

Under risk-aversion, the idea that the degree of risk-aversion could matter for the success of cooperative agreements was first developed by Endres and Ohl (2003) in a simple prisoners' dilemma type of game. They argue that cooperation may reduce the variance of payoffs, which makes cooperation more attractive for risk-averse players than free-riding. Whether this conclusion carries over to a richer setting with coalition formation is analysed in *Boucher and Bramoullé (2010, Chapter 41)* and Finus, Pintassilgo and Ulph (2014). For instance, regarding the second stage, for a given coalition an increase in uncertainty, that is, an increase of the variance of the damage parameter, but keeping the expected value constant, increases signatories' abatement efforts. The same holds for an increase in risk-aversion. Regarding the first stage, increasing uncertainty and risk tends to decrease the equilibrium coalition size. As a tendency, learning is more beneficial the more risk-averse players are, as long as risk-aversion stays below a threshold. Above this threshold, with no or partial learning, all players fully cooperate because of the fear of possible very high damages, which may not be the case under full learning.

An experimental analysis of the effect of risk and uncertainty on the costs and benefits of public good provision is conducted in *Gangadharan and Nemes (2009, Chapter 44)*. They also allow subjects to invest, either in order to reduce the degree of uncertainty or to influence the probabilities of the return on public good provision. They find that uncertainty and risk about the benefits from public good provision reduce equilibrium contributions and that subjects make use of the possibility of reducing uncertainty if this is not too costly.

FEATURE 2(D): PAYOFF FUNCTION

Another interesting departure emerges from giving up the assumption of continuous payoff functions. *Barrett (2013, Chapter 42)* considers the possibility that the damage function from global emissions may have a kink. The kink constitutes a threshold at which damages may become very large. He shows that if the threshold is known, this transforms the cooperation problem into a simple coordination game, with full cooperation being the outcome, provided the damage costs above the threshold are relatively high compared to the cost of avoiding them. Barrett also shows that if the impact above the threshold is unknown, this result still holds, though not if the threshold itself is unknown, despite that the risk of high damages may be very high. Essentially, an uncertain threshold makes the expected damage function continuous, with all the problems known from the standard model. Experimentally, *Tavoni et al. (2011, Chapter 43)* test a similar setting, though players have asymmetric endowments and know the threshold, as well as the expected loss in case the threshold of total contributions to a public good is not reached. They find that inequality reduces the prospects of reaching the target but that communication increases the success dramatically. Successful groups tend to eliminate inequality over the course of the game, with rich players signalling willingness to redistribute early on.

FEATURE 2(E): PAYOFF STRUCTURE

The payoff structure is also an important feature for determining the outcome of coalition formation. Considering a dynamic instead of a static payoff structure does not necessarily change the fundamental incentive structure in a coalition game if membership is fixed (as defined in the first stage, see Table 1), as the partition function may still possess the same standard properties as a static payoff function. This is evident, for instance, from Kwon (2006), in the fishery context, or from *Eyckmans and Finus (2006, Chapter 36)* and Rubio and Casino (2005), in a global emission game. A more interesting analysis emerges in a game with flexible membership. Interestingly, like for a static payoff structure, the two prototype stability concepts have been considered, the core in *Germain et al. (2003, Chapter 47)* and internal and external stability in *Rubio and Ulph (2007, Chapter 45)*. *Germain et al. (2003, Chapter 47)* is an extension of the idea developed in the static context in *Chander and Tulkens (1997, Chapter 5)* to a difference model with a terminal time T (which can be finite or infinite). In the dynamic context, an imputation must lie in the core along the entire time path. The structure of the transfer formula is similar to the Chander–Tulkens transfer rule, though the actual transfers paid or received change over time. The striking thing is that an analytical proof can be obtained under mild conditions for open and closed loop Nash equilibrium strategies, namely that the core is non-empty. This is not possible in Rubio and Ulph's model, and hence they conduct a large number of simulations. *Rubio and Ulph (2007, Chapter 45)* consider a difference game with an infinite horizon, where in each period identical countries decide whether or not to become a member of a coalition. Their simulations predict that membership decreases over time, which is not in line with the actual record of most IEAs (see, for instance, Finus 2003), suggesting that further modifications of their model will be needed. Without doubt such extensions will be difficult as dynamic membership introduces substantial complications in their setting. Partly because of this difficulty, an alternative approach was followed by *Breton, Sbragia and Zaccour (2010, Chapter 49)*. Instead of assuming that in each period countries solve a membership game, as in *Rubio and Ulph (2007, Chapter 45)*, the authors analyse a model where countries gradually reach a stable IEA (in terms of internal and external stability), based on the assumption that the proportion of signatory countries evolves over time following the discrete-time replicator dynamics.[12] This implies that as long as the levels of the stock of pollution are unstable, the coalition size is not determined by a membership game (that is, it does not necessarily satisfy the internal and external stability condition). It also implies that the payoffs of signatories and non-signatories are not identical throughout the game. Hence, the replicator dynamics is used to determine the membership decision. Their numerical simulations show that in this framework non-signatories will gradually join the coalition, a behaviour frequently observed in actual IEAs. They also show that once the equilibrium is reached (including the condition that the coalition has to be internally and externally stable), partial cooperation is stable from this instance onwards.

De Zeeuw (2008, Chapter 46) uses the concept of farsightedness to analyse a difference game with an infinite time horizon in which the steady state implies no emission irrespective of the degree of cooperation but where under cooperation the zero emissions target is reached faster and with lower costs than in the absence of cooperation. This setting is very different from most other papers discussed in this introduction. Therefore, it would be interesting to know whether the results obtained hold in a more general setting. The paper starts by showing that in a one-shot version of the model with static payoffs, farsighted stability implies larger stable coalitions than those obtained under internal and external stability. For the dynamic model, the concept of dynamic farsighted stability[13] is introduced, implying that a coalition is stable if deviations do not pay, assuming that the deviator will be an outsider with respect to the smaller stable coalition that emerges in the next period. Again, dynamics introduces a complexity, which requires one to resort to simulations. The author focuses on a set of simulations with eight countries. In contrast to the conclusion under static farsighted stability, the numerical results for dynamic farsighted stability show that countries can only coordinate on a large stable coalition if the relative weight attached to the damage costs as compared to the abatement costs is small. If this relative weight is small, the level of emissions is slowly adjusted downwards along the equilibrium path and hence the incentive structure in the game changes only slowly over time, while the opposite is true if the relative weight is large. Thus, the gains from free-riding are small with a small relative weight attached to damage costs, sustaining larger coalitions than if the reverse is true.

Weikard, Dellink and van Ierland (2010, Chapter 48) consider a sequence of stage games in which coalitions will form. In each stage, the decision is based on discounted payoffs from this stage until the final stage. Within a stage game, players behave like in a standard membership model by deciding first on membership and then on abatement strategies. Abatement strategies within the coalition are chosen cooperatively and non-members choose their abatement non-cooperatively, based on the discounted payoffs until the end of the game. The number of stages is varied and called rounds. What is now different to the standard model is that in earlier stages, coalitions can be enforced by using renegotiation-proof threat strategies. If, say, several coalitions are stable in the last round, then players can choose the worst Pareto-undominated outcome as a threat in the round preceding the last round in order to enforce membership. In this way, they work their way backwards to round 1, the initial round, when the game starts. Thus, their model combines elements of a standard membership model and a standard compliance model, using renegotiation-proof punishment strategies to enforce membership. Despite the fact that they assume a stock pollutant, they can solve the game backwards, as marginal damages are assumed to be constant and hence do not depend on the level of the stock. Their simulations are based on the STACO model, which is a calibrated climate model with twelve regions. They focus on the impact of the number of renegotiation rounds on the stability of coalitions. For instance, if only one renegotiation round is considered, they find that China and the US form the unique stable coalition, while larger coalitions emerge if there are more renegotiation rounds. It appears that in their model an increase of renegotiation rounds increases the threat options of punishment, allowing to enforce larger stable coalitions, although the grand coalition is never stable. Unfortunately, it is not possible to conclude how general this result is.

The discussion of the previous papers highlights the difficulties encountered in analysing analytically non-cooperative coalition formation games in a dynamic context. *Rubio and Ulph (2007, Chapter 45)*, *Breton, Sbragia and Zaccour (2010, Chapter 49)*, *Weikard, Dellink and van Ierland (2010, Chapter 48)* and *de Zeeuw (2008, Chapter 46)* all rely on simulations, which implies that it is not possible to evaluate the generality of their results. In addition, these models make strong additional assumptions, like membership decisions based on replicator dynamics (*Breton, Sbragia and Zaccour 2010, Chapter 49*) or phasing out of emissions irrespective of the degree of cooperative behaviour (*de Zeeuw 2008, Chapter 46*).

FEATURE 2(F): OTHER PAYOFF COMPONENTS

There was always a temptation to consider other payoff components or a modified payoff function to obtain more optimistic results. Already *Carraro and Siniscalco (1993, Chapter 3)* consider the possibility that some coalition members are committed to cooperation, hence ignoring their free-rider incentives, and use this slack of enforcement power to pay outsiders to enlarge their coalition. Other attempts were undertaken by Jeppesen and Andersen (1998) and Hoel and Schneider (1997), who add a term to the payoff of non-signatories that captures the disutility of remaining an outsider. Not really surprising, larger coalitions will form, which follows almost by assumption.[14] A more interesting idea is pursued in *Lange and Vogt (2003, Chapter 33)*, with similar ideas considered in Gruning and Peters (2010) and Lange (2006).[15] They consider ERC preferences (equity, reciprocity and competition) where utility of a country is the weighted sum of the absolute material payoff function and the relative material payoff function. The relative material payoff function obtains its maximum at the average payoff of all countries. Any deviation from the arithmetic mean is interpreted as inequality and counts as disutility. For the assumption of ex-ante symmetric countries, they show that if relative payoffs receive a sufficiently high weight compared to absolute payoffs, the grand coalition emerges as an equilibrium coalition. For asymmetric countries, predictions are much more difficult and depend on the particular assumptions. In this case, equity preferences do not always lead to better outcomes, which is also confirmed by the theoretical papers by Gruning and Peters (2010) and Lange (2006).[16] The experimental evidence presented in *Kosfeld, Okada and Riedl (2009, Chapter 34)* suggests that social preferences by participants (that is, inequality aversion in the sense of Fehr and Schmidt 1999) do play an important role. They find larger stable coalitions than predicted by standard preferences. Interestingly, their setting combines membership with compliance as non-compliance of members is sanctioned if an agreement forms.

Negotiation Models

Most of the membership models (as well as the compliance models discussed below) abstract from the negotiating process. This is the focus of negotiation models, which have mainly looked at global public goods in general, with climate change being a particular application. One of the main issues of this line of research is the burden-sharing rule, which emerges from negotiations between countries and groups of countries pursuing their self-interests.

Particular emphasis is given to (a) the role of asymmetric information (as countries may hide information during the negotiations; Rotillon and Tazdaït 1996, *Caparrós, Péreau and*

Tazdaït 2004, Chapter 21, Jakob and Lessmann 2012, Espínola-Arredondo and Muñoz-García 2011 and Helm and Wirl 2014), (b) the role of delegation (as international negotiations are carried out by delegates and countries' representatives; Segendorff 1998, Buchholz, Haupt and Peters 2005, Roelfsema 2007 and Harstad 2008), (c) the consequences of the ratification process needed to enforce an international agreement (Courtois and Tazdaït 2008), (d) the possibility of countries forming groups during the negotiation process to gain bargaining power (*Caparrós and Péreau 2013, Chapter 22*), (e) the impact of a unilateral commitment by one of the agents involved in the negotiation process (Hoel 1991, Brandt 2004 or Konrad and Thum 2014) and (f) the influence of expectations regarding a future bargaining process about investment decisions in abatement technologies (Beccherle and Tirole 2011, Urpelainen 2012 and *Harstad 2012, Chapter 23*).

Although a simple bargaining process à la Rubinstein predicts that, under perfect information, an agreement is reached immediately, most of the papers mentioned above yield rather pessimistic results. For example, introducing asymmetric information implies that an agreement will be delayed as countries use offers and counter-offers during the bargaining process to gain information about the characteristics/interests of other parties (Rotillon and Tazdaït 1996 and *Caparrós, Péreau and Tazdaït 2004, Chapter 21*). In a simple model, in which only one variable is private information (which can only take on two possible values), the delay is modest as each country (player) needs only one offer to determine with certainty the type of its opponent (that is, it suffices to make an offer above the demands of a weak player and below that of a strong player). However, in reality there are many variables, and each may take on many possible states of the world. Hence, in order to determine the types of opponent players, a player would need to know the responses to many different offers (a process that increases the delay if each player can only make one offer at a time). Delays are also frequent if not only two but several players are involved in the negotiations, as shown in *Caparrós, Péreau and Tazdaït (2004, Chapter 21)* where negotiations take place between a 'Northern coalition' and two 'Southern coalitions' and the South can have 'low' or 'high' minimum transfer demands. They distinguish between a complementary relationship (both Southern coalitions are needed for an effective agreement) and a substitutable relationship (an agreement with one Southern coalition suffices to lead to an effective agreement). Complementarity gives Southern coalitions significant bargaining power and encourages them to negotiate separately, while substitutability reduces the bargaining power of the Southern coalitions and encourages them to negotiate together. This holds for complete information and asymmetric information, with asymmetric information further reinforcing the position of complementary coalitions and further weakening that of substitutable players. The reason is that complementary coalitions are more reluctant to reveal that they have low transfer demands, as they know that the North will always need to come back with another offer if they reject the current one. In contrast, substitutable coalitions are more willing to disclose this information, as they know that one Southern coalition is enough for the North to strike an effective deal. As the authors point out, this implies that the Southern coalitions have a great incentive to be seen as complementary. From a Southern perspective, an optimal negotiation path looks as follows: first, sign a general agreement (such as the Kyoto Protocol) which makes the Southern groups complementary by setting a minimum participation clause that de facto requires the key players to ratify the agreement and, second, start a discussion about the details of a treaty (like the Marrakech Accords in the case of the Kyoto Protocol) where the separation in two complementary groups gives the Southern coalitions strong bargaining power in order influence the negotiations on the amount of the transfers (or any other type of benefits) in their favour. Of course, one may argue that the same reasoning can be applied to the North, which should also prefer to negotiate in two separate groups. Whether this is the case is the question investigated in *Caparrós and Péreau (2013, Chapter 22)*, discussed below, as in *Caparrós, Péreau and Tazdaït (2004, Chapter 21)* this possibility is ruled out by assumption.

Before turning to this question, it is important to clarify the term 'coalitions', as referred to in the previous paragraph. 'Coalitions' means here institutions conducting negotiations on behalf of its members but are not viewed as the final outcome of a bargaining process (like in membership models) but as the starting point and as a vehicle of negotiations. 'Coalitions', as they are commonly referred to in studies of international relations, have played a significant role in climate change negotiations (Hampson and Hart 1995). During all the negotiation rounds,[17] countries have grouped into 'coalitions', such as the Umbrella Group (the US and similarly minded countries during the nineties), the Group of 77 and China (a large group of developing countries), the Alliance of Small Island States and, more recently, the Rainforest Coalition (a group including almost all countries with rainforests, except Brazil). To avoid confusion about the meaning of the term 'coalitions', *Caparrós and Péreau (2013, Chapter 22)* refer to these groups of countries which have decided to negotiate together as a group, as 'negotiation-coalitions'. The authors analyse whether there is a natural tendency for the negotiations to be carried out among negotiation-coalitions, distinguishing between the incentives that arise from efficiency gains and those originating from the incentives to increase the bargaining power in negotiations. Their analysis shows that if bargaining power encourages the formation of a negotiation-coalition on one side of the table, the tendency will go in the opposite direction on the other side of the table. That is, if the South has incentives to form a unique negotiation-coalition for bargaining power reasons, the North will not be interested in doing the same among its group members (and vice versa). By discussing the more realistic values for the different parameters of their model, the authors come to the conclusion that the most likely configuration is a unified North and a separated South (and hence the discussion of *Caparrós, Péreau and Tazdaït 2014, Chapter 21* above remains valid).

Beccherle and Tirole (2011) and Urpelainen (2012) show that countries have an incentive to underinvest in clean technologies before negotiations start in order to improve upon their threat points, giving them more bargaining power. *Harstad (2012, Chapter 23)* further analyses this issue in a differential game with n agents, where countries can sign contracts on emissions but, for most of the analysis, not on abatement technology investments. Contracts can be 'complete', where countries commit to their emissions and their investments in abatement technologies, or 'incomplete', where countries can only sign contracts on their levels of emissions. Contracts are assumed to be efficient and determined, for example, using the Nash bargaining solution. The author analyses first a framework without renegotiation, where the default if negotiations fail is no agreement, and then a framework with renegotiations, where countries can renegotiate after (part of) the investment is sunk and where the threat point of the Nash bargaining solution is the previous interim agreement. Results show that a sequence of short-term incomplete contracts without renegotiation makes everybody worse off, as countries invest less in abatement technologies when they anticipate future negotiations (due to the hold-up problem: if a country has a large stock of abatement technology, it will be required to abate more in any efficient agreement). The first-best solution can be achieved either by signing a complete contract (emissions and investment levels) without renegotiations, or via long-term incomplete contracts on emissions with renegotiations. Renegotiations solve the hold-up problem because countries that underinvested have a weaker bargaining position during the renegotiation phase and, anticipating this, countries invest more initially.

The results in these papers imply that negotiations should either proceed very fast in order to prevent strategic investment decisions (an option lost a long time ago in the context of climate change), or should target long-term agreements, as this can solve the hold-up problem.

Another important issue that yields rather pessimistic results is the delegation of international negotiations to countries' representatives. Segendorff (1998) analyses strategic delegation in a model where two countries bargain over the provision of a transboundary public good. In each country there is a principal who has the option to delegate the task of negotiations to an agent (where the principal can be seen as a decisive voter). The analysis assumes a two-stage game: first, both principals simultaneously choose agents and, second, agents bargain over the provision of the public good (where the second stage is solved using the Nash bargaining solution). The author considers a weak and a strong delegation model, the difference being that in the former the breakdown allocation is done by the principal while in the latter the agent also decides upon the breakdown allocation. Principals choose agents that have lower preferences for the public good than their own because this lowers the reservation utility and increases their bargaining power. However, as both principals adopt this strategy, delegation makes at least one of the principals worse off compared to self-representation (where the principal negotiates). When delegation is strong, this negative effect can even more than offset the gains from reaching an agreement. Buchholz, Haupt and Peters (2005) confirm this qualitative result in a model where voting is explicitly introduced and the median voter elects his preferred politician. The key intuition at work continues to be that sending a delegate with lower than own environmental preferences to the negotiation table helps to obtain a large share of the surplus from cooperation.[18] A similar result is obtained when analysing the impact of ratification requirements. Courtois and Tazdaït (2008) show that a strong domestic constraint, that is, a parliament that is reluctant to ratify an agreement, increases the bargaining power of a player.

Unilateral action also does not seem to be a very promising strategy to increase cooperation (Hoel 1991), unless it can be used as a signal that abatement costs are lower than expected, as this may encourage countries to cooperate which would normally abstain from a treaty due to the fear of high abatement costs (Brandt 2004). However, as Brandt shows, this mechanism only works if abatement costs between countries are strongly correlated. In contrast, conditional pledges are probably a better way of pushing joint policies forward, as the recent analysis by Helm and Wirl (2014) shows. These authors apply a principal–agent model under asymmetric information to North–South climate negotiations, assuming that the principal has all the bargaining power and the contract is binding. The asymmetry of information is related to the private valuation of the damages by the South. The analysis shows that the optimal contract implies that the principal should condition its own emission reductions on those of the agent (reciprocal relationship or matching), such that it reduces emissions more substantially if the agent also does so.

Taken together, the various results obtained from negotiation models yield rather pessimistic results. Further research is needed to analyse how negotiations could be structured to favour positive outcomes. Long-term agreements seem to be the most obvious solution, although enforceability problems may become more prevalent the longer the time horizon of an agreement is, especially in a world where governments constantly change and do not always share the views of the government involved in the original negotiations (as the example of the US government and the Kyoto Protocol illustrates).

Compliance Models

Compliance models typically abstract from the process of coalition formation.[19] Roughly speaking, for a given set of players, typically the grand coalition, it is tested whether the discounted payoff from compliance is larger than the discounted payoff from taking a free ride and subsequent punishment. Crucial factors as to whether compliance can be enforced are the punishment options and the credibility of sanctions, which are somehow related. The credibility of sanctions is closely related to the definition of an equilibrium, where the demands on the credibility of sanctions increases along the following sequence: Nash, subgame perfect, weakly renegotiation-proof, strongly renegotiation-proof and strong subgame-perfect equilibrium (see Table 3.3, p.126, Finus 2003). The higher the demands on credibility, the more difficult it becomes to enforce compliance. Whereas *Folmer, van Mouche and Ragland (1993, Chapter 12)*, Folmer and van Mouche (1994), *Hannesson (1997, Chapter 10)* and Mäler (1994) consider only subgame-perfect strategies, *Barrett (1994, Chapter 2)* already considers weakly and strongly renegotiation-proof and even strong subgame-perfect equilibrium strategies (though the latter is not mentioned). In particular, the concept of renegotiation-proof equilibrium captures an essential obstacle of enforcing compliance, namely that those conducting the punishment should not be worse off than during cooperation (Mohr 1988). Hence, simple trigger strategies, which are normally subgame perfect, where all players revert to the non-cooperative equilibrium once non-compliance is detected until the end of the game (finite time horizon) or until infinity (infinite time horizon), can normally not be used for punishment. More sophisticated punishment strategies must be considered where the culprit must show repentance to allow the punishers not to be worse off during the punishment phase than during cooperation. At the same time, punishment cannot last too long such that the culprit finds it attractive to accept punishment, which is rewarded with a return to cooperation instead of continuous punishment. Tit-for-tat strategies capture this spirit of forgiveness. The concept of renegotiation-proof equilibrium is used, for instance, in the papers by Asheim and Holtsmark (2009), *Asheim et al. (2006, Chapter 37)*, *Barrett (2002, Chapter 19)*, Endres and Finus (1998), *Finus and Rundshagen (1998, Chapter 18)*, *Finus and Tjøtta (2003, Chapter 11)* and Froyn and Hovi (2008). The fact that enforcement is much more difficult if sanctions have to be renegotiation-proof is evident, for instance, by comparing the papers by Mäler (1994) and *Finus and Tjøtta (2003, Chapter 11)*. In the former paper, a socially optimal solution can be enforced by a subgame-perfect trigger strategy, whereas in the latter paper, this is not possible using renegotiation-proof strategies. The grand coalition is only renegotiation-proof for substantially less ambitious emission reduction targets than those in the social optimum. This is also evident from the theoretical papers by *Barrett (2002, Chapter 19)* and *Finus and Rundshagen (1998, Chapter 18)*. But also in *Hannesson's (1997, Chapter 10)* fishing supergame using subgame-perfect trigger strategies, the number of agents which an agreement can sustain may be quite low for reasonable discount rates. A similar analysis for a renewable resource is conducted in *Tarui et al. (2008, Chapter 16)*, though in their paper punishment is divided in two phases. In *Hannesson (1997, Chapter 10)* and *Tarui et al. (2008, Chapter 16)* the analysis is more complicated than in a repeated game as stocks change over time and hence the punishment options are not the same along the entire time path. A similar idea with changing punishment options over time is developed in Chapter 12 of Finus (2001).

The role of different trigger strategies in sustaining equilibria that improve over the business as usual (BAU) emission scenario is also analysed in *Dutta and Radner (2009, Chapter 17)* in a differential game with an infinite time horizon, where a set of countries pollute and the stock of greenhouse gases accumulates over time. Although they present some basic numerical simulations, *Dutta and Radner (2009, Chapter 17)* provide an analytical solution of their model and characterise the non-cooperative Markov-perfect equilibria (MPE) of their game. In a third-best setting, they analyse the Pareto optimal outcomes among the equilibria that can be sustained by a threat of reverting to the BAU scenario. These equilibria improve welfare for all discount rates, but only for high discount rates can the social optimum be reached using such kinds of punishment strategies. In a second-best setting, they study the Pareto optimal outcomes among the equilibria that can be sustained by any threat, focusing on the reversion to the worst equilibrium. The qualitative results of the third-best analysis still hold, as emissions are always lower than under the BAU scenario but again only identical to the global optimum for high discount rates. The interesting feature of all the equilibria just described is that they are particularly simple, as steady-state emissions are constant and independent from the initial state of the system. As the authors point out, the implication is that there is no such as thing as starting 'too late' on emission control, as the same steady state will be reached regardless of the initial situation (as long as the initial GHG level is below the steady-state level). Nevertheless, the authors complete the analysis by showing that there are in fact MPEs in which the current level of GHG matters.

Another interesting feature in the context of credible punishment is the introduction of a deposit-refund system, where coalition members deposit money into a fund at the beginning of the game, receive interest payments as long as they comply, but lose parts or all of their deposit once they take a free-ride. In the case of free-riding, the deposit may be used to compensate compliant players. This allows conducting a renegotiation-proof punishment as compliant countries can be made better off than during cooperation, provided the deposit is large enough. This idea has been proposed in Finus (2002) and has been formally developed in *Gerber and Wichardt (2009, Chapter 13)*, further developed in Gerber and Wichardt (2013), applied in Hovi et al. (2012) to the case of climate change and experimentally tested in *Cherry and McEvoy (2013, Chapter 14)*, who confirm that a deposit system can improve the prospects of cooperation.

A slightly different idea is pursued in *McEvoy and Stranlund (2009, Chapter 15)*, who combine elements of membership and compliance models. They analyse an IEA in which parties to the agreement finance an independent monitor who audits the compliance performance of the members of an agreement. These audits reveal instances of non-compliance so that they can be sanctioned. They find that costly monitoring of compliance limits the circumstances under which international cooperation is beneficial, but when an IEA forms, it will often involve greater participation than an IEA that does not require costly monitoring.

Consequently, costly monitoring of IEAs can produce higher environmental quality. Moreover, under certain conditions, aggregate welfare is higher when IEAs require costly monitoring.

Finally, *Folmer, van Mouche and Ragland (1993, Chapter 12)* and Folmer and van Mouche (1994) consider issue linkage in a compliance model. Issue linkage has two implications. First, it works like a compensation scheme in the presence of ex-ante asymmetric players if payoffs obtained from the different issues constitute a kind of mirror image. Low payoffs in one game can be compensated by higher payoffs in the other game. As individual rationality is a necessary condition for sustaining cooperation in a repeated game through subgame-perfect strategies, issue linkage increases the chances for cooperation. Second, and closely related to this, the strategy space of possible punishment options increases through issue linkage as the number of Nash equilibria in a constituent game increases (Finus 2001). The fact that such positive results do not only hold for subgame perfect but also renegotiation-proof punishment strategies has been demonstrated by Cesar and de Zeeuw (1997).

Notes

1. For a regional externality, like the depletion of tuna stocks in the North Atlantic, all fishing nations with direct (coastal states) or indirect access (distant water fishing nations) access should be part of an agreement.
2. We suggest two guiding principles for possible extensions. On the one hand, in the light of environmental externalities, cooperation is a positive sum game, that is, the aggregate net benefit from cooperation is positive and is particularly large the stronger externalities across players are and the more players are involved in an externality problem. Cooperation can be beneficial to all players provided appropriate compensation measures (for example, monetary or in-kind transfers) are used. On the other hand, players pursue their self-interests, that is, players try to maximise their share of the global gain from cooperation, or, to put it differently, minimise their contribution to a cooperative arrangement. Ideally, this is captured through a non-cooperative negotiation and bargaining process, preceding an agreement. This does not preclude including social norms and preferences in an analysis, it only means that models should not assume in an ad hoc manner that negotiations are always conducted in a cooperative spirit or that countries' representatives ignore the opportunity to free-ride. Ideally, such behaviour should follow endogenously from the model itself.
3. It is this interpretation which allows one to relate the γ-core in the context of the characteristic function to the framework of the partition function. See more on this below.
4. The assumptions regarding those rules are discussed below in the context of the second stage of membership models.
5. An extensive comparison between internal and external stability and the core is provided in Finus (2001, 2003) and Tulkens (1998).
6. Applications to IEAs are found in Finus and Rundshagen (2006b) and Finus, Rundshagen and Eyckmans (2014). Ray and Vohra (2001) consider a general public good provision game.
7. A Nash equilibrium considers only single (one-shot) deviations whereas a strong Nash equilibrium allows for any group (one-shot) deviations. Group deviations are also considered in a coalition-proof Nash equilibrium, but (initial) deviations are only deemed feasible if either no further deviations by some of the original deviators occur or those further deviations are also beneficial to the original deviators. Hence, the idea of one-shot deviations is given up. A similar idea is captured by an equilibrium binding agreement, though the starting point is the grand coalition and deviations and subsequent deviations mean that partition can only become finer. That is, if a group or group of players splits up from the grand coalition into sub-coalitions, only players among a coalition can split up further. The most sophisticated concept is that of a farsighted equilibrium which does not impose any restrictions on deviations and considers all possible chain reactions after an initial

deviation. For practical purposes, this concept often turns out to be too complicated. Farsightedness has been applied in Diamantoudi and Sartzekatis (2002), Ecchia and Mariotti (1997), Eyckmans (2001), Osmani and Tol (2009) and *de Zeeuw (2008, Chapter 46)*.

8. Under open membership, players can freely join other coalitions or become singletons and this does not trigger any reaction by other players. Under exclusive membership, players cannot freely join other coalitions and if a player leaves a coalition to become a singleton, the coalition to which he belonged or even other coalitions may break apart, depending on the exact definition of exclusivity and consensus. See Finus and Rundshagen (2006a, 2009) for details.
9. Issue linkage in the context of compliance models is associated with quite different driving forces discussed below under compliance models.
10. A critical review of these papers and an agenda for future research is found in Finus (2003), section 3.3.5.
11. This is not only true in the context of environmental issues but also more generally. See, for instance, the surveys by Bloch (1997) and Yi (1997 and 2003).
12. The authors also assume that signatory countries punish non-signatory countries, at a cost borne by the signatories.
13. Farsightedness, as introduced by Chwe (1994), is a concept that allows for multiple chain reactions to each deviation, but assumes that there are no time preferences: 'players care only about the end outcome and not how many moves it takes to get there.' The concept of dynamic farsightedness in *de Zeeuw (2008, Chapter 46)* assumes that players care about the entire path and discount the future.
14. A detailed critical analysis is provided in Finus (2003).
15. An interesting analysis about the effect of different equity criteria for the success of coalition formation is conducted in Pol, Weikard and van Ierland (2012).
16. One other possibility to extend the standard payoff function is suggested by the concept of ancillary benefits, also called co- or secondary benefits. Investing in greenhouse gas (GHG) emission reduction does not only generate a public benefit, but also a private benefit. See Finus and Rübbelke (2013).
17. In the official website of the United Nations Framework Convention on Climate Change, one can find the description of the different groupings of countries during the negotiations. See http://unfccc.int/parties_and_observers/parties/negotiating_groups/items/2714.php. See also www.rainforestcoalition.org.
18. Nevertheless, Roelfsema (2007) obtains the opposite result, namely that a median voter that cares sufficiently for the environment has an incentive to delegate negotiations to an agent that cares even more for the environment, in a model where there is no bargaining stage. This highlights that the key for the results discussed in the main text is the bargaining stage.
19. This is different in *Finus and Rundshagen (1998, Chapter 18)* where this process is explicitly modelled as a bargaining process.

References

Aakre, S. (2013), 'The Nature, Design and Feasibility of Robust Climate Agreements', PhD Dissertation, University of Oslo.

Aakre, S. and J. Hovi (2010), 'Emission Trading: Participation Enforcement Determines the Need for Compliance Enforcement', *European Union Politics*, **11** (3), 427–45.

Altamirano-Cabrera, J.C. and M. Finus (2006), 'Permit Trading and Stability of International Climate Agreements', *Journal of Applied Economics*, **9** (1), 19–48.

Altamirano-Cabrera, J.C., M. Finus and R. Dellink (2008), 'Do Abatement Quotas Lead to More Successful Climate Coalitions?', *Manchester School*, **76** (1), 104–29.

Asheim, G.B. and B. Holtsmark (2009), 'Renegotiation-Proof Climate Agreements with Full Participation: Conditions for Pareto-Efficiency', *Environment and Resource Economics*, **43**, 519–33.

Barrett, S. (1990), 'The Problem of Global Environmental Protection', *Oxford Review of Economic Policy*, **6** (1), 68–79.
Barrett, S. (2001), 'International Cooperation for Sale', *European Economic Review*, **45** (10), 1835–50.
Barrett, S. (2003), *Environment and Statecraft: The Strategy of Environmental Treaty-Making*, New York: Oxford University Press.
Barrett, S. and R. Stavins (2003), 'Increasing Participation and Compliance in International Climate Change Agreements', *International Environmental Agreements: Politics, Law, and Economics*, **3** (4), 349–76.
Beccherle, J. and J. Tirole (2011), 'Regional Initiatives and the Cost of Delaying Binding Climate Change Agreements', *Journal of Public Economics*, **95**, 1339–48.
Bloch, F. (1996), 'Sequential Formation of Coalitions in Games with Externalities and Fixed Payoff Division', *Games and Economic Behavior*, **14** (1), 90–123.
Bloch, F. (1997), 'Non-Cooperative Models of Coalition Formation in Games with Spillovers', in C. Carraro and D. Siniscalco (eds), *New Directions in the Economic Theory of the Environment*, Cambridge, UK: Cambridge University Press, Chapter 10, 311–52.
Botteon, M. and C. Carraro (1997), 'Burden-Sharing and Coalition Stability in Environmental Negotiations with Asymmetric Countries', in C. Carraro (ed.), *International Environmental Negotiations: Strategic Policy Issues*, Cheltenham, UK and Northampton, MA: Edward Elgar, Chapter 3, 26–55.
Botteon, M. and C. Carraro (1998), 'Strategies for Environmental Negotiations: Issue Linkage with Heterogeneous Countries', in N. Hanley and H. Folmer (eds), *Game Theory and the Environment*, Cheltenham, UK and Northampton, MA: Edward Elgar, Chapter 9, 180–200.
Brandt, U.S. (2004), 'Unilateral Actions, the Case of International Environmental Problems', *Resource and Energy Economics*, **26**, 373–91.
Buchholz, W., A. Haupt and W. Peters (2005), 'International Environmental Agreements and Strategic Voting', *Scandinavian Journal of Economics*, **107** (1), 175–95.
Calvo, E. and S.J. Rubio (2012), 'Dynamic Models of International Environmental Agreements: A Differential Game Approach', *International Review of Environmental and Resource Economics*, **6**, 289–339.
Caparrós, A., J.-C. Péreau and T. Tazdaït (2003), 'Coalition et Accords Environnementaux Internationaux', *Revue Française de Economie*, **2** (18), 199–232.
Carraro, C. and C. Marchiori (2003), 'Stable Coalitions', in C. Carraro (ed.), *The Endogenous Formation of Economic Coalitions*, Cheltenham, UK and Northampton, MA: Edward Elgar, Chapter 5, 156–98.
Carraro, C. and C. Marchiori (2004), 'Endogenous Strategic Issue Linkage in International Negotiations', in C. Carraro and V. Fragnelli (eds), *Game Practice and the Environment*, Cheltenham, UK and Northampton, MA: Edward Elgar, Chapter 3, 65–86.
Carraro, C. and D. Siniscalco (1997), 'R&D Cooperation and the Stability of International Environmental Agreements', in C. Carraro (ed.), *International Environmental Negotiations: Strategic Policy Issues*, Cheltenham, UK and Northampton, MA: Edward Elgar, Chapter 5, 71–96.
Cesar, H. and A. de Zeeuw (1997), 'Issue Linkage in Global Environmental Problems', in A. Xepapadeas (ed.), *Economic Policy for the Environment and Natural Resources: Techniques for the Management and Control of Pollution*, Cheltenham, UK and Northampton, MA: Edward Elgar, Chapter 7, 158–73.
Chander, P. and H. Tulkens (1992), 'Theoretical Foundations of Negotiations and Cost Sharing in Transfrontier Pollution Problems', *European Economic Review*, **36** (2–3), 388–99.
Chander, P. and H. Tulkens (1995), 'A Core-Theoretic Solution for the Design of Cooperative Agreements on Transfrontier Pollution', *International Tax and Public Finance*, **2** (2), 279–93.
Chwe, Michael Suk-Young (1994), 'Farsighted Coalitional Stability', *Journal of Economic Theory*, **63** (2), 299–325.
Conconi, P. and C. Perroni (2002), 'Issue Linkage and Issue Tie-in in International Negotiations', *Journal of International Economics*, **57**, 423–47.
Courtois, P. and T. Tazdaït (2008), 'Accord Climatique: Concessions et Ratifications', *Revue Économique*, **59**, 719–35.
Dannenberg, A. (2011), 'Coalition Formation and Voting in Public Goods Games', *Strategic Behavior and the Environment*, **1**, 1–23.

Dannenberg, A., B. Sturm and C. Vogt (2010), 'Do Equity Preferences Matter for Climate Negotiators? An Experimental Investigation', *Environment and Resource Economics*, **47**, 91–109.
Dellink, R. and M. Finus (2012), 'Uncertainty and Climate Treaties: Does Ignorance Pay?', *Resource and Energy Economics*, **34** (4), 565–84.
Dellink, R., M. Finus and N. Olieman (2008), 'The Stability Likelihood of an International Climate Agreement', *Environmental and Resource Economics*, **39** (4), 357–77.
Diamantoudi, E. and E.S. Sartzetakis (2002), 'International Environmental Agreements – The Role of Foresight', CORE Discussion Paper No. 2002/61, Center for Operations Research and Econometrics, Université Catholique de Louvain, Louvain-la-Neuve, Belgium. www.core.ucl.ac.be/services/psfiles/dp02/dp2002-61.pdf.
Ecchia, G. and M. Mariotti (1997), 'The Stability of International Environmental Coalitions with Farsighted Countries: Some Theoretical Observations', in C. Carraro (ed.), *International Environmental Negotiations: Strategic Policy Issues*, Cheltenham, UK and Northampton, MA: Edward Elgar, Chapter 10, 172–92.
Eichner, T. and R. Pethig (2013), 'Self-Enforcing Environmental Agreements and International Trade', *Journal of Public Economics*, **102**, 37–50.
Endres, A. and M. Finus (1998), 'Renegotiation-Proof Equilibria in a Bargaining Game over Global Emission Reductions – Does the Instrumental Framework Matter?', in N. Hanley and H. Folmer (eds), *Game Theory and the Global Environment*, Cheltenham, UK and Northampton, MA: Edward Elgar, Chapter 7, 135–64.
Endres, A. and C. Ohl (2003), 'International Environmental Cooperation with Risk Aversion', *International Journal of Sustainable Development*, **6** (3), 378–92.
Espínola-Arredondo, A. and F. Muñoz-García (2011), 'Free-Riding in International Environmental Agreements: A Signaling Approach to Non-Enforceable Treaties', *Journal of Theoretical Politics*, **23** (1), 111–34.
Eyckmans, J. (2001), 'On the Farsighted Stability of the Kyoto Protocol', Working Paper Series, Faculty of Economics and Applied Economic Sciences, University of Leuven, No. 2001–03.
Eyckmans, J. and M. Finus (2004), 'An Almost Ideal Sharing Scheme for Coalition Games with Externalities', Working Paper No. 155.2004, Fondazione Eni Enrico Mattei, Italy.
Eyckmans, J., M. Finus and L. Mallozzi (2012), 'A New Class of Welfare Maximizing Sharing Rules for Partition Function Games with Externalities', Bath Economics Research Paper 6–2012.
Eyckmans, J. and H. Tulkens (2003), 'Simulating Coalitionally Stable Burden Sharing Agreements for the Climate Change Problem', *Resource and Energy Economics*, **25** (4), 299–327.
Fehr, E. and K.M. Schmidt (1999), 'A Theory of Fairness, Competition, and Cooperation', *Quarterly Journal of Economics*, **114** (3), 817–68.
Finus, M. (2001), *Game Theory and International Environmental Cooperation*, Cheltenham, UK and Northampton, MA: Edward Elgar.
Finus, M. (2002), 'Game Theory and International Environmental Cooperation: Any Practical Application?', in C. Böhringer, M. Finus and C. Vogt (eds), *Controlling Global Warming: Perspectives from Economics, Game Theory and Public Choice*, Cheltenham, UK and Northampton, MA: Edward Elgar, Chapter 2, 9–104.
Finus, M. (2003), 'Stability and Design of International Environmental Agreements: The Case of Global and Transboundary Pollution', in H. Folmer and T. Tietenberg (eds), *International Yearbook of Environmental and Resource Economics 2003/4*, Cheltenham, UK and Northampton, MA: Edward Elgar, Chapter 3, 82–158.
Finus, M. (2008), 'Game Theoretic Research on the Design of International Environmental Agreements: Insights, Critical Remarks and Future Challenges', *International Review of Environmental and Resource Economics*, **2** (1), 29–67.
Finus, M., J.C. Altamirano-Cabrera and E.C. van Ierland (2005), 'The Effect of Membership Rules and Voting Schemes on the Success of International Climate Agreements', *Public Choice*, **125** (1), 95–127.
Finus, M. and S. Maus (2008), 'Modesty May Pay!', *Journal of Public Economic Theory*, **10** (5), 801–26.
Finus, M. and P. Pintassilgo (2012), 'International Environmental Agreements under Uncertainty: Does

the Veil of Uncertainty Help?', *Oxford Economic Papers*, **64**, 736–64.

Finus, M., P. Pintassilgo and A. Ulph (2014), 'International Environmental Agreements with Uncertainty, Learning and Risk Aversion', Cesifo Working Paper No. 4589 Category 9: Resource and Environment Economics.

Finus, M. and D.T.G. Rübbelke (2013), 'Public Good Provision and Ancillary Benefits: The Case of Climate Agreements', *Environmental and Resource Economics*, **56**, 211–22.

Finus, M. and B. Rundshagen (2003), 'Endogenous Coalition Formation in Global Pollution Control: A Partition Function Approach', in C. Carraro (ed.), *The Endogenous Formation of Economic Coalitions*, Cheltenham, UK and Northampton, MA, USA: Edward Elgar, Chapter 6, 199–243.

Finus, M. and B. Rundshagen (2006a), 'A Micro Foundation of Core Stability in Positive-Externality Coalition Games', *Journal of Institutional and Theoretical Economics*, **162** (2), 329–46.

Finus, M. and B. Rundshagen (2006b), 'Participation in International Environmental Agreements: The Role of Timing and Regulation', *Natural Resource Modeling*, **19** (2), 165–200.

Finus, M. and B. Rundshagen (2009), 'Membership Rules and Stability of Coalition Structures in Positive Externality Games', *Social Choice and Welfare*, **32**, 389–406.

Finus, M., B. Rundshagen and J. Eyckmans (2014), 'Simulating a Sequential Coalition Formation Process for the Climate Change Problem: First Come, but Second Served?', *Annals of Operations Research*, **220** (1), 5–23.

Finus, M., M.E. Sáiz and E.M.T. Hendrix (2009), 'An Empirical Test of New Developments in Coalition Theory for the Design of International Environmental Agreements', *Environment and Development Economics*, **14** (1), 117–37.

Folmer, H. and P. van Mouche (1994), 'Interconnected Games and International Environmental Problems II', *Annals of Operations Research*, **54**, 97–117.

Folmer, H. and P. van Mouche (2002), 'The Acid Rain Game: A Formal and Mathematically Rigorous Analysis', in B. Kriström, P. Dasgupta and K.-G. Löfgren (eds), *Economic Theory for the Environment: Essays in Honour of Karl-Göran Mäler*, Cheltenham, UK and Northampton, MA: Edward Elgar, Chapter 9, 138–61.

Froyn, C.B. and J. Hovi (2008), 'A Climate Agreement with Full Participation', *Economics Letters*, **99**, 317–19.

Fuentes-Albero, C. and S.J. Rubio (2010), 'Can International Environmental Cooperation Be Bought?', *European Journal of Operational Research*, **202**, 255–64.

Gerber, A. and P.C. Wichardt (2013), 'On the Private Provision of Intertemporal Public Goods with Stock Effects', *Environmental and Resource Economics*, **55** (2), 245–55.

Gruning, C. and W. Peters (2010), 'Can Justice and Fairness Enlarge International Environmental Agreements?', *Games*, **1**, 137–58.

Hampson, O. and M. Hart (1995), *Multilateral Negotiations: Lessons from Arms Control, Trade, and the Environment*, Baltimore and London: The Johns Hopkins University Press.

Harstad, B. (2008), 'Do Side Payments Help? Collective Decisions and Strategic Delegation', *Journal of the European Economic Association*, **6** (2–3), 468–77.

Helm, C. and F. Wirl (2014), 'The Principal–Agent Model with Multilateral Externalities: An Application to Climate Agreements', *Journal of Environmental Economics and Management*, DOI: 10.1016/j.jeem.2013.11.006.

Hoel, M. (1991), 'Global Environmental Problems: The Effects of Unilateral Actions Taken by One Country', *Journal of Environmental Economics and Management*, **20**, 55–70.

Hoel, M. and K. Schneider (1997), 'Incentives to Participate in an International Environmental Agreement', *Environmental and Resource Economics*, **9** (2), 153–70.

Hovi, J., M. Greaker, C. Hagem and B. Holtsmark (2012), 'A Credible Compliance Enforcement System for the Climate Regime', *Climate Policy*, **12** (6), 741–54.

Jakob, M. and K. Lessmann (2012), 'Signaling in International Environmental Agreements: The Case of Early and Delayed Action', *International Environmental Agreements: Politics, Law and Economics*, **12** (4), 309–25.

Jeppesen, T. and P. Andersen (1998), 'Commitment and Fairness in Environmental Games', in N. Hanley and H. Folmer (eds), *Game Theory and the Environment*, Cheltenham, UK and Northampton, MA: Edward Elgar, Chapter 4, 65–83.

Jørgensen, S., G. Martín-Herrán and G. Zaccour (2010), 'Dynamic Games in the Economics and Management of Pollution', *Environmental Modeling and Assessment*, **15** (6), 433–67.
Kaitala, V. and M. Lindroos (1998), 'Sharing the Benefits of Cooperation in High Seas Fisheries: A Characteristic Function Game Approach', *Natural Resource Modeling*, **11** (4), 275–99.
Kaitala, V., K.-G. Mäler and H. Tulkens (1995), 'The Acid Rain Game as a Resource Allocation Process with an Application to the International Cooperation among Finland, Russia and Estonia', *Scandinavian Journal of Economics*, **97** (2), 325–43.
Kallbekken, S. and J. Hovi (2007), 'The Price of Non-Compliance with the Kyoto Protocol: The Remarkable Case of Norway', *International Environmental Agreements: Politics, Law and Economics*, **7** (1), 1–15.
Katsoulacos, Y. (1997), 'R&D Spillovers, Cooperation, Subsidies and International Agreements', in C. Carraro (ed.), *International Environmental Negotiations: Strategic Policy Issues*, Cheltenham, UK and Northampton, MA: Edward Elgar, Chapter 6, 97–109.
Kolstad, C.D. (2007), 'Systematic Uncertainty in Self-Enforcing International Environmental Agreements', *Journal of Environmental Economics and Management*, **53** (1), 68–79.
Kolstad, C.D. (2010), 'Equity, Heterogeneity and International Environmental Agreements', *B.E. Journal of Economic Analysis and Policy*, **10** (2), Article 3.
Kolstad, C.D. and A. Ulph (2008), 'Learning and International Environmental Agreements', *Climatic Change*, **89** (1–2), 125–41.
Kolstad, C.D. and A. Ulph (2011), 'Uncertainty, Learning and Heterogeneity in International Environmental Agreements', *Environmental and Resource Economics*, **50**, 389–403.
Konrad, K.A. and M. Thum (2014), 'Climate Policy Negotiations with Incomplete Information', *Economica*, DOI: 10.1111/ecca.12065.
Kroeze-Gil, J. and H. Folmer (1998), 'Linking Environmental and Non-Environmental Problems in an International Setting: The Interconnected Games Approach', in N. Hanley and H. Folmer (eds), *Game Theory and the Environment*, Cheltenham, UK and Northampton, MA: Edward Elgar, Chapter 8, 165–80.
Kwon, O. (2006), 'Partial International Coordination in the Great Fish War', *Environmental and Resource Economics*, **33** (4), 463–83.
Lange, A. (2006), 'The Impact of Equity-Preferences on the Stability of International Environmental Agreements', *Environmental and Resource Economics*, **34** (2), 247–67.
Mäler, K.-G. (1990), 'International Environmental Problems', *Oxford Review of Economic Policy*, **6** (1), 80–108.
Mäler, K.-G. (1994), 'Acid Rain in Europe: A Dynamic Perspective on the Use of Economic Incentives', in E.C. van Ierland (ed.), *Developments in Environmental Economics, Vol. 4: International Environmental Economics – Theories, Models and Applications to Climate Change, International Trade and Acidification*, North-Holland, Amsterdam: Elsevier, Chapter 15, 351–72.
McEvoy, D.M., J.J. Murphy, J.M. Spraggon and J.K. Stranlund (2010), 'The Problem of Maintaining Compliance within Stable Coalitions: Experimental Evidence', *Oxford Economic Papers*, **63**, 475–98.
McEvoy, D.M. and J.K. Stranlund (2010), 'Costly Enforcement of Voluntary Environmental Agreements', *Environment and Resource Economics*, **47**, 45–63.
McGinty, M. (2011), 'A Risk-Dominant Allocation: Maximizing Coalition Stability', *Journal of Public Economic Theory*, **13** (2), 311–25.
Milinski, M., T. Rohl and J. Marotzke (2011), 'Cooperative Interaction of Rich and Poor Can Be Catalyzed by Intermediate Climate Targets', *Climatic Change*, **109**, 807–14.
Milinski, M., R.D. Sommerfeld, H.-J. Krambeck, F.A. Reed and J. Marotzke (2008), 'The Collective-Risk Social Dilemma and the Prevention of Simulated Dangerous Climate Change', *Proceedings of the National Academy of Sciences of the USA*, **105** (7), 2291–4.
Mohr, E. (1988), 'On the Incredibility of Perfect Threats in Repeated Games: Note', *International Economic Review*, **29** (3), 551–5.
Nagashima, M., H.P. Weikard, K. de Bruin and R. Dellink (2011), 'International Climate Agreements under Induced Technological Change', *Metroeconomica*, **62** (4), 612–34.
Osmani, D. and R. Tol (2009), 'Toward Farsightedly Stable International Environmental Agreements',

Journal of Public Economic Theory, **11** (3), 455–92.
Pethig, R. (ed.) (1992), *Conflicts and Cooperation in Managing Environmental Resources, Microeconomic Studies*, Berlin: Springer.
Pintassilgo, P. (2003), 'A Coalition Approach to the Management of High Seas Fisheries in the Presence of Externalities', *Natural Resource Modeling*, **16** (2), 175–97.
Pintassilgo, P., M. Finus, M. Lindroos and G. Munro (2010), 'Stability and Success of Regional Fisheries Management Organizations', *Environmental and Resource Economics*, **46**, 377–402.
Pintassilgo, P. and M. Lindroos (2008), 'Coalition Formation in Straddling Stock Fisheries: A Partition Function Approach', *International Game Theory Review*, **10** (3), 303–17.
Pol, T.D. van der, H.P. Weikard and E.C. van Ierland (2012), 'Can Altruism Stabilize International Environmental Agreements?', *Ecological Economics*, **81**, 112–20.
Ray, D. and R. Vohra (2001), 'Coalitional Power and Public Goods', *Journal of Political Economy*, **109** (6), 1355–84.
Roelfsema, H. (2007), 'Strategic Delegation of Environmental Policy Making', *Journal of Environmental Economics and Management*, **53**, 270–75.
Rotillon, G. and Tazdaït, T. (1996), 'International Bargaining in the Presence of Global Environmental Change', *Environmental and Resource Economics*, **8**, 293–314.
Rubio, S.J. and B. Casino (2005), 'Self-Enforcing International Environmental Agreements with a Stock Pollutant', *Spanish Economic Review*, **7** (2), 89–109.
Segendorff, B. (1998), 'Delegation and Threat in Bargaining', *Games and Economic Behavior*, **23**, 266–83.
Sumaila, U.R. (1999), 'A Review of Game-Theoretic Models of Fishing', *Marine Policy*, **23** (1), 1–10.
Tulkens, H. (1998), 'Cooperation versus Free-Riding in International Environmental Affairs: Two Approaches', in N. Hanley and H. Folmer (eds), *Game Theory and the Environment*, Cheltenham, UK and Northampton, MA: Edward Elgar, Chapter 2, 30–44.
Ulph, A. (1998), 'Learning about Global Warming?', in N. Hanley and H. Folmer (eds), *Game Theory and the Environment*, Cheltenham, UK and Northampton, MA: Edward Elgar, Chapter 13, 255–86.
Ulph, A. (2004), 'Stable International Environmental Agreements with a Stock Pollutant, Uncertainty and Learning', *Journal of Risk and Uncertainty*, **29** (1), 53–73.
Urpelainen, J. (2012), 'Technology Investment, Bargaining, and International Environmental Agreements', *International Environmental Agreements*, **12**, 145–63.
Wagner, U.J. (2001), 'The Design of Stable International Environmental Agreements: Economic Theory and Political Economy', *Journal of Economic Surveys*, **15** (3), 377–411.
Weikard, H.-P., M. Finus and J.C. Altamirano-Cabrera (2006), 'The Impact of Surplus Sharing on the Stability of International Climate Agreements', *Oxford Economic Papers*, **58** (2), 209–32.
Yi, S.-S. (1997), 'Stable Coalition Structures with Externalities', *Games and Economic Behavior*, **20** (2), 201–37.
Yi, S.-S. (2003), 'Endogenous Formation of Economic Coalitions: A Survey of the Partition Function Approach', in C. Carraro (ed.), *The Endogenous Formation of Economic Coalitions*, Cheltenham, UK and Northampton, MA: Edward Elgar, Chapter 3, 80–127.

Part I
Foundations

[1]

Chapter 12

THE ACID RAIN GAME*

KARL GÖRAN MÄLER
Stockholm School of Economics, Box 6501, 113 83 Stockholm (Sweden)

1 INTRODUCTION

During the last couple of years, the problems connected with emissions of sulfur- and nitrogen oxides and the resulting damages on eco-systems have come more and more in the focus. The death of forests in central Europe and Scandinavia have been quite important news items and rightly so, because if the European eco-systems will change so drastically there will be a substantial reduction in timber supply, that will effect not only nature conservationists but also everyday man because of the economic consequences in a conventional sense. Thus, the problems of acid rains seem to be a very important issue for economists to analyse. Moreover, it offers a fascinating multitude of intellectual challenges, one being that the information on causes and effects is very uncertain, another being that it concerns the use of a common property resource in a very asymmetric way, a third being that it is about a game with incomplete information and with many players so that problems with incompatible incentives will be at the heart, and finally, the parties involved are different nations with no agreed rules of the game. In addition to these challenges, there are of course a multitude of equally challenging empirical problems.

The objective of this paper is to give a skeleton of an analytical model in which some of the above problems, and in particular the problems of international cooperation, will be analysed. However, in view of the complexities of the interactions between national

* This chapter is an outgrow of a research project on the economics of acid rains at the Stockholm School of Economics, financed by the Swedish Environment Protection Board. I am grateful to Lars Bergman for many rewarding discussions on the topics covered in this chapter and to Clas Olsson for excellent research assistance. I would also like to express my gratitude to Partha Dasgupta for teaching me game theory and in particular the concept of a strong equilibrium. Finally I would like to thank two anonymous referees as well as Henk Folmer and Ekko van Ierland for many valuable suggestions which have improved the presentation very much. They are, of course, absolved from remaining errors and mistakes.

emissions and the resulting environmental damage, we will make a few, rather serious simplifications. Our basic model will concentrate on national sulfur emissions in different countries in Europe. Thus, we will not look into the very relevant issue of spatial distribution of the emissions in each country. Nor will we look into the spatial distribution of damages within each country. Finally, we will neglect the effects of nitrogen oxide emissions and the formation of ozone, although it seems clear that these factors are perhaps as much to be blamed as sulfur emissions for the observed damages to forests. Moreover, it should be stressed that this paper is one in a series of different papers, analysing different aspects of international cooperation in the environmental field.

Thus we will discuss the following scenario. Each European country is emitting sulfur oxides, the amount of which depends on energy consumption, sulfur content in the burnt fuels, the combustion technology and stackgas cleaning technology. By switching to fuels with lower sulfur content, by changing combustion technology, by reducing energy consumption etc. it is possible to reduce the emissions of sulfur, but at a cost. Thus we will postulate a cost function for reducing the emissions. The expected control cost function for country i will be denoted $C_i(E_i)$ where E_i denotes the emission in country i. This function is, of course, decreasing in E_i, as the cost of abatement increases with abatement. We will also assume C_i to be strictly convex (although in some simulation it will be linear or piecewise linear). Moreover, we will assume it to be sufficiently smooth.

The emissions will have local effects due to the resulting ambient concentrations of sulfur oxides. These local effects are health effects, corrosion to materials, damage to vegetation etc. We will for the discussion in this paper assume that we have a monetary damage function for these local effects so that we can form the net control cost function by subtracting the local damage from the control cost function.[1] We will in the sequel reinterpret C_i as the expected net control cost function. The cost functions actually used in the simulations presented in this paper are, however, gross, i.e. do not include local damage costs.

The sulfur emitted will be transported by winds in the atmosphere and will also be transformed by chemical processes from sulfur oxides to sulphates. Ultimately, the

[1] This construction is not completely correct as the local damage from increased ambient concentrations of sulfur oxides is not directly related to the emissions but also to the height of the stack, the location of the stack etc. We will, however, neglect those factors here.

sulphates will be removed from the atmosphere by direct "dry" deposition or by rains-"wet" deposition. It turns out that there exists at least one meteorological model that describes this transportation and transformation and is simple enough to be used for an economic analysis of the acid rain problem and is accepted by most European countries as giving a fair description of the actual processes, namely the EMEP model. The EMEP model is based on a grid by which Europe is divided into about 700 squares. The model assumes sulfur emissions in these squares and by using observations on actual winds etc. the air package above a square is followed as it moves from one square to another. The model predicts the chemical changes within the airpackage and the removal of sulfur from the package in the form of deposition. With steady state climatic conditions, the model reduces to a transfer matrix with a dimension equal to about 700 times 700. An element in the matrix gives the amount of sulfur deposition in one square, following the emission of one ton in another square. However, for the purpose of this study, the matrix is aggregated into a country times country matrix, i.e. with a dimension 28 times 28, so that a typical element in the matrix describes the deposition in one country that is due to the emission of one ton in another country.[2] However, Iceland has been excluded, so the actual matrix is of order 27 times 27. Finally, as the contribution to the acid rain problem from Luxemburg is negligible, Luxemburg has been deleted from all tables containing results from the calculations.

The transfer is denoted by the matrix A. If E is the vector of emission levels in the European countries and if Q is the vector of sulfur deposition in the countries, the steady state model for the transport of sulfur is simply $Q = AE$.

The deposition of sulfur gives rise to various environmental damages. The first environmental problems from acid rains to be observed were acidification of surface water. It was noted, mainly in Scandinavia, that the Ph-level of lakes and streams was falling, and in some cases to such low levels that further life of traditional species was inhibited.[3] It soon became quite clear that even the ground water was affected and that in the long run, the growth of the forests would be reduced. These effects are now being observed both in Scandinavia and central Europe.

---

2 See Binmore and Dasgupta (1986).

3 One of the early reports on environmental damage from acid rains is to be found in the Swedish case study to the U.N. conference on human environment in Stockholm 1972.

However, the experts on forest ecology are not in agreement on the factors behind the damages.[4] There seems to be three explanations that have been put forward. According to the first, the forest damages are due to acidification of and the resulting chemical changes in the soil. The acidification of the soil is then explained by the deposition of sulfur from the atmosphere. Another explanation gives the blame to increased ambient concentrations of nitrogen and sulfur oxides. These concentrations are very high close to the points of emissions but fall very rapidly with the distance. One of the main sources of nitrogen oxides is the automobile, and it is possible in many countries to watch dying trees close to motorways. If this would be the only explanation of forest damages, then the international aspects of the problem would be negligible and it would be almost completely a domestic issue. (Recently, research results have, however, indicated that NO_x may also be transported by winds for long distances.)

Finally, there is the "stress theory", which in a way integrates the two previous ones, namely that the defense to extra stress the trees can provide is reduced by pollution. Thus the explanation would consist of many different factors. It should be clear from above that there is no simple answer to the question: what are the environmental consequences from acid rains? In fact, the uncertainty is so great that one could imagine that acid rains have nothing to do with the death of forest and also that they are the main and only culprit. (It should be added that the damage from acid rains to surface and ground water is much more firmly established.)

Because of this uncertainty, we should then talk about an expected damage function of deposition of sulfur. We, therefore, posit the existence of an expected damage function for each country which relates the deposition of sulfur to a monetary measure of damage. Assume that the deposition in country i is Q_i. Then we will denote the expected damage cost function (or simply damage function) by $D_i(Q_i)$. For each country we thus have an expected net control cost function and an expected damage cost function. We will assume that these functions as well as the EMEP matrix that relates the emissions from one country to deposition in all other countries, are known by all countries.

The net control cost functions, the EMEP matrix and the damage functions define a non-cooperative game, in which the strategy for a country consists of choosing an emission level and the payoff is equal to minus the sum of the net control cost and the

[4] For a summary of different hypotheses explaining the death of forests, see Hinrichson (1986).

damage cost (except for an uninteresting constant). This game is quite similar to the ones analysed for common property resources, except that those games generally deal with a symmetric situation while this acid rains game deals with a highly asymmetric case. The reason this 'acid rain game' is asymmetric is that the matrix A is non-symmetric because of the prevalent wind directions. Emissions in some countries are more harmful than in others simply because of their locations. In section 2, different non-cooperative solutions to this game are defined and analysed and compared with the full cooperative solution (i.e. the solution that would emerge if all countries would cooperate and maximize their joint payoff, assuming that utility is transferable). In the following sections different cooperative equilibrium concepts are defined, discussed and illustrated by simulations.

2 NON-COOPERATIVE EQUILIBRIA AND THE FULL COOPERATIVE SOLUTION

The purpose of the simulations to be presented in the next section is to get a feeling for the gains from cooperation the European countries could expect. Ideally, one would therefore be interested in the core and related equilibrium concepts of the acid rain game. In section 6 these concepts are defined and discussed in more detail. In this section the interest is focused on the "full cooperative solution". This was defined in the previous section as simply that vector E of emission levels that minimizes the expected total cost, i.e.

$$\min \Sigma_i \left(C_i(E_i) + D_i(Q_i)\right)$$

$$\text{subject to } Q = AE.$$

This solution concept implicitly requires transferable utility, i.e. that gains in one country can be transferred to other countries in order to achieve another distribution of gains and losses. As both cost and damage functions are measured in monetary units we thus assume that utility is linear in income and transferable between countries (which incidentally requires that the current exchange rates are equilibrium rates).

The reason the solution is defined as the minimum of expected cost (expected instead of simply cost) is that in general both damage- and control costs are uncertain. However, we shall in the sequel disregard this kind of uncertainty. Thus, the full cooperative solution could as well be defined without the expectation operator.[5] There is another kind

[5] The expectation operator has been included because in some companion papers the role of uncertainty is the focus of analysis.

of uncertainty which we must look into, however. This has to do with the information one country may have about costs and damage in other countries. In general, the control costs and environmental damage in one particular country is known only to that country (under the present assumptions). The lack of information is, in fact, so great that total emissions of sulfur in one country are unknown to other countries and the total deposition of sulfur in one country can only be estimated with the aid of the EMEP-model given the assumptions about the pattern of emissions. In this report, this aspect will be neglected. Instead, in a subsequent paper, this aspect will be the main problem to be discussed.

The full cooperative solution is a special case of the "Pareto-efficient" outcome. A vector $(E,Q) = \{E_1, E_2, \ldots, E_n, Q_1, \ldots, Q_n\}$ is said to be Pareto efficient if there does not exist another feasible vector (E', Q') such that the total cost (control- and damage cost) for each country at these alternative emission and deposition levels is less than or equal to the total cost at the (E, Q) level. If the feasible set is convex (which it is with our assumptions on the control- and damage cost functions and the linearity of the transport models), then all Pareto-efficient outcomes can be characterized as vectors (E, Q) that minimize

$$\Sigma_i\ \alpha_i\{C_i(E_i) + D_i(Q_i)\}$$

for a certain selection of $\alpha_1, \ldots, \alpha_n$, with $\alpha_i > 0$, $i = 1, \ldots, n$.

If (E', Q') is any allocation of emissions and depositions that is feasible (i.e. Q' = AE'), then P(E', Q') denotes the set of feasible allocations of emissions and depositions that Pareto dominates (E', Q'), i.e.

$$P(E', Q') = \{(E, Q);\ C_i(E_i) + D_i(Q_i) \le C_i(E_i) + D_i(Q_i),\ Q = AE\}$$

A subset of the Pareto efficient allocations is thus a subset of the boundary of P(E', Q'). Note that for a given (E', Q'), the full cooperative solution need not be a member of P(E', Q'). If that is not the case when (E', Q') is the initial situation, then the full cooperative solution will not be obtained unless there are sidepayments among the countries.

We can define (at least) two different non-cooperative equilibrium concepts.

i) Dominant equilibrium (E^*_k, Q^*_k), $Q^*_k = A\ E^*_k$, $k = 1,\ldots n$ is a dominant equilibrium if for all E, Q such that Q = AE it is true that for all $k = 1,\ldots n$

$$C_k(E^*_k) + D_k(Q^*_k) < C_k(E_k) + D_k(Q_k)$$

Thus, a dominant equilibrium is characterized by a set of strategies E^*_k such that irrespective of what other countries do, it would not be beneficial for country k to change its strategy.

It is very easy to see that in general no dominant equilibrium will exist and we will give a proof in connection with the discussion of Nash equilibrium. However, under one special condition which is the basis for the simulations discussed later, a dominant equilibrium will exist. This condition is that the expected damage function is linear so that the marginal damage is constant. If that is the case the cost minimization emission in a country is determined by the condition that marginal abatement cost equals the marginal damage times the proportion of the emission that will be deposited in the own country. As the marginal damage is independent of the emissions in other countries, it follows that the cost minimizing emission is independent of the emissions in other countries and thus constitutes a dominant strategy.

ii) Nash equilibrium. A Nash equilibrium is defined as a pair (E^*, Q^*) such that

$$C_k(E_k^*) + D_k(Q_k^*) \leq C_k(E_k') + D_k(Q_k'),\ k = 1,\ldots n$$

where Q' is defined by

$$Q' = AE'$$

$$E_l' = E_l^* \qquad l = 1, \ldots, n, l \neq k$$

Thus a Nash equilibrium is characterized by the condition that if all other countries are emitting their Nash equilibrium quantities, then it is optimal for the remaining country to do that too.

The Nash equilibrium concept has a very solid base in economic theory and the intuitive reasons behind the concept should be clear.[6] It is based on the idea that the countries are rational. That means that they, if they have the necessary information, can calculate the optimal behavior of all other countries (including the other's optimal response to the amount emitted in the own country) and therefore the own optimal response. This rationality is based on one crucial assumption, however, namely that each country has complete information on the others' emissions, control costs, depositions and damage costs. As was noted above, such is not the case and the Nash equilibrium concept (at least as it has been defined here) cannot be rigorously defended. However, in spite of this we will for the rest of this section disregard this criticism and continue as if the game under consideration is a game with complete certainty.

---

6 See Binmore and Dasgupta (1986).

In order to be able to prove the existence of a Nash equilibrium we will make the following assumptions:

a) The control cost- and damage functions are twice continuously differentiable;

b) $D'_k(Q_k)$ (here and in the sequel the symbol ' attached to a function denotes the derivative) goes to infinity with Q_k (that is, we are looking at the other extreme compared with the assumption of constant marginal damage made above);

c) C'_k goes to zero with E_k;

d) all elements a_{kk} of matrix A are different from zero for all countries k.

The best reply of country k given that the other countries emit E_l, $l = 1, \ldots, n$, $l \neq k$, is given by the function $\phi(E)$ defined as the emission that minimizes the total cost in country k given the emissions in the other countries.

It is easily seen that for each $k = 1, \ldots, n$ there exist $\overset{o}{E}_k$ such that there exists an E' within the cube defined by the origin and $\overset{o}{E}_k$, $k = 1, \ldots, n$ in n + 1 space, such that $C_k(E'_k) + D_k(AE')$ is less than the corresponding costs for points outside the cube. Thus we can restrict ourselves to this compact set when searching for a Nash equilibrium. But then the feasibility sets of all players are compact and the payoff functions are continuous and there exists a Nash equilibrium.

It is easily seen that under the assumptions made, the best reply functions are implicitly defined by

$$dC_i/dE_i + a_{ii}\, dD_i/dQ_i = 0, \; i = 1, 2, \ldots, n.$$

The Jacobian of this system of equations is given by

$$\begin{bmatrix} C''_1(E_1) + a^2_{11}\, D''(Q_1) & \ldots a_{11}a_{1n}D''(Q_1) \\ \ldots & \\ a_{nn}\, a_{n1}\, D''(Q_n) & \ldots C''_n(E_n) + a^2_{nn}\, D''(Q_n) \end{bmatrix}$$

where the symbol " denotes the second derivative.

The sum of the off-diagonal elements in row i is equal to

$$\sum_{j \neq i} a_{ii}a_{ij}\, D''(Q_i)$$

and the diagonal element is

$$C_i''(E_i) + a^2_{ii}D''(Q_i)$$

The Jacobian has therefore a dominant diagonal if

$$C_i'' + a_{ii}D'' (a_{ii}D'' (a_{ij}) > 0.$$
$$J \neq i$$

Scrutiny of the EMEP model reveals that the term within parenthesis is positive for most European countries. In the cases where this term within parenthesis is less than zero, it has a rather small magnitude (in the order of .1). Multiplied with a_{ii} the second term will be small and it seems reasonable to assume that the Jacobian has a dominant diagonal. This means, however, that the Nash equilibrium is unique according to well-known theorems.[7] Thus, it seems reasonable to assume that there exists a unique Nash equilibrium in the European acid rain game as it has been formulated in this section.

What are the differences between the full cooperative solution and the Nash solution? This question is interesting from the following points of view. A natural hypothesis would be that the present situation can be characterized as a Nash equilibrium. A comparison between the Nash equilibrium and the full cooperative solution would then indicate first the gains from cooperation but second, and more importantly, the allocation of the necessary reductions of sulfur oxide emissions among different countries. However, the assumption that the present situation can be characterized as a Nash equilibrium may appear to be far removed from reality. In fact, there are a number of international agreements on reductions of sulfur emissions in Europe and it is questionable whether these agreements can be interpreted as part of a Nash equilibrium. However, so far no transnational payments have been involved in these agreements, which could be interpreted by saying that the present situation must belong to the set $P(E^n,Q^n)$ of allocations that dominate the Nash equilibrium (E^n,Q^n). If it can be shown that the full cooperative solution does not belong to the set $P(E^n,Q^n)$, then one can conclude that some kind of side payments are called for when the full cooperative solution between the European countries is looked for.

3 SIMULATION RESULTS

In order to enable numerical simulations of the Nash equilibrium, the set of allocations that dominates that equilibrium and the full cooperative solution, the following strategy has been followed:

a) Control cost functions were guesstimated on the basis of some plots produced by the Acid Rains Project at IIASA, Laxenburg.[8] These cost functions were taken to be quadratic

---

[7] See for example James Friedman (1986), Chapter 5.3, Theorem 2.6.

[8] Amann, M., and G. Kornai (1987).

(with the exception of German Democratic Republic). The IIASA cost functions have the drawback that they do not include fuel substitution or switch to fuels with lower sulfur content and that they assume exogenous given energy demands. In particular, the cost functions are estimated on the basis of the expected energy demands for the year 2000. In spite of this, the functions have been applied in this paper to the energy consumption pattern 1984. Moreover, based on the information from IIASA, maximum amounts of pollution control have been assumed for the different countries. This grossly overstates the cost of control of sulfur emissions.

b) The damage cost functions were assumed to be linear, so that the marginal damage cost is constant, independent of the amount of deposition. If the initial situation is a Nash equilibrium then the absolute value of the marginal damage cost times the appropriate diagonal element in the EMEP matrix must be equal to the marginal control cost i.e.

$$-dC_i/dE_i = a_{ii}dD_i/dQ_i.$$

By using this necessary condition for a Nash equilibrium, the damage cost function can be calibrated such that the marginal damage cost is equal to the marginal control cost divided by a_{ii}. In particular, this means that the damage cost function represents the evaluation of the damage that the respective governments make today. It thus corresponds to what is usually called revealed preferences. It is, however, important to understand that this does not necessarily mean that the damage cost function estimated by conventional methods would be equal to the assumed damage cost function. The approach taken here is simply to assume that the damage evaluation revealed by actual policy decisions is used for the simulations.

In the discussion on acid rains in Europe, it is often claimed that some countries are using too low estimates of the damage. Therefore, in some simulations it has been assumed that the marginal damages in GDR and Czechoslovakia are 50 percent higher than what the corresponding control cost would generate, and for Poland, the damage cost has been assumed to be 100 percent higher. However, these adjustments turned out to be of minor importance, and these simulations are not reproduced in this paper.

It is obvious that changed information will change the perceived damage and therefore the chosen strategy also. Our numerical illustrations therefore only represent the current perception of damage. As the following will show, this does not really matter so long as the perceptions change uniformly over countries.

c) This calibration thus yields a damage cost function for each country that can be used to calculate the set of Pareto efficient outcomes, the full cooperative solution and other solution concepts such as the core. However, if the true damage cost function is convex instead of linear, which seems probable, then this calibration will yield an underestimate of the true damage cost function. In particular, one may end up with overestimates of the gains from cooperation, as the benefits from reductions in sulfur deposition will be overstated. It is therefore very important to bear this bias in mind when the following results are interpreted. Moreover, although this assumption of constant marginal damage enables explicit calculations, it also removes one important and interesting connection between the European countries. With constant marginal damage, the best strategy choice in one country is independent of what other countries do, and the Nash equilibrium turns out to be a dominant equilibrium.

d) The way the game is set up implies that the utility can be measured in monetary terms and that the cost (damage- and control cost) figures have the same meaning for all countries involved. This requires that the exchange rate is an equilibrium one and that the cost of capital (freely transferable across national borders) is the same in all countries. We know for sure that this assumption is not correct, partly because some of the European countries experience different economic systems than others.

e) All simulations have been carried out with GAMS – General Algebraic Modelling System, a software developed at the World Bank.[9] The data in the matrix of transport coefficients and the initial emissions (1984) and the assumed maximal emission reductions are taken from Lehmhaus, Saltbones and Eliasson (1986).

4 FULL COOPERATIVE SOLUTION WITH SIDEPAYMENTS (Fcs)

We will start by looking at the "full cooperative solution" when sidepayments between the different countries are possible. This means that the net benefits for some countries may turn out to be negative but these countries can be compensated by cash payments if the total European benefits are positive. The results are given in Table 1.

The calculations show that the total emissions in Europe would be reduced by about 40 percent in the full cooperative solution compared with the present situation (emissions 1984), that almost all countries would gain from the full cooperative solution and that a few countries – UK, Italy and Spain – would lose from participating in the cooperation.

---

[9] GAMS was developed at the World Bank by David Kendrick and Alexander Meeraus. It is now marketed by The Scientific Press, 507 Seaport Court, Redwood City, CA 94062, USA.

242

(The losses that Finland and Luxemburg would experience are negligible). Spain's loss is also almost negligible, while Italy would experience a moderate loss and the UK a substantial loss. Obviously, the UK would have no incentives to participate in organized cooperation to reduce the sulfur emissions in line with the full cooperation.

These results do not depend crucially on the assumed control cost functions. If we have assumed too high a value for the marginal control cost for a particular country, this

TABLE 1

Net benefits from the full cooperative solution.

	Emission control 1000 ton SO_2	Percentage reduction	Benefits mill. D Mark
ALB	10	42	22
AUS	31	21	324
BEL	112	36	191
BUL	179	36	28
CZE	1219	75	152
DEN	130	86	119
FIN	25	14	–2
FRA	104	10	879
GDR	1040	80	11
FRG	1183	86	328
GRE	303	86	52
HUN	635	77	5
IRE	27	38	71
ITA	634	33	–84
NET	105	62	565
NOR	3	6	272
POL	560	27	599
POR	15	19	10
ROM	83	83	420
SPA	231	14	–29
SWE	6	4	606
SWI	10	23	192
TUR	299	62	0
USSR	107	2	1510
UK	1494	81	–336
YUG	465	79	346
TOT	9011	39	6248

would to a certain extent be compensated by too high a value for the marginal damage cost, because of the way we calibrate the damage cost function. In fact, sensitivity analysis yields the expected result that it is the EMEP matrix that is crucial as long as we are willing to make the assumption that the damage cost function is linear.

A few things should be noted about the figures in this table (some of which also apply to later tables). First, a few countries are required to abate their emissions of sulfur up to the maximum amount. In this group are the UK, Czechoslovakia, East and West Germany and a few others. In view of the rather arbitrary upper limits on emission reductions that have been imposed on the solutions, it can be concluded that these countries would have to reduce their emissions further given a more realistic cost of abatement function. Furthermore, it should be observed that the total emission reduction required is about forty percent, which is more than the thirty percent agreed upon in the "30-club".[10]

One should also note that the abatement requirements vary very much among the countries. In general, the Scandinavian countries are required to abate less than the central European countries. One reason for this is that abatement in the Scandinavian countries has already driven up the marginal control cost so that it would be cheaper from an European perspective to reduce the emissions elsewhere. But, perhaps a more important factor is that Scandinavia is downstream or "downwind" in the EMEP model relative the rest of Europe. Moreover, the Mediterranian countries are at average, not required to do as much abatement as the countries in Central Europe. This is so because the damage in Africa and Middle East from their emissions is not included in the analysis.

Finally, a few countries will end up with net losses. It would not be rational for any country to sign a binding agreement by which they would expect losses. Therefore, we can conclude that even if binding agreements between countries on emission reductions could be made, we should not expect a full cooperative solution to result, unless the strategy space for each country is expanded to include sidepayments. Without sidepayments, we would not expect the UK, Italy and Spain to agree on an emission control plan that would result in non-negligible losses.

[10] The "30-club" is a group of European and North American countries that have agreed to reduce their emissions by at least 30 percent.

244

5 PARETO DOMINANT OUTCOMES

Assume that sidepayments are not feasible. What would be a "good" outcome if the countries would cooperate? One such outcome would be an agreement on emission reductions that would minimize the total European damage- and emission costs but that would leave no country worse off. That outcome (the Pareto dominating outcome – Pdo) is given in Table 2.

TABLE 2

Net benefits when no country is made worse off (Pdo).

	Emission control 1000 ton SO_2	Percentage reduction	Benefits mill. D Mark
ALB	11	42	22
AUS	33	22	314
BEL	123	40	96
BUL	181	36	27
CZE	1219	75	148
DEN	130	86	119
FIN	24	14	8
FRA	119	12	696
GDR	1026	79	0
FRG	1183	86	242
GRE	303	86	52
HUN	635	77	2
IRE	57	82	6
ITA	508	27	0
NET	119	70	464
NOR	3	6	225
POL	571	28	565
POR	26	32	2
ROM	83	83	416
SPA	116	7	0
SWE	6	3	549
SWI	10	24	173
TUR	299	62	0
USSR	444	7	1437
UK	747	40	0
YUG	465	79	329
TOT	8440	37	5892

The total benefits are reduced by about 6 percent compared to the full cooperative solution with sidepayments if sidepayments are not allowed. The total emission control in Europe would also be about 6 percent smaller in the Pdo than in the full cooperative solution. The main difference between the two solutions is the level of emission control in the UK. In the Pdo the UK is required to abate 40 percent of the initial emissions while in the full cooperative solution it is required to abate 81 percent. However, when side payments are not feasible, it is not certain that the countries would agree on minimizing the total European damage- and control costs. The main reason for choosing the minimization of the total damage- and control cost is that this would yield a maximal surplus which could be distributed among the countries in some way and thereby secure

TABLE 3
Maximum potential gains from individual minimization; million D mark.

Countries	USSR	Sweden	UK	Poland	Germany
ALB	17	16	17	17	17
AUS	115	246	97	76	39
BEL					
BUL		11			
CZE	253	209	232	96	62
DEN	131	122	129	130	
FIN					3
FRA		687	629		
GDR	24	6			300
FRG	477	330	224	407	890
GRE	57	56	57	57	57
HUN	20	11	19		2
IRE	10	7		10	1
ITA		49			
NET	477	434	335	477	146
NOR	224	131		222	
POL		190		864	
POR					
ROM	455	440	452	388	423
SPA		2	6		
SWE		608			
SWI	176	135	146	176	125
USSR	1737	1492		1276	1360
UK			96		
YUG	386	345	368	349	316

for each country a maximum payoff. With no sidepayments, the interests of the countries are much more in conflict with each other. In order to analyse this issue, the following simulations were carried out. The optimal allocation of emission control was calculated when the objective was to minimize the total damage and control cost in the USSR, Sweden, the UK, Poland and the two Germanies combined, respectively. Each of these calculations then shows the maximum potential gain each of these countries could expect from participating in the cooperation. The reason the Federal Republic of Germany and the German Democratic Republic are combined is simply that they are very similar with respect to the atmospheric transport model. The results are given in Table 3.

Table 3 is quite revealing. It is quite clear that Sweden should have a very strong interest in reducing the total emissions. When the Swedish net benefits are maximized, she gains only 11 percent compared with the Pdo and only 0.3 percent compared with the full cooperative solution. The important role the UK is playing is also clear. If the British benefits are maximized, the net benefits in many countries would drop to zero. It is also clear from the table that Czechoslovakia would gain quite a lot, irrespective of which country is maximizing. If the Germans would be successful and more or less dictate the emission control strategy, Poland would suffer because it would then be forced to reduce its emissions to such an extent that its own net benefits would fall to zero. Anyhow, the table shows that the countries have quite disparate wishes in negotiations on voluntary restrictions of emissions.

6 COALITION FORMATION

Instead of either paying some countries for reducing their emissions (with side payments) or in some other way making concessions (for example by making extra big emission control efforts) in order to achieve an agreement involving all European countries, some countries could try to form coalitions to find out whether they could do better than on their own. In theory, this should be analysed with the aid of cooperative game theory, but as has already been pointed out, the sheer number of countries makes it almost impossible to calculate the characteristic function and the equilibrium concepts that are based on it.

However, a few analytical results can be derived. Assume that the vector E of national emissions is a candidate for an agreement. That vector is blocked by a coalition M of countries if there exist emission levels E'_i for the countries in the coalition such that no country in the coalition is worse off and at least one is better off, irrespective of what the other countries do. A vector E is said to belong to the core if it cannot be blocked.[11] An allocation of emissions among countries that is in the core has thus a certain stability. No coalition can do better for itself than it can with a vector in the core, because the countries outside the coalition can "revenge" by making certain changes in their emissions. However, this concept is not terribly interesting because all Pareto efficient allocations will belong to the core. The reason is that a coalition trying to block a vector in the core can be met by big increases of emissions from the countries outside the coalition. Only if the coalition consists of countries that do not import any sulfur from other European countries, would it be able to block a Pareto efficient allocation. The only such coalition would consist of Iceland alone, a not very interesting case. Moreover, the only way countries outside a coalition M can prevent M from blocking is by increasing emissions and thereby increasing their own damage cost. This threat is therefore hardly credible.

A more interesting equilibrium concept is the strong equilibrium.[12] This concept is based on a more restricted assumption of what countries outside the coalition will do. In particular, a coalition can upset a vector E if there exists E'_i, for all i in the coalition, such that none of the countries in the coalition is worse off with E'_i, given that the countries outside the coalition will not find it to their advantage to change their behavior (i.e. their emissions). In terms of the model in this section, this means that a coalition could gain by playing the noncooperative game against the coalition of all other countries. It is shown in the Appendix that if the EMEP matrix were symmetric, the set of emission allocations that are strong equilibria would be empty if the number of countries is sufficiently large. However, it is possible to extend that argument to an asymmetric EMEP matrix as long as all countries are damaged by acid rain and they also contribute to the rains. Thus, in the European acid rain game, one should therefore not expect to find strong equilibria. Any Pareto efficient allocation of emissions can therefore be upset by a coalition that can do better on its own compared with a complete European agreement.

However, in spite of this rather negative result, quantitative studies of the economic

---

[11] This concept and the following one – the strong equilibrium – is defined and discussed in most textbooks on game theory. For a lucid discussion see Luce and Raiffa (1964). The concept of strong equilibrium is discussed in Dasgupta and Heal (1981), Chapter 2.

[12] See Dasgupta, Heal, op.cit.

248

TABLE 4

Net benefits with coalition formation.

Coalition members	Emission control 1000 ton SO_2	Percentage reduction	Benefits mill. D Mark
ALB	9	37	22
AUS	29	20	277
BEL	106	35	50
BUL	127	34	33
CZE	1219	75	125
DEN	124	82	127
FRA	91	9	466
GDR	1040	80	–47
FRG	1183	86	78
GRE	303	86	51
HUN	635	77	–9
IRE	11	15	–1
NET	99	58	400
NOR	3	6	175
POL	560	27	544
POR	5	6	0
ROM	83	83	398
SWE	5	3	478
SWI	8	19	95
TUR	299	62	
USSR			1377
YUG	465	79	253
TOT	9011	39[1)]	6002
Non-coalition members			
FIN			5
ITA			148
SPA			5
UK			87
TOT			246
EUROPE		40	

1) average for all countries

incentives for various coalitions may yield further insights. In this section, one particular coalition will be studied, namely the coalition of all countries that are not making a negative net benefit in the full cooperative solution. Thus, we will look at the coalition consisting of all countries except Finland, Italy, Spain and the UK. These countries, not in the coalition, are assumed to maximize their net benefits. As the marginal damage cost does not depend on the emissions in other countries, it follows that they will carry out their Nash-strategies, i.e. they will stick to their initial emissions, whatever the coalition decides to do. The result is shown in Table 4.

The total net benefits accruing to the coalition is less than what the coalition could have obtained by having cooperation with all countries and compensating those countries that would have experienced negative net benefits. The net benefit to the coalition would in that case have been 6248 million D-Mark after sidepayments of the order of 451 million D-Mark had been made. Moreover, the emission control in Europe would be significantly lower with the non-coalition countries outside an agreement.

However, Table 4 shows that both Italy and the UK have strong incentives to stay out of any agreement. By staying outside and sticking to the Nash-strategies, both countries can gain significantly. This is probably what can be seen today on the scene of international negotiations on emission control, at least for the role played by the UK. Thus, although this analysis is far from complete, in that it does not study the coalition formation systematically, it gives some insights into the possibility of international bargaining.

7 SUMMARY

In this paper, an attempt has been made to accomplish two objectives: to create an analytical framework for problems of international cooperation on transboundary pollution and to present a first round of estimates of the incentives different European countries may have to participate in such cooperation on controlling sulfur emissions.

The framework chosen has been the theory of cooperative games. The formulation of the conflicts between different European countries as variable sum game has certain advantages. First of all, it points out and identifies the strategic aspects of the behavior of the different countries at the negotiating table. It also identifies the kind of gains that a cooperative outcome would imply. Finally it gives a means of quantifying the net benefits to the countries from participating in European cooperation, More specifically, the present situation was assumed to represent a Nash non-cooperative equilibrium, in which each country optimizes its own net benefit, taking the strategies and payoffs of the other countries into account. By assuming that the damage cost function is linear in the deposition of sulfur and by using some crude estimates of the cost of controlling sulfur emissions it is possible to calibrate the damage cost function with the help of the EMEP atmospheric transport model in such a way that the present situation represents a Nash equilibrium in the corresponding model of the game. Having thus calibrated the damage cost function, calculations of the gains for the different countries from different cooperative solutions can easily be made. The common conclusion from almost all simulations was that there is a need for international transfers in order to motivate all countries to participate. Thus some countries should be bribed to reduce their emissions. Only if the cooperative agreement is such that not all possibilities of mutual gain are exploited it will be unnecessary to make such transfers. However, even the full cooperative solution in which the countries agree to reduce their emissions in such a way as to minimize the total European damage- and control cost is not stable for coalition formation. It was shown that the set of strong equilibria, i.e. agreements on reductions such that no coalition of countries could do better on their own, given that the other countries minimize their costs, is probably empty. Thus, it may very well be so that for every possible agreement, there exists a coalition that could upset that agreement. The reason the conclusion is rather vague is due to the fact that it was shown for a symmetric game of managing a common property resource that if the number of players is great enough, there is no strong equilibrium. However, we have no indication on the precise meaning of "great enough". Moreover, the European acid rain game is highly asymmetric.

Nevertheless, as long as countries are mainly concerned with their own welfare, it seems that international transfers are necessary to support European wide agreements on emission reductions. In this paper, these transfers have been assumed to be cash payments. However, other kinds of transfer of wealth or command over resources are possible.

APPENDIX

Assume there is a common property resource which is exploited by n agents. Let the benefit to agent i from exploiting the resource be $B^i(x_i,\sum_j x_j)$, where x_i is the amount of exploitation on part of agent i. B^i is assumed to be increasing in its first argument and decreasing in its second argument. In terms of the acid rain game, B^i could be interpreted as the negative of the damage and control cost in country i, x_i emission in country i and $\sum x_j$ the deposition of sulfur in country i (thus the EMEP matrix would simply consist of "ones" in all cells).

Let us assume that the benefit functions are the same for all agents. That makes the game completely symmetrical and each equilibrium concept will give the same value of x_i for all agents.

The Nash equilibrium is defined as x^N defined by

$$B_1(x^N,nx^N) + B_2(x^N,nx^N) = 0$$

where B_1 and B_2 are the partial derivatives with respect to the first and second argument resp.

The Pareto efficient equilibrium $\hat{x}$ is defined from

$$B_1(\hat{x},n\hat{x}) + nB_2(\hat{x},n\hat{x}) = 0.$$

where subscripts denote derivatives.

We will show that for n great enough, there does not exist a strong equilibrium. Let M be a coalition of agents that tries to upset $\hat{x}$. Let M have m members and let the complementary coalition have $s = n - m$ members.

Assume that the members of M chooses x_m in such a way that

$$B(x_m, mx_m + sx_s) = \max_x B(x, mx_m + sx_s).$$

In a similar way, the complementary coalition will choose x_s in such a way that

$$B(x_s, mx_m + sx_s) = \max_x B(x, mx_m + sx).$$

If n is sufficiently greater than m, x_s will be close to $\hat{x}$ and it follows that

$$B(x_m, mx_m + sx_s) \geq B(\hat{x}, m\hat{x} + sx) \simeq B(\hat{x}, m\hat{x} + s\hat{x}) = B(\hat{x}, n\hat{x}).$$

Thus if the number of agents is sufficiently large, it is possible to find a coalition that would upset the Pareto efficient solution.

REFERENCES

Amann M. and G. Kornai, 1987. Cost Functions for Controlling SO_2 emissions in Europe. Working Paper May 1987, WP-87-065, IIASA, A-2361 Laxenburg, Austria.

252

Air pollution across national boundaries, 1972. The impact on the environment of sulfur in air and precipitation. Sweden's case study for the United Nations Conference on the human environment, Stockholm.

Binmore K. and P. Dasgupta, eds, 1986. Economic Organizations as Games, Basil Blackwell Ltd.

Dasgupta P. and G. Heal, 1981. Economic Theory and Exhaustible Resources, Cambridge.

Eliasson A. and J. Saltbones, 1983. Modelling of long-range transport of sulphur over Europe: a two-year model run and some model experiments. Atmos. Environ. 17 1457-1473.

Friedman J., 1986. Game Theory with Application to Economics, New York.

Hinrichson D., 1986. Multiple Pollutants and forest Decline, AMBIO, Vol. XV, No 5.

Lehmhaus J., J. Saltbones, and A. Eliasson, 1986. A modified Sulphur Budget for Europe 1980, EMEP/MSC-W Report 1/86, Norwegian Meteorological Institute, Oslo.

Luce D., H. Raiffa, 1964, Games and Decisions, New York 1964.

The Norwegian Meteorological Institute, 1987. Sulphur Budgets for Europe for 1979, 1980, 1981, 1982, 1983, 1984 and 1985. EMEP/MSC-W Not 4/87.

[2]

Oxford Economic Papers 46 (1994), 878–894

SELF-ENFORCING INTERNATIONAL ENVIRONMENTAL AGREEMENTS

By SCOTT BARRETT

Faculty of Economics, London Business School, Sussex Place, Regent's Park, London, NW1 4SA

Centre for Social and Economic Research on the Global Environment

1. Introduction

It is well-known that collective well-being can be increased if all countries cooperate in managing shared environmental resources like the climate and ozone layer, but that if this improved situation is attained, every country will earn even higher returns by free-riding on the virtuous behavior of the remaining cooperators. However, the fact that countries can do better if cooperation can be sustained suggests that countries have incentives to develop institutions which can punish free riding. Such institutions do in fact exist in the form of international environmental agreements (IEAs). Indeed, more than 100 IEAs are currently in force, the most recent of which include the Montreal Protocol on Substances that Deplete the Ozone Layer and the Framework Convention on Climate Change. The essential feature of IEAs is that they must be self-enforcing. No country can be forced to sign an IEA, and signatories to an IEA can always withdraw from the agreement. If IEAs can improve the management of shared environmental resources, they must make it attractive for countries to want to sign, and to want to carry out the terms of the agreement. This paper employs some concepts from game theory to explore the properties of self-enforcing IEAs.

Two different approaches are considered. The first is a model in which the number of signatories, the terms of the agreement, and the actions of non-signatories are all determined endogenously. This model assumes that signatories maximize their collective net benefits, recognizing how their choice of treaty obligations affects the actions of nonsignatories, while nonsignatories each choose their actions on the assumption that other countries will not respond. The mechanism that sustains the IEA is as follows. When a country joins the IEA, the other signatories increase their abatement levels, and hence reward the country for acceding to the agreement; when a country withdraws, the remaining signatories reduce their abatement levels, and hence punish the country for withdrawing from the agreement. These punishments and rewards are credible, because the signatories always maximize their collective net benefits. However, the magnitude of the punishments and rewards may not be sufficiently large to sustain a self-enforcing IEA consisting of many countries. Depending on the functional specification, Section 2 of this paper shows that the self-enforcing IEA may not exist, or it may consist of no more than two or three countries, or it may support up to as many countries as share the resource (depending in this last case on parameter values).

The other approach is to model the IEA as an infinitely repeated game. This approach has the attraction of increasing credible punishments compared with the other model, but it has the drawback of not solving endogenously for the number of signatories, the terms of the agreement, and the actions of non-signatories. Instead, this model can be employed to calculate the maximum number of countries that could support the full cooperative outcome. While the folk theorem guarantees that the full cooperative outcome can be sustained by any number of countries provided the rate of discount is sufficiently small, we shall see in Section 3 that this result may not hold for a self-enforcing IEA.[1] The reason is that the punishments that sustain the full cooperative outcome may be vulnerable to renegotiation, and a self-enforcing IEA must clearly be renegotiation-proof. As with the previous model, the ability of the infinitely-repeated game to sustain the full cooperative outcome depends on the functional specification and parameter values.

These different models yield one consistent and striking result. The first model shows that for global environmental problems, an IEA may achieve a high degree of cooperation, but only when the difference between global net benefits under the noncooperative and full cooperative outcomes is small. When this difference is large, a self-enforcing IEA cannot support a large number of signatory countries.[2] The second model shows that the maximum number of countries that can sustain the full cooperative outcome may be large, but only when the difference between global net benefits under the noncooperative and full cooperative outcomes is small. When this difference is large, the full cooperative outcome cannot be sustained by a self-enforcing IEA.

By necessity, the analysis imposes a number of restrictive assumptions, and the above result must be seen in the light of these. The most important assumptions are: (i) that all countries are identical; (ii) that each country's net benefit function is known by all countries, and known to be known by all countries; (iii) that the choice instrument is restricted to pollution abatement; (iv) that abatement levels are instantly and costlessly observable; (v) that the pollutant does not accumulate in the environment; and (iv) that the cost functions are independent. Further research is needed to learn whether the basic conclusion of this paper would be upset by changes in these assumptions.[3]

[1] Of course, the folk theorem only asserts that any feasible individually rational payoffs can be supported as a Nash equilibrium if the game is repeated infinitely often and players are sufficiently patient. Hence, the full cooperative outcome is only one among many supergame equilibria.

[2] Unless, of course, the agreement merely codifies the noncooperative outcome.

[3] A number of studies have considered these issues, but not within the context of the models presented in this paper. Black *et al.* (1993) consider information asymmetries in a model where the agreement specifies the minimum number of countries that must sign the agreement for it to come into force. With complete information, the solution to their model would trivially require that all countries sign the agreement. Such an agreement would not, however, be (*ex post*) self-enforcing. Barrett (1992) considers the implications for reaching a negotiated agreement when countries are different and may select different policy instruments (abatement levels, pollution taxes, and tradeable emission entitlements), but this study does not require that the agreement that is reached

2. The model of an IEA

2.1. *Linear marginal abatement benefits and costs*

Consider a world of $i = 1, \ldots, N$ identical countries, each of which emits a pollutant that damages a shared environmental resource. i's current abatement benefits are assumed to depend on current total abatement as follows

$$B_i(Q) = b(aQ - Q^2/2)N \quad (1)$$

where $B_i(Q)$ is i's abatement benefit, a and b are positive parameters, and Q is global abatement. According to (1), the marginal benefit of the first unit of abatement for i is ab/N, and that of the ath is zero.[4] The parameter b is the slope of the global marginal benefit function (found by summing (1) over all N countries, and differentiating with respect to Q).

Each country's abatement costs are assumed to depend on its own abatement level and no one else's. For country i, the abatement cost function is assumed to be given by

$$C_i(q_i) = cq_i^2/2 \quad (2)$$

where $C_i(q_i)$ is i's abatement cost and q_i is i's abatement (in eq. (1), $Q = \Sigma_i q_i$). The parameter c represents the slope of each country's marginal abatement cost curve.

2.1.1. *The gains to full cooperation.* Denote i's net benefits by π_i (i.e., $\pi_i = B_i(Q) - C_i(q_i)$), and denote global net benefits by Π (i.e., $\Pi = \Sigma_i \pi_i$). The full cooperative outcome is found by choosing Q to maximize Π. The solution requires setting each country's marginal cost of abatement (MC_i) equal to the global marginal benefit of abatement (MB), and is illustrated in Fig. 1. The full cooperative abatement levels are

$$Q_c = aN/(N + \gamma), \qquad q_c = a/(N + \gamma) \quad (3)$$

where Q_c is global abatement and q_c is an individual country's abatement under the full cooperative outcome, and $\gamma \equiv c/b$.

The noncooperative or open access outcome can be computed by invoking the usual Cournot conjecture that every country chooses its abatement level

be sustainable by a self-enforcing IEA. Carraro and Siniscalco (1991) consider a model in which signatories to an IEA may choose side payments to induce nonsignatories to accede to the IEA, but for signatories to want to make such side payments, it is essential that these countries commit to certain abatement levels, and such a commitment is inconsistent with the notion of self-enforcement. Mäler (1989) and Hoel (1990) analyze cooperative solutions to the control of persistent pollutants, but which are not subject to a self-enforcement constraint. Finally, Heal (1992) considers the case where each country's abatement costs depend on the abatement actions of other countries, as well as its own, but evaluates only the Nash noncooperative equilibrium.

[4] The problem specified here does not impose an upper bound on each country's abatement effort. Clearly, emissions must be at least as great as a units, since the marginal benefit of abating a units is zero. To solve the problem we can assume that emissions exceed a by a sufficient margin so that the maximal abatement constraint is never binding.

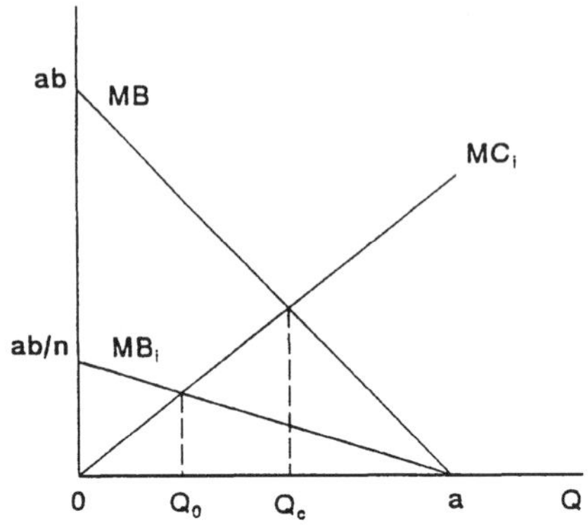

FIG. 1. The non-cooperative and full-cooperative outcomes

on the belief that the abatement levels of all other countries are given. That is, each country chooses q_i to maximize π_i, taking all $q_j, j \neq i$, as fixed. The solution requires setting each country's own marginal benefit of abatement (denoted MB_i in Fig. 1) equal to its own marginal cost of abatement. Denoting Q_o global abatement and q_o any individual country's abatement under the noncooperative or open access outcome, we find

$$Q_o = a/(1+\gamma), \qquad q_o = a/N(1+\gamma) \tag{4}$$

We see immediately that $Q_c > Q_o$.[5] Each country is better off at the full cooperative solution, but none has an incentive to choose q_c unilaterally.

It is clear from Fig. 1 that the difference between global net benefits under the full cooperative and non-cooperative outcomes depends on the slopes of the marginal abatement cost and benefit curves. Let Π_c and Π_o denote global net benefits under full cooperation and open access, respectively. Then we have:

Lemma 1 Let $\theta(N) = [\sqrt{(N^2 + 8N)} - N]/4$. Then (i) $\Pi_c - \Pi_o$ increases in b when $\gamma > \theta(N)$ and decreases in b when $\gamma < \theta(N)$ for c given; (ii) $\Pi_c - \Pi_o$ increases in c for γ fixed; and (iii) θ is monotonic and approaches a value of 1 as N becomes very large.

Proof See Appendix.

Lemma 1 tells us that the gains to cooperation ($\Pi_c - \Pi_o$) are larger the closer γ is to $\theta(N)$ and the larger is c. For global environmental problems, N exceeds 100 and $\theta \approx 1$. When c is small and b is large, unilateral abatement is substantial, and the gains to cooperation are relatively small. When c is large and b is small, countries will not abate their emissions by much, even if cooperation is full. When $c \approx b$ and both parameter values are small, $Q_c - Q_o$ will be large, but the gains to cooperation will be small. When $c \approx b$ and both

[5] This is a well known result. See, for example, chapter 2 of Dasgupta (1982).

parameters are large, both $Q_c - Q_o$ and $\Pi_c - \Pi_o$ will be large. Clearly, IEAs can achieve the most when these last conditions hold.

2.1.2. *The self-enforcing IEA.* Suppose that a number of countries negotiate an IEA. Clearly, the number of signatories, the terms of the IEA, and the abatement levels of nonsignatories must be determined jointly. It seems reasonable to suppose that nonsignatories will choose their abatement levels taking as given the abatement decisions of all other countries. Let α denote the proportion of countries that sign the IEA. There are then αN signatories and $(1-\alpha)N$ nonsignatories. Let Q_n denote the abatement of all nonsignatories, and let q_n denote the abatement of any individual nonsignatory. Since all nonsignatories are identical, $Q_n = (1-\alpha)Nq_n$. The reaction functions of nonsignatories are given by

$$Q_n(\alpha, Q_s) = (1-\alpha)(a - Q_s)/(\gamma + 1 - \alpha) \tag{5}$$

Signatories are assumed to choose Q_s to maximize their collective net benefits, with q_s being identical for all signatory countries, subject to (5).[6] The solution is

$$Q_s^*(\alpha) = a\alpha^2 N\gamma/[(\gamma + 1 - \alpha)^2 + \alpha^2 N\gamma] \tag{6}$$

Substituting (6) into (5) yields

$$Q_n^*(\alpha) = a(1-\alpha)(\gamma + 1 - \alpha)/[(\gamma + 1 - \alpha)^2 + \alpha^2 N\gamma] \tag{7}$$

It is easily verified that the full cooperative ($\alpha = 1$) and noncooperative ($\alpha = 0$) solutions, given by eqs (3) and (4) respectively, are special cases of (6) and (7).

The remaining problem is to determine $\alpha^* N$, or the number of signatories to the self-enforcing IEA. In determining α^*, it will prove helpful to invoke a concept of stability developed for the analysis of cartels (d'Aspremont *et al.* 1983). Let π_s be the net benefit of every signatory and π_n that of every nonsignatory.

Definition An IEA consisting of αN signatories is self-enforcing if

$$\pi_n(\alpha - 1/N) \leqslant \pi_s(\alpha) \quad \text{and} \quad \pi_n(\alpha) \geqslant \pi_s(\alpha + 1/N) \tag{8}$$

If (8) holds, no signatory will want to withdraw unilaterally from the agreement; such a withdrawal would reduce the signatory's abatement level, and hence its costs, but its defection from the IEA would weaken the agreement, and the resulting loss in benefits would more than offset the reduction in costs gained by the withdrawal. Similarly, no nonsignatory, acting alone, would want

[6] The assumption that the q_s are identical is easily justified. If the problem of signatories agreeing on individual country obligations is modelled as a cooperative game, then it can be shown that if all signatories are identical, the Nash bargaining solution will require that each country undertake the same level of abatement.

TABLE 1
Stability analysis for hypothetical example

α	q_s	q_n	π_s	π_n	Q	Π
0.0	—	8.0	—	472.0	80.0	4,720.0
0.1	1.9	8.5	476.8	468.1	78.7	4,690.0
0.2	4.2	8.7	474.0	466.6	78.2	4,681.2
0.3	6.7	8.4	472.3	468.9	78.9	4,699.4
0.4*	8.9*	7.6*	472.2*	474.9*	81.1*	4,738.1*
0.5	10.5	6.3	473.7	482.5	84.2	4,781.2
0.6	11.3	4.9	476.4	489.4	87.7	4,816.0
0.7	11.5	3.6	479.5	494.3	91.0	4,839.8
0.8	11.1	2.5	482.7	497.3	93.8	4,855.9
0.9	10.5	1.6	485.4	498.8	95.9	4,867.8
1.0	9.8	—	487.8	—	97.6	4,878.0

to accede to the IEA; although the recruit's benefits would rise, its abatement costs would rise even more.[7]

The concept is best understood by considering an example. Let $N = 10$, $a = 100$, $b = 1$, and $c = 0.25$; and denote global net benefits by $\Pi(\alpha)$; i.e. $\Pi(\alpha) = \alpha N\pi_s + (1 - \alpha)N\pi_n$. Table 1 shows the net benefit and abatement levels corresponding to each possible α, and Fig. 2 illustrates the example. To solve for α^*, start at $\alpha = 0$ and compare $\pi_n(0)$ with $\pi_s(0.1)$. Clearly, $476.8 > 472.0$, and hence it will play a nonsignatory to accede to the agreement. Now compare $\pi_n(0.1)$ with $\pi_s(0.2)$. Again, $474.0 > 468.1$, and it will pay a nonsignatory to cooperate. Continuing in this way one finds that nonsignatories always do better by acceding when $\alpha < 0.4$. Likewise, starting at $\alpha = 1.0$, one finds that signatories always do better by withdrawing from the agreement whenever $\alpha > 0.4$. Hence, an IEA consisting of four signatories is the only self-enforcing IEA for this problem.[8]

[7] The concept of self-enforcement defined by (8) was first employed by Barrett (1989), but has also been used by Carraro and Siniscalco (1991) and Hoel (1992). The essential feature of (8) is that it does not consider whether coalitions of two or more countries would want to accede to or withdraw from the IEA. Heal (1992) defines a related concept which he calls the minimum critical coalition. This is a group of countries S, which (i) adopts an abatement policy yielding each member of S positive net benefits, and (ii) does not contain a proper subset of countries which yields its members positive net benefits.

[8] Notice that $\pi_s(\alpha^* + 1/N) - \pi_s(\alpha^*) > \pi_n(\alpha^*) - \pi_s(\alpha^* + 1/N)$ in this example. This implies that signatories to a self-enforcing IEA have an incentive to offer a side payment to nonsignatories to join the agreement. The difficulty is that the agreement to offer a side payment is not self-enforcing. While the signatories to the self-enforcing IEA have a collective incentive to offer a side payment to a nonsignatory, each of the original signatories would do better still by withdrawing from the agreement since $\pi_n(\alpha^*) > \pi_s(\alpha^* + 1/N)$ (by definition).

Carraro and Siniscalco (1991) propose a way out of this dilemma. They show that if the signatories to the self-enforcing IEA can commit themselves to cooperate whatever the magnitude of α, then the total number of signatories can be increased (although the full cooperative outcome may still be unattainable). The snag in this proposal, as noted in footnote 2, is that the commitment is not credible; the expanded agreement is not self-enforcing. If commitment could be made credible, then the full cooperative outcome would be easy to sustain: every country would simply commit to choose the full cooperative abatement level q_c.

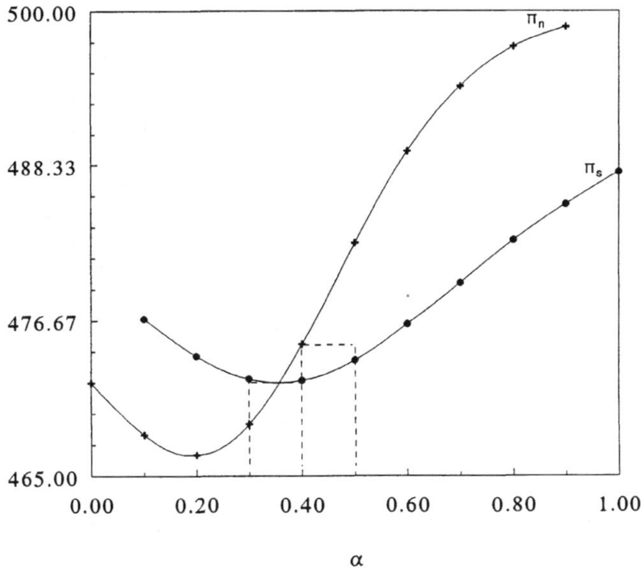

FIG. 2. Graphical analysis of self-enforcing IEA

Unfortunately, a full characterization of the solution cannot be obtained analytically for this functional specification. However, simulations reveal a very simple and compelling relationship between γ and α^*, and between γ and the ability of an IEA to close the gap between the full cooperative and non-cooperative outcomes.

2.1.3. *Simulations.* Table 2 presents solutions to α^*N for various values of b and c when $N = 100$.[9] It is clear from the simulations that the number of signatories to the self-enforcing IEA increases as γ decreases. It is encouraging to find that a self-enforcing IEA can consist of many signatories, if only when γ is small. However, Lemma 1 demonstrated that if γ is small, then the difference between global net benefits under the full cooperative and noncooperative outcomes will also be small. Hence, IEAs signed by many countries do not increase global net benefits by very much compared with the non-cooperative outcome. Indeed, Tables 3 and 4, which give the total abatement and net benefit figures corresponding to a selection of nine cells from Table 2, confirm that the self-enforcing IEA increases global abatement and global net benefits by very little when γ is small. These results suggest that IEAs signed by a large number of countries have very little effect.

Table 2 shows that when γ is large, few countries will sign the self-enforcing

[9] The value taken by the parameter a does not affect the results in Table 2.

TABLE 2
Number of signatories out of 100 for various values of c and b

	c										
	0.01	*0.05*	*0.10*	*0.25*	*0.50*	*1.00*	*2.00*	*5.00*	*10.00*	*20.00*	*100.00*
0.01	3	3	3	2	2	2	2	2	2	2	2
0.05	6	3	3	3	3	2	2	2	2	2	2
0.10	11	4	3	3	3	3	2	2	2	2	2
0.25	21	6	4	3	3	3	3	2	2	2	2
0.50	35	11	6	4	3	3	3	3	2	2	2
1.00	51	18	11	5	4	3	3	3	3	2	2
2.00	68	30	18	9	5	4	3	3	3	3	2
5.00	84	51	35	18	11	6	4	3	3	3	2
10.00	92	68	51	30	18	11	6	4	3	3	3
20.00	96	81	68	46	30	18	11	5	4	3	3
100.00	100	96	92	81	68	51	35	18	11	6	3

TABLE 3
*Global abatement under the noncooperative, IEA, and full co-operative outcomes**

		c		
		0.01	*1*	*100*
	0.01	500.0	9.9	0.1
		503.9	10.1	0.1
		990.1	500.0	9.9
b	1	990.1	500.0	90.9
		990.2	503.9	95.2
		999.9	990.1	909.1
	100	999.9	990.1	500.0
		1,000.0	990.2	503.9
		1,000.0	999.9	990.1

* Assumes $N = 100$ and $a = 1{,}000$. The top number in each cell is Q_o, the middle $Q(\alpha^*)$, and the bottom Q_c.

IEA. Since IEAs typically specify a minimum number of signatories,[10] this means that an agreement is unlikely to be reached when γ is large. But Tables 3 and 4 indicate that the difference in global abatement and global net benefits under the full cooperative and noncooperative outcomes is trivial in absolute terms when γ is large. Hence, failure to reach agreement is of no great consequence when γ is large.

[10] For example, the Framework Convention on Climate Change will not come into force until 50 countries have ratified the agreement. See Black *et al.* (1993) for an analysis showing why countries might want to specify a minimum number of signatories.

TABLE 4
*Global net benefits under the noncooperative, IEA, and full cooperative outcomes**

		c		
		0.01	*1*	*100*
	0.01	3,738 3,757 4,950	98 100 2,500	1 1 50
b	1	499,902 499,903 499,950	373,750 375,659 495,050	9,803 9,990 250,000
	100	49.999,949 49,999,950 49,999,950	49,990,197 49,990,293 49,995,000	37,375,000 37,565,851 49,504,950

* Assumes $N = 100$ and $a = 1{,}000$. The top number in each cell is Π_0, the middle $\Pi(\alpha^*)$, and the bottom Π_c.

The situation is very different where c and b are both large. As Lemma 1 showed, it is here that the difference between global net benefits under the noncooperative and full cooperative outcomes is large. Table 4 confirms this. But from Table 2 we see that few countries are likely to sign an IEA for such a pollutant.

Proposition 1 For global environmental problems characterized by eqs (1) and (2), the self-enforcing IEA will be signed by a lot of countries, each undertaking substantial abatement, when γ is small, but under these circumstances the IEA increases global net benefits by very little compared with the noncooperative outcome. Cooperation would increase net benefits substantially when c and b are both large, but under these circumstances the self-enforcing IEA cannot sustain a large number of signatories.

Table 5 shows the magnitude of free riding for these simulations. Signatories abate more than they would in the absence of an IEA, while nonsignatories abate less. When $\gamma = 1$, only three countries cooperate. Every signatory abates significantly more than under the noncooperative outcome, and every non-signatory just a little less. But because there are so many more nonsignatories in this case, the free-rider effect is huge in total and the IEA has little overall impact. When $\gamma = 0.01$, there are 51 signatories. In this case the free-rider effect is small—but so too is the additional abatement of signatories. Again, the self-enforcing IEA has little overall impact. Finally, Table 6 shows the net benefit levels corresponding to the abatement levels in Table 5. All countries receive higher net benefits when there exists a self-enforcing IEA compared with the noncooperative outcome, but nonsignatories take a free ride on signatories.

It must be emphasized that Proposition 1 applies to environmental problems

TABLE 5
*Country abatement under the noncooperative, IEA, and full cooperative outcomes**

		c		
		0.01	*1*	*100*
	0.01	5.00	0.10	—
		7.55	0.20	—
		4.96	0.10	—
		9.90	5.00	0.10
b	1	9.90	5.00	0.10
		10.00	7.55	0.20
		9.80	4.96	0.10
		10.00	9.90	5.00
	100	10.00	9.90	5.00
		10.00	10.00	7.55
		—	9.80	4.96
		10.00	10.00	9.90

* Assumes $N = 100$ and $a = 1{,}000$. The top number in each cell is q_o, the second $q_s(\alpha^*)$, the third $q_n(\alpha^*)$, and the bottom q_c.

TABLE 6
*Country net benefits under the noncooperative, IEA, and full cooperative outcomes**

		c		
		0.01	*1*	*100*
	0.01	37.38	0.98	0.01
		37.41	0.99	0.01
		37.57	1·00	0.01
		49.50	25.00	0.50
b	1	4,999.02	3,737.50	98.03
		4,999.02	3,740.84	98.50
		4,999.04	3,757.07	99.93
		4,999.50	4,950.50	2,500.00
	100	499,999.50	499,901.97	373,750.00
		499,999.50	499,901.98	374,083.96
		—	499,903.92	375,707.21
		499,999.50	499,950.00	495,049.50

* Assumes $N = 100$ and $a = 1{,}000$. The top number of each cell is π_o, the second $\pi_s(\alpha)$, the third $\pi_n(\alpha^*)$, and the bottom π_c.

for which N is large. As N becomes small, the effect of withdrawal or accession by any country on the abatement level of the other cooperators increases, and hence free riding can be punished more effectively. To take an example, suppose $N = 5$, $a = 100$ and $c = b = 1$. Then the self-enforcing IEA consists of three countries with $Q_o = 50$, $Q(\alpha^*) = 63$ and $Q_c = 83$; and with $\Pi_0 = 3{,}500$, $\Pi(\alpha^*) = 3{,}869$, and $\Pi_c = 4{,}162$. In this case, the self-enforcing IEA is able to close the gap between the noncooperative outcomes quite substantially.

2.2. *Constant marginal benefits and linear marginal costs*

Suppose that maginal benefits are constant; i.e.

$$B_i(Q) = \omega Q \tag{9}$$

Then, if the abatement cost function is given by e.g. (2), the self-enforcing IEA can be solved analytically.

Proposition 2 For the model specification given by eqs (2) and (9), the self-enforcing IEA consists of 2 countries when $N = 2$ and 3 countries when $N \geqslant 3$.

Proof See Appendix.

2.3. *Constant marginal benefits and logarithmic marginal costs*

Suppose now that abatement benefits continue to be given by eq. (9) but that total abatement costs for country i are now given by

$$C_i(q_i) = x\sigma[(1 - q_i/x)\ln(1 - q_i/x) + q_i/x] \tag{10}$$

The marginal abatement cost function is then

$$MC_i(q_i) = -\sigma \ln(1 - q_i/x) \tag{11}$$

If the parameter x is taken to equal emissions in the absence of any abatement effort, q_i/x is percentage abatement. In that case, the marginal abatement cost function (11) is the same form as was estimated by Nordhaus (1990) for US carbon dioxide emissions.[11] Since Nordhaus's model also assumes constant marginal benefits, eqs (9) and (10) describe Nordhaus's model of the greenhouse effect. However, Nordhaus does not disaggregate to the individual country level. Here it is assumed that all countries are identical and have the same cost and benefit functions.[12] The solution to this problem, like the previous one, can be derived analytically.

[11] The cost function specification is described in footnote 10 of Nordhaus (1990), p. 19.

[12] Since Nordhaus (1990) calculates a global marginal benefit, the parameter ω in eq. (9) can be taken to equal each country's (identical) share of the global marginal benefit.

Lemma 2 For the model specification given by eqs (9) and (10),

(i) $\pi_n(0) = \pi_s(1/N) = \pi_n(1/N)$
(ii) $\pi_s(2/N) > \pi_n(1/N)$
(iii) $\pi_n(\alpha) > \pi_s(\alpha + 1/N) \forall \alpha \geqslant 2/N$

Proof See Appendix.

Lemma 2 implies:

Proposition 3 For the model specification given by eqs (9) and (10), the self-enforcing IEA consists of 2 countries for $N \geqslant 2$.

2.4. *Linear marginal benefits and constant marginal costs*

Suppose now that country i's benefit function is given by (1) and its cost function by

$$C_i(q_i) = dq_i \tag{12}$$

Then the self-enforcing IEA may be solved easily.

Proposition 4 For the model specification given by eqs (1) and (12), a self-enforcing IEA consisting of 2 or more countries does not exist.

Proof See Appendix.

With this specification, any increase in Q_s is entirely offset by a reduction in Q_n. There is therefore nothing to be gained by cooperating.

Clearly, the results obtained using the specification given by eqs (1) and (2) are not generally robust. According to the specifications examined in this section, the number of signatories may be limited to just a few whatever parameter values characterise the problem being investigated.[13] However, the results of all specifications considered here are consistent in one respect: when N is large, an agreement achieving substantial emission reductions over and above the non-cooperative levels is not self-enforcing.

3. IEAs as repeated games

The preceding model is unable to sustain the full cooperative outcome in general because its punishments, though credible, are small. It is well known that the full cooperative outcome of the one-shot game can be sustained as a sub-game perfect equilibrium of the infinitely repeated game if the rate of discount is sufficiently small. However, this section will show that repetition may not allow IEAs to sustain the full cooperative outcome, even for arbitrarily small discount rates, because the punishments that sustain cooperation in infinitely repeated games may be vulnerable to renegotiation, and such punishments could not be supported by a self-enforcing IEA.

[13] d'Aspremont *et al.* (1983), Hoel (1992), and Carraro and Siniscalco (1991) also solve for models yielding two or three cooperators.

Let π_i now denote country i's average payoff in the infinitely repeated game, and consider the following functional specification

$$\pi_i = b(aQ - Q^2/2)/N - dq_i - (ab - dN)^2/2bN \quad (13)$$

where the constant term ensures that each country is guaranteed a payoff of at least zero. Proposition 4 tells us that specification (13) will not allow a stable IEA to improve on the noncooperative outcome. We shall see that this is not necessarily so when the game is infinitely repeated.

The full cooperative outcome yields a global payoff of $d(N-1) \times (2ab - d(N+1))/2b$. Hence, the set of feasible, individually rational payoffs is

$$V^* = \{v \mid v_i \geqslant 0, \textstyle\sum v_i \leqslant d(N-1)(2ab - d(N+1))/2b\}$$

Following Farrell and Maskin (1989), a payoff vector v may be said to be (weakly) renegotiation proof (for discount rates near zero) if abatement levels q_j^j, q_k^j can be chosen to punish j for cheating, such that

$$\max_{q_j} \{(b/N)[a(q_j + (N-1)q_k^j) - (q_j + (N-1)q_k^j)^2/2] - dq_j - (ab - dN)^2/2bN\} \leqslant v_j \quad (14)$$

$$(b/N)[a(q_j^j + (N-1)q_k^j) - (q_j^j + (N-1)q_k^j)^2/2] - dq_k^j - (ab - dN)^2/2bN \geqslant v_k \quad (15)$$

Condition (14) ensures that any one country cannot gain by cheating on the agreement.[14] Condition (15) ensures that countries have no incentive to renegotiate. For this problem, it also makes sense to require that the payoff vector v be Pareto efficient in the punishment phase.[15] If we further assume that j chooses for q_j^j its minimax abatement level, $q_j^j = (ab - dN)/b$, then $q_k^j = \mathrm{d}/b$. Substituting, we require for every country i

$$\mathrm{d}[2abN - \mathrm{d}(N+1)^2]/2bN \geqslant v_i \geqslant \mathrm{d}^2(N-1)/b \quad (16)$$

Under the full cooperative outcome, every country receives a payoff of $\mathrm{d}(N-1)(2ab - \mathrm{d}(N+1))/2bN$. Substituting this value into (16), we find:

Proposition 5 For the specification given by eq. (13), the full cooperative outcome can be sustained as a self-enforcing IEA for a common discount rate close to zero if the number of countries N does not exceed

$$\bar{N} = \min(ab/d - 1, 2ab/3d - \tfrac{1}{3})$$

It is easy to see that $\bar{N}$ is increasing in a and b, and decreasing in d. $\bar{N}$ is calculated for various values of b and d in Table 7.[16] The number of countries that can sustain the full cooperative outcome varies substantially, depending

[14] As elsewhere in the paper, I am assuming that countries do not form coalitions.
[15] Farrell and Maskin (1989) refer to such an equilibrium as being strongly perfect.
[16] The table considers parameter values which yield only interior solutions.

TABLE 7
*Illustration of maximum number of countries that can sustain the full cooperative outcome**

a	b	d	$\bar{N}$	$(\Pi_c - \Pi_o)/(N-1)^2$
100	1	0.99	67	0.49
100	1	0.75	88	0.28
100	1	0.65	102	0.21
100	2	1.98	67	0.98
100	2	1.50	88	0.56
100	2	1.30	102	0.42
1,000	1	9.9	67	49
1,000	1	7.5	88	28
1,000	1	6.5	102	21
1,000	2	19.8	67	98
1,000	2	15.0	88	56
1,000	2	13.0	102	42

* Assumes $N = 100$, to ensure interior solutions.

on the parameter values, and is higher the larger is b and the smaller is d. It can be shown that $\Pi_c - \Pi_o = d^2(N-1)^2/2b$ for specification (13). The last column in Table 7 indicates that when $\bar{N}$ is large, the difference $\Pi_c - \Pi_o$ is small, and that when $\bar{N}$ is small, the difference $\Pi_c - \Pi_o$ may be very large. These results are reminiscent of those obtained in Section 2.1. Clearly, even infinite repetition may not allow the IEA to sustain the full cooperative outcome when IEAs must be self-enforcing.[17]

4. Conclusions

This paper has shown that self-enforcing international environmental agreements (IEAs), which establish rules for managing shared environmental resources, may not be able to improve substantially upon the noncooperative outcome. Two different modelling approaches support this conclusion. The model of a self-enforcing IEA, which solves jointly for the number of signatories, the terms of the IEA, and the actions of nonsignatories, shows that, depending on the functional specification, a self-enforcing IEA may not exist, or it may not be able to sustain more than two or three signatory countries, in which case the IEA cannot increase global net benefits substantially when the number of countries that share the resource is very large. For one functional specification, the self-enforcing IEA can sustain a large number of signatories—up to as many as share the resource—but only when the difference in net benefits between the noncooperative and full cooperative outcomes is very small. When this difference is very large, the self-enforcing IEA can sustain only a small number of signatories, and hence can only marginally improve upon the noncooperative

[17] Barrett (1994) carries out a similar analysis for the conservation of biological diversity, a global public good, but using a different functional form. Though this model can only be evaluted using simulation, the analysis arrives at a similar conclusion.

outcome when the environmental resource is shared globally. The other model, which takes the IEA to be an equilibrium to an infinitely repeated game, but one which is renegotiation-proof, shows that the full cooperative outcome can be sustained by a large number of countries, but only when the difference in global net benefits between the noncooperative and full cooperative outcomes is small. When this difference is large, the full cooperative outcome can be sustained by only a few countries, or possibly by none at all.

Despite the arguably harsh assumptions that underpin these models, it is tempting to consider an actual IEA in the light of this analysis. The Montreal Protocol on Substances that Deplete the Ozone Layer, originally negotiated in 1987, has now been signed by about 80 countries, and in its original form committed each signatory to cut its production and consumption of the so-called hard CFCs by 50%.[18] This agreement is understandably considered to be a substantial achievement of international cooperation.[19] However, when seen in the context of this paper, an analysis of ozone depletion by the US Environmental Protection Agency (1988) suggests that the Montreal Protocol may not have increased global net benefits substantially compared with the noncooperative outcome. Assuming 'substantial levels of participation by other nations', the EPA estimates that the US would receive a benefit, primarily in the form of a reduction in the incidence of skin cancers, of $3,575bn, and incur a cost of $21bn, if it acceded to the treaty. If the US cut its production and consumption of CFCs by 50% unilaterally, the cost to the US would remain unchanged, but the benefit to the US would fall to only $1,373bn. While one would ideally like to know the marginal costs and benefits, these total benefit and cost figures strongly suggest that the US had an incentive to carry out the terms of the Montreal Protocol unilaterally. There is no compelling reason to believe that these benefit and cost estimates would be wildly different for most other countries in relative terms, and so, if the EPA's figures are broadly correct and consistent with the beliefs of the Montreal Protocol negotiators, it seems that this agreement may not have achieved as much as is commonly claimed. That conclusion is consistent with the analysis presented in this paper.

[18] The agreement has since been amended, and strengthened in a number of respects.
[19] But see the analysis by Bohm (1990).

REFERENCES

D'ASPREMONT, C. A., JACQUEMIN,, J., GABSZEWEIZ, J., and WEYMARK, J. A. (1983). 'On the Stability of Collusive Price Leadership', *Canadian Journal of Economics*, **16**, 17–25.

BARRETT, S. (1989). 'On the Nature and Significance of International Environmental Agreements', mimeo, London Business School.

BARRETT, S. (1992). 'International Environmental Agreements as Games', in R. Pethig (ed.), *Conflict and Cooperation in Managing Environmental Resources*, Springer-Verlag, Berlin.

BARRETT, S. (1994). 'The Biodiversity Supergame', *Environmental and Resource Economics*, **4**, 111–22.

BLACK, J., LEVI, M. D., and DE MEZA, D. (1992). 'Creating a Good Atmosphere: Minimum Participation for Tackling the "Greenhouse Effect"', *Economica*, **60**, 281–93.

BOHM, P. (1990). 'Efficiency Aspects of Imperfect Treaties on Global Public Bads: Lessons from the Montreal Protocol', mimeo, Department of Economics, University of Stockholm.

CARRARO, C. and SINISCALCO, D. (1991). 'Strategies for the International Protection of the Environment', mimeo, Fondazione E.N.I. 'Enrico Mattei'.

DASGUPTA, P. (1982). *Control of Resources*,, Harvard University Press, Cambridge, MA.

FARRELL, J. and MASKIN, E. (1989). 'Renegotiation in Repeated Games', *Games and Economic Behaviour*, **1**, 327–60.

HEAL, G. (1992). 'International Negotiations on Emission Control', *Structural Change and Economic Dynamics*, forthcoming.

HOEL, M. (1990). 'CO_2 and the Greenhouse Effect: A Game Theoretic Exploration', mimeo, Department of Economics, University of Oslo.

HOEL, M. (1992). 'International Environment Conventions: The Case of Uniform Reductions of Emissions', *Environmental and Resource Economics*, **2**, 141–59.

MÄLER, K.-G. (1989). 'The Acid Rain Game, 2', mimeo, Department of Economics, Stockholm School of Economics.

NORDHAUS, W. D. (1990). 'To Slow or Not to Slow: The Economics of the Greenhouse Effect', mimeo, Department of Economics, Yale University.

US ENVIRONMENTAL PROTECTION AGENCY (1988). *Regulatory Impact Analysis: Protection of Stratospheric Ozone*, Washington, D.C.

APPENDIX

Proof of Lemma 1 (i) The potential gains to cooperation are

$$\Pi_c - \Pi_o = a^2c\gamma(N-1)^2/2N(N+\gamma)(\gamma+1)^2 \tag{A.1}$$

Differentiating with respect to γ while holding c constant gives

$$\partial(\Pi_c - \Pi_o)/\partial\gamma \gtreqless 0 \quad \text{if} \quad (2\gamma^2 + \gamma N - N) \lesseqgtr 0 \tag{A.2}$$

This quadratic equation has one positive solution, $\gamma = \theta(N)$.

(ii) From (A.1) we see immediately that $\partial(\Pi_c - \Pi_o)/\partial c\,|_{\gamma=\bar{\gamma}} > 0$.

(iii) Notice that $\theta(N) = [\sqrt{(N^2 + 8N)} - N]/4$ can be rewritten

$$(1-\theta)/(2\theta^2) = 1/N \tag{A.3}$$

As $N \to \infty$, $1/N \to 0$, and hence $\theta \to 1$. □

Proof of Proposition 2 The solutions to the appropriate optimization problems are

$$q_n = \omega/c, \qquad q_s = \omega\alpha N/c$$

Substituting, we obtain

$$\pi_n(\alpha) = \omega^2 N(1 - \alpha + \alpha^2 N - 1/2N)/c \tag{A.4a}$$

$$\pi_s(\alpha) = \omega^2 N(1 - \alpha + \alpha^2 N/2)/c \tag{A.4b}$$

$$\pi_n(\alpha - 1/N) = \omega^2 N(1 - 3\alpha + 3/2N + \alpha^2 N)/c \tag{A.46c}$$

$$\pi_s(\alpha + 1/N) = \omega^2 N(1 - 1/2N + \alpha^2 N/2)/c \tag{A.4d}$$

Substituting (A.4a) and (A.4d) into (8) yields $\alpha \geqslant 2/N$. Substituting (A.4b) and (A.4c) into (8) and solving for the resulting quadratic equation yields $3/N \geqslant \alpha \geqslant 1/N$. Since we require that both of these conditions hold, the self-enforcing IEA consists of $3 \geqslant \alpha N \geqslant 2$ countries. Clearly, if $N = 2$, then both countries will sign the IEA. Upon substitution, it is easy to show that $\pi_s(3/N) = \pi_n(2/N)$. If it is assumed that an additional country will joint the IEA when $\pi_s(\alpha + 1/N) = \pi_n(\alpha)$, then the self-enforcing IEA will consist of three countries whenever $N \geqslant 3$. □

Proof of Lemma 2 (i) The solutions to the appropriate optimization problems are

$$q_n = a(1 - e^{-\omega/\sigma}), \qquad q_s = a(1 - e^{-\omega\alpha N/\sigma})$$

Substituting , we obtain

$$\pi_n(\alpha) = [\omega N - \sigma - \alpha\omega N\, e^{-\omega\alpha N/\sigma} + e^{-\omega/\sigma}(\alpha\omega N - \omega N + \sigma + \omega)] \tag{A.5a}$$

$$\pi_s(\alpha) = a[\omega N - \sigma - \omega N(1 - \alpha)\, e^{-\omega/\sigma} + \sigma\, e^{-\omega\alpha N/\sigma}] \tag{A.5b}$$

Part (i) of Lemma 2 can be proved by substituting the appropriate values for α into eqs (A.5a) and (A.5b).

(ii) Setting $\alpha = 2/N$ in (A.5b) and $\alpha = 1/N$ in (A.5a), and substituting, the inequality in (ii) implies

$$\omega/\sigma > 1 - e^{-\omega/\sigma} \tag{A.6}$$

Let $z = \omega/\sigma$. When $z = 0$, both sides of inequality (A.6) equal zero (and hence the inequality does not hold at $z = 0$). The LHS of inequality (A.6) rises with z at rate 1, while the RHS rises at rate e^{-z}. The LHS of (A.6) therefore rises more rapidly with z than the RHS. Since both sides of (A.6) equal zero when $z = 0$, the inquality holds $\forall z > 0$.

(iii) Upon substitution, the inequality in (ii) implies

$$-\alpha\omega N\, e^{-\omega\alpha N/\sigma} + \sigma\, e^{-\omega/\sigma} > \sigma\, e^{-\omega/\sigma}\, e^{-\omega\alpha N/\sigma} \tag{A.7}$$

$\forall \alpha \geqslant 2/N$.

First determine that (A.7) holds for $\alpha = 2/N$. Upon substituting, one obtains

$$1 - e^{-2\omega/\sigma} > 2w\, e^{-\omega/\sigma}/\sigma \tag{A.8}$$

Again, let $z = \omega/\sigma$. When $z = 0$, both sides of (A.8) equal zero. The derivative of the LHS is $2\,e^{-2z}$, while that of the RHS is $2\,e^{-z}(1 - x)$. When $\alpha = 2/N$, the derivative of the LHS of (A.8) exceeds that of the RHS provided $z > 1 - e^{-z}$. But this condition is the same as (A.6), which has already been shown to hold. Hence (iii) holds for $\alpha = 2/N$. Now take the derivative of both sides of (A.7) with respect to α. The derivative of the LHS of (A.7) exceeds the derivative of the RHS if

$$\alpha\omega N/\sigma > 1 - e^{-\omega/\sigma} \tag{A.9}$$

We know (A.9) holds for $\alpha = 1/N$. Hence, it must hold for $\alpha > 1/N$. Since (A.7) holds for $\alpha = 2/N$, it must therefore also hold for all $\alpha \geqslant 2N$. □

Proof of Proposition 4 Each non-signatory will equate its own marginal benefit with its own constant marginal cost, yielding the non-signatory reaction function

$$Q_n = a - dN/b - Q_s \tag{A.10}$$

Signatories will equate their combined marginal benefit with their individual marginal costs, yielding the reaction function

$$Q_s = a - d/\alpha b - Q_n \tag{A.11}$$

Solving (A.10) and (A.11) yields $\alpha N = 1$. From (A.10) we see that $Q_n + Q_s = a - dN/b$, which is the same abatement level achieved when $Q_s = 0$. □

[3]

Journal of Public Economics 52 (1993) 309–328. North-Holland

Strategies for the international protection of the environment

Carlo Carraro

Department of Economics, University of Udine, Via Tomadini 30, 33 100 Udine, Italy and Fondazione Eni 'Enrico Mattei', Milano, Italy

Domenico Siniscalco*

Faculty of Economics, University of Torino, Pza Albarello 8, 10 122 Torino, Italy and Fondazione Eni 'Enrico Mattei', Milan, Italy

Received July 1991, final version received June 1992

This paper analyses profitability and stability of international agreements to protect the environment in the presence of trans-frontier or global pollution. Each country decides whether or not to coordinate its strategy with other countries. A coalition is formed when conditions of profitability and stability (no free-riding) are satisfied. It is shown that such coalitions exist; that they tend to involve a fraction of negotiating countries; and that the number of signatory countries can be increased by means of self-financed transfers. However, expanding coalitions requires some form of commitment. Such schemes of commitment and transfers can even lead to cooperation by all countries.

1. Introduction

A large amount of pollutants are discharged in the atmosphere and water systems, as a result of human activity in each country. The emissions often affect other countries, as well as the global environment. In economic terms, each polluting country benefits from using the environment as a receptacle for emissions but, at the same time, is also damaged by environmental deterioration. While the benefit is related to domestic emissions only, the damage is related to both domestic and foreign emissions which negatively affect the environment. Hence, a problem of international externalities arises which, in the present institutional setting, can be solved only by voluntary agreements among sovereign countries. Such agreements have been quite common in

Correspondence to: D. Siniscalco, Fondazione Eni Enrico Mattei, Via S. Sofia 27, I-20122 Milano, Italy.

*We are grateful to K.J. Arrow, C. D'Aspremont, M. Hoel, K.G. Maler, P. Michel and two anonymous referees, for helpful comments and suggestions on a previous version of this paper. The usual disclaimers apply.

recent years,[1] and they seem to share some features: they are often characterized by cooperative behaviour among the individual countries involved; they usually have only a sub-group of the negotiating countries as signatories (partial cooperation); and they tend to use various forms of transfers, typically to the developing countries, as a key instrument for increasing the number of signatories.

The existing literature on the protection of the international environment does not sufficiently convey the characteristics of international agreements mentioned above. The traditional contributions on trans-national commons describe countries' environmental interaction as a one-shot Prisoner's Dilemma, where free-riding inevitably leads to the 'tragedy of commons' [for a discussion, see Ostrom (1990)]. In more recent works on the subject, the repetition of the Prisoner's Dilemma, under appropriate assumptions, can enlarge the set of equilibrium outcomes, and characterize situations where all countries cooperate [for a discussion, see Maler (1989), Barrett (1992)]. Partial cooperation and the role of transfers, however, are not usually considered in either approach.

This paper presents a general framework to analyse the profitability and stability of international agreements to protect the environment in the presence of trans-frontier or global pollution. In our analysis, international negotiations are modelled as games in which sovereign countries bargain over emission control. With respect to emission control, countries may choose to act either cooperatively or non-cooperatively. In the former case, cooperative agreements can involve a sub-group of countries, whose number can be expanded by means of 'self-financed' welfare transfers. The framework re-interprets results which were already presented in the recent environmental literature with reference to specific cases [e.g. Maler (1989), Newbery (1991), Barrett (1991, 1992), Hoel (1992), Kaitala et al. (1992)]. More importantly, it provides new results on the emergence of partial cooperation, transfers, and the role of commitment.

The main conclusions of the paper can be summarized as follows:

(i) the strategic interaction among countries in a common environment does not necessarily lead to the 'tragedy of commons', but there is a wide range of possible voluntary agreements to control emissions;

(ii) beyond non-cooperative emission control, there exist partial cooperative agreements among sub-groups of countries (coalitions) which are profitable and stable;

(iii) gains from partial cooperation can be used to expand existing coalitions by inducing other countries to cooperate using self-financed welfare transfers. To sustain broader coalitions by means of transfers, however, a minimum

[1]Experts quote more than 150 international agreements which have been signed to protect the environment in various regions. The protocols on CFC, and the ongoing negotiation on global warming provide examples of agreements at the global level.

degree of commitment must be introduced into the game, thus changing its rules. The various forms of commitment proposed in the paper are less demanding than a commitment of cooperation by all players, since they may involve only a fraction of the cooperating countries. Such schemes of partial commitment and transfers can even lead to cooperation by all countries.

The paper is organized as follows. In section 2, the general framework is introduced and some kinds of agreements which can lead to pollution control are defined. Results on stable coalitions, the role of transfers and commitment are also provided. In section 3, the main results and implications of the proposed framework are discussed. Finally, in section 4, some extensions of the model are proposed, together with the scope for further work.

2. The analytical framework

2.1. Players, payoffs, strategies

Consider n countries ($n \geqq 2$) that interact in a common environment, and bargain over emission control of a specific pollutant. Each country i benefits from using the environment as a factor of production and as a receptable for emissions. The welfare of each country, however, is negatively affected both by its own emissions x_i and by other countries' emissions x_{-i}, where x_{-i} is the vector $(x_1, \ldots, x_{i-1}, x_{i+1}, \ldots, x_n)$.

Country i's benefit and damage can be represented in the welfare function $P_i(x) = B_i(x_i) - D_i(x_i, x_{-i})$, where $B_i(x_i)$ denotes benefits arising from the use of the environment for production and consumption activities; $D_i(x_i, x_{-i})$ denotes damages (welfare losses) resulting from pollution emissions;[2] and $x \equiv (x_1 \ldots x_n) = (x_i, x_{-i})$ is the vector of all countries' emissions.

Let us consider the benefit function $B_i(x_i)$. It is assumed that a reduction in pollution, which can be achieved through domestic environmental policies, is costly and reduces benefits. The benefit function, which depends on abatement costs, is related to the level of technology, the economic structure, the level of development, and the endowment of resources of a country. By technology we mean more than the mechanical process of turning inputs into outputs; we mean useful knowledge and experience, institutions and organizational structures, norms and values that govern the processes of production and exchange.

The damage function $D_i(x_i, x_{-i})$ depends on a country's perception of the effects of emissions of a given pollutant, as well as on the evaluation of such effects. Consequently, the damage function is based on a subjective evaluation of environmental goods. The specific functional form of $D_i(\cdot)$ can be

[2]The parameters of the damage function may include the ith row of the transportation matrix $A = \{\alpha_{ij}\}$, where α_{ij} is the share of the country j's emissions affecting country i.

determined using appropriate models of environmental impact evaluation. In such models, an index usually summarizes both the measurement of the relevant physical damages and their evaluation.

Countries consider one pollutant at a time. Let δ_i be the maximum level of pollution emissions for country i. It is computed by maximizing environmental benefits $B_i(x_i)$ without taking into account the associated costs evaluated through the function $D_i(\cdot)$. Moreover, δ_i can be seen as a measure of a country's size and level of development. The 'emission game' between the n countries is defined by a triple (N, S, P), and by appropriate rules: as usual, $N=\{1\ldots n\}$ is the set of players, $S=S_1\times\cdots\times S_n$, where $S_i=[0,\delta_i]$, is the strategy space, and $P=(P_1(x)\ldots P_n(x))$ is the payoff vector. Complete information is assumed. Problems arising from asymmetric information will briefly be mentioned in section 4.

By taking into account reciprocal externalities, a country may decide whether or not to cooperate with other countries in order to reduce total emissions. At this stage, we assume that cooperative agreements are not binding. The decision regarding whether or not to cooperate is the outcome of a 'metagame' in which each country anticipates the choice (cooperative or non-cooperative) of the other countries, and the relative outcomes in terms of emission levels.[3]

Let us begin by analysing the outcomes of the game under alternative strategic combinations. First, we assume that countries play simultaneously and non-cooperatively. Thus, country i's optimal level of emissions is determined by equating marginal benefits with marginal costs, given the emission levels set by the other countries. The solution of the system of first-order conditions determines the Nash equilibria of the game. For simplicity's sake, we assume the equilibrium to be unique. The Nash equilibrium of the non-cooperative game can also be determined by computing the fixed point of countries' best-reply functions. Let $R_i(x)$, $x=(x_1\ldots x_n)$, the country i's best-reply function, where $R_i(x)=\{x_i\colon P_i(x_i,x_{-i})\geqq P_i(s_i,x_{-i})$, for all $s_i\in S_i\}$. The non-cooperative equilibrium x^0 is defined by $x^0=R(x^0)$, where $R(x)=(R_1(x)\ldots R_n(x))$.

Alternatively, countries can decide to set emissions cooperatively. In this case, we assume that a bargaining process takes place in order to achieve a

[3]We restrict our analysis to one-shot games. In terms of additional equilibrium outcomes, analysing repeated games would be fruitful only if appropriate trigger or stick/carrot strategies could sustain cooperation as an equilibrium outcome. The level of emissions, however, can hardly be conceived as a trigger variable which can be increased strategically in response to other countries' defection. Some reasons are the following: firstly, emission reduction, in cases such as CO_2 or CFC, involves substantial and irreversible investments. Secondly, expanding emissions as a retaliation could generate environmental damage primarily to the triggering country. Finally, an increase in emissions can hardly be used as a selective punishment. Other effective punishments (e.g. trade protectionism) could be even more costly for the triggering country and therefore, not credible. For these reasons, we believe that trigger or stick/carrot strategies are not particularly helpful in sustaining environmental cooperation.

Pareto optimal outcome. The bargaining process may lead to the formation of a coalition among j countries, where j goes from 2 (the smallest feasible coalition) to n (when all countries set emissions by taking into account reciprocal externalities). We define 'full cooperation' as a situation in which a coalition formed by n countries emerges, while a 'partial cooperation' is a coalition formed by $2 \leq j < n$ countries. In this work, we determine the cooperative outcome of the game by using the Nash bargaining solution.[4] Moreover, we use the non-cooperative equilibrium $x^0 = (x_1^0 \dots x_n^0)$ as the threat point of the bargaining process.[5] Stated more formally, j countries will act cooperatively when they set emissions maximizing the joint product of the difference between $P_j(x)$ and P_i^0, where P_i^0 denotes the non-cooperative welfare.

Before setting emission levels, each country must therefore decide whether or not to act cooperatively. This decision can be modelled by defining a 'metagame' in which countries choose between the cooperative and the non-cooperative strategy by anticipating the outcomes of the related emission game.[6]

2.2. *Profitability and stability of coalitions*

Let $P_i(j)$ be country i's welfare when it decides to cooperate, and $j-1$ when other countries also cooperate; whereas $Q_i(j)$ is its welfare when country i does not join the coalition formed by j countries. Moreover, let j stand for the set of cooperating countries, and J^0 denote the set of countries that play non-cooperatively. Let us suppose, for simplicity, that all countries are symmetric, i.e. the welfare function is not country specific. We do not therefore index the welfare functions P and Q and their parameters.

A minimum requirement must be met for an environmental coalition to be formed: the welfare of each country signing the cooperative agreement must be larger than the non-cooperative welfare. In other words, country i gains from joining the coalition, with respect to its position when no countries cooperate, if $P(j) > P^0$. This leads to:

Definition 1. A coalition formed by j players is *profitable* if $P(j) > P^0$ for all countries belonging to J.

[4]This assumption is not crucial, because, in the rest of the analysis, all countries will be assumed to have the same benefit and damage functions. As was pointed out by a referee, in this case any bargaining solution in the literature would give the same result.

[5]In a two-player game, this means interpreting Rubinstein's alternating offers model as a model in which players face a risk that, if the agreement is delayed, then the opportunity they hope to exploit it jointly may be lost [see Binmore et al. (1986)].

[6]Most environmental studies model this 'metagame' as a one-shot Prisoner's dilemma, in which non-cooperation is the dominant strategy.

Of course, this only represents a minimum requirement that may not suffice to induce many countries to sign a cooperative agreement. The main problem preventing the formation of any coalition is the possibility of free-riding by some countries. Free-riding can be explained as follows: since one country can profit from the reduction of emissions by cooperating countries, it has an incentive to let other countries to sign the cooperative agreement. If all countries are symmetric, no cooperation takes place. In other words, the 'metagame' in which countries choose between cooperation and non-cooperation is represented as a Prisoner's dilemma. As we will see, however, this representation of countries' strategic choice may not be appropriate.

The problem can also be stated more formally. For each country, the crucial comparison is between $P(j)$, the country's payoff for belonging to the j-coalition, and $Q(j-1)$, the country's payoff when it exists the coalition, and lets other $j-1$ countries sign the cooperative agreement. Let $Q(j-1)-P(j)$ be a country's incentive to defect from a coalition formed by j players, whereas $P(j+1)-Q(j)$ is the incentive for a non-cooperating country to join a j-coalition [which, consequently, becomes a $(j+1)$-coalition]. Thus, a stable coalition can be defined as follows:

Definition 2. A coalition formed by j players is *stable* if there is no incentive to defect, i.e. $Q(j-1)-P(j)<0$, for all countries belonging to J, and there is no incentive to broaden the coalition, i.e. $P(j+1)-Q(j)<0$, for all countries belonging to J^0.

This definition corresponds to that of cartel stability presented in the oligopoly literature [see D'Aspremont and Gabszewicz (1986), a similar definiton is also used in Barrett (1991)].

It has been shown that under fairly general conditions stable coalitions exist [see Donsimoni et al. (1986)]. However, this does not satisfactorily address the problem of protecting international commons because, as has been demonstrated both in the oligopoly and in the environmental literature [see, for example, D'Aspremont et al. (1983), D'Aspremont and Gabszewicz (1986), Carraro and Siniscalco (1991), Hoel (1992)], stable coalitions are generally formed by $j \leqq n$ players, where j is a small number, regardless of n.[7] This leads us to the following question: can the j cooperating countries expand the coalition through self-financed welfare transfers to the remaining players?

Given the previously stated rules of the game, the answer is no. This is demonstrated by the following proposition:

[7]More satisfactory results are presented in Barrett (1991), who shows that, under certain conditions, and given a specific functional form for the welfare function, large stable coalitions exist.

Proposition 1. Suppose no countries can commit to the cooperative strategy. Then, in this case, no self-financed transfer T from the j cooperating countries to the other non-cooperating countries can successfully enlarge the original coalition.

Proof. For the transfer to be self-financed, it cannot be larger than the gain that the j players obtain from moving to a $(j+1)$-coalition. Furthermore, in order to add one player to a j-coalition, the transfer T must be larger than the loss incurred by the $j+1$ player by entering it. These two conditions yield:

$$[P(j+1)-P(j)] \geqq T > Q(j)-P(j+1). \tag{1}$$

This condition makes it possible to self-finance an enlarged coalition. However, is this broadened coalition stable? The $j+1$ player does not defect if the transfer is larger than $Q(j)-P(j+1)$. However, by definition of stable coalition, $P(j+1)<Q(j)$: the j players of the original coalition have therefore an incentive to defect; their incentive being greater because of the transfer made to the $j+1$ player. Hence, the $(j+1)$-coalition is unstable. Q.E.D.

This leads to the following conclusion: welfare transfers from countries belonging to a stable coalition to non-cooperating countries cannot be used to expand the initial coalition, unless the rules of the game are changed. There are various rules that can lead to the formation of larger stable coalitions. We will focus our attention on the role of commitment. If all countries were committed to cooperation, obviously no free-riding would exist. We show, however, that there are several, less demanding forms of commitment which, if associated with appropriate welfare transfers, can lead to large stable coalitions. There are two crucial elements in our analysis: partial commitments (only a subset of the n countries commit to cooperation), and welfare transfers (bribing).

2.3. Expanding coalitions

Hereafter, we shall analyse the implications of four types of commitment, that could serve as possible blueprints for environmental cooperation. Of course, other types of institutional mechanisms could be proposed as well. The four types of commitments are:

(i) Only the j countries belonging to the stable coalition commit to cooperation (stable coalition commitment).

(ii) The j countries are committed to cooperation and any new signatory, as soon as it enters the expanded coalition, must commit to cooperation as well (sequential commitment).

(iii) The number of committed countries is such that appropriate transfers can induce all other countries to cooperate (full-cooperation minimum commitment).

(iv) A subset of non-cooperating countries commits to transfer welfare in order to induce the remaining non-signatories to cooperate, and to guarantee the stability of the resulting coalition (external commitment).

As has already been shown, it is necessary to impose constraints on the amount of transfers allowed: if no restriction were imposed, all non-signatories could be bribed. Therefore, we assume:

(i) transfers must be self-financed, i.e. the total transfer T must be lower than the gain that the committed countries obtain from expanding the coalition;

(ii) the move to a larger coalition must be Pareto-improving, i.e. all countries must be better off than in the situation preceeding the coalition expansion, and better off with respect to the case of no-cooperation (the larger coalition must be profitable); and

(iii) committed countries choose the transfer in order to maximize the number of signatories (given the above two constraints).[8]

Given these restrictions, the following question can be asked: Under what conditions can partial commitments and transfers expand the initial j-coalition? Some answers are provided by the following propositions:

Proposition 2 (Stable coalition commitment). *Suppose the j countries belonging to the stable coalition are committed to cooperation. If $P(j+s)>P(j)$ and $Q(j+s)>Q(j)$ for all positive s, $s \leqq n-j-1$, then at most r countries can be induced to join the initial coalition, where r is the largest integer satisfying:*

$$r<j[P(j+r)-P(j)]/[Q(j+r-1)-P(j+r)]. \tag{2}$$

Proof. The initial j countries can use their gains resulting from broadening the coalition in order to finance other countries' cooperation. This gain is equal to $j[P(j+r)-P(j)]$, and it is positive if $P(j+r)>P(j)$. In order for transfers to be self-financed, they must be greater than the incentive to defect for the r countries that have to enter the coalition. This incentive is equal to $r[Q(j+r-1)-P(j+r)]$. Hence:

$$j[P(j+r)-P(j)]>r[Q(j+r-1)-P(j+r)]. \tag{2'}$$

Moreover, the maximum transfer $j[P(j+r)-P(j)]$ must be larger than the

[8]Notice that transfers enter a country's payoff in an additive way. Otherwise, each cooperating country would maximize its payoff with respect to both emissions and transfers. This would be the case if transfers were carried out using policy instruments which interact with other economic variables, and with emissions in particular. Such an extension of the paper is discussed in the concluding section.

total loss suffered by entering countries; this loss is equal to $r[Q(j)-P(j+r)]$. This condition can be expressed as follows:

$$j[P(j+r)-P(j)]>r[Q(j)-Q(j+r-1))+(Q(j+r-1)-P(j+r))]. \quad (3)$$

Notice that $[Q(j)-Q(j+r-1)]\leqq 0$ and $[Q(j+r-1)-P(j+r]>0$. Hence, (2′) implies (3). Since (2) implies (2′), the proposition is proved. The newly entered countries have no incentive to defect, and benefit from joining the coalition; moreover, the initial cooperating countries benefit from expanding the coalition, and are committed to cooperation. Therefore, the new equilibrium constitutes a Pareto improvement. Q.E.D.

The meaning of Proposition 2 is the following: starting from a j-coalition, the commitment of its members induces, through appropriate transfers, other r countries to join the initial coalition; notice that r is larger, the larger the gain attained from expanding the coalition, and the lower the incentive to defect from it.

Proposition 3 (Sequential commitment). *Suppose that a stable coalition formed by j countries exists. If each country, when entering the coalition, commits to cooperation, and if* $P(j+s)>P(j+s-1)$ *for all* $1\leqq s\leqq r$, *then at most r countries can be induced to join the initial coalition if*

$$j+s>[Q(j+s-1)-P(j+s-1)]/[(j+s)-P(j+s-1)] \quad (4)$$

for all $1\leqq s\leqq r$, *and, when* $j+r<n$,

$$j+r<[Q(j+r)-P(j+r+1)]/[P(j+r+1)-P(j+r)]. \quad (4')$$

Proof. Suppose that j countries form a stable coalition, and are committed to cooperation. For one more country to be induced to cooperate, the transfer must be lower than or equal to the total gain derived from moving to a $(j+1)$-coalition, and larger than the loss incurred by the entering country. This gives

$$j[P(j+1)-P(j)]>Q(j)-P(j+1)>0. \quad (5)$$

Suppose this condition is satisfied and that the $(j+1)$th country, when entering the coalition, commits to cooperation. An additional country can be induced to join the coalition if

$$(j+1)[P(j+2)-P(j+1)]>Q(j+1)-P(j+2)>0. \quad (6)$$

By iterating this reasoning, one can say that r countries can be induced to join the initial coalition through appropriate transfers, and the sequential commitment, if

$$(j+s-1)[P(j+s)-P(j+s-1)]>Q(j+s-1)-P(j+s)>0 \quad (7)$$

for all $1 \leqq s \leqq r$. If this condition is satisfied for all s such that $j<s \leqq n-j$, then all countries will join the coalition. Thus, full cooperation can be achieved. Otherwise, the coalition cannot be expanded further if

$$(j+r)[P(j+r+1)-P(j+r)]<Q(j+r)-P(j+r+1). \quad (7')$$

Full cooperation cannot always be achieved because there may exist values of r for which the gain from further broadening the coalition is lower than the loss incurred by the entering country, i.e. the transfer would not be sufficient to induce one more country to join the coalition.

Eq. (7) can be re-written as

$$j+s>[Q(j+s-1)-P(j+s-1)]/[P(j+s)-P(j+s-1)] \quad (8)$$

for all $1 \leqq s \leqq r$, and

$$j+r<[Q(j+r)-P(j+r+1)]/[P(j+r+1)+P(j+r)]. \quad (8')$$

It should be noted that, when the $(j+r)$-coalition is formed, all countries gain from broadening the coalition. Moreover, given the assumption regarding commitment, the coalition is stable. Q.E.D.

Proposition 3 has two implications. First, as in the case of stable coalition commitment, the expanded coalition is larger, the larger the gain from moving to a wider coalition, and the lower the incentive to defect from it. Secondly, this form of commitment is more demanding than the previous one, since it eliminates the problem of guaranteeing the coalition stability (by the sequential commitment assumption). Hence, unless (4′) holds for some $r<n-j$, full cooperation can be achieved.

Proposition 4 (Full-cooperation minimum commitment). *If* $Q(n-1)>Q(i)$ *for all positive* $i<n-1$, *at least a fraction*

$$i/n>[Q(n-1)-P(n)]/[Q(n-1)-P(i)] \quad (9)$$

of the n countries must be committed to a cooperative strategy for all n countries to cooperate.

Proof. Let us assume that i countries are committed to cooperation regardless of the size of the coalition that is formed. Suppose the i countries offer transfers to induce the other $(n-i)$ countries to join the coalition. The total transfer T must be less than or equal to the gain that the i players achieve from entering the n-coalition. Moreover, this transfer should compensate the $(n-i)$ players for the loss from joining the coalition, and should also offset their incentive to defect from the n-coalition. In order to compensate the $(n-i)$ players for the loss from joining the n-coalition, the following condition must be met:

$$i(P(n)-P(i)) \geqq T > (n-i)(Q(i)-P(n)). \tag{10}$$

This condition ensures that the enlarged coalition is self-financed. It can be re-written as

$$i[Q(i)-P(i)] > n[Q(i)-P(n)]. \tag{10'}$$

In order to offset the incentive to deviate from the n-coalition, the gain $P(n)$ plus the transfer $i(P(n)-P(i))/(n-i)$ must be larger than the defector's welfare $Q(n-1)$, i.e. rearranging the equation

$$i(Q(n-1)-P(i)) > n(Q(n-1)-P(n)), \tag{11}$$

which is equivalent to (9). Notice that both sides of the equation are positive. Eq. (11) can be re-written as

$$i[(Q(n-1)-Q(i))+(Q(i)-P(i))] > n[(Q(n-1)-Q(i))+(Q(i)-P(n))]. \tag{11'}$$

Let us show that (11′) implies (10′). Assume that (11′) holds as an equality. Thus, we can solve it with respect to $Q(i)-P(i)$. This expression can then be substituted into eq. (10′), yielding $(n-i)[Q(n-1)-Q(i)]>0$, which is satisfied for all positive $i<n-1$. Hence, condition (9) guarantees that both the financing condition (10) and the no-defection condition (11) are satisfied. As a consequence, the $(n-i)$ players that join the initial coalition have no incentive to defect. The initial i players are instead committed to co-operation. Finally, the move to the n-coalition is a Pareto improvement. Q.E.D.

Proposition 4 states that there exists a minimum number of countries that should commit to cooperation for all the remaining countries to be induced to cooperate using appropriate transfers. The minimum number of committed cooperating countries decreases as the gain from moving from the i-coalition to the n-coalition increases, and as the incentive to defect from the latter decreases.

The last case we would like to explore considers how an environmental coalition can be expanded using external commitment. Non-cooperating countries gain when the cooperative agreement is expanded (because they receive less emissions). Consequently, a subset of these countries could find it profitable to induce other non-signatories to enter the coalition, and to secure its stability. Suppose, then, that a fraction of non-cooperating countries commits to finance environmental cooperation (emission reduction in other countries). What is the largest number of countries that can be induced to form a stable coalition?

Proposition 5 (External commitment). *Suppose that j countries cooperate, and that $n-j-r$ non-cooperating countries commit to transfer welfare both to induce r non-signatories to cooperate, and to guarantee the stability of the resulting coalition. If $P(j+s)>P(j)$ and $Q(j+s)>Q(j)$ for all positive $s \leqq n-j-1$, a stable $(j+r)$-coalition can be formed if*

$$(j+r)/n < 1/(1+\theta), \tag{12}$$

where $\theta = [Q(j+r-1)-P(j+r)]/[Q(j+r)-Q(j)]$.

Proof. Assume there exists a stable j-coalition. The $(n-j-r)$ countries who do not join the coalition gain from financing, through appropriate transfers, a larger coalition, if $Q(j+r)$, their payoff when the $(j+r)$-coalition is formed, less $(Q(j+r-1)-P(j+r))(j+r)/(n-j-r)$, the transfer to the $(j+r)$ cooperating countries, is larger than $Q(j)$, their payoff before broadening the coalition. This is true if (12) holds. Moreover, $Q(j+r)-(Q(j+r-1)-P(j+r))(j+r)/(n-j-r)$ must be larger than $P(j+r+1)$, i.e. no additional countries want to join the coalition. This is true if

$$(n-j-r)[Q(j+r)-P(j+r+1)] > (j+r)[Q(j+r-1)-P(j+r)], \tag{13}$$

which can be written as

$$n[Q(j+r)-P(j+r+1)] > (j+r)[Q(j+r)-P(j+r)+Q(j+r-1) - P(j+r+1)]. \tag{13'}$$

Comparing (12) and (13′), it is easy to see that (12) implies (13′) [and therefore (13)].

Finally, we have to prove that the players in the $(j+r)$-coalition have no incentive to defect. This is true if $P(j+r)$, the welfare when the $(j+r)$-coalition is formed, plus $Q(j+r-1)-P(j+r)$, the transfer each cooperating country receives, is not lower than $Q(j+r-1)$, the welfare that each cooperating country would receive by defecting from the coalition. This implies $Q(j+r-1) \geqq Q(j+r-1)$, which obviously holds (we assume that whenever a country is indifferent between cooperation and defection, it cooperates). Moreover, $P(j+r)+[Q(j+r-1)-P(j+r)]$ is larger than $P(j)$, the welfare that countries in the stable coalition received before its expansion, because $P(j+r)>P(j)$ by assumption and $Q(j+r-1)-P(j+r)>0$ by the stability condition; it is also larger than $Q(j)$, the welfare that countries entering the coalition received before, because $Q(j+r-1)>Q(j)$ by assumption.

As a consequence, all players in the $(j+r)$-coalition do not defect, all players outside the coalition do not want to join it, and the move to a $(j+r)$-coalition constitutes a Pareto improvement. Q.E.D.

The conclusion reached is the following: r additional countries can be induced to cooperate, and the $(j+r)$-coalition is stable if the remaining $(n-j-r)$ non-cooperating countries commit to carry on appropriate transfers to the $(j+r)$ cooperating countries. The dimension of the resulting coalition increases as the incentive to defect from a $(j+r)$-coalition decreases, and as the gain that non-cooperators achieve from moving to a $(j+r)$-coalition increases.

3. Results and applications to environmental agreements

3.1. Some general comments

The framework proposed in the previous section aims at explaining the emergence of environmental cooperation without the help of trigger or stick/carrot mechanisms. Our approach was chosen because, to our knowledge, no international negotiation to protect the environment has ever used pollution as a triggering variable, and because partial cooperation and transfers seem to be common features of many recent agreements in this field.

The crucial steps of the analysis are two: we first characterize stable coalitions, and give some formal conditions for their existence (subsection 3.2). We then use such stable coalitions as a starting point for wider coalitions, expanded by means of transfers and commitments (subsection 3.3). The existence of small stable coalitions is a result already obtained in the most recent environmental literature. Subsection 3.3 on expanded coalitions provides some new results.

The whole framework deserves some general comment, at this stage. If stable coalitions exist, the 'metagame' in which countries decide whether or not to cooperate is not a Prisoner's dilemma. Assume that a j-stable coalition is formed: by definition, no incentive to defect exists $(Q(j-1)<P(j))$. All countries, however, have an additional incentive not to cooperate. Since non-cooperating countries gain from the others' cooperative behaviour, each country has an incentive to let other countries form the coalition $(Q(j)>P(j))$. This is not a Prisoner's dilemma because the situation in which

Table 1

		Country h	
		C	N
Country i	C	$P(j+1), P(j+1)$	$P(j), Q(j)$
	N	$Q(j), P(j)$	$Q(j-1), Q(j-1)$

one group of countries cooperates and the others do not is an equilibrium of the 'metagame'. This is shown by the 2×2 example in table 1.

In this table, C and N denote the cooperative and non-cooperative strategies, respectively, and the payoff pairs indicate countries' welfare, as defined in the previous section. Table 1 represents a situation in which $j-1$ countries cooperate. A stable coalition is formed by j countries. Countries i and h are the marginal countries with respect to the stable coalition. Both i and h have an incentive to join the coalition (by definition of stability). Country i's most preferred outcome is the one in which it lets h cooperate. However, if country h does not act cooperatively, country i will choose to do so, in order to belong to the stable coalition (by definition of stability). Formally, this is implied by the following inequalities: $Q(j) > P(j+1) > P(j) > Q(j-1)$. The first and the last inequalities are implied by the stability of the j-coalition; $P(j+1) - P(j)$ holds by assumption (see subsections 2.2 and 2.3). Hence, non-cooperation is not the dominant strategy. This game is known as a *chicken game* (a game belonging to the class of coordination games). There are two equilibria (N, C) and (C, N), but all countries have an incentive to let the others cooperate.[9] The game has no dominant strategy; countries' attempts to choose non-cooperation, in order to let the others cooperate, may lead to the worst possible outcome (N, N). The cooperative outcome (C, C) is not Pareto optimal.

A stable coalition can be expanded by transfers to non-cooperating countries, provided some form of commitment takes place. The intuition behind this result is simple. Welfare transfers to non-cooperating countries decrease by T the payoff of the countries belonging to the j-coalition, preserving profitability (transfers are self-financed), but creating instability. The instability has to be dealt with by some form of commitment. As we

[9]The impasse can be solved by the introduction of asymmetries into the game. If countries have different preferences, technology or environmental endowment, it is possible to determine which countries are likely to form a coalition. For example, in the case of external commitment, countries with higher abatement costs are likely to finance emission reductions in countries with lower abatement costs, who therefore form the coalition. In the case of stable coalition commitment, countries in which environmental policy is part of a package of coordinated policies, or large countries that heavily affect the global environment, are more likely to commit themselves to cooperation, thus attracting other cooperators. If the game were repeated, countries with higher discount rate would be more likely to form a coalition.

anticipated in the introduction, the various forms of commitment we analyse are less demanding than the commitment by all players, assumed in cooperative games. In our framework, the commitment of only a fraction of the n countries can ensure the stability of wider coalitions, and can even lead to full cooperation.

Notice, moreover, that welfare transfers are indeed impossible and/or inefficient if they are based only on emission reductions. The simplest instrument to transfer wealth is probably cash. There are, however, other appropriate instruments, such as trade or debt policy and technology transfers.

3.2. Stable coalitions and best-reply functions

Over and above the previous comments, a number of specific questions deserve answers: Under what economic conditions does a stable coalition exist? How many countries belong to the stable coalition? Which form of commitment is likely to increase the number of countries belonging to the stable coalition? What is the size of the largest coalition in each different case?

To answer such questions, it is necessary to specify a particular form for the benefit and damage functions of the different countries. Under such restriction, the environmental literature does provide some helpful insights into stable coalitions: for example, Barrett (1991), Carraro and Siniscalco (1991) and Hoel (1992) show that stable coalitions for the protection of the environment exist under reasonable specifications of the benefit and damage functions. The same conclusion is reached in the cartel stability literature [see D'Aspremont et al. (1983) and Schmalensee (1987), for example]. More general results on the existence of stable coalitions in oligopolistic markets can be found in D'Aspremont and Gabszewicz (1986) and Donsimoni et al. (1986). These results prove that there are cases in which the metagame describing countries' interaction in environmental negotiations is not a Prisoner's dilemma. These works also show that stable coalitions are generally formed by a subset of all players of the game, and that this subset is often small. Only Barrett (1991), using a specific example, provides numerical simulations in which the number of cooperating countries approximates n.

In addition, there is one result which appears in most of the works on the international environment, but is seldom discussed: the pattern of interdependence among countries, as described by the slope of their best-reply function, is crucial for understanding the effectiveness of cooperative and non-cooperative emission control. The reason being that the more negative the slope of the best-reply functions, the larger the incentive to deviate from any coalition [i.e. $Q(j-1)-P(j)$ in the game]. Drawing from results proved in

Carraro and Siniscalco (1991), we would like to clarify the role played by the slope of countries' best-reply function.

Let us first consider non-cooperative emission control. Countries that interact in a common environment with mutual externalities set their emissions by equating their own marginal benefit to marginal damage, given the emissions set by the other countries. In this context, country *i*'s actual emissions are generally lower than emissions δ_i, which maximize its benefit function. Moreover, non-coopeative emissions are increasingly reduced as the slope of the best-reply functions become increasingly negative. The literature shows this is the case when the impact of foreign emissions for country *i* is high, the perceived damage is high, and the benefit (abatement cost) is low [on this point see Barrett (1991) and Carraro and Siniscalco (1991)].[10]

The reason that lies behind this result is intuitively simple. The best-reply functions reflect, inter alia, the marginal damage produced by foreign countries' emissions. If this marginal damage to country *i* is relevant, then the best non-cooperative response of country *i* to an expansion of foreign emission is an emission reduction. This reduction will be greater, the lower the benefit of domestic emissions, and vice versa. The difference between δ_i and the actual non-coopertive emissions of country *i* is therefore positively related to the slope of country *i*'s best-reply function.

Let us now consider cooperation. In this case, countries bargain over emission levels in order to achieve an optimal aggregate outcome, taking into account reciprocal externalities. As is well known, cooperation among all countries is profitable and optimal, but it is intrinsically undermined by free-riding. However, if the best-reply functions are orthogonal or near orthogonal, there is some scope for partial cooperation in the form of agreements among small groups of countries, which can be profitable and stable. The reason, again, has to do with free-riding behaviour, as reflected by the best-reply function of a country which does not belong to the partial coalition. If this best-reply function is negatively sloped, the non-cooperating country will expand its emissions if the coalition restricts them, offsetting the effort of the cooperating countries. If, on the contrary, the best-reply functions are orthogonal or near orthogonal, the free-rider will simply enjoy the cleaner environment without paying for it, but will not offset the emission reduction by the cooperating countries.

Similarly, with negatively sloped best-reply functions, the number of countries is also crucial for the existence of stable coalitions. If n is large, a stable coalition is unlikely to exist, as a cooperative contraction by few

[10]Notice that high perceived damage, low abatement cost, and high impact of foreign emissions are only necessary conditions for the negative slope of the best-reply functions. Separable damage functions imply orthogonal best-reply functions, whatever the other parameters.

countries is offset by the reactions of many others [see Barrett (1991) and Carraro and Siniscalco (1991)].

The above considerations suggest that, in environmental agreements, there is a sort of trade-off. When the best-reply functions are negatively sloped there is a high degree of interdependence. Likewise, non-cooperative emission control can lead to substantial emission reduction. But if one or more countries unilaterally or cooperatively reduces emissions, this contraction is offset by an expansion by the non-cooperating countries. This kind of interaction undermines all kinds of cooperation, as the free-riding behaviour implies a substantial loss for countries who wish to cooperate. With an orthogonal or near-orthogonal best-reply function the situation is somehow reversed. Non-cooperative emission control leads to small emission reductions, but the scope for cooperation is now greater: if a number of countries cooperatively reduce their emissions, this reduction is not offset by free-riders, who simply enjoy a better environment but do not directly damage countries which cooperate. In this case, then, stable coalitions exist.

3.3. Transfers and commitments

The result that stable coalitions are formed by a subset of the n countries, and that this subset is often small, led us to consider strategies for expanding stable coalitions. Coalition expansions can be achieved through welfare transfers and some form of commitment. How much can a stable coalition be expanded? Which type of commitment leads to full cooperation?

Proposition 1 in section 2 holds for any form of the benefit and damage function: it shows that any attempt to expand a stable coalition by means of transfers is flawed without some form of commitment. This point, which may be relevant to the environmental policy debate, has already been discussed in section 2 and subsection 3.1.

We have attempted, therefore, to explore various forms of commitment in order to sustain expanded coalitions.

The 'stable coalition commitment' (Proposition 2) is the first to be discussed. Let us suppose a stable coalition exists. If all members of the coalition commit to cooperation, they can use the gains resulting from moving to a larger coalition to bribe other countries. How many countries can be induced to enter the coalition using this strategy?

The existing environmental literature is not very helpful in addressing these issues, as they are relatively new. One specific example we provide elsewhere [Carraro and Siniscalco (1991)], assuming symmetric countries and linear-quadratic benefit and damage functions, shows that the answer depends once more on the slope of the best-reply functions. With orthogonal (or near-orthogonal) best-reply functions, a stable coalition of three countries can induce four other countries to cooperate, irrespective of n. When the best-reply functions become negatively sloped, the possibility of bribing other countries gradually decreases to zero.

Besides the example mentioned above, we believe that this type of commitment can be relevant to capture some features of the current negotiation on global warming: why should a group of countries which already cooperate (say EC countries) commit and transfer 'new and additional resources' to other countries (say the LDCs)?

Sequential commitment, discussed in Proposition 3, is possibly less realistic, but shows how it is possible to reach wider coalitions. Countries belonging to a stable coalition commit to cooperation and bribe other countries. As any of the latter countries joins the coalition, it must commit in order to further expand the coalition. In this way, it is easy to show that, with orthogonal (or near-orthogonal) best-reply functions, sequential commitment leads to full cooperation, as the conditions on the stability of the coalition itself are removed by this form of commitment.

Full cooperation can also be achieved if about 60 percent of the n countries commit to cooperation (we are referring here to Proposition 4), while the external commitment of Proposition 5, again with orthogonal or near-orthogonal best-reply functions, can induce about 70 percent of the n countries to cooperate. The latter case can be helpful in understanding the recent proposal by some industrial countries (e.g. the Scandinavian countries) to subsidize environmental programmes in other countries [e.g. Eastern Europe; see Kaitala et al. (1992)].

As the above discussion has already pointed out, if the best-reply functions are negatively sloped, all the results are generally less favourable. In particular, sequential commitment does not lead to full cooperation, and the impact of each type of commitment decreases as the slope of the best-reply functions increases (in absolute value).

As examples can provide only anecdotes and special results, we believe that a better understanding of environmental negotiations can only come from serious applied work. Only empirical work can justify alternative specifications of countries' interdependence. Only empirical work, moreover, can support an intuition we submit: while in the traditional case of common-property goods (fisheries, pastures, forests, etc.) the payoff functions give rise to non-orthogonal best-reply functions, in the case of some global pollutants, e.g. CO_2 or CFC, the best-reply functions are probably orthogonal (or near orthogonal).[11]

[11]Indeed, we can hardly think of any countries that expand their own CO_2 or CFC emissions in response to other countries' reductions. As the above discussion has suggested, this leaves room for cooperative agreements.

4. Conclusions and scope for further work

In the next few years the intenational protection of the environment will increasingly rely on international agreements, although they often involve substantial difficulties. To what extent can the proposed analysis be useful, and how can it be extended?

The analytical framework we proposed is highly simplified and the results obtained must be interpreted with great caution. However, given the difficulties and failures of many attempts to reach global agreements, it shows a promising route for research and policy analysis.

Firstly, the framework and the results show that it is rather sterile to study optimal agreements among all countries, if such agreements are profitable but intrinsically unstable. The structure of such agreements can be useful as a benchmark, but it is very unlikely to be realized in practice. Secondly, they show that there is a full range of possible agreements among sovereign countries to protect the international environment. There are cases in which an effective protection can be obtained non-cooperatively. In other cases, an effective environmental protection can be reached through partial co-operation and transfers.

Coming to the scope for further work, a number of relevant issues have still to be addressed. First, it would be useful to provide a sort of taxonomy relating the various pollutants to appropriate damage functions. Only then would it be possible to contextualize policy analysis, obtaining meaningful results for each case. Secondly, it would be useful to re-appraise the instruments to implement cooperation. Emissions, in many cases, are very difficult to monitor. The various economic instruments needed to implement an agreement, therefore, must be designed in a way that prevents cheating. So far, the literature has compared the various agreements in terms of efficiency, i.e. maximum profitability. Our analysis proposes another criterion: an instrument must be efficient, but it must also be effective in preventing or discouraging free-riding. In other words, it must also be designed to promote the stability of the agreements.

Finally, three extensions should be attempted:

(i) Asymmetric information should be introduced. Countries' preferences can hardly be observed. If we remove the assumption of complete information, each country that is induced to enter a coalition would be tempted to overstate the damage and claim for greater incentives. The solution to this problem is to embody an appropriate information or self-selection premium in the transfer to each country that enters the coalition.

(ii) It would be relevant to introduce some asymmetries into the game (by assuming for example different damage and abatement costs across countries), in order to evaluate coalition profitability and stability when countries have different incentives to join it.

(iii) The benefit function should account for the interaction between environmental variables and the policy instruments designed to carry on transfers. This prevents the analysis of environmental policy as such, but implies its integration into a wider analysis which accounts for other economic variables in each country's payoff.

A last point concerns institutions. In the present setting there is not an institution that has the authority to impose supernational regulations on countries and regions. Stable and expanded coalitions can be seen as a first step towards such institutions, as the member countries would find it costly to individually perform the transfer, the monitoring, and the enforcement activities which are usually associated with the management of a cooperative agreement.

References

Barrett, S., 1991, The paradox of international environmental agreements, Mimeo. (London Business School).

Barrett, S., 1992, International environmental agreements as games, in: R. Pethig, ed., Conflicts and cooperation in managing environmental resources (Springer-Verlag, Berlin) 18–33.

Binmore, K., A. Rubinstein and A. Wolinski, 1986, The Nash bargaining solution in economic modelling, Rand Journal of Economics 17, 176–188.

Bohm, P., 1990, Efficiency aspects of imperfect treaties on global public bads: Lessons from the Montreal protocol, Mimeo.

Carraro, C. and D. Siniscalco, 1991, Transfers and commitments in international negotiations, paper prepared for the ESF task force 3 on environmental economics, in: K.G. Maler, ed., International environmental problems: An economic perspective (Kluwer Academic Publishers, Dordrecht) forthcoming.

D'Aspremont, C.A. and J.J. Gabszewicz, 1986, On the stability of collusion, in: G.F. Matthewson and J.E. Stiglitz, eds., New developments in the analysis of market structure (Macmillan, New York) 243–264.

D'Aspremont, C.A., A. Jacquemin, J.J. Gabszewicz and J. Weymark, 1983, On the stability of collusive price leadership, Canadian Journal of Economics 16, 17–25.

Donsimoni, M.P., N.S. Economides and H.M. Polemarchakis, 1986, Stable cartels, International Economic Review 27, 317–327.

Hardin, G., 1968, The tragedy of commons, Science 162, 1243–1248.

Hardin, G. and J. Baden, 1977, Managing the commons (Freeman and Co., New York) 294.

Hoel, M., 1992, International environment conventions: The case of uniform reductions of emissions, Environmental and Resource Economics 2, 141–160.

Kaitala, V., M. Pojola and O. Tahvonen, 1992, Trans-boundary air pollution and soil acidification: A dynamic analysis of acid rain game between Finland and the USSR, Environmental Resource Economic 2, 161–182.

Maler, K.G., 1989, The acid rain game, in: H. Folmer and E. Ireland, eds., Valuation methods and policy making in environmental economics (Elsevier, New York) 188–205.

Newbery, D.M., 1991, Acid rains, Economic Policy 1, 42–88.

Nitze, W.A., 1990, The greenhouse effect: Formulating a convention (The Royal Institute of International Affairs, London) 66.

Ostrom, E., 1990, Governing the commons (Cambridge University Press, Cambridge).

Schmalensee, R., 1987, Competitive advantage and collusive optima, International Journal of Industrial Organisation 5, 351–367.

[4]

International Environment Conventions: The Case of Uniform Reductions of Emissions*

MICHAEL HOEL
University of Oslo, P.O. Box 1095, Blindern, N-0317 Oslo 3, Norway.

Abstract. Several serious environmental problems have a global character. International cooperation to reduce emissions for this type of problems often takes the form of an agreement among the cooperating countries to cut back emissions by a uniform percent rate compared with some base year. This type of agreements has two disadvantages. In the first place, it is well known from environmental economics that equal percentage reductions of emissions from different sources usually gives an inefficient outcome, in the sense that the same environmental goals could be achieved at lower costs through a different distribution of emission reductions. A second problem with agreements of equal percentage reductions is that not all countries will find it in their interest to participate in such agreements. In the paper, it is assumed that the set of countries which participate in an agreement is endogenously determined, with a country participating in an agreement provided that this makes the country better off than it would have been in a situation without any agreement. The agreement among the participating countries is assumed to be a uniform percentage reduction of their emissions. The countries have different opinions about what this uniform percentage should be. In the paper, it is assumed that the outcome is determined by the median country of the participating countries. The assumptions above lead to a particular equilibrium, in which some but not all countries cooperate. The equilibrium reduction of emissions for the cooperating countries is also derived. This equilibrium compared with the first best optimum within the context of simple numerical example.

Key words. International environment conventions, uniform reductions of emissions.

1. Introduction

Several serious environmental problems have a global character. Examples include the effect of emissions of CFC on the ozone layer, and the effect of emissions of carbon dioxide on the global climate. An important feature of these two examples is that it is only global emissions which affect the environment. Another important feature is that consequences of environmental changes may be quite different for different countries.

International cooperation often takes the form of an agreement among the cooperating countries to cut back emissions by a uniform percent rate compared with some base year. This type of agreements has two disadvantages. In the first place, it is well known from environmental economics that

* Presented at the conference "Environmental Cooperation and Policy in the Single European Market", Venice, April 16—20, 1990. The paper is part of the research project "Energy and Society" at the Centre for Research in Economics and Business Administration (SNF), Oslo. I am grateful to Ignazio Musu and Henk Folmer for useful comments on an earlier version of the paper.

Environmental and Resource Economics **2**: 141—159, 1992.

equal percentage reductions of emissions from different sources usually gives an inefficient outcome, in the sense that the same environmental goals could be achieved at lower costs through a different distribution of emission reductions. This source of inefficiency might in fact be quite large: Tietenberg (1985) reports that for 17 types of emissions to air and water, a move from rigid regulations of the type "uniform percentage reductions" to the cost minimizing way of achieving the same environmental goal would reduce costs by 7—95%, with a cost reduction exceeding 50% for 10 of the 17 cases.

A second problem with agreements of equal percentage reductions is that not all countries will find it in their interest to participate in such agreements. A likely minimum requirement for a country to participate in an agreement is that the country is better off under the agreement than it is without any international agreement. The reason for such a requirement is the following. As long as there is no international law to force countries to participate in an agreement, each country can choose to be a free rider outside the agreement instead of participating in the agreement. If the country stands outside the agreement it can enjoy (almost) the same benefits of reduced emissions as if it participates in the agreement, while it doesn't bear any of the costs of reducing emissions. An important motive for a country to participate in an agreement instead of being a free rider is that by being a free rider it increases the risk of the whole agreement breaking down. This motive for participating in the agreement is stronger the more the country has to lose from the agreement breaking down. Obviously, a country which doesn't lose anything from the agreement breaking down has no incentive to participate in the agreement, and it will therefore instead choose to be a free rider.

It is this second issue which is the main topic of the present paper. In order to focus on this issue, it is assumed that all countries have the same costs of reducing emissions. A first best optimum is therefore characterized by all countries reducing their emissions proportionately. However, the countries are assumed to differ with respect to their evaluation of environmental changes. For the case of the greenhouse effect, the reason for this could be that the consequences of climatic changes differ significantly between countries. For instance, some island and coastal countries (the Maldives, Bangladesh) are very exposed to a rising sea level due to global warming. On the other hand, one could argue that parts of, e.g., USSR and Canada would benefit from a warmer climate.

When countries have different environmental cost functions, the first best optimum (maximizing the sum of social welfare) might make some countries worse off than they are in a non-cooperative equilibrium. This is discussed and illustrated with a particular numerical example in Sections 2 and 3. Unless side payments are allowed, this first best optimum is therefore not feasible, cf. the discussion above. In Section 4 we therefore consider the constrained social optimum, in which no country is worse off under the agreement than in the non-cooperative equilibrium. Under this constrained social optimum, emission reductions differ between countries, with larger reductions for the countries most concerned about the environment than under the first best optimum. For the particular numerical example considered, the welfare loss relative to the first best optimum is quite small.

Sections 5—7 consider the case in which all countries participating in the agreement have equal percentage reductions of emissions. In Section 5, it is shown that the stricter the international agreement is (i.e., the smaller are the emissions for each participating country), the fewer countries will participate in the agreement. The uniform emission reduction which maximizes the sum of social welfare for *all* countries is derived in Section 6, and it is shown that emission levels for the participating countries are higher than they are under the first best optimum. Since non-participating countries don't reduce their emissions, total emissions are therefore also higher than they are in the first best optimum. For the numerical example considered, total emissions are 45% higher than they are in the first best optimum.

There is no reason to believe that the participating countries will try to maximize the sum of social welfare over *all* countries. Instead, it is assumed in Section 7 that each participating country is only concerned with its own social welfare. Since the participating countries have different environmetnal cost functions, they also have different opinions about how much emissions should be reduced. In Section 7 it is assumed that the participating countries agree upon the emission reduction which is the most preferred reduction for the median country (among the participating countries). For the numerical example considered, total emissions in this case are only about 3% higher than the level derived in Section 6.

Section 8 gives an analysis of a somewhat stricter requirement for participating in an agreement than the requirement discussed above. Following Barrett (1989), a two stage game is considered, in which the countries decide whether or not to join the agreement in the first stage. The emission levels of the participating countries are decided in the second stage of the game. With this setup, a country will find it better to participate in the agreement than to be a free rider only if the agreement is so much better when it participates that this environmental benefit exceeds the costs of the country participating. An equilibrium with some countries cooperating may exist also in this case. However, for the numerical example considered, only two countries will cooperate, and there will hardly be any difference between this equilibrium and the non-cooperative equilibrium.

Some concluding remarks are given in Section 9.

2. The Model

There are N countries, which have equal revenue functions $R(v_j)$, where v_j measures emissions from country j. It is assumed that $R(0) = 0$, $R'(v) > 0$ and $R''(v) < 0$ for $v < \bar{v}$ and $R'(\bar{v}) = 0$. In other words, $R(v_j)$ measures revenue addition relative to the case of no emissions, and $v_j = \bar{v}$ is the optimal amount of emissions from all countries in the absence of environmental considerations.

Total emissions are

$$V = \sum_i v_i \tag{2.1}$$

and country j's environmental damage function is assumed to be equal to $m_j V$, where m_j is a non-negative constant parameter which measures the marginal environmental cost of country j. It is assumed that countries are numbered so that

$$m_1 \geqslant m_2 \geqslant \ldots \geqslant m_N \geqslant 0. \tag{2.2}$$

In other words, the countries with the lowest numbers are the countries which care most about the environment.

We define the sum of the marginal environmental costs by

$$M = \sum_i m_i > 0. \tag{2.3}$$

In the absence of any international cooperation, emissions are assumed to be determined as the Nash equilibrium of the non-cooperative game in which all countries choose their emissions simultaneously.[1] This Nash equilibrium is characterized by each country j choosing v_j so that $R(v_j) - m_j V$ is maximized, taking all other v_i $(i \neq j)$ as given. This gives

$$R'(v_j) = m_j \quad j = 1, \ldots, N. \tag{2.4}$$

The assumption that each country has constant marginal environmental costs implies that each country's optimal emission level is independent of the emission levels chosen by other countries. When N is large, each m_j will be small relative to M, and we shall assume that m_j is so close to zero that $R'(v_j) = 0$, i.e., $v_j = \bar{v}$ is a good approximation to (2.4).

Throughout the paper, we shall illustrate the general results with a specific example. In this example it is assumed that

$$R(v) = v - \tfrac{1}{2} v^2, \tag{2.5}$$

i.e.,

$$R'(v) = 1 - v \tag{2.6}$$

implying that $\bar{v} = 1$. Moreover, the marginal environmental costs are distributed uniformly between $1/N$ and 0, so that

$$m_j = \frac{N-j}{N-1} \frac{1}{N}. \tag{2.7}$$

If country j could choose equal emission levels for all countries including itself, it would choose v_j so that $R(v_j) - m_j N v_j$ was maximized, giving $R'(v_j) = m_j N$. With the present example, this implies (cf. (2.6) and (2.7))

$$v_j = \frac{j-1}{N-1}. \qquad (2.8)$$

In other words, the larger j is, the larger is the preferred uniform emission level, with $v_1 = 0$ and $v_N = 1$. For the median country, i.e., $j = (N+1)/2$, the preferred uniform emission level is equal to 0.5.

Total marginal environmental costs for this example follow from (2.3) and (2.7)

$$M = \sum_i m_i = \frac{1}{(N-1)N}\left(N^2 - \sum_i i\right) = 0.5. \qquad (2.9)$$

The non-cooperative equilibrium is given by (2.4), which in the present example gives (cf. (2.6) and (2.7))

$$v_j = 1 - \frac{1}{N}\,\frac{N-j}{N-1} \geqslant 1 - \frac{1}{N}. \qquad (2.10)$$

For, e.g., $N = 100$, all countries thus choose emission levels in the interval [0.99, 1]. The approximation $v_j = 1$ for the non-cooperative case is therefore a good approximation in this example.

3. The First-Best Optimum

The sum of net benefits for all countries is

$$B = \sum_i [R(v_i) - m_i V]. \qquad (3.1)$$

where V is given by (2.1). A first best social optimum follows from maximizing (3.1) with respect to all v_j, which gives

$$R'(v_j) = M, \quad j = 1, \ldots, N, \qquad (3.2)$$

where M is the sum of marginal environmental costs. In other words, marginal revenues are equal for all countries, and equal to the sum over all countries of the marginal environmental costs of the emissions. Since we have assumed that all countries have equal revenue functions, (3.2) implies that optimal emissions are equal for all countries.

If side payments are allowed, the total net benefit from cooperating may be distributed in any way between the countries.[2] In particular, all countries can be made better off than they are without any cooperation.

Without side payments, some "low m countries" might be worse off under the first best optimum than they are without any cooperation. This is certainly true if $m_N = 0$: In this case country N doesn't care about the environment, and any reduction of v_N below $\bar{v}$ makes this country worse off.

With our numerical example (3.2) may be written

$$1 - v_j = M. \tag{3.2}$$

Since $M = 0.5$ (cf. (2.9)), we thus have $v_j = 0.5$ in this case. In this case the first best optimum thus implies a 50% reduction of emissions for all countries (compared with the non-cooperative case). The optimal emission level is in this case equal to the median countrys most preferred uniform emission level.

It is straightforward to verify that

$$\begin{aligned} v_j &= 0.5 \Rightarrow B = 0.125N, \\ v_j &= 1 \Rightarrow B = 0. \end{aligned} \tag{3.3}$$

The total net gain relative to the non-cooperative case is thus $0.125N$. Denote the corresponding net gain for country j by b_j. We have

$$b_j = [0.5 - 0.5 \cdot (0.5)^2 - m_j N \cdot 0.5] - [1 - 0.5 - m_j N],$$

i.e., using (2.7)

$$b_j = 0.5 \frac{N-j}{N-1} - 0.125. \tag{3.4}$$

From (3.4) it follows that

$$b_j \geqslant 0 \quad \text{iff} \quad j \leqslant 0.75N + 0.25. \tag{3.5}$$

With, e.g., $N = 100$, the 75 countries with highest marginal environmental costs are thus better off under the first best optimum than without any cooperation. The 25 countries with the lowest environmental concern prefer the non-cooperative outcome to the first best optimum.

4. Constrained Social Optimum

Consider the case in which B is maximized subject to

$$b_j = [R(v_j) - m_j V] - [R(\bar{v}) - m_j N] \geqslant 0, \tag{4.1}$$

i.e., that no countries should be worse off under the social optimum than under the non-cooperative equilibrium. The Lagrangian corresponding to this optimization problem is

$$L = \sum_i [R(v_i) - m_i V] - \sum_i \lambda_i b_i, \tag{4.2}$$

where V and b_i are given by (2.1) and (4.1), respectively. The first order conditions are

$$(1+\lambda_j)R'(v_j) = \sum_i (1+\lambda_i)m_i. \tag{4.3}$$

Using $\Sigma_i m_i = M$ it is clear that

$$\sum_i (1+\lambda_j)m_i = M + \sum_i \lambda_i m_i = (1+\lambda_j)M - \lambda_j \sum_i m_i + \sum_i \lambda_i m_i$$
$$= (1+\lambda_j)M + \sum_i (\lambda_i - \lambda_j)m_i.$$

Inserting this into (4.3) gives

$$R'(v_j) = M + \frac{\sum_i (\lambda_i - \lambda_j)m_i}{1+\lambda_j}, \tag{4.3'}$$

where $\lambda_j \geqslant 0$ and $\lambda_j b_j = 0$.

Consider first the "high m countries", i.e., the countries for which the constraint (4.1) in not binding in equilibrium. For there countries $\lambda_j = 0$. As long as $\lambda_i > 0$ for sufficiently large values of i, the second term on the r.h.s. of (4.3′) is positive for countries with $\lambda_j = 0$. These countries therefore have lower emissions in the constrained optimum than they have in the first best optimum.

The countries with the lowest m_j-values have the highest λ_j-values. For these countries the term $\Sigma_i(\lambda_i - \lambda_j)m_i$ is therefore negative, i.e., they have higher emissions than under the first-best optimum. If $m_N = 0$, it follows directly from (4.1) that $v_N = \bar{v}$, i.e., country N has no reduction in its emissions compared with the non-cooperative case.

Figure 1 illustrates the case of a constrained optimum when $m_j - m_{j-1}$ is negative and sufficiently small for all j and $m_N > 0$. In the first best optimum, all countries have emission levels equal to v^0. In the constrained optimum, emissions from country j are given by v_j^c.

In the specific example considered previously the first best optimum gave emissions equal to 0.5 from all countries, and total net benefits of cooperating equaled 12.5 for $N = 100$. For the constrained optimum, numerical calculations show that 71 countries remain unconstrained (for $N = 100$), and that the emissions from these countries are 0.49. The emissions from unconstrained countries are thus slightly lower than they are in the first best optimum. Since emission levels for most of the 29 constrained countries are above 0.5, total emissions in the present case equal 53.8, compared with 50.0 in the first best optimum. Total social gains from cooperating are equal to 11.86, i.e., only 5% lower than the total social gains under the first best optimum.

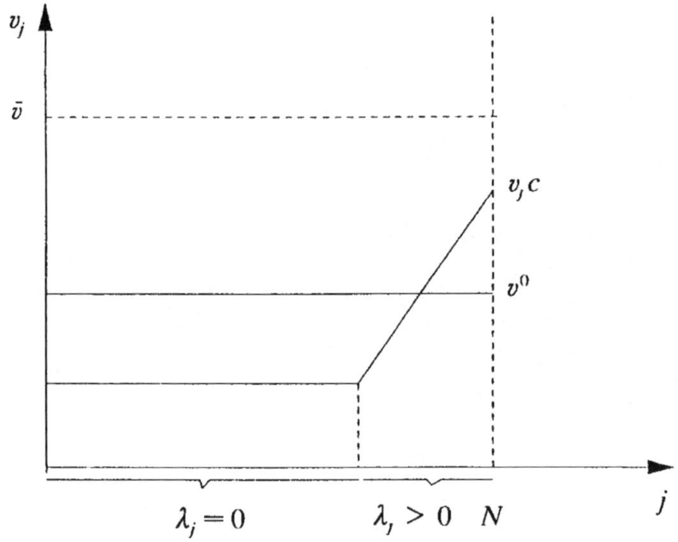

Fig. 1.

5. Participation in an Agreement with Uniform Emissions

The numerical example in the end of the previous section indicates that the lack of side payments does not necessarily imply a large total welfare loss, as long as emissions are allowed to differ between countries. However, if emissions are restricted to be equal for all cooperating countries, the loss might be considerably larger. The reason is that the requirement that no country is worse off under an agreement than without might make several countries prefer not to join the agreement. This section considers the relationship between the emission level agreed upon and the number of participating countries.

Using v to denote the uniform emission level under an agreement, and n as the number of participating countries, we have

$$V = nv + (N - n)\bar{v}. \tag{5.1}$$

The requirement $b_j \geqslant 0$ may be written as

$$R(v) - m_j V \geqslant R(\bar{v}) - m_j N\bar{v}$$

or, using (5.1)

$$nm_j \cdot (\bar{v} - v) \geqslant R(\bar{v}) - R(v) \tag{5.2}$$

or

$$nm_j \geqslant \frac{R(\bar{v}) - R(v)}{\bar{v} - v} \equiv g(v). \tag{5.2'}$$

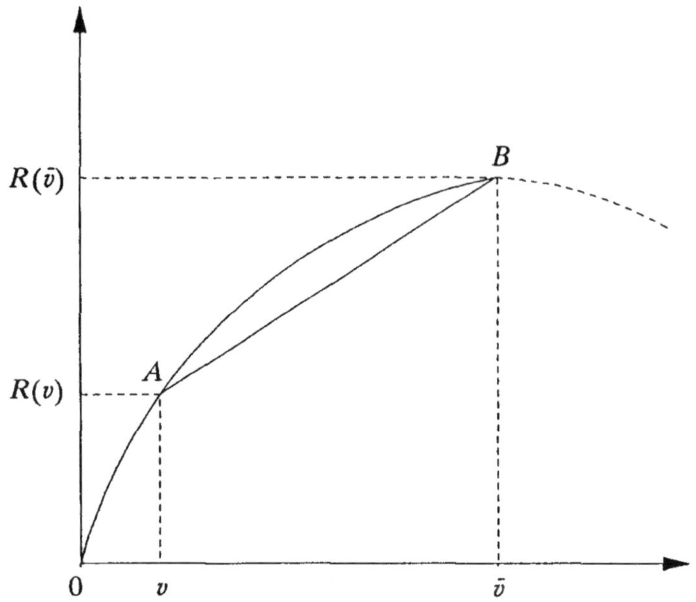

Fig. 2.

From Figure 2 we see that $g(v)$ is given by the slope of the line AB, i.e.,

$$\begin{aligned} &g(\bar{v}) = 0 \\ &g'(v) < 0 \\ &g(v) < R'(v). \end{aligned} \tag{5.3}$$

If (5.2) holds for $j \leqslant n$ for some agreement v', it therefore also holds for an alternative agreement $v'' > v'$.

The marginal country n satisfying (5.2) is given by

$$n(v) = \max\{n \mid nm_n \geqslant g(v)\}. \tag{5.4}$$

Unless the sequence $m_1, m_2, \ldots$ declines very rapidly, nm_n is an increasing function of n for small values of n. We shall assume that nm_n increases monotonically until n reaches a value $\hat{n} \leqslant N$, after which nm_n decreases monotonically if $\hat{n} < N$. Assume that $nm_n < g(0)$ for all n. In other words, for tight enough emission controls, there does not exist any group of cooperating countries which are all better off with the emission agreement than without. The maximal value $\hat{v}$ of v which satisfies (5.2′) for some value of n is thus positive, and given by

$$g(\hat{v}) = \hat{n}m_{\hat{n}} = \max_{n \in [0, N]} \{nm_n\}. \tag{5.5}$$

If $v < \hat{v}$, i.e., $g(v) > g(\hat{v})$, $nm_n < g(v)$ no matter what n is. In other words, for emission levels below $\hat{v}$, there does not exist any group of cooperating countries which are all better off with the emission agreement than without.

If $v = \hat{v}$, $\hat{n}$ countries are at least as well off under an agreement with $v = \hat{v}$ for the $\hat{n}$ participating countries and $v = 1$ for the $N - \hat{n}$ remaining countries as they are with no agreement.

If $\hat{n} = N$, any agreement with $v \geqslant \hat{v}$ makes all countries better off (or equally well off) than they are in the non-cooperative case. No agreement with $v < \hat{v}$ is possible as long as countries which don't gain from the agreement choose emission levels equal to $\bar{v}$.

If $\hat{n} < N$, which seems most plausible, the countries $1, \ldots, \hat{n}$ will want to participate in an agreement giving $v = \hat{v}$ for the participating countries. For higher emission levels, more countries will want to participate, and *all* countries will be better off under the agreement if v is so high that $Nm_N \geqslant g(v)$. Since $g(v) > 0$ for $v < \bar{v}$, there will only exist emission levels below $\bar{v}$ satisfying $Nm_N \geqslant g(v)$ if $m_N > 0$. In other words, all countries can be better off by agreeing to reduce emissions only if all countries care positively about the environment.

The relationship between n and v is illustrated in Figure 3 for the case in which $\hat{n} < N$ and $m_N > 0$. For $v < \hat{v}$, i.e., $g(v) > g(\hat{v})$, no group of countries will find it in their interest to participate in an agreement. For $v \geqslant \hat{v}$, i.e., $g(v) \leqslant g(\hat{v})$, the maximal number of countries participating in the

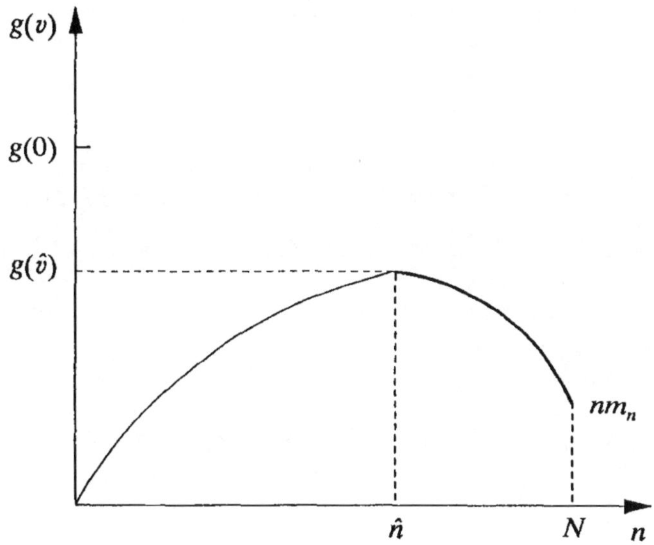

Fig. 3.

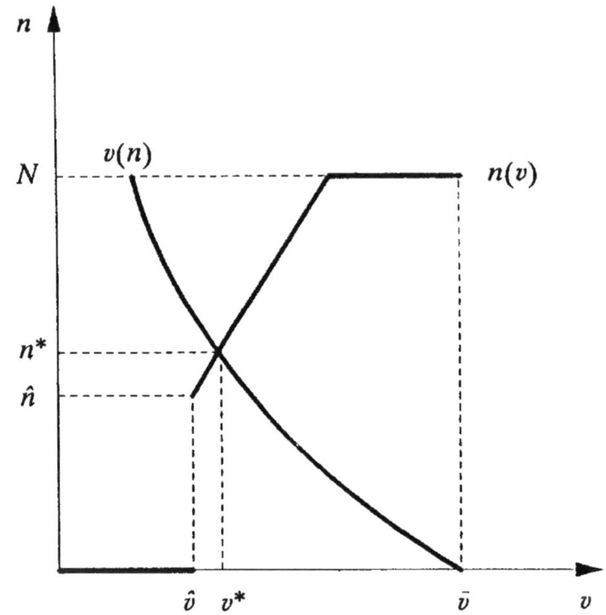

Fig. 4.

agreement is given by the heavily drawn part of the nm_n curve in Figure 3. In the subsequent sections, we shall denote this relationship between v and n by $n(v)$. We thus have $n(v) = 0$ for $v < \hat{v}$, $n(\hat{v}) = \hat{n}$, and $n'(v) > 0$ for $v > \hat{v}$ if $n(v) < N$. This function is given by the heavily drawn discontinuous curve in Figure 4.

For our specific example the revenue functions are given by $R(v) = v - v^2/2$. From (5.2′) it is therefore clear that the function $g(v)$ in this case is

$$g(v) = \frac{0.5 - v + v^2/2}{1 - v} = \frac{1 - 2v + v^2}{2(1 - v)} = \frac{1 - v}{2}. \tag{5.6}$$

Moreover, from (2.7) we find

$$nm_n = \frac{n(N - n)}{N - 1}\,\frac{1}{N}. \tag{5.7}$$

Solving (5.5) for this case thus gives $\hat{n} = N/2$. For N large it follows from (5.5)—(5.7) that

$$\hat{v} = 1 - \frac{N}{2(N - 1)} \approx 0.5. \tag{5.8}$$

In other words, proposals of equal emissions below $v = 0.5$ will not be supported by any countries, while a proposal of $v = 0.5$ will be supported by 50% of the countries.

For $v > \hat{v}$ it follows from (5.5)–(5.7) that the relationship between n and v is given by

$$v = 1 - \frac{2n(N-n)}{N(N-1)} \tag{5.9}$$

or, solving for n

$$n(v) = \frac{N}{2}\left\{1 + \left[1 - 2(1-v)\,\frac{N-1}{N}\right]^{1/2}\right\} \tag{5.10}$$

or

$$\frac{n(v)}{N} \approx 0.5[1 + (2v-1)^{1/2}]. \tag{5.10'}$$

In other words, $n(0.5) \approx 0.5N$, $n(1) = N$, and $n''(v) < 0$ for $v > \hat{v}$.

Total emissions are equal to

$$V = n(v)v + [N - n(v)]\bar{v}. \tag{5.11}$$

6. Socially Optimal Uniform Reduction

If one restricts oneself to uniform reductions, and participation in the agreement is given by $n(v)$ as explained in the previous section, total welfare is given by

$$B = n(v)R(v) + (N - n(v))R(\bar{v}) - M[n(v)v + (N - n(v))\bar{v}]. \tag{6.1}$$

The level of emissions which maximizes B is given by

$$\begin{aligned} &n(v)R'(v) + [R(v) - R(\bar{v})]n'(v) - \\ &\quad M(v - \bar{v})n'(v) - Mn(v) = 0. \end{aligned} \tag{6.2}$$

Some manipulation of (6.2) gives, using the definition (5.2) of $g(v)$

$$R'(v) = AM, \tag{6.3}$$

where

$$A = \frac{1 - (\bar{v} - v)\dfrac{n'(v)}{n(v)}}{1 - \dfrac{g(v)}{R'(v)}(\bar{v} - v)\dfrac{n'(v)}{n(v)}} < 1 \tag{6.4}$$

since $n'(v) > 0$ and $g(v) < R'(v)$ (cf. (5.3)). Comparing (6.3) with the first best optimum (3.2), it is thus clear that emissions from the participating countries are higher than under the first best optimum. Since non-participating countries do not reduce their emissions in the present case, total emissions are also higher in the present case than in the first best optimum.

In our specific example, $R'(v) = 1 - v$ and $M = 0.5$, so that (6.3)—(6.4) may be written as

$$1 - v = \frac{1 - (1 - v)\dfrac{n'(v)}{n(v)}}{1 - \dfrac{1}{2}(1 - v)\dfrac{n'(v)}{n(v)}} \cdot 0.5. \tag{6.5}$$

Solving (6.5) for $n'(v)$, we find

$$n'(v) = n \cdot \frac{2v - 1}{v(1 - v)}. \tag{6.6}$$

Inserting v and $1 - v$ from (5.9) gives, after some manupulation

$$n'(v) = \frac{N(N-1)}{2(N-n)} \cdot \frac{N(N-1) - 4n(N-n)}{N(N-1) - 2n(N-n)}. \tag{6.7}$$

From (5.9) we can also derive

$$n'(v) = \left(\frac{dv}{dn}\right)^{-1} = \frac{N(N-1)}{4n - 2N}. \tag{6.8}$$

Eliminating $n'(v)$ from (6.7) and (6.8) we can solve for n (for any given N). For $N = 100$ we find $n = 77$, which inserted into (5.9) gives $v = 0.64$. Total emissions and total welfare follow from (5.11) and (6.1), giving $V = 72.3$ and $B = 8.85$. Compared with the first best optimum, total emissions have thus increased by 45%, while the net benefits of an agreement have dropped from 12.5 to 8.85, i.e., a reduction equal to 29%.

7. The Most Preferred Uniform Reduction of the Median Country

There is no reason to believe that the n countries participating in an agreement will choose the emission level which maximizes total welfare, as derived in Section 6. In the public choice literature, it is often assumed that it is the preferences of the median voter which determine the outcome in situations of this type (see, e.g., Mueller, 1979). In the present context, this means that the n countries which cooperate agree upon the emission level which is the median value of the most preferred emission levels of the n

countries. For any given value of n, the most preferred emission level of country j ($\leqslant n$) is the value of v which maximizes

$$b_j = R(v) - m_j V$$

or, inserting V from (5.13)

$$b_j = R(v) - m_j[nv + (N-n)\bar{v}]. \tag{7.1}$$

The value of v which maximizes b_j is given by

$$R'(v) = m_j n. \tag{7.2}$$

Since $m_j n < m_j N$, the most preferred emission level of country j is higher than the most preferred emission level for the case in which *all* countries have equal emissions.

Since m_j is declining in j, the most preferred v-value, given by (7.2), is lower the lower is j. Among the participating countries, $j = (n+1)/2$ is the median country. This country's most preferred emission level is therefore equal to

$$R'(v) = m_{(n+1)/2}n \tag{7.3}$$

which gives v as a function of n, denoted by $v(n)$.[3]

Since $(n+1)/2 \approx n/2$, we have $nm_{(n+1)/2} \approx 2(n/2)m_{n/2}$. For $n \leqslant \hat{n}$ the function $(n/2)m_{n/2}$ is increasing in n. From (7.3) and $R''(v) < 0$ it therefore follows that $v'(n) < 0$ for $n \leqslant \hat{n}$, as illustrated in Figure 4. (In Figure 4, it is assumed that $v'(n) < 0$ for *all* n. This will be the case if $\hat{n} \geqslant N/2$, since $\hat{n} \geqslant N/2$ implies that $(n/2)m_{n/2}$ is rising in n for all n up to N.)

The intersection between the curves $n(v)$ and $v(n)$ gives the equilibrium pair (n^*, v^*): When n^* countries cooperate, they will agree upon v^* as their common emission level. With this emission level, the cooperating countries are better off than they are without any agreement, while each of the $(N - n^*)$ remaining countries would be worse off participating in an agreement with emission level v^* than they would be in the non-cooperative equilibrium.

Since the curve $n(v)$ is discontinous, the curves $n(v)$ and $v(n)$ need not have any point of intersection. If $v(\hat{n}) < \hat{v}$, we get the situation described in Figure 5. The Appendix gives an example of such a situation. With this situation, there will be no equilibrium with countries cooperating. Whatever number of countries one tentatively lets cooperate, they agree upon an emission level which either implies that some of these countries would have been better off in the non-cooperative equilibrium or that some of the non-cooperating countries would have been better off participating in the agreement than they would have been in the non-cooperative equilibrium.

When Figure 5 is valid, the non-cooperative equilibrium is the only possible equilibrium: When no countries cooperate, no country can be in a

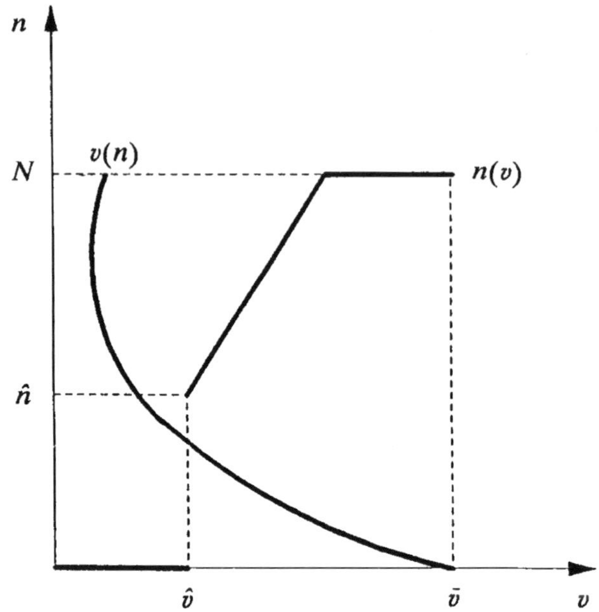

Fig. 5.

position where it would have been better off joining an agreement than it is under the non-cooperative equilibrium.

For our specific example with $R'(v) = 1 - v$, we can insert for $m_{(n+1)/2}$ from (2.7) in (7.3). After same manipulation this yields

$$v(n) = 1 - \frac{(2N - n - 1)n}{2N(N - 1)} \tag{7.4}$$

which is a declining function of n.

The intersection between the $n(v)$- and $v(n)$-curve is found by equating the right hand sides of (5.9) and (7.4), which gives

$$2n^*(N - n^*) = \frac{n^*}{2}(2N - n^* - 1),$$

i.e.,

$$n^* = \frac{2N + 1}{3}. \tag{7.5}$$

Solving for $N = 100$ we thus find $n = 67$, which inserted into (7.4) gives $v = 0.55$. Total emissions and total welfare follow from (5.13) and (6.1), giving

$V = 70.1$ and $B = 8.28$. In this example the total emissions and total welfare are thus quite close to the values which maximize social welfare (derived in Section 6).

8. The Free Rider Problem

Throughout the paper, it has been assumed that a sufficient condition for a country to participate in an international agreement is that the country is better off under the agreement than it is under the non-cooperative outcome. As was pointed out in Section 1, however, this condition is a *minimum* requirement for participation. Even if a country is better off participating in an agreement than it would be without any agreement, it might be even better off if the other countries cooperated while it itself stood outside the agreement. Given the decision of other countries of whether or not to cooperate, a country will find it better to participate in the agreement than to be a free rider only if the agreement is so much better when it participates that this environmental benefit exceeds the costs of the country participating. As Barrett (1989) has shown, one may have an equilibrium with some countries cooperating also in this case. Barrett considers a two stage game, in which the countries simultaneously decide whether or not to join the agreement in the first stage. The emission level of the participating countries is decided in the second stage of the game.[4]

The problem of choosing the appropriate emission level in the second stage of the game was treated in Section 7. One possible choice of emission level at this stage of the game is the most preferred emission level of the median country (among the participating countries). In this case the emission level is given by (7.3), which defines the function $v(n)$ from Section 7.

In the first stage of the game, the countries decide whether or not to cooperate with other countries. If any countries cooperate, it will obviously be the countries with largest concern for the environment, i.e., the countries with the lowest index numbers. Assume that n countries cooperate (i.e., countries 1, 2, . . . , n). Country n is better off cooperating than being a free rider if

$$R(\bar{v}) - R(v(n)) \leqslant m_n[(\bar{v} + (n-1)v(n-1)) - nv(n)]. \tag{8.1}$$

The l.h.s. of (8.1) is the revenue increase for country n of being a free rider instead of cooperating with the other $n - 1$ countries. The terms in the square brackets in the r.h.s. of (8.1) give the increase in emissions if country n decides to be a free rider instead of cooperating: If it cooperates, emissions from the n countries are $nv(n)$, while they are $\bar{v} + (n-1)v(n-1)$ if country n chooses to be a free rider. The emissions from the remaining $N - n$ countries are $\bar{v}$ whatever country n chooses to do.

If there are n countries cooperating in equilibrium, country $n + 1$ must be better off being a free rider than it is by cooperating with the n first countries, i.e.,

$$R(\bar{v}) - R(v(n+1)) \geqslant m_{n+1}[(\bar{v} + nv(n)) - (n+1)v(n+1)]. \quad (8.2)$$

The group of countries cooperating in equilibrium is thus 1, 2, . . . , n; with n determined by the conditions (8.1) and (8.2). One can of course not rule out the possibility of $n = 1$, i.e., that no countries cooperate.

Using our numerical example, we can insert (2.5) and (2.7) into (7.3) and (8.1)—(8.2). Solving for $N = 100$, we find that $n = 2$.[5] In other words, only two countries cooperate in equilibrium. Moreover, these two countries only reduce their emissions from 0.99 to 0.98. This equilibrium is therefore for all practical purposes equal to the non-cooperative equilibrium in which each country's emission level is given by (2.4). It is also interesting to note that the number of participating countries is almost independent of the total number of countries: Two countries cooperate for $N \geqslant 5$, while there is no cooperation if $N < 5$.[6]

9. Conclusions

Table I below summarizes some of the results for the numerical example considered. It is clear from this table that the welfare loss associated with an agreement of the type "equal percentage reductions" is quite substantial (compared with a first best optimum or the constrained optimum making all countries better off). In considering this welfare loss, one should also bear in

Table I. A First best optimum (Section 3); B Constrained optimum (Section 4); C Maximum social welfare given equal emissions from participating countries (Section 6); D Equal emissions from all participating countries, emission level chosen by median country among participating countries (Section 7)

	A	B	C	D
Percent participation	100	100	77	67
Percent emission reduction by participating country	50	51[a]	36	45
Percent reduction of total emission	50	46	28	30
Welfare relative to no agreement (in % of max)	100	95	71	66

[a] The 71 "unconstrained" countries reduce their emissions by 51%. The remaining countries reduce their emissions so much that they are equally well off under the agreement as they are in the non-cooperative equilibrium. This implies a reduction of emissions from these countries by less than 51%.

mind that the model is constructed so that uniform percentage emission reductions is optimal if side payment can be used so that all countries are willing to participate in the agreement.

Appendix

Consider the revenue function

$$R(v) = \begin{cases} v^{0.2} & \text{for } v \leqslant 1 \\ 1 & \text{for } v > 1, \end{cases} \tag{A.1}$$

i.e., $\bar{v} = 1$ and $R(\bar{v}) = 1$. For this case we have

$$R'(v) = 0.2v^{-0.8} \tag{A.2}$$

and

$$g(v) \equiv \frac{R(\bar{v}) - R(v)}{\bar{v} - v} = \frac{1 - v^{0.2}}{1 - v}. \tag{A.3}$$

For the m-distribution given by (2.7), the value $\hat{v}$ (defined by (5.5)) follows from (A.3), (2.7) and (5.5). Straightforward calculations give

$$\hat{v} = 0.55. \tag{A.4}$$

The function $v(n)$ follows from (7.3), which after inserting from (A.2) and (2.7) gives

$$0.2\, v^{-0.8} = \frac{n}{N(N-1)} \left(N - \frac{n+1}{2} \right). \tag{A.5}$$

This gives v as a declining function of n. For $N = 100$ we find

$$\begin{aligned} v(1) &= 1, \\ v(50) &= 0.45, \\ v(100) &= 0.32. \end{aligned} \tag{A.6}$$

In particular, it follows from (A.4) and (A.6) that $v(\hat{n}) = v(50) < \hat{v}$. In other words, we have the situation which is illustrated in Figure 5.

Notes

1 See, e.g., Hoel (1991, 1992) and van der Ploeg and de Zeeuw (1991) for discussions of Nash equilibria in similar international environmental games.

2 The total net benefit from cooperating is equal to the difference between the value of B from (3.1) when all v_j are optimal and the value of B when all $v_j = \bar{v}_j$, which holds in the non-cooperative case.

[3] Instead of assuming that the median country chooses the emission level, one could assume that the emission level is chosen so that the sum of welfare for the participating countries is maximized. This will generally give a different condition than (7.3), and therefore also a different function $v(n)$. However, when the m-values are uniformly distributed as assumed in our specific example (cf. (2.7)), it is easily seen that these two different assumptions lead to the same condition (7.3).

[4] Barrett assumes that all countries are equal, so that an agreement will require all participating countries to have the same emission level.

[5] In the numerical calculation, the true non-cooperative emission levels, given by (2.4), are used instead of the approximation $\bar{v}$.

[6] For very large values of N, we get $v = 3$: $N = 10^8$ gives $v = 3$, while $N = 10^7$ gives $v = 2$.

References

Barrett, S. (1989), 'On the Nature and Significance of International Environmental Agreements', mimeo, London Business School.

Hoel, M. (1991), 'Global Environmental Problems: The Effects of Unilateral Actions Taken by One Country', *Journal of Environmental Economics and Management* **20**, 55—70.

Hoel, M. (1992), 'Emission Taxes in a Dynamic Game of CO_2 Emissions', in R. Pethig (ed.), *Conflict and Cooperation in Managing Environmental Resources*, Springer Publishing House.

Mueller, D. C. (1979), *Public Choice*, Cambridge University Press.

van der Ploeg, F. and de Zeeuw, A. (1991), 'International Aspects of Pollution Control', *Environmental and Resource Economics* **2**, 117—139, this issue.

Tietenberg, T. (1985), *Emission Trading: An Exercise in Reforming Pollution Policy*, Resources for the Future, Washington D.C.

[5]

International Journal of Game Theory (1997) 26: 379–401

Game Theory

The Core of an Economy with Multilateral Environmental Externalities

PARKASH CHANDER

Indian Statistical Institute and California Institute of Technology, Sansanwal 7, New Delhi 110016, India

HENRY TULKENS

CORE, Université Catholique de Louvain, 34 Voie du Roman Pays, B-1348 Louvain-la-Neuve, Belgium

Abstract: When environmental externalities are international – i.e. transfrontier – they most often are multilateral and embody public good characteristics. Improving upon inefficient laissez-faire equilibria requires voluntary cooperation for which the game-theoretic core concept provides optimal outcomes that have interesting properties against free riding. To define the core, however, the characteristic function of the game associated with the economy (which specifies the payoff achievable by each possible coalition of players – here, the countries) must also reflect in each case the behavior of the players which are not members of the coalition. This has been for a long time a disputed issue in the theory of the core of economies with externalities.

Among the several assumptions that can be made as to this behaviour, a plausible one is defined in this paper, for which it is shown that the core of the game is nonempty. The proof is constructive in the sense that it exhibits a strategy (specifying an explicit coordinated abatement policy and including financial transfers) that has the desired property of nondomination by any proper coalition of countries, given the assumed behavior of the other countries. This strategy is also shown to have an equilibrium interpretation in the economic model.

1 Introduction

We deal in this paper with an economy with several agents, whose productive activities generate "multilateral" externalities, i.e., externalities that each one of them can both generate and be a recipient of. We have in mind externalitites that are detrimental for the recipients. We also call these externalities "environmental" because we assume that they exhibit public goods (actually public "bads") characteristics in the sense that when they are generated, they affect all agents in the economy. We should also call them "additive" because we further assume that what is received by any recipient is simply the sum of what is emitted by the various generators. For such an economy we call upon the concept of the *core* of a cooperative game associated with it, in order to identify joint actions that would (*a*) improve upon inefficient laissez-faire equilibria, (*b*) achieve Pareto efficiency, and (*c*) induce this outcome in such a way that it deters both individual and coalitional free riding behaviours.

0020–7276/97/3/379–401$2.50 © 1997 Physica-Verlag, Heidelberg

Being motivated by an interest of long standing[1] for the sources of cooperation between countries on issues of transfrontier pollution (a currently prominent example of multilateral environmental externalities), we shall interpret our model below as one of an international economy, where the countries are also the players in the games associated with it.[2] While other interpretations are conceivable, in our present interpretation the core outcome we are seeking for may be seen as a "strategically stable" result of negotiations on joint pollution abatement.

To define the core, the characteristic function of the game – which specifies the payoff achievable by each possible coalition of players – must also reflect, in the case of games with externalities, the behavior of the players which are *not* members to the coalition. This is a disputed issue, and alternative assumptions in this respect lead to alternative core concepts such as, *e.g.* the α- and β-cores. These are studied and contrasted to each other in various externalities contexts by SCARF 1971, STARRETT 1972 and LAFFONT 1977 (chapter V).[3] However, when using one of these assumptions in analyzing international environmental negotiations by means of his acid rain game, (actually one implying a particular version of the α-core, as we shall see below), MÄLER 1989 comes to the conclusion that the core concept is useless because it results in too large a set of outcomes, actually the whole set of Pareto optima.

In this paper, we adopt an alternative assumption on the behavior of players outside each coalition that to our knowledge has not been explored in the literature, and that we think to be a more plausible one in the context of environmental externalities. We use this assumption to formulate what we call the γ-characteristic function of the game, and for the ensuing γ-core concept we prove nonemptiness. The proof is constructive in the sense that it exhibits a joint strategy (*i.e.* in our pollution interpretation an explicit coordinated abatement policy accompanied by financial transfers) that has the desired property of nondomination by any proper coalition of the players. Furthermore, this strategy is shown to have an equilibrium interpretation which is conceptually analogous to that of *ratio equilibrium* (KANEKO 1977; MAS COLELL and SILVESTRE 1989) in the context of economies with public goods.

Our analysis assumes full information throughout: this is because our primary concern is with a satisfactory definition and an existence proof – which are lacking – of the concept of core allocation for the kind of economy we are

---

[1] A first version of the international environmental model alluded to here was given in TULKENS 1979, and elaborated upon from a game theoretic point of view in CHANDER and TULKENS 1991, 1992a and 1992b.

[2] The "acid rain game" of MÄLER 1989 is a similar international environmental model, to which we apply, in CHANDER and TULKENS 1995, the concepts and results developed

[3] The issue also arises in games associated with economies with public goods, as in FOLEY 1970, CHAMPSAUR 1975, MOULIN 1987 and CHANDER 1993.

considering. This should provide firm ground for further work where other information structures are allowed for.

2 The Model of an Economy with Multilateral Environmental Externalities and its Efficiency Characterization

The agents of the economy (countries in our international interpretation) are denoted by the index i, with $N = \{i | i = 1, 2, \ldots, n\}$. Three categories of commodities are considered:

(i) a standard private good, whose quantities are denoted by $x \geq 0$ if they are consumed, and by $y \geq 0$ if they are produced;
(ii) pollutant discharges, the quantities of which are denoted by $p \geq 0$; and
(iii) ambient pollutant quantities, denoted by $z \leq 0$.

Each agent i's preferences are represented by a utility function $u_i(x_i, z)$ satisfying:

Assumption 1: $u_i(x_i \cdot z) = x_i + v_i(z)$ *i.e.* quasi-linearity; and

Assumption 2: $v_i(z)$ concave, differentiable and such that $dv_i/dz \equiv \pi_i(z) > 0$ for all $z \leq 0$.

With each agent i there is furthermore associated a technology, described by the production function $y_i = g_i(p_i)$, satisfying:

Assumption 3: $g_i(p_i)$ strictly concave and differentiable over an interval; and

Assumption 4: There exists $p_i^0 > 0$ such that

$$\frac{dy_i}{dp_i} \equiv \gamma_i(p_i) \begin{cases} > 0 \text{ if } p_i < p_i^0 & \text{(i)} \\ = 0 \text{ if } p_i \geq p_i^0 & \text{(ii)} \\ = \infty \text{ if } p_i = 0. & \text{(iii)} \end{cases}$$

Inputs, which are not explicitly mentioned in the production functions, are subsumed in the functional symbols g_i. Although this amounts to treat them as fixed in the analysis that follows, our results will not rest in an essential way on that assumption, which is made here mostly for expositional convenience.

All possible behaviors within the economy, in terms of the consumption, production and pollutant discharge decisions taken by its agents, as well as in terms of the the resulting values of ambient pollutant are formally described by feasible states:

Definition 1: Feasible states of the economy (or "allocations") are vectors

$$(x, p, z) \equiv (x_1, \ldots, x_n; p_1, \ldots, p_n; z)$$

such that

$$\sum_{i \in N} x_i \leq \sum_{i \in N} g_i(p_i) \tag{1}$$

and

$$z = -\sum_{i \in N} p_i. \tag{2}$$

Among feasible states,

Definition 2: A Pareto efficient state of the economy is a feasible state (x, p, z) such that there exists no other feasible state (x', p', z') for which $u_i(x_i', z') \geq u_i(x_i, z)$ for all $i \in N$ with strict inequality for at least one i.

To characterize efficient states, the usual first order conditions take in this case the form of the following system of equalities:

$$\sum_{j \in N} \pi_j(z) = \gamma_i(p_i), \quad i = 1, 2, \ldots, n. \tag{3}$$

Henceforth, we shall always write π_N for $\Sigma_{j \in N} \pi_j$. Existence of (several) Pareto efficient states easily follows from our assumptions and, due to Assumption 4(iii) in particular, one has $p_i > 0$ for all i in any efficient state.

A further property of efficient states in our model, important for our purposes, is the following:

Proposition 1: Assumptions 1–3 imply that in all Pareto efficient states, the vector of emission levels $(p_1, \ldots, p_n)$ is the same.

Proof: Suppose not, and let $(\bar{x}, \bar{p}, \bar{z})$ and $(\tilde{x}, \tilde{p}, \tilde{z})$ be two states, both Pareto efficient, with $\bar{p}_i \neq \tilde{p}_i$ for some i. On the one hand, if $\tilde{z} = \bar{z}$, then $\pi_N(\tilde{z}) = \pi_N(\bar{z})$ implies by (3) that $\gamma(\bar{p}_i) = \gamma_i(\tilde{p}_i)\ \forall i \in N$; but this, by strict concavity of all functions $g_i(p_i)$, is only possible if $\tilde{p}_i = \bar{p}_i\, \forall i$. If, on the other hand, $\tilde{z} \neq \bar{z}$, say $\tilde{z} < \bar{z}$ without loss of generality, the fact that $\pi_N(\tilde{z}) \geq \pi_N(\bar{z})$ now implies by (3) that $\gamma_i(\tilde{p}_i) \geq \gamma_i(\bar{p}_i)\, \forall i$, which in turn implies by concavity that $\tilde{p}_i \leq \bar{p}_i\, \forall i$. However, since $\tilde{z} = -\Sigma_{i \in N} \tilde{p}_i$ and $\bar{z} = -\Sigma_{i \in N} \bar{p}_i$, one would have $\tilde{z} \geq \bar{z}$, which is a contradiction.

3 Noncooperative Games Associated with the Economy

In this section we examine various noncooperative equilibrium concepts – the Nash equilibrium, the strong Nash equilibrium and the coalition proof Nash equilibrium – that might be called upon to describe the state of the economy in the absence of cooperation. We conclude that in the case of detrimental environmental externalities studied here only the Nash equilibrium is appropriate for that purpose.

We consider now each agent i of the economy as a player in an n-person noncooperative game. To that effect, let

$$T_i = \{(x_i, p_i) | 0 \leq p_i \leq p_i^0; 0 \leq x_i \leq g_i(p_i^0)\}, \quad i \in N$$

be the strategy set of each player i, and

$$T(S) = \{(x_i, p_i)_{i \in S} | 0 \leq p_i \leq p_i^0 \forall_{i \in S} \text{ and } 0 \leq \sum_{i \in S} x_i \leq \sum_{i \in S} g_i(p_i^0)\}$$

be the space of joint strategies of players in S. Clearly, $T(S) \supset X_{i \in S} T_i$. Let T denote the space of joint strategies of all players, i.e., $T \equiv T(N)$. Any joint strategy choice $[(x_1, p_1), \ldots, (x_n, p_n)] \in T$ of the players induces a feasible state (x, p, z) of the economy if $z = -\sum p_i$.

Then, if for each $i = 1, \ldots, n$ and any $[(x_1, p_1), \ldots, (x_n, p_n)] \in T$ we choose $u_i(x_i, z) = x_i + v_i(z)$ with $z = -\Sigma_{j \in N} p_j$ as the payoff for player i and write $u = (u_1, \ldots, u_n)$, we have defined a noncooperative game $[N, T, u]$, associated with the economy.

Definition 3: For the noncooperative game $[N, T, u]$, the joint strategy choice $[(\bar{x}_1, \bar{p}_1), \ldots, (\bar{x}_n, \bar{p}_n)]$ is a Nash equilibrium if for all $i \in N$,

$(\bar{x}_i, \bar{p}_i)$ maximizes $x_i + v_i(z)$

subject to

$$x_i \leq g_i(p_i)$$

and

$$p_i + z = -\sum_{\substack{j \in N \\ j \neq i}} \bar{p}_j.$$

Definition 4: For the economy, a disagreement equilibrium is the state $(\bar{x}_1, \ldots, \bar{x}_n; \bar{p}_1, \ldots, \bar{p}_n; \bar{z})$ with $\bar{z} = -\Sigma_{i \in N} \bar{p}_i$ induced by the Nash equilibrium $[(\bar{x}_1, \bar{p}_1), \ldots, (\bar{x}_n, \bar{p}_n)]$ of the game $[N, T, u]$.

To characterize a Nash equilibrium or a disagreement equilibrium, the first order conditions of the maximization problem in Definition 3 yield the well known system of equalities:

$$\pi_i(\bar{z}) = \gamma_i(\bar{p}_i), \quad i = 1, \ldots, n. \tag{4}$$

From the fact that the system (4) differs from (3), the standard statement is derived that a disagreement equilibrium is not an efficient state of the economy. Furthermore, it will prove useful to have the following two properties established:

Proposition 2: For the game $[N, T, u]$ there exists a Nash equilibrium; this equilibrium is unique.

Proof: The existence of a Nash equilibrium follows from standard theorems (see, e.g., FRIEDMAN 1990), since each player's strategy set is compact and convex, and each player's payoff function is concave, and therefore continuous, and bounded. All these conditions are obviously met in our model. To prove uniqueness of the equilibrium $[(\bar{x}_1, \bar{p}_1), \ldots, (\bar{x}_n, \bar{p}_n)]$, suppose contrary to the assertion that there exists another Nash equilibrium

$$[(\hat{x}_1, \hat{p}_1), \ldots, (\hat{x}_n, \hat{p}_n)] \neq [(\bar{x}_1, \bar{p}_1), \ldots, (\bar{x}_n, \bar{p}_n)].$$

Without loss of generality assume that $\Sigma\hat{p}_i \leq \Sigma\bar{p}_i$, entailing $\hat{z} = -\Sigma\hat{p}_i \geq \bar{z} = -\Sigma\bar{p}_i$. From the characterization of the disagreement equilibrium and the concavity of the functions $v_i(z)$, we would then have $\pi_i(\hat{z}) \leq \pi_i(\bar{z})$ as well as $\gamma_i(\hat{p}_i) \leq \gamma_i(\bar{p}_i)$ for each $i \in N$. But given the concavity of the production functions, this last inequality would imply that $\hat{p}_i \geq \bar{p}_i$ for each $i \in N$, which contradicts the assumption that $\Sigma\hat{p}_i \leq \Sigma\bar{p}_i$, and $[(\hat{x}_1, \hat{p}_1), \ldots, (\hat{x}_n, \hat{p}_n)] \neq [(\bar{x}_1, \bar{p}_1), \ldots, (\bar{x}_n, \bar{p}_n)]$. □

Turning to the economy, the existence and uniqueness of a disagreement equilibrium state are also established by Proposition 2.

Let us now examine the relevance for our externalities model of two alternative equilibrium concepts which are offered by the noncooperative game theoretic literature, namely the **strong Nash equilibrium** and the coalition proof Nash equilibrium.

Definition 5: For the noncooperative game $[N, T, u]$, the joint strategy choice $[(\tilde{\bar{x}}_1, \tilde{\bar{p}}_1), \ldots, (\tilde{\bar{x}}_n, \tilde{\bar{p}}_n)]$ is a strong Nash equilibrium if for all $S \subseteq N$,

$$[(\tilde{\bar{x}}_i, \tilde{\bar{p}}_i)_{i \in S}] \text{ maximizes } \sum_{i \in S} [x_i + v_i(z)] \tag{5.a}$$

$$\text{subject to } \sum_{i \in S} x_i \leq \sum_{i \in S} g_i(p_i) \tag{5.b}$$

$$\text{and} \sum_{i \in S} p_i + z = \begin{cases} 0 & \text{if } S = N \\ -\sum_{j \in N \setminus S} \tilde{\tilde{p}}_j & \text{if } S \neq N. \end{cases} \tag{5.c}$$

For the analysis of economies with detrimental environmental externalities, this concept is unfortunately of no use because of the following result.

Proposition 3: For the game $[N, T, u]$, there does not exist a strong Nash equilibrium.

Proof: Definition 5 implies both that a strong Nash equilibrium is also a Nash equilibrium, and that it induces a Pareto efficient state of the economy. However, as was observed earlier, the Nash equilibrium cannot induce such a Pareto efficient state. It follows from this contradiction that a strong Nash equilibrium cannot exist for the game $[N, T, u]$.[4] □

Having noted that strong Nash equilibria almost never exist in general. BERNHEIM, PELEG, and WHINSTON (1987) comment that the strong Nash concept is "too strong". They therefore propose as an alternative the **coalition-proof Nash equilibrium.** For the externalities model we are dealing with here, however, this refinement cannot bring much conceptual progress for the following reason: a coalition-proof Nash equilibrium is always also a Nash equilibrium, and in our $[N, T, u]$ game the Nash equilibrium is unique; therefore, any coalition-proof Nash equilibrium – if one exists at all – cannot be different from the Nash equilibrium characterized above.

The coalition proof Nash equilibrium concept might be used, however, not for describing the state of the economy in the absence of cooperation, as we have done so far, but instead for predicting and/or characterizing the outcome of negotiations. In this perspective, the fact that a coalition-proof equilibrium is not a Pareto optimum in general prevents us to use it, as we are searching for efficient outcomes. Moreover the concept of coalition-proof Nash equilibrium involves, as in the case of strong Nash equilibrium, the assumption that when a coalition deviates, it takes as given the strategies of the complement – an assumption that we do not find justified, as we shall argue in the following section.

4 Cooperative Games Associated with the Economy

4.1 Inappropriateness of the α-Characteristic Function

Turning now to the cooperative part of our analysis, let us associate, with every coalition S, the number $w(S)$, called the "worth" of the coalition S and defined as

[4] MÄLER (1989) shows that for his "acid rain game" also there exists no strong Nash equilibrium.

the highest aggregate payoff $\Sigma_{i \in S} u_i$ that the members of the coalition can achieve using some strategy. Thus, the pair $[N, w(\cdot)]$ consisting of the players set N and the characteristic function $w(\cdot)$ defines a cooperative game (with transferable utilities) associated with our economy.

Stated in full, the characteristic function reads

$$w(S) = \max_{\{(x_i, p_i)_{i \in S}\}} \sum_{i \in S} [x_i + v_i(z)] \tag{6}$$

For the relation between this function and our economy to be meaningful – i.e. to correspond to feasible states, the variable z should satisfy condition (2). However, this equality involves strategic choices made by players who are *not* members of S. Thus, the worth of a coalition in our game is not only a function of actions taken by its members, but also of actions of players outside the coalition.

This typical feature of cooperative games associated with economies with externalities and/or public goods requires the characteristic function to specify explicitly what the actions are both of the members of S and of the other players. A familiar way to get around this problem has been to assume that the players outside the coalition adopt those strategies that are *least favorable* to the coalition. That is, the characteristic function is defined as

$$w^{\alpha}(S) = \max_{\{(x_i, p_i)_{i \in S}\}} \sum_{i \in S} [x_i + v_i(z)] \tag{7.a}$$

$$\text{subject to } \sum_{i \in S} x_i \leq \sum_{i \in S} g_i(p_i) \tag{7.b}$$

$$\sum_{i \in S} p_i + z \begin{cases} = -\sum_{j \in N \setminus S} p_j^0 & \text{if } S \neq N \\ = 0 & \text{if } S = N. \end{cases} \tag{7.c}$$

Games with a characteristic function of this type are used in most of the studies dealing with the core of economies with public goods (see, e.g., FOLEY 1970, CHAMPSAUR 1975, MOULIN 1987 and CHANDER 1993). This is also the case with the game used by SCARF 1971 in his study of cores of economies with externalities: his "α-characteristic function" is similar to the one defined above[5]. And more recently, pushing this "pessimistic" view to the

[5] LAFFONT 1977 shows that the α-and the β-characteristic functions are identical for cooperative games associated with economies exhibiting detrimental externalities of the type dealt with here. Hence, all our subsequent developments on α-cores will also be valid for β-cores.

extreme, MÄLER 1989 assumes in his analysis of the acid rain game that there is no upper bound such as our p_i^0 for the individual pollutant discharges; thus, coalitions can be hurt by up to infinite amounts of pollutants emitted by players outside the coalition.

This extreme form of the assumption behind the α-characteristic function suggests that it is actually ill-adapted to the issues at stake with environmental externalities. Indeed there are at least two grounds for considering that assuming such strategies on the part of the players outside the coalition is questionable. On the one hand, these strategies may not maximize the individual payoffs of these players. On the other hand, they rest on the presumption that when a coalition forms, its payoff is what it would get when members of the complementary coalition act so as to minimax this coalition's payoff. But why should they behave in this extreme fashion, and why should the coalition formed presume that they would follow such strategies?

Searching for alternative assumptions in the spirit of the strong Nash equilibrium or of the coalition proof Nash equilibrium is not satisfactory either: they assume for their part that when a coalition deviates from the equilibrium it takes as given the strategies of its complement at that equilibrium. But as before, this may not maximize the payoff of the complement's members; and in addition, why should their strategies remain unchanged when S forms?

4.2 Equilibrium with Respect to a Coalition

These considerations lead us to introduce here a concept in which it is assumed that when a coalition forms it neither takes as given the strategies of its complement, nor does it assume that the complement would follow minimax strategies: instead, it looks forward to the best reply payoff corresponding to the equilibrium that its actions would induce. More specifically, we propose to assume that when S forms players outside S do not take particular coalitional actions against S, but adopt only individually best reply strategies. This results in a Nash equilibrium between S and the remaining players, with the members of S thus playing their joint best response to the individual strategies of the others.[6] In terms of our economic model and of its associated game, this is formalized as follows.

Definition 6: For the noncooperative game $[N, T, u]$, given a coalition $S \subset N$, a Nash equilibrium with respect to S is the joint strategy choice $[(\tilde{x}_1, \tilde{p}_1), \ldots,$

[6] Our analysis might be extended to the case when $N \backslash S$ may also form, but we do not pursue this possibility in the present paper. It may be noted however that if the players outside S form one or more non-singleton coalitions then the worth of S is generally higher. That players outside S act individually is therefore equivalent to granting S a certain degree of pessimism.

$(\tilde{x}_n, \tilde{p}_n)] \in T$ where

(i) $[(\tilde{x}_i, \tilde{p}_i)_{i \in S}]$ maximizes $\sum_{i \in S} [x_i + v_i(z)]$ (9.a)

subject to $\sum_{i \in S} x_i \leq \sum_{i \in S} g_i(p_i)$ (9.b)

and $\sum_{i \in S} p_i + z = - \sum_{j \in N \setminus S} \tilde{p}_j$; (9.c)

(ii) $\forall j \in N \setminus S$, $(\tilde{x}_j, \tilde{p}_j)$ maximizes $x_j + v_j(z)$ (9.d)

subject to $x_j \leq g_j(p_j)$ (9.e)

and $p_j + z = - \sum_{\substack{i \in N \\ i \neq j}} \tilde{p}_j$. (9.f)

As suggested by our choice of terminology, this equilibrium amounts to a Nash equilibrium between the coalition S acting as one individual player and the other players acting alone.

Definition 7: For the economy, given a coalition $S \subset N$, a partial agreement equilibrium with respect to S is the state $(\tilde{x}_1, \ldots, \tilde{x}_n; \tilde{p}_1, \ldots, \tilde{p}_n; \tilde{z})$ with $\tilde{z} = -\Sigma_{i \in N} \tilde{p}_i$ induced by $[(\tilde{x}_1, \tilde{p}_1), \ldots, (\tilde{x}_n, \tilde{p}_n)]$, the Nash equilibrium with respect to S of Definition 6.

In terms of first order conditions, a partial agreement equilibrium with respect to S so defined is characterized by the system of n equalities:

$$\sum_{j \in S} \pi_j(\tilde{z}) = \gamma_i(\tilde{p}_i), \quad i \in S$$

and

$$\pi_j(\tilde{z}) = \gamma_j(\tilde{p}_j), \quad j \in N \setminus S.$$

Note that inequality (9.b) in Definition 6 allows for some members of S possibly to consume more, or less, than what they produce. The strategy thus includes the possibility of transfers of private goods among the members of the coalition. The inequality also implies that the algebraic sum of these transfers be zero within the coalition. Further properties of a partial agreement equilibrium with respect to a coalition S that are important for the sequel are collected hereafter.

Proposition 4: For any coalition $S \subset N$,

(i) there exists a partial agreement equilibrium with respect to S;
(ii) the individual emission levels corresponding to this equilibrium are unique;
(iii) the individual emission levels of the players outside of S are not lower than those at the disagreement equilibrium,
(iv) although the total emissions are not higher.

Proof: The existence proof follows from similar arguments as in Proposition 2. The uniqueness of individual emission levels follows from similar arguments as in Proposition 1. We therefore only formally prove here the remaining two parts of the proposition.

Let $\tilde{z} = -\sum \tilde{p}_i$ and $\bar{z} = -\sum \bar{p}_i$, where $(\tilde{p}_1, \ldots, \tilde{p}_n)$ and $(\bar{p}_1, \ldots, \bar{p}_n)$ are the emission levels corresponding to a partial agreement equilibrium with respect to S and to the disagreement equilibrium, respectively. We first show that $\tilde{z} \geq \bar{z}$.

Suppose contrary to the assertion that $\tilde{z} < \bar{z}$. We must then have $\pi_i(\tilde{z}) \geq \pi_i(\bar{z})$ for each $i \in N$. From the characterizations of a partial agreement equilibrium and of the disagreement equilibrium it follows that

$$\sum_{j \in S} \pi_j(\tilde{z}) = \gamma_i(\tilde{p}_i) \geq \gamma_i(\bar{p}_i) = \pi_i(\bar{z}), \quad \forall i \in S$$

and

$$\pi_j(\tilde{z}) = \gamma_j(\tilde{p}_j) \geq \gamma_j(\bar{p}_j) = \pi_j(\bar{z}), \quad \forall j \in N \backslash S.$$

From the strict concavity of each function g_i, it follows that $\tilde{p}_i \leq \bar{p}_i$ for each $i \in N$. But this contradicts our supposition that $\tilde{z} < \bar{z}$. Hence, we must have $\tilde{z} \geq \bar{z}$.

Finally, since $\tilde{z} \geq \bar{z}, \pi_j(\tilde{z}) \leq \pi_j(\bar{z})$ for each $j \in N \backslash S$. The inequalities above and strict concavity of g_j imply that $\tilde{p}_j \geq \bar{p}_j$ for each $j \in N \backslash S$. □

Note that in view of Assumption 2, the total emissions corresponding to a partial agreement equilibrium with respect to a coalition of two or more players are strictly lower than those corresponding to the disagreement equilibrium. This means that as compared to the disagreement equilibrium, the players outside a coalition of two or more players are strictly better-off at a partial agreement equilibrium with respect to that coalition – which is actually a form of free riding on the part of those players.

4.3 The γ-Characteristic Function

We are now equipped to introduce a new characteristic function, that embodies the behavioral assumptions we found desirable at the beginning of the previous subsection. We call it the "partial agreement characteristic function" or, for short, the γ-characteristic function, and denote it by w^{γ}. It is defined by:

$$w^{\gamma}(S) = \max_{\{(x_i,p_i)_{i\in S}\}} \sum_{i\in S}[x_i + v_i(z)] \tag{10.a}$$

$$\text{subject to } \sum_{i\in S} x_i \leq \sum_{i\in S} g_i(p_i) \tag{10.b}$$

$$\text{and } \sum_{i\in S} p_i + z = -\sum_{j\in N\setminus S} p_j, \tag{10.c}$$

$$\text{where } \forall j \in N\setminus S, (x_j, p_j)\text{maximizes } x_j + v_j(z) \tag{10.d}$$

$$\text{subject to } x_j \leq g_j(p_j) \tag{10.e}$$

$$\text{and } p_j + z = -\sum_{\substack{i\in N\\ i\neq j}} p_i. \tag{10.f}$$

The worth of coalition S is thus determined by an equilibrium concept, namely that of partial agreement equilibrium with respect to S. It is not assumed that the players outside the coalition S do the worst, as with the α-characteristic function; nor is it assumed, as in the concepts of strong and coalition-proof Nash equilibria, that they do not react to the actions of S.

Let us note however that for each S, the value assigned by the partial agreement characteristic function w^{γ} is at least as much as that assigned by the α-characteristic function, *i.e.* $w^{\gamma}(S) \geq w^{\alpha}(S)$ for each $S \subset N$. In fact, examples can be constructed such that $w^{\gamma}(S) > w^{\alpha}(S)$ for all $S \subset N$. This implies that the core of the game $[N, w^{\gamma}]$, *i.e.* the "γ-core", is, if nonempty, contained in the "α-core", and possibly smaller. Moreover, if we had assumed that when a coalition S forms, the complementary coalition $N\setminus S$ also forms and chooses the best response strategy for its members given what S does, then the core of the so-defined game would most likely be smaller than that of the game $[N, w^{\gamma}]$, and even perhaps empty in which case a stable strategy for the grand coalition N would be impossible to find.[7]

[7] This remark points to the work of CARRARO and SINISCALCO 1993 who in a model with identical agents assume that when S forms and achieves the aggregate payoff $w(S)$, if some $i \in S$ leaves S, the coalition $S\setminus\{i\}$ remains formed.

5 A Strategy in the γ-Core with Linear Pay-off Functions

Let us recall:

Definition 8: For any cooperative game $[N, w]$, a strategy of the all player coalition N is said to belong to the core of the game if for any subset $S \subset N$ the payoff it yields to the members of S is larger than $w(S)$, *i.e.* the payoff that S can achieve by itself.

Emptiness *vs.* nonemptiness of the core typically depends upon the form of the characteristic function, which reflects the power of each coalition in the game. In the presence of externalities, this power is crucially affected by the assumed behavior of the players outside the coalition. Thus, with the α-characteristic function, coalitions are weakened by the presumed minimax behavior, letting hope that the corresponding α-core be nonempty. This is indeed the case in SCARF 1971, as well as in the version given by LAFFONT 1977 of the SHAPLEY and SHUBIK 1969 "garbage game",[8] and also in MÄLER's 1989 acid rain game.

The former two results, however, do not bear on an economy with externalities of the environmental type we are dealing with here, and Mäler's argument is only an informal one. Furthermore, no results are available, to our knowledge, on cores of games with the characteristic function w^γ we have introduced above.

To find out whether or not the core of a cooperative game is nonempty various approach can be used. Two qualitative ones are offered by SHAPLEY 1967 and SHAPLEY 1971 who respectively show nonemptiness if the game is balanced or convex. None of these approaches proved succesful in our case. We therefore turn to another, constructive, approach, which is to exhibit a strategy for which we can show that it satisfies Definition 8.

Specifically, let $(x^*, p^*, z^*) = (x_1^*, \ldots, x_n^*; p_1^*, \ldots, p_n^*; z^*)$ be the Pareto efficient state defined by

$$\begin{aligned} x_i^* &= g_i(\bar{p}_i) - \frac{\pi_j^*}{\pi_N^*}\left(\sum_{j\in N} g_j(\bar{p}) - \sum_{j\in N} g_j(p_j^*)\right), \quad i \in N, \\ z^* &= -\sum p_i^*, \end{aligned} \tag{11}$$

They show that then it may be better for i to leave S. As this advantage grows with the size of coalitions, they conclude that only small coalitions can prevail, and that N will never form. More recently, in a model where coalition formation is endogeneous, RAY and VOHRA 1996 also conclude that in general N cannot form.

[8] This is also the case in the many studies of economies with public goods alluded to above of FOLEY 1970, CHAMPSAUR 1975, MOULIN 1987 and CHANDER 1993, where the unfavorable behavior consists in producing no public good at all; these have typically large cores.

where $\pi_i^* = \pi_i(z^*)$ for each $i \in N$ and $(p_1^*, \ldots, p_n^*)$ and $(\bar{p}_1, \ldots, \bar{p}_n)$ are the (unique) individual emission levels corresponding to the Pareto efficient states and to the disagreement equilibrium, respectively.[9] By Pareto efficiency $\pi_N^* = \gamma_i(p_i^*)$ for each $i \in N$.

As can be easily seen, this state of the economy[10] implies that for each $i \in N$,

$$(x_i^*, p_i^*, z^*) \text{ maximizes } x_i + v_i(z)$$

subject to

$$x_i \leq g_i(\bar{p}_i) - \frac{\pi_i^*}{\pi_N^*}\left(\sum_{j \in N} g_j(\bar{p}_j) - \sum_{\substack{j \in N \\ j \neq i}} g_j(p_j^*) - g_i(p_i)\right),$$

$$p_i + z = -\sum_{j \neq i} p_j^*$$

and

$$0 \leq p_i \leq p_i^0.$$

which means that the state $(x_1^*, \ldots, x_n^*; p_1^*, \ldots, p_n^*; z^*)$ is an equilibrium concept. As it can be given an interpretation analogous to that of the ratio equilibrium (see KANEKO 1977, and MAS-COLELL and SILVESTRE 1989), we shall refer to it as the ratio equilibrium with respect to the disagreement equilibrium.[11]

We now prove the nonemptiness of the γ-core by showing that the ratio equilibrium with respect to the disagreement equilibrium belongs to the core of the game $[N, w^\gamma]$.

We first consider a special case of our general model, namely the one where it is assumed that the payoff functions are linear, i.e.:

Assumption 1': $u_i(x_i, z) = x_i + \bar{\pi}_i z, \quad \bar{\pi}_i > 0.$

---

[9] CHANDER 1993 analyzes an instantaneous analog of this cost sharing rule in a public good context.

[10] Expression (11) implies transfers $T_i (> 0$ if received, < 0 if paid) of the private good, such that $\forall i \in N, x_i^* = g_i(p_i^*) + T_i$, where

$$T_i = -(g_i(p_i^*) - g_i(\bar{p}_i)) + \frac{\pi_i^*}{\pi_N^*}\left(\sum_{j \in N} g_j(p_j^*) - \sum_{j \in N} g_j(\bar{p}_j)\right),$$

with the property that $\Sigma_{i \in N} T_i = 0$

[11] Note that one can also consider the Pareto efficient state defined by:

$$x_j^* = g_i(p_i^0) - \frac{\pi_i^*}{\pi_N^*}\left(\sum_{j \in N} g_j(p_j^0) - \sum_{j \in N} g_j(p_j^*)\right), i \in N.$$

But we are unable to establish the same properties for this Pareto efficient state. It seems that the reference point matters.

This is actually the case for which MÄLER 1989 proves the nonexistence of a strong Nash equilibrium. In SHAPLEY and SHUBIK's 1969 garbage game also, the payoff functions are linear (unlike here, however, their externalities are directional and involve no diseconomies of scale).

As can be easily seen from the characterizations of the Nash equilibrium and of a partial agreement equilibrium, an important consequence of the linearity assumption is that in a partial agreement equilibrium with respect to a coalition, the emission levels of the players outside the coalition are the same as in the Nash equilibrium. In fact, under the linearity assumption the Nash equilibrium is a dominant strategy equilibrium.

Theorem 1: Under the linearity Assumption 1′, the strategy $[(x_1^*, p_1^*), \ldots, (x_n^*, p_n^*)]$ of the grand coalition N that induces the ratio equilibrium (x^*, p^*, z^*) of the economy belongs to the core of the game $[N, w^\gamma]$.

Proof: Suppose contrary to the assertion, that the strategy inducing $(x_1^*, \ldots, x_n^*; p_1^*, \ldots, p_n^*; z^*)$ is not in the core of the game $[N, w^\gamma]$. Then there exists a coalition $S \subset N$ and a strategy for S inducing the feasible state $(\tilde{x}_1, \ldots, \tilde{x}_n; \tilde{p}_1, \ldots, \tilde{p}_n; \tilde{z})$ such that $(\tilde{x}_1, \ldots, \tilde{x}_n; \tilde{p}_1, \ldots, \tilde{p}_n; \tilde{z})$ is a partial agreement equilibrium with respect to S, and $\tilde{x}_i + \bar{\pi}_i \tilde{z} > x_i^* + \bar{\pi}_i z^*$ for all $i \in S$. From the characterization of a partial agreement equilibrium with respect to a coalition, it follows that $\tilde{p}_i = \bar{p}_i$ for all $i \in N \backslash S$, that $\tilde{p}_i \geq p_i^*$ for all $i \in S$, and that $\Sigma_{i \in S} \tilde{x}_i = \Sigma_{i \in S} g_i(\tilde{p}_i)$.

Consider now the alternative efficient state $(\hat{x}_1, \ldots, \hat{x}_n; p_1^*, \ldots, p_n^*; z^*)$, defined as:

$$
\begin{aligned}
\hat{x}_i &= g_i(\tilde{p}_i) - \frac{\bar{\pi}_i}{\bar{\pi}_N}\left(\sum_{i \in N} g_i(\tilde{p}_i) - \sum_{i \in N} g_i(p_i^*)\right), i \in N, \\
z^* &= -\sum_{i \in N} p_i^*.
\end{aligned}
\tag{12}
$$

We show below that, as far as the members of S are concerned, one has

$$
\sum_{i \in S} \hat{x}_i + \sum_{i \in S} \bar{\pi}_i z^* > \sum_{i \in S} \tilde{x}_i + \sum_{i \in S} \bar{\pi}_i \tilde{z},
\tag{13}
$$

which implies

$$
\sum_{i \in S} \hat{x}_i + \sum_{i \in S} \bar{\pi}_i z^* > \sum_{i \in S} x_i^* + \sum_{i \in S} \bar{\pi}_i z^*
\tag{14}
$$

if S is able to achieve $\tilde{x}_i + \bar{\pi}_i \tilde{z} > x_i^* + \bar{\pi}_i z^*$ for all $i \in S$ as it is supposed to do.

We further show that as far as the other players are concerned,

$$
\hat{x}_i + \bar{\pi}_i z^* \geq x_i^* + \bar{\pi}_i z^* \text{ for all } i \in N \backslash S.
\tag{15}
$$

As together the inequalities (14) and (15) imply that the state $(\hat{x}_1, \ldots, \hat{x}_n; p_1^*, \ldots, p_n^*; z^*)$ is Pareto superior to the Pareto efficient allocation $(x_1^*, \ldots, x_n^*; p_1^*, \ldots, p_n^*; z^*)$, we get an impossibility. Proving (13) and (15) will thus establish the theorem.

To show (13), the definition (12) allows one to write

$$\begin{aligned}\sum_{i\in S}\hat{x}_i + \sum_{i\in S}\bar{\pi}_i z^* &= \sum_{i\in S} g_i(\tilde{p}_i) - \frac{\sum_{i\in S}\bar{\pi}_i}{\bar{\pi}_N}\left(\sum_{i\in N} g_i(\tilde{p}_i) - \sum_{i\in N} g_i(p_i^*)\right) + \sum_{i\in S}\bar{\pi}_i z^* \\ &= \sum_{i\in S}\tilde{x}_i - \frac{\sum_{i\in S}\bar{\pi}_i}{\bar{\pi}_N}\left(\sum_{i\in N} g_i(\tilde{p}_i) - \sum_{i\in N} g_i(p_i^*)\right) + \sum_{i\in S}\bar{\pi}_i z^* - \sum_{i\in S}\bar{\pi}_i\tilde{z} + \sum_{i\in S}\bar{\pi}_i\tilde{z} \\ &= \sum_{i\in S}\tilde{x}_i + \frac{\sum_{i\in S}\bar{\pi}_i}{\bar{\pi}_N}\left[\bar{\pi}_N(z^* - \tilde{z}) - \left(\sum_{i\in N} g_i(\tilde{p}_i) - \sum_{i\in N} g_i(p_i^*)\right)\right] + \sum_{i\in S}\bar{\pi}_i\tilde{z}.\end{aligned} \tag{16}$$

From the respective characterizations of a Pareto efficient state, and of a partial agreement equilibrium, we have for all $i \in N$, $\bar{\pi}_N = \gamma_i(p_i^*)$ and $\tilde{p}_i \geq p_i^*$. Hence, the strict concavity of each function g_i implies

$$\bar{\pi}_N > \frac{g_i(\tilde{p}_i) - g_i(p_i^*)}{\tilde{p}_i - p_i^*}, \quad \text{for all } i \in N,$$

which in turn implies that

$$\bar{\pi}_N(z^* - \tilde{z}) > \sum_{i\in N} g_i(\tilde{p}_i) - \sum_{i\in N} g_i(p_i^*).$$

Then (13) follows from (16).

On the other hand, from the respective characterizations of the disagreement equilibrium and of a partial agreement equilibrium with respect to a coalition, we have $\tilde{p}_i \leq \bar{p}_i$, for all $i \in N$, $\tilde{p}_i = \bar{p}_i$ for all $i \in N\backslash S$, and thus $\Sigma_{i\in N} g_i(\tilde{p}_i) \leq \Sigma_{i\in N} g_i(\bar{p}_i)$. Therefore,

$$\begin{aligned}\hat{x}_i &= g_i(\tilde{p}_i) - \frac{\bar{\pi}_i}{\bar{\pi}_N}\left(\sum_{i\in N} g_i(\tilde{p}_i) - \sum_{i\in N} g_i(p_i^*)\right) \\ &\geq g_i(\bar{p}_i) - \frac{\bar{\pi}_i}{\bar{\pi}_N}\left(\sum_{i\in N} g_i(\bar{p}_i) - \sum_{i\in N} g_i(p_i^*)\right) = x_i^* \quad \text{for all } i \in N\backslash S.\end{aligned}$$

These inequalities establish (15). □

Note that the domination used in the proof of Theorem 1 is more demanding than the usual one in that it requires any dominating payoff vector to correspond to an equilibrium strategy combination. Thus, the α-core includes the γ-core, just as the coalition proof Nash equilibria include the strong Nash equilibria if they are both nonempty. In other words, the γ-core seems to satisfy a consistency requirement.[12]

6 γ-Core Property of the Strategy Under Non-linear Payoff Functions

Doing away with the linearity of payoff functions makes the interaction between the strategic variables more complex. To see this, note that the Nash equilibrium is no longer a dominant strategy equilibrium. Moreover, it need not be true that $\tilde{p}_i \leq \bar{p}_i$ for all $i \in S$, where $\tilde{p}_i, i \in N$, are the emission levels corresponding to the partial agreement equilibrium with respect to S. As these inequalities play a crucial role in the proof of Theorem 1, we do extend our result to the more general case by imposing a condition which is sufficient to ensure that these inequalities continue to hold. Consider the following:

Assumption 1″: For all $S \subset N, S \neq N, |S| \geq 2, \Sigma_{i \in S}\pi_i(z^*) \geq \pi_j(\bar{z}), j \in S$, where $\bar{z}$ and z^* correspond to the disagreement equilibrium and a Pareto efficient state, respectively.

Clearly, this assumption is satisfied when the payoff functions are linear. In words, the assumption says that the (now non-constant) marginal utilities of z should not fall "too much" between the disagreement equilibrium and the Pareto optimum. It covers a large class of quadratic utility functions, among others.

Proposition 5: Under Assumption 1″, the emission level of each player in the coalition of a partial agreement equilibrium is not higher than the one corresponding to the Nash equilibrium.

Proof: Let $S \subset N$ be some coalition, and let $((\tilde{x}_1, \tilde{p}_1), \ldots, (\tilde{x}_n, \tilde{p}_n))$ be a partial agreement equilibrium with respect to S. Using an argument which is analogous to that in the proof of Proposition 4, we first show that $\Sigma_{i \in N}\tilde{p}_i \geq \Sigma_{i \in N}p_i^*$, where $p_i^*, i \in N$, are the emission levels corresponding to a Pareto optimum.

Indeed, suppose $\Sigma_{i \in N}\tilde{p}_i < \Sigma_{i \in N}p_i^*$ instead, implying $\tilde{z} > z^*$. By concavity of the functions $v_i(z)$, one would have $\pi_j(\tilde{z}) \leq \pi_j(z^*) \forall j \in N$, and from this, together with the first order characterizations of a partial agreement

[12] We are grateful to the referee who made that point.

equilibrium and of an optimum, it would follow that, for each $j \in S$.

$$\sum_{k \in S} \pi_k(\tilde{z}) = \gamma_j(\tilde{p}_j) \leq \sum_{k \in S} \pi_k(z^*) \leq \sum_{k \in N} \pi_k(z^*) = \gamma_j(p_j^*).$$

and for each $j \in N \backslash S$,

$$\pi_j(\tilde{z}) = \gamma_j(\tilde{p}_j) \leq \pi_j(z^*) \leq \gamma_j(p_j^*).$$

Since $\gamma_j(p_j)$ is decreasing in p_j, these two expressions imply that $\tilde{p}_j \geq p_j^* \forall j \in N$, contradicting the supposition.

Now, since $\tilde{z} \leq z^*$, concavity again implies that $\Sigma_{i \in S}\pi_i(\tilde{z}) \geq \Sigma_{i \in S}\pi_i(z^*)$ for all $S \subset N$. From the characterizations of the Nash equilibrium and of the partial agreement equilibrium with respect to S, it then follows that $\gamma_j(\tilde{p}_j) \geq \gamma_j(\bar{p}_j)$ for all $j \in S$, that is $\tilde{p}_i \leq \bar{p}_j$ for all $j \in S$.

We can now state: □

Theorem 2: Under Assumption 1″, the strategy $[(x_1^*, p_1^*), \ldots, (x_n^*, p_n^*)]$ of the grand coalition N that induces the ratio equilibrium (x^*, p^*, z^*) of the economy belongs to the core of the game $[N, w^\gamma]$.

Proof: Suppose, contrary to the assertion, that the strategy inducing the above state (x^*, p^*, z^*) is not in the core of the game $[N, w^\gamma]$. Then there exists a coalition $S \subset N, S \neq N$, and a Nash equilibrium with respect to S, $[(\tilde{x}_1, \tilde{p}_1), \ldots, (\tilde{x}_n, \tilde{p}_n)]$, such that

$$\tilde{x}_i + v_i(\tilde{z}) > x_i^* + v_i(z^*) \text{ for all } i \in S, \tag{17}$$

where $\tilde{z} = -\Sigma_{i \in N}\tilde{p}_i$.

Define a new Pareto efficient state of the economy $(\hat{x}_1, \ldots, \hat{x}_n; p_1^*, \ldots, p_n^*; z^*)$ as follows:

$$\hat{x}_i = g_i(\tilde{p}_i) - \frac{\pi_i^*}{\pi_N^*}\left(\sum_{j \in N} g_j(\tilde{p}_j) - \sum_{j \in N} g_j(p_j^*)\right), i = 1, \ldots, n.$$

We claim that inequality (17) implies that

$$\sum_{i \in S} \hat{x}_i + \sum_{i \in S} v_i(z^*) > \sum_{i \in S} \tilde{x}_i + \sum_{i \in S} v_i(\tilde{z}) \tag{18}$$

and

$$\sum_{i \in N \backslash S} \hat{x}_i > \sum_{i \in N \backslash S} x_i^*. \tag{19}$$

As these two inequalities together with (17) clearly contradict the Pareto efficiency of $(x_1^*, \ldots, x_n^*; p_1^*, \ldots, p_n^*; z^*)$, our theorem is proved if we establish them.

We first prove (18). By the definition of the new allocation,

$$\sum_{i \in S} \hat{x}_i = \sum_{i \in S} g_i(\tilde{p}_i) - \left(\sum_{i \in S} \pi_i^*/\pi_N^*\right)\left(\sum_{j \in N} g_j(\tilde{p}_j) - \sum_{j \in N} g_j(p_j^*)\right)$$
$$\geq \sum_{i \in S} \tilde{x}_i - \left(\sum_{i \in S} \pi_i^*/\pi_N^*\right)\left(\pi_N^*\left(\sum_{j \in N} \tilde{p}_j - \sum_{j \in N} p_j^*\right)\right),$$

using the concavity of the functions g_i, as well as the Pareto efficiency condition (3); but the last expression is equal to:

$$\sum_{i \in S} \tilde{x}_i - \left(\sum_{i \in S} \pi_i^*/\pi_N^*\right)\pi_N^*(z^* - \tilde{z}).$$

Thus,

$$\sum_{i \in S} \hat{x}_i + \sum_{i \in S} \pi_i^* z^* \geq \sum_{i \in S} \tilde{x}_i + \sum_{i \in S} \pi_i^* \tilde{z},$$

that is

$$\sum_{i \in S} \hat{x}_i + \sum_{i \in S} v_i(z^*) \geq \sum_{i \in S} \tilde{x}_i + \sum_{i \in S} v_i(\tilde{z}) + \left(\sum_{i \in S} v_i(z^*) - \sum_{i \in S} v_i(\tilde{z}) - \sum_{i \in S} \pi_i^*(z^* - \tilde{z})\right)$$
$$\geq \sum_{i \in S} \tilde{x}_i + \sum_{i \in S} v_i(\tilde{z}),$$

(*i.e.* (18)), since from concavity $v_i(z^*) - v_i(\tilde{z}) \geq \pi_i^*(z^* - \tilde{z})$ for all i.

Next we prove inequality (19). By definition,

$$\sum_{i \in N \setminus S} \hat{x}_i = \sum_{i \in N \setminus S} g_i(\tilde{p}_i) - \left(\sum_{i \in N \setminus S} \pi_i^*/\pi_N^*\right)\left(\sum_{j \in N} g_j(\tilde{p}_j) - \sum_{j \in N} g_j(p_j^*)\right)$$
$$= \sum_{i \in N \setminus S} g_i(\bar{p}_i) - \left(\sum_{i \in N \setminus S} \pi_i^*/\pi_N^*\right)\left(\sum_{j \in N} g_j(\bar{p}_j) - \sum_{j \in N} g_j(p_j^*)\right)$$
$$+ \left(\sum_{i \in N \setminus S} g_i(\tilde{p}_i) - \sum_{i \in N \setminus S} g_i(\bar{p}_i)\right)$$
$$+ \left(\sum_{i \in N \setminus S} \pi_i^*/\pi_N^*\right)\left(\sum_{j \in N} g_j(\bar{p}_j) - \sum_{j \in N} g_j(\tilde{p}_j)\right)$$

$$= \sum_{i\in N\setminus S} x_i^* + \left(\sum_{i\in N\setminus S} g_i(\tilde{p}_i) - \sum_{i\in N\setminus S} g_i(\bar{p}_i)\right)$$
$$+ \left(\sum_{i\in N\setminus S} \pi_i^*/\pi_N^*\right)\left(\sum_{j\in N} g_j(\bar{p}_j) - \sum_{j\in N} g_j(\tilde{p}_j)\right)$$
$$= \sum_{i\in N\setminus S} x_i^* + \left[\left(\sum_{i\in N\setminus S} g_i(\tilde{p}_i) - \sum_{i\in N\setminus S} g_i(\bar{p}_i)\right)\right.$$
$$\left. - \left(\sum_{i\in N\setminus S} \pi_i^*/\pi_N^*\right)\left(\sum_{i\in N\setminus S} g_i(\tilde{p}_i) - \sum_{i\in N\setminus S} g_i(\bar{p}_i)\right)\right]$$
$$- \left(\sum_{i\in N\setminus S} \pi_i^*/\pi_N^*\right)\left(\sum_{i\in S} g_i(\tilde{p}_i) - \sum_{i\in S} g_i(\bar{p}_i)\right).$$

As Propositions 4 and 5 imply that $\tilde{p}_i \geq \bar{p}_i$ for all $i \in N\setminus S$ and $\tilde{p}_i \leq \bar{p}_i$ for all $i \in S$, we have $\Sigma_{i\in N\setminus S}\hat{x}_i > \Sigma_{i\in N\setminus}x_i^*$, *i.e.* (19). □

We had referred earlier to an international environmental externalities literature that concludes negatively on the issue of full cooperation. These authors (MÄLER 1989, BARRETT 1990, CARRARO and SINISCALCO 1993) all establish their claim under the alternative assumption of identical players–thus, identical countries. It is therefore of interest that the opposite result can be obtained here under the same assumption, as we now show.

Assumption 5: For all $i,j \in N, u_i = u_j$, i.e., the utility[13] functions of the countries are identical.

Proposition 6: Proposition 5 holds under Assumption 5 instead of Assumption 1″.

Proof: When the payoff functions are identical, it follows from the characterizations of the Nash equilibrium and of the partial agreement equilibrium with respect to S that for all $i,j \in S$, $\gamma_i(\bar{p}_i) = \gamma_j(\bar{p}_j)$ and $\gamma_i(\tilde{p}_i) = \gamma_j(\tilde{p}_j)$. Since $\Sigma_{i\in S}\tilde{p}_i \leq \Sigma_{i\in S}\bar{p}_i$ by Proposition 4, it follows that $\gamma_i(\tilde{p}_i) \geq \gamma_i(\bar{p}_i)$ for all $i \in S$, that is, $\tilde{p}_i \leq \bar{p}_i$ for all $i \in S$. □

Corollary: Theorem 2 holds under Assumption 5 instead of Assumption 1″.

[13] and not the production functions, as assumed in earlier versions of this paper.

Proof: Identical to the proof of Theorem 2. □

7 Concluding Remarks

For the cooperative game associated with an economy with detrimental environmental externalities, we have exhibited a strategy in the γ-core, that is, Pareto optimal joint actions of the all-players set N such that no coalition $S \subset N$ can do better for its members. The essence of our contribution lies (i) in identifying this fully cooperative strategy and (ii) in showing that to deter deviating behavior of any $S \subset N$ against it – that is, to deter free riding by any S, the breaking up of the players not in S into singletons acting rationally is sufficient.

We interpret our result as an argument supporting the view that, on logical grounds, full cooperation and efficiency can prevail in economies of this type, in spite of the fact that neither a strong Nash equilibrium nor a Pareto efficient coalition-proof Nash equilibrium can be shown to exist for them.

The proposed allocation has an equilibrium interpretation, nevertheless: it is analogous to that of ratio and therefore of Lindahl equilibrium in economies with public goods. For such economies, it is known that the ratio equilibrium belongs to the core (actually, the α-core). Since the γ-core is a smaller set of allocations, our result establishes an additional property of the ratio equilibrium.

Note also that the constructive nature of our result has the virtue of allowing one to compute the γ-core allocation in applied problems, as is done, for instance, in GERMAIN, TOINT and TULKENS 1996 for the international acid rains model of MÄLER's 1989.

Finally, while the restrictive assumptions made have permitted to introduce and illustrate the basic concepts in a more transparent way, extending our analysis to a wider class of utility functions and other information structures are obvious directions for further research.

Acknowledgements: Acknowledgements are due to Karl Göran Mäler for numerous fruitful discussions and his hospitality at the Beijer International Institute for Ecological Economics, Royal Swedish Academy of Sciences, Stockholm, during May-June 1992, as well as to Francis Bloch and Jacques Drèze for their comments on earlier versions of this paper. We also benefitted from discussions at seminar presentations made at CORE, the Copenhagen Business School, the Séminaire Cournot (Paris I), the 1994 FEEM-IMOP workshop "Environment: Policy and Market Structure" in Athens, and the 1994 Maastricht meeting of the Econometric Society. This work was initially stimulated by the European Science Foundation programme "Environment, Science and Society". The first author is grateful to California Institute of Technology for providing an adequate environment for completing this research. The second author thanks the Fonds de la Recherche Fondamentale Collective, Brussels (convention $n°2.4589.92$) and the Commission of the European Communities (DG XII: Project "Environmental Policy, International Agreements and International Trade") for their support.

This text presents research results of the Belgian Program on Interuniversity Poles of Attraction initiated by the Belgian State, Prime Minister's Office, Science Policy Programming. The scientific responsibility is assumed by its authors.

8 References

Barrett S (1990) The problem of global environmental protection. Oxford Review of Economic Policy 6 (1): 68–79

Bernheim B.D, Peleg B, Whinston MD (1987) Coalition proof Nash equilibria I concepts. Journal of Economic Theory 42: 1–12

Carraro C, Siniscalco D (1993) Strategies for the international protection of the environment. Journal of Public Economics 52: 309–328

Champsaur (1975) How to share the cost of a public good. International Journal of Game Theory 4: 113–129

Chander P (1993) Dynamic procedures and incentives in public good economies. Econometrica 61: 1341–1354

Chander P, Tulkens H (1991) Strategically stable cost sharing in an economic-ecological negotiation process. CORE Discussion Paper n°9135, revised October 1992, to appear in: Mäler KG (Ed.) International environmental problems. A volume sponsored by the European Science Foundation, Kluwer, forthcoming

Chander P, Tulkens H (1992a) Theoretical foundations of negotiations and cost sharing in transfrontier pollution problems. European Economic Review 36 (2/3): 288–299

Chander P, Tulkens H (1992b) Aspects stratégiques des négociations internationales sur les pollutions transfrontières et du partage des coûts de l'épuration. Revue Economique 43 (4): 769–781

Chander P, Tulkens H (1995) A core-theoretical solution for the design of cooperative agreements on transfrontier pollution. International Tax and Public Finance 2 (2): 279–294

Foley D (1970) Lindahl solution and the core of an economy with public goods. Econometrica 38: 66–72

Friedman J (1990) Game theory with applications to economics, second edition. Oxford University Press

Germain M, Toint P. Tulkens H (1996) Calcul économique itératif pour les négociations internationales sur les pluies acides entre la Finlande. la Russie et l'Estonie. Annales d'Economie et de Statistique (Paris) 43: 101–127

Kaneko M (1977) The ratio equilibria and the core of the voting game $G(N, W)$ in a public goods economy. Econometrica 45: 1589–1594

Laffont JJ (1977) Effets externes et théorie économique. Monographies du Séminaire d'Econométrie, Editions du Centre national de la Recherche Scientifique (CNRS), Paris

Mäler KG (1989) The acid rain game. In: Folmer H, Van Ierland E (Eds.) Valuation methods and policy making in environmental economics. Elsevier, Amsterdam: 231–252

Mas-Colell A, Silvestre J (1989) Cost sharing equilibria: A Lindahlian approach. Journal of Economic Theory 47: 329–256

Moulin H (1987) Egalitarian-equivalent cost sharing of a public good. Econometrica 55: 963–976

Ray D, Vohra R (1993) Equilibrium binding agreements. Working Paper 21, Department of Economics, Boston University, forthcoming in Journal of Economic Theory

Scarf H (1971) On the existence of a cooperative solution for a general class of N-person games. Journal of Economic Theory 3: 169–181

Shapley L (1967) On balanced sets and cores. International Journal of Game Theory Quarterly 14: 453–460

Shapley L (1971) Cores of convex games. International Journal of Game Theory 1: 11–26
Shapley L, Shubik M (1969) On the core of an economic system with externalities. American Economic Review LIX: 678–684
Starrett D (1972) A note on externalities and the core. Econometrica 41 (1): 179–183
Tulkens H (1979) An economic model of international negotiations relating to transfrontier pollution. In: Krippendorff K (Ed.) Communication and control in society. Gordon and Breach Science Publishers, New York: 199–212

Received September 1995
Revised version September 1996

Part II
Technical Advances

[6]

Stable International Environmental Agreements: An Analytical Approach

EFFROSYNI DIAMANTOUDI
Concordia University

EFTICHIOS S. SARTZETAKIS
University of Macedonia

Abstract

In this paper we examine the formation of international environmental agreements (IEAs). We show that the welfare of the signatories does *not* increase monotonically with respect to the number of signatories. We provide an analytical solution of the leadership model. In particular, we find that if the number of countries is greater than four then there exists a unique stable IEA with either two, three, or four signatories. Furthermore, we show that the welfare of the signatories is almost at its lowest level when the IEA is *stable*. While in our model each country's choice variable is emissions, we extend our results to the case where the choice variable is abatement efforts.

1. Introduction

Some of the most important environmental problems urgently calling for solution are problems related to transboundary pollution. Environmental problems such as ozone depletion, climate change, and marine pollution have been the focus of intense negotiations at the international level over the past two decades. Given the high priority environmental problems receive at the policy front, it is not surprising that there is a growing effort

---

Effrosyni Diamantoudi, Department of Economics, Concordia University, 1455 De Maisonneuve Blvd West, Montreal Quebec H3G 1M8, Canada (ediamant@alcor.concordia.ca). Eftichios S. Sartzetakis, Department of Accounting and Finance, University of Macedonia (esartz@uom.gr).

The authors would like to thank Licun Xue, Scott Barrett and Carlo Carraro and participants of the CREE 2001 conference for their suggestions. We are also indebted to two referees for their insightful comments.

Received March 4, 2003; Accepted September 10, 2004.

Journal of Public Economic Theory, 8 (2), 2006, pp. 247–263.

to analyze international environmental agreements (IEAs) at the theoretical front. A significant part of the literature on IEAs utilizes game theory to model the formation of a single coalition that reduces pollution. There are two main directions in which the IEA literature has been developed over the last 15 years. The first argues that the formation of an IEA resembles the voluntary provision of a public good (with externalities) and formalizes countries' behavior as a cooperative game. It shows that an IEA ratified by all countries is stable (Chander and Tulkens 1995, 1997). The second direction uses the tools of noncooperative game theory to model the formation of an IEA. The latter is the direction we follow in this paper.

The noncooperative approach examines both the case where *all* countries (members or not of the IEA) make their decisions simultaneously (Carraro and Siniscalco 1993), and the case where the countries that have ratified the IEA (signatories) act as a *leader*, whose decision precedes the decision of the countries that remain outside the IEA (Barrett 1994). The simultaneous case has been resolved by De Cara and Rotillon (2001), Finus and Rundshagen (2001), and Rubio and Casino (2001) leading to the conclusion that, when cost and benefit functions are quadratic, the stable IEA will involve no more than two countries. In the leadership approach the most important contributions are based so far on simulations. While the size of the stable IEA remains unknown, simulations in Barrett (1994) suggest that a stable IEA may include a large number of countries, even the grand coalition. We believe that the leadership model, in which an individual country that decides unilaterally will wait to observe the decision of a coalition whose emission will influence significantly global pollution, is compelling enough to warrant further investigation. Thus, in this paper we adopt the leadership model which we solve analytically and our results complement the simulation results in Barrett (1994).

In particular, each country's welfare (or payoff) is expressed as the difference between the benefits from the country's emissions and the damages from the aggregate emissions. In the leadership literature it is assumed that, in the first stage, countries signing the IEA form a coalition and behave *cooperatively* by maximizing the coalition's aggregate welfare and in the second stage, the countries that do not participate in the agreement observe the results of the agreement and behave *noncooperatively* by maximizing their individual welfare. Naturally, when the coalition (leader) maximizes its welfare in the first stage, it foresees and takes into account the nonsignatories' (followers) behavior. Due to the lack of supra-national authorities that could enforce nonbinding agreements, IEAs have to be self-enforcing in the sense that they are immune to deviation by the countries involved. An IEA is considered to be stable if none of its signatories has an incentive to withdraw (this aspect of stability is known as *internal stability*) and none of the nonsignatories has an incentive to further participate in the agreement (this aspect of stability is known as *external stability*).[1] Such a coalitional stability notion was originally introduced by D'Aspremont et al. (1983) in the study of stable cartels in a price leadership model. However, our model and D' Aspremont et al.'s (1983) model differ significantly: (i) while in our model nonmembers behave strategically in theirs they behave as price takers, (ii) unlike the cartel formation case, in the IEA case members' welfare *does not* increase monotonically with respect to the size of the coalition. We study the problem of deriving the size of a stable IEA in a model very similar to Barrett (1994) with the main difference being the choice variable: in our model countries choose emission levels whereas in his they choose abatement efforts.

The main contribution of this paper is the complete analytical solution of the coalition formation model with quadratic benefit and damage functions. We find that a stable coalition consists of either two, three, or four members if the total number of countries is greater than four. Furthermore, we show that the welfare level of the signatories is very close to its lowest value when the IEA is stable. Our results corroborate the outcome of the static models by anticipating very little participation in an IEA. In fact, the predicted size of a stable IEA is so small in both static and dynamic models that it contradicts empirical observations[2] and establishes the need for an alternative approach to modeling countries' behavior in international environmental negotiations. Along this vain, Hoel and Schneider (1997) propose a *simultaneous* model similar to that of Carraro and Siniscalco (1993) with one major difference: they introduce nonenvironmental costs incurred by the nonsignatories.[3] In particular, in the primitive simultaneous model, when a country exits the coalition, there are two forces in effect: (i) the increase in its own emissions that results in higher benefits (e.g., due to cheaper production), and (ii) the increase in total emission levels that results in higher damages (e.g., due to an increase in global environmental pollution). If the increase in benefits exceeds the increase in damages the country has indeed an incentive to exit. Hoel and Schneider (1997) introduce an additional cost incurred by the exiting country representing nonenvironmental costs such as political ones. Naturally, if this additional cost is high enough it may reverse the original

[1]The reader should take note that the notions of *internal* and *external stability* introduced in this paper are completely different from those introduced by von Neumann and Morgenstern (1944) within the concept of the (abstract) stable set.

[2]For example, from the 194 members of the United Nations General Assembly, 184 have ratified the Montreal Protocol, 158 the Basel Convention, 164 the Convention on International Trade in Endangered Species and although the Kyoto Protocol has not come into effect yet, it has been ratified by 119 countries, 29 of which belong to the Annex I countries.

[3]Additionally the damage function is linear as opposed to quadratic but this makes their analysis simpler without disturbing the results. It is shown in De Cara and Rotillon (2001) and Finus and Rundshagen (2001) that this simpler version without nonenvironmental costs results also in very small coalitions.

incentives, inducing thus the country to stay in the coalition. Due to this effect, Hoel and Schneider (1997) are able to support larger coalitions, including the grand coalition.

Our work parallels that of Konishi and Lin (1999) in terms of the coalition formation analysis employed. However, in Konishi and Lin (1999) the primitive model is cartel formation with Cournot fringe whereas in this paper it is IEA formation. While the two models share many common features, among which free-riding incentives by the coalition members, there are nevertheless, significant differences. As we show in this paper, an IEA can never contain more than four countries whereas a cartel, as Konishi and Lin (1999) show, may include a larger number of firms.

Our results, severely restricting the size of stable coalitions, complement Barrett's (1994) suggestion that stable IEAs could consist of any large number of countries by relating the size of a stable coalition to the domain of the choice variable. We convert our model's choice variable from emission levels to abatement efforts making, thus, our model directly comparable to his framework. In doing so, we formulate the link between the two approaches and show that our results survive such a conversion. Assuming that abatement cannot exceed the current flow of emissions we show that Barrett's (1994) model yields stable coalitions consisting of no more than four members. It is only when abatement can exceed the current flow of emissions, such as in the case of a stock pollutant whose stock could be technologically and economically viable to reduce, that Barrett's (1994) model could support stable IEAs consisting of more than four countries. The proofs of all the results presented in the paper are delineated in the Appendix.

2. The Model

We assume that there exist n identical countries, $N = \{1, \ldots, n\}$. Production and consumption in each country i generates emissions $e_i \geq 0$ of a global pollutant as an output. The term *global* pollutant indicates that we assume pollution to be a public bad and that individual emissions impose negative externalities on all other countries. Similarly, in Section 4 where the model is specified in terms of abatement effort, individual abatement effort is assumed to be a public good. The social welfare of country i, w_i, is expressed as the net between the total benefits from country i's emissions, $B_i(e_i)$, and the damages $D_i(E)$ from the aggregate emissions, E, including country i's emissions. Since countries are assumed to be identical we henceforth drop the subscripts from the functions. As each country i's emission level increases, its benefits $B(e_i)$ increase as well. We consider the following quadratic benefit function for each country $i \in N$, $B(e_i) = b[ae_i - \frac{1}{2}e_i^2]$, where a and b are positive parameters. Country i's damages from pollution depend on aggregate pollution, E, where $E = \sum_{i \in N} e_i$. We assume a quadratic damage function for each country $i \in N$, of the following form $D(E) = \frac{1}{2}c(E)^2$, where c is a positive parameter.[4]

With these specifications, each country i's welfare function becomes

$$w(e_i) = b\left[ae_i - \frac{1}{2}e_i^2\right] - \frac{c}{2}\left(\sum_{i \in N} e_i\right)^2. \quad (1)$$

2.1. The (Pure) Noncooperative Case

In the noncooperative case each country chooses its emission level taking the other countries' emissions as given. That is, country i behaves in a typical Cournot fashion maximizing Equation (1). The first-order condition of the above maximization problem yields country i's emission reaction function, $e_i = \frac{ba - c\sum_{j \neq i} e_j}{b + c}$.

Since we have assumed complete symmetry, all countries generate the same level of emission at the equilibrium, denoted by e_{nc}. The solution of the reaction functions' system yields

$$e_{nc} = \frac{a}{1 + \gamma n}, \quad (2)$$

where $\gamma = \frac{c}{b}$. Consequently, the aggregate emission level under the (purely) noncooperative case is, $E_{nc} = ne_{nc} = \frac{na}{1+\gamma n}$.

2.2. Full Cooperation

Under full cooperation, the grand coalition maximizes the joint welfare. The first-order condition yields the aggregate emission level, $E_c = \frac{an}{\gamma n^2 + 1}$. Since each country contributes $\frac{1}{n}$ of the total emissions, the per country emission level, e_c, is

$$e_c = \frac{E_c}{n} = \frac{a}{\gamma n^2 + 1}. \quad (3)$$

It is easily verifiable that each country emits less and is better off in the case of full cooperation than under noncooperation, that is, $e_c < e_n$ and $w_c > w_n$.

However, in this one stage, purely simultaneous framework each country has an incentive to cheat on the agreement and free-ride on the emission

[4] An alternative form of the damage function is also used in the literature, see, for example, Barrett (1994). According to their functional form, each country's damages are a share of aggregate emissions, that is, $D(E) = \frac{1}{2n}c(E)^2$. The difference between the two forms is a difference in parameter specification and it does not affect the results. The full analysis using this alternative functional form is available to the interested reader upon request.

reduction achieved by the countries complying with the agreement. In what follows we examine the two-stage framework where the incentive to free ride on the coalition's cooperating efforts may be offset by the adjustment of the coalition's emissions upon a member's deviation. The equilibrium number of countries participating in an IEA, is derived by applying the notions of internal and external stability of a coalition as was originally developed by D'Aspremont et al. (1983) and extended to IEAs by Carraro and Siniscalco (1993) and Barrett (1994).

2.3. Coalition Formation

Assume that a set $S \subset N$ of countries sign an agreement and $N \backslash S$ do not. Let the size of coalition S be $|S| = s$, the total emission generated by the coalition be E_s, while each member of the coalition emits e_s such that $E_s = se_s$. In a similar manner, each nonsignatory country emits e_{ns}, yielding a total emission level $E_{ns} = (n - s)e_{ns}$.

The nonsignatories behave noncooperatively after having observed the choice of signatories. Their maximization problem results to a best response function of the form presented earlier. However, now only $n - s$ countries stay outside of the emission reduction agreement emitting e_{ns}, while the rest s countries emit in total E_s, that is, $\sum_{i \in N} e_i = (n - s)e_{ns} + se_s$. Substituting this into the reaction function yields each nonsignatory country's emissions $e_{ns} = \frac{a - \gamma se_s}{1 + \gamma(n - s)}$ as a function of the signatory countries' emission e_s. The aggregate nonsignatory emission level is $E_{ns} = \frac{(a - \gamma se_s)(n - s)}{1 + \gamma(n - s)}$.

Signatories choose their emission level by maximizing their collective welfare while taking into account the behavior of nonsignatories. That is, signatories choose e_s by solving the following maximization problem,

$$\max_{e_s} s[B(e_s) - D(se_s + (n - s)e_{ns}(e_s))].$$

The first-order condition yields the emission of the signatories,

$$e_s = a\left[1 - \frac{\gamma sn}{\Psi}\right], \tag{4}$$

where $\Psi = X^2 + \gamma s^2$ and $X = 1 + \gamma(n - s)$. The aggregate emission level by the signatories is $E_s = sa[1 - \frac{\gamma sn}{\Psi}]$. Substituting the value of e_s into the reaction function of nonsignatories yields,

$$e_{ns} = e_s + \frac{a\gamma n(s - X)}{\Psi}. \tag{5}$$

The total emission level by nonsignatories is $E_{ns} = (n - s)[e_s + \frac{a\gamma n(s - X)}{\Psi}]$.

The full-cooperative and the pure noncooperative solutions can be derived as special cases of the above solution. That is, when $s = n$, the problem reduces to the full cooperative solution and $e_s = e_c$, while when $s = 0$, it reduces to the pure noncooperative solution, and, $e_{ns} = e_{nc}$.

The aggregate emission level $E = E_{ns} + E_s$ is

$$E = \frac{naX}{\Psi}. \tag{6}$$

Unlike the previous two cases where $e_{nc} > 0$ and $e_c > 0$ always hold, in the coalition formation case we have to restrict the parameters of the model in order to guarantee that our solutions are interior, that is, we need to restrict the parameters so that $e_s > 0$ and $e_{ns} > 0$. The following proposition establishes the necessary conditions for interior solutions.

PROPOSITION 1: *$e_s > 0$ if and only if $\gamma < \frac{4}{n(n-4)}$ and $n > 4$, $e_{ns} > 0$ if $\gamma < \frac{4}{n(n-4)}$ and $n > 4$.*

The intuitive explanation behind these conditions is that for emissions to be positive it must be that the relative impact of damages to benefits is not very high (recall that $\gamma = c/b$). Although such a restriction may seem benign at first, it is of great importance since it is this condition that restricts the size of the stable coalition to 2, 3, or 4 countries as we formally show in Section 3.

Despite its importance, this condition has been overlooked so far, simply because the model is most commonly defined in terms of abatement efforts rather than in terms of emissions (the prominent example is the work of Barrett 1994). In Section 4 we convert our model's choice variable to abatement effort and, while establishing the direct link between the two models, we extend the constraint to the converted model as well, validating, thus, the immunity of our results to the selection of choice variable.

The last step in fully formulating our model is the determination of the welfare level of signatories and nonsignatories for any given s. This is done by simply substituting the emission levels e_s, e_{ns} and E with their equilibrium values from Equations (4–6), respectively, into the corresponding welfare functions. We denote the indirect welfare function of the signatories by ω_s while that of the nonsignatories by ω_{ns}, which take the following form:

$$\omega_s = ba^2\left[\frac{1}{2} - \frac{n^2\gamma}{2\Psi}\right], \quad \text{and} \quad \omega_{ns} = ba^2\left[\frac{1}{2} - \frac{n^2\gamma X^2(1+\gamma)}{2\Psi^2}\right]. \tag{7}$$

The properties of these indirect welfare functions are established in Proposition 2.

PROPOSITION 2: *Consider the indirect welfare functions of signatory and nonsignatory countries, $\omega_s(s)$ and $\omega_{ns}(s)$, respectively, and let $z^{\min} = \frac{1+\gamma n}{1+\gamma}$. Then,*

(i) $z^{\min} = \arg\min_{s\in\Re\cap[0,n]} \omega_s(s)$,

(ii) *$\omega_s(s)$ increases in s if $s > z^{\min}$ and it decreases in s if $s < z^{\min}$,*

(iii) *$\omega_{ns}(s) \lesseqgtr \omega_s(s)$ for all $s \lesseqgtr z^{\min}$.*

(iv) *If, moreover, $z^{\min}$ is an integer then the two indirect welfare levels are equal at $s = z^{\min}$, that is, $\omega_{ns}(s^{\min}) = \omega_s(s^{\min})$.*

We would like to point out that these indirect welfare functions do not exhibit the same properties with those in D' Aspremont et al. (1983). While in the latter paper the welfare functions are monotonically increasing, in our analysis there exist situations (with sufficiently small coalitions), where a country is better off as a member of the coalition than it is outside the coalition and as the coalition grows its members' welfare drops. The difference stems from the fact that in the price leadership model the fringe behaves nonstrategically, i.e., its members behave as price-takers, not conceptualizing the impact of their actions on the market price. In contrast, in the IEAs case the nonsignatories behave strategically by explicitly taking into account the negative effect their individual emissions have on their welfare via global pollution. Not surprisingly, the same observation has been made in Konishi and Lin (1999).

3. The Size of Stable IEAs

We now proceed with the determination of the size of the stable IEA, denoted by s^*, using the internal and external stability conditions. Recall that the internal stability condition ensures that if a country were to defect unilaterally, its gains from free riding would be outweighed by the adjustment (due to its defection) of the emission levels of the remaining members of the IEA. The external stability condition ensures that no other nonsignatory country finds it beneficial to unilaterally join the IEA. Formally, the internal and external stability conditions are

$$\omega_s(s^*) \geq \omega_{ns}(s^* - 1) \quad \text{and} \quad \omega_s(s^* + 1) \leq \omega_{ns}(s^*) \, ,$$

respectively. The following proposition establishes all the possible sizes of the unique stable IEA.

PROPOSITION 3: *For $n > 4$, there exists a unique stable IEA whose size s^* is such that $s^* \in \{2, 3, 4\}$.*

We illustrate the results presented in Proposition 3 by considering a numerical example that leads to $s^* = 3$. We assume $n = 10$, $a = 10$, $b = 6$ and $c = 0.39999$, which result in $\gamma = 0.066665$. Observe that $\gamma < \frac{4}{n(n-4)} \Leftrightarrow 0.066665 < 0.066667$ satisfying the interior solution constraint.

In both Figures 1 and 2 $\omega_s(s)$ is depicted by the solid line, $\omega_{ns}(s)$ by the crossed line and $\omega_{ns}(s - 1)$ by the dashed line. All three indirect welfare functions are plotted against different coalition sizes s. While Figure 1 plots the functions for all possible values of $s = 0, \ldots, 10$, Figure 2 focuses on the values of interest, that is, $s = 1, \ldots, 4$. Observe that coalition $s^* = 3$ is internally stable, i.e., $\omega_s(s^*) > \omega_{ns}(s^* - 1)$ since the dashed curve is below the solid curve. Moreover, $s^* = 3$ is externally stable, i.e., $\omega_s(s^* + 1) < \omega_{ns}(s^*)$ since $s^* + 1$ is after the intersection of the dashed and the solid curves. Therefore, the coalition of size $s^* = 3$ is stable.

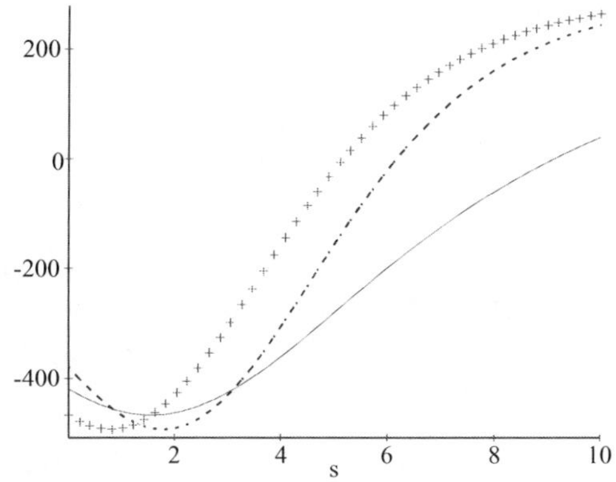

Figure 1: Functions for all possible values of $s = 0, \ldots, 10$

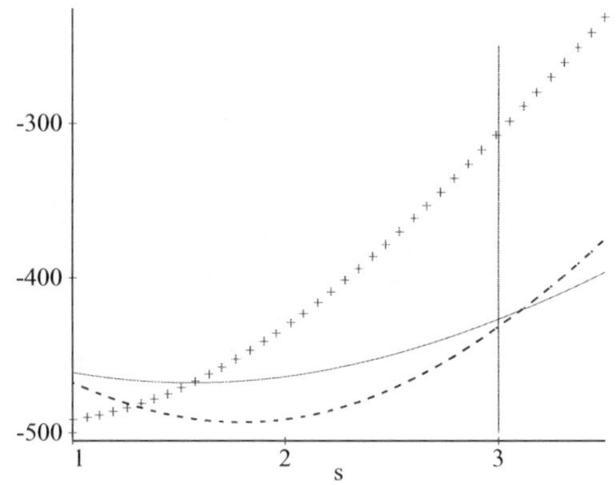

Figure 2: Values of interest, $s = 1, \ldots, 4$

Remark 1: An important observation stemming from the above analysis is that the size of the stable coalition is slightly larger than that for which the welfare of the signatories is at its minimum.

Closer to our results, Rubio and Casino (2001) have suggested that a coalition consisting of two countries is the only stable coalition, but, their result is derived by constraining the indirect welfare levels to be positive. Such a constraint is unjustified since welfare functions are invariant to positive monotonic transforms and hence their cardinal values are insignificant.

4. Emissions versus Abatement

As we mentioned in the previous section, our result in Proposition 3 regarding the size of the stable coalition complements that of Barrett (1994) where the same type of quadratic benefits and costs functions are used. Although the main difference between Barrett's (1994) and our model is the choice variable, abatement effort and emissions, respectively, the difference in the results is not due to the choice variable but rather to the restrictions imposed on the choice variable. As we show in this section the two models are equivalent and all results carry over, as long as abatement does not exceed the flow of emission. The difference arises in the case of a stock pollutant, whose stock could be technologically and economically viable to reduce.[5] In such a case, it could be possible to abate more than the current flow of emission, that is, it could be possible to have a negative net flow of emission. Then, Proposition 3 does not hold anymore eliminating, thus, the difference between our present work and Barrett's (1994) contribution. Lemma 1, however, applies to both models and the stable IEA remains the largest integer below $z^{\min}+1$, while the pessimistic observation, outlined by Remark 1, that a stable IEA is (almost) the least rewarding to its members is still in effect. That is, when the feasibility constraint is not binding we can have large coalitions whose members attain close to the lowest, in terms of coalition size, net welfare.

In the rest of this section we illustrate that the two models are directly comparable when abatement is defined as a reduction in the flow of emission. Barrett (1994) assumes that countries derive benefits from aggregate abatement Q, with country i's benefits given by $B_i(Q)=\frac{\hat{b}}{n}(\hat{a}Q-\frac{1}{2}Q^2)$. Each country's costs depend on its own abatement, that is, $C_i(q_i)=\frac{\hat{c}}{2}q_i^2$, where $\hat{b}$, $\hat{a}$, and $\hat{c}$ are parameters and n denotes the number of countries.[6] Within this framework, it is asserted in Barrett (1994, Proposition 1) that stable IEAs can be signed by a large number of countries for low values of $\hat{\gamma}=\frac{\hat{c}}{\hat{b}}$, that is, when the importance of own abatement costs is small relative to the benefits derived from aggregate abatement. Although the model can be solved in a manner parallel to ours, the goal here is to derive the abatement model from the emission model, establishing the equivalence between them.[7]

If, as discussed earlier, abatement effort is defined as a reduction in the flow of emissions then abatement is meaningful only in the presence of emissions, and thus, the maximum level of abatement is constrained by the maximum uncontrolled flow of emissions. In other words, the abatement model is derived from the emission model. Denote by $\bar{E}$ the

[5]For example, in the case of carbon dioxide serious consideration is given recently to the technological option of geological and oceanic storage, usually referred to as carbon sequestration. Although currently at the experimental stage, its pottential is explored both at national and international level. UNFCCC has invited IPCC to prepare a Special Report on the subject, and the results of the first workshop were presented in the 20th Session of IPCC, see IPCC-XX/Doc.19 (10.II.2003).

[6]We added hats in the symbols b, a, and c to distinguish them from those we have already used.

[7]We can provide the full solution to the interested reader on request.

uncontrolled, aggregate emissions level, that is, the level of emissions associated with zero abatement, and by E the controlled emissions level we derived in the previous section. According to the above definition of abatement the domain of Q, as captured by $B_i(Q)$, should be derived from the emissions model that independently determines the level of uncontrolled emissions. That is, each country's uncontrolled level of emissions is derived directly from its benefit function $B_i(e_i)$ and it is $\bar{e} = a$, and thus, $\bar{E} = na$. By extension, country specific and aggregate abatements are then defined as $q_i = \bar{e} - e_i$, and $Q = \bar{E} - E = na - E$, respectively. Substituting these definitions into country i's welfare function defined in terms of abatement yields, $w_i = \frac{\hat{b}}{n}[\hat{a}(na - E) - \frac{1}{2}(na - E)^2] - \frac{\hat{c}}{2}(a - e_i)^2$. This expression can take the following form which facilitates direct comparison with the welfare function specified in terms of emissions in Equation (1).

$$w_i = \hat{c}\left[ae_i - \frac{1}{2}e_i^2\right] - \frac{\hat{b}}{2n}E^2 + \frac{\hat{b}}{n}(na - \hat{a})E + \left[\hat{b}\hat{a}a - \frac{\hat{b}na^2}{2} - \frac{\hat{c}a^2}{2}\right]. \quad (8)$$

By setting $\hat{c} = b$, $\hat{b} = nc$ and $\hat{a} = na$, Equation (8) reduces to $w_i = b[ae_i - \frac{1}{2}e_i^2] - \frac{c}{2}E^2 + \frac{cna^2}{2}(n - \frac{1}{\gamma n})$, where γ has been defined in Section 2 as $\gamma = \frac{c}{b}$. Note that the last term is just a constant that only scales welfare levels and does not affect the solution of the problem. Therefore, the same solution is derived whether we specify welfare in terms of emissions, that is, $w_i = b[ae_i - \frac{1}{2}e_i^2] - \frac{c}{2}E^2$, or in terms of abatement, that is, $w_i = \frac{\hat{b}}{n}[\hat{a}Q - \frac{1}{2}Q^2] - \frac{\hat{c}}{2}q_i^2$, as long as $\hat{c} = b$, $\hat{b} = nc$, $\hat{a} = na$, and $\hat{\gamma} = \frac{\hat{c}}{\hat{b}} = \frac{1}{\gamma n}$. For example, one can derive the abatement level of signatory countries using Equation (4) in Section 2 ($e_s = a - \frac{a\gamma sn}{\Psi}$), simply by recalling the definition of abatement, that is, $e_s = \bar{e} - q_s$ which implies that $q_s = \frac{a\gamma sn}{\Psi}$.[8]

Using the above equivalence between the two models we can now support the derived abatement model specification with the necessary constraints from the primary emission model. Recall that Proposition 2 provides the necessary conditions to ensure that the choice variables are positive, that is, $e_s \geq 0$ and $e_{ns} \geq 0$. These constraints though, imply the following conditions for the corresponding abatement levels, $q_s \leq a$, and $q_{ns} \leq a$. Note that the latter constraints are equivalent with those stemming from the benefit function $B_i(Q)$, that is, $Q \leq \hat{a}$ which implies $q \leq \frac{\hat{a}}{n} = \frac{an}{n} = a$. Since the parameters $\hat{a}$, $\hat{b}$, and $\hat{c}$ are directly derived from the emission model, they carry over the constraints imposed on a, b, and c, namely, $\gamma < \frac{4}{n(n-4)} \Longleftrightarrow \frac{c}{b} < \frac{4}{n(n-4)}$. Replacing c and b yields $\frac{\hat{b}/n}{\hat{c}} < \frac{4}{n(n-4)}$, which is equivalent to $\hat{\gamma} = \frac{\hat{c}}{\hat{b}} > \frac{n-4}{4}$.

[8]Simple parameter transformation using the definitions in the beginning of the paragraph yields $q_s = \frac{\hat{a}\alpha\hat{\gamma}}{(\hat{\gamma}+1-\alpha)^2 + \alpha^2 n\hat{\gamma}}$, which if multiplied by $n\alpha$ yields the total abatement level of signatory countries, given in equation (6), p. 882, Barrett (1994).

If these conditions are taken into account, it is immediate that the admissible sizes of a stable coalition reduce to 2, 3, and 4 as was the case in Section 3. To illustrate the equivalence between the two models consider the first example constructed in Barrett (1994). The parameters' values are $n = 10$, $\hat{a} = 100$, $\hat{b} = 1$, and $\hat{c} = 0.25$, implying $\hat{\gamma} = \frac{\hat{c}}{\hat{b}} = 0.25$, and the stable coalition allegedly consists of four countries. However, the chosen values of $\hat{b}$ and $\hat{c}$ clearly violate the maximum abatement constraint established earlier, requiring that $\hat{\gamma} > 1.5$. The violation of the maximum abatement constraint is evident from the data presented in Barrett (1994, table 1), since the abatement of signatory countries exceeds the corresponding uncontrolled level of emissions $\bar{e} = \frac{\hat{a}}{n} = 10$. That is, each signatory abates more than it can ever emit. In this case, restricting $\hat{\gamma} > 1.5$ yields stable coalitions consisting of either two or three countries depending on the value of $\hat{\gamma}$. In general, restricting the value of $\hat{\gamma}$ to the admissible range, we find that the stable coalition consists of either two, three, or four countries, depending on how close the value of $\hat{\gamma}$ is to its lower bound.

5. Conclusions

The present paper studies the size of stable coalitions that ratify IEAs concerning transboundary environmental problems. A coalition is considered stable when no signatories wish to withdraw while no more countries wish to participate. Within this framework we show that, contrary to the general perception in the literature, the welfare levels of both the signatories and the nonsignatories do *not* monotonically increase in the size of the coalition. Furthermore, in the case of small coalitions, signatories are better off than nonsignatories while as the coalition grows sufficiently the opposite is true.

We find that the size of the stable coalition is not only very small, but it is also invariant to the value of the model's parameters. Moreover, it is very close to the worst, in terms of the members' welfare, coalition size.

All these problematic features of a stable coalition suggest that there exists a caveat in the model. One explanation of the results is that when each country acts it does not foresee the disappointing outcome in which it may end up. Instead, it myopically concentrates on its own action ignoring the reactions of others. In a companion to this paper we study stable IEAs when countries behave in a more sophisticated manner and are forward looking.

There are, however, other venues one can explore. Asymmetry among countries has not yet been studied while in the real world it is widespread. For example, not all countries possess identical technologies, leading thus, to varying abating costs. Similarly, the (perceived) impact of environmental damages differs from country to country, hence the Damage function can vary as well. Such asymmetries can be incorporated in our model by indexing parameter γ by country. Then, a coalition will be characterized not only by its size but also by the identity of those in it. Spatial topology is another dimension that can be added to the basic model when regional pollution problems are studied. Emissions from a given country may affect only its neighbors instead of all the countries. A network will be more appropriate in modeling such a situation.

Lastly, in the present work it is assumed that there is only one IEA (hence one coalition). Although it is a natural assumption, it would be very interesting to examine whether it is also the outcome of a model that allows *ex ante* many coalitions to form. There are several works that model endogenous coalition formation, for example, Block (1996) and Ray and Vohra (1999).

Appendix

Although in our model s is a nonnegative integer smaller than n, for the ease of exposition and calculations in the proofs we assume that s is a real number taking values from $[0, n]$. When necessary, at the end of some proofs we convert s back to being an integer.

Proof of Proposition 1: From Equation (4) we know that $e_s = a[1 - \frac{\gamma s n}{\Psi}]$. Hence $e_s > 0 \Leftrightarrow [1+\gamma(n-s)]^2 - \gamma s(n-s) > 0$. Let $A(s) = [1+\gamma(n-s)]^2 - \gamma s(n-s) = 1+\gamma(n-s)[\gamma(n-s)-(s-2)]$ and consider $\underline{s} = \arg\min_s A(s) = \frac{2\gamma n+2+n}{2\gamma+2}$. For $A(s) > 0$ for all s it suffices that $A(\underline{s}) > 0$. Observe that since $(n-\underline{s}) = \frac{n-2}{2\gamma+2}$ and $(\underline{s}-2) = \frac{(n-2)(2\gamma+1)}{2\gamma+2}$ we have $A(\underline{s}) = \frac{4\gamma n-\gamma n^2+4}{4\gamma+4}$. Then $A(\underline{s}) > 0 \Leftrightarrow 4\gamma n - \gamma n^2 + 4 > 0 \Leftrightarrow \gamma < \frac{4}{n(n-4)}$ and the latter is true from our hypothesis.

From Equation (5) we know that $e_{ns} = e_s + \frac{a\gamma n(s-X)}{\Psi} = a[1 - \frac{\gamma s n}{\Psi}] + \frac{a\gamma n(s-X)}{\Psi}$. For $e_{ns} > 0$ it suffices that $[1+\gamma(n-s)](1-\gamma s) + \gamma s^2 > 0$. Let $\Phi(s) = [1+\gamma(n-s)](1-\gamma s)+\gamma s^2$ and consider $\bar{s} = \arg\min_s \Phi(s) = \frac{\gamma n+2}{2\gamma+2}$. For $\Phi(s) > 0$ for all s it suffices that $\Phi(\bar{s}) > 0$. Observe that since $1+\gamma(n-\bar{s}) = \frac{\gamma n(\gamma+2)+2}{2\gamma+2}$ and $(1-\gamma\bar{s}) = \frac{2-\gamma^2 n}{2\gamma+2}$ we have $\Phi(\bar{s}) = [\frac{\gamma n(\gamma+2)+2}{2\gamma+2}][\frac{2-\gamma^2 n}{2\gamma+2}] + \frac{\gamma(\gamma n+2)^2}{(2\gamma+2)^2}$. Note that for $\Phi(\bar{s}) > 0$ it suffices that $\frac{2-\gamma^2 n}{2\gamma+2} > 0 \Leftrightarrow \gamma < \sqrt{\frac{2}{n}}$. But we already know from our hypothesis that $\gamma < \frac{4}{n(n-4)}$ and since $\frac{4}{n(n-4)} < \sqrt{\frac{2}{n}}$ for all $n \geq 6$ it is indeed the case that $\gamma < \sqrt{\frac{2}{n}}$ if $n \geq 6$. Moreover, when $n = 5$ we have $\Phi(\bar{s}) = -\frac{1}{4}\frac{25\gamma^3-20\gamma-4}{\gamma+1}$. For $\Phi(\bar{s}) > 0$ it suffices that $25\gamma^3 - 20\gamma - 4 < 0$ which is true since $\gamma < \frac{4}{5}$. ■

Proof of Proposition 2:

1–2. Observe that $\frac{\partial\omega_s}{\partial s} = \frac{ba^2\gamma^2 n^2}{\Psi^2}(s-X)$. Thus, $\frac{\partial\omega_s}{\partial s}|_{s=z^{\min}} = 0 \Leftrightarrow z^{\min} = \frac{1+\gamma n}{1+\gamma}$. Since $\frac{\partial^2\omega_s}{\partial s^2} > 0$ for all γ and n the first order condition is sufficient. Moreover, observe that $\frac{\partial\omega_s}{\partial s} \lesseqgtr 0$ if $s \lesseqgtr X \Leftrightarrow s \lesseqgtr z^{\min}$.

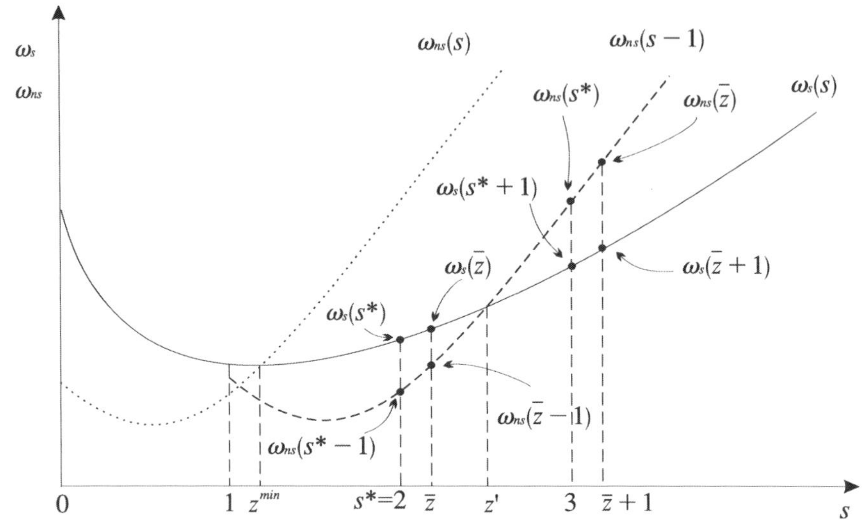

Figure A1: Proposition 3 analysis illustration

3–4. Combining the expressions in (7), the welfare of nonsignatory countries can be expressed as a function of signatories' welfare as follows: $\omega_{ns} = \omega_s + \frac{ba^2\gamma^2 n^2}{2\Psi^2}(X + s)(s - X)$. Then it is obvious that $\omega_{ns} \lesseqgtr \omega_s$, for $s \lesseqgtr X \Leftrightarrow s \lesseqgtr z^{\min}$. If, moreover, $z^{\min}$ is an integer, then when $s = z^{\min} \Leftrightarrow s = X$ and $\omega_{ns}(z^{\min}) = \omega_s(z^{\min})$. ■

Proof of Proposition 3: Unfortunately, allowing s to take noninteger values and then setting $\omega_s(z') = \omega_{ns}(z' - 1)$ where $z' \in [0, n]$ does not provide an analytical solution for z' due to computational limitations and the model has remained unsolved. Fortunately, it is not z' that we are interested in, per se. Instead, it is the largest integer $s^* \leq z'$ that we are looking for as we formally explain in the Stability section.

We are able to bypass the difficulties of solving the complicated polynomial that results from $\omega_s(z') = \omega_{ns}(z' - 1)$ by "guessing" some value $\bar{z}$, that satisfies both stability conditions, not necessarily with equality. Then we adjust it to the appropriate integer.

Stability: To illustrate our analysis we use Figure A1. The curve $\omega_s(s)$ denotes the welfare of the signatories for a size of coalition s, while curves $\omega_{ns}(s)$ and $\omega_{ns}(s - 1)$ denote the welfare of the nonsignatories when the size of coalition is s and $s - 1$, respectively. By its definition $z^{\min}$ is such that $\omega_s(z^{\min}) = \omega_{ns}(z^{\min})$. We now define $\bar{z} := z^{\min} + 1$, and by Lemma 1 we deduce that $\bar{z}$ satisfies the internal and external stability conditions: $\omega_s(\bar{z}) \geq \omega_{ns}(\bar{z} - 1)$ and $\omega_s(\bar{z} + 1) \geq \omega_{ns}(\bar{z})$, respectively.

Let z' be the smallest s such that $\omega_s(z') = \omega_{ns}(z' - 1)$. It is straightforward to show[9] that $\omega_s(z^{\min}) > \omega_{ns}(s - 1)$ for all $s < \bar{z}$. Then, from the internal and external stability of $\bar{z}$ we can conclude that $\bar{z} < z' < \bar{z} + 1$ and hence $\omega_s(s) > \omega_{ns}(s - 1)$ for all $s < z'$. Let $\lfloor x \rfloor$ denote the largest integer that is less than or equal to (if x is an integer itself) x. Then, the size of the stable coalition is $s^* = \lfloor z' \rfloor$. The internal stability of s^* $(\omega_s(s^*) \geq \omega_{ns}(s^* - 1))$ is satisfied due to the fact that z' is, by definition, the first intersection between $\omega_s(s)$ and $\omega_{ns}(s - 1)$, and since $\omega_s(\bar{z}) > \omega_{ns}(\bar{z} - 1)$ and $\bar{z} < z'$ we have $\omega_s(s^*) \geq \omega_{ns}(s^* - 1)$. Similarly, the external stability of s^*, i.e., $\omega_{ns}(s^*) \geq \omega_s(s^* + 1)$, is satisfied since $\omega_{ns}(s - 1) > \omega_s(s)$ for all $s > z'$ (as we illustrate below, under the Uniqueness section) and $s^* + 1 > z'$.

Recall that $z^{\min} = \frac{\gamma n + 1}{\gamma + 1}$, rearranging the expression yields $\gamma = \frac{z^{\min} - 1}{n - z^{\min}}$. We know that $0 < \gamma < \frac{4}{n(n-4)}$, thus, $0 < \frac{z^{\min} - 1}{n - z^{\min}} < \frac{4}{n(n-4)}$. From $0 < \frac{z^{\min} - 1}{n - z^{\min}}$ we get that $z^{\min} > 1$. From $\frac{z^{\min} - 1}{n - z^{\min}} < \frac{4}{n(n-4)}$ we get that $z^{\min} < \frac{n^2}{n^2 - 4n + 4} < 2$ if $n > 6$. Therefore, $1 < z^{\min} < 2$, and by extension $2 < \bar{z} < 3$, and $3 < \bar{z} + 1 < 4$, hence $2 < z' < 4$.

Since we know that $2 < z' < 4$ we can conclude that if $z' < 3$ then $s^* = 2$ (this is the case depicted in Figure A1), whereas if $3 \leq z'$ then $s^* = 3$.

Moreover, $1 < z^{\min} < 3$ if $4 < n \leq 6$, hence $2 < \bar{z} < 4$ and $3 < \bar{z} + 1 < 5$, and thus $2 < z' < 5$. Then, the size of the stable coalition s^* can take the values

$$\begin{aligned} s^* &= 2 \text{ if } z' < 3 \\ s^* &= 3 \text{ if } z' < 4 \\ s^* &= 4 \text{ if } z' \geq 4, \end{aligned}$$

if $4 < n < 6$. In the special case where $n = 6$ the possibility of $s^* = 4$ is ruled out below when we show the uniqueness of s^*.

Uniqueness: We have already argued above that $\omega_s(s) > \omega_{ns}(s - 1)$ for all $s < z'$. Thus, all coalitions $s < s^*$ are externally unstable since $\omega_s(s + 1) > \omega_{ns}(s)$. In order to show that s^* is the only size of a stable IEA it suffices to show that all coalitions of size $s > z'$ are internally unstable, i.e., $\omega_{ns}(s - 1) > \omega_s(s)$, for all $n > 4$.

Using the expressions in (7) we derive that

$$\omega_{ns}(s-1) - \omega_s(s)$$
$$= \frac{ba^2 n^2 \gamma}{2} \left[\frac{\Psi^2(s-1) - \Psi(s)\Psi(s-1) + \Psi(s)\gamma(s-1)^2 - \Psi(s)\gamma X^2(s-1)}{\Psi(s)\Psi^2(s-1)} \right].$$

To show that $\omega_{ns}(s - 1) - \omega_s(s) > 0$ for all $s > z'$ suffices to show that $\Xi(s) = \Psi^2(s - 1) - \Psi(s)\Psi(s - 1) + \Psi(s)\gamma(s - 1)^2 - \Psi(s)\gamma X^2(s - 1) > 0$. Substituting all the relevant values, the expression can be further simplified to the following rather long polynomial:

[9] The calculations are available upon request.

$$\begin{aligned}\Xi(s) = \gamma(&-8\gamma ns + 3 - 4s - 12\gamma^2 sn - 2\gamma^2 ns^3 + 2\gamma^3 ns + 2\gamma ns^2\\ &+\gamma^3 + 5\gamma^2 + 8\gamma^2 n - 12\gamma^2 s + 9\gamma^2 s^2 + 15\gamma s^2 + 8\gamma - 18\gamma s\\ &+6\gamma n - 2\gamma^3 s - 2\gamma^3 s^2 + 2\gamma^3 n - \gamma^3 n^2 + 2\gamma^2 n^2 - 6\gamma^4 ns^2\\ &+4\gamma^3 s^3 - 6\gamma s^3 - 2\gamma^3 n^3 - \gamma^4 s^2 + 2\gamma^4 s^3 - \gamma^4 s^4 + \gamma^2 s^4\\ &-4\gamma^2 s^3 - \gamma^4 n^2 - 2\gamma^4 n^3 - \gamma^4 n^4 - 8\gamma^3 ns^2 - \gamma^3 s^4 + 4\gamma^4 n^3 s\\ &+\gamma s^4 + 6\gamma^4 n^2 s + 2\gamma^3 s^3 n - \gamma^3 s^2 n^2 + 2\gamma^4 ns - 4\gamma^2 n^2 s\\ &+8\gamma^2 ns^2 - 6\gamma^4 n^2 s^2 + \gamma^2 n^2 s^2 + 4\gamma^4 ns^3 + 6\gamma^3 n^2 s + s^2).\end{aligned}$$

We know that $\omega_{ns}(s-1) = \omega_s(s)$ at $s = z'$ for all $n > 4$. We proceed by showing that $\Phi'(s) = \frac{d\Phi}{ds} > 0$ for all $s \geq \bar{z}$ and for all $n > 4$, where $\Phi(s) = \frac{1}{\gamma}\Xi(s)$. To do that we show that it is positive at its lowest value, i.e., $\Phi'(\tilde{s}) > 0$ where $\tilde{s} = \arg\min_{s\geq\bar{z}} \Phi'(s)$. We argue that $\tilde{s} = \bar{z}$ since $\frac{d\Phi'(s)}{ds} = \frac{d^2\Phi(s)}{ds^2} > 0$. The calculations are omitted due to their length and are available upon request.

LEMMA 1: *Consider $\bar{z}$ such that $\bar{z} = z^{\min} + 1$, then $\bar{z}$ satisfies the internal and external stability conditions.*

Proof:

Internal stability: From Proposition 1 we know that $\omega_s(z^{\min}) = \omega_{ns}(z^{\min})$ and that $\omega_s(s)$ increases in s if $s > z^{\min}$. Then, $\omega_s(z^{\min}+1) > \omega_s(z^{\min})$, thus, $\omega_s(z^{\min}+1) > \omega_{ns}(z^{\min})$ which is equivalent to the internal stability condition $\omega_s(\bar{z}) > \omega_{ns}(\bar{z}-1)$.

External stability: External stability is shown by substituting $\bar{z} = \frac{\gamma n+1}{\gamma+1} + 1$ into the external stability condition $\omega_{ns}(\bar{z}) > \omega_s(\bar{z}+1)$. The inequality reduces to $\gamma\frac{2\gamma^2 n^3 + (-3\gamma^2 + 4\gamma - \gamma^3)n^2 + (8\gamma^3 + 2\gamma + 14\gamma^2 + 2)n + 6 - \gamma^2 - 4\gamma^4 - 11\gamma^3 + 14\gamma}{(\gamma+1)^3} \geq 0$. It suffices to show that the following inequality holds:

$$\begin{bmatrix} 2\gamma^2 n^3 + (4\gamma - \gamma^3 - 3\gamma^2)n^2 \\ +(2 + 14\gamma^2 + 8\gamma^3 + 2\gamma)n \\ +6 + 14\gamma - \gamma^2 - 4\gamma^4 - 11\gamma^3 \end{bmatrix} \geq 0.$$

Observe that $4\gamma - \gamma^3 - 3\gamma^2 \geq 0$ for $\gamma \leq 1$, while $6 + 14\gamma - \gamma^2 - 4\gamma^4 - 11\gamma^3 \geq 0$ for $\gamma < 1.0937$. Therefore, the external stability condition is satisfied since $\gamma < \frac{4}{n(n-4)}$ and $n > 4$ imply that $\gamma < 1$. ■

References

BARRETT, S. (1994) Self-enforcing international environmental agreements, *Oxford Economic Papers* **46**, 878–894.

BLOCK, F. (1996) Sequential formation of coalitions in games with externalities and fixed payoff division, *Games and Economic Behavior* **38**, 201–230.

CARRARO, C., and D. SINISCALCO (1993) Strategies for the international protection of the environment, *Journal of Public Economics* **52**, 309–328.

CHANDER, P., and H. TULKENS (1995) A core-theoretic solution for the design of cooperative agreements on transfrontier pollution, *International Tax and Public Finance* **2**, 279–293.

CHANDER, P., and H. TULKENS (1997) The core of an economy with multilateral environmental externalities, *International Journal of Game Theory* **26**, 379–401.

D'ASPREMONT, C. A., J. JACQUEMIN, J. GABSZEWEIZ, and J. A. WEYMARK (1983) On the stability of collusive price leadership, *Canadian Journal of Economics* **16**, 17–25.

DE CARA, S., and G. ROTILLON (2001) Multi greenhouse gas international agreements, mimeo.

FINUS, M., and B. RUNDSHAGEN (2001) Endogenous coalition formation global pollution control, Working Paper, FEEM, Nota di Lavoro 43.2001.

HOEL, M., and K. SCHNEIDER (1997) Incentives to participate in an international environmental agreement, *Environmental and Resource Economics* **9**, 153–170.

KONISHI, H., and P. LIN (1999) Stable cartels with a cournot fringe in a symmetric oligopoly, *Keio Economics Studies* **36**, 1–10.

RAY, D., and R. VOHRA (1999) A theory of endogenous coalition structures, *Games and Economic Behavior* **26**, 286–336.

RUBIO, J. S., and B. CASINO (2001) International cooperation in pollution control, Mimeo.

VON NEUMANN, J., and O. MORGENSTERN (1944) *Theory of Games and Economic Behavior*. Princeton, NJ: Princeton University Press.

[7]

Oxford Economic Papers 58 (2006), 233–263 **233**
doi:10.1093/oep/gpl002

Self-enforcing international environmental agreements revisited

By Santiago J. Rubio* and Alistair Ulph†

*Department of Economic Analysis, University of Valencia, Edificio Departamental Oriental, Avda. de los Naranjos s/n, 46022 Valencia, Spain; e-mail: Santiago.Rubio@uv.es

†Institute of Science and Technology UMIST, University of Manchester, UK

Barrett's (1994) paper on transboundary pollution abatement shows that if the signatories of an international environmental agreement act in a Stackelberg fashion, then, depending on parameter values, a self-enforcing IEA can have any number of signatories between two and the grand coalition. Barrett obtains this result using numerical simulations in a pollution abatement model where he is not constraining emissions to be non-negative. Recent attempts to use analytical approaches and to explicitly recognize the non-negativity constraints have suggested that the number of signatories of a stable IEA may be very small. The way such papers have dealt with non-negativity constraints is to restrict parameter values to ensure interior solutions for emissions. We argue that a more appropriate approach is to use Kuhn-Tucker conditions to derive the equilibrium of the emissions game. When this is done we show, analytically, that the key results from Barrett's paper are maintained. Finally, we explain why his main conclusion is correct although his analysis can implicitly imply negative emissions.

JEL classification: C72, D62, F02, Q20.

1. Introduction

Over the last two decades, one of the factors driving an increased sense of interdependence between countries is the need to tackle global environmental problems such as climate change, ozone depletion, and loss of biological diversity, amongst others. Tackling such problems requires some form of agreement between countries, and the Framework Convention on Climate Change, the Montreal Protocol on Substances that Deplete the Ozone Layer, and the Convention on Biodiversity are important examples of such International Environmental Agreements (IEAs). However, the very different experiences of these agreements illustrates the crucial importance of understanding how to design agreements which give countries incentives to both join and abide by such agreements. Economists have emphasized two important features: agreements must be profitable, which means that there must be potential gains to all signatory countries; more importantly, in the absence of any international authority, agreements must be self-enforcing, i.e. there must be incentives for countries acting in their own self-interest to want to join or stay in an agreement.

One of the earliest definitions of a self-enforcing agreement was the concept of a stable IEA, which means that no individual signatory country has any incentive to leave the IEA, and no non-signatory country has an incentive to join, given the membership decisions of all other countries.[1] Models based on this concept include Carraro and Siniscalco (1991, 1993), Hoel (1992), Barrett (1994), Na and Shin (1998), amongst many others. Carraro and Siniscalco (1991) and Hoel (1992) have shown that if signatory countries act in Cournot fashion with respect to non-signatories, then a stable IEA consists of three countries when marginal environmental damage is constant (i.e., when the countries' best-replay functions are orthogonal), and of two countries when marginal damage increases with emissions (i.e., when the best-replay functions have a negative slope), in both cases irrespective of the number of countries affected.[2] In the paper published by Barrett in 1994, which has had a great influence in subsequent literature on IEA, it is shown that for a model of abatement with quadratic costs and benefits where the countries in the agreement act in a Stackelberg fashion, an IEA may achieve a high degree of cooperation, but only when the gains of cooperation are small. When the gains are large, a self-enforcing IEA cannot support a large number of signatories (see Proposition 1 on page 886).[3] The rationale for the difference in outcomes between Cournot and Stackelberg models is that if one country were to leave the IEA, with Cournot behaviour, the non-signatories expand their emissions and the remaining signatory countries partially accommodate this by

---

[1] There are a number of other concepts of what makes an agreement self-enforcing. Chander and Tulkens (1995, 1997) draw on cooperative game concepts. See Tulkens (1998) for a systematic comparison of these two approaches. Other concepts, such as far-sightedness, have been developed, for instance, by Ecchia and Mariotti (1997, 1998), Ray and Vohra (2001), and Diamantoudi and Sartzetakis (2002b); see Finus (2001) and Wagner (2001) for excellent overviews, and, more recently, Barrett (2003) for a broad presentation of the strategy of environmental treaty-making.

[2] Carraro and Siniscalco (1991, 1993) also show that the number of signatory countries can be increased by means of self-financed transfers. However, expanding coalitions requires some form of commitment. Petrakis and Xepapadeas (1996) extend this result to the case in which the countries are not identical using an emissions game with orthogonal best-replay functions like the one studied by Hoel (1992). Hoel and Schneider (1997) point out that the prospect of receiving a transfer tends to reduce the incentive a country might have to commit itself to cooperation so that if the disincentive is strong, total emissions will be higher with side payments. More recently, Barrett (2001) has shown that with strong asymmetry side payments become the vehicle for increasing participation in a cooperative agreement.

[3] The results obtained so far in the literature as regards participation vary widely. The number of signatories depends on the specification of the net benefit function; for some specifications, it depends on the parameter values; and, for some specifications, it depends on the Cournot/Stackelberg assumption. For a more comprehensive and technical discussion of the importance of specification, see Finus (2001) and Barrett (2003, Table 7.1). In this paper we focus on Barrett's model (quadratic costs and benefits in combination with the Stackelberg assumption).

reducing their emissions. On the other hand with Stackelberg behaviour, if a signatory were to leave the IEA the remaining signatories would expand their emissions. Thus the incentives to leave an IEA are greater with Cournot behaviour than with Stackelberg.

Since this model has become something of a workhorse tool to study IEAs, it is important that its properties are well understood. In this respect we think that the early analysis developed by Barret (1994) could be improved in two ways. First, it relied on numerical simulations to derive his Proposition 1.[4] Second, it assumed that aggregate emissions are large enough to avoid the maximal abatement constraint binding (see footnote 4 in Barrett's paper). However, it can be shown that this assumption on aggregate emissions does not guarantee non-negative emissions for each individual country.[5] Recent papers have attempted to generalize Barrett's numerical analysis to evaluate the robustness of his results. Finus (2001) presents an analytical generalization of Barrett's results and shows that the higher the level of environmental damages, the greater the size of the stable IEA. Finus assumes interior solutions but it is easy to show that if emissions are not constrained, high environmental damages imply in fact negative emissions. Diamantoudi and Sartzetakis (2002a) and Rubio and Casino (2001) also use analytical approaches, but recognize the need to ensure that emissions are non-negative.[6] They reach even more pessimistic conclusions—that even with Stackelberg behaviour the number of signatories of a stable IEA will be small—no greater than four.

However the way Diamantoudi and Sartzetakis deal with the non-negativity constraint is to compute an interior solution and then restrict parameter values

[4] To be fair to Barrett's (1994) paper it should be pointed out that three additional models with different specifications of the net benefit function are analytically solved in the paper, although these other models have received less attention in subsequent literature.

[5] This assumption is a potential problem for the quadratic specification studied by Barrett (1994) but it is not necessarily a problem for other specifications as, for instance, the one studied by Carraro and Siniscalco (1991). These authors develop an analytical solution for a symmetric Cournot equilibrium. In their model they assume an environmental damage function that is quadratic with respect to local emissions but linear with respect to imported emissions, i.e., environmental damage depends on the product between local emissions and total emissions that affect the country. As a result of this specification the solution of the game always yields interior solutions. In this paper, we focus on a global environmental problem so that we assume that the environmental damage is a quadratic function of the aggregate emissions.

[6] A referee pointed out that sequestration of carbon dioxide could be viewed as a form of negative emissions. However, this arises in the context of a stock pollutant, so that in one period it may be possible that reductions in the stock of greenhouse gases through carbon sequestration might exceed any positive flow of new emissions of greenhouse gases. Another referee suggested that some mechanisms such as the clean development mechanism proposed in the framework of the Kyoto Protocol could explain negative emissions. Countries can indeed cut emissions in other countries and one could imagine that the overall abatement is greater than the country's emissions. However, even in this extreme case by definition the aggregate abatement cannot be greater than aggregate emissions so that a non-negativity constraint on aggregate emissions should be taken into account in this case as well.

to ensure that the resulting emissions are always strictly positive, see their Proposition 1. It is not surprising that this restriction on parameters restricts the number of signatories in a stable IEA. Rubio and Casino go further and restrict parameters to ensure that payoffs are non-negative, which is difficult to justify. We argue that neither of these approaches is appropriate. In this paper we also use an analytical approach and deal with the non-negativity constraint by simply imposing it directly on the choice of emissions by both signatory and non-signatory countries and using Kuhn-Tucker conditions to derive the equilibrium of the game. Thus for some parameter values, the emission game will result in corner solutions.

Our findings for the Stackelberg equilibrium of the emissions game show that the number of countries in a stable IEA is directly related to the level of marginal environmental damage so that when marginal environmental damage is low enough a stable IEA consists of at most three countries while when marginal environmental damage is large enough the unique stable IEA is the grand coalition.[7] The rationale for this kind of relationship is given by the fact that interdependence among the countries occurs through the environmental damage function. Thus when the marginal environmental damage is relatively high, the countries in the agreement choose emission levels which induce the non-signatories to select low emissions, making exit from the agreement unprofitable. With lower marginal environmental damage cost these effects are weakened, so that some countries find it profitable to leave the agreement, i.e. the free-riding cannot be avoided by Stackelberg leadership. Finally, we clarify the previous results in the literature which have been derived assuming interior solutions. According to our results, restricting parameter values to guarantee interior solutions is a sufficient condition to get stable IEAs with a small number of signatories but it is not a necessary condition. In this paper we show that a stable IEA with a small number of countries can involve a corner solution. In other words, our results establish that what is really a necessary and sufficient condition to get stable IEAs with a small number of signatories is that the marginal environmental damage is low.

Thus in this paper we have shown that the results derived for the model of stable IEAs in the paper by Barrett (1994), which used numerical calculations on the linear-quadratic version of the model and ignored the issue of non-negative emissions, carry through when derived analytically in a model which seriously takes the need to ensure that emissions are non-negative. The reason why considering non-negative emissions does not change the main results of the

[7] We have also shown analytically that when the environmental damage is a quadratic function of the aggregate emissions and non-negativity constraints are taken into account, the previous results obtained by Carraro and Siniscalco (1991) for the Cournot equilibrium are unaffected. See Rubio and Ulph's (2002a) working paper.

literature is that the definition of a stable agreement depends on the sign of the difference between payoffs to signatories and non-signatories as the number of signatories varies, as we shall show. Taking into account the need for emissions to be non-negative obviously changes the value of these payoffs, but not the sign of differences in payoffs. Thus this paper not only analytically derives results for the Barrett (1994) model using an appropriate treatment of non-negative emission constraints, but disproves the claim that taking into account such constraints makes a significant difference to known results.

Finally, we would like to clarify three issues. Firstly, we want to explicitly recognize that our results have been obtained assuming that all countries are identical, as in Barrett's (1994) paper. This assumption, although restrictive, allows us to get an analytical solution of the game and thus to advance the analysis of the stability of IEAs.[8] Secondly, we want to highlight that it is important, at least for one case, to extend the analysis in order to consider the possibility of zero emissions not only from a mathematical point of view but also from an economic point of view. We are considering the case of non essential emissions. In a partial equilibrium analysis, like the one we develop in this paper, emissions are non essential when the marginal benefit from emissions is positive but finite for zero emissions. For this type of emissions, if the marginal environmental damage is great enough it makes sense to consider the possibility that emissions can be completely eliminated. This means that there exists a substitute so that if the damage is great enough, emissions of a particular pollutant can be zero. This is the case studied in this paper, where we assume that the benefit from emissions is given by a quadratic-linear function so that the marginal benefit is positive for zero emissions, but finite. Finally, following the approach adopted by Carraro and Siniscalco (1993), Barrett (1994), and Chander and Tulkens (1997) among others, we focus our analysis on the case where only one IEA is formed and the only question remaining is the size of the self-enforcing agreement. This approach is justified in this paper because our aim is to review Barrett's (1994) numerical analysis in order to analytically evaluate the robustness of his results when the non-negativity constraints are taken into account. Nevertheless, this assumption may be reasonable for global environmental problems, such as the climate change problem, for which IEAs are usually unique and launched by the United Nations. In the last analysis, it could be interpreted as an institutional constraint.

In Section 2 we present the basic model for an international emissions game and introduce the definition of a stable international environmental agreement. In Section 3 we derive the stability results for Stackelberg behaviour. Section 4 concludes.

[8] Although countries are symmetric, the equilibrium is asymmetric with signatories and non-signatories polluting in different amounts.

2. An international emissions game

2.1 The basic model

In this section we present the basic linear-quadratic model of an international emissions game.[9] There are N identical countries, $i=1,\dots,N$. We define $q_i \geq 0$ as the level of emissions generated by country i, $Q_i \equiv \sum_{k\neq i} q_k$ the total emissions generated by all countries other than i, and $Q=\sum_k q_k = Q_i + q_i$ as the total emissions generated by all N countries. Each country derives a gross benefit from its emissions and also suffers environmental damage, which depends on the global level of emissions. Thus each country has a net benefit(payoff) function:

$$\pi(q_i, Q_i) \equiv \alpha q_i - \frac{\beta}{2} q_i^2 - \frac{\gamma}{2}(q_i + Q_i)^2.$$

We assume that $\alpha > 0$, $\beta > 0$ and $\gamma > 0$. It should be clear that w.l.o.g. we can normalize one of the parameters and we choose to normalize by setting $\gamma = 1$. To emphasize this normalization we rewrite the net benefit function as:

$$\pi(q_i, Q_i) \equiv a q_i - \frac{b}{2} q_i^2 - \frac{1}{2}(q_i + Q_i)^2. \quad (1)$$

We shall think of b as β/γ—the ratio of the (absolute) slope of the marginal benefit curve and the slope of the marginal damage cost curve, so a low value of b is to be interpreted as a (relatively) high marginal damage cost.[10]

2.2 Stable international environmental agreements

We model an International Environmental Agreement as a three-stage model, in which in the first stage (the Membership Game) each country decides whether or not to join an IEA, in the second stage the signatories determine their emissions, and in the third stage the non-signatories determine their emissions. Thus, the countries first play a simultaneous game, the Membership Game, and given the outcome of this game they play a sequential game that we call the Emissions Game. We describe each game briefly, in reverse order.

2.2.1 *The Emissions Game* Suppose that, as the outcome of the first-stage game, there are n signatory countries (a typical signatory being denoted by s) and $N-n$ non-signatory countries (a typical non-signatory being denoted by f,

[9] In this model, as in the early papers discussed in the introduction, we deal with a flow pollutant. For analysis using a stock pollutant, see Rubio and Ulph (2002b, 2003).

[10] Notice that while we work with an emissions game, it can be shown that this is equivalent to the model of an abatement game developed by Barrett (1994). See Appendix 1.

for fringe or free-rider). Each non-signatory country takes the emissions of all other countries as given and chooses its emissions to maximize its own net benefits. Using symmetry, this will define a non-signatory reaction function linking the emissions of a typical non-signatory country to the emissions of a typical signatory country. Signatory countries choose emissions to maximize the aggregate payoff of the n signatories, recognizing how this will affect emissions of non-signatories.

Two issues need further clarification. First, there is the question of the timing of the players' actions. Following Barrett (1994) we argue that membership of an IEA acts as a form of commitment device, which we model by thinking of signatories setting their emissions before non-signatories. In this case signatories can calculate what emissions non-signatories will choose (in terms of the non-signatory reaction function), and choose their own emissions to maximize aggregate net benefit. This is the Stackelberg model.[11]

The second issue is that emissions by signatories and non-signatories have to be non-negative and we simply impose this introducing a non-negativity constraint in the decision problems of signatories and non-signatories. Thus for some parameter values the emissions of either a signatory or non-signatory could be zero (a corner solution).

The outcome of the emissions game then, is that for any number of signatories n we can define the equilibrium payoffs to signatory and non-signatory countries: $\pi^s(n)$, $\pi^f(n)$.

2.2.2 *The Membership Game* We follow Hoel (1992), Carraro and Siniscalco (1993), Barrett (1994), and others in saying that an IEA is self-enforcing if it is stable, where the concept of stability is borrowed from the literature on cartel stability (d'Aspremont et al., 1983). For $2 \leq n \leq N$ we define $\Delta(n) = \pi^s(n) - \pi^f(n-1)$; thus:

Definition 1 An IEA with n signatories is stable if it satisfies the conditions: $\Delta(n) \geq 0$, i.e. $\pi^s(n) \geq \pi^f(n-1)$ (internal stability condition); $\Delta(n+1) \leq 0$, i.e. $\pi^f(n) \geq \pi^s(n+1)$ (external stability condition).

Internal stability simply means that any signatory country is at least as well off staying in the IEA as it is quitting, assuming that all other countries do not change their membership decisions. In a similar way external stability requires that any non-signatory is at least as well off remaining a non-signatory as it will be being a signatory, again assuming that other countries do not change their membership decisions.

[11] We would like to explicitly acknowledge that this notion of equilibrium for the emissions game is the Stackelberg version of the 'Partial Agreement Nash Equilibrium (PANE) with respect to a coalition' defined by Chander and Tulkens (1997).

We can also think of a stable IEA as a Nash equilibrium of a simultaneous open membership game where the strategies for each country are to sign or not to sign.[12] A country takes the membership decisions of all other countries as given. Suppose these have resulted in a membership of m, $0 \leq m \leq N-1$. Then the payoffs to a country are $\pi^s(m+1)$ if it signs and $\pi^f(m)$ if it does not. So it will join if $\pi^s(m+1) \geq \pi^f(m)$ and otherwise not join. For an IEA with n^* members to constitute a Nash equilibrium of the Membership Game, it must have paid each signatory to sign, so $\pi^s(n^*) \geq \pi^f(n^*-1)$. Similarly it must have paid each non-signatory not to join, so $\pi^f(n^*) \geq \pi^s(n^*+1)$. These are just the conditions for Internal and External Stability.

In this next section we analyse stable IEAs for the Stackelberg model with non-negative emissions.

3. Stable Stackelberg IEAs with non-negative emissions

In this section we analyse stable IEAs when signatory countries act collectively as a Stackelberg leader and emissions are restricted to be non-negative. We begin with the emissions game.

3.1 Stackelberg emissions game with non-negative emissions

Suppose there are n signatories and $N-n$ non-signatories. A non-signatory country k takes Q_{fk} as given and chooses q_{fk} to solve:

$$\max_{q_{fk}\geq 0} \pi_k^f = aq_{fk} - \frac{b}{2}q_{fk}^2 - \frac{1}{2}(q_{fk} + Q_{fk})^2,$$

where q_{fk} stands for the level of emissions generated by non-signatory k, $k=1,\ldots,N-n$, and Q_{fk} for the total emissions generated by all countries other than non-signatory k.

The first order condition is:

$$\frac{\partial \pi_k^f}{\partial q_{fk}} = a - bq_{fk} - (q_{fk} + Q_{fk}) \leq 0, \quad q_{fk} \geq 0, \quad q_{fk}\frac{\partial \pi_k^f}{\partial q_{fk}} = 0. \tag{2}$$

(2) defines the non-signatory reaction function for a country k allowing for the fact that emissions must be non-negative, so part of the reaction function has $q_{fk}=0$. Now signatories are assumed to coordinate in order to

[12] In open membership games, any player is free to join or leave a coalition. In our case each country chooses one of the two possible strategies and the agreement is formed by all players who choose to sign.

maximize their collective net benefits taking into account the reaction function of the followers:

$$\max_{q_{s1},\dots,q_{sn},q_{f1},\dots,q_{f,N-n}\geq 0} \Pi^s = \sum_{i=1}^{n} \pi_i^s = \sum_{i=1}^{n}\left[aq_{si} - \frac{b}{2}q_{si}^2 - \frac{1}{2}(q_{si}+Q_{si})^2\right],$$

$$s.t. \quad -a + bq_{fk} + (q_{fk} + Q_{fk}) \geq 0,$$

where q_{si} stands for the level of emissions generated by signatory i, $i=1,\dots,n$, and Q_{si} for the total emissions generated by all countries other than signatory i.

Under the assumption of symmetry $q_{f1}=\dots=q_{f(N-n)}=q_f$, $q_{s1}=\dots=q_{sn}=q_s$ and $Q=nq_s+(N-n)q_f$, so that the previous optimization problem reduces to:

$$\max_{q_s\geq 0,\ q_f\geq 0} n\pi_s = n[aq_s - \frac{b}{2}q_s^2 - \frac{1}{2}(nq_s+(N-n)q_f)^2]$$

$$s.t. \quad -a + bq_f + nq_s + (N-n)q_f \geq 0, \tag{3}$$

The Lagrange function for the problem is

$$L = n\left[aq_s - \frac{b}{2}q_s^2 - \frac{1}{2}\left(nq_s+(N-n)q_f\right)^2\right] + \lambda\left(-a + bq_f + nq_s + (N-n)q_f\right),$$

and the KTCs are

$$\frac{\partial L}{\partial q_s} = n\left[a - bq_s - n(nq_s+(N-n)q_f) + \lambda\right] \leq 0, \tag{4}$$

$$q_s \geq 0, \quad q_s\frac{\partial L}{\partial q_s} = 0,$$

$$\frac{\partial L}{\partial q_f} = -n(N-n)(nq_s+(N-n)q_f) + \lambda(b+N-n) \leq 0, \tag{5}$$

$$q_f \geq 0, \quad q_f\frac{\partial L}{\partial q_f} = 0,$$

$$\frac{\partial L}{\partial \lambda} = -a + bq_f + nq_s + (N-n)q_f \geq 0, \tag{6}$$

$$\lambda \geq 0, \quad \lambda\frac{\partial L}{\partial \lambda} = 0.$$

Equilibrium in the emissions game involves solving (4)–(6) simultaneously, taking non-negativity constraints into account. In principle there are three possibilities:
(i) *Interior Solution* ($q_s > 0$, $q_f > 0$). From (4)–(6) we obtain:

$$q_s = \frac{a\left[b^2 - (N-n)(n-2)b + (N-n)^2\right]}{b\left[(b+N-n)^2 + bn^2\right]}, \tag{7}$$

$$q_f = \frac{a\left[b^2 + (N+n^2-2n)b - (N-n)n\right]}{b\left[(b+N-n)^2 + bn^2\right]}, \tag{8}$$

so that $q_s > 0$, $q_f > 0$ iff

$$g(b,n) = b^2 - (N-n)(n-2)b + (N-n)^2 > 0,$$
$$h(b,n) = b^2 + (N+n^2-2n)b - (N-n)n > 0.$$

(ii) *Signatory Corner Solution* ($q_s = 0$, $q_f > 0$). From (4)–(6) this requires:

$$q_f = \frac{a}{b+N-n}, \quad g(b,n) \leq 0. \tag{9}$$

(iii) *Non-Signatory Corner Solution* ($q_s > 0$, $q_f = 0$). From (4)–(6) this requires:

$$q_s = \frac{a}{n}, \quad h(b,n) \leq 0. \tag{10}$$

It is easy to show that for the Stackelberg equilibrium a solution $q_s = 0$, $q_f = 0$ does not satisfy the KTCs. Given the emissions for each kind of solution the net benefits can be obtained by substitution.
(i) *Interior Solution*:

$$\pi^s(n) = \frac{a^2}{2b}\left\{1 - \frac{N^2 b}{(b+N-n)^2 + bn^2}\right\}, \tag{11}$$

$$\pi^f(n) = \frac{a^2}{2b}\left\{1 - \frac{(b+1)N^2(b+N-n)^2}{\left[(b+N-n)^2 + bn^2\right]^2}\right\}. \tag{12}$$

(ii) *Signatory Corner Solution*:

$$\pi^s(n) = -\frac{a^2(N-n)^2}{2(b+N-n)^2}, \quad \pi^f(n) = \frac{a^2[b - (N-n)(N-n-2)]}{2(b+N-n)^2}. \tag{13}$$

(iii) *Non-Signatory Corner Solution*:

$$\pi^s(n) = -\frac{a^2(b + n(n-2))}{2n^2}, \quad \pi^f(n) = -\frac{a^2}{2}. \tag{14}$$

Now note that for the interior solution the full-cooperative level of emissions is given by eq. (7) for $n = N$ and that the full-noncooperative Cournot level of emissions is given by eq. (8) for $n = 0$. In these two cases we have an interior solution. For this reason we focus on $n = 1, 2, \ldots, N-1$ in the rest of this section.

We now want to determine more precisely for which parameter values the three different solutions occur. This analysis is developed in the following pages and the final results are presented in Fig. 3. The results clearly depend on the signs of $g(b, n)$ and $h(b, n)$. In Fig. 3, $b_1(n)$ is the function implicitly defined by $h(b, n) = 0$ and $b_2(n)$ and $b_3(n)$ are the functions implicitly defined by $g(b, n) = 0$, so that these functions allow us to determine for which values (b, n), $g(b, n)$ and $h(b, n)$ yield negative values or zero and then to know when the solution of the problem is a corner solution.[13] The next proposition fixes n and considers for which values of b we get each of the three solutions.

Proposition 1 For any n, there exists a unique solution to the Stackelberg emissions game with non-negative emissions as follows: (i) for $n \in \{1, 2, 3\}$, there exists $b_1(n)$ defined above such that for $b \leq b_1(n)$ the equilibrium is the non-signatory corner solution while for $b > b_1(n)$ the equilibrium is the interior solution. (ii) For $n = 4$, $g(N-4, 4) = 0$ and $b_1(4) < N-4$ so that: (a) for $b \leq b_1(4)$ the equilibrium is the non-signatory corner solution, (b) for $b_1(4) < b < N-4$ the solution is the interior solution, (c) for $b = N-4$ the equilibrium is the signatory corner solution, (d) for $b > N-4$ the equilibrium is again the interior solution. (iii) For $n \in (4, N)$, $b_1(n) < b_2(n) < b_3(n)$ so that : (a) for $b \leq b_1(n)$ the equilibrium is the non-signatories corner solution, (b) for $b \in (b_1(n), b_2(n))$ the equilibrium is the interior solution, (c) for $b \in [b_2(n), b_3(n)]$ the equilibrium is the signatories corner solution, and (d) for $b > b_3(n)$ the solution is the interior solution.

Proof See Appendix 2.

This Proposition determines, for a given value of n, the critical values of b that yield one type of solution or another. For instance, we have shown that if b is big enough, the Stackelberg equilibrium is an interior solution for any value of n. Notice that the three kinds of solutions that the problem

[13] Notice that $g(b, n)=0$ defines two functions because for a given n the equation has two positive real solutions: $b_2(n) < b_3(n)$. However, $h(b, n) = 0$ only has a positive real solution that defines $b_1(n)$.

may present can be explained by resorting to the marginal conditions defined by the f.o.cs. of the problem. Next, we explain the logic behind the non-signatories corner solutions since, as will be seen in Section 3.2, this kind of solution supports high levels of cooperation. To do this we first rewrite the optimization problem of signatories using the reaction function of non-signatories as an equality

$$\max_{q_s \geq 0} n\pi_s = n\left[aq_s - \frac{b}{2}q_s^2 - \frac{1}{2}\left(\frac{(N-n)a + nbq_s}{b+N-n}\right)^2\right].$$

For this problem the f.o.c. can be written as

$$a - bq_s = \frac{nb(N-n)a}{(b+N-n)^2} + \frac{n^2b^2}{(b+N-n)^2}q_s,$$

where the l.h.s. stands for the marginal benefit of emissions and the r.h.s for the marginal environmental damage. Now we can use this condition to explain why when b is low enough the solution is that non-signatories cut their emissions completely. By first considering the limit case of $b=0$ we see that the problem becomes linear and the solution would be $q_s=+\infty$ and $q_f=-\infty$ as (7) and (8) establish. Thus, when b is low enough the optimization problem is 'quasi-linear' and both the marginal benefit curve and the marginal environmental damage curve are very 'flat'. Under these conditions the marginal benefit of emissions slowly decreases and the marginal environmental damage also increases slowly resulting in large emissions for the signatories. Thus as for $b=0$ the emissions tend to infinite, there exists a critical value for b given by $b_1(n)$ such that if $b \leq b_1(n)$ the signatories' emissions are large enough to induce the non-signatories to choose zero emissions. At this point the reader should take into account that the signatories enjoy a strategic advantage and that emissions are strategic substitutes, which explains why zero emissions can be the optimal policy for the non-signatories if signatories' emissions are large enough.[14]

[14] One referee sees this solution as empirically irrelevant and he thinks that is explained by the fact that signatories act as a leader group. Although it is true that this assumption plays a key role in the derivation of this solution as we have just explained, two more things are required: a marginal benefit from emissions that is positive but finite for zero emissions, and a low enough marginal cost, i.e. a low value of b. Given these three conditions we show that the non-signatory corner solution is one of the outcomes of the model proposed by Barrett (1994) when it is assumed that emissions can be zero. At this point, we would like to clarify that the two main conclusions of the paper, Propositions 4 and 5, do not depend on the existence of this kind of solutions. Nevertheless, we have decided to include it to offer to the reader a complete characterization of the outcomes that can be obtained from Barrett's (1994) model without any additional constraint on parameter values.

In order to complete the analysis of the parameter space we now characterize functions $b_1(n)$, $b_2(n)$ and $b_3(n)$ which allow us to determine the shape of the contours $g(b,n)=h(b,n)=0$ of the functions $g(b,n)$ and $h(b,n)$. For function $b_1(n)$, $b_1(0)=b_1(N)=0$ and it is easy to show that this function presents a unique extreme in the interior of the interval $(1,N-1)$ which is a maximum.[15] Then there exists, at least, one integer, $\hat{n}$, in that interval that maximizes $b_1(n)$ so that $b_1(\hat{n})$ is the maximum value of the function given by an integer in domain $\{1,2,\ldots,N-1\}$.[16] Given this behaviour of function $b_1(n)$ we are able to establish for a given b what the solution of the emissions game is for non-signatories and for different values of n. For $b \leq b_1(\hat{n})$, define $n_1 \leq n_2$ as positive real roots of $b=b_1(n)$ then we find that:

Lemma 1 (i) If $b>b_1(\hat{n})$, for all n the equilibrium will be an interior solution. (ii) If $b \leq \hat{b}_1(\hat{n})$, there exist n_1, n_2 defined above such that the equilibrium will be an interior solution for n such that $n \notin [n_1,n_2]$ and a corner solution for n when $n \in [n_1,n_2]$.
Proof See Appendix 3.

Moreover, as $b_1(N-1)$ is the minimum value of $b_1(n)$ for $n=1,2,\ldots,N-1$ we can also establish that:

Corollary 1 If $b \in (0,b_1(N-1)]$, $n_1<1$ and $n_2 \geq N-1$ so the only interior solutions are $n=0$ and $n=N$. In other words, the equilibrium is a corner solution for non-signatories for all n.
Proof See Appendix 4.

We illustrate these results in Fig. 1.[17]

The kind of solution for signatories emissions depends on whether b belongs to the interval $[b_2(n),b_3(n)]$. It can be shown that $b_2(n)$ is a strictly convex, decreasing function defined in the interval $[4,N]$ with $b_2(4)=N-4$ and $b_2(N)=0$. Then for $n=4,5,\ldots,N-1$ the maximum value of the function is $N-4$ and the minimum value is $b_2(N-1)$.

On the other hand, $b_3(n)$ is a strictly concave function with a maximum value equal to $N(N-4)/4$ in interval $(4,N-1)$ with $b_3(4)=N-4$ and $b_3(N)=0$.

---

[15] In order to study the behaviour of $b_1(n)$ in that interval we assume that n is a real number and once we know the properties of $b_1(n)$ then we are able to characterize the values of $b_1(n)$ with respect to n but now with n restricted to be an integer number. The same approach is followed to study $b_2(n)$ and $b_3(n)$.

[16] Function $b_1(n)$ has a unique maximum, $\tilde{n}$, in the real interval $(1,N-1)$. If the maximum is an integer, the function has the same maximum in the domain $\{1,2,\ldots,N-1\}$ that we call $\hat{n}$. If $\tilde{n}$ is not an integer and the function is not symmetric then $\hat{n}$ is the closest integer to $\tilde{n}$ that yields the maximum value for $b_1(n)$. If the function is symmetric we could have two maximums $\hat{n}_1$ and $\hat{n}_2$ such that $\hat{n}_2=\hat{n}_1+1$ provided that $\tilde{n}$ is in the centre of interval $(\hat{n}_1,\hat{n}_2)$.

[17] In order to simplify the graphical representation we assume that the maximum value of real function $b_1(n)$ is the integer number $\hat{n}$.

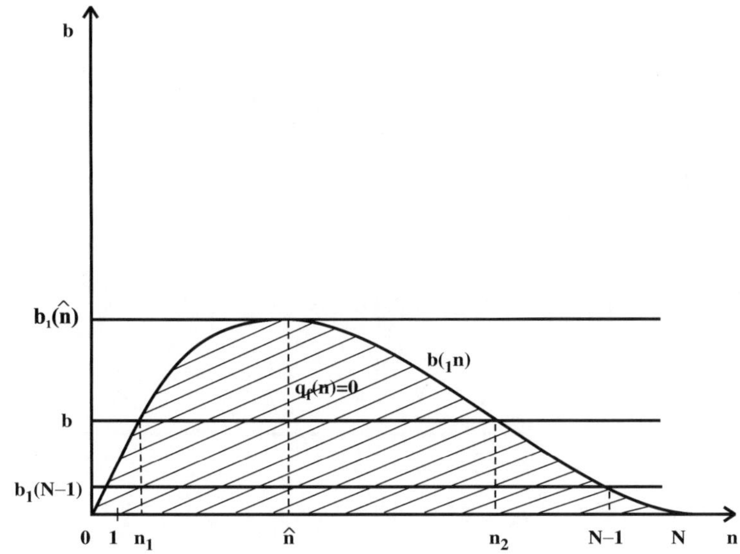

Fig. 1. Corner solutions for non-signatories

Then there exists, at least, one integer, n^*, that maximizes $b_3(n)$ so that $b_1(n^*)$ is the maximum value of the function given by an integer in domain $\{4, 5, \ldots, N-1\}$ and therefore the function increases for $4 \leq n < n^*$ and decreases afterwards.[18] See Fig. 2. For this behaviour of functions $b_2(n)$ and $b_3(n)$ we are able to establish for a given b what the solution of the emissions game is for signatories and for different values of n. For $b \in [b_2(N-1), N-4]$, define n_3 as the unique, positive real root of $b = b_2(n)$ and $n_4 > n_3$ as the unique positive real root of $b = b_3(n)$. For $b \in (N-4, b_3(n^*)]$, define $n_3 \leq n_4$ as the positive real roots of $b = b_3(n)$. Thus:

Lemma 2 (i) If $b > b_3(n^*)$, for all n the equilibrium will be an interior solution. (ii) If $b \in [b_2(N-1), b_3(n^*)]$, there exist n_3, n_4 defined above depending on whether b is greater or less than $N-4$ such that the equilibrium will be an interior solution for n when $n \notin [n_3, n_4]$ and a signatory corner solution for n when $n \in [n_3, n_4]$. (iii) If $b \in (0, b_2(N-1))$, then for all n the equilibrium will be an interior solution.
Proof See Appendix 5.

[18] Function $b_3(n)$ has a unique maximum, $\tilde{n}$, in the real interval $(4, N-1)$ for $N > 5$. If the maximum is an integer, the function has the same maximum in the domain $\{4, \ldots, N-1\}$ that we call n^*. If $\tilde{n}$ is not an integer and the function is not symmetric then n^* is the closest integer to $\tilde{n}$ that yields the maximum value for $b_3(n)$. If the function is symmetric we could have two maximums n_1^* and n_2^* such that $n_2^* = n_1^* + 1$ provided that $\tilde{n}$ is in the centre of interval (n_1^*, n_2^*).

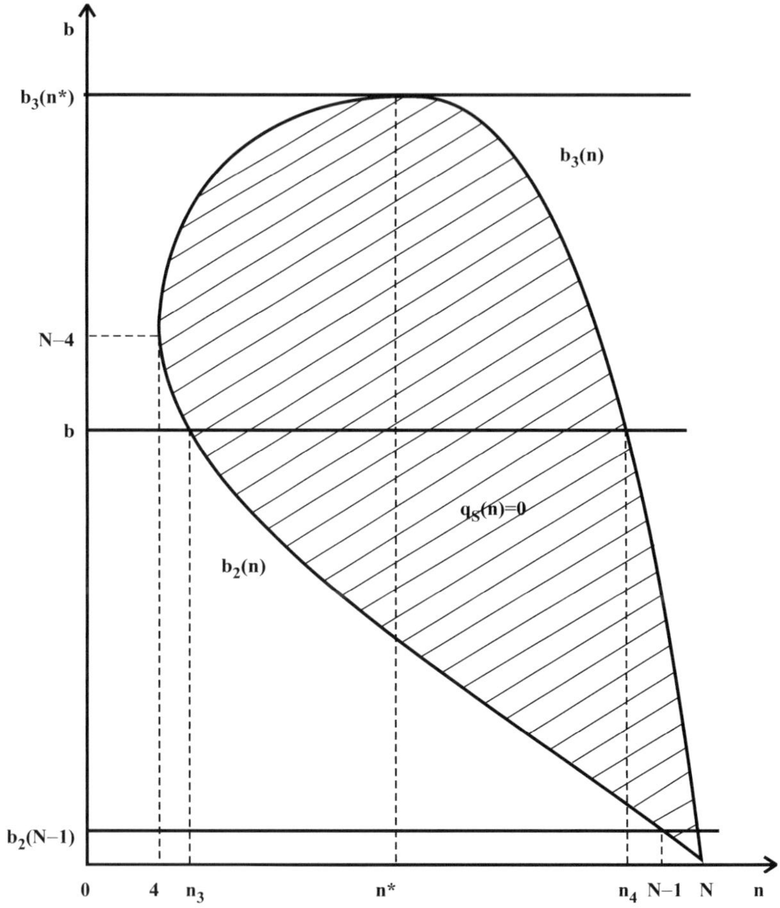

Fig. 2. Corner solutions for signatories

We illustrate this Lemma in Fig. 2.[19]

Now we can completely characterize the parameter space. The next Proposition summarizes Lemmas 1 and 2.

Proposition 2 (i) If $b > b_3(n^*)$ we have that for all n, the equilibrium will be an interior solution for signatories and non-signatories. (ii) If $b \in (b_1(\hat{n}), b_3(n^*)]$, the equilibrium will be an interior solution for non-signatories for all n, however for signatories there exist $\mathbf{n}_3$, $\mathbf{n}_4$ defined above such that the equilibrium will be an interior solution for n when $n \notin [n_3, n_4]$ and a corner solution when $n \in [n_3, n_4]$.

[19] In order to simplify the graphical representation, we assume that the maximum value of real function $b_3(n)$ is given by the integer n^*.

(iii) If $b \in [b_2(N-1), b_1(\hat{n})]$, there exist n_1, n_2, n_3 and n_4 defined above such that the equilibrium will be an interior solution for non-signatories for n when $n \notin [n_1, n_2]$ and a non-signatory corner solution for n when $n \in [n_1, n_2]$. Moreover the equilibrium will be an interior solution for signatories for n when $n \notin [n_3, n_4]$ and a signatory corner solution for n when $n \in [n_3, n_4]$. (iv) If $b \in (b_1(N-1), b_2(N-1))$, the equilibrium will be an interior solution for signatories for all n and a corner solution for non-signatories also for all n except for $n = N-1$. (v) Finally, if $b \in (0, b_1(N-1)]$, the equilibrium will be an interior solution for signatories for all n and a corner solution for non-signatories also for all n.
Proof See Appendix 6.

We illustrate this proposition in Fig. 3.[20]

As can be seen, the parameter space is divided into five regions which are defined by four critical values of b. The first one is the maximum value of $b_3(n)$. Above this value the equilibrium will be an interior solution for signatories and non-signatories. Under this value the corner solutions appear. The second critical value is the maximum value of $b_1(n)$. Between these two critical values the equilibrium will be a corner solution for signatories for some values of n. The third critical value is the minimum value of $b_2(n)$, i.e. $b_2(N-1)$. In this third region the type of equilibrium depends on n. For low values of n we will find a corner solution for non-signatories. However for high values of n the signatories will choose zero emissions. Finally, the last critical value is $b_1(N-1)$ which defines a transition region almost identical to the bottom region for which the equilibrium will be an interior solution for signatories and a corner solution for non-signatories.

Thus for any parameters N and b, Proposition 2 indicates, for any number of signatories n, what type of solution there is to the Stackelberg emissions game, with corner solutions to take account of non-negative emissions constraints. Next, we conduct the stability analysis using these results and the corresponding outputs and equilibrium payoff functions for signatories and non-signatories.

3.2 Membership game for Stackelberg model with non-negative emissions

In this section we show that the scope of the international cooperation for controlling an environmental problem critically depends on the level of the marginal

[20] In order to simplify the graphical representation, we assume that the maximum values of real functions $b_1(n)$ and $b_3(n)$ are given by the integers $\hat{n}$ and n^*. Notice that as $b_2(n)$ is a decreasing, strictly convex function and $b_2(N) = b_1(N) = 0$, $b_1(n)$ must be also a strictly convex function for large enough values of n. The previous results apply for $N > 5$ although with minimal changes they are also valid for $N = 5$.

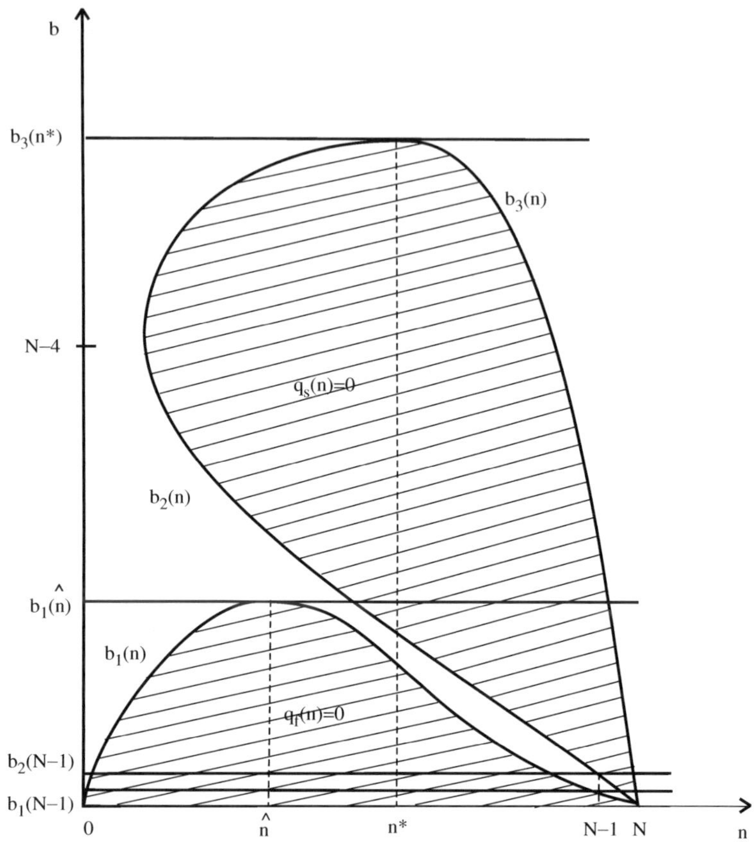

Fig. 3. Parameter space

environmental damage. We begin analysing the stability for (relatively) high marginal damages, i.e., for low values of *b*.

Proposition 3 If $b \leq b_1(N-1)$ the unique stable IEA of the Stackelberg model with non-negative emissions is the grand coalition.

Proof To conclude that the grand coalition is self-enforcing it is only necessary to check if the internal stability condition is satisfied, i.e. if $\Delta(N) = \pi^s(N) - \pi^f(N-1) \geq 0$ according to Definition 1. From (11) we know that the net benefits of the full-cooperative equilibrium of the emission game are $\pi^s(N) = a^2/2(b+N^2)$. Moreover, Proposition 2 establishes that for this range of parameter values the equilibrium of the emissions game is the non-signatories corner solution for $n<N$ but for this kind of solution the net benefits of the non-signatories are negative since their emissions are zero, see (14). The result is that the internal stability condition is satisfied: $\Delta(N)>0$. □

When b is low enough to induce the non-signatories to cut their emissions completely, the grand-coalition, i.e. the full-cooperative equilibrium, is self-enforcing since no country has incentives to exit from the agreement.

Next we show that the grand coalition cannot be a stable agreement for lower values of damage costs.

Proposition 4 If $b \in [b_2(N-2), b_2(4) = N-4]$, there exists an upper bound given by the smallest integer no less than n_3 for the number of countries that belong to a self-enforcing IEA. This upper bound decreases when b increases.
Proof See Appendix 7.

This result establishes that the scope of cooperation is very sensitive to changes in the level of marginal environmental damage. So that we have to expect that a reduction in the marginal damage leads to a reduction in the level of cooperation reached by a self-enforcing IEA. The explanation for this kind of relationship is given by the fact that the interdependence among the countries occurs through the damage function. Thus, when the marginal environmental damage is relatively high (a low b), the leadership of the countries in the agreement is strong and the signatories choose emission levels which induce non-signatories to select low values of emission, making exit from the agreement unprofitable. These effects are weakened as environmental damage costs get smaller.

Finally, we focus on the scope of cooperation when $b > N-4$.

Proposition 5 If $b > N-4$ and $N > 5$ the maximum level of cooperation that can be achieved by a self-enforcing IEA is three.
Proof See Appendix 8.

Taking into account that $N(N-4)/4 > N-4$ when $N > 5$ and that $b > N(N-4)/4$ is a sufficient condition to have an interior solution for signatories and non-signatories we can conclude that:

Corollary 2 If $b > N(N-4)/4$ then the Stackelberg equilibrium is an interior solution for signatories and non-signatories and the maximum level of cooperation that can be achieved by a self-enforcing IEA is three.

It is also easy to show that for a large enough value of b the stable level of cooperation is two. These conclusions clarify the previous results in the literature which have been derived assuming that there are interior solutions. According to our results, restricting parameter values to guarantee interior solutions is a sufficient condition to get stable IEAs with a small number of signatories but it is not a necessary condition. We have obtained that it is enough with $b > N-4$ to have a maximum of three countries in an IEA. But between $N-4$ and $N(N-4)/4$ the Stackelberg equilibrium is a corner solution for different values of n depending on the value of b. This means that what is necessary and sufficient to get a small degree of cooperation is a high value of b and not interior solutions for signatories and non-signatories.

Thus we have shown that even if we take non-negative emission constraints seriously, the Stackelberg model can have stable IEAs as large as the grand coalition and as low as a bilateral agreement, depending on the value of the marginal environmental damage.

Finally, although we have shown that, allowing for non-negative emission constraints, it is still possible to get the grand coalition as a stable IEA, it could still be the case that imposing non-negative emissions has a significant effect on the size of a stable IEA in the sense that for any particular set of parameter values the size of IEA is significantly smaller than would be calculated if one simply ignored the constraints. To test this we have taken values of $a = 1000$, values of $N = 10, 20, 150$, and 1500 values of b. For each set of parameter values we calculated the size of the stable IEA imposing non-negative emission constraints and without imposing such constraints. Three points emerged: (i) first, we confirmed that for all N, by varying b, the maximum size of stable IEA obtained was the grand coalition, whether or not the non-negative emission constraints were imposed; (ii) for any set of parameter values, the size of the stable IEA with the non-negative emissions constraints imposed was never greater than the size of stable IEA when the constraints were ignored; (iii) but crucially, the maximum difference in the size of stable IEA between restricted and unrestricted emissions is 1. We illustrate this in Table 1 by showing, for a range of values of N, the average size of stable IEA (averaged over different values of b) with and without the constraints. As can be seen, the difference in average size by imposing the constraints is tiny as one would expect given that the maximum difference is 1. This means that for a lot of cases there is no difference in the level of cooperation with and without constraints. One has to go to the third decimal place to detect a difference in average size.

4. Conclusions

In this paper we have provided analytical proofs of the main results of the linear-quadratic version of the widely used model of stable IEAs introduced by

Table 1 Average size of stable IEA with and without constraints

N	With constraints	Without constraints
10	6.268	6.272
30	16.677	16.680
50	27.148	27.151
70	37.629	37.631
90	48.110	48.111
110	58.589	58.590
130	69.068	69.069
150	79.545	79.546

Carraro and Siniscalco (1991) and Barrett (1994). Moreover we have shown, analytically, that these results are robust to the introduction of constraints that emissions must be non-negative. While such constraints significantly complicate the analysis, they leave the main findings of the original literature almost completely unaffected. Since it is clearly right that such non-negative constraints should be taken into account, it is important to know the original results are robust and this is one of the contributions of our paper. Another contribution is that our results qualify the claim by Diamantoudi and Sartzetakis (2002a) that imposing non-negative emissions constraints significantly reduces the size of the stable IEAs. This paper argues that this claim is only right when emissions are restricted to be strictly positive. However, if countries can completely cut their emissions the size of the stable IEAs depends directly on the marginal environmental damage, so that a high degree of cooperation can be reached if the marginal environmental damage is big enough. Thus, this paper gives an analytical support to Barrett's conclusion on the scope of cooperation that can be achieved by a self-enforcing IEA.

Of course, there are many other respects in which the original models of stable IEAs need to be extended—richer concepts of self-enforcing agreements, asymmetric countries, concepts of fairness, dealing with stock pollutants, allowing for uncertainty and learning—and the authors of the original papers and many others have made important contributions to addressing these extensions (see again Finus, 2001, for an excellent survey). We have also addressed some of these extensions (Rubio and Ulph, 2002b, 2003; Ulph, 2002, 2004). However, since the basic model continues to attract interest, it is important to ensure that its properties rest on thorough analysis, and this paper contributes to this purpose.

Acknowledgements

Research of Santiago J. Rubio was supported by the Ministerio de Ciencia y Tecnología under grant BEC2000-1432 and by Fundación BBVA. We also acknowledge the financial support from the Instituto Valenciano de Investigaciones Económicas. Finally, we would like to thank Begona Casino for her help with the figures and two anonymous referees for their useful comments. Regarding any remaining inadequacies, the usual caveat applies.

References

Barrett, S. (1994) Self-enforcing international environmental agreements, *Oxford Economic Papers*, **46**, 878-94.

Barrett, S. (2001) International cooperation for sale, *European Economic Review*, **45**, 1835–50.

Barrett, S. (2003) *Environment and Statecraft*, Oxford University Press, Oxford.

Carraro, C. and Siniscalco, D. (1991) Strategies for the international protection of the environment, Discussion Paper 568, Centre for Economic Policy Research, London.

Carraro, C. and Siniscalco, D. (1993) Strategies for the international protection of the environment, *Journal of Public Economics*, **52**, 309–28.

Chander, P. and Tulkens, H. (1995) A core-theoretic solution for the design of cooperative agreements on transfrontier pollution, *International Tax and Public Finance*, **2**, 279–93.

Chander, P. and Tulkens, H. (1997) The core of an economy with multilateral environmental externalities, *International Journal of Game Theory*, **26**, 379–401.

D'Aspremont, C.A., Jacquemin, J., Gadszeweiz, J., and Weymark, J. A. (1983) On the stability of collusive price leadership, *Canadian Journal of Economics*, **16**, 17–25.

Diamantoudi, E. and Sartzetakis, E. S. (2002a) Stable international environmental agreements: an analytical approach, Nota Di Lavoro 7.2002, Fondazione Eni Enrico Mattei, Milan.

Diamantoudi, E. and Sartzetakis, E.S. (2002b) International environmental agreements: the role of foresight, mimeo, Department of Economics, University of Aarhus.

Ecchia, G. and Mariotti, M. (1997) The stability of international environmental coalitions with farsighted countries: some theoretical observations, in C. Carraro (ed.) *International Environmental Negotiations*, Edward Elgar, Cheltenham.

Ecchia, G. and Mariotti, M. (1998) Coalition formation in international environmental agreements and the role of institutions, *European Economic Review*, **42**, 573–82.

Finus, M. (2001) *Game theory and international environmental cooperation*, Edward Elgar, Cheltenham.

Hoel, M. (1992) International environmental conventions: the case of uniform reductions of emissions, *Environmental and Resource Economics*, **2**, 141–59.

Hoel, M. and Schneider, K. (1997) Incentives to participate in an international environmental agreement, *Environmental and Resource Economics*, **9**, 153–70.

Na, S.L. and Shin, H.S. (1998) International environmental agreements under uncertainty, *Oxford Economic Papers*, **50**, 173–85.

Petrakis, E. and Xepapadeas, A. (1996) Environmental consciousness and moral hazard in international agreements to protect the environment, *Journal of Public Economics*, **60**, 95–110.

Ray, D. and Vohra, R. (2001) Coalition power and public goods, *Journal of Political Economy*, **109**, 1355–84.

Rubio, S. and Casino, B. (2001) International cooperation in pollution control, Working Paper WP-AD 2001-21, Instituto Valenciano de Investigaciones Económicas, Valencia.

Rubio, S. and Ulph, A. (2002a) Leadership and self-enforcing environmental agreements with non-negative emissions, Discussion Paper in Economics and Econometrics 0211, Department of Economics, University of Southampton.

Rubio, S. and Ulph, A. (2002b) A simple dynamic model of international environmental agreements with a stock pollutant, Discussion Paper in Economics and Econometrics 0209, Department of Economics, University of Southampton.

Rubio, S. and Ulph, A. (2003) An infinite-horizon model of dynamic membership of international environmental agreements, Working Paper No. 57.2003, Fondazione Eni Enrico Mattei, Milan.

Tulkens, H. (1998) Cooperation versus free-riding in international environmental affairs: two approaches, in N. Hanley and H. Folmer (eds), *Game theory and the environment*, Edward Elgar, Cheltenham, ch 2.

Ulph, A. (2002) Environmental agreements with uncertainty and learning — the case of stock-dependent unit costs, Discussion Paper in Economics and Econometrics 0208, Department of Economics, University of Southampton.

Ulph, A. (2004) Stable international environmental agreements with a stock pollutant, uncertainty and learning, *Journal of Risk and Uncertainty*, **29**, 53–73.

Wagner, U.J. (2001) The design of stable international environmental agreements: economic theory and political economy, *Journal of Economic Surveys*, **15**, 377–411.

Appendix 1

On the correspondence between the model developed in the paper and in Barrett (1994)

It is easy to establish the correspondence between the two models through a variable change. Basically, we follow the procedure proposed by Diamantoudi and Sartzetakis (2002a, see Section 4) offering here a justification for the application of this procedure. If we distinguish between current emissions, e_i, and potential or unabated emissions, $\bar{e}$ abatement can be defined as $q_i = \bar{e} - e_i$.[21] As we are going to explain how to convert Barrett's (1994) abatement model into an emission model, in this appendix we have decided to use the same notation as the one used by Barrett to represent the abatement. Thus $q = \sum_i q_i = \sum_i (\bar{e} - e_i) = N\bar{e} - E$, where $E = \sum_i e_i$. Now substituting q_i and Q in expressions (1) and (2) in Barrett's paper we obtain

$$B_i(E) = \frac{b}{N}\left(a(N\bar{e} - E) - \frac{(N\bar{e} - E)^2}{2}\right), \quad C_i(e_i) = \frac{c}{2}(\bar{e} - e_i)^2$$

that yields

$$\pi_i = c\bar{e}e_i - \frac{c}{2}e_i^2 + \frac{b(N\bar{e} - a)}{N}E - \frac{b}{2N}E^2 + \pi_i(\bar{e}), \tag{15}$$

where $N\bar{e} - a \geq 0$ according to footnote 4 in Barrett's paper: '. . .Clearly, emissions must be at least as great as a units, since the marginal benefit of abating a units is zero. . .'. However, $N\bar{e} - a > 0$ does not make sense in (15) since it implies that the aggregate emissions yield benefits in the interval $(0, 2(N\bar{e} - a))$. Thus it seems clear that potential emissions in each country must be equal to a/N. In this case (15) is

$$\pi_i = \frac{ca}{N}e_i - \frac{c}{2}e_i^2 - \frac{b}{2N}E^2 + \pi_i\left(\frac{a}{N}\right), \tag{16}$$

that corresponds to the standard quadratic model of emissions where $\pi_i(a/N)$ is a constant. (16) is equivalent to the net benefit function in Barrett's paper in the sense that the same results can be obtained using this function instead of the net benefit function used by Barrett.

[21] The potential emissions are the same for all countries since we are assuming that all the countries are identical, i.e., $\bar{e}$ is a parameter of the model.

For instance, if the full cooperative solution is calculated the result is $e_c = \gamma a/(\gamma + N)$ where $\gamma = c/b$, which yields the following level of abatement

$$q_c = \frac{a}{N} - e_c = \frac{a}{\gamma + N},$$

see (3) in Barrett's paper. The same result is also obtained for the full non-cooperative solution. Now the relationship between our model and Barrett's model can be easily clarified. We rewrite (16) as

$$\pi_i = \frac{b_B}{N}\left[\frac{c_B a_B}{b_B} e_i - \frac{N c_B}{2 b_B} e_i^2 - \frac{1}{2}E^2\right] + \pi_i\left(\frac{a}{N}\right), \tag{17}$$

where subscript B denotes the parameters that appear in Barrett's paper. Then we can define a and b in our paper as

$$a = \frac{c_B a_B}{b_B} = \gamma_B a_B, \quad b = \frac{c_B N}{b_B} = \gamma_B N, \tag{18}$$

and rewrite (17) as

$$\pi_i = \frac{b_B}{N}\left[a e_i - \frac{b}{2} e_i^2 - \frac{1}{2}E^2\right] + \pi_i\left(\frac{a}{N}\right). \tag{19}$$

This is not exactly our net benefit function (1) but our results also apply to this model (Barrett's model).[22] The reason is pretty obvious. Since $\pi_i(a/N)$ is a constant and b_B/N is multiplying, the first order conditions obtained in the paper are the same first order conditions as the ones that can be derived using (26). Moreover, as the stability conditions are defined as a difference, $\pi_i(a/N)$ disappears from the difference and b_B/N does not affect the sign so that we are going to find the same results for the stability analysis. This is a detailed explanation of the correspondence between the model developed in our paper and the one studied by Barrett (1994).

Appendix 2

Proof of Proposition 1

First, we show that there does not exist any value of b such that $g(b, n), h(b, n) \leq 0$ for a given value of n. For a given n $h(b, n)$ is a strictly convex function with respect to b with a minimum for a negative value of b and an intersection point with the vertical axis that is also negative. This implies that $h(b, n) = 0$ only has a positive real solution given by

$$b_1(n) = \frac{1}{2}\left\{-(N + n^2 - 2n) + \left(n^4 - 4n^3 + 2Nn^2 + N^2\right)^{1/2}\right\}, \tag{20}$$

so that $h(b, n)$ will be negative or zero if $b \in (0, b_1(n)]$ and strictly positive if $b > b_1(n)$.

[22] Notice that e_i stands for emissions in this appendix and q_i for emissions in the main text. In order to obtain an exact correspondence between the two models, including the level of net benefits, it should be rewritten the emission model in terms of the abatement using the change of variable proposed in this Appendix.

On the other hand, function $g(b,n)$ is strictly convex with respect to b and presents a minimum for $b=(N-n)(n-2)/2$. For this minimum the value of the function is $n(N-n)^2(4-n)/4$ which implies that $g(b,n)>0$ for $n=1$, 2, 3 and $b>0$. Then we can conclude that there does not exist any value of b such that $g(b,n)$, $h(b,n)\le 0$ for $n=1$, 2, 3 since $g(b,n)$ is always positive. For $n=4$, $g(b,4)=0$ for $b=N-4$ and positive for $b\neq N-4$ but as $b=N-4$ is larger than $b_1(4)$ if $N>4$, $g(b,n)$, $h(b,n)$ cannot be negative or zero at the same time for $n=4$.[23] For $n>4$, $g(b,n)=0$ has two positive real solutions

$$b_2(n)=\frac{1}{2}\{(N-n)[n-2-(n^2-4n)^{1/2}]\}, \tag{21}$$

$$b_3(n)=\frac{1}{2}\{(N-n)[n-2+(n^2-4n)^{1/2}]\} \tag{22}$$

such that $b_2(n)<b_3(n)$ and $g(b,n)$ will be negative or zero if $b\in[b_2(n),b_3(n)]$.

Next, we show that $b_1(n)$ is lower than $b_2(n)$. Let us suppose now that $b_1(n)\ge b_2(n)$ which yields

$$\begin{aligned} &-(N+n^2-2n)+(n^4-4n^3+2Nn^2+N^2)^{1/2}\\ &\ge (N-n)(n-2)-(N-n)(n^2-4n)^{1/2}>0, \end{aligned} \tag{23}$$

simplifying terms we obtain

$$(n^4-4n^3+2Nn^2+N^2)^{1/2}\ge N(n-1)-(N-n)(n^2-4n)^{1/2}>0.$$

Then squaring and simplifying terms again we get

$$n(n^2-(3+N)n+3N)\ge -(n-1)(N-n)(n^2-4n)^{1/2},$$

where the left-hand side of the inequality is negative for $n\in(4,N)$, then multiplying by -1 we obtain

$$0<-n(n^2-(3+N)n+3N)\le (n-1)(N-n)(n^2-4n)^{1/2}.$$

Finally, squaring again and simplifying terms we get a contradiction

$$4n(n^2-2Nn+N^2)\le 0,$$

since $n^2-2Nn+N^2$ is positive for $n<N$. Consequently, we can establish that $b_1(n)<b_2(n)$ for all $n>4$ and then $g(b,n)$ and $h(b,n)$ cannot be negative or zero at the same time. This means that for each combination (b,n) only one of the three solutions characterized in the paper applies.

Now according to the sign of $h(b,n)$ and $g(b,n)$ it can be established for each combination (b,n) which kind of solution the problem has. For $n=1$, 2, 3 we have two kinds of solutions. For $b\le b_1(n)$ we have that $h(b,n)\le 0$ and the equilibrium yields $q_f=0$ and $q_s>0$ whereas for

[23] We assume that N is large enough to satisfy $N>4$.

$b > b_1(n)$ $h(b,n) > 0$ and the equilibrium is the interior solution since $g(b,n)$ is also strictly positive. For $n = 4$ we find the three kinds of solutions but the signatory corner solution only for $b = N - 4$ since only for $(b = N - 4, n = 4)$ $g(b,n)$ is zero. Finally, as we can order the critical values of b: $b_1(n) < b_2(n) < b_3(n)$, if $b > b_3(n)$, $h(b,n)$ and $g(b,n)$ are strictly positive and the equilibrium yields $q_s > 0$, $q_f > 0$, if $b \in [b_2(n), b_3(n)]$ $g(b,n)$ is negative or zero and the equilibrium yields $q_s = 0$, $q_f > 0$, if $b \in (b_1(n), b_2(n))$ $g(b,n)$ and $h(b,n)$ are strictly positive again and then the equilibrium yields $q_s > 0$, $q_f > 0$, and finally if $b \leq b_1(n)$ $g(b,n)$ is strictly positive and $h(b,n)$ is negative or zero and the equilibrium then yields $q_s > 0$, $q_f = 0$. The graphical representation of this result appears in Fig. 3.

Appendix 3

Proof of Lemma 1

The results in Lemma 1 are shown from the properties of $b_1(n)$. Thus, what we show first is that $b_1(0) = b_1(N) = 0$ and that $b_1(n)$ presents a unique extreme in the interval $(1, N-1)$ which is a maximum.

By substitution it is easy to check that $b_1(0) = b_1(N) = 0$. To show that the unique extreme of $b_1(n)$ is a maximum we use the inverse function of $b_1(n)$. In order to obtain this function we rewrite $h(b,n)$ as

$$h(b,n) = (b+1)n^2 - (2b+N)n + (b+N)b,$$

and then from $h(b,n) = 0$ we get:

$$\mathrm{n}^+(\mathrm{b}) = \frac{2b + N + (N^2 - 4b^3 - 4b^2)^{1/2}}{2(b+1)}, \quad \mathrm{n}^-(\mathrm{b}) = \frac{2b + N - (N^2 - 4b^3 - 4b^2)^{1/2}}{2(b+1)}, \quad (24)$$

which yield real values for $N^2 - 4b^3 - 4b^2 \geq 0$. Thus these two functions yield real values in the interval $[0, \tilde{b}]$ where $\tilde{b}$ is the unique positive solution of equation $N^2 - 4b^3 - 4b^2 = 0$. Moreover, the sign of their first derivatives is:[24]

$$\frac{dn^+}{db} = -\frac{2b^3 + 6b^2 + 4b + N^2 + (N-2)(N^2 - 4b^3 - 4b^2)^{1/2}}{2(b+1)^2(N^2 - 4b^3 - 4b^2)^{1/2}} < 0,$$

$$\frac{dn^-}{db} = \frac{2b^3 + 6b^2 + 4b + N^2 - (N-2)(N^2 - 4b^3 - 4b^2)^{1/2}}{2(b+1)^2(N^2 - 4b^3 - 4b^2)^{1/2}} > 0,$$

and

$$n^+(\tilde{b}) = n^-(\tilde{b}) = \frac{2\tilde{b} + N}{2(\tilde{b}+1)} = \tilde{n},$$

so that it can be concluded that $n^-(b)$ is the inverse function of $b_1(n)$ in interval $[0, \tilde{n}]$ and that it is increasing in that interval and also that $n^+(b)$ is the inverse function of $b_1(n)$ in interval $[\tilde{n}, N]$ and that it is decreasing in that interval. Thus function $b_1(n)$ is

[24] It is easy to show that the numerator of dn^-/db is positive.

increasing in interval $[0, \tilde{n})$ and decreasing in interval $(\tilde{n}, N]$ and therefore it presents a maximum at $n = \tilde{n}$.

Notice that

$$\lim_{b \to \tilde{b}^-} \frac{dn^+}{db_1} = -\infty; \qquad \lim_{b \to \tilde{b}^-} \frac{dn^-}{db_1} = +\infty,$$

which implies that $db_1/dn = 0$ when $n = \tilde{n}$. Finally, it is straightforward to show that $\tilde{n} \in [1, N-1]$. Thus given the properties of $b_1(n)$ we can determine for any b the kind of solutions depending on the value of n. We explain this below.[25]

According to footnote 16 in the main text we define $\hat{n}$ as the integer that maximizes $b_1(n)$ in the domain $\{1, 2, \ldots, N-1\}$. Thus if $b > b_1(\hat{n})$, $b > b_1(n)$ for all n and according to Proposition 1 the equilibrium is the interior solution for all n. If $b < b_1(\hat{n})$ equation $b = b_1(n)$ has two positive real roots that we call n_1, n_2 such that $n_1 < n_2$. Thus as $b_1(n)$ is first increasing and afterwards decreasing $b > b_1(n)$ when $n < n_1$ and $n > n_2$ and accordingly the equilibrium is the interior solution in those two cases. However, when $n \in [n_1, n_2]$, $b \leq b_1(n)$ and Proposition 1 establishes that the equilibrium is the corner solution. Finally, for $b = b_1(\hat{n})$ we find three possibilities. If $\hat{n} = \tilde{n}$ equation $b_1(\hat{n}) = b_1(n)$ yields $n_1 = n_2 = \hat{n}$ and the equilibrium is the interior solution for all n except for $\hat{n}$. If $\hat{n} \neq \tilde{n}$, because ñ is not an integer and the function is not symmetric, equation $b_1(\hat{n}) = b_1(n)$ yields $n_1 < n_2$ where one of these two values is by definition $\hat{n}$, the closest integer to $\tilde{n}$, see footnote 16. Thus given the behaviour of the function around $\tilde{n}$ the distance between n_1 and n_2 must be less than the unity and the equilibrium is again the interior solution for all n except for $\hat{n}$. If the function is symmetric it could present two maximum $\hat{n}_1 = n_1$ and $\hat{n}_2 = n_2$ provided that ñ is in the centre of interval $(\hat{n}_1, \hat{n}_2)$, and the equilibrium would be the interior solution for all n except for $\hat{n}_1$ and $\hat{n}_2$.[26]

Appendix 4

Proof of Corollary 1

This result is immediate from Lemma 1 provided that $b_1(n-1)$ is the minimum value of $b_1(n)$ for $n = 1, 2, \ldots, N-1$ since in this case $n_1 < 1$ and $n_2 \geq N-1$. As $b_1(n)$ is first increasing for $n < \tilde{n}$ and afterwards decreasing, $b_1(N-1)$ is the minimum value if $b_1(N-1)$ is lower than $b_1(1)$.

First we calculate these two values:

$$b_1(1) = \frac{1}{2}\left\{-(N-1) + (N^2 + 2N - 3)^{1/2}\right\},$$
$$b_1(N-1) = \frac{1}{2}\left\{-(N^2 - 3N + 3) + (N^4 - 6N^3 + 15N^2 - 14N + 5)^{1/2}\right\}.$$

Let us suppose that $b_1(1) \leq b_1(N-1)$ which yields

$$-(N-1) + (N^2 + 2N - 3)^{1/2} \leq -(N^2 - 3N + 3) + (N^4 - 6N^3 + 15N^2 - 14N + 5)^{1/2},$$

[25] Figure 1 can help the reader to follow our argument.

[26] Notice that symmetry is a necessary condition to have two maximums but not a sufficient condition.

simplifying terms we obtain

$$0 < N^2 - 4N + 4 + (N^2 + 2N - 3)^{1/2} \leq (N^4 - 6N^3 + 15N^2 - 14N + 5)^{1/2}.$$

Then squaring and simplifying terms again we get

$$0 < 2(N^2 - 4N + 4)(N^2 + 2N - 3)^{1/2} \leq 2N^3 - 10N^2 + 16N - 8.$$

Finally, squaring again and simplifying terms we get a contradiction

$$0 \leq -16N^5 + 144N^4 - 512N^3 + 896N^2 - 768N + 256,$$

since the right-hand side of the inequality is negative for $N \geq 3$. Thus we can conclude that $b_1(1) > b_1(N-1)$ and from Lemma 1we obtain Corollary 1.

Appendix 5

Proof of Lemma 2

As for Lemma 1, the proof of Lemma 2 derives from the properties of the functions $b_2(n)$ and $b_3(n)$. By substitution in (21) we obtain that $b_2(4) = N - 4$ and $b_2(N) = 0$. Remember that signatories' emissions are always positive for $n \in \{1, 2, 3\}$ and $b > 0$. See Appendix 2.

On the other hand, if we take the first derivative of $b_2(n)$ we obtain

$$\frac{db_2}{dn} = \frac{1}{2}\left\{N - 2n + 2 + \frac{2n^2 - (6 + N)n + 2N}{(n^2 - 4n)^{1/2}}\right\},$$

that presents the following limits:

$$\lim_{n \to 4} \frac{db_2}{dn} = -\infty,$$

$$\lim_{n \to N} \frac{db_2}{dn} = \frac{1}{2}\left\{-(N - 2) + \frac{N(N - 4)}{(N^2 - 4N)^{1/2}}\right\} < 0 \quad \text{for } N > 4.$$

Moreover, its second derivative is:

$$\frac{d^2 b_2}{dn^2} = \frac{1}{2}\left(-2 + \frac{2n^3 - 12n^2 + 12n + 4N}{(n^2 - 4n)^{1/2}}\right).$$

Let us suppose that this second derivative is negative or zero. This implies that

$$0 < 2n^3 - 12n^2 + 12n + 4N \leq 2(n^2 - 4n)^{1/2},$$

squaring and reordering terms we obtain the following inequality

$$4n^6 - 48n^5 + 192n^4 + (16N - 288)n^3 - (96N - 140)n^2 + (96N + 16)n + 16N^2 \leq 0.$$

It is easy but tedious to show that the left-hand side of this inequality is positive for $n \geq 4$ thus yielding a contradiction. So that we can conclude that $d^2b_2/dn^2 > 0$ which allows us to establish that $b_2(n)$ is a decreasing, strictly convex function in interval $[4, N]$.

Next, we study the properties of function $b_3(n)$. By substitution in (22) we obtain that $b_3(4) = N - 4$ and $b_3(N) = 0$. Moreover, it is easy to show that $b_3(n)$ is a strictly concave function with a maximum in interval $(4, N-1)$ equal to $N^2/2(N-2)$.[27] So that for $n \in (4, N^2/2(N-2))$ $b_2(n)$ increases and for $n \in (N^2/2(N-2), N)$ decreases. The maximum value for the function is $b_3(N^2/2(N-2)) = N(N-4)/4 > b_2(4) = N - 4$. Thus given the properties of $b_2(n)$ and $b_3(n)$, we can determine for any b the kind of solutions for signatories depending on the value of n. Below we explain this.[28]

According to footnote 18 in the main text we define n^* as the integer that maximizes $b_1(n)$ in the domain $\{4, \ldots, n-1\}$. Then if $b > b_3(n^*)$, $b > b_3(n)$ for all n and according to Proposition 1 the equilibrium is the interior solution for all n. If $b \in (N-4, b_3(n^*))$ equation $b = b_3(n)$ has two positive real roots that we call n_3, n_4 such that $n_3 < n_4$. Then as $b_3(n)$ is first increasing and afterwards decreasing $b > b_3(n)$ when $n < n_3$ and $n > n_4$ and accordingly the equilibrium is the interior solution in these two cases. However, when $n \in [n_3, n_4]$ $b \leq b_1(n)$ and Proposition 1 establishes that the equilibrium is the corner solution. If $b \in [b_2(N-1), N-4]$, $b = b_3(n)$ has a unique positive solution n_4 that along with n_3 obtained from $b = b_2(n)$ defines an interval $[n_3, n_4]$ for which $b \in [b_2(n), b_3(n)]$, then according to Proposition 1 the equilibrium is the corner solution when n belongs to that interval and the interior solution if that is not the case. If $b \in (0, b_2(N-1))$, $b < b_2(n)$ for all n and according to Proposition 1 the equilibrium is the interior solution for all n. Finally, for $b = b_3(n^*)$ we find three possibilities. If $n^* = \tilde{n}$ equation $b_3(n^*) = b_3(n)$ yields $n_3 = n_4 = n^*$ and the equilibrium is the interior solution for all n except for n^*. If $n^* \neq \tilde{n}$, because $\tilde{n}$ is not an integer and the function is not symmetric, equation $b_3(n^*) = b_3(n)$ yields $n_3 < n_4$ where one of these two values by definition is n^*, the closest integer to $\tilde{n}$, see footnote 18. Thus, given the behaviour of the function around $\tilde{n}$, the distance between n_3 and n_4 must be less than the unity and the equilibrium is again the interior solution for all n except for n^*. If the function is symmetric it could present two maximum $n_1^* = n_1$ and $n_2^* = n_2$ provided that $\tilde{n}$ is in the centre of interval (n_1^*, n_2^*), and the equilibrium would be the interior solution for all n except for n_1^* and n_2^*.

Appendix 6

Proof of Proposition 2

Points (i)–(iii) and (v) follow from Lemmas 1 and 2 and Corollary 1 provided $b_1(\hat{n}) < b_3(n^*)$. Point (iv) only occurs if $b_1(1) > b_2(N-1)$ and $b_1(N-2) > b_2(N-1)$ since function $b_1(n)$ is first increasing and afterwards decreasing and $b_1(N-1) < b_2(N-1)$ according to Proposition 1. First we show that $b_1(\hat{n}) < b_3(n^*)$ for $N > 5$. From the proofs of Proposition 1 and Lemma 2 we know that $b_1(4) < b_2(4) = b_3(4) = N - 4 < b_3(n^*)$.

On the other hand, $b_1(n)$ can be increasing or decreasing at $n = 4$ depending on the number of countries, N. Let us suppose that

$$\frac{db_1(4)}{dn} = -3 + \frac{16 + 4N}{(32N + N^2)^{1/2}} \geq 0.$$

[27] In order to have an interval with more than one point we assume that $N > 5$.

[28] Figure 2 can help the reader to follow our argument.

This implies that

$$16 + 4N \geq 3(32N + N^2)^{1/2},$$

that squaring and simplifying yields

$$256 - 160N + 7N^2 \geq 0.$$

It is easy to show that the left-hand side of the inequality is negative for $N \in (5, 21]$. So that we can conclude that $db_1(4)/dn < 0$, which means that $\hat{n} \leq 4$. The previous result also allows us to establish that for $N > 21$, the first derivative of $b_1(n)$ at $n = 4$ is positive, which means that $\hat{n} \geq 4$.

Next, we suppose that $N \in (5, 21]$. In this case, $\hat{n}$ is an integer in the domain $\{1, 2, 3, 4\}$. If $\hat{n} = 4$, we already know that $b_1(4) < N - 4 < b_3(n^*)$ and it is established that $b_1(\hat{n}) < b_3(n^*)$. If $\hat{n}$ were an integer different from 4, it is easy to show that $b_1(n) < N - 4$ for $n = \{1, 2, 3\}$ so that we can also conclude that $b_1(\hat{n}) < b_3(n^*)$ for this values of n.

Next, we suppose that $N > 21$. In this case, $\hat{n} \geq 4$. For $\hat{n} = 4$ the same argument as the one we have just used is applied. For $\hat{n} > 4$, we know from Proposition 1 that $b_1(\hat{n}) < b_2(\hat{n})$ and from the proof of Lemma 2 that $b_2(\hat{n}) < N - 4 < b_3(n^*)$ so that we find that $b_1(\hat{n}) < b_3(n^*)$ as we wanted to establish.

Finally, we show that $b_1(1) > b_2(N-1)b$ and $b_1(N-2) > b_2(N-1)$. Let us suppose first that $b_1(1) \leq b_2(N-1)$. This implies that

$$0 < \left((N-1)^2 + 4(N-1)\right)^{1/2} \leq 2N - 4 - \left((N-1)^2 - 4(N-1)\right)^{1/2}.$$

Squaring yields

$$2(2N-4)\left((N-1)^2 - 4(N-1)\right)^{1/2} \leq (2N-4)^2 - 8(N-1),$$

squaring again, simplifying and reordering terms we obtain the following inequality

$$32N^3 - 240N^2 + 448N - 256 \leq 0.$$

This inequality yields a contradiction for $N > 5$ so that we can conclude that $b_2(N-1) < b_1(1)$.

Next, we suppose that $b_1(N-2) \leq b_2(N-1)$. This implies that

$$0 < (N^4 - 10N^3 + 41N^2 - 72N + 48)^{1/2} \leq -4N + N^2 + 5 - (N^2 - 6N + 5)^{1/2}.$$

Squaring yields

$$2(-4N + N^2 + 5)(N^2 - 6N + 5)^{1/2} \leq 2N^3 - 14N^2 + 26N - 18.$$

Squaring again, simplifying and reordering terms we obtain the following inequality

$$16N^4 - 144N^3 + 400N^2 - 464N + 176 \leq 0$$

This inequality yields a contradiction fro $N > 5$, so that we can conclude that $b_2(N-1) < b_1(N-2)$.

Appendix 7

Proof of Proposition 4

For $b \in [b_2(N-2), N-4]$, $n_4 \in (N-1, N)$ so that we have a signatory corner solution for $n \in [n_3, n_4]$. This is a consequence of the fact that $b_3(N-1) > b_2(4) = N-4$.[29] Now, given a value of b we select n such that n and $n-1 \in [n_3, n_4]$ and we check if the internal stability condition can be satisfied. In this case both n and $n-1$ are signatory corner solutions, so that according to (13) from the main text the internal stability condition is given by

$$\begin{aligned}\Delta(n) &= -\frac{a^2(N-n)^2}{2(b+N-n)^2} - \frac{a^2[b-(N-n+1)(N-n-1)]}{2(b+N-n+1)^2} \\ &= -\frac{a^2[b^3 + L(n)b^2 + M(n)b + P(n)]}{2(b+N-n)^2(b+N-n+1)^2},\end{aligned}$$

where

$$L(n) = 2(N-n) + 1 > 0, \quad M(n) = 3(N-n)^2 + 2(N-n) > 0$$

$$P(n) = 2(N-n)^3 + 2(N-n)^2 > 0$$

So, $\Delta(n)$ is negative and, in fact, it is negative for all $b > 0$. Consequently, if there exists a self-enforcing IEA the number of countries in the agreement must be equal to or less than n_3. In fact, as it is not guaranteed that n_3 is an integer, the upper bound for the number of countries in a self-enforcing IEA must be defined as the smallest integer no less than n_3. Finally, from Lemma 2 we know that $b_2(n)$ is a strictly convex, decreasing function defined in interval [4, N], then as n_3 is defined as the unique, positive real root of equation $b = b_2(n)$, we can conclude that n_3 decreases when b increases and that consequently, the smallest integer no less than n_3 also decreases.

Appendix 8

Proof of Proposition 5

According to Proposition 4, if $b = N-4$ the maximum number of countries in a self-enforcing IEA is four, then the question to answer now is whether an agreement of four countries can be self-enforcing for $b > N-4$. For $b > N-4$ and $n = 4$, according to Proposition 1, the equilibrium is the interior solution for signatories and non-signatories. Thus using (11) and (12) from the main text we get

$$\Delta(4) = -\frac{a^2N^2k(b)}{2\left[(b+N-3)^2 + 9b\right]^2\left[(b+N-4)^2 + 16b\right]},$$

[29] This is very easy to show, so we omit it.

where

$$k(b) = 3b^4 + (6N + 48)b^3 + (2N^2 - 40N + 101)b^2 \\ - (2N^3 - 7N^2 - 10N + 39)b - (N^4 - 14N^3 + 73N^2 - 168N + 144).$$

It can be easily shown that $k''(b)$ is a strictly convex function with a minimum at $b = (8 + N)/2$. For this minimum, the second derivative is positive for all N. This implies that $k'(b)$ is increasing for all $b > 0$ and consequently is also increasing for $b > N - 4$. Then as $k'(N-4) = 252N^3 - 1212N^2 - 3408N + 16716$ is positive for $N > 5$ we can conclude that the first derivative is positive for $b > N - 4$, which implies that $k(b)$ is increasing for $b > N - 4$. Finally, as $f(N-4) = 8N^4 - 99N^3 + 362N^2 - 207N - 676$ is positive for $N > 5$, $k(b)$ is positive for $b > N - 4$ so that $\Delta(4)$ is negative and the internal stability condition is not satisfied. This shows that $n = 4$ cannot be the Nash equilibrium of the membership game.

[8]

Journal of Environmental Economics and Management 65 (2013) 326–344

Contents lists available at SciVerse ScienceDirect

Journal of Environmental Economics and Management

journal homepage: www.elsevier.com/locate/jeem

Participation games and international environmental agreements: A non-parametric model

Larry Karp [a,b], Leo Simon [a,c,*]

[a] Department of Agricultural and Resource Economics, University of California, Berkeley, CA 94720-3310, United States
[b] The Ragnar Frisch Center for Economic Research, Norway
[c] The Giannini Foundation of Agricultural Economics, United States

ARTICLE INFO

Article history:
Received 24 March 2011
Available online 15 September 2012

Keywords:
Stable coalitions
Participation game
International environmental agreement
Climate agreement
Trans-boundary pollution
Investment spillovers

ABSTRACT

We examine the size of stable coalitions in a participation game that has been used to model international environmental agreements, cartel formation, R&D spillovers, and monetary policy. The literature to date has relied on parametric examples; based on these examples, a consensus has emerged that in this kind of game, the equilibrium coalition size is small, except possibly when the potential benefits of cooperation are also small. In this paper, we develop a non-parametric approach to the problem, and demonstrate that the conventional wisdom is not robust. In a general setting, we identify conditions under which the equilibrium coalition size can be large even when potential gains are large. Contrary to previously examined leading special cases, we show that reductions in marginal abatement costs in an international environmental game can increase equilibrium membership, and we provide a measure of the smallest reduction in costs needed to support a coalition of arbitrary size.

1. Introduction

There is a rapidly expanding literature on the prospects for international environmental agreements (IEAs) to combat trans-boundary pollution problems such as the depletion of the ozone layer, the proliferation of greenhouse gases and, more generally, climate change. Much of this literature is based on a particular participation game. Due to the difficulty of analyzing this game, virtually all research on the topic specifies particular functional forms for the costs and benefits of abatement. Based on this research, the unchallenged consensus opinion among researchers in this field is that IEAs are ineffective precisely when the potential gains from cooperation are large [5,26,27,33]. Moreover, in leading special cases, an innovation – for example, a reduction in abatement costs – that increases the potential gains from cooperation will leave unchanged or reduce equilibrium participation. In these cases, a cost reduction that one would expect to be welfare enhancing has no effect on, or even reduces the fraction of the welfare gains from cooperation that could potentially be realized through the formation of an IEA. The existing theory on IEAs is thus pessimistic: sovereign nations cannot easily be induced to provide public goods; voluntary, self-enforcing agreements are not effective in providing public goods, and plausible remedies, such as investments that reduce abatement costs, will not help and might even be counterproductive. This pessimism spills into policy advice: it has been suggested that efforts devoted to promoting a climate agreement

* Corresponding author at: Department of Agricultural and Resource Economics, University of California, Berkeley, CA 94720-3310, United States. Fax: +1 510 643 8911.
E-mail addresses: karp@berkeley.edu (L. Karp), leosimon@berkeley.edu (L. Simon).

0095-0696/$ - see front matter © 2012 Elsevier Inc. All rights reserved.
http://dx.doi.org/10.1016/j.jeem.2012.09.002

requiring signatories to reduce greenhouse gas emissions are bound to be unsuccessful. A corollary, discussed by Barrett [5, Chap. 15] and Stiglitz [43], is that an IEA cannot be successful in the absence of some kind of external punishment.

We show that the basis for this pessimism is fragile, in that it relies on specific functional forms. All of the forms considered in the IEA literature share a common property: the marginal abatement cost function is convex at positive levels of abatement. This property is not required by classical micro-theory, which assumes only that the cost function itself is convex. Indeed, as we illustrate below in Section 3.2, when the marginal cost encompasses more than one abatement technology, it is typically not convex. When this excessively stringent condition is relaxed, the pessimistic results in the literature can easily be overturned. In this paper, we avoid the straightjacket of specific functional forms, focusing instead on a non-parametric expression for the function – we call it the *joiner's gain function* – that determines the equilibrium level of participation. When the marginal benefits from abatement are linear, there is a simple decomposition of this function that leads to a quite general theory of IEAs, and provides intuition that helps in analyzing the case where marginal benefits of abatement are concave. We show that if marginal abatement costs are weakly convex and benefits are linear, then the equilibrium size of the IEA never exceeds three countries. To this extent, the basis for the literature's pessimism is more general than previously thought. However, when abatement costs are convex but *marginal* abatement costs are strictly concave, the equilibrium coalition has at least three members; in some cases most or all potential members join. Thus, for some classically admissible cost functions – although not the ones previously analyzed in the literature – the pessimistic conclusion is overturned.

When marginal benefits are linear, our decomposition of the joiner's gain function is a powerful analytic tool. We use it to determine, for any integer up to the total number of potential IEA signatories, the type and minimal size of the cost reduction needed to support an equilibrium with at least this many members, in circumstances where under the original cost function an IEA of this size would not be stable. It is thus possible to capture up to 100% of the potential gains from cooperation that are available under the appropriately specified, reduced cost function. We then show that arbitrarily large coalitions can also be sustained when the benefit of abatement is concave.

In most of this paper – except for Section 3.2 – we restrict attention to cost reductions that globally weakly decrease marginal abatement costs. Such reductions necessarily increase the equilibrium level of abatement by an IEA with a given number of members. This restriction limits the number of special cases that we need to address, allowing us to focus on the most familiar scenario. By contrast, some recent papers have considered technological innovations that reduce total abatement costs, while increasing marginal abatement costs over some range of abatement levels. (See for example [1,2,7,22,23].) Under these conditions, a cost reduction may reduce the equilibrium level of abatement for an IEA of a given size.

Our results are policy-relevant for two reasons. First, as noted above, our results show that the theoretical basis for pessimism about self-enforcing IEAs may have been exaggerated. In particular, our work challenges those who would advise policymakers that a successful IEA requires a punishment device. Second, using the techniques that we develop, it is relatively straightforward to determine the effect on participation incentives of a cost reducing technological innovation.

The literature on IEAs to which we contribute is based on a two-stage game developed in D'Aspremont et al. [14].[1] In the first stage agents have a binary decision, whether to join or stay out of the coalition; in the second stage, agents take an action (abatement, sales, etc.). Members of the coalition internalize the effect of their actions on other coalition members. The outcome at the participation stage is a non-cooperative Nash equilibrium, and the outcome at the second ("abatement") stage is either a Nash or Stackelberg equilibrium.

While our research was motivated by the difficulty of sustaining IEAs, this two-stage game has been used to study a variety of problems. Because the essential features of the model arise in several fields, the relevance of our results extends well beyond the setting of this paper. In an IO setting, Donsimoni et al. [20] provide conditions under which the size of the stable coalition is uniquely determined. De Bondt [15] and Katz [35] use the D'Aspremont et al. [14] definition of coalition stability to study collusion in R&D games with spillovers. Thoron [44] shows that stable coalitions and Nash equilibrium outcomes to the binary action participation game are equivalent. Kohler [36] uses the definition of stability to study cooperation in monetary policy. Escrihuela-Villar [24] and Bos and Harrington [9] use the same definition of stability in a setting where the post-participation outcome is a non-cooperative equilibrium to a repeated game, rather than the result of optimization by the coalition. Bloch and Dutta [8] explain the relation between the particular definition of coalition stability used in these papers, and other prominent definitions.

Early applications of this model to IEA formation include [3,11,30]. Barrett [4] and Finus and Maus [28] show that coalitions with modest ambitions may be more successful than those that try to fully internalize damages. Using functional forms that are standard for this literature, Barrett [6] presents a model in which cost-reducing investments can improve the outcome under a stable coalition if there are increasing returns to scale (IRTS) to the adoption decision. With IRTS, the participation game becomes a coordination game, and there may exist a non-cooperative equilibrium in which all nations participate. As Heal and Kunreuther [29] suggest, a coordination game can also be induced by technological spillovers or trade policy. They provide conditions under which there is a tipping point in the participation game: if a sufficient

[1] An alternative to the participation game that we use is based on games of coalition formation, such as [12,21,41,46]. These models invoke a more sophisticated interpretation of rationality, in which agents understand how their provisional decision to join or leave a coalition would affect other agents' participation decisions; nations are "farsighted". Diamantoudi and Sartzetakis [17], Eyckmans [25], de Zeeuw [16], and Osmani and Tol [40] apply this notion to IEA models. An entirely distinct line of research models climate policy using a cooperative game.

number of countries join, it is individually rational for all countries to join. Hoel and De Zeeuw [31] show that cost-reducing investments (without IRTS) can improve the performance of an IEA if investment can reduce abatement costs sufficiently that a dominant strategy for countries is to abate at the socially optimal level. These examples of coordination games or dominant strategies contrast sharply with the examples cited above in which cost reductions weaken incentives to participate. Taken together, the two sets of examples may suggest that only special kinds of cost reductions, e.g. those associated with IRTS or dominant strategies, promote participation. We show that unless marginal costs are required to be convex, this conjecture is false. Dixit and Olson [19] and Hong and Karp [32] study the game when agents use mixed strategies at the participation stage. Ulph [45], Kolstad [37], Kolstad and Ulph [38] and Karp [34] examine the effect on participation and welfare when agents anticipate learning about costs or damages after the participation stage, before they choose abatement. Kosfeld et al. [39], Burger and Kolstad [10] and Dannenberg et al. [13] find moderate support in laboratory settings for the prediction of low participation in the simplest game; they also examine extensions of that game.

Section 2 introduces some terminology relating to our joiner's gain function and its decomposition. Section 3 focuses on three parametric examples designed to introduce the main themes of the paper. In Section 4, we consider arbitrary convex abatement costs, but require that marginal benefits from abatement are linear. In Section 5, we assume that the abatement benefit function is strictly concave. Section 6 concludes. The Appendix contains the proofs of all Remarks and Propositions.

2. Preliminaries

We use "s" (for signatories) subscripts to denote signatories to the agreement and "f" (for freeloaders) subscripts to denote non-signatories. Superscripts indicate the number of signatories in the agreement. Signatories delegate their abatement decisions to the coalition, which chooses abatement to maximize coalition welfare. Each non-signatory ignores the public good nature of abatement and maximizes its individual welfare. Each of the N countries is identical, so the model predicts equilibrium participation and abatement, but not the identity of participants. There are two variants of the second stage, a Cournot variant in which signatories and non-signatories choose abatement levels simultaneously, and a Stackelberg variant in which the signatories choose first. This paper considers the Cournot equilibrium. An online addendum studies the Stackelberg case.

Abatement is a public good. Each country derives a benefit $B(Q)$ from the global level of abatement, Q, and incurs a cost based on its individual abatement level, q. The individual abatement cost function is convex, denoted by $C(q)$. Because $B'(0)=1$, non-signatories choose an abatement level of zero whenever $MC(0)>1$; we examine simple examples of this case in Sections 3.1 and 3.2. In the remainder of this paper, we assume that $MC(0)\leq 1$.

Although the number of signatories, n, has a natural interpretation only when n is an integer, all of the variables in the paper are mathematically well-defined for any non-negative real value of n. Accordingly, we define all functions below on the non-negative reals. For $r\in\mathbb{R}_+$, we denote by $\lceil r\rceil$ the smallest integer weakly greater than r and by $\lfloor r\rfloor$ the greatest integer weakly less than r.

Suppose that in the first stage of the participation game, $r\in\mathbb{R}_+$ countries choose to participate in an IEA. Let q_s^r and q_f^r denote the abatement levels of signatories and non-signatories, respectively, and let $Q^r=rq_s^r+(N-r)q_f^r$ denote aggregate abatement. The objective functions for signatories and non-signatories are, respectively,

$$U^s=rB(Q^r)-C(q_s^r) \tag{1a}$$

$$U^f=B(Q^r)-C(q_f^r) \tag{1b}$$

In an interior equilibrium of the Cournot variant of the abatement stage, the abatement levels q_s^r and q_f^r for signatories and non-signatories, respectively, are simultaneously determined as the solutions to the first order conditions (2a) and (2b) below

$$0=r\frac{\partial B(Q^r)}{\partial q_s^r}-C'(q_s^r) \tag{2a}$$

$$0=\frac{\partial B(Q^r)}{\partial q_f^r}-C'(q_f^r) \tag{2b}$$

For $r\in\mathbb{R}_+$, we let $g(r)$ denote the "joiner's gain" function, representing the increment to a non-signatory's payoff if it becomes the rth member of an IEA

$$g(r)=\begin{cases}[B(Q^r)-C(q_s^r)]-[B(Q^{r-1})-C(q_f^{r-1})] & \text{if } r\leq N\\ -1 & \text{if } r>N\end{cases} \tag{3}$$

For $r\leq N$, the first term in square brackets is the net (private) benefit obtained by a signatory to an r-member IEA; the second term is the net benefit to not joining: if it did not join, the non-signatory would benefit from aggregate abatement Q^{r-1} and incur a cost $C(q_f^{r-1})$. Therefore, the difference between the two terms in square brackets is indeed the net gain to a potential signatory of becoming the rth member of an IEA. We assume that a country indifferent between joining an IEA or not chooses to join. Breaking ties in this way, an IEA with an integer $n \geq 2$ members[2] is a *stable equilibrium* iff

$$g(n) \geq 0 > g(n+1). \tag{4}$$

The first inequality is known as the "internal stability condition"—a member of an IEA with n members has no incentive to leave—while the second is called the "external stability condition"—a non-member strictly prefers not to join an IEA that has n members. Note that the external stability condition is vacuously satisfied if $n=N$ because in this case there are no non-members; for this reason we have set $g(n)=-1$ for $n>N$.

To solve the model, the natural approach is to compute the real-valued roots of the equation $g(\cdot)=0$, and then check each root to see if the integers on either side of it satisfy (4). For a *non-integer* value[3] of $r \in \mathbb{R}_+$ s.t. $g(r)=0$, we say that r is a *stable root of* g if

$$g(\lfloor r \rfloor) \geq 0 > g(\lceil r \rceil) \tag{4'}$$

Having made this point, we shall for the remainder of the paper focus on integer-valued solutions to the model, and index solutions with n rather than r.

In Sections 2–4, we assume that benefits are linear in abatement, and choose units so that each country's benefit $B(Q)$ is equal to Q. Linearity simplifies the analysis considerably, because each non-signatory has a dominant abatement strategy: in particular, its abatement choice is independent of both the size of, and the collective abatement decision made by the IEA; hence under this assumption the Cournot and Stackelberg variants are equivalent. In Section 5, we generalize the model to allow for benefits that are concave in Q.

When there are n signatories and $B(Q)=Q$, each signatory in an interior equilibrium abates at the level $q_s^n > 0$, where q_s^n solves $n=MC(q)$, while each non-signatory abates at $q_f > 0$, where q_f solves $1=MC(q)$, i.e., q_f is independent of n. In this case, we can rewrite the joiner's gain function (3) as

$$\begin{aligned} g(n) &= nq_s^n + (N-n)q_f - C(q_s^n) - [(n-1)q_s^{n-1} + (N-n+1)q_f - C(q_f)] \\ &= n(q_s^n - q_s^{n-1}) + (q_s^{n-1} - q_f) - [C(q_s^n) - C(q_s^{n-1}) + C(q_s^{n-1}) - C(q_f)] \end{aligned} \tag{5}$$

For heuristic reasons, it is helpful to write

$$g(n) = CR(n) - UA(n) \quad \text{where in the linear case} \tag{6a}$$

$$UA(n) = C(q_s^{n-1}) - C(q_f) - (q_s^{n-1} - q_f) > 0 \tag{6b}$$

$$CR(n) = n(q_s^n - q_s^{n-1}) - [C(q_s^n) - C(q_s^{n-1})] > 0 \tag{6c}$$

UA (unilateral action) is the utility loss to a non-signatory if it increases its abatement from q_f to the level q_s^{n-1} produced by the signatories to an $(n-1)$-member IEA, while other countries maintain the same abatement levels; because the non-signatory's abatement level q_f is individually optimal, the cost increment of this unilateral step is greater than the benefit increment. CR (collective response) is the utility gain to the signatory when all signatories to the augmented coalition respond to the additional member by increasing abatement from q_s^{n-1} to q_s^n; because q_s^n is collectively optimal for the augmented coalition, this second increment in aggregate abatement is greater than the second increment in cost.

3. Some special cases

To motivate the research orientation of this paper, we present in this section three parametric special cases. These serve as a tutorial for readers unfamiliar with the topic, and illustrate some of the broad conclusions of the literature reviewed above, as well as showing that these conclusions are not robust.

Our first example, drawn from Barrett [5], has been widely discussed in the literature cited above. It illustrates the pessimistic conventional wisdom that technical innovations that reduce abatement costs tend to reduce, rather than increase equilibrium participation. The second example, based on one in Baker and Adu-Bonnah [1], generalizes the first but yields quite different conclusions. In particular, it illustrates how non-convexities in the marginal cost function can arise in a natural way through technological innovation, and how these non-convexities can induce stable IEAs with arbitrarily many members. The themes of our third special case, based on Barrett [3], mimic those of our second, but in a piecewise linear setting. In both the second and third cases, we illustrate the usefulness of our decomposition (6) of the joiner's gain function.

---

[2] For many specifications of the model, (4) can be satisfied for $n=1$. Because in this case there is no distinction between signatories and non-signatories, such solutions are clearly uninteresting; we ignore them throughout the paper.

[3] Because $\lfloor r \rfloor = \lceil r \rceil$ for any integer n, (4') will be violated whenever the root of g is an integer. A necessary and sufficient condition for *any* real-valued root r to be stable is that $g(\lfloor r \rfloor) \geq 0 > g(\lceil r+\varepsilon \rceil)$ for sufficiently small $\varepsilon > 0$.

3.1. Constant marginal costs with capacity constraints

In our first example, benefits are linear in aggregate abatement, while marginal costs are constant up to a capacity constraint, normalized to 1. A country's cost of abatement level q is

$$C(q) = \begin{cases} cq & \text{for } 0 \le q \le 1 \\ \infty & \text{for } q > 1 \end{cases} \tag{7}$$

We choose units so that each country's benefit is equal to the level of aggregate abatement, Q. As the marginal benefit of abatement is constant at 1, this model is interesting only when $c > 1$, so that no country, acting individually, has an incentive to abate.

In this model, signatories to an IEA of size n will abate to the capacity level of 1 iff $n \ge \lceil c \rceil$; if $n < \lceil c \rceil$, signatories abate zero. Non-signatories always abate zero. To obtain the joiner's gain function if this model, we substitute these properties into (5)

$$g(n) = \begin{cases} 0 & \text{if } n < \lceil c \rceil \\ n-c & \text{if } n = \lceil c \rceil \\ 1-c & \text{if } n > \lceil c \rceil \end{cases} \tag{8}$$

An IEA with $n > \lceil c \rceil$ is not internally stable, because if one member were to leave, the remaining IEA would continue to abate at capacity. Our tie-breaking assumption (see p. 7 and Eq. (4)) implies that IEAs with $n < \lceil c \rceil$ are not externally stable. The unique stable equilibrium, $n = \lceil c \rceil$, results in a positive level of abatement.

A striking property of this specification is that equilibrium coalition size weakly increases with abatement costs. Moreover, equilibrium global welfare equals[4] $(N-c)\lceil c \rceil$, which is a "saw-toothed" function of c: when c reaches any integer n, participation increases discontinuously while costs increase continuously, leading to a jump in equilibrium welfare; as c increases through the interval $(n,n+1)$, participation remains unchanged but costs increase, so that global welfare declines.

One might expect technological innovations that reduce the cost of abatement to increase global welfare, but the model with cost specification (7) delivers the opposite result: it implies that IEAs with large numbers of signatories are sustainable only if abatement is very costly, and that efforts to increase abatement efficiency may be counter-productive. We focus on cost-reducing technological enhancements that *can* increase both equilibrium participation and aggregate welfare.

3.2. Piecewise constant marginal costs

Our second example, based on Baker and Adu-Bonnah [1], generalizes our first and illustrates that a cost-reducing innovation might either reduce equilibrium membership (as in Section 3.1) or increase it. We assume that the only source of emissions is electricity generation, and that countries can produce electricity using any combination of coal, gas or solar technologies. By choice of units, a country that uses all coal produces one unit of emissions. A country that uses all solar produces no emissions; one that uses all gas produces $1-x$ units, with $x < 1$. The marginal cost of switching from coal to solar is $c > 1$ and the marginal cost of switching from coal to gas is $a > x$. These inequalities imply that it is not individually rational for a country to use either solar or gas. Gas will be an efficient means of reducing emissions iff $a/x < c$; if $a/x \ge c$ the model below is in effect no different from the one in Section 3.1, in which the unique equilibrium is $\lceil c \rceil$.

Now consider a technological innovation which reduces a/x to below c, moving gas onto the efficiency frontier. To avoid dealing with some distracting special cases, we assume that

$$a < 1 \quad \text{and} \quad \lceil c-a/1-x \rceil > \lceil a/x \rceil + 1 \tag{9}$$

Fig. 1 graphs the marginal cost of abatement both when $a/x \ge c$ (the heavy solid line) and when $a/x < c$ (the dashed step function). To verify its shape, note that $\alpha < x$ units of abatement can be achieved by converting a fraction α/x of the coal generators to gas, at a constant marginal cost of a/x; to abate $\alpha \ge x$, some fraction $y(\alpha)$ of the coal plants must be converted to solar; the minimum fraction required solves $y(\alpha)+x(1-y(\alpha)) = \alpha$, implying that[5] the marginal cost of increasing α beyond x is $c-a/1-x$.

[4] In equilibrium, $\lceil c \rceil$ countries each abate one unit at a cost of c, so that each of N countries benefit from $Q = \lceil c \rceil$ units of abatement. Hence, global welfare is $N\lceil c \rceil - \lceil c \rceil c = (N-c)\lceil c \rceil$.

[5] From the equation defining $y(\alpha)$, $dy(\alpha)/d\alpha = 1/1-x$. Letting $C(\alpha) = a(1-y(\alpha))+cy(\alpha)$ denote the total cost of abating α, we have $dC(\alpha)/d\alpha = (c-a)dy(\alpha)/\alpha = c-a/1-x$.

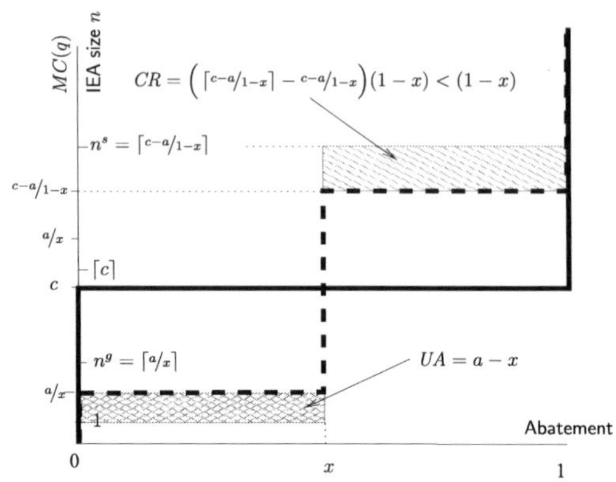

Fig. 1. An example with piecewise constant marginal cost.

An n-member IEA uses whichever generation technology is most profitable; it is straightforward to verify that this method is

$$\begin{cases} \text{coal} & \text{if } n < \lceil a/x \rceil \\ \text{gas} & \text{if } \lceil a/x \rceil \le n < \lceil c-a/1-x \rceil \\ \text{solar} & \text{if } \lceil c-a/1-x \rceil \le n \end{cases}$$

Taking as given the IEA's technology decision rule, we can now define $g : \mathbb{N} \to \mathbb{R}$, the joiner's gain function

$$g(n) = \begin{cases} 0 & \text{if } n < \lceil a/x \rceil \\ nx-a & \text{if } n = \lceil a/x \rceil \\ x-a & \text{if } n \in (\lceil a/x \rceil + 1, \lceil c-a/1-x \rceil - 1] \\ n-(n-1)x-c & \text{if } n = \lceil c-a/1-x \rceil \\ 1-c & \text{if } n > \lceil c-a/1-x \rceil \end{cases} \tag{10}$$

The only two candidates for stable equilibrium are $\lceil a/x \rceil$ and $\lceil c-a/1-x \rceil$, the two sizes at which switching to a new technology first becomes profitable. At all other values of n exceeding $\lceil a/x \rceil$, $g(\cdot) < 0$; values of $n < \lceil a/x \rceil$ are not externally stable because of our tie-breaking assumption. For $n \in (\lceil a/x \rceil + 1, \lceil c-a/1-x \rceil - 1]$, gas technology is profitable even with $n-1$ members; hence the addition of an additional member would increase abatement by only x units, while costing the joiner $a > x$ units. (Inequality (9) ensures that the interval $(\lceil a/x \rceil + 1, \lceil c-a/1-x \rceil - 1]$ is non-empty.) Moreover, $g(\lceil a/x \rceil)$ is unambiguously positive, so an IEA with $\lceil a/x \rceil$ members is always a stable equilibrium. The sign of $g(\lceil c-a/1-x \rceil)$ is ambiguous; $n^s := \lceil c-a/1-x \rceil$ satisfies (4) iff

$$g(n^s) = n^s - (n^s - 1)x - c \ge 0 \tag{11}$$

An important characteristic of the post-innovation marginal cost function is that it is concave over the interval $[x,1)$. Without this property, it would not be the case that for $n \ge 3$, $CR(n)$ could be arbitrarily large relative to $UA(n)$. The role that locally concavity plays in supporting large agreements is discussed heuristically in Section 3.3 and proved formally in Proposition 1.

We now decompose $g(n^s)$. Using the notation introduced on p. 6

$$q_f = 0,\ q_s^{n^s-1} = x \text{ and } q_s^{n^s} = 1, \quad \text{while } C(q_s^{n^s-1}) = a \text{ and } C(q_s^{n^s}) = c \tag{12a}$$

From (6), the internal stability condition (11) can be written as $CR(n^s) - UA(n^s)$, where

$$UA(n^s) = a - x \tag{12b}$$

$$CR(n^s) = \left\lceil \frac{c-a}{1-x} \right\rceil (1-x) - (c-a) = \left(\left\lceil \frac{c-a}{1-x} \right\rceil - \frac{c-a}{1-x} \right)(1-x) < (1-x) \tag{12c}$$

Eqs. (12b) and (12c) imply that $g(n^s) < 1-a$, so a necessary condition for $g(\lceil c-a/1-x \rceil) \ge 0$ (stability of n^s) is that $a < 1$. If this property holds – we assume it in (9) – there exists $x < 1 < a/x < c$ such that $g(\lceil c-a/1-x \rceil) \ge 0$ so that a solar IEA is stable. Whether this inequality is satisfied, however, depends in a fragile way on parameter values. To analyze this dependence, we fix $a \in (0,1)$ and define CX(a) to be the set of (c,x) values which, paired with a, support a solar IEA

$$\text{CX}(a) = \{(c,x) \in \mathbb{R}^2 : x < 1 < a/x < c \text{ and } g(\lceil c-a/1-x \rceil) \ge 0\} \tag{13}$$

Define AX(c) and AC(x) analogously to CX(a). Heuristically, if CX(a)$\subsetneq$CX(a'), for all $0<a'<a<1$, then the likelihood of a stable solar IEA is strictly monotone decreasing in the average cost of gas generation. On the other hand, if there exists $c\neq c'$ such that AX(c) and AX(c') are *not nested*[6] then the likelihood of a stable solar IEA is not monotonically related to c.

Remark 1.

(a) For all $x<a'<a<1$, CX(a)$\subsetneq$CX(a');
(b) For all $1<c'$, there exists $c\approx c'$ s.t. AX(c') and AX(c) are not nested;
(c) For all $0<x'<1$, there exists $x\approx x'$ s.t. AC(x') and AC(x) are not nested.

Part (a) is at the first sight paradoxical. Its implication is that when the gas technology is more expensive, it is *more* rather than less difficult to sustain a solar IEA. Upon reflection, however, the paradox is easily resolved: the difficulty of sustaining a solar IEA depends on the incremental benefit that a non-signatory obtains from participating in it, relative to free-riding on the gas IEA that would prevail if she did not participate. This incremental benefit is weakly *lower*, the *less* efficient is the gas technology. The reason is that the non-signatory does not care about the cost of abatement under the gas IEA, while the benefit she obtains from free-riding on the gas IEA weakly increases with its cost. To see this, recall first our observation on p. 10 that in the example presented in Section 3.1, global abatement increases weakly with abatement costs; the same is true for the gas IEA: the more expensive is gas, the more members are required in order to sustain a gas IEA, and hence the more abatement results from the agreement; true, the cost per IEA member is greater also, but as noted above, the non-signatory contemplating joining the solar IEA does not take this cost into account. Finally, the greater is the level of abatement under the gas IEA, the smaller is the *incremental* benefit of switching to a solar IEA. To summarize, the higher is the cost of abatement under the gas technology, the smaller is the incremental benefit to a non-signatory– the difference between total abatement under solar and total abatement under gas – of joining the solar IEA.

Why does not the same monotonicity argument above apply also to a decrease in x? Clearly, a decrease in x also weakly increases the number of members – $n^g:=\lceil a/x\rceil$ – needed in order to sustain a gas IEA. Unlike an increase in a, however, a decrease in x affects in a non-monotonic way the benefit to a free-rider of the gas IEA. Indeed, as x decreases holding other parameters constant, the average benefit to a non-signatory of a gas IEA declines so long as $\lceil a/x\rceil$ remains constant, but jumps up discontinuously once the integer threshold is crossed. To summarize, the impact of a small change in x on the incremental benefit to a non-signatory of joining a solar IEA may be positive or negative, depending on initial conditions.

The parameter c has a quite different impact than either a or x on the sustainability of a solar IEA. Just as the cost to signatories of abatement in the gas IEA increases monotonically in a, so the cost of a solar IEA increases monotonically in c. But a non-signatory considering joining such an agreement weighs the *net* benefit of solar against the *gross* benefit of gas. This tradeoff depends on c in a non-monotonic way: from (11), the difference is $\lceil c-a/1-x\rceil-c-x/1-x$, which cycles with c in a "saw-tooth" pattern between $x-a/1-x<0$ and $1-a/1-x>0$.

The above example illustrates a number of key aspects of this paper. First, it complements the point we made on p. 2: when an abatement cost function encompasses multiple abatement technologies, there are good reasons to expect that marginal abatement costs will not be convex. Second, it demonstrates how easy it is, once this widely adopted but not well-justified convexity assumption is relaxed, to construct models in which technological innovations both globally reduce abatement costs and also increase equilibrium participation. Third, it highlights in a particularly dramatic way a point reiterated in our conclusion: the stability of high-participation IEAs may depend sensitively on parameter values.

3.3. Linear marginal costs

We begin this subsection with a review of the model with linear marginal costs and benefits. Using a simple diagramatic argument, we confirm Barrett's [3] result that for this specification, the unique stable equilibrium is $n=3$, regardless of the slope of the marginal cost function. A change in the slope of the marginal cost function changes the equilibrium level of abatement, leaving unchanged both the equilibrium number of signatories and the fraction of potential global welfare gains realized in the game. We then show that by introducing a kink in the marginal cost function, flattening it north-east of the kink so that marginal costs are now concave, we can support any integer $3<n\leq N$ as a second stable equilibrium; $n=3$ remains as an equilibrium. Thus, we show in this second context that technological innovations that reduce costs *are* compatible with increased, indeed even 100%, participation. In Fig. 2, we again decompose the joiner's gain function $g(n)$ into two components, $CR(n)$ and $UA(n)$ (see (5)–(6c)). The cost function is $C(q)=q^2/2M$, for some $M\in\mathbb{R}$, so that $C'(q)=q/M$. The upward sloping curve represents both the marginal cost curve $MC(q)$ and its inverse, $S(p)$. Recall that $q_s^{n^s}$ solves $n=MC(q)$, while q_f solves $1=MC(q)$. From (6b), $UA(n)$ is the area under the marginal cost curve between q_f and q_s^{n-1}, minus the area of the rectangle with boundaries q_f and q_s^{n-1} and height 1. The difference is the area of the cross-hatched triangle labeled $UA(n)$. From (6c), $CR(n)$ is the area of the rectangle with boundaries q_s^{n-1} and $q_s^{n^s}$ and height n minus the area under the marginal cost curve between q_s^{n-1} and $q_s^{n^s}$. The difference is the area of the hatched triangle labeled $CR(n^s)$. This second triangle has unit height, regardless of n; the height of the first

[6] Sets A and B are said to be not nested if $A\not\subseteq B$ and $B\not\subseteq A$.

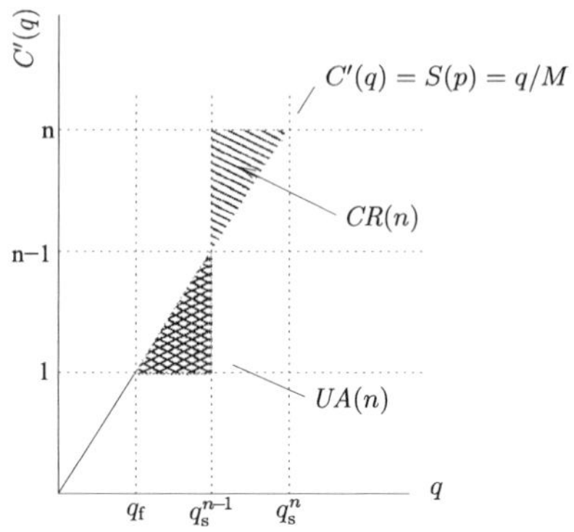

Fig. 2. Linear benefits and linear marginal costs.

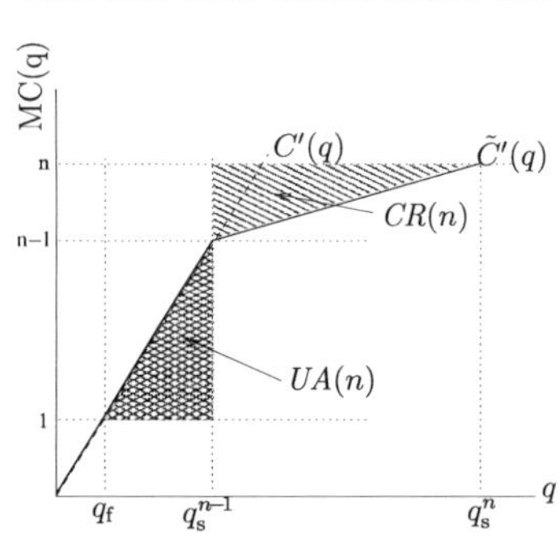

Fig. 3. Linear benefits and kinked marginal costs.

triangle is $n-2$. Because the two triangles are similar, the area $CR(n) \geq UA(n)$ iff $n \leq 3$. The equilibrium condition (4) is that $CR(n) \geq UA(n)$ while $CR(n+1) < UA(n+1)$. We have thus established.

Remark 2. When marginal benefit and marginal cost are both linear, the unique stable equilibrium size is $n=3$.

Moreover:

Remark 3. When three countries form an IEA, the equilibrium global gain is $6M(N-2)$; this is a fraction $12(N-2)/N(N^2-2N+1)$ of the total potential gain that could be achieved if all N countries joined the agreement. This fraction is independent of the cost parameter M and decreases with N for $N>3$.

Although an increase in M lowers total and marginal costs and increases global welfare, technology enhancements that reduce M have no effect on the fraction of the potential welfare gain that is achieved in equilibrium. In Section 4, Proposition 1 establishes that an equilibrium size exceeding $n=3$ cannot be sustained if the *marginal* cost function is convex, i.e., if $C'''(\cdot) \geq 0$. Membership can, however, exceed $n=3$ if this restriction is relaxed, allowing the marginal cost function to be locally concave beyond $q=q_s^{n-1}$. (We reiterate that conventional economic theory requires only that the *cost* function is convex; no restrictions are imposed on marginal costs except that they be non-decreasing.) Indeed, we have

Remark 4. Assume that initially, marginal benefit and marginal cost are both linear. For any $3<\overline{n} \leq N$, a stable equilibrium of size $\overline{n}$ can be implemented by a cost-reducing innovation resulting in a piece-wise linear marginal cost curve.

Fig. 3 illustrates Remark 4. In the figure, the modified cost function is $\tilde{C}(\cdot)$, whose derivative is piecewise linear and agrees with $C'(\cdot)$ for $q \leq q_s^{n-1} = (\overline{n}-1)M$, but is flatter than $C'(\cdot)$ for larger values of q. Specifically, let

$$\tilde{C}'(q) = \begin{cases} q/M & \text{if } q \leq (\overline{n}-1)M \\ c_0 + q/K & \text{if } q > (\overline{n}-1) \end{cases}$$

where $K > M$ and $c_0 = (\overline{n}-1)(1-M/K)$. ($c_0$ is chosen so that $\tilde{C}'(\cdot)$ is continuous at $q = q_s^{n-1} = (\overline{n}-1)M$.) The new marginal cost curve $\tilde{C}'(\cdot)$ is depicted in Fig. 3 by a heavy solid line; the original one by a dashed line. Note that the area of $CR(n)$ just exceeds the area of $UA(n)$: the shorter height of the first triangle is now more than offset by its longer base. The cost reduction from $C(\cdot)$ to $\tilde{C}(\cdot)$ leaves individual abatement unchanged at M if no IEA is implemented; in this case the reduction has no effect on global welfare. The cost reduction can however increase participation in an IEA. Indeed, if the reduction is sufficiently great – specifically, if $K \geq M(N-2)^2$ (see the proof of Remark 4) – an agreement to which all N countries are signatories is stable, and all of the potential gains from cooperation under this cost structure are realized.

4. Convex costs and linear benefits

In this section we generalize the model to incorporate arbitrary convex cost functions, while maintaining the restriction that benefits are linear in abatement. We begin by showing that if the *marginal* cost function is initially strictly convex in abatement, then an IEA with more than two members is unsustainable. Once this restriction is relaxed (maintaining convexity of the *cost* function), an IEA of any size can be sustained. Our approach is to fix $n \leq N$ and a given strictly convex cost function for which n is not a stable equilibrium, then lower this function to the point at which a stable equilibrium of size *at least n* exists. When benefits are linear in abatement, there are countless ways of attaining this goal: we identify the "smallest possible" reduction in the cost function that can accomplish it. This result is formalized in Proposition 2. Our motivation for this exercise is that cost-reducing technological progress is more difficult to achieve, the greater is the extent to which costs must be reduced. Our "smallest possible" reduction can be interpreted as the least resource-intensive way to implement a coalition with at least n members.

The cost reduction we construct on p. 7 below (see (28) and (29)) implements a coalition with exactly n members if and only if $g(n+1) \leq g(n) \leq 0$ under the initial cost function. If instead $g(n+1) \geq g(n)$, this reduced cost function will satisfy only the internal, but not the external stability condition at n (see (4)). In this second case, however, both internal and external stability will "eventually" be satisfied

if n is internally stable, there exists $n' \geq n$ such that n' is a stable equilibrium (14)

To confirm this claim we consider the two cases, first where n is externally stable and then when it is not. If n is externally stable, then by definition n is stable. If $\overline{n}$ is not externally stable, let $\underline{\mathcal{N}} = \{n' \in (n+1, N] \cap \mathbb{N} : \tilde{g}(n') < 0\}$, where $\tilde{g}(\cdot)$ is the joiner's gain function corresponding to the reduced cost function. If $\underline{\mathcal{N}}$ is non-empty, let $\underline{n}$ be its smallest element. In this case, because $\tilde{g}(\underline{n}-1) > 0 > \tilde{g}(\underline{n})$, $n' = \underline{n}-1$ is both internally and externally stable. On the other hand, if $\underline{\mathcal{N}}$ is empty, then $\tilde{g}(N) \geq 0$ and N will be both internally and externally stable, because the external stability condition is satisfied vacuously at N. A straightforward extension of the methodology we develop in this section shows that it is possible to implement a coalition with exactly n members even when $g(n+1) > g(n)$.

For the rest of this section, we fix a differentiable, strictly convex cost function C, and let $S(\cdot)$ denote the inverse of $MC(\cdot)$. When the benefit function is linear, the functions $UA(n)$ and $CR(n)$ defined above ((6b) and (6c)) can be rewritten in a particularly convenient form

$$UA(n) = C(q_s^{n-1}) - C(q_f) - (q_s^{n-1} - q_f) = (n-2)q_s^{n-1} - \int_1^{n-1} S(p)\,dp \tag{15a}$$

$$CR(n) = n(q_s^{\overline{n}} - q_s^{n-1}) - [C(q_s^{\overline{n}}) - C(q_s^{n-1})] = \int_{n-1}^{n} S(p)\,dp - q_s^{n-1} \tag{15b}$$

4.1. Strictly convex and strictly concave marginal costs

Fig. 2 provides geometric intuition for why the formulations in (6) and (15) are equivalent provided benefits are linear. In our discussion of the figure on p. 17, the cross-hatched area $UA(n)$ was identified as the area under the marginal cost curve between q_f and q_s^{n-1}, minus the area of the rectangle with vertical boundaries q_f and q_s^{n-1} and height 1. As Fig. 2 illustrates, this area is also equal to the area of the rectangle bounded vertically by 0 and q_s^{n-1}, and horizontally by 1 and $n-1$, minus the trapezoid to the left of $S(p)$ between 1 and $n-1$. The area of the difference between the trapezoid and the rectangle is given by the right-hand side of (15a). The hatched area $CR(n)$ was identified on p. 17 as equal to the area of the rectangle with vertical boundaries q_s^{n-1} and $q_s^{\overline{n}}$ and height n minus the area under the marginal cost curve between q_s^{n-1} and $q_s^{\overline{n}}$. It is also equal to the trapezoid to the left of the curve $S(p)$ between $n-1$ and n, minus the area of the rectangle between 0 and q_s^{n-1} with height one. The area of the difference between this second rectangle and smaller trapezoid is equal to the right-hand side of (15b).

We use (15) to simplify expression (5) for the joiner's gain function $g(n) = CR(n) - UA(n)$

$$\begin{aligned} g(n) &= \int_{n-1}^{n} S(p)\,dp - q_{in}^{n-1} - \left[(n-1-1)q_{in}^{n-1} - \int_1^{n-1} S(p)\,dp\right] \\ &= \int_1^{n} S(p)\,dp - (n-1)q_{in}^{n-1} = \int_1^{n} S(p)\,dp - (n-1)S(n-1) \end{aligned} \tag{16}$$

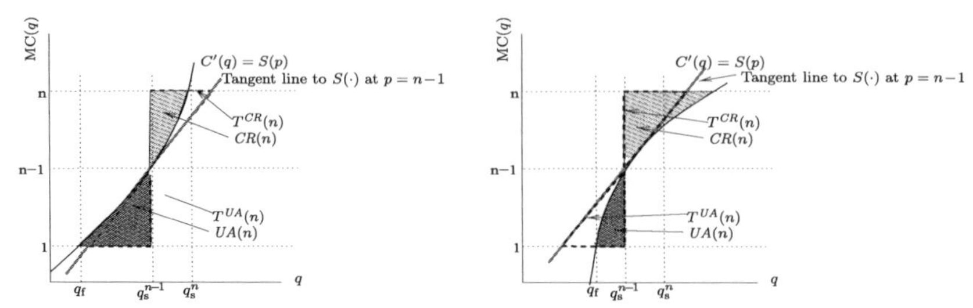

Fig. 4. Intuition for Proposition 1.

Using expression (16), Proposition 1 establishes some relationships between the curvature of the marginal cost function, and the number of members in a stable coalition.[7]

Proposition 1. *If the marginal cost function is strictly convex, so that S(p) is strictly concave, then the largest stable equilibrium is less than or equal to 2. If the marginal cost is strictly concave, the smallest stable equilibrium is greater than or equal to 3.*

The left panel of Fig. 4 provides intuition for the first part of Proposition 1. Because $MC'(\cdot)$ is strictly convex (i.e., $S(\cdot)$ is strictly concave), its graph lies everywhere except at $p=n-1$ strictly to the left of the line that is tangent to $S(\cdot)$ at $p=n-1$. Therefore, the hatched region $CR(n)$ is strictly contained in the triangle $T^{CR}(n)$, indicated by dashed lines, which is bounded below by $n-1$ and above by n, and so has height 1. Because $T^{CR}(n)$ is similar to the triangle $T^{UA}(n)$, marked by dotted lines, with height $n-2$, the area of $T^{UA}(n)$ weakly exceeds that of $T^{CR}(n)$ for all $n \geq 3$. Moreover, the cross-hatched region $UA(n)$ strictly contains $T^{UA}(n)$. It follows that for all $n \geq 3$, $UA(n) > CR(n)$, verifying that $g(n) < 0$, for every integer $n > 2$. The first part of Proposition 1 now follows from (4). A parallel argument, illustrated by the right panel of Fig. 4, establishes that when marginal costs are strictly concave, $UA(3) > CR(3)$, so $n=3$ is internally stable. By (14), there exists $n \geq 3$ such that n is a stable equilibrium, establishing the second part of the proposition.

The intuition underlying Proposition 1 can be extended to provide insight into the relationship between the internal and external stability conditions. Say that a function $f : \mathbb{R}_+ \to \mathbb{R}_+$ is *convex beyond x* if $f(\cdot)$ is convex on the interval $[x,\infty]$. It will be clear by inspection of Fig. 4 that if for some $\overline{n}$, the marginal cost curve is convex beyond $MC^{-1}(\overline{n})$, then as n increases beyond $\overline{n}$, the breadth of $CR(n)$ at its base will either remain constant or decrease. Its height, however, will remain constant at 1. On the other hand, both the height and breadth of $UA(n)$ will increase with n. It follows immediately that

$$\begin{aligned}&\text{for any } \overline{n} \in \mathbb{N}, \text{ if marginal cost is convex beyond } MC^{-1}(\overline{n}), \text{ then}\\&\quad \text{the joiner's gain function } g(\cdot) \text{ is strictly decreasing beyond } \overline{n}\end{aligned} \tag{17}$$

An immediate consequence of (17) is that if marginal cost is convex beyond $MC^{-1}(\overline{n})$, then the external stability requirement is satisfied beyond $\overline{n}$ whenever the internal stability requirement is satisfied (i.e., for $n \geq \overline{n}$, $g(n) \leq 0$ implies $g(n+1) < 0$). Summarizing

$$\text{if } MC \text{ is convex beyond } MC^{-1}(\overline{n}), \text{ then } n \geq \overline{n} \text{ is stable if } g(n) = 0 \tag{18}$$

In Proposition 2 below, we reduce a given initial cost function in a way that leaves the original cost function unchanged beyond $\overline{n}$, while the joiner's modified gain function is zero at $\overline{n}$. It follows from (18) that if the marginal cost were originally convex, then our construction is sufficient to ensure that $\overline{n}$ is stable. This fact is significant because in the literature on participation games to date, costs are almost invariably parameterized so that marginal costs are convex. Indeed, we are unaware of *any* paper in the related literature in which this property is violated.

4.2. Admissible cost reductions

We now consider the effect on the equilibrium size of an IEA of a technological improvement that lowers the abatement cost function. We model this kind of improvement by adding to an initially specified inverse marginal cost function $S(\cdot)$ a non-negative function that is positive on an open subset of its domain. Observe that a (weak) increase in $S(\cdot)$, for all p, implies a (weak) decrease in marginal costs, for all q.

Let $\varepsilon(p) \geq 0$ denote a *cost reduction function*. The new (lower) inverse marginal cost function is denoted by $\tilde{S}(\cdot) = S(\cdot) + \varepsilon(\cdot)$. In an n-member IEA, the per-member equilibrium level of abatement is $S(n)$ before the cost reduction,

[7] Using a convex marginal cost function which had previously been used to estimate the costs of greenhouse gas abatement, Barrett's [3] Proposition 4 shows that the equilibrium IEA consists of two members.

and $\tilde{S}(n)$ after it. Thus, $\varepsilon(n)$ is the amount by which the cost reduction increases per-member abatement for an n-member IEA. Let $\tilde{g}(\cdot)$ denote the joiner's gain function under the modified cost structure $\tilde{S}(\cdot)$.

We say that a cost reduction function $\varepsilon(\cdot)$ is *admissible* if it satisfies certain properties. To streamline the exposition, we allow $\varepsilon(\cdot)$ to be discontinuous, in which case there will be p-values at which the optimal abatement choice is not uniquely defined.[8] To resolve indeterminacies, we assume

If multiple abatement levels satisfy the first order condition (2)
each non-signatory chooses the *lowest* optimal abatement level (19a)

each signatory chooses the *highest* optimal abatement level (19b)

From the exposition which follows, it will be clear that we could have restricted $\varepsilon(\cdot)$ to be continuous, and thus eliminated all indeterminacies. This would have required more notation, increased the complexity of proofs, and made our heuristic explanations more cumbersome, without changing any of our formal results. We now define the conditions for admissibility:

Definition 1. For given $S(\cdot)$, $\varepsilon(p)$ is an admissible cost reduction function if

$$\varepsilon(p) \geq 0 \quad \text{with strict inequality on some open subset of } \mathbb{R} \tag{20a}$$

$$S(\cdot)+\varepsilon(\cdot) \text{ is a non-decreasing function of } p \tag{20b}$$

Inequality (20a) implies that for any given level of membership, the cost reduction never decreases abatement, and strictly increases it for some levels; inequality (20b) states that marginal cost after the reduction is non-decreasing, so that costs remain convex after the cost reduction. For convenience, we add a third restriction

$$\varepsilon(p) = 0 \quad \text{for all } p \in (N+1, \infty) \tag{20c}$$

Because the largest possible IEA has N members, q_s^n never exceeds $S(N)$. Hence marginal cost reductions at levels of $q > S(N)$ can have no impact on equilibrium outcomes, so restriction (20c) is without loss of generality. Replacing $S(\cdot)$ with $\tilde{S}(\cdot)$ in (16), we see that the change in the joiner's gain function at n due to the cost reduction $\varepsilon(\cdot)$ is

$$\Delta^{\varepsilon} g(n) = \tilde{g}(n) - g(n) = \int_1^n \varepsilon(p)\, dp - (n-1)\varepsilon(n-1) \tag{21a}$$

Similarly, the change in $g(\cdot)$ at $n+1$ is

$$\Delta^{\varepsilon} g(n+1) = \int_1^{n+1} \varepsilon(p)\, dp - n\varepsilon(n) \tag{21b}$$

$$\Delta^{\varepsilon} g(n+1) = \Delta^{\varepsilon} g(n) - n\varepsilon(n) + (n-1)\varepsilon(n-1) + \int_n^{n+1} \varepsilon(p)\, dp \tag{21c}$$

For any $3 \leq n \leq N$, we identify in Section 4.3 the "smallest possible" or "most efficient" reduction in abatement costs such that an IEA with at least n members is stable. Because we consider cost reductions over an interval, there are many alternative definitions of "smallest possible". We adopt a natural specification, evaluating $\varepsilon(\cdot)$'s that satisfy conditions (20) according to the following norm[9]:

$$\|\varepsilon\| = \int_0^{N+1} \varepsilon(p)\, dp \tag{22}$$

that is, we take the integral over the change in inverse marginal costs from 0 to $N+1$. For $\varepsilon(\cdot)$'s satisfying restriction (20c), this norm is equivalent to one that computes the integral of marginal costs (cf. inverse marginal costs) over an interval of q's containing the set on which the new and the old marginal cost curves (cf. the inverse marginal cost curves) are permitted to differ.

4.3. *Cost decreases that increase equilibrium membership*

Fix $\overline{n} > 2$ and a marginal cost function whose inverse is S. The integer $\overline{n}$ may be unstable under S because it fails the test either for internal stability—i.e., $g(\overline{n}) < 0$ or for external stability—i.e., $g(\overline{n}+1) \geq 0$. The pessimistic results in the literature to date are all consequences of the fact that for the functional forms which that literature considers, it is the internal rather

[8] A function $\varepsilon(p)$ such that for some p, $\lim_{\delta \searrow 0} \varepsilon(p-\delta) = \underline{q} < \overline{q} = \lim_{\delta \searrow 0} \varepsilon(p+\delta)$ is consistent with a cost function $C(q)$ which is affine on the interval $[\underline{q}, \overline{q}]$, i.e., marginal cost is flat on this interval. In this case, all of the q's in $[\underline{q}, \overline{q}]$ will be equi-profitable. Strictly, speaking, $S(p)+\varepsilon(p)$ is not the inverse of any marginal cost function, since viewed as a function of q rather than p, $\tilde{S}(q)+\varepsilon(q)$ is not one-to-one.

[9] Our conditions (20) for admissibility confine us to considering only cost reductions that weakly reduce marginal costs globally. This is a very strong restriction. For example, the cost reduction we consider in Section 3.2 involves an increase in marginal costs over some region. If we had imposed a weaker admissibility criterion, which admitted the possibility that cost reduction could locally increase marginal costs, thus relaxing a self-imposed constraint, we could have supported a given membership level with a cost reduction that was much smaller under our metric.

than the external condition that is violated for large n. Accordingly, to limit the number of cases we need to consider we assume that under the original cost structure, the internal stability condition fails at $\overline{n}$, i.e., $g(\overline{n})$ is negative.

Our goal, then, is to shift $g(\cdot)$ up at $\overline{n}$. From (5), $g(\cdot)|_{\overline{n}}$ decreases with $q_s^{\overline{n}-1}$; moreover, $q_s^{\overline{n}-1}$ increases as $MC(\cdot)|_{q_s^{\overline{n}-1}}$ shifts down, so that a cost modification which lowers $MC(\cdot)$ on a neighborhood of $q_s^{\overline{n}-1}$ will, holding all else constant, shift $g(\cdot)|_{\overline{n}}$ even further down. On the other hand, Eq. (5) also reveals that $g(\cdot)$ shifts up at $\overline{n}$ as the cost of producing $q_s^{\overline{n}}$ decreases. Accordingly, the cost modification we construct below leaves $MC(\cdot)$ unchanged at $q_s^{\overline{n}-1}$ itself, while lowering it on an interval to the left of $q_s^{\overline{n}-1}$, thus lowering the cost of producing $q_s^{\overline{n}}$, while $q_s^{\overline{n}-1}$ remains constant. As Fig. 4 illustrates and Proposition 1 formalizes, this reduction must be sufficiently large that the modified marginal cost curve must on some interval to the left of $q_s^{\overline{n}-1}$ lie below the tangent line to $MC(\cdot)$ at $q_s^{\overline{n}-1}$ – in other words the modified marginal cost curve must be locally concave on some interval – otherwise $g(\cdot)$ will necessarily be negative for all $n \geq 3$. (As we note on p. 29, we could also increase $g(\cdot)$ at $\overline{n}$ to some extent by lowering marginal cost between $q_s^{\overline{n}-1}$ and $q_s^{\overline{n}}$; there is, however, no efficiency gain to this alternative approach.)

Before proceeding with our construction, we note some key inequalities and introduce some terminology. Given a cost reduction function $\varepsilon(\cdot)$, a necessary and sufficient condition for $\overline{n}$ to be stable given the reduced cost structure $\tilde{S}(\cdot) = S(\cdot) + \varepsilon(\cdot)$ is that

$$\tilde{g}(\overline{n}) = g(\overline{n}) + \Delta^{\varepsilon} g(\overline{n}) \geq 0 > g(\overline{n}+1) + \Delta^{\varepsilon} g(\overline{n}+1) = \tilde{g}(\overline{n}+1) \tag{23}$$

It follows from (23) that $\overline{n}$ will be stable given a cost reduction $\varepsilon(\cdot)$ iff

$$\Delta^{\varepsilon} g(\overline{n}) \geq -g(\overline{n}) \geq 0 \tag{24a}$$

$$g(\overline{n}+1) + \Delta^{\varepsilon} g(\overline{n}+1) < 0 \tag{24b}$$

We established at the beginning of Section 4 that if (24a) is satisfied at $\overline{n}$, then there exists $n' \geq \overline{n}$ such that n' is stable. Accordingly we say that

$\varepsilon(\cdot)$ *implements at least* $\overline{n}$ if $\varepsilon(\cdot)$ satisfies (20) and if $\Delta^{\varepsilon} g(\overline{n})$ satisfies (24a) (25a)

$\varepsilon(\cdot)$ *implements exactly* $\overline{n}$ if $\varepsilon(\cdot)$ satisfies both (24a) and (24b) (25b)

That is, to implement at least $\overline{n}$, it is necessary only to satisfy internal stability at $\overline{n}$; to implement exactly $\overline{n}$, external stability must be satisfied as well. The cost reduction we construct below satisfies internal stability with equality, i.e., sets $\Delta^{\varepsilon} g(\overline{n}) = -g(\overline{n})$. As noted on p. 22 (see (18)), if the initial cost function belongs to a class that has been considered in the related literature to date and internal stability is satisfied with equality, then external stability is satisfied as well. For such cost functions, the distinction between "at least" and "exactly" is moot.

There are countless admissible cost reductions that implement at least $\overline{n}$. We identify the minimal value of $\|\varepsilon\|$, taken over the set of all such reductions. When satisfying internal stability with equality does *not* imply external stability, there will not exist a minimal value of $\|\varepsilon\|$ that implements exactly $\overline{n}$. Accordingly, we define the

minimal cost reduction for at least $\overline{n}$ as $\epsilon_{\min}^{\overline{n}} = \min\{\|\varepsilon\| : \varepsilon(\cdot) \text{ implements at least } \overline{n}\}$ (26a)

infimal cost reduction for exactly $\overline{n}$ as $\epsilon_{\inf}^{\overline{n}} = \inf\{\|\varepsilon\| : \varepsilon(\cdot) \text{ implements exactly } \overline{n}\}$ (26b)

Finally, say that

$\varepsilon(\cdot)$ is *efficient for at least* $\overline{n}$ if $\varepsilon(\cdot)$ implements at least $\overline{n}$ and $\|\varepsilon\| = \epsilon_{\min}^{\overline{n}}$ (27a)

$\varepsilon(\cdot)$ is *efficient for exactly* $\overline{n}$ if $\varepsilon(\cdot)$ implements exactly $\overline{n}$ and $\|\varepsilon\| = \epsilon_{\inf}^{\overline{n}}$ (27b)

A cost reduction that is efficient for exactly $\overline{n}$ exists only if satisfying internal stability with equality implies external stability.

The construction below identifies an $\varepsilon(\cdot)$ that is efficient for at least $\overline{n}$. First, for each $\lambda_1 \in [0,1)$, define

$$\varepsilon^{\lambda_1}(p) = \begin{cases} \lambda_1(S(\overline{n}-1) - S(p)) & \text{if } p \in [1, \overline{n}-1) \\ 0 & \text{otherwise} \end{cases} \tag{28}$$

For $\lambda_1 \approx 2/3$, the graph of $\varepsilon^{\lambda_1}(\cdot)$ is depicted in the left panel of Fig. 5 by heavy dashed lines. It is vertical until $p=1$, at which point it jumps horizontally, then slopes negatively until $p = (\overline{n}-1)$, beyond which it is again vertical. In the right panel of the figure, $\varepsilon^{\lambda_1}(\cdot)$ is added to $S(\cdot)$ to form $\tilde{S}^{\lambda_1}(\cdot)$, which by construction agrees with $S(\cdot)$ on $[0,1) \cup [\overline{n}-1, \infty)$, is discontinuous at $p=1$, then rises more steeply than $S(\cdot)$ over the interval $(1, \overline{n}-1]$. Note that as λ_1 approaches 1, $\lim_{\delta \searrow 0} \tilde{S}^{\lambda_1}(1+\delta)$ approaches $q_s^{\overline{n}-1}$, while $S^{\lambda_1}(\cdot)$ rises increasingly steeply on $(1, \overline{n}-1]$. In the left panel, $UA(\overline{n})$ is larger than $CR(\overline{n})$, so that $g(\overline{n}) \equiv CR(\overline{n}) - UA(\overline{n}) < 0$. In the right panel, the modified region $\tilde{UA}^{\lambda_1}(\overline{n})$ (the region above $p=1$, below $p = \overline{n}-1$, right of $\tilde{S}^{\lambda_1}(\cdot)$ and left of $q_s^{\overline{n}-1}$) is considerably smaller. The location of q_f remains unchanged, in accordance with our assumption (19a) about how indeterminacies are resolved. It will be clear from the figure that when $\lambda_1 = 0$, $\tilde{S}(\cdot)$ is identical to $S(\cdot)$, so that the area

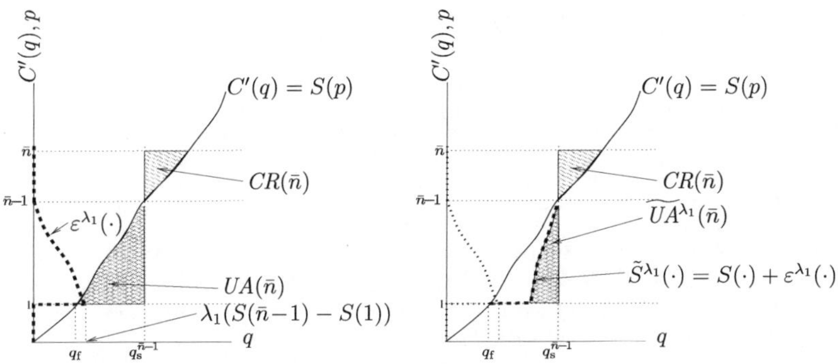

Fig. 5. Shrinking $UA(\overline{n})$ increases $\tilde{g}(\overline{n})$.

of $\tilde{UA}^{\lambda_1}(\overline{n})$ exceeds that of $CR(\overline{n})$, while as λ_1 approaches 1, the area of $\tilde{UA}^{\lambda_1}(\overline{n})$ shrinks to zero and is hence eventually smaller than that of $CR(\overline{n})$. Since the area $\tilde{UA}^{\lambda_1}(\overline{n})$ changes continuously with λ_1 it follows from the intermediate value theorem that

$$\text{there exists } \lambda_1^* \in (0,1) \text{ s.t. the areas of } \tilde{UA}^{\lambda_1^*}(\overline{n}) \text{ and } CR(\overline{n}) \text{ are equal} \tag{29}$$

Hence for the function $\tilde{S}^{\lambda_1^*}$, $\tilde{g}^{\lambda_1^*}(\overline{n}) = 0$. From (21a), $\Delta^{\lambda_1^*}g(\overline{n}) = \tilde{g}^{\lambda_1^*}(\overline{n}) - g(\overline{n})$, so that $\Delta^{\lambda_1^*}g(\overline{n}) = -g(\overline{n})$. Moreover, by construction, $\varepsilon^{\lambda_1^*}(\overline{n}-1) = 0$. Hence from (21a) and (22), $\|\varepsilon^{\lambda_1^*}\| = -g(\overline{n})$, so that $\epsilon^{\overline{n}}_{\text{min}}$, the minimal cost reduction for at least $\overline{n}$ is at most $-g(\overline{n})$. Part 1 of Proposition 2 establishes that in fact $\epsilon^{\overline{n}}_{\text{min}} = -g(\overline{n})$.

A property of the construction (28) is that $\varepsilon^{\lambda_1}(p) = 0$, for all $p \geq \overline{n}-1$. Hence from (21c), $\Delta^{\lambda_1^*}g(\overline{n}) = \Delta^{\lambda_1^*}g(\overline{n}+1)$. Since from (23), $\tilde{g}^{\lambda_1^*}(\overline{n}+1) = g(\overline{n}+1) + \Delta^{\lambda_1^*}g(\overline{n}+1)$ while $\tilde{g}^{\lambda_1^*}(\overline{n}) = 0 = g(\overline{n}) + \Delta^{\lambda_1^*}g(\overline{n})$, we have that

$$\tilde{g}^{\lambda_1^*}(\overline{n}+1) = g(\overline{n}+1) - g(\overline{n}) \tag{30}$$

It now follows from (24b) and (30) that if $g(\overline{n}+1) < g(\overline{n})$, then $\overline{n}$ is stable under $\varepsilon^{\lambda_1^*}$.

The construction above is by no means unique: if the heavy dashed line that bounds $\tilde{UA}^{\lambda_1}(\overline{n})$ on the left were replaced by *any* other non-decreasing line such that the area of the corresponding region $UA(\overline{n})$ were equal to that of $\tilde{UA}^{\lambda_1}(\overline{n})$, then the cost reduction function associated with that line would also be efficient for at least $\overline{n}$. There is also a second class of $\varepsilon(\cdot)$'s that have this property: functions in this class are positive on $(\overline{n}-1,\overline{n})$ and thus increase the area of $CR(\overline{n})$ rather than simply reducing the area of $UA(\overline{n})$. There is, however, a limit in general on how much can be accomplished by increasing $CR(\overline{n})$ without also reducing $UA(\overline{n})$; for example, if $S(\cdot)$ were nearly affine on $(\overline{n}-1,\overline{n})$, then one could not do much better than double the area of $CR(\overline{n})$, which, if the initial difference between $UA(\overline{n})$ and $CR(\overline{n})$ were sufficiently large, would be alone insufficient to equalize the two areas. By contrast, regardless of the initial relative sizes of $UA(\overline{n})$ and $CR(\overline{n})$, the area of $UA(\overline{n})$ can always be reduce sufficiently to equalize them.

Proposition 2 below summarizes and extends the preceding graphical analysis. The extensions are proved in the Appendix. The result focuses exclusively on conditions under which the *internal* stability condition (24a) is satisfied. As we observed on p. 19 (statement (14)), if the internal stability condition is satisfied at $\overline{n}$, then both the internal and external stability conditions will be satisfied for some $n' \geq \overline{n}$. For this reason, the first three parts of Proposition 2 relate to properties that hold for "at least $\overline{n}$". The fourth part notes that when the external stability constraint is slack – i.e., when $g(\overline{n}+1) < g(\overline{n})$ – then a property will hold for at least $\overline{n}$ if and only if it holds for exactly $\overline{n}$.

Proposition 2. *Let C be a twice differentiable, strictly convex cost function and let B be a linear benefit function. For* $1 < \overline{n} \leq N$, *assume that* $g(\overline{n}) < 0$ *so that* $\overline{n}$ *is not stable. Then*

(1) *the minimal cost reduction for at least* $\overline{n}$ *is* $-g(\overline{n})$.
(2) $\exists \lambda_1^* \in (0,1)$ *s.t. the cost reduction* $\varepsilon^{\lambda_1^*}$ *defined by* (28) *is efficient for at least* $\overline{n}$.
(3) $\varepsilon(\cdot)$ *is efficient for at least* $\overline{n}$ *iff* $\int_1^{\overline{n}} \varepsilon(p)\,dp = -g(\overline{n})$ *and* $\varepsilon([0,1) \cup \{\overline{n}-1\} \cup [\overline{n},\infty]) = 0$.
(4) *if* $g(\overline{n}+1) < g(\overline{n})$ *then* $\varepsilon(\cdot)$ *is efficient for at least* $\overline{n}$ *iff* $\varepsilon(\cdot)$ *is efficient for exactly* $\overline{n}$.

An immediate corollary of part (1) of Proposition 2 is the intuitive result that if $g(\cdot)$ is monotone decreasing in n over an interval, then larger cost reductions are required in order to implement larger coalitions in that interval. Note from part (3) that efficient cost reductions necessarily leave unchanged the original cost function beyond $\overline{n}$. Thus, if the initial *marginal* cost function is convex, then from (17) and part (4), $\varepsilon(\cdot)$ is efficient for exactly $\overline{n}$ iff it is efficient for at least $\overline{n}$. As we have emphasized above, marginal costs are indeed convex for the functions that have been considered in the literature on participation games to date, so that part (4) applies to these functions.

Now suppose that $g(\overline{n}+1) \geq g(\overline{n})$. As we noted on p. 19, cost reductions that are efficient for at least $\overline{n}$ will now violate the external stability constraint. It is possible to implement exactly $\overline{n}$ in this case, but costs must be reduced still further. The required construction is similar to our first reduction (28), although the analysis is less "clean" in several respects. First, we were able to satisfy the internal stability condition at $\overline{n}$ while holding constant the abatement levels q_f, $q_s^{\overline{n}-1}$ and $q_s^{\overline{n}}$; to satisfy external stability, on the other hand, it is now necessary to increase $q_s^{\overline{n}}$. (We need a cost reduction $\varepsilon(\cdot)$ s.t. $\Delta^{\varepsilon}g(\overline{n}+1) < 0$. From (21b), it is now necessary that $\varepsilon(\overline{n}) > 0$, since $\Delta^{\varepsilon}g(\overline{n}+1)$ increases with $\varepsilon(p)$ for all other values of p.) Second, because of the strict inequality in the external stability condition (24b), a minimal cost reduction no longer exists. Third, while it is possible to identify an infimal cost reduction, a closed-form expression for this infimum does not exist. For completeness, Proposition 3 summarizes most of what can be said when $g(\overline{n}+1) \geq g(\overline{n})$ under the original cost function. The proof, which is rather technical, is included in the online addendum.

Proposition 3. *Let C be a twice differentiable, strictly convex cost function and let B be a linear benefit function. For* $1 < \overline{n} \leq N$, *assume that* $g(\overline{n}) \leq g(\overline{n}+1) < 0$, *so that* $\overline{n}$ *is not stable. Then*

(1) *there exists a cost reduction that implements exactly* $\overline{n}$,
(2) *if* $g(\overline{n}+1) = g(\overline{n})$ *(resp.* $g(\overline{n}+1) > g(\overline{n})$*), then* $\epsilon_{\inf}^{\overline{n}} = -g(\overline{n})$ *(resp.* $\epsilon_{\inf}^{\overline{n}} > -g(\overline{n})$*).*

Propositions 2 and 3 highlight the importance of the relative magnitudes of $g(\overline{n})$ and $g(\overline{n}+1)$, under the assumption that both are negative. If $g(\overline{n}) > g(\overline{n}+1)$ the external stability constraint is slack, in the sense that internal stability guarantees external stability. If $g(\overline{n}) < g(\overline{n}+1)$, the external stability condition is binding. In the latter case, some cost reduction on an open interval above $\overline{n}$ is required in order to satisfy the external stability condition.

5. General costs, concave abatement benefits

In this section we assume that the benefit function $B(\cdot)$ is increasing and strictly concave in aggregate benefits. We continue to assume, as in Section 4, that the abatement cost function C is initially strictly convex. To avoid dealing with nonnegativity constraints (see for example [18,42]) we assume

$$C'(0) < B'(Nq^*) \quad \text{where } q^* \text{ is the socially optimal level of abatement} \tag{31}$$

Since signatory abatement levels are bounded above by q^* in any Cournot or Stackelberg equilibrium, this assumption guarantees that an interior solution to the non-signatory's first order condition (2b) exists.

When we relax the assumption of linear marginal benefits, our problem becomes more complex for two related reasons. First, we can no longer normalize a non-signatory's marginal benefit function to unity; nor can we normalize a signatory's marginal benefit function to the number of members of the IEA. A consequence is that our simplification (5) of expression (3) for the joiner's gain function is no longer valid; also invalid is our decomposition (6), derived from the simplification (5), of the joiner's gain function into *UA* and *CR*. Second, a non-signatory's optimization problem is no longer independent of the decisions made by signatories, so that for each n, the abatement choices made by signatories and non-signatories must be determined simultaneously. Moreover, if an additional member joins an IEA, so that total abatement by signatories increases, abatement levels by non-signatories will decrease in response. (Accordingly, we now denote non-signatory abatement when there are n signatories by q_f^n rather than q_f.)

Because of these complexities, the cost-reduction required to implement an IEA with at least n members is less straightforward than the one we constructed to prove Proposition 2. Moreover, in this context, it would be difficult to characterize the class of efficient cost reductions. Accordingly the scope of Proposition 4, presented in Section 5 below, is narrower than its counterpart, Proposition 2, for linear benefits. We have not attempted to identify the further cost-reduction that would be required to implement exactly n, so we have no counterpart to Proposition 3. Another consequence of the additional complexity is that the Cournot and Stackelberg versions of our game no longer yield identical results. In the remainder of this section, we restrict our attention to the Cournot case. In an online addendum, we establish that given a benefit and initial cost function, a cost-reduction that implements at least n as a Cournot equilibrium will also implement at least n as a Stackelberg equilibrium.

Fix a strictly increasing marginal cost function C', a strictly decreasing marginal benefit function B', and $1 < \overline{n} \leq N$. Fig. 6 illustrates a reduced marginal cost function $\tilde{C}'$ which implements at least $\overline{n}$ as a stable equilibrium. The original marginal cost curve C' is indicated by a thin solid line and the modified curve $\tilde{C}'$ by a heavy dotted line. The two curves overlap near the origin, and beyond $q_s^{\overline{n}-1}$. The region where they differ consists of a segment that is arbitrarily close to horizontal, followed by one that is arbitrarily close to vertical.

We show that under the modified marginal cost function $\tilde{C}'$, the joiner's gain function $\tilde{g}(\cdot)$ satisfies the internal stability condition (24a). From (14), both internal and external stability hold for some $n' \geq \overline{n}$. The key to our construction is to ensure that for both $\overline{n}-1$ and $\overline{n}$, the optimal levels of abatement for non-signatories, respectively $q_f^{\overline{n}-1}$ and $q_f^{\overline{n}}$, are virtually the same as, but slightly smaller than $q_s^{\overline{n}-1}$; we impose this by making modified marginal costs virtually vertical on the interval $(q_f^{\overline{n}}, q_s^{\overline{n}-1})$ which contains $q_f^{\overline{n}-1}$.

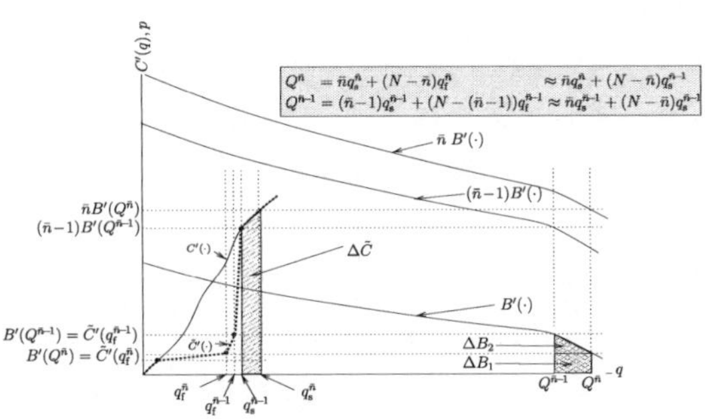

Fig. 6. Implementing $\overline{n}$ with concave benefits.

Because $q_s^{\overline{n}-1}$ and $q_f^{\overline{n}-1}$ are arbitrarily close together under this construction, aggregate abatement with $\overline{n}-1$ signatories is $Q^{\overline{n}-1} \approx Nq_s^{\overline{n}-1}$. Similarly, because $q_s^{\overline{n}-1}$ and $q_f^{\overline{n}}$ are arbitrarily close together, aggregate abatement with $\overline{n}$ signatories is $Q^{\overline{n}} \approx \overline{n}q_s^{\overline{n}} + (N-\overline{n})q_s^{\overline{n}-1}$. Thus, $Q^{\overline{n}} - Q^{\overline{n}-1} \approx \overline{n}(q_s^{\overline{n}} - q_s^{\overline{n}-1})$. This property simplifies our analysis because, by construction, we have eliminated the complication that non-signatory abatement depend on $\overline{n}$.

A closely related consequence of our construction is that when there are $\overline{n}-1$ signatories under the modified cost structure, it costs a signatory virtually the same to produce $q_s^{\overline{n}-1}$ as it costs a non-signatory to produce $q_f^{\overline{n}-1}$. This implies that in the joiner's gain function $\tilde{g}(\cdot)$ (see expression (3)), the term $\tilde{C}(q_f^{\overline{n}-1})$ can be replaced by $\tilde{C}(q_s^{\overline{n}-1})$, so that $\tilde{g}(\overline{n}) \approx \Delta B - \Delta\tilde{C}$, where ΔB is the per-signatory gain from increased abatements, and $\Delta\tilde{C}$ is the per-signatory incremental cost, when the IEA acquires an additional signatory. To establish that $\tilde{g}(\overline{n})$ is positive, we now only need to show that ΔB exceeds $\Delta\tilde{C}$.

The magnitude of $\Delta\tilde{C}$ (the tall thin hatched trapezoid in Fig. 6) is strictly less than the area of the rectangle that encloses it with height $\overline{n}B'(Q^{\overline{n}})$ and width $(q_s^{\overline{n}} - q_s^{\overline{n}-1})$. ΔB is the short wide cross-hatched trapezoid to the right of the figure, the union of the rectangle ΔB_1 and triangular-shaped ΔB_2. The height of ΔB_1 is $B'(Q^{\overline{n}})$, i.e., $1/\overline{n}$ times the height of rectangle that contains $\Delta\tilde{C}$. The width of ΔB_1 is $Q^{\overline{n}} - Q^{\overline{n}-1} \approx \overline{n}(q_s^{\overline{n}} - q_s^{\overline{n}-1})$ i.e., $\overline{n}$ times the width of the rectangle that contains $\Delta\tilde{C}$. Thus, the area of ΔB_1 strictly exceeds that of $\Delta\tilde{C}$. Moreover, the trapezoid ΔB also includes ΔB_2 which has positive area. Thus $g(\overline{n}) = \Delta B - \Delta\tilde{C} > 0$, verifying that $\overline{n}$ is internally stable. This informal argument is made precise in Proposition 4 below.

Proposition 4. *Assume that the benefit function B and initial cost function C are both twice continuously differentiable, that B is strictly concave and C is strictly convex. Then for any* $1 < \overline{n} \le N$, *there exists an admissible cost reduction such that under the modified cost function* $\tilde{C}$, *an IEA with at least* $\overline{n}$ *members is stable.*

6. Conclusion

An important strand of the theory of IEAs focuses on the non-cooperative Nash equilibria of the participation game introduced by Aspremont et al. [14]. The literature to date, which has focused exclusively on parametric examples, has been highly pessimistic about the prospects for international environmental cooperation. Its conclusion is that the equilibrium size of a stable IEA is small except when the potential gains from cooperation are also small. Moreover, although reductions in abatement costs necessarily increase the potential gains from cooperation, analysis of leading special cases has shown that such reductions leave unchanged or reduce equilibrium membership.

This paper approaches the problem from a non-parametric perspective. We introduce a novel decomposition of the gains to participation. For linear marginal benefits and general convex abatement costs, we use this decomposition to show that regardless of the potential gains from cooperation, if *marginal* abatement costs are convex then the equilibrium coalition size cannot exceed three. Maintaining convexity of marginal abatement costs, reductions in abatement costs cannot increase membership. In this respect, the results obtained in the previous literature are robust.

In an important respect, however, these previous results are fragile. If marginal costs are strictly concave, the equilibrium coalition size is at least 3. More generally, if marginal abatement costs are locally concave over a critical range of abatement levels, there are no restrictions on the size of the equilibrium coalition. Indeed, the non-cooperative Nash equilibrium can realize up to 100% of the potential gains from cooperation, even when these potential gains are large. Moreover, we show that cost reductions *can* increase equilibrium membership and, when marginal benefits are linear, we derive a lower bound on the magnitude of the cost reduction needed to induce an arbitrary level of cooperation. Finally, we show that when benefits are strictly concave in abatement, it is still possible to support agreements among an arbitrarily large fraction of potential member countries.

Our analysis highlights the danger of drawing sweeping conclusions, however plausible, from parametric examples. In particular, our results undermine the apparently robust basis for pessimism that has become almost conventional wisdom in the field. On the other hand, one should not interpret our results as sufficient grounds for optimism about the prospects for real-world global agreements, since the kinds of cost functions we show to be consistent with such agreements might be difficult to design in practice.

Acknowledgments

The authors thank Megan Stevenson for excellent research assistance. Rolf Golembek, two anonymous referees and an associate editor, and seminar participants at Berkeley, Banff, Rome and Venice provided many helpful comments.

Appendix A. Proofs

Proof of Remark 1. From (6) and (12), we have, after rearranging

$$g(\lceil c-a/1-x\rceil) = CR(\lceil c-a/1-x\rceil) - UA(\lceil c-a/1-x\rceil) = (1-x)\left(\underbrace{\lceil c-a/1-x\rceil}_{\text{Term L}} - \underbrace{c-x/1-x}_{\text{Term R}}\right) \tag{32}$$

(a) Fix $1 < a' < a$ and $(c,x) \in CX(a)$. Since Term L is weakly decreasing in a on $(x,1)$, $\lceil c-a'/1-x\rceil \geq \lceil c-a/1-a\rceil \geq c-x/1-x$, so that $(c,x) \in CX(a')$. To prove that $CX(a)$ is a strict subset of $CX(a')$, pick (c,x) such that $c-a/1-x = \lfloor c-x/1-x\rfloor < c-x/1-x < \lceil c-x/1-x\rceil$. Clearly $\lceil c-a/1-x\rceil = \lfloor c-x/1-x\rfloor$ so that $(c,x) \notin CX(a)$. But since $a' < a$, $\lceil c-a'/1-x\rceil \geq \lfloor c-x/1-x\rfloor + 1 > c-x/1-x$, so that $(c,x) \in CX(a')$. This proves that $CX(a) \subsetneq CX(a')$.

(b) Fix $1 < c' < c$. Note that Term L is weakly, while Term R is strictly increasing in c.

  (i) First pick (a,x) such that $c'-a/1-x = \lfloor c'-x/1-x\rfloor < c'-x/1-x < \lceil c'-x/1-x\rceil$. Clearly $\lceil c'-a/1-x\rceil = \lfloor c'-x/1-x\rfloor$ so that $(a,x) \notin AX(c')$. Now pick c larger than c' but sufficiently close that $c-x/1-x < \lceil c-x/1-x\rceil = \lceil c'-x/1-x\rceil$ and hence $\lceil c-a/1-x\rceil \geq \lfloor c-x/1-x\rfloor + 1 > c-x/1-x$, so that $(a,x) \in AX(c) \backslash AX(c')$.

  (ii) Now pick (a,x) such that $\lceil c'-a/1-x\rceil = c'-x/1-x$ so that $(a,x) \in AX(c')$. Since $a > x$, we can pick c larger than c' but sufficiently close that $\lceil c-a/1-x\rceil = \lceil c'-a/1-x\rceil < c-x/1-x$, so that $(a,x) \in AX(c') \backslash AX(c)$. Hence $AX(c') \ominus AX(c) \neq \emptyset$.

(c) Since, once again, Term L is weakly, while Term R is strictly increasing in x, we can prove this part with an argument analogous to that which proved part (b). □

Proof of Remark 2. Let $MC(q) = q/M$. Because marginal costs are linear, $q_s^{\overline{n}} - q_s^{\overline{n}-1} = M$ which implies the area of $CR(n) = M/2$ and $(n-1-1)/(q_s^{\overline{n}-1} - q_f^{\overline{n}}) = 1/M$, so $q_s^{\overline{n}-1} - q_f^{\overline{n}} = M(n-2)$ which implies that the area of $UA(n) = (M/2)(n-2)^2$. The root of $g(n) = 0$ satisfies $g(n) = M/2 - (M/2)(n-2)^2 = 0$ or $1-(n-2)^2 = 0$ so the solutions are 3,1. Because we have $g'(n) = -M(n-2) < 0$ for $n > 2$ we know that the root $n=3$ is stable, i.e., $g(2) > g(3) = 0 > g(4)$. This confirms that for linear marginal costs, $n=3$ is the unique non-trivial IEA size. □

Proof of Remark 3. The equilibrium level of aggregate abatement with the coalition is $3(3M)+(N-3)M = M(6+N)$, so the global benefit of abatement is $MN(6+N)$. The aggregate equilibrium cost of abatement is $(N-3)(M/2)+3(9M/2) = \frac{1}{2}M(N+24)$, so global equilibrium net benefits are $MN(6+N) - \frac{1}{2}M(N+24) = \frac{1}{2}M(11N+2N^2-24)$. In the absence of a coalition, each nation abates at M, so global net benefits are $N(NM-M/2)$. The global gain due to the formation of the coalition is therefore $\frac{1}{2}M(11N+2N^2-24) - N(NM-M/2) = 6M(N-2)$. If every nation were to join the coalition, aggregate welfare would be $N(N^2M - N^2M/2) = \frac{1}{2}N^3M$. The actual gain relative to the potential gain, i.e., the fraction of the potential gain from cooperation actually achieved, is

$$\frac{6M(N-2)}{\frac{1}{2}N^3M - N(NM-\frac{M}{2})} = 12\frac{N-2}{N(N^2-2N+1)} \qquad \square$$

Proof of Remark 4. We solve for K such that $\overline{n}$ is a stable equilibrium. As in the proof of Remark 2, the area of $UA(\overline{n}) = (M/2)(\overline{n}-2)^2$, while the area of $CR(\overline{n})$ is now $(q_s^{\overline{n}} - q_f^{\overline{n}})/2 = K/2$. Hence $CR(\overline{n}) = UA(\overline{n})$ if $K = (\overline{n}-2)^2M$. Moreover, because for $n > \overline{n}-1$, $g(n) = (M/2)((\overline{n}-2)^2-(n-2)^2)$, $q'(\cdot) < 0$ on $(\overline{n}-1,\infty)$. Thus $g(\overline{n}) \geq 0 > g(\overline{n}+1)$, verifying that condition (4) for $\overline{n}$ to be a stable equilibrium is satisfied. □

Proof of Proposition 1. Let $n=2$ and let the tangent to the marginal cost curve, evaluated at q_s^2, be $MC'(q_s^2) = 1/M$. The area of triangles $T^{CR}(3)$ and $T^{UA}(3)$ both equal $M/2$. Since by strict convexity, $MC(q)$ lies strictly above the tangent line through q_s^2, for all $q \neq q_s^2$, $CR(3) < M/2 < UA(3)$. Therefore, $g(3) = CR(3) - UA(3) < 0$. Now assume $n > 2$. From (16)

$$\frac{dg(n)}{dn} = S(n) - S(n-1) - (n-1)\frac{dS(n-1)}{dn}$$

But since $S(\cdot)$ is strictly concave, $S(n) < S(n-1) + dS(n-1)/dn$. Hence

$$\frac{dg(n)}{dn} < S(n-1) + \frac{dS(n-1)}{dn} - S(n-1) - (n-1)\frac{dS(n-1)}{dn} = -(n-2)\frac{dS(n-1)}{dn}$$

which is negative for $n > 2$. Since $g(3) < 0$ and $g(n)$ is strictly decreasing for $n \geq 3$, there can be no integer $n > 2$ such that $g(n) \geq 0$. Therefore, there are no stable equilibria with n greater than 2. A proof of the second part of the proposition is sketched in the text. □

Proof of Proposition 2. Part (4) has been proved in the text, by the line immediately following (30). We also proved in the text that $\varepsilon^{\lambda_1^*}$ implements at least $\overline{n}$ and that $\|\varepsilon^{\lambda_1^*}\| = -g(\overline{n})$. To prove that $-g(\overline{n})$ is the minimal cost reduction for at least $\overline{n}$ (part (1)), it suffices to prove that $\|\varepsilon\| \geq -g(\overline{n})$, for any admissible cost reduction $\varepsilon(\cdot)$ that implements at least $\overline{n}$. It will then follow immediately that $\varepsilon^{\lambda_1^*}$ is efficient for at least $\overline{n}$ (part (2)).

Assume now that $\varepsilon(\cdot)$ implements at least $\overline{n}$. Substituting (21a) into (22), we obtain

$$\|\varepsilon\| = \Delta^{\varepsilon} g(\overline{n}) + (\overline{n}-1)\varepsilon(\overline{n}-1) + \int_0^1 \varepsilon(p)\,dp + \int_{\overline{n}}^{N+1} \varepsilon(p)\,dp \underbrace{\geq}_{\text{from (24a)}} -g(\overline{n}) + (\overline{n}-1)\varepsilon(\overline{n}-1) + \int_0^1 \varepsilon(p)\,dp + \int_{\overline{n}}^{N+1} \varepsilon(p)\,dp \quad (33)$$

Since admissibility requires $\varepsilon(\cdot) \geq 0$, (33) implies that $\|\varepsilon\| \geq -g(\overline{n})$, which is what we needed to prove. This completes the proof of parts (1) and hence (2).

To prove part (3), we first establish sufficiency. Assume that $\int_1^{\overline{n}} \varepsilon(p)\,dp = -g(\overline{n})$ and that $\varepsilon([0,1) \cup \{\overline{n}-1\} \cup [\overline{n},\infty]) = 0$. Since $\varepsilon(\overline{n}-1) = 0$, (21a) implies $\Delta^{\varepsilon} g(\overline{n}) = -g(\overline{n})$, so that from (24a), ε implements at least $\overline{n}$. Since $\varepsilon([0,1) \cup [\overline{n},\infty]) = 0$, (22) implies $\|\varepsilon\| = -g(\overline{n})$. It now follows that $\varepsilon(\cdot)$ is efficient for at least $\overline{n}$.

We now prove necessity. Assume that $\varepsilon(\cdot)$ is efficient for at least $\overline{n}$, so that from part (1) of the proposition, $\|\varepsilon\| = -g(\overline{n})$. This, together with (33) implies

$$0 \geq (\overline{n}-1)\varepsilon(\overline{n}-1) + \int_0^1 \varepsilon(p)\,dp + \int_{\overline{n}}^{N+1} \varepsilon(p)\,dp \quad (34)$$

But since $\varepsilon(\cdot)$ is non-negative, (34) implies

$$\varepsilon(\overline{n}-1) = \varepsilon((0,1)) = \varepsilon((\overline{n},N+1)) = 0 \quad (35)$$

while

$$\text{If } \varepsilon(\cdot) \text{ satisfies (20) and } \varepsilon(p) > 0 \text{ then } \exists \delta > 0 \text{ s.t. } \varepsilon(\cdot) > 0 \text{ on } (p, p+\delta) \quad (36)$$

Eq. (36) restates statement A-4 in the online addendum, where it is proved. Eqs. (35) and (36) imply $\varepsilon(0) = \varepsilon(\overline{n}) = 0$, verifying that $\varepsilon([0,1) \cup \{\overline{n}-1\} \cup [\overline{n},\infty]) = 0$. Hence $-g(\overline{n}) = \|\varepsilon\| \equiv \int_0^{N+1} \varepsilon(p)\,dp = \int_1^{\overline{n}} \varepsilon(p)\,dp$. □

Proof of Proposition 4. Let C be an initial strictly convex cost function. We distinguish below between "script Q's" (denoted $\mathcal{Q}$) and regular Q's: The $\mathcal{Q}^r$'s are solutions to certain equations relating B' and C', while the Q^r's, as usual, denote aggregate abatements, i.e., $Q^r = rq_s^r + (N-r)q_f^r$. Having defined our $\mathcal{Q}^r$'s, we will construct a modified cost function $\tilde{C}$ with the property that for each r, aggregate abatement Q^r with r signatories equals $\mathcal{Q}^r$.

Now fix $\overline{n} \leq N$ and $\delta \geq 0$. Let $\mathcal{Q}^{\overline{n}-1}(\delta) = Nq_s^{\overline{n}-1}(\delta) - (N-(\overline{n}-1))\delta$, and let $\mathcal{Q}^{\overline{n}}(\delta) = Nq_s^{\overline{n}}(\delta) - 2(N-\overline{n})\delta$, where

$$q_s^{\overline{n}-1}(\delta) \text{ is defined by the condition } C'(q_s^{\overline{n}-1}(\delta)) = (\overline{n}-1)B'(\mathcal{Q}^{\overline{n}-1}(\delta)) \quad (37a)$$

$$q_s^{\overline{n}}(\delta) \text{ is defined by the condition } C'(q_s^{\overline{n}}(\delta)) = \overline{n}B'(\mathcal{Q}^{\overline{n}}(\delta)) \quad (37b)$$

Thus,

$$\mathcal{Q}^{\overline{n}}(\delta) - \mathcal{Q}^{\overline{n}-1}(\delta) = N(q_s^{\overline{n}}(\delta) - q_s^{\overline{n}-1}(\delta)) - \delta(N-(\overline{n}-1)) \quad (38)$$

Note that for some $\epsilon > 0$, $q_s^{\overline{n}}(0) - q_s^{\overline{n}-1}(0) \geq 2\epsilon$, since otherwise, $\mathcal{Q}^{\overline{n}}(0) - \mathcal{Q}^{\overline{n}-1}(0) \leq 0$, and, since B is strictly concave, $\overline{n}B'(\mathcal{Q}^{\overline{n}}(0))/(\overline{n}-1)B'(\mathcal{Q}^{\overline{n}-1}(0)) \geq \overline{n}/(\overline{n}-1)$ while, since C is strictly convex, $C'(q_s^{\overline{n}}(0))/C'(q_s^{\overline{n}-1}(0)) < 1$, in which case equalities (37a) and (37b) could not be satisfied simultaneously. Hence by continuity, there exists $\overline{\delta}$ s.t. if $\delta < \overline{\delta}$, then $q_s^{\overline{n}}(\delta) - q_s^{\overline{n}-1}(\delta) \geq \epsilon$ and $\mathcal{Q}^{\overline{n}}(\delta) - \mathcal{Q}^{\overline{n}-1}(\delta) > 0$.

Fig. 6 illustrates a modified cost function $\tilde{C}(\cdot|\delta) \leq C(\cdot)$, for $0 < \delta < \overline{\delta}$. We first identify two intervals on which C and $\tilde{C}(\cdot|\delta)$ agree, and define $\tilde{C}(\cdot|\delta)$ at some critical points in between

$$\tilde{C}'(q|\delta) = \begin{cases} C'(q) & \text{if } q \leq (C')^{-1}(B'(\mathcal{Q}^{\overline{n}}(\delta)) - \delta) \\ B'(\mathcal{Q}^{\overline{n}}(\delta)) & \text{if } q = q_s^{\overline{n}-1} \end{cases} (\delta) - 2\delta B'(\mathcal{Q}^{\overline{n}-1}(\delta)) \text{if } q = q_s^{\overline{n}-1}(\delta) - \delta C'(q) \text{if } q \geq q_s^{\overline{n}-1}(\delta). \quad (39)$$

For the remainder of the proof we will suppress the argument δ when referring to the functions that depend on it. It is straightforward to verify that $\tilde{C}'(\cdot)$ strictly increases with q over the restricted domain for which it is has so far been defined. For the remaining intervals, define $\tilde{C}'(\cdot)$ so that the entire function is strictly increasing and differentiable. Observe also that since C' is strictly increasing, if $\delta > 0$ is sufficiently close to zero, then $\tilde{C}'(\cdot)$ will lie strictly below $C(\cdot)$ on the interval $((C')^{-1}(B'(\mathcal{Q}^{\overline{n}})-\delta),q_s^{\overline{n}-1})$.

By construction, under the cost function $\tilde{C}$, a signatory to an IEA with $\overline{n}-1$ members chooses $q_s^{\overline{n}-1}$ while a non-signatory chooses $q_f^{\overline{n}-1}=q_f^{\overline{n}-1}-\delta$. To verify this, note that

$$Q^{\overline{n}-1} \equiv (\overline{n}-1)q_s^{\overline{n}-1}+(N-(\overline{n}-1))q_f^{\overline{n}-1} = Nq_s^{\overline{n}-1}-(N-(\overline{n}-1))\delta \equiv \mathcal{Q}^{\overline{n}-1} \tag{40}$$

so that

$$\tilde{C}'(q_f^{\overline{n}-1}) \equiv \tilde{C}'(q_s^{\overline{n}-1}-\delta) \underbrace{=}_{\text{from (39)}} B'(Q^{\overline{n}-1})$$

$$\tilde{C}'(q_s^{\overline{n}-1}) \underbrace{=}_{\text{from (39)}} C'(q_s^{\overline{n}-1}) \underbrace{=}_{\text{from (37a)}} (\overline{n}-1)B'(\mathcal{Q}^{\overline{n}-1}) \underbrace{=}_{\text{from (40)}} (\overline{n}-1)B'(Q^{\overline{n}-1})$$

Similarly, a signatory to an IEA with $\overline{n}$ members chooses $q_s^{\overline{n}}$ while a non-signatory chooses $q_f^{\overline{n}}=q_s^{\overline{n}-1}-2\delta$, since

$$Q^{\overline{n}} = \overline{n}q_s^{\overline{n}}+(N-\overline{n})q_f^{\overline{n}} = Nq_s^{\overline{n}}-2(N-\overline{n})\delta = \mathcal{Q}^{\overline{n}} \tag{41}$$

so that

$$\tilde{C}'(q_f^{\overline{n}}) \equiv \tilde{C}'(q_s^{\overline{n}-1}-2\delta) \underbrace{=}_{\text{from (39)}} B'(Q^{\overline{n}})$$

$$\tilde{C}'(q_s^{\overline{n}}) \underbrace{=}_{\text{from (39)}} C'(q_s^{\overline{n}}) \underbrace{=}_{\text{from (37b)}} (\overline{n})B'(\mathcal{Q}^{\overline{n}}) \underbrace{=}_{\text{from (40)}} \overline{n}B'(Q^{\overline{n}})$$

We now show that $g(\overline{n})>0$. In the chain of relationships below, we use approximation signs ($\approx$) to indicate that the difference between the right and left hand sides of the approximation sign is $O(\delta)$.[10] Because δ can be chosen to be arbitrarily small, the approximations can be treated for our purposes as effectively equalities. From (3)

$$g(\overline{n}) = B(Q^{\overline{n}})-\tilde{C}(q_s^{\overline{n}})-[B(Q^{\overline{n}-1})-\tilde{C}(q_f^{\overline{n}-1})] = B(Q^{\overline{n}})-\left(\tilde{C}(q_s^{\overline{n}-1})+\int_{q_s^{\overline{n}-1}}^{q_s^{\overline{n}}}\tilde{C}'(q)\,dq\right)-[B(Q^{\overline{n}-1})-\tilde{C}(q_f^{\overline{n}-1})]$$

Because $\tilde{C}'$ is strictly increasing

$$g(\overline{n}) > B(Q^{\overline{n}})-(\tilde{C}(q_s^{\overline{n}-1})+\tilde{C}'(q_s^{\overline{n}})(q_s^{\overline{n}}-q_s^{\overline{n}-1}))-[B(Q^{\overline{n}-1})-\tilde{C}(q_f^{\overline{n}-1})]$$

Since $\tilde{C}(q_f^{\overline{n}-1}) \approx \tilde{C}(q_s^{\overline{n}-1})$

$$g(\overline{n}) \approx B(Q^{\overline{n}})-\tilde{C}'(q_s^{\overline{n}})(q_s^{\overline{n}}-q_s^{\overline{n}-1})-B(Q^{\overline{n}-1}) = \int_{Q^{\overline{n}-1}}^{Q^{\overline{n}}} B'(q)\,dq-\tilde{C}'(q_s^{\overline{n}})(q_s^{\overline{n}}-q_s^{\overline{n}-1}).$$

Since $B'<0$ and, from (38), $Q^{\overline{n}}-Q^{\overline{n}-1}=N(q_s^{\overline{n}}-q_s^{\overline{n}-1})-\delta(N-(\overline{n}-1))>0$,

$$g(\overline{n}) > B'(Q^{\overline{n}})(Nq_s^{\overline{n}}-Nq_s^{\overline{n}}-\delta(N-(\overline{n}-1)))-\tilde{C}'(q_s^{\overline{n}})(q_s^{\overline{n}}-q_s^{\overline{n}-1}) \approx B'(Q^{\overline{n}})(Nq_s^{\overline{n}}-Nq_s^{\overline{n}})-\tilde{C}'(q_s^{\overline{n}})(q_s^{\overline{n}}-q_s^{\overline{n}-1})$$
$$= (NB'(Q^{\overline{n}})-\tilde{C}'(q_s^{\overline{n}}))(q_s^{\overline{n}}-q_s^{\overline{n}-1}) \geq (\overline{n}B'(Q^{\overline{n}})-\tilde{C}'(q_s^{\overline{n}}))(q_s^{\overline{n}}-q_s^{\overline{n}-1}) \underbrace{=}_{\text{from (37b)}} 0 \quad \square$$

References

[1] E. Baker, K. Adu-Bonnah, Investment in risky R&D programs in the face of climate uncertainty, Energy Economics 30 (2) (2008) 465–486.
[2] E. Baker, L. Clarke, E. Shittu, Technical change and the marginal cost of abatement, Energy Economics 30 (2008) 2799–2816.
[3] S. Barrett, Self-enforcing international environmental agreements, Oxford Economic Papers 46 (1994) 878–894.
[4] S. Barrett, Consensus treaties, Journal of Institutional and Theoretical Politics 158 (2002) 519–541.
[5] S. Barrett, Environment and Statecraft, Oxford University Press, 2003.
[6] S. Barrett, Climate treaties and breakthrough technologies, American Economic Review 96 (2006) 22–25.
[7] Y. Bauman, M. Lee, K. Seeley, Does technological innovation really reduce marginal abatements costs? Environmental Resource Economics 40 (2008) 507–527.

[10] *Big O notation*: We write $x(\delta)-y(\delta)=O(\delta)$, if there exists $\alpha \in \mathbb{R}$ such that as δ approaches zero, $|x(\delta)-y(\delta)|<\alpha\delta$.

[8] F. Bloch, B. Dutta, Formation of networks and coalitions, in: J. Benhabib, A. Bisin, M. Jackson (Eds.), Handbook of Social Economics, 2008.
[9] I. Bos, J. Harrington Jr., Endogenous cartel formation with heterogeneous firms, The RAND Journal of Economics 41 (1) (2010) 92–117.
[10] N. Burger, C. Kolstad, Voluntary Public Goods Provision: Coalition Formation and Uncertainty, NBER Working Papers 15543, 2009.
[11] C. Carraro, D. Siniscalco, Strategies for the international protection of the environment, Journal of Public Economics 52 (1993) 309–328.
[12] M.S. Chwe, Farsighted coalitional stability, Journal of Economic Theory 63 (1994) 299–325.
[13] A. Dannenberg, A. Lange, B. Sturm, On the Formation of Coalitions to Provide Public Goods—Experimental Evidence from the Lab, Working paper, 2009.
[14] C. D'Aspremont, A. Jacquemin, J.J. Gabszewicz, J.A. Weymark, On the stability of collusive price leadership, The Canadian Journal of Economics 16 (1) (1983) 17–25.
[15] R. De Bondt, Spillovers and innovative activities, International Journal of Industrial Organization 15 (1) (1997) 1–28.
[16] A. de Zeeuw, Dynamic effects on the stability of international environmental agreements, Journal of Environmental Economics and Management 55 (2) (2008) 163–174.
[17] E. Diamantoudi, E. Sartzetakis, International Environmental Agreements—The Role of Foresight, University of Aarhus Working Paper 2002-10, 2002.
[18] E. Diamantoudi, E. Sartzetakis, Stable international environmental agreements: an analytical approach, Journal of Public Economic Theory 8 (2) (2006) 247–263.
[19] A. Dixit, M. Olson, Does voluntary participation undermine the Coase Theorem, Journal of Public Economics 76 (2000) 309–335.
[20] M. Donsimoni, N. Economides, H. Polemarchakis, Stable cartels, International Economic Review 27 (2) (1986) 317–327.
[21] G. Ecchia, M. Mariotti, Coalition formation in international environmental agreements and the role of institutions, European Economic Review 42 (3–5) (1998) 573–582.
[22] A. Enderes, C. Ohl, International environmental cooperation with risk aversion, International Journal of Sustainable Development 6 (2003) 378–392.
[23] A. Endres, T. Friehe, Incentives to diffuse advanced abatement technology under environmental liability law, Journal of Environmental Economics and Management 62 (1) (2011) 30–40.
[24] M. Escrihuela-Villar, A note on cartel stability and endogenous sequencing with tacit collusion, Journal of Economics 96 (2) (2009) 137–147.
[25] J. Eyckmans, On the Farsighted Stability of the Kyoto Protocol, CLIMNEG Working Paper 40, CORE, Universite Catholique de Louvain, 2001.
[26] M. Finus, Game Theory and International Environmental Cooperation, Edward Elgar, Cheltenham, 2001.
[27] M. Finus, 3. Stability and Design of International Environmental Agreements: The Case of Transboundary Pollution, The International Yearbook of Environmental and Resource Economics 2003/2004: A Survey of Current Issues, 2003, p. 82.
[28] M. Finus, S. Maus, Modesty may pay! Journal of Public Economic Theory 10 (2008) 801–826.
[29] G. Heal, H. Kunreuther, Tipping Climate Negotiations, NBER Working Paper 16954, 2011.
[30] M. Hoel, International environmental conventions: the case of uniform reductions of emissions, Environmental and Resource Economics 2 (1992) 141–159.
[31] M. Hoel, A. De Zeeuw, Can a focus on breakthrough technologies improve the performance of international environmental agreements, Environmental and Resource Economics (2010) 1–12.
[32] F. Hong, L. Karp, International environmental agreements with mixed strategies and investment, Journal of Public Economics 96 (2012) 685–697.
[33] A. Ioannidis, A. Papandreou, E. Sartzetakis, International Environmental Agreements: A Literature Review, Cahiers de Recherche du GREEN 00-08, Université Laval, Québec, 2000.
[34] L. Karp, The effect of learning on membership and welfare in an International Environmental Agreement, Climatic Change 110 (2012) 499–505.
[35] M. Katz, An analysis of cooperative research and development, The RAND Journal of Economics 17 (4) (1986) 527–543.
[36] M. Kohler, Coalition formation in international monetary policy games, Journal of International Economics 56 (2) (2002) 371–385.
[37] C.D. Kolstad, Systematic uncertainty in self-enforcing international environmental agreements, Journal of Environmental Economics and Management 53 (2007) 68–79.
[38] C.D. Kolstad, A. Ulph, Learning and international environmental agreements, Climatic Change 89 (2008) 125–141.
[39] M. Kosfeld, A. Okada, A. Riedl, Institution formation in public goods games, American Economic Review 99 (2009) 1335–1355.
[40] D. Osmani, R. Tol, Toward farsightedly stable international environmental agreements, Journal of Public Economic Theory 11 (2009) 455–492.
[41] D. Ray, R. Vohra, Coalitional power and public goods, Journal of Political Economy 109 (6) (2001) 1355–1382.
[42] S.J. Rubio, A. Ulph, Self-enforcing international environmental agreements revisited, Oxford Economic Papers 58 (2006) 233–263.
[43] J.E. Stiglitz, A New Agenda for Global Warming, Economists' Voice, 2006, pp. 1–4.
[44] S. Thoron, Formation of a coalition-proof stable cartel, Canadian Journal of Economics 31 (1) (1998) 63–76.
[45] A. Ulph, Stable international environmental agreements with a stock pollutant, uncertainty and learning, Journal of Risk and Uncertainty 29 (2004) 53–73.
[46] L. Xue, Coalitional stability under perfect foresight, Economic Theory 11 (1998) 603–627.

[9]

Int J Game Theory (2001) 30:141–146

International Journal of Game Theory

On the existence of a cooperative solution for a coalitional game with externalities**

Carsten Helm*

School of Business and Economics, Humboldt-University Berlin, Spandauer Str. 1, D-10178 Berlin, Germany (e-mail: helm@wiwi.hu-berlin.de)

Received June 2000/Revised version March 2001

Abstract. This paper analyzes a game in coalitional form that is derived from a simple economy with multilateral externalities. Following Chander and Tulkens (1997) we assume that agents react to a blocking coalition by choosing individual best reply strategies. A non-empty core of this game is established by showing that the game is balanced. The proof relies only on standard convexity assumptions and, therefore, substantially generalizes the results in Chander and Tulkens (1997).

Key words: coalitional games, γ-core, multilateral externalities, balanced games

1. Introduction

We analyze a game in coalitional form that is derived from a simple economy with multilateral externalities. Due to those externalities, the payoff of a coalition that blocks global cooperation crucially depends on the reaction of the other players. Often the standard assumption that those external agents act to minimize the coalitional payoff seems inappropriate (Rosenthal 1971).

In the context of transboundary environmental problems, for example, increasing own emissions so as to minimize the payoff of a defecting coalition (α-characteristic function) would inflict substantial damages upon the punishing agents themselves. Therefore, Chander and Tulkens (1997) have proposed the alternative blocking rule that external agents choose their action individually as a best reply to the coalitional action. This leads to the 'partial agreement characteristic function' or, for short, the γ-characteristic function.

** I would like to thank Henry Tulkens, Parkash Chander, Dominique Demougin, Bertrand Koebel, Bengt-Arne Wickström and two anonymous referees for useful comments and suggestions.

Based on this behavioral assumption about coalition external agents, we prove that a coalitional game with multilateral externalities is balanced and, therefore, has a non-empty core. As this approach requires only standard convexity assumptions about payoff functions, it substantially generalizes the results in Chander and Tulkens (1997).

2. The model and some definitions

A coalitional game with transferable utility (N, v) consists of a finite set N of players, indexed by $i, j = \{1, 2, \ldots, n\}$, and a characteristic function v that specifies the worth $v(S)$ of every non-empty coalition $S \subseteq N$. We consider a simple economy with multilateral externalities. A unique private consumption good is produced at the nonnegative level $x_i \geq 0$ from the quantity of input $e_i \geq 0$. The technology is described by a production function $x_i = f_i(e_i)$, which is assumed to be increasing, differentiable and concave in e_i. Each agent's preferences are represented by a quasilinear utility function $u_i(x_i, e) = x_i - d_i(e)$. Here, $d_i(e)$ is i's disutility as a positive, increasing, differentiable and convex function of the level of the externality $e \equiv \sum_{i \in N} e_i$. This setting is general enough to comprise a large class of environmental pollution problems. For concreteness, we refer to $d_i(e)$ as agent i's damages caused by overall emissions e.

Agents outside of a blocking coalition choose their individual best reply to the action of all other agents. This behavioral assumption underlies the *γ-characteristic function* (Chander and Tulkens 1997), which is defined by

$$v^{\gamma}(S) = \max_{\{(x_i, e_i)_{i \in S}\}} \sum_{i \in S} u_i(x_i, e) \tag{1}$$

$$\text{subject to} \sum_{i \in S} x_i \leq \sum_{i \in S} f_i(e_i) \quad \text{and} \quad e = \sum_{i \in S} e_i + \sum_{i \in N \backslash S} e_i, \tag{2}$$

$$\text{where for all } i \in N \backslash S, \quad (x_i, e_i) \text{ maximizes } u_i(x_i, e) \tag{3}$$

$$\text{subject to } x_i \leq f_i(e_i) \quad \text{and} \quad e = e_i + \sum_{j \in N \backslash \{i\}} e_j. \tag{4}$$

A payoff vector $(\pi_i)_{i \in N}$ lies in the *γ-core* if

$$\sum_{i \in N} \pi_i = v(N), \quad \text{and for all } S \subset N : v^{\gamma}(S) \leq \sum_{i \in S} \pi_i. \tag{5}$$

The first part establishes a feasibility condition and the second part a stability condition for the grand coalition N to form.

One way to prove a non-empty core is to show that the underlying game is convex (Shapley 1971). However, this is not the case for our model (Eyckmans 1997). Chander and Tulkens (1997) have taken the alternative route of pointing out a specific allocation in the core. In particular, they proposed a version of the ratio equilibrium (Kaneko 1977), which would allocate the benefits of emissions such that

$$x_i^* = f_i(e_i^N) - \frac{d_i'(e^*)}{\sum_{i \in N} d_i'(e^*)} \sum_{i \in N} (f_i(e_i^N) - f_i(e_i^*)) \quad \text{for all } i \in N, \tag{6}$$

where superscripts * and ^{N} denote Pareto-efficient respectively Nash equilibrium levels. Accordingly, each agent gets the goods he produces in the Nash equilibrium minus a weighted share of the difference between the Nash production level and the Pareto-efficient production level.

However, agents with relatively high marginal damages would require substantial emission reductions in the efficient solution. Therefore, the other agents may have little interest to cooperate with them and form a defecting coalition, at least if there are sufficient gains from cooperation within this coalition. Furthermore, given the high marginal damages of the coalition external agents, their best reply strategy will involve only low emissions and, therefore, not provide a serious threat to the defecting coalition.[1]

To exclude such cases, Chander and Tulkens (1997) assume that for all $S \subset N$, $|S| \geq 2 : \sum_{j \in S} d_j'(e^*) \geq d_i'(e^N)$, $i \in S$. Using the Bondareva-Shapley theorem, we will now show that a non-empty γ-core exists even for those situations where the ratio equilibrium fails.

3. Balancedness of a coalitional game with multilateral externalities

The Bondareva-Shapley theorem elaborates on the following concept of balanced weights. Let $\mathscr{C}$ denote the set of all coalitions that can be formed and $\mathscr{C}_i = \{S \in \mathscr{C} : i \in S\}$ the subset of those coalitions of which i is a member. A vector $(\delta_S)_{S \in \mathscr{C}} : \delta_S \in [0, 1]$ for all $S \in \mathscr{C}$ is a *balanced collection of weights* if for all $i \in N : \sum_{S \in \mathscr{C}_i} \delta_S = 1$.

Proposition 1. *(Bondareva 1963; Shapley 1967) A coalitional game with transferable payoff (N, v) has a non-empty core if and only if for every balanced collection of weights $\sum_{S \in \mathscr{C}} \delta_S v(S) \leq v(N)$.*

Accordingly, for the core to be non-empty there must be no feasible pattern of coalition formation that yields a higher total payoff than the grand coalition can achieve. In showing that this condition is satisfied for the above economic-ecological model, the following lemma will prove useful. We denote an optimal input vector for coalition S by $(e_i^S)_{i \in S}$, and the reply of coalition external agents by $(\tilde{e}_i^S)_{i \in N \setminus S}$. Hence, total emissions in a 'partial agreement equilibrium' are given by $e^P \equiv \sum_{i \in S} e_i^S + \sum_{i \in N \setminus S} \tilde{e}_i^S$.

Lemma 1. *For the coalitional game with multilateral externalities (N, v^γ),*

$$\sum_{S \in \mathscr{C} \setminus \mathscr{C}_i} \delta_S \sum_{j \in S} e_j^S \leq \sum_{S \in \mathscr{C}_i} \delta_S \sum_{j \in N \setminus S} \tilde{e}_j^S. \tag{7}$$

Proof: Chander and Tulkens (1997, 389) have shown that $e^P \leq e^N$ and $\tilde{e}_j^S \geq$

[1] A simple example is the following three country model with quadratic benefit and damage functions: $f_1(e_1) = 14e_1 - 2e_1^2$, $d_1(e) = 4e^2$, $f_2(e_2) = e_2 - 0.075e_2^2$, $d_2(e) = 0.05e^2$, $f_3(e_3) = 6e_3 - 0.5e_3^2$, $d_3(e) = 0.95e^2$. Solving the model yields $v^\gamma(\{2, 3\}) = 6 > \sum_{i=2}^{3} u_i(x_i^*, e^*) = 3.64$, hence the coalition $S = \{2, 3\}$ has an objection. Note that in the efficient solution only country 1 has positive emissions ($e_1^* = 1$), in the Nash equilibrium only country 2 ($e_2^N = 4$), and in the partial agreement equilibrium only country 3 ($e_3^S = 2$).

e_j^N for all $j \in N \backslash S$ (see Appendix). Intuitively, external agents free-ride on the emission reductions achieved by cooperation within coalition S. It follows immediately that $\sum_{j \in S} e_j^S \leq \sum_{j \in S} e_j^N$. Hence

$$\sum_{S \in \mathscr{C} \backslash \mathscr{C}_i} \delta_S \sum_{j \in S} e_j^S \leq \sum_{S \in \mathscr{C} \backslash \mathscr{C}_i} \delta_S \sum_{j \in S} e_j^N \tag{8}$$

$$= \sum_{S \in \mathscr{C}} \delta_S \sum_{j \in S} e_j^N - \sum_{S \in \mathscr{C}_i} \delta_S \sum_{j \in S} e_j^N \tag{9}$$

$$= \sum_{j \in N} e_j^N - \sum_{S \in \mathscr{C}_i} \delta_S \sum_{j \in S} e_j^N \tag{10}$$

$$= \sum_{S \in \mathscr{C}_i} \delta_S \left(\sum_{j \in N} e_j^N - \sum_{j \in S} e_j^N \right) \tag{11}$$

$$= \sum_{S \in \mathscr{C}_i} \delta_S \sum_{j \in N \backslash S} e_j^N \tag{12}$$

$$\leq \sum_{S \in \mathscr{C}_i} \delta_S \sum_{j \in N \backslash S} \tilde{e}_j^S. \quad \square$$

Proposition 2. *The coalitional game with multilateral externalities (N, v^γ) is balanced and, therefore, has a non-empty γ-core.*

Proof: For each $i \in N$ let $e_i^{\mathscr{C}} = \sum_{S \in \mathscr{C}_i} \delta_S e_i^S$ denote i's total emissions from emitting in each coalition of which it is a member the optimal level e_i^S, weighted by the associated δ_S. By concavity of the production function,

$$f_i(e_i^{\mathscr{C}}) = f_i\left(\sum_{S \in \mathscr{C}_i} \delta_S e_i^S \right) \geq \sum_{S \in \mathscr{C}_i} \delta_S f_i(e_i^S). \tag{13}$$

Turning to the damage function one gets

$$d_i\left(\sum_{j \in N} e_j^{\mathscr{C}} \right) = d_i\left(\sum_{j \in N} \sum_{S \in \mathscr{C}_j} \delta_S e_j^S \right) = d_i\left(\sum_{S \in \mathscr{C}} \delta_S \sum_{j \in S} e_j^S \right) \tag{14}$$

$$= d_i\left(\sum_{S \in \mathscr{C}_i} \delta_S \sum_{j \in S} e_j^S + \sum_{S \in \mathscr{C} \backslash \mathscr{C}_i} \delta_S \sum_{j \in S} e_j^S \right) \tag{15}$$

$$\leq d_i\left(\sum_{S \in \mathscr{C}_i} \delta_S \sum_{j \in S} e_j^S + \sum_{S \in \mathscr{C}_i} \delta_S \sum_{j \in N \backslash S} \tilde{e}_j^S \right) \tag{16}$$

$$= d_i\left(\sum_{S \in \mathscr{C}_i} \delta_S \left(\sum_{j \in S} e_j^S + \sum_{j \in N \backslash S} \tilde{e}_j^S \right) \right) \tag{17}$$

$$\leq \sum_{S \in \mathscr{C}_i} \delta_S d_i\left(\sum_{j \in S} e_j^S + \sum_{j \in N \backslash S} \tilde{e}_j^S \right). \tag{18}$$

Here, (14) follows from the fact that j's total emissions from passing through all coalitions of which it is a member summed over all agents are equivalent to the emissions of each individual coalition S summed over all coalitions. Lemma 1 is used in the step leading to (16), and (18) follows from convexity of the damage function.

Finally, by cohesiveness of the coalitional game, what means that the worth of the grand coalition is at least as large as the sum of the worths of the members of any partition of N, and the previous steps

$$v(N) \geq \sum_{i \in N} f_i(e_i^{\mathscr{C}}) - \sum_{i \in N} d_i\Big(\sum_{j \in N} e_j^{\mathscr{C}}\Big) \tag{19}$$

$$\geq \sum_{i \in N} \sum_{S \in \mathscr{C}_i} \delta_S f_i(e_i^S) - \sum_{i \in N} \sum_{S \in \mathscr{C}_i} \delta_S d_i\Big(\sum_{j \in S} e_j^S + \sum_{j \in N \backslash S} \tilde{e}_j^S\Big) \tag{20}$$

$$= \sum_{S \in \mathscr{C}} \delta_S \sum_{i \in S} f_i(e_i^S) - \sum_{S \in \mathscr{C}} \delta_S \sum_{i \in S} d_i\Big(\sum_{j \in S} e_j^S + \sum_{j \in N \backslash S} \tilde{e}_j^S\Big) \tag{21}$$

$$= \sum_{S \in \mathscr{C}} \delta_S \Big(\sum_{i \in S} f_i(e_i^S) - \sum_{i \in S} d_i\Big(\sum_{j \in S} e_j^S + \sum_{j \in N \backslash S} \tilde{e}_j^S\Big)\Big) \tag{22}$$

$$= \sum_{S \in \mathscr{C}} \delta_S v^{\gamma}(S), \tag{23}$$

where the last step follows from the definition of the γ-characteristic function. □

4. Concluding remarks

Potential areas for future research are generalizations of the above economic-ecological model and, in particular, a systematic analysis of a wider range of alternative conjectures about blocking rules. One such alternative, where agents $N \backslash S$ form coalitions and choose coalitional (rather than individual) best-reply strategies, has been analyzed by Funaki and Yamato (1999). As they show for a simple game in which agents have identical production functions, this further weakening of the threat to blocking coalitions may lead to an empty core. However, consistency seems to require that also the stability of $N \backslash S$ itself – now being a coalition – needs to be addressed (Ray and Vohra 1997).

5. Appendix

Proof of $\tilde{e}_i^S \geq e_i^N \; \forall i \in N \backslash S$ *and* $e^P \leq e^N$ *(Chander and Tulkens 1997)*

By contradiction, assume that $e^P > e^N$. Convexity of damage functions then implies $d_i'(e^P) \geq d_i'(e^N)$. Together with the conditions for the partial agreement equilibrium and the Nash equilibrium one gets

$$f_i'(e_i^S) = \sum_{j \in S} d_j'(e^P) \geq d_i'(e^P) \geq d_i'(e^N) = f_i'(e_i^N) \; \forall i \in S, \quad \text{and} \tag{24}$$

$$f_i'(\tilde{e}_i^S) = d_i'(e^P) \geq d_i'(e^N) = f_i'(e_i^N) \; \forall i \in N \backslash S. \tag{25}$$

By concavity of production functions this would imply that emissions in the partial agreement equilibrium are lower than those in the Nash equilibrium for all $i \in N$ – a contradiction. Hence $e^P \leq e^N$, and therefore $d_i'(e^P) \leq d_i'(e^N)$. Reversing the sign in (25) accordingly yields $\tilde{e}_i^S \geq e_i^N \; \forall i \in N \backslash S$.

References

Bondareva ON (1963) Some applications of linear programming methods to the theory of cooperative games [in Russian]. Problemy Kibernetiki 10:119–139
Chander P, Tulkens H (1997) A core of an economy with multilateral environmental externalities. International Journal of Game Theory 26:379–401
Eyckmans J (1997) On the incentives of nations to join international environmental agreements. Ph.D. thesis, Katholieke Universiteit Leuven
Funaki Y, Yamato T (1999) The core of an economy with a common pool resource: A partition function form approach. International Journal of Game Theory 28(2):157–171
Kaneko M (1977) The ratio equilibria and the core of the voting game $G(N, W)$ in a public goods economy. Econometrica 45(7):1589–1594
Ray D, Vohra R (1997) Equilibrium binding agreements. Journal of Economic Theory 73:30–78
Rosenthal RW (1971) External economies and cores. Journal of Economic Theory 3:182–188
Shapley LS (1967) On balanced sets and cores. Naval Research Logistics Quarterly 14:453–460
Shapley LS (1971) Cores of convex games. International Journal of Game Theory 1:11–26

Part III
Compliance

[10]

JOURNAL OF ENVIRONMENTAL ECONOMICS AND MANAGEMENT **32**, 309–322 (1997)
ARTICLE NO. EE970971

Fishing as a Supergame

Rögnvaldur Hannesson

The Norwegian School of Economics and Business Administration, Helleveien 30, N-5035 Bergen-Sandviken, Norway

Received February 6, 1996; revised May 15, 1996

This paper considers how cooperative solutions to games of sharing fish resources can be supported by threat strategies. With highly mobile fish stocks, the number of agents compatible with a cooperative self-enforcing solution is not very high for reasonable values of the discount rate, but sensitive to changes in the discount rate and costs and to cost heterogeneity. With migrating stocks, where growth and reproduction depend on how much all agents leave behind after harvesting, the likelihood of a cooperative, self-enforcing equilibrium is increased. With a dominant player and a competitive fringe the rents and optimum stock level of the dominant player fall quickly as the share of the competitive fringe increases. © 1997 Academic Press

1. INTRODUCTION

Exploitation of renewable resources such as fish stocks shared by a limited number of agents involves strategic choices. Should the agents cooperate and maximize their aggregate returns, or is cooperation a futile exercise undertaken by the naive, who will be outsmarted by the realistic? There exists by now a voluminous literature on this subject. Much of this literature was inspired by the international disputes over fishing limits in the 1970s and the deliberations of the third United Nations Conference on the Law of the Sea 1973–1982 which endorsed the 200-mile exclusive economic zone. Two early papers in this genre are Munro [11] and Levhari and Mirman [9]. Recently Fischer and Mirman [2, 3] have extended the analysis to interacting species. Three of these papers compare Nash equilibria and global optima while Munro is concerned with bargaining solutions in cooperative games. All these papers deal with the case of two agents, which is true of most of the literature on game theory and fisheries; few authors have considered explicitly the importance of the number of agents for obtaining a cooperative solution. An exception is Clark [1] who does not, however, consider the implications of threat strategies in repeated games. Hämäläinen *et al.* [6] do so, but again limit themselves to the case of two agents.

The motivation to explore the importance of the number of agents is provided by the fact that many important fish stocks enclosed by the 200-mile limit are shared by two or more coastal states. What is the likelihood that they will cooperate in setting the rate of exploitation of the stocks and how does it depend on the number of states sharing a stock? A further motivation is provided by the straddling of some fish stocks outside the 200-mile limit where they are accessible by fleets of any nationality. What are the chances of cooperation under those circumstances, or more to the point perhaps, how will the strategy of the coastal state(s) controlling the main part of a stock be affected by this competition?

0095-0696/97 $25.00

This latter case is a topical one. In 1993 the UN convened a conference on the exploitation of straddling and highly migratory fish stocks. The conference was concluded in August 1995 with an agreement that authorizes regional organizations to manage fisheries outside the 200-mile limit. All who fish in a given area will have to abide by the rules agreed by the relevant organization. It is unclear, however, what if anything limits the membership of such organizations; the agreement says only that member nations must have a "real interest" in fishing in the area. Neither is it clear how decisions will be taken, whether this will be by majority vote, a qualified majority, or a consensus. Last but not least, it is unclear what sanctions can and will be applied to those who do not cooperate.

In this paper we consider how critical the number of agents sharing a fish stock is for realizing the cooperative (globally optimal) solution. The problem is formulated as a repeated game of an infinite duration (a supergame). We begin by considering a game of N identical agents. Then we look into the importance of cost differences among agents. In both these formulations agents are assumed to exploit a truly common stock; i.e., the density of the stock will be the same for all. We then consider a somewhat different problem which may be more relevant for countries sharing a fish stock. Each agent is assumed to exploit a certain portion of the stock being accessible for exploitation only in that agent's territory but nevertheless linked to the remainder of the stock through a common growth function. The cooperative solution is defined as the global optimum, while the noncooperative solution is the one where each agent realizes his individual optimum without taking into account the stock growth externality. Finally we look at a case meant to reflect the straddling stock situation where a dominant agent maximizes his profits, taking into account a fringe of exploiters who have access to a portion of the stock outside his territory. The focus here is on the efficiency losses vis-à-vis the global optimum.

2. THE BASIC MODEL

Assume that the growth of the fish stock is determined by how much is left behind after harvesting; i.e., the stock at the beginning of period t is a function $G(S_{t-1})$, where S_{t-1} is the stock left behind after harvesting in period $t-1$ (the size of the stock left behind will be referred to as the abandonment level). Ignoring natural mortality of the stock while it is being fished, the amount caught in period t will be $G(S_{t-1}) - S_t$. At a given price (p) the revenue (R) obtained in period t will be

$$R_t = p[G(S_{t-1}) - S_t]. \tag{1}$$

Assume that the marginal cost of fish is inversely proportional to the size of the stock at any point in time.[1] The total cost (C) in period t will then be

$$C_t = \int_{S_t}^{G(S_{t-1})} \frac{c}{x}\, dx = c[\ln G(S_{t-1}) - \ln S_t], \tag{2}$$

[1] This cost function obtains if the cost per unit of fishing effort is constant and the catch per unit of effort is proportionate to the size of the exploited stock. The latter obtains if the stock is always evenly distributed over a given area. While popular and not unreasonable, this is obviously a special case. The emphasis here is on obtaining numerical results, which makes it necessary to use a simple but not unreasonable cost function. For a further discussion, see [7].

where c is a cost parameter. Since the quantity caught (Q) is $Q = G - S$, with G being given at the beginning of each period, this function has the usual properties $C_Q = -C_S > 0$ and $C_{QQ} = -C_{SS} > 0$ (subscripts denote derivatives).

The present value (V) of fishing rent ($R - C$), for an infinite time horizon, is

$$V = \sum_{t=0}^{\infty} \delta^t \{p[G(S_{t-1}) - S_t] - c[\ln G(S_{t-1}) - \ln S_t]\}. \qquad (3)$$

Maximizing V with respect to S_t gives the first order condition[2]

$$-(p - c/S_t^0) + \delta[p - c/G(S_t^0)]G'(S_t^0) = 0, \qquad (4)$$

where $\delta = 1/(1 + r)$ is the discount factor, r being the discount rate, with G' denoting the first derivative of G and S^0 the optimum value of S.

As an illustration in numerical calculations below we shall use the discrete variant of the logistic growth function

$$G(S) = S[1 + a(1 - S/K)], \qquad (5)$$

where a and K are parameters (intrinsic growth rate and carrying capacity, respectively). This gives

$$G'(S) = 1 + a(1 - 2S/K). \qquad (6)$$

3. COOPERATIVE EQUILIBRIUM: IDENTICAL AGENTS

Suppose there are N identical agents who share a fish stock. Suppose further that they plan to harvest the stock for an indefinite period of time. If they cooperate in realizing the optimal solution (which is identical from everybody's perspective) each will get $1/N$th of the total profits in each period. If one of them deviates from the optimal solution he will get more, as long as the deviation has not been discovered and punished. Assume that deviation would be detected after one period and that the other agents then would retaliate by fishing down the stock in each period until further depletion becomes unprofitable, i.e., until the marginal cost of fish caught has risen to equal the price (cf. Eq. (2)). This is in fact the best strategy they could follow, as long as the deviating agent depletes the stock to the level where fishing becomes unprofitable (an alternative trigger strategy will be commented upon below). In view of the above cost and revenue functions (Eqs. (1) and (2)), the abandonment level of the stock (S^*) would then be

$$S^* = c/p. \qquad (7)$$

[2] The same first order condition also obtains for a finite time horizon except for the last period where the optimal abandonment level is $S^* = c/p$ when the stock beyond the horizon has no value. If the initial stock is less than the optimal stock it will be necessary to leave it unfished for one or more periods, until $G(S_{t-1}) > S^0$.

For an infinite time horizon, the present value of the cooperative strategy (V^0) for a typical agent is

$$V^0 = \frac{\pi^0}{N} \frac{1}{1-\delta}, \tag{8}$$

where

$$\pi^0 = p[G(S^0) - S^0] - c[\ln G(S^0) - \ln S^0] \tag{9}$$

and S^0 is the abandonment level of the stock along the optimal (cooperative) stationary path.

The present value of the payoff for an agent that deviates from the cooperative solution and is then punished by all other agents playing noncooperatively forever is

$$V^d = \frac{\pi^0}{N} + \pi^d + \frac{\pi^*}{N} \frac{\delta}{1-\delta}, \tag{10}$$

where

$$\pi^d = p(S^0 - S^*) - c(\ln S^0 - \ln S^*) \tag{11}$$

and

$$\pi^* = p[G(S^*) - S^*] - c[\ln G(S^*) - \ln S^*]. \tag{12}$$

In the first period, the defector gets the same profit as in the cooperative solution, as all other participants play cooperatively, and in addition he gets the profit of driving the stock down to the noncooperative abandonment level. In the second and all later periods he will be punished by all other agents playing noncooperatively and gets only the profit obtained in the noncooperative solution (cf. Eq. (7)).

If defection is not profitable, $V^0 > V^d$, which implies

$$N < \frac{\delta}{1-\delta} \frac{\pi^0 - \pi^*}{\pi^d}. \tag{13}$$

As $\delta \to 1$ the right-hand side of (13) approaches infinity and defection will never be profitable; the losses from being punished will always outweigh the temporary gains of defecting. For a positive discount rate ($\delta < 1$) the temporary gains of defecting may outweigh the long term loss of playing noncooperatively rather than cooperatively. How likely this is depends on N, the number of players. The gains from defecting accrue to the defector while the losses from playing noncooperatively rather than cooperatively are shared by all participants. The temptation of defecting therefore becomes greater the more participants there are. Specifying the parameters of the growth equation $G(S)$ makes it possible to find the critical value of N for alternative values of the discount factor. A selection of results is presented in Fig. 1.

The values of the parameters a, c, and δ in Fig. 1 are meant to reflect a realistic range. The parameter a is the maximum relative rate of growth (cf. Eq. (6)). Figure 1 shows solutions for two alternative values of a, 20 and 50%. The ratio c/S shows the marginal cost of fish (cf. Eq. (2)). The maximum size of the stock is K (cf. Eq.

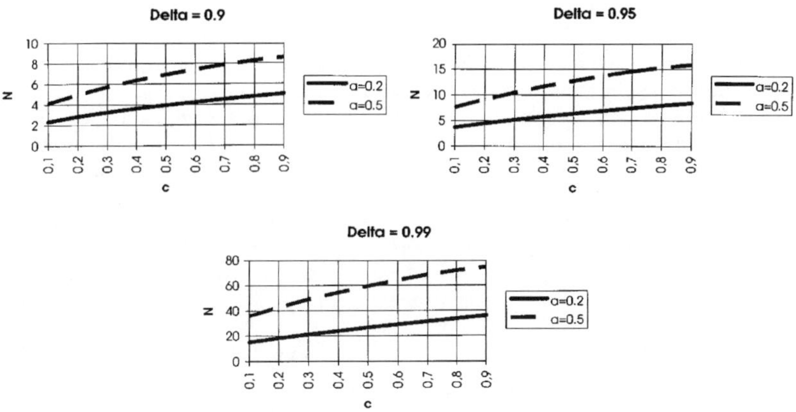

FIG. 1. Relationship between the maximum number of players in a cooperative solution and the minimum marginal cost of fish.

(5)), so c/K can be interpreted as the minimum marginal cost of fish, prevailing at the beginning of exploitation of a pristine stock. Since K has been set equal to 1, c shows the lowest marginal cost of fish as a fraction of the pristine stock. The values of c in Fig. 1 range from 0.1 to 0.9. A value of 1 would imply that fishing would never be profitable while a value of 0 would imply no cost of fishing. Finally, the three values of δ in Fig. 1 imply discount rates of approximately 1, 5, and 11%.

Figure 1 indicates that N, the maximum number of participants in the fishery compatible with a self-enforcing cooperative solution, will be highly sensitive to the discount factor and the cost of fishing. As already mentioned, a higher discount rate (lower value of δ) makes the cooperative solution less likely. For $c = 0.1$ and $a = 0.2$ N is only 2 when $\delta = 0.9$, but rises to 3 and 15, respectively, as δ increases to 0.95 and 0.99. A higher marginal cost of fish makes it more likely that the cooperative solution will prevail; the abandonment level of the stock in the noncooperative solution will be higher and the gains from defection smaller. This is readily seen from the figure; as c increases from 0.1 to 0.9 N approximately doubles. Finally, raising a from 0.2 to 0.5 approximately doubles N for any given c.

If we interpret N as the number of countries which share a resource and have the necessary control over their fishing fleets, the results in Fig. 1 are not too discouraging. Whenever stocks are fully contained within the 200-mile zone but migrate between different national zones, the number of countries with access rights is usually highly restricted. Quite often there are just two countries sharing a stock, like Canada and the United States on George's Bank, while several countries share the stocks of the North Sea. A discount rate of the order of 5–10% appears realistic and would accommodate a cooperative solution among a few countries.

For the case of stocks located outside the 200-mile zone the conclusion is much more pessimistic. The number of potential exploiters of such stocks is high, as witnessed by the fact that some boat owners use flags of convenience when fishing on the high seas. Up to now this number has in fact been indefinite, as international law has not recognized any mechanism to limit access to fishing outside the 200-mile zone. It remains to be seen whether the agreement reached by the UN conference on highly migratory and straddling fish stocks will change this.

The results in Fig. 1 in part corroborate a tendency which Mesterton-Gibbons [10] has termed "comedy of the commons," namely that cooperation can arise spontaneously and be self-enforcing if the commons are sufficiently unproductive. In Fig. 1 we see that the number of participants that can be accommodated in a self-enforcing cooperative solution increases with the cost of exploitation. However, the more productive stock (the one with an intrinsic growth rate of 0.5) accommodates more participants in a self-enforcing cooperative solution than the less productive stock, which would seem to contradict the comedy of the commons. The reason why the more productive stock accommodates more participants in a self-enforcing cooperative equilibrium is that the losses from playing noncooperatively are greater, being associated with foregone potential growth, while the gains from defecting are due to fishing down the stock in one period.

It is possible to give an alternative interpretation of the results in Fig. 1. These results were derived on the assumption that the punishment of a defector would go on forever. This is unnecessarily heavy handed. The defector might promise to mend his ways, and all participants might be well advised to revert to the cooperative solution. However, in order to be an effective deterrent the punishment must go on for a sufficiently long time to make the cooperative strategy more attractive for a potential defector. To accomplish this the $N - 1$ participants might threaten to play the noncooperative strategy for a finite number of periods. We may ask what is the minimum number of periods (T) in which the noncooperative strategy must be played in order to deter a potential defector. Instead of (10), define $V^d(T)$ as[3]

$$V^d(T) = \frac{\pi^0}{N} + \pi^d + \sum_{t=1}^{T} \delta^t \frac{\pi^*}{N} + \frac{\delta^{T+1}}{1-\delta}\frac{\pi^0}{N}. \tag{10'}$$

An effective punishment requires $V^0 > V^d(T)$, which implies

$$T > \frac{\ln F}{\ln \delta}, \tag{14}$$

where

$$F = 1 - \frac{1-\delta}{\delta}\frac{N\pi^d}{\pi^0 - \pi^*}. \tag{15}$$

As $F \to 0$, the right-hand side of (14) approaches infinity. The value of N which gives $F = 0$ is precisely the same as would make (13) an equality. Hence it is possible to interpret the figures in Fig. 1 as the number of participants that will make it possible to deter a defector by threatening to play the noncooperative strategy for a finite number of periods.

[3] I am grateful to an anonymous referee for pointing this out. Punishment strategies that last for a finite period are discussed in [12], Chapter 8.6.

4. COOPERATIVE EQUILIBRIUM WHEN COSTS DIFFER AMONG AGENTS

The analysis so far has been based on the assumption that all agents have identical cost functions. However, what if some agents are more efficient than others? Such agents might be tempted to "undercut" high cost agents and fish down the stock to a level where the high cost agents are barred from entry.[4] The incentive to do so will be strongest when there are many high cost agents and few low cost agents. Here we shall look at an example with one low cost agent and $N - 1$ high cost agents, with cost parameters c_l and c_h, respectively, but otherwise identical cost functions.

The low cost agent will be able to fish down the stock to a lower abandonment level than the high cost agents without incurring losses. In the noncooperative solution two cases may arise. (i) The low cost agent depletes the stock to the level c_l/p, but since the stock at the beginning of each period is $G(c_l/p)$, the high cost agents can still do some profitable fishing if $G(c_l/p) > c_h/p$. (ii) The cost difference is so great that $G(c_l/p) < c_h/p$. In that case the low cost agent can undercut the high cost agents and exclude them from the fishery altogether. It would not, however, be profitable for the low cost agent to leave behind a smaller stock than that which gives $G(S^*) = c_h/p$ (i.e., exactly undercuts the high cost agents), unless the optimum stock for the low cost agent alone is less than this.

Hence the abandonment level of the stock in a noncooperative equilibrium will be

$$S^* = \max\left[\frac{c_l}{p}, G^{-1}\left(\frac{c_h}{p}\right)\right]. \tag{16}$$

The profit of the low cost agent in a noncooperative play may be split into two parts. The first is the profit of fishing down the stock from $G(S^*)$ to c_h/p. In this phase (which may be nonexistent) the low cost agent gets only $1/N$th of the catch. In the second phase the low cost agent drives the stock further down to S^* and has all this catch to himself. The two parts of the profit of the low cost agent thus are

$$\pi_1^* = p[G(S^*) - c_h/p] - c_l[\ln G(S^*) - \ln(c_h/p)] \tag{17a}$$

$$\pi_2^* = p[c_h/p - S^*] - c_l[\ln(c_h/p) - \ln S^*]. \tag{17b}$$

The gain from defection will be greater than before, because some and possibly all of the catch in the noncooperative equilibrium does not have to be shared with all agents but will be taken by the low cost agent alone. Equation (10) becomes

$$V^d = \frac{\pi^0}{N} + \pi^d + \left[\frac{\pi_1^*}{N} + \pi_2^*\right]\frac{\delta}{1-\delta} \tag{10''}$$

and (13) becomes

$$N < \frac{\delta(\pi^0 - \pi_1^*)}{\delta\pi_2^* + (1-\delta)\pi^d}. \tag{13''}$$

[4] This is what happens in the closed loop solution derived by Clark [1].

The question arises as to what would be the optimum stock level in the cooperative solution, since the high cost agents and the low cost agent will not agree on this. A global optimum would entail fishing by the low cost agent only, with side payments to the high cost agents. If this is not possible, the agents might agree on some mutually acceptable abandonment level of the stock that allows both low cost and high cost agents to make a profit. Here it will be assumed that the cooperative solution is determined with reference to the cost of the low cost agent, as this will minimize the likelihood that the low cost agent will want to defect. Note that this solution does not impose losses on the high cost agents as long as $pS^0 > c_h$.

As before, we can calculate the number of participants accommodated by a self-enforcing cooperative equilibrium (N). A sample of such results is shown in Fig. 2 and compared to the case with cost homogeneity. The dashed curves in Fig. 2 show N when all agents have the same cost (same as in Fig. 1). The solid curves show what happens when one agent has a lower cost parameter (c) than the rest. As the cost of one agent starts to fall below that of the rest N drops quickly, particularly when the cost is high. Note that when the cost is high N is also high, and the losses from ending up in a noncooperative solution are low compared to the gains from defecting, since the losses are spread among all players. This is why it takes only a slight difference in costs to produce a steep fall in N when N is high initially.

The effect of cost heterogeneity is seen to be substantial; it does not take a great difference in costs to reduce N to a number not much higher than two, irrespective of the discount rate, cost level, or productivity of the fish stock. There may be reason, therefore, for less optimism than expressed above for the likelihood of a cooperative solution emerging among a limited number of countries sharing a fish stock.

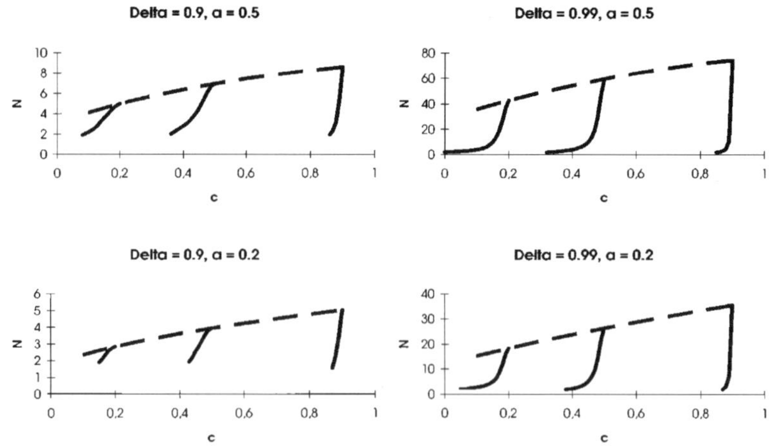

FIG. 2. How the minimum number of players in a cooperative solution depends on the cost difference between a low cost player and high cost players. Dashed line, equal cost for all players.

5. COOPERATIVE EQUILIBRIUM WITH LIMITED MIGRATION OF STOCKS ACROSS BOUNDARIES

In the above examples it was implicitly assumed that the agents fish simultaneously on the same stock or, what amounts to the same thing, that each agent fishes in his own area but that the fish redistribute themselves instantaneously between the areas so that the distribution is always the same in all areas. A more realistic example could be where each agent fishes in his own area and the fish migrate only slowly between areas or visit these areas sequentially with a seasonal pattern.[5] The growth of the stock could still be dependent on the aggregate size of the stock, because of a seasonal pattern of breeding migration or because the eggs and larvae are distributed over the entire habitat of the stock irrespective of where they are spawned.

Here we shall look at a stylized version of this. Let the stock be measured as density, i.e., units of fish per unit area (tonnes per square km, for example). The reason for taking this approach is that the marginal cost of fish depends on the density of the stock and thus indirectly on the size of the stock, provided that the area it occupies does not shrink in proportion to the depletion of the stock. To make the comparison with previous cases easier we define the unit area as the entire area that the stock occupies, assumed always to be of the same size. Assume that the area is divided among N agents. In each period we start with a stock (G) that is uniformly distributed over the area. The amount available for agent i will be $\gamma_i G$, where γ_i is agent i's share of the area where the stock is located.[6] The fish are assumed not to migrate between the agents' subareas during each fishing period, so each agent has full control over the abandonment level of the stock in his area ($\gamma_i S_i$). After each fishing period the stock grows and redistributes itself randomly over the entire area. This leads to the growth function

$$G(\Sigma\gamma_i S_{i,t-1}),$$

where

$$\Sigma\gamma_i = 1. \tag{18}$$

The growth of the stock depends on what all agents leave behind ($\Sigma\gamma_i S_i$), and all agents start fishing a stock of the same density at the beginning of each period $[G(\Sigma\gamma_i S_{i,t-1})]$.

The globally optimal solution will be the same as derived above (Eq. (4)). For this solution to be realized, each agent would have to leave behind a stock compatible with the global optimum, and he would get a share in the total profits equal to his share of the area where the stock is located. If, on the other hand, each agent optimizes for himself, the optimum stock to be left behind in period t

[5] The case of sequential fishing is dealt with in [8] and will not be considered here.

[6] The assumption of uniform distribution is unnecessarily strong. All that is required is that the stock always redistribute itself in the same way after each fishing period. The share parameters γ_i will then be constant.

will be given by maximizing a modified version of (3):

$$V_i = \sum_{t=0}^{\infty} \delta^t \gamma_i \{ p[G(\Sigma \gamma_j S_{j,t-1}) - S_{i,t}] - c[\ln G(\Sigma \gamma_j S_{j,t-1}) - \ln S_{i,t}] \}. \quad (19)$$

Note that the profit function is multiplied by γ_i because the stock is measured as density, with the unit area being equal to the entire habitat of the stock.

The first order condition for maximum is

$$-[p - c/G(\Sigma \gamma_j S_j^0)] + \delta \gamma_i G'(\Sigma \gamma_j S_j^0)[p - c/S_i^0] = 0. \quad (20)$$

We shall now define "defection" as optimization of each agent for himself, i.e., choosing the abandonment level which is optimal according to Eq. (20). We denote this by S^* and reserve the notation S^0 for the globally optimal abandonment level. We continue to assume that defection will be discovered after the period in which it occurs. In the period after defection occurs, other agents adjust their harvesting to attain the abandonment level (S^*), so in this case the punishment strategy does not begin to bite until two periods after defection begins. The stock returning at the beginning of the period after defection will be $G(S^0\gamma_{-i} + S^*\gamma_i)$, where the notation "$-i$" means all agents except agent i, the defecting agent. The condition for defection not being profitable is

$$\frac{\pi^0}{1-\delta} > \pi_1^d + \delta \pi_2^d + \frac{\delta^2}{1-\delta}\pi^*, \quad (21)$$

where

$$\pi_i^0 = \gamma_i \{ p[G(\Sigma \gamma_j S_j^0) - S_i^0] - c[\ln G(\Sigma \gamma_j S_j^0) - \ln S_i^0] \} \quad (22a)$$

$$\pi_{1,i}^d = \gamma_i \{ p[G(\Sigma \gamma_j S_j^0) - S_i^0] - c[\ln G(\Sigma \gamma_j S_j^0) - \ln S_i^0] \}$$
$$+ \gamma_i \{ p(S_i^0 - S_i^*) - c(\ln S_i^0 - \ln S_i^*) \} \quad (22b)$$

$$\pi_{2,i}^d = \gamma_i \{ p[G(\Sigma S_j^0 \gamma_{-i} + S_i^* \gamma_i) - S_i^*] - c[\ln G(\Sigma S_j^0 \gamma_{-i} + S_i^* \gamma_i) \text{-} \ln S_i^*] \} \quad (22c)$$

$$\pi_i^* = \gamma_i \{ p[G(\Sigma \gamma_j S_j^*) - S_i^*] - c[\ln G(\Sigma \gamma_j S_j^*) - \ln S_i^*] \}. \quad (22d)$$

The expression on the left-hand side of (21) is the present value of playing cooperatively while the expression on the right-hand side is the present value of defecting. In the period when defection occurs the defector gets the extra benefit of reducing the density of the stock in his area to the level S^* (Eq. (22b)). This benefit is, of course, less than in previous cases where the defector could reduce the overall density of the stock, due to its instantaneous migration. At the end of this period the defector is found out and everyone depletes his stock to density S^* from then onward, but the stock emerging at the beginning of the period after defection has not yet reached the long term equilibrium (Eq. (22c)). The noncooperative equilibrium is reached in the second period after defection (Eq. (22d)).

The maximum number of players that can be accommodated in a cooperative equilibrium is defined implicitly by (21). Table I shows some solutions for the case with identical agents ($\gamma_i = 1/N$). In general it is much more likely that a coopera-

TABLE I

Maximum Number of Agents (N) Compatible with Sustaining the Global Optimum as a Cooperative Self-Enforcing Equilibrium ($a = 0.2$)

	$c = 0.1$					$c = 0.05$				
		Global optimum		Individual optimum			Global optimum		Individual optimum	
δ	N	S^0	$G^0 - S^0$	S^*	$G^* - S^*$	N	S^0	$G^0 - S^0$	S^*	$G^* - S^*$
0.8	4	0.212	0.0335	0.105	0.0187	2	0.133	0.0231	0.057	0.0108
0.85	7	0.267	0.0391	0.102	0.0184	2	0.195	0.0314	0.058	0.0109
0.9	∞	0.347	0.0453	0.100	0.0180	11	0.291	0.0413	0.051	0.0097

tive solution will emerge in this case than in the case with instantaneous redistribution of fish analyzed in Section 4. It was noted in Section 4 that N increases with c, a, and δ. Here we get $N = \infty$ for $c = 0.1$, $\delta = 0.9$, and $a = 0.2$. N falls quickly, however, when c or δ are reduced below these values; reducing δ to 0.85 (18% rate of discount) or c to 0.05 (lowest marginal cost of fish equal to 5% of the pristine stock) reduces N to 7 and 11, respectively. Even with this highly restricted migration, a cooperative solution is not guaranteed.[7]

Table I also shows figures that indicate the waste associated with agents choosing the individually optimal rather than the globally optimal solution (note that the pristine stock and its density are equal to 1). Superscript "o" refers to the global optimum while "*" refers to the individual optimum; if the number of participants were to increase by one the equilibrium solution would go from the global optimum to one very close to the individual optimum for the number of agents shown in the table. The individually optimal stock level is in all cases very much lower than the globally optimal level and in fact not much greater than the abandonment level with free access (c). The amount harvested is equal to the difference $G - S$. In the two cases shown in Table I the amount harvested in the noncooperative solution (the individual optimum) is only a quarter to one-half of the cooperative (globally optimal) harvest.

6. STRADDLING STOCKS

The final case we shall consider is one where a single agent has a high degree of control over a stock which is partly accessible to an indefinite number of agents. The case we have in mind is the one where a fish stock is largely contained within the 200-mile zone but straddles into the high seas. There are many real world cases of this kind, e.g., the groundfish stocks of the Grand Banks of Newfoundland, the Alaska pollack which is shared between the United States and Russia and straddles an area popularly known as the Donut Hole, and the Arcto-Norwegian cod which is shared by Norway and Russia and straddles an area similarly known as the Loophole.

[7] If the other players discover the defection immediately and start their punishment strategy in the same period as the defection occurs, their stock abandonment level in the period of defection will be the same as the defector's. The first term on the right-hand side of (21) will be divided by N, the second drops out and the last term will be multiplied by δ instead of δ^2. Expression (21) becomes a variation around an optimal path and is always negative. Hence defection will never be profitable and the cooperative equilibrium will be compatible with any number of agents.

We shall model this in a manner similar to the previous section. A certain fraction of the stock is supposed to be located within the "dominant agent's" area and the remainder outside and accessible to an indefinite number of agents. We ignore transboundary migration during the phase of exploitation; the dominant agent can choose the abandonment level of the stock within his area, while the remaining agents are assumed not to be interested in a cooperative solution, their number being indefinite, and hence choose an abandonment level for the remainder of the stock equal to the break-even marginal cost of fish ($p = c/S^*$). The transboundary externality is captured by letting the growth of the stock depend on total abandonment in the two areas, like in Eq. (18) above.

The dominant agent is assumed to choose an optimal abandonment level for his part of the stock, taking into account that the "fringe" agents choose the abandonment level $S^* = c/p$ for "their" part of the stock. The optimal steady-state abandonment level of the stock for the dominant agent will be given by a modified version of Eq. (20). With α being the share of the stock controlled by the dominant agent, and letting S and S^* denote the abandonment level of the dominant agent and the fringe agents, respectively, the objective function of the dominant agent becomes

$$V = \sum_{t=0}^{\infty} \delta^t \alpha \{ p[G(\alpha S_{t-1} + (1-\alpha)S^*) - S_t] - c[\ln G(\alpha S_{t-1} + (1-\alpha)S^*) - \ln S_t] \}, \qquad (19')$$

with the optimum abandonment level being given by

$$-(p - c/S^0) + \delta\alpha G'(\alpha S^0 + (1-\alpha)S^*)[p - c/G(\alpha S^0 + (1-\alpha)S^*)] = 0. \qquad (20')$$

A set of optimum solutions for the dominant agent are shown in Table II. The solution for $\alpha = 1$ is, of course, identical to the globally optimal solution. A striking feature of the solutions shown in Table II is how quickly the total present value and the optimum abandonment level of the stock fall with the dominant agent's share of the total stock. A dominant agent who controls 95% of the stock chooses an optimum stock level which is only 75% of the globally optimal level, and the total present value of profits (fringe plus dominant agent) is only 67% of the global optimum. This indicates that only a minor straddling of fish stocks into

TABLE II

Optimum Stock Level, Growth, and Present Value (PV) of Profits for Different Stock Shares of the Dominant Agent ($a = 0.2$, $c = 0.1$, and $\delta = 0.9$)

α	S^0 (dominant agent)	G	PV dominant agent	PV fringe	Sum PV
1	0.347	0.392	0.330	0	0.330
0.95	0.260	0.291	0.179	0.042	0.221
0.9	0.198	0.219	0.097	0.040	0.138
0.8	0.144	0.158	0.039	0.025	0.064
0.5	0.111	0.125	0.010	0.013	0.023
0.1	0.102	0.118	0.001	0.01	0.015

the high seas may result in substantial losses of efficiency. As the dominant agent's share of the stock is reduced further, the optimum abandonment level of the stock falls rapidly and so does the present value of aggregate profits; when the dominant agent controls one-half of the stock, the optimum abandonment level is only slightly above the free access abandonment level (0.1), and the aggregate profits are only about 5% of the global maximum.

Another interesting feature of the results in Table II is that beyond a certain point the fringe is not better off by having access to a larger part of the stock; the fringe's profit actually decreases slightly as its share of the stock increases from 5 to 10%. The reason is that the dominant agent maintains a large and productive stock as long as he has a high degree of control and some of the benefits of a large stock spills over to the fringe due to the growth externality.

7. CONCLUSION

The above analysis indicates that the number of agents that will cooperate in setting the exploitation rate for a shared fish stock is quite limited, but probably not much lower than the number of countries typically sharing a fish stock contained within 200 miles, and quite possibly higher. Cost heterogeneity will greatly limit this number, but in repeated games with an indefinite horizon low cost agents will not necessarily outfish high cost agents as in Clark's closed loop differential game [1]. The number of agents compatible with a self-enforcing cooperative solution is highly sensitive to the discount rate.

For stocks that are contained within separate national fishing zones but the growth of which is interrelated, depending on the aggregate stock size, the number of agents compatible with a self-enforcing cooperative equilibrium is much higher, but becomes very sensitive to fishing costs and the discount rate as these fall below a certain value. It is not at all unlikely that this critical number of agents will be comparable with or higher than the number of countries sharing stocks that are fully contained within 200 miles. Letting one of these zones represent the high seas with an indefinite number of agents, with one agent acting as a "leader" in the remaining zone, it appears, however, that even an ostensibly insignificant "fringe" may lead to significant losses of efficiency.

REFERENCES

1. C. W. Clark, Restricted access to common-property fishery resources: A game-theoretic analysis, *in* "Dynamic Optimization and Mathematical Economics" (P.-T. Liu, Ed.), pp. 117–132, Plenum, New York (1980).
2. R. D. Fischer and L. J. Mirman, Strategic dynamic interaction. Fish wars. *J. Econom. Dynamics Control* **16**, 267–287 (1992).
3. R. D. Fischer and L. J. Mirman, The compleat fish wars: Biological and dynamic interactions, *J. Environ. Econom. Management* **30**, 34–42 (1996).
4. D. Fudenberg and J. Tirole, "Game Theory," MIT Press, Cambridge, MA (1991).
5. R. Gibbons, "A Primer in Game Theory," Harvester Wheatsheaf, London (1992).
6. R. P. Hämäläinen, V. Kaitala, and A. Haurie, Bargaining on whales: A differential game model with Pareto optimal equilibria, *Oper. Res. Lett.* **3**, 5–11 (1984).
7. R. Hannesson, "Bioeconomic Analysis of Fisheries," Fishing News Books, Oxford (1993).

8. R. Hannesson, Sequential fishing: Cooperative and non-cooperative equilibria, *Natural Resource Modeling* **9**, 51–59 (1995).
9. D. Levhari and L. J. Mirman, The great fish war: An example using a dynamic Cournot–Nash solution, *Bell J. Econom.* **11**, 322–334 (1980).
10. M. Mesterton-Gibbons, Game–theoretic resource modeling, *Natural Resource Modeling* **7**, 93–147 (1993).
11. G. R. Munro, The optimal management of transboundary renewable resources, *Canad. J. Econom.* **12**, 355–376 (1979).
12. M. J. Osborne and A. Rubinstein, "A Course in Game Theory," MIT Press, Cambridge, MA (1994).
13. J. Vislie, On the optimal management of transboundary renewable resources: A comment on Munro's paper, *Canad. J. Econom.* **2**, 870–875 (1987).

[11]

ELSEVIER

Journal of Public Economics 87 (2003) 2031–2048

JOURNAL OF PUBLIC ECONOMICS

www.elsevier.com/locate/econbase

The Oslo Protocol on sulfur reduction: the great leap forward?

Michael Finus[a,*], Sigve Tjøtta[b]

[a]*Department of Economics, Institute of Economic Theory, University of Hagen, Profilstr. 8, 58084 Hagen, Germany*

[b]*Department of Economics, University of Bergen, Fosswinkelsgt. 6, N-5007 Bergen, Norway*

Received 15 December 2000; received in revised form 8 October 2001; accepted 7 January 2002

Abstract

The increasing number of international environmental agreements (IEAs), and the fact that under some agreements emissions have been reduced beyond agreed targets, are frequently viewed as success stories of international cooperation. We argue that success can only be measured if abatement targets under an IEA are compared with estimated abatement levels in the absence of a treaty, and are evaluated in terms of costs and benefits. We analyze the Oslo Protocol on sulfur reduction, showing that this IEA is not the great leap forward when compared with the calculated Nash equilibrium and the social optimum. However, we also demonstrate that under the Oslo Protocol, more ambitious targets could not have been realized due to strong free rider incentives.

Keywords: International environmental agreements; Sulfur reduction; Self-enforcing and effective treaties

JEL classification: Q72; D62; D74; Q20; Q28

1. Introduction

During the last 30 years more than 100 international environmental agreements (IEAs) have been signed. Politicians, the media, and the public have all seen this

*Corresponding author. Fax: +49-2331-987-302.

E-mail address: michael.finus@fernuni-hagen.de (M. Finus).

0047-2727/02/$ – see front matter
doi:10.1016/S0047-2727(02)00042-7

as a success. However, a signed IEA need not infer success. Two points are important: (1) information is needed on what would have happened without a treaty; (2) abatement targets have to be evaluated in terms of costs and benefits. Thus, we consider an IEA to be effective if it leads to a global welfare gain compared to a no-treaty situation.

Political scientists have done the bulk of research on the efficacy of the provision of public goods in the context of IEAs (e.g., Chayes and Chayes, 1993, 1995; Haas et al., 1993; Victor et al., 1998). However, their counterfactual analysis is often based on ad hoc assumptions rather than on a theory of the behavior of governments, and they are vague on commenting the success of IEAs. For instance, Wettestadt (1996, p. 67) summarizes the results of several studies on the Helsinki Protocol on the reduction of sulfur in Europe, which was signed in 1985, as follows: "There seems to be a shared feeling of more success than failure."

In contrast, the list of efficacy analyses of IEAs conducted by economists is very short. Though there are many cost benefit analyses on various environmental problems (e.g., acid rain, see Mäler, 1989; Kaitala et al., 1991; or global warming, see Falkinger et al., 1996; Fankhauser and Kverndokk, 1996), they usually do not evaluate the actual targets of an IEA and do not provide information on points 1 or 2. An exception is Murdoch and Sandler (1997a,b), who conducted econometric analyses on the Helsinki Protocol and the Montreal Protocol (on the reduction of CFCs), suggesting that agreed emission ceilings under both agreements are more in line with non-cooperative than with cooperative behavior of governments.

In this paper we analyze the successor to the Helsinki Protocol, 'The Protocol on the Further Reduction of Sulfur Emissions', signed in Oslo in 1994. In the spirit of Murdoch and Sandler, our aim is to evaluate the efficacy of this IEA. Unlike Murdoch and Sandler, our model allows evaluation of the various abatement levels also in terms of net benefits. We find that abatement targets and global welfare under the Oslo Protocol are substantially lower than in a socially optimal contract and even lower than in non-cooperative Nash equilibrium strategy. Given this negative result, we ask whether more ambitious abatement targets could have been realized in a self-enforcing IEA. We show that the socially optimal abatement level is not self-enforcing and that the maximal self-enforcing abatement level is only slightly higher than that of the Nash equilibrium. This suggests that low abatement targets under the Oslo Protocol may be due to free rider problems.

Methodologically, our approach to the acid rain problem is inspired by Mäler (1989, 1994). Like Mäler, we use the revealed preference approach to calibrate the damage cost functions of countries and we test the stability of various treaty designs in a supergame framework. Our analysis differs from Mäler's in three respects. First, Mäler calibrates constant marginal damage cost functions, which implies that the optimal emission strategy of a country is independent of the strategy of neighboring countries (orthogonal best reply functions). In contrast, we assume linear marginal damage cost functions, which imply that optimal strategies depend mutually on each other (non-orthogonal reaction functions), provided there is transboundary pollution. Second, whereas Mäler (1994) studies stability by applying the concept of a subgame-perfect equilibrium, we use the concept of a renegotiation-proof equilibrium, which emphasizes more the credibility of threat strategies. Third, Mäler assumes that transfers can be used to balance asymmetric payoffs so that stability of a treaty is easy to obtain. In contrast, we rule out transfers since the Oslo Protocol has no provision for transfers and they have hardly been used in past IEAs. We believe that our assumptions better reflect the interaction of countries in the context of transboundary pollution. They imply that it is more difficult to control for free riding in our than in Mäler's model.

As Mäler (1994) points out, the analysis of stability in a supergame framework rests on the assumptions of an infinite time horizon and flow pollutants. The first assumption may be justified by the fact that IEAs typically do not specify the termination of a treaty, and therefore the end of the game is not known with certainty. That is, there is a probability that the game will continue (see Gibbons, 1992, p. 90; Osborne and Rubinstein, 1994, p. 135). The second assumption is made since differential and difference games, which allow us to model emissions as stock pollutants, only deficiently allow studying stability aspects (free riding versus compliance). In differential games, either open-loop strategies or closed-loop Markov strategies are usually assumed for computational simplicity. Whereas the former is essentially a one-shot game, the latter conditions strategies only on actions of the immediate previous stage (Mäler and de Zeeuw, 1998). For difference games there are not many general results available to study the stability of an IEA, as they are difficult to solve (see de Zeeuw and van der Ploeg, 1991). We are aware that our choice of a supergame is far from an ideal assumption, however, it is the best one, given the focus of this study.

In the following, we provide background information on the sulfur reduction agreements and motivate our research questions. In Section 3, we introduce the model and report on its calibration in Section 4. Simulation results are presented in Section 5, and Section 6 concludes with a short summary of the main results and points to future research questions.

2. Historical background, hypotheses and research questions

2.1. From Geneva to Gothenburg

The 'Framework Convention on Long-Range Transboundary Air Pollution' was signed in 1979 by 33 countries in Geneva, expressing concern about the harmful effects of transboundary pollution. Though this convention only constitutes a declaration of good will, it served as a basis for follow-up protocols. In 1984, 22 states signed the monitoring and evaluation protocol 'EMEP', which requires that signatories report on their emissions to the treaty secretariat. Shortly after, three protocols followed, regulating sulfur emissions. 'The Protocol on the Reduction of

Sulfur Emissions and their Transboundary Fluxes by at least 30 Percent', signed in Helsinki in 1985, called on all participants to reduce annual emissions uniformly by 30% based on annual emissions in 1980, to be accomplished by 1993. It came into force in 1987. At present (20 August 2001), 22 countries are parties to this agreement of which 19 are original signatories (see Table 1).[1]

The focus of this paper, 'The Protocol on Further Reduction of Sulfur Emissions', was signed in Oslo in 1994, and came into force in 1998. There are 28 signatories, 23 of which have ratified the treaty so far (see Table 1). The Oslo Protocol is one of the first IEAs specifying differentiated abatement targets for signatories, expressed as annual emission ceilings for the years 2000, 2005, and 2010. Since only emission ceilings for the year 2000 are completely specified, we assume that this is the target year of the Oslo Protocol.

Finally, 'The Protocol to Abate Acidification, Eutrophication and Ground-Level Ozone' was signed by 31 states in Gothenburg in 1999, though it has not yet entered into force. This protocol is an extension of the Helsinki and Oslo Protocols (sulfur), as well as the Sofia (nitrogen oxides) and Geneva Protocols (volatile organic compounds), calling on further reduction of these pollutants.

2.2. The Helsinki Protocol: empirical data and research hypotheses

Table 1 reveals that many signatories of the Helsinki Protocol (e.g., Austria, Belgium, France, and Norway) achieved the reduction target already in 1985. Also, two of the non-signatories (Ireland and Spain) reduced annual emissions by more than 30% already by 1985.

All signatories met the target in 1993, and most countries' reductions are well above 30%. Also some non-signatories (e.g., Estonia and Lithuania) reduced annual emissions by more than 30%. Overall, in 1993, signatories reduced annual emissions by 55%. However, also non-signatories reduced annual emissions by 31%, exceeding the target level of the Helsinki Protocol.

These observations suggest that reductions are not the result of the Helsinki Protocol but merely the result of a non-cooperative abatement policy. Admittedly, signatories reduced annual emissions on average more than non-signatories. However, such differences can also be explained by differences in environmental damages and abatement costs between signatories and non-signatories if countries follow a non-cooperative strategy. Murdoch and Sandler (1997a) support this interpretation, suggesting that emissions under the Helsinki Protocol are in line with non-cooperative behavior.

Thus, our hypothesis is: countries sign agreements that codify emission ceilings equal to or even smaller than those that would follow from non-cooperative

[1]Liechtenstein and Canada, which are also parties to the Helsinki agreement, are not listed in Table 1. Liechtenstein has been omitted because of its marginal share of total emissions. Canada is not listed because of lacking data.

Table 1
Historical emissions and status of signature and ratification[a]

Country (1)	Emission 1980 (2)	Emission 1985 (3)	Emission 1993 (4)	Reduction 1980–85 (%) (5)	Reduction 1980–93 (%) (6)	Signature/ ratification (7)
Austria	397	195	71	51	82	Ge,H,O,(Go)
Belarus	740	690	382	7	48	Ge,H
Belgium	828	400	294	52	64	Ge,H,O,(Go)
Bulgaria	2050	2314	1426	−13	30	Ge,H,(O),(Go)
Croatia	150	165	114	−10	24	)Ge(,O,(Go)
Czech Republic	2257	2277	1419	−1	37	)Ge(,)H(,O,(Go)
Denmark	450	339	156	25	65	Ge,H,O,(Go)
Estonia	239	239	145	0	39	)Ge(,)H(
Finland	584	382	124	35	79	Ge,H,O,(Go)
France	3338	1470	1121	56	66	Ge,H,O,(Go)
Germany	7514	7732	3153	−3	58	Ge,H,O,(Go)
Greece	400	500	545	−25	−36	Ge,O,(Go)
Hungary	1633	1404	756	14	54	Ge,H,(O),(Go)
Ireland	222	140	157	37	29	Ge,O,(Go)
Italy	3800	1733	1490	54	61	Ge,H,O,(Go)
Latvia	57	57	44	0	23	)Ge(,(Go)
Lithuania	311	304	125	2	60	)Ge(
Luxembourg	24	17	15	29	38	Ge,H,O,Go
Netherlands	490	261	157	47	68	Ge,H,O,(Go)
Norway	140	97	35	31	75	Ge,H,O,(Go)
Poland	4100	4300	2725	−5	34	Ge,(O),(Go)
Portugal	266	198	300	26	−13	Ge,(Go)
Rep. of Moldova	308	282	128	8	58	)Ge(,(Go)
Romania	1055	1255	928	−19	12	Ge,(Go)
Russian Federation	7161	6191	3456	14	52	Ge,H,(O),
Slovakia	780	613	325	21	58	)Ge(,)H(,O,(Go)
Slovenia	234	241	183	−3	22	)Ge(,O,(Go)
Spain	3319	2190	2071	34	38	Ge,O,(Go)
Sweden	508	266	101	48	80	Ge,H,O,(Go)
Switzerland	116	76	34	34	71	Ge,H,O,(Go)
Turkey	860	322	354	63	59	Ge
Ukraine	3849	3463	2194	10	43	Ge,H,(O)
UK	4913	3766	3185	23	35	Ge,O,(Go)
Yugoslavia	406	478	401	−18	1	)Ge(
Other countries	1527	1527	1539	0	−1	
Natural sources	4318	4318	4409	0	−2	
Total	59 345	50 202	34 061	15	43	

[a] Source: Emissions: EMEP Status Report, 1997, Part 1, Table 3.5. Status of signature and ratification as of 20 August 2001: various homepages UNEP. Legend: Emissions (columns 2–4) are expressed in 1000 tons SO_2 SO_2/year. Reductions (columns 5 and 6) are expressed as percentage reduction with respect to 1980 annual emissions. () means that a country is a signatory but has not ratified the agreement yet.) (indicates that a country was a non-signatory but succeeded later; Ge = 1979 Geneva Framework Convention; H = 1985 Helsinki Protocol; O = 1994 Oslo Protocol; and Go = 1999 Gothenburg Protocol. Other countries comprise Africa, Albania, Bosnia, Cyprus, Georgia, Iceland, Kazakhstan, Liechtenstein, The FYR Macedonia. Natural sources comprise The Baltic Sea, The Mediterranean Sea, The North Sea, Remaining N.E. Atlantic, Natural Oceanic, and Volcanic.

behavior. As mentioned in Section 1, we test this hypothesis for the Oslo Protocol and find strong support for it. Therefore, in a second step, we analyze the possibilities of implementing a self-enforcing agreement with more ambitious abatement targets.

3. Model

3.1. The Nash Equilibrium and social optimum

Let $E = (E_1, E_2, \ldots, E_N)$ be a SO_2-emission vector of a group of N countries. Furthermore, let $Q = (Q_1, Q_2, \ldots, Q_N)$ be a vector of sulfur depositions in these N countries. The deposition of receptor i is given by $Q_i = \Sigma_j^N a_{ji}E_j + \overline{Q}_i$ where $\overline{Q}_i$ are background depositions from natural sources and emissions from countries not covered by our study. The coefficient a_{ji} is the portion of total emissions released by emitter j, which is deposited at receptor i. It mainly reflects annual wind patterns and other natural factors. Damages from emissions in country i, D_i, depend on depositions, and are represented by the convex function $D_i = D_i(Q_i)$. Benefits from emissions are given by the concave benefit function $B_i = B_i(E_i)$. Taken together, net benefits in country i are given by

$$\Pi_i = B_i(E_i) - D_i(Q_i). \tag{1}$$

If the N countries behave non-cooperatively, each country maximizes (1) with respect to its own emissions, taking emissions from other countries and background emissions as given. From the first order conditions, the Nash equilibrium emission vector $E^{\mathrm{N}} = (E_1^{\mathrm{N}}, \ldots, E_N^{\mathrm{N}})$ follows. In contrast, a socially optimal solution would require that each country not only considers its own damages, but also those in the other countries when choosing its emission level. From the first order conditions, emissions in the social optimum $E^{\mathrm{S}} = (E_1^{\mathrm{S}}, E_2^{\mathrm{S}}, \ldots, E_N^{\mathrm{S}})$ follow.[2] Total emissions are lower in the social optimum than in the non-cooperative equilibrium, i.e., $\Sigma_{i=1}^{\mathrm{N}} E_i^{\mathrm{S}} < \Sigma_{i=1}^{\mathrm{N}} E_i^{\mathrm{N}}$, though on a country basis $E_i^{\mathrm{N}} < E_i^{\mathrm{S}}$ may hold.[3]

3.2. Stability of an agreement: assumptions

Despite the fact that full (social optimum) or partial cooperative agreements are associated with global welfare gains, cooperation proves difficult in reality. Due to the lack of a supranational enforcement authority, *two constraints* constitute a

[2] Appendix A shows how the Nash equilibrium and the social optimum are computed for the simulations on which we report in Section 5.

[3] See Folmer and van Mouche (2000) for a proof.

major obstacle for successful international cooperation. The *individual rationality constraint* requires every country to reach a net benefit (payoff) above a certain threshold level to take part in an agreement. In a supergame framework, this benchmark is given by the minimax payoff, which follows from the various versions of the Folk theorem (Fudenberg and Maskin, 1986).[4] Though one obvious instrument to guarantee individual rationality is transfers, we rule them out since the Oslo Protocol has no provision for transfers and empirical evidence indicates that governments in the past have shown a great reluctance to implement transfer schemes.[5] The *stability constraint* requires that the free rider incentive can be controlled by deterrent threat strategies. Whether this condition holds is tested by applying the concept of weakly renegotiation-proof equilibrium.[6] Thus, if we subsequently write about a stable emission vector, we mean a vector that is individually rational and constitutes a weakly renegotiation-proof equilibrium (WRPE) in a supergame.

3.3. Stability of an agreement: concept

A self-enforcing IEA requires that signatories are able to stabilize an agreement by appropriate threat and punishment strategies. The harsher the punishment at the disposal of countries, the easier it is to deter free riding. However, punishment has to be credible in order to be effective. We apply Farrell and Maskin's (1989) stability concept of a weakly renegotiation proof equilibrium (WRPE): only those emission vectors are considered to be stable which can be supported by threat strategies that leave the punishers at least as well off when carrying out the punishment than if all countries comply with the treaty obligations. This gives the punishers an incentive to actually carry out the punishment if needed and consequently the threat is credible (Mohr, 1988).

More concretely, in a supergame an emission vector $E^* = (E_1^*, \ldots, E_N^*)$ is a WRPE if (a) it is individually rational for all countries, and if and only if the payoff stream (b) to each country from sticking to E^* over the entire period of the game is at least as high as (c) when taking a free ride in one period and

---

[4]In the context of the transboundary emission game, the minimax payoff is given by $\min_{E_{-i}} \max_{E_i} \Pi_i(E_i, E_{-i})$. All countries except i will choose their highest emission level $E_{-i}^{\max}$ to punish country i. Country i replies by choosing its best reply, i.e., $\Pi_i^{\min\max} = \Pi_i(E_i(E_{-i}^{\max}), E_{-i}^{\max})$.

[5]This observation has initiated a series of papers on issue linkage as a measure to guarantee individual rationality (Cesar and de Zeeuw, 1996 and Folmer et al., 1993). The literature suggests that the lack of transfers is due to the problem of implementing self-enforcing transfers between donors and recipients and within the group of donors (Finus, 2002). To consider these problems would complicate the analysis substantially, to ignore them, as in Mäler (1994), solves stability problems almost by assumption.

[6]Other stability concepts applied to the analysis of IEAs comprise the concept of internal and external stability (Barrett, 1994; Carraro and Siniscalco, 1993) and the γ-core (Chander and Tulkens, 1995).

subsequently being punished for a period of time. Moreover, each country's payoff stream (d) when conducting the punishment must be at least as high as (e) when sticking to E^*. In order to make the payoff streams under (b) and (c) as well as under (d) and (e) comparable, payoffs received in each period over the entire period of the game have to be discounted by a discount factor $\delta = p/(1+r)$, $0 \leq \delta \leq 1$, to time $t = 0$, where p is the probability that the game continuous and r is the discount rate (time preference rate). The lower the discount factor, the more attractive is free riding and the more difficult it is to establish a self-enforcing agreement. Similarly, the lower the discount factor, the less attractive it is to conduct the punishment. Therefore, in order not to overestimate stability problems, we make the (optimistic) assumption of discount factors close to one. This assumption has the nice property that a WRPE can be characterized in terms of only two conditions.[7] An emission vector E^* is a WRPE if there *exists* for every country i a punish vector $E^i = (E^i_i, E^i_{-i})$ (the superscript denotes the country being punished) such that

$$\Pi_i(E^*) \geq \Pi_i(E_i(E^i_{-i}), E^i_{-i}), \tag{2}$$

and

$$\Pi_j(E^*) \leq \Pi_j(E^i_i, E^i_{-i}) \quad \text{for all } j \in N, j \neq i \tag{3}$$

can be satisfied where $\Pi_i(E^*) \geq \Pi_i^{\min\max}$ for all $i \in N$. Π_i is the average discounted payoff country i receives in each period and $E_i(E^i_{-i})$ denotes country i's best (stage game) reply to punishment E^i_{-i}. Inequality (2) ensures that a potential free rider i can be punished by all other countries through punishment E^i_{-i} when deviating from the agreed strategy profile E^*. Inequalities (3) secure that this punishment is also credible. The details of the computations are provided in Appendix A.

4. Data and calibration

Since benefits from emissions can also be interpreted as the opportunity cost of reducing emission, we use cost functions from the program RAINS developed by IIASA. Since these functions are piecewise linear functions, we approximate them with the following quadratic functions

$$B_i = b_{i1}E_i - \tfrac{1}{2}b_{i2}E_i^2. \tag{4}$$

We estimate the parameters of each country's benefit function by ordinary least squares, in a similar fashion as Førsund and Nævdal (1998). We chose the

[7] See Farrell and Maskin (1989). A discussion and derivation is provided in Finus and Rundshagen (1998a,b).

'no-control strategy' scenario as Førsund and Nævdal (1998), and Klaassen et al. (1994), which assumes that countries emit according to the business as usual path. We chose 1995 as the base year, which allows us to use the most recent data.

The RAINS cost functions assume an exogenous energy demand and a limited input substitution, and are therefore only defined up to a certain emission reduction. The corresponding lower bound of emission is denoted by $E_i^{\min}$. For the upper bound of emission we require $E_i \leq E_i^{\max} = b_{i1}/b_{i2}$ to ensure $B_i' \geq 0$. Note that $E_i^N < E_i^{\max}$, implies that $\Pi_i^N > \Pi_i^{\min\ \max}$ for all i holds.

The transportation matrix was computed on the basis of the data generated by the resource-receptor multi-layer model developed by the Norwegian Meteorological Institute in Oslo.[8] The transportation matrix reflects the average relations over several years, so that the same matrix can be used for the calibration year 1995 and the target year 2000 of the Oslo Protocol.

For the damage cost function we assume a quadratic function of the form

$$D_i = \tfrac{1}{2} d_i Q_i^2. \tag{5}$$

To determine the parameter d_i, we calibrate the model by using the revealed preference approach as used by Mäler (1989). That is, we assume that before the Oslo agreement came into force, countries had been in the Nash equilibrium.[9] This assumption rests on our finding in Section 2 that the Helsinki Protocol was a non-binding agreement. From the first order conditions in the Nash equilibrium (assuming an interior solution)

$$b_{i1} - b_{i2}E_i = a_{ii} d_i Q_i \tag{6}$$

we derive the parameter d_i by inserting E_i^{95} and Q_i^{95} in (6) and solving for d_i. Eq. (6) implicitly defines a country's best reply function: optimal emissions in country i depend negatively on foreign emissions as long as $a_{ii} \neq 0$ and $a_{ij} \neq 0$ for at least one j. In contrast, a constant marginal damage cost function, as assumed by Mäler, would imply that optimal emissions in country i are independent of foreign emissions (orthogonal reaction functions). Since a country's optimal deviation is guided by its best reply function, the free rider incentive is higher for non-orthogonal than for orthogonal best reply functions.

Since the target year of the Oslo Protocol is 2000, we update the damage cost functions to this year. We assume that environmental quality is a normal good where the demand for environmental quality increases with income. Therefore, we

[8] EMEP-Status Report 1/97, Part 1, Ch. 6, Table 6.3.

[9] That is, we implicitly assume a game of complete information. This seems a good approximation for abatement costs since they are regularly published for instance by the International Institute of Applied Systems Analysis. It seems at least a fair approximation for the evaluation of environmental damages since governments have access to records of past emission levels, official statements, press releases, and they regularly exchange opinions during negotiations leading to an IEA.

raise the parameter d_i by 2.5% annually, reflecting the average growth rate of GDP over the period 1995–2000. We also conducted a sensitivity analysis with an annual increase of damages of 1 and 4%, but found all qualitative results on which we report in Section 5 unchanged.[10] The cost parameters are not updated since the RAINS data for the year 2000 suggest that control costs remain almost constant. All input data for the calibration are given in Appendix B, except the transportation matrix, which is available upon request from the authors.

5. Results

In this section we discuss emissions, net benefits, and stability of different scenarios: Oslo Protocol, Nash equilibrium, social optimum, and maximal stable emission reduction. Results are reported as annual figures in the year 2000 in Table 2. In a repeated game framework, this implies that these emissions and differences in net benefits occur every year after 2000 under the respective scenarios.

5.1. The Oslo Protocol: the great leap forward?

Total annual emissions in the Nash equilibrium are lower than those implied by the Oslo Protocol. This protocol is only binding for seven countries with respect to the Nash equilibrium. Italic numbers in the second column indicate these countries. For most of these seven countries, the difference between Oslo and the Nash equilibrium is relatively small. Thus, it seems that the Oslo Protocol constitutes only a weak constraint for this group of countries. Even for non-signatories, Nash equilibrium emissions are lower than those predicted by EMEP.

The total annual net benefit under the Oslo Protocol is ECU 3606 million lower than in the Nash equilibrium. On average, a country would suffer an annual loss of net benefits of 26%. However, the welfare implications are unevenly distributed among countries, a fact revealed by a relatively large standard deviation of 149%. For instance, signatories gain slightly an average of 3%, whereas non-signatories would suffer an average loss of net benefits of 116%. It is also evident that for many countries for which the Oslo Protocol is not binding (e.g., Finland, Hungary, and Norway) net benefits under the Oslo Protocol are *lower* than in the Nash equilibrium. However, total emissions under the Oslo Protocol are not only higher than in the Nash equilibrium, but also substantially higher than in the social optimum. Moreover, Oslo emission ceilings imply a total annual loss of ECU 6869 million in net benefits compared to the social optimum, exceeding those compared to the Nash equilibrium substantially.

[10] Results are available upon request from the authors.

Table 2
Emissions and net benefits under the Oslo-Agreement, Nash Equilibrium, Social Optimum, and Maximum Stable Emission Reduction in Year 2000[a]

Country (1)	E_i^O (2)	E_i^N (3)	E_i^S (4)	E_i^M (5)	$\Pi_i^O - \Pi_i^N$ (6)	$\Pi_i^S - \Pi_i^N$ (7)	$\Pi_i^M - \Pi_i^N$ (8)
Austria	78	62	40	61	25	390	23
Belgium	248	237	145	233	139	135	12
Bulgaria	*1374*	1470	794	1436	−20	−118	1
Croatia	133	58	34	57	−128	139	8
Czech Rep.	1128	1058	755	1043	254	202	21
Denmark	*90*	139	75	136	66	128	10
Finland	116	79	70	78	−190	366	20
France	*868*	979	792	970	4	51	7
Germany	*1300*	2949	2182	2912	−1865	−116	16
Greece	595	547	546	547	−14	136	7
Hungary	898	691	540	683	−138	31	6
Ireland	*155*	165	146	164	−1	3	0
Italy	*1330*	1392	1230	1384	−43	97	7
Luxembourg	10	8	6	8	20	31	2
Netherlands	*106*	139	86	137	47	128	10
Norway	34	32	30	32	−29	457	25
Poland	2583	2326	1206	2271	25	−494	1
Russian Fed.	4440	2935	2705	2924	−1007	321	19
Slovakia	337	232	152	228	−107	121	9
Slovenia	130	118	72	115	−22	−11	2
Spain	2143	2027	1580	2005	−3	−76	0
Sweden	100	85	84	85	249	675	38
Switzerland	60	34	20	33	−2	−6	0
Ukraine	2310	1618	1373	1606	−250	121	9
UK	2449	2344	1653	2309	5	−240	0
Belarus	456	271	206	268	−131	90	6
Estonia	(110)	104	22	100	−34	54	6
Latvia	(38)	35	28	34	−120	403	24
Lithuania	(164)	106	64	104	−30	50	5
Portugal	304	271	266	270	−6	7	0
Rep. of Moldova	(140)	57	52	57	−141	130	8
Romania	(912)	895	612	881	−55	20	5
Yugoslavia	(680)	428	215	417	−106	36	5
Rest	6305	6305	6305	6305			
Total	32 124	30 196	24 088	29 891	−3606	3263	318

[a] All figures are annual data in the year 2000. Emissions, E_1, in columns 2–5 are expressed in 1000 tons SO_2/year and differences in net benefits between scenario A and B, $\Pi_i^A - \Pi_i^B$, in columns 6–8, are expressed in million ECU/year where net benefits Π_i^A, (Π_i^B) are benefits minus damage costs from emissions under scenario A (B). Superscripts O, N, S and M stand for Oslo Protocol, Nash equilibrium, social optimum and maximum stable emission reduction (as described in Section 5.3). Signatories of the Oslo Protocol are indicated by bold types. The signatories Canada, Liechtenstein, and the European Union are not listed. Belarus and Portugal (bold and italics) are listed in the agreement but are not signatories. For signatories emissions reflect emission ceilings of Annex II of the Oslo Protocol. For non-signatories are projected emissions from EMEP Status Report, 1997, Part 1, Table 3.5. Projected emissions are indicated by brackets; italic emissions imply that the Oslo agreement is binding compared to Nash equilibrium emissions. 'Rest' comprises Africa, Albania, Bosnia, Cyprus, Georgia, Iceland, Kazakhstan, Liechtenstein, The FYR Macedonia, Turkey, and natural sources. For all scenarios, annual emissions of 'Rest' are assumed at the 1995 level.

2042 M. Finus, S. Tjøtta / Journal of Public Economics 87 (2003) 2031–2048

Seen together, the evidence shows a similar pattern, as discussed in Section 2 for the Helsinki Protocol, and is in line with findings of Murdoch and Sandler (1997a). This allows for the following conclusions: (1) emission ceilings under the Oslo Protocol are not in line with cooperative behavior by countries. (2) The Oslo Protocol is not 'the great leap forward', neither in terms of emission reduction nor in terms of net benefits. (3) We expect that most signatories will reduce their emissions more than required in the Oslo Protocol. That is, emission patterns in the year 2000 will resemble those in the predicted Nash equilibrium more than emission ceilings under the Oslo Protocol.[11] (4) The fact that signatories gain on average under the Oslo Protocol indicates that the decision to accede to a treaty is a rational choice.

Based on the evidence and interpretation above, a natural conjecture is that under the Oslo Protocol stability should not be a problem. A check whether the Oslo Protocol constitutes a WRPE according to the procedure outlined in Section 3.3 and Appendix A, confirms this.

5.2. The Social Optimum: a stable great leap forward?

Given our negative conclusion with respect to the efficacy of the Oslo Protocol, we analyze whether a more ambitious treaty could have been realized under the constraint that this treaty is stable. From a normative point of view, an obvious candidate for such a test is the social optimum. Since we expect emissions in 2000 to be close to the Nash equilibrium, we benchmark the social optimum with the Nash equilibrium.

From Table 2 it is evident that total annual emissions in the social optimum are substantially lower than in the Nash equilibrium. In terms of 1993 emission levels (Table 1), the social optimum would call for a 29% reduction of total annual emissions in the year 2000, exceeding the 11% reduction in the Nash equilibrium substantially. The overall annual net benefit gain in the social optimum compared to the Nash equilibrium is ECU 3263 million, and the average annual gain to a country is 56%. Thus, from a global point of view, a socially optimal solution should clearly be advocated.

However, the gains in net benefits are rather unevenly distributed among countries with a standard deviation of 99%, though there is no marked difference between signatories (average gain of 53%) and non-signatories (average gain of 67%). Nevertheless, some countries would lose in the social optimum compared to

[11]An obvious way to verify the third conclusion is to compare the predicted Nash equilibrium emissions with the actual emissions in 2000. Unfortunately, however, this data is not yet available (EMEP, 2001). Nevertheless, the conclusion can be supported by the fact that signatories reduce annual aggregate emissions by 15% from 1993 to 2000 under the non-cooperative scenario. In contrast, non-signatories (including countries not listed in Tables 1 and 2) reduce total annual emissions by 7%. A similar relation between signatories (55%) and non-signatories (33%) was observed for actual emission reductions under the Helsinki Protocol. This suggests that our prediction for the Oslo Protocol is in line with the observations under the Helsinki Protocol, as discussed in Section 2.2.

the calculated Nash equilibrium. This could happen if a country has relatively low marginal benefits from emissions per unit of caused damages, which therefore is required to contribute much to the socially optimal agreement, and/or has relatively low environmental damages.

If $\Pi_i^N = \Pi_i^{\min\max}$, it were immediately evident from Table 2 that a socially optimal solution is not a WRPE, since the individual rationality constraint would be violated. However, also for $\Pi_i^N > \Pi_i^{\min\max}$ and discount factors of all countries close to one, our test reveals that the social optimum does not constitute a weakly renegotiation proof equilibrium (WRPE) in a supergame. This result differs from Mäler (1994) who found the social optimum to be a subgame-perfect equilibrium (SPE). Comparing both analyses, four important differences are identified: (1) the free rider incentive is greater in our model since our specification of the damage cost functions implies non-orthogonal reaction functions, whereas Mäler's specification implies orthogonal best reply functions. (2) Mäler assumes unlimited transfers to compensate losers and that transfers can be implemented, whereas we rule out transfers. (3) We apply the stronger concept of WRPE which allows for less severe punishments than the concept of SPE as used by Mäler. (4) Mäler tests stability assuming a trigger strategy where countries return to Nash equilibrium emissions for the rest of the game after a country has deviated. Hence, the emission space is restricted to $E_i \le E_i^N$, which implies for individual rationality $\Pi_i(E^*) \ge \Pi_i^N$ for all $i \in N$. In contrast, we assume a wider emission space, i.e., $E_i \le E_i^{\max}$, which implies $\Pi_i(E^*) \ge \Pi_i^{\min\max}$ for all $i \in N$ where $\Pi_i^N > \Pi_i^{\min\max}$. Assumptions 1, 2, and 3 imply that instability is less of a problem in Mäler's than in our model. The opposite holds for assumption 4. Thus, it seems that the effects of assumptions 1, 2 and 3 dominate the effect of assumption 4.

5.3. Maximal stable emission reductions: a pragmatic leap forward?

An obvious question, which arises from the 'negative' result above, is: What is the maximum emission reduction that could be stabilized as a WRPE? To answer this question, we search along an emission path starting from the Nash equilibrium, which is stable by definition, towards the non-stable but efficient social optimum:

$$E_i = \gamma E_i^N + (1-\gamma)E_i^S \quad \text{for all } i \text{ and } 0 \le \gamma \le 1. \tag{7}$$

We find that emission vectors for which $0.95 \le \gamma \le 1$ holds are WRPE, and those for which $0 \le \gamma < 0.95$ holds are not.[12] For $\gamma = 0.95$ annual emissions and associated net benefits are listed in Table 2.

[12] More precisely, we performed a line search to check for weakly renegotiation-proofness and found that $\gamma = 1.00$, 0.975 and 0.95 are WRPE and $\gamma = 0.94$, 0.925, 0.90, 0.75, 0.50 . . . 0.00 are not. Therefore, strictly speaking, $\gamma = 0.95$ does not necessarily reflect the most efficient WRPE along the line given by (7). However, due to computational limitations, a complete line search is not possible.

It is evident that for $\gamma = 0.95$ all countries gain compared to the Nash equilibrium. However, if the maximal WRPE emission reduction were implemented, total annual emissions would only be slightly lower than those in Nash equilibrium, and substantially higher than those in the social optimum. In terms of emission levels in 1993, the maximal WRPE emission reduction is 12%, whereas in the Nash equilibrium and social optimum emission reductions are 11 and 29%, respectively. On average, a country gains 5% compared to net benefit in the Nash equilibrium. Given this small gain, the standard deviation of 9.4% is relatively large, which suggests that the main stability problem is due to the asymmetries of the transboundary pollution problem. In other words, the chances of implementing a self-enforcing agreement which calls upon reducing annual emissions by substantially more than what would follow from pursuing a non-cooperative strategy, are not very high in the acid rain context.

6. Summary and conclusion

We calibrate a numerical model for 33 European countries suffering from sulfur emission using the revealed preference approach. In a first step we showed that abatement targets and annual net benefits under the Oslo Protocol are substantially lower than in the social optimum and even lower than in the Nash equilibrium. Therefore, we concluded that this protocol is by no means 'the great leap forward', and conjectured that many signatories will overfulfil the terms of this treaty in their own self-interest. In a second step we tested whether the socially optimal contract would be stable in the sense of a weakly renegotiation-proof equilibrium, and we computed the maximal stable emission reduction. It turned out that the social optimum is not stable, and that the maximal stable emission reduction only slightly exceeds that in the Nash equilibrium due to large asymmetries reminiscent of the acid rain problem. Therefore, it would be interesting in future research to check stability of other forms of line search and to analyze whether the formation of subcoalitions could achieve more than the grand coalition. Moreover, efficacy of other international environmental agreements may be tested. A first attempt in this direction is Murdoch and Sandler (1997b) who confirm our hypothesis for the Montreal Protocol that IEAs are non-binding treaties. Of course, this conclusion leaves us with a puzzling question: Why do states sign agreements specifying abatement obligations which they will meet or even overfulfil in their own interest anyway? We have no final answer yet. Nevertheless, we suggest the following possible explanations: first, environmentally conscious voters are rationally badly informed (given their marginal influence on election outcomes and the cost to obtain thorough background information about IEAs), whereas industry possesses private information about abatement costs and, given the importance of this interest group, holds a higher stake in the political system. Consequently, governments trying to capture political support have an incentive to accede to an IEA, to take on a relatively low abatement responsibility, and to later overfulfil their abatement targets, selling accession, and overfulfillment as good deeds to 'green voters', without imposing high cost on industry. Second, overfulfillment of an agreement may give governments slack with respect to other policy issues where compliance is more difficult. Thus the acid rain game may be part of a larger game where issues are linked and concessions are traded.[13]

Acknowledgements

The first author gratefully acknowledges the receipt of grant No. II 69982 by the Volkswagen-Foundation. He particularly wishes to thank Alfred Endres and Bianca Rundshagen for many fruitful discussions. Both authors have benefited from grants by the Norwegian Research Council (Ruhrgas) and would like to thank three anonymous referees and co-editor Lans Bovenberg for very constructive comments and suggestions, and Frank Brockmeier, Dimitrios Gotsis, and Frank Lobigs for research assistance.

Appendix A

The *social optimum* is computed by solving the following program.

$$\begin{aligned}&\text{Find}\quad (E_1,E_2,\dots,E_N)\quad\text{and}\quad (Q_1,Q_2,\dots,Q_N)\\&\text{which maximize}\quad \sum_{i=1}^{N} B_i(E_i)-D_i(Q_i)\quad \text{for } i=1,2,\dots,N\\&\text{subject to}\quad Q_i\geq\sum_{j=1}^{N} a_{ji}E_j\geq 0\quad\text{and}\quad E_i^{\max}\geq E_i\geq E_i^{\min}\\&\text{for } i=1,2,\dots,N.\end{aligned}\tag{P1}$$

The *Nash equilibrium* is calculated by solving the following program.

$$\begin{aligned}&\text{Find}\quad (E_1,E_2,\dots,E_N),(Q_1,Q_2,\dots,Q_N),(S_1^+,S_2^+,\dots,S_N^+)\\&\text{and}\quad (S_1^-,S_2^-,\dots,S_N^-)\\&\text{which minimize}\quad \sum_{i=1}^{N} S_i^+ + S_i^-\\&\text{subject to}\quad Q_i=\sum_{j=1}^{N} a_{ji}E_j\geq 0, B_i'(E_i)+S_i^+ - S_i^- = a_{ii}D_i'(Q_i),\\&\qquad E_i^{\max}\geq E_i\geq E_i^{\min},\\&\qquad S_i^+\geq 0\ \text{and}\ S_i^-\geq 0\quad \text{for } i=1,2,\dots,N\end{aligned}\tag{P2}$$

[13] This possibility was suggested by one of the referees. For references on issue linkage, see Footnote 5.

2046 M. Finus, S. Tjøtta / Journal of Public Economics 87 (2003) 2031–2048

where S_i^+ and S_i^- are slack variables. *Weakly renegotiation-proofness* of an emission vector E^* is checked by solving for *each* country i the following program:

$$
\begin{aligned}
&\text{Find vectors} \quad (S_1, S_2, \ldots, S_N) \quad \text{and} \quad (E_1, E_2, \ldots, E_N) \\
&\text{which minimize} \quad \sum_{j=1}^{N} S_j \\
&\text{subject to} \quad \Pi_i(E^*) + S_i \geq \Pi_i(E_i(E^i_{-1}), E^i_{-i}) \\
&\qquad \Pi_j(E^*) - S_j \leq \Pi_j(E^i_i, E^i_{-i}) \quad \text{for all } j \neq i. \\
&\qquad S_j \geq 0, E_j^{\max} \geq E_j \geq E_j^{\min} \quad \text{for all } j = 1, 2, \ldots, N
\end{aligned}
\tag{P3}
$$

where S_i are slack variables. We define Z^i as the optimal value of program (P3). If $Z^i = 0$ for all countries i, we say that E^* is a WRPE otherwise E^* is not. The best reply of country i is denoted $E_i(E^i_{-i})$.

Appendix B

Input data

Country	Emission 1995	$E_i^{\min}$	$E_i^{\max}$	b_1	b_2
Austria	74	16	221	3.70	0.016745
Belarus	324	14	309	2.77	0.008980
Belgium	294	52	416	4.58	0.011010
Bulgaria	1497	110	1761	0.93	0.000530
Croatia	63	21	111	2.79	0.025040
Czech Rep.	1091	196	1400	3.74	0.002667
Denmark	150	20	265	3.97	0.014990
Estonia	110	16	166	1.92	0.011543
Finland	96	48	294	3.24	0.011040
France	989	125	1076	4.88	0.004538
Germany	2995	419	3410	4.21	0.001236
Greece	556	31	640	4.19	0.006540
Hungary	699	314	778	4.51	0.005800
Ireland	166	21	177	3.04	0.017140
Italy	1437	126	1856	3.80	0.002050
Latvia	38	18	69	4.60	0.066300
Lithuania	107	18	119	4.22	0.035476
Luxembourg	8	3	10	7.89	0.788600
Netherlands	147	32	224	4.51	0.020125
Norway	35	18	65	7.25	0.111040
Poland	2337	340	2428	2.00	0.000825
Portugal	272	18	284	3.49	0.012288
Rep. of Moldova	59	16	79	3.15	0.040000
Romania	912	91	1073	1.69	0.001579
Russian Fed.	2983	455	3410	3.58	0.001050

Slovakia	238	192	292	3.04	0.010400
Slovenia	119	8	133	5.34	0.040060
Spain	2061	142	2354	1.77	0.000750
Sweden	94	52	183	5.74	0.031390
Switzerland	34	2	35	3.16	0.090200
Ukraine	1639	308	1825	2.35	0.001290
UK	2365	212	2541	2.57	0.001010
Yugoslavia	462	29	837	1.54	0.001840
Rest	6305				
Total	30 724				

Emissions are taken from EMEP Status Report, 1997, Part 1, Table 3.5, and are expressed in 1000 tons SO_2 per annum. 'Rest' comprises Africa, Albania, Bosnia, Cyprus, Georgia, Iceland, Kazakhstan, Liechtenstein, The FYR Macedonia and Turkey as well as natural sources. $E_i^{\min}$ = minimum emission beyond which reduction is not possible; $E_i^{\max}$ = maximum emission at which $B_i' = 0$. b_1 and b_2 are estimated parameters from the cost data produced by RAINS (version 7.2) for the 'no-control strategy' in 1995, as described in Section 4.

References

Barrett, S., 1994. Self-enforcing international environmental agreements. Oxford Economic Papers 46, 878–884.

Carraro, C., Siniscalco, D., 1993. Strategies for the international protection of the environment. Journal of Public Economics 52, 309–328.

Cesar, H., de Zeeuw, A., 1996. Linkage in global environmental problems. In: Xepapadeas, A. (Ed.), The Economic Policy for the Environment and Natural Resources. Edward Elgar, Cheltenham, pp. 157–173.

Chander, P., Tulkens, H., 1995. A Core-theoretic solution for the design of cooperative agreements on transfrontier pollution. International Tax and Public Finance 2, 279–293.

Chayes, A.H., Chayes, A., 1993. On compliance. International Organization 47, 175–205.

Chayes, A.H., Chayes, A., 1995. In: The New Sovereignty. Harvard University Press, Cambridge (USA) and London.

EMEP, 1997, Status Report 1997, Oslo.

EMEP, 2001, Status Report 2001, Oslo.

Falkinger, J., Hackl, F., Pruckner, G.J., 1996. A Fair mechanism for efficient reduction of global CO_2-emissions. Finanzarchiv 53, 308–331.

Fankhauser, S., Kverndokk, S., 1996. The global warming game—Simulations of a CO_2 reduction agreement. Resource and Energy Economics 18, 83–102.

Farrell, J., Maskin, E., 1989. Renegotiation in repeated games. Games and Economic Behavior 1, 327–360.

Finus, M., 2002. Game theory and international environmental cooperation: any practical application. In: Böhringer, C., Finus, M., Vogt, C. (Eds.), Controlling Global Warming: Perspectives from Economics, Game Theory and Public Choice, Chapter 2, Edward Elgar, Cheltenham, UK, in press.

Finus, M., Rundshagen, B., 1998a. Toward a positive theory of coalition formation and endogenous instrumental choice in global pollution control. Public Choice 96, 145–186.

Finus, M., Rundshagen, B., 1998b. Renegotiation-proof equilibria in a global emission game when players are impatient. Environmental and Resource Economics 12, 275–306.

Folmer, H., van Mouche, P., Ragland, S., 1993. Interconnected games and international environmental problems. Environmental and Resource Economics 3, 313–335.

Folmer, H., van Mouche, P., 2000. The acid rain game. A mathematically rigorous analysis. In: Dasgupta, P., Kristroem, B., Loefgren, K.-G. (Eds.), Festschrift in Honor of Karl Göran Mäler. Edward Elgar, Cheltenham, in press.

Førsund, F.R., Nævdal, E., 1998. Efficiency gains under exchange-rate emission trading. Environmental and Resource Economics 12, 403–423.

Fudenberg, D., Maskin, E., 1986. The folk theorem in repeated games with discounting or with incomplete information. Econometrica 54, 533–554.

Gibbons, R., 1992. In: A Primer in Game Theory. Harvester Wheatsheaf, New York.

Haas, P.M., Keohane, R.O., Levy, M.A. (Eds.), 1993. Institutions for the Earth. MIT Press, Cambridge (USA) and London.

Kaitala, V., Pohjola, M., Tahvonen, O., 1991. An analysis of SO_2 negotiations between Finland and the Soviet Union. Finnish Economic Papers 4, 104–112.

Klaassen, G.A., Førsund, F.R., Amann, M., 1994. Emission trading in Europe with an exchange rate. Journal of Environmental and Resource Economics 4, 305–330.

Mäler, K.-G., 1989. The acid rain game. In: Folmer, H., van Ierland, E. (Eds.), Valuation Methods and Policy Making in Environmental Economics. Elsevier, Amsterdam, pp. 231–252, Chapter 12.

Mäler, K.-G., 1994. Acid rain in Europe: a dynamic perspective on the use of economic incentives. In: van Ierland, E. (Ed.), International Environmental Economics. Developments in Environmental Economics 4. Elsevier, Amsterdam, pp. 351–372, Chapter 15.

Mäler, K.-G., de Zeeuw, A., 1998. The acid differential game. Environmental and Resource Economics 12, 167–184.

Mohr, E., 1988. On the incredibility of perfect threats in repeated games: note. International Economic Review 29, 551–555.

Murdoch, J.C., Sandler, T., 1997a. Voluntary cutbacks and pretreaty behavior: The Helsinki protocol and sulfur emissions. Public Finance Review 25, 139–162.

Murdoch, J.C., Sandler, T., 1997b. The voluntary provision of a pure public good: The case of reduced CFC emissions and the Montreal protocol. Journal of Public Economics 63, 331–349.

Osborne, M.J., Rubinstein, A., 1994. In: A Course in Game Theory. MIT Press, Cambridge (USA) and London.

Victor, D.G., Raustiala, K., Skolnikoff, E.B. (Eds.), 1998. The Implementation and Effectiveness of International Environmental Commitments. MIT Press, Cambridge (USA) and London.

Wettestad, J. 1996. Acid Lessons? Assessing and Explaining LRTAP Implementation and Effectiveness, IIASA Working Paper, WP-96-18, Laxenburg.

de Zeeuw, A.J., van der Ploeg, F., 1991. Difference games and policy evaluation: A conceptual framework. Oxford Economic Papers 43, 612–636.

[12]

Interconnected Games and International Environmental Problems

HENK FOLMER, PIERRE v. MOUCHE and SHANNON RAGLAND*

Abstract. The purpose of this paper is to introduce the concept of interconnected games and to show its relevance for modeling international environmental problems. It is argued that an interconnected game approach to international environmental problems may enhance cooperation and provide an alternative to the use of financial side payments to induce countries to cooperate. Two types of interconnected games are distinguished in this paper, i.e. direct sum games and tensor games. In the former all the constituting isolated games are games in strategic form and in the latter they are repeated games. In both cases the interconnected game can be interpreted as a multiple objective game, but only the setting where a trade-off is made for the vector-payoffs is considered. In addition to the formal definition of these types of interconnected games, some elementary results concerning Nash equilibria of such games are derived.

Key words. Game theory, repeated games, multiple objective games, tensor games, environmental economics, economic modeling.

1. Introduction

Several environmental problems have international dimensions because their impacts cannot be confined to the country of origin. This attribute is most obvious in the case of global environmental problems, such as ozone depletion and climate change. It also applies to various other kinds of environmental problems, such as acid rain and the pollution of many international rivers and seas.

This interdependency implies that coordination among the countries involved is necessary for effective and efficient solutions to these environmental problems (see for example Mäler, 1989, Folmer, 1993). However, an essential feature of these problems is the absence of an institution with the international jurisdiction to enforce agreements. Thus, any agreement must be voluntary as well as multilateral. It should be observed, that international institutions exist, such as the United Nations, which have been authorised to coordinate international environmental policy. These institutions, however, cannot enforce agreements, so any member can refuse to execute a policy.

Given the characteristics of transboundary pollution, (non-cooperative) game theoretic models have an intuitive appeal. Static as well as dynamic game theoretic models have been used in the economic literature pertaining to transboundary pollution because they can address the environmental interdependencies and the institutional setting (see for example Mäler, 1989,

Environmental and Resource Economics **3**: 313—335, 1993.

1991, van der Ploeg and de Zeeuw, 1992, Kaitala *et al.*, 1992, Hoel 1992a, 1992b). This literature demonstrates theoretically as well as through simulations that there are substantial gains associated with efficient international coordination. It should be noticed that static game theoretic approaches to transboundary pollution problems may be unrealistic because most environmental problems are dynamic and even possess structural time dependence.[1]

The application of game theoretical models to subjects of international concern, such as arms races, transboundary pollution, trade and common property resources has enhanced our understanding of these forms of interaction a great deal. However, to the extent that the models used thus far have all focussed on an isolated topic, they are unable to explain some types of observed behavior. For instance, isolated models studying transboundary pollution suggest that countries might coordinate their actions such that better outcomes, in terms of lower pollution levels, are possible as long as no country is made worse off under the new equilibrium. Consequently, such models are at a loss to explain why some countries join international pollution conventions that appear, within the context of the specific pollution issue, to reduce their welfare.[2] Such actions can only be interpreted as irrational behavior in an isolated model framework.

The problem of concluding optimal agreements with respect to international environmental problems has been at the limelight in the relevant literature for some years now. The basic problem is that these agreements will not necessarily be preferred to the status quo by all countries. The reason is that some of the relevant countries will be made worse off by the agreement and will therefore choose to not participate. A country may be made worse off because its total net benefit from the agreement is negative (see, amongst others, Mäler, 1989). But even if its total net benefit is positive, it may be more profitable to behave as a free rider rather than to participate in the agreement (Barrett, 1990).[3]

In response to the dilemma of negative net benefits it is often suggested that countries who would receive positive net benefits if the agreement is concluded may provide incentives in the form of financial side payments to those whose net benefit would otherwise be negative.[4] A well-known example of the use of side payments is the Rhine Treaty in which inter alia the Netherlands has agreed to compensate France for the reduction of emissions. The potential use of side payments in the context of abatement of acidification in Europe has been discussed by Mäler, 1989 whose simulations of an SO_2 reduction model reveal that Finland, Italy, Spain, and the United Kingdom would experience losses from the full cooperative solution and therefore must be compensated to induce their cooperation. Shaw, 1988 and Kaitala *et al.*, 1992 similarly illustrate via simulations the advantages of allowing side payments. Theoretical treatments of side payments for two-country models can be found in Mäler, 1990 and Markusen, 1975.

Although side payments of the direct transfer type make more efficient solutions possible, there is a good deal of resistance to their implementation. Theoretically they signify application of a victim pays principle rather than a polluter pays principle which is contrary to international policy prescriptions.[5] Moreover, the anticipation of such side payments may provide an incentive to initially minimise abatement policies even below non-cooperative levels thereby extracting larger side payments. This will be particularly relevant when the initial cooperators have imperfect information about the preferences for environmental quality and abatement costs in the non-cooperating countries. Finally Mäler, 1990 suggests two other reasons why such side payments are rarely seen. The first is their prejudicing effect of characterising the victim country as a 'weak negotiator'. Secondly, such conspicuous financial side payments are perhaps unnecessary since the countries in question are typically engaged in several areas of negotiation; a condition which enables the countries to exchange concessions in fields of relative strength. One may speak here of 'side payments in kind'. It is in particular in this scenario where we hope interconnected games prove to be of use.

An interconnected approach to an international environmental problem becomes relevant if the countries in question are also involved in other problems in which the net-benefits of cooperation are (in some sense) reversed. For instance, a country who suffers from transboundary pollution may be an important trade partner of the polluting country and could thereby offer trade concessions to induce the latter to cooperate on environmental issues. Using a two-country example, it might be the case that country 1 is located downstream or downwind of country 2 and therefore suffers from the latter's pollution. Thus, country 1 would like to see country 2 clean up its emissions. However, at the same time, country 1 limits imports of some good from country 2; a practice which country 2 would like to see discontinued. Traditionally one might expect two separate negotiations to evolve; one addressing the environmental dispute and another the trade issue. If this is the case, any 'improvement' in either area will be accompanied by some form of financial side payment. Clearly, if one assumes that the status quo is characterised by each nation maximising its own welfare, then any voluntary reductions in transboundary pollution or unilateral changes in trade policy would make the acting country worse off. Consequently, such policy changes must be compensated via side payments. However, the countries, being aware of the asymmetry of their predicament, could also base negotiations (implicitly or explicitly) upon some exchange of concessions across the two arenas. Country 1 increases imports from country 2, and in exchange the latter reduces the amount of transboundary pollution generated within its borders.[6]

The linkage between different problems may also take the form of a threat to withdraw some extant advantage. For example, trade restrictions are often used to influence some aspect of the targeted nation's behavior other than its trade policies. Recent examples include trade sanctions against South Africa and Iraq.[7] In this case, threats other than the ones that the conventional game theoretic models would predict are utilised. Since the latter are limited to a single issue, all actions including threats must pertain to that issue.

Linking environmental problems to (non-)environmental problems may become more likely, the greater the interdependency between the relevant countries becomes. This applies in particular to institutions like the European Community, which experience a strong increase in domains of cooperation (Folmer and Howe, 1991). However, it can also be observed in other frameworks. For instance, in the development economics literature there is a growing tendency to relate debt forgiveness or interest rate subsidies to resource saving (Hansen, 1991).

The neglect of possible linkages between various problems substantially reduces the usefulness of numerous modelling approaches, including game theory. The purpose of this paper is therefore to formalise the linkages between several problems in a game theoretical framework. The presentation will mainly be at an abstract level. However, the simple example in section 5 will serve to illustrate the various definitions and results as well as the relevance of interconnected games for international environmental problems.

In a couple of instances, modeling techniques used in the literature are comparable in form or spirit to interconnected games. In his study of rational choice in politics, Tsebelis, 1990 develops 'nested games' to convey the importance of context when analyzing political decision-making. For example, individual voting decisions may seem irrational on the face, but they appear rational when viewed in a larger context of say party politics. Bernheim and Winston, 1990 essentially utilize an interconnected game structure in their analysis of the impacts of multi-market contact on oligopolistic behavior. The present paper, however, deals with the notion of interconnected modeling in a formal game theoretic framework. To the authors' knowledge this has not yet been done.

The following observations apply. First, this paper deals with interconnected games in the case where the constituting isolated games are all of the same type. Second, two types of interconnected games are distinguished here and formalised, viz. direct sum games and tensor games. In the former the constituting isolated games are games in strategic form; in the latter they are repeated games with discounting.[8] Third, both types of interconnected games can be interpreted as a multiple objective game, because only the setting where a trade-off is made using linear weights is considered here. Fourth, it should be noticed that in modeling real-world problems our particular construction is useful only if the constituting isolated games are strategy and payoff independent. We speak of 'strategy dependency' if a strategy choice of a player in one constituting isolated game directly restricts the possible strategy choices of players in the other constituting isolated games and we speak of 'payoff dependency' if such a choice directly influences payoffs in other constituting isolated games. Fifth, the interconnected games that we consider don't take into account the possibility of structural time dependence, which is an important feature in applied dynamic modeling of environmental problems. Such extensions are possible but for the sake of convenience are not considered in this paper. Sixth, often one adds a topological structure to the strategy sets and a topological and geometrical structure to the payoff functions. However, we have choosen not to do so in this paper, because we want to focus merely on the fundamental structure of interconnected games. Finally, note that the strategies used throughout this paper are pure strategies.

A central point in the analysis of interconnected games is to compare the new results for interconnected games with those of their constituting isolated games. In this regard the topic of Nash equilibria is studied. Concerning direct sum games, we shall prove that there exists a bijection between the set of Nash equilibria of a direct sum game and the cartesian product of the Nash equilibria of its constituting isolated games. Concerning tensor games, we shall prove that there exists an injection of the cartesian product of the sets of Nash equilibria of the constituting isolated games of a tensor game into the set of Nash equilibria of this tensor game. The above bijective result concerning direct sum games means that direct sum games are in some sense trivial. However, they should be discussed since they serve as the stage game for tensor games.

Unlike direct sum games, tensor games are not trivial constructions. The possible argument that a tensor game can be interpreted as a repeated game and therefore is a superfluous concept does not hold. This is already reflected by the lack of existence of a (canonical) bijection between the set of Nash-equilibria of a tensor game and the cartesian products of the sets of Nash-equilibria of its constituting isolated games. It can also be made clear by referring to the analogy between games in strategic form and repeated games. A repeated game could be considered as a game in strategic form but nobody would claim that the framework of repeated games is therefore superfluous. In both the case of the repeated and the tensor game an additional structure is introduced. A repeated game has a dynamic structure, which a game in strategic form does not have. Similarly, the feature which distinguishes a tensor game from a direct sum game is its dynamic structure. Moreover its multi objective structure distinguishes it from repeated games.

The results obtained in this paper lead us very carefully to the theme 'interconnection may sustain more cooperation' for tensor games similar to the theme 'repetition enables cooperation' for repeated games. Hence, interconnection may serve as an additional explanation of cooperation.

The further organisation of this paper is as follows. The formal definitions of direct sum and tensor games are respectively given in sections 2 and 3. In section 4 we discuss the relationship between Nash equilibria for these types of interconnected games and those of their constituting isolated games. The theoretical results are illustrated by a simple example for the two-player case, in section 5. Conclusions are given in section 6 and formal proofs are given in an appendix.

2. Direct Sum Games

We shall first introduce notation for games in strategic form because the direct sum games considered in this paper are constructed out of a given collection (static) games in strategic form. For details on games in strategic form the reader is referred to for example Friedman (1991).[9]

- A game in strategic form is made up of non-empty (strategy) set X^j $(1 \leqslant j \leqslant N)$ and of (payoff) functions f^j: $X^1 \times \cdots \times X^N \to \mathbb{R}$ $(1 \leqslant j \leqslant N)$. Such a game will denoted by

$$\boldsymbol{\Gamma} = \langle X^1, \ldots, X^N; f^1, \ldots, f^N \rangle.$$

- No topological assumptions will be made for the strategy set X^j and no smoothness and geometric assumptions will be made for the payoff functions f^j. However, to avoid some technicalities later on we shall always assume that each f^j is a bounded function.
- The convention of representing cartesian products and vectors (referring to all N players together) by boldface type will be used here. For example $\boldsymbol{x}$ stands for the (row) vector $(x^1, \ldots, x^N)$ and $\boldsymbol{X}$ stands for $X^1 \times \cdots \times X^N$.[10]
- We shall refer to elements of $\boldsymbol{X} = \Pi_{j=1}^{N} X^j$ as multi-strategies. Moreover, we define $X^{\hat{\imath}} := \Pi_{j=1, j \neq i}^{N} X^j$. Sometimes it is convenient to identify $\boldsymbol{X}$ with $X^i \times X^{\hat{\imath}}$ and to write according $\boldsymbol{x} \in \boldsymbol{X}$ as $\boldsymbol{x} = (x^i, x^{\hat{\imath}})$.

Direct sum games will be constructed out of M given games in strategic form with the same N players in each game. In this context we shall speak of M objectives. In connection with possible trade-offs between the M objectives we introduce a linear function α: $\mathbb{R}^M \to \mathbb{R}$ that is strictly increasing in each of its M variables. α will be called a weight.[11] If α^j is a weight of player j for all j, then we call $\boldsymbol{\alpha}: = (\alpha^1, \ldots, \alpha^N)$ a multi-weight. When α^j is a weight, there (uniquely) exist positive ${}_k\alpha^j$ $(1 \leqslant k \leqslant M)$ such that $\alpha^j(w_1, \ldots, w_M) = \Sigma_{k=1}^{M} {}_k\alpha^j w_k$.

We now present the formal definition of a direct sum game:

DEFINITION 2.1 Given games in strategic form

$${}_k\boldsymbol{\Gamma} = \langle {}_kX^1, \ldots, {}_kX^N; {}_kf^1, \ldots, {}_kf^N \rangle \; (1 \leqslant k \leqslant M)$$

and a multi-weight $\boldsymbol{\alpha}$. We define the trade-off direct sum game

$$(\oplus_{k=1}^{M} {}_k\boldsymbol{\Gamma})_{\alpha}$$

as the game in strategic form

$$\left\langle \prod_{k=1}^{M} {}_kX^1, \ldots, \prod_{k=1}^{M} {}_kX^N; f^1_{\alpha^1}, \ldots, f^N_{\alpha^N} \right\rangle$$

where the payoff $f^i_{\alpha^i}$ is defined as

$$f^i_{\alpha^i}\left(\begin{pmatrix} {}_1x^1 \\ \vdots \\ {}_Mx^1 \end{pmatrix}, \ldots, \begin{pmatrix} {}_1x^N \\ \vdots \\ {}_Mx^N \end{pmatrix} \right) := \sum_{k=1}^{M} {}_k\alpha^i \, {}_kf^i({}_kx^1, \ldots, {}_kx^N).$$

◇

For reasons of convenience we shall use the simpler notation $\oplus \boldsymbol{\Gamma}$, instead of $(\oplus_{k=1}^{M} {}_k\boldsymbol{\Gamma})_\alpha$. We call ${}_1\boldsymbol{\Gamma}, \ldots, {}_M\boldsymbol{\Gamma}$ the constituting isolated games of $\oplus \boldsymbol{\Gamma}$.

The following remarks apply. First, the α^j represent the trade-offs between the vector payoffs of the M constituting isolated games. Hence, we are dealing with trade-off multiple objective games (Zhao, 1991). Secondly, in the case of two players, 1 and 2, who play two bimatrix games $\boldsymbol{A} = (A^1, A^2)$ and $\boldsymbol{B} = (B^1, B^2)$ and a given multi-weight $\boldsymbol{\alpha}$, the trade-off direct sum game of $\boldsymbol{A}$ and $\boldsymbol{B}$ is given by the bimatrix game $(A^1 \circledS_{\alpha^1} B^1, A^2 \circledS_{\alpha^2} B^2)$, where for two matrices X, Y and $\alpha\colon \mathbb{R}^2 \to \mathbb{R}$ the matrix $X \circledS_\alpha Y$ is obtained from the tensor product of X and Y by replacing all products xy by $\alpha(x, y)$. For an example see section 5.

3. Tensor Games

Tensor games have a multiple objective and a dynamic structure. A tensor game is constructed out of a given collection of repeated games. First we introduce notation for repeated games with discounting and perfect monitoring. Again we refer for more details to Friedman (1991).

- A repeated game is made up of a game in strategic form

  $$\boldsymbol{\Gamma} = \langle X^1, \ldots, X^N; f^1, \ldots, f^N \rangle,$$

  called the stage game, a number T (positive or ∞) and a number $\delta \in [0, 1]$. Such a game will be denoted by $\langle \boldsymbol{\Gamma}; T; \delta \rangle$. T is the number of repetitions and δ is a (player independent) discount factor. Given a repeated game, the strategy set $[X^i]$ equals the collection of sequences of mappings[12] $\sigma^i = (\sigma^i_t)_{0 \leqslant t < T}$ with $\sigma^i_t\colon \Pi_{\tau=0}^{t-1} \Pi_{j=1}^{N} X^j \to X^i$ and the payoff function $[f^i]$ is defined as

  $$[f^i](\boldsymbol{\sigma}) := \sum_{t=0}^{T-1} \delta^t f^i(\boldsymbol{a}_t(\boldsymbol{\sigma})),$$

with the outcome path $a^j_t(\boldsymbol{\sigma}) \in X^j$ ($0 \leqslant t < T$) inductively defined by $a^j_0(\boldsymbol{\sigma}) := \sigma^j_0$ and $a^j_t(\boldsymbol{\sigma}) := \sigma^j_t(\boldsymbol{a}_0(\boldsymbol{\sigma}), \boldsymbol{a}_1(\boldsymbol{\sigma}), \ldots, \boldsymbol{a}_{t-1}(\boldsymbol{\sigma}))$ ($1 \leqslant t < T$). When $T = \infty$, we always suppose to avoid convergence problems that $\delta < 1$.

Definition 3.1 formally defines tensor games.

DEFINITION 3.1 Given repeated games $\langle {}_k\boldsymbol{\Gamma}; T; \delta \rangle$ ($1 \leqslant k \leqslant M$) and a multi-weight $\boldsymbol{\alpha}$. The (trade-off) tensor game $(\otimes^M_{k=1} \langle {}_k\boldsymbol{\Gamma}; T; \delta \rangle)_\alpha$ is defined as the repeated game

$$\langle (\oplus^M_{k=1}\, {}_k\boldsymbol{\Gamma})_\alpha; T; \delta \rangle.$$

◇

We also use the simpler notation $\otimes \boldsymbol{\Gamma}$. We call $\langle {}_1\boldsymbol{\Gamma}; T; \delta \rangle, \ldots, \langle {}_M\boldsymbol{\Gamma}; T; \delta \rangle$ the constituting isolated games of $\otimes \boldsymbol{\Gamma}$.

The direct sum game $(\oplus^M_{k=1}\, {}_k\boldsymbol{\Gamma})_\alpha$ serves as the stage game for the tensor game $\otimes \boldsymbol{\Gamma}$. Hence, the strategy set for player i in $\otimes \boldsymbol{\Gamma}$ equals $[\Pi^M_{k=1}\, {}_kX^i]$ and his payoff function in this game is denoted by $[f^i_{\alpha^i}]$ and is given by (denoting multi-strategies of the tensor game by $\boldsymbol{\Sigma}$)

$$[f^i_{\alpha^i}]\,(\boldsymbol{\Sigma}) = \sum_{t=0}^{T-1} \delta^t f^i_{\alpha^i}(A^1_t(\boldsymbol{\Sigma}), \ldots, A^N_t(\boldsymbol{\Sigma})) \tag{1}$$

where $A^j_t(\boldsymbol{\Sigma})$ ($0 \leqslant t < T$) denotes now the outcome path for player j in the tensor game. More explicitly (denoting components referring to the objectives by a pre-subscript):

$$[f^i_{\alpha^i}]\,(\boldsymbol{\Sigma}) = \sum_{t=0}^{T-1} \delta^t \sum_{k=1}^{M} {}_k\alpha^i\, {}_kf^i({}_k(A^1_t(\boldsymbol{\Sigma})), \ldots, {}_k(A^N_t(\boldsymbol{\Sigma}))).$$

In the case where α^i is completely additive, i.e. $\alpha^i(w_1, \ldots, w_M) = \Sigma^M_{k=1} w_k$, (1) becomes

$$[f^i_{\alpha^i}]\,(\boldsymbol{\Sigma}) = \sum_{t=0}^{T-1} \delta^t \sum_{k=1}^{M} {}_kf^i({}_k(A^1_t(\boldsymbol{\Sigma})), \ldots, {}_k(A^N_t(\boldsymbol{\Sigma}))).$$

Roughly speaking, a strategy for a player in $\otimes \boldsymbol{\Gamma}$ consists of rules for each period t that determine what he will play at period t as a function of the history of the game, i.e. all previous actions.

4. Nash Equilibria of Direct Sum and Tensor Games

This section deals with Nash equilibria for direct sum and tensor games and the relationship between these equilibria and those of its constituting isolated games. For direct sum games this relationship will be simple, whereas for tensor games it will be more complicated. First we shall deal with direct sum games and then with tensor games. It should be observed that all the proofs of the theorems and propositions are given in the appendix. Before proceeding we introduce the following notation.

- A pre-subscript $*$ is used to indicate aggregation across the pre-subscripts denoting the M objectives. For example ${}_*x^i = \begin{pmatrix} {}_1x^i \\ \vdots \\ {}_Mx^i \end{pmatrix}$.

The following definition is more basic. Consider a direct sum game $\oplus \boldsymbol{\Gamma}$. Given ${}_k\boldsymbol{x} \in {}_k\boldsymbol{X}$ $(1 \leqslant k \leqslant M)$, we define the multi-strategy $\sqsubset \boldsymbol{x} \sqsupset \in \Pi_{k=1}^{M}\, {}_kX^1 \times \cdots \times \Pi_{k=1}^{M}\, {}_kX^N$ by

$$\sqsubset \boldsymbol{x} \sqsupset := ({}_*x^1, \ldots, {}_*x^N).$$

Notice that

$$f^i_{\alpha^i}(\sqsubset \boldsymbol{x} \sqsupset) = \sum_{k=1}^{M} {}_k\alpha^i\, {}_kf^i({}_k\boldsymbol{x}). \tag{2}$$

DIRECT SUM GAMES: A direct sum game is a trivial construction in the sense that the analysis of a direct sum game normally boils down to the analysis of its constituting isolated games. In this context we mention the following Theorem 4.1.

THEOREM 4.1 Consider a direct sum game $(\oplus_{k=1}^{M}\, {}_k\boldsymbol{\Gamma})_\alpha$. Let ${}_k\boldsymbol{x} \in {}_k\boldsymbol{X}$ $(1 \leqslant k \leqslant M)$ and consider $\sqsubset \boldsymbol{x} \sqsupset$. Then, $\sqsubset \boldsymbol{x} \sqsupset$ is a Nash equilibrium of $(\oplus_{k=1}^{M}\, {}_k\boldsymbol{\Gamma})_\alpha$ if and only if ${}_k\boldsymbol{x}$ is a Nash equilibrium of ${}_k\boldsymbol{\Gamma}$ $(1 \leqslant k \leqslant M)$. Moreover, for each Nash equilibrium Y of $(\oplus_{k=1}^{M}\, \boldsymbol{\Gamma})_\alpha$ there exist (unique) Nash-equilibria ${}_k\boldsymbol{x}$ of ${}_k\boldsymbol{\Gamma}$ $(1 \leqslant k \leqslant M)$ such that $Y = \sqsubset \boldsymbol{x} \sqsupset$. ◇

This theorem says that for each Nash-equilibrium of the direct sum game there exist corresponding Nash-equilibria of the constituting isolated games and vice verse.

TENSOR GAMES: Here the situation becomes much more interesting, because the analysis of a tensor game only partially boils down to the analysis of its constituting isolated games. This is formalized in Theorem 4.2 which is the counterpart of Theorem 4.1. Before going into detail, some definitions and propositions are presented. We begin with the notion of a disconnected strategy. Such strategies refer to actions where for each period t and for each objective k the action taken as a function of the history of the game up to period t only depends on the history of objective k and not on the histories of other objectives. We call such a strategy a disconnected strategy. More precisely:

DEFINITION 4.1 Consider a tensor game $\otimes \boldsymbol{\Gamma}$. We call a strategy $\Sigma^i \in [\Pi_{k=1}^{M} {}_kX^i]$ of player i a disconnected strategy if for each k and t (with $1 \leqslant k \leqslant M$ and $1 \leqslant t < T$)

$$_k(\Sigma_t^i(\boldsymbol{h}_0, \boldsymbol{h}_1, \ldots, \boldsymbol{h}_{t-1})) = {}_k(\Sigma_t^i(\boldsymbol{g}_0, \boldsymbol{g}_1, \ldots, \boldsymbol{g}_{t-1})),$$

for all t-period histories $(\boldsymbol{h}_0, \boldsymbol{h}_1, \ldots, \boldsymbol{h}_{t-1}), (\boldsymbol{g}_0, \boldsymbol{g}_1, \ldots, \boldsymbol{g}_{t-1})$ satisfying

$$_kh_\tau^j = {}_kg_\tau^j \,(1 \leqslant j \leqslant N, 0 \leqslant \tau \leqslant t-1).$$

We denote the set of disconnected strategies of player i by DIS^i. ◇

Intuitively this means that the strategies of player i of the isolated games also belong to the strategy set of player i of the tensor game.

Notice that $T = 1$ as well as $M = 1$ imply that $DIS^i = [\Pi_{k=1}^{M} {}_kX^i]$. Given repeated games $\langle {}_k\boldsymbol{\Gamma}; T; \delta \rangle$ $(1 \leqslant k \leqslant M)$, there exists a natural way to identify disconnected strategies in the strategy sets of $\otimes \boldsymbol{\Gamma}$: Given ${}_k\sigma^i \in [{}_kX^i]$ $(1 \leqslant k \leqslant M)$, we define $\Sigma^i = (\Sigma_t^i)_{0 \leqslant t < T} \in [\Pi_{k=1}^{M} {}_kX^i]$, with

$$\Sigma_t^i: \prod_{\tau=0}^{t-1} \left(\prod_{j=1}^{N} ({}_1X^j \times \cdots \times {}_MX^j) \right) \to {}_1X^i \times \cdots \times {}_MX^i,$$

by[13]

$$\Sigma_t^i \left(\left(\begin{pmatrix} {}_1h_0^1 \\ \vdots \\ {}_Mh_0^1 \end{pmatrix}, \ldots, \begin{pmatrix} {}_1h_0^N \\ \vdots \\ {}_Mh_0^N \end{pmatrix} \right), \ldots, \left(\begin{pmatrix} {}_1h_{t-1}^1 \\ \vdots \\ {}_Mh_{t-1}^1 \end{pmatrix}, \ldots, \begin{pmatrix} {}_1h_{t-1}^N \\ \vdots \\ {}_Mh_{t-1}^N \end{pmatrix} \right) \right) =$$

$$\begin{pmatrix} {}_1\sigma_t^i(({}_1h_0^1, \ldots, {}_1h_0^N), \ldots, ({}_1h_{t-1}^1, \ldots, {}_1h_{t-1}^N)) \\ \vdots \\ {}_M\sigma_t^i(({}_1h_0^1, \ldots, {}_1h_0^N), \ldots, ({}_Mh_{t-1}^1, \ldots, {}_Mh_{t-1}^N)) \end{pmatrix}.$$

It is evident that $\Sigma^i \in DIS^i$. We shall usually denote Σ^i by $\langle \sigma \rangle^i$. Furthermore, for given ${}_k\boldsymbol{\sigma}$ $(1 \leqslant k \leqslant M)$, we can define in a natural way, via the above definition, $\langle \sigma \rangle^i$ $(1 \leqslant i \leqslant N)$, i.e. $\langle \boldsymbol{\sigma} \rangle$. Notice that $\langle \boldsymbol{\sigma} \rangle = {}_1\boldsymbol{\sigma}$ if $M = 1$.

PROPOSITION 4.1 For all ${}_k\boldsymbol{\sigma} \in \Pi_{j=1}^{N}[{}_kX^j]$ $(1 \leqslant k \leqslant M)$ one has

$$A_t^i(\langle \boldsymbol{\sigma} \rangle) = \begin{pmatrix} {}_1a_t^i({}_1\boldsymbol{\sigma}) \\ \vdots \\ {}_Ma_t^i({}_M\boldsymbol{\sigma}) \end{pmatrix} \quad (1 \leqslant i \leqslant N, 0 \leqslant t < T). \tag{3}$$

Moreover, if α is a weight one has

$$[f^i_\alpha](\langle \sigma \rangle) = \alpha([_1f^i](_1\sigma), \ldots, [_Mf^i](_M\sigma)) = \sum_{k=1}^{M} {}_k\alpha[_kf^i](_k\sigma). \qquad (4)$$

◇

Proposition 4.1 says that in the case of disconnected strategies the outcome path and payoff functions of the tensor game factorize.

THEOREM 4.2 Let ${}_k\sigma \in \Pi_{j=1}^N [_kX^j]$ a multi-strategy of $\langle {}_k\Gamma; T; \delta \rangle$ $(1 \leqslant k \leqslant M)$ and consider $\langle \sigma \rangle$. Then $\langle \sigma \rangle$ is a Nash equilibrium of $\otimes\Gamma$ if and only if ${}_k\sigma$ is a Nash equilibrium of $\langle {}_k\Gamma; T; \delta \rangle$ $(1 \leqslant k \leqslant M)$. ◇

We shall call Nash equilibria as they appear in Theorem 4.2 simple, because they are trivial from an interconnected point of view because they correspond to Nash-equilibria of the constituting isolated games. More precisely:

DEFINITION 4.2 We call a Nash equilibrium Σ of a tensor game $\otimes\Gamma$ simple if there exist Nash equilibria ${}_k\sigma$ of $\langle {}_k\Gamma; T; \delta \rangle$ $(1 \leqslant k \leqslant M)$ such that $\Sigma = \langle \sigma \rangle$. ◇

We now turn to Nash equilibria of a tensor game which are not simple. In this context we shall make use of what we call Friedman-trigger strategies.[14]

Given a repeated game $\langle \Gamma; T; \delta \rangle$, we call $\sigma^i = (\sigma^i_t)_{0 \leqslant t < T} \in [X^i]$ a trigger strategy for player i if there exists $p \in X$ and $n^i \in X^i$, such that $\sigma^i_0 = p^i$, $\sigma^i_t(p, \ldots, p) = p^i$ and $\sigma^i_t(x_0, \ldots, x_{t-1}) = n^i$ if $(x_0, \ldots, x_{t-1}) \neq (p, \ldots, p)$ $(1 \leqslant t < T)$. We also denote such an σ^i by $\langle\langle p, n^i \rangle\rangle$. Moreover, if $\langle\langle p, n^j \rangle\rangle$ is a trigger strategies for player j $(1 \leqslant j \leqslant N)$, then we denote the corresponding multi-strategy by $\langle\langle p, n \rangle\rangle$. We shall call a trigger multi-strategy $\langle\langle p, n \rangle\rangle$ where n is a Nash equilibrium of Γ and $f^j(p) > f^j(n)$ $(1 \leqslant j \leqslant N)$ a Friedman-trigger multi-strategy. Hence, in the case of a tensor game, $\langle\langle P, N \rangle\rangle$ is a Friedman-trigger multi-strategy means that $P, N \in \Pi_{k=1}^M {}_kX^1 \times \cdots \times \Pi_{k=1}^M {}_kX^N$, N is a Nash equilibrium of $\oplus\Gamma$ and $f^j_{\alpha'}(P) > f^j_{\alpha'}(N)$ $(1 \leqslant j \leqslant N)$.

In connection with the following theorem we denote by ${}_k\phi^i$ the best reply payoff ${}_k\phi^i: {}_kX^{\hat{\imath}} \to \mathbb{R}$ for player i in ${}_k\Gamma$.

THEOREM 4.3 Let $\Sigma = \langle\langle P, N \rangle\rangle$ be a Friedman-trigger multi-strategy of a tensor game $(\otimes_{k=1}^M < {}_k\Gamma; T; \delta \rangle)_\alpha$. Then

1. If $T \geqslant 2$ and $M \geqslant 2$, then under mild conditions Σ is not disconnected (for example if there exists a k with ${}_kP^i \neq {}_kN^i$ and a $k' \neq k$ such that the strategy set ${}_{k'}X^i$ contains at least two elements).

2. Suppose $T = \infty$. $\boldsymbol{\Sigma}$ is a Nash equilibrium of $(\otimes_{k=1}^{M} < {}_k\boldsymbol{\Gamma}; T; \delta\rangle)_{\boldsymbol{\alpha}}$ if and only if for all $1 \leqslant j \leqslant N$

$$\delta\left(\sum_{k=1}^{M} {}_k\alpha^j({}_k\phi^j({}_kP^j) - {}_kf^j({}_k\boldsymbol{N}))\right) \geqslant$$

$$\sum_{k=1}^{M} {}_k\alpha^j({}_k\phi^j({}_kP^j) - {}_kf^j({}_k\boldsymbol{P})). \tag{5}$$

3. Suppose $T = \infty$. There exists a discount factor δ_0 that satisfies (5) (with $\delta = \delta_0$). If δ_0 is such, then for all discount factors η with $\eta \geqslant \delta_0$, $\boldsymbol{\Sigma}$ is a Nash equilibrium of $(\otimes_{k=1}^{M} \langle {}_k\boldsymbol{\Gamma}; T; \eta\rangle)_{\boldsymbol{\alpha}}$. ◇

Theorem 4.3 shows that Friedman-trigger multi-strategies allow players to achieve 'cooperative outcomes' in games that are non-cooperative, as the players' strategies contain possibilities to punish players who are observed to deviate.[15] However, tensor games do not only contain provisions to punish in the same (constituting isolated) game, as in the case of repeated games, but also in different constituting isolated games. This is the major distinguishing feature of tensor games.

For tensor games various types of trigger strategies other than the one mentioned in Theorem 4.3 exist. They are more interesting and may be more credible from an interconnected point of view. An example of such an strategy is considered in the example given in the next section.

Concerning the existence of Friedman-trigger strategies for tensor games we mention the following result.

PROPOSITION 4.2 Given trigger multi-strategies $\langle\langle {}_k\boldsymbol{p}, {}_k\boldsymbol{n}\rangle\rangle$ of $\langle {}_k\boldsymbol{\Gamma}; \boldsymbol{T}; \boldsymbol{\delta}\rangle$ $(1 \leqslant k \leqslant M)$. Consider the trigger multi-strategy $\boldsymbol{\Sigma} := \langle\langle \sqsubset \boldsymbol{p} \sqsupset, \sqsubset \boldsymbol{n} \sqsupset\rangle\rangle$ of $\otimes\boldsymbol{\Gamma}$. Then

1. $\boldsymbol{\Sigma}$ is a Friedman-trigger multi-strategy of $\otimes\boldsymbol{\Gamma} \Leftrightarrow {}_k\boldsymbol{n}$ is a Nash equilibrium of ${}_k\boldsymbol{\Gamma}$ and $\Sigma_{k=1}^{M} {}_k\alpha^j\ {}_kf^j({}_k\boldsymbol{p}) > \Sigma_{k=1}^{M} {}_k\alpha^j\ {}_kf^j({}_k\boldsymbol{n})$ $(1 \leqslant k \leqslant M)$.
2. A sufficient condition for $\boldsymbol{\Sigma}$ to be a Friedman-trigger multi-strategy of $\otimes\boldsymbol{\Gamma}$ is that $\langle\langle {}_k\boldsymbol{p}, {}_k\boldsymbol{n}\rangle\rangle$ is a Friedman-trigger multi-strategy of $\langle {}_k\boldsymbol{\Gamma}; T; \delta\rangle$ $(1 \leqslant k \leqslant M)$. ◇

Notice that from the above results it indeed follows that 'in general' a tensor game has non-simple Nash equilibria (for high enough discount factor).

5. An Example

The foregoing is illustrated by means of an artificial example comprising two countries, denoted by 1 and 2, and two repeated games, a transboundary pollution game $\langle {}_1\boldsymbol{\Gamma};\ T;\ \delta\rangle$ and a trade game $\langle {}_2\boldsymbol{\Gamma};\ T;\ \delta\rangle$. The stage games of these games are given by the following 2×2 bi-matrix games

$$ {}_1\boldsymbol{\Gamma} = \begin{pmatrix} 3-\varepsilon^1, 1 & -2-\varepsilon^1, 2 \\ 5, -1 & 0, 0 \end{pmatrix}, \quad {}_2\boldsymbol{\Gamma} = \begin{pmatrix} 2, 2-\varepsilon^2 & -1, 5 \\ 3, -3-\varepsilon^2 & 0, 0 \end{pmatrix}. $$

Here $0 < \varepsilon^1 < 3$ and $0 < \varepsilon^2 < 2$.

Both ${}_1\boldsymbol{\Gamma}$ and ${}_2\boldsymbol{\Gamma}$ have the typical properties of the classic prisonners' dilemma game. Each country has a strictly dominant strategy (namely, the second one). The associated Nash equilibrium is not Pareto-optimal (in the weak sense) and each country's Nash equilibrium payoff (0) equals its minmax payoff.[16] As usual for 2×2-bimatrix prisonner dilemma games we refer to the dominant strategy also as the 'defection strategy' D and to the other strategy also as the 'cooperation strategy' C.

The interpretation of the stage game ${}_1\boldsymbol{\Gamma}$ is now as follows: ${}_1\boldsymbol{\Gamma}$ deals with an uni-directional pollutant generated by country 2 and affecting country 1. Country 2 chooses whether to reduce its pollution emission (C), or not (D) and country 1 chooses whether to make a side payment (C) of the amount 2, or not (D). More precisely, if both countries defect, then the status quo prevails, there is neither a reduction in transboundary pollution nor a side payment, and each country receives 0 for its payoff. If country 2 reduces pollution, but no side payment is made, the cost of cooperation to country 2 is -1, while country 1 receives a net benefit of 5. Moving then to the mutual cooperation cell of the bimatrix, country 1 makes a side payment of the amount 2 and country 2's payoff becomes 1. Note, however, that country 1's payoff is $3 - \varepsilon^1$ instead of 3. The ε^1 represents some loss associated with financial side payments, e.g., it might represent the 'psychic cost of being labeled a weak negotiator'.[17] It is not difficult to imagine that such losses occur when nations apparently succumb to openly bribing their neighbours. Practically, in terms of the payoffs, one may suppose that these psychic losses incorporate reputational damages that would arise in the event of financial side-payments.[18] Finally, if country 2 does not reduce its pollution, but country 1 makes the side payment the payoffs are $-2 - \varepsilon^1$ and 2.[19]

Game ${}_2\boldsymbol{\Gamma}$ is in some sense symmetrical to ${}_1\boldsymbol{\Gamma}$. For ${}_2\boldsymbol{\Gamma}$ mutual cooperation now requires country 1 to remove some import restriction and country 2 to make a side payment of the amount 3. In this game ε^2 might again represent the utility loss associated with country 2's side payment.

Consider now the repeated games $\langle {}_1\boldsymbol{\Gamma};\ T;\ \delta\rangle$ and $\langle {}_2\boldsymbol{\Gamma};\ T;\ \delta\rangle$. The interpretation of these games is now as usual derived from the interpretation of the stage games: In each of T periods each country has two actions from which to choose. These actions, selected simultaneously in each period, represent some type of activity that can be observed on an annual basis, e.g., allowing imports to rise from 100,000 to 175,000 tons per year, reducing annual emissions from 1,000 tons to 800 tons, or making an annual side payment. The duration of the repeated games is determined by, T, the number of times the stage games are repeated. It is assumed that this number is the same in both repeated games. In the case of a single shot game, i.e., $T = 1$, a strategy is identical to selecting one of the two possible actions; if $T > 1$ the strategies will be more complex since they must specify a fully contingent plan of play, i.e. for each period they must specify an action for every conceivable history of the game. We assume perfect monitoring in each period, and that the discount factor δ is identical across players and games. Period t payoffs are received in the beginning of period t.

To examine how the capability to exchange concessions across the games might alter the outcome vis a vis isolated negotiations, one must consider both scenarios. First consider which outcomes may be obtained when the games are played independently. What reasonable expectations might one have regarding the outcomes of these games? To answer this question it is convenient to distinguish between T finite and $T = \infty$.

Consider the game $\langle {}_1\boldsymbol{\Gamma}; T; \delta \rangle$. In the case of finite T, it is well known that playing a Nash equilibrium of $\langle {}_1\boldsymbol{\Gamma}; T; \delta \rangle$ leads in every period to the unique Nash equilibrium of the stage game, i.e. to the outcome, that is country 1 plays D and country 2 plays D. Thus, in this case each Nash equilibrium leads to an average payoff of 0. In the case of $T = \infty$ however, Fact A.1 (Appendix) implies for this case that the Friedman-trigger multi-strategy $\langle\langle (C, C), (D, D) \rangle\rangle$ is a Nash equilibrium of $\langle {}_1\boldsymbol{\Gamma}; T; \delta \rangle$ for $\delta \geqslant (2 + \varepsilon^1)/5$.[20] This Nash equilibrium gives an average payoff of $3 - \varepsilon^1$ for country 1 and 1 for country 2. Notice that these average payoffs are >0. However, there does not exist a Nash equilibrium leading to the outcome (C, D) in every period, which is interesting from the point of view that it would give the highest possible value, 4, for the sum of the average payoffs across countries. This is due to the fact that such outcomes path would imply that one of the countries be paid in every period less than his minmax payoff. For $\langle {}_2\boldsymbol{\Gamma}; T; \delta \rangle$ analogous considerations and conclusions hold.

Now let us consider what happens when we interconnect the two games. Before doing so some details concerning the interpretation of the interconnection must be first addressed. Suppose that the interconnected game comprises simultaneous play of the two games such that each country plays a strategy containing actions for both games instead of one strategy for each game. Payoffs for the interconnected game are supposed to be unweighted sums of the payoffs from the constituting isolated games. More precisely, at each period t $(0 \leqslant t < T)$ each country receives a total payoff equal to the sum of the payoffs from games 1 and 2 at this period. Thus, using the terminology introduced, we consider the tensor game $\otimes\boldsymbol{\Gamma}$ of $\langle {}_1\boldsymbol{\Gamma}; T; \delta \rangle$, $\langle {}_2\boldsymbol{\Gamma}; T; \delta \rangle$ with completely additive weights.

The outcome path of the tensor game which maximises the sum of the payoffs, $[f^1] + [f^2]$, would be a play in every period of $\binom{D}{C}$ for country 1 and of $\binom{C}{D}$ for country 2. This sum corresponds to a total average payoff 8, whereas mutual cooperation in every period can only give $8 - \varepsilon^1 - \varepsilon^2$. Consider the following multi-strategy $\boldsymbol{\Sigma}$: Country 1 plays $\binom{D}{C}$ in period 0 and in every period thereafter as long as country 2 has not previously played D in game 1; otherwise it plays $\binom{D}{D}$. Country 2 plays $\binom{C}{D}$ in period 0 and in every subsequent period as long as country 1 has not played D in game 2; otherwise it plays $\binom{D}{D}$. Of course, $\boldsymbol{\Sigma}$ leads in each period to the in the above sense optimal outcome $\binom{D}{C}$ for country 1 and $\binom{C}{D}$ for country 2. We notice that $\boldsymbol{\Sigma}$ is not a trigger multi-strategy, as defined in the preceding section and that therefore Theorem 4.3 cannot be applied. However it is not so difficult to see that if $T = \infty$, $\boldsymbol{\Sigma}$ is a Nash equilibrium of the tensor game $\otimes \boldsymbol{\Gamma}$ as long as $\delta > 1/5$.

Notice that, according to the theory in section 2, we have (using the Ⓢ operation of section 2) for the stage game of $\otimes \boldsymbol{\Gamma}$ the bi-matrix game

$$\begin{pmatrix} 5-\varepsilon^1, 3-\varepsilon^2 & 2-\varepsilon^1, 6 & -\varepsilon^1, 4-\varepsilon^2 & -3-\varepsilon^1, 7 \\ 6-\varepsilon^1, -2-\varepsilon^2 & 3-\varepsilon^1, 1 & 1-\varepsilon^1, -1-\varepsilon^2 & -2-\varepsilon^1, 2 \\ 7, 1-\varepsilon^2 & 4, 4 & 2, 2-\varepsilon^2 & -1, 5 \\ 8, -4-\varepsilon^2 & 5, -1 & 3, -3-\varepsilon^2 & 0, 0 \end{pmatrix}$$

and that the fourth strategy of country 1 and the fourth strategy of country 2 form the only Nash equilibrium of this game in accordance with Theorem 4.1.

The above example demonstrates the potential advantages of allowing countries to 'trade' concessions across issues under negotiation and describes what circumstances best lend themselves to such a trade. The innovation of the interconnected game framework is that countries can now condition their choice of actions in the environmental area to outcomes previously observed in the trade arena, and vice versa. This capability enriches the set of Nash equilibria and maximises the sum across countries of the obtainable payoff values.

6. Conclusions

In this paper we introduced the concept of interconnected games. This concept is a very obvious and potentially important one. We formalised this concept for the special case where each constituting isolated game is a repeated game and for the case where each constituting isolated game is a game in strategic form; the former one certainly being the most interesting.

The aim of this paper was primarily to introduce the above concept, to derive some elementary results and to discuss the relevance of interconnected games for modeling international environmental problems. For convenience various restricting assumptions were made such as the trade-off assumption and the absence of structural time dependence. It is quite well possible to relax these assumptions. Also the very elementary analysis given here should in any case be extended to more specific aspects of interconnection. Particularly, the analysis of Nash equilibria should be continued and refinements such as subgame perfection should be considered. Avenues of future work include: Interconnection of games other than games in strategic form or repeated games, trade-offs versus no trade-offs, folk theorems, the addition of a topological structure to the strategy sets and of a topological (and geometrical) structure to the payoff functions, appropriate solution concepts, cooperative solutions, the effects of informational refinements such as asymmetric, incomplete or uncertain information, stability of equilibria, potential power asymmetry (with respect to negotiations). A central point in the analysis of interconnected games should always be how the results for interconnected games relate to the (corresponding) results for the constituting isolated games.

Similar to the theme 'repetition enables cooperation' for repeated games, 'interconnection may sustain more cooperation' may be a theme for tensor games. Hence, interconnection may serve as an additional explanation of cooperation. In terms of a specific international environmental game, for example this means that by linking this environmental game to a non-environmental game, or by linking it to another environmental game, countries which without the linkage would have no incentive to cooperate may be stimulated to cooperate. In this respect interconection may provide an alternative to side payments which are often recommended in the context of international environmental policy. In this context it is important that the net benefits in the several areas of negotiation are reversed.

One may expect that interconnected games may not only be interesting for modeling of international environmental problems but even for modeling of economic problems in general. Also it may be interesting for modeling in other sciences. Of special (policy) interest might be the analysis of seemingly irrational behavior as in the case of countries joining international pollution conventions that appear, within the context of a specific pollution issue, to reduce their welfare.

A. Proofs

The notation used in the main text will be used in the following proofs. Moreover ${}_*X^j := \Pi_{k=1}^{M}\, {}_kX^j$.

Proof: [THEOREM 4.1] Let ${}_k\boldsymbol{x}$ be a Nash equilibrium of ${}_k\boldsymbol{\Gamma}$. We shall first show that $\sqsubset\boldsymbol{x}\sqsupset$ is a Nash equilibrium of the direct sum game, i.e. that $({}_*x^1, \ldots, {}_*x^N)$ is a Nash equilibrium of $\oplus\boldsymbol{\Gamma}$. Well, for all i and ${}_*y \in {}_*X^i$ we have $f^i_{\alpha^i}({}_*y, \sqsubset x\sqsupset^{\hat{i}}) = \alpha^i({}_1f^i({}_1y, {}_1x^{\hat{i}}), \ldots, {}_Mf^i({}_My, {}_Mx^{\hat{i}}))$. Because ${}_kf^i({}_ky, {}_kx^{\hat{i}}) \leqslant {}_kf^i({}_k\boldsymbol{x})$ for all k, we find $f^i_{\alpha^i}({}_*y, \sqsubset x\sqsupset^{\hat{i}}) \leqslant \alpha^i({}_1f^i({}_1\boldsymbol{x}), \ldots, {}_Mf^i({}_M\boldsymbol{x})) = f^i_{\alpha^i}({}_*x^i, \sqsubset x\sqsupset^{\hat{i}})$. This proves the if part.

To prove the only if part we suppose $\sqsubset\boldsymbol{x}\sqsupset = ({}_*x^1, \ldots, {}_*x^N)$ is a Nash equilibrium of $\oplus \boldsymbol{\Gamma}$. We have to prove that ${}_k\boldsymbol{x}$ is a Nash equilibrium of ${}_k\boldsymbol{\Gamma}$ $(1 \leqslant k \leqslant M)$. Fix k, i and $z \in {}_kX^i$. For all ${}_*y \in {}_*X^i$ we have $f^i_{\alpha^i}(\Box\boldsymbol{x}\Box) \geqslant f^i_{\alpha^i}({}_*y, \Box\boldsymbol{x}\Box^{\hat{i}})$. In particular this is true for ${}_*y \in {}_*X^i$ defined by ${}_ky = z$, ${}_ly = {}_lx^i (l \neq k)$. For this ${}_*y$ we have thus $\alpha^i({}_1f^i({}_1\boldsymbol{x}), \ldots, {}_Mf^i({}_M\boldsymbol{x})) \geqslant \alpha^i({}_1f^i({}_1y, {}_1x^{\hat{i}}), \ldots, {}_Mf^i({}_My, {}_Mx^{\hat{i}}))$. Because in this inequality ${}_lf^i({}_l\boldsymbol{x}) = {}_lf^i({}_ly, {}_lx^{\hat{i}})$ $(l \neq k)$, we obtain, using that α^i is strictly increasing in each variable, that ${}_kf^i({}_k\boldsymbol{x}) \geqslant {}_kf^i({}_ky, {}_kx^{\hat{i}})$. Thus ${}_kf^i({}_k\boldsymbol{x}) \geqslant {}_kf^i(z, {}_kx^{\hat{i}})$.

The last statement is a consequence of the above and the following remark: If $\boldsymbol{Y}$ is a multi-strategy of the direct sum game, then there exists a (even unique) multi-strategy ${}_k\boldsymbol{x}$ of ${}_k\boldsymbol{\Gamma}$ $(1 \leqslant k \leqslant M)$ such that $\boldsymbol{Y} = \Box\boldsymbol{x}\Box$. □

In the sequel the notion of best reply payoff function is needed. Consider a direct sum game $\oplus\boldsymbol{\Gamma}$. The best reply payoff function for player i is defined as the function $\Phi^i: \Pi_{j \neq i}(\Pi_{k=1}^M {}_kX^j) \to \mathbb{R}$ given by

$$\Phi^i(z) := \sup_{y \in \Pi_{k=1}^M {}_kX^i} f^i_{\alpha^i}(y, z).$$

Such expressions can usually be expressed in terms of objects of the constituting isolated games. For example, if ${}_k\phi^i$ denotes the best reply payoff ${}_k\phi^i: {}_kX^{\hat{i}} \to \mathbb{R}$ for player i in ${}_k\boldsymbol{\Gamma}$, then the following result holds:

LEMMA A.1 Fix i. Then for all $z \in \Pi_{j \neq i}(\Pi_{k=1}^M {}_kX^j)$ we have

$$\Phi^i(z) = \sum_{k=1}^{M} {}_k\alpha^i \; {}_k\phi^i({}_kz).$$

◇

Proof: $\Phi^i(z) = \sup_{y \in {}_*X^i} f^i_{\alpha^i}(y, z) = \sup_{y \in {}_*X^i} \Sigma_{k=1}^M {}_k\alpha^i \; {}_kf^i({}_ky, {}_kz) = \Sigma_{k=1}^M {}_k\alpha^i \sup_{{}_ky \in {}_kX^i} {}_kf^i({}_ky, {}_kz) = \Sigma_{k=1}^M {}_k\phi^i({}_kz)$. □

In connection with the sequel we introduce the following notation.

- Notation: If $y = \left(\begin{pmatrix} {}_1x^1 \\ \vdots \\ {}_Mx^1 \end{pmatrix}, \ldots, \begin{pmatrix} {}_1x^N \\ \vdots \\ {}_Mx^N \end{pmatrix} \right) \in \Pi_{j=1}^N(\Pi_{k=1}^M {}_kX^j)$, then

$pr_ky = ({}_kx^1, \ldots, {}_kx^N)$.

LEMMA A.2 Let ${}_k\boldsymbol{\sigma}$ be a multi-strategy of $\langle {}_k\boldsymbol{\Gamma}; T; \delta \rangle$ $(1 \leqslant k \leqslant M)$. Consider $\langle \boldsymbol{\sigma} \rangle$. Fix i and consider a multi-strategy $\boldsymbol{\Sigma} = (\Lambda, \langle \sigma \rangle^{\hat{i}})$ of $\otimes\boldsymbol{\Gamma}$. Then, for each k, there exists an open loop strategy ${}_k\lambda^i$ of player i in $\langle {}_k\boldsymbol{\Gamma}; T; \delta \rangle$ such that with ${}_k\boldsymbol{\tau} := ({}_k\lambda^i, {}_k\sigma^{\hat{i}}) \in \Pi_{j=1}^N[{}_kX^j]$ we have for $\langle \boldsymbol{\tau} \rangle$

$$A^j_t(\langle \boldsymbol{\tau} \rangle) = A^j_t(\boldsymbol{\Sigma}) \text{ for all } j, t. \tag{6}$$

◇

Proof: Define ${}_k\lambda^i = ({}_k\lambda^i_t)_{0 \leqslant t < T}$ as the open loop strategy where at each period t, for all possible t-period histories, ${}_k(A^i_t(\boldsymbol{\Sigma}))$ is played. Then (6) is true for $j = i$ and all t because, using (3), we have $A^i_t(\langle \boldsymbol{\tau} \rangle) = \langle \tau \rangle^i_t(\mathbf{A}_0(\langle \boldsymbol{\tau} \rangle), \ldots, \mathbf{A}_{t-1}(\langle \boldsymbol{\tau} \rangle)) =$

$$\begin{pmatrix} {}_1\tau^i_t(pr_1\mathbf{A}_0(\langle \boldsymbol{\tau} \rangle), \ldots, pr_1\mathbf{A}_{t-1}(\langle \boldsymbol{\tau} \rangle)) \\ \vdots \\ {}_M\tau^i_t(pr_M\mathbf{A}_0(\langle \boldsymbol{\tau} \rangle), \ldots, pr_M\mathbf{A}_{t-1}(\langle \boldsymbol{\tau} \rangle)) \end{pmatrix} =$$

$$\begin{pmatrix} {}_1\lambda^i_t(pr_1\mathbf{A}_0(\langle \boldsymbol{\tau} \rangle), \ldots, pr_1\mathbf{A}_{t-1}(\langle \boldsymbol{\tau} \rangle)) \\ \vdots \\ {}_M\lambda^i_t(pr_M\mathbf{A}_0(\langle \boldsymbol{\tau} \rangle), \ldots, pr_M\mathbf{A}_{t-1}(\langle \boldsymbol{\tau} \rangle)) \end{pmatrix} = \begin{pmatrix} {}_1(A^i_t(\boldsymbol{\Sigma})) \\ \vdots \\ {}_M(A^i_t(\boldsymbol{\Sigma})) \end{pmatrix} = A^i_t(\boldsymbol{\Sigma}).$$

For $j \neq i$ and $t = 0$ the statement is true because $A^j_0(\langle \boldsymbol{\tau} \rangle) = \langle \tau \rangle^j_0 = \begin{pmatrix} {}_1\tau^j_0 \\ \vdots \\ {}_M\tau^j_0 \end{pmatrix} = \begin{pmatrix} {}_1\sigma^j_0 \\ \vdots \\ {}_M\sigma^j_0 \end{pmatrix} = \langle \sigma \rangle^j_0 = \Sigma^j_0 = A^j_0(\boldsymbol{\Sigma})$. We shall prove now by induction on t that the statement is true for all j. For $t = 0$ this has already been done. Now suppose the statement is true for all j for all t with $0 \leqslant t \leqslant L$ for some L. We shall prove that it is true now for $t = L + 1$:

$$A^j_{L+1}(\langle \boldsymbol{\tau} \rangle) = \langle \boldsymbol{\tau} \rangle^j_{L+1}(\mathbf{A}_0(\langle \boldsymbol{\tau} \rangle), \ldots, \mathbf{A}_L(\langle \boldsymbol{\tau} \rangle)) =$$

$$\langle \tau \rangle^j_{L+1}(\begin{pmatrix} pr_1\mathbf{A}_0(\langle \boldsymbol{\tau} \rangle) \\ \vdots \\ pr_M\mathbf{A}_0(\langle \boldsymbol{\tau} \rangle) \end{pmatrix}, \ldots, \begin{pmatrix} pr_1\mathbf{A}_L(\langle \boldsymbol{\tau} \rangle) \\ \vdots \\ pr_M\mathbf{A}_L(\langle \boldsymbol{\tau} \rangle) \end{pmatrix}).$$

Because of Proposition 4.1 this equals

$$\langle \tau \rangle^j_{L+1}(\begin{pmatrix} {}_1a_0({}_1\boldsymbol{\tau}) \\ \vdots \\ {}_Ma_0({}_M\boldsymbol{\tau}) \end{pmatrix}, \ldots, \begin{pmatrix} {}_1a_L({}_1\boldsymbol{\tau}) \\ \vdots \\ {}_Ma_L({}_M\boldsymbol{\tau}) \end{pmatrix}) =$$

$$\begin{pmatrix} {}_1\tau^j_{L+1}({}_1\boldsymbol{a}_0({}_1\boldsymbol{\tau}), \ldots, {}_1\boldsymbol{a}_L({}_1\boldsymbol{\tau})) \\ \vdots \\ {}_M\tau^j_{L+1}({}_M\boldsymbol{a}_0({}_M\boldsymbol{\tau}), \ldots, {}_M\boldsymbol{a}_L({}_M\boldsymbol{\tau})) \end{pmatrix} = \begin{pmatrix} {}_1\sigma^j_{L+1}({}_1\boldsymbol{a}_0({}_1\boldsymbol{\tau}), \ldots, {}_1\boldsymbol{a}_L({}_1\boldsymbol{\tau})) \\ \vdots \\ {}_M\sigma^j_{L+1}({}_M\boldsymbol{a}_0({}_M\boldsymbol{\tau}), \ldots, {}_M\boldsymbol{a}_L({}_M\boldsymbol{\tau})) \end{pmatrix} =$$

$$\begin{pmatrix} {}_1\sigma^j_{L+1}(pr_1\mathbf{A}_0(\langle \boldsymbol{\tau} \rangle), \ldots, pr_1\mathbf{A}_L(\langle \boldsymbol{\tau} \rangle)) \\ \vdots \\ {}_M\sigma^j_{L+1}(pr_M\mathbf{A}_0(\langle \boldsymbol{\tau} \rangle), \ldots, pr_M\mathbf{A}_L(\langle \boldsymbol{\tau} \rangle)) \end{pmatrix} = \langle \sigma \rangle^j_{L+1}(\mathbf{A}_0(\langle \boldsymbol{\tau} \rangle), \ldots, \mathbf{A}_L(\langle \boldsymbol{\tau} \rangle)).$$

Because of the induction assumption this becomes $\langle\sigma\rangle^{j}_{L+1}(\mathbf{A}_0(\boldsymbol{\Sigma}), \ldots, \mathbf{A}_L(\boldsymbol{\Sigma})) = \Sigma^{j}_{t}(\mathbf{A}_0(\boldsymbol{\Sigma}), \ldots, \mathbf{A}_L(\boldsymbol{\Sigma})) = \mathbf{A}^{j}_{t}(\boldsymbol{\Sigma})$. □

Proof: [PROPOSITION 4.1] Here is a proof by induction on t. The statement is

true for $t = 0$, namely: $A^{\iota}_0(\langle\boldsymbol{\sigma}\rangle) = \langle\sigma\rangle^{\iota}_0 = \begin{pmatrix} {}_1\sigma^{\iota}_0 \\ \vdots \\ {}_M\sigma^{\iota}_0 \end{pmatrix} = \begin{pmatrix} {}_1a^{\iota}_0({}_1\boldsymbol{\sigma}) \\ \vdots \\ {}_Ma^{\iota}_0({}_M\boldsymbol{\sigma}) \end{pmatrix}$.

Suppose now that the statement is true for all periods 0, 1, ..., t for a certain t with $0 \leqslant t < T-1$. We shall prove that the statement is then also true for period $t+1$:

$$A^{\iota}_{t+1}(\langle\boldsymbol{\sigma}\rangle) = \langle\sigma\rangle^{\iota}_{t+1}(\mathbf{A}_0(\langle\boldsymbol{\sigma}\rangle), \ldots, \mathbf{A}_t(\langle\boldsymbol{\sigma}\rangle)) =$$
$$\langle\sigma^{\iota}\rangle_{t+1}((A^1_0(\langle\boldsymbol{\sigma}\rangle), \ldots, A^N_0(\langle\boldsymbol{\sigma}\rangle)), \ldots, (A^1_t(\langle\boldsymbol{\sigma}\rangle), \ldots, A^N_t(\langle\boldsymbol{\sigma}\rangle))).$$

Because of the induction assumption this equals

$$\langle\sigma\rangle^{\iota}_{t+1}((\begin{pmatrix} {}_1a^1_0({}_1\boldsymbol{\sigma}) \\ \vdots \\ {}_Ma^1_0({}_M\boldsymbol{\sigma}) \end{pmatrix}, \ldots, \begin{pmatrix} {}_1a^N_0({}_1\boldsymbol{\sigma}) \\ \vdots \\ {}_Ma^N_0({}_M\boldsymbol{\sigma}) \end{pmatrix}), \ldots, (\begin{pmatrix} {}_1a^1_t({}_1\sigma) \\ \vdots \\ {}_Ma^1_t({}_M\sigma) \end{pmatrix}, \ldots, \begin{pmatrix} {}_1a^N_t({}_1\boldsymbol{\sigma}) \\ \vdots \\ {}_Ma^N_t({}_M\boldsymbol{\sigma}) \end{pmatrix})).$$

Because of (3) this in turn equals $\begin{pmatrix} {}_1\sigma^{\iota}_{t+1}({}_1\mathbf{a}_0({}_1\boldsymbol{\sigma}), \ldots, {}_1\mathbf{a}_t({}_1\boldsymbol{\sigma})) \\ \vdots \\ {}_M\sigma^{\iota}_{t+1}({}_M\mathbf{a}_0({}_M\boldsymbol{\sigma}), \ldots, {}_M\mathbf{a}_t({}_M\boldsymbol{\sigma})) \end{pmatrix}$ and this

is just the definition of $\begin{pmatrix} {}_1a^{\iota}_{t+1}({}_1\boldsymbol{\sigma}) \\ \vdots \\ {}_Ma^{\iota}_{t+1}({}_M\boldsymbol{\sigma}) \end{pmatrix}$.

If α is a weight, then by the above result we obtain $[f^{\iota}_{\alpha}](\langle\boldsymbol{\sigma}\rangle) = \Sigma^{T-1}_{t=0}\ \delta^t\ \Sigma^M_{k=1}\ {}_k\alpha_k f^t({}_k(A^1_t(\langle\boldsymbol{\sigma}\rangle)), \ldots, {}_k(A^N_t(\langle\boldsymbol{\sigma}\rangle))) = \Sigma^M_{k=1}\ {}_k\alpha\ \Sigma^{T-1}_{t=0}\ \delta^t\ {}_kf^t({}_ka^1_t({}_k\boldsymbol{\sigma}), \ldots, {}_ka^N_t({}_k\boldsymbol{\sigma})) = \Sigma^M_{k=1}\ {}_k\alpha[{}_kf]\,({}_k\boldsymbol{\sigma}))$. □

Proof: [THEOREM 4.2] Let each ${}_k\boldsymbol{\sigma}$ be a Nash equilibrium of $\langle {}_k\boldsymbol{\Gamma};\ T;\ \delta\rangle$. We have to show that $\langle\boldsymbol{\sigma}\rangle$ is a Nash equilibrium of $\otimes\boldsymbol{\Gamma}$. Fix a player i. Consider a multi-strategy $\boldsymbol{\Sigma} = (\Lambda, \langle\sigma\rangle^{\hat{\iota}})$ with $\Lambda \in [\Pi^M_{k=1}\ {}_kX^{\iota}]$. We have to show that $[f^{\iota}_{\alpha^{\iota}}](\boldsymbol{\Sigma}) \leqslant [f^{\iota}_{\alpha^{\iota}}](\langle\boldsymbol{\sigma}\rangle)$. Therefore, we take, using Lemma A.2 open loop strategies ${}_k\lambda^{\iota} \in [{}_kX^{\iota}]$ $(1 \leqslant k \leqslant M)$ such that with ${}_k\boldsymbol{\tau} := ({}_k\lambda^{\iota}, {}_k\sigma^{\hat{\iota}})$ one has $A^j_t(\langle\tau\rangle) = A^j_t(\boldsymbol{\Sigma})$ for all j, t. Thus $\mathbf{A}_t(\boldsymbol{\Sigma}) = \mathbf{A}_t(\langle\boldsymbol{\tau}\rangle)$ and therefore $[f^{\iota}_{\alpha^{\iota}}](\boldsymbol{\Sigma}) = \Sigma^{T-1}_{t=0}\ \delta^t\alpha^{\iota}({}_1f^{\iota}(pr_1\mathbf{A}_t(\boldsymbol{\Sigma})), \ldots, {}_Mf^{\iota}(pr_M\mathbf{A}_t(\boldsymbol{\Sigma}))) = [f^{\iota}_{\alpha^{\iota}}](\langle\boldsymbol{\tau}\rangle)$. Using (4), we obtain $[f^{\iota}_{\alpha^{\iota}}](\boldsymbol{\Sigma}) = \alpha^{\iota}([{}_1f^{\iota}]({}_1\boldsymbol{\tau}), \ldots, [{}_Mf^{\iota}]({}_M\boldsymbol{\tau}))$. Because ${}_k\boldsymbol{\sigma}$ is a Nash equilibrium of $\langle {}_k\boldsymbol{\Gamma};\ T;\ \delta\rangle$, we have $[{}_kf^{\iota}]({}_k\boldsymbol{\tau}) \leqslant [{}_kf^{\iota}]({}_k\boldsymbol{\sigma})$. Because α^{ι} is increasing in each variable we obtain $[f^{i}_{\alpha^i}](\boldsymbol{\Sigma}) \leqslant \alpha^{i}([{}_1f^{i}]({}_1\boldsymbol{\sigma}), \ldots, [{}_Mf^{i}]({}_M\boldsymbol{\sigma})) = [f^{i}_{\alpha^i}](\langle\boldsymbol{\sigma}\rangle)$. This proves the if part.

To prove the only if part suppose that $\langle\boldsymbol{\sigma}\rangle$ is a Nash equilibrium of $\otimes\boldsymbol{\Gamma}$. We have to prove that ${}_k\boldsymbol{\sigma}$ is a Nash equilibrium of $\langle {}_k\boldsymbol{\Gamma}; T; \delta\rangle$ for all k. Fix k, i and $\xi = (\xi_t)_{0 \leqslant t < T} \in [{}_kX^i]$. Because $\langle\boldsymbol{\sigma}\rangle$ is a Nash equilibrium, we have for all $\eta \in [\Pi^{M}_{k=1}\, {}_kX^i]$ that $[f^{i}_{\alpha^i}](\langle\boldsymbol{\sigma}\rangle) \geqslant [f^{i}_{\alpha^i}](\eta, \langle\sigma\rangle^{\hat{i}})$. Thus in particular with ${}_l\rho^i = {}_l\sigma^i$ $(l \neq k)$, ${}_k\rho^i = \xi$, we have for $\langle\rho\rangle^i \in [\Pi^{M}_{k=1}\, {}_kX^i]$ that $[f^{i}_{\alpha^i}](\langle\boldsymbol{\sigma}\rangle) \geqslant [f^{i}_{\alpha^i}](\langle\rho\rangle^i, \langle\sigma\rangle^{\hat{i}})$. Denoting $(\langle\rho\rangle^i, \langle\sigma\rangle^{\hat{i}})$ by $\langle\boldsymbol{\gamma}\rangle$, we have $\alpha^{i}([{}_1f^{i}]({}_1\boldsymbol{\sigma}), \ldots, [{}_Mf^{i}]({}_M\boldsymbol{\sigma})) \geqslant \alpha^{i}([{}_1f^{i}]({}_1\boldsymbol{\gamma}), \ldots, [{}_Mf^{i}]({}_M\boldsymbol{\gamma}))$. Because ${}_l\boldsymbol{\gamma} = {}_l\boldsymbol{\sigma}$ $(l \neq k)$, we obtain, using that α^i is strictly increasing in each variable, $[{}_kf^{i}]({}_k\boldsymbol{\sigma}) \geqslant [{}_kf^{i}]({}_k\boldsymbol{\gamma})$. That is $[{}_kf^{i}]({}_k\boldsymbol{\sigma}) \geqslant [{}_kf^{i}](\xi, {}_k\sigma^{\hat{i}})$. □

The following Fact A.1 (see for example Friedman (1991)) describes the 'power of (infinite) repetition' in a quantitative way:

FACT A.1 Let $\boldsymbol{\sigma} = \langle\langle \boldsymbol{p}, \boldsymbol{n}\rangle\rangle$ be a Friedman-trigger multi-strategy of a repeated game $\langle\boldsymbol{\Gamma}; \infty; \delta\rangle$. Then

1. $\boldsymbol{\sigma}$ is a Nash equilibrium of $\langle\boldsymbol{\Gamma}; \infty; \delta\rangle$ if and only if for all $1 \leqslant j \leqslant N$
$$\delta(\phi^{j}(p^{\hat{j}}) - f^{j}(\boldsymbol{n})) \geqslant \phi^{j}(p^{\hat{j}}) - f^{j}(\boldsymbol{p}). \tag{7}$$
2. There exists a discount factor δ_0 that satisfies (7) (with $\delta = \delta_0$). If δ_0 is such, then for all discount factors η with $\eta \geqslant \delta_0$, $\boldsymbol{\sigma}$ is a Nash equilibrium of $\langle\boldsymbol{\Gamma}; \infty; \eta\rangle$. ◇

The intuitive reason for the existence of such Nash equilibria is that players may have an incentive to cooperate, since they can be punished in the future if they try to improve their current period reward.[21]

The following result shows that trigger strategies can be disconnected only in some degenerate cases.

LEMMA A.3 Consider a tensor game with $T \geqslant 2$ and $M \geqslant 2$. Let $\langle\langle \boldsymbol{P}, N^i\rangle\rangle \in [\Pi^{M}_{k=1}\, {}_kX^i]$ be a trigger strategy for player i. Under mild conditions (for example if there exists a k with ${}_kP^i \neq {}_kN^i$ and a $k' \neq k$ such that the strategy set ${}_{k'}X^i$ contains at least two elements) this strategy is not disconnected. ◇

Proof: Here is the proof under the restrictions mentioned above. Denote $\langle\langle \boldsymbol{P}, N^i\rangle\rangle$ by Σ^i. We shall give 1-period histories $\boldsymbol{h}_0$ and $\boldsymbol{g}_0$ with ${}_kh^{j}_0 = {}_kg^{j}_0$ $(1 \leqslant j \leqslant N)$ but ${}_k(\Sigma^{i}_1(\boldsymbol{h}_0)) \neq {}_k(\Sigma^{i}_1(\boldsymbol{g}_0))$.

Well, take $\boldsymbol{h}_0 = \boldsymbol{P}$, fix $\xi \in {}_{k'}X^i$ with $\xi \neq {}_{k'}P^i$ and defining $\boldsymbol{g}_0$ by ${}_{k'}g^{i}_0 = \xi$ and ${}_lg^{j}_0 = {}_lh^{j}_0$ in all other cases. Then, indeed ${}_kh^{j}_0 = {}_kg^{j}_0$ $(1 \leqslant j \leqslant N)$ and because $\boldsymbol{g}_0 \neq \boldsymbol{P}$ it follows ${}_k(\Sigma^{i}_1(\boldsymbol{h}_0)) = {}_k(\Sigma^{i}_1(\boldsymbol{P})) = {}_kP^i \neq {}_kN^i = {}_k(\Sigma^{i}_1(\boldsymbol{g}_0))$. □

Proof: [THEOREM 4.3]

1. This is a consequence of Lemma A.3
2. From Fact A.1 we have that $\langle\langle \boldsymbol{P}, \boldsymbol{N}\rangle\rangle$ is a Nash equilibrium if and only if $\delta(\Phi^j(P^j) - \alpha^j({}_1f^j({}_1N), \ldots, {}_Mf^j({}_M\boldsymbol{N}))) \geqslant \Phi^j(P^j) - \alpha^j({}_1f^j({}_1\boldsymbol{P}), \ldots, {}_Mf^j({}_M\boldsymbol{P}))$ $(1 \leqslant j \leqslant N)$. Because of Lemma A.1 we have $\Phi^j(P^j) = \alpha^j({}_1\phi^j({}_1P^j), \ldots, {}_M\phi^j({}_MP^j))$. Finally the linearity of the weights implies the desired inequality.
3. This is a consequence of the strict inequalities ${}_kf^j({}_k\boldsymbol{P}) > {}_kf^j({}_k\boldsymbol{N})$ and 2. □

Proof: [PROPOSITION 4.2]

1. $\boldsymbol{\Sigma}$ is a Friedman-trigger multi-strategy if and only if $[\boldsymbol{n}]$ is a Nash equilibrium of $\oplus \boldsymbol{\Gamma}$ and $f^j_{\alpha^j}(\sqsubset \boldsymbol{p} \sqsupset) > f^j_{\alpha^j}(\sqsubset \boldsymbol{n} \sqsupset)$ $(1 \leqslant j \leqslant N)$, thus, by Theorem 4.1 and (2), if and only if ${}_k\boldsymbol{n}$ is a Nash equilibrium and $\Sigma_{k=1}^M {}_k\alpha^j\ {}_kf^j({}_k\boldsymbol{p}) > \Sigma_{k=1}^M {}_k\alpha^j\ {}_kf^j({}_k\boldsymbol{n})$ $(1 \leqslant k \leqslant M)$.
2. Suppose each $\langle\langle {}_k\boldsymbol{p}, {}_k\boldsymbol{n}\rangle\rangle$ is a Friedman-trigger multi-strategy. Then $\langle\langle \sqsubset \boldsymbol{p} \sqsupset, \sqsubset \boldsymbol{n} \sqsupset\rangle\rangle$ is a Friedman-trigger strategy of the tensor game. Indeed: Because of Theorem 4.1, $[\boldsymbol{n}]$ is a Nash equilibrium of $\oplus \boldsymbol{\Gamma}$ and for each j we have $f^j_{\alpha^j}(\sqsubset \boldsymbol{p} \sqsupset) = \Sigma_{k=1}^M {}_k\alpha^j\ {}_kf^j({}_k\boldsymbol{p}) > \Sigma_{k=1}^M {}_k\alpha^j\ {}_kf^j({}_k\boldsymbol{n}) = f^j_{\alpha^j}(\sqsubset \boldsymbol{n} \sqsupset)$. □

Notes

* Folmer and v. Mouche: Landbouwuniversiteit Wageningen, Postbus 8130, 6700 EW Wageningen, The Netherlands; Ragland: University of Colorado at Boulder, Boulder, Colorado 80309-0256, U.S.A. An earlier version of this paper was presented at the EAERE conference in Stockholm, June 1991. The authors appreciate comments made by conference participants and journal referees.

[1] The approach adopted in this paper with tensor games is clearly dynamic, but without structural time dependence.

[2] Bohm (1990, pp. 12—13) suggests that some countries might have chosen to be signatories to the Montreal Protocol on CFC's even though abatement costs exceed benefits: 'a reason for joining may be that countries simply want to be part of a cooperative movement in order to benefit from its side-effects (e.g., avoiding risks of losing partners for other forms of international cooperation). The side-effects of joining the Montreal Protocol could conceivably be greater than the net abatement costs, i.e., the excess of abatement costs over environmental benefits.' Mäler (1990, p. 87) notes two examples wherein 'one country may want to make concessions in order to improve friendly neighborhood relations and thereby achieve advantages in other areas of mutual interest'. The first is the Columbia River Treaty (1961) between Canada and the United States, and the second is a 1973 agreement regarding desalinisation of the Colorado River between Mexico and the United States. Both agreements required the United States to incur large costs while Mexico and Canada enjoyed resulting benefits; an outcome that traditional modeling approaches would not have predicted.

[3] Barrett (1990) argues that an important incentive to cooperate follows from the risk that the

agreement would not be effected if the country decided to be a free rider. The more a country has to lose should the agreement not materialise, the stronger this incentive will be. One might add that if the participants are allowed to react to the free riding behavior, as they would in a repeated game context, the incentive to free riding is also reduced.

[4] Often the literature treats these side payments as direct cash transfers or as indirect cash transfers via a fund which specifies contributions and reimbursements according to some rule (see for example Hoel, 1992a).

[5] For instance the Treaty of Rome, as amended by the Single European Act, which underlies the European Community states that community environmental policy shall be based on the principle that 'the polluter should pay' (article 130r(2)).

[6] In section 5 an example similar to this is considered in more detail.

[7] As an example in the environmental arena, the United States threatened Japan with trade sanctions on imports of fish and other wildlife imports (worth more than $353 million per year) if it continued to import shells of the endangered hawksbill turtle. These shells were made into eyeglass frames, and other items by a cottage industry worth $125 million annually. Two days before the sanctions were to go into effect, Japan announced a plan to phase out imports of the endangered hawksbill turtle. ('Japan Cuts Imports of Rare Sea Turtle', International Herald Tribune, May 18—19, 1991, pp. 1, 14.)

[8] For an explanation of the terminology see section 2. Of course, the class of tensor games contains the class of direct sum games.

[9] For a less technical book see for example Holler and Illing (1991).

[10] The accompanying subscripts, parentheses, etcetera will not be in boldface type.

[11] Strictly speaking we are dealing with linear strict weights.

[12] In this definition σ_0^i should be interpreted as an element of X^i.

[13] Σ_0^i should be interpreted as nothing else than an element of ${}_1X^i \times \cdots \times {}_MX^i$.

[14] Various more complicated types of trigger strategy exists for a repeated game. However, for our purposes Friedman-trigger strategies are sufficient.

[15] Notice the similarity between (5) and the last formula on page 3 in Bernheim and Winston (1990).

[16] See for example Friedman (1991). Also for some well known results to be presented in this section we refer to this book.

[17] Rather than subtracting ε^1 from country 1's payoff, one might add an ε^1 term to country 2's payoff. The effects will be similar, but the interpretation would now reflect Bohm's (1990, pp. 12—13) suggestion that some signatories to the Montreal Protocol to reduce CFCs anticipated positive side-effects to signing the Protocol.

[18] For more aspects of reputation in this context see Freedman (1991) and the references therein.

[19] This method of treating a side payment as a strategy might seem highly artificial. Recall, however, that country 1 cannot commit itself to make a side payment in the event that country 2 reduces the amount of pollution. Therefore, the side payment was explicitly incorporated in the game.

[20] Remember that $\langle\langle(C, C), (D, D)\rangle\rangle$ means: play C in the first period and in each subsequent period if the opponent has not previously played D; otherwise play D forever.

[21] This can be seen as follows: $\langle\langle \boldsymbol{p}, \boldsymbol{n} \rangle\rangle$ means that players agree to play $\boldsymbol{p}$ in each period, which gives each player a higher payoff than at the Nash equilibrium $\boldsymbol{n}$. As soon as some player j violates the agreement (to obtain some short-term advantage), all other players will play n^j. So, if a player j tries to obtain a short-term advantage at the cost of the other players, he will be punished by the other players in the sense that from the next period on he will obtain (only) $f^j(\boldsymbol{n})$. So all players will respect the agreement if for all players the loss due to the retaliation strategy $\boldsymbol{n}$ exceeds the maximal attainable gain from a one-period defection. This condition will be the easier satisfied, the less are discounted future payoffs, i.e., the closer the discount factor δ is to 1.

References

Barrett, S. (1990), 'The Problem of Global Environmental Protection', *Oxford Review of Economic Policy* **6**(1), 68–79.

Bernheim, B. and M. Whinston (1990), 'Multimarket Contact and Collusive Behavior', *Rand Journal of Economics* **21**(1), 1–26.

Bohm, P. (1990), 'Efficiency Issues and the Montreal Protocol on CFC's', *The World Bank, Environment Working Paper,* Washington D.C.

Folmer, H. and C. Howe (1991), 'Environmental Problems and Policy in the Single European Market', *Environmental Resource Economics* **1**(1), 17–42.

Folmer, H. (1993), *The Need for Cooperation on International Environmental Problems,* Mimeo, University of Wageningen, Wageningen.

Friedman, J. (1991, 1986), *Game Theory with Applications to Economics,* Oxford University Press, Oxford.

Hansen, S. (1989), 'Debt for Nature Swaps: Overview and Discussion of Key Issues', *Ecological Economics* **1**, 77–93.

Hoel, M. (1992a), 'Emission Taxes in a Dynamic Game of CO_2 Emissions', in R. Pethig, ed., *Conflict and Cooperation in Managing Environmental Resources.* Springer, Berlin.

Hoel, M. (1992b), 'International Environment Conventions: The Case of Uniform Reductions of Emissions', *Environment and Resource Economics* **2**(2), 141–159.

Holler, M. and G. Illing (1991). *Einführung in die Spieltheorie.* Springer-Verlag, Berlin.

Kaitala, V., M. Pohjola and O. Tahvonen (1992), 'Transboundary Air Pollution and Soil Acidification: A Dynamic Analysis of an Acid Rain Game between Finland and the USSR', *Environmental and Resource Economics* **2**(2), 161–168.

Mäler, K. (1989), 'The Acid Rain Game', in H. Folmer and E. van Ierland, eds., *Valuation Methods and Policy Making in Environmental Economics.* Elsevier, Amsterdam, 231–252.

Mäler, K. (1990), 'International Environmental Problems', *Oxford Review of Economic Policy* **6**(1), 80–108.

Mäler, K. (1991), *The Acid Rain Game II.* Mimeo. Stockholm School of Economics, Stockholm.

Markusen, J. (1975), 'Cooperative Control of International Pollution and Common Property Resources', *The Quarterly Journal of Economics* **89**, 618–632.

van der Ploeg, F. and A. de Zeeuw (1992), 'International Aspects of Pollution Control', *Environmental and Resource Economics* **2**(2), 117–139.

Shaw, R. (1988), Transboundary Acidification in Europe and the Benefits of International Cooperation. Paper presented at the conference "Pollution Knows No Frontiers", Bulgaria 16–20 October 1988.

Tsebelis, G. (1990), *Nested Games.* University of California Press, Berkeley.

Zhao, J. (1991), 'The Equilibria of a Multiple Objective Games', *International Journal of Game Theory* **20**, 171–182.

[13]

Journal of Public Economics 93 (2009) 429–439

Contents lists available at ScienceDirect

Journal of Public Economics

journal homepage: www.elsevier.com/locate/econbase

Providing public goods in the absence of strong institutions

Anke Gerber[a,*], Philipp C. Wichardt[b]

[a] Department of Economics, University of Hamburg, Germany
[b] Department of Economics, University of Bonn, Germany

ARTICLE INFO

Article history:
Received 15 January 2008
Received in revised form 15 October 2008
Accepted 16 October 2008
Available online 30 October 2008

JEL classification:
C72
D61

Keywords:
Public goods
Cooperation
Institutions
Climate-change treaties

ABSTRACT

This paper proposes a simple two-stage mechanism to establish positive contributions to public goods in the absence of powerful institutions to provide the public good and to sanction free-riders. In this mechanism players commit to the public good by paying a deposit prior to the contribution stage. If there is universal commitment, deposits are immediately refunded whenever a player contributes her specified share to the public good. If there is no universal commitment, all deposits are refunded and the standard game is played. For suitable deposits, prior commitment and full ex post contributions are supported as a subgame-perfect Nash equilibrium for the resulting game. As the mechanism obviates the need for any ex post prosecution of free-riders, it is particularly suited for situations where players do not submit to a common authority as in the case of international agreements.

1. Introduction

One of the most prominent examples for the failure of markets concerns the provision of public goods. In this case, the conflict of interest between the socially desirable and individually optimal contribution to the public good commonly prevents the implementation of Pareto optimal solutions — not only theoretically but also empirically (e.g. Fehr and Gächter, 2000). Moreover, due to the often immense welfare gains at stake (e.g. clean water/air), the question of how to establish high contribution rates in public goods games is a key issue in economic policy.

An "easy" way out of such social dilemmas are sanctioning institutions that can enforce the desired contributions to public goods (e.g. a reliable jurisdiction). In fact, already casual evidence suggests that, once individual deviations from previously agreed contribution rates can be appropriately punished, many public goods can be — and indeed are — established at a level close to the social optimum (e.g. public transport, health care, quiet sleeping hours at night). Evidence from laboratory experiments further supports this observation (e.g. Falkinger et al., 2000; Fehr and Gächter, 2000); more recent studies even show that individuals, when facing the choice between a social environment with and without sanctioning possibilities, choose the environment with sanctioning (Gürek et al., 2006; see also Kosfeld et al., forthcoming). Moreover, upon being placed in the desired environment, most players indeed fully contribute to the public good and punish free-riders (Gürek et al., 2006). And, of course, inverting the role of the institution, an effect similar to that of sanctions can be achieved through the implementation of some rewarding scheme, given that the institution is provided with the required funds; rewards may, for example, take the form of matching grants as in Boadway et al., (1989). In either case, however, it is necessary that the players submit to some common authority which has the right/power to enforce some kind of punishment or rewarding scheme.

* Corresponding author. Department of Economics, University of Hamburg, Von-Melle-Park 5, D-20146 Hamburg, Germany.
E-mail addresses: anke.gerber@wiso.uni-hamburg.de (A. Gerber), philipp.wichardt@uni-bonn.de (P.C. Wichardt).

0047-2727/$ – see front matter
doi:10.1016/j.jpubeco.2008.10.006

Alternatively, the successful provision of a public good is also possible if the good can be provided by some central institution. This of course requires that the institution is in a position not only to produce the public good but also to obtain the necessary resources from the agents. Thus, the institution either needs the power to tax, or else it must be possible to condition the provision of the public good on the contributions of the agents; in the latter case, considered by Bagnoli and Lipman (1989), the public good is provided if the sum of the contributions is sufficiently large.[1]

In all these cases, however, the provision of the public good eventually hinges on the existence of an institution which is strong in at least one of the following senses: (i) the institution has the power to implement transfers between the agents, i.e. it has to be able to impose (monetary) punishments and/or rewards; or (ii) it has the ability/technology to produce the public good itself and to raise the necessary funds, e.g. because it also has the power to tax. Clearly, such conditions are likely to be satisfied in case of local public goods, where governmental institutions exist to back up the enforcement of the desired contributions and/or where the public good is directly provided by the government (education, street-light, national security etc.). Yet, not least due to the effects of globalisation, an increasing number of public goods do not belong to this category.

Prominent examples for public goods which have to be provided in the absence of strong institutions can be found in connection with environmental issues such as global warming. In fact, questions about the implementation of such public goods have attracted considerable scientific attention in recent years (e.g. Carraro, 1999; Bosello et al., 2003; Dutta and Radner, 2004). In the case of global warming, for example, the public good "clean air" has to be provided in a decentralised way so that each country has to take measures in order to reduce greenhouse gas emissions. At the same time effective sanctioning of free-riders is difficult to establish since there is no institution like a world government which can punish sovereign countries. Accordingly, it is rather unsurprising that we observe a major compliance problem with climate-change treaties like the Kyoto Protocol.[2]

In the present paper, we take up this issue and propose a simple mechanism aimed to establish the provision of a public good in an environment without strong institutions; i.e. the agents cannot be forced to contribute to the public good (e.g. by taxes), effective ex post sanctioning is difficult, if not impossible to enforce, and the public good has to be provided in a decentralised way by the agents themselves. For our analysis, we take as given some agreement on individual contributions to the public good; i.e. we are not concerned with the agents' incentives to actually reach such an agreement but only with its eventual implementation. The considered agreement can, for example, be thought of as the result of some unspecified cooperative bargaining process among the agents. It may be characterised by a number of desirable properties like efficiency, individual rationality, equity etc. Like the Kyoto protocol, however, the agreement is not enforceable due to the absence of sufficiently strong institutions. Hence, it will be implemented only if all agents find it in their self-interest to do so.

The idea of our mechanism is to allow players to take an action, prior to the contribution stage, which renders it a dominant strategy to comply with the agreement. More specifically, we consider a 2-stage modification of a general public goods game. In stage 1, players can choose to pay a deposit to a neutral institution. If at the end of stage 1 everyone has paid the required deposit, then in stage 2 the public goods game is played and deposits are refunded to those who contribute to the public good. If some player has not paid the deposit, all deposits are refunded (potentially deducting a small fee to sustain the institution administering deposits) and the standard public goods game is played in stage 2. With these modifications, universal commitment in the form of paying the deposit and full contributions to the public good now can be rationalised as a subgame-perfect Nash equilibrium of the resulting game. Moreover, the corresponding equilibrium even is in stage-wise weakly dominant strategies. Thus, the mechanism does not require that players know each other's preferences but only that the social planner who designs the mechanism has perfect information about the agents' preferences so as to determine the adequate contributions and deposits.[3] Furthermore, as players essentially execute their own punishment (pay the deposit), the neutral institution considered here only has to resist demands to repay forfeited deposits. This, however, appears far easier to enforce than collecting fines from free-riders ex post.

The rest of the paper is structured as follows. In Section 2, we introduce and analyse our mechanism. We start with an introductory example, then present the general analysis and finally discuss extensions to multi-period public goods problems and issues concerning renegotiation-proofness. Section 3 concludes.

2. Model and results

This section is divided into three parts: an introductory example (Section 2.1), the general analysis (Section 2.2), and a discussion of possible extensions (Section 2.3).

[1] Yet another way many public goods problems are "solved" in practice is via social norms which prescribe the desired behaviour and which are followed by the players in order to avoid mental distress such as feelings of guilt (e.g. Charness and Dufwenberg, 2006) or conflicts with ones identity (e.g. Wichardt, 2008; see also Akerlof and Kranton, 2000, 2005). In such cases, however, there is no conflict between individual incentives and aggregate welfare which is the topic of our paper.

[2] The Kyoto Protocol, in fact, nicely demonstrates the problems that arise in the absence of strong institutions. The current enforcement mechanism of the Kyoto Protocol simply "punishes" a party who has exceeded the assigned amount of emissions by imposing an additional emission deduction of 30% in the next period and by suspending the Party's eligibility to make transfers under emission trading until the Party is reinstated (cf. "http://unfccc.int/kyoto_protocol/compliance/items/2875.php"). The form of punishment used here, however, obviously suffers from a similar compliance problem as the original agreement and hence cannot be considered an effective threat.

[3] Clearly, if the agents themselves negotiate the individual contributions to the public good and determine deposits afterwards, they must have perfect information about each others' preferences.

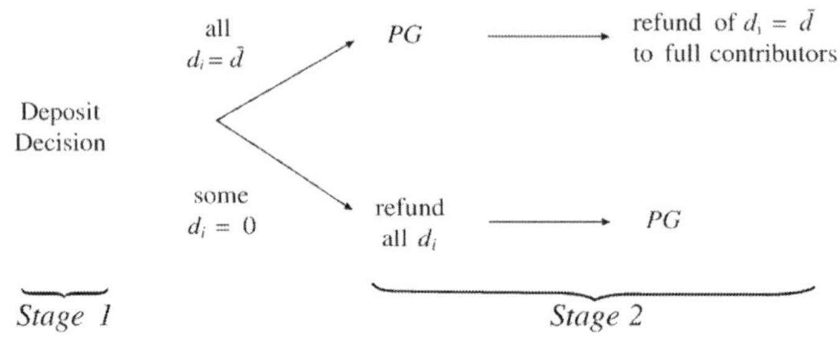

Fig. 1. Graphical illustration of the 2-stage game $\widehat{PG}$. If in stage 1 all players pay the required deposit, PG is played and deposits are refunded to those who contributed $\bar{c}$; otherwise all deposits are refunded first and then PG is played.

2.1. Introductory example

The purpose of the subsequent example is to illustrate the main features of the proposed mechanism for the simple case of a symmetric linear public goods problem. We concentrate on the main aspects of the argument and hence defer any discussion of our assumptions to the next subsection where the general case is analysed.

Consider the following linear, symmetric public goods game denoted by PG. There are two players. Each of them is equipped with an initial endowment e and has to choose a contribution $c_i \in [0,e]$, $i=1,2$, to some public good. The players' preferences over contribution profiles (c_1,c_2) are represented by the utility function $\pi_i(c_1,c_2)=e-c_i+a(c_1+c_2)$, $i=1,2$, with $\frac{1}{2}<a<1$.

Given the specification of PG, it is a strictly dominant strategy for both players to choose $c_i=0$. Accordingly, the unique Nash-equilibrium of PG is given by $c_1^0=c_2^0=0$. However, the outcome corresponding to $c_1^0=c_2^0=0$ is Pareto dominated by the full contribution outcome, i.e. $c_1=c_2=e=:\bar{c}$, because $\pi_i(\bar{c},\bar{c})=e-\bar{c}+2a\bar{c}=2ae>e=\pi_i(0,0)$.[4] The question, thus, is whether and how full contribution can be implemented.

In the sequel, we show that full contributions, $\bar{c}$, by both players can be implemented as a subgame-perfect equilibrium in a 2-stage variation of PG, denoted by $\widehat{PG}$, which is structured as follows: In stage 1, both players simultaneously decide whether to transfer a refundable deposit $\bar{d}>0$ to a neutral institution, e.g. a trustee. The size of the deposit $\bar{d}$ will be determined later. At the end of stage 1, after all players i have chosen $d_i \in \{0, \bar{d}\}$, the profile of deposits (d_1,d_2) is revealed to both players. If both players have deposited $\bar{d}$, then stage 2 directly continues with PG and deposits are refunded to those players who contributed the desired amount, i.e. $\bar{c}$, to the public good. If $d_i=0$ for at least one player i, then, in stage 2, all deposits are refunded immediately and PG is played after that; see Fig. 1 for an illustration.

The stage game of $\widehat{PG}$ that results in case some player i has deposited $d_i=0$ is denoted by PG^0; the stage game that results in case $d_1=d_2=\bar{d}$ is denoted by PG^*. Because PG^0 starts by (automatically) refunding all deposits, the payoff structure of PG^0 is identical to that of PG. Hence, zero contribution is the dominant strategy for both players so that the unique Nash equilibrium of PG^0 again is given by $c_1=c_2=0$. By contrast, the players' payoffs in PG^*, where refunds are conditional on full contribution to the public good, are given by:

$$\pi_i^*(c_1,c_2)=\begin{cases} e-\bar{d}-c_i+a(c_1+c_2), & \text{if } c_i \neq \bar{c} \\ e-c_i+a(c_1+c_2), & \text{if } c_i=\bar{c}. \end{cases}$$

From this it follows immediately that player i's dominant strategy in PG^* is $c_i=\bar{c}$, if deposits are sufficiently large, i.e. if $\bar{d}>(1-a)\bar{c}$.[5] Accordingly, for any profile (d_1,d_2) of deposits paid in stage 1, there is a unique Nash equilibrium in strictly dominant strategies for the resulting subgame at stage 2 and player i's strictly dominant strategy is given by

$$\hat{c}_i^*(d_1,d_2)=\begin{cases} \bar{c}, & \text{if } d_1=d_2=\bar{d} \\ 0, & \text{else.} \end{cases}$$

Intuitively, by paying the deposit $\bar{d}$ in stage 1, players effectively commit to full contributions in the public goods game by implementing a deterrent punishment for themselves. The commitment is conditional, though, as the threat of punishment is only activated in case all players committed to the public good in this way. However, once it is activated, no further steps to execute the punishment are necessary as deposits are already paid and hence the credibility of the (conditional) commitment is not an issue.

Finally, if play continues with the strategy profile $(\hat{c}_1^*,\hat{c}_2^*)$ at stage 2, then $d_i=\bar{d}$ is a weakly dominant strategy for player i, $i=1,2$, at stage 1. To see this, consider the case of player 1. If player 2 deposits $\bar{d}$, then $d_1=\bar{d}$ is strictly better than $d_1=0$. This follows because $\pi_i(\bar{c},\bar{c})>\pi_i(0,0)$ and $(\hat{c}_1^*,\hat{c}_2^*)$ implements $c_1=c_2=\bar{c}$ if $d_1=d_2=\bar{d}$ but only $c_1=c_2=0$ if $d_1=0$. If instead player 2 chooses $d_2=0$,

[4] As maximal contributions need not be Pareto optimal in the general case analysed in Section 2.2, we already use the more general notation, i.e. $\bar{c}$ instead of e, here. Later $\bar{c}$ will refer to the desired contribution.

[5] Observe that $\bar{d}>(1-a)\bar{c}$ for $\bar{d}$ sufficiently close to e.

then the outcome that is implemented by $(\hat{c}_1^*, \hat{c}_2^*)$ in stage 2 is independent of d_1, namely $c_1 = c_2 = 0$, so that player 1 is indifferent between $d_1 = 0$ and $d_1 = \bar{d}$. By symmetry the same argument holds for player 2. Accordingly, $(d_i, \hat{c}_i)_{i=1,2}$, with $d_i = \bar{d}$ and $\hat{c}_i = \hat{c}_i^*$ for $i = 1,2$, is the unique subgame-perfect Nash equilibrium in stage-wise (weakly) dominant strategies.[6] Moreover, in this equilibrium both players contribute the desired amount $\bar{c}$ to the public good.

2.2. General analysis

We now consider a standard public goods problem with n players referred to as consumers. Each consumer i, $i = 1, \ldots, n$, has a preference relation over consumption of a private and a public good. These preferences are represented by a utility function $U_i : \mathbb{R}_+^2 \to \mathbb{R}$, where $U_i(x_i, z)$ is i's utility from consuming x_i units of the private good and z units of the public good. We assume that U_i is strictly increasing:

Assumption A. For all i, it holds that

$$\frac{\partial U_i(x_i, z)}{\partial x_i} > 0 \quad \text{and} \quad \frac{\partial U_i(x_i, z)}{\partial z} > 0.$$

Each consumer i has an initial endowment $e_i > 0$ of the private good while the initial level of the public good is zero. The public good is produced in a decentralised way, i.e. for all i there exists a differentiable production function f_i: $[0, e_i] \to \mathbb{R}_+$ with $f_i(0) = 0, f_i' > 0$, and $f_i'' < 0$. If consumer i contributes $c_i \in [0, e_i]$, $i = 1, \ldots, n$, the realised amount of the public good is

$$z = F\left(\sum_{i=1}^{n} f_i(c_i)\right)$$

for some differentiable function F: $\mathbb{R}_+ \to \mathbb{R}_+$ with $F' > 0$, and consumer i's utility is given by:

$$\pi_i(c_1, \ldots, c_n) := U_i\left(e_i - c_i, F\left(\sum_{j=1}^{n} f_i(c_j)\right)\right)$$

Remark 1. From a formal point of view the differentiation between the global and the individual production functions, i.e. between F and the f_i, is not necessary. Yet, in view of applications, it is convenient to separate between these functions. For example, in the context of global warming, one may want to differentiate between technological progress, that reduces the costs for the abatement of greenhouse gas emissions in some country, and natural changes in the decomposition of greenhouse gases in the atmosphere. In this case, f_i can be thought of as the country's technology for reducing greenhouse gas emissions, while $F(\Sigma_i q_i)$ can be interpreted as the "quality" of the atmosphere if each country i has reduced its greenhouse gas emissions by q_i.[7]

Finally, we assume that positive contributions are always strictly dominated by zero contributions. As we will discuss later (see Section 2.3) this assumption can be relaxed. Nevertheless, at the present stage it is useful as it best captures the essence of the public goods problem at issue.

Assumption B. For all $(c_1, \ldots, c_n) \in \times_{j=1}^{n} [0, e_j]$, it holds that

$$\frac{\frac{\partial U_i(x_i, z)}{\partial x_i}}{\frac{\partial U_i(x_i, z)}{\partial z}} > f_i'(c_i) F'(z),$$

for all x_i and for all $i = 1, \ldots, n$, where $z = F(\Sigma_j f_j(c_j))$.

Hence, we assume that for each consumer i the marginal rate of substitution of the private for the public good is larger than i's marginal rate of transformation in the public good's production. Under Assumption B it follows that $\partial\pi_i(c_1, \ldots, c_n)/\partial c_i < 0$ for all i and all contribution profiles $(c_1, \ldots, c_n)$.

Remark 2. The linear public goods problem presented as an introductory example satisfies Assumptions A and B. In this case, $U_i(x_i, z) = x_i + az$ and $f_i(c_i) = c_i$ for some positive constant a with $a < 1$, and F is given by $F(q) = q$.

The resulting public goods problem is modelled as an n-player game PG, where each player i has a strategy set $S_i = [0, e_i]$ and payoff function $\pi_i : \times_{j=1}^{n} S_j \to \mathbb{R}$. Under Assumption B the following Proposition is immediate:

Proposition 1. *The strategy profile c^0 with $c_i^0 = 0$ for all i is the unique Nash equilibrium for PG. Moreover, $c_i^0 = 0$ is a strictly dominant strategy for all i.*

As, in general, the no contribution equilibrium of Proposition 1 does not lead to an efficient outcome, there commonly is a conflict between individual incentives and aggregate welfare. The mechanism proposed below remedies this deficiency. More

[6] Of course, there also exists another subgame-perfect equilibrium, where both players $i = 1,2$, choose $d_i = 0$ in stage 1 and then play $\hat{c}_i^*$ in stage 2. However, since $d_i = 0$ is weakly dominated by $d_i = \bar{d}$ if play continues with $(\hat{c}_1^*, \hat{c}_2^*)$ at stage 2, this equilibrium seems less plausible than the full contribution equilibrium characterised first.

[7] Observe that the quality of the atmosphere need not be a linear function of the aggregate reduction in greenhouse gas emissions.

specifically, it is designed to implement any contribution profile $\bar{c}=(\bar{c}_1,\ldots,\bar{c}_n)\in\times_{j=1}^{n}(0,e_j]$ that strictly Pareto dominates the no contribution profile, i.e. any $\bar{c}$ that satisfies the following condition:

PDOM. For all $i=1,\ldots,n$, $\pi_i(\bar{c}_1,\ldots,\bar{c}_n)>\pi_i(0,\ldots,0)$.

Beyond Condition PDOM, no further assumptions on $\bar{c}$ are necessary. In particular, it is not required that $\bar{c}$ yields a Pareto efficient allocation.

Remark 3. As mentioned before, the vector of contributions $\bar{c}$ can, for example, be thought of as the result of some cooperative bargaining process among the players which cannot be enforced, e.g. due to a lack of strong institutions. In case of the linear public goods problem discussed in the introductory example Condition PDOM is satisfied for $a>1/2$.

In order to implement $\bar{c}$, consider the following 2-stage modification of PG:

Stage 1. All players simultaneously choose to pay a deposit $d_i\in\{0,\bar{d}_i\}$ with $\bar{d}_i>0$ to a neutral institution, e.g. a trustee; later, $\bar{d}_i$ will be chosen such that contributing $\bar{c}_i$ in the public goods game PG is a dominant strategy for player i. At the end of stage 1, the profile of all deposits paid, $d=(d_1,\ldots,d_n)$, is revealed to the players.

Stage 2. The interaction in this stage depends on d in the following way. If at the end of stage 1 we have $d_i=0$ for some i, then all deposits are refunded and the public goods game PG is played. If, however, at the end of stage 1 we have $d_i=\bar{d}_i$ for all i, then a public goods game PG* is played for which payoffs are as follows:

$$\pi_i^*(c_1,\ldots,c_n)=\begin{cases}U_i(e_i-c_i-d_i,F(\Sigma_j f_j(c_j))), & \text{if } c_i\neq\bar{c}_i\\ U_i(e_i-c_i,F(\Sigma_j f_j(c_j))), & \text{if } c_i=\bar{c}_i\end{cases}$$

Thus, if each player i has decided to pay the deposit $\bar{d}_i$, these deposits are only refunded to those players who contribute to the public good in the desired way, i.e. for whom $c_i=\bar{c}_i$. For the moment we simply assume that the neutral institution keeps any forfeited deposits; a discussion of alternative ways to deal with non-refunded deposits is provided in Section 2.3.1.

The 2-stage game defined above is denoted by $\widehat{PG}$. Player i's payoff in $\widehat{PG}$ is given by:

$$\Pi_i[(d_1,c_1),\ldots,(d_n,c_n)]=\begin{cases}\pi_i^*(c_1,\ldots,c_n), & \text{if } d_j=\bar{d}_j \text{ for all } j\\ \pi_i(c_1,\ldots,c_n), & \text{else}\end{cases}$$

A strategy for player i in $\widehat{PG}$ is given by a tuple $(d_i,\hat{c}_i)$, where $d_i\in\{0,\bar{d}_i\}$ and $\hat{c}_i$: $\{0,\bar{d}_1\}\times\ldots\times\{0,\bar{d}_n\}\rightarrow[0,e_i]$.

Let $\bar{d}_i\leq e_i$ be such that

$$U_i\left(e_i-\bar{c}_i,F\left(\sum_{j\neq i}f_j(c_j)+f_i(\bar{c}_i)\right)\right)>U_i\left(e_i-\bar{d}_i,F\left(\sum_{j\neq i}f_j(c_j)\right)\right) \tag{1}$$

for all $c_j\in[0,e_j]\ j\neq i$.[8] Since U_i,f_i and F are strictly increasing and $\bar{c}_i>0$, it follows that inequality (1) is satisfied for $\bar{d}_i=\bar{c}_i$. Hence, by continuity it is also satisfied for any $\bar{d}_i$ in a sufficiently small neighbourhood of $\bar{c}_i$. The following result, then, is readily established:

Proposition 2. *Let Assumptions A and B be satisfied. Then, for any profile $\bar{c}$ of contributions to the public goods game PG that satisfy Condition PDOM, the strategy profile $(\bar{d}_i,\hat{c}_i^*)_{i=1,\ldots,n}$ with*

$$\hat{c}_i^*(d)=\begin{cases}\bar{c}_i, & \text{if } d_j=\bar{d}_j \text{ for all } j\\ 0, & \text{if } d_j=0 \text{ for some } j\end{cases}$$

is a subgame-perfect Nash equilibrium of $\widehat{PG}$ if $\bar{d}_i$ satisfies inequality (1) for all i. Moreover, for all i, $\hat{c}_i^*$ is a strictly dominant strategy at stage 2 and $\bar{d}_i$ is a weakly dominant strategy at stage 1, if play continues with the strategy profile $\hat{c}^*$ at stage 2.[9]

Proof. Consider stage 2 first. If $d_j=0$ for some j, then PG is played in stage 2. In this case, $\hat{c}_i^*(d)=0$ for all i is the unique Nash equilibrium and it is in strictly dominant strategies (cf. Proposition 1). If $d_j=\bar{d}_j$ for all j, then $c_i=\bar{c}_i$ is a strictly dominant strategy for player i in PG*. To see this, let c_{-i} be an arbitrary profile of contributions for all players except i. By Eq. (1) we have

$$U_i\left(e_i-\bar{c}_i,F\left(\sum_{j\neq i}f_j(c_j)+f_i(\bar{c}_i)\right)\right)>U_i\left(e_i-\bar{d}_i,F\left(\sum_{j\neq i}f_j(c_j)\right)\right)$$
$$\Leftrightarrow\pi_i^*(\bar{c}_i,c_{-i})>\pi_i^*(0,c_{-i})$$
$$\Leftrightarrow\pi_i^*(\bar{c}_i,c_{-i})>\pi_i^*(c_i,c_{-i}) \quad \text{for all } c_i\neq\bar{c}_i$$

where the last equivalence follows from Assumption B. Hence, $\hat{c}_i^*(d)$ as given in the statement of the proposition is the unique Nash equilibrium in stage 2 and it is in strictly dominant strategies. What remains to be shown is that $\bar{d}_i$ is a weakly dominant strategy for player i at stage 1 given that play continues with $\hat{c}^*$. Consider first the case where $d_j=0$ for some $j\neq i$. Then,

[8] Choosing $\bar{d}_i$ so that inequality (1) is satisfied, of course, requires knowledge of the players' utility functions.

[9] $\bar{d}_i$ is only weakly dominant since player i is indifferent between $d_i=\bar{d}_i$ and $d_i=0$ whenever $dj=0$ for some $j\neq i$, in which case all deposits are refunded.

independent of player i's choice of d_i all deposits are refunded and the game continues with the play of PG at stage 2. Hence,

$$\begin{aligned}\Pi_i[(\bar{d}_i,\hat{c}_i^*(\bar{d}_i,d_{-i})),(d_{-i},\hat{c}_{-i}^*(\bar{d}_i,d_{-i}))] &= \pi_i(\hat{c}^*(\bar{d}_i,d_{-i}))\\ &= U_i(e_i,0)\\ &= \pi_i(\hat{c}^*(0,d_{-i}))\\ &= \Pi_i[(0,\hat{c}_i^*(0,d_{-i})),(d_{-i},c_{-i}^*(0,d_{-i}))].\end{aligned}$$

Next, consider the case where $d_j=\bar{d}_j$, for all $j\neq i$. Then, by Condition PDOM

$$\begin{aligned}\Pi_i[(\bar{d}_i,\hat{c}_i^*(\bar{d}_i,\bar{d}_{-i})),(\bar{d}_{-i},\hat{c}_{-i}^*(\bar{d}_i,\bar{d}_{-i}))] &= \pi_i^*(\hat{c}^*(\bar{d}_i,\bar{d}_{-i}))\\ &= \pi_i^*(\bar{c}_1,\ldots,\bar{c}_n)\\ &= \pi_i(\bar{c}_1,\ldots\bar{c}_n)\\ &> \pi_i(0,\ldots.0)\\ &= \pi_i(\hat{c}^*(0.\bar{d}_{-i}))\\ &= \Pi_i[(0,\hat{c}_i^*(0,\bar{d}_{-i})),(\bar{d}_{-i},\hat{c}_{-i}^*(0.\bar{d}_{-i}))]\end{aligned}$$

As we have seen in the proof of Proposition 2, for any deposit profile d there is a unique equilibrium $\hat{c}^*(d)$ at stage 2 of $\widehat{PG}$. This is not true for stage 1 since $\bar{d}_i$ is only weakly dominant for player i. Any tuple $(d,\hat{c}^*)$, with $d_i=0$ for at least two i, also constitutes a subgame-perfect Nash equilibrium of $\widehat{PG}$. In such an equilibrium, there is no contribution to the public good at stage 2. However, $d_i=0$ is weakly dominated by $d_i=\bar{d}_i$ if play continues with $\hat{c}^*$. In fact, $(\bar{d},\hat{c}^*)$ is the unique subgame-perfect Nash equilibrium in stage-wise weakly dominant strategies.

As a last step, we show that the above result remains to hold if the institution collecting and administering the deposits is assumed to be costly, and if players, when paying their deposits, have to bear a small share of this cost. To see this, consider following modification of $\widehat{PG}$, denoted by $\widetilde{PG}$. Different from the previous case, assume now that a fraction $\varepsilon>0$ of any deposit made is kept for the maintenance of the respective institution. Thus, if at the end of stage 1 we have $d_i=0$ for some i, then in stage 2 the game PG is played except that now player i's payoff function is given by:

$$\tilde{\pi}_i(c_1,\ldots,c_n)=\begin{cases}U_i(e_i-\varepsilon\bar{d}_i-c_i,F(\Sigma_jf_j(c_j))), & \text{if } d_i=\bar{d}_i\\ U_i(e_i-c_i,F(\Sigma_jf_j(c_j))), & \text{if } d_i=0\end{cases}$$

If at the end of stage 1 we have $d_i=\bar{d}_i$ for all i, then again PG* is played with the following modification of the players' payoff functions:

$$\tilde{\pi}_i^*(c_1,\ldots,c_n)=\begin{cases}U_i(e_i-\bar{d}_i-c_i,F(\Sigma_jf_j(c_j))), & \text{if } c_i\neq\bar{c}_i\\ U_i(e_i-\varepsilon\bar{d}_i-c_i,F(\Sigma_jf_j(c_j))), & \text{if } c_i=\bar{c}_i\end{cases}$$

Hence, player i's payoff in $\widetilde{PG}$ is

$$\tilde{\Pi}_i[(d_1,c_1),\ldots,(d_n,c_n)]=\begin{cases}\tilde{\pi}_i^*(c_1,\ldots,c_n), & \text{if } d_j=\bar{d}_j \text{ for all } j\\ \tilde{\pi}_i(c_1,\ldots c_n), & \text{else.}\end{cases}$$

Proposition 3 below states that also under these modifications everyone paying the deposit in stage 1 and full contributions to the public good in stage 2 still can be implemented as a subgame-perfect Nash equilibrium of $\widetilde{PG}$.

Proposition 3. *Let* Assumptions A and B *be satisfied and let* $\bar{c}$ *be any contribution profile that satisfies Condition PDOM. Furthermore, assume that* $\bar{d}_i$ *satisfies Eq.* (1) *for all i. Then, there exists* $\bar{\varepsilon}>0$ *such that for* $0<\varepsilon<\bar{\varepsilon}$, $\widetilde{PG}$ *has exactly two subgame-perfect Nash equilibria* $(d^*,\hat{c}^*)$. *In one equilibrium,* $d_i^*=\bar{d}_i$ *for all i in stage 1; in the other equilibrium* $d_i^*=0$ *for all i in stage 1. In both equilibria* $\hat{c}^*$ *is as given in* Proposition 2. *Both equilibria are strict.*

The proof immediately follows from the fact that by continuity Condition PDOM implies that

$$U_i\left(e_i-\varepsilon\bar{d}_i-\bar{c}_i,F\left(\sum_j f_j(\bar{c}_j)\right)\right) > U_i(e_i,0)$$

and that inequality (1) implies that

$$U_i\left(e_i - \varepsilon \bar{d}_i - \bar{c}_i, F\left(\sum_{j \neq i} f_j(c_j) + f_i(\bar{c}_i)\right)\right) > U_i\left(e_i - \bar{d}_i, F\left(\sum_{j \neq i} f_j(c_j)\right)\right)$$

for all i if ε is sufficiently small.

2.3. Discussion and extensions

In the following, we provide a discussion of several extensions of the preceding model. To begin with, we propose some alternatives to the use of forfeited deposits off the equilibrium path. We then show how the proposed mechanism can be applied to public goods games that do not have a unique equilibrium in dominant strategies. Afterwards we demonstrate how the mechanism applies in cases where the provision of the public good necessitates contributions over more than one period so that intertemporal aspects have to be taken into account. Finally, we address issues concerning incentives to renegotiate and argue that our mechanism indeed is renegotiation-proof in the sense of Farrell and Maskin (1989).

2.3.1. Forfeited deposits

In our game, whenever all players have paid the required deposits in stage 1, these deposits are only refunded to those players who contribute the desired amount to the public good in stage 2. As we have seen, in equilibrium all players contribute the desired amount and hence all deposits are refunded. However, off the equilibrium path some deposits are not refunded and the question is what happens to these forfeited deposits. So far, we have assumed that these deposits are kept by the neutral institution without specifying their further use.

In our view there are at least three alternatives, which all support the equilibrium outcome we have derived before. The first is an inefficient "burning" of the forfeited deposits. The second is to add forfeited deposits to the contributors' refunds, e.g. according to the relative size of their contribution $\bar{c}_i$ to the public good compared to sum of all full contributions $\bar{c}_j$ that were actually made and to simply refund all deposits in case of insufficient contributions by all players ($c_i \neq \bar{c}_i$ for all i).[10] The third is to keep the forfeited deposits within the respective fund and to save them for (efficient) use on issues that also are of common interest to all players but that lie outside the public goods problem considered here, e.g. in the context of international agreements, to finance first aid in case of catastrophes such as droughts. The advantage of the second alternative, of course, is that it technically remains within the boundaries of the strategic problem at hand. A social planner, who has to solve multiple allocation problems, may find the third option slightly more appealing, though.

2.3.2. Public goods game without dominant strategies

So far, we have assumed that for each player i zero contribution, i.e. $c_i = 0$, is a strictly dominant strategy in PG (Assumption B). However, as mentioned before, this assumption was made for expositional purposes only. In fact, the desired contributions can still be implemented as a subgame-perfect Nash equilibrium of our extended game if we abandon Assumption B and, instead, apply a minor adjustment to the corresponding condition PDOM on the Pareto dominating strategy profile. To this end let ξ_i denote a mixed strategy of player i. We then require that the targeted contribution profile $\bar{c} = (\bar{c}_1, \ldots, \bar{c}_n)$, Pareto dominates all Nash equilibria $(\xi_1^0, \ldots, \xi_n^0)$ of *PG* including mixed equilibria.[11] Hence, we replace condition PDOM by the following condition PDOM′, where, slightly abusing notation, π_i now refers to expected payoffs:

PDOM′: For all $i = 1, \ldots, n$ and any Nash equilibrium $\xi^0 = (\xi_1^0, \ldots, \xi_n^0)$ of *PG*, it holds that $\pi_i(\bar{\xi}_1, \ldots, \bar{\xi}_n) > \pi_i(\xi_1^0, \ldots, \xi_n^0)$, where $\bar{\xi}_i$ denotes the mixed strategy of player i that places probability one on $\bar{c}_i, i = 1, \ldots, n$; i.e. the contribution profile $\bar{c}$ Pareto dominates the Nash equilibrium ξ^0.

Proposition 2 then becomes:

Proposition 2′. *Let* Assumption A *be satisfied. Then, for any contribution profile* $(\bar{c}_1, \ldots, \bar{c}_n)$ *that satisfies PDOM′ and any Nash equilibrium* $(\xi_1^0, \ldots, \xi_n^0)$ *of PG, the strategy profile* $(\bar{d}_i, \hat{\xi}_i^*)_{i=1,\ldots,n}$ *with*

$$\hat{\xi}_i^*(d) = \begin{cases} \xi_i^0, & \text{if } d_j = 0 \text{ for some } j \\ \bar{\xi}_i, & \text{if } d_j = \bar{d}_j \text{ for all } j \end{cases}$$

is a subgame-perfect Nash equilibrium of $\widehat{PG}$ *if* $\bar{d}_i$ *satisfies inequality* (1) *for all i. Moreover, for all i,* $\bar{d}_i$ *is a weakly dominant strategy at stage 1, if play continues with the strategy profile* $\hat{\xi}^*$ *at stage 2.*

The proof is analogous to Proposition 2 except that in case $d_j = 0$ for some j playing $\hat{\xi}_i^*(d) = \xi_i^0$ is not a dominant strategy for player i at stage 2. Instead, in this case, $\hat{\xi}^*(d)$ with $\hat{\xi}_i^*(d) = \xi_i^0$ for all i constitutes "only" a Nash equilibrium of the subgame in stage 2.

2.3.3. Intertemporal aspects and renegotiation-proofness

The subsequent discussion is divided into two parts. To begin with we outline how our mechanism can be applied in case the underlying public goods game extends over more than one period. In the second part, we consider the possibility of renegotiations and argue that the players' strategies implemented by the proposed mechanism are renegotiation-proof in the sense of Farrell and Maskin (1989).

[10] The strategy profile specified in Proposition 2 remains a subgame-perfect equilibrium under the described refunding scheme because the best any player i can do in stage 2 still is to play $\bar{c}_i$, given $d_j = \bar{d}_j$ for all j in stage 1 and given $c_j = \bar{c}_j$ for all $j \neq i$; otherwise player i would lose $\bar{d}_i$ which under this scheme is redistributed to players $j \neq i$.

[11] Observe that PG has at least one Nash equilibrium in mixed strategies since the strategy sets are compact and the payoff functions are continuous.

A repeated public goods problem. For the preceding argument, we have assumed that the public good to be provided requires contributions only in one period. Yet, many public goods, in particular non-durable ones, necessitate repeated contributions over a longer horizon. A valid question to be asked, therefore, is whether the mechanism proposed in Section 2.2 extends to the case of repeated contributions to a public good. In the sequel, we address this issue and demonstrate by means of a simplified example how our mechanism can indeed be adapted to public goods that require repeated contributions by all players.[12]

Intuitively, the main idea again is to first collect deposits and to then refund in each period a share of these deposits to those who contributed to the public good in that period. Collecting deposits once at the beginning is preferable to a repeated collection at the start of each period as it saves transaction costs[13]; also, it implies a credible long run commitment in case everyone has paid the deposit.[14] If the deposits collected at the beginning are sufficiently high to cover the necessary refunds in each period, initial commitment, i.e. full deposit payments, and provision of the desired per period contributions to the public good again can be implemented as a subgame-perfect Nash equilibrium.

Consider an n-player version of the linear public goods problem PG discussed in Section 2.1, where for all $i=1,\dots, n$, $U_i(x_i,z)=x_i+az$, with $\frac{1}{n}<a<1$, $f_i(c_i)=c_i$ and $F(q)=q$. Suppose that there are $T\geq 1$ periods and that players discount future payoffs with a per period discount factor $\delta=\frac{1}{1+r}$, where r is the one-period riskless interest rate. In all periods $t=1,\dots, T$, each player i receives an endowment $e>0$ of the private good that she can either consume, invest into a riskless asset or spend on the production of the public good. For simplicity, we assume that the public good is non-durable, i.e. there is no stock of the public good that can be transferred across periods; the generalisation to the case of durable public goods, i.e. cases with "stock-effects," is straightforward.

As players' utility functions are linear in the private good and because $\delta=\frac{1}{1+r}$, players are indifferent between immediate consumption and investment into the riskless asset. To further simplify the exposition, we therefore assume that all players immediately consume any fraction of period-t-endowments that is not spent on the public good. Hence, we obtain a repeated game, PG^T, where in each period t player i chooses a contribution $c_i^t\in[0,e]$ to the public good and player i's payoff in period t is

$$\pi_i^t(c_1^t,\dots,c_n^t)=e-c_i^t+a\sum_{j=1}^{n}c_j^t.$$

Given any sequence of contribution profiles $(c^1,\dots, c^T)$, player i's overall payoff then is given by

$$\sum_{t=1}^{T}\delta^{t-1}\pi_i^t(c_1^t,\dots,c_n^t).$$

It is an immediate consequence of the above specification of PG^T, that the T-period public goods game has a unique subgame-perfect equilibrium in dominant strategies, namely zero contributions in all periods by all players, i.e. $c_i^t=0$ for all t and all i. Because $an>1$ by assumption, this equilibrium is Pareto dominated by any sequence of contribution profiles $(c^1,\dots, c^T)$ with $c_i^t=\bar{c}>0$ for all t and all i. The purpose of the subsequent discussion is to show how this contribution profile can be supported as a subgame-perfect equilibrium in a T-period version of our mechanism, denoted by $\widehat{PG}^T$.

As indicated above, the basic idea of the T-period variation of our mechanism again is to collect deposits whose conditional refund renders full desired contribution to the public good, $\bar{c}$, a dominant strategy for all players. Hence, similar to the one-period case, the mechanism starts with an additional deposit stage preceding the contribution stage of period 1. As before, if all players have paid the required deposit, part of the deposit is refunded each period and the remainder is invested into the riskless asset; the details are explicated below. Compared to the previous 1-period case, however, two modifications are necessary. First of all, deposits now have to provide incentives for T periods instead of 1. And secondly, in case deposit payments are incomplete and all deposits are refunded immediately, we assume that the next period again starts with a new deposit stage. This is done in order to allow for an implementation of the desired outcome with a one period delay.

Following this procedure and again assuming $\bar{d}>\bar{c}(1-a)$, the deposits required at the beginning of a period t are given by

$$\bar{D}^t=\bar{d}\sum_{\tau=t}^{T}\left(\frac{1}{1+r}\right)^{\tau-t}. \tag{2}$$

[12] It has come to our attention that Gersbach and Winkler (2007) in a recent working paper propose a related approach to an intertemporal public goods problem albeit in the specific context of global warming. Moreover, although their analysis uses a refunding mechanism which is similar to ours, it does not address some issues which we believe are essential to this problem. Most importantly, their paper limits itself to a 2-period game with homogeneous players and proposes a mechanism that relies on taxes and their redistribution through a refunding scheme. Yet, it does not discuss the consequences of this redistribution for the players' initial incentives to participate in the refunding scheme, although this aspect is crucial in the more realistic case of heterogeneous players; in our model the participation decision corresponds to the decision to pay the required deposit in stage 1 of the game.

[13] International treaties, e.g. on greenhouse gas reductions, commonly have to be approved by national parliaments. And treaties that require repeated payments would usually imply repeated voting, repeated control of the deposits made etc., all of which is time-consuming and costly.

[14] Recall that there is also an inefficient subgame-perfect Nash equilibrium of the two-stage deposit-contribution game where at least two players do not pay the required deposit and there is no contribution to the public good at stage 2. Therefore, if we simply repeat the two-stage game, there is always the risk that players play the inefficient equilibrium at some point in the future even if they have played the efficient full-contribution equilibrium in the past. If, however, deposits are collected once at the beginning, then a successful completion of the deposit stage implies that full contributions to the public good are a strictly dominant strategy for all players in all future periods.

Given this, consider an arbitrary period t. We distinguish two cases: (a) there is a deposit stage in period t and (b) there is none. Case (a) occurs in period 1 and in any period $t>1$, unless the deposit stage was successfully completed in some period $\tau<t$; below we will give a definition of the successful completion of the deposit stage. To begin with, consider the case where there is a deposit stage in period t. In this case, each player first chooses a deposit $D_i^t \in \{0,\bar{D}^t\}$.[15] Once all deposit choices are made, the resulting profile of deposits is revealed to all players and the game enters the contribution stage. In the contribution stage, players have to decide on their contribution to the public good. If $D_i^t=0$ for some i, then all deposits are refunded immediately and afterwards the game PG is played, where each player i chooses $c_i^t \in [0,e]$. Hence, player i's period t payoff is given by

$$\pi_i^t(c_1^t,\dots,c_n^t) = e - c_i^t + a\sum_{j=1}^{n} c_j^t.$$

If instead, $D_i^t=\bar{D}^t$ for all i, we say that *the deposit stage was successfully completed in period t*. In this case, the game PG is played and afterwards, each player i who contributes $c_i^t=\bar{c}$ to the public good in period t obtains a partial refund $\bar{d}$ out of her deposit $\bar{D}^t$. Accordingly, player i's period t payoff is given by

$$\pi_i^{*t}(c_1^t,\dots,c_n^t) = \begin{cases} e-\bar{D}^t - c_i^t + a\sum_{j=1}^{n} c_j^t, & \text{if } c_i^t \neq \bar{c} \\ e-\bar{D}^t + \bar{d} - c_i^t + a\sum_{j=1}^{n} c_j^t, & \text{if } c_i^t = \bar{c} \end{cases}$$

Moreover, whatever remains of player i's deposit $\bar{D}^t$ (which is $\bar{D}^t$, if $c_i^t \neq \bar{c}$, and $\bar{D}^t-\bar{d}$, if $c_i^t=\bar{c}$) is transferred to an account that bears the riskless interest rate. Thus, by construction, the value of player i's account in period $t+1$ is at least $\bar{D}^{t+1}$. Accordingly, once the deposit stage has been successfully completed in some period t, no further deposit stages in later periods are necessary.

Finally, if there is no deposit stage in period $t>1$ (recall that there is always is a deposit stage in period 1), then the game immediately enters the contribution stage for that period and payoffs are

$$\pi_i^{*t}(c_1^t,\dots,c_n^t) = \begin{cases} e - c_i^t + a\sum_{j=1}^{n} c_j^t, & \text{if } c_i^t \neq \bar{c} \\ e + \bar{d} - c_i^t + a\sum_{j=1}^{n} c_j^t, & \text{if } c_i^t = \bar{c} \end{cases}$$

Proposition 4 below generalises our previous results for the one period case to the present T-period setting. It shows that all players paying the required deposit in period 1 and full contributions to the public good in all later periods again is the outcome of a subgame-perfect equilibrium of $\widehat{PG}^T$.

Proposition 4. *Let PG be the linear public goods game specified above and let $\bar{d}>(1-a)\bar{c}$. Then the following strategies support the contribution $\bar{c}$ by all players in all periods as a subgame-perfect Nash equilibrium of $\widehat{PG}^T$. For any period t, $1\le t\le T$:*

(i) If there is a deposit stage in period t, each player i pays the full deposit, i.e. $D_i^{*t}=\bar{D}^t$, for all $i=1,\dots,n$; then,
  1. if $D_i^{*t}\neq\bar{D}^t$ for some i, each player i, $i=1,\dots,n$, chooses $c_i^{*t}=0$ in the contribution stage of period t.
  2. if $D_i^{*t}=\bar{D}^t$ for all i, each player i, $i=1,\dots,n$, chooses $c_i^{*t}=\bar{c}$ in the contribution stage of period t.

(ii) If there is no deposit stage in period t, i.e. the deposit stage of some period $\tau<t$ was successfully completed, then each player i, $i=1,\dots,n$, chooses $c_i^{*t}=\bar{c}$ in the contribution stage of period t.

For all i and all t, c_i^{*t} is a strictly dominant strategy in the contribution stage of period t. Moreover, for all i, if there is a deposit stage in period t, then D_i^{*t} is a weakly dominant strategy for player i, if play continues with the strategy profile $(c_1^{*\tau},\dots,c_n^{*\tau})$ in the contribution stages of all subsequent periods $\tau\ge t$.

The proof of Proposition 4 is a straightforward generalisation of the proof of Proposition 2, so we only give a sketch of the argument.

Sketch of the Proof. As before, we first note that once all players have paid the required deposits in some period t^*, i.e. there was a deposit stage in t^* and $D_i^{t^*}=\bar{D}^{t^*}$ for all i, then full contribution to the public good in all subsequent periods, i.e. $c_i^t=\bar{c}$ for all $t\ge t^*$, is a strictly dominant strategy for all players; again this essentially follows from the fact that $\bar{d}>(1-a)\bar{c}$. Similarly, $c_i^t=0$ again is the optimal choice for all contribution stages that have not been preceded by a successfully completed deposit stage. Thus, it remains to be shown that paying the full deposit whenever there is a deposit stage in some period t^* is a weakly dominant strategy, if play continues with $(c_1^{*t},\dots,c_n^{*t})$ in the contribution stages of all subsequent periods $t\ge t^*$. To see this, consider period T first. If there is a deposit stage in period T, then $\bar{D}^T=\bar{d}$ is a weakly dominant strategy for any player, if play continues with $(c_1^{*T},\dots,c_n^{*T})$; this follows directly from Proposition 2. The argument, then, can be iterated back in order to cover all periods $t<T$, i.e. if there is a

[15] Regarding the payment of the deposit, we assume that players can borrow $\bar{D}^t$ at a per period interest rate of r.

deposit stage in period $t<T$, then $\bar{D}^t$ is a weakly dominant strategy for any player if subsequent play follows the equilibrium strategies as specified in the proposition. This is due to the fact that paying the deposit has the same consequences as not paying it if some other player does not pay the required deposit; in this case, all deposits are refunded immediately. If, however, all other players pay the required deposits, then paying the deposit is optimal as in that case deposits can be used to induce the desired contributions to the public good already in period t and, by construction,

$$\sum_{\tau=t}^{T} \delta^{\tau-t}\left[e-\bar{c}+\bar{d}+an\bar{c}\right]-\bar{D}^{t} > e+\sum_{\tau=t+1}^{T} \delta^{\tau-t}\left[e-\bar{c}+\bar{d}+an\bar{c}\right]-\delta\bar{D}^{t+1}.$$

□

Remark 4. Regarding the use of forfeited deposits, the alternatives discussed in Section 2.3.1 are also feasible here. As before, though, the exact details of the solution are of minor importance and can be adapted to the problem at hand.

Renegotiation-proofness. Subgame-perfection ensures that the equilibrium strategies are individually credible, i.e. no player has an incentive to change her strategy in any subgame on or off the equilibrium path, it may nevertheless be that the punishment is not optimal for the group as a whole, since there exists another continuation equilibrium that strictly Pareto-dominates the "punishment equilibrium." For example, if equilibrium strategies prescribe the punishment of deviators in a way that hurts all players, no punishment and simply proceeding with the earlier equilibrium path strictly Pareto-dominates the execution of the punishment.[16] Accordingly, in this case there is an incentive for the group of all players to renegotiate so that the initial threat to punish is incredible. A subgame-perfect equilibrium that is robust to such renegotiations is called *renegotiation-proof* (cf. Farrell and Maskin, 1989).

Naturally, renegotiation-proofness plays an important role in the context of international environmental agreements and their enforcement (Barrett, 1994; see also Finus and Rundshagen, 1998; Froyn and Hovi, 2008). In the context of our mechanism, however, it is not an issue as all players effectively take care of their own punishment, namely by paying the required deposit to a central agency; in case of later insufficient contributions by some player, the punishment then is carried out by the agency simply by keeping the respective deposit. Thus, there is no costly punishment on the part of the non-deviators, which usually gives rise to potential Pareto-improvements through renegotiations. Moreover, in any subgame the equilibrium strategies, where full deposits are paid at the earliest opportunity and full contributions are paid in all contribution stages from then on, strictly Pareto-dominate any other subgame-perfect equilibrium. Hence, there is no scope for improvements, neither on nor off the equilibrium path.[17]

3. Concluding remarks

In many situations, like in the case of global warming, there is a general consensus among the affected parties that the provision of a particular public good is desirable. Yet, at the same time everyone knows that, ex post, there is a strong incentive to free ride on the contributions of others if there is no institution which can enforce the desired contributions. Hence, low contribution rates are what commonly results.

The mechanism proposed in this paper dissolves the tension between the socially desirable and the individually rational through the introduction of an initial commitment stage. In this stage, players can voluntarily commit to making the desired contribution to the public good by paying an adequate deposit. If all players commit to the public good, deposits are only refunded to those who later make the full desired contribution; otherwise all deposits are refunded immediately. Thus, in effect, commitment to the public good implements a self-sanctioning scheme which, if activated, renders full contributions to the public good a strictly dominant strategy.

The main advantage of the initial commitment stage is that, similar to the public goods decision, general commitment again is socially desirable. Yet, different from the contribution decision in the public goods game, commitment now is also individually rational. This is because the respective self-sanctioning scheme is activated only if all players choose to commit and commitment has no consequences otherwise; contributions to the public good itself, by contrast, are irreversible once made. Accordingly, free-riding is not an issue in the commitment stage.

Moreover, the proposed mechanism does not require the presence or establishment of powerful institutions to implement the desired contributions to the public good. The reason is that no ex post punishment of free-riders is required because of the conditional ex ante self-sanctioning involved in the procedure. All that is needed is an independent institution that collects deposits, monitors the players' contributions and refunds deposits to those who have made the desired contributions to the public good. Hence, the mechanism is particularly suited to implement agreements in situations where there is no common authority for all players that can enforce the punishment of free-riders, e.g. in the case of international agreements such as the Kyoto protocol.

[16] For a specific example consider an infinitely repeated Prisoner's Dilemma game and assume that both players play grim-trigger (i.e. cooperate as long as the opponent does, defect until infinity in case of a deviation of the opponent). In this case, given a single deviation, players are better off by coordinating again on cooperation rather than executing the punishment.

[17] Of course, an inefficient use of forfeited deposits may induce a desire to renegotiate the mechanism itself. This, however, does not concern the players' strategy choices and, hence, is not at issue in the context of renegotiation-proofness in the sense defined above. Nevertheless, requiring an efficient use of forfeited deposits certainly not only renders the mechanism more appealing from a theoretical perspective but also is more plausible in view of applications (cf. Section 2.3.1).

Finally, regarding practical applications, we believe that these are both possible and promising. In fact, whenever international agreements are negotiated, a possible way to complement other incentives to comply with the agreement, e.g. due to reputation effects, would be to add some monetary deposit, which then is used in the way described above; the deposit could, e.g., be payable to the World Bank. Moreover, given the complementary character of the deposits, we would speculate that the size of the deposits necessary to alleviate the compliance problem is considerably smaller than what the preceding discussion — which neglects external aspects such as reputation — might suggest. Yet, to ultimately judge in how far the implementation of such a mechanism is possible, given the diversity of national legal systems and the different national interest groups, is beyond the scope of this paper. Nevertheless, our results suggest that from an economic perspective it might be worthwhile to explore what is politically feasible in this respect.

Acknowledgements

We are grateful to Sylvain Béal, Patrick Schmitz, Avner Shaked and three anonymous referees for helpful comments. Anke Gerber acknowledges support from the center of excellence "Integrated Climate System Analysis and Prediction" (CliSAP) at the University of Hamburg. Philipp Wichardt gratefully acknowledges financial support from the German Research Foundation (DFG) through GRK 629 and SFB/TR 15 at the University of Bonn. This paper was originally circulated under the title "Providing Public Goods without Strong Sanctioning Institutions" (IEW-Working Paper No. 303, University of Zurich, 2006).

References

Akerlof, G., Kranton, R., 2000. Economics and identity. Quarterly Journal of Economics 115, 715–753.
Akerlof, G., Kranton, R., 2005. Identity and the economics of organizations. Journal of Economic Perspectives 19, 9–32.
Bagnoli, M., Lipman, B.L., 1989. Provision of public goods: fully implementing the core through private contributions. The Review of Economic Studies 56, 583–601.
Barrett, S., 1994. The biodiversity supergame. Environmental and Resource Economics 4, 111–122.
Boadway, R., Pestieau, P., Wildasin, D., 1989. Tax-transfer policies and the voluntary provision of public goods. Journal of Public Economics 39, 157–176.
Bosello, F., Buchner, B., Carraro, C., 2003. Equity, development, and climate change control. Journal of the European Economic Association 1, 601–611.
Carraro, C., 1999. International Environmental Agreements on Climate Change. Kluwer Academic Publishers, Dordrecht. The Netherlands.
Charness, G., Dufwenberg, M., 2006. Promises and partnership. Econometrica 74, 1579–1601.
Dutta, P., Radner, R., 2004. Self-enforcing climate change treaties. Proceedings of the National Academy of Sciences 101, 5174–5179.
Falkinger, J., Fehr, E., Gächter, S., Winter-Ebmer, R., 2000. A simple mechanism for the efficient provision of public goods: experimental evidence. The American Economic Review 90, 247–264.
Farrell, J., Maskin, E., 1989. Renegotiation in repeated games. Games and Economic Behavior 1, 327–360.
Fehr, E., Gächter, S., 2000. Cooperation and punishment in public goods experiments. The American Economic Review 90, 980–994.
Finus, M., Rundshagen, B., 1998. Renegotiation-proof equilibria in a global emission game when players are impatient. Environmental and Resource Economics 12, 275–306.
Froyn, C., Hovi, J., 2008. A climate agreement with full participation. Economics Letters 99, 317–319.
Gersbach, H., Winkler, R., 2007. On the Design of Global Refunding and Climate Change. CER-ETH - Center of Economic Research at ETH Zurich, Woking Paper No. 07/69.
Gürek, O., Irlenbusch, B., Rockenbach, B., 2006. The competitive advantage of sanctioning institutions. Science 312, 108–111.
Kosfeld, M., Okada, A., Riedl, A. forthcoming. Institution Formation in Public Goods Games. The American Economic Review.
Wichardt, P., 2008. Why and How Identity Should Influence Utility. SSRN eLibrary.

[14]

Environ Resource Econ (2013) 54:63–77
DOI 10.1007/s10640-012-9581-3

Enforcing Compliance with Environmental Agreements in the Absence of Strong Institutions: An Experimental Analysis

Todd L. Cherry · David M. McEvoy

Accepted: 24 July 2012 / Published online: 10 August 2012

Abstract This paper uses laboratory experiments to evaluate the performance of a deposit-refund mechanism used to enforce compliance with voluntary public-good commitments made in the absence of strong regulatory institutions. With this mechanism agents decide whether to join an agreement and pay a deposit prior to making their contribution decisions. If an agreement receives sufficient membership to form, members then make their contribution decisions and compliant members are refunded their deposits. If an agreement does not form, then deposits are immediately refunded and a standard voluntary contribution game is played. We find that the deposit-refund mechanism achieves nearly full efficiency when agreements require full participation, but is far less effective, and in some cases disruptive, when agreements require only partial participation. As the mechanism does not require the existence of strong sanctioning institutions, it is particularly suited for enforcing compliance with international environmental agreements.

Keywords Enforcing compliance · Environmental agreements · Experimental economics · Deposit refund · International environmental agreements

JEL Classification D02 · Q50 · H41 · H87 · C90

T. L. Cherry
Economics Department, Appalachian State University, 3095 Raley Hall, Boone, NC 28608, USA
e-mail: cherrytl@appstate.edu

T. L. Cherry
CICERO Center for International Climate and Environmental Research,
Oslo, P.O. Box 1129, Blindern, 0318 Oslo, Norway

D. M. McEvoy (✉)
Economics Department, Appalachian State University, 3108 Raley Hall, Boone, NC 28608, USA
e-mail: mcevoydm@appstate.edu

1 Introduction

The management of many environmental and natural resources gives rise to *social dilemma* situations in which individually rational decisions are not collectively rational. Public goods (e.g., protection of ozone layer, preservation of biological diversity) tend to be underprovided and common-pool-resources (e.g., fisheries, forest resources) tend to be overexploited. Many social dilemma situations may be easily resolved given the existence of strong regulatory institutions. For example, a well-functioning government could provide efficient levels of public goods if it can adequately collect taxes and use the revenue to produce public goods (e.g., public transportation). Likewise, common-pool resources may be managed efficiently given an effective regulatory institution and credible enforcement provisions (e.g., federal regulation of commercial fisheries). In many cases, however, these types of social dilemmas exist in the absence of strong regulatory institutions. The problem of managing greenhouse-gas emissions, or transboundary environmental problems in general (e.g., managing CFC emissions, trade in endangered species, biodiversity preservation), falls into this category. Sovereignty requires that transboundary environmental public goods are provided in a decentralized fashion without an overarching global government that can heavy-handedly regulate the management of shared resources.[1]

In the absence of traditional regulatory approaches, transboundary resources are typically managed through voluntary institutions called international environmental agreements (IEAs). While there are many obstacles in enacting effective IEAs, there are two potentially formidable challenges. First, participation with IEAs is voluntary, meaning nations are free to decide whether to join an agreement. This presents a challenge because commitments made under an agreement, by definition, restrict a member's use of a shared environmental or natural resource, and there are often strong incentives to free ride. The second challenge is that compliance with the terms of an agreement needs to be enforced. Provisions for enforcement must be included as part of the agreement and cannot depend on the presence of a strong regulatory body. Furthermore, the enforcement provisions need to be effective in the sense that they will be severe enough to deter noncompliance and credible in the sense that they will actually be imposed. While a great deal of research has addressed the first challenge posed—the one of voluntary participation—less attention has been paid to the fact that compliance with voluntary agreements needs to be enforced. In this paper we address the issue of enforcing compliance with IEAs by examining one promising instrument that exhibits the necessary features for effective enforcement. The instrument explored here—a deposit-refund mechanism—has recently been proposed in the theoretical literature (Gerber and Wichardt 2009) and the reported results are encouraging. Building on their study, we empirically test the theoretical predictions of alternative deposit-refund schemes using experimental methods.

While some research does address the compliance issue directly (e.g., Barrett 1994, 1997, 2008; Finus and Rundshagen 1998; Barrett and Stavins 2003; Hovi et al. 2007; Finus 2008; McEvoy and Stranlund 2009; McEvoy et al. 2011) much of the economics literature avoids the problem of maintaining compliance by designing models, and experiments, in which agreement members are required to comply with their commitments (e.g., Hoel 1992; Carraro and Siniscalco 1993, 1998; Hoel and Schneider 1997; Rubio and Ulph 2006, 2007; Kolstad 2007; McGinty 2007; Carraro et al. 2009; Kosfeld et al. 2009; Breton et al. 2010; Dannenberg

[1] In a similar spirit, many small fisheries are managed in a decentralized way in which a collection of fishermen is given local control of managing their shared resource (e.g., Chilean abalone fishing communities, Maine lobster fishery).

et al. 2010; McEvoy 2010).[2,3] In other words, these studies do not give agreement members the opportunity to violate the treaty. This is not a trivial issue. Ignoring the possibility of noncompliance abstracts from a fundamental problem for the effectiveness of cooperative agreements; that is, the opportunity to violate commitments could have significant impacts on the management of shared resources. In fact, many attribute the impending failure of the Kyoto Protocol to the lack of credible enforcement of compliance. Barrett (2003, p. 360) argues, "…if the negotiators had reflected on the need for enforcement and on the difficulty of devising an effective enforcement mechanism earlier in the process, they may have negotiated a different kind of treaty—one that sustained more cooperation."[4]

Motivated by the need to explore enforcement strategies within environmental agreements in the absence of strong institutions, we adapt the general model of Gerber and Wichardt (2009) and utilize a set of laboratory experiments to test the effectiveness of different deposit-refund schemes. While the theoretical approach is useful for predicting how the deposit-refund system can facilitate cooperation, an experimental approach allows for its empirical test in the absence of naturally occurring data.[5] While deposit-refund mechanisms are used in a number of simple settings (e.g., return of recyclable solid waste, environmental performance bonds, etc.), we analyze a deposit-refund game within an agreement formation context.[6,7] The mechanism is straightforward. A group of players individually decide whether or not to join an agreement, in which members to the agreement commit to contributing to the public good.[8] Upon joining the agreement, each member is required to pay a deposit to a neutral, third-party institution. The agreement enters into force if enough players join

2 Barrett (1994) and Finus and Rundshagen (1998) examine enforcement in IEAs using renegotiation-proof punishment strategies with which agreement members can punish noncompliance by reducing their provision in the future. We take a different approach to enforcing IEAs by examining a static game (no future punishment possibilities) in the absence of a strong institution that can credibly impose financial penalties on noncompliant members.

3 Kosfeld et al. (2009) theoretically examine a coalition formation game given the threat of sanctions imposed in response to noncompliance by an exogenous institution. Their experimental tests, however, preclude the possibility of noncompliance by making commitments binding. McEvoy and Stranlund (2009) theoretically examine an enforcement strategy within a static game in which IEA members finance a third-party that has the power to impose financial sanctions on noncompliant members. The mechanism is tested experimentally in McEvoy et al. (2011) with a coalition formation game that includes costly and incomplete monitoring for noncompliance. They find compliance rates of roughly 70 percent when coalitions form. Their experiments differ from ours in that enforcement costs directly affect the minimum participation requirements in their design. Also, if a coalition does not form in their experiment then no player contributes to the public good and nonmembers, by default, never contribute to the public good.

4 Although compliance with the Kyoto Protocol cannot be formally evaluated until the end of the first commitment period (December 2012), many countries are currently on pace to significantly exceed their quotas (Tourney and Fujiwara 2010).

5 The use of laboratory experiments in the evaluation of public policies is well established (Plott 1987; Shogren and Hurley 1999; Cason and Plott 1996; Stranlund et al. 2011), and experiments are particularly well suited to evaluate policies that have not been implemented in the field, as in our case. This research follows a number of recent experimental studies used to analyze the effectiveness of international environmental institutions (Kosfeld et al. 2009; McEvoy et al. 2011; Tavoni et al. 2011).

6 See Fullerton and Wu (1998) for an analysis of deposit-refund mechanisms applied to environmental policies.

7 A similar deposit-refund mechanism for enforcing compliance with international climate agreements was briefly mentioned in Finus (2008).

8 In our simple game, the deposit-refund mechanism is designed to motivate compliance with public-good commitments (e.g., emissions abatement). However, deposit-refunds may also be tied to indirect measures of cooperative behavior. For example, deposits could be refunded if member countries achieve a target Renewable Portfolio Standard or target investment level in research and development.

(and pay deposits) to satisfy a minimum membership requirement.[9] If the membership requirement is not satisfied, then no deposits are paid and no agreement enters into force. If the membership requirement is satisfied, then the agreement enters into force and members then decide whether or not to comply with their commitments (i.e., whether to contribute to the public good). After these decisions are made, the third-party institution pays back the deposit if the member is found compliant. Otherwise, the noncompliant agreement member is not refunded their deposit. The deposit-refund scheme does not require strong institutions in the sense that the neutral third party cannot force members to contribute (e.g. through taxation) and cannot engage in ex post sanctioning (e.g., issuing fines in response to noncompliance). Rather, the neutral institution simply has to withhold the initial deposit.[10]

In our setting the deposit-refund mechanism can be likened to performance bonds that are levied in order to ensure fulfillment of environmental obligations. An example is the Surface Mining Control and Reclamation Act (1977) which requires firms to purchase bonds before attaining the permits to begin a mining project. The bond amounts are refunded if the mined land is adequately reclaimed. Performance bonds are also used in the management of timber, oil and gas industries. While deposit-refunds/performance bonds, to our knowledge, have not been included as compliance provisions in environmental agreements, the fact that similar mechanisms are used in other environmental policy dimensions suggests that such mechanisms may be effective.[11]

Our experimental design consists of three treatments. The first is a simple voluntary contributions game which is used as our baseline. The baseline captures the unilateral management of shared environmental and natural resources within social dilemma situations. The second treatment is an agreement formation game with a deposit-refund mechanism that requires full participation. In this treatment all players need to join the agreement before it enters into force. If full participation is not achieved, deposits are returned and players revert to the same voluntary contributions game as in the baseline treatment. The third and final treatment is an agreement formation game with a deposit-refund mechanism that only requires partial participation before entry into force. This treatment is motivated by many international agreements that require less than full participation, most notably the Kyoto Protocol.[12]

[9] Minimum membership requirements are standard in treaties that address international environmental issues. For example, the Kyoto Protocol to the United Nations Framework Convention on Climate Change required ratification by at least 55 parties prior to its entry into force in February 2005. Similarly, the Montreal Protocol on Substances that Deplete the Ozone Layer required at least 11 countries to ratify it before it entered into force in 1989. While many treaties require only a subset of affected parties to ratify for entry into force, some require all parties to join. For example, the Convention for the Protection of the Marine Environment of the North-East Atlantic required accession of all negotiating parties.

[10] It is common for IEAs to empower third-party institutions as part of the agreement process. Examples include the Compliance Committee established under the Marrakesh Accords as part of the Kyoto Protocol and the Multilateral Fund established under the Montreal Protocol.

[11] A different type of deposit-refund system is included as part of the Stability and Growth Pact (SGP) of the European Economic and Monetary Union as a mechanism to motivate compliance with honoring debt limits. Under the SGP, a member country incurring a deficit greater than three percent of its GDP is required to pay a deposit (ranging between 0.2 and 0.5 % of GDP) to the European Union Council. The deposit is to be refunded if the non-compliant country reduces its debt below the three percent threshold within two years. Otherwise, the country could lose its deposit. This deposit-refund scheme, however, is very different from the one we propose. Deposits made under the SGP are to be paid *after* a country is found noncompliant, rather than before. The deposits are therefore effectively fines imposed in response to noncompliance that would be repaid in the event that the violator returns to being in compliance. The mechanism has never actually been applied in practice.

[12] While some international agreements require ratification by all parties before entering into force, most (especially global treaties) require a minimum less than the total number of negotiating countries (see Barrett 2003, pp. 158–194).

We find that when full participation is required under a deposit-refund mechanism, the agreement formation process is almost 100 percent efficient. Agreements enter into force in nearly every trial and compliance is full. This is a dramatic increase in efficiency compared to the baseline case (i.e., the unilateral management case). Specifically, we observe a 33 percentage point increase in efficiency under voluntary agreements with a deposit-refund scheme that require full participation compared to unilateral management. Interestingly, agreements with deposit refunds that require only partial participation do not improve upon the unilateral management situation. Agreements formed infrequently in this scenario, but when agreements did form, average contributions were higher compared to those under unilateral management. Surprisingly, when agreements failed to form, contributions were dramatically lower than under the baseline treatment (26 vs. 60 % efficiency). This suggests that behavior in the first stage affects cooperative behavior in the second stage. Players contributed far less to the public good when an agreement failed compared to when there was no option for an agreement to form.

Our research is also related to the extensive literature on threshold public good experiments with money-back guarantees (Van de Kragt et al. 1983; Dawes et al. 1986; Isaac et al. 1989; Rapoport and Eshed-Levy 1989; Erev and Rapoport 1990; Bagnoli and McKee 1991; Suleiman and Rapoport 1992; Rapoport and Suleiman 1993; Marks and Croson 1998, 1999; Cadsby and Maynes 1999; McEvoy 2010; Rauchdobler et al. 2010). In these experiments, subjects make individual contribution decisions, and if the threshold level of contributions (or contributors) is satisfied, then the public good is provided. Otherwise, contributions are returned. The threshold, like the minimum participation requirement in our experiments, acts as a commitment device. Our experiment differs in an important way from these experiments by giving subjects the opportunity to violate their commitments in a second stage.

2 Experimental Design

Our *baseline* treatment is a familiar discrete choice public goods game. Players are in groups of five. Each subject is endowed with \$13 and she must decide whether or not to contribute \$8 from her endowment to a public account.[13] Every \$1 contributed to the public account returns \$0.50 to each person in the group (i.e., the marginal per-capita return is 0.5). Players make their contribution decisions simultaneously. Therefore, a player's payoff function is

$$\pi_i = 13 + 4(q_i + q_{-i}) - 8q_i, \tag{1}$$

where q_i equals one if player i contributes \$8 to the public account and zero if player i contributes zero, and q_{-i} is the sum of the binary contribution decisions by the other four players. In a non-cooperative Nash equilibrium, no contributions are made to the public account. This is confirmed by subtracting a player's payoff when contributing zero to the public good from their payoff when they contribute \$8, $\pi_i(q_i = 0) - \pi_i(q_i = 1) = \$ - 4$. However, achieving the social optimum requires that all five players in the group contribute \$8 to the public account. In this case all players earn \$25.

In contrast to the theoretical prediction of no contributions, years of experimental research informs us that some subjects *are* expected to contribute to public goods in social dilemma situations such as this. A common result in a repeated game setting is that contributions start

[13] In the experiment we abstract away from the environmental and natural resources context and have subjects making decisions whether to contribute to a generic public account.

in the range of 40–60 percent of the maximum and decrease as the experiment progresses (Ledyard 1995; Chaudhuri 2011).

In our second treatment, *full participation*, we give subjects the opportunity to form an agreement with the goal of increasing contributions to the public good.[14] The agreement works as follows: In the first stage, subjects decide simultaneously whether to join an agreement. An agreement is said to *form* if and only if all five players join. When an agreement does form, its members enter a non-binding commitment to contribute \$8 to the public account. In addition, joining an agreement requires that players pay an upfront deposit of d as part of the membership process. If an agreement forms after the first-stage decisions are made, then in the second stage its members decide individually whether to comply with their commitment and contribute \$8 to the public account. If a member does contribute \$8 to the public account in the second stage (i.e., she complies) then she is refunded her deposit of d. Otherwise, if she does not contribute \$8 she is not refunded her deposit. Finally, if at least one player fails to join the agreement in the first stage then no agreement forms. In this case, all deposits are instantly refunded to those players that joined the agreement and all players revert back to the baseline game in stage two.

Therefore, if an agreement forms in the first stage of the game, individuals face the following payoff functions during stage two

$$\begin{aligned}\pi^c &= 9 + 4q_{-i}\\ \pi^{nc} &= 13 + 4q_{-i} - d,\end{aligned} \tag{2}$$

where the superscripts c and nc indicate compliance and noncompliance by members to an agreement, respectively. Recall, that if at least one player does not join, then no agreement forms and all players face the same payoff function in [1] during stage two.

With this agreement mechanism, members earn higher payoffs complying provided that $d > 4$. Indeed, as long as $d > 4$ the efficient solution can be implemented as part of a subgame-perfect Nash equilibrium; meaning, all five players join the agreement in the first stage and then contribute \$8 to the public account in the second stage. We set the deposit amount at \$5 for our experiments.[15] The solution is straightforward and is found using backward induction. While in the second stage, there are two possible scenarios that players may face. Consider first the scenario in which no agreement forms in the first stage. In this case the game-theoretic prediction, like the baseline game, is that no player contributes to the public account and each player earns \$13. The other possible scenario in stage two is that an agreement has formed at the end of stage one. In this case each member decides between complying with the agreement and contributing their \$8 or not complying and keeping their \$8. A comparison of the payoff functions in equation [2] illustrates that in this subgame all agreement members are expected to comply with their commitments. When compliance is full the value of the public account is \$20, and since each player is refunded their \$5 deposit they earn \$25. Given the expected earnings in stage two, joining the agreement in stage one is a weakly dominant strategy for each player.[16]

Our third and final treatment, *partial participation*, examines the deposit-refund mechanism when full participation is not required. This treatment proceeds like the second

[14] In the experiments we referred to agreements as *clubs*.

[15] Noncompliance is detected with certainty in this game. An obvious extension to this research would be to relax this feature and allow for incomplete monitoring for compliance.

[16] There also exists another subgame-perfect Nash equilibrium in which no agreement forms in stage one and all players choose not to contribute to the public account in stage two. However, since not joining the agreement in stage one is weakly dominated by joining, the efficient equilibrium is more compelling.

Table 1 Expected earnings for agreement members and nonmembers

	Agreement size		
	3	4	5
Member	\$17	\$21	\$25
Nonmember	\$25	\$29	–

treatment, but now an agreement will form if at least three players join in stage one. Therefore, when agreements form, they can range from having three, four or five members. This means that in cases in which participation is less than full, members and nonmembers can coexist. Members, as before, commit to contributing \$8 to the public account and must pay an initial deposit of \$5. Nonmembers, in contrast, do not commit to contributing \$8 and also do not pay the \$5 deposit. Rather, nonmembers simply play the baseline game.

We again solve the game using backward induction. When an agreement fails to form, as in this second treatment, players are expected to earn \$13. On the other hand, from the comparison of the payoff functions in equation [2] and a deposit amount of \$5, when agreements do form we expect members to comply with their commitments and contribute \$8 to the public account (the cost of noncompliance (\$5) is greater than the benefit (\$4)). Nonmembers, if they exist, are expected not to contribute the public account and earn \$13 plus the value of the public account. Therefore, nonmembers will earn strictly more than members (\$8 more) and hence there is a financial incentive to not join the agreement and to free ride on the contributions to the public account. Table 1 shows the *expected* earnings for members and nonmembers given the three possible agreement sizes.

A comparison of the payoffs from Table 1 illustrates that while nonmembers always earn more than agreement members, all players are better off when an agreement forms compared to having no agreement at all (recall, players expect to earn \$13 when an agreement does not form). Hence players have two conflicting incentives; on one hand the incentive to free ride encourages nonparticipation, and on the other hand the incentive to ensure an agreement forms encourages participation. This tension allows for two subgame-perfect Nash equilibria in this game. An equilibrium exists in which no agreement forms (less than three players join in the first stage) and no contributions are made to the public account in the second stage. In addition, an agreement of size three can exist in equilibrium in which the members fully comply with their commitments and contribute to the public good.[17] In short, we expect this third treatment to be weakly more efficient than the baseline treatment and less efficient than the full participation treatment.

Participants were recruited from the general student body at Appalachian State University. For each of the three treatments, we conducted a session with 15 subjects. The subjects were placed in groups of five and the groups were reshuffled for each of 10 rounds. For each session, all of the rules and parameters remained constant over the ten rounds. Subjects were informed that they would only be paid for one round of play and that round would be chosen

[17] To see why profitable agreements of sizes four or five cannot be part of a subgame-perfect Nash equilibrium in the partial participation treatment, consider the case in which an agreement of size five forms in stage one. In stage two, all five players comply with the agreement and each \$25. However, if one player were to leave the agreement in stage one, then the four remaining players would comply in stage two each earning \$21 and the single free rider would earn \$33. Clearly, leaving the agreement is payoff enhancing for the defector. The same is true for an agreement of size four, since one player can leave the agreement and earn more money defecting. However, with an agreement of size three, if one player leaves then the agreement collapses and each earn \$13 which is lower than their payoff of \$17 if the player stays in the agreement.

at random after all of the rounds were completed.[18] To mitigate end-game effects, the subjects were not told the exact number of rounds. The experiments were hand run and lasted roughly one hour for the baseline treatment and 1.5 h for the deposit-refund treatments. The experiment yielded 450 individual-level observations, 150 for each treatment.

In the baseline treatment, subjects were provided with a *contribution decision sheet* at the start of each round that contained a table displaying all of their earnings possibilities as well as check boxes that subjects used to indicate their decisions. Subjects had to check whether or not to contribute $8 to the public account. After all subjects in the room were finished making their decisions, the sheets were collected and the contributions were tallied. At that point the results for the round were recorded onto the subjects' decisions sheets and handed back. At the end of each round subjects were informed (via their decision sheet) of how many players in their group contributed $8 to the public account and their individual earnings. Subjects were also provided with a record sheet on which they recorded the results for each round of play.

In the other two treatments, the contribution decision was preceded by an agreement decision. An *agreement decision sheet* was used by subjects to indicate their decision whether or not to join the agreement. This sheet also contained tables that displayed all of their earnings possibilities. After all of the participants made their agreement decision, the sheets were collected, the decisions were tabulated and the results were recorded on the individual decision sheets. The sheets were then returned to the subjects. At that point each subject was informed of how many subjects joined the agreement, whether or not it formed and whether or not a deposit of $5 was deducted. Then subjects moved to the contribution decision, as described in the baseline. Using the contribution decision sheet, subjects indicated whether or not they contribute $8 to the public account. Once subjects were finished making their contribution decisions, the sheets were collected, the results tabulated and recorded. Subjects were then informed of how many group members contributed to the public account, whether their deposit was returned and their earnings for the round. Subjects recorded their round earnings on their record sheet and the next round began.

3 Results

Figure 1 illustrates the percentage of players that contributed to the public good by treatment and round. Numbers from the baseline setting initially reveal the familiar pattern of positive, though less than socially optimal, contributions that decay over time (Ledyard 1995; Chaudhuri 2011), but in later rounds, contributions exhibit a more stable pattern that is often observed in the closely related threshold public good setting (e.g., Croson and Marks 2000). Initially, 73 percent of players contribute, but the number falls to about 50 percent during the session and ends at 60 percent in the final rounds. Table 2 summarizes the contribution levels across all rounds for each treatment. The average contribution level in the baseline treatment was 60 percent.

Result 1 A deposit-refund mechanism that requires full participation leads to nearly efficient public-good provision.

Consistent with theoretical predictions, 97.3 percent of subjects (146/150) chose to join the agreement, which led to agreements forming in 90 percent of trials (27/30). As shown in

[18] Subjects were informed in advance that a bingo ball cage would be used to select the binding round. This method ensured that each round was equally likely to be binding and therefore equally important in determining payoffs.

Table 2 Percentage contributing by treatment

	Pooled		
	Mean	SD	N
Baseline	0.600	0.491	150
Partial	0.553	0.499	150
Full	0.933	0.250	150
Total	0.696	0.461	450

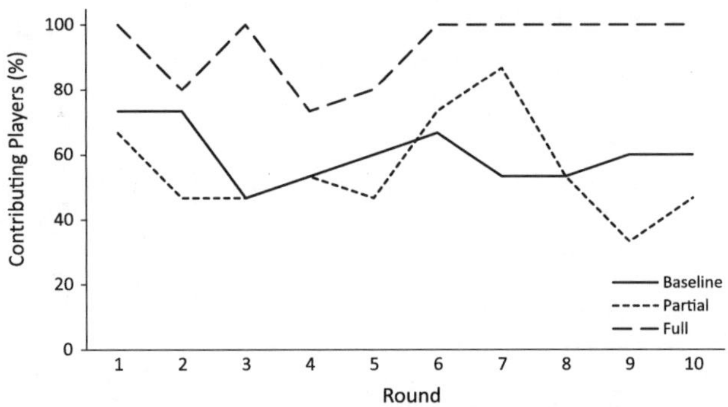

Fig. 1 Percentage contributing by treatment over rounds

Fig. 1, the decision to join is universal (100 %) in seven of the ten rounds, and such cooperation is sustained over the final five rounds. Moreover, when agreements formed subjects were perfectly compliant with their commitment to contribute to the public good. The combination of a high percentage of agreement formation and perfect compliance lead to an overall contribution level of 93.3 percent (Table 2). Therefore we find support for the theoretical prediction from section two, and from Gerber and Wichardt (2009), that a deposit-refund mechanism that requires full participation will lead to efficient public-good provision.

Result 2 A deposit-refund mechanism that requires only partial participation does not improve upon average baseline public-good contribution levels.

When only partial participation was required, 54.7 percent of subjects (82/150) joined the agreement with agreements forming in only 53.3 percent of trials (16/30).[19] Overall, 55.3 percent of subjects (83/150) contributed to the public good (Table 2), which is statistically equivalent to contributions in the baseline setting (55.3 vs. 60.0 %, $p = 0.414$).[20] When agreements did form, they had an average of 3.69 members (out of five). These summary results indicate that the incentive to free ride in the partial participation setting caused low

[19] Although most subjects varied their participation decision over the course of the experiment, two subjects did not change their strategy, one being a 'strong cooperator' (always joined) and the other being a 'strong free rider' (never joined). To examine whether participation decisions were made conditional on the success/failure of a the previous round's agreement, we estimate a probit model regressing the decision to join an agreement on whether an agreement formed in the previous round while including subject fixed effects. We see very little evidence of condition participatory behavior as indicated by an insignificant coefficient for whether the previous agreement formed ($p = 0.178$) and only two significant subject-specific coefficients.

[20] Reported p-values are from Mann-Whitney rank-sum tests.

Table 3 Percentage contributing by treatment and membership decision

	Partial			Full		
	Mean	SD	N	Mean	SD	N
Agreement formed	0.763	0.428	80	1.000	0.000	135
Members	0.983	0.130	59	1.000	0.000	135
Nonmembers	0.143	0.358	21	–	–	0
No agreement	0.314	0.467	70	0.333	0.488	15
Joined	0.261	0.449	23	0.454	0.522	11
Did not join	0.340	0.479	47	0.000	0.000	4
Total	0.553	0.499	150	0.933	0.250	150

levels of agreement formation, and less than full participation when agreements did form. However, the deposit-refund scenario that requires only partial participation was effective at achieving nearly full compliance from its members (98.3 percent). This implies that the deposit-refund mechanism, in both the partial and full participation cases, effectively facilitates contributions from the members when an agreement forms.

We examine the data in more detail by stratifying contribution decisions by whether the agreement formed and whether group members signed up to join the agreement. The numbers, as reported in Table 3, reveal key elements behind the overall performance measures of the deposit-refund mechanisms.

Result 3 When agreements form, a deposit-refund mechanism that requires only partial participation causes an increase in contributions relative to the baseline.

Considering only the partial participation case, we see from Table 3 that contribution levels when an agreement formed were more than double the levels observed when an agreement does not form; 76.3 vs. 31.4 percent. A comparison of the contribution rates under the deposit-refund mechanism when agreements formed and the baseline case shows a significant increase in public-good provision (76.3 vs. 60.0 %, $p = 0.014$). When agreements form, most of the public-good contributions are provided by members; 98.3 percent of members contribute and 14.3 percent of nonmembers contribute. Note further that the percentage of nonmembers contributing in this scenario is significantly lower than the percentage of contributors in the baseline (14.3 vs. 60.0 %, $p = 0.000$). Therefore free-riding is exacerbated among nonmembers when partial agreements form.

Result 4 When agreements fail to form, deposit-refund mechanisms cause a decrease in public-good contributions relative to the baseline.

An interesting finding is the stark difference in behavior when agreements failed to form and the baseline. Recall, if an agreement does not form, the group and its members remain in the baseline public-good setting. Thus, at the time of the contribution decision, the only difference between the deposit-refund treatments and the baseline is a previously failed attempt to form an agreement. The data suggest that this failure matters a great deal. In the baseline treatment, 60.0 percent of members contributed to the public good, which is nearly twice the levels

Table 4 Probit models: determinants of contribution decision

	Model 1	Model 2	Model 3
Constant	0.637**	0.536*	0.762**
	(0.314)	(0.329)	(0.347)
D-R full	1.316***	1.783***	2.510***
	(0.319)	(0.314)	(0.407)
D-R partial	−0.117	0.669*	2.286***
	(0.325)	(0.136)	(0.498)
D-R & agreement failed	–	−1.620***	−3.118***
		(0.293)	(0.436)
D-R partial & agreement formed & did not join	–	–	−3.709***
			(0.642)
Chi-square	92.93***	93.26***	143.85***
N	450	450	450

For each probit probability model, the dependent variable is the subject's binary decision whether to contribute to the public good. Estimates use robust standard errors (in parentheses) and clustering at the individual level. Models control for period-specific fixed-effects. *,** and *** indicate significance at the 10, 5 and 1 % levels

observed in the full and partial participation treatments (33.3 and 31.4 %, respectively).[21] It appears the failure of forming an agreement can undermine cooperation.

We confirm these results with a conditional analysis by estimating a series of probit models of individual contribution decisions. In each model, the dependent variable is the subject's decision to contribute to the public good (1 if contribute; 0 otherwise). Estimates control for period-specific fixed-effects and take into account that individual subjects make repeated decisions by employing robust standard errors and clustering at the individual level.[22,23] The results are reported in Table 4. We begin with a basic model in which the dependent variable is estimated as a function of the two deposit-refund treatments (omitting the baseline). The results are consistent with our initial findings; that is, a deposit-refund scheme that requires full participation increases a subject's likelihood of contributing to the public good relative to the baseline (marginal effect = 0.333), while contributions are not significantly different with a partial participation scheme.

The second model adds an additional independent variable that interacts the presence of a deposit-refund mechanism (full or partial) with the failure of an agreement forming. The estimate for the interaction variable indicates how the failure to form an agreement influences the likelihood of contributing to the public good, while controlling for individual treatment effects. Results reveal a negative and highly significant relationship (marginal effect = −0.566), indicating that subjects are less likely to contribute to the public good after a failed attempt to form an agreement, as compared to the absence of any opportunity to form an agreement. This indicates that while successful agreement formation significantly improves cooperation, an unsuccessful effort might lead to worse outcomes in comparison to not pursuing an agreement. The final model adds an interaction variable to isolate players

[21] A comparison between those that joined and those that did not join a failed agreement (combining data from both treatments with deposit-refund mechanisms) suggests both player types respond to failed coalitions in similar ways - 32.4 percent contributions by joiners vs. 31.4 percent contributions by non-joiners ($p = 0.925$).

[22] Joint-hypotheses tests on period-specific fixed-effects are significant for the first and third models ($p = 0.018$ and $p = 0.005$), and insignificant for the second model ($p = 0.197$).

[23] Results were robust to logit and linear probability specifications.

that do not join a partial agreement to disentangle the effect of a successfully formed partial agreement on members and nonmembers. Estimates show that the new interaction term is significant and negative (marginal effect = −0.878) while the previously insignificant coefficient for the partial participation term is now significant and positive (marginal effect = 0.364). Results show the formation of a partial agreement leads to greater contributions from members while also exacerbating free-riding from nonmembers.

4 Conclusion

Without a global governmental system that can provide direct regulation of resource use, countries are left to manage shared resources voluntarily through environmental agreements. While a decent amount of research has focused on the challenge of getting nations to voluntarily participate in these agreements, very little research has examined the problem of maintaining compliance with the commitments of the members. This paper uses lab experiments to evaluate the performance of one promising mechanism—a deposit-refund mechanism—in which players pay a deposit upon joining an agreement and are later refunded the deposit if they are not in violation of their commitments. Our results provide an empirical test of the deposit-refund mechanism recently proposed by Gerber and Wichardt 2009, but adapted here within an agreement formation game to provide a public good.

Our results show that the deposit-refund mechanism has differential impacts, depending on whether full or partial participation is required. When the participation of all players is required before an agreement enters into force, the mechanism achieves near-perfect efficiency levels. That is, agreements form in almost every trial and members are perfectly compliant when agreements do form (all deposits are refunded). The result confirms the encouraging theoretical prediction of Gerber and Wichardt (2009) that the deposit-refund mechanism will achieve significant efficiency gains relative to a voluntary contribution mechanism. However, our research explores an important variation of the deposit-refund system not analyzed by Gerber and Wichardt. That is, we show that when less than full participation is required for an agreement to enter into force—the situation that characterizes most global-environmental agreements (e.g., Kyoto Protocol)—the deposit-refund mechanism does not increase average efficiency measures. Agreements form only about half of the time. When agreements do form, the deposit-refund mechanism is successful at motivating full compliance of its members and therefore a subset of the players act more cooperatively relative to the baseline case. At the same time, however, a deposit-refund mechanism that requires only partial participation exacerbates free-riding behavior in two important ways. First, when agreements form the nonmembers are less cooperative compared to players in the absence of such a mechanism. Second, when agreements fail to form players act significantly less cooperative than when they had no opportunity to join an agreement. Therefore it appears that a previously failed attempt at forming an agreement can undermine cooperation.

Our experiments, like the model proposed by Gerber and Wichardt, necessarily abstract from many of the practical implications of implementing a deposit-refund system within an international environmental agreement. However, the basic mechanism is simple and its most critical element is that parties empower a neutral third-party institution with the power to withhold financial deposits in cases of noncompliance. The financial institution would serve as an escrow account in which monies will be disbursed given the fulfillment of contractual emissions abatement responsibilities. Since third-party financial institutions have been implemented within other IEAs (e.g., Multilateral Fund under the Montreal Protocol), this challenge appears to be achievable. In the model we test, players face predefined emissions abatement commitments, deposit amounts and a minimum number of countries required to ratify for entry into force. Apart from paying deposits, stage one in our experiments can be likened to Annex I countries deciding whether to ratify the Kyoto Protocol given defined emissions limits (relative to their 1990 levels) and a minimum participation constraint (at least 55 of the parties needed to ratify).[24]

In our experiments, players have identical payoff functions. In reality, countries involved in IEAs typically have differential commitments as members, and most importantly, different incentives to violate those commitments. Since the up-front deposit paid by each member must weakly exceed the financial incentive to cheat, heterogeneous members should also face different deposit amounts. To provide a rough idea of the size of these deposits, a report by the Energy Information Administration (1998) compiled cost estimates for the United States to comply with the Kyoto Protocol (a seven percent reduction in 1990 level emissions) that ranged from \$52 billion to \$437 billion (0.5–4.2 % of 2010 GDP). Clearly, deposit amounts for large greenhouse-gas emitting nations are non-trivial. Moreover, uncertainty in the benefit/cost functions or the realities of incomplete (or inaccurate monitoring) will likely play an important role in determining optimal deposit sizes.

Our overarching results suggest that a deposit-refund mechanism can be very effective when agreements require full participation, but is less effective, and potentially disruptive, when agreements require only partial participation. While these results help inform the design of effective institution formation, there are some obvious extensions that should be carried out to explore other relevant issues. As highlighted previously, a useful extension would be to introduce heterogeneity into the players' payoff functions and commitment levels, which would translate into requiring differential deposit amounts. While this is likely not a problem theoretically, it is unclear how this will impact the effectiveness of the mechanism. If parties exhibit other-regarding preferences (like inequality aversion), which is sometimes demonstrated to be the case (Charness and Rabin 2002), having differential commitments and deposits could undermine the positive features of the mechanism. Other useful extensions include introducing uncertainty by making the benefits of cooperation uncertain and/or by relaxing the assumption of perfect and complete monitoring for compliance.

References

Bagnoli M, McKee M (1991) Voluntary Contributions Games: Efficient Private Provision of Public Goods. Econ Inquiry 29(2):351–366

Barrett S (1994) Self-enforcing international environmental agreements. Oxf Econ Pap 46(1):878–894

Barrett S (1997) The strategy of trade sanctions in international environmental agreements. Resour Energy Econ 19(1):345–361

Barrett S (2003) Environment and statecraft: the strategy of environmental treaty-making. Oxford University Press, Oxford and New York

Barrett S (2008) Climate treaties and the imperative of enforcement. Oxf Rev Econ Policy 24(2):239–258

Barrett S, Stavins R (2003) Increasing participation and compliance in international climate Change agreements. Int Environ Agreem Polit Law Econ 3:349–376

Breton M, Sbragia L, Zaccour G (2010) A dynamic model for international environmental agreements. Environ Resour Econ 45(1):25–48

Cadsby CB, Maynes E (1999) Voluntary provision of threshold public goods with continuous contributions: experimental evidence. J Pub Econ 71(1):53–73

Carraro C, Siniscalco D (1993) Strategies for the international protection of the environment. J Pub Econ 52(3):309–328

24 Under the Kyoto Protocol, the parties also needed to account for at least 55 percent of 1990 green-house-gas emissions.

Carraro C, Siniscalco D (1998) International environmental agreements: incentives and political Economy. Eur Econ Rev 42(3–5):561–572

Carraro C, Marchiori C, Oreffice S (2009) Endogenous minimum participation in international environmental treaties. Environ Resour Econ 42(3):411–425

Cason TN, Plott CR (1996) EPA's new emissions trading mechanism: a laboratory evaluation. J Environ Econ Manag 30(2):133–160

Charness G, Rabin M (2002) Understanding social preferences with simple tests. Q J Econ 117(3):817–869

Chaudhuri A (2011) Sustaining cooperation in laboratory public goods experiments: a selective Survey of the Literature. Exp Econ 14(1):47–83

Croson R, Marks M (2000) Step returns in threshold public goods: a meta- and experimental analysis. Exp Econ 2:239–259

Dannenberg A, Lange A, Sturm B (2010) On the formation of coalitions to provide public goods—experimental evidence from the lab. National bureau of economic research working paper no. 15967

Dawes RM, Orbell JM, Simmons RT, Vande Kragt AJC (1986) Organizing groups for collective action. Am Polit Sci Rev 80(4):1171–1185

Energy Information Administration (1998) Impacts of the Kyoto Protocol on United States energy markets and economic activity. Office of integrated analysis and forecasting, United States Department of Energy, Washington, DC

Erev I, Rapoport A (1990) Provision of step-level public goods: the sequential contribution mechanism. J Confl Resolut 34(3):401–425

Finus M, Rundshagen B (1998) Toward a Positive Theory of Coalition Formation and Endogenous Instrument Choice in Global Pollution Control. Public Choice 96(1–2):145–186

Finus M (2008) The enforcement mechanisms of the Kyoto Protocol: flawed or promising concepts. Lett Spatial Resour Sci 1(1):13–25

Fullerton D, Wu W (1998) Policies for green design. J Environ Econ Manag 36(2):131–148

Gerber A, Wichardt P (2009) Providing public goods in the absence of strong institutions. J Pub Econ 93 (3–4):429–439

Hoel M (1992) International environmental conventions: the case of uniform reductions of emissions. Environ Resour Econ 2(2):141–159

Hoel M, Schneider K (1997) Incentives to participate in an international environmental treaty. Environ Resour Econ 9(2):153–170

Hovi J, Froyn CB, Bang G (2007) Enforcing the Kyoto Protocol: can punitive consequences restore compliance?. Rev Int Stud 33(3):435–449

Isaac MR, Schmidtz D, Walker JM (1989) The assurance problem in a laboratory market. Pub Choice 62(3):217–236

Kolstad C (2007) Systematic uncertainty in self-enforcing international environmental agreements. J Environ Econ Manag 53(1):68–79

Kosfeld M, Okada A, Riedl A (2009) Institution formation in public goods games. Am Econ Rev 99(4): 1335–1355

Ledyard J (1995) Public good: a survey of experimental results. In: Kagel J, Roth A (eds) The handbook of experimental economics. Princeton University Press, Princeton pp 111–194

Marks M, Croson R (1998) Alternative rebate rules in the provision of a threshold public good: an experimental investigation.
J Pub Econ 67(2):195–220

Marks M, Croson R (1999) The Effect of incomplete information in a threshold public goods experiment. J Pub Econ 67(2):195–220

McEvoy D (2010) Not it: opting out of voluntary coalitions that provide a public good. Pub Choice 142(1–2):9–23

McEvoy D, Murphy JJ, Spraggon JM, Stranlund JK (2011) The problem of maintaining compliance with stable coalitions: experimental evidence. Oxf Econ Pap 63(3):475–498

McEvoy D, Stranlund J (2009) Self-enforcing international environmental agreements with costly monitoring for compliance. Environ Resour Econ 42:491–508

McGinty M (2007) International environmental agreements among asymmetric nations. Oxf Econ Pap 59(1):45–62

Plott C (1987) Dimensions of parallelism: some policy applications of experimental methods. In: Alvin Roth (ed) Laboratory experimentation in economics: six points of view. Cambridge University Press, Cambridge

Rapoport A, Eshed-Levy D (1989) Provision of step-Level public goods: effects of greed and fear of being gypped. Organ Behav Hum Decis Process 44(3):325–344

Rapoport A, Suleiman R (1993) Incremental contribution in step-level public goods games with asymmetric players. Organ Behav Hum Decis Process 55(2):171–194
Rauchdobler J, Sausgruber R, Tyran J (2010) Voting on thresholds for public goods: experimental evidence. Pub Finance Anal 66(1):34–64
Rubio SJ, Ulph A (2006) Self-enforcing international environmental agreements revisited. Oxf Econ Pap 58(2):233–263
Rubio SJ, Ulph A (2007) An infinite-horizon model of dynamic membership of international environmental agreements. J Environ Econ Manag 54(3):296–310
Shogren JF, Hurley TM (1999) Experiments in environmental economics. In: van den Bergh JCJM (ed) Handbook of environmental and resource economics. Edward Elgar, Cheltenham
Stranlund JK, Murphy JJ, Spraggon JM (2011) An experimental analysis of compliance in dynamic emissions markets. J Environ Econ Manag 62(3):414–429
Suleiman R, Rapoport A (1992) Provision of step-level public goods with continuous contribution. J Behav Decis Mak 5(2):133–153
Tavoni A, Dannenberg A, Kallis G, Löschel A (2011) Inequality, communication and the avoidance of disastrous climate change in a public goods game. Proc Natl Acad Sci 108(29):11825–11829
Tourney D, Fujiwara N (2010) National commitments, compliance and the future of the Kyoto Protocol. Centre Eur Policy Stud 226:1–228
Van de Kragt AJC, Orbell JM, Dawes RM (1983) The minimal contributing set as a solution to public goods problems. Am Polit Sci Rev 77(1):112–122

[15]

Environ Resource Econ (2009) 42:491–508
DOI 10.1007/s10640-008-9220-1

Self-enforcing International Environmental Agreements with Costly Monitoring for Compliance

David M. McEvoy · John K. Stranlund

Accepted: 7 June 2008 / Published online: 29 June 2008

Abstract Theoretical analyses of international environmental agreements (IEAs) have often employed the concept of *self-enforcing agreements* to predict the number of parties to such an agreement. The term *self-enforcing*, however, is a bit misleading. The concept refers to the stability of cooperative agreements, not to enforcing compliance with these agreements once they are in place. In this paper we analyze an IEA game in which parties to an agreement finance an independent monitor who audits the compliance performance of the members of an agreement. These audits reveal instances of noncompliance so they can be sanctioned. We find that costly monitoring of compliance limits the circumstances under which international cooperation to protect the environment is worthwhile, but when IEAs do form they will often involve greater participation than IEAs that do not require costly monitoring. Consequently, costly monitoring of IEAs can produce higher international environmental quality. Moreover, under certain conditions, aggregate welfare is higher when IEAs require costly monitoring.

Keywords International environmental agreements · Self-enforcing agreements · Compliance · Monitoring

JEL Classification Q5 · H41 · C72 · F53

D. M. McEvoy (✉)
Economics Department, Appalachian State University, Raley Hall,
416 Howard Street, ASU Box 32051, Boone, NC 28608-2051, USA
e-mail: mcevoydm@appstate.edu

J. K. Stranlund
Department of Resource Economics, University of Massachusetts-Amherst,
80 Campus Center Way, 214 Stockbridge Hall, Amherst, MA 01003, USA
e-mail: stranlund@resecon.umass.edu

1 Introduction

International environmental agreements (IEAs) made between sovereign nations seeking to manage shared environmental and natural resources are susceptible to two sources of free riding. First, because participation with IEAs is voluntary, countries can decide at will whether to become a party to an agreement. If an IEA only requires a subset of countries to join before entering into force, as is typically the case, incentives exist for some countries to stay out of the agreement and free ride off the provision of the cooperating others. Second, if compliance with the terms of the agreement is not properly enforced, parties to an IEA will have an incentive to violate the terms of the agreement and free ride off those countries that do comply. Previous theoretical analyses of IEAs have focused on the impact the first form of free riding has on the effectiveness of IEAs, while avoiding altogether the possibility that member countries may not fully comply with their commitments. In this paper, we deal directly with the issue of noncompliance within IEAs by constructing and analyzing a game in which parties to an IEA finance an independent monitor who audits, with some endogenous probability, the compliance performance of individual members.

All effective international agreements must include costly compliance monitoring.[1] For example, the Implementation Committee of the Montreal Protocol is responsible for gathering compliance information about countries that are suspected to be in violation of the Protocol (Benedick 1998).[2] Likewise, the Compliance Committee, one of the institutions constructed to enforce compliance with the Kyoto Protocol, is designed to monitor the compliance behavior of signatory countries and then report their findings to the Enforcement Branch, which has the power to impose sanctions (UNFCC 2002). The Convention on International Trade in Endangered Species (CITES) makes a similar provision for monitoring for compliance through the Standing Committee, its own management body responsible for enforcement.[3] The enforcement provisions included in CITES and the Montreal and Kyoto Protocols allow their respective management bodies to monitor a party's behavior only when it is suspected to be violating the treaty. In contrast, the International Whaling Commission, the management body appointed under the International Convention for the Regulation of Whaling, established the International Observers Scheme to monitor signatories' behavior. This policy required member countries to each appoint observers and finance monitoring activities on a rotating basis.[4] Although the design of monitoring and enforcement institutions differ across international agreements, each share the common characteristic that the cost of monitoring activities must be bourn by the parties themselves.

Despite the fact that costly monitoring for compliance must be part of any international agreement, theoretical analyses of IEAs usually assume that compliance can be enforced without cost. In the literature IEAs are often analyzed using the equilibrium concept of a *self-enforcing agreement*.[5] A self-enforcing agreement made between agents is defined as a single coalition of cooperating agents from which no member of the coalition wishes to

[1] This point is argued by Heister et al. (1997) who provide a qualitative analysis of why international environmental agreements need to be enforced and the available strategies for enforcing compliance.

[2] Protocol on Substances that Deplete the Ozone Layer, Montreal, September 1987; Amended 1990, 1992, 1995, 1997, 1999.

[3] Convention on International Trade in Endangered Species, Washington, March 1973; Amended 1979.

[4] The International Observers Scheme was in effect from 1972 until the moratorium on whaling was imposed in 1982 (Oberthur 1998).

[5] Other stability concepts have been employed in the analysis of IEAs, including 'core' solutions from cooperative game theory (Chandler and Tulkens 1995, 1997), renegotiation-proof equilibria (Barrett 1994; Finus and Rundshagen 1998) and farsighted stability in the context of repeated games (Osmani and Tol 2007).

withdraw (the coalition is *internally stable*) and no non-member wishes to join (the coalition is *externally stable*).[6] This definition of coalitional stability was first proposed by D'Aspremont et al. (1983) for their analysis of cartel stability. Barrett (1994) appears to have been the first to apply it to the analysis of international environmental relations. However, the term *self-enforcing* is a bit misleading, because it does not address the problem of enforcing the compliance of the members of a cooperative agreement. Most authors who use this equilibrium concept simply assume that parties to a self-enforcing IEA comply fully with the terms of the agreement (e.g. Hoel 1992; Carraro and Siniscalco 1993, 1998; Barrett 1994, 2003; Kolstad 2007).[7] Barrett (1994) and Finus and Rundshagen (1998) have examined enforcement mechanisms for IEAs. They focus on renegotiation-proof punishment strategies with which parties to an agreement sanction violators by jointly reducing their provision of international environmental quality in the future.

We take a different approach to the study of compliance and enforcement of IEAs that is motivated by the costly monitoring activities that are part of established international agreements. We maintain the concept of a self-enforcing agreement to determine the equilibrium number of parties to an IEA. However, to counteract the incentive that parties have to violate the terms of an IEA, they finance and empower an independent body with the power to monitor their compliance behavior. Parties to an IEA that are monitored and deemed in violation of the agreement incur a costly sanction. We consider two types of sanctions—one in which an enforcement body imposes a financial sanction on detected violations, and the other in which detected violators, as well as those who choose not to participate in an IEA, face a costly deterioration of their reputation as cooperative members of the international community.[8]

Our efforts yield several new results. First, the range of international environmental problems within which a welfare-improving self-enforcing IEA can form is smaller when monitoring is necessary and costly. This follows because monitoring increases the cost of cooperation, which in turn limits the set of situations under which a treaty can actually increase international welfare. Second, when an IEA is expected to form, it will often have more members

[6] Models of self-enforcing IEAs have included provisions for trade sanctions (Barrett 1997a,b), side payments (Carraro and Siniscalco 1993; Carraro and Botteon 1997; Hoel and Schneider 1997), issue linkage (Folmer et al. 1993; Carraro and Siniscalco 1997; Botteon and Carraro 1998) and minimum participation requirements (Barrett 1998a, 2003; Carraro et al. 2003). For a review of many of these extensions see Wagner (2001).

[7] There is ample evidence that many existing IEAs experience significant problems with noncompliance. For example, CITES has witnessed hundreds of infractions every year since its inception in 1979 (Carraro and Siniscalco 1999; Finus 2004). Similarly, customs officers throughout the world frequently intercept flows of CFC products even though these substances have been banned under the Montreal Protocol since 1991 (Finus 2004). Although the performance of the Kyoto Protocol cannot formally be evaluated until the end of the first commitment period (2008–2012), many countries are currently on pace to significantly exceed their quotas (Friends of the Earth 2006). Although it is universally accepted that some level of noncompliance with IEAs exists, there remains debate as to the magnitude and consequences of noncompliance. On one side of the debate, Chayes and Chayes (1991) argue that high levels of compliance are often observed in IEAs without formal enforcement mechanisms and therefore enforcement is not a real concern. However, Downs et al. (1996) argue that high levels of compliance may result because the lack of enforcement motivates parties to form shallow agreements to begin with.

[8] Because our model is one of enforcing compliance within stable coalitions it is related to the vast literature on the economics of law enforcement. For a review of this literature see Polinsky and Shavell (2000). There is also a large literature on the economics of enforcing environmental policies. Recent contributions on enforcing emissions standards, tradable emissions permits, and emissions taxes include Arguedas (2007), Stranlund (2007), and Macho-Stadler and Perez-Castrillo (2006), respectively. The fundamental difference between our work and the law enforcement literature is that we examine enforcement within a model of endogenous coalition formation while the typical model of enforcement involves a government enforcing a rule that it has imposed.

when the parties must pay for compliance monitoring. This occurs because the additional cost of being a party to an IEA must be offset with an increase in the level of environmental quality, which is realized through an increase in the number of parties to the agreement. As a result, costly monitoring of IEAs can produce higher environmental quality. In addition, provided that the number of potential parties to an agreement is sufficiently large, social welfare can be greater under an IEA that is costly to monitor. Finally, we find that our principal results are accentuated when compliance monitoring is not perfectly accurate. That is, monitoring errors further reduce the set of situations in which multilateral cooperation is welfare enhancing and thus limit the opportunities for an IEA to form. However, when an agreement does form, monitoring inaccuracy can generate higher participation and environmental quality.

The rest of this paper proceeds as follows. After presenting a basic model of a self-enforcing IEA in Sect. 2. In Sect. 3 we extend the model to allow for noncompliance and assume that parties to a self-enforcing IEA finance an independent monitor who checks on their compliance choices and applies a financial sanction in cases of detected noncompliance. Section 4 contains the principal results of our work. In this section we derive the consequences of costly monitoring of IEAs on the value of international cooperation, the circumstances under which a welfare-improving IEA will form, and the self-enforcing number of members of an IEA. In Sect. 5 we analyze the effects of costly monitoring on global welfare. In Sect. 6 we consider an alternative model in which IEA violators, as well as those who do not participate in the agreement, suffer a costly loss in reputation. We conclude in Sect. 7.

2 The Basic Model of a Self-enforcing IEA

In this section, we present a standard model of a self-enforcing IEA—one that does not require enforcement—to review the structure of these games and to provide a baseline for determining the effects of costly compliance monitoring on the outcomes of these games. For simplicity, throughout our analysis countries make a discrete choice whether to abate their emissions. Although a model with continuous abatement choice would allow for more general results, the discrete choice model has been adopted in a number of previous studies because of its tractability. Thus, following Barrett (2003); Ulph (2004) and Kolstad (2007), consider a situation where N identical countries each emit a uniformly mixed transboundary pollutant. Country i's welfare is

$$w_i = A + b(q_i + q_{-i}) - cq_i, \tag{1}$$

where q_i is equal to one if i abates its emissions and zero if it does not, q_{-i} is the sum of the abatement decisions made by all other countries, b is the constant benefit of abatement by i or any other country, c is the constant individual cost of abatement, and A is a positive constant. Assume that $b < c$ so that no country will unilaterally abate its emissions, but that $Nb > c$ so that the countries' joint welfare will be maximized when they all abate their emissions.

Recognizing the benefits from agreeing to control their emissions, countries have the incentive to form an IEA to do so. As in Ulph (2004) and Kolstad (2007), the formation of an IEA is modeled as a two-stage game. In the first stage, each country must decide independently whether to become a party to an IEA. Assume that countries have complete information regarding each others' welfare functions and that they can form only a single cooperative coalition. In the second stage, all countries choose whether to abate their emissions. Countries that become members of an IEA in the first stage commit to decisions in the second stage that maximize their joint welfare, given the decisions of the non-member nations. In this section we assume perfect compliance with the agreement so that members of the IEA do not have to be monitored for compliance. While the parties to an IEA carry through on their commitments to maximize their joint welfare in the second stage, non-member countries maximize their individual welfare. Since the individual cost of abatement outweighs the benefit for each country (i.e., $b < c$), all countries that are not members of an IEA will decide to not abate their emissions in the second stage.

Given the decision of all non-member countries to not abate their emissions, the members of an IEA will abate their emissions in the second stage as long as the coalition is *profitable* in the sense that each member of the coalition is at least as well off as when all countries choose not to control their emissions. Let s denote the number of parties to an IEA who all agree to abate their emissions. Moreover, let $w^p(s)$ denote the welfare of each of the parties to the IEA and let $w^{np}(s)$ denote the welfare of each of the countries that are not party to the agreement. Throughout the paper the superscript p signals that the country in question is a party to an IEA, while the superscript np signals that the country is not a party to the agreement. From (1) we have:

$$w^p(s) = A + bs - c;$$
$$w^{np}(s) = A + bs. \tag{2}$$

Since $b < c$, $Nb > c$, and $w^p(s)$ is increasing in s, there exist coalition sizes that are strictly greater than one and weakly less than N that are profitable. The smallest of these profitable coalitions is

$$s^{nc}_{\min} = \min\left\{s|w^p(s) \geq w^{np}(0)\right\} = \min\{s|s \geq c/b\}. \tag{3}$$

Throughout, the superscript nc denotes values when the decisions of the members of an IEA do not require costly monitoring (i.e., compliance is assumed perfect). If $s \geq s^{nc}_{\min}$ from the first stage of the game, the members of the coalition will agree to abate their emissions in the second stage of the game. If $s < s^{nc}_{\min}$, the members of the IEA maximize their joint welfare by not abating their emissions in the second stage.

The subgame perfect equilibrium of the game is called a *self-enforcing agreement* in the IEA literature. Intuitively, a self-enforcing agreement is one for which no party to the agreement wants to leave the agreement and no non-party wishes to join. More formally:

Definition 1 An IEA consisting of s countries that comply with the agreement without the need for costly monitoring is self-enforcing if and only if:

$$(i)\ w^p(s) \geq w^{np}(s-1) \text{ for } s > s^{nc}_{\min}, \text{ or } w^p(s^{nc}_{\min}) \geq w^{np}(0) \text{ for } s = s^{nc}_{\min}$$
$$(ii)\ w^{np}(s) \geq w^p(s+1). \tag{4}$$

Requirement (i) is that the agreement is *internally stable* in the sense that no party has an incentive to leave a self-enforcing agreement. Requirement (ii) is that the agreement is *externally stable* in the sense that no non-party wishes to join the IEA. The only internally stable coalition is the one with $s = s^{nc}_{\min}$ members. To see why, note that for $s > s^{nc}_{\min}$, (2) reveals that $w^p(s) \geq w^{np}(s-1)$ implies $b \geq c$, which violates our assumption that $b < c$. On the other hand, an IEA with $s^{nc}_{\min}$ members is internally stable, because if one member of the coalition defected then the remaining members would find it unprofitable to abate their emissions. The defector's welfare (as well as every other country's welfare) would then be $w^{np}(0)$, which is weakly less than $w^p(s^{nc}_{\min})$. Finally, an IEA with $s^{nc}_{\min}$ members is also externally stable. Using requirement (ii) and the welfare functions (2), $w^{np}(s^{nc}_{\min}) \geq w^p(s^{nc}_{\min}+1)$ implies $b \leq c$, which is satisfied because $b < c$. Since an agreement consisting of $s^{nc}_{\min}$ members is the only internally and externally stable coalition size, it is the equilibrium number of members of a self-enforcing IEA.

Proposition 1 *Let the equilibrium number of members of a self-enforcing agreement that does not require costly monitoring be* s^{nc}. *Then,* s^{nc} *is the smallest profitable coalition; that is,*

$$s^{nc} = s^{nc}_{\min} = \min\{s|s \geq c/b\}. \tag{5}$$

The equilibrium number of members of an IEA that does not require costly monitoring is unique, strictly greater than 1, and weakly less than N. Note that [5] indicates that participation with an IEA is increasing in the cost of abatement, c, and decreasing in the individual benefit of abatement, b. Thus, Barrett's (1994, 2003) claim that international cooperation to protect the environment will be greatest when it is needed least is easily verified.

It is clear that the concept of a self-enforcing agreement applies to the stability of a cooperating coalition, not to the coalition members' decisions to comply with the terms of the agreement once they have joined. Indeed, if countries are able to observe each others' compliance decisions perfectly and without cost, then a member country has no incentive to join an agreement and then violate its requirements. If it did so all the other participating nations would observe this violation and would realize that they would be worse off if they continued to abate their emissions. Thus, under perfect information a member of a self-enforcing IEA would not violate the terms of the agreement because this would cause the agreement to collapse.

In reality, however, nations are not able to observe the compliance behavior of others so easily. Under imperfect information about individual nations' compliance behavior, a country may be motivated to join an agreement and then decide to violate its terms. If the other participating countries do not observe this act of noncompliance, they will not automatically respond by refusing to abate their emissions. The violator may then be able to escape the cost of compliance while enjoying the benefit of cooperation of those that continue to abate their emissions.

To demonstrate the incentive parties have to violate the terms of the IEA under imperfect information regarding compliance behavior, assume that a member of an IEA with s members can fail to abate its emissions without being detected. If the party chooses to be noncompliant and the other member countries continue to abate their emissions because they can not observe this violation, the violator's payoff is $w^{np}(s-1) = A + b(s-1)$. If it instead remained compliant its payoff is $w^{p}(s) = A + bs - c$. Subtracting $w^{np}(s-1)$ from $w^{p}(s)$ yields $b - c$, which is negative by assumption indicating that member countries always have an incentive to violate an agreement as long as they believe that the other members are unable to observe their violation. However, with homogenous countries, each should expect that all the others will have the same motivation to violate an IEA as they do—if one member of an agreement has an incentive to violate an agreement, then all members do. Thus, under asymmetric information about countries' abatement decisions and without monitoring to check on these decisions, no self-enforcing agreement exists (except for the trivial case of $s = 0$). This motivates the implementation of a monitoring mechanism to check on the compliance decisions of the members of an IEA and a sanction to punish detected violators.

3 Self-enforcing IEAs when Compliance is Costly to Monitor

We now modify the basic model of a self-enforcing IEA to give its members the opportunity to decide whether to comply with the terms of their agreement. We assume that members cannot observe each other's compliance decisions and, therefore, each of them has an incentive to violate the requirements of the agreement in the absence of adequate monitoring. To counteract this incentive the parties of an IEA finance an independent monitor who audits their compliance decisions. A fixed sanction is applied when violations are detected.

As before the model is analyzed as a stage game, but now the game has four stages. The first stage (*membership stage*) proceeds exactly as before; that is, each country chooses independently whether to join the IEA. In the second stage (*treaty stage*), members of the IEA jointly agree on whether to abate their emissions and, if they do agree to abate their emissions, each of them is required to contribute funds to the independent monitor. If IEA members jointly decide not to control their emissions in stage two, they do not fund the monitor, an effective IEA does not form, and the game ends. (Throughout, we refer to an *effective* IEA as one that actually leads to emissions control). If an effective IEA forms in stage two, all countries make their abatement decisions independently in stage three (*abatement stage*). Finally, in the fourth stage (*enforcement stage*) the monitor randomly audits the abatement decisions of member countries with the funding provided to it in the second stage, and a sanction is applied when the monitor finds a violation. Since the treaty, abatement, and enforcement stages are new with this study, we now describe them in detail. Because the game is solved by backward induction, we start with the last stage.

3.1 Enforcement Stage

If the game reaches this stage, an IEA with s members has formed, each member has agreed to abate their emissions, each member has contributed x dollars to fund the monitor, and all countries have made their abatement decisions. In the enforcement stage, the monitor randomly audits the abatement decisions of the IEA members with a probability that is an increasing function of the amount of funding the members contribute. Each additional dollar of funding allows the number of random audits to increase by $\alpha > 0$; that is, α is the constant marginal productivity of resources devoted to monitoring. If s parties to an agreement each provide x to fund the monitor, then the number of random audits conducted is $sx\alpha$, and the probability that any party is audited is

$$\pi = sx\alpha/s = x\alpha \qquad (6)$$

We will demonstrate shortly that an effective IEA will involve a monitoring probability in the half-closed interval $(0, 1]$.

Monitoring of the countries' compliance behavior, however, may be subject to a host of possible errors, including errors due to erroneous or missing data, reporting errors, and errors in evaluating available performance data. Given an audit, let $\rho^1 \in [0, 1)$ denote the probability that a compliant country is judged to be noncompliant (a Type I error), and let $\rho^2 \in [0, 1)$ denote the probability that a noncompliant country is judged to be compliant (a Type II error).[9] These probabilities are common knowledge.

In practice, there are a variety of ways noncompliant countries may be sanctioned. Much of our analysis in this regard is motivated by the enforcement mechanism of the Kyoto Protocol established under the Marrakesh Accords. These established an enforcement body (called the Compliance Committee) with the power to sanction noncompliant parties by reducing their greenhouse gas emissions quota for the next commitment period. Therefore, we assume that a party that is revealed to be noncompliant by the monitor incurs a known exogenous sanction of f. Others have criticized the enforcement strategy of the Kyoto Protocol as being non-credible (e.g. Barrett 2003), because a party to an IEA who decides not to abate its

[9] See Kaplow and Shavell (1994) for the effects of monitoring inaccuracy in a standard model of optimal law enforcement.

emissions could just as easily choose not to pay the sanction if its violation is discovered. Moreover, a party that is found to be in violation of the requirements of an IEA could simply leave the agreement.

In Sect. 6 we consider another type of sanction that cannot be avoided so easily. There is strong agreement in the economic and political science literature that a party's willingness to jointly manage transnational resources has value beyond a single treaty (Keohane 1986; Chayes and Chayes 1991; Simmons 1998; Barrett 2003). Complying with international agreements builds a country's reputation as a cooperating member of the international community, and that reputation allows it to be a part of other beneficial agreements. Chayes and Chayes (1991, p. 320) introduce the flip-side of this argument: "A reputation for unreliability cannot be confined to the area of activity in which it is earned. It is inevitable that a state's defection from treaty rules will generate repercussions and linkages throughout the network of its relationships with others in the community. And these more diffuse responses can be calibrated more finely than formal sanctions." We would add that if failing to comply with an IEA deteriorates a country's reputation, then not participating with an agreement in first place is also likely to be damaging. In Sect. 6 we model sanctions for agreement violations and for non-participation as costly reputation effects, and show that our qualitative conclusions about the impacts of costly monitoring on self-enforcing IEAs are largely preserved with this approach.[10]

3.2 Abatement Stage

At this point in the game the members of an IEA have agreed to abate their emissions and have funded the monitor. In the abatement stage both members of an IEA and non-members independently choose whether to control their emissions. Since the individual cost of abatement outweighs the individual benefit, each non-member will decide not to abate their emissions in this stage. Member countries, however, make this decision by comparing the expected cost of not complying with the agreement and the benefit of noncompliance.

Assume that the parties to an IEA are risk neutral and that they comply if they are at least indifferent between compliance and noncompliance. Then, given an effective agreement consisting of s parties, in the abatement stage an individual country will comply if its expected welfare from doing so is not less than its expected welfare from noncompliance. Parties to the agreement cannot observe the others' compliance choices, so one party's violation of its commitment to abate its emissions will go unnoticed unless it is discovered by the monitor. Given the probability of an audit π, the probabilities of Type I and Type II monitoring errors ρ^1 and ρ^2, and the sanction f, it is straightforward to show that a party's expected welfare from compliance is $w^p(s) - x - \pi\rho^1 f$, while its expected welfare from noncompliance is $w^{np}(s-1) - x - \pi(1-\rho^2)f$, where recall that $w^p(s)$ and $w^{np}(s-1)$ are defined by (2). Note that $\pi\rho^1 f$ is a compliant party's expected sanction when it is falsely judged to be noncompliant and $\pi(1-\rho^1)f$ is a noncompliant party's expected sanction from a correct determination that it is noncompliant.[11] A party to an IEA complies with the terms of the

[10] A third possible form the sanction for noncompliance could take is a restriction on the 'club' benefits parties receive from belonging to an IEA. For example, the International Monetary Fund does not allow noncompliant members to access the pool of funds jointly provided by the signatories to the agreement (Chayes and Chayes 1991). This arrangement is similar to posting a performance bond when ratifying an agreement, which is paid back to the party if they comply with the agreement.

[11] One may wonder why a compliant country that is incorrectly determined to be noncompliant (Type I error) would passively accept the sanction. At the very least, a compliant country that faces a sanction for noncompliance will attempt to prove that it has, in fact, satisfied the terms of the agreement. Under the Kyoto Protocol,

agreement if and only if

$$[w^p(s) - x - \pi\rho^1 f] - [w^{np}(s-1) - x - \pi(1-\rho^2)f]$$
$$= \pi f(1-\rho^1-\rho^2) - (c-b) \geq 0 \tag{7}$$

Clearly, $\pi f(1-\rho^1-\rho^2) \geq (c-b)$ is a necessary condition for an effective IEA. If this condition did not hold, no party to an IEA would comply with its requirements and an effective agreement would not form. Note further that no amount of monitoring will be sufficient to induce compliance if $f(1-\rho^1-\rho^2) < (c-b)$. This indicates that low sanctions or severe monitoring errors can prevent the formation of an effective IEA. For the remainder of this analysis we assume that $f(1-\rho^1-\rho^2) \geq (c-b)$ so that an effective IEA can form. Since $c > b$, this condition clearly requires $\rho^1 + \rho^2 < 1$.

3.3 Treaty Stage

In this stage members of an IEA agree to abate their emissions and to fund the monitor provided that these decisions maximize their joint welfare. Because we assume that each country's abatement decision is not directly observable by other countries, members of an IEA cannot credibly commit to abatement. This, of course, is the reason for monitoring. Each country that wishes to join an effective agreement is required to contribute funds to the monitor. These payments are perfectly observed, so a country is not able to promise to make the payment and then fail to do so.

Each party's contribution to the monitor, x, is endogenous, so we determine this value first. If an effective agreement is to form, each party would like to contribute as little as possible while providing the monitor with sufficient resources to maintain compliance with the agreement. This requires a payment x so that $\pi f(1-\rho^1-\rho^2)-(c-b) \geq 0$ binds. Since $\pi = x\alpha$ from (6), the contribution to the monitor of an IEA that is required of all parties to the agreement is

$$x = (c-b)/(\alpha f(1-\rho^1-\rho^2)). \tag{8}$$

Note that this payment decreases with the size of the sanction, f, and the marginal productivity of resources devoted to monitoring, α, but is increasing in the probabilities the monitor commits a Type I or Type II error, ρ^1 and ρ^2, and the gain from noncompliance, $c-b$. Note further that $\pi = x\alpha = (c-b)/(f(1-\rho^1-\rho^2))$ is the equilibrium probability the monitor audits any member of an effective IEA in the enforcement stage of the game. Our assumptions that $c > b$, $\rho^1 + \rho^2 < 1$, and $f(1-\rho^1-\rho^2) \geq (c-b)$ guarantee that $\pi \in (0, 1]$.

Parties to an IEA will fund the monitor and jointly agree to abate their emissions only if they will be at least as well off as without an agreement. That is, again, abatement must be profitable for the members of the agreement. Given s parties to an agreement that each expects to earn $w^p(s) - x - \pi\rho^1 f = A + bs - c - x - \pi\rho^1 f$ if they abate their emissions, and an individual country's welfare in the absence of an IEA, A, the minimum size profitable coalition is the smallest s such that $A + bs - c - x - \pi\rho^1 f \geq A$. Substituting for x from (8) and rearranging terms yields the smallest profitable coalition when an IEA requires costly

Footnote 11 continued
for example, if the enforcement branch of the Compliance Committee finds a party to be in violation, that party is entitled to a hearing at which it can present additional evidence and expert testimony. Further, if a party is still deemed noncompliant after the appeal process, it is given the chance to submit a final written testimony to the enforcement branch (UNFCCC 2002). This process appears to be designed to limit the probability of a Type I error.

monitoring:

$$s^c_{\min} = \min\left\{s|s \geq \frac{c + \pi\rho^1 f}{b} + \frac{c-b}{b\alpha f(1-\rho^1-\rho^2)}\right\}. \tag{9}$$

(The superscript 'c' is used to denote values when compliance with an IEA requires costly monitoring). If $s \geq s^c_{\min}$ from the first stage of the game (the membership stage), the members of the coalition will agree to abate their emissions in the treaty stage of the game. Furthermore, each member pays x from (8) to the monitor in this stage and will comply with the terms of the agreement in stage three (the abatement stage). On the other hand, if $s < s^c_{\min}$ from the first stage, the members of the IEA maximize their joint welfare by not abating their emissions and not funding the monitor. Clearly, if $s < s^c_{\min}$ an effective IEA does not form and the game concludes.

4 Properties of Self-enforcing IEAs that Require Costly Monitoring

The subgame perfect equilibrium of the stage game we have just described is either a self-enforcing agreement under which all members of the agreement abate their emissions, or an effective IEA does not form. For an IEA to form, profitable coalitions must exist, which they will as long as $N \geq s^c_{\min}$; that is, using (9), $N \geq (c+\pi\rho^1 f)/b+(c-b)/(b\alpha f(1-\rho^1-\rho^2))$. In the model without costly monitoring of Sect. 2, profitable coalitions always exist because we assumed at the outset that $N > c/b$. However, since $c > b$ and $\rho^1+\rho^2 < 1$, c/b is strictly less than $(c + \pi\rho^1 f)/b + (c - b)/(b\alpha f(1 - \rho^1 - \rho^2))$. This implies that while profitable coalitions always exist when monitoring is not costly, they may not exist when parties to an IEA must fund a monitor to maintain compliance with the agreement. Therefore, we have our first conclusion about the impact of costly monitoring on self-enforcing environmental agreements:

Proposition 2 *The circumstances under which an effective IEA will form are diminished by the need for costly compliance monitoring.*

That $N \geq (c + \pi\rho^1 f)/b + (c - b)/(b\alpha f(1 - \rho^1 - \rho^2))$ is required for an effective IEA to form is simply a statement about whether international cooperation to limit a global pollutant can increase aggregate welfare when we take account of the costs of compliance monitoring. Note that $(c + \pi\rho^1 f)/b + (c - b)/(b\alpha f(1 - \rho^1 - \rho^2))$ is decreasing in the marginal productivity of monitoring resources, α, and the size of the sanction, f. Decreasing either of these parameters increases the payment cooperators pay to monitor an agreement, leading to a decrease in the set of opportunities for an effective IEA. Note also that $(c + \pi\rho^1 f)/b + (c - b)/(b\alpha f(1 - \rho^1 - \rho^2))$ is increasing in the probability the monitor commits either a Type I or Type II error, indicating that the presence of monitoring errors further reduces the circumstances under which an effective IEA will form. Finally, $(c+\pi\rho^1 f)/b+(c-b)/(b\alpha f(1-\rho^1-\rho^2))$ is increasing in c and decreasing in b; hence, the set of circumstances under which an effective IEA will form is smaller when abatement costs are higher and when the individual benefit from abatement is lower.

As with the IEA model without costly monitoring in Sect. 2, a self-enforcing IEA that requires costly monitoring is the coalition size that is both internally and externally stable. And like the model in Sect. 2, it is easy to demonstrate that the only internally and externally stable coalition is the smallest profitable coalition given by (9), provided that profitable coalitions actually exist. Therefore:

Proposition 3 *Let s^c be the equilibrium number of members of a self-enforcing agreement when monitoring for compliance with the agreement is costly. Then,*

$$s^c = s^c_{\min} = \min\left\{s|s \geq (c + \pi\rho^1 f)/b + (c - b)/(b\alpha f(1 - \rho^1 - \rho^2))\right\}, \quad (10)$$

provided that $N \geq s^c_{\min}$. If $N < s^c_{\min}$, there are no profitable coalitions and an effective self-enforcing agreement will not form.

Comparing Proposition 3 to Proposition 1 allows us to determine how costly monitoring affects the equilibrium number of members of an IEA. When an agreement with costly monitoring forms, (10) indicates that the equilibrium size of the coalition s^c is the least s for which $s \geq (c + \pi\rho^1 f)/b + (c - b)/(b\alpha f(1 - \rho^1 - \rho^2))$. When an IEA does not require costly monitoring, (5) indicates that the size of the coalition s^{nc} is the least s for which $s \geq c/b$. Recall that $1 - \rho^1 - \rho^2 > 0$ is required for an agreement with costly monitoring to form. Therefore, $(c + \pi\rho^1 f)/b + (c - b)/(b\alpha f(1 - \rho^1 - \rho^2)) > c/b$ implies $s^c \geq s^{nc}$, and our next proposition.

Proposition 4 *If an effective IEA that is costly to monitor forms, the number of parties to the agreement will be no less, and will typically be greater, than if the IEA did not require costly monitoring.*

The intuition behind Proposition 4 is straightforward. Since funding the monitor of an IEA is an additional cost of joining the agreement, more countries are required to participate and abate their emissions to make the agreement profitable. This produces the seemingly paradoxical result that costly monitoring of IEAs can produce higher environmental quality.

Generally speaking, participation in an IEA increases with the costs of participation and decreases with the benefit that participation provides to all the countries. Thus, participation with an effective IEA that is costly to monitor increases with a country's abatement cost, c, and decreases with the individual benefit of some country's abatement, b. More importantly for the purposes of this paper, any change in an enforcement parameter that increases the costs of monitoring will result in greater participation, as long as the IEA remains viable. Thus, participation and environmental quality increase with decreases in the marginal productivity of monitoring resources, α, and the sanction for noncompliance, f. Moreover, because monitoring inaccuracy increases the costs of maintaining compliance with an effective IEA, participation and environmental quality increase with monitoring inaccuracy as long as this does not prevent the formation of an IEA.

5 Costly Monitoring of IEAs and Aggregate Welfare

In this section we examine the differences in aggregate welfare under self-enforcing IEAs that are costly to monitor and under those for which compliance can be enforced without cost. Costly monitoring produces two countervailing effects on aggregate welfare. The first effect is the cost of monitoring itself which, holding participation with an IEA constant, diminishes aggregate welfare. However, costly monitoring can produce greater participation and higher environmental quality, which increases aggregate welfare. We demonstrate that it is possible that the latter effect dominates the former so that costly monitoring of international cooperation actually produces greater aggregate welfare.

Given s^c parties to an agreement that each earn $w^p(s^c) - x - \pi\rho^1 f = A + bs^c - c - x - \pi\rho^1 f$, and $N - s^c$ free-riding countries that each earn $w^{np}(s^c) = A + bs^c$, aggregate welfare when cooperation requires costly monitoring is

$$\begin{aligned} W^c(s^c) &= s^c(A + bs^c - c - x - \pi\rho^1 f) + (N - s^c)(A + bs^c) \\ &= NA + s^c(Nb - c - x - \pi\rho^1 f). \end{aligned} \tag{11}$$

Recall that the superscript 'c' identifies variables and functions when compliance with an IEA requires costly monitoring. For convenience, let us assume that s^c is continuous. Then, from (10), $s^c = (c + \pi\rho^1 f)\,/(b + (c - b)/\,/(b\alpha f(1 - \rho^1 - \rho^2))$ as long as it is not greater than N. Substitute this and $x = (c - b)/(\alpha f(1 - \rho^1 - \rho^2))$ from (8) into $W^c(s^c)$ to obtain

$$\begin{aligned} W^c(s^c) = NA &+ \left(\frac{c + \pi\rho^1 f}{b} + \frac{c - b}{b\alpha f(1 - \rho^1 - \rho^2)}\right) \\ &\times \left(Nb - c - \frac{c - b}{\alpha f(1 - \rho^1 - \rho^2)} - \pi\rho^1 f\right) \end{aligned} \tag{12}$$

Aggregate welfare when a self-enforcing IEA does not require costly monitoring is $W^{nc}(s^{nc}) = NA + s^{nc}(Nb - c)$. Allowing s^{nc} to be continuous, substitute $s^{nc} = c/b$ from (5) into $W^{nc}(s^{nc})$ to obtain

$$W^{nc}(s^{nc}) = NA + (c/b)\,(Nb - c)\,. \tag{13}$$

Now subtract (13) from (12):

$$\begin{aligned} W^c(s^c) - W^{nc}(s^{nc}) &= \left(\frac{c - b}{b\alpha f(1 - \rho^1 - \rho^2)} + \frac{\pi\rho^1 f}{b}\right) \\ &\times \left[N - \left(\frac{2c + \pi\rho^1 f}{b} + \frac{c - b}{b\alpha f(1 - \rho^1 - \rho^2)}\right)\right]. \end{aligned}$$

Since the first term of $W^c(s^c) - W^{nc}(s^{nc})$ is positive by assumption, the sign of $W^c(s^c) - W_{nc}(s^{nc})$ is equal to the sign of the term in hard brackets, which provides our next proposition.

Proposition 5 *Costly monitoring of an IEA results in higher aggregate welfare if and only if*

$$N > \frac{2c + \pi\rho^1 f}{b} + \frac{c - b}{b\alpha f(1 - \rho^1 - \rho^2)}. \tag{14}$$

Proposition 5 indicates that costly monitoring of an effective IEA produces higher aggregate welfare when the number of potential parties to the agreement is large enough.

Figure 1 illustrates how $W^c(s^c)$ and $W^{nc}(s^{nc})$ vary with the number of potential parties to an IEA and with the accuracy of compliance monitoring. To draw this graph we have assumed, without loss of generality, that $A = 0$. Aggregate welfare when an IEA does not require costly monitoring, $W^{nc}(s^{nc})$, is zero for $N \leq c/b$. For $N > c/b$, $W^{nc}(s^{nc})$ is positive and Eq. (13) indicates that it increases linearly at rate c. The size of a self-enforcing IEA does not change as N increases; it remains constant at c/b. However, aggregate welfare increases with N because increasing N means we are increasing the number of free-riding countries, each of which benefits from the abatement efforts of the c/b parties to the agreement.

The dashed function in Fig. 1 denoted $W^c(s^c|\rho^1 = \rho^2 = 0)$ is aggregate welfare for a self-enforcing IEA that is costly to monitor and monitoring is perfectly accurate. The dotted function, $W^c(s^c|\rho^1 + \rho^2 > 0)$, is aggregate welfare for a self-enforcing IEA that is costly to monitor and monitoring is inaccurate. Note that $W^c(s^c)$, with and without monitoring errors, is equal to zero for a larger range of N than $W^{nc}(s^{nc})$. This follows because cooperative

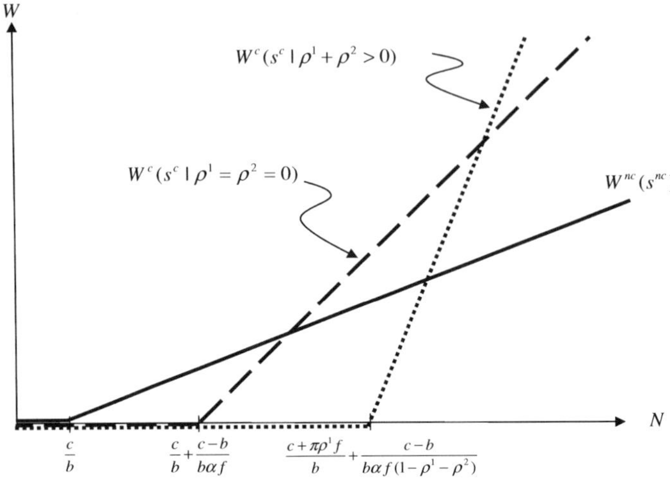

Fig. 1 A comparison of aggregate welfare under self-enforcing international environmental agreements that require costly monitoring and aggregate welfare under agreements that do not require costly monitoring

abatement efforts that require costly monitoring increase aggregate welfare under a smaller set of circumstances than when cooperation does not require monitoring (Proposition 2).

If monitoring is perfectly accurate and $N \geq c/b + (c-b)/(b\alpha f)$, the coalition $s^c = c/b + (c-b)(b\alpha f)$ forms and aggregate welfare increases with N at rate $c + (c-b)/(\alpha f)$. For $N \geq c/b - (c-b)/(b\alpha f)$, the relationship between $W^c(s^c|\rho^1 = \rho^2 = 0)$ and $W^{nc}(s^{nc})$ depends on the positive welfare effect of higher abatement when an IEA must be monitored (because $s^c > s^{nc}$ from Proposition 4) and the negative welfare effect of the costs of monitoring. The level of N where $W^c(s^c|\rho^1 = \rho^2 = 0)$ and $W^{nc}(s^{nc})$ intersect is $2c/b + (c-b)/(b\alpha f)$. Therefore, for $N \in [c/b + (c-b)/(b\alpha f), 2c/b + (c-b)/(b\alpha f))$, the monitoring-cost effect dominates the higher-abatement effect so that aggregate welfare is lower when the equilibrium IEA requires costly monitoring. However, when N exceeds $2c/b + (c-b)/(b\alpha f)$ the higher-abatement effect dominates the monitoring-cost effect so that aggregate welfare is higher when compliance requires costly monitoring.

Monitoring errors further reduce the set of circumstance under which an effective IEA will form. For a large enough N, it is straightforward to show that $W^{nc}(s^{nc})$ and $W^c(s^c|\rho^1 + \rho^2 > 0)$ intersect at $N = (2c + \pi\rho^1 f)/b + (c-b)/(b\alpha f(1-\rho^1-\rho^2))$. For N greater than this value, costly and inaccurate monitoring of an IEA leads to higher aggregate welfare than if the IEA did not require monitoring. Moreover, Eq. (12) indicates that when $W^c(s^c|\rho^1 + \rho^2 > 0)$ is positive it rises more quickly than $W^c(s^c|\rho^1 = \rho^2 = 0)$. Therefore, monitoring inaccuracy results in higher aggregate welfare than under perfectly accurate monitoring if N is large enough.[12] The reason for this counterintuitive result is the now familiar monitoring cost/abatement tradeoff. Monitoring inaccuracy reduces aggregate welfare because it leads to higher monitoring costs, but inaccuracy also increases aggregate welfare because it leads to higher abatement. The latter effect dominates when the number of potential parties to an IEA is large enough.

[12] It is straightforward to show that $W^c(s^c|\rho^1 + \rho^2 > 0)$ and $W^c(s^c|\rho^1 = \rho^2 = 0)$ intersect at $N = (2c + \pi\rho^1 f)/b + (c-b)(2-\rho^1-\rho^2)/(b\alpha f(1-\rho^1-\rho^2))$.

6 Self-enforcing IEAs with Costly Monitoring and Reputation Effects

Up to this point we have assumed that an enforcement body of an IEA can impose a sanction on noncompliant members. Given the possibility that a violator may simply refuse to pay a sanction, which we discussed in Sect. 3, we now examine the consequences of an alternative model that includes reputation sanctions that cannot be avoided. Modeling costly reputation effects that result from violating the terms of an IEA forces us to also consider the possibility that countries that refuse to join an IEA in the first place may also suffer a deterioration of their reputation as cooperative members of the international community. Therefore, we now let f denote a sanction that is the value of a member country's reputation loss when it violates an IEA, and let g be the value of a country's reputation loss when it refuses to join an IEA. Since it is unreasonable to assume that all countries suffer a loss of reputation when no country joins an IEA, assume that $g = 0$ when an IEA does not form.[13]

The noncompliance reputation loss, f, enters our costly monitoring model in the same way as before. Thus, all changes in our results are due solely to the inclusion of the non-participation reputation loss, g. Moreover, as we demonstrate below, the primary effect of this non-participation sanction is that it can be large enough to induce full participation with an IEA that does not require costly monitoring. It is only in this case that there are substantive changes in the results we have obtained thus far. On the other hand, when a model of costless monitoring and a non-participation sanction predicts less than full participation, all of our qualitative results about the impacts of costly monitoring on the levels of IEA participation, environmental quality, and aggregate welfare continue to hold.

We start by revising the welfare functions in (2) to include the non-participation sanction g. For $s \geq 1$, the new welfare functions when compliance with an IEA does not require costly monitoring are now:

$$w^p(s) = A + bs - c;$$
$$w^{np}(s) = A + bs - g. \tag{15}$$

As before, $w^p(0) = w^{np}(0) = A$. Note that

$$w^p(s) - w^{np}(s-1) = \begin{cases} b + g - c, & \text{for } s > 1 \\ b - c < 0, & \text{for } s = 1. \end{cases}$$

Suppose for now that $b + g - c \geq 0$. Then $w^p(s) \geq w^{np}(s-1)$ for all $s > 1$, indicating that if more than one country joins an IEA, then all other countries will join as well. By assumption, no country will unilaterally abate its emissions because $b - c < 0$; however, since $Nb > c$, each country is better off if they all form an IEA to abate their emissions than if none do. Therefore, when the reputation loss from not participating with an IEA is large enough so that $b + g - c \geq 0$, a self-enforcing IEA that does not require costly monitoring for compliance will involve all countries.

Suppose now that the non-participation loss is not so large; that is, suppose that $b+g-c < 0$. As with the model of Sect. 2, since $b < c$, $Nb > c$, and $w^p(s)$ is increasing in s, there exist coalition sizes that are strictly greater than one and weakly less that N that are profitable. Furthermore, using (15) and noting that $w^{np}(0) = A$, the smallest of these profitable

[13] We thank an anonymous reviewer for suggesting that we investigate the impacts on our model of non-participation and noncompliance reputation sanctions. We should note that these sanctions need not be reputation losses, but could take on other forms. For example, the Montreal Protocol deters non-participation by imposing trade restrictions on ozone-depleting chemicals between parties and non-parties. These same trade restrictions can also be used to punish noncompliant parties (see Barrett 1998b, p. 36).

coalitions is $s_{\min}^{nc} = \min\{s|w^p(s) \geq w^{np}(0)\} = \min\{s|s \geq c/b\}$, which is identical to the minimum sized profitable coalition in the model of Sect. 2 (see Eq. (3)). Moreover, this is the only internally and externally stable coalition. Hence, when $b + g - c < 0$, Proposition 1 continues to hold so that the equilibrium number of members of an IEA that does not require costly monitoring is $s^{nc} = \min\{s|s \geq c/b\}$. The only modification to Proposition 1 needed when non-participation produces a reputation loss is to state that $s^{nc} = N$ when $b+g-c \geq 0$.

Let us now determine how reputation sanctions affect the self-enforcing number of members of an IEA that requires costly compliance monitoring. As detailed in Sect. 3, the game entails four stages that proceed in exactly the same way, except that now if a member is determined to be noncompliant, the monitor makes this information public and the violator suffers a reputation loss, the value of which is f. Despite this re-interpretation of f, it enters the model exactly as in the model of Sect. 3. Therefore, a party of an IEA complies with its terms if and only if $\pi f(1-\rho^1-\rho^2) \geq c-b$ (Eq. (7)), and provided that $f(1-\rho^1-\rho^2) \geq (c-b)$, each member country must contribute $x = (c-b)/(\alpha f(1-\rho^1-\rho^2))$ to the monitor to maintain compliance with the agreement (Eq. (8)). Moreover, the minimum size profitable coalition is the same $s_{\min}^c$ that is specified in Eq. (9), and profitable coalitions exist if $N \geq s_{\min}^c$.

The main effect of modeling reputation sanctions for both violations and non-participation is that the non-participation sanction may be large enough to produce full compliance with an IEA, just as in the model without costly compliance monitoring. From Sect. 3, a party's expected welfare from complying with an IEA with s members that requires costly monitoring is $A + bs - c - \pi\rho^1 f - x$. If instead the country decides not to be a member of the IEA it suffers the reputation loss of g and earns $A + b(s-1) - g$ (provided that $s > 1$). The IEA obtains full participation if $A + bs - c - \pi\rho^1 f - x \geq A + b(s-1) - g$, which reduces to $b + g - c - (\pi\rho^1 f + x) \geq 0$. Recall that without the need for costly monitoring, full participation is realized when $b + g - c \geq 0$. Since $\pi\rho^1 f + x > 0$, if the non-participation reputation loss is large enough to promote full participation with costly compliance monitoring, then it is large enough to promote full participation when costly monitoring is not needed. Therefore, when $b + g - c - (\pi\rho^1 f + x) \geq 0$, the self-enforcing IEA involves all countries whether compliance monitoring is costly or not.

However, in all cases in which $(b + g) - c - (\pi\rho^1 f + x) < 0$, the non-participation loss is not sufficient to induce full compliance with an IEA. Under these circumstances, it is straightforward to demonstrate that Propositions 2 and 3 of Sect. 4 are unchanged. That is, costly monitoring limits the circumstance under which an effective IEA forms, but when one does form its membership is s^c, as specified in Eq. (10). However, because the non-participation reputation loss can produce full participation with an IEA, we must modify Proposition 4 when this sanction is present. There are two circumstances under which an effective IEA that requires costly monitoring does not involve higher participation than if it did not require costly monitoring. One is the case in which the non-participation reputation loss is so large that an IEA involves full participation whether compliance monitoring is costly or not. Recall that this is the case when $b + g - c - (\pi\rho^1 f + x) \geq 0$. The other case is when $b + g - c - (\pi\rho^1 f + x) < 0$, but $b + g - c < 0$. In this case, an IEA would involve full participation if it did not require costly monitoring, but there would be less than full participation if it does require costly compliance monitoring. When both noncompliance and non-participation incur reputation losses, costly compliance monitoring of an effective IEA produces greater participation only when the non-participation reputation loss is low enough so that $b + g - c < 0$.

The fact that a loss for not participating with an IEA can produce full participation also causes us to modify our conclusions in Sect. 5 about the welfare effects of costly monitoring. There we noted that because costly compliance monitoring can increase the number of participants and environmental quality, it can also produce higher aggregate welfare as well because the increase in environmental quality offsets the costs of monitoring under some circumstances. With both noncompliance and non-participation sanctions, this result continues to hold as long as participation is less than complete when costly monitoring is not needed. However, when a non-participation sanction is large enough to induce full compliance when monitoring is not needed, participation and environmental quality cannot be greater when costly monitoring is required. In this case costly monitoring unequivocally reduces aggregate welfare because the costs of monitoring an IEA will never be offset by an increase in environmental quality.

7 Concluding Remarks

We have analyzed games of self-enforcing IEAs when parties to such agreements finance an independent institution to monitor compliance with the IEA. We have shown that costly monitoring limits the circumstances under which international cooperation to control a transboundary pollutant will increase aggregate welfare. However, when an effective IEA does form, participation with the agreement and environmental quality can be greater than when an IEA does not require costly monitoring. In fact, under some circumstances, costly monitoring of an IEA is associated with higher aggregate welfare as well -not for the parties to an IEA, however, but for the countries that choose to free ride on the agreement.

Most of our results are robust to the problem of inaccurate monitoring and to different types of sanctions. We demonstrate that monitoring inaccuracy can further reduce the set of international environmental problems that can be addressed with an environmental agreement, but when an agreement is worthwhile, monitoring inaccuracy can produce higher participation and environmental quality. We also examined two types of sanctions—one that is imposed on noncompliant parties by an enforcement body, while the other involves reputation losses that are incurred by both noncompliance with an agreement and by not participating with the agreement in the first place. The qualitative results of both specifications are the same, except when the non-participation loss is so large that it induces full participation with an agreement.

Our results have important implications for multilateral management of environmental externalities. Assuming away monitoring and enforcement problems related to IEAs artificially enlarges the scope of mutually beneficial agreements between countries, and may artificially reduce the number of participating countries that are necessary to make these agreements worthwhile. In other words, for some international environmental problems, monitoring costs either require more members for an IEA to come into effect or they render mutual cooperation inefficient.

The models developed in this paper can and should be extended to include additional features. An obvious extension would be to recognize the real-world heterogeneity of countries involved in international environmental relations. Other features of international cooperation such as side payments, issue linkage, and minimum participation requirements that others have addressed, can and should be examined in the context of costly monitoring of these agreements. Certainly there are many extensions that should be addressed to gain a more complete picture of how costly monitoring affects voluntary coalition formation to protect the international environment.

Acknowledgements The authors acknowledge funding for this project from the Center for Public Policy and Administration, University of Massachusetts, Amherst, and the Cooperative State Research Extension, Education Service, U. S. Department of Agriculture, Massachusetts Agricultural Experiment Station, and the

Department of Resource Economics under Project No. MAS00871. The authors gratefully acknowledge the helpful comments of James Murphy, John Spraggon, Erin Baker, and Tim Perri. We would also like to thank Michael Rauscher and two anonymous reviewers for their helpful comments.

References

Arguedas C (2007) To comply or not to comply? Pollution standard setting under costly monitoring and sanctioning. Forthcoming in Environ Resour Econ

Barrett S (1994) Self-Enforcing International Environmental Agreements. Oxf Econ Pap 46(1):878–894

Barrett S (1997a) The strategy of trade sanctions in international environmental agreements. Resour Energy Econ 19(1):345–361. doi:10.1016/S0928-7655(97)00016-X

Barrett S (1997b) Heterogeneous international environmental agreements. In: Carraro Carlo (ed) International Environmental Negotiations. Edward Elgar, Cheltenham, UK and Northampton, MA

Barrett S (1998a) On the theory and diplomacy of environmental treaty making. Environ Resour Econ 11(1):317–333. doi:10.1023/A:1008243528330

Barrett S (1998b) Political economy of the Kyoto protocol. Oxf Rev Econ Policy 14(4):20–39. doi:10.1093/oxrep/14.4.20

Barrett S (2003) Environment and statecraft: the strategy of environmental treaty-making. Oxford University Press, Oxford and New York

Benedick RE (1998) Ozone diplomacy: new directions in safeguarding the planet, enlarged edition. Harvard University Press, Cambridge, MA and London, England

Botteon M, Carraro C (1998) Strategies for environmental negotiations: issue linkage with heterogeneous countries. In: Hanley N, Folmer H (eds) Game Theory and the Environment. Edward Elgar, Cheltenham, UK and Northampton, MA

Carraro C, Siniscalco D (1993) Strategies for the international protection of the environment. J Public Econ 52(3):309–328. doi:10.1016/0047-2727(93)90037-T

Carraro C, Botteon M (1997) Burden sharing and coalition stability in environmental negotiations with asymmetric countries. In: Carraro C (ed) International environmental negotiations. Edward Elgar, Cheltenham, UK and Northampton, MA

Carraro C, Siniscalco D (1997) R&D Cooperation and the stability of international environmental agreements. In: Carraro C (ed) International environmental negotiations. Edward Elgar, Cheltenham, UK and Northampton, MA, pp 1–2

Carraro C, Siniscalco D (1998) International environmental agreements: incentives and political economy. Eur Econ Rev 42(3–5):561–572. doi:10.1016/S0014-2921(97)00118-9

Carraro C, Marchiori C, Oreffice S (2003) Endogenous minimum participation in international environmental treaties. Nota Di Lavaro 113.2003. Fondazione Eni Enrico Mattei Working Paper

Chandler P, Tulkens H (1995) A core-theoretic solution for the design of cooperative agreements on transfrontier pollution. Int Tax Public Finance 2(2):279–293. doi:10.1007/BF00877502

Chandler P, Tulkens H (1997) The core of an economy with multilateral environmental externalities. Int J Game Theory 26(3):379–401

Chayes A, Chayes AH (1991) Compliance without enforcement: state behavior under regulatory treaties. Negotiation J 7(1):311–330. doi:10.1111/j.1571-9979.1991.tb00625.x

D'Aspremont C, Jacquemin A, Gabszewicz JJ, Weymark JA (1983) On the stability of collusive price leadership. Can J Econ 16(1):17–25. doi:10.2307/134972

Downs GW, Rocke DM, Barsoom PM (1996) Is the good news about compliance good news about cooperation? Int Organ 50(3):379–340

Finus M (2004) International cooperation to resolve international pollution problems. Fondazione Eni Enrico Mattei Working Paper

Finus M, Rundshagen B (1998) Toward a positive theory of coalition formation and endogenous instrument choice in global pollution control. Public Choice 96(1–2):145–186. doi:10.1023/A:1005016623915

Folmer H, van Mouche P, Ragland S (1993) Interconnected games and international environmental problems. Environ Resour Econ 3(4):313–335. doi:10.1007/BF00418815

Friends of the Earth (2006) Climate treaty one year old, but emissions still rising. Press Release, February 15, 2006 http://www.foe.co.uk/resource/press_releases/climate_treaty_one_year_ol_14022006.html

Heister J, Mohr E, Stahler F, Stoll P-T, Wolfum R (1997) Strategies to enforce compliance with an international CO_2 Treaty. Int Environ Aff 9(1):22–53

Hoel M (1992) International environment conventions: the case of uniform reductions of emissions. Environ Resour Econ 2(2):141–159

Hoel M, Schneider K (1997) Incentives to participate in an international environmental treaty. Environ Resour Econ 9(2):153–170

Kaplow L, Shavell S (1994) Accuracy in the determination of liability. J Law Econ 37(1):1–16. doi:10.1086/467304

Keohane RO (1986) Reciprocity in international relations. Int Organ 40(1):1–27

Kolstad C (2007) Systematic uncertainty in self-enforcing international environmental agreements. J Environ Econ Manage 53(1):68–79. doi:10.1016/j.jeem.2006.08.001

Macho-Stadler I, Perez-Castrillo D (2006) Optimal enforcement policy and firm's emissions and compliance with environmental taxes. J Environ Econ Manage 51(1):110–131. doi:10.1016/j.jeem.2005.06.001

Oberthur S (1998) The International convention for the regulation of whaling: from over-exploitation to total prohibition. Yearbook of international co-operation on environment and development. Earthscan Publishers, London

Osmani D, Tol RSJ (2007) Toward farsightedly stable international environmental agreements: part one. Hamburg University Working Papers, FNU-140

Polinsky AM, Shavell S (2000) The economic theory of public enforcement of law. J Econ Lit 38(1):45–76

Simmons BA (1998) Compliance with international agreements. Annu Rev Polit Sci 1(1):75–93. doi:10.1146/annurev.polisci.1.1.75

Stranlund JK (2007) The regulatory choice of noncompliance in emissions trading programs. Environ Resour Econ 38(1):99–117. doi:10.1007/s10640-006-9058-3

Swanson T, Johnston S (1999) Global environmental problems and international environmental agreements: the economics of international institution building. Edward Elgar, Cheltenham, UK and Northampton, MA

Ulph A (2004) Stable international environmental agreements with a stock pollutant, uncertainty and learning. J Risk Uncertain 29(1):53–73. doi:10.1023/B:RISK.0000031445.13939.e4

United Nations Framework Convention on Climate Change (2002) Report on the conference of the parties: VII session. Part Two: actions taken by the conference of the parties. Marrakesh, January 2002

Wagner UJ (2001) The design of stable international environmental agreements: economic theory and political economy. J Econ Surv 15(3):377–411. doi:10.1111/1467-6419.00143

[16]

ELSEVIER

Available online at www.sciencedirect.com

ScienceDirect

Journal of Environmental Economics and Management 55 (2008) 37–51

JOURNAL OF ENVIRONMENTAL ECONOMICS AND MANAGEMENT

www.elsevier.com/locate/jeem

Cooperation in the commons with unobservable actions

Nori Tarui[a], Charles F. Mason[b], Stephen Polasky[c,*], Greg Ellis[d]

[a] *Department of Economics, University of Hawaii, USA*
[b] *Department of Economics and Finance, University of Wyoming, USA*
[c] *Department of Applied Economics, University of Minnesota, 1994 Buford Avenue, Saint Paul, MN 55108 USA*
[d] *Department of Economics, University of Washington, USA*

Received 2 February 2006
Available online 14 September 2007

Abstract

We model a dynamic common property resource game with unobservable actions and non-linear stock-dependent costs. We propose a strategy profile that generates a worst perfect equilibrium in the punishment phase, thereby supporting cooperation under the widest set of conditions. We show under what set of parameter values for the discount rate, resource growth rate, harvest price, and the number of resource users, this strategy supports cooperation in the commons as a subgame perfect equilibrium. The strategy profile that we propose, which involves harsh punishment after a defection followed by forgiveness, is consistent with human behavior observed in experiments and common property resource case studies.

Keywords: Common property resources; Renewable resources; Cooperation; Dynamic games; Worst perfect equilibrium; Subgame perfection; Unobservable actions

1. Introduction

Since Gordon [19], economists have known that individuals have incentives for excessive exploitation of common property resources. These incentives for excessive exploitation lead to the "tragedy of the commons" [23]. There are two potential externalities that lead to misguided incentives. Increased efforts by harvesters can impose crowding costs on other harvesters; this is at its core a static externality [5]. This situation is quite similar to that facing individuals contributing to a public good. Indeed, one can think of the decision to reduce harvesting as a sort of contribution to the public good of "less crowding." There may also be a dynamic externality: larger current harvests reduce future resource stocks, which in turn may increase future harvest costs or constrain future harvest levels [30]. With either static or dynamic externalities, privately optimal behavior leads to socially excessive harvesting of a common property resource, yielding smaller welfare flows to society.

*Corresponding author. Fax: +1 612 625 2729
E-mail address: polasky@umn.edu (S. Polasky).

0095-0696/$ - see front matter © 2007 Elsevier Inc. All rights reserved.
doi:10.1016/j.jeem.2007.08.001

Despite the grim predictions of tragedy, a variety of researchers have found that many actual common property harvesting regimes manage common property resources in a reasonably efficient manner, often for long periods of time, even with many agents involved (for reviews see [14, 31, 32]). The typical successful management regime has some means of limiting access to the commons and some means of punishment for over-harvesting. Access may be restricted to members of a particular community or group. Community members are responsible for monitoring and enforcement. Punishment can involve some type of loss of privilege, either temporary or permanent, or, for major offenses, banishment from the group.

A large number of experimental studies have investigated the common property problem, as well as the closely related public goods provision problem [27]. While many studies have found tendencies towards privately optimal behavior, several studies point to the possibility of more socially efficient outcomes. In one class of studies, subjects are able to identify specific players as defectors. Several papers show that when defectors can be singled out for punishment more socially efficient outcomes result [6, 15, 34]. However, the observation of increased social efficiency does not rely solely on the ability to punish specific individuals who defect. Mason and Philips [29] study a game in which punishments cannot be tailored to the individual, but they still find that more socially efficient outcomes obtain. In a second class of studies, socially attractive outcomes occur even though subjects only know aggregate behavior [7, 9, 21, 25, 33]. In the majority of these studies, more cooperative outcomes are enhanced by communication. While several of these papers place subjects in a finitely repeated game, some do not [29, 34, 38]. These various empirical observations underscore the importance of understanding, from a conceptual point of view, how cooperative arrangements might take shape.

There has long been interest in examining the theoretical underpinnings of cooperative behavior. A rich literature in applied game theory has developed over the last two decades, evaluating the conditions under which equilibrium supports socially desirable outcomes in the presence of unilateral short-term incentives to deviate [4, 8, 11–13, 22, 26, 35, 37]. The basic thrust of this literature has been to show that cooperation in the commons can be supported as a subgame perfect equilibrium, under certain conditions, through the use of strategies that include credible future punishment for deviations from the cooperative outcome. Dutta [11] proves that cooperation can be sustained for sufficiently patient players, an extension of the folk theorem result for repeated games. Several papers analyze dynamic resource games for arbitrary discount factors where payoffs are independent of stock [4, 8, 13, 37].

While assuming stock independence is analytically convenient, this approach eliminates dynamic cost externalities important for a wide range of real-world common property resources[1]. It also makes extraction to zero stock (extinction) profitable, which allows rapid extraction to extinction to be a credible punishment. Our analysis, like Hannesson [22], Laukkanen [26], and Polasky et al. [35], explicitly considers the role of non-linear stock-dependent costs. We follow Clark [10], and others, in using a discrete-time model in which the marginal cost of harvest declines with increasing stock. When marginal cost declines in stock, there can be a strictly positive stock level that generates zero profits and below which further harvest generates negative marginal profit. In this case, credible punishment strategies are more complex, typically involving an on-going path of harvesting rather than rapid extinction.

Another issue we address is imperfect monitoring of harvests. Punishing individual defectors is possible only if the resource users can identify who cheated. Polasky et al. [35] analyze a game with costless monitoring and individual punishment schemes. However, monitoring is often prohibitively costly in the context of natural resource use. Ostrom [31] suggests that, though monitoring is an important factor behind cooperative use of commons, cooperation is observed in some settings with imperfect monitoring, a finding in the experimental literature as well. In this paper, we assume that the resource stock level is observable but individual players' actions are not observable. When the resource stock falls below the cooperative level, what can be inferred is that *someone* cheated, but not *who* cheated. Therefore, punishment must be symmetric, targeting the whole group, rather than asymmetric, targeting a particular individual who cheated. Abreu et al. [3] analyzed cooperation under imperfect monitoring in repeated games. This paper addresses cooperation under imperfect monitoring in a dynamic game where the stage game evolves endogenously given how the resource was used in previous periods.

[1] Dutta [12] assumes that payoffs are linear in stock.

Two prior papers, Hannesson [22] and Laukkanen [26], have addressed the issue of supporting cooperation in a common property resource with imperfect monitoring and non-linear stock-dependent costs. Laukkanen [26] analyzed cooperation in a two-player game in which fish stocks migrate between exclusive harvest zones so that players move sequentially. Cooperative solutions are supported by the threat to revert to non-cooperative Nash equilibrium if cheating is suspected. Hannesson [22] uses a model similar to ours in that N players move simultaneously in a discrete-time common property harvest game. Hannesson uses a simple punishment strategy in which players draw down the stock to the zero profit level of stock, where price equals marginal cost of harvest, in every period following a defection.

In contrast, we use a two-part punishment scheme, with heavy penalties in the first phase followed by eventual recovery of the stock to the cooperative level in the second phase. With a two-part punishment scheme, using the worst perfect equilibrium (i.e., a subgame perfect equilibrium with the lowest possible payoffs) following a defection offers the strongest possible incentive not to defect in the first place. In general, cooperation can be supported under a wider set of conditions by reversion to a worst perfect equilibrium rather than reversion to Nash equilibrium [1, 2, 35]. Moreover, with a two-part punishment scheme it need not be the case that cooperation is harder to support as the group size increases. We present a numerical simulation in Section 5 that suggests cooperation could be easier to support with four players than with three. Our conjecture is that increases in group size have two conflicting effects. On the one hand, increases in group size reduce the share of the cooperative pie that any one player is allocated; all else equal this would tend to make cooperation harder to support. On the other hand, the set of potential punishers who must respond to an individual player that has cheated is larger when the group is larger. All else equal, when the number of agents is larger, it is less likely that a single agent can obtain positive profits from deviating during the first phase of the punishment. As such, larger punishments become credible, which tends to make cooperation easier to support. The numerical results we present below show that the combined impact of these two conflicting effects will be to make cooperation easier to support when group size is increased starting from relatively small values—at least in the context of our numerical example. Ultimately, however, the former effect dominates, and cooperation becomes more difficult to support as group size rises. These qualitative results are consistent with the experimental finding in Mason and Philips [29], who found more cooperative behavior with intermediate group sizes than with slightly smaller or slightly larger groups.

In the next section, we describe the dynamic common property resource game with unobservable actions and non-linear stock-dependent costs. We characterize the cooperative outcome in Section 3. In Section 4, we analyze strategies designed to support cooperation as a subgame perfect equilibrium. We propose a two-part punishment scheme and demonstrate that, this two-part scheme can constitute a worst perfect equilibrium once a player has cheated, which then supports cooperation as a subgame perfect equilibrium under the widest possible set of circumstances. Section 5 contains a numerical illustration. Section 6 contains concluding comments.

2. The game

We consider a discrete-time dynamic game with $N \geqslant 3$ players, indexed by i, who jointly harvest a common property renewable resource.[2] In every period t ($= 0, 1, \dots$), each player i chooses a harvest level, $h_{it} \geqslant 0$. We assume that players choose harvest strategies in period t simultaneously. Define total harvest in period t as $h_t = \Sigma_{i=1}^{N} h_{it}$. Let s_t equal the amount of stock at the beginning of period t and z_t equal the stock at the end of period t (escapement), with $z_t = s_t - h_t$. Between the end of period t and the start of period $t+1$, stock grows according to a biological growth function: $s_{t+1} = g(z_t)$. We assume that the biological growth function, g(.), is a positive and strictly concave function for stock sizes between 0 and a positive carrying capacity (K): $g(0) = 0, g(K) = K,$ and $s < g(s), g''(s) < 0$ for all $s \in (0, K)$. The initial stock $s_0 \in (0, K]$ is given.

We assume that players observe stock at the beginning (s_t) and end (z_t) of each period so that they know aggregate harvest (h_t), but that there is imperfect monitoring of harvests: player i cannot observe h_{jt} for $j \neq i$. Each player's action in period t can depend on the initial level of stock in period t (s_t) and the history of

[2] Note that with $N = 2$, the other player's harvest can be inferred from knowledge of one's own harvest and total harvest.

harvests through period $t-1$. Player i's strategy in period t specifies i's action conditional on the information the player has in period t. We consider only pure strategies within a given period.

We assume that unit harvest cost depends upon stock size, $c(s)$. The unit harvest cost function $c(s)$ is twice continuously differentiable, with $c(s)>0$, $c'(s)<0$, $c''(s)>0$. We also assume that unit costs rise sufficiently rapidly as stock gets small so that $\int_0^s c(\omega)\,dw=\infty$ for any positive stock. Let $p>0$ be the exogenous price of harvest. For this to be an economically interesting problem, it must be worth harvesting at some stock level, and so we assume that $p>c(K)$. Define the zero profit level of stock as $\underline{s}$: $p-c(\underline{s})=0$.

In any period, we assume that player i receives a share of total returns equal to i's share of total harvest. Thus, player i's return in period t is given by

$$r_i(h_{1t},\ldots,h_{Nt},s_t)=\begin{cases}\frac{h_{it}}{h_t}\int_{s_t-h_t}^{s_t}[p-c(\omega)]\,d\omega, & \text{for } h_t\leqslant s_t\\ -\infty, & \text{for } h_t>s_t.\end{cases},$$

In essence, the assumption of equal sharing of returns is equivalent to assuming that the share of harvest for each player is constant throughout the time period. Alternatively, one could assume that all players harvest at the same rate during the time that they are active but that players harvesting more stock during the time period harvest for a longer amount of time. This alternative approach would weight harvest costs more heavily toward players that harvest greater amounts, i.e., that players with smaller shares earn higher profit per unit of harvest.

Each player's payoff for the entire game is given by the discounted sum of the period returns. We assume there is a common discount factor δ, $0<\delta<1$, by which payoffs are discounted from one period to the next.

3. Characterizing cooperative outcome

The joint-rent maximization problem is given by

$$\max_{\{h_t\}_{t=0}^{\infty}}\sum_{t=0}^{\infty}\delta^t\int_{s_t-h_t}^{s_t}[p-c(\omega)]\,d\omega$$

$$\text{s.t.}\quad s_{t+1}=g(s_t-h_t),\quad h_t\geqslant 0,\quad s_t\geqslant 0,\quad t=0,1,\ldots\text{given}\quad s_0\in(0,K].$$

We can rewrite this problem as a dynamic programming problem focusing on escapement rather than harvest as the choice variable. The Bellman equation for this formulation is

$$V(s_t)=\max_{0\leqslant z_t\leqslant s_t}\int_{z_t}^{s_t}[p-c(\omega)]\,d\omega+\delta V(g(z_t)),$$

An optimal solution to the Bellman equation, which characterizes optimal escapement (z^*) in every period with positive harvest, must satisfy

$$-[p-c(z^*)]+\delta g'(z^*)V'(g(z^*))=0,$$

where $V'(g(z^*))=[p-c(g(z^*))]$. Such z^* exists under the assumptions on c and g. We maintain the following assumption throughout this paper.

Assumption 1. $(\partial/\partial s)(-(p-c(s))+\delta g'(s)[p-c(g(s))])<0$ for all s.

Under Assumption 1 there is a unique optimal escapement. The optimal harvest policy, h^*, is

$$h^*(s)=\begin{cases}s-z^* & \text{if}\quad s-z^*\geqslant 0,\\ 0 & \text{otherwise.}\end{cases}$$

For cases where $s_0 \geqslant z^*$, the value function is

$$\bar{V}(s_0) = \int_{z^*}^{s_0} [p - c(\omega)]\, d\omega + \delta \bar{V}(g(z^*))$$
$$= \int_{z^*}^{s_0} [p - c(\omega)]\, d\omega + \frac{\delta}{1-\delta} \int_{z^*}^{g(z^*)} [p - c(\omega)]\, d\omega = w(s_0) + \frac{\delta}{1-\delta} w(g(z^*)),$$

where $w(s) \equiv \int_{z^*}^{s} [p - c(\omega)]\, d\omega$ for $s \geqslant z^*$. When $s_0 < z^*$, the optimal solution requires no harvest until stock has increased above z^*. Define $T(s_0)$ by

$$T(s_0) \equiv \min\{t | s_t \geqslant z^*,\ h_l = 0 \quad \text{for all } l \geqslant 0,\ s_l = g(s_{l-1}) \quad \text{for all } l \geqslant 1\},$$

i.e. the minimum number of iterations of the growth function $g(.)$ with $s_t = g(g(\ldots g(s_0)\ldots)) \geqslant z^*$. Then the value function is given by

$$V(s_0) = \begin{cases} \bar{V}(s_0) & \text{if} \quad s_0 \geqslant z^*, \\ \delta^{T(s_0)} \bar{V}(s_{T(s_0)}) & \text{if} \quad s_0 < z^*, \end{cases}$$

where $s_{T(s_0)}$ is the stock size after $T(s_0)$ periods of growth starting from stock size s_0.

We assume that the cooperative solution among N players involves an equal division of harvest among the players: $h_i^*(s_0) \equiv h^*(s_0)/N$, for all i. Let $v = V/N$ be the individual payoff under cooperation.

4. Characterizing a subgame perfect strategy profile

4.1. A two-part punishment scheme to support cooperation

Here we propose a history-dependent strategy profile σ^* that supports a cooperative outcome and uses a two-part punishment scheme that generates a worst perfect equilibrium once a player has defected from the cooperative outcome. Reversion to a worst perfect equilibrium when a player cheats yields the greatest possible punishment, which in turn gives the greatest incentive not to cheat. Note that a player can always set $h_{it} = 0$ for all t yielding a payoff of 0. An equilibrium punishment strategy cannot force a player to earn present value negative payoffs, though losses in some periods are possible if offset by positive payoffs in other periods.

4.1.1. Strategy profile σ^*

Phase I: In period 0, each player i plays $h_i^*(s_0) \equiv h^*(s_0)/N$. If $z_0 < s_0 - h^*(s_0)$, with resulting stock $s_1 = g(z_0)$, then move to phase II(s_t) in period 1. In period $t = 1,2,\ldots$, each player i plays $h_i^*(s_t) \equiv h^*(s_t)/N$ as long as $z_{t-1} \geqslant s_{t-1} - h^*(s_{t-1})$ (that is, as long as the stock is equal to or above the level induced by h^*). If $z_{t-1} < s_{t-1} - h^*(s_{t-1})$, with resulting stock $s_t = g(z_{t-1})$, then move to phase II(s_t) in period t.

Phase II(s_t): Each player i plays action $h_i^{\mathrm{P}}(s_t)$ defined below, resulting in escapement of $z_t^{\mathrm{P}}(s_t)$. If escapement in period t equals $z_t^{\mathrm{P}}(s_t)$, then move back to phase I. If escapement in period t does not equal $z_t^{\mathrm{P}}(s_t)$, with resulting stock s_{t+1}, then begin phase II(s_{t+1}) in the following period $t+1$.

Once a player over-harvests the resource, all players conduct phase-II actions where everyone over-extracts the resource to a point where all players incur one-period negative profits. Each player has an incentive to absorb losses in this period because then the resource stock is allowed to recover, players resume cooperation, and each player earns positive returns. Not carrying out the punishment means that phase II will start again in the following period. These two parts of punishment (severe punishment, recovery) can be combined so that each player earns present value returns of zero (the worst perfect equilibrium payoff) evaluated from the beginning of the punishment phase.

We define the punishment strategy starting in period τ to be consistent with the worst perfect equilibrium starting from the punishment phase. The lowest present value payoff (minmax value) for this game is 0. Define $h_i^P(s_\tau) \equiv h^P(s_\tau)/N$ such that a player will earn zero present value profit by playing this action in period τ followed by playing $h_i^*(s)$ thereafter, assuming that the other players play according to the equilibrium strategy:

$$\frac{1}{N}\int_{s_\tau - h^P(s_\tau)}^{s_\tau} [p - c(\omega)]\,d\omega + \delta v\big(g(z_\tau^P(s_\tau))\big) = 0, \tag{1}$$

where $z_\tau^P(s_\tau) = s_\tau - h^P(s_\tau)$, and $v\big(g(z_\tau^P(s_\tau))\big)$ is a player's present value of profit from the cooperative solution starting with $g(z_\tau^P(s_\tau))$ amount of stock. Under the assumptions on the cost function, c, there will exist a phase-II harvest $h^P(s_\tau)$ that satisfies (1) for any stock level s_τ. The first term in Eq. (1) is equal to

$$\frac{1}{N}\int_{s_\tau - [(N-1)/N]h^P(s_\tau)}^{s_\tau} [p - c(\omega)]\,d\omega + \frac{1}{N}\int_{s_\tau - h^P(s_\tau)}^{s_\tau - [(N-1)/N]h^P(s_\tau)} [p - c(\omega)]\,d\omega \tag{2}$$

This expression must be negative for Eq. (1) to be satisfied.

We find conditions under which no player is better off by any one-shot deviation from phases I and II at any stock level. Because each player's period-wise return is bounded from above and the discount rate is positive, the principle of optimality for discounted dynamic programming applies. Therefore, in order to prove that σ^* is subgame perfect, it is sufficient to show that any one-shot deviation cannot be payoff-improving for any player [18]. Because this is a dynamic game, we need to verify that no player has an incentive to deviate from the prescribed strategy in any phase and under any possible stock level.[3]

Given stock s, each player who follows cooperation in phase I earns $v(s)$. If player i deviates from $h_i^*(s)$, then the maximum the player can obtain is (given zero continuation payoff upon detection of cheating)

$$\pi(s) \equiv \max_{h_i \geqslant 0} r_i(h_i, h_{-i}^*(s), s) = \max_{h_i \geqslant 0} \frac{h_i}{\sum_{j\neq i} h_j^*(s) + h_i} \int_{s - \sum_{j\neq i} h_j^*(s) - h_i}^{s} [p - c(\omega)]\,d\omega,$$

where $h_{-i}^*(s) = (h_1^*(s), \ldots, h_{i-1}^*(s), h_{i+1}^*(s), \ldots, h_N^*(s))$.

We now show that σ^* is a subgame perfect equilibrium under certain conditions.

Assumption 2. $v(K) \geqslant \pi(K)$, where K is the carrying capacity of the renewable resource.

Assumption 2 states that the present value payoffs from following cooperative strategy must be greater than or equal to the maximum one period deviation payoffs when stock at the beginning of the period equals the carrying capacity, K. In what follows we show that no player can gain by deviating in phase I given any current stock level, when Assumption 2 holds. Then we show that no player can gain by deviating in phase II given any stock level.

4.2. Sustaining cooperation in phase I

In this section we show that, under Assumption 2, no player can gain by deviating in phase I at any initial stock level, $s \in (0, K]$. Beginning with stock s, each player earns $v(s)$ by playing σ^*. If player i deviates in phase I, then the player earns at most $\pi(s)$ in the present period. Starting in phase II, player i earns present value payoffs of 0. Hence, player i has no incentive to deviate in phase I if the following condition holds:

$$v(s) \geqslant \pi(s) \quad \text{for all} \quad s. \tag{3}$$

The following lemmas will show that $v(K) \geqslant \pi(K)$ is sufficient for the above inequality to hold.

[3] See Dutta [11, 12] and Polasky et al. [35] for a similar analysis in a dynamic game context.

Let $h^{\mathrm{D}}(s)$ be the harvest that maximizes the payoff upon deviation when stock size is s where the other players choose $h^*_{-i}(s)$. Lemma 1 characterizes $h^{\mathrm{D}}(s)$.

Lemma 1. The optimal deviation $h^{\mathrm{D}}(s)$ satisfies

$$h^{\mathrm{D}}(s)\begin{cases} = 0 & \textit{if} \quad s \in (0, \underline{s}], \\ = s - \sum_{j \neq i} h_j^*(s) - \underline{s} & \textit{if} \quad s \in (\underline{s}, z^*], \\ > s - \sum_{j \neq i} h_j^*(s) - \underline{s} & \textit{if} \quad s \in (z^*, K]. \end{cases}$$

See Appendix A for the proof. Lemma 1 states that the optimal deviation in period t given $s > \underline{s}$ is to extract the resource to a point at or below $\underline{s}$. The intuition for the result is that at $\underline{s}$ the unit rent is zero. With further extraction, the total return decreases but the player's share of total returns increases. If other players are harvesting, it is therefore optimal for a deviator to harvest so that the resulting resource stock is less than $\underline{s}$. Lemma 1 implies the following about the slopes of π and v.

Lemma 2. $\pi'(s) > v'(s)$ for all $s \in [z^*, K)$.

See Appendix A for the proof. Lemma 2 implies that π decreases faster than v as stock decreases from K to some level $s \geqslant z^*$. This implies that $v(s) \geqslant \pi(s)$ for all $s \in [z^*, K)$ if $v(K) \geqslant \pi(K)$.

For $v(K) \geqslant \pi(K)$ to imply inequality (3), it remains to show that $v(s) \geqslant \pi(s)$ for all $s \in (\underline{s}, z^*)$ whenever $v(z^*) \geqslant \pi(z^*)$. Lemma 3 addresses this last point.

Lemma 3. $v''(s) - \pi''(s) < 0$ holds for any $s \in (\underline{s}, z^*)$.

See Appendix A for the proof.

Lemma 4. If $v(K) \geq \pi(K)$, then $v(s) \geq \pi(s)$ for all $s \in [s, K)$.

Proof. It follows from Lemma 3, $v'(\underline{s}) > \pi'(\underline{s})$ and $v'(z^*) < \pi'(z^*)$ that there exists unique $\hat{s} \in (\underline{s}, z^*)$ such that

(i) $\pi'(\hat{s}) = v'(\hat{s})$,
(ii) $\pi'(s) > v'(s)$ if $\hat{s} < s \leqslant z^*$, and
(iii) $\pi'(s) < v'(s)$ if $\underline{s} \leqslant s < \hat{s}$.

Together with the fact that $v(\underline{s}) > \pi(\underline{s})$, the assumption that $v(K) \geqslant \pi(K)$, and Lemma 2, we conclude that $v\ (s) = \pi\ (s)$ for all $s \leqslant K$. □

Since, $v(s) \geqslant \pi(s)$for all s, no player would find it profitable to deviate from phase I.

Fig. 1 describes an example where $v(K) \geqslant \pi(K)$ holds. The figure is based on a unit cost function $c(s) = 1/s$ and a logistic growth function $g(s) = s + rs(1 - s/K)$ where $r < 1$ is the intrinsic growth rate of the resource. As in Lemma 4, $v(K) \geqslant \pi(K)$ implies $v(s) \geqslant \pi(s)$ for all $0 < s \leqslant K$.

4.3. Supporting cooperation in phase II

Now we show that players cannot gain by deviating in phase II(s) for any post-deviation stock level s. In order to do so we need one additional condition to be satisfied.

Assumption 3. $P(s_\tau) \equiv \int_{s_\tau - [(N-1)/N]h^{\mathrm{P}}(s_\tau)}^{s_\tau} [p - c(\omega)]\, d\omega \leqslant 0$ for all $s_\tau > \underline{s}$.

The expression $P(S_\tau)$ is the combined profit to all players when all players but one play the punishment strategy and the other player harvests nothing. This expression is the first term in expression (2) given above. When Assumption 3 holds, a player cannot deviate from the punishment strategy and earn positive profits during the harsh punishment phase.

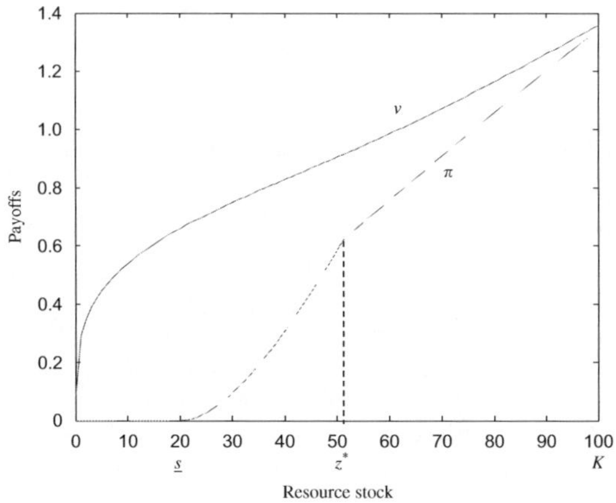

Fig. 1. Payoffs upon cooperation (v) and deviation (π). This figure is based on an example with $r = 0.5$, $K = 100$, $\delta = 0.9$, $p = 0.05$, and $N = 4$. In this case, we have $\underline{s} = 20$ and $z^* = 51.3$. Function v is the payoff upon cooperation in phase I, and function π represents the maximum payoff upon deviation from phase I.

Lemma 5. Under Assumption 3, no player has an incentive to deviate in phase II(s) for all s.

Proof. By following the equilibrium strategy in phase II(s), a player earns 0 present value returns. In the first period of phase II(s), the maximum payoff player i can earn upon deviation is given by

$$\max_{h_i \geqslant 0} \frac{h_i}{h_i + \sum_{j \neq i} h_j^{\mathrm{P}}(s)} \int_{s-h_i-\sum_{j \neq i} h_j^{\mathrm{P}}(s)}^{s} [p - c(\omega)]\, d\omega$$

Under Assumption 3, $\int_{s-\sum_{j\neq i} h_j^{\mathrm{P}}(s)}^{s} [p - c(\omega)]\, d\omega \leqslant 0$ for all $s > \underline{s}$ (and is certainly negative for $s < \underline{s}$). Therefore, there is no positive h_i that can generate a positive return when other players play according to the punishment strategy: $h_j = h_j^{\mathrm{P}}$ for $j \neq i$. A deviation against the punishment triggers the start of a new punishment phase in which all players earn present value profits of zero. Therefore, there is no profitable deviation in the severe punishment phase (the period following a defection from the cooperative strategy). If there is no deviation in the first period of phase II(s), players then revert back to phase I, for which we have already demonstrated that no profitable deviations exist. Hence, under Assumption 3 ($P(s) \leq 0$ for all $s > \underline{s}$), there are no profitable unilateral deviations for any level of stock. □

Assumption 3 is likely to hold when N is not too small or growth of the stock is not too large. Because the optimal deviation drives escapement below s, as long as stock growth is not too fast, stock may be either lower than s or else not far above. With large N, any one player's share of harvest is small making it unlikely that even if this player harvests nothing that escapement will be above s.

4.4. Summary

Lemmas 1–5 imply that there is no profitable deviation against strategy s^* in either phase I or II. We summarize the previous analysis in the following proposition.

Proposition 1. *Under Assumptions 1, 2, and 3, σ^* is a subgame perfect equilibrium and supports a cooperative (first best) outcome.*

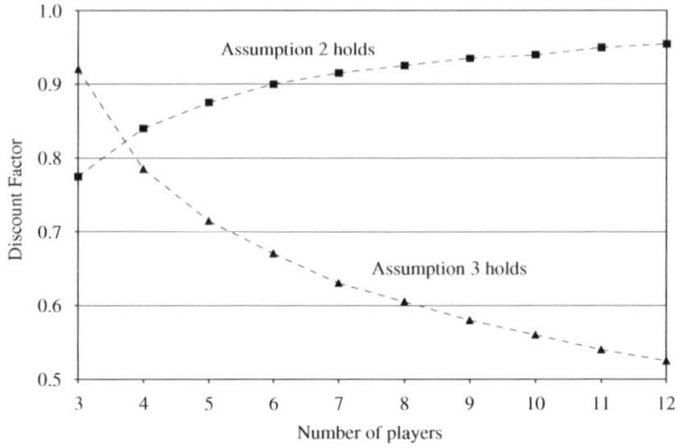

Fig. 2. Tradeoff between Assumptions 2 and 3. This figure is based on an example with $r = 0.5$, $K = 100$, and $p = 0.05$. Assumption 2 holds in the area above the curve with square markers. Assumption 3 holds in the area above the curve with triangle markers.

5. Numerical illustration

We use a numerical example to illustrate the results of the previous section. We assume a unit cost function $c(s) = 1/s$ and a logistic growth function $g(s) = s + rs(1 - s/K)$ where $r < 1$. We describe when the sufficient conditions for Proposition 1 hold.

Fig. 2 shows the combinations for the number of players, N, and the discount factor, δ, for which Assumptions 2 and 3 hold. Assumption 2 states that the present value of the cooperative strategy exceeds the value of deviation for a player when the stock level begins at the unexploited carrying capacity level of stock, K. In the figure, any combination of N and δ above (below) the line with square markers satisfies (does not satisfy) Assumption 2. Assumption 3 states that a player cannot earn positive profits during the punishment phase ($P(s) \leqslant 0$). Any combination of N and δ above (below) the line with triangle markers satisfies (does not satisfy) Assumption 3. For $N \geqslant 4$, Assumption 2 is the more binding constraint (i.e., holds for a smaller range of discount factors). As shown in the figure, Assumption 2 holds for a smaller range of discount factors as N grows. In other words, it becomes more difficult to support cooperation as N grows. This is the standard finding in the literature. On the other hand, for small N, ($N = 3$ in the case shown in Fig. 2), Assumption 3 is the more binding constraint. Because Assumption 3 holds for a wider range of values of discount factors as N gets larger, it may be possible to support cooperation as a subgame perfect equilibrium outcome for a larger N even though it is not possible to do so for smaller N. This counter-intuitive result may occur because having more players makes it easier to ensure that no player can profitably deviate against the punishment strategy, which is necessary to ensure that the strategies constitute a subgame perfect equilibrium.

Fig. 3 draws the value of function $P(s)$ defined in Assumption 3 given optimal phase-I deviations. This value must be non-positive in order for the phase-II strategy to be a best response for each player. With $\delta = 0.9$, Assumption 3 holds with $N = 4$ but not with $N = 3$.

The implication is that the range of potential punishments increases when group size increases from 3 to 4. This increase in N reduces the degree to which any one individual must increase harvest to inflict the punishment, which tends to lower the cost associated with following through with the punishment. As such, larger punishments become credible, which tends to make cooperation easier to support. But increases in group size raise the potential gains from defection: as N grows, each player is accorded a smaller share of the cooperative pie, which increases the potential one-time gain from defection. Our simulation results indicate that, for the particular example we analyze, this effect becomes more important than the enhanced credibility of punishments for values of N at or above 4. On balance, we see a non-monotonic effect of group size upon cooperative tendencies. It is interesting to compare this result to the findings in Mason and Philips [29]. In experiments based on a common-property resource with groups of 2, 3, 4, and 5 players, they found the most

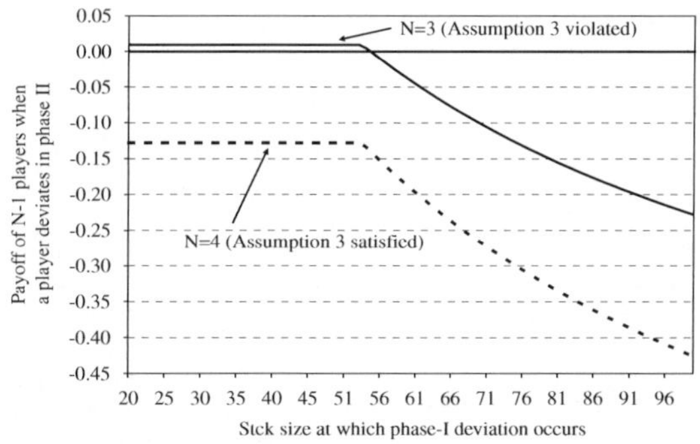

Fig. 3. Assumption 3 holds with $N = 4$, but not with $N = 3$. This figure is based on an example with $r = 0.8$, $K = 100$, $\delta = 0.9$, and $p = 0.05$.

cooperative behavior occurred with an intermediate group size of 3, rather than at the lowest group size (2) or highest group size (5).

For a smaller intrinsic resource growth rate r, Assumption 3 holds with lower discount factors but Assumption 2 holds only with higher discount factors. In Polasky et al. [35] with observable actions, an increase in resource growth rate was favorable to supporting a cooperative outcome. With unobservable actions, players have a stronger incentive to deviate from phase II when the resource grows faster.

If Assumption 3 does not hold and $P(s)>0$ for some stock levels, then each player may be better off by deviating in phase II of σ^*. In this case, the worst perfect equilibrium payoff will be positive. Hence, by deviating in phase I a player can receive a positive one-shot return as well as positive present-value returns starting from the next period. Phase II must be redefined so that the payoffs upon deviation in phase II is lower than the positive worst perfect equilibrium payoff $w>0$. The condition for phase I then becomes $v(s) \geqslant \pi(s) + \delta w$ for all s.

6. Discussion

We analyzed cooperation in a common-property resource among N harvesters in a discrete-time model with non-linear stock-dependent costs and imperfect monitoring of resource users' harvests. Given that individual harvests are unobservable, punishment to deter over-harvesting involves symmetric punishment of all resource users. This paper proposed a strategy profile with symmetric two-part punishment where, upon someone's deviation from cooperation, every resource user over-harvests the resource for one period, followed by reestablishment of cooperative harvesting. This two-part punishment scheme can be designed so that players receive the worst perfect equilibrium payoffs following a defection, thereby inflicting the greatest possible punishment, which in turn supports cooperation under the widest set of conditions. Using this approach we found conditions under which cooperation can be supported as a subgame perfect equilibrium.

In the worst perfect equilibrium, payoffs are driven as low as possible after a defection making a player indifferent between following the two-part punishment strategy and further defection. In some sense, a player has nothing else to lose by defecting against the punishment strategy. One concern then with worst perfect equilibrium is that it involves too harsh punishment and gives insufficient incentive for players to carry out the punishment strategy. An alternative strategy would be to inflict lesser punishment, which would give greater incentive to carry out the punishment following a defection. The weaker the punishment, however, the lower is its deterrence value in preventing cheating. In reality, there will likely be tradeoffs between providing deterrence to cheating and providing incentives to carry out punishment should someone cheat.

Unlike the case of observable actions [35], cooperation is not necessarily supportable when the number of resource users is small, when the resource grows faster, or when the players have a larger discount factor. When each player's action is observable to others the players can apply a punishment specific to the player who deviated from cooperation. In this case, the players who did not cheat can receive positive payoffs during the punishment phase, with only the player who cheated subject to the worst perfect equilibrium payoffs. With unobserved actions, punishment is symmetric and applied to all players. To be a worst perfect equilibrium, a strategy must ensure that no player can earn positive profits during the punishment phase, a constraint that can be harder to satisfy with fewer players.

The simple punishment strategy proposed in this paper, severe group punishment following a defection followed by forgiveness and a return to cooperation, has several attractive features. First, as we have emphasized through the paper, such a strategy can be designed to generate the strongest possible punishment (worst perfect equilibrium), which gives the greatest possible deterrent to defection.

Second, such a strategy is consistent with empirical evidence on how people behave in common property and public goods games. In such situations, people appear to behave in a reciprocal manner, meaning they are nice to those who cooperate but quite willing to inflict punishment on those who fail to be cooperative, even when such punishment is costly to themselves [15–17]. Ostrom [32] summarizes typical behavior observed in common property and public goods games as players being "conditional cooperators" and "willing punishers." In much of the empirical literature, the identity of the cheating player is known and punishments fall directly on that player. In our case, players will know that someone cheated but not who cheated and punishment falls on the group rather than the individual, which may weaken some of the willingness to inflict costly punishment. It seems likely that people who are supposed to cooperate in harvesting a common property resource would be willing to exact some revenge even when not directed to a particular guilty party upon learning that someone in the community has cheated. Whether they would go as far as we have posited in specifying the worst-perfect equilibrium strategy is questionable. However, even lesser punishments willingly inflicted immediately following cheating may be sufficient in some circumstances to deter defection from the cooperative strategy.

Finally, the two-part punishment strategy, unlike a simple single-part punishment strategy such as a simple trigger strategy like that used in Hannesson [22], is relatively immune to renegotiation. The only period in which payoffs are not on the Pareto frontier is the period immediately following a defection, when players probably are willing to inflict costly punishment. As long as the group "takes its medicine" by going through with the punishment, cooperation is reestablished and payoffs are once again on the Pareto frontier, and hence, not subject to renegotiation pressure. In contrast, a simple single-part punishment strategy involves dooming players to suboptimal payoffs forever following a defection. This fact raises questions about why players would not at some future time find in their best interests to forgive and forget, i.e., renegotiate to get back to the Pareto frontier. But, of course, if they do so, this will call into question the deterrence value of the single-part punishment, which in turn may cause cooperation to fail.

In this paper, we assumed deterministic resource transitions and perfect monitoring of stock. In fisheries, and other renewable resources, unexpected fluctuations in stock due to random environmental events are common as is imprecise measurement of stock levels [36,39]. One can extend previous studies on repeated oligopoly game with random demand [2,20] to the common property resource situation with stochastic growth or imprecise measurement. While analyzing an N-player model with uncertain stock would present formidable technical challenges, it is clearly an interesting direction for future work.

Acknowledgments

We thank Joe Herriges, two anonymous referees, Katerina Sherstyuk, and workshop participants at the University of Hawaii for helpful suggestions.

Appendix A

A.1. Uniqueness of the optimal harvesting rule

The first best outcome is given by a solution to the following problem.

$$\max_{\{h_t\}_{t=0}^{\infty}} \sum_{t=0}^{\infty} \delta^t \int_{s_t-h_t}^{s_t} [p - c(\omega)]\, d\omega$$

$$\text{s.t.} \quad s_{t+1} = g(s_t - h_t), \quad t = 0, 1, \ldots, \quad h_t \geqslant 0, \quad t = 0, 1, \ldots \text{ given} \quad s_0 \in (0, K].$$

Letting $z_t = s_t - h_t$, the first-order condition is

$$-[p - c(z_t)] + \delta g'(z_t)[P - c(g(z_t))] = 0 \quad \text{if} \quad z_t \leqslant s_t,$$

and

$$z_t = s_t \quad \text{if} \quad z^* > s_t \quad \text{for all } t.$$

That is, the optimal recruitment given the beginning-of-period-stock s_t is independent of s_t. The existence of a solution z^* follows from

$$-[p - c(\underline{s})] + \delta g'(\underline{s})[p - c(g(\underline{s}))] > 0 \quad (g' > 0, \quad g(s) > s \quad \text{and} \quad c' < 0),$$

$$-[p - c(K)] + \delta g'(K)[p - c(g(K)] < 0 \quad (g'(K) < 1, g(K)) = K \text{ and } \delta < 1)$$

and the continuity of c and g. A sufficient condition for the uniqueness of the solution z^* is that the second-order derivative is negative:

$$(*) \quad c'(z) + \delta g''(z)[p - c(g(z))] - \delta c'(g(z))[g'(z)]^2 < 0 \quad \text{for all} \quad z \in (\underline{s}, K).$$

The first two terms are negative and the third term is positive, so the sign is indeterminate. This condition holds, for example, if $c(s) = 1/s$ and $g(s) = s^{\alpha}$ with $0 < \alpha < 1$. In this case, the first-order condition is $-[p - 1/z_t] + \delta\alpha z_t^{\alpha-1}[p - 1/(z_t^{\alpha})] = -p + (1 - \delta\alpha)z_t^{-1} + p\delta\alpha z_t^{\alpha-1} = 0$and the second-order derivative is $-(1 - \delta\alpha)z_t^{-2} + p\delta\alpha(\alpha - 1)z_t^{\alpha-2} < 0$ for all z_t. Condition (*) also holds if $c(s) = 1/s$ and $g(s) = s + rs(1 - s/K)$ with $r \in (0, 1)$ and $K > 0$. In this case, the first-order condition is $-p + ((1 - \delta)/z_t) + \delta p\big(1 + r - (2rz_t/K)\big) + (\delta r/(K(1 + r) - rz_t)) = 0$ and the second-order condition is

$$-\frac{1-\delta}{z_t^2} - \frac{\delta r\left[2pK\big(1 + r - (rz_t/K)\big)^2 - r\right]}{[K(1 + r) - rz_t]^2} < 0,$$

which always holds because $p > c(K) = 1/K$ (and hence the expression inside the square bracket is positive).

A.2. Proof of Lemma 1

The first-order condition for $h^{\mathrm{D}}(s)$ is

$$\frac{\partial}{\partial h_i} r_i(h_i, h_{-i}^*(s), s) = \frac{\sum_{j\neq i} h_j^*(s)}{(h_i + \sum_{j\neq i} h_j^*(s))^2} \int_{s-\sum_{j\neq i} h_j^*(s)-h_i}^{s} [p - c(\omega)]\, d\omega + \frac{h_i}{(h_i + \sum_{j\neq i} h_j^*(s))} [p - c(s - \sum_{j\neq i} h_j^*(s) - h_i)] \leqslant 0.$$

Clearly the optimal $h^{\mathrm{D}}(s)$ is zero when the stock level is below $\underline{s}$. When $s \in (\underline{s}, z^*]$, we have $h_j^*(s) = 0$ for all $j \neq i$ and hence the above derivative is equal to zero when $s - \sum_{j\neq i} h_j^*(s) - h_i = s - h_i = \underline{s}$. If $s \in (z^*, K]$, then $h_j^*(s) > 0$ for all j and hence the above derivative is positive for all h_i such that $s - \sum_{j\neq i} h_j^*(s) - h_i \geqslant \underline{s}$. This means that the optimal unilateral deviation for player i is to harvest so that the stock level is lower than $\underline{s}$.

A.3. Proof of Lemma 2

For $s \in [z^*, K)$, we have $v'(s) = (1/N)[p - c(s)]$ and

$$\pi'(s) = \frac{h^D(s)\left[H^*_{-i}(s) + h^D(s)\right] - h^D(s)\left[H^{*\prime}_{-i}(s) + h^{D\prime}(s)\right]}{\left[H^*_{-i}(s) + h^D(s)\right]^2} \int_{s-H^*_{-i}(s)-h^D(s)}^{s} [p - c(\omega)]\, d\omega$$
$$+ \frac{h^D(s)}{H^*_{-i}(s) + h^D(s)} \left[p - c(s) - \left[\left(p - c(s - H^*_{-i}(s) - h^D(s))\right)\left(1 - H^{*\prime}_{-i}(s) - h^{D\prime}(s)\right)\right]\right]$$
$$= \frac{N-1}{N} \frac{h^D(s)}{\left[H^*_{-i}(s) + h^D(s)\right]^2} \int_{s-H^*_{-i}(s)-h^D(s)}^{s} [p - c(\omega)]\, d\omega$$
$$+ \frac{h^D(s)}{H^*_{-i}(s) + h^D(s)} \left[p - c(s) - \left[\left(p - c(s - H^*_{-i}(s) - h^D(s))\right)\frac{1}{N}\right]\right],$$

where $H^*_{-i}(s) \equiv \sum_{j \neq i} h^*_j(s)$ and $h^D(s)$ maximizes the payoff upon deviation when stock size is s. The second equality follows from the envelope theorem and $H^{*\prime}_{-i}(s) = (N-1)/N$ for all $s \in (z^*, K)$. In the last expression, the first term is positive because $\int_{s-H^*_{-i}(s)-h^D(s)}^{s} [p - c(\omega)]\, d\omega > 0$. In the second term, we have $p - c(s - H^*_{-i}(s) - h^D(s)) < 0$ and $(h^D(s)/(H^*_{-i}(s) + h^D(s))) > 1/N$ by Lemma 1. Hence, $\pi'(s)$ is greater than $(1/N)[p - c(s)]$. These facts imply $\pi'(s) > v'(s)$ for $s \in [z^*, K)$.

A.4. Proof of Lemma 3

For all $s \in (\underline{s}, z^*)$ we have $\pi(s) = \int_{\underline{s}}^{s} [p - c(w)]\, dw$. The functional form of v on $(\underline{s}, z^*)$ depends on the value $T(\underline{s})$, that is, the number of periods the players should choose zero harvest on the efficient path. We will show that $v'(s) - \pi'(s)$ is strictly decreasing for an arbitrary value of $T(\underline{s})$. First, consider stock level $s < z^*$ which is sufficiently close to z^* such that $g(s) > z^*$. We have $T(s) = 1$ and hence $v(s) = (\delta/N) \int_{z^*}^{g(s)} [p - c(w)]\, dw + \delta^2 v(g(z^*))$. Then

$$v'(s) - \pi'(s) = -(p - c(s)) + \frac{\delta}{N}[p - c(g(s))]g'(s)$$
$$= -\frac{1}{N}(p - c(s)) + \frac{\delta}{N}[p - c(g(s))]g'(s) - \frac{N-1}{N}(p - c(s))$$
$$= \frac{1}{N} f(s) - \frac{N-1}{N}(p - c(s)),$$

where $f(s) \equiv -(p - c(s)) + \delta g'(s)[p - c(g(s))]$. Because $f' < 0$ by Assumption 1 and $c' < 0$, we have

$$v''(s) - \pi''(s) = \frac{1}{N} f'(s) + \frac{N-1}{N} c''(s) < 0.$$

Second, suppose $T(\underline{s}) > 1$. Then, for stock size s such that $g(s) < z^* < g(g(s))$ (i.e. $T(s) = 2$), efficiency requires players to choose zero harvest for two periods. Hence,

$$v'(s) = \frac{\delta^2}{N}[p - c(g(g(s)))]g'(g(s))g'(s)$$

and we have

$$v'(s) - \pi'(s) = -(p - c(s)) + \frac{\delta}{N}[p - c(g(s))]g'(s) - \frac{\delta}{N}[p - c(g(s))]g'(s) + \frac{\delta^2}{N}[p - c(g(g(s)))]g'(g(s))g'(s)$$
$$= \frac{1}{N} f(s) - \frac{N-1}{N}(p - c(s)) + \frac{\delta}{N} g'(s)[-[p - c(g(s))] + [p - c(g(g(s)))]g'(g(s))]$$
$$= \frac{1}{N} f(s) - \frac{N-1}{N}(p - c(s)) + \frac{\delta}{N} g'(s) f(g(s)).$$

Hence,

$$v''(s) - \pi''(s) = \frac{1}{N}f'(s) + \frac{N-1}{N}c'(s) + \frac{\delta}{N}\left[g''(s)f(g(s)) + g'(s)f'(g(s))g'(s)\right] < 0$$

because $g' > 0, g'' < 0, f' < 0$, and $f(s) > 0$ for all $s < z^*$.
Similarly, suppose $T(\underline{s}) > 2$ and consider stock $s > \underline{s}$ such that $g(g(s)) < z^* < g(g(g(s)))$. We have

$$\begin{aligned} v'(s) - \pi'(s) &= -(p - c(s)) + \frac{\delta}{N}[p - c(g(s))]g'(s) - \frac{\delta}{N}[p - c(g(s))]g'(s) + \frac{\delta^2}{N}[p - c(g(g(s)))]g'(g(s))g'(s) \\ &= -\frac{\delta^2}{N}[p - c(g(g(s)))]g'(g(s))g'(s) + \frac{\delta^3}{N}[p - c(g(g(g(s))))]g'(g(g(s)))g'(g(s))g'(s) \\ &= \frac{1}{N}f(s) - \frac{N-1}{N}(p - c(s)) + \frac{\delta}{N}g'(s)f(g(s)) + \frac{\delta^2}{N}g'(s)g'(g(s))f(g(g(s))). \end{aligned}$$

It then follows that

$$v''(s) - \pi''(s) = \frac{1}{N}f' + \frac{N-1}{N}c' + \frac{\delta}{N}(g''f + g'f'g') + \frac{\delta^2}{N}\left[(g''g' + g'g''g')f + g'g'f'\right] < 0$$

(the arguments of functions f and g are omitted). The same argument holds for any $s \in (\underline{s}, z^*)$ given any value of $T(\underline{s})$. Hence, $v''(s) - \pi''(s) < 0$ holds for any $s \in (\underline{s}, z^*)$.

References

[1] D. Abreu, On the theory of infinitely repeated games with discounting, Econometrica 56 (1988) 383–396.
[2] D. Abreu, D. Pearce, E.S. Stacchetti, Optimal cartel equilibria with imperfect monitoring, J. Econ. Theory 39 (1986) 251–269.
[3] D. Abreu, D. Pearce, E.S. Stachetti, Toward a theory of discounted repeated games with imperfect monitoring, Econometrica 58 (1990) 1041–1063.
[4] J. Benhabib, R. Radner, The joint exploitation of a productive asset: a game-theoretic approach, Econ. Theory 2 (1992) 155–190.
[5] G. Brown, An optimal program for managing common property resources with congestion externalities, J. Polit. Econ. 82 (1974) 163–173.
[6] M. Casari, C.R. Plott, Decentralized management of common property resources: experiments with a centuries-old institution, J. Econ. Behav. Organ. 51 (2003) 217–247.
[7] T.N. Cason, F.U. Khan, A laboratory study of voluntary public goods provision with imperfect monitoring and communication, J. Devel. Econ. 58 (1999) 533–552.
[8] J. Cave, Long-term competition in a dynamic game: the cold fish war, RAND J. Econ. 18 (1987) 596–610.
[9] J.M. Chermak, K. Krause, Individual response, information, and intergenerational common pool problems, J. Environ. Econ. Manage. 43 (2002) 47–70.
[10] C.W. Clark, Profit maximization and the extinction of animal species, J. Polit. Econ. 81 (4) (1973) 950–961.
[11] P.K. Dutta, A folk theorem for stochastic games, J Econ. Theory 66 (1995) 1–32.
[12] P.K. Dutta, Collusion, discounting and dynamic games, J. Econ. Theory 66 (1995) 289–306.
[13] P.K. Dutta, R.K. Sundaram, The tragedy of the commons?, Econ. Theory 3 (1993) 413–426.
[14] D. Feeny, S. Hanna A. F. McEvoy, Questioning the assumptions of the 'tragedy of the commons' model of fisheries, Land Econ. 72 (1996) 187–205.
[15] E. Fehr, S. Gachter, Cooperation and punishment in public goods experiments, Amer. Econ. Rev. 90 (2000) 980–994.
[16] E. Fehr, S. Gachter, Fairness and retaliation: the economics of reciprocity, J. Econ. Perspect. 14 (3) (2000) 159–181.
[17] E. Fehr, S. Gachter, Altruistic punishment in humans, Nature 415 (2002) 137–140.
[18] D. Fudenberg, J. Tirole, Game Theory, MIT Press, Cambridge, MA, 1991.
[19] H.S. Gordon, The economic theory of a common-property resource: the fishery, J. Polit Econ. 62 (1954) 124–142.
[20] E.J. Green, R.H. Porter, Noncooperative collusion under imperfect price information, Econometrica 52 (1984) 87–100.
[21] S. Hackett, E. Schlager, J. Walker, The role of communication in resolving commons dilemmas: experimental evidence with heterogeneous appropriators, J. Environ. Econ. Manage. 27 (1994) 99–126.
[22] R. Hannesson, Fishing as a supergame, J. Environ. Econ. Manage. 32 (1997) 309–322.
[23] G. Hardin, The tragedy of the commons, Science 162 (1968) 1243–1248.
[25] R.M. Isaac, J.M. Walker, Communication and free-riding: the voluntary contribution mechanism, Econ. Inquiry 26 (1988) 585–605.
[26] M. Laukkanen, Cooperative and non-cooperative harvesting in a stochastic sequential fishery, J. Environ. Econ. Manage. 45 (2003) 454–473.
[27] J.O. Ledyard, Public good: a survey of experimental research, in: J.H. Kagel, A.E. Roth (Eds.), The Handbook of Experimental Economics, Princeton University Press, Princeton, NJ, 1995.

[29] C.F. Mason, O.R. Philips, Mitigating the tragedy of the commons through cooperation: an experimental evaluation, J. Environ. Econ. Manage. 34 (1997) 148–172.
[30] C.F. Mason, S. Polasky, The optimal number of firms in the commons: a dynamic approach, Can. J. Econ. 30 (1997) 1143–1160.
[31] E. Ostrom, Governing the Commons, Cambridge University Press, Cambridge, UK, 1990.
[32] E. Ostrom, Collective action and the evolution of social norms, J. Econ. Perspect. 14 (3) (2000) 137–158.
[33] E. Ostrom, R. Gardner, J. Walker, Rules, Games, and Common Pool Resources, University of Michigan Press, Ann Arbor, 1994.
[34] E. Ostrom, J.M. Walker, R. Gardner, Covenants with and without a sword: self-governance is possible, Amer. Polit. Sci. Rev. 86 (1992) 404–417.
[35] S. Polasky, N. Tarui, G.M. Ellis, C.F. Mason, Cooperation in the commons, Econ. Theory 29 (2006) 71–88.
[36] J. Roughgarden, F. Smith, Why fisheries collapse and what to do about it?, Proceed. Nat. Acad. Sci. 93 (1996) 5078–5083.
[37] A. Rustichini, Second best equilibria for games of joint exploitation of a productive asset, Econ. Theory 2 (1992) 191–196.
[38] A. Sadrieh, H.A.A. Verbon, Inequality, cooperation, and growth: an experimental study, Europ. Econ. Rev. 50 (5) (2005) 1197–1222.
[39] G. Sethi, C. Costello, A. Fisher, M. Hanemann, L. Karp, Fishery management under multiple uncertainty, J. Environ. Econ. Manage. 50 (2005) 300–318.

[17]

Journal of Economic Behavior & Organization 71 (2009) 187–209

Contents lists available at ScienceDirect

Journal of Economic Behavior & Organization

journal homepage: www.elsevier.com/locate/econbase

A strategic analysis of global warming: Theory and some numbers☆

Prajit K. Dutta[a,*], Roy Radner[b]

[a] Departmentof Economics, Columbia University, New York, NY 10027, United States
[b] Stern School of Business, New York University, New York, NY 10012, United States

ARTICLE INFO

Article history:
Received 23 March 2007
Received in revised form 20 January 2009
Accepted 23 January 2009
Available online 26 February 2009

JEL classification:
C73
Q54

Keywords:
Global warming
Dynamic games
Treaty
Calibration

ABSTRACT

We model the global warming process as a dynamic commons game in which the players are countries, their actions at each date produce emissions of greenhouse gases, and the state variable is the current stock of greenhouse gases. The theoretical analysis is complemented by a calibration exercise. The first set of results establishes theoretically, and then with illustrative numbers, the over-emissions due to a "tragedy of the commons." The power of simple sanctions to lower emissions and increase welfare is then examined as is the effect of cost asymmetry. Finally, a complete theoretical charactrization is provided for the best equilibrium, and it is shown that it has a very simple structure; it involves a constant emission rate through time.

1. Introduction

The dramatic rise of the world's population in the last three centuries, coupled with an even more dramatic acceleration of economic development in many parts of the world, has led to a transformation of the natural environment by humans that is unprecedented in scale. In particular, on account of the greenhouse effect, *global warming* has emerged as a central problem, unrivalled in its potential for harm to life as we know it on planet Earth. Seemingly the consequences are everywhere: melting and break-up of the world's ice-belts in the Arctic and the Antarctic; heat-waves that set all-time temperature highs in Western Europe and sub-Saharan Africa; storms increased in frequency and ferocity including Hurricane Katrina, typhoons in Japan, and flooding in Mumbai; and immediately graspable economic consequences, such as the shut-down of part of the New York subway due to a heat wave, and the startling images from Al Gore's movie "An Inconvenient Truth" showing the possible disappearance of major parts of coast-lines in China, India, the United States, and throughtout the globe.[1] Here are three additional facts:

1. Eleven of the last 12 years (1995–2006) have been amongst the 12 warmest years in the instrumental record of global surface temperatures (since 1850). The 100 year linear trend (1906–2005) of 0.74 °C is larger than the corresponding trend of 0.6 °C given in the Third Assessment Report of the IPCC for 1901–2000.[2]

☆ The first version of this paper was written in 2000.

* Corresponding author.
E-mail address: pkd1@columbia.edu (P.K. Dutta).

[1] Global Warming and its consequences have been, of course, widely written about. Al Gore's recent book and movie "An Inconvenient Truth" is an excellent, easy-to-read examination of the issues. More detailed analysis can be found in Hansen (2006). Two authoritative recent treatments are the Stern Review on the Economics of Climate Change, October 2006 and the IPCC Synthesis Report, November 2007.

[2] From "Climate Change", IPCC (2007).

0167-2681/$ – see front matter © 2009 Elsevier Ltd. All rights reserved.
doi:10.1016/j.jebo.2009.01.013

188 P.K. Dutta, R. Radner / Journal of Economic Behavior & Organization 71 (2009) 187–209

2. The Intergovernmental Panel on Climate Change (IPCC) predicts that if we go on with "Business as Usual", by 2100 global sea levels will probably have risen by 9–88 cm and average temperatures by between 1.5 and 5.5 °C.
   Various factors contribute to global warming, but the major one is an increase in greenhouse gases (GHGs), primarily carbon dioxide, so-called because they are transparent to incoming shortwave solar radiation but trap outgoing longwave infrared radiation. Increased carbon emissions due to the burning of fossil fuel is commonly cited as the principal immediate cause of global warming. The "natural levels" of GHGs (e.g., pre-1800 levels) raised the earth's average temperature by some 33 °C, from −18 to +15. (Cline 1992, p. 4.).[3]
3. Before the Industrial Revolution, atmospheric CO_2 concentrations were about 270–280 parts per million (ppm). They now stand at almost 380 ppm and have been rising at about 1.5 ppm annually.

The IPCC Synthesis Report (2007, p. 257) says "Warming of the climate system is unequivocal, as is now evident from observations of increases in global average air and ocean temperatures, widespread melting of snow and ice, and rising global average sea level."

It is clear that addressing the global warming problem will require the coordinated efforts of the world's nations. In the absence of an international government, that coordination will have to be achieved by way of an international environmental treaty. For a treaty to be implemented, it will have to align the incentives of the signatories by way of rewards for cutting greenhouse emissions and punishments for not doing so. For an adequate analysis of this problem one needs a dynamic and fully strategic approach. A natural methodology for this is the theory of Nash equilibria of dynamic games, which we adopt in this paper. In other words, our goal is to characterize the equilibria of a well-defined dynamic global warming game, as well as the globally (Pareto) optimal outcomes (which are typically not sutstainable as equilibria of the game). We then supplement the theoretical analysis with a calibration exercise that yields numerical estimates of equibrium and optimal trajectories of the system.

The present paper is part of an ongoing research project in which we have addressed various aspects of the global warming problem. We describe some of the other results of this project, as well as goals for further research, in Section 8. While our approach is not entirely new, we believe that we have carried the analysis farther than what is currently available in the published literature; for some relevant bibliographic notes, see Section 9.

The rest of the introduction provides a brief discussion of background issues regarding global warming as well as the model and results of this paper.

1.1. Background and principal issues

Although there is considerable uncertainty about the exact costs of global warming, the two principal sources will be a rise in the sea-level and climate changes. The former may wash away low-lying coastal areas such as Bangladesh and the Netherlands. Climate changes are more difficult to predict: tropical countries will become more arid and less productive agriculturally; there will be an increased likelihood of hurricanes, fires and forest loss; and there will be the unpredictable consequences of damage to the natural habitat of many living organisms. On the other hand, emission abatement imposes its own costs. Higher emissions are typically associated with greater GDP and consumer amenities (via increased energy usage). Reducing emissions will require many or all of the following costly activities: cutbacks in energy production, switches to alternative modes of production, investment in more energy-efficient equipment, investment in R&D to generate alternative sources of energy, and so on.

The principal features of the global warming problem are

- *The Global Common*: although the sources of carbon buildup are localized, it is the total stock of GHGs in the global environment that will determine the amount of warming.
- *Near-irreversibility*: since the stock of greenhouse gases depletes slowly, the effect of current emissions can be felt into the distant future.
- *Asymmetry*: some regions will suffer more than others.
- *Non-linearity*: the costs can be very non-linear; a rise in one degree may have little effect, but a rise in several degrees may be catastrophic.

The theoretical framework that accommodates all of these features is an *asymmetric dynamic commons* model with the global stock of greenhouse gases as the (common) state variable. We study just such a model with all the above characteristics *except* non-linearity and, additionally, with two other features:

- *A Strategic Model*: although the players (countries) are relatively numerous, there are some very large players and some blocks of like-minded countries, such as the U.S., Western Europe, China, and Japan. That warrants a strategic analysis. Furthermore a lesson of history is that the only good international treaties are those that are incentive-compatible and

[3] The IPCC (2007) report declared unequivocally "most of the observed warming over the last 50 years is likely to have been due to the increase in greenhouse gas concentrations."

self-enforcing for the signatories. (Indeed an objection to the Kyoto Protocol, although not the one commonly cited, is that it was not incentive-compatible; more on this in Section 8.)

- *A Simple Linear Model*: the model we study embeds a simple linear description of how global climate change affects the physical and economic world; it is assumed that emissions add linearly to the stock of GHGs and the stock of GHGs affect country costs linearly. (Technically, we use a "dynamic affine model"; see Sobel, 1990.) In particular that implies that, say, a 20 percent increase in the stock of GHGs will cause a 20 percent increase in the costs of climate change.

The linearity of the model is undoubtedly restrictive in several ways. It implies that our model is unable to analyze catastrophes or certain kinds of feedback effects running back from climate change to economic costs. For instance, some scientists have argued that a doubling of pre-industrial GHG levels will lead to a 3–5 °C temperature increase, causing an annual loss of 5 percent of GDP, but any increase beyond that could cause up to a 20 percent loss in GDP (Stern, 2006 and less explicitly, IPCC Synthesis Report 2007). Similarly, many scientists have pointed out that there could be several potential non-linear feedback effects: the shutting down of the Atlantic Ocean thermohaline circulation, the release of methane from the melting of permafrost, the melting of polar ice-caps, to name a few.

We make the linearity assumption for three reasons. First, for a theoretical exercise to have any chance of informing policy-makers in even one country, let alone 200, its conclusions have to be simple, and the linear model delivers that. (Indeed all the proposals at Kyoto or follow-up conferences have involved simple across-the-board cuts in emissions.) Second, it allows us to derive closed-form solutions for global Pareto optimal outcomes and some equilibria and to calibrate our model and derive some numerical guideposts.[4] Third, there is little consensus on what is the correct form of non-linearity in costs. Partly the problem stems from the fact that some costs are not going to be felt for another 50–100 years, and forecasting the nature of costs on that horizon length is at best a hazardous exercise.[5] Hence, instead of postulating one of many possible non-linear cost functions, all of which may turn out to be incorrect for the long-run, we opt instead to work with a cost function that may be thought of as a linear approximation to any number of actual non-linear specifications. To the extent that the world might remain in the linear segment for the next 50 or 100 years, and since discounting makes this the more relevant time-frame, our results are approximations of what a fully non-linear model would deliver. One can also think of our exercise as suggesting a lower bound on the economic consequences of global warming since catastrophic consequences are not analyzed.[6]

Although a full discussion of the relevant literature may be found in Section 9, it is worth remarking here on two influential recent publications: the Stern Review (Stern, 2006) and the IPCC Synthesis Report (IPCC 2007). The Stern Review empirically examines two contrasting scenarios. The first looks at the economic costs of "Business as Usual", that is, the costs due to climate change that the world's economies would face if GHG emissions continued to increase at current projections.[7] Even without examining the non-linear feedback effects discussed above, it comes up with an estimate that BAU will cause a 5 percent drop in world GDP per year starting around 2050, and it argues that this cost could be significantly higher, upto 20 percent, once non-linearities are accounted for. The second scenario examines the costs associated with cutting emissions in such a fashion that the stock of GHG stabilizes around 550 parts per million (ppm).[8] It concludes that the costs to mitigation and adaptation will be around 1 percent of GDP per year. Hence, it delivers a call for international coordination on mitigation and adaptation (Executive Summary, especially pp. x–xiii.).

The IPCC Report is similar in its conclusions about mitigation costs although it focuses more on the range of possibilities. It considers stabilization by year 2050 in a band ranging from 435 to 710 parts per million (ppm) and concludes that these will cost in range between 5.5 percent of world GDP through −1 percent (Summary for Policymakers, pp. 23–24.) Again, when coupled with the 5 percent, and possibly higher, GDP reduction from inaction, this too is a call for urgent international coordination on reducing GHG emission levels.

Many scholars have pointed out that the focus on 2× pre-industrial CO_2 levels of 550 parts per million (ppm), and the pathways to that steady-state, are essentially arbitrarily chosen and that the right approach is to examine what is the globally optimal GHG dynamic. One way to view the contribution of this project, modest as it is, that we do just that: we examine the GPO and then the best equilibrium emission strategies. Furthermore, we take seriously the idea that these emission patterns will be followed only if the nations of the world wish to do so, something, again, that the Stern and IPCC Reports ignore.

---

[4] The calibration and extensive numerical calculations for a model with approximately 180 countries have been done by Dr. Sangwon Park. Space limitations prevent us from presenting the details here, but they can be found in (found in Park, 2004).

[5] This point has been made forcefully by, amongst others, Thomas Schelling. See, for example, Schelling (2002).

[6] One other restriction implied by linearity is that our model is unable to deliver complex dynamics including chaos. For an example of such behavior, please see Chen (1997) as well as the broader discussion in Rosser (2001).

[7] These costs include health costs, declining crop yields, flood and storm damage, loss of ecosystems, and so on. The Review uses an integrated assessment model to impute economic costs and includes a sensitivity analysis.

[8] Emission cuts are projected to arise from a combination of greater efficiency in energy usage, a switch to cleaner technologies such as nuclear power and hydro-electricity, and the generation of new technologies through R&D.

1.2. The results

The present paper has five main parts: Benchmark Model, Baseline Global Pareto optima and "Business As Usual" Equilibria, Better Equilibria (including second and third-best), Asymmetry, and the Greenhouse Trap.

Model (Section 2): in the model, in each period each country adds to the global stock of greenhouse gases (GHGs) by its emission of those gases. The existing stock depreciates at a constant rate over time. Emissions arise from the production and consumption of energy, and other productive activities, and hence higher emissions are associated with, at least up to a point, higher benefits. The incremental cost of global warming is directly proportional to the amount of GHGs in the earth's atmosphere. Both the benefits as well as the costs vary across countries. The model has a relatively small set of parameters that guide it, and drawing on estimates for those parameters that are available in the literature, we are also able to calibrate the model.[9]

Baseline results (Section 3): we first show that *Global Pareto optimal* (GPO) emission profiles involve a constant emission of GHGs by each country (i.e., constant across GHG levels, and hence through time).[10] Next we characterize a Markov Perfect equilibrium (MPE) that is qualitatively similar to the GPO solutions; it too involves a vector a^* of constant emission rates by each country. Not surprisingly, this equilibrium, which we call *Business as Usual (BAU)*, exhibits the tragedy of the commons; a^* is strictly higher than the GPO vectors. The numerical estimates for this over-emission are illustrative. For base-line parameter values, the GPO is between 8 and 10 percent lower than the BAU level for developed economies and between 28 and 34 percent lower for developing economies and China. (Contrast this with the Kyoto-sanctioned 5 percent cut for developed economies.) Corresponding welfare loss in per period terms ranges from 0.1 percent for base-line costs to 2 percent for higher cost estimates.

Better equilibria/treaties (Sections 4 and 6): we then turn to the main question(s) of this paper: can countries sign a Kyoto-like agreement that slows down the pace of global warming *and* is honored by all signatories?

In Section 4 we start with the *third-best*: Nash or BAU reversion *trigger strategy equilibria* since these are the simplest to implement. The two main theoretical results here show that for all discount factors and all countries, one can do strictly better; the highest welfare level is strictly higher than in the BAU, and the minimum incentive-compatible emission levels are strictly lower than the BAU levels. Second, if discount factors are high enough, then, in fact, the GPO emission levels are themselves sustainable. For the base-line parameter values, the minimum sustainable emissions are lower than the GPO levels, for all regions except Eastern Europe; the greatest discrepancies arise for Russia and Ukraine where the GPO emission levels are, respectively, 39 percent and 40 percent lower than the BAU emission levels but the maximum sustainable cuts from BAU levels are only 20 percent and 19 percent, respectively.

Second-best: in Section 6 we characterize the full set of subgame perfect equilibria (SPE). We show that the *best equilibrium* has a remarkably simple structure; it too has a constant emission vector along the equilibrium path (at a rate strictly lower than the BAU rate a^*). The sanctions that support such a low emission level utilize the *worst equilibrium* for each player. That equilibrium is asymmetric (even in a symmetric game) and has an interesting two-stage structure. In the first stage, and for exactly one period, all countries other than the deviant country emit a "very high" amount, in excess of the BAU level. This behavior punishes all countries. For the sanctioning countries, the compensation is that in the second stage play proceeds to an equilibrium they prefer (and one that the sanctioned country does not). This is the equilibrium that maximizes a weighted sum of (equilibrium) payoffs, with zero weight on the sanctioned country. In such an equilibrium, the country with zero weight typically has to settle for very low emissions. This then is the long-term cost that a country suffers by breaking an agreement; its emission "quota" is permanently reduced *in an incentive-compatible fashion.*

The effect of *asymmetry* is studied in Section 5 by analyzing trigger strategy equilibria. Since there is no obvious measure of asymmetry with many countries, we restrict attention to two countries, or regions, and measure asymmetry by the difference in (marginal) costs of GHG. We show that increasing asymmetry leads to strict welfare loss (i.e., a reduction in the total payoffs of the two countries). Again we provide numerical analogs of these results by focussing on China and the United States.

Greenhouse traps: in Section 7 we consider emission strategies that are Markovian but not necessarily constant, and we demonstrate the existence of MPE that have simultaneously "basins of tragic attraction", levels of GHGs that are inoptimally high so that once the system reaches them there are no incentives to escape, as well as "basins of low GHG buildup" (regions of carbon buildup that are no more than what a global planner would recommend). In other words, we show that there are MPE that simultaneously have characteristics of "bad" and "good" equilibria; where the system ends up depends then on where it started.

Section 8 discusses extensions of the model, including the incorporation of technological change, while Section 9 includes a literature review. Proofs are gathered in Section 10.

[9] The model has been generalized in Dutta and Radner (2004, 2006) and Dutta and Radner (in preparation) to account, respectively, for technical change, population growth, and endogenous capital stock dynamics. See Section 8 for a further discussion.

[10] The GPO emission is constant on account of the linearity in the damage-cost function; that, coupled with the constant depreciation rate, implies that the marginal valuation of a unit of emission is independent of the existing stock of GHG.

2. A simple model

In this section we present a simplified model to illustrate the basic strategic ideas. In this model there is no population growth and no possibility of changing the emissions producing technologies in each country.[11] However, the countries may differ in their "sizes," their emissions technologies, and their preferences.

There are I countries. The emission of (a scalar index of) greenhouse gases during period t by country i is denoted by $a_i(t)$. (Time is discrete, with $t = 0, 1, 2, \ldots$, ad inf.) Let $A(t)$ denote the global (total) emission during period t;

$$A(t) = \sum_{i=1}^{I} a_i(t). \tag{1}$$

The total (global) stock of greenhouse gases (GHGs) at the beginning of period t is denoted by $g(t)$. The law of motion for the total GHG is

$$g(t+1) = A(t) + \sigma g(t), \tag{2}$$

where σ is a given parameter ($0 < \sigma < 1$). We may interpret $(1 - \sigma)$ as the fraction of the beginning-of-period stock of GHG that is dissipated from the atmosphere during the period. The "surviving" stock, $\sigma g(t)$, is augmented by the quantity of global emissions, $A(t)$, during the same period.

Suppose that the utility of country i in period t is

$$v_i(t) = h_i[a_i(t)] - c_i g(t). \tag{3}$$

The function h_i represents, for example what country i's gross national product would be at different levels of its own emissions, holding the global level of GHG constant. This function reflects the costs and benefits of producing and using energy as well as the costs and benefits of other activities that have an impact on the emissions of GHGs (e.g., the extent of forestation). For a given population there will be an optimal level of energy use, forestation, and so on, and hence an optimal level of emissions. It therefore seems natural to assume that h_i is a strictly concave C^2 function that reaches a maximum and then decreases thereafter.[12]

The parameter $c_i > 0$ represents the marginal cost to the country of increasing the global stock of GHG. Of course, it is not the stock of GHG itself that is costly, but the associated climatic conditions. As discussed above, in a more general model the cost would be non-linear. Similarly, the separability of the payoff function is a simplification; in a more general model, a country's GNP level might itself depend on the GHG stock level.[13]

The total payoff (utility) for country i is

$$v_i = \sum_{t=0}^{\infty} \delta^t v_i(t). \tag{4}$$

For the sake of simplicity, we have taken the discount factor, δ, to be the same for all countries.

At time t, a history of play is an enumeration of past GHG levels up to and including the current one as well as past emissions for all countries. A strategy for country i at time t, call it π_{it}, is a measurable map from the set of histories to the set of emissions. A *strategy* for the whole game, denoted by π_i, is a sequence of such strategies: $\pi_i \equiv [\pi_{i0}, \pi_{i1}, \ldots, \pi_{it}, \ldots]$.[14] Let π denote a profile of strategies, one for each country: $\pi = (\pi_1, \pi_2, \ldots, \pi_I)$. A particularly simple strategy for country i is to condition her action at period t only on the state at that period; such a strategy is called a *Markovian strategy.*

Each profile π, and the initial level of GHG, g_0, defines in an obvious manner country i's payoff in period t; denote this payoff by $v_i(t; \pi, g_0)$. Thus associated with each profile is a total discounted payoff for each player:

$$v_i(\pi, g_0) \equiv \sum_{t=0}^{\infty} \delta^t v_i(t; \pi, g_0).$$

As usual, a profile of strategies π^* forms a *Nash Equilibrium* (or just an *equilibrium*) from the initial GHG level g_0 if no player can increase her total discounted expected payoff by *unilaterally* changing her strategy. A profile π^* forms a *subgame perfect Equilibrium* (SPE) if after every history the continuation strategies of π^* form a Nash equilibrium. If a profile of Markovian strategies forms a Nash equilibrium from every initial state then it is called a *Markov Perfect Equilibrium* (MPE); note that an MPE is also an SPE.

Note that if there is imperfect monitoring then player i's history will contain only her own past actions. In this paper we shall restrict attention to the perfect monitoring case, although our results on Markovian strategies will apply to the case of

[11] Population growth is studied in Dutta and Radner (2006) while certain kinds of technological changes are allowed in Dutta and Radner (2004).

[12] However, none of the results depend on h having a finite argmax.

[13] We thank a referee for this suggestion.

[14] Without loss of generality we restrict attention to pure strategies.

imperfect monitoring, as well. This is because, by definition, Markovian strategies only condition on the observable part of history (i.e., the total stock of GHGs).

2.1. An illustrative example

The theoretical results of this paper will be illustrated throughout by use of a calibrated model using parameter values that have been considered in the literature. The aim is thereby to provide some numerical orders of magnitude for the results stated here. In this subsection we will briefly describe the numerical analysis, and in subsequent sections we will present the numerical analogs immediately following the appropriate theoretical results.

The numerical example is taken from a much more extensive analysis by Sangwon Park. In his model, 184 countries are grouped for calibration purposes into eight regions: the United States, Western Europe, Other High Income, Eastern Europe, Middle Income, Lower Middle Income, China, and Lower Income. (This grouping enables Park to utilize the data and estimates provided by Nordhaus and Boyer (2000).) The base year is 1998 ($t = 0$). Referring to Eq. (3), the utility of country i net of warming costs, its gross national product after subtracting the amount of capital needed to maintain the capital stock at its 1998 level, h_i, is given by

$$h_i[a_i(t)] = \phi_i K_i^{\gamma_i} L_i^{1-\gamma_i-\beta_i}\left[\frac{a_i(t)}{f_i}\right]^{\beta_i} - p_i\left[\frac{a_i(t)}{f_i}\right]$$

where K_i and L_i are the capital and labor inputs, respectively; f_i is the emission factor of country i and p_i is the price of "energy" (and, recall, a_i is the emission of country i). To explain the last two terms in greater detail, imagine that energy is a proxy for a scalar index of emissions-producing inputs, measured in coal–equivalent metric tons. Then for every unit of energy there are associated f_i units of emissions. Countries differ in their emission factors: countries with cleaner technologies have lower f_i than those with "dirtier" technologies. The Greek letters are parameters. These parameters as well as the price of energy, p_i, were the same for all countries in the same region (that is what defines a region) while the damage coefficient c_i is based on a combination of country and group-specific data and is drawn from Fankhauser (1995). All three production inputs – capital, labor, energy – are country-specific.

All countries have the same discount factor, δ, which for sensitivity analysis has been varied between 0.97 and 0.995. These would seem to bracket the values commonly discussed in the literature, motivated by different interpretations of δ as reflecting "social values" or the returns on investment. In what follows utility is measured in 1990 U.S. dollars and emissions are measured in gigatons of carbon.

3. Two benchmarks

In this section we characterize two benchmarks: the global Pareto optima, and a simple Markov Perfect Equilibrium, called "Business As Usual".

3.1. Global pareto optima

Let $x = (x_i)$ be a vector of positive numbers, one for each country. A *Global Pareto Optimum (GPO)* corresponding to x is a profile of strategies that maximizes the weighted sum of country payoffs,

$$v = \sum_i x_i v_i, \tag{5}$$

which we shall call *global welfare*. Without loss of generality, we may take the weights, x_i, to sum to I.

Theorem 1. *Given a vector of weights, x, let $\hat{V}(g)$ be the maximum attainable global welfare starting with an initial GHG stock equal to g; then*

$$\hat{V}(g) = \widehat{u} - wg, \tag{6}$$

$$w = \frac{1}{1-\delta\sigma}\sum_i x_i c_i,$$

$$\widehat{u} = \frac{\sum_i x_i h_i(\hat{a}_i) - \delta w\hat{A}}{1-\delta}, \quad \hat{A} = \sum_i \hat{a}_i,$$

and $\hat{a}_i$ is determined by

$$x_i h_i'(\hat{a}_i) = \delta w. \tag{7}$$

(It is assumed that this last equation has a solution.) Furthermore, in the GPO solution, country i would be required to use a constant emission equal to $\hat{a}_i$ *in all periods and after all histories.*

Theorem 1 states that, independently of the level of GHG, g, each country should emit an amount $\hat{a}_i$. The fact that the optimal emission is constant follows from the linearity of the model in g. Notice that on account of the linearity in the gas buildup equation Eq. (2) a unit of emission in period t can be analyzed in isolation as a surviving unit of size σ in period $t+1$, σ^2 in period $t+2$, σ^3 in period $t+2$, and so on. On account of the linearity in cost, these surviving units add $(\sum_i x_i c_i)\times\delta\sigma$ in period $t+1$, $(\sum_i x_i c_i)\times(\delta\sigma)^2$ in period $t+2$, and so on; therefore, the marginal lifetime cost is

$$\frac{1}{1-\delta\sigma}\sum_i x_i c_i,$$

or w_i, and that marginal cost is independent of g.

3.2. A Markov-perfect equilibrium: "Business as Usual"

This MPE shares the feature that the equilibrium emission rate of each country is constant in time, and it is the unique MPE with this property. We shall call it the "Business as Usual" equilibrium. Note that in this equilibrium each country takes account of the incremental damage to itself caused by an incremental increase in its emission rate, but does not take account of the damage caused to other countries. [In some contexts the phrase "Business as Usual" is used to describe a situation in which each country does not even take account of the damage to itself.]

Theorem 2. ("Business as Usual" Equilibrium). *Let g be the initial stock of GHG. For each country I, let* $a_i{}^*$ *be determined by*

$$h_i'(a_i^*) = \delta w_i, \tag{8}$$

$$w_i = \frac{c_i}{1-\delta\sigma},$$

and let its strategy be to use a constant emission equal to $a_i{}^*$ *in each period; then this strategy profile is a MPE, and country i's corresponding payoff is*

$$V_i^*(g) = u_i^* - w_i g, \tag{9}$$

$$u_i^* = \frac{h_i(a_i^*) - \delta w_i A^*}{1-\delta}, \quad A^* = \sum_j a_j^*.$$

The intuition for the existence of an MPE with constant emissions is similar to the analogous result for the GPO solution. As long as other countries do not make their emissions contingent on the level of GHGs, country i has a constant marginal lifetime cost to emissions; this follows from the linearity of the model. Notice that on account of the linearity in the transition (gas buildup) equation – Eq. (2) – a unit of emission in period t can be analyzed in isolation as a surviving unit of size σ in period $t+1$, σ^2 in period $t+2$, σ^3 in period $t+2$, and so on. On account of the linearity in cost, these surviving units add $c_i \times \delta\sigma$ in period $t+1$, $c_i \times (\delta\sigma)^2$ in period $t+2$, and so on, i.e., the marginal lifetime cost to country i on account of its own emission is $(c_i/(1-\delta\sigma))$ or w_i. And that marginal cost is independent of g. There are, however, other MPEs in which countries do condition on g, and some of them will be described in Section 7.

3.3. Comparison of the GPO and business as usual

The preceding results enable us to compare the emissions in the GPO with those in the Business as Usual MPE:

$$\text{GPO}: h_i'(\hat{a}_i) = \frac{\delta\sum_j x_j c_j}{x_i(1-\delta\sigma)},$$
$$\text{BAU}: h_i'(a_i^*) = \frac{\delta c_i}{1-\delta\sigma}. \tag{10}$$

Since

$$x_i c_i < \sum_j x_j c_j,$$

194 P.K. Dutta, R. Radner / Journal of Economic Behavior & Organization 71 (2009) 187–209

it follows that

$$\frac{\delta c_i}{1-\delta\sigma} < \frac{\delta \sum_j x_j c_j}{x_i(1-\delta\sigma)}.$$

Since h_i is concave, it follows that

$$a_i^* > \hat{a}_i. \tag{11}$$

Note that this inequality holds except in the trivial case in which all welfare weights are zero (except one). In other words, there is a tragedy of the commons whenever there is some externality to emissions. In turn, all this follows from Eq. (8), which says that in the BAU equilibrium each country only considers its own marginal cost and ignores the cost imposed on other countries on account of its emissions; in the GPO solution that additional cost is, of course, accounted for. It follows that the GPO is strictly Pareto superior to the MPE for an open set of welfare weights x_i (and leads to a strictly lower steady-state GHG level for all welfare weights).

3.4. Numerical results on the GPO and BAU

In this subsection we present numerical results on the emission levels and welfare values (especially in a comparative sense) of the GPO and BAU solutions. As indicated above, the results are derived by using numerical estimates for parameter values that have been used elsewhere by Fankhauser (1995) and Nordhaus and Boyer (2000). We start with the benchmark case in which $\delta = 0.97$ and cost coefficients c_i are as given in Fankhauser. This case has been calibrated so that the BAU matches available data and estimates for 1998. The benchmark GPO then uses the same parameter values and initial conditions. (Throughout, in computing GPO trajectories, we have taken the country-welfare weights x_i to be equal to 1. The consequences of using other welfare weights have been explored in Park.) (Table 1).

The most striking results are (a) BAU emission levels are most significantly higher (relative to GPO) for Eastern Europe, presumably on account of Soviet-era energy technologies; (b) the next significant departures are for China and Lower Middle Income countries (that includes India), and this is particularly worrisome if the major growth spurt of the next 50 years comes from those countries, which will then move them way past Eastern Europe in over-emission; and (c) that the United States and Western Europe account for over 40 percent of current total emissions and, hence, a cut-back to GPO levels in those regions alone can significantly narrow the overall GPO–BAU gap.

The differences between GPO and BAU are magnified if either of the two basic parameters are increased. Consider an increase in the discount factor to 0.995 (reflecting the idea that environmental discount factors ought to be closer to 1 for inter-generational equity reasons); this is Table 2. The main findings while qualitatively similar (Eastern Europe is the

Table 1
Benchmark case ($\delta = 0.97$, cost = Fankhauser c_i, Year = 1998).

Region/emissions	BAU (Gtc)	GPO (Gtc)	% Difference ((BAU − GPO)/BAU)
United States	1.50	1.36	9
Western Europe	0.86	0.79	8
Other high income	0.59	0.53	9
Eastern Europe	0.74	0.45	39
Middle income (MI)	0.41	0.36	12
Lower MI	0.58	0.41	30
China	0.85	0.56	34
Lower income	0.66	0.48	28
Total	6.18	4.93	20

Table 2
Sensitivity analysis I ($\delta = 0.995$, cost = Fankhauser c_i, Year = 1998).

Region/emissions	BAU (Gtc)	GPO (Gtc)	% Difference ((BAU − GPO)/BAU)
United States	1.50	1.15	23
Western Europe	0.86	0.69	20
Other high income	0.59	0.46	22
Eastern Europe	0.74	0.27	64
Middle income (MI)	0.41	0.29	28
Lower MI	0.58	0.27	54
China	0.85	0.32	62
Lower income	0.66	0.32	52
Total	6.18	3.76	39

Table 3
Value loss when $\delta = 0.97$.

Value loss/cost parameter	FC	2XFC	3XFC	4XFC	5XFC
(GPO – BAU)/BAU in 1998 dollars	0.11%	0.4%	0.82%	1.34%	1.95%

Table 4
Value loss when $\delta = 0.995$.

Value loss/cost parameter	FC	2XFC	3XFC	4XFC	5XFC
(GPO – BAU)/BAU in 1998 dollars	0.68%	2.14%	4.06%	6.37%	9.02%

greatest problem) show that China and Lower Income Countries have almost as much over-emissions. This could pose a huge problem down the line if these economies continue to grow strongly.[15]

We now turn to value comparisons; that is how much is the welfare loss due to over-emission and its consequent increase in global temperatures? The welfare analysis takes account of the infinite future. In particular, what is computed is the present discounted average value of GNP less carbon costs (i.e., $(1-\delta)V_i$)so that when δ is close to 1, the discounted average value is close to the long-term yearly average. Note that if 2 percent of a country's GNP were invested every year, the rate of growth of its economy would be dramatically increased.

Table 3 provides value loss estimates for $\delta = 0.97$ while Table 4 provides the parallel estimates for $\delta = 0.995$. The carbon costs parameter is drawn from Fankhauser (1995), denoted FC, and the sensitivity analysis is to vary them between his estimates (=FC) and five times his estimates (=5XFC):

4. Other equilibria I: BAU sanctions

In this section we analyze a "third-best" problem: maximize a weighted sum of country utilities when the emission policies are incentive-compatible under the threat of BAU emissions. Recall that an equilibrium is *second best* if its outcome is Pareto optimal among the outcomes that can be sustained by some(subgame-perfect) equilibrium. The set of second best equilibria will be characterized in Section 6.2. For some given subset $\mathcal{E}$ of equilibria, an equilibrium is third best relative to $\mathcal{E}$ if it is Pareto optimal among the outcomes that can be sustained by some equilibrium in $\mathcal{E}$. In the present section, $\mathcal{E}$ is the set of equilbria that are sustained by a threat of reverting to the BAU in case there is a defection.

There are three main results. First, for all discount factors, the third-best solution qualitatively mirrors the BAU and GPO solutions; there is a constant emission level a_i' that country i emits, independently of the stock of GHGs. Second, the set of sustainable emissions includes the BAU level but is strictly larger than that. In particular, the minimum incentive-compatible emission level is strictly lower than the BAU level, and that is true for every country. Hence, BAU sanctions are strictly welfare improving for all countries and all discount factors. Third, if discount factors are high enough, then, in fact, the GPO emission levels are themselves the third-best solution. Finally, we provide empirical illustration for these results.

4.1. BAU Sanctions: third-best and minimum emissions

Let $x = (x_i)$ be a vector of positive numbers, one for each country. A *Third-Best Optimum (TBO)* corresponding to x is a profile of "norm" strategies that maximizes the weighted sum of country payoffs,

$$v = \sum_i x_i v_i, \tag{12}$$

subject to BAU reversion (i.e., subject to the constraint, detailed below, that should any country i not follow the norm, all countries would switch to BAU emissions vector $a^* = (a_1^*, a_2^*, \ldots, a_I^*)$ starting in from the following period). As before, and without loss of generality, we may take the weights, x_i, to sum to I. The following result characterizes the TBO:

Proposition 1. *There exists a vector $\tilde{a}$ of constant emission levels $\tilde{a}_i$ such that on the equilibrium path country i's TBO strategy is to use a constant emission equal to $\tilde{a}_i$ in all periods, where $\tilde{a}_i'$ satisfies the incentive constraint:*

$$\text{for every } i, \quad h_i(\widetilde{a}_i) - \delta w_i\left(\widetilde{a}_i + \delta\sum_{j\neq i}\widetilde{a}_j\right) \geq h_i(a_i^*) - \delta w_i\left[a_i^* + \delta\sum_{j\neq i}a_j^*\right].$$

[15] A change in the second parameter, raising the cost estimates to 5× Fankhauser (reflecting possible catastrophic costs) increases the range of over-emissions to 56 percent (Western Europe) to 90 percent (Eastern Europe).

Table 5
Minimum emissions under BAU sanctions; $\delta = 0.97$ and Fankhauser costs.

Region/emissions	BAU (Gtc)	GPO (Gtc)	MIN (Gtc)
United States	1.50	1.36	1.22
Western Europe	0.86	0.79	0.67
Other high income	0.59	0.53	0.45
Eastern Europe	0.74	0.45	0.54
Middle income (MI)	0.41	0.36	0.28
Lower MI	0.58	0.41	0.32
China	0.85	0.56	0.45
Lower income	0.66	0.48	0.43
Total	6.18	4.93	4.36

It might help to sketch the derivation of the simple incentive constraint above. Consider any constant emission level vector $\widetilde{a} = \widetilde{a}_1, \widetilde{a}_2, \ldots, \widetilde{a}_I$. It can be easily shown that this emission level is sustainable as an equilibrium norm provided

$$\text{for every } i \text{ and } a_i, \quad h_i(\widetilde{a}_i) - c_i g + \delta\left[\widetilde{u}_i - w_i\left(\sigma g + \sum_j \widetilde{a}_j\right)\right] \geq h_i(a_i) - c_i g + \delta\left[u_i^* - w_i\left(\sigma g + a_i + \sum_{j \neq i} \widetilde{a}_j\right)\right], \tag{13}$$

where $\widetilde{u}_i$ and u_i^* are the present discounted values to emission policies that emit at constant rate-vectors of, respectively, $\widetilde{a}$ and a^* into the infinite future; recall that in the event of a deviation, the threat is to revert to the BAU emissions vector a^* forever. In turn, the above equation can be simplified to

$$\text{for every } i \text{ and } a_i, \quad h_i(\widetilde{a}_i) - \delta w_i \widetilde{a}_i + \delta \widetilde{u}_i \geq h_i(a_i) - \delta w_i a_i + \delta u_i^*.$$

From Eq. (8), the characterization of the BAU emission level, it follows that the deviation profit (the RHS of the equation above) is maximized at a_i^*. Using that fact and the definitions of $\widetilde{u}_i$ and u_i^*,[16] the incentive constraint can be rewritten in the form above.

It is immediate that the BAU emission policy is sustainable by the threat of BAU reversion (of course!) since the inequality is trivially satisfied when $\widetilde{a}_i = a_i^*$. What is also not very difficult to show is that the GPO emission policy also becomes sustainable at a high enough δ. Formally, we have

Proposition 2. *(a) The welfare that is achievable under the threat of BAU emissions is at least as high as u^*.*

(b) Suppose that the GPO solution under equal country weighting, ($x_i = x_j$ forall i, j) Pareto-dominates the BAU solution for all high $\delta \geq \delta'$; then there is a cut-off discount factor $\widetilde{\delta} \in (\delta', 1)$ such that, above it, the GPO emission policy is sustainable as an equibrium norm.

Consider the minimum emission problem:

$$Min \sum_i \widetilde{a}_i$$

over all incentive-compatible emission levels (i.e., over all $\widetilde{a}$ that satisfy the incentive constraint above). The next proposition characterizes the minimum emissions.

Proposition 3. *(a) There is a unique emission level $\underline{a}$ that is the minimum emission level sustainable by the threat of BAU reversion.*

(b) For all δ, the minimum emission level is strictly lower than the BAU level a^* (i.e., $\underline{a} << a^*$).

4.2. BAU sanctions: numerical evidence on minimum emissions

In this subsection we present empirical evidence on the emission cuts that are sustainable under the threat of BAU sanctions. We show that this is, for most regions, even lower than the GPO levels. There is one significant exception to this: Eastern Europe. We also present some country-specific evidence. Again countries from the ex-Soviet Bloc are the ones where significant cuts appear most difficult to implement; for Russia and Ukraine the GPO emission levels are, respectively, 39 percent and 40 percent lower than the BAU emission levels, but the maximum sustainable cuts are only 20 percent and 19 percent, respectively.

In Tables 5 and 6 below, MIN refers to the minimum emission level $\underline{a}$.

[16] Recall (see Eq. (9)) $u_i^* = (h_i(a_i^*) - \delta w_i \sum a_j^*/1 - \delta)$ and $\widetilde{u}_i = (h_i(\widetilde{a}_i) - \delta w_i \sum \widetilde{a}_j/1 - \delta)$. By extension, u^* is the vector $(u_1^*, u_2^*, \ldots, u_I^*)$ and $\widetilde{u}$ is the vector $(\widetilde{u}_1, \widetilde{u}_2, \ldots, \widetilde{u}_I)$.

Table 6
A comparison of minimum emissions and GPO emissions for selected countries.

Country/emissions	(MIN – GPO)/BAU (%)
UnitedStates	–9
Brazil	–31
China	–21
India	–9
Korea DPR	+22
Poland	+5
Russia	+17
Ukraine	+21

5. Asymmetry

One critical feature of the climate change problem is that countries differ greatly in the consequences they are likely to suffer. Whereas low-lying places such as Bangladesh may see parts of their countries washed away by rising sea-waters or equatorial countries may become uninhabitably hot, northern regions of the world such as parts of North America and northern Europe will bear limited costs and may even benefit from longer growing seasons and lower heating costs in the winter. In this section we analyze: how asymmetry (in costs) limits the extent of sustainable cuts. Are they optimal cuts? What are the welfare consequences of asymmetry? As in the last section, we will limit attention to equilibria that are sustainable by the threat of reversion to the BAU equilibrium. For this section, we will focus on

- "*Two countries*, (i.e., $I = 2$)."

We study two countries (or two regions) because it is easier to define asymmetry in terms of a single-dimensional parameter (namely the difference in the costs c_1 and c_2) and it is less obvious what we mean by an "increase" in asymmetry when we have I costs.[17]

5.1. Incentives and welfare

No matter what emissions policy we consider, when country i's marginal damage cost of global warming, c_i, increases, its total and marginal lifetime cost increases. As we saw in Section 3, the emission in any period t can be treated in isolation, and costs $c_i/(1-\delta\sigma)$ per unit of emission. Hence when c_i increases, *everything else being equal*, country i should be willing to cut emissions since benefits are unchanged and the marginal cost is higher (and conversely for a decrease in costs).

The problem though is that everything else does not actually remain unchanged when costs change. In particular, the sanction (i.e., the BAU solution) itself changes with the cost. In principle then the comparative statics of incentives are unclear; whereas a higher cost makes cheating on a treaty less profitable, it also makes the sanction to be suffered less severe. In this subsection we show that nevertheless the first effect always triumphs; a higher-cost country will agree to every cut that a lower cost country agrees to, and possibly more.

Consider any constant emissions policy with emissions at rate $\widetilde{a}$. Let $IC(c)$ denote the emission vectors that are incentive-compatible for both countries under the threat of BAU reversion, that is,

$$IC(c) = \{\widetilde{a}_1, \widetilde{a}_2 \geq 0 : h_i(\widetilde{a}_i) - \delta w_i(\widetilde{a}_i + \delta\widetilde{a}_j) \geq h_i(a_i^*) - \delta w_i(a_i^* + \delta a_j^*), \quad i = 1, 2, \quad i \neq j\}.$$

By extension, let $IC_i(c)$ denote the projection of that set (i.e., denote all emissions $\widetilde{a}_i$ of country i that are incentive-compatible given some incentive-compatible emission $\widetilde{a}_j$). The next result shows that as country i's cost increases, and j's cost decreases, its set of incentive-compatible emissions, $IC_i(c)$, gets larger as well. Conversely, the other country's set of incentive-compatible emissions, $IC_j(c)$, gets smaller.

Theorem 3. *Consider an emission vector $\widetilde{a}$ less than the BAU vector a^*: if $(\widetilde{a}_1, \widetilde{a}_2)$ is in $IC_i(c)$, that is, is incentive compatible (under costs c_1 and c_2) then it is still incentive-compatible for country i at $c_i + e$ and $c_j - e$, for $e > 0$. Conversely, $IC_j(c_i + e, c_j - e)$ is a subset of $IC_j(c)$.*

Recall the symmetric third-best problem:

$$Max\, V_1(g; a) + V_2(g; a), \tag{14}$$

where $V_i(g; a)$ is country i's lifetime payoff from emission policy $a(.)$, and this emission policy is incentive compatible under the threat of BAU reversion. Denote cost asymmetry by a single-dimensional parameter $d = c_1 - c_2$. Without loss of generality, let us consider cost changes in which c_1 increases while c_2 decreases, so that d is positive and increasing and let us start from

[17] Of course, the same analysis would work if the entire set I of countries were divisible into two groups, provided within each group every country had the same benefits and costs.

Table 7
Third-best emission levels as a function of cost difference *d*.

d	US (GtC)	China (GtC)
0.2	1.46400	0.77206
0.3	1.46369	0.77201
0.4	1.46208	0.77586
0.5	1.46048	0.77971
0.6	1.45601	0.79646
0.7	1.45332	0.81295

$d = 0$. In order to focus on the effect of incentive changes alone, let us keep the total cost, $c_1 + c_2$ constant since the total is the only way costs enter the global welfare function.

Theorem 4. *Suppose that the GPO is sustainable as a third-best equilibrium when costs are equal, $d = 0$, and suppose that $c_1 + c_2$ remains constant. Overall welfare, as measured by the sum of country payoffs are non-increasing in cost asymmetry d. It remains unchanged for small asymmetry, until a cut-off asymmetry level $\widehat{d}$, and thereafter strictly decreases.*

5.2. Asymmetry: some numbers

In this subsection the two countries are the US (country 1) and China (country 2), whose actual costs are quite similar to each other. Table 7 illustrates the *third-best* as it changes with asymmetry *d*.

6. Other equilibria II: a complete characterization of SPE

In this section we will characterize all possible subgame perfect equilibria (SPE) of the model. We will do that in two steps. In Section 6.1 we will characterize the SPE *payoff* correspondence. This characterization will be then used in Section 6.2 to study the *extreme equilibria*, the best and the worst equilibria. Although the theoretical analysis is complete, numerical estimates of extreme equilibria, and especially the worst equilibrium, are more difficult to compute. That exercise is left for future research.

6.1. The equilibrium payoff correspondence

We will show that the SPE payoff correspondence has a surprising simplicity; the *set* of equilibrium payoffs at a level g is a simple linear translate of the set of equilibrium payoffs from some benchmark level, say, $g = 0$. Consequently, it will be seen that the *set* of emission levels that can arise in equilibrium from level g is identical to those that can arise from equilibrium play at a GHG level of 0. Note that the fact that the set of equilibrium possibilities is invariant to the level of g is perfectly consistent with the possibility that, in a particular equilibrium, emission levels vary with g. (Indeed we will see some of those equilibria in Section 7.) However, the invariance property will make for a particularly simple characterization of the best and worst equilibria.

Let $\Xi(g)$ denote the *set* of equilibrium payoff vectors with initial state g (i.e., each element of $\Xi(g)$ is the payoff to *some* SPE starting from g).

Theorem 5. *The equilibrium payoff correspondence Ξ is linear; there is a compact set $U \subset \Re^I$ such that for every initial state g*

$$\Xi(g) = U - \{w_1 g, w_2 g, \ldots, w_I g\}$$

where $w_i = c_i/(1 - \sigma\delta)$, $i = 1, \ldots, I$. In particular, consider any SPE, any period t and any history of play up until t. Then the payoff vector for the continuation strategies must necessarily be of the form

$$v - (w_1 g_t, w_2 g_t, \ldots, w_I g_t)$$

where $v \in U$ (and g_t is the state at period t).

The theorem is proved by way of a bootstrap argument. We presume that a (candidate) payoff set has this invariance and show that the linear structure of the model confirms the conjecture. Consequently, we generate another candidate payoff set that is also state-invariant. Then we look for a fixed point of that operator.[18]

[18] In other words, we employ a generalized version of the Abreu et al. (1990) operator to generate the SPE correspondence. We need to generalize the APS argument since that was formulated for repeated games alone.

6.2. Extreme equilibria

We will now use the result of the previous subsection to characterize the best (and the worst) equilibria in the global climate change game. Consider the *second-best problem* (from initial state g and for a given vector of welfare weights $x = (x_i; i = 1, \ldots, I)$), the problem of maximizing a weighted sum of *equilibrium payoffs*:

$$\max \sum_{i=1}^{I} x_i V_i(g),\ V(g) \in \Xi(g).$$

Note that we consider all possible equilibria, so, we consider equilibria that choose to condition on current and past GHG levels as well as equilibria that do not. The result states that the best equilibrium *need not* condition on GHG levels.

Theorem 6. *There exists a constant emission level $\bar{a} \equiv \bar{a}_1, \bar{a}_2, \ldots, \bar{a}_I$ such that no matter what the initial level of GHG, the second-best policy is to emit at the constant rate $\bar{a}$. In the event of a deviation from this constant emissions policy by country i, play proceeds to i's worst equilibrium. Furthermore, the second-best emission rate is always strictly lower than the BAU rate* (i.e., $\bar{a} < a^*$). *Above a critical discount factor* (*less than* 1), *the second-best rate coincides with the GPO emission rate $\widehat{a}$.*

The theorem is attractive for three reasons. First, it says that the best possible equilibrium behavior is no more complicated than BAU behavior, so there is no argument for delaying a treaty (to cut emissions) merely because the status quo is simple. Second, the cut required to implement the second-best policy is an across-the-board cut; independently of anything else, country i should cut its emissions by the amount $a_i^* - \bar{a}_i$.[19] Third, the second-best is exactly realized at high discount factors rather than asymptotically approached as the discount factor tends to 1.

Sanctions will be required if countries break with the second-best policy,[20] and without loss of generality we can restrict attention to the worst such sanction. We turn now to a characterization of this worst equilibrium (for, say, country i). One definition will be useful for this purpose.

Definition 1. *An i-less second-best equilibrium is the solution to a second-best problem in which the welfare weight of i is set equal to zero* (i.e., $x_i = 0$).

By the previous theorem, every such problem has a solution in which on the equilibrium path, emissions are a constant. Denote that emission level $a(x_{-i})$.

Theorem 7. *There exists a "high" emission level $\bar{\bar{a}}(i)$(with $\sum_{j \neq i} \bar{\bar{a}}_j(i) > \sum_{j \neq i} a_j^*$) and an i-less second-best equilibrium $a(x_{-i})$ such that at the country i's worst equilibrium*:

1. *Each country emits at rate $\bar{\bar{a}}_j(i)$ for one period (no matter what g is), $j = 1, \ldots, I$.*
2. *From the second period onwards, each country emits at the constant rate $a_j(x_{-i}), j = 1, \ldots, I$.*

If any country k deviates at either stages 1 *or* 2, *play switches to k's worst equilibrium starting with the very next period after the deviation.*

Put another way, for every country i, a sanction is made up of two emission rates, $\bar{\bar{a}}(i)$ and $a(x_{-i})$. The former imposes immediate costs on country i. The way it does so is by increasing the emission levels of countries $j \neq i$. The effect of this is a temporary increase in incremental GHG, but due to the irreversibility of gas accumulation, a permanent increase in country i's costs, enough of an increase to wipe out any immediate gains that the country might have obtained from the deviation. Of course this additional emission also increases country j's costs. For the punishing countries, however, this increase is offset by the subsequent permanent change, the switch to the emission vector $a(x_{-i})$, which permanently increases their quota at the expense of country i's.

The fact that there is a temporary loosening of environmental regulations as part of environmental sanctions is reminiscent of GATT rules where tariffs can be temporarily imposed by countries that seek to punish illegitimate trade practices on the part of others.

7. Greenhouse trap

In every equilibrium that we have studied so far (BAU, the third and second-best), each country emits at a constant rate regardless of GHG level. Hence, the dynamics of every such equilibrium is also simple; at a constant cumulative rate A, the stock of greenhouse gases (GHGs) grow to a steady-state of $(A/(1 - \sigma))$. In particular, there is no such thing as starting "too

[19] Our model operates at the aggregative level alone and, in particular, we do not address the issue of how national governments will implement cuts that they agree to (in the national interest). However, it seems quite likely that an across-the-board cut will be easier to implement (and will be perceived to be fairer to all) than one which is sensitively tied to levels of GHG.

[20] A major criticism of the Kyoto accord is that it did not incorporate sanctions and hence would never be carried out. For details, see Barrett (2003, Chapter 15) and the further discussion in Section 8.

200 P.K. Dutta, R. Radner / Journal of Economic Behavior & Organization 71 (2009) 187–209

late" on emission control. As long as the current GHG level is less then the target steady-state, a successful effort to bring emissions down to A will guarantee that the world will not zoom past that steady-state. Put yet another way, the current GHG level has no long-term implication.

In this section we demonstrate the richness of the model by identifying some MPEs in which current GHG levels matter, MPEs in which there is a so-called "greenhouse trap". If the world starts below some critical level of GHG, say $\tilde{g}$, then it grows no farther than that level. However if the system starts above $\tilde{g}$ (or somehow crosses into the higher region) then greenhouse gases are trapped into growing and eventually grow to the BAU steady-state.

The key to these equilibria is a richer interaction between the (Markov) emission levels of country i and the rest of the world. In particular, we will consider emission policies that are Markovian but not constant, say an emission policy (vector) such as $a(g)$. Now country i has an incentive, everything else being equal, to emit in such a fashion that the global stock of GHG grows towards a region where $a_{-i}(.)$ are lower. Of course every country has such an incentive, so each country will wish to drive g_t towards a region where emission levels are low for the other countries. Hence, the conjecture for the group as a whole is that there will be MPEs with regions of "abnormally" low emissions, and every country will (a) have an incentive to stay in such regions once the system gets there and (b) have an incentive to participate in pushing the system towards such regions.

The conjecture is almost correct. We will show that there are indeed such equilibria (in fact there are many) each of which has such a "good" region of low emissions. Furthermore, from *most* (but not all) initial GHG levels outside this good region, countries will have an incentive to drive the system to the good region. The one additional complication is that in order for there to be a good region there must also be some "bad" region (of high emissions) from which countries do *not* find their way to the good region. The presence of a bad region, and the associated spectre of landing there, is what keeps the countries honest in the good region.[21]

To keep the exposition simple, we are going to present results only for the symmetric case.[22] Accordingly, when we speak of the Pareto optimal solution we will refer to the symmetric solution, and to avoid clutter we will simply denote that solution by $\widehat{a}$ (with associated steady state $\widehat{g}$). We shall present two results of increasing generality.

Consider the following symmetric Markovian strategy $a(.)$: if the GHG level is below the Pareto optimal steady state $\widehat{g}$, emissions take the game immediately to that state. On the other hand, if the GHG level exceeds $\widehat{g}$, then emissions are at the (high) BAU level of a^*. In other words,

$$a(g) = \frac{\widehat{g} - \sigma g}{I}, \quad g \leq \widehat{g}$$
$$= a^*, \quad g > \widehat{g}.$$

In the terminology of the immediately preceding discussion, the region below the Pareto optimal steady state $\widehat{g}$ is the "good" region of (relatively) low emission levels whereas the region above is the "bad" region of high emissions. Consider the following condition ("L" for "large"), which says that the BAU emission level is sufficiently larger than the GPO level.

Condition L:

$$\frac{a^*}{\widehat{a}} > \max\left(\frac{I}{I-1}, \frac{1}{1-\sigma}\right),$$

where (recall) I is the number of players and σ is the persistence of CO_2 in the earth's atmosphere.[23]

Theorem 8. *Suppose that Condition L holds. Then there is a cut-off value of the discount factor, say $\widehat{\delta}$, such that $a(.)$ is an MPE for all $\delta \geq \widehat{\delta}$. In such an equilibrium, the GHG level converges in one period to the Pareto optimal steady state $\widehat{g}$ if the initial level is below $\widehat{g}$ whereas it converges asymptotically to the BAU steady state g^* if the initial level is above $\widehat{g}$.*

The reader might wonder how useful this last theorem is if we suspect that the world is currently already past the Pareto optimal steady state. Are there MPE that have steady states less than the BAU steady state of g^* (but higher than the desired but unattainable steady state of $\widehat{g}$)?

We will now demonstrate the existence of an MPE that differs in three ways from the one above. First, in addition to the good and bad regions, there will be a "latent good" region, a region of low GHG levels from which the stock will gradually grow until it reaches the good region (whereupon it will stay there). Second, the good region (with emissions lower than the BAU level a^*) will extend beyond the Pareto optimal steady state $\widehat{g}$; in fact it will extend quite close to the BAU steady

[21] Readers familiar with repeated games will note the obvious connection with the idea of history-dependent punishments. The point to note though is that these are not history-dependent equilibria since they only depend on the GHG level and not on past emissions; hence they are more difficult to construct. They are also more sparing in terms of informational requirements since they do not require i to condition on (or even know) past emission levels of the other countries.

[22] The first of the two results is easily generalized to the asymmetric case.

[23] Since $(I/I-1)$ approaches 1 for I large, Condition L really boils down to $(a^*/\widehat{a}) \geq (1/1-\sigma)$ whenever we have a large number of players. This condition will always hold under standard asymptotic conditions on h. To see this, note that simple algebra shows that $(h'(\widehat{a})/h'(a^*)) = I$, and hence $(a^*/\widehat{a})$ is large whenever I is large.

state g^*. (So no matter where the world is today, however close to the worst possibility, we can still put the brakes on in an incentive-compatible way!) Third, we will considerably weaken Condition L.

Let $\tilde{g}$ be any GHG level that is higher than $\hat{g}$ but no more than $((I-1)/I)g^*$; $\tilde{g}$ will be our candidate low steady state. We will consider Markov strategies in which stocks from a left neighborhood of $\tilde{g}$ (to be defined shortly) come in one step to $\tilde{g}$ and stocks above $\tilde{g}$ exhibit BAU behavior:

$$\begin{aligned}\tilde{a}(g) &= a^*, \quad g > \tilde{g} \\ &= \frac{\tilde{g} - \sigma g}{I}, \quad \tilde{g}_1 \le g \le \tilde{g}\end{aligned} \tag{15}$$

where

$$\tilde{g}_1 \equiv \frac{I-1}{I}\tilde{g}.$$

For stocks lower than $\tilde{g}_1$, the "latent good" region, the emission levels will be such that GHG levels grow (gradually) until they get into the $[\tilde{g}_1, \tilde{g}]$ region. These emission levels cannot, however, be solved for in closed form. Instead we will employ a fixed point argument to show the following.

Theorem 9. *Suppose that*

$$\frac{a^*}{\tilde{a}} > \frac{I}{I-1}.$$

Then there is a cut-off value of the discount factor, say $\tilde{\delta}$, such that for all $\delta \ge \tilde{\delta}$, there is a MPE $\tilde{a}(.)$ whose behavior above $\tilde{g}_1$ is as given above. Below $\tilde{g}_1$, the stock grows although it remains below $\tilde{g}$, that is,

$$\sigma g + I\tilde{a}(g) \in (g, \tilde{g}), \text{for all} g < \tilde{g}.$$

In such an equilibrium, the GHG level converges to the steady state $\tilde{g}$ if the initial level is below it whereas it converges asymptotically to the BAU steady state g^* if the initial level is above $\tilde{g}$.

8. Discussion and extensions

In terms of conclusions, our findings are similar in terms of welfare loss to those of other authors, including those who focus on the first-best issue. Nordhaus and Boyer calculate a United States loss from global climate change of the order of 0.5 percent of GDP (for doubling of pre-industrial levels); other estimates have been in the 1–2 percent range. This agrees with our findings of a decrease in global welfare (respectively, US welfare) in the range of 0.6–2 percent (respectively, 0.3–1 percent) for different discount rates with our benchmark damage cost coefficients. Where we differ is in the size of the emission cuts that our model predicts. Whereas Kyoto had called for cuts in the 10 percent range relative to 1990 levels and some of the literature has proposed that even that is too much to be globally optimal (see Nordhaus and Boyer), our numerical computations ask for much deeper emission cuts. We conjecture that the dynamic element of our model (that gases can persist in the atmosphere for a 100 years or more, which is in line with the known evidence) make the deeper cuts optimal. Furthermore, our results for both potential welfare increases and emission cuts are quite sensitive to increases in the damage cost coefficients.

The Kyoto Protocol appears not to be the basis of a self-enforcing treaty, principally because it does not build in any effective sanctions that would be applied if countries fail to meet their targets. This contrasts with GATT and its attendant institutional structure, the World Trade Organization (WTO). All that is said in the Kyoto protocol (in Article 18) is that procedures and mechanisms for compliance should be determined by the parties at their first meeting and should include "an indicative list of consequences". Subsequently, at the Hague November 2000 meeting, the most popular proposal (which came from the Dutch Environment Minister Jan Pronk) was that countries would face an escalating series of target reductions in the future if they failed to comply in the current stage. A watered-down version of this proposal was adopted in Bonn in March, 2001, after the United States had pulled out of the treaty, yet even this version had several problems; in principle, countries could postpone retribution indefinitely, and the base from which the enhanced reduction would be required would be worked out in the future, and so on (Barrett, 2003, Chapter 15). One contribution of this paper is to show that fairly simple sanctions (BAU reversion) can be quite effective, and even the most severe sanctions (the worst equilibria) are fairly simple.

A key feature of our model is that (1) the one-period payoff for each country is additively separable in GDP and the damage cost, and (2) the (incremental) damage cost is linear in the current stock of greenhouse gas. This property of the model allows us to get closed-form solutions for the Business as Usual and Pareto optimal solutions, characterize the equilibrium payoff correspondence, and investigate other special classes of equilibria. It also facilitates the calibration of the model, the numerical calculation of various trajectories, and sensitivity analyses. The disadvantage is that it results in a number of cases in unrealistic "bang-bang" strategies, that is, strategies in which the emission rates are constant after the first period. This aspect of the results needs to be taken "with a grain of salt." In a more realistic model, one would expect that these strategies would display a more gradual adaptation to the current levels of greenhous gas. Our conjecture is that the analysis of the affine model yields reasonable approximations to equilibrium and optimal trajectories, except in the very short run. However, precise tests of this conjecture will have to await future research.

In Dutta and Radner (2006) we generalized our current model to allow for population change and demonstrate qualitatively similar theoretical results. In Dutta and Radner (2004) we allowed for simple technological change and presented some theoretical and numerical results on the GPO and BAU solutions. In Dutta and Radner (in preparation) we incorporate capital accumulation (and technical change). The main question that we hope to address in that model is under what conditions does the prevention of global warming slow down the rate of economic growth. A second question that we hope to look at is: (how) does asymmetry in the current level of economic development affect sustainability of agreements about emission cuts. We have some preliminary results on the first question but not a complete solution. Finally, we hope to develop and analyze a "complete" model that incorporates all of the above features.

9. Bibliographic notes

The literature on (symmetric) dynamic commons games is exceedingly rich and goes back over 25 years. The earliest model was that of Levhari and Mirman (1980) who studied a particular functional representation of the neo-classical growth model with the novel twist that the capital stock could be "expropriated" by multiple players. Subsequently several authors (Sundaram, 1989; Sobel, 1990; Benhabib and Radner, 1992; Rustichini, 1992; Dutta and Sundaram, 1992, 1993; Sorger, 1998) studied this model in great generality and established several interesting properties relating to existence of equilibria, welfare consequences, and dynamic paths.[24]

Another variant of that model has been studied by Tornell and Velasco (1992) and, subsequently, Long and Sorger (2006). In that model, the public capital stock grows linearly at a fixed rate of growth. The expropriated capital need not be consumed immediately but can instead be made to grow privately at a rate lower that the public growth rate. The effect of such expropriation (or capital outflows) is examined by the various authors.

More recently in a series of papers by Dockner and his co-authors, the growth model has been directly applied to environmental problems including the problem of global warming. The paper closest to the current one is Dockner et al. (1996). It studies a model of global warming that has some broad similarities to the one we have studied here. In particular, the transition equation is identical in the two models. What is different is that they impose linearity in the emissions payoff function h (whereas we have assumed it to be strictly concave) while their cost to g is strictly convex (as opposed to ours which is linear).

The consequence of linearity in the benefit function h is that the GPO and BAU solutions have a "most rapid approach" property; if $(1-\sigma)g$, the depreciated stock in the next period, is less than a most preferred g^*, it is optimal to jump the system to g^*. Else it is optimal to wait for depreciation to bring the stock down to g^*. In other words, linearity in benefits implies a "one-shot" move to a desired level of gas g^*, which is thereafter maintained, while linearity in cost (as in our model) implies a constant emission rate.[25] The authors then show that a trigger strategy can realize the GPO for high enough discount factors. The paper also shows that there might be other MPE, and it gives an example of an MPE with complex dynamics. That MPE in this environmental model is identical, as the authors acknowledge, to a complex dynamics example first provided by Dutta and Sundaram (1993) in the context of the growth model.

Where the two papers truly differ in terms of analysis is that the Dockner at al. paper has a) no discussion of the effects of asymmetry or b) a full characterization of all subgame perfect equilibria. Additionally, that paper does not attempt to calibrate its model to generate illustrative numbers.[26]

A large volume of literature exists that directly focuses on the economics of climate change. A central question there is to determine the level of emissions that is globally optimal. An excellent example of this is Nordhaus and Boyer. Several of those papers, including the Nordhaus and Boyer paper, analyze only the "competitive" model, not taking strategic considerations fully into account.[27] A smaller volume of literature emphasizes the need for treaties to be self-enforcing, presenting a strategic analysis of the problem (see Barrett, 2003 and Finus, 2001). Where we depart from that literature is in the dynamic modelling; we allow GHGs to accumulate and stay in the environment for a (possibly long) period of time. By contrast the Barrett and Finus studies

[24] Some of these papers allow asymmetry; however, none of them analyzes the effect of asymmetries. One significant exception is the recent paper of Long and Sorger that explicitly considers asymmetry in appropriation costs within the Tornell and Velasco model. They show that increasing asymmetry lowers equilibrium rate of growth, a result broadly consistent with the one shown in this paper that asymmetry is deletrious.

[25] What is unclear in their model is why the multiple players would have the same target steady-state g^*. It would appear natural that, with asymmetric payoffs, each player would have a different steady-state. The existence of a MRAP equilibrium would consequently appear problematical. The authors impose a condition that implies that there is not too much asymmetry.

[26] There are is one other interesting and relevant paper (Dockner and Nishimura, 1999) that focuses on complex dynamics example and shows that it emerges in a couple of related models as well (for example, a model in which every country's pollution affects one immediate neighbor).

[27] To be fair, Nordhaus and Boyer (2000) and Nordhaus and Yang (1996) do consider strategic models but restrict themselves to open-loop strategies.

restrict themselves to purely repeated games, which implies that the state variable, gas stock, remains constant over time.

10. Proofs

10.1. Global pareto optima and BAU equilibrium

Proof of Theorem 1. We shall show by dynamic programming arguments that the Pareto-optimal value function is of the form $\widehat{V} = \sum_{i=1}^{I} x_i[\widehat{u}_i - w_i g]$. We need to be able to find the constants $\widehat{u}_i$ to satisfy

$$\sum_{i=1}^{I} x_i[\widehat{u}_i - w_i g] = Max_{a_1,\ldots,a_I} \sum_{i=1}^{I} x_i \left[h_i(a_i) - c_i g + \delta \left(\widehat{u}_i - w_i \left(\sigma g + \sum_{j=1}^{I} a_j \right) \right) \right]. \tag{16}$$

Collecting terms that need maximization we can reduce the equation above to

$$\sum_{i=1}^{I} x_i \widehat{u}_i = Max_{a_1,\ldots,a_I} \sum_{i=1}^{I} x_i [h_i(a_i) - \delta w_i \sum_{j=1}^{I} a_j] + \delta \sum_{i=1}^{I} x_i \widehat{u}_i. \tag{17}$$

It is clear that the solution to this system is the same for all g; call this (first-best) solution $\widehat{a}_i$. Elementary algebra reveals that

$$\widehat{u}_i = \frac{h_i(\widehat{a}_i) - \delta w_i \sum_{j=1}^{I} \widehat{a}_j}{1-\delta} \quad \text{and} \quad w_i = \frac{c_i}{1-\delta\sigma}.$$

It is also obvious that $x_i h_i'(\hat{a}_i) = \delta w$, where $w = \sum_{i=1}^{I} x_i w_i$. □

Proof of Theorem 2. Consider any constant emission strategy for players other than i; call these constants $\bar{a}_j$. We shall now show by dynamic programming arguments that the best response value function is of the form $\overline{u}_i - w_i g$, where

$$\overline{u}_i = \frac{h_i(a_i^*) - \delta w_i(a_i^* + \sum_{j \neq i} \bar{a}_j)}{1-\delta}.$$

We therefore need to show that

$$\overline{u}_i - w_i g = Max_{a_i} \left[h_i(a_i) - c_i g + \delta(\overline{u}_i - w_i) \left(\sigma g + a_i + \sum_{j \neq i} \bar{a}_j \right) \right]. \tag{18}$$

Collecting the terms that need maximization we can reduce the above equation to

$$\overline{u}_i = Max_{a_i}[h(a_i) - \delta w_i a_i] + \delta \overline{u}_i. \tag{19}$$

It is clear that the solution to this system is the same for all g and equals a_i^* where $h_i'(a_i^*) = \delta w_i$. □

10.2. Equilibria sustained by BAU sanctions

Proof of Proposition 1. The proof follows the lines of that for the analogous result on the second-best along the lines of the proof of Theorems 6 and 7(see below). Evidently, equilibria that are sustained by the threat of BAU reversion are SPE, and hence their values belong to the SPE payoff correspondence that we identify in Theorem 6- $U - \{w_1 g, w_2 g, \ldots, w_I g\}$ where U is a compact subset of $\Re^I$. Define therefore the set of incentive-compatible emission vectors a (associated with the threat of BAU reversion) as

$$A = \{a \in \Re_+^I : h_i(a_i) + \delta(u_i' - w_i a_i) \geq h(\widetilde{a}_i) + \delta(u_i^* - w_i \widetilde{a}_i), \forall \widetilde{a}_i, i, \text{for some} u' \in U\} \tag{20}$$

where

$$u_i^* = \frac{h_i(a_i^*) - \delta w_i \sum_{j=1}^{I} a_j^*}{1-\delta}$$

is the BAU equilibrium payoff.

The third-best equilibrium is therefore found by solving the following problem:

$$Max_{a_i,u_i}\sum_i x_i v_i = Max_{a_i,u_i}\sum_{i=1}^{I} x_i\left[h_i(a_i)+\delta\left[u_i - w_i\sum_{k=1}^{I} a_k\right]\right],$$

subject to the incentive compatibility constraint Eq. (20). Since the constraint is independent of g, and so is the objective function, it follows that the solution is as well. The proposition is proved. □

Proof of Proposition 2. The only part left to prove is part (b). Note that as $\delta \uparrow 1$, the incentive constraint, Eq. (20), becomes

$$h_i(\widetilde{a}_i) - w_i(1)\left(\widetilde{a}_i + \sum_{j\neq i}\widetilde{a}_j\right) \geq h_i(a_i^*) - w_i(1)\left(a_i^* + \sum_{j\neq i} a_j^*\right), \forall i, \tag{21}$$

where

$$w_i(1) = \lim_{\delta\uparrow 1} w_i(\delta) = \lim_{\delta\uparrow 1}\frac{c_i}{1-\sigma\delta} = \frac{c_i}{1-\sigma}.$$

The expression on the LHS, $h_i(\widetilde{a}_i) - w_i(1)\sum_j \widetilde{a}_j$, is precisely the long-run average (LRA) value from a policy that emits at the constant rate $\widetilde{a}$. In particular then, if we pick the candidate $\widetilde{a}$ to be the GPO emission under the LRA criterion with equal weights for all countries, the incentive constraint Eq. (21) is satisfied by the definition of GPO, at $\delta = 1$. Add to that (a) that the LRA GPO is the limit of discounted average GPO values (see Dutta, 1991) and (b) the assumption that the equal weight GPO solution Pareto-dominates the BAU value for all high discount factors. The proof is complete. □

Proof of Proposition 3. We shall first prove part (b). Start then at the BAU solution a^* and consider a reduction of every country's emission by an amount ε. We shall show that the emission vector $a^* - \varepsilon$ is sustainable by the BAU sanction for appropriately small but positive ε. Note that the LHS of the incentive constraint, Eq. (20), becomes

$$h_i(a_i^* - \varepsilon) - \delta w_i\left(a_i^* - \varepsilon + \delta\sum_{j\neq i}(a_j^* - \varepsilon)\right),$$

and that can be rewritten as

$$h_i(a_i^*) - \delta w_i\left(a_i^* + \delta\sum_{j\neq i}(a_j^* - \varepsilon)\right) - [h_i'(z_i^*) - \delta w_i]\varepsilon$$

for some $z_i^* \in (a_i^* - \varepsilon, a_i^*)$. That in turn equals

$$h_i(a_i^*) - \delta w_i\left(a_i^* + \delta\sum_{j\neq i} a_j^*\right) + \{\delta^2 w_i(I-1) - [h_i'(z_i^*) - \delta w_i]\}\varepsilon$$

As $\varepsilon \downarrow 0$, $z_i^* \to a_i^*$. Since $h_i'(a_i^*) - \delta w_i = 0$, it follows that $\delta^2 w_i(I-1) - [h_i'(z_i^*) - \delta w_i] > 0$ for small ε. In particular, the incentive constraint

$$h_i(a_i^* - \varepsilon) - \delta w_i\left(a_i^* - \varepsilon + \delta\sum_{j\neq i}(a_j^* - \varepsilon)\right) \geq h_i(a_i^*) - \delta w_i\left(a_i^* + \delta\sum_{j\neq i} a_j^*\right), \forall i. \tag{22}$$

holds. The proof of part (b) is complete.

At the minimum incentive-compatible emission level, every IC constraint must hold with equality. Consider instead the alternative case; suppose the constraint is a strict inequality for country *j*. In that case we can further reduce that country's emission without violating its constraint. However lowering *j*'s emission only helps satisfy every other country's constraint. Hence, we will have lowered total emissions without violating incentive constraints, which is a contradiction.

Rewriting the incentive constraints as equalities yields

$$h_i(a_i) - \delta w_i\left(a_i + \delta\sum_{j\neq i} a_j\right) = \Psi_i, \forall i$$

where $\Psi_i = h_i(a_i^*) - \delta w_i(a_i^* + \delta\sum_{j\neq i}a_j^*)$. Let us suppose that $\underline{a}$ is a solution to the emission minimization problem and let $\underline{A} = \sum_i \underline{a}_i$. Rewriting the immediately preceding equation we have

$$h_i(\underline{a}_i) - (1-\delta)w_i\underline{a}_i = \Psi_i + \delta^2 w_i\underline{A}.$$

Given the strict concavity of h, it is evident that $\underline{a}_i$ is uniquely defined by the above equation. The proof is complete. □

10.3. Asymmetry

Proposition 1 above greatly simplifies the incentive analysis for the third-best problem since we can now restrict ourselves to constant emissions policies and the consequent incentive constraint:

$$h_i(a_i) - \delta w_i(a_i + \delta a_j) \geq h_i(a_i^*) - \delta w_i(a_i^* + \delta a_j^*),\ i,\quad \text{and for } j \neq i. \tag{23}$$

Proof of Theorem 3. Consider an increase in w_1 (to w_1') and a simultaneous decrease in w_2. Denote $S_i(a)$ as the incentive slack of country i; therefore,

$$S_i(a) = h_i(a_i) - \delta w_i(a_i + \delta a_j) - [h_i(a_i^*) - \delta w_i(a_i^* + \delta a_j^*)].$$

Similarly, denote the incentive slack under the new costs S_i'. Evidently, for country 1, the difference in the two slacks, $S_1'(a) - S_1(a)$, equals

$$\{h_1(a_1) - \delta w_1'(a_1 + \delta a_2) - [h_1(a_1) - \delta w_1(a_1 + \delta a_2)]\} - \{h_1(a_1'*) - \delta w_1'(a_1'* + \delta a_2'*) - [h_1(a_1^*) - \delta w_1(a_1^* + \delta a_2^*)]\}$$

where $a'*$ is the BAU emission level for costs w'. Call the term in the first (curly) bracket, $\{\}_1$ and the second one $\{\}_2$. Simple algebra yields

$$\{\}_1 = \delta(w_1 - w_1')[a_1 + \delta a_2]. \tag{24}$$

Note that

$$h_1(a_1'*) - \delta w_1(a_1'*) < h_1(a_1^*) - \delta w_1(a_1^*),$$

by the definition of a_1^*. Furthermore, $a_2'* > a_2^*$ since country 2's costs have gone down. Combining those two facts we have that

$$\begin{aligned}\{\}_2 &< h_1(a_1'*) - \delta w_1'(a_1'* + \delta a_2'*) - [h_1(a_1'*) - \delta w_1(a_1'* + \delta a_2'*)],\\ &= \delta(w_1 - w_1')[a_1'* + \delta a_2'*].\end{aligned} \tag{25}$$

Since $a_1'* + \delta a_2'* \geq a_1 + \delta a_2$, it further follows $\{\}_2 < \{\}_1$; therefore, $S_1'(a) - S_1(a) = \{\}_1 - \{\}_2 > 0$.

To prove the result for country 2, we mimic the above argument except in reverse. The theorem is proved. □

Proof of Theorem 4. Consider starting from the symmetric situation where $c_1 = c_2$ or, equivalently, the symmetric situation where the lifetime costs are equal:

$$w_1 = \frac{c_1}{1-\delta\sigma} = \frac{c_2}{1-\delta\sigma} = w_2.$$

Now consider varying the costs in a manner that the costs of country 1 "South" are increased while those of country 2 "North" are decreased. Let us measure the asymmetry in costs by the difference $d = c_1 - c_2$ or, equivalently, the difference in lifetime costs:

$$\frac{d}{1-\delta\sigma} = \theta = w_1 - w_2.$$

In other words, we shall consider an increase in asymmetry as exemplified by an increase in d (or θ). Furthermore, to ensure that the only effect is that of an increase in asymmetry, we shall hold constant the sum of the two costs $w_1 + w_2$.

By hypothesis, the GPO is third-best when $w_1 = w_2$.

Even when the cost difference becomes positive, the GPO continues to be incentive-compatible for both countries as long as d is small. However, beyond a cut-off value, it is no longer incentive-compatible for the low cost country 2 "North" (see Theorem 3). At that point, in the third-best solution only one of the incentive constraints is binding. Furthermore, and again from Theorem 3 above, that has to be the incentive constraint for country 2 (the lower-cost country).

Consider two different θ^1 and θ^2, say $\theta^1 > \theta^2$ and let $a^1 = (a_1^1, a_2^1)$ and $a^2 = (a_1^2, a_2^2)$ be the corresponding third-best emission vectors. Since the incentive constraint for country 1 "South" is slack, it follows that if θ^1 is chosen sufficiently close to θ^2, then a_1^1 is also incentive compatible at θ^2. By Theorem 3 above, country 2's incentive region shrinks as its costs decrease (i.e., its emission under θ^1, a_2^1, must also have been incentive compatible at θ^2). Putting the two together, the emission vector a^1 is incentive compatible at θ^2, but the choice made for that parameter is a^2. That implies that total welfare is higher at a^2 than at a^1 at the parameter θ^2, but since the total welfare function is unchanged as the cost difference changes (since $w_1 + w_2$ is constant), it follows that total welfare is lower at θ^1 compared to θ^2. □

10.4. All equilibria

Proof of Theorem 5. We shall employ the generalization of the Dynamic Programming Bellman equation that was introduced by Abreu et al. to study the set of SPE in repeated games. Let us start with the conjecture that the correspondence of SPE payoffs, V, has the structure claimed in the result (i.e., is of the form $U - \{w_1 g, w_2 g, \ldots, w_I g\}$ where U is a compact subset of $\Re^I$). Let $\underline{u}^i$ denote the vector in U that gives country i the worst payoff,

$$\underline{u}^i = \min\{u_i | u \in U\},$$

and denote i's payoff in that minimum as $\underline{u}^i_i$. Define the set of incentive-compatible emission vectors a (associated with this candidate equilibrium payoff set) as

$$A = \{a \in \Re^I_+ : h_i(a_i) + \delta(u'_i - w_i a_i) \geq h(\tilde{a}_i) + \delta(\underline{u}^i_i - w_i \tilde{a}_i), \forall \tilde{a}_i, i, \text{for some} u' \in U\}. \tag{26}$$

Now define the APS operator BV (that takes the correspondence V into another correspondence) as follows:

$$BV = \Big\{v(.) : \exists \text{selections} a(.) \text{and} u(.,.) \text{satisfying} a(g) \in A, \text{and} u(g', g) \in U, \forall g', g.$$

$$v_i(g) = h_i(a_i) - c_i g + \delta\left(u_i\left(\sigma g + \sum_{j=1}^{I} a_j(g), g\right) - w_i\left(\sigma g + \sum_{j=1}^{I} a_j(g)\right)\right)\Big\}.$$

Note that $v(.)$ can be rewritten as

$$v_i(g) = \left[h_i(a_i) + \delta\left(u_i\left(\sigma g + \sum_{j=1}^{I} a_j(g), g\right) - w_i \sum_{j=1}^{I} a_j(g)\right)\right] - w_i g$$

Note from the incentive set A that exactly the same set of actions satisfy the incentive constraint at every value of g. By hypothesis the "continuation payoffs" $u(g', g) \in U$, a set that is independent of g as well. It follows that BV is a correspondence of the same form as V (i.e., $BV = U' - \{w_1 g, w_2 g, \ldots, w_I g\}$ for some U' that is a subset of $\Re^I$). Standard arguments using the Maximum theorem can be employed to show that U' must be a compact set as well.

Start then with a set U_0 (or correspondence V_0) that is compact and "large" (one that contains all feasible payoffs in the game). Define U_1 via the APS operator as above; U_1 is compact as well. It must also be non-empty because it must contain the BAU payoff u^*. By definition, $U_1 \subset U_0$. Define U_n recursively in this fashion. It is easy to check that the operator is monotone, and hence U_n is a decreasing sequence of non-empty compact sets. Associated with each is the correspondence $V_n = U_n - \{w_1 g, w_2 g, \ldots, w_I g\}$. Let U be the intersection of these U_n sets. APS type arguments then show that $BV = V$ and that V contains all SPE payoffs. The theorem is proved. □

Proof of Theorem 6. Given the previous theorem, the second-best equilibrium is found by solving the following problem:

$$Max_{a, u(.,g) \in U} \sum_{i=1}^{I} x_i \left[h_i(a_i) + \delta\left[u_i\left(\sigma g + \sum_{k=1}^{I} a_k, g\right) - w_i \sum_{k=1}^{I} a_k\right]\right]$$

where, for every g, $u(\sigma g + \sum_{k=1}^{I} a_k, g) \in U$, and the emissions have to satisfy the incentive compatibility constraint

$$h_j(a_j) + \delta\left(u_j\left(\sigma g + \sum_{k=1}^{I} a_j, g\right) - w_j a_j\right) \geq h_j(\tilde{a}_j) + \delta(\underline{u}^j_j - w_j \tilde{a}_j), \forall \tilde{a}_j, j.$$

Since the constraint is independent of g, as is the objective function, it follows that the solution is as well. □

Proof of Theorem 7. In light of our previous results, the worst equilibrium for player i, if the state is g, is a solution to the following problem:

$$Min_{a, u(.,g) \in U} h_i(a_i) + \delta\left[u_i\left(\sigma g + \sum_{k=1}^{I} a_k, g\right) - w_i \sum_{k=1}^{I} a_k\right] \tag{27}$$

$$s.t. h_j(a_j) + \delta\left(u_j\left(\sigma g + \sum_{k=1}^{I} a_j, g\right) - w_j a_j\right) \geq h_j(\tilde{a}_j) + \delta(\underline{u}^j_j - w_j \tilde{a}_j), \forall \tilde{a}_j, j. \tag{28}$$

Since $u(.,g) \in U$ (a state independent set) it follows that the solution to the problem is going to be state independent as well. Furthermore, by arguments that we have seen before, it is clear that if player j is going to deviate (and expects the worst equilibrium for her to be the continuation regardless of how much she deviated by), then her optimal deviation is a_j^*. In other words, Eqs. (27) and (28) can be simplified to yield the following:

$$\underline{u}_i = Min_{a,\widetilde{u} \in U} h_i(a_i) + \delta \left[\widetilde{u}_i - w_i \sum_{k=1}^{I} a_k \right] \tag{29}$$

$$s.t. h_j(a_j) + \delta(\widetilde{u}_j - w_j a_j) \geq h_j(a_j^*) + \delta(\underline{u}_j^j - w_j a_j^*), \forall j. \tag{30}$$

Note that it must be the case that the incentive constraint for player i, whose worst equilibrium we are analyzing, is binding in the above problem. If not, we can perturb a_i in such a manner (without changing $a_j, j \neq i$) such that $h_i(a_i) - \delta w_i a_i$ is lowered *and* player *i's* incentive constraint continues to be satisfied. Since the other's actions do not change, their incentives are untouched and we lower the value of the minimand in Eq. (29). In light of that observation, we can rewrite the problem as

$$Max_{a,\widetilde{u} \in U} \sum_{k \neq i} a_k$$

$$s.t. h_j(a_j) + \delta(\widetilde{u}_j - w_j a_j) \geq h_j(a_j^*) + \delta(\underline{u}_j^j - w_j a_j^*), \forall j \tag{31}$$

It immediately follows that in the solution, $\bar{\bar{a}}(i)$, $\sum_{j \neq i} \bar{\bar{a}}_j(i) \geq \sum_{j \neq i} a_j^*$. We know however that there are equilibrium payoffs that are strictly higher than the BAU payoffs (for example, the third-best payoffs are strictly higher). Hence from the incentive constraints it follows that there are emission vectors for the other players that are incentive compatible and whose sum is strictly higher than the BAU emissions.

A little reflection also shows that the optimization problem of Eq. (31) must be equivalent to some i – *less* second problem. After all the higher is the equilibrium payoff $\widetilde{u}_j$, the easier it is to get country j to emit a high amount this period (in order to punish i). From the result for second-best problems, it then follows that from the second period onwards, the emission must be at some constant level $a(x_{-i})$. The theorem is proved. □

10.5. Greenhouse trap

Proof of Theorem 8. Suppose that the initial GHG level g_0 is above the Pareto optimal steady state $\widehat{g}$. In that case, country i faces a best response problem that is identical to the one it would face if the rest of the world emitted a^* at *all* levels of GHG. Note that even if i drops her emission to zero, the GHG level would still remain above $\widehat{g}$; after all, $\sigma g + (I-1)a^* \geq \sigma g + I\widehat{a} > \widehat{g}$. (Recall Condition L: $(I-1)a^* \geq I\widehat{a}$ and that $(\sigma\widehat{g} + I\widehat{a} = \widehat{g})$. It follows therefore that, if $g > \widehat{g}$, the best response for i is to play the BAU level a^*.

Suppose instead that $g \leq \widehat{g}$. There are two possibilities for *i's* emission: either the emission level is chosen to be so high that the next period's GHG stock is above $\widehat{g}$, or it is not. Consider the two cases separately.

Case 1 Emissions such that $g_{t+1} > \widehat{g}$, that is, $a_i > a(g_t)$.

In this case, the lifetime payoffs are $h(a_i) - cg_t + \delta[u^* - wg_{t+1}]$, where u^* is the BAU payoff, $u^* = (h(a^*) - \delta wIa^*)/1 - \delta$. Substituting for the relevant expressions, the optimal deviation in this case is then determined by

$$Max_{a_i > a(g)} h(a_i) - cg + \delta \left[\frac{h(a^*) - \delta wIa^*}{1-\delta} - w(\sigma g + (I-1)a(g) + a_i) \right].$$

Collecting only the terms that involve a_i, the above expression immediately simplifies to

$$Max_{a_i > a(g)} h(a_i) - \delta w a_i,$$

and its solution is, in fact, the BAU emission level a^*.[28] Hence, the best deviation payoff in this case reduces to

$$u^* - wg + \delta w(I-1)(a^* - a(g)).$$

Case 2 Emissions such that $g_{t+1} \leq \widehat{g}$, that is, $a_i \leq a(g_t)$.

It will be more convenient to think of emission levels in this case as $a(g) - \epsilon, \epsilon > 0$. Since we only need to consider one-shot deviations, and because lowering emissions unilaterally will simply delay by one period the convergence to the steady state $\widehat{g}$, it follows that the optimal deviation can be derived from the following exercise:

$$Max_{\epsilon \geq 0} h(a(g) - \epsilon) + \delta \left[h\left(\widehat{a} + \frac{\sigma\epsilon}{I}\right) + c\epsilon \right]$$

[28] Another way of saying what we just did is that *if* country i is going to push the game into the BAU region, then she might as well take the unilaterally optimal action a^* to do so.

where we have used the following facts: (i) an ϵ cut-back by country i leaves the state of the system at $\widehat{g}-\epsilon$ in the next period, and (ii) that requires an additional $(\sigma\epsilon/I)$ emission by each country to bring the GHG level two periods hence up to the Pareto optimal steady state of $\widehat{g}$. It is straightforward to verify that the maximand above is concave. A sufficient condition for the maximizer to be $\epsilon=0$ (which is what we want to show) is that the derivative at that point be non-positive, or,

$$\frac{\sigma\delta}{I}h'(\widehat{a}) \le h'(a(g)) - \delta c.$$

By the second part of Condition L, $a^* \ge a(g)$, and hence by the concavity of h, a sufficient condition for the last inequality is

$$\frac{\sigma\delta}{I}h'(\widehat{a}) \le h'(a^*) - \delta c.$$

Note that $h'(\widehat{a}) = (\delta Ic/1-\sigma\delta)$ and that $h'(a^*) = (\delta c/1-\sigma\delta)$. After making the appropriate substitutions, it is easy to see that the two sides of the inequality are actually equal to each other (and to $\delta c(\sigma\delta/1-\sigma\delta)$). Hence, the optimal deviation is not to deviate.

Combining the two cases we have the following: $a(g)$ is a best response (to $a(g)$) provided it yields higher lifetime payoffs than a^*. Note that the lifetime payoffs to $a(g)$ are

$$h[a(g)] - wg - \delta wIa(g) + \delta\widehat{u}.$$

Elementary algebra says that the lifetime payoffs to $a(g)$ are higher if and only if

$$h[a(g)] - \delta wIa(g) + \delta\widehat{u} \ge h(a^*) - \delta wIa^* + \delta u^* + \delta w(I-1)(a^* - a(g)).$$

Since $h(a) - \delta wIa$ is a concave function that is maximized at $\widehat{a}$ and since $a(g) \in [\widehat{a}, a^*]$, it follows that $h[a(g)] - \delta wIa(g) \ge h(a^*) - \delta wIa^*$. Hence a sufficient condition for the last inequality to hold is that

$$\widehat{u} \ge u^* + \delta w(I-1)(a^* - a(g)).$$

In turn the second term in the right-hand side is strictly smaller than $\delta wI(a^* - \widehat{a})$, so a sufficient condition for the inequality is

$$\widehat{u} \ge u^* + \delta wI(a^* - \widehat{a}).$$

From the definitions, it follows that the condition is equivalent to

$$h(\widehat{a}) - \delta wI\widehat{a} \ge h(a^*) - \delta wIa^* + (1-\delta)\delta wI(a^* - \widehat{a}).$$

As $\delta \uparrow 1$, the inequality converges to

$$h(\widehat{a}) - w(1)I\widehat{a} \ge h(a^*) - w(1)Ia^*,$$

where the present undiscounted cost is $w(1) = c/(1-\sigma)$, and $\widehat{a}$ and a^* are respectively the solutions to $Max\, h(a) - \delta Iw(1)a$ and $\text{Max}\, h(a) - \delta w(1)a$ (and hence $\widehat{a} \ne a^*$). By definition, the last inequality is actually a strict inequality. It follows therefore that the sufficient condition for $a(g)$ to be a (strictly) better response than a^* holds for all δ appropriately close to 1. □

Proof of Theorem 9. The proof may be found in Dutta and Radner (2007). □

Acknowledgements

This paperhas benefitted from the comments of seminar audiences at the University of Michigan, University of Rochester, Boston University, CalTech, UCLA, Columbia University, The Stockholm School of Economics and New York University as well as the thoughtful comments of a referee and the Editor of this journal. Of course, all errors are our very own.

References

Abreu, D., Pearce, D., Stachetti, E., 1990. Towards a general theory of discounted repeated games with discounting. Econometrica 58, 1041–1065.
Barrett, S., 2003. Environment and Statecraft: The Strategy of Environmental Treaty-Making. Oxford University Press, Oxford.
Benhabib, J., Radner, R., 1992. Joint exploitation of a productive asset. Economic Theory 2, 155–190.
Chen, Z., 1997. Can economic activity lead to climate chaos? An economic analysis of global warming. Canadian Journal of Economics 30, 349–366.
Cline, W., 1992. The Economics of Global Warming. Institute for International Economics, Washington, DC.
Dockner, E., Long, N., Sorger, G., 1996. Analysis of Nash equilibria in a class of capital accumulation games. Journal of Economic Dynamics and Control 20, 1209–1235.
Dockner, E., Nishimura, K., 1999. Transboundary pollution in a dynamic game model. Japanese Economic Review 50, 443–456.
Dutta, P., 1991. What do discounted optima converge to? A theory of discount rate asymptotics in economic models. Journal of Economic Theory 55, 65–94.
Dutta, P., Radner, R., 2004. Self-enforcing climate change treaties. Proceedings of the National Academy of Sciences, USA 101–14, 5174–5179.
Dutta, P., Radner, R., 2006. Population growth and technological change in a global warming model. Economic Theory 29, 251–270.
Dutta, P., Radner, R., in preparation. Choosing cleaner technologies: global warming and technological change.
Dutta, P., Radner, R., 2007. A strategic model of global warming: theory and some numbers, mimeo, Columbia University.

Dutta, P., Sundaram, R., 1992. Markovian equilibrium in a class of stochastic games: existence theorems for discounted and undiscounted models. Economic Theory 2, 197–214.
Dutta, P., Sundaram, R., 1993. How different can strategic models be?: Journal of Economic Theory 60, 42–61.
Fankhauser, S., 1995. Valuing Climate Change, The Economics of the Greenhouse. Economic and Social Research Council, London.
Finus, M., 2001. Game Theory and International Environmental Cooperation. Edward Elgar, Cheltenham.
Hansen, J., 2006. The Threat to the Planet, New York Review of Books, July 13, 2006.
Inter-governmental panel on climate change, 2007. Climate change, The synthesis report. Geneva, Switzerland, IPCC. Available at http://www.ipcc.ch/ipccreports/assessments-reports.htm (accessed 05.12.07).
Levhari, D., Mirman, L., 1980. The great fish war: an example using a dynamic Cournot-Nash solution. Bell Journal of Economics 11, 322–334.
Long, N., Sorger, G., 2006. Insecure property rights and growth: the role of appropriation costs, wealth effects and heterogenity. Economic Theory 28, 513–529.
Nordhaus, W., Yang, Z., 1996. A regional dynamic general equilibrium model of alternative climate-change strategies. American Economic Review 86, 741–765.
Nordhaus, W., Boyer, J., 2000. Warming the World: Economic Models of Global Warming. MIT Press, Cambridge, MA.
Park, S, 2004. Calibrating a model of global warming. Columbia University Economics Department, unpublished.
Rosser, B., 2001. Complex ecologic-economic dynamics and environmental policy. Ecological Economics 37, 23–37.
Rustichini, A., 1992. Second-best equilibria for games of joint exploitation of a productive asset. Economic Theory 2, 191–196.
Schelling, T., 2002. Greenhouse effect, in: The Concise Encyclopedia of Economics. Available at http://www.econlib.org/LIBRARY/Enc/GreenhouseEffect.html (accessed 05.12.07).
Sobel, M., 1990. Myopic solutions of affine dynamic models. Operations Research 38, 847–853.
Sorger, G., 1998. Markov-perfect Nash equilibria in a class of resource games. Economic Theory 11, 79–100.
Stern, N., 2006. Review on the Economics of Climate Change, HM Treasury, London, UK, http://www.sternreview.org.uk (accessed 05.12.07).
Sundaram, R., 1989. Perfect equilibrium in a class of symmetric dynamic games. Journal of Economic Theory 47, 153–177.
Tornell, A., Velasco, A., 1992. The tragedy of the commons and economic growth: why does capital flow from poor to rich countries? Journal of Political Economy 100, 1208–1231.

Part IV
Negotiations, Second-Best Designs and Institutions

[18]

Public Choice **96:** 145–186, 1998.

Toward a positive theory of coalition formation and endogenous instrumental choice in global pollution control *

MICHAEL FINUS & BIANCA RUNDSHAGEN
Department of Economics, University of Hagen, Profilstr. 8, D-58084 Hagen, Germany

Accepted 18 November 1997

Abstract. The paper analyzes the coalition formation process in a global emission game with asymmetric countries where the number of signatories, the abatement target and the policy instrument are chosen simultaneously. Exemplarily, a uniform emission reduction quota and an effluent charge are considered. Stability is analyzed in a supergame framework by applying the concept of coalition-proofness. The analysis also considers the impact of impatient agents, restricted and simple punishment profiles. Two main results may be mentioned: First, paradoxically, IEAs achieve only little (if signed at all) if the externality problem is distinct. Second, our model helps to explain the frequent appearance of emission quotas in international pollution control despite the recommendations by economists to use market-based instruments like effluent charges.

1. Introduction

The literature on coalition formation in international pollution control has grown immensely over the recent years. Two basic strands may be distinguished: The first strand estimates the costs and benefits of various abatement targets under different cost allocation rules. Bohm and Larsen (1993), Fankhauser and Kverndokk (1996), Kverndokk (1993), Welsch (1992, 1995), and Tahvonen (1994) identify the winners and losers of a hypothetical agreement to reduce "greenhouse gases" whereas Kaitala, Pohjola and Tahvonen (1992), Mäler (1989) and Newbery (1990) do this in the case of transboundary fluxes. In this literature, the rationale for a group of countries to form an IEA is based on whether the group as a whole receives more than in the status quo. Moreover, individual rationality is mostly checked by demanding that each country must receive more in an IEA than if it does not participate. Though these conditions might be seen as necessary for an IEA to come

* The authors gratefully acknowledge the receipt of grant No. II 69982, "Cooperative Solutions for International Environmental Problems" by the Volkswagen-Foundation. Both authors benefited from several discussions and comments. In particular, they like to thank Alfred Endres, Bernd Genser, Ekko van Ierland, Jörn Martiensen, Ernst Mohr and Carsten Schmidt. The comments of an anonymous referee are also gratefully acknowledged.

into force, they are certainly not sufficient. In particular, one has to check whether the omnipresent free-rider incentive in an international context can be controlled by appropriate measures.

This is the focus of the second strand of literature which studies within a theoretical framework the conditions of how many countries will accede to an IEA and what is needed to broaden a coalition. Except Chander and Tulkens (1992, 1995), all papers employ stability concepts of non-cooperative game theory. The reason is that IEAs must be self-enforcing because no binding commitments are possible.[1] Basically, the models of Barrett (1994b), Bauer (1992), Carraro and Siniscalco (1992, 1993) and Hoel (1992) employ a stability concept borrowed from the oligopoly literature (e.g., d'Aspermont, Jacquemin, Gabszeweiz and Weymark, 1983) where a coalition is thought to be stable if no country wants to accede to the coalition (external stability) and no country wants to leave the coalition (internal stability). The equilibrium coalition size is determined as an adjustment process of signatories and non-signatories. This is particularly evident in Barrett (1994a, b), where signatories behave as Stackelberg leaders. However, also in the other papers listed above some kind of conjectural variation is assumed.[2] For instance, internal stability is checked by comparing the payoff of a signatory if it takes part in an IEA with the payoff if it leaves *and* the remaining signatories and non-signatories adjust to the new situation. Thus, implicitly, it is assumed that once a signatory accedes to an IEA it complies with its obligation (see Barrett, 1995). This seems a rather optimistic assumption.

Though the above-mentioned papers do not control free-riding in a strict sense, most models come to the pessimistic conclusion that only two or three countries will form a stable coalition. Only Barrett (1994b) finds a coalition of up to N countries in the case of linear marginal abatement and damage costs, symmetric countries, and assuming the signatories to be Stackelberg leaders. For the supergame version of his model the result is confirmed.

With respect to the time dimension the question arises whether an adjustment process should not be better modeled as a dynamic game which comprises of more than two stages. Particularly, since most IEAs are of an open-end character in that they do not specify the termination of the contract, accession to and compliance of an IEA may be better analyzed as an infinite supergame (see Barrett, 1994a; Dockner and Van Long, 1993; Endres and Finus, 1998). This is the framework we will choose in this paper.

The assumption of symmetric countries is typical for most models and is due to computational ease. Of the early papers only Hoel (1992) and Bauer (1992) make an exception. Hoel assumes countries to have the same abatement costs but different environmental damage costs. Bauer models an inversely proportional relation between abatement and damage costs. To offset the complication of asymmetric countries Hoel assumes constant marginal damage costs and Bauer constant marginal benefits from abatement. This amounts to orthogonal reaction functions. However, this assumption does not really address the question of interaction between countries suffering from a global pollutant.

Recent papers modeling asymmetric countries are Barrett (1997) and Botteon and Carraro (1997). Whereas Barrett distinguishes only two types of countries for the linear marginal abatement and damage cost case, Botteon and Carraro use estimates from Musgrave (1994) of five world regions for the abatement and damage cost functions. Both papers confirm that for constant marginal benefits from abatement at the most three countries will form a stable IEA. For linear marginal abatement and damage costs Barrett again confirms his finding of the identical country case that up to N countries might form a coalition, but will not improve very much upon the status quo. We find similar results as Barrett, assuming linear marginal abatement and cost functions. As Hoel (1992) we let damage costs vary between countries according to a uniform distribution, but assume identical abatement costs. The latter assumption ensures that the model is still tractable.

Whereas for symmetric countries the agreement on an abatement target and the cost allocation among signatories is trivial, this is less straightforward in the case of asymmetric countries. Usually it is assumed that in case a grand coalition forms the socially optimal abatement target is realized. In case a subcoalition signs an agreement, signatories choose that abatement target which maximizes aggregate welfare of its members. For example, Barrett (1997) uses the Shapley value and Botteon and Carraro (1997) employ the Shapley value and the Nash bargaining solution (allowing for transfers).

Of course, these concepts of cooperative game theory (and the underlying assumptions) have some normative appeal and possess some axiomatic properties. However, these concepts beg the question whether they really reflect what is going on in international politics. In particular, they rely on three crucial assumptions which are usually not met in reality.

First, it is (implicitly) assumed that there is some cooperative spirit when it comes to choose an abatement target and to allocate abatement burdens. Second, the cooperative bargaining concepts mostly rely on the possibility of unlimited transfers.[3] Third, complete information is an essential prerequisite in order to determine the solution of these concepts. Additionally, in the theoretical literature, the choice of the instrument to achieve a particular abatement target is not considered.

In contrast, we suspect that potential signatories hold different views regarding a joint abatement target, though they may generally agree that something should be done to improve environmental quality. Moreover, we expect that governments controversially discuss how abatement duties should be allocated among countries acceding to an IEA. Of course, these allocations crucially depend on the particular design of an agreement and the policy instrument chosen to achieve an abatement target.

The assumption of transfers has a sound justification in welfare economics. It is well known that transfers can foster cooperation and enhance efficiency by compensating countries which suffer from a particular policy. However, in reality they have hardly been observed. The few IEAs which have a provision for transfers, are mostly very unspecific regarding transfer obligations. In fact, it seems safe to claim that the actual transfers which have been conducted so far are negligible compared to (expected) abatement and damage costs. The same is true with respect to technical assistance to developing countries which has been more of a lip-service than actual policy.[4]

The assumption that all information is available to determine a bargaining outcome may also not hold in reality. In particular, one should expect that governments try to influence a decision in its favor and therefore will not reveal their environmental preferences. Depending on the decision process tactical maneuvers may be successful or can be neutralized.

Since we are mainly concerned with a *positive analysis* of coalition formation and instrumental choice in international pollution control, we give up some of the restrictive assumptions mentioned above and introduce some additional features into the model which are typical for many IEAs. Of course, not all of these changes are as well axiomatically founded as the bargaining concepts of cooperative game theory.

We assume that country's representatives meet in order to discuss joint abatement measures. Two policy regimes to achieve an abatement target are considered: a uniform emission reduction quota and a uniform emission tax. In a uniform quota regime each country must reduce its emissions by the same percentage from some base year. In a uniform tax regime the same tax rate is applied in each country. For both policy regimes we do not consider transfer payments. For the tax regime we therefore assume that the revenues raised remain in the country of origin. This seems to be suggestive because it is unlikely that countries will hand over their tax sovereignty to an IEA secretariat (Bohm, 1994: 190).

Within each regime the political representatives put forward a proposal. Since the proposals will typically differ and a government cannot be forced into an agreement, we assume that governments settle for the "lowest common denominator". That is, in the quota regime the lowest percentage emission reduction is finally agreed upon by the signatories. In the tax regime this is the lowest tax proposal. This decision rule is frequently applied in international organizations, e.g., European Union and United Nations Organizations (Lenschow, 1996). The stability analysis in Sections 2 and 3 and some other properties discussed below will illustrate the rationale behind this (inefficient) decision rule. In particular, complete information is not necessary to determine the bargaining outcome.

The quota has been chosen because it is part of many IEAs. For example, the Montreal Protocol specifies an emission reduction of CFCs by 20% based on 1989 emission levels to be accomplished by 1998. Another example is the Helsinki Protocol which suggests a reduction of long-range transboundary air pollutants from 1980 levels by 30% by 1993. The emission tax has been chosen because it is frequently recommended by economists as an efficient mean to tackle environmental problems. In the case of a "pure" global environmental problem the restriction to a uniform tax solution does not cause efficiency losses.

Uniform solutions are a typical feature of many IEAs and might be due to several reasons.[5] First, in public the notion prevails that uniform solutions are apparently "fair". Second, to agree on differentiated solutions takes time and is therefore associated with higher transaction costs. In contrast, uniform solutions constitute some kind of focal point in the sense of Schelling (1960) on which bargaining partners feel relatively easy to agree.

The model presented below is able to explain and to depict two phenomena of existing IEAs: First, the model captures the fact that the regulation of most pollutants has not been successful in the past though there are some rare exceptions, e.g., the Montreal Protocol. Either reduction levels have been low or only few countries have ratified an agreement. Second, the model helps to explain why an effluent charge has not been part of any IEA, though economists have recommended this instrument (among others) as an efficient mean to tackle international pollution. We present arguments which allow the conjecture that it might be rational to use an inefficient emission quota in a second-best world.

Thus, we are in the tradition of the polit-economic literature analyzing (in the national context) why market-based instruments have hardly been used in environmental policy (e.g., Buchanan and Tullock, 1975; Dewees, 1983; Pashigan, 1985; Ursprung, 1992) and environmental quality is often below the Pareto-efficient level (e.g., Endres and Finus, 1997b; Hahn, 1990). For simplicity, we only look at governments' decisions at an aggregate level, and do not include national interest groups in the analysis as for instance, Hillman and Ursprung (1994) do in the context of trade protection and environmental policy.

150

In what follows we first derive the stage game equilibria of the emission game if a grand coalition forms (Section 2.1), and then check for stability in a dynamic context applying the concept of coalition-proofness (2.2). In subsection 2.2.1 the concept is laid out. Then a stability analysis of the bargaining outcomes in the two policy regimes and the social optimum is conducted for a discount factor close to one (2.2.2), but also for more impatient players (2.2.3), and for sub-optimal punishment profiles (2.2.4). Section 3 deals with the formation of sub-coalitions. First, the formation process is described (3.1), then the equilibrium coalition size is determined when the policy regime is exogenously given (3.2), and, finally, when also the policy choice is endogenously determined (3.3). Section 4 winds up with a summary and suggests some possible future research.

2. The grand coalition

2.1. *Statics*

2.1.1. *The basic model and some benchmarks*

Assume N countries suffering from a global pollutant E with $E = \sum_{i=1}^{N} E_i$, where the subscript i denotes the country and E_i are emissions originating from country i. Let damages from emissions in country i depend on global emissions, $D_i(E)$, and increase in E, $D_i' > 0$, at an increasing rate, $D_i'' > 0$. Further assume that countries differ with respect to the perception of environmental damages. In particular let damages in a country be given by

$$D_i(E) = i \cdot \frac{c}{2} \cdot E^2, c > 0, i \in \{1, \ldots, N\} \tag{1}$$

where c is a cost parameter. This implies that damages are (discrete) uniformly distributed. Countries with a higher index i suffer more from pollution than those with a lower index number.

However, countries do not only suffer from pollution. Emissions are also an input in the production and consumption of goods from which benefits, $B_i(E_i)$, are derived. It seems plausible to assume that these benefits increase in emissions in some range at a decreasing rate, $B_i'' < 0$. To depict such a relation, we assume benefits in each country to be given by

$$B_i(E_i) = b\left(dE_i - \frac{1}{2}E_i^2\right), b > 0, d > 0, E_i \leq 2d, \tag{2}$$

where b and d are benefit parameters, $E_i \leq 2d$ is necessary to ensure $B_i > 0$ (with $B_i' > 0$ for $E_i \leq d$), and defines the upper boundary of emission space in our pollution game. Thus, the emission strategy space is given by $E_i \in [0, 2d]$. Taken together, net benefits of country i, NB_i, are given by

$$NB_i = B_i(E_i) - D_i(E). \tag{3}$$

Equations (1) to (3) define the normal form of the stage game, $G = (I, \Psi, P)$, with I the set of players, Ψ the set of strategies, and P the set of payoffs in the game. In the Nash equilibrium of the stage game each country equates marginal benefits to marginal damages ($B_i'(E_i^N) = D_i'(E^N)$), taking the emissions of the other countries as given. This leads to equilibrium emissions

$$E_i^N = \frac{d(2\gamma + N^2 + N - 2iN)}{2\gamma + N^2 + N}, E^N = \frac{2dN\gamma}{2\gamma + N^2 + N} \tag{4}$$

with associated net benefits

$$NB_i^N = \frac{bd^2(4b^2 + 4bcN^2 + 4bcN - 4bciN^2 + c^2N^4 + 2c^2N^4 + 2c^2N^3 + c^2N^4 - 4c^2i^2N^2)}{2(2b + cN^2 + cN)^2},$$

$$NB^N = \frac{bd^2N(12b^2 + 12bcN + 6bcN^2 - 6bcN^3 + c^2N^2 - c^2N^4)}{6(2b + cN^2 + cN)^2}, \tag{5}$$

where the superscript N stands for "Nash equilibrium" and $\gamma = b/c$. γ can be interpreted as benefit–cost parameter ratio. In the later analysis it turns out that all qualitative results depend only on this ratio, independently of the absolute values of b and c.

From Equation (4) it is evident that countries with a lower index choose higher emissions in the Nash equilibrium because they perceive damages as less severe. In order to ensure positive emissions we have to require

$$NNC_1 : \gamma \geq \frac{N^2 - N}{2}, \tag{6}$$

where NNC stands for "non-negativity constraint". However, net benefits in a country, NB_i^N, as well as aggregate net benefits, NB^N, may well be negative in the Nash equilibrium depending on the values of the parameters.

In order to contrast the non-cooperative outcome with what would be required from a global point of view, we compute emission rates and net benefits in the social optimum.

$$E_i^S = \frac{2d\gamma}{2\gamma + N^3 + N^2}, E^S = \frac{2dN\gamma}{2\gamma + N^3 + N^2} \tag{7}$$

$$NB_i^S = \frac{2b^2d^2(b + cN^3 + cN^2 - icN^2)}{(2b + cN^3 + cN^2)^2}, NB^S = \frac{b^2d^2N}{2b + cN^3 + cN^2} \tag{8}$$

152

The superscript S stands for "social optimum". Casual inspection of Equations (4) and (7) confirms a standard result: In the Nash equilibrium aggregate emissions are too high from a social point of view. Basic algebra also reveals that global welfare in the Nash equilibrium is lower than in the social optimum, i.e., $NB^N < NB^S$ holds. Moreover, we find $\partial(NB_i^S - NB_i^N)/\partial i > 0$ and $NB_i^S - NB_i^N <> 0$. That is, countries benefit differently from a socially optimal abatement policy. In fact, countries which perceive environmental damages as less severe may lose in the social optimum compared to the Nash equilibrium. The exact relation depends on the parameter values. Independently of these values, however, $NB_1^S - NB_1^N < 0$ and $NB_N^S - NB_N^N > 0$ hold. This preliminary result already hints at problems associated with the stability of an IEA specifying socially optimal abatement targets.

Of course, in a static context the social optimum is not stable in any case since in equilibrium $B_i'(E_i^S) > D_i'(E^S)\forall i$ holds. Each country is tempted to take a free-ride on the other countries abatement efforts. But also in a dynamic context the uneven distribution of the gains from joint abatement may jeopardize the stability of an IEA as will become apparent in the sequel.

In order to benchmark the results obtained below, we analyze the relative difference between aggregate emissions in the Nash equilibrium and in the social optimum with respect to two important parameters of the model.

$$\Delta E = \frac{(E^N - E^S)}{E^N}; \quad \frac{\partial \Delta E}{\partial N} > 0; \quad \frac{\partial \Delta E}{\partial \gamma} < 0. \tag{9}$$

The larger the number of countries which suffer from the global externality and the lower the benefit–cost ratio of emissions, the more important it is to reach a cooperative agreement. The former result is typical for the provision of public goods: The externality is distinct if many countries suffer from the global pollutant. The latter result indicates that abatement is particularly attractive if environmental damages are very high compared to opportunity costs of abatement. In contrast, if opportunity costs of abatement are very high, only moderate emission reductions (expressed as percentage reduction compared to the Nash equilibrium) are advisable from a global perspective.[6]

2.1.2. *The quota and tax regime*

Whereas in the previous sub-section we determined two benchmark cases, a non-cooperative and a full-cooperative solution, we are now concerned with more real-world situations. We assume that delegates of countries meet in order to discuss joint abatement targets which improve upon the non-cooperative situation.[7] Hence, the Nash equilibrium is assumed to be the starting point in the negotiations. Of course, the proposals of delegates will depend crucially on the abatement allocation rules. These in turn depend on the policy regimes of which two are considered: a uniform emission reduction quota and a uniform effluent tax.

Under the quota regime each country has to reduce its emissions by the same percentage R_i^Q from the initial situation. With this information, and the assumption that all countries participate in the agreement, a government's proposal, $R_i^Q (0 \leq R_i^Q \leq 1)$, is derived from

$$\max_{R_i^Q} \left[B_i((1 - R_i^Q) \cdot E_i^N) - D_i(1 - R_i^Q) \cdot E^N) \right] \tag{10}$$

to which the solution is (using [1] and [2])

$$R_i^Q = \frac{2iN(2N\gamma - 2\gamma - N^2 - N + 2iN)}{4\gamma^2 + 4N^2\gamma + 4N\gamma + 8iN\gamma + 4iN^2\gamma + N^4 + 2N^3 + N^2 - 4iN^3 - 4iN^2 + 4i^2N^2} \tag{11}$$

Using the non-negativity constraint NNC_1 it can be shown that $\partial R_i^Q / \partial i > 0$ holds. That is, the higher a government perceives environmental damages, the higher is its reduction proposal. Since the proposals differ among countries, an agreement must be reached which of the N proposals will be adopted. Two basic solutions come to mind: Either all countries seek an agreement which is acceptable to all delegates or only a sub-group of countries strives for an agreement among themselves. The latter solution might be particularly attractive for countries with high environmental damages which feel that an agreement among all countries leads to a too lax joint environmental policy. In this section, however, we want to confine the analysis to the first solution, and treat the second possibility in Section 3.

However, also within the two "categories of solutions" several agreement procedures are possible. To narrow the analysis further down, we focus on the "lowest common denominator decision rule", henceforth abbreviated LCD-decision rule. As has been argued in the introduction, its popularity in international politics is most likely due to the fact that no country can be forced into an agreement. Moreover, this agreement procedure has the property that "truthtelling" is a dominant strategy.[8] That is, a country has no incentive to conceal its true preferences and to choose its proposal strategically in order to influence the bargaining outcome in its favor. This seems particularly important in international negotiations where it is difficult to verify country specific abatement costs, but even more environmental preferences, and where free-riding poses a serious problem.

Of course, a decision rule where the median country's proposal is adopted would also foster truthtelling, but – as will become apparent later – would often specify too high abatement targets. For the particular model specification of (1) to (3) it is true that the median country's proposal corresponds to

154

the "socially optimal emission reduction", given a uniform emission quota is applied, i.e., $R^Q_{(N+1)/2} = R^{Q*}$ (see (13) below). Results presented in subsection 2.2.2 show that only in a few cases such a high reduction target is stable.

If the LCD-decision rule is applied and since $\partial R^Q_i/\partial i > 0$ holds, country 1 is the bottleneck in the negotiations and the abatement target agreed upon is given by

$$R^Q_1 = \frac{2N(2N\gamma - 2\gamma - N^2 + N)}{4\gamma^2 + 8N^2\gamma - 4N\gamma + N^4 - 2N^3 + N^2}. \tag{12}$$

Computing country specific net benefits reveals that $NB^Q_i \geq NB^N_i$ holds for all countries. That is, all countries gain from the emission reduction proposed by country 1. At an aggregate level the following relations can be established: $NB^S > NB^Q > NB^N$ and $E^S < E^Q < E^N$. Hence, a quota agreement improves upon the non-cooperative situation, though countries agree only on the lowest common denominator. However, aggregate welfare under the quota regime is lower than in the social optimum. This is due to the institutional restriction of a uniform solution (which leads to cost inefficiencies) and the agreement procedure LCD.

If we impose only the institutional restriction on an abatement policy and calculate the "socially optimal emission reduction quota", we find

$$R^{Q*} = \frac{N(N^2 - 1)(6\gamma + N)}{12\gamma^2 + 6\gamma N^3 + 6\gamma N^2 + N^4 - N^2}. \tag{13}$$

This emission reduction R^{Q*} may serve as benchmark in the later analysis in Section 2.2.

Note that even though an IEA specifies moderate abatement targets according to the LCD-decision rule, such an agreement would not be stable in a *static* context. In fact, independently of the reduction level agreed upon by the N countries, $B_i'(R^Q) > D_i'(R^Q)$ holds.

Let us now turn to an IEA in which an effluent tax is used to reduce global pollution. In particular, we assume that it is understood by all countries that a uniform tax rate t is imposed in each country and that receipts of the revenues remain in each country.

To determine the proposals of the N countries it is necessary that each country knows besides its own net benefit function the marginal benefit functions of the other N-1 countries. This information is the basis to compute the adjustment process of the other countries to a given tax rate. Hence, the information requirements under a tax regime and a quota regime are different (Endres, 1996). In the latter regime it suffices for a government to have knowledge of present emissions (and of course of its own net benefit function).[9] In the former regime this information is not important but instead knowledge on the opportunity cost of abatement of all participants is necessary.

Whereas the information on current emissions will be most likely public available, one should suspect that knowledge on other countries opportunity cost of abatement is more difficult to obtain. However, also in the tax regime complete information is not necessary because knowledge on the other countries perceived damages is irrelevant (and will probably not be available in reality) for country i to compute its optimal tax proposal t_i.[10]

To derive the proposal of a country i we use the information that in equilibrium all countries reduce emissions until marginal opportunity costs of abatement are equal to the tax rate, $B_i'(E_i^T(t)) = t$. Here, the superscript T stands for "tax regime". In contrast to the quota regime, in the tax regime cost efficiency is "automatically" ensured.

Using (2), and rearranging terms, we have $E_i^T = d - t/b$ in equilibrium. Inserting this information in (3), a country's proposal, t_i, is derived from

$$\max_{t_i} \left[B_i(E_i^t(t_i)) - D_i(\sum_{j=1}^{N} E_j^T(t_i)) \right] \tag{14}$$

to which the solution is

$$t_i = \frac{ibcdN^2}{b + icN^2}, \tag{15}$$

and where $\partial t_i / \partial_i > 0$ holds. That is, a country's tax proposal increases in perceived environmental damages. Consequently, if countries agree on the lowest common denominator, country 1 is the bottleneck in the negotiations. Therefore, the tax rate laid down in the treaty is

$$t_1 = \frac{bcdN^2}{b + cN^2}, \tag{16}$$

or, alternatively, expressed as relative emission reduction from the Nash equilibrium, R_1^T,

$$R_1^T = \frac{N^2 - N}{2(\gamma + N^2)}. \tag{17}$$

Comparing net benefits of country i under the tax regime (if the LCD-decision rule is applied) with the Nash equilibrium reveals that all countries benefit compared to the initial situation, except country 1, i.e., $NB_i^T \geq NB_i^N \forall i, i \neq 1$. At first thought this seems at odds with the assumption that country 1 determines the terms of the agreement. However, at second thought it is apparent that in the tax regime in which all countries contribute to a joint abatement policy, countries with low index numbers carry a *relatively* high abatement burden, considering their immediate interest. Though the allocation of abatement burdens in such a regime is cost efficient, it differs much from the initial reduction level in the Nash equilibrium. In contrast, in the quota regime all countries carry the same relative abatement burden which is more favorable for countries with a low index. In sub-coalitions, which we discuss in Section 3, the problem of an uneven distribution of the net benefits is less pronounced in the tax regime and all signatories gain from an agreement. Nevertheless, the instability of agreements is still a problem.

At an aggregate level, we find $NB^S > NB^T > NB^N$ and $E^S < E^T < E^N$. As in the quota regime, a tax agreement improves upon the initial situation but is a second-best solution from a global point of view. In contrast to the quota regime, the instrument as such is not responsible for this second-best result because abatement is conducted efficiently. Only the agreement on the "lowest common denominator" is responsible for this sub-optimality.

Again, in order to have a benchmark in the analysis below, we compute the socially optimal effluent tax (Pigouvian tax) which is given by

$$t^* = \frac{bcdN^2(N+1)}{2b + cN^3 + cN^2}, \tag{18}$$

or expressed as a relative reduction from the Nash equilibrium.

$$R^{T*} = \frac{N(N^2-1)}{2\gamma + N^3 + N^2}, \tag{19}$$

where $R^{T*} > R_1^T$ and $R^{T*} = R^S$, with $R^S = \Delta E$ as defined in Equation (9). That is, a uniform Pigouvian tax is efficient and is higher than the tax rate proposed by country 1. Note that, as in the quota regime, for our model specification the median country's proposal is equal to the socially optimal tax rate, i.e., $R^T_{(N+1)/2} = R^{T*}$. Moreover, $B_i'(R^T) > D_i'(R^T)\ \forall i$ holds for $R^T \geq R^{T_1}$, implying that the tax agreement is also not stable in a static context.

A preliminary comparison of the stage equilibria of the two policy instruments reveals that $E^T > (<,=) E^Q$ holds if $\gamma > (<,=) N(N+1)/2$. That is, for a given number of countries N, global emissions resulting from a bargaining solution in the quota regime are lower than in the tax regime if abatement is relatively expensive compared to perceived environmental damages (γ is big). With respect to global welfare a comparison is less straightforward. Nevertheless, it turns out that a sufficient condition for $NB^T < NB^Q$ is $\gamma > 2N^2$ and for $NB^T > NB^Q$ is $\gamma < N(N+1)/2$. This implies that if abatement is perceived to be relatively expensive compared to environmental damages aggregate welfare of the bargaining outcome in the quota regime is higher than in the tax regime. However, for a final evaluation of the two policy instruments, stability must be checked in a dynamic context.

2.2. *Dynamics*

2.2.1. *Stability concept*

In the previous sub-sections it became apparent that a grand coalition agreeing on the lowest common denominator can improve upon the initial situation, though socially optimal emission reductions and welfare levels are not obtained. Moreover, it was demonstrated that in a static world an incentive to defect always prevails which causes an IEA to be potentially unstable. This was true regardless of the policy regime. Therefore, in the subsequent sub-sections, we investigate whether the stage game equilibria are stable in a supergame framework.

It is well known from game theory that the chances for cooperation are higher in such a framework. For example, the folk-theorem asserts that every individual rational payoff vector (giving each player more than her minimax payoff) can be sustained in the long-run through subgame-perfect strategies provided the discount factor is close to one (Fudenberg and Maskin, 1986).[11] In contrast to Nash strategies, subgame-perfect strategies include only *credible* threats to deter cheating. Simply speaking, credibility means that to punish a defector must constitute a Nash equilibrium of the remaining game, at each stage defection might occur.

Though the concept of subgame-perfectness has its merits, it misses an important point. It allows threat strategies to be called credible which might give a punisher less than he received under the agreement. However, such strategies imply a strong incentive for the punisher not to carry out the punishment and to treat bygones as bygones. Consequently, the threat loses it credibility and therefore cannot deter cheating (Mohr, 1988).

A concept which rules out such non-credible threats is that of renegotiation-proof equilibria which is due to Farrell (1983) and Bernheim and Ray (1985). Since then, many different versions of this concept emerged in the literature, e.g., Asheim (1991), Benoît and Krishna (1993), Bergin and MacLeod (1993) and Ray (1994). In the analysis below, we employ the weakly-renegotiation-proof conditions proposed by Farell and Maskin (1989).[12] The authors defined the concept for only two players, but, as we will argue below, an extension to N players seems to be justified in the context of a global emission game.[13] Accordingly, an equilibrium satisfying the conditions specified below for N players (N > 2) may be called weakly coalition-proof, abbreviated WCP (see, e.g., Bernheim, Peleg and Whinston, 1987).

Let $\Gamma^{T=\infty}(\delta) = (I, \Phi, P)$ denote the dynamic game which is played infinitely, with I the set of players, Φ the set of strategies and P the set of payoffs

158

in the game. Here the payoff P_i can be interpreted as discounted average payoff $NB_{ia} = (1-\delta)\cdot NB_i^{\infty} = \sum_{t=0}^{\infty} \delta^t NB_i$. Further let $E^* = (E_1^*, \ldots, E_N^*)$ be the emission tuple of an agreement and $E^i = (E_1^i, \ldots, E_i^i, \ldots, E_N^i)$ a punishment emission vector to punish country i. In line with Abreu's simple strategy profiles (see Abreu, 1988), (N + 1) strategy vectors (E^* and E^i $\forall i$) are sufficient to check for stability.

For this stability check it is helpful to divide a game into two *phases*. In the *normal phase* all countries comply with the treaty and E^* is played. In case a country i defects in some period the other countries punish country i in the subsequent periods by choosing a punishment emission level E_{-i}^i, where $-i$ denotes all countries except i and $E_{-i}^i = \sum_{j \neq i} E_j^i$. Within this *punishment phase* two sub-cases may be distinguished: 1) Country i accepts the punishment (*repentance phase*) which has been previously specified regarding extent and duration. If the play follows this route all countries return to the normal phase after the punishment time. 2) Alternatively, country i chooses its best reply $E_i(E_{-i}^i)$ to the punishment. In this case the punishment is continued infinitely (*retaliation phase*).

In order for E^* to be a WCP-equilibrium the following conditions have to be satisfied for all countries i and j (see Farrell and Maskin, 1989, in the two player case):

$$NB_i^P(E_i^i, E_{-i}^i) \geq NB_i^D(E_i(E_{-i}^i), E_{-i}^i) \tag{20}$$

$$(1-\delta_i)NB_i^{max}(E_i^*(E_{-i}^*), E_{-i}^*) + \delta_i NB_i^P(E_i^i, E_{-i}^i) \leq NB_i^*(E_i^*, E_{-i}^*) \tag{21}$$

$$NB_j^*(E^*) \leq NB_j^R(E^i)\ \forall j \neq i \tag{22}$$

with

$$NB_i^P = (1-\delta_i^{t_i})NB_i^R + \delta_i^{t_i} NB_i^* \text{ and } 0 \leq \delta < 1. \tag{23}$$

Equation (20) requires that the discounted payoff of a potential defector country i if it is punished but shows repentence, NB_i^P, must be at least as high as if defection is continued, NB_i^D. This gives a (potential) cheater an incentive to show repentance and to meet a prerequisite to reestablish cooperation (normal phase).

Equation (21) requires that the discounted payoff from deviating from the normal phase, NB_i^{max} (best available deviation of i to a given emission rate E_{-i}^*) and accepting the punishment for the agreed time must be lower (or equal) than if complying always.

Equation (22) ensures that the punishment is designed such that no punisher has an incentive to renegotiate on the agreed punishment. Hence, the payoff during the repentance phase, NB_j^R, must be at least as high as in the normal phase. Especially this last condition distinguishes a renegotiation-proof equilibrium from a subgame-perfect equilibrium in many games.

Equation (23) defines the continuation punishment payoff, NB_i^P, which is the weighted average payoff of the retaliation phase, NB_i^R, and the normal phase, NB_i^*. t_i is the punishment time.

In an N-country world the inequality system above implies that $N(N+1)$ conditions have to be satisfied simultaneously in order for E^* to be a WCP equilibrium. Usually, in N-player games, things are (even) more complicated because not only deviations by one single player have to be considered but also deviations of various groups of players. Basically, one has to check for all possible coalition formations of (potential) deviators. Accordingly, an excessively large number of conditions has to be satisfied for a WCP-equilibrium. Hence, in games where it is beneficial to some players to deviate jointly, a coalition-proof equilibrium can only be determined for very simple and/or finite games which mostly lack an economic interpretation (see, e.g., Bernheim and Whinston, 1987).

However, in the global emission game, a country planning to defect has no incentive to do so jointly because the defection payoff is higher if other countries comply. But also on the side of the punishers there is no incentive to form a coalition when breaching the punishment obligations. If during the punishment phase a country among the punishers were to emit more (or less) than requested to punish the defector, it would not ask other countries to follow suit. The punishment constitutes a public good to control an agreement and therefore free-riding is most attractive to a country if conducted by itself. Therefore, if a punisher does not meet its obligations, he is treated as a defector and what has been said above applies.

2.2.2. *Stability analysis in the case of* $\delta \rightarrow 1$

In this sub-section we test whether emission reductions in the social optimum and those specified in the two policy regimes can be sustained in the long-run using credible threat strategies in the sense defined above.

In the case of δ_i close to 1 ($\delta_i^{t_i} \rightarrow 1$) this test is simplified because (20) and (21) reduce to[14]

$$\begin{aligned} &NB_i^*(E_i^*, E_{-i}^*) > NB_i^D(E^i(E_{-i}^i), E_i^i) \text{ if } E_i^* \neq E_i(E_{-i}^*), \\ &NB_i^*(E_i^*, E_{-i}^* \geq NB_i^D(E_i(E_{-i}^i), E_{-i}^i) \text{ if } E_i^* = E_i(E_{-i}^*) \end{aligned} \qquad (24)$$

This implies that an emission vector E^* is a WCP-equilibrium if punishment vectors exist which satisfy (22) and (24) for all i and j, $i \neq j$. Unfortunately,

160

it is impossible to conduct this check analytically in the case of N > 2.[15] Instead, one has to rely on simulations. However, this procedure seems to be justified because all simulation results presented below are quite robust and show a clear pattern. Table 1 tabulates some of these results which take NNC_1 in consideration. The algorithm used to conduct the stability test is described in Appendix 1.

Table 1. Stability analysis of the grand coalition for $\delta \to 1^a$

N	γ	Quota					Tax				
		R^{Q*} (%)	R^{Q_1} (%)	$R^{Q_{max}}$ (%)	DO^{Q_1} (%)	$DO^{Q_{max}}$ (%)	R^{T*} (%)	R^{T_1} (%)	$R^{T_{max}}$ (%)	DO^{T_1} (%)	$DO^{T_{max}}$ (%)
5	50	***48.40***	19.60	42.00	64.0	97.4	***48.00***	13.30	23.30	49.6	74.4
5	200	21.90	8.00	21.90	59.3	98.9	***21.80***	4.40	18.10	37.8	97.1
5	500	10.50	3.70	10.50	56.9	98.7	10.40	1.90	10.40	34.2	100.0
5	1000	5.60	1.90	5.60	55.9	98.7	5.60	0.98	5.60	32.9	100.0
5	2500	2.30	0.80	2.30	55.3	98.7	2.30	0.40	2.30	32.1	100.0
5	5000	1.20	0.40	1.20	55.0	98.6	1.18	0.20	1.18	31.8	100.0
5	10.000	0.60	0.20	0.60	54.9	98.6	0.60	0.10	0.60	31.7	100.0
5	50.000	0.12	0.04	0.12	54.8	98.6	0.12	0.02	0.12	31.5	100.0
5	100.000	0.06	0.02	0.06	54.7	98.6	0.06	0.01	0.06	31.5	100.0
10	50	***83.00***	28.90	44.00	57.4	79.0	***82.50***	***30.00***	-5.00	***61.1***	-8.1
10	200	***66.20***	21.90	28.30	55.2	67.1	***66.00***	***15.00***	9.40	***41.0***	27.4
10	500	***47.20***	12.80	19.00	46.9	64.2	***47.10***	7.50	10.50	29.8	40.1
10	1000	***32.00***	7.50	14.00	41.4	68.2	***31.90***	4.10	9.50	24.4	50.8
10	2500	***16.20***	3.30	9.20	36.8	81.0	***16.20***	1.70	7.30	20.5	69.7
10	5000	***8.90***	1.70	6.00	35.0	89.0	***8.90***	0.90	5.50	19.1	85.5
10	10.000	***4.69***	0.88	4.20	34.0	98.6	***4.69***	0.45	3.73	18.4	95.8
10	50.000	0.98	0.18	0.98	33.2	99.7	0.98	0.09	0.98	17.8	100.0
10	100.000	0.49	0.09	0.49	33.1	99.7	0.49	0.04	0.49	17.7	100.0
20	200	***90.80***	***31.10***	31.00	***56.8***	56.6	***90.70***	***31.70***	-31.20	***58.4***	-77.4
20	500	***85.00***	***27.50***	23.60	***54.3***	47.8	***84.90***	***21.10***	-6.00	***44.0***	-13.9
20	1000	***76.80***	***20.70***	18.00	***46.7***	41.8	***76.70***	***13.60***	1.50	***32.5***	4.3
20	2500	***59.60***	11.50	12.20	34.8	36.7	***59.60***	***6.55***	4.95	***21.0***	16.1
20	5000	***43.40***	6.60	8.80	27.9	36.4	***43.40***	3.52	5.06	15.7	22.1
20	10.000	***28.10***	3.50	6.00	23.5	38.1	***28.10***	1.80	4.40	12.7	28.9
20	50.000	***7.40***	0.75	2.80	19.3	61.1	***7.40***	0.38	2.46	10.1	55.6
20	100.000	***3.80***	0.38	1.90	18.7	74.6	***3.80***	0.19	1.70	9.7	69.1

Table 1. Continued

N	γ	Quota					Tax				
		R^{Q*} (%)	R^{Q_1} (%)	$R^{Q_{max}}$ (%)	DO^{Q_1} (%)	$DO^{Q_{max}}$ (%)	R^{T*} (%)	R^{T_1} (%)	$R^{T_{max}}$ (%)	DO^{T_1} (%)	$DO^{T_{max}}$ (%)
50	2500	***94.30***	***30.10***	16.00	***53.7***	31.0	***94.30***	***24.50***	−23.5	***45.4***	−55.5
50	5000	***90.90***	***23.80***	12.00	***45.5***	24.7	***90.90***	***16.30***	−8.8	***32.8***	−20.1
50	10.000	***84.7***	***16.20***	9.40	***34.6***	21.0	***84.70***	***9.80***	−1.88	***21.9***	−4.4
50	50.000	***54.90***	***4.45***	4.30	***15.6***	15.4	***54.90***	***2.33***	1.99	***8.3***	7.2
50	100.000	***38.20***	2.33	3.10	11.9	15.6	***38.20***	1.20	1.92	5.0	9.8
100	5000	***98.00***	***32.90***	14.00	***55.8***	26.5	***98.00***	***33.00***	−64.80	***56.1***	−175.0
100	10.000	***97.10***	***30.40***	11.00	***52.9***	21.4	***97.10***	***24.80***	−30.60	***44.6***	−72.8
100	50.000	***90.10***	***14.10***	5.00	***28.8***	12.9	***90.10***	***8.30***	−3.30	***17.5***	−7.5
100	100.000	***82.60***	***8.20***	4.30	***18.9***	10.1	***82.60***	***4.50***	−0.43	***10.6***	−1.0
100	300.000	***62.10***	***3.10***	2.50	***9.7***	7.9	***62.10***	***1.60***	0.90	***5.1***	3.0

[a] R^{Q*} = socially optimal emission reduction in the quota regime, R^{Q_1} = emission reduction in the quota regime according to the LCD-decision rule, $R^{Q_{max}}$ = maximal emission reduction in the quota regime which can be backed as a WCP equilibrium, $DO^{Q_{max}}$ = degree of optimality for $R^{Q_{max}}$, DO^{Q_1} = degree of optimality for R^{Q_1}, R^{T*} = R^S = socially optimal emission reduction, R^{T_1} = emission reduction in the tax regime according to the LCD decision rule, $R^{T_{max}}$ = maximal emission reduction in the tax regime which can be backed WCP, $DO^{T_{max}}$ = degree of optimality for $R^{T_{max}}$, DO^{T_1} = degree of optimality for R^{T_1}

In Table 1 emission reductions and welfare indicators are provided for the social optimum, the tax and the quota regime. Emission reductions are expressed as percentage reduction compared to the initial situation. The degree of optimality is defined as

$$DO^h = \frac{NB^h - NB^N}{NB^S - NB^N} \cdot 100 \qquad (25)$$

where h stands for the policy regime and the particular solution, e.g., $h = T_1$ or $h = Q_{max}$. Put differently, DO^h indicates how close a solution comes to the social optimum. By definition, the degree of optimality in the social optimum is 100%. Since the tax is an efficient instrument and therefore the emission reduction under a Pigouvian tax, R^{T*}, corresponds to the socially optimal emission reduction, R^S, $DO^{T*} = DO^S = 100\%$ holds.

For the two policy regimes we provide information on the socially optimal emission reduction (R^{Q*} and R^{T*}), emission reductions if country 1 determines the terms of the agreement (R^{Q_1} and R^{T_1}) and the maximal emission reductions which can be backed by WCP threat strategies (R^Q_{max} and R^T_{max}).

162

The maximal emission reduction is provided to evaluate the two policy instruments independently of the decision rule chosen among the negotiators. The entries in bold italics indicate that a solution is not stable in sense of WCP. In other words, these solutions specify an abatement target above R^T_{max} or R^Q_{max}.

From Table 1 four main results may be derived:

1) Whenever the potential gains from cooperation would be great (N big and γ small, see Equation (9) stability poses a problem. This holds general for "optimal" and second-best solutions.
2) The socially optimal solution ($R^S = R^{T*}$) is only stable in a few cases where N is small and γ is big. These are those parameter constellations for which abatement targets are rather low. The same holds true for a socially optimal emission quota (R^{Q*}). Hence, whenever a joint abatement policy is most attractive from a global point of view, "optimal solutions" are not stable. Recalling that the median country proposal in the tax regime is identical to R^{T*} and in the quota regime to R^{Q*}, it is evident that such proposals have hardly a chance of realization. Therefore, it does not seem sensible to consider the median country proposal any further.
3) Whenever the tax agreement is stable under the LCD-decision rule, then this is also true for the quota regime but the degree of optimality and global emission reductions are higher.
4) Considering the maximal backable emission reduction the degree of optimality in the tax regime is higher than in the quota regime for the "non-critical" parameter values (N small and γ big). In Table 1 this relation holds for $N = 5$, $\gamma \geq 500$ and $N = 10$, $\gamma \geq 50.000$. For all other parameter values the opposite holds.

The results summarized in 3) and 4) can be taken as a first indication that in a second-best world, the choice of a quota regime to tackle global emissions might be rational. Of course, governments are not concerned with global efficiency. However, according to Table 1, the typical efficiency argument in favor of the tax regime does not hold anymore when confronted with a second-best environment.

It seems important to note that for a discount factor close to one a grand coalition is only stable if the potential gains from cooperation are low. This result is independent of the decision rule and the policy regime. This is in line with results obtained by Barrett (1994a and b). In particular for global environmental problems where N is big, a stable grand coalition seems very unlikely even if politicians are almost perfectly patient. But even if a grand coalition were stable, reduction targets agreed upon would be very low. This result is confirmed by the historical evidence of IEAs signed so far.

2.2.3. *Stability analysis for* $\delta << 1$

In the previous sub-section the stability analysis was confined to a discount factor close to one. This is the standard assumption in the literature. The reason is simple. On the one hand, the theoretical literature is only concerned whether there are WCP-equilibria in a game. On the other hand, the applied literature faces the problem than once departing from simple games, a solution is difficult to obtain.

In reality, however, political representatives are hardly (almost) perfectly patient. On the contrary, in democracies elections are held regularly and therefore short-term success is important for politicians (Hahn, 1987: 300)). Therefore, it seems obvious to extent the analysis to impatient negotiators. In particular, we want to find out whether there are differences regarding the "patience requirement" in the two policy regimes. Of course, such an analysis is only interesting for those parameter constellations discussed in the previous section where stability is possible for $\delta \rightarrow 1$. In order to reduce the data set further we restrict the analysis to $N \leq 20$. For a given emission tuple E^*, the minimum discount factor requirement of country i, δ_i^{min}, satisfies the $N + 1$ inequalities given by (20) to (22) and makes at least one inequality binding. To determine, δ_i^{min} an algorithm is used, similar to the one used in the case of $\delta \rightarrow 1$. The details are given in Appendix 2. In Table 2 simulation results are presented for selected parameter constellations.

Table 2. Discount rate requirement for country 1 using the LCD-decision rule[a]

		γ								
		50	200	500	1000	2500	5000	10000	50000	100000
N =5	Quota	16	10	9	9	8	8	8	8	8
	Tax	7	5	5	5	5	5	5	5	5
N = 10	Quota	5.4	1.9	2.0	1.7	1.6	1.5	1.5	1.4	1.4
	Tax	–	–	1.0	1.2	1.1	1.1	1.1	1.1	1.1
N = 20	Quota		–	–	–	0.1	0.3	0.3	0.3	0.3
	Tax		–	–	–	–	0.2	0.2	0.2	0.2

[a]The discount rate r_1^{max} has been calculated on the 1% significance level for $N = 5$ and on the 0.1% level for $N = 10$ and $N = 20$. "–" means not stable for $r_1 \rightarrow 0$. The blank spaces indicate a non-defined parameter constellation (NNC_1).

For convenience, δ_i^{min} is expressed as maximum interest rate, r_i^{max}, where $\delta_i^{min} = 1/(1 + r_i^{max})$ holds.[16] Therefore, r_i^{max} is the maximal discount rate for which a country i can be deterred from taking a free-ride by the other N-1 countries. If the actual discount rate r_i is higher than r_i^{max}, then the agreement is not stable.

It turns out that country 1 has always the highest interest rate requirement. Hence, in Table 2, attention has been restricted to r_1^{max} only. From this it follows that $r_1^{max} \geq r_i$ is a necessary and sufficient condition for country 1 ($i = 1$) not to take a free-ride (provided all other countries comply with the agreement), but it is only a sufficient condition for country i, $i \neq 1$.

Two main results can be derived from Table 2:

1) Only if the number of countries suffering from the global externality is small, a stable agreement can be expected. Already in the case of ten countries and $\gamma \geq 200$ the discount rate must be less than 2%. For N = 20 countries must be even more patient, i.e., $r_1^{max} < 0.4$. The result backs the assertion above that a grand coalition is very unlikely to form if a pollution problem is of real global character.
2) The discount rate requirement for country 1 is less restrictive under the quota than under the tax regime. If countries with a low index are not only less interested in emission reductions than their neighbors, but also have a higher time preference rate, this implies that "critical" countries find it easier to comply with a uniform emission reduction quota than with a uniform effluent charge.

The first result, again, suggests to investigate the formation of sub-coalitions. This will be done in Section 3 below. The second result can be interpreted as an additional explanation why the quota regime has been so popular in international pollution control, though it causes inefficiencies.

2.2.4. *Stability analysis for restricted punishment profiles*

Up to now it has been assumed that if countries punish a defector they do this optimally (see Appendices 1 and 2). In particular, it has been assumed that the punishment emission vectors are chosen in such a way that the WCP conditions in Equation (22) are satisfied "most easily". This implies that the increase in emissions compared to the normal phase, which is necessary to punish a defector i, is allocated optimally among the group of punishers.

However, in case of many countries such an "optimal coordination" might be difficult to obtain. Especially in the light of (possible) high transaction costs, it seems plausible to assume that punishment strategies are chosen which follow a simple pattern. In addition, to determine an optimal allocation of punishment duties requires that governments reveal their true preferences.

Table 3. Discount rate requirement for restricted punishment profiles and N = 5[a]

		γ						
		50	200	500	1000	5000	10000	100000
Quota	no restriction	16	10	9	9	8	8	8
	P_1	13	9	8	8	7	7	7
	P_2	14	10	8	8	8	7	7
Tax	no restriction	7	5	5	5	5	5	5
	P_1	6	5	5	5	5	5	5
	P_2	6	5	5	5	5	5	5

[a]The discount rate r_1^{max} has been calculated on the 1% significance level.

If governments keep this information private for strategic reasons, one has to reckon with strategically manipulated answers. A simple punishment profile, which does not require complete information, might therefore be used in reality. Moreover, in reality it takes time to coordinate an optimal punishment action. This renders the "deterrence potential" to be rather low for a potential defector. In contrast, if countries agree on a simple punishment strategy, an immediate reaction has a higher chance of realization.

If simple punishment profiles are used to remedy the problems mentioned above, uniform "punishment agreements" seem to be plausible focus points. In the analysis below, two punishment versions are considered: $P_1 : E_j^i = (1+k)E_j^*$ and $P_2 : E_j^i = E_j^* + k$. These versions imply that in the punishment phase each country $j \neq i$ increases its emissions either proportionally (P_1) or absolutely (P_2) by the same amount k compared to the normal phase. Table 3 shows the changes involved when using these strategy profiles. The algorithm to derive the results is outlined in Appendix 3.

Of course, even though simple punishment strategies might reduce transaction costs (which are not modeled) and possess some other advantages, they increase the discount rate requirements (because they are not optimal). Therefore, there exists a dilemma: On the one hand, if many countries participate in an IEA simple strategy profiles are most likely employed to make coordination easier. On the other hand, in this case the discount rate requirements are already very high in the absence of simple strategy profiles. Again, the result underlines our previous claim that for global externalities a grand coalition is very unlikely to form.

Comparing the two punishment profiles P_1 and P_2, we find no difference in the tax regime. This is due to the specification of the abatement cost function as given in Equation (2) for which equilibrium emissions in the tax regime

166

are equal for all countries ($E_i^* = E_1^{T_1} = E_2^{T_1} = \ldots = E_N^{T_1}$). In contrast, for the quota regime P_2 is less restrictive than P_1. The reason is that countries with a higher index suffer particularly from an increase in emissions in the punishment phase. Hence, an optimal punishment allocation requires that, if total emissions of the punishers have to be increased by the amount K compared to the normal phase to punish a country i, countries with higher index numbers should receive a relatively big share of K. If their portions are not sufficiently big, then Equation (22) is violated.

Applying the punishment profile P_2, countries with a high index receive a relatively high share of K. Not so however if punishment profile P_1 is employed. Because emission levels of high index countries are relatively low in the normal phase, a proportional allocation of punishment duties is not conducive for stability. Interestingly, even though it seems to be suggestive in the case of the quota regime to apply the punishment profile P_1, since it uses the same allocation rule as in the normal phase, this leads to a higher discount rate requirement than if punishment profile P_2 is applied. The result indicates that the choice of the punishment profile may have an impact on the stability of an agreement. Hence, for a successful IEA it is not only important which policy instrument is chosen to achieve an envisaged abatement target, but also to agree on the punishment duties in case defection occurs.

In reality, also other restrictions on the punishment profile may apply. For example, there might be an upper ceiling on the level of the punishment. A reason could be that sometimes governments negotiate more than one issue at the same time. Consequently, they punish defection only modestly in order not to jeopardize relations in other sensitive areas of interest. Moreover, one may argue that to increase punishment emissions above non-cooperative abatement levels is difficult once abatement measures have been implemented.

Generally, the smaller a ceiling s, below which punishment emissions must lie, the higher is the discount rate requirement for each country (the lower is $r^{max}{}_i$). Since a demonstration of this result is given elsewhere, we skip it here.[17]

If the punishment ceiling is given by $s = E_i^N$ (recall E_i^N is the initial emission level), then a grand coalition is not stable under a uniform tax regime, no matter how low the tax rate might be. In contrast, under a quota regime a grand coalition can be stable provided discount factors are sufficiently close to one. The proof of this result is given in Appendix 4. Once more, the result underlines the possible advantages of a quota regime in a second best world.

3. The sub-coalition

3.1. *The formation process*

So far we assumed that if countries decide to jointly improve environmental quality they seek to form a coalition of all countries. It became apparent that there are many obstacles which render the probability of a stable grand coalition to be rather low. This result is perfectly in line with existing environmental conventions in which only some of N countries participate if they come into force at all. For example, the Convention for the Regulation of Whaling has been signed by 36 countries and the Antarctic Treaty by 26 countries even though all states should have an interest in the problem. The Rio declaration might be viewed as an exception. However, it is more an intention of good will rather than an IEA with precise abatement targets.

Basically, environmentally conscious countries face a dilemma: On the one hand, they like to see many countries participating in an IEA to improve environmental quality and to share the abatement burden with them. On the other hand, due to the LCD decision rule, if too many countries join an IEA the abatement target which can be established within an IEA might be very low. Moreover, it has been demonstrated that less environmentally concerned countries are also the critical ones regarding stability. Therefore, one should suspect that those countries which have a high interest in an international environmental policy seek partners to join them, but look only for those with similar interests. This procedure seems even more likely when taking into account that coordination in international politics is associated with transaction costs.

In the following we do not model transaction costs explicitly because its effects are rather obvious (transaction cost increase as a function of the number of signatories), but its impact should be kept in mind when interpreting the results below.

From a theoretical point of view, basically, all kinds of sub-coalitions could form.[18] To narrow the analysis down, we assume that countries with higher environmental damages will form a coalition among themselves (see also Botteon and Carraro, 1997 and Hoel, 1992). Therefore, in the context of our model the set of non-signatories is given by $\Omega^{NS} = \{1, 2. \dots, N_0\}$, and the set of signatories by $\Omega^{S} = \{N_0 + 1, \dots, N\}$. Within the coalition three issues have to be decided simultaneously: abatement target, instrumental choice and coalition size.

3.1.1. *Abatement target*

Among the signatories we assume that the LCD-decision rule is used to determine the abatement target. As in the case of a grand coalition it turns out

168

that for a given coalition size it is the country with the smallest index which makes the lowest proposal. Therefore, it is this country's government which determines the terms off the agreement (see Equations (26) and (27) below).

For example, if N = 10 and $\Omega^S = \{7, 8, 9, 10\}$, then country 7 is the bottleneck country and its proposal is laid down in the treaty.

For the non-signatories we assume that they choose their initial non-cooperative emission level, $E^N{}_i$. Of course, there is an incentive for non-signatories to increase their emissions as a reaction to the higher abatement efforts by signatories. Such leakage effects have been described and modeled by Bohm and Larsen (1993), Golombek, Hagem and Hoel (1995) and others in a static context. These effects cannot be ruled out *a priori*. Therefore, in a dynamic setting we check whether "free-riding" by non-signatories can be controlled by signatories through suitable punishments.

From this it follows that the stability of an IEA comprises two components:

1) External stability: It has to be checked whether it is possible for a coalition to control emissions of non-signatories using WCP strategies. That is, non-signatories are not allowed to increase their emissions above the initial level, E_i^N, given in Equation (4).
2) Internal stability: It has to be investigated whether signatories can control their agreement internally using WCP strategies.

Due to lack of space we restrict the analysis to $\delta \rightarrow 1$. A detailed description of the stability test is given in Appendix 5.

3.1.2. *Instrumental choice*

Potential signatories decide the choice of the instrument employed in the IEA according to the majority decision rule (MDR). E.g., if N = 10, $\Omega^S = \{7, 8, 9, 10\}$ and preferences are given by 7: Q, 8: Q: 9: Q and 10: T this coalition of four countries would employ an emission quota. The same applies for any preliminary coalition which is constructed during the coalition formation process.

3.1.3. *Coalition size*

We assume that the coalition formation process starts with country N (the most environmentally conscious country) approaching country N-1 in order to discuss a joint environmental policy. In this first round each country makes a proposal regarding an emission reduction. According to the LCD decision rule only the smaller of both suggestions will be realized. This is country N-1's proposal. Since one can show that country N always prefers to cooperate with country N-1, even though country N's proposal is not accepted, the LCD assumption seems indeed plausible.[19]

For this preliminary sub-coalition the associated welfare levels of country N and N-1 can be computed. They serve as a benchmark to decide whether a third country should be asked to join. In the second round of the negotiation process country N-2 is asked to submit its proposal if country N-2, N-1 and N would form a coalition. For this new constellation also country N and N-1 will make an offer. Again, the parties have to agree on the smallest offer which is submitted by the country with the lowest index number, i.e., country N-2. Provided country N and N-1 benefit if N-2 joins (and determines the condition of the treaty), then these countries offer country N-2 to participate in their preliminary agreement. In case both countries were to lose if country N-2 joins, these countries would form a coalition only among themselves. If country N and N-1 hold different views regarding an extension of its preliminary coalition the coalition formation process stops. This is so because otherwise the country which is against the enlargement can threaten to boycott the coalition. Hence, the coalition size is decided according to a unanimity decision rule (UDR) and formation process stops once a country votes against an enlargement.[20] In order to keep track of the simultaneous decision process, we proceed in two steps. In the first step we determine the equilibrium number of signatories assuming that the instrumental choice is exogenously given (Section 3.2). In the second step, this is done if the instrumental choice is endogenously determined within the negotiation process leading to an IEA (Section 3.3). Results are discussed in Section 3.4.

3.2. *The equilibrium number of signatories in the two policy regimes*

To determine the equilibrium coalition size, we first have to compute the abatement proposal of a potential signatory j. In the quota regime, we use two pieces of information. First, country j knows that country 1 to N_0 choose their initial emission level $E^N{}_i$. Therefore, emissions of all non-signatories are given by $E^{NS} = \sum_{i=1}^{N0} E_i^N$. Second, if country $N_0 + 1$ to N comply with country j's proposal, $R^Q{}_j$, then signatories' emissions will be $E^S = (1 - R_j^Q) \cdot \sum_{i=N0+1}^{N} E_i^N$. Consequently, aggregate emissions are given by $E^Q = E^{NS} + E^S$. Inserting this information in Equations (1) to (3), it is straightforward to compute a signatory j's proposal to be

$$
\begin{aligned}
R_j &= \frac{2\gamma jN(2N\gamma - 2N_0\gamma + NN_0^2 - N^2N_0 - N - N^2 + 2jN - 2\gamma}{A} \\
A: &= 4\gamma^3 - 8j\gamma^2N - 4j\gamma N^2 + \gamma N^2 - 4j\gamma N^3 + 4\gamma^2N + 2\gamma N^3 - 8j\gamma^2NN_0 \\
&+ 4j\gamma^2N^2 - 4j\gamma N^3N_0 + 4j\gamma^2N_0^2 - 2jN^3N_0^3 - 4j\gamma NN_0^3 \\
&+ 4j^2\gamma N^2 + 8j\gamma N^2N_0^2 + jN^4N_0^2 + \gamma N^4 + 4\gamma^2N^2 + jN^2N_0^4, \qquad (26)
\end{aligned}
$$

170

where $\partial R_j/\partial j > 0$ holds. That is, the signatory with the lowest index number is the bottleneck country. According to our notation this is country $N_0 + 1$.

Using a similar procedure, we compute a signatory's tax proposal:

$$t_j = \frac{bcdjN(N - N_0)(2b + N^2c - N_0^2c + Nc - N_0c)}{(2b + N^2c + Nc)(b + N^2jc + jN_0^2c - 2jN_0Nc)} \tag{27}$$

Again, $\partial t_j/\partial j > 0$ holds, implying that the proposal of the signatory with the lowest index is accepted within the IEA.

In order to rule out corner solutions, we have to ensure $t_{N0+1} < bd$ for any values of N_0 and N. It turns out that a sufficient condition is

$$NNC_2 : \gamma > N^3/6 \tag{28}$$

which we assume to hold in the remainder.

To gain a deeper understanding of the formation process, a typical example is shown in Table 4.

We start by considering the formation process in the quota regime. From Table 4 it is evident that up to a coalition size of eight countries – including country 3 to 10 – all potential signatories favor to enlarge the coalition. If eight countries participate in the IEA, only three countries are in favor of an enlargement. Therefore, according to the UDR, the equilibrium coalition size comprises eight countries (italic entries).

Up to a coalition size of five countries the reduction level proposed by country $N_0 + 1$, R_{N0+1}, increases. In this range ambitious abatement targets are proposed by country $N_0 + 1$ because the marginal benefits from a joint emission reduction are high. Then, for a coalition size above five countries, the proposal of country $N_0 + 1$ decreases. Nevertheless, up to a coalition size of eight countries, all potential signatories benefit from an enlargement. In this range, the benefit to potential signatories from an additional contributor is stronger than the (negative) effect of a lower joint abatement target. If more than eight countries were to join, then the reduction target would become too low for countries with high environmental damages. A majority of countries is against an enlargement.

In contrast, in the tax regime the equilibrium coalition size is not eight but seven countries. In this example country 4's tax proposal, $t_4 = 1665$, cannot be backed by WCP strategies. Therefore, we assume that the marginal stable rate, $t_{max} = 1650$, will be applied in the agreement. Since countries 5 to 10 prefer a coalition of seven countries including country 4 to a coalition of only six countries, though only the low tax rate t_{max} is applied, one should expect that country 4 is asked to join.

For other parameter constellations Table 5 summarizes the relevant information which will be discussed in Section 3.4.

Table 4. The formation process of a sub-coalition[a]

Coalition size N^a		1	2	3	4	5	6	7	*8*	9	10
N_0+1		10	9	8	7	6	5	4	***3***	2	1
Quota:	R_{N^0+1}	0	29.3%	45.0%	51.3%	51.9%	49.1%	**44.1%**	***37.5%***	29.1%	17.8%
	R_{max}	0	32.5%	53.5%	65.0%	69.9%	67.0%	**54.0%**	***44.0%***	34.0%	24.0%
	R^Q	0	5.1%	11.9%	18.5%	23.8%	27.5%	**29.3%**	***29.0%***	25.8%	17.8%
	NB_{10}	-21900	-20810	-18190	-15080	-12160	-9810	**-8200**	***-7550***	-8280	-1630
	NB_9		-17300	-15140	-12440	-9830	-7660	**-6140**	***-5460***	-6030	-8950
	NB_8			-12100	-9810	-7500	-5520	**-4090**	***-380***	-3780	-6270
	NB_7				-7180	-5170	-3390	**-2040**	***-1320***	-1550	-3620
	NB_6					-2850	-1260	**0**	***740***	670	-980
	NB_5						870	**2030**	***2790***	2870	1650
	NB_4							**4060**	***4830***	5070	4250
	NB_3								***6850***	7250	6840
	NB_2									9420	9420
	NB_1										11980
	Enlargem.										
	yes/no	1:0	2:0	3:0	4:0	5:0	6:0	**7:0**	***3:5***	0:9	–

(*continued overleaf*)

Table 4. Conanued.

Coalition size N[a]	1	2	3	4	5	6	7	8	9	10
N_0+1	10	9	8	7	6	5	4	3	2	1
Tax: T_{N0+1}	845	**1444**	**1783**	**1918**	**1915**	**1822**	*1665*	1451	1166	750
t_{max}	84.5	**1515**	**1958**	**2201**	**2299**	**2010**	*1650*	1340	1050	730
R^T	0	**5.1%**	**12.1%**	**18.9%**	**24.4%**	**28.1%**	*29.2%*	25.0%	19.3%	10.5%
NB_{10}	−21900	**−20660**	**−17890**	**−14600**	**−11500**	**−8980**	*−7420*	−810.0	−10110	−14520
NB_9		**−17440**	**−15130**	**−12250**	**−9460**	**−7140**	*−5630*	−6090	−7780	−11650
NB_8			**−12370**	**−9910**	**−7420**	**−5290**	*−3850*	−4080	−5440	−8790
NB_7				**−7560**	**−5380**	**−3450**	*−2060*	−2070	−3130	−5930
NB_6					**−3350**	**−1600**	*−270*	−60	−800	−3060
NB_5						**240**	*1520*	1950	1530	−200
NB_4							*3310*	3960	3860	2660
NB_3								5970	6180	5520
NB_2									8510	8390
NB_1										11250
Enlargem. yes/no	1:0	**2:0**	**3:0**	**4:0**	**5:0**	**6:0**	*3:4*	2:6	0:9	-
Countries in favor of a quota		9	8	7–8	6–7	5–7	**4–7**	**3–10**	2–10	1–10
Countries in favor of a tax		**10**	**9–10**	**9–10**	**8–10**	**8–10**	8–10			
Instrument chosen		T	T	T	T	T	Q	Q	Q	Q

[a]The example assumes N = 10, b = 300, c = 1 and d = 10.

Table 5. Equilibrium size of a sub-coalition for $\delta \rightarrow 1$[a]

N	γ		Coalition size	R^{SC} (%)	R^{GC} (%)	DO^{SC} (%)	DO^{GC} (%)	$DO^{GC_{max}}$ (%)	Instrumental choice
5	50	Q	4	22.9	19.6	65.1	64.0	97.4	T
		T	**4**	22.0	13.3	64.2	49.6	74.4	
	200	Q	**4**	9.6	8.0	60.4	59.3	98.9	Q
		T	4	7.8	4.4	53.2	37.8	97.1	
	500	Q	**4**	4.4	3.7	58.0	56.9	98.7	Q
		T	4	3.4	1.9	49.4	34.2	100.0	
	1000	Q	**4**	2.3	1.9	57.0	55.9	98.7	Q
		T	4	1.7	1.0	47.9	32.9	100.0	
	2500	Q	**4**	0.9	0.8	56.4	55.3	98.7	Q
		T	4	0.7	0.4	47.0	32.1	100.0	
	5000	Q	**4**	0.5	0.4	56.1	55.0	98.6	Q
		T	4	0.4	0.2	46.6	31.8	100.0	
	10000	Q	**4**	0.2	0.2	56.0	54.9	98.6	Q
		T	4	0.2	0.1	46.5	31.7	100.0	
	50000	Q	**5**	0.04	0.04	54.9	54.8	98.6	Q
		T	4	0.04	0.02	46.3	31.5	100.0	
	100000	Q	**5**	0.02	0.02	54.7	54.7	98.6	Q
		T	4	0.02	0.01	46.2	31.5	100.0	
10	200	Q	**8**	34.6	21.9	72.9	55.2	67.1	Q
		T	7	31.7	9.4	67.1	27.4	27.4	
	500	Q	**8**	21.7	12.8	66.2	46.9	64.2	Q
		T	8	20.2	7.5	63.4	29.8	40.1	
	1000	Q	**8**	13.2	7.5	61.0	41.4	68.2	Q
		T	8	11.7	4.1	56.3	24.4	50.8	
	2500	Q	**8**	6.1	3.3	56.2	36.8	81.0	Q
		T	8	5.2	1.7	50.4	20.5	69.7	
	5000	Q	**8**	3.2	1.7	54.2	35.0	89.0	Q
		T	8	2.7	0.9	48.1	19.1	85.5	
	10000	Q	**8**	1.6	0.88	53.1	34.0	98.6	Q
		T	8	1.4	0.45	46.8	18.4	95.8	
	50000	Q	**8**	0.3	0.18	52.2	33.2	99.7	Q
		T	8	0.3	0.09	45.7	17.8	100.0	
	100000	Q	**8**	0.2	0.09	52.1	33.1	99.7	Q
		T	8	0.1	0.04	45.6	17.7	100.0	

(*continued overleaf*)

174

Table 5. Continued

N	γ		Coalition size	R^{SC} (%)	R^{GC} (%)	DO^{SC} (%)	DO^{GC} (%)	DO^{GC}_{max} (%)	Instrumental choice
20	2500	Q	**14**	24.6	11.5	61.5	34.8	36.7	Q
		T	12	22.1	5.0	55.4	16.1	16.1	
	5000	Q	**14**	18.7	6.6	61.9	27.9	36.4	Q
		T	13	16.7	3.5	56.5	15.7	22.1	
	10000	Q	**15**	10.8	3.5	57.7	23.5	38.1	Q
		T	15	10.2	1.8	55.3	12.7	28.9	
	50000	Q	**15**	2.4	0.75	51.1	19.3	61.6	Q
		T	15	2.2	0.38	48.1	10.1	55.6	
	100000	Q	**15**	1.2	0.38	50.1	18.7	74.6	Q
		T	15	1.1	0.19	47.1	9.7	69.1	

R^{SC} = global emission reduction of the sub-coalition; R^{GC} = global emission reduction of the grand coalition if country 1's proposal is applied, or, if this is not stable, the maximal stable emission reduction (see Table 1); DO^{SC} = degree of optimality for R^{SC}; DO^{GC} = degree of optimality for R^{GC}; DO^{GC}_{max} = degree of optimality of the grand coalition if R^{max} is realized.

3.3. *The endogenous determination of instrumental choice*

So far we treated the policy regime as exogenously given and derived the equilibrium coalition size for each regime. Now, we also consider the choice of the policy regime within the negotiation process. Of course, this choice will also affect the coalition size and the abatement target agreed upon in an IEA.

Generally, countries may hold different views regarding the instrument which is applied in an IEA. As a tendency, countries with low index numbers prefer the emission quota because for a given reduction level they have to carry a lower reduction burden than under the tax regime.[22] In contrast, countries with high index numbers prefer a tax because this leads to an emission allocation which favors them.

However, since the instrumental issue is linked to the other two issues, countries with high index numbers might also favor a quota if this implies that more countries participate in an IEA and/or the abatement target agreed upon is higher than in a tax regime. To gain more insight in the forces we use again Table 4 to illustrate a typical negotiation process.

From the previous discussion we know that for this example the equilibrium coalition size comprises seven countries in the case of the tax regime, and eight countries in the case of the quota regime. For each possible coalition size the countries which are in favor of a particular instrument are listed in the second and third last row.

For example, if country 9 and 10 form a coalition, country 9 prefers a quota regime ($-17300 > -17440$) and country 10 a tax regime ($-20660 > -20810$). In case of such a draw, we assume the preliminary coalition to settle for the tax. This "conservative" assumption seems appropriate because it turns out below that for most parameter constellations the quota regime will finally be chosen among the signatories. Generally, however, we assume that the majority decision rule (MDR) is applied with respect to this issue. The outcome of the decision process is indicated for each stage by the bold numbers and listed in the last row in Table 4.

However, in the preliminary coalition comprising countries 9 and 10 the instrumental issue is not really important because both countries unanimously prefer if one more country joins their "club". This is true, irrespectively of the policy regime. Consequently, the coalition is enlarged. Only if six countries have joined the coalition the instrumental choice becomes important: Three countries prefer a tax and three a quota regime. If one more country joins, four signatories prefer a quota and three a tax regime. According to the MDR, signatories agree upon the quota. Then, the coalition is extended to eight countries of which all unanimously prefer the quota and the extension of the coalition ceases.

Of course, we also have to consider possible strategic linkages between issues. For example, country 10 might think of blocking an enlargement at an early stage of the formation process in order to force the decision process in favor of the tax regime. In the example, country 10 could veto an extension from six to seven countries in order to have a majority of countries favoring the tax. However, comparing net benefits of country 10 in a coalition of six countries applying a tax rate with a coalition of eight countries applying a quota, country 10 prefers a coalition of eight countries. Therefore, country 10 will not make use of its veto right at this early stage.

According to the procedure described above, the endogenous policy choice has been determined for other parameter constellations. The results are presented in the last column of Table 5. The equilibrium number of signatories (if all three issues are decided simultaneously) has been marked bold in the column displaying the "Coalition Size".

3.4. *Results*

From the results presented in Table 5 four main conclusions can be drawn:

176

1) The relative coalition size (N*/N) is not very sensitive to changes in the parameter values, but decreases in N and increases in γ. This implies that the number of signatories is lower for the "critical" parameter values (N big and γ small).
2) The degree of optimality of the sub-coalition, DO^{SC}, is higher than of the grand coalition, DO^{GC}, if the LCD-decision rule is applied. The difference is particularly distinct for the critical parameter values.
3) For the critical parameter values (N big and γ small) the degree of optimality of the sub-coalition, DO^{SC}, is higher than of the grand coalition which chooses the maximal backable emission reduction target, $DO^{GC_{max}}$ (e.g., N = 20 and $\gamma = 2500$).
4) In most cases the quota regime emerges from the decision process. Only if N is small a tax regime may be chosen among signatories (e.g., N = 5 and $\gamma = 50$).

The first result confirms the findings for the grand coalition that whenever the gains from cooperation would be great, an IEA achieves relatively little.

The second and third result underlines the rationale behind the formation of sub-coalitions. Even though the signatories are not concerned with global welfare but only with welfare in their countries, agreements based on compromises (LCD-decision rule) might be more efficient if signed among a small group of relatively homogenous countries instead if a grand coalition is formed. In the light of possible high transaction costs, this argument becomes even more momentum.

The rationale for the fourth result is the following: According to Equation (1) a low value of N implies a low variance in damages between countries. Together with Equations (2) and (3) this implies also a low variance in net benefits. Therefore, for small N the allocation of net benefits from a joint abatement policy is also under a tax regime relatively even. Consequently, a majority of countries votes for a tax in order to capture the efficiency gains associated with this instrument.

In contrast, for large N interests between countries are heterogeneous which leads to an uneven distribution of net benefits from emission reductions under a uniform tax. This in turn causes stability problems which can only be partially compensated by low reduction levels.

In contrast, under a quota regime abatement is based on initial emission levels which already reflect the preferences of countries to some extent. Hence, the abatement burden is allocated more in line with the interests of countries. This also allows that higher abatement targets are approved within an IEA and/or a higher stability.

Since there exists an incentive for the initiator governments (countries with a high index) that other countries contribute to a joint environmental policy, a quota regime becomes particularly attractive if the global externality problem is distinct which is the case if N is large (see Equation (9)).

The findings help to explain why in reality no tax has been observed so far as part of IEAs, though this instrument has been advocated frequently by economists as an efficient instrument to tackle global externalities. According to the simulations a tax will only find acceptance among a small number of countries and if abatement costs are relatively moderate compared to perceived environmental damages.[22]

4. Summary and conclusion

The coalition formation process of asymmetric countries has been studied. In this process three issues are decided: abatement target, instrumental choice and coalition size. Exemplarily, a uniform emission reduction quota and a uniform tax have been considered where the abatement target is decided according to the lowest common denominator. Additionally, the social optimum has been determined as a benchmark value. Within a dynamic framework stability of a grand coalition (Section 2) and of sub-coalitions (Section 3) have been analyzed, applying the concept of coalition-proofness.

For the grand coalition it was shown that a socially agreement is not stable if many countries are affected by the externality problem and if perceived environmental damages are relatively high compared to abatement costs (critical parameter values). It turned out that these are those parameter constellations where a socially optimal agreement could achieve much compared to the non-cooperative equilibrium. Also in the two policy regimes a stable agreement is difficult to obtain for these critical values. However, since these policy regimes specify more moderate abatement targets, instability is less problematic than in a socially optimal treaty.

It was demonstrated that especially for those countries which have only a minor interest in emission reductions, political representatives have to be very patient. This is particularly true for externality problems of real global character. It became apparent that there exists a dilemma: On the one hand, there are many arguments which back the conjecture that simple punishment profiles are particularly used if many countries suffer from a global externality. On the other hand, it was shown that simple punishment profiles are sub-optimal and increase the discount rate requirements which are already very high if many countries suffer from pollution and optimal punishment profiles are used.

178

Due to the stability problems of a grand coalition the formation process of sub-coalitions was considered in Section 3. In this process the equilibrium size of the sub-coalition, the policy instrument and the environmental standard were determined. It became apparent that an agreement among a sub-coalition might be more efficient than of a grand coalition. Moreover, the tax will only emerge from the bargaining process if only few countries are involved in an externality problem. For global pollutants, where the number of countries suffering from an externality is large, governments agree on a quota regime. This finding contributed to explain why a uniform emission quota is encountered in many IEAs, though it is an inefficient instrument.

Admittedly, the results which we obtained were derived from a stylized model and relied to a large extent on simulations. Therefore, final conclusions have to be drawn with caution. However, as was demonstrated, the general findings are quite robust regarding a variation of the assumptions. In particular, the results are very much in line with the past experience in international pollution control.

Regarding future research we propose three main extensions: First, our analysis could be extended to cover other instruments, e.g., tradable permits (see Endres and Finus, 1996a, 1998). It would be interesting to find out whether the quota regime will also emerge from the negotiation process leading to an IEA if other instruments are among the choice set. Second, the coalition formation process could be studied if the environmental issue is linked to other policy issues, e.g., trade (Barrett, 1996), or cooperation in R&D (Carraro and Siniscalco, 1995). This seems to be suggestive because issue linkage plays a substantial role in actual international policy (Ragland, Bennett and Yolles, 1996). Third, the design of existing IEAs could be investigated by combining the two approaches described in the introduction: empirical estimates of costs and benefits from pollution control and theoretical concepts of coalition formation and stability control. If the models of public choice are added to such an analysis this would have to be regarded as a major step to understand the construction of present IEAs and those which have failed to come into force.

Notes

1. The papers of Black, Levi and de Meza (1992), partially Carraro and Siniscalco (1992: Section 4; 1993: Section 2.3) and Heal (1993), though they do not apply stability concepts of cooperative game theory, assume some commitment by countries. This is, however, not compatible with the notion of self-enforcement of non-cooperative game theory.
2. The non-zero conjectural variation assumption has played some role in the literature on the provision of public goods. See, e.g., Cornes and Sandler (1984, 1985) and Guttman (1987, 1991).

3. For example Chander and Tulkens (1996) apply the core-concept to international pollution control showing that with an appropriate transfer scheme a stable coalition of all countries could form.
4. The question why in reality countries show such a reluctance to transfers has not yet been answered in the literature to our knowledge. However, this phenomenon initiated a series of papers on interconnected games. See Cesar and de Zeeuw (1994), Folmer, von Mouche and Ragland (1993) and Ragland, Bennett and Yolles (1996).
5. Besides the above-mentioned agreements also HELCOM for the protection of the Baltic Sea, or the International Convention for the Prevention of Pollution of the Sea by Oil specify uniform duties for signatories. See also Hoel (1991, 1992) and Barrett (1992) on this point.
6. With respect to welfare a similar comparison can be conducted. Defining $\Delta NB = (NB^S - NB^N)/NB^N$ we find $\partial \Delta NB/\partial \gamma < 0$ and $\partial \Delta NB/\partial N > 0$. For a sensible interpretation, however, we have to require net benefits in the Nash equilibrium to be positive.
7. The bargaining setting is borrowed from Endres (1996, 1997).
8. For a proof of this claim and an extensive discussion of other properties of this bargaining solution see Endres and Finus (1996a).
9. According to our assumption present emissions are given by the emission vector in the Nash equilibrium of the stage game. Usually, complete information is necessary to determine this equilibrium. In a wider interpretation the Nash equilibrium may be viewed as the result of a dynamic adjustment process (cobweb process). Then, each country only needs to have knowledge on its own reaction function and complete information is not necessary.
10. However, we do not model incomplete information explicitly.
11. For more than two players the payoff space has to be of "full dimensionality". See Fudenberg and Maskin (1986).
12. There exists also a strong version of the concept requiring that a renegotiation-proof equilibrium is undominated by any other equilibrium. This amounts to require that an equilibrium must lie on the Pareto-frontier. Except the social optimum, none of the pragmatic solutions satisfies this condition. Since we are mainly concerned with a positive analysis of the formation of IEAs we feel that this requirement is unnecessarily too strong.
13. For the case of $\delta \rightarrow 1$ such an extension can be found in Barrett (1994b).
14. See Finus and Rundshagen (1998) Appendix B for a proof.
15. See Endres and Finus (1996a) for an analytical check in the case of two countries. However, for $\delta << 1$, a general solution cannot be obtained even for two countries. See Finus and Rundshagen (1998).
16. Hence, $\delta_i \rightarrow 1$ is equivalent to $r_i \rightarrow 0$.
17. See Finus and Rundshagen (1998) in the case of two countries.
18. This is particularly true in a supergame in which the number of potential equilibria is large. For a similar argument in a static context, see Barrett (1997).
19. This result is apparent when observing Table 4. A general proof of this result is available upon request by the authors.
20. Alternatively, one could also consider a majority decision rule (MDR) with respect to this issue. However, it turns out that the equilibrium coalition size is quite robust with respect to the decision rule. See the working paper version.
21. This result has been demonstrated in the case of the grand coalition. For the sub-coalition it is also true that the country with the lowest index, i.e., $N_0 + 1$, prefers the quota instead of the tax regime. The proof is simple and follows by construction.

180

22. The results also hold if the coalition size is decided according to a MDR. See the working paper version.
23. For the appropriateness of this assumption in the case of $\delta \to 1$ see Finus and Rundshagen (1998).

References

Abreu, D. (1988). On the theory of infinitely repeated games with discounting. *Econometrica* 56: 383–396.

Asheim, G.B. (1991). Extending renegotiation-proofness to infinite games. *Games and Economic Behavior* 3: 278–294.

Barrett, S. (1992). International environmental agreements as games. In R. Pethig (Ed.), *Conflicts and cooperation in managing environmental resources. Microeconomics Studies*, Ch. 1: 11–37. Berlin et al: Springer Verlag.

Barrett, S. (1994a). The biodiversity supergame. *Environmental and Resource Economics* 4: 111–122.

Barrett, S. (1994b). Self-enforcing international environmental agreements. *Oxford Economic Papers* 46: 878–894.

Barrett, S. (1995). Toward a theory of international environmental cooperation. Discussion Paper 60.95. Milan: Fondazione Eni Enrico Mattei.

Barrett, S. (1997). The strategy of trade sanctions in international environmental agreements. *Resource and Energy Economics* 19: 345–361.

Barrett, S. (1997). Heterogeneous international agreements. In C. Carraro (Ed.), *International environmental negotiations: Strategic policy issues*, 9–25. Cheltenham: Edward Elgar.

Bauer, A. (1992). International cooperation over greenhouse gas abatement. Mimeo. Seminar für empirische Wirtschaftsforschung, University Munich.

Benoît, J.-P. and Krishna, V. (1993). Renegotiations in finitely repeated games. *Econometrica* 61: 303–323.

Bergin, J. and MacLeod, W.B. (1993). Efficiency and renegotiations in repeated games. *Journal of Economic Theory* 61: 42–73.

Bernheim, D., Peleg, B. and Whinston, M.D. (1987). Coalition-proof Nash equilibria: I. Concepts. *Journal of Economic Theory* 42: 1–12.

Bernheim, D. and Ray, D. (1985). Pareto-perfect Nash equilibria. Mimeo. Stanford University.

Bernheim, D. and Whinston, M.D. (1987). Coalition-proof Nash equilibria: II. Applications. *Journal of Economic Theory* 42: 13–29.

Black, J., Levi, M.D. and Meza, D. de (1992). Creating a good atmosphere: Minimum participation for tackling the "greenhouse effect". *Economica* 60: 281–293.

Bohm, P. (1994). Making carbon emission quota agreements more efficient: Joint implementation versus quota tradability. In G. Klaassen and Finn R. Forsund, *Economic instruments for air pollution control*, Ch. 9: 187–208. Dordrecht et al.: Kluwer.

Bohm, P. and Larsen, B. (1993). Fairness in a tradeable-permit treaty for carbon emissions reductions in Europe and the Former Soviet Union. *Environmental and Resource Economics* 4: 219–239.

Botteon, M. and Carraro, C. (1997). Burden-sharing and coalition stability in environmental negotiations with asymmetric countries. In C. Carraro (Ed.), *International environmental negotiations: Strategic policy issues*, 26–55. Cheltenham: Edward Elgar.

Buchanan, J.M. and Tullock, G. (1975). Polluters' profits and political response: Direct controls versus taxes. *American Economic Review* 65: 139–147.

Carraro, C. and Siniscalco, D. (1992). The international dimension of environmental policy. *European Economic Review* 36: 379–387.

Carraro, C. and Siniscalco, D. (1993). Strategies for the international protection of the environment. *Journal of Public Economics* 52: 309–328.

Carraro, C. and Siniscalco, D. (1997). R&D cooperation and the stability of international environmental agreements. In C. Carraro (Ed.), *International environmental negotiations*: Strategic policy issues, 71–96. Cheltenham: Edward Elgar.

Cesar, H. and de Zeeuw, A. (1996). Issue linkage in global environmental problems. In A. Xepapadeas (Ed.), *The economic policy for the environment and natural resources*, 157–173.

Chander, P. and Tulkens, H. (1992). Theoretical foundations of negotiations and cost sharing in transfrontier pollution problems. *European Economic Review* 36: 388–398.

Chander, P. and Tulkens, H. (1995). A core-theoretic solution for the design of cooperative agreements on transfrontier pollution. *International Tax and Public Finance* 2: 279–293.

Chander, P. and Tulkens, H. (1996). The core of an economy with multilateral environmental externalities. *International Journal of Game Theory* 26: 379–402.

Cornes, R. and Sandler, T. (1984). The theory of public goods: Non-Nash behaviour. *Journal of Public Economics* 23: 367–379.

Cornes, R. and Sandler, T. (1985). On the consistency of conjectures with public goods. *Journal of Public Economics* 27: 125–129.

D'Aspermont, C., Gabszeweiz, J.J. and Weymark, J.A. (1983). On the stability of collusive price leadership. *Canadian Journal of Economics* 16: 17–25.

Dewees, D.D. (1983). Instrument choice in environmental policy. *Economic Inquiry* 21: 53–71.

Dockner, E.J. and Van Long, N. (1993). International pollution control: Cooperative versus noncooperative strategies. *Journal of Environmental Economics and Management* 24: 13–29.

Endres, A. (1996). Designing a greenhouse treaty: Some economic problems. In E. Eide and R. van den Bergh (Eds.), *Law and economics of the environment*, 201–224. Oslo: Juridisk.

Endres, A. (1997). Increasing environmental awareness to protect the global commons: A Curmudgeon's view. *Kyklos* 50: 3–27.

Endres, A. and Finus, M. (1996a). International environmental agreements: How the policy instrument affects equilibrium emissions and welfare. Working Paper No. 231 (revised version). University of Hagen, Hagen.

Endres, A. and Finus, M. (1996b). Umweltpolitische Zielbestimmung im Spannungsfeld gesellschaftlicher Interessengruppen: Theorie und Empirie. In H. Siebert (Ed.), *Elemente einer rationalen Umweltpolitik. Expertisen zur umweltpolitischen Neuorientierung*, 35–133. Tübingen: J.C.B. Mohr.

Endres, A. and Finus, M. (1998). Renegotiation-proof equilibria in a bargaining game over global emission reductions: Does the instrumental framework matter? To appear in N. Hanley and H. Folmer (Eds.), *Game theory and the global environment*. Cheltenham: Edgar Elgar.

Fankhauser, S. and Kverndokk, S. (1996). The global warning game: Simulations of a CO_2 reduction agreement. *Resource & Energy Economics* 18: 83–102.

Farrell, J. (1983). Credible repeated game equilibrium. Unpublished manuscript.

Farrell, J. and Maskin, E. (1989). Renegotiation in repeated games. *Games and Economic Behavior* 1: 327–360.

182

Finus, M. and Rundshagen, B. (1998). Renegotiation-proof equilibria in a global emission game when players are impatient. *Environmental and Resource Economics.* (forthcoming)

Folmer, H. von Mouche, P. and Ragland, S. (1993). Interconnected games and international environmental problems. *Environmental and Resource Economics* 3: 313–335.

Fudenberg, D. and Maskin, E. (1986). The folk theorem in repeated games with discounting or with incomplete information. *Econometrica* 54: 533–554.

Golombek, G., Hagem, C. and Hoel, M. (1995). Efficient incomplete international agreements. *Resource and Energy Economics* 17: 25–46.

Guttman, J.M. (1987). A non-Cournot model of voluntary collective action. *Economica* 54: 1–19.

Guttman, J.M. (1991). Voluntary collective action. In A. Hillman (Ed.), *Markets and Politicians*, Ch. 2: 27–41. Boston: Kluwer.

Hahn, R.W. (1987). Jobs and environmental quality: Some implications for instrument choice. *Policy Sciences* 20: 289–306.

Hahn, R.W. (1990). The political economy of environmental regulation: Towards a unifying framework. *Public Choice* 65: 21–47.

Heal, G. (1993). Formation of international environmental agreements. Discussion Paper 28.93. Series: Economics, Energy, Environment. Milan: Fondazione Eni Enrico Mattei.

Hillman, A.L. and Ursprung, H.W. (1994). Greens, supergreens, and international trade policy: Environmental concerns and protectionism. In C. Carraro (Ed.), *The international dimension of environmental policy, 75–108. Dordrecht: Kluwer.*

Hoel, M. (1991). Global environmental problems: The effects of unilateral actions taken by one country. *Journal of Environmental Economics and Management* 20: 55–70.

Hoel, M. (1992). International environment conventions: The case of uniform reductions of emissions. *Environmental and Resource Economics* 2: 141–159.

Kaitala, W., Pohjola, M. and Tahvonen, O. (1992). Transboundary air pollution and soil acidification: A dynamic analysis of an acid rain game between Finland and the USSR. *Environmental and Resource Economics* 2: 161–181.

Kverndokk, S. (1993). Global CO_2 agreements: A cost-effective approach. *Energy Journal* 14: 91–112.

Lenschow, A. (1996). Der umweltpolitische Entscheidungsprozeß in der Europäischen Union am Beispiel der Klimapolitik. In H.G. Brauch (Ed.), *Klimapolitik*, Ch. 7: 89–104. Berlin et al.: Springer-Verlag.

Mäler, K.-G. (1989). The acid rain game. In H. Folmer and E. van Ierland (Eds.), *Valuation methods and policy making in environmental economics*, Ch. 12: 231–252. Amsterdam: Elsevier.

Mohr, E. (1988). On the incredibility of perfect threats in repeated games: Note. *International Economic Review* 29: 551–555.

Musgrave, P. (1995). Pure global externalities: International efficiency and equity. In: L. Bovenberg and S. Cnossen (Eds.), *Public economics and the environment in an imperfect world*. Boston: Kluwer.

Newbery, D. (1990). Acid rain. *Economic Policy* 5: 287–346.

Nordhaus, W.D. (1993). Rolling the "DICE": An optimal transition path for controlling greenhouse gases. *Resource and Energy Economics* 15: 27–50.

Pashigian, P. (1985). Environmental regulation: Whose self-interests are being protected? *Economic Inquiry* 23: 551–584.

Ragland, S.E., Bennett, L.L. and Yolles, P. (1996). Unidirectorial problems with international river basins: A game theoretic approach. Paper presented at the European Meeting of Resource and Environmental Economists, Lisbon, 26–29 June, 1996.

Schelling, T.C. (1960). *The strategy of conflict.* Cambridge: Harvard University Press.
Ray, D. (1994). Internally renegotiation-proof equilibrium sets: Limit behavior with low discounting. *Games and Economic Behavior* 6: 162–177.
Tahvonen, O. (1994). Carbon dioxide abatement as a differential game. *European Journal of Political Economy* 10: 685–705.
Ursprung, H.W. (1992). The political economy of environmental decision-making. Discussion Paper II-176. Faculty of Economics, University of Konstanz, Konstanz.
Welsch, H. (1992). Inequality aspects of alternative CO_2 agreement designs. *OPEC Review* 16: 23–36.
Welsch, H. (1995). Incentives for forty-five countries to join various forms of carbon reduction agreements. *Resource and Energy Economics* 17: 213–237.

184

Appendix

Appendix 1: Stability test of the grand coalition for $\delta \to 1$

For the test whether an emission tuple $E^* = (E_1^*, E_2^*, \ldots, E_N^*)$ is weakly coalition-proof the following algorithm is used.

A. Set $i = 1$.
B. Determine the amount K by which all countries j, $j \neq i$, have to increase their emissions during the punishment of country i in order to fulfill Equation (24).
   B.I Determine the reaction function of country i, $E_i(E_{-i})$.
   B.II Insert $E^i_{-i} = K + E^*_{-i}$ in Equation (24) and solve for $K > K_{min}$. Define $K : K_{min} + \epsilon$, with ϵ a sufficiently small positive number.
C. Check whether Equation (22) can be satisfied for all countries j if they jointly increase emissions by the amount K.
   C.I Set $E^i_i = 0$ in Equation (22) to determine the outer boundaries of the WCP emission space.[23] Hence, $E^i = K + E^*_{-i}$, where K has been determined in step B.II.
   C.II For all $j \neq i$ let $E^i_j := E^*_j + k_j$ with k_j the amount by which country j increases its emission to punish i. Further, define $S_{min} := 0$ and $S_{max} := 0$.
   C.III Repeat for all $j \neq i$:
   Insert E^i in Equation (22). Solve Equation (22) for k_j to have $k_j \in [k_{min}, k_{max}]$ or $k_j \in \oslash$. In the first case define $S_{min} := S_{min} + k_{min}$ and $S_{max} := S_{max} + k_{max}$. In the second case stop C.III and define $S_{min} := 0$ and $S_{max} := 0$.
   C.IV If, and only if $K \in [S_{min}, S_{max}]$, then country i can be punished by WCP strategies. If this is the case continue with C.V. Otherwise stop.
   C.V Define $i := i + 1$. Repeat all steps starting from B. If the algorithm runs for all i, $i \in \{1, \ldots, N\}$, then E^* is a WCP-equilibrium.

Appendix 2: Determination of δ_i^{min} for unrestricted punishment profiles

Assume E^i to be given. In order to satisfy Equation (21) for small δ_is, NB_i^P should be small. This is equivalent to demand δ_i^t to be small. From Equation (20), however, we have

$$(1 - \delta^t)NB_i^R + \delta^t NB_i^* \geq NB_i^D \Leftrightarrow \delta_i^t \geq \frac{NB_i^D - NB_i^R}{NB_i^* - NB_i^R} \tag{A1}$$

from which the optimal punishment time

$$t^* = \left[\frac{\log\left(\frac{NB_i^D - NB_i^R}{NB_i^* - NB_i^R}\right)}{\log\delta_i} \right] \text{ or approximated } t_a^* := \frac{\log\left(\frac{NB_i^D - NB_i^R}{NB_i^* - NB_i^R}\right)}{\log\delta_i} \quad \text{(A2)}$$

follows. Because $t^* \geq 1$ must hold, $\delta_i^t \leq \delta_i$ follows and we have

$$\delta_i \geq \frac{NB_i^D - NB_i^R}{NB_i^* - NB_i^R}. \quad (20')$$

Thus, for $t = t_a^*$ $NB_i^P = NB_i^D$ holds, and Equation (21) reduces to

$$(1 - \delta_i)NB_i^{max} + \delta NB_i^D \leq NB_i^*. \quad (21')$$

Hence, for a country i the determination of δ_i^{min}, which is the smallest δ_i that satisfies Equations (20′), (21′) and (22), proceeds as follows:

A. Set $\delta_i = 1$.
B. Set $\delta_i := \delta_i - s$, with s a suitable step size (e.g., $s = 0.01$ or $s = 0.001$).
C. Determine the amount K_{min}^a by which all countries j, $j \neq i$, have to increase their emissions during the punishment of i in order to satisfy Equation (21′).
   C.I Determine the best deviation in the normal phase of country i, $E_i(E_{-i}^*)$, to have NB_i^{max}.
   C.II Insert $E_{-i}^i = K + E_{-i}^*$ in Equation (21′) and solve for $K \geq K_{min}^a$.
D. Check whether there exist E_i^i and $K \geq K_{min}^a$, so that Equations (20′) and (22) can be satisfied jointly.
   D.I Define $E_i^i := 0$ in Equations (20′) and (22).
   D.II Solve Equation (20′) for $K \geq K_{min}^b$. Define $K := \max\{K_{min}^a, K_{min}^b\}$.
   D.III Let $E_j^i := E_j^* + k_j$ with k_j the amount by which country j increases its emission to punish i. Moreover, let $E^i = E_i^i + E_{-i}^i = E_i^i + E_{-i}^* + K$ and define $S_{min} := 0$ and $S_{max} := 0$.
   D.IV Repeat for all $j \neq i$:
   Evaluate E^i and insert this in Equation (22). Solve Equation (22) for k_j to have $k_j \in [k_{min}, k_{max}]$ or $k_j \in \emptyset$. In the first case $S_{min} := S_{min} + k_{min}$ and $S_{max} := S_{max} + k_{max}$. In the second case stop D.IV and define $S_{min} := 0$, $S_{max} := 0$.
   D.V a) If $K \in [S_{min}, S_{max}]$, then i can be punished using WCP-strategies. In this case start again with B in order to test if δ_i may be further reduced.
   b) If $K \notin [S_{min}, S_{max}]$ and $E_i^i < E_i^*$ define $E_i^i := E_i^i + h$, with h a suitable step size. Go back to D.II.
   c) If $K \notin [S_{min}, S_{max}]$ and $E_i^i \geq E_i^*$, a punishment using WCP-strategies is not possible. Then, $\delta_i^{min} := \delta_i + s$. Stop algorithm.

186

Appendix 3: Determination of δ_i^{min} for restricted punishment profiles

The algorithm proceeds similar as formulated in Appendix 2, however, with the following simplification. Replace D.III and D.IV by

D.III′ Let $P_1 : E_j^i := (1 + k)E_j^*$ or $P_2 : E_j^i := E_j^* + k$, with k determined from $E^i(k) = E_{-i}^* + K$.
DIV′ Evaluate E^i and test if Equation (22) is true for all $j \neq i$. (This case corresponds to D.V a).)

Appendix 4: Restricted punishment strategies (Nash)

Preposition: If the punishment strategy is restricted to $E_j^i \leq E_i^N$, then a tax rate, t_1, proposed by country 1 for the grand coalition, cannot be backed by WCP-strategies, even though $\delta \rightarrow 1$.

Proof: If $E_j^i \leq E_i^N$, then $E_{-i}^i \leq E_{-i}^N$. Moreover, $\partial(NB_i^D(E_i(E_{-i}^i), E_{-i}^i))/\partial E_{-i}^i < 0$. Then from Equation (24) we have $NB_i^* \geq NB_i^D(E_i(E_{-i}^i), E_{-i}^i) \geq NB_i^D(E_i(E_{-i}^N), E_{-i}^N) = NB_i^N(E_i^N, E_{-i}^N)$. On the other hand, from sub-section 2.1.2 we know that for t_1 $NB_1^T < NB_1^N$ holds, which completes the proof.

Appendix 5: Stability test of a sub-coalition for $\delta \rightarrow 1$

The test proceeds along the same lines as laid out in Appendix 1. First, the amount K is determined by which the signatories have to increase their emissions to punish a non-signatory to satisfy Equation (24). Then, it is checked whether the signatories can provide such a K under the restriction that Equation (22) is met for each signatory. Accordingly, the same procedure is applied for the case that a signatory breaches the contract.

[19]

Consensus Treaties

by

SCOTT BARRETT *

Treaties seeking to supply transnational public goods are usually modeled as a game in which cooperation is incomplete, with some countries participating and some not. This paper shows that, if the usual rationality concept is weakened somewhat, countries will often do better by negotiating a consensus treaty in which provision of the public good is chosen to maximize the collective payoff of all countries subject to participation in the treaty being full. The theory developed here thus explains when, and why, it may be better to negotiate a treaty that is "broad but shallow" rather than "narrow but deep." (JEL: C 7, H 41, K 33, Q 2)

1 *Introduction*

Nation states are sovereign. There is no World Government that can coerce states into supplying transnational public goods like global climate change mitigation, no authority that can enforce agreements between states to supply these goods. Of course, states can, and do, enter into such agreements. But the choice to participate in an international treaty is voluntary. Successful international cooperation thus depends on whether free riding can be deterred (SANDLER [1997]).

It has become standard to model the constraint of sovereignty as affecting only the level of participation in international agreements. Moreover, a basic finding of the literature is that participation is typically low (see HOEL [1992], CARRARO AND SINISCALCO [1993], and BARRETT [1994]). That is, few countries participate in a treaty seeking to supply a global public good. In real agreements, by contrast, participation levels are sometimes high. The Framework Convention on Climate Change, for example, has been ratified by 186 countries. Moreover, this agreement was negotiated with the aim of supporting the maximum effort to protect the climate, subject to sustaining full participation. According to a Swedish diplomat (KJELLEN [1994, p. 151]) who played a leading role in these talks, "a major objective of the negotiation was to have all the big players sign the convention during the Earth Summit at Rio."

* I am grateful to Parkash Chander, the editor of this journal, and two anonymous referees for comments on an earlier draft. I am also grateful to participants at the NBER's Summer Institute Public Economics Workshop and Harvard University's Environmental Economics seminar for helpful comments.

Journal of Institutional and Theoretical Economics
JITE 158 (2002), 529–547 © 2002 Mohr Siebeck – ISSN 0932-4569

But does agreement by consensus imply that the first best cooperative outcome is being sustained? The diplomat just quoted readily blames the feebleness of the final treaty on the overriding desire for consensus. "If we were to reach consensus," he says, "there could be no commitments on targets and timetables in the final agreed text" (p. 151). This hints at a trade-off. Countries can reach a consensus around a weak agreement – one that attracts full participation but demands that relatively little of the public good be supplied. Alternatively, countries can also negotiate a more potent but incomplete agreement – one in which every participant supplies relatively more of the public good but participation is less than full. When negotiating the Framework Convention on Climate Change, countries chose the consensus option.

At the negotiations on the Montreal Protocol on Substances that Deplete the Ozone Layer, countries chose differently. At one point, the United States proposed that the treaty should not enter into force until countries accounting for at least 90 percent of global consumption of ozone-depleting substances had ratified the agreement. According to Richard BENEDICK [1998, p. 89], the chief US negotiator in Montreal, "When the United States proposed this percentage at Montreal, the reaction was almost universally negative; only the Soviet Union supported 90 percent. Many observers feared that such a requirement could hold the treaty hostage to Japan or the Soviet Union, which might then weaken the protocol by extracting other concessions as the price for adherence." In the end, the negotiators agreed that the treaty should enter into force after being ratified by at least 11 countries, together accounting for at least two-thirds of global consumption.[1]

The usual approach to modeling international environmental agreements cannot accommodate this apparent trade-off between the depth and breadth of cooperation. This approach (see HOEL [1992]; CARRARO AND SINISCALCO [1993]; and BARRETT [1992], [1994], [1999], [2001]) assumes that the cooperating countries take their participation level as given and choose a provision level so as that maximizes their collective payoff both on and off the equilibrium path. In this paper I refer to this kind of agreement as being *strongly collectively rational* (SCR). For any given problem, an SCR treaty is unique and typically incomplete; some countries cooperate and some free ride.

To sustain a consensus treaty requires a different approach. One possibility, following HOEL AND SCHNEIDER [1997], is to assume that every country pays a penalty for not participating – a penalty that increases in the number of countries that do participate. Obviously, if the penalty were large enough, a consensus could always be sustained. However, in this setting there is no tension between the depth and breadth of cooperation. Moreover, the penalty for not conforming is in a sense self-imposed and essentially psychic since imposition of the penalty costs the participating countries nothing.

[1] Using a combination of trade restrictions and side payments, the Montreal Protocol has sustained both broad and deep cooperation; see BARRETT [1997], [2001].

In this paper I develop a different approach. Rather than alter preferences, I modify the rationality assumption. In particular, I consider the alternative of a *weakly collectively rational* (WCR) treaty, a concept modeled after FARRELL AND MASKIN'S [1989] notion of a weakly renegotiation-proof equilibrium.[2] WCR treaties can support varying levels of participation (they can always support a consensus treaty) and so provide the means for evaluating the trade-off between treaties that are "narrow but deep" and "broad but shallow."

The WCR concept provides a tool by which the "punishment dilemma" in repeated games can be avoided. Cooperation in a prisoners' dilemma game must be sustained by the credible threat to punish deviations, and there is an intimate connection between the credibility of such a threat and the assumed rationality of the players. It is customary to assume that players are individually rational (IR), and that equilibria must be subgame perfect. The folk theorems, for example, tell us that the famous Grim strategy can sustain cooperation in an infinitely repeated prisoners' dilemma provided that the players are sufficiently patient. However, the threat to punish is only credible in a limited sense. Though it is in each player's interests to play Grim given that all others play Grim, both players (in a 2-player game) would always do better by renegotiating their way out of a punishment phase. Given that treaties are negotiated – openly, explicitly, and deliberately – and given that countries cannot commit to participating in and complying with a treaty, it is implausible in my view to believe that countries will fail to anticipate the need to renegotiate. Put differently, a theory seeking to explain how countries come to cooperate must also explain how countries come to punish deviations from cooperation. It seems to me to be inconsistent to assume that countries can cooperate only in cooperative phases and not also in punishment phases. Indeed, in actual practice, deviations from a cooperative outcome almost invariably provoke a collective response, beginning with the convening of a meeting of the aggrieved parties to discuss the actions they should take against the deviant. For all of these reasons, treaties to supply transnational public goods must be renegotiation-proof.

To avoid renegotiation, the countries responsible for punishing a defection must get at least as high a per-period payoff in a punishment phase as they get in a cooperative phase. Assuming that each player's payoff increases in the amount of the public good supplied by others but decreases in the amount supplied by itself, the implications are obvious. To dull the incentives to renegotiate, the country being punished must supply more of the public good in a punishment phase compared to the cooperative phase and/or the countries carrying out the punishment must supply less of the public good in a punishment phase compared to the cooperative phase.

These choices cannot be made arbitrarily. They must be rational. And it can be shown that the choice of a rationality concept matters. Depending on whether the

[2] FINUS AND RUNDSHAGEN [1998] also use the concept of a renegotiation-proof equilibrium to study transnational pollution games, but they limit their analysis to two-player situations. See also ENDRES AND FINUS [1998].

SCR or WCR concepts are used, qualitatively different kinds of agreements can be sustained.

For an SCR treaty, as mentioned earlier, the cooperating countries maximize their collective payoff both on and off the equilibrium path (that is, in the cooperative and punishment phases). The problem for cooperation is that when many countries are called upon to punish a deviation, maximization of this group's collective payoff will typically demand weak punishments. The countries called upon to punish will hurt themselves if they lower their provision levels very much, but by not lowering their provision levels by much they fail to deter free riding. This is why participation is limited in SCR treaties.

WCR treaties are renegotiation-proof in a different sense. For a WCR treaty, provision levels can be varied, both on and off the equilibrium path. To sustain a high level of cooperation requires stiff punishments, and yet these will typically not be credible when participation is high. To maintain credibility, punishments must be made smaller. But for these punishments to deter deviations, cooperative provision levels must also be reduced. Cooperation among a larger number of countries can therefore be sustained provided that the cooperative provision levels are reduced. Greater breadth (participation) is attained at the expense of lower depth (per-country provision).

An example best illustrates the difference between the concepts. Suppose that a country withdraws from an agreement seeking to supply a public good, or fails to accede to this agreement in the first place, or ratifies the treaty but fails to comply with it. Suppose, too, that the countries that are parties to the agreement (or that remain as parties and comply with its terms) are required by the treaty to punish the deviant. If the countries that are required to punish the deviant can only choose between imposing the prescribed punishment or not, and they cannot do better collectively except by imposing the punishment, then the treaty is WCR. If, however, these countries can choose to impose different punishments than required by the treaty, but cannot gain collectively by doing so, then the treaty is SCR. As suggested by this example, all SCR treaties are WCR but the reverse is not true. SCR treaties are more credible.

Though this added credibility is to be desired, it comes at the cost of reduced flexibility. The experimental literature shows that cooperation is strongly influenced by concerns for fairness. The obvious attraction of a consensus agreement is that it is a symmetric outcome. It is thus envy-free in the sense that no country would prefer to swap places with another country (in an incomplete treaty, parties would prefer to swap places with non-parties). If there is a widely shared belief that all countries ought to supply the public good, then why should deviations from this belief take a back seat to credibility? Experimental evidence also suggests that, when given the opportunity to punish a free rider, cooperating players will take it, even if imposing a punishment is self-damaging (FEHR AND GÄCHTER [2000]).[3] Players

[3] As pointed out by a referee, other evidence, such as AXELROD's [1984] tournament, suggests that players may gain by being more forgiving.

do not like being made a "sucker." In general, the SCR concept does not provide the opportunity for either of these motivations to be expressed. They are, however, easily accommodated by the WCR assumption. In a sense, the WCR assumption is a kind of compromise between the IR and SCR assumptions.

The tradeoff between depth and breadth which is examined by the WCR concept has implications for welfare that are ambiguous. Depending on the circumstances, countries could be better off with a "broad but shallow" treaty or with a "narrow but deep" treaty. Though concerns for fairness may favor a consensus, it is a stretch to believe that fairness will dominate all other considerations, including efficiency (just as it is wrong to let credibility dominate fairness). Recall that the strategy of treaty negotiation varies. Some treaties seek to reach a consensus. Others do not. The reason for this variation is a puzzle, but in this paper I shall provide an explanation: under some circumstances, the countries that negotiate a treaty will prefer a consensus – not (only) for reasons of fairness but because a consensus treaty makes them better off.

The remainder of the paper is organized as follows. Section 2 provides a general formulation of a WCR treaty. Sections 3 and 4 apply the concept to two alternative specifications. Both of these assume that benefits increase linearly in the provision of a transnational public good and that costs are independent so that the non-cooperating countries have dominant strategies. However, in Section 3 I assume that costs are linear whereas in Section 4 I assume that costs are quadratic. These differences, it turns out, matter to treaty formation. In the former case, countries will negotiate a consensus treaty rather than an incomplete treaty only if a small fraction of countries would participate in an incomplete treaty. If marginal costs are increasing, however, then countries will always do better by negotiating by consensus. Section 5 uses the theory to question the logic behind the enforcement mechanisms incorporated in the recently negotiated Kyoto Protocol. The concluding section summarizes the paper's main conclusions.

2 *Concepts*

It is usual to model a treaty for the supply of a transnational public good as a stage game in which the decisions to participate in the treaty and to supply a particular level of the public good are made sequentially (see HOEL [1992]; CARRARO AND SINISCALCO [1993]; and BARRETT [1992], [1994], [2001]). However, I have shown elsewhere (BARRETT [1999]) that cooperation can also be modeled as a repeated game and that the equilibria for the two formulations correspond under the assumption of strong collective rationality. Here I adopt the repeated game structure since this is more appropriate to the analysis of alternative rationality assumptions.

To allow the depth of cooperation to vary, I take choices to be continuous. To allow the breadth of cooperation to vary, I assume that the number of countries is *at least* two (our main interest shall be with cases where $N \geq 3$). Consistent with the previous literature, I assume that the supply of the public good is determined by the summation of per-country provision levels.[4] Let Q_i be the set of feasible provision levels for i, where i is some member of A. If, in some period, country i supplies a provision level $q_i \in Q_i$, and every other member of A, indexed by $j(j \neq i)$ supplies a provision level $q_j \in Q_j$, then i's payoff in this period is given by $\pi_i(q_i, q_{-i})$, where $q_{-i} = \sum\limits_{j \in A, j \neq i} q_j$.

The underlying game is taken to be a symmetric prisoners' dilemma, defined as a game for which there exist two feasible strategies (vectors), q^o and q^c, such that: (i) $\pi_i(q_i^o, q_{-i}) > \pi_i(q_i, q_{-i})\ \forall q_i \neq q_i^o,\ \forall q_{-i},\ \forall i$ (i.e., q_i^o is dominant);[5] and (ii) $\pi_i(q_i^c, q_{-i}^c) > \pi_i(q_i^o, q_{-i}^o)\ \forall i$ (i.e., q_i^o is Pareto inefficient). I shall further take q^c to denote the full cooperative outcome; that is, $\sum_i \pi_i(q_i^c, q_{-i}^c) > \sum_i \pi_i(q_i, q_{-i})\ \forall q_i \neq q_i^c$, $\forall q_{-i} \neq q_{-i}^c$. For brevity, the Nash equilibrium payoff shall be denoted π^o and the full cooperative payoff π^c.

Following FARRELL AND MASKIN [1989], I assume throughout that countries are patient (discount factors are set equal to one). Doing so simplifies the mathematics. It also adds clarity to the results. As is well known, cooperation is harder to support the more impatient are the players. Here I show that full cooperation may not be supportable, even if countries do not discount the future at all.

A treaty is an agreement among a set S of countries, where S is a subset of A, the set of all countries with an interest in a transnational public good. There are N countries in total, so that $A = \{1, ..., N\}$. A treaty specifies for each member of S a strategy. This tells each member the level of provision that should be chosen in each period. Following entry into force, this strategy depends (possibly) on the entire history of play. All countries in A are symmetric, and so all members of S are symmetric. I assume that a treaty consisting of symmetric countries is itself symmetric and so prescribes that its parties play symmetric strategies.[6]

Consider now the definition of a WCR treaty. In the cooperative phase of such an agreement, each party earns a per-period payoff of π_s^w, where $\pi^o \leq \pi_s^w \leq \pi^c$. Suppose, however, that party j cheats. Then a punishment phase will be triggered. In a punishment phase, the treaty tells the deviant, j, to play q_j^j and for every other party to play q_{-j}^j. As noted in the introduction, q_j^j will normally be at least as large as the cooperative provision level; q_{-j}^j, by contrast, will be no greater than the cooperative provision level. In Section 4 I illustrate the role played by q_j^j. In general, the larger is q_j^j, the larger can be q_{-j}^j; and the larger is q_{-j}^j, the greater is

4 For a discussion of other kinds of problems, see SANDLER [1997] and ARCE M. [2000]. Of course, the renegotiation-proof concept can be applied to different kinds of problems; for 2-player examples, see FARRELL AND MASKIN [1989].

5 For analyses where this assumption does not hold, see HEAL [1993], BARRETT [1994], and BUCHHOLZ, HASLBECK, AND SANDLER [1998].

6 The assumption about symmetric strategies is not always needed; see the appendix to BARRETT [1999]. The assumption about symmetric countries is common in the literature. For an analysis showing how and why this assumption may matter, see BARRETT [2001].

the number of countries that can sustain full cooperation (smaller punishments are more credible).

Formally:

Definition 1: A treaty comprising $S = \{1, ..., k\}$ signatories is weakly collectively rational (WCR) if, for discount factors equal to one, and invoking the punishments (q_j^j, q_{-j}^j) to punish party j for cheating, it sustains the per-period payoff $\pi_s^w = \sup_{\pi_s \in \Pi_s^w} \pi_s$ for each member of S, where $\Pi_s^w = \{\pi_s : \pi_s \text{ satisfies (1) and (2)}\}$, with

(1) $$\max_{q_j \in Q_j} \pi_j(q_j, q_{-j}^j) \leq \pi_s$$

and

(2) $$\pi_i(q_j^j, q_{-j}^j) \geq \pi_s \; \forall i, \quad j \in S, \quad i \neq j.$$

Condition (1) ensures that party j can do no better than to re-establish cooperation, given that a previous deviation has occurred. If cooperation is re-established, j will get π_s^w every period. If cooperation is not re-established, j will get a payoff no greater than π_s^w every period. Hence, j cannot lose and may possibly gain by re-establishing cooperation. Of course, if j plays q_j^j in the punishment phase and if the other parties play q_{-j}^j, j will get a per-period payoff that is smaller than π_s^w (this is needed to ensure that j has no incentive to deviate in the first place). However, j would rather that this happen because, without discounting, it is the long run that matters, and in the long run j can be sure of getting a higher payoff by re-establishing cooperation.

Condition (2) guarantees that no party $i, i \neq j$, can gain by overlooking the deviation, reverting to a cooperative phase without punishing j. Moreover, by the assumption of symmetry, (2) implies also that all of the aggrieved countries cannot do better collectively than by imposing the threatened punishment. So these countries have no incentive to renegotiate.

This definition of a WCR treaty takes the level of participation (and, hence, non-participation, since N is fixed) as given. It also takes the behavior of non-signatories as given. But, of course, by assumption, the Nash equilibrium of the one-shot game is in dominant strategies and so the behavior of non-signatories will be unaffected by the treaty; that is, the behavior of non-signatories *is* given. Importantly, as we vary the level of participation (that is, the breadth of cooperation), k, π_s^w will change (as will the provision of the public good). This allows us to evaluate the tradeoff between the depth and breadth of cooperation.

Assuming that players are symmetric and patient and have linear payoffs, whether the constraint of self-enforcement bites depends on the number of players. It turns out that, in a 2-player prisoners' dilemma, we can have our cake and eat it. That is, for discount factors sufficiently large, full cooperation can be sustained by a strongly renegotiation-proof agreement (FARRELL AND MASKIN [1989], VAN DAMME [1989]). This is an agreement in which the deviant plays a penance strategy, increasing its provision level in a punishment phase – punishing itself, essentially, for deviating – while the other player decreases its provision in the punishment phase and thus punishes the deviant. However, to explore the trade-off between the depth and the breadth of cooperation, we plainly need to consider situations involving $N > 2$ players. Extension of the FARRELL AND MASKIN [1989] concept to the $N > 2$ case is non-trivial, however, not least because coalitions may form. The extension employed here rules out the arbitrary formation of coalitions; it considers the interests of one group of countries as against the interests of individual countries acting unilaterally. Moreover, if there are a multiple of equilibria that can be sustained by a group of given size, the approach followed here assumes that the best of these (from the perspective of the group) will be selected. It is because these assumptions are special and emphasize the collective aspect of cooperation that I refer to such treaties as being collectively rational rather than simply renegotiation-proof.

By contrast, a WCR treaty will exist for any level of participation – that is, for any $k \geq 2$. To sustain a greater breadth (participation level, k), parties only need decrease the depth of cooperation (that is, the cooperative provision level). At the very least, since the non-cooperative outcome is weakly renegotiation-proof, countries can always support an outcome at least as good as the non-cooperative outcome. For the two examples illustrated later in the paper, they can always do strictly better.

How will participation be determined? Since all N countries have an interest in the outcome of a treaty negotiation, all will participate in the negotiation process. Superficially, then, it seems plausible that in the treaty negotiation stage, negotiators will write a treaty that maximizes each country's *ex ante* payoff – the payoff each can expect to get before countries have chosen whether to be a signatory or a non-signatory. However, such a treaty may not seem so appealing *ex post*, and the negotiators should be able to anticipate this. If a consensus treaty is negotiated (that is, a treaty for which $S = A$), each country will get the same payoff *ex post* that it expects to get *ex ante*, but if an incomplete treaty is negotiated, signatories will get a lower payoff and non-signatories a higher payoff *ex post*. That a subset of countries should free ride may seem unfair, and experiments have shown that fairness can be important determinant of an equilibrium (see, for example, OSTROM [1998]). From an *ex post* perspective, a consensus treaty is especially compelling because it represents a symmetric equilibrium to a symmetric game and is thus "focal."[7]

However, fairness should not be the only consideration. To sustain a consensus, a treaty may need to be watered down so much that little is achieved overall. From an *ex ante* perspective, it may be better to negotiate a treaty in which much more is achieved, even if only by a subset of countries. So it seems that countries should be able to choose *ex ante* between having a consensus or an incomplete treaty.

Like a consensus treaty, an incomplete treaty must be compelling *ex post*. Arguably, the most appealing incomplete treaty is one that is collectively rational *on*

[7] RUNGE [1984, p. 161] also argues in favor of symmetry, claiming that no country "wants to be the odd man out." However, he assumes that countries have "fairminded" preferences, making the provision of public goods a coordination game. In this paper, countries are shown to prefer a consensus even though they care only about self-interest.

the equilibrium path. This, you will remember, is how cooperative provision levels are calculated by HOEL [1992], CARRARO AND SINISCALCO [1993], and BARRETT [1992], [1994], [1999], [2001]. Signatories to such a treaty may resent the non-signatories, but at least they can take comfort in the knowledge that their treaty yields them the highest welfare, given the level of participation. I refer to this kind of treaty as an incomplete, focal treaty.

Formally, a consensus treaty, an incomplete treaty, and an incomplete, focal treaty may be defined as:

Definition 2: A WCR consensus treaty is a WCR treaty for which $S = A$. A WCR incomplete treaty is a WCR treaty for which S contains at least two countries and the set $N \backslash S$ is non-empty.

Definition 3: A WCR incomplete, focal treaty is a WCR incomplete treaty comprising the largest number of signatories satisfying (1) and (2) with $\pi_s = \pi_s^c$, where π_s^c is the payoff each signatory gets when the provision by every signatory j is the solution to $\max_{q_j \in Q_j} \sum_{i \in S} \pi_i(q_j, q_{-j})$ for all $i, j \in S$.

With these concepts, we can now determine which of the two kinds of treaty will be preferred *ex ante* – that is, in the negotiation stage.

3 Constant Marginal Costs

For this specification, the payoffs and feasible actions are given by

$$\text{(3)} \quad \pi_i = bQ - cq_i, \quad \text{where } Q = \sum_{j \in A} q_j, q_i \in [0, 1], \quad \text{and } bN > c > b > 0.$$

π_i is country i's payoff, the difference between its gross benefit of public good provision and its own cost of supply; Q is the quantity of the public good provided; q_i is country i's supply of the public good, which can vary between 0 and 1; N is the total number of countries that can both supply and benefit from the public good; and b and c, respectively, are the constant marginal benefits and costs of public good supply to i.

The restrictions on the parameter values coupled with the payoff specification mean that the problem being studied here is a linear, N-player prisoners' dilemma. If we instead assumed $b > c$, then all countries would supply the public good unilaterally. If we assumed $c > bN$, then countries would not supply the public good even if they cooperated fully. In neither of these cases would an agreement be needed. Assuming $bN > c > b$ introduces a conflict between what countries will choose to do unilaterally and what they would prefer that all countries do collectively. It is the tension that this literature has rightly focused on. The important benchmark solutions are given in Proposition 1.

Proposition 1: For model (3), the Nash equilibrium of the one-shot game is $q_i^o = 0$, with $\pi_i^o = 0$; the full cooperative outcome is $q_i^c = 1$, with $\pi_i^c = bN - c$ for all $i \in A$.

The proof is easy. Because payoffs are linear, and given that $c > b$, each country's payoff decreases in its own abatement. Hence, each country can do no better than to play the minimal abatement level (zero). The aggregate payoff for all countries increases in each country's abatement under the assumption that $bN > c$. Hence, full cooperation requires that each country play the maximal abatement level (one). The payoffs are found by substitution.

3.1 Sustaining Full Cooperation

If game (3) is repeated infinitely often, then we know from the folk theorems that full cooperation can always be sustained by an IR treaty if every country's discount factor is sufficiently large. For collectively rational treaties, however, it will not always be possible to sustain full cooperation. Upon setting $\pi_s = \pi^c$ in (1) and (2), we obtain (see the Appendix):

Proposition 2: For the infinitely repeated version of game (3), and for discount factors close to 1, the full cooperative outcome can only be sustained by a WCR treaty if $N \leq \overline{N}$, where $\overline{N}$ is the largest integer not greater than $(b + c)/b$.

Notice that, since $c > b$ by assumption, full cooperation can always be sustained if $N = 2$ (see also FARRELL AND MASKIN [1989], VAN DAMME [1989]). However, if $N \geq 3$, then it may not be possible to sustain full cooperation by means of collectively rational punishments.[8] This is worth underlining because the international relations literature relies almost without exception on the analysis of 2×2 models.

3.2 Consensus Treaties

When $N \leq \overline{N}$, we can safely assume that a WCR treaty will sustain full cooperation. But suppose that N exceeds this critical value. Then a consensus treaty can only sustain a strictly smaller payoff than π^c in a cooperative phase. We know that WCR consensus treaties will always exist – since at the very least a consensus treaty can sustain the Nash equilibrium of the one-shot game. Proposition 3 (proved in the Appendix) yields a more remarkable result: a WCR consensus treaty can sustain a payoff strictly greater than π^o, even as N gets very large.

Proposition 3: Let $N > \overline{N}$. Then, for the infinitely repeated version of game (3), and for discount factors close to 1, a WCR consensus treaty sustains a per-period payoff of $\pi_S^w(N) = b^2(N - 1)/c$.

If the per-period payoff that each country realizes in a cooperative phase is lowered sufficiently, a WCR consensus agreement can be sustained for any N. This is because lowering the payoff to cooperation reduces the incentive to renegotiate, making the threat to punish more credible. More surprisingly, Proposition 3 shows that a WCR

[8] This is also demonstrated in BARRETT [1999], where it is shown that a SCR treaty also can sustain full cooperation if and only if N is no greater than $(b + c)/b$.

consensus treaty can sustain a cooperative phase payoff that exceeds π^o (recall that, for model (3), $\pi^o = 0$). Indeed, π_s^w is strictly increasing in the number of signatories (N in a consensus treaty)! This might seem surprising but recall from (3) that π^c is also increasing in N. We should expect π_s^w to fall relative to π^c and this can be easily confirmed. To be precise, it can be shown that, for $N > \overline{N}$, π_s^w/π^c is decreasing in N and approaches b/c as $N \to \infty$.

Finally, notice that, for $N > \overline{N}$, $\overline{N}$ will be the number of signatories to the incomplete, focal treaty. This follows immediately from Definition 3 and the observation that non-signatories will play q^o irrespective of the level of participation or the provision achieved by the incomplete treaty.

3.3 Consensus v. Incomplete, Focal Treaties

As noted previously, the earlier literature finds that the self-enforcing agreement will normally be incomplete, with some countries participating and some not. This equilibrium arises from the assumption that treaties must be SCR. If treaties need only be WCR, however, then a variety of agreements can be sustained. For the reasons noted previously, the most compelling of these are the consensus treaty and the incomplete, focal agreement. For model (3), the incomplete, focal agreement derived from the WCR assumption is identical to the equilibrium agreement derived from the SCR concept. Hence, a direct comparison can be made between the kinds of agreements that are sustained by the approach taken in the earlier literature and the alternative of a consensus treaty proposed here. Below I derive the conditions under which the consensus option will be preferred at the negotiation stage.

Signatories to a WCR consensus treaty will each earn a per-period payoff of $\pi_s^w(N)$ *ex post* – that is, after the treaty has been negotiated and enters into force. *Ex post*, signatories to an WCR incomplete, focal treaty consisting of $\overline{N}$ parties will each get $\pi_s^w(\overline{N}; N)$, whereas non-signatories will get $\pi_n^w(\overline{N}; N)$, but no country will know *ex ante* whether it will be a signatory or a non-signatory. The *expected* payoff to any country of having an incomplete, focal agreement will be $\hat{\pi} = [\overline{N}\pi_s + (N - \overline{N})\pi_n]/N$. From an *ex ante* perspective each country will prefer to negotiate by consensus provided $\pi_S^w(N) \geq \hat{\pi}(\overline{N})$. A little algebra (see the Appendix) can prove the following:

Proposition 4: Let $N > \overline{N}$. Then, from an *ex ante* perspective, if treaties need only be WCR, all countries will prefer a consensus treaty to an incomplete, focal treaty provided $N \geq c^2/b^2$.

It is worth noting that, whichever type of treaty is actually negotiated, no country will have an incentive to deviate *ex post* provided conditions (1) and (2) are satisfied. Both kinds of treaty are self-enforcing (in the WCR sense).

Consider an example. Suppose $N = 99$, $b = 2$, and $c = 15$. Then an incomplete, focal treaty will consist of 8 signatories – and from an *ex ante* perspective all countries will prefer to negotiate by consensus. However, if $b = 2$, and $c = 131$, then an incomplete, focal treaty will consist of 66 signatories – and, from an *ex ante* perspective, all countries will prefer this treaty to a consensus. A consensus treaty will be preferred only when an incomplete, focal treaty would attract participation by a small fraction of countries.

Note that the trade-off between depth and breadth is present in both of these situations. When $N = 99$, $b = 2$, and $c = 15$, the 8 signatories supply $q = 1$ each, whereas parties to the consensus option supply only $q = 0.14$ each. When $b = 2$, and $c = 131$, the 66 parties to the incomplete, focal treaty supply $q = 1$ each, whereas the parties to the consensus treaty supply $q = 0.045$ each.

4 *Increasing Marginal Costs*

Assume now that payoffs are given by

(4) $$\pi_i = bQ - cq_i^2/2, q_i \in [0, q^{max}], \text{ and } b, c > 0.$$

Model (4), employed previously by HOEL [1992], CARRARO AND SINISCALCO [1993], and BARRETT [1994], assumes that the marginal cost of provision increases linearly. Assuming that the feasibility constraints admit an interior solution, we get:

Proposition 1a: For (4), the Nash equilibrium of the one-shot game is $q_i^o = b/c$, with $\pi_i^o = b^2(2N-1)/2c$; the full cooperative outcome is $q_i^c = bN/c$, with $\pi_i^c = b^2N^2/2c$ for all $i \in N$.

The proof is easy. With abatement levels being chosen independently, the first order conditions require $b - cq_i = 0$. Since the second order condition for a maximum is satisfied, the non-cooperative outcome is $q_i^o = b/c$. If countries cooperate fully, by contrast, the first order condition requires $bN - cq_i = 0$. Since the second order condition for a maximum is once again satisfied, full cooperation requires $q_i^c = bN/c$. The payoffs corresponding to these two outcomes are found by simple substitution.

Model (4) thus yields another prisoners' dilemma. The important difference between the two models is that the non-cooperative and full cooperative provision levels for this specification are determined by the parameter values rather than by the feasibility constraints, which do not bind by assumption.

And, yet, when countries cooperate, the results can be sensitive to the feasibility constraints. The ability to sustain cooperation by means of a WCR treaty depends on choice of q_j^j. The higher j can set q_j^j, the higher the other signatories can set q_{-j}^j, making more credible the threat by these countries to punish j for deviating from their agreement. In the analysis that follows I take it that $q^{max} = bN/c$. This assumption is made for convenience. I shall indicate as I proceed how the results depend on this assumption.

Proposition 2a: For the infinitely repeated version of model (4), and for discount factors close to 1, the maximum number of countries that can sustain full cooperation by means of a WCR treaty is $\overline{N} = 5$, provided $q_i \in [0, bN/c]$.

This result (proved in the Appendix) is qualitatively different from Proposition 2. For model (4), the maximum number of countries capable of sustaining full cooperation by means of a WCR treaty is independent of parameter values. Moreover, it can be shown that a SCR treaty for model (4) can only sustain full cooperation for a smaller N (specifically, $N = 3$; for a proof in a one-shot setting, see HOEL [1992], CARRARO AND SINISCALCO [1993], and BARRETT [1994]). In addition, and as explained above, the details of Proposition 2a depend on the feasibility constraints. If q_j^j can be increased, constraint (2) will be relaxed. Proposition 2a assumes $q^{max} = bN/c$. Let us suppose instead that $q^{max} = (1 + \alpha)bN/c$ with $\alpha \geq 0$ being a scale parameter indicating the maximum feasible abatement level (as explained later, α can also be interpreted as specifying the "penalty" to deviating). It can be shown that the maximum number of countries capable of sustaining the full cooperative outcome by means of a WCR treaty is $N \leq 3 + 4\alpha + 2\sqrt{1 + 6\alpha + 4\alpha^2}$. If $\alpha = 0$, as Proposition 2a assumes, then no more than five countries can sustain the full cooperative outcome by means of a WCR treaty. If $\alpha = 1$ – if feasible abatement can be twice the full cooperative level – then 13 countries can sustain the full cooperative outcome by such a treaty. And if $\alpha = 9$ – if the maximum feasible abatement level is ten times the full cooperative level – then 77 countries can sustain the full cooperative outcome using a WCR consensus treaty. The smaller is q^{max} relative to q^c, the smaller is the number of countries that can sustain full cooperation.

Proposition 3a: Assume $q_i \in [0, bN/c]$. Then, for the infinitely repeated version of model (4), for $N > \overline{N}$, and for discount factors close to 1, a WCR consensus treaty sustains a per-period payoff of $\pi_s^w = b^2[2(N-1)\sqrt{2N-1} + 1]/2c$.

(The proof is given in the Appendix.) Of course, if the feasibility constraints were relaxed, a greater payoff could be sustained by a WCR, consensus treaty. More interesting is the result that π_s^w is increasing in N. However, it is easy to show that, for $N > \overline{N}$, π_s^w/π^c is decreasing in N and approaches zero as $N \rightarrow \infty$ (for any upper bound on provision).

It only remains now to determine the conditions under which countries will prefer a consensus treaty to an incomplete, focal treaty – that is, an incomplete treaty that is collectively rational on the equilibrium path. For model (3), the equilibrium number of participants in an incomplete focal treaty is the same for both the SCR and WCR concepts. For model (4), the corresponding participation rates are not equal but they are close. As noted above, for $N \geq 5$, the SCR treaty consists of 3 countries while the WCR treaty consists of 5. Here, however, we obtain a stronger result as regards preference for a consensus treaty.

Proposition 4a: Let $q_i \in [0, bN/c]$. Then, from an *ex ante* perspective, all countries will always prefer a WCR consensus treaty to a WCR incomplete, focal treaty.

Notice that this last result (proved in the Appendix) holds for *any* N. Why, from an *ex ante* perspective, is a consensus treaty always attractive for model (4) but not for model (3)? The essential reason is that, with marginal abatement costs increasing, the total cost of sustaining any particular level of public good provision can always be reduced by expanding participation and redistributing the burden of provision. For model (4), there remains a trade-off between the breadth and depth of cooperation in the sense that participating countries each supply more of the public good in an incomplete, focal treaty than in a consensus treaty for "large" N. However, from an *ex ante* perspective, welfare considerations favor the consensus option.

5 Commentary on the Kyoto Protocol

The models analyzed here, though simple, illuminate important aspects of the Kyoto Protocol. Here I comment briefly on just two.

Though I have emphasized that a consensus treaty is "fair," this is in the context of symmetric countries. In the real world, countries are decidedly asymmetric, and in 1995 it was decided that, on grounds of fairness, the industrialized countries only should be subject to quantitative emission ceilings in a new climate protocol. On its own, differential emission limits drive a wedge between the marginal costs of industrialized and non-industrialized countries. In doing so, pressure for a consensus agreement should build. But, as noted above, the efficiency advantage of having a consensus clashes here with the equity disadvantage. The negotiators reconciled this conflict by including a "clean development mechanism" in the Kyoto Protocol. This essentially gives the industrialized countries credit for reducing emissions in the non-industrialized countries. The problem is that such a mechanism is likely to entail substantial transaction costs. If so, pressure may build for a redesign – perhaps to allow full-blown trading in emission entitlements among all countries. Equity concerns could then be addressed by the allocation of the initial distribution of entitlements.

However, the efficacy of quantitative emission limits, with or without trading, requires enforcement, and the Kyoto Protocol incorporates a flawed enforcement mechanism. The original agreement, negotiated in December 1997 did not specify any consequences for non-compliance. Later, however, in meetings held in Bonn and Marrakesh in 2001, a compliance mechanism was agreed. This requires that, if a party were found to have exceeded its allowed emissions in the first control period (2008–2012) by, say, X tons, then it would be required to reduce its emissions by an additional $1.3X$ tons in the *next* control period (presumably, 2013–2017). This so-called penalty rate is equivalent to choosing $\alpha = 0.3$ in the analysis of Section 4.

As shown in Section 4, however, setting $\alpha > 0$ is not sufficient for enforcement. Enforcement requires that the *other* parties punish the deviant. Indeed, the reason for setting $\alpha > 0$ is to reduce the cost to the other parties of doing this. The Bonn–Marrakesh agreement, however, does not require that the other parties change their behavior at all in a punishment phase. It relies entirely on self-inflicted punishment.

Indeed, the problem is even worse than this for several reasons. First, the self-imposed punishment can be delayed – potentially, forever. According to the agreement, a country that fails to punish itself in the second control period (for a failure to comply in the first control period) is required to punish itself in the third period, with the penalty increased (though the future value of α would need to be decided in a future amendment). That is, a country is not only not punished by others for failing to comply. It is also not punished by others for failing to punish *itself* for failing to comply. Second, the allowed levels of emissions in future control periods have yet to be negotiated and are thus endogenous to the process. Though the intention is plainly for future emission levels to be negotiated before the start of the previous control period, there is no mechanism within the treaty to ensure that this will be done. In any event, a country that expects compliance to be difficult can simply insist on easier future emission constraints. Third, a country that finds compliance difficult, or fears it might fail to comply, may simply withdraw from the treaty. Parties are allowed to do this, and there is little incentive in the treaty for them not to do so. Finally, the Kyoto Protocol prohibits compliance mechanisms having "binding consequences" except as agreed by amendment. An amendment is essentially a new treaty, and so the compliance mechanism agreed in Bonn/Marrakesh cannot actually be enforced as regards the parties to the Kyoto Protocol.

6 *Conclusions*

This paper looks at international cooperation problems in a different way than has been the tradition in the literature. The previous literature has assumed that treaties are strongly collectively rational (SCR). In this paper I assume that treaties are instead weakly collectively rational (WCR). SCR treaties are typically able to sustain only incomplete treaties. By contrast, as shown in this paper, WCR treaties can always support a consensus – and a non-trivial one at that. For two specifications of payoffs, I show that a WCR consensus treaty strictly improves on the non-cooperative outcome.

Though the obvious attraction of a consensus treaty is that it is symmetric and thus, in a sense, fair, this paper shows that a consensus treaty has an added advantage: it can sustain more cooperation than an incomplete treaty. With increasing marginal costs, an incomplete treaty is almost begging to be renegotiated. Even if the overall level of abatement were not increased, exchange between parties and non-parties could make everyone better off. The natural way to exploit this potential for cost savings, given the symmetry of the players, would be for them to negotiate a consensus treaty from the start. That, in doing so, countries can sustain even more cooperation is a bonus.

The WCR concept is also more compatible with real world observations. The prediction of the earlier literature that agreements will typically be incomplete sits rather awkwardly next to the facts. The earlier literature (see HOEL [1992], CARRARO AND SINISCALCO [1993], BARRETT [1994]) finds that, if payoffs conform to model (4), the self-enforcing treaty will consist of just three signatories – irrespective of how large N is (that is, for *any* $N \geq 3$). However, it is very rare for a treaty to attract a small number of parties relative to N, and no treaty seeking to supply a global public good has as little as three parties. The results presented here for WCR treaties are more consistent with the facts.

Finally, this paper also shows that the earlier literature misleads when it implies that full participation is synonymous with full cooperation. A failure to cooperate fully has different guises. It can be manifest in under-provision as well as non-participation.

Appendix

Proof of Proposition 2: Upon substituting, (1) becomes

(A1) $$\max_{q_j} \left\{ b\left(q_{-j}^j + q_j\right) - cq_j \right\} \le bN - c.$$

Inequality (A1) can be rewritten

(A1a) $$bq_{-j}^j \le bN - c.$$

Condition (2) requires

(A2) $$b\left[q_{-j}^j + q_j^j\right] - cq_{-j}^j/(N-1) \ge bN - c.$$

It is natural to take $q_j^j = 1$. Then (A2) becomes

(A2a) $$[b - c/(N-1)]\, q_{-j}^j \ge b(N-1) - c.$$

Suppose $b(N-1) > c$. Then (A2a) requires $q_{-j}^j = (N-1)$, and this implies, by (A1a), that $b - c \ge 0$, which is false by assumption. Suppose instead that $b(N-1) - c \le 0$. Then (A2a) holds for *any* feasible q_{-j}^j. However, (A1a) will *not* hold for any q_{-j}^j. Together, (A1a) and (A2a) require $bq_{-j}^j \le bN - c \le b$ or $0 \le q_{-j}^j \le (bN-c)/b < (N-1)$ and $N \le (b+c)/b$. *Q.E.D.*

Proof of Proposition 3: Condition (1) requires

(A3) $$\max_{q_j} \left\{ b\left(q_{-j}^j + q_j\right) - cq_j \right\} \le \pi_s$$

or

(A3a) $$bq_{-j}^j \le \pi_s.$$

Condition (2) requires

(A4) $$b\left(q_{-j}^j + q_j^j\right) - cq_{-j}^j/(N-1) \ge \pi_s.$$

Letting $q_j^j = 1$, (A4) becomes

(A4a) $$[b - c/(N-1)]q_{-j}^j \ge \pi_s - b.$$

Together, (A3a) and (A4a) require

(A5) $$\frac{\pi_s}{b} \ge q_{-j}^j \ge \frac{\pi_s - b}{[b - c/(N-1)]}$$

or $b^2(N-1)/c \ge \pi_s$. *Q.E.D.*

Proof of Proposition 4: We know that $\pi_s^{w}(N) = b^2(N-1)/c$. We also know that $\pi_s^{w}(\overline{N}; N) = b\overline{N} - c$ and $\pi_n^{w}(\overline{N}; N) = b\overline{N}$. Every country will prefer a consensus agreement to an incomplete, focal agreement if $\pi_s^{w}(N) \geq \hat{\pi}(\overline{N})$. Upon substituting, we get

(A6) $$\frac{b^2(N-1)}{c} \geq \frac{[(b\overline{N}-c)\overline{N} + b\overline{N}(N-\overline{N})]}{N}.$$

The right hand side of (A6) is strictly increasing in $\overline{N}$. We know from Proposition 2 that $\overline{N} \leq (b+c)/b$. Hence (A6) will be a sufficient condition if we set $\overline{N} = (b+c)/b$. Upon doing so, and rearranging, (A6) becomes

(A6a) $$b^3N^2 - b\,[b^2 + c(b+c)]N + c^2(b+c) \geq 0.$$

The left hand side of this weak inequality is a quadratic. Let N^{min} and N^{max} denote the two values of N for which the quadratic term equals zero. By inspection, the inequality holds for $N^{min} \geq N \geq N^{max}$. Solving the quadratic yields $N^{min} = (b+c)/b$ and $N^{max} = c^2/b^2$. Since $\overline{N}$ is the largest integer less than or equal to $(c+b)/b$, and since N^{min} and N^{max} need not be integers, it follows that $N^{min} \geq \overline{N}$. But the proposition holds for $N > \overline{N}$. Hence, $\hat{\pi} \geq \pi_s^{w}$ only holds for $N > \overline{N}$ if $N \geq N^{max}$ or $N \geq c^2/b^2$. *Q.E.D.*

Proof of Proposition 2a: Substituting into (1) yields

(A7) $$\max_{q_j} \left\{ b\left(q_{-j}^{j} + q_j\right) - cq_j^2/2 \right\} \leq b^2N^2/2c.$$

Solving (A7) gives

(A7a) $$q_{-j}^{j} \leq \frac{b(N^2-1)}{2c}.$$

By (2),

(A8) $$b\left(q_{-j}^{j} + q_j^{j}\right) - \frac{c\left[q_{-j}^{j}/(N-1)\right]^2}{2} \geq \frac{b^2N^2}{2c}.$$

Let $q_j^j = bN/c$. Then (A8) becomes

(A9) $$q_{-j}^{j}\left[b - \frac{cq_{-j}^{j}}{2(N-1)^2}\right] \geq \frac{b^2N(N-2)}{2c}.$$

Upon solving this quadratic, one obtains $q_i^j \in [b(N-2)/c, bN/c]$ for $i \neq j$. Of course, (A8) must also be satisfied (that is, $q_i^j \leq b(N+1)/2c$), and one can show that $bN/c > b(N+1)/2c$. (Suppose instead that $bN/c \leq b(N+1)/2c$. Then we have $N \leq 1$, and by assumption we require $N \geq 2$). Hence, (A7a) and (A9) together require

(A10) $$\frac{b(N-2)}{c} \leq q_i^j \leq \frac{b(N+1)}{2c}$$

or $N \leq 5$. *Q.E.D.*

Proof of Proposition 3a: We require

(A11) $$\max_{q_j} \{b(q^j_{-j} + q_j) - cq_j^2/2\} \leq \pi^w_s$$

or

(A11a) $$bq^j_{-j} + \frac{b^2}{2c} \leq \pi^w_s.$$

We also require

(A12) $$b(q^j_{-j} + q^j_j) - c[q^j_{-j}/(N-1)]^2/2 \geq \pi^w_s.$$

Setting $q^j_j = q^c = bN/c$, (A12) becomes

(A12a) $$q^j_{-j}[b - cq^j_{-j}/2(N-1)] + b^2N/c \geq \pi^w_s.$$

Together, (A11a) and (A12a) require

$$bq^j_{-j} + \frac{b^2}{2c} \leq \pi^w_s \leq q^j_{-j}\left[b - \frac{cq^j_{-j}}{2(N-1)}\right] + \frac{b^2N}{c},$$

which implies

(A13) $$q^j_i \leq b\sqrt{2N-1}/c$$

for $i \neq j$. The right hand side of (A13) is less than bN/c for $N \geq 2$. Letting $q^j_i = b\sqrt{2N-1}/c$ and substituting into (A11a) yields $\pi^w_s \leq b^2[2(N-1)\sqrt{2N-1}+1]/2c$. *Q.E.D.*

Proof of Proposition 4a: We know from Proposition 2a that countries will negotiate a consensus agreement that sustains full cooperation for $N \leq 5$. Hence, we need only show that countries will also want to negotiate a consensus agreement for $N > 5$. Every country will prefer a consensus agreement to an incomplete, focal agreement for $N > 5$ if $\pi^w_s(N) \geq \hat{\pi}(5; N)$. Upon substituting for $\hat{\pi}$ and π^w_s, we get $\pi^w_s(N) - \hat{\pi}(5; N) = N(N-1)\sqrt{2N-1} - N^2 - 19N + 60$. Tedious algebra then gives $\pi^w_s(N) - \hat{\pi}(5; N) = (N-5)\Omega(N)$, where $\Omega(N) = N^4 + 2N^3 - 7N^2 - 156N + 360$. It is easy to show that $\Omega(N)$ is positive and increasing for $N \geq 5$. Hence, $\pi^w_s(N) > \hat{\pi}(5; N)$ for $N > 5$. *Q.E.D.*

References

ARCE M., D. G. [2000], "Leadership and the Aggregation of International Collective Action," *Oxford Economic Papers*, 53, 114–137.

AXELROD, R. [1984], *The Evolution of Cooperation*, Basic Books: New York.

BARRETT, S. [1992], "International Environmental Agreements as Games," pp. 11–35 in: R. Pethig (ed.), *Conflict and Cooperation in Managing Environmental Resources*, Springer-Verlag: Berlin.

-- [1994], "Self-Enforcing International Environmental Agreements," *Oxford Economic Papers*, 46, 878–894.

-- [1997], "The Strategy of Trade Sanctions in International Environmental Agreements," *Resource and Energy Economics*, 19, 345–361.

-- [1999], "A Theory of Full International Cooperation," *Journal of Theoretical Politics*, 11, 519–541.
-- [2001], "International Cooperation for Sale," *European Economic Review*, 45, 1835–1850.
BENEDICK, R. E. [1998], *Ozone Diplomacy*, Harvard University: Cambridge, MA.
BUCHHOLZ, W., C. HASLBECK, AND T. SANDLER [1998], "When does Partial Co-Operation Pay?" *Finanzarchiv*, 55, 1–20.
CARRARO, C., AND D. SINISCALCO [1993], "Strategies for the International Protection of the Environment," *Journal of Public Economics*, 52, 309–328.
ENDRES, A., AND M. FINUS [1998], "Renegotiation-Proof Equilibria in a Bargaining Game over Global Emission Reductions: Does the Instrumental Framework Matter?" pp. 135–164 in: N. Hanley and H. Folmer (eds.), *Game Theory and the Global Environment*, Edward Elgar: Cheltenham.
FARRELL, J., AND E. MASKIN [1989], "Renegotiation in Repeated Games," *Games and Economic Behavior*, 1, 327–360.
FEHR, E., AND S. GÄCHTER [2000], "Cooperation and Punishment in Public Goods Experiments," *American Economic Review*, 90, 980–994.
FINUS, M., AND B. RUNDSHAGEN [1998], "Renegotiation-Proof Equilibria in a Global Emission Game When Players are Impatient," *Environmental and Resource Economics*, 12, 275–306.
HEAL, G. [1993], "Formation of International Environmental Agreements," pp. 301–322 in: C. Carraro (ed.), *Trade, Innovation, Environment*, Kluwer: Dordrecht.
HOEL, M. [1992], "International Environment Conventions: The Case of Uniform Reductions of Emissions," *Environmental and Resource Economics*, 2, 141–159.
-- AND K. SCHNEIDER [1997], "Incentives to Participate in an International Environmental Agreement," *Environmental and Resource Economics*, 9, 153–170.
KJELLEN, B. [1994], "A Personal Assessment," pp. 149–174 in: I. M. Mintzer and J. A. Leonard (eds.), *Negotiating Climate Change: The Inside Story of the Rio Convention*, Cambridge University: Cambridge.
OSTROM, E. [1998], "A Behavioral Approach to the Rational Choice Theory of Collective Action," *American Political Science Review*, 92, 1–22.
RUNGE, C. F. [1984], "Institutions and the Free Rider: The Assurance Problem in Collective Action," *Journal of Politics*, 46, 154–181.
SANDLER, T. [1997], *Global Challenges*, Cambridge University: Cambridge.
VAN DAMME, E. [1989], "Renegotiation-Proof Equilibria in Repeated Prisoners' Dilemma," *Journal of Economic Theory*, 46, 206–217.

Scott Barrett
Paul H. Nitze School of Advanced International Studies
Johns Hopkins University
1619 Massachusetts Avenue, NW
Washington, DC 20036
USA
E-mail: sbarrett@jhu.edu

[20]

Public Choice (2012) 151:565–584
DOI 10.1007/s11127-010-9759-6

Environmental cooperation: ratifying second-best agreements

Pierre Courtois · Guillaume Haeringer

Received: 4 May 2010 / Accepted: 21 December 2010 / Published online: 15 January 2011

Abstract As an alternative to the environmental cartel approach, we assume that an international environmental agreement aims simply at providing a collective response to a perceived threat. Given this less demanding concept of cooperation and considering that most treaties become enforceable only after ratification by a sufficient number of participants, we examine the set of self-enforceable agreements. This set contains first-best but also second-best agreements that do not maximize the collective welfare of members but meet environmental and/or participative requirements. We study the properties of this set and discuss admissible values of targets and thresholds that favour economics over environmental objectives and *vice versa*.

Keywords International environmental agreement · Social welfare · Abatement bound · Self-enforcement · Ratification threshold

1 Introduction

The number of major international environmental agreements (IEA) that have come into force has grown significantly over the last 50 years.[1] Biodiversity, long-range transboundary air pollution, ozone depleting substances, and climate change are all being reg-

[1] Several online projects have been developed which list those agreements, describe their content, and analyse their performance. Two such are the IEA database project developed by the University of Oregon and the Multilaterals project developed by the Fletcher School at Tufts University. The IEA database lists 1 039 multilateral and 1 538 bilateral environmental agreements.

P. Courtois (✉)
Dept. of Economics, INRA, Montpellier, France
e-mail: courtois@supagro.inra.fr

G. Haeringer
Dept. of Economics, Universitat Autonoma de Barcelona, Bellaterra, Spain

ulated by major multilateral treaties.[2] Considering the possible costs incurred and the incentives to free ride on collective efforts, the enactment of those IEA may seem surprising. However, the price of cooperation and the incentives to free ride are not necessarily high. Agreed environmental targets may rely on a combination of considerations that go beyond economic efficiency and aim, for example, at meeting participation or environmental quality objectives. Moreover, after an IEA has been signed, the requirements for its ratification must be fulfilled before it becomes enforceable.[3]

The literature on non-cooperative IEA that builds on the work of Hoel (1992), Carraro and Siniscalco (1993) and Barrett (1994) pays scant attention to these two specificities.[4] By equating agreements to cartels maximizing countries' collective welfare, the main focus of this literature is on the stability of first-best agreements.[5] However, as Black et al. (1993) and Okada (1993) show, the ratification threshold rule controls free riding and may ensure stability. Defining an IEA as a cartel of countries aimed at maximizing collective welfare is also a questionable representation of the outcome of international environmental negotiations (Vaubel 1986). Collective welfare is a cooperative objective while negotiation resembles a non-cooperative process (Caparros et al. 2004). Moreover, consensus and environmental performance are not necessarily at odds with economic efficiency and the goal of environmental cooperation may go beyond welfare considerations (Frey 1997; Vaubel et al. 2007). Prescriptions assigned by recent IEA illustrate the predominance of non-optimal equal reduction formulae that are justified on the basis of environmental considerations. The Montreal Protocol (1988) stipulates a gradual eradication of chlorofluorocarbons from production processes, which Mostapha Tolba, the Executive Director of the negotiations, justifies on the basis of environmental rather than economic concerns (Benedick 1998). Article 2 of the United Nations Framework Convention on Climate Change states that the commitment to a 5.2% reduction in CO2 undertaken by Kyoto members is aimed at stabilizing greenhouse gas concentrations in the atmosphere at a level that would prevent dangerous anthropogenic disturbance to the climate system. Analysing agreed targets within the trans-boundary air pollution and ozone layer protocols, Dietrich (1995) and Patt (1999) confirm those statements, arguing that critical loads[6] and uniform targets were chosen for the purpose of providing a consensual response to the potential environmental threat. Grubb et al. (1999) and Böhringer and Vogt (2004) argue that agreed policies should be seen as the lowest common denominator in order to maximize acceptance.

Two key questions arise. First, if we assume that an IEA may prescribe any profitable collective responses to a perceived threat, what are the economic and environmental properties of the set of enforceable agreements? In other words, if we allow for second-best targets, what can we expect to achieve economically and/or environmentally? Second, considering

[2]In this paper we use the terms protocol, agreement and treaty interchangeably.

[3]Rutz (2002) reviews over 100 IEA's and shows that about 98% of agreements comprise a ratification procedure as stipulated by the Vienna Convention (1969). For an overview of ratification thresholds see, e.g., Article 25 of the Kyoto Protocol (1997), Article 16 of the Ozone Layer Montreal Protocol (1987), Article 15 of the Oslo Protocol on Further Reduction of Sulphur Emissions (1994), the NOX Protocol (1988) and the VOC Protocol (1991).

[4]For surveys of this extensive literature, see Barrett (2003, 2007) and Finus (2001, 2008).

[5]Note that this approach differs from the purely public deterrence model of alliances or with the joint product model of alliances reviewed in Cornes and Sandler (1996) or in Sandler and Hartley (2001).

[6]Critical loads are generally thought of as threshold concentration pollutants that can be tolerated without unacceptable environmental impacts. They were first used in the protection of aquatic ecosystems from eutrophying pollutants, in the 1980s.

this less demanding concept of cooperation, what can we say about participation? The assumption of super-additivity of the public good technology usually equates full cooperation with economic efficiency, but in this case why do ratification clauses not prescribe unanimity? It is unclear also why one particular threshold is chosen in preference to some other and, more generally, what is the exact role of this threshold in treaty-making.

Some answers have been put forward by the literature. Barrett (2003, 2007) questions the choice of target. In studying the diversity of aggregation technologies, he examines the range of policy objectives that could be pursued, showing that welfare maximization might call for eradication, or for small-scale or large-scale control. Complementary to this approach, we are more interested in questioning the goal at the origin of the choice of a target. Barrett (2002) and Finus and Maus (2008) study second-best policies as a means of achieving greater participation. Their questions are related to ours, but their setting is different. They examine cooperation from the angle of joint welfare, set the free rider incentive as the cornerstone of their approach, and study the problem as if there were no ratification clause. In contrast, several authors analyse minimum participation rules, but all focus on the first-best solution of environmental cartels.[7] Currarini and Tulkens (2004) study the allocation rule to design an efficient agreement, given that the ratification threshold constraint must be met; Rutz (2002) analyses the IEA game in a non-cooperative setting and shows that ratification overcomes the stability issue; Kohnz (2005) looks at ratification design from the angle of contract theory and assumes asymmetrical information between agents; Harstad (2006) examines the minimum participation rule in EU policy-making and considers uncertainty as an explanation for non-unanimity. Finally, Weikard et al. (2009) analyse the minimum participation rule when countries are heterogeneous and Carraro et al. (2009) endogenize the threshold, considering a take-it-or-leave-it offer, and show that a full-participation rule is necessarily the optimal outcome.

Complementary to these contributions, we examine the set of second-best agreements, based on their ratification clauses. In line with the Vienna Convention on Treaties (1969), we assume that IEA are simply the collective consent of states to be bound [Article 2a], expressed by means of ratification [Article 14(a)]. We thus exclude the idea that an IEA is equivalent to a cartel, and suppose that an agreement can prescribe any environmental policy as soon as it becomes more stringent than the business-as-usual policy. According to this view, an IEA can aim at either maximizing the social welfare of its participants and reaching a specific environmental goal or simply maximizing participation at the lowest common denominator. We assume that a country has to comply with the prescriptions of an agreement only if it has ratified it and only if the number of the other countries that have done so is at least as high as the ratification threshold. This allows us to characterize the set of all IEA that can be enforced.

Formally, considering an agreement as a minimal environmental policy target and a ratification threshold, we consider a two-stage game in which countries decide first whether to ratify or not, and second about the choice of environmental policy, i.e., the abatement level. We then study the subgame perfect equilibrium of this game, which allows us to describe the complete set of agreements that are self-enforceable.

The paper is organized as follows. Section 2 presents the model. In Sect. 3 we describe the ratification game and the minimum requirements for an agreement to come into force.

[7] Note that Black et al. (1993) is an application of the literature on the voluntary contribution to public goods which studies how threshold levels can have ambiguous effects on contribution levels and, therefore, on the efficiency of the outcome (Bagnoli and Lipman 1989, 1992; Palfrey and Rosenthal 1984; Nitzan and Romano 1990; Suleiman 1997; McBride 2006 and Dixit and Olson 2000, among many others).

In Sect. 4 we analyse the interplay between countries' welfare and the total abatement level and discuss participation levels. Section 5 concludes. Most of the proof is relegated to the Appendix.

2 The model

2.1 Preliminaries: the abatement game

Following Barrett (1994), Finus and Maus (2008) and many papers in the IEA literature, we consider a (finite) set of n identical countries,[8] $N = \{1, \ldots, n\}$, where each country i has to choose an abatement level of pollutant, q_i. For the sake of simplicity, we assume that for each country $i \in N$, the range of possible abatement levels is defined by $X_i = [0, \infty)$. We denote by $\mathbf{q}$ the vector of abatement levels, i.e., $\mathbf{q} = (q_i)_{i \in N}$, and by Q the aggregate abatement level, $Q = \sum_{j \in N} q_j$.

We assume that abatement is a public good with congestion.[9] In other words, abatement allows for global environmental damage to be avoided, and therefore benefits each country symmetrically, but when abatement levels become drastic they affect negatively the functioning of the international economy. When the aggregate abatement level is Q, country i gets $B_i(Q) = aQ - \frac{1}{2}Q^2$ where a is a positive parameter. Abatement is also individually costly in the sense that each country pays the cost of its own abatement effort. The greater the level of a country's abatement, the higher the marginal cost will be. The cost function of country i is convex and following Barrett (1994) we assume $C_i(q_i) = \frac{c}{2}q_i^2$ where c is a positive parameter.

For a given vector of abatement the net payoff of country i is then given by the following equation,[10]

$$u_i(\mathbf{q}) = B_i(Q) - C_i(q_i) = aQ - \frac{1}{2}Q^2 - \frac{c}{2}q_i^2. \tag{1}$$

As usual, we write q_{-i} to denote the $(n-1)$-dimensional vector $(q_h)_{h \in N \setminus \{i\}}$. The abatement game has a unique equilibrium, in which each country chooses the same abatement level, q_0,

$$q_0 = \frac{a}{n+c}. \tag{2}$$

This yields the following utility level

$$u_0 = u_i(\mathbf{q}_0) = \frac{a^2(n^2 + 2nc - c)}{2(n+c)^2}. \tag{3}$$

where $\mathbf{q}_0 = (q_0, \ldots, q_0)$

[8] We are aware that this is a strong assumption but it is a necessary condition for analytical simplicity. General heterogeneous cost and benefit functions are theoretically implementable in our setting but add considerable complexity to the interpretation of results.

[9] Congestion here refers to the decreasing part of the function, meaning that over-abatement is harmful to the economy. This is a common assumption in climate change impact modelling, which usually assumes that high levels of abatement are inconsistent with economic growth (see e.g., Nordhaus 2007). Note, however, that most of our results carry over if we model total abatement as a public good without congestion.

[10] Alternatively, we could consider a model in which countries choose their level of emissions of a pollutant. Diamantoudi and Sartzetakis (2006) show that these two approaches are equivalent for this class of payoffs.

2.2 The restricted abatement game

Traditionally, the literature on environmental agreements defines an IEA as a set of countries that jointly choose abatement levels that maximize their collective welfare. This contrasts with most international treaties where an IEA is more an environmental target, usually grounded in an effect-based philosophy, and an enforcement mechanism, i.e., a ratification procedure. In the paper, we consider the latter approach.

In order to illustrate what happens in the absence of enforcement procedures when we do not equate IEA to cartels, we start assuming that an IEA is simply a set of countries that collectively choose abatement levels superior to their business-as-usual policies. Restricting our analysis to the target of an IEA and not considering the enforcement mechanism allows us to focus on some basic characteristics of IEA, such as countries' best responses, and IEA effectiveness.

Because we assume that countries are symmetrical, imposing the same lower bound α on the abatement levels of countries participating in the IEA is a natural assumption. We consider only values of α such that $\alpha > q_0$.[11]

The focus is on whether countries have incentives to follow the recommendations of the IEA and therefore to join the treaty. From a strategic point of view, participation in an IEA consists of an alteration to the strategy set in the abatement game, which we refer to as the restricted abatement game. More precisely, the possible levels of abatement of a country i participating in an IEA will be X_i^{IEA},

$$X_i^{\text{IEA}}(\alpha) = [\alpha, \infty),$$

and if it chooses not to participate, its attainable abatement levels are unchanged, i.e., $X_i = [0, \infty)$. In order to avoid any confusion, we omit the term α in the strategy set of a member country of the IEA and write X_i^{IEA} rather than $X_i^{\text{IEA}}(\alpha)$.

While strategy sets are affected if there is an IEA (for participant countries only), payoffs are not. Note that from a formal point of view, the situation with an IEA defines a different game to the situation with no IEA described at the beginning of this section.[12] To keep things simple we simplify the notations and use (1) to denote the payoffs from the games with and without an IEA.

Since participation in an IEA bounds the choices available to a country, it affects its strategic behavior. Note that the payoff function for each country $i \in N$ is continuous in $\mathbf{q}$ and strictly concave in q_i. It follows that each country's best-reply is continuous and single valued. For a non-participant country, its best reply can be defined as follows,

$$BR_i(q_{-i}) = \{q_i \in X_i : u_i(q_i, q_{-i}) \geq u_i(q_i', q_{-i}) \text{ for all } q_i' \in X_i\}.$$

Because q_i must belong to X_i for all $i \in N$, we thus have

$$BR_i(q_{-i}) = \max\left\{\frac{a - \sum_{j \neq i} q_j}{1+c}; 0\right\}.$$

[11] As explained in the Introduction, we consider the choice of the value α to be a secondary issue. The objective in this paper is to characterize the different levels of α that allow for the existence of a stable IEA. We use our characterization of the set of admissible values of α to give some insights about which value is more likely to be chosen, depending on the objectives of the signatories of the IEA, i.e., maximizing the total abatement level or maximizing the collective welfare of the signatories to the IEA.

[12] The main difference of the case without an IEA is the domain of the payoff functions. Without an IEA, the domain of each country's payoff function is $\prod_{i \in N} X_i$ while with an IEA the domain becomes $\prod_{i \in S} X_i^{\text{IEA}} \times \prod_{i \in N \setminus S} X_i$.

Consider now the case of a country, say i, participating in the IEA. In this case the best-reply, denoted br_i, is defined as follows,[13]

$$br_i(q_{-i}) = \{q_i \in X_i^{\mathrm{IEA}} : u_i(q_i, q_{-i}) \geq u_i(q_i', q_{-i}) \text{ for all } q_i' \in q_i\}.$$

The function br_i can easily be characterized from the function BR_i and the bound α.[14]

Lemma 1 *Let u_i be continuous and strictly quasi-concave in q_i for all $i \in N$. Let $X_i = [0, \infty)$ and $X_i^{\mathrm{IEA}} = [\alpha, \infty)$. Then,*

$$br_i(q_{-i}) = \begin{cases} \alpha & \text{if } BR_i(q_{-i}) < \alpha, \\ BR_i(q_{-i}) & \text{if } \alpha \leq BR_i(q_{-i}). \end{cases}$$

Lemma 1 states that whenever a country's best response is to choose an abatement level higher than α, then the country is free to do so. However, if the best response consists of choosing an abatement level lower than α, then the country chooses an abatement level equal to α. Note that this result may not hold if the payoff functions are not strictly quasi-concave. Because we are analysing a model in which the countries choose an abatement level, only a minimal abatement level is relevant. See Bade et al. (2009) for a statement of this result when a country has a minimal and a maximal bound.

Given the existence of an IEA, a crucial question is how many countries will follow the IEA's recommendation. To answer this, we first need to characterize the equilibria of the restricted abatement games, for any number of countries participating in the IEA. Denoting signatory countries by subscript s and non-signatories countries by the subscript ns, we can propose the following.

Proposition 1 *Let $\alpha > q_0$ and let S be a coalition of countries that follow the IEA's recommendation, i.e., restrict their attainable abatement levels to $[\alpha, \infty)$, with $s = \sharp S$. There is a unique equilibrium in the abatement game in which each signatory country chooses an abatement level equal to $q_s(s) = \alpha$ and each non-signatory country chooses $q_{ns}(s)$ where,*

$$q_{ns}(s) = \begin{cases} \frac{a - s\alpha}{c + n - s} & \text{if } s \leq \frac{a}{\alpha}, \\ 0 & \text{otherwise.} \end{cases} \tag{4}$$

By Proposition 1, all countries participating (respectively, not participating) in an IEA can be treated symmetrically. It follows that what matters when computing the payoffs of a country is the size of the IEA and whether it participates in the agreement or not. For simplicity, we will write $u(s, 1)$ to denote the payoff of a country participating in an IEA with s countries, and $u(s, 0)$ for the payoff for a country not participating in an IEA which includes s countries. When there is no IEA the payoff of a country can be denoted simply as u_0, and we write $u(n)$ to denote a country's payoff if the IEA involves all countries.

13 Note that since the domain of the best-reply br_i depends on the set of strategies of all other countries and thus, on the set of countries participating in the IEA, we should write $br_i^{X^{\mathrm{IEA}}}$ instead of br_i.

14 Bade et al. (2009) provide a characterization of the restricted best reply whenever the original and the constrained strategy sets are both compact and convex subsets of the real line (which includes e.g., the case where an IEA also imposes an upper bound) and the payoff functions are strictly quasi-concave.

Perhaps the most basic question to answer relates to the effectiveness of the IEA, that is, whether it increases the total abatement level. When an IEA is established, all of the countries following its recommendations will abate to a greater extent than would have been the case without the IEA ($\alpha > q_0$ by assumption). However, according to Proposition 1, countries not participating in the IEA will abate $q_{ns}(s)$ which is always below q_0. Nevertheless, we can show that as long as the abatement level of each country conforms to an equilibrium (i.e., abatement levels are given by Proposition 1), an IEA always increases the total abatement level. This result holds, irrespective of the stability of the IEA. To see this, let α be a minimal abatement level imposed by the IEA, and suppose that there are t countries following the IEA's recommendation. Suppose first that $t < a/\alpha$. According to Proposition 1, the total abatement will be $Q(t)$, where

$$Q(t) = (n-t)q_{ns}(t) + t\alpha = \frac{a(n-t) + \alpha tc}{c+n-t}.$$

When there is no IEA, the total abatement level is $Q(0)$,

$$Q(0) = nq(0) = \frac{na}{n+c}.$$

It suffices then to compare both $Q(t)$ and $Q(0)$. We obtain,

$$\frac{a(n-t) + \alpha tc}{c+n-t} > \frac{an}{c+n} \Leftrightarrow \alpha > \frac{a}{c+n}.$$

By assumption the second inequality is always satisfied. If on the contrary, we have $t \geq a/\alpha$ then $q_{ns}(t) = 0$, and thus $Q(t) \geq a$. Because $an > a/(n+c)$, we obtain the following result,

Proposition 2 *For any IEA, total abatement is greater than would have been obtained without the IEA.*

3 The IEA game

3.1 The game

We showed in the previous section that given a minimal abatement level α and a number of countries committed to abating to a level not less than α, abatement levels and payoffs are uniquely determined. What still has to be known to explain the forging of an IEA is how many countries will commit to a minimal bound on their abatement levels.

To find this, we consider the following two-stage game, with perfect information between each stage. The first stage consists of the ratification stage, where countries choose simultaneously between two actions: R, for ratifying, and NR, for not ratifying. In the second stage, all countries simultaneously choose an abatement level.

Ratification of an IEA by a country can be interpreted here as a conditional commitment by the country to participate in the IEA. The conditionality comes from the presence of a ratification threshold, t, which consists of the minimal number of countries required to ratify the agreement for the IEA to come into force. It is the combination of each country's decision in the first stage and the ratification threshold that determines which restricted abatement game is played in the second stage. If i is a country choosing NR in the first stage then its second stage action is $X_i = [0, \infty)$. Suppose now that country i chooses R in the first stage, and denote by T the set of all countries, including i, that choose R in the first stage. The action set of country i in the second stage is defined as follows:

$$X_i = \begin{cases} X_i^{\text{IEA}} = [\alpha, \infty) & \text{if } \sharp T \geq t, \\ X_i = [0, \infty) & \text{otherwise.} \end{cases}$$

Ratification of the agreement by a country is then a binding decision because if a country has ratified the IEA it cannot choose an abatement level below α. However, this decision is conditional on the fact that at least $t - 1$ other countries also ratify the agreement. We can then say that the agreement enters into force if it has been ratified by at least t countries. The formal definition of an IEA follows:

Definition 1 An International Environmental Agreement (α, t) is a proposal to put a lower bound $\alpha > q_0$ on countries' abatement levels, and a ratification threshold t.

3.2 Stability

Given the present framework, a natural equilibrium concept is subgame perfection. Because for any number of countries participating in the IEA equilibrium in the restricted abatement game always exists and because countries' first-stage action sets are finite, subgame perfect equilibrium always exists.

Traditionally, the literature focuses mostly on the stability concept originally introduced by d'Aspremont et al. (1983) in the literature on cartels. This concept is based on a combination of two stability requirements, respectively internal and external stability. According to this concept, a coalition S is said to be *internally stable* if no country in S has an incentive to leave the coalition, and to be *externally stable* if no country outside S has an incentive to join the coalition. A coalition S is stable if it is both internally and externally stable. Subgame perfection in our framework turns out to be equivalent to internal and external stability if the coalition is defined as the set of countries participating in the IEA. Indeed, in a subgame perfect equilibrium, no ratifying country wants to change its first-stage action by deciding not to ratify. Similarly, a non-ratifying country has no incentive to change its first-stage action by ratifying the IEA. In other words, the choice of ratifying *versus* not ratifying translates into a choice of staying in or staying out of the coalition of countries respecting the IEA's recommendations.

Note, however, that the existence of a subgame perfect equilibrium (and therefore a stable coalition) is not sufficient to ensure that the IEA will come into force: it needs to be ratified by at least t countries, i.e. the ratification threshold requirement. For this reason, a stable coalition (or stable IEA) always refers to a group of countries choosing R in the first stage and being larger than or equal to the ratification threshold t. Stability is then defined as follow,

Definition 2 An IEA (α, t) of size $s \geq t$ is stable if:

- $u(s, 1) > u(s - 1, 0)$ and $u(s + 1, 1) < u(s + 1, 0)$;
- at least t countries fulfill the ratification requirement.

3.3 Stability without ratification threshold

We consider first the case of no ratification threshold included in the protocol. Note that the absence of a ratification clause is equivalent to the case where the ratification threshold is set to 1, since a country's decision to follow the IEA's recommendation is binding, regardless of the decisions of the other countries. We show first that it is not possible to have a stable IEA including all countries.

Proposition 3 *If $t = 1$, for any $\alpha > q_0$, there is no stable IEA with n members.*

Since an IEA induces participants to choose a non-Nash abatement level, there is always one country that will want to deviate. By Proposition 1 the abatement level chosen by non-deviating participant will be constrained (i.e., $\geq \alpha$), the deviating country will be the only one able to fully adjust its abatement level, and achieve its best-reply. Remark that the result also holds for any number of countries ratifying the IEA. If there are some countries that are not participating in the IEA, the abatement levels of these countries will change if one participating country decides to withdraw from the IEA. In this case, the abatement level chosen by a country i withdrawing from the IEA will be the outcome of a new equilibrium.

Proposition 4 *If $t = 1$, for any $s \in \{1, \dots, n\}$, there is no stable IEA with s countries.*

Without a stabilization mechanism, Proposition 4 states that there is no hope of obtaining a stable IEA. This suggests that an international agreement should not only consist of an abatement level, but also include a mechanism regulating enforceability. This result contrasts with a regular finding in the IEA literature, that there exists a parameter space for at least small coalitions to be stable. The main difference lies in the fact that in our approach cooperation does not allow for yielding extra gains from cartel formation. We assume that countries participating in an IEA do not consider the joint payoff of the coalition but only their own payoff, causing countries to prefer playing according to their best reply. In the next section we show that imposing a ratification threshold greater than 1 is a natural mechanism to ensure the coming into force of a stable IEA.

3.4 Stability with ratification threshold

We now consider the case of a non-trivial ratification threshold, i.e., $t \geq 2$. We first look at conditions on t for the agreement to be stable. To do so, we examine whether the set $Z(\alpha) = \{s \leq n \mid u(s, 1) - u_0 \geq 0\}$ is not empty. This translates in studying conditions for a country to ratify an IEA when it has been ratified by $t - 1$ other countries. It turns out that there is never over-ratification.

Proposition 5 *Suppose that (α, t) is a stable IEA. There is no equilibrium in which more than t countries ratify the IEA.*

This result is a direct consequence of Proposition 4. More precisely, suppose that t' countries ratify the IEA and $t' > t$, where t is the ratification threshold. Consider now a ratifying country. Holding the strategy of the other countries fixed, this country receives a payoff equal to $u(t', 1)$. If it decides not to ratify, it receives a payoff equal to $u(t' - 1, 0)$. According to Proposition 4, we have $u(t', 1) < u(t' - 1, 0)$: whenever there is over-ratification some countries have an incentive to withdraw from the agreement. On the contrary, when there are just t countries ratifying, by withdrawing a member of the t-coalition switches from payoff $u(t, 1)$ to u_0, the Nash equilibrium payoff of the abatement game without IEA. In other words, if the country opts not to ratify, the number of ratifiers will be below the threshold t and no IEA will come into force. The next result gives the necessary and sufficient condition for the minimal abatement level α and ratification threshold t to ensure the existence of a stable IEA.

Proposition 6 *A stable IEA* (α, t) *is self-enforceable if and only if:*

- $\alpha \in (q_0, \overline{\alpha}]$, *where* $\overline{\alpha} = q_0\sqrt{1+c}$;
- $t \in [\underline{t}(\alpha), \overline{t}(\alpha)]$ *where* $\underline{t}(\alpha)$ *and* $\overline{t}(\alpha)$ *are the solution of the following programs,*

$$\underline{t}(\alpha) = arg\min_{t\in\{0,\dots,n\}} u(t, 1) - u_0$$

$$such\ that \quad u(t, 1) - u_0 \geq 0,$$

$$\overline{t}(\alpha) = arg\max_{t\in\{0,\dots,n\}} u(t, 1) - u_0$$

$$such\ that \quad u(t, 1) - u_0 \geq 0.$$

We say that a minimal abatement level α is *admissible* if $\alpha \in (q_0, q_0\sqrt{1+c}]$. If α does not lie within this interval, emission reduction is too high for an IEA to be profitable for its members whatever is t.[15] Similarly, we say that a ratification threshold t is *admissible* if there exists an abatement level α such that $t \in Z(\alpha)$ making the agreement profitable. We deduce that self-enforceability is conditioned by admissibility requirements and the conjunction of admissible abatement level and admissible ratification thresholds let us delineate the set of admissible IEA.

Lemma 2 *If* $t(\alpha)$ *is an admissible threshold, there are* $n!/(n-t)!t!$ *distinct stable IEA that may enter into force.*

It follows that for any $\alpha \in (q_0, \overline{\alpha}]$, it exists a subset of self-enforceable IEA of size $(\overline{t}(\alpha) - \underline{t}(\alpha))!$ We depict the set of self-enforceable treaties in Fig. 1; positive values on each of the curves representing a subset of self-enforcing IEA for a given α.

We deduce that for any $\alpha \in]q_0, \overline{\alpha}]$, a large set of acceptable IEA (α, t) may be enforced. Observe that in Fig. 1, $\overline{\alpha} > \alpha_1 > \alpha_2 > \alpha_3$, the less demanding the IEA in terms of target, the larger the set of admissible cooperative coalition will be. For example, when $\alpha = \alpha_1$, there are $(\overline{t} - \underline{t})!$ IEA (α, t) that are profitable and stable for t countries, with $t \in [\underline{t}(\alpha), \overline{t}(\alpha)]$. The lower the admissible α, the larger the interval $[\underline{t}(\alpha), \overline{t}(\alpha)]$ and the number of admissible agreements there will be. Conversely, at the highest admissible target $\alpha = \overline{\alpha}$, only one coalition made of $t = \overline{t} = \underline{t}$ countries can be enforced. Note that for any $\alpha \in]q_0, \overline{\alpha}[$, there is more than one enforceable treaty, each with a different number of signatory countries. The reverse is true and we observe that there exists $\alpha \in]q_0, \overline{\alpha}]$ such that an IEA (α, t) is enforceable with $t \in [2, n]$. An extreme case is when t is set to n. Then, we obtain the result that any Pareto improvement is attainable in a stable IEA involving *all* countries.[16]

So why do most agreements not fix an abatement level that maximizes the collective welfare by setting a ratification threshold equal to n? First, it can be assumed that requiring

[15]Note that the maximal value of abatement given by Proposition 6 does not, in principle, yield a stable IEA. The reason is that the bound $\overline{\alpha} = q_0\sqrt{1+c}$ may not be sufficiently high to ensure that there is an *integer* such that $u(t, 1) \geq u(0)$—although we know that when considering the trivial abatement level $\alpha = q_0$ there is a stable IEA for any ratification threshold t. For the sake of simplicity, in the remainder of the paper, we consider that $\overline{\alpha}$ is such that there is an integer t such that $u(t, 1) > u(0)$.

[16]By *any* Pareto improvement we mean any choice of α such that all countries choosing the abatement level α makes them better off compared to the Nash equilibrium level q_0.

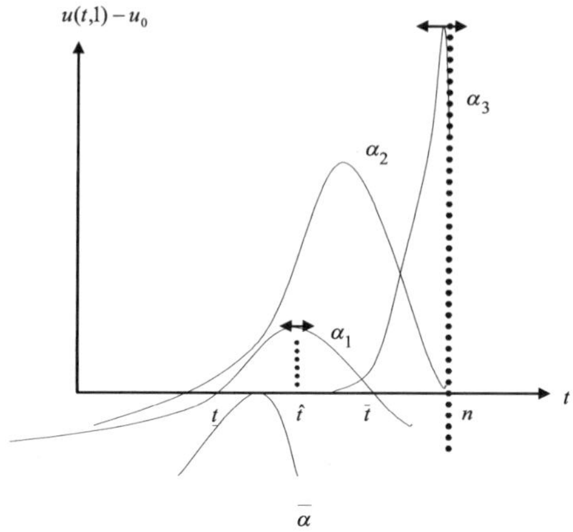

Fig. 1 The set of self-enforceable IEA

unanimous ratification would be too demanding, thereby decreasing the likelihood that the IEA comes into force by giving the power to any country to block the collective effort. Second, Proposition 4 shows that there are strong incentives to free ride. That is, for any abatement level, every country would prefer to have a ratification threshold strictly below n and be one of the countries not ratifying. Third, an environmental goal may be attained without unanimity and by disregarding efficiency; a low threshold may be easier to bargain over. Hence, although a stable IEA involving all countries is theoretically plausible, there are good reasons to explain that negotiation may also lead to a threshold lower than n.

4 Welfare *versus* environmental quality

We start by considering the question of welfare. Let t^* be an admissible ratification threshold. By Proposition 5, any stable IEA contains $s^* = t^*$ countries. From the stability condition of the IEA, we have that,

$$u(t^*, 1) \geq u_0.$$

All countries forming part of the IEA are better off compared to the situation without the IEA. Furthermore, since the benefit of abatement is shared by all countries and only the cost of abatement is country-specific, we have,

$$\forall t = 1, \ldots, n, \quad u(t, 0) > u(t, 1).$$

Combining this two previous observations we then have the following result.

Proposition 7 *A stable IEA is always welfare-improving for all countries, for any admissible ratification threshold.*

We now focus on the interplay between the welfare of signatories to an IEA, the ratification threshold and the equilibrium total abatement level. While an IEA benefits the environment and the welfare of all countries, it turns out that maximizing environmental impact (i.e. minimizing damage) is often not equivalent to maximization of the signatories' welfare.

We characterize first the agreements that maximize the total abatement level. Let α be an admissible minimal abatement level. It follows that for any threshold $t \in [\underline{t}(\alpha), \bar{t}(\alpha)]$ there is an equilibrium in which only s countries ratify, yielding total abatement of $Q(s)$,

$$Q(s) = (n-s)\frac{a - s\alpha}{c + n - s} + s\alpha.$$

To facilitate the analysis, let $g(s)$ be the differentiable mapping from $\mathbb{R}$ to $\mathbb{R}$ such that $g(s) = (n-s)\frac{a-s\alpha}{c+n-s} + s\alpha$.[17] By differentiating we obtain,

$$g'(s) = \frac{c(\alpha(n+c) - a)}{(-c - n + s)^2}.$$

Since $\alpha > a/(n+c)$, $g'(s) > 0$ for any s. We have the following proposition.

Proposition 8 *Let α be an admissible level of abatement. The total abatement is maximized when the ratification threshold is set to $\bar{t}(\alpha)$.*

For a given target α, total abatement is maximised when the participation attain the highest admissible number. Note that in the set of IEA $(\alpha, \bar{t}(\alpha))$ maximizing total abatement, one of them is environmentally preferable. We denote it as $\overline{\text{IEA}}$, and it is such that α is the highest admissible target given $\bar{t} = n$. To see this, remark that the mapping $g(s)$ is always increasing in s and in α, with $\alpha \in (q_0, q_0\sqrt{1+c}]$. Graphically, this corresponds to the IEA $(\alpha_2, \bar{t})$ depicted in Fig. 1.

If the negotiated threshold is chosen so as to maximize the environmental impact, that is, set to $\bar{t}(\alpha)$, the signatories' welfare is very close to their welfare without IEA.[18] It turns out that for a given abatement level α, the ratification threshold that maximizes the welfare of signatories (at equilibrium, and provided that the IEA comes into force) is not necessary the same as the threshold maximizing total abatement.

Proposition 9 *Let α be an admissible level of abatement. The welfare of ratifying countries is maximized when the ratification threshold is set to $\hat{t}(\alpha)$, where*

$$\hat{t}(\alpha) = \begin{cases} \frac{a}{\alpha} & \textit{if } n \geq \frac{c}{\sqrt{1+c}-1}, \\ n & \textit{otherwise}. \end{cases}$$

Welfare maximizing IEA$(\alpha, \hat{t}(\alpha))$ are depicted in Fig. 1. They are at the top of each curve delineating profitability levels. We deduce that for any target α, such that $\alpha =]q_0, \bar{\alpha}]$, there is an optimal ratification threshold $\hat{t}(\alpha)$ such that IEA $(\alpha, \hat{t}(\alpha))$ maximizes the participants' welfare. Among those IEA, one is economically preferable and allows for members to achieve their highest possible welfare. We denote it as $\widehat{\text{IEA}}$ and it is such that $\hat{t}(\alpha) = \frac{a}{\alpha} = n$. Graphically, this corresponds to the IEA(α_3, n) depicted in Fig. 1. Collective effort is optimal and the burden is shared equally among all members.

The combination of Propositions 8 and 9 shows that given a minimal abatement level α, the maximization of the welfare of the signatories is often not at odds with the maximization

[17] Since $Q(s)$ is defined over integers, we need to define a new function which coincides with $Q(s)$, that is differentiable. Another, *albeit* more tedious method, would be to compute $Q(s) - Q(s-1)$.

[18] Recall that by definition, $\bar{t}(\alpha)$ is the highest integer such that $u(\bar{t}(\alpha), 1) \geq u_0$.

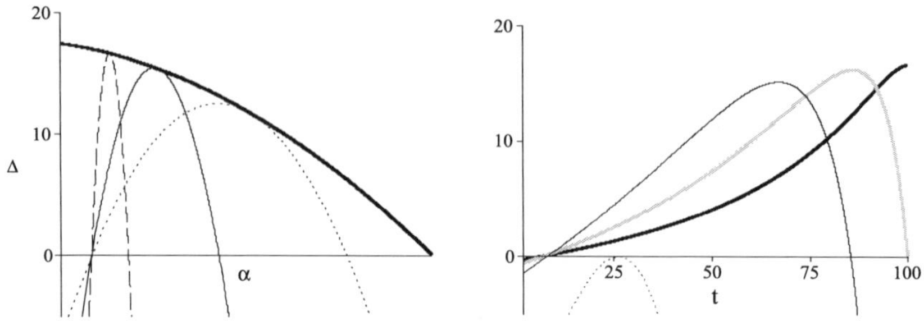

Fig. 2 Example $n = 100$, $a = 35$, $c = 20$

of the total abatement. This is the case whenever $n < \frac{c}{\sqrt{1+c}-1}$. Indeed, from the definition of $\bar{t}(\alpha)$, it can only be the case when $\hat{t}(\alpha) = n$. Whenever n is too large, $\hat{t}(\alpha) \neq n$, the maximal global abatement level cannot be a corollary of the maximization of the signatories' welfare. In fact, having an agreement that maximizes both the welfare and the quality of the environment constitutes an exception rather than a rule. This may be true if and only if α is such that $\hat{t}(\alpha) = \bar{t}(\alpha) = n$.

Note that both $\overline{\text{IEA}}$ and $\widehat{\text{IEA}}$ prescribe a ratification threshold $t = n$. However, the design of these two IEA usually differ in the abatement target, with the first prescribing a higher α than the second. We illustrate it with an example by setting $n = 100$, $a = 35$, $c = 20$.

Again, we examine profitability ($\Delta = u(t, 1) - u_0$) according to the prescriptions (α, t) assigned. The graph at the right-hand side is very similar to Fig. 1. Each curve depicts the relationship between profitability and participation levels for a given minimum abatement level α. Positive values represent the subset of IEA (α, t) that may be enforced. Observe that the higher is α, the smaller is the set of enforceable treaties. The highest admissible abatement target is $\overline{\alpha} = 1.33$ and is depicted in dots. This abatement target leads to a single enforceable IEA made of $n = 28$ countries and performing a total abatement $Q = 29.53$. Observe in bold the two subsets containing $\overline{\text{IEA}}$ and $\widehat{\text{IEA}}$. Pertaining to the grey curve, $\overline{\text{IEA}}$ assigns to the $n = 100$ countries forming it, a minimum abatement $\alpha = 0.406$, yielding a total abatement $Q = 40.68$ and a profitability close to zero. Conversely, pertaining to the black curve, $\widehat{\text{IEA}}$ assigns to the $n = 100$ countries forming it, a minimum abatement $\alpha = \frac{a}{n} = 0.35$ close to q_0 but allowing for yielding high marginal benefits. Those numbers have no tangible significance *per se* but they highlight the discrepancy between economic and environmental objectives, the gap between the two being all the greater when countries have a high incentive to free ride. The left-hand graph is complementary to the right-hand graph. It illustrates the subset of acceptable treaties when participation level t is set. Each curve depicts the relationship between profitability and abatement levels for a given ratification threshold t. For example, the curve depicted in dots represents the relationship between Δ and α when $t = 45$. Note that usually, there are several admissible values of α rendering the entry into force of a treaty made of t countries possible (i.e., $\Delta \geq 0$). Most of those treaties are second-best. Observe that the subsets depicted in thin lines are strictly concave and hump shaped. We deduce that the higher t is, the fewer and the lower the admissible abatement levels α will be. Focus now on the bold curve and call $\widehat{\alpha}$ the abatement level that maximises profitability for any t set. We deduce from Proposition 9 that $\widehat{\alpha}(t) = a/t$ allowing for representing the profitability Δ for any value $\widehat{\alpha}(t)$. This bold curve represents the subset of welfare maximizing IEA, and we observe that since a/α is decreasing in α, the higher the minimal abatement level required by the IEA is, the less likely it will be that the maximization of the signatories' welfare coincides with maximization of total abatement. Yet, whenever $\bar{t}(\alpha) \neq \hat{t}(\alpha)$, we obtain the surprising result that whenever the ratification threshold is set to maximize the signatories' welfare, total abatement in equilibrium is independent of the minimal level α. To see this, note that $\bar{t}(\alpha) \neq \hat{t}(\alpha)$ implies that $\hat{t}(\alpha) = a/\alpha$. Setting the ratification threshold to be equal to $\hat{t}(\alpha)$, total abatement in equilibrium is,

$$\frac{a(n - \hat{t}(\alpha)) + \alpha\hat{t}(\alpha)c}{c + n - \hat{t}(\alpha)} = \frac{a(n - \hat{t}(\alpha) + c)}{c + n - \hat{t}(\alpha)} = a.$$

If c is high enough, $\overline{\alpha}$ is such that there are as many stable and welfare-maximizing IEA as there are integers between a/q_0 and $a/\overline{\alpha}$. The difference between each of them is the distribution of the burden. Since global abatement of the coalition is unchanged, a low α would allow large coalition to be stable while a high α making bringing the entry into force difficult.

5 Conclusion

Far from being damaging to cooperation, ratification procedures are a necessary mechanism to overcome free riding. Since it is always worthwhile for a country to be the outsider to an existing IEA, the ratification threshold rule binds countries at the margin to join the agreement, making the size of the IEA equal to the threshold. While this allows accurate predictions about the number of countries participating in the IEA, it highlights the fragility of such agreements. Any country ratifying an IEA becomes pivotal, meaning that its non-ratification is sufficient to bring down the entire IEA. This is certainly the main reason why this threshold never requires full participation.

We show that a large typology of IEA can come into force, with or without consensus. By cooperation, we assume that countries commit simply to a binding environmental target which does not need to maximize the joint welfare of the cooperative coalition. In this respect, our approach is purely non-cooperative since countries may care only about their own welfare, although that does not prevent the ex-post introduction of efficiency mechanisms, such as joint implementation or tradeable permits, which can only Pareto improve our results. Given this less demanding concept of cooperation, we portray the set of IEA that may enter into force. According to the abatement bound and the threshold rule that have been established, the IEA that comes into force adheres strictly to that bound and that threshold. In other words, abatements equal the bound, and size equals the threshold. It follows that the bound and the threshold negotiated fully characterize the IEA that eventually comes into force. We describe a set of welfare-maximizing IEA as well as a set of environmentally maximizing IEA, and show that they are usually not at odds. We define the design of the most preferable IEA in terms of abatement level and of individual members' welfare. The two prescribe a threshold ensuring complete participation, but generally differ in terms of the targets assigned.

Further research on this topic should include: (1) introducing heterogeneity into the model, and (2) studying the threshold more deeply by focusing on countries' bargaining powers. Considering heterogeneity is not an easy task since it would likely involve the multiplicity of equilibria in our setting. A solution would be to consider the various partitions that could be enforced, but this would render the interpretation of results difficult. Considering bargaining power, as in Caparros et al. (2004), is more feasible and would translate into endogenizing the ratification threshold, as in Carraro et al. (2009). From this perspective, we could examine an IEA design where the ratification rule constitutes a strategic tool for the most powerful countries to impose the participation ratios that best fit their expectations.

Acknowledgements The authors want to thank seminar audiences at University of Adelaid, Autonoma de Barcelona, Melbourne, Marseille, New South Wales, Sydney and Toulouse, as well as helpful comments and suggestions from Guy Meunier, Philippe Quirion, Ludovic Renou, two anonymous referees and the Editor William Shughart II. The usual disclaimer applies.

Appendix

Proof of Proposition 1 Let S be the set of countries restricted by the IEA. Suppose first that (s,α) is such that $a - s\alpha \geq 0$. We first show that for all $i \in S$, the unique equilibrium is such that $q_i = \alpha$. Let $\hat{q} = (\hat{q}_h)_{h\in N}$ be an equilibrium of the abatement game with s countries signing the IEA and suppose there is a country participating to the IEA, say i, such that $\hat{q}_i \neq \alpha$. If follows by Lemma 1 that $\hat{q}_i = BR_i(\hat{q}_{-i})$. Let j denote any country not participating to the IEA, i.e., $\hat{q}_j = BR_j(\hat{q}_{-j})$. Let $\hat{Q}_{-ij} = \sum_{h\neq i,j} \hat{q}_h$. We then have

$$\hat{q}_i = \frac{a - \hat{Q}_{-ij} - \hat{q}_j}{1+c}, \tag{5}$$

$$\hat{q}_j = \frac{a - \hat{Q}_{-ij} - \hat{q}_i}{1+c}. \tag{6}$$

Solving (5) and (6) we then find that $\hat{q}_i = \hat{q}_j$. Repeating this for all countries not participating to the IEA, and because $\hat{q}_i \geq \alpha$, we have

$$\hat{q}_h \geq \alpha, \quad \forall h \in N. \tag{7}$$

Let j be any country not participating to the IEA. Because $\hat{q}_{-j} \geq (n-1)\alpha$, we have

$$\hat{q}_j = \frac{a - \hat{q}_{-j}}{1+c} < \frac{a-(n-1)\alpha}{1+c}. \tag{8}$$

Moreover, observe that $\alpha > a/(n+c)$ implies that

$$\alpha > \frac{a-(n-1)\alpha}{1+c}. \tag{9}$$

Combining (8) and (9) we obtain that country j's best reply is to choose an abatement level strictly lower than α, a contradiction with (7).

Hence, all countries participating to the IEA choose an abatement level equal to α. The abatement game reduces then to a game with $n - s$ players, where the payoffs functions are given by

$$\forall i \in N_-, u_i(q) = \left(a(s\alpha + Q_{ns}) - \frac{1}{2}(s\alpha + Q_{ns})^2\right) - \frac{c}{2}q_i^2, \tag{10}$$

where N_- is the set of countries not participating to the IEA, and $Q_{ns} = \sum_{h\in N_-} q_h$. It is easy to see that this game admits a unique Nash equilibrium in which all countries (not participating to the IEA) would choose an abatement level equal to $q_{ns}(s) = (a - s\alpha)/(c + n - s)$. Because $s\alpha < a$, $q_{ns}(s) \in X_i$, for all $i \in S$.

Suppose now that (s,α) is such that $a - s\alpha < 0$. It follows that $q_{ns}(s) = 0$. We claim that in this case the unique equilibrium is then q^* such that $q_i^* = \alpha$ for all $i \in S$ and $q_i^* = 0$ if

$i \notin S$. Because $q_i \in [\alpha, \infty)$ for $i \in S$, we have $\sum_{i \in S} q_i \geq s\alpha$, and thus $Q_{-j} = \sum_{i \in N \setminus j} q_i \geq s\alpha$ for all $j \notin S$. Recall that the best reply of a country $j \notin S$ is $q_j^* = \max\{(a - Q_{-j})/(1+c); 0\}$, which implies that $q_j^* = 0$. Consider now $i \in S$, and suppose that $q_i^* \neq \alpha$. If follows by Lemma 1 that $q_i^* = BR_i(q_{-i}^*)$. Since for all $j \in S$ we have $q_j \geq \alpha$, country i's original best reply is bounded, i.e., $BR_i(q_{-i}^*) \leq (a - (s-1)\alpha)/(1+c)$. To show that we must have $q_i^* = \alpha$ it suffices to show that $BR_i(q_{-i}^*) \leq \alpha$, and thus that $(a - (s-1)\alpha)/(1+c) < \alpha$. This inequality is equivalent to $\alpha > a/(c+s)$. Since $\alpha > q_0 = a/(c+n)$ and $s \leq n$, the result follows. □

Proof of Proposition 3 Let S be the coalition of countries constrained by the IEA. By Proposition 1 the abatement level chosen by each participant is equal to α. Suppose now that one country, say i, decides to withdraw from the IEA. Again by Proposition 1, the abatement level chosen by the remaining participants will be equal to α, and that of country i will be equal to $(a - (n-1)\alpha)/(1+c)$, i.e., $BR_i((n-1)\alpha)$. Since $(q_h = \alpha)_{h \in N}$ is not a Nash equilibrium of the (unrestricted) abatement game, we have $u_i(BR_i((n-1)\alpha), (n-1)\alpha) > u_i(\alpha, (n-1)\alpha)$. That is, country i is strictly better off withdrawing from the IEA. □

Proof of Proposition 4 If there is no stable IEA with one or more countries then it must be that the IEA game has a unique subgame perfect equilibrium in which all countries choose NR in the first stage. Hence, it suffices to show that for any $\alpha > q_0$ and $s = 1, \ldots, n$, $u(s, 1) < u(s-1, 0)$.

Consider first the case where s and α are such that $s\alpha < a$. From Proposition 1 it follows that $q_{ns}(s) > 0$. Let $\Delta(\alpha, s) = u(s-1, 0) - u(s, 1)$, which yields,

$$\Delta(\alpha, s) = -\frac{c}{2(c+n-s)^2(1+c+n-s)^2}(\alpha c - a + \alpha n)(A + \alpha B), \tag{11}$$

where

$$A = a(n^2 - 2sn - c^2 - c + s^2),$$
$$B = n^3 + 3cn^2 - 4n^2 s + 2n^2 - 6cns + 3cn + 3c^2 n + 5s^2 n$$
$$- 4sn + c^2 + 3s^2 c - 2sc + 2s^2 - 2c^2 s + c^3 - 2s^3.$$

Observe that Δ is a polynomial in α of degree 2, whose roots are

$$r_1 = \frac{a}{c+n}, \qquad r_2 = -\frac{A}{B}.$$

Computing $r_1 - r_2$ we obtain,

$$r_1 - r_2 = \frac{A(c+n) + aB}{(c+n)B} = \frac{(s-n)(s-c-n)(s-c-n-1)}{(c+n)B}.$$

Because $s \leq n$, $r_1 > r_2$ if and only if $B < 0$. Moreover, the coefficient of $\alpha 2$ in Δ is equal to $B(c+n)$, implying that Δ is convex (resp. concave) if $B > 0$ (resp. $B < 0$). Since $\alpha > r_1$, $\Delta(\alpha, s) > 0$ for any s whenever $B > 0$. If $B < 0$ then $r_1 < r_2$ and $\Delta(\alpha, s) > 0$ for any $\alpha \in (r_1, r_2)$, which completes the proof of the proposition.

Suppose now that $s\alpha > a$, and let i be a country that does not ratify the IEA. Let $Q(s) = (n-s)q_{ns}(s) + (s-1)\alpha$ and $Q(n-s) = (n-s-1)q_{ns}(s-1) + (s-1)\alpha$. From Proposition 1 $q_s(s) = q_s(s-1) = \alpha$, and $q_{ns}(s-1) = 0$ or $q_{ns}(s-1) = (a - s\alpha)/(c+n-s)$.

Suppose first that $q_{ns}(s-1)=0$. Hence, it must be the case that $(s-1)\alpha \geq a$, which implies that $Q(s) > q(s-1) \geq \alpha$. Therefore, we have $aQ(s-1) - \frac{1}{2}Q(s-1)^2 > aQ(s) - \frac{1}{2}Q(s)^2$. Adding the cost of abatement, $q_{ns}(s-1)$ and α respectively, we obtain $aQ(s-1) - \frac{1}{2}Q(s-1)^2 > aQ(s) - \frac{1}{2}Q(s)^2 - \frac{c}{2}\alpha^2$, which is tantamount to $u(s-1,0) > u(s,1)$, the desired result.

Consider now the case when $q_{ns}(s-1) > 0$. It follows that $(s-1)\alpha < a$, and thus $s \in [a/\alpha, a/\alpha+1]$. We now show that $u(s-1,0)$ is minimized when $s = a/\alpha$. First, observe that the total abatement $Q(s-1) < a$ if and only if $(s-1)\alpha < a$. Hence, $Q(s-1)$ is increasing in s and takes its lowest value when $s = a/\alpha$. The abatement of a country not ratifying $q_{ns}(s-1)$ is also a decreasing function of s, which implies that $c/2q_{ns}(s-1)^2$ is maximized when $s-1 = a/\alpha - 1$. Therefore, $u(s-1,0) \geq u(a/\alpha, 0)$. Furthermore, $s > a/\alpha$ implies that $Q(s) \geq a$, and thus we have $u(s,1) \leq u(\alpha,1) = \frac{1}{2}(a^2-\alpha^2)$. Computing $\Delta = u(s-1,0) - u(s,1)$, we obtain

$$\Delta \geq \frac{c\alpha^2(a^2 - \alpha(2a(n+c+1) - \alpha((c+n)^2+c+2n)))}{2(a-c\alpha-n\alpha-\alpha)^2}.$$

To show that $\Delta > 0$, it suffices to show that the following holds true,

$$\alpha > \frac{2a(n+c+1)}{(c+n)^2+c+2n}. \tag{12}$$

Consider now the following inequality,

$$\frac{a}{c+n} > \frac{2a(n+c+1)}{(c+n)^2+c+2n}. \tag{13}$$

Because $\alpha > a/(c+n)$, (13) being true implies that (12) is true as well. Now, it is readily verified that (13) always holds true (the above inequality simplifies to $n > 0$), and thus we have $u(s-1,0) > u(s,1)$, which completes the proof. □

Proof of Proposition 6 Let $\Delta = u(s,1) - u_0$. Suppose first that $q_{ns}(s) \neq 0$, i.e., non-ratifying countries are not constrained. Since $\alpha > q_0$, it is convenient to pose $\alpha = \beta q_0$ with β being a parameter strictly greater than 1. Simplifying we then obtain:

$$\Delta = -\frac{a^2c}{2(c+n-s)^2(n+c)^2}(-X+Y+\beta(\beta X - Y)) \tag{14}$$

where

$$X = 2nc + cs^2 - 2sc + n^2 + s^2 - 2ns + c^2,$$
$$Y = 2c^2s + 2cns.$$

Consider now the function $f(s): \mathbb{R} \to \mathbb{R}$ with $f(s) = -X + Y + \beta(\beta X - Y)$. Observe that the sign of $f(s)$ is the opposite of the sign of Δ. Furthermore, the mapping $f(s)$ is a polynomial of degree 2, such that the coefficient of s_2 is equal to $(1+c)(\beta^2-1)$, which implies that $\Delta > 0$ whenever $s \in (s_1, s_2)$ where s_1 and s_2 are the two roots of $f(s)$,

$$s_1 = \frac{(1+c+\beta-\sqrt{c+c^2-c\beta^2})(n+c)}{\beta c+\beta+c+1}, \tag{15}$$

$$s_2 = \frac{(1+c+\beta+\sqrt{c+c^2-c\beta^2})(n+c)}{\beta c+\beta+c+1}, \tag{16}$$

$s_1, s_2 \in \mathbb{R}$ if and only if $\beta \leq \sqrt{1+c}$, the desired result.

Suppose now that $q_{ns}(s) = 0$, i.e., $s > a/\alpha$. In this case the total abatement is equal to a if s countries ratify, which gives the following payoff,

$$u(s, 1) = \frac{1}{2}(a^2 - \alpha^2).$$

Substracting u_0 to $u(s, 1)$ we obtain

$$\Delta = \frac{c(ca^2 - \alpha^2 n^2 - 2c\alpha^2 n - c^2\alpha^2 + a^2)}{2(n+c)^2},$$

Δ is positive if and only if $\alpha \in (-q_0\sqrt{1+c}, q_0\sqrt{1+c})$, which completes the proof of the proposition. □

Proof of Proposition 9 Consider the difference $u(s, 1) - u_0$ given by (14). Abusing notation, let $d(s)$ be the differentiable mapping from $[0, n]$ to $\mathbb{R}$ such that $d(s) = -\frac{a^2 c}{2(c+n-s)^2(n+c)^2} f(s)$, where $f(s)$ is the polynomial function defined in the proof of Proposition 6. Differentiating by s we obtain,

$$d'(s) = -\frac{c^2(a - n\alpha - c\alpha)(a - s\alpha)}{(c+n-s)^3},$$

$d'(s)$ is thus strictly positive if and only if $s < \frac{a}{\alpha}$. Given that $d(s)$ is continuous, the function reaches its maximum when $s = \frac{a}{\alpha}$.

It remains to show that there exists several admissible values of α such that $a/\alpha \leq n$ or, equivalently, $\alpha \geq a/n$. From Proposition 6, α is admissible only if $\alpha \leq \frac{a}{c+n}\sqrt{1+c}$. Hence, if suffices to show that

$$\frac{a}{c+n}\sqrt{1+c} \geq a/n.$$

Simplifying we obtain the following condition,

$$n \geq \frac{c}{\sqrt{1+c} - 1}. \tag{17}$$

Hence, whenever (17) holds true, the threshold maximizing the welfare is equal to $\frac{a}{\alpha}$. Otherwise, the threshold maximizing the welfare is equal to n. □

References

d'Aspremont, C., Jacquemin, A., Gabszewicz, J., & Weymark, J. (1983). On the stability of collusive price leadership. *Canadian Journal of Economics*, *16*, 17–25.

Bade, S., Haeringer, G., & Renou, L. (2009). More strategies, more Nash equilibria. *Journal of Economic Theory*, *144*, 1817–1831.

Bagnoli, M., & Lipman, B. L. (1989). Provision of public goods: fully implementing the core through private contributions. *Review of Economic Studies*, *56*, 583–601.

Bagnoli, M., & Lipman, B. L. (1992). Private provision of public goods can be efficient. *Public Choice*, *74*, 59–78.

Barrett, S. (1994). Self-enforcing international environmental agreements. *Oxford Economic Papers*, *46*, 878–894.

Barrett, S. (2002). Consensus treaties. *Journal of Institutional and Theoretical Economics*, *158*, 529–547.

Barrett, S. (2003). *Environment and statecraft: the strategy of environmental treaty making*. New York: Oxford University Press.

Barrett, S. (2007). *Why cooperate? The incentive to supply global public goods*. Oxford: Oxford University Press.

Benedick, R. E. (1998). *Ozone diplomacy: new directions in safeguarding the planet* (2nd edn.) Cambridge: Harvard University Press.

Black, J., Levi, M. D., & de Meza, D. (1993). Creating a good atmosphere: minimum participation for tackling the greenhouse effect. *Economica*, *60*, 281–293.

Böhringer, C., & Vogt, C. (2004). The dismantling of breakthrough: The Kyoto protocol as symbolic policy. *European Journal of Political Economy*, *20*, 597–618.

Caparros, A., Péreau, J. C., & Tazdaït, T. (2004). North-South climate change negotiations: a sequential game with asymmetric information. *Public Choice*, *121*, 455–480.

Carraro, C., & Siniscalco, D. (1993). Strategies for the international protection of the environment. *Journal of Public Economics*, *52*, 309–328.

Carraro, C., Marchiori, C., & Oreffice, S. (2009). Endogenous minimum participation in international environmental treaties. *Environmental & Resource Economics*, *42*, 411–425.

Cornes, R., & Sandler, T. (1996). *The theory of externalities, public goods, and club goods* (2nd ed.). Cambridge: Cambridge Univ. Press.

Currarini, S., & Tulkens, H. (2004). Stable international agreements on transfontier pollution with ratification constraints. In C. Carraro & V. Fragnelli (Eds.), *Game practice and the environment* (pp. 9–36). Cheltenham: Edward Elgar Publishing

Diamantoudi, E., & Sartzetakis, E. (2006). Stable international environmental agreements: an analytical approach. *Journal of Public Economic Theory*, *8*(2), 247–263.

Dietrich, W. F. (1995). *The challenge of selecting environmental policy goals: case studies regarding the use of critical levels*, CSIA Discussion Paper 95-05, Kennedy School of Government, Harvard University.

Dixit, A., & Olson, M. (2000). Does voluntary participation undermine the Coase theorem? *Journal of Public Economics*, *76*(3), 309–335.

Finus, M. (2001). *Game theory and international environmental cooperation*. Cheltenham: Edward Elgar.

Finus, M. (2008). Game theoretic research on the design of international environmental agreements: insights, critical remarks and future challenges. *International Review of Environmental and Resource Economics*, *2*, 1–39.

Finus, M., & Maus, S. (2008). Modesty may pay! *Journal of Public Economic Theory*, *10*(5), 801–826.

Frey, B. S. (1997). The public choice of international organizations. In D. Mueller (Ed.), *Perspectives on public choice* (pp. 106–123). Cambridge: Cambridge University Press.

Grubb, M., Vrolijk, C., & Brack, D. (1999). *The Kyoto Protocol, a guide and assessment*. London: Royal Institute of International Affairs.

Harstad, B. (2006). Flexible integration? Mandatory and minimum participation rules. *The Scandinavian Journal of Economics*, *108*(4), 683–702.

Hoel, M. (1992). International environment conventions: the case of uniform reductions of emissions. *Environmental & Resource Economics*, *2*, 141–159.

Kohnz, S. (2005). *Ratification quotas in international contracts: an example of emission reduction*. Mimeo, Yale University.

McBride, M. (2006). Discrete public goods under threshold uncertainty. *Journal of Public Economics*, *90*(6–7), 1181–1199.

Nitzan, S., & Romano, R. (1990). Private provision of a discrete public good with uncertain cost. *Journal of Public Economics*, *42*, 357–370.

Nordhaus, W. (2007). *The challenge of global warming, economics model and environmental policy in the DICE 2007 Model*. Mimeo, Yale University.

Okada, O. (1993). The possibility of cooperation in an n-person Prisoners' dilemma with institutional arrangements. *Public Choice*, *77*(3), 629–656.

Palfrey, T., & Rosenthal, H. (1984). Participation and the provision of discrete public goods: a strategic analysis. *Journal of Public Economics*, *24*, 171–193.

Patt, A. (1999). Separating analysis from politics: acid rain in Europe. *Review of Policy Research*, *16*(34), 104–137.

Rutz, S. (2002). *International environmental agreements: much ado about nothing*? Ph.D thesis. Aachen: Shaker Verlag.
Sandler, T., & Hartley, K. (2001). Economics of alliances: the lessons for collective action. *Journal of Economic Literature*, *39*, 869–896.
Suleiman, R. (1997). Provision of step-level public goods under uncertainty: a theoretical analysis. *Rationality and Society*, *9*, 163–187.
Vaubel, R. (1986). A public choice approach to international organization. *Public Choice*, *51*, 39–58.
Vaubel, R., Dreher, A., & Soylu, U. (2007). Staff growth in international organizations: a principal-agent problem? An empirical analysis. *Public Choice*, *133*, 275–295.
Weikard, H. P., Wangler, L., & Freytag, A. (2009). *Minimum participation rules with heterogeneous countries*. Jena Economic Research Papers in Economics 2009-077.

[21]

Public Choice **121:** 455–480, 2004.

North-South climate change negotiations: A sequential game with asymmetric information *

A. CAPARRÓS[1], J.-C. PÉREAU[2] & T. TAZDAÏT[3]
[1]*Department of Economics, Institute of Economics and Geography (IEG), Spanish Council for Scientific Research (CSIC), 28006 Madrid, Spain; E-mail: acaparros@ieg.csic.es;* [2]*OEP, University of Marne-la-Vallée, Cité Descartes, Marne la Vallée Cedex 2, France; E-mail: pereau@univ-mlv.fr;* [3]*CNRS-EHESS-CIRED Jardin Tropical, 94736 Nogent sur Marne Cedex, France; E-mail: tazdait@centre-cired.fr*

Accepted 12 September 2003

Abstract. This article determines the conditions under which the Southern countries should act together, or separately, while negotiating with the North about climate change policy and about the conditions for future Southern engagement. The paper models the international negotiations with complete and with asymmetric information in a dynamic framework. Results show that, depending on their characteristics, the different players can obtain benefits delaying the moment of the agreement.

1. Introduction

In 1992 virtually all the countries of the world signed, and ratified, the United Nations Framework Convention on Climate Change whose ultimate objective is to achieve the 'stabilization of greenhouse gas concentration in the atmosphere at a level that would prevent dangerous anthropogenic inference with the climate system'. Five years later, the Kyoto Protocol was signed, engaging the countries included in Annex I (OECD and economies in transition) to reduce their overall emission by five per cent in 2012 compared to 1990. This agreement was further discussed and refined, and in 2001, almost all the countries in the world (with the only significant absence of the United States of America) agreed on the Marrakech Accords, which are expected to enable enforcement of the Kyoto Protocol (this requires that countries accounting for 55% of Annex I carbon dioxide emissions ratify it[1]).

Although it is fair to acknowledge that the industrialized countries (IC) are the main historic, and present, causers of anthropogenic climate change,

* We wish to thank comments and suggestions by an anonymous referee and by the participants at the EAERE2003 conference in Bilbao (Spain) and the AFSE2003 conference in Paris (France). Alejandro Caparrós acknowledges the financial support from the European Commission (EEP Research Training Network).

the solution of the problem will need the implication of the developing countries (DC) in a more comprehensive way than the one proposed in the Kyoto Protocol (based on projects). This is especially true if we assume that the DC will be able to develop as fast as they need. In fact, the incorporation of the DC is the main issue on the negotiation agenda and the absence of binding targets for the DC was one of the main arguments used by President Bush to reject the Kyoto Protocol.

In the next negotiation rounds we shall see the North (the Annex-I countries which have ratified the Kyoto Protocol) negotiating with the South (Non-Annex I countries) the most convenient way to bring them on board. The North, led by the European Union, has recognized its historical responsibility (in official texts but also accepting to be the first one to reduce) and should be ready to compensate the South in some way for accepting limitations to their urgently needed development. This compensation could take the form of monetary transfers but could also be based on other type of advantages.

Until now the DC have talked in the international climate change arena with a single voice, the so-called "G77 and China" group. Nevertheless, until now, the DC, although Parties of the negotiations, have not been affected by binding reductions. Thus, they have been less involved in the "hard part" of the negotiations. In the new context, the Southern countries have to decide whether they continue to talk with a single voice or whether they split up into two or more groups (the G77 and China is a highly heterogeneous group, including small countries highly vulnerable to climate change and countries like Saudi Arabia strongly interested in oil exports). Talking together gives bargaining power, but under some circumstances, it may be more convenient to split up to take advantage of internal differences. Rotillon, Tazdaït and Zeghni (1996) analysed this issue with complete information and particular benefit functions. Nevertheless, official statistics in the South are rare and incomplete and real knowledge about the abatement costs and options in the South are not at all available to the North. On the contrary, due to the high amount of existing studies, and due to the implementation of the Kyoto Protocol, the South will have almost complete information about the cost and benefit functions of the North. Hence, we discuss this issue in the context of asymmetric information and with general functional forms.

Finally, if the Southern countries decide to split up into two groups (e.g.), these groups can be *complementary* or *substitutable*. We call them *complementary* if the North needs to sign an agreement with both groups to be effective. *Substitutable* implies that an agreement with one of the groups is enough to produce positive benefits. At first glance, the Southern groups should be considered substitutable in the case of climate change since if one group, e.g. representing 50% of the Southern emissions, accepts to reduce its future emissions (or to increase them less) this will have a positive effect on climate change. Nevertheless, the opposite situation is also possible. As described above, in 1997 the Kyoto Protocol was signed, but stating that it will only come into force if countries representing 55% of Annex I emissions ratify it. This gave Russia and other countries a lot of bargaining power since they became complementary, especially after the withdrawal of the United States. This situation might be repeated if an initial Protocol states that the DC will only be obliged to reduce their emission patterns if DC representing 55% (e.g.) of Non-Annex I emissions ratify it.[2] Hence, we discuss in our model the circumstances under which the South is interested in pushing for such a situation.

Another relevant characteristic of international negotiations, and in particular of climate change negotiations, is that they take a great deal of time to conclude. In principle, if negotiators were rational, we should expect them to finish in one single negotiation round. Nevertheless, the delay in the negotiations can be used, and has been used in the past, to take advantage and to gain better knowledge of the characteristics of the opponent. Therefore, we model the negotiations in a dynamic framework to analyse the circumstances under which the Parties benefit from delaying the agreement.

Game theory literature on cooperation in the context of climate change has followed two approaches,[3] although both of them assimilate cooperation and coordination of environmental policies. On the one hand, we have what Tulkens (1998) called the thesis of the "small stable coalitions" (Carraro and Siniscalco, 1993; Barret, 1994; Péreau and Tazdaït, 2001; among others). On the other hand, we have the thesis of the "grand stable coalition" (Chander and Tulkens, 1997; Egteren and Tang, 1997; Funaki and Yamato, 1999; Helm, 2001; among others). According to the first approach, only a small number of countries will accept to sign an agreement (the stability concept used comes from the analysis of cartels in industrial economics). In this framework, a coalition including all countries is not sustainable since the benefits arising from cooperation are a public good (distributed to all countries), while the costs are borne only by the countries forming the coalition. On the contrary, in the second approach all countries are interested in cooperation (taken individually or in sub-groups). According to Chander and Tulkens (1995, 1997) an agreement should not only be based on pareto optimality to be effective, but should also prevent any sub-group of countries to be interested in secession. Hence, the objective is to set up a burden-sharing rule able to favour the cooperation of all. However, these two approaches have in common that the negotiating process has been completely ignored, as well as significant characteristics like information asymmetry, countries heterogeneity,[4] or even the possibilities of renegotiation. In the non-cooperative games evoked by Carraro and Siniscalco, negotiation is modelled as simultaneous discourse, not leaving any space to dialogue (i.e. the possibility of carrying out counter-propositions). In the models of endogenous coalition formation (Chander and Tulkens, 1995; 1997), cooperation is the result of the direct intervention of a third party: a supranational authority acting as a central planner. This central planner is assumed to be good-willed, to defend the global interests and to have enforcement power; somehow irrealistic assumptions in the context of real climate change negotiations.

To overcome these limitations, we analyse future climate change negotiations in a dynamic framework, with asymmetric information, with heterogeneous coalitions and explicitly including the possibility to make counter-propositions (remark that bargaining games are strictly competitive games, since this may be useful for the interpretation of our results). In technical terms, we develop a sequential game with asymmetric information in a bilateral, as well as in a multilateral framework. Therefore, we deal with hidden knowledge issues and signalling theory; and look for pooling and separating equilibria. This model is set out using general functional forms and could therefore be applied to most of the current and future discussions between the North and the South. It applies, more generally, to all circumstances where negotiations take place between, on the one hand, one player whose characteristics are well known and who is ready to pay the other players to obtain something, and, on the other hand, one or two players whose characteristics are not known.

The article is organised as follows. In Section 2 we present the general model. In Section 3, we analyse the outcomes of the game with complete information. The study of the negotiation with incomplete information is proposed in Section 4. Finally, Section 5 highlights the main conclusions.

2. The model

We analyse a bargaining game between the North (Annex I countries which have ratified the Kyoto Protocol), which is assumed to be neutral to risk, and a group of developing countries (e.g. the G77 and China). The negotiation is about the value of the transfers that the North (N) will grant to countries of the South (S) willing to cooperate and commit themselves. The Southern countries are assumed to be identical and C in number. To assume identical countries could be seen as a limitation (see above); however, it is important to recall that we are assuming that all the countries of the Southern coalition talk with a single voice. Therefore, there is no problem in considering that a country like China has more votes (parts of C) in the internal vote (in fact, C could also refer to the amount of emissions as in the example described in the introduction). We call $0 < \delta_i < 1$ the discount rate of agent i, i = N, S . The monetary transfer (or other type of advantage) granted by the North is noted τ. The benefit for the North is the difference between the abatement reached ($B(C) = B$), and the cost of the transfers to the C countries of the Southern coalition (τC):

$$\Pi(\tau, t) = \delta_N^t[B - \tau C] \tag{1}$$

The utility function of the Southern coalition is:

$$U(\tau, t) = \delta_S^t[(\tau - b)C] \tag{2}$$

We assume that the minimum amount of transfers asked for by the Southern coalition (b) is decided in an internal vote among the coalition (as it occurs in the international negotiations, where G77 and China have private meetings where they decide their position). With this amount of transfers b the utility is zero. We also define the "reservation transfer", as the maximum transfer that the North can accept. This amount is given by the benefits that the North obtains from the reductions in the South (we assume that the negotiation is about the amount of transfers for a given level of Southern effort). The maximum transfer level is B/C, and the benefit for the North would be zero. That is, the negotiation area is bounded by b and B/C. Thus, for $B/C < b$, the agreement set is empty. To make the agreement possible, we assume $B > Cb$. All the elements of the negotiation are known, except the minimum amount of transfers that is acceptable to the Southern group. This amount can have two possible values b^- and b^+ (i.e. $b \in \{b^-, b^+\}$), where $b^- < b^+ < B/C$. The Northern negotiator has an *a priori* probability function for these values: $p_N(b^+) = p^+$ and $p_N(b^-) = p^- = 1 - p^+$.

The players propose offers alternatively, which refer to the amount of transfers for a given effort. The negotiation goes on with no time limit. We call the game where the North (respectively the South) makes the first offer J_N^∞ (J_S^∞). Let us analyse the structure of the J_N^∞ game.

The North proposes an amount of transfers τ_N. The South accepts (Yes) or refuses (No) the proposition. If the South accepts, an agreement is signed in the first period. The objective functions of the agents are:

$$\Pi(\tau_N, 0) = B - \tau_N C \quad \text{and} \quad U(\tau_N, 0) = (\tau_N - b)C$$

If the South rejects the offer, the negotiation continues with a counter-offer from the South. If the North accepts this new offer, an agreement is signed in the second period, with an amount of transfers τ_S. Thus:

$$\Pi(\tau_S, 1) = \delta_N[B - \tau_S C] \quad \text{and} \quad U(\tau_S, 1) = \delta_S[(\tau_S - b)C]$$

460

Otherwise, the negotiation starts again with a Northern offer τ_N. In the case of an agreement:

$$\Pi(\tau_N, 2) = \delta_N^2[B - \tau_N C] \text{ and } U(\tau_N, 2) = \delta_S^2[(\tau_N - b)C]$$

And so on. In the case of perpetual disagreement we get:

$$\Pi(\tau_0, \infty) < \Pi(\tau, t) \text{ and } U(\tau_0, \infty) < U(\tau, t) \text{ for the final t.}$$

We assume that both agents have a given preference for the present. Hence, for two moments in time t and s, with $t < s$, we have:

$$\Pi(\tau, t) \geq \Pi(\tau, s) \text{ and } U(\tau, t) \geq U(\tau, s)$$

with $\tau = b^-, b^+$. The inequality will be strict if $\Pi > 0$ (respectively $U > 0$).

Since the acceptation of an offer closes the negotiation, a history in time t, h^t, describes a possible evolution of the negotiations from the beginning at date 0 until date t (including the behaviour in t). More precisely, it gives the sequence of propositions leading to a disagreement, if there were propositions, knowing that the North revises its beliefs taking into account the history. Call H the set of all the possible histories and H^t the set of all the possible histories with a duration t. A pure strategy for i is a function f_i where:

$$f_i^t : H_i^{t-1} \rightarrow [b^-, B/C] \cup \{Yes, No\}$$

specifies after each history h_i^{t-1} ($\in H_i^{t-1}$) the behaviour of i at time t. In particular, it tells us if after h_i^{t-1}, i will propose a new offer, or refuse or accept the offer of j. Hence, a history is formed either by a proposition between b^- and B/C, or by a refusal or an acceptation. This last choice can only appear at t. The result of the negotiation will also depend upon the beliefs of the North. The resolution of the dynamic game with incomplete information is based on the concept of sequential equilibrium (Kreps and Wilson, 1982). By definition, a set of strategies and a family of beliefs form a sequential equilibrium if the following three conditions hold: (i) the strategies of both players are optimal in all the points of the game, given the beliefs of the North, (ii) For any information set reached the beliefs of the North are given by the Bayes' law, and (iii) the beliefs must be the limit of, at least, a sequence of reasonable beliefs (i.e. for a sequence of possible deviation).[5]

3. Complete information

3.1. *Bilateral negotiation*

Following a similar resolution method to the one proposed by Rubinstein (1982), we calculate the unique perfect equilibrium with complete information to which both players agree immediately in the first negotiation round (see Hovi (2001) for an alternative application of this methodology to the Clean Development Mechanism). The minimum amount of transfers b requested by S is supposed to be common knowledge. Let $\tau^*(N, S)$ be the equilibrium transfer when the North opens the negotiations and $\tau^*(S, N)$ the equilibrium transfer when it was proposed by the South.[6] We obtain:

$$\tau^*(N, S) = \frac{bC(1 - \delta_S) + \delta_S(1 - \delta_N)B}{(1 - \delta_N\delta_S)C} \tag{3}$$

and

$$\tau^*(S, N) = \frac{B(1 - \delta_N) + bC\delta_N(1 - \delta_S)}{(1 - \delta_N\delta_S)C} \tag{4}$$

Concerning the position of the players in the game, we can formulate the following corollary. It says that N (respectively S) prefers to play the game when it opens the negotiations (J_N^∞, respectively J_S^∞), instead of the opposite when it responds (J_S^∞, respectively J_N^∞).

Corollary 1. If J_N^∞ and J_S^∞ are two games with perfect information, being b common knowledge, we obtain the following two expressions:

$$J_N^\infty \succ_N J_S^\infty$$
$$J_S^\infty \succ_S J_N^\infty$$

Proof. We can see that:

$$\tau^*(S, N) > \tau^*(N, S) \Longleftrightarrow B > Cb$$

Since by definition $B > Cb$, the inequality always holds.

S wishes to obtain the highest amount of transfers and prefers the situation where $\tau = \tau^*(S, N)$, i.e. the amount of transfers that it obtains in the game where it talks first. On the contrary, N wishes to obtain the lowest amount of transfers and prefers the situation where $\tau = \tau^*(N, S)$, i.e. the amount of transfers that it obtains in the game where it talks first. □

462

3.2. *Multilateral negotiation*

Let us now consider the case where the South is organised in two separate groups, S_i, $i = 1, 2$, representing C_i countries, with $C_1 + C_2 = C$. The benefit obtained by the North from the abatement of group i is noted B_i, $i = 1, 2$. The aim is to study the union and disunity behaviours of the groups in complete information according to their degree of complementarity and substitutability. We assume that within each group, the members vote to determine their minimum level of transfer claims. Let b_i $(i = 1, 2)$ be the transfer requested by the group i (resulting from the internal vote within each group) and b the transfer requested by the groups while acting together (resulting from the internal vote within the group formed by both sub-groups).

The bargaining process has the same characteristics as previously described. The game consists in alternative offers and the North makes the initial ones (one offer to each group). If one group accepts the offer while the other refuses, the negotiation with the former group is finished; going on with the latter group. If both groups simultaneously refuse (accept), the negotiation keeps going on (ends) with both groups. This process goes on until an agreement is reached.

Let $0 < \delta_S < 1$ be the fixed discount rate of the groups (joined or separated). We have $\forall i = 1, 2$:

$$\tau^*(N, S_i) = \frac{b_i C_i (1 - \delta_S) + \delta_S (1 - \delta_N) B_i}{(1 - \delta_N \delta_S) C_i} \tag{5}$$

Both groups will accept to form a unique alliance if, by their cohesion, they reach a more satisfactory level than they would have reached acting separately:

$$U_1 + U_2 < U \tag{6}$$

where U_i is the utility of the groups $(\forall i = 1, 2)$. Condition (6) involves:

$$bC - b_1 C_1 - b_2 C_2 < B - B_1 - B_2 \tag{7}$$

Let us define $V_i = (B - bC) - (B_j - b_j C_j)$, $i \neq j$, the maximum contribution of group i to the benefit of the North. Condition (7) is rewritten as follows:

$$V_1 + V_2 > B - bC \tag{8}$$

The two Southern groups should act together when the North benefits from negotiating separately with each of them. This only happens if the distribution rule among the members of the Southern group allows them to obtain more together than what they would obtain separately. The North is only motivated to favour two bilateral negotiations if it intends to propose a low level of transfers. In this configuration, by standing separately, the bargaining power of the South is lower, as their threat to refuse any agreement is less credible. If one group tries to use that threat, the North will propose a slightly more interesting offer to the other group. Being confronted with two distinct groups, it could propose a low level of transfers and take advantage of the disunity and lead both Southern groups to accept its offer. To foil that strategy, the two groups have to act together.

If condition (8) is not respected, the Southern groups should participate separately in the negotiations. In this case, the North is inclined to propose a high level of transfers from the beginning of the negotiation. Since the North proposes a high level of transfers, its intention is to conclude an agreement as soon as possible . Each group is therefore incited to take advantage of that situation by accepting the overbid game. The separation of the groups accentuates this behaviour and increases their bargaining power. Hence, we obtain the following proposition:

Proposition 1.

1. When $V_1 + V_2 > B - bC$, a unique subgame perfect equilibrium exists in which the groups act jointly.
2. When $V_1 + V_2 < B - bC$, a unique subgame perfect equilibrium exists in which the groups act separately.

The outcome of the negotiations depends widely on the degree of commitment of the North. It defines the whole process of negotiations by making the first offer. If the Southern countries anticipate a moderate commitment, they will be united during the negotiations, since that will give them significant bargaining power. Conversely, if they anticipate a high commitment, they will be dissociated into two separate groups.

We will now analyse the behaviour of the Southern groups according to their degree of substitutability and complementarity. The groups are said complementary when the two groups have to sign the agreement so that benefits can arise (see introduction). They are called substitutable when the agreement can be reached with only one of the groups. If we accept perfect complementarity, an agreement signed by only one group has no impact. Therefore, $\forall C_i$, $i = 1, 2$, we have $V_1 = V_2 = 0$. Hence:

$$V_1 + V_2 < B - bC \tag{9}$$

Since relation (9) always holds for the complementary groups, we assume that they will act separately during negotiations. In this way both can unilaterally compromise the negotiations in case of disagreement. On the other hand, when the groups are substitutable, they had better act together to preserve their bargaining power. The following expression is found inverting equation (6):

$$V_1 + V_2 > B - bC \tag{10}$$

According to the North, the inequality (10) means that the maximum additional contribution to the Northern payoff of the two united groups is low. Thus, an agreement with one of the groups is enough. We obtain the following proposition:

Proposition 2.

1. When the two groups are perfectly complementary, they will act separately.
2. When the two groups are perfectly substitutable, they will act jointly.

When the Southern groups are complementary; the North would like to negotiate with the united groups. Conversely, the Southern groups will adopt the asymmetric behaviour. Being complementary, they are aware of being the driving forces and that incites them to become two distinct groups. Such behaviour gives them bargaining power and allows them to request a higher amount of transfers.

When the groups are substitutable; the North needs to conclude an agreement with only one of them. Conversely, the Southern groups are incited to act united to preserve their bargaining power. Nevertheless, that alliance is fragile since one of the groups can betray its partner accepting a slightly lower level of transfers, which guarantees the signing of an agreement.

4. One-sided incomplete information

4.1. *Bilateral negotiation*

We now analyse the game in which the North opens the negotiations and does not know the amount of transfers requested by the Southern countries. We might justify this situation as follows: the North has complete, easily accessible statistics, as well as a high amount of studies estimating the costs of climate change policy in the North. In addition, the North (formed by the IC which have ratified the Kyoto Protocol) will have started its climate change policy and revealed its reduction potentials. On the contrary, in most of the Southern countries statistics are incomplete and the knowledge about the cost of alternative climate change policies is not well known. Therefore, the South has better information about the amount of transfers that it can request compared to the information available to the North about the minimum level of transfers (or other benefits) that the South would accept.

To simplify, we assume that the amount of transfers requested by the Southern group can take two values: b^- and b^+, with $b^- < b^+ < B/C$. Either it is low, and we will call the group type b^-, or it is high and the group will be of type b^+. According to Harsanyi (1967), this game is formalised by introducing a new player, Nature, which selects the group b^+ (respectively b^-) as the Northern opponent with probability p^+ (respectively $1 - p^+$). The distribution of probability is common knowledge.

Since we use the concept of sequential equilibrium, we have to specify the form of the beliefs of the North. We will define pessimistic beliefs[7] (Rubinstein, 1988). Hence, each time an event of probability zero appears, the North considers that it is facing b^+.

The relation $b^- < b^+ < B/C$ means that the North is in a position of strength faced with b^-, while it is in a position of weakness faced with b^+ (who has stronger transfer demands). Within the game in complete information, the equilibrium transfer issued from the negotiation with b^+ is higher or equal to the equilibrium transfer in the negotiation with b^-. Hence:

$$\tau^*(N, S^+) \geq \tau^*(N, S^-) \tag{11}$$

With one-sided incomplete information, b^- is interested in persuading the North that it has strong demands. Therefore, its strategy consists in adopting behaviour similar to b^+. Lemmas (1) to (3) show this result.

Lemma 1. If in any sequential equilibrium of the game J_N^∞, two types of Southern groups propose a counter-offer, they will make the same one. In this case, b^- is interested in suggesting a counter-offer identical to the one proposed by b^+.

Lemma 2. In any sequential equilibrium of J_N^∞:

a) If b^- refuses a proposition τ_N, b^+ is interested in doing as well.
b) Conversely, if b^+ accepts a proposition τ_N, it is in the interest of b^- to do as well.

Lemma 3. In any sequential equilibrium of J_N^∞, a group, whatever its type, will accept any offer τ_N that satisfies:

$$\tau_N \geq (1 - \delta_S)b + \delta_S \tau^*(S^+, N) \tag{12}$$

As the range of possibilities of b^- is larger than those of b^+, there are levels of transfer τ_N likely to be accepted by the first and refused by the second. The following lemma (4) confirms that.

Lemma 4. In any sequential equilibrium of J_N^∞, if:

$$(\tau^*(S^+, N) - b^-)\delta_S + b^- \leq \tau_N \tag{13}$$

b^- accepts the proposition τ_N while b^+ rejects it and offers $\tau^*(S^+, N)$, which will be accepted by N.

Lemmas (1) to (4) lead us to propositions (3) to (5) for the one-sided incomplete information game.

Proposition 3. In any sequential equilibrium of the game J_N^∞ where the North is characterised by pessimistic beliefs and where equation (13) is satisfied, we obtain the following results:

1. If:
$$p^+ < \frac{C[\tau^*(N, S^+) - \tau_N]}{B(1 - \delta_N) + C[\delta_N \tau^*(S^+, N) - \tau_N]} = g$$
The North proposes the level of transfers τ_N. b^- accepts it while b^+ refuses it to make a counter-offer $\tau^*(S^+, N)$, accepted by the North.
2. If $p^+ > g$, the North proposes the level of transfers $\tau^*(N, S^+)$ which is immediately accepted by the Southern groups whatever their type.
3. If $p^+ = g$, two cases are conceivable:
– Either the North proposes the transfer $\tau^*(N, S^+)$ which is immediately accepted by the Southern group, whatever its type.
– Or it proposes the transfer τ_N that satisfies (13), but it is refused by b^+. The counter-offer of b^+ is then equal to $\tau^*(S^+, N)$ and is accepted by the North.

The results of this proposition show that the asymmetry of information totally favours b^- as it is incited to adopt opportunistic behaviour. The North can however determine the real characteristic of its opponent. For that, it just has to propose a transfer τ_N that satisfies (13), which will be a sufficient incentive for b^- to accept and give up the option of making a counter-offer. Nevertheless, since $\tau_N > \tau^*(N, S^-)$, b^-, although forced to reveal its characteristics, benefits from its opportunism. Let us add that the situation of the North is even more unfavourable if equation (13) does not hold. The following proposition illustrates this situation.

Proposition 4. In any sequential equilibrium of the game J_N^∞ where the

North is characterised by pessimistic beliefs and where the relation (13) does not hold, if the North opens the negotiations proposing $\tau^*(N, S^-)$, it will be refused by the Southern groups whatever their type.

However, the asymmetry of information does not have a great impact on the date of signing of an agreement. The following proposition shows that an agreement will be rapidly obtained.

Proposition 5. In any sequential equilibrium of the game J_N^∞ characterised by an asymmetry favourable to the Southern groups, the negotiations will never last more than two periods.

This proposition is similar to the one developed by Osborne and Rubinstein (1990). The idea is as follows: Suppose an agreement is reached at the period n ($n \in \aleph^*$) on a proposition of the Southern group. Lemma (1) defines the behaviour of b^-. Its strategy consists in imitating the behaviour of b^+ by making the same counter-offer. Thus, whatever the type of group, the counter-offers will be $\tau^*(S^+, N)$. In addition, the North is not interested in prolonging the negotiations until the period n, since it can accept the transfer $\tau^*(S^+, N)$ at the second period, obtaining higher earnings due to its preference for the present. Hence, if a proposition τ_N is refused by both types of Southern groups, the agreement can only emerge at the following period.

On the contrary, if the offer of first period τ_N is accepted by b^-, the North can determine the real characteristics of its opponent. Therefore the North knows that a counter-offer τ_S can only come from a b^+. Since it cannot increase its gain by refusing, the North will accept τ_S. Thus, the negotiation will end at the second period. Finally, if the offer τ_N is a sufficient incentive to be accepted by the two types of union, i.e. $\tau_N \geq \tau^*(N, S^+)$, the agreement will take place at the first period. However, remark that this result is only valid for a situation where the incomplete information is reduced to two possibilities. If the problem gets more complicated, the agreement will no longer be reached at the second period.

Propositions (3) and (4) show that with asymmetry of information, two categories of agreements are likely to be signed. We can associate a sequential equilibrium to each of these types of agreements.

On the one hand, an agreement is possible where b^+ accepts the equilibrium transfer obtained from the game without asymmetry of information. The equilibrium transfer is $\tau^*(S^+, N)$ and b^- accepts the offer τ_N defined by (13). We call this sequential equilibrium 'separating'. The North is certain to determine the characteristic of the Southern group. By proposing the transfer

τ_N, only b^- will accept it. In case of rejection, the North will suppose that its opponent is b^+.

On the other hand, an agreement concluded on a transfer corresponding to b^+, is highly profitable for b^-. This latter group obtains more than what it expected in complete information. We find here a 'pooling' sequential equilibrium. The agreement doesn't reflect the type of the contracting parties.

The importance of this second case depends strongly on the beliefs of the North. There is a monotonous relation between the value p^+ and the possible emergence of a pooling sequential equilibrium. The higher (lower) the initial belief p^+ of the North, the higher (lower) are the chances to meet an agreement with a transfer $\tau^*(N, S^+)$ from the first negotiation round. Therefore, if the Southern group was able to emit in a pre-game phase a signal not corresponding to its type, the asymmetry could quickly favour them. Keeping personal information plays an essential role in the bargaining process.

Proposition (5) shows that time takes a new strategic dimension in the presence of asymmetry of information. In this case, time favours the Southern groups. We have seen that the strategy of b^- is to adopt opportunistic behaviour, by posting demands identical to the ones requested by b^+. This strategy allows it to emit a signal not corresponding to its own type. Furthermore, even if the negotiations continue, the North is not able to discover the real characteristic of the Southern group. Hence, to maintain the advantage granted by the asymmetry of information, b^- has to persevere and not change its strategy. Therefore, the North is reluctant to prolong the negotiations, since it will not be able to discover the real characteristics of its opponent. Taking into account the temporal dimension and its asymmetry of information, the North is eager to conclude an agreement quickly. Nevertheless, this configuration favours b^-. If the initial offer of the North is judged unsatisfactory, it will make a positive profit, equal to $\tau^*(S^+, N) - \tau_N$, since its counter-offer is higher than the one it would have requested in the absence of asymmetry of information. On the contrary, b^+ will receive a transfer corresponding to its characteristics (whatever the initial situation).

This proposition also shows that asymmetry of information could prolong the negotiations, although an agreement is likely to be achieved at the second period. This delay is a consequence of asymmetry of information, and corresponds to the time needed by the North to determine the Southern group's type.

4.2. *Multilateral negotiation*

We now analyse a negotiation between the North and two distinct Southern groups S_1 and S_2. When the North is confronted with the group S_i $(i = 1, 2)$ the associated distribution of probability is written $P_{N_i}(b^+) = p_i^+$

and $P_{N_i}(b^-) = 1 - p_i^+$. In extension to proposition (3), we have:

$$g_i = \frac{C_i[\tau^*(N, S_i^+) - \tau_{Ni}]}{B_i(1-\delta_N) + C_i[\delta_N \tau^*(S_i^+, N) - \tau^*(N, S_i^-)]} \quad i = 1, 2 \tag{14}$$

Since we focus on configurations involving a delay during the negotiation, we exclude the case where $p_i^+ > g_i$, $(i = 1, 2)$. When $p_i^+ < g_i$ $(i = 1, 2)$ three cases are conceivable according to the Southern groups' characteristics.

In the first configuration the group j is of type b_j^-, and the group k is of type b_k^+, $j \neq k$. Two sub-cases have to be considered. If both groups act together, they can claim either a transfer b^-, and obtain an agreement during the first stage, or a transfer b^+, which will extend the negotiation to the second period.

Sub-case 1. If acting together b_j^- and b_k^+ request b^-, they are interested in separating themselves if and only if:

$$U_j(\tau_{Nj}, 0) + U_k(\tau^*(S_k^+, N), 1) > U(\tau_N, 0) \tag{15}$$

That involves:

$$b_j^- < \frac{(1-\delta_N\delta_S)(Cb^- - C_k b_k^+) - \delta_N(1-\delta_S)(Cb^+ - C_j b_j^+ - C_k b_k^+)}{(1-\delta_N\delta_S)C_j} - \frac{(1-\delta_N)(B - B_j - B_k)}{(1-\delta_N\delta_S)C_j} = \alpha_1 \tag{16}$$

Sub-case 2. If acting together b_j^- and b_k^+ request b^+, they are interested in separating themselves if and only if:

$$U_j(\tau_{Nj}, 0) + U_k(\tau^*(S_k^+, N), 1) > U(\tau^*(S^+, N), 1) \tag{17}$$

Solving:

$$b_j^- < \frac{(1-\delta_N\delta_S)(Cb^+ - \delta_S C_k b_k^+) - (1-\delta_S)(Cb^+ - \delta_N\delta_S C_j b_j^+ - \delta_N\delta_S C_k b_k^+}{\delta_S(1-\delta_N\delta_S)C_j} - \frac{(1-\delta_N)(B - B_j - B_k)}{(1-\delta_N\delta_S)C_j} = \alpha_2 \quad (18$$

Since it can easily be shown that $\alpha_1 < \alpha_2$, for $b_j^- < \alpha_1$ whatever the claim of the union, the Southern groups favour disunity. For sub-case 1, b_k^+ is confronted with a delay while the union would lead to an agreement from the first stage. That delay represents the necessary time for b_k^+ to separate itself from b_j^- and reveal its demands to the North. For the latter, these demands reveal the characteristics of b_k^+ only if it has previously determined the type of b_j^-. b_k^+ would prefer disunity if it would obtain a higher benefit by showing its type. That requires that b_j^- concludes an agreement on a low level of transfers in the first period.

For sub-case 2, the negotiation delay is less costly for b_k^+ than in the previous case, since the delay would also occur acting together. Two simultaneous interpretations of the delay are possible. In the case of united groups, the delay represents the time needed by the North to infer the type of coalition it is confronted by. For the case of disunity, it describes the necessary time for b_k^+ to be dissociated from b_j^-. The sub-cases (1) and (2) correspond to a situation where the groups are complementary. This configuration reinforces the bargaining power of b_k^+. When it realizes that b_j^- wishes to demand a low transfer, it is interested in extending the negotiation to obtain a higher level of transfers in the second period.

When the inequalities (16) and (18) are inverted, we end up with a situation where the Southern groups are substitutable and act together. For the symmetric sub-case (1), despite union, the negotiation goes on. The common action does not give a sufficiently dissuasive bargaining power to end the negotiation in the first round. The bargaining power of the union is compensated by the presence of asymmetry of information, since the North uses the delay to determine the characteristics of the Southern group. Although the agreement is signed with the transfer $\tau^*(S^+, N)$, the union is confronted with a negotiation cost.

In the second configuration, we assume that the Southern group j is of type b_j^+ and the group k of type b_k^+. In this case the negotiation will end in the second period even if there is no union. Disunity is favourable to the Southern groups if:

$$U_j(\tau^*(S_j^+, N), 1) + U_k(\tau^*(S_k^+, N), 1) > U(\tau^*(S^+, N), 1) \quad (19)$$

Solving:

$$b_k^+ < \frac{B_j + B_k - B + Cb^+ - C_j b_j^+}{C_k} \quad (20)$$

Equation (20) corresponds to the condition of complementarity given by proposition (1). It shows that each group is interested in acting separately when their respective demands, although high, are not enough to take a relevant part of the benefit. Despite their complementarity, each group has room to manoeuver until a certain level. If this level is exceeded, the situation leads the groups to unite to obtain the bargaining power allowing them to reach an agreement based on a high level of transfers. This situation is associated to substitutability between the groups. In both cases, we find the initial interpretation of the negotiation delay. The delay becomes profitable to the North, it helps it to infer the characteristics of each group or the union of groups.

Finally, in the third configuration, we suppose that the group j is of type b_j^- and the group k of type b_k^-. The delay can take place when the groups act jointly and require b^+. Disunity is profitable if:

$$U_j(\tau_{Nj}, 0) + U_k(\tau_{Nk}, 0) > U(\tau^*(S^+, N), 1) \quad (21)$$

Solving:

$$b_k^- < \frac{(1-\delta_N\delta_S)(Cb^+ - C_jb_j^-) - \delta_N(1-\delta_S)(Cb^+ - C_jb_j^+ - C_kb_k^+)}{(1-\delta_N)C_k} - \frac{(B - B_j - B_k)}{C_k} \quad (22)$$

The Southern groups will favour disunity within a complementarity framework when their minima individual demands are low. That will incite them to accept the offer of the North. Acting together, their transfer demands are increased due to their strong bargaining power. Therefore, they are interested in extending the negotiation. Thus, the delay can be interpreted as the means to affirm high demands. The negotiation costs can be compensated by the anticipated higher benefit in terms of transfers. Hence, the groups characterised by low minimum demands are willing to claim more by being united.

We now focus on a situation in which the beliefs of the North lead to a separating equilibrium with one of the Southern groups and to a pooling equilibrium with the other, i.e. $p_j^+ > g_j$ and $p_k^+ < g_k$, $j \neq k$. In this situation, group j (whatever its characteristics) will accept the offer $\tau^*(N, S_j^+)$ at the first round. Its utility is written:

$$U_j(\tau^*(N, S_j^+), 0) = [\tau^*(N, S_j^+) - b_j^-]C_j \text{ if } S_j \text{ is of type } b_j^- \quad (23)$$

$$U_j(\tau^*(N, S_j^+), 0) = [\tau^*(N, S_j^+) - b_j^+]C_j \text{ if } S_j \text{ is of type } b_j^+ \quad (24)$$

The utility of group k is:

$$U_k(\tau_{Nk}, 0) = [\tau_{Nk} - b_k^-]C_k \quad \text{if } S_k \text{ is of type } b_k^- \quad (25)$$

$$U_k(\tau^*(S_k^+, N), 1) = \delta_S[\tau^*(S_k^+, N) - b_k^+]C_k \quad \text{if } S_k \text{ is of type } b_k^+ \quad (26)$$

The situations (23)+(25) and (23)+(26) are included in the cases previously discussed. For the remaining situations, we limit ourselves to those in which the negotiation delay appears. The situation (24)+(25) leads to the following relation:

$$U_j(\tau^*(N, S_j^+), 0) + U_k(\tau_{Nk}, 0) > U(\tau^*(S^+, N), 1) \quad (27)$$

472

For

$$b_k^- < \frac{(1-\delta_N\delta_S)(\delta_S Cb^+ - C_j b_j^+) - (1-\delta_S)(\delta_N\delta_S Cb^+ - C_j b_j^+ - \delta_N\delta_S C_k b_k^+)}{\delta_S(1-\delta_N\delta_S)C_k} - \frac{(1-\delta_N)(B - B_j - B_k)}{(1-\delta_N\delta_S)C_k} \tag{28}$$

b_j^+ and b_k^- will tackle the negotiations separately. When the Southern groups are complementary and their individual demands high, they are willing to act separately. On the other hand, when they are substitutable, they will act jointly and force the negotiation delay because the beliefs of the North lead it to make a low opening proposition.

Let us now analyse the situation (24)-(26). Two cases are conceivable according to the response of the union to the initial proposition of the North:

$$U_j(\tau^*(N, S_j^+), 0) + U_k(\tau^*(S_k^+, N), 1) > U(\tau^*(N, S^+), 0) \tag{29}$$

or

$$U_j(\tau^*(N, S_j^+), 0) + U_k(\tau^*(S_k^+, N), 1) > U(\tau^*(S^+, N), 1) \tag{30}$$

These two configurations lead to the same condition, namely equation (20). Thus, although b_k^+ bears a negotiation delay, disunity is still favoured. In the first case, the complementarity allows b_k^+ to affirm its demands by refusing the initial offer of the North. Therefore the delay allows b_k^+ to obtain what it expected. However, when they are substitutable, b_k^+ is interested in joining b_j^+ to increase its bargaining power and not to bear a delay. In the second case, the Southern groups also favour disunity instead of union. Moreover, complementarity favours b_j^+ since it is induced to sign an agreement as of the first period, whilst, being united or acting separately, b_k^+ is still confronted with a delay.

To resume, the different configurations analysed lead to the following results: in the presence of asymmetric information, when the Southern groups form a union (when they are substitutable) the delay favours the North. That delay corresponds to the time necessary for it to acquire the relevant information concerning the demands of the groups. On the other hand, when the Southern groups act separately (complementarity situation), the delay has several interpretations. If one of the two Southern groups is of type b^+, while the other is of type b^-, the delay corresponds to the time necessary for b^+ to dissociate from b^-. When both groups are of type b^+, the delay is used by one and/or the other group to affirm its transfer demands.

The comparison of equilibrium in complete and incomplete information in multilateral negotiations game gives the following proposition:

Proposition 6. When both groups are of type b_i^+, the conditions of complementarity-substitutability in presence of asymmetric information are identical to those obtained with complete information. On the contrary, when at least one of the groups is of type b_i^-, the conditions of complementarity (substitution) are wider (more restrictive).

Within a multilateral negotiation with incomplete information, the existence of information asymmetry reinforces the bargaining power of complementary groups and reduces the power of substitutable groups. As with complete information, the complementarity, requiring the agreement of both groups, gives them significant bargaining power compelling the North to propose a high level of transfers. In this context, the one and/or the other Southern group will take advantage of its private information to increase its bargaining power and obtain an agreement on a high transfer level.[8] The advantages of complementarity are therefore reinforced by the lack of information of the North. Asymmetry of information makes the conditions of complementarity obtained with complete information less restrictive. On the contrary, in the presence of substitutable groups, the North is interested in acquiring information on the characteristics of the groups, in order to oppose them by searching an agreement with only one of them. The possible race downwards leads the Southern groups to reveal their private information reducing their bargaining power. The drawbacks of substitutability are reinforced by informational asymmetry since, if the Southern groups request high transfers, the negotiations cannot be concluded in the first period. Thus, the conditions of substitutability are more restrictive than with complete information.

Considering these results, the South will be interested in repeating the pattern of the Kyoto Protocol. That is, the relatively rapid signing of a general agreement which makes the Southern groups complementary, followed by a discussion about the details (the Marrakech Accords in the case of the Kyoto Protocol) where the separation in two complementary groups will give them strong bargaining power to tackle the negotiations on the precise amount of the transfers.

Moreover, if both Southern groups have high demands, either complementary of substitute, they will never accept less than what they would obtain in complete information. Likewise, the North will never offer more than the maximum amount of transfers that it would have proposed with complete information. As an agreement is only reached with a transfer level corresponding to the demands of the Southern groups, the presence of asymmetry of information does not benefit them. Under these circumstances, the conditions of complementarity-substitutability are identical with complete and incomplete information.

5. Conclusion

The objective of this paper has been to determine the different forms that the international agreements to enlarge the Kyoto Protocol to include the developing countries might take. The main characteristics of the negotiation are the presence of asymmetry of information about the capacities of the South to reduce their emissions and the possibility to make the negotiations last more than one round. Although we have analysed the climate change negotiations, the framework developed is wide enough to represent most of the international negotiations between the North and the South, as long as they deal with an issue where the North is willing to transfer money (or other economic advantages) in exchange of some commitment by the Southern countries; and as long as available data about the North are more accurate than data about the South.

In a bilateral framework, we have shown that asymmetry of information is not always beneficial for the developing countries coalition. Industrial countries will be interested in continuing negotiations and may delay the moment of an agreement to obtain information about the characteristics of the Southern coalition. Compared to the outcome with complete information, the developing countries will pay a negotiation cost due to the delay.

In a multilateral framework, the information asymmetry reinforces the negotiation power of the complementary coalitions, which will generally act separately, and reduces that of the substitutable coalitions, which will generally act together. Hence, the developing countries are interested in forcing a situation where they are complementary by setting, at an early stage (pre-game) of the negotiations, a high minimum level of ratifications for the agreement to come into force (as it was done for the Kyoto Protocol). If the southern coalitions manage to become complementary, they will act separately but the agreement will not be reached in the first negotiation period. The delay can now be interpreted as the time needed by the coalition with strong demands to dissociate from the coalition with weak demands.

Notes

1. As well as 55 countries, including Annex I as well as Non-Annex I countries.
2. In fact, this kind of clause is common in international agreements.
3. For a survey see Finus (2001).
4. Most of the work with heterogeneous countries is based on numerical simulations (Barrett, 1997; Botteon and Carraro, 1997, 1998) and hence does not provide general results. The theoretical work available has been developed within the highly restricted framework of countries characterised by specific welfare functions (Hoel, 1992; Heal, 1994).

5. We could also adopt a perfect bayesian equilibrium concept. However, as Kreps and Wilson (1982) show, this concept has the drawback that it defines a great amount of solutions, while the sequential equilibrium permits to restrain them.
6. These transfer equilibria are derived from a backward induction reasoning (Rubinstein, 1982).
7. We retain that type of beliefs among the family of possible beliefs (optimistic, passive or continued) since they are the most economically relevant within this model.
8. We assume that when they act jointly, the Southern groups request a high level of transfers.

References

Barrett, S. (1994). Self enforcing international environmental agreements. *Oxford Economic Papers* 46: 878–894.

Barrett, S. (1997). Heterogeneous international environmental agreements. In C. Carraro (Ed.), *International environmental negotiations. Strategic policy issues.* Cheltenham: Edward Elgar.

Botteon, M. and Carraro, C. (1997). Burden-sharing and coalition stability in environmental negotiations with asymmetric countries. In C. Carraro (Ed.), *International environmental negotiations. Strategic policy issues.* Cheltenham: Edward Elgar.

Botteon, M. and Carraro, C. (1998). Environmental coalitions with heterogeneous countries: Burden-sharing and carbon leakage. In A. Ulph (Ed.), *Environmental policy, international agreements and international trade.* Oxford: Oxford University Press.

Carraro, C. and Siniscalco, D. (1993). Strategies for the international protection of the environment. *Journal of Public Economics* 2: 309–328.

Chander, P. and Tulkens, H. (1995). A core-theoretic for the design of cooperative agreements on transfrontier pollution. *International Tax and Public Finance* 2: 279–294.

Chander, P. and Tulkens, H. (1997). The core of an economy with multilateral environmental externalities. *International Journal of Game Theory* 26: 379–401.

Egteren, H. and Tang, J. (1997). Maximum victim benefit: A fair division process in transboundary pollution problems. *Environmental and Resources Economics* 10: 363–386.

Finus, M. (2001). *Game theory and international environmental cooperation.* Cheltenham: Edward Elgar.

Funaki, Y. and Yamato, T. (1999). The core of an economy with a common pool resource: A partition function approach. *International Journal of Game Theory* 28: 157–171.

Harsanyi J. (1967–68). Games with incomplete information played by Bayesian players. *Management Science* 14: 159–182, 320–334, 486–502.

Heal, G. (1994). Formation of international agreements. In C. Carraro (Ed.), *Trade, innovation, environment.* Dordrecht: Kluwer Academic Press.

Helm, C. (2001). On the existence of a cooperative solution for a coalitional game with externalities. *International Journal of Game Theory* 30: 141–146.

Hoel, M. (1992). International environment convention: The case of uniform reductions of emissions.*Environmental and Resource Economics* 2: 141–159.

Hovi, J. (2001). Decentralized enforcement, sequential bargaining, and the clean development mechanism. *CICERO Working Paper* 12–2001.

Kreps, D. and Wilson, R. (1982). Sequential equilibria. *Econometrica* 50: 863–984.

Osborne, M. and Rubinstein, A. (1990). *Bargaining and market.* San Diego: Academic Press.

476

Péreau, J.-C. and Tazdaït, T. (2001). Cooperation and unilateral commitment in the presence of global environmental problems. *Environmental and Resource Economics* 20: 225–239.
Rotillon, G., Tazdaït, T. and Zeghni S. (1996). Bilateral or multilateral bargaining in the face of global environmental change? *Ecological Economics* 18: 177–187.
Rubinstein, A. (1982). Perfect equilibrium in a bargaining model. *Econometrica* 50: 97–109.
Rubinstein, A. (1985). A bargaining model with incomplete information about time preferences. *Econometrica* 53: 1151–1172.
Rubinstein, A. (1988). Choice of conjectures in a bargaining game with incomplete information. In A.E. Roth (Ed.), *Game theoretic models of bargaining*. Cambridge: Cambridge University Press.
Tulkens, H. (1998). Cooperative vs. free riding in international environmental affairs: Two approaches. In N. Hanley and H. Folmer (Eds.), *Game theory and the environment*. London: Edward Elgar.

Appendix

Proof of Lemma 1

See Rubinstein (1985). Nevertheless, the general principle of the demonstration is as follows. Let us consider the history for which b^+ and b^- reject the offer of N and suggest different offers in return. Let $\tau^*(S^+, N)$ be the offer of b^+ and τ_S^- the one of b^-, with:

$$B/C \geq \tau^*(S^+, N) > \tau_S^- > \tau^*(S^-, N) \tag{31}$$

After the Southern group has made its proposition, N is able to determine the characteristics of its opponent.

- Behaviour of N towards b^+. Faced with a proposition $\tau^*(S^+, N)$, N identifies its opponent as being b^+. Knowing that it could not propose less than $\tau^*(N, S^+)$ at the next period, it will accept the counter-offer. By offering the transfer $\tau^*(N, S^+)$, its benefit is not increased compared to that which it would obtain accepting $\tau^*(S^+, N)$ a period later.
- Behaviour of N towards b^-. Two cases are possible:

Observing τ_S^- satisfying (31) and therefore strictly lower to $\tau^*(S^+, N)$, N knows that it is confronted with b^-. It just has to propose in the following period $\tau_S^- - \epsilon > \tau^*(S^-, N)$, with $\epsilon \to 0$ in a way that $\Pi(\tau_S^-, t) < \Pi(\tau_S^- - \epsilon, t + 1)$. That strategy is profitable as it can improve its benefit. On the other hand, it deteriorates simultaneously the utility of b^-.

Suppose now that N accepts τ_S^-. b^- is then incited to deviate as it knows that N will also accept $\tau^*(S^+, N)$. Since both offers τ_S^- and $\tau^*(S^+, N)$ can be accepted, b^- is interested in requesting $\tau^*(S^+, N)$. □

Proof of Lemma 2

For each of these two cases, we consider the game at any date t where the Southern group is the proposer.

- b^- makes an offer τ_S different from τ_N if the utility associated to its proposition τ_S is strictly higher to the one obtained with τ_N. However, the lemma 1 stipulates that the counter-offer of b^- is equal to $\tau^*(S^+, N)$. The condition is then written:

$$U(\tau_N, t - 1) \leq U(\tau^*(S^+, N), t)$$

or

$$C(\tau_N - b^-)\delta_S^{t-1} \leq C(\tau^*(S^+, N) - b^-)\delta_S^t$$

or

$$\tau_N - [(1 - \delta_S)b^- + b^+\delta_S] \leq (\tau^*(S^+, N) - b^+)\delta_S$$

- Since $b^+ \geq (1-\delta_S)b^- + b^+\delta_S$, we obtain: $C(\tau_N - b^+) \leq C(\tau^*(S^+, N) - b^+)\delta_S$. Thus, when b^- refuses τ_N, b^+ does as much.
- Suppose b^+ accepts a proposition τ_N. Hence:

$$U(\tau_N, t-1) \geq U(\tau^*(S^+, N), t)$$

or

$$\tau_N - [b^-\delta_S + (1-\delta_S)b^+] \geq (\tau^*(S^+, N) - b^-)\delta_S$$

Since $b^- \leq b^-\delta_S + (1-\delta_S)b^+$, we have $C(\tau_N - b^-) \geq C(\tau^*(S^+, N) - b^-)\delta_S$. □

Proof of lemma 3

Suppose the Southern group does not accept τ_N. The lemma (1) tells us that the two types of groups will suggest the same counter-offer τ_S, with $\tau_S \geq b^+$. However, the highest proposition the Southern group can make is $\tau^*(S^+, N)$. Hence, for any transfer τ_N satisfying the relation $\tau_N \geq (1-\delta_S)b + \delta_S\tau^*(S^+, N)$,we have $\tau_N - b \geq (\tau^*(S^+, N) - b)\delta_S$. Therefore, the Southern group will accept the transfer τ_N.

Let us specify the expression of the offer τ_N according to the type of the Southern group. If the group is of type b^-, we have, for $b = b^-$: $\tau_N^- \geq (1-\delta_S)b^- + \delta_S\tau^*(S^+, N)$ and for $b = b^+$: $\tau_N^+ \geq (1-\delta_S)b^+ + \delta_S\tau^*(S^+, N) = \tau^*(N, S^+)$.

Thus, although b^- is incited to behave like b^+ (according to lemma 1 and 2), it will have more room to manoeuvre than b^+ (according to lemma 3). I.e., its set of satisfactory propositions is wider than the one of b^+:

$$\begin{aligned}\tau_N^- \in A &= [(1-\delta_S)b^- + \delta_S\tau^*(S^+, N), B/C] \\ \tau_N^+ \in B &= [(1-\delta_S)b^+ + \delta_S\tau^*(S^+, N), B/C]\end{aligned}$$

and $B \subset A$. For $\delta_S \to 1$, we have $\tau = B/C$. The behaviour of the Southern groups b^- and b^+ coincides when b^- is patient as they request both the highest transfer that N can accept. Patience allows b^- to claim the requests of b^+. On the other hand, if b^- is impatient, its transfer demand becomes its reference transfer for the negotiation. □

Proof of lemma 4

We successively present the behaviours of b^-, b^+ and N.

- Behaviour of b^-:

b^- will accept τ_N if that transfer level incites it, at the next period, to renounce earnings associated to a proposition $\tau^*(S^+, N)$. It is enough that:

$$U(\tau_N, t-1) \geq U(\tau^*(S^+, N), t)$$

or

$$\tau_N - b^- \geq (\tau^*(S^+, N) - b^-)\delta_S$$

Due to its preference for the present, we have:

$$\tau_N \geq (\tau^*(S^+, N) - b^-)\delta_S + b^- \tag{32}$$

– Behaviour of b^+:
If b^+ prefers τ_S to τ_N, the following inequality has to hold:

$$U(\tau_N, t-1) < U(\tau^*(S^+, N), t) \tag{33}$$

or

$$\tau_N < \tau^*(N, S^+) \tag{34}$$

– Behaviour of two types of groups:
According to equation (32), b^- is not interested in refusing the offer τ_N since it will not increase its payoffs by a counter-proposition $\tau^*(S^+, N)$. On the contrary, according to (33), b^+ refuses τ_N since it can enhance its utility by proposing $\tau^*(S^+, N)$. Thus, for an offer τ_N satisfying the condition (13), b^- has to accept it while b^+ can refuse it to propose afterwards $\tau^*(S^+, N)$.
– Behaviour of N:
If b^- accepts the offer of τ_N while b^+ rejects it to propose afterwards $\tau^*(S^+, N)$, N is able to determine the type of its opponent. In fact, it knows that for a level of transfers τ_N, satisfying (33), b^- should not reject it (while b^+ should). N, knowing the type of group it is negotiating with, will accept the offer $\tau^*(S^+, N)$. □

Proof of Proposition 3

- If N opens the negotiations on an offer $\tau^*(N, S^+)$, its expected payoff is:

$$\begin{aligned} E\Pi^+ &= p^+[B - \tau^*(N, S^+)C] + (1 - p^+)[B - \tau^*(N, S^+)C] \\ &= B - \tau^*(N, S^+)C \end{aligned} \tag{35}$$

  On the other hand, if it suggests the pure strategy τ_N satisfying (13), its expected payoff becomes:

$$E\Pi^- = (1 - p^+)[B - \tau_N C] + p^+\delta_N[B - \tau^*(S^+, N)C]$$

  According to lemma (4), the offer will be accepted by b^- and rejected by b^+. The latter will propose the transfer $\tau^*(S^+, N)$, which will be accepted by N in the second period. The strategy τ_N is optimal for N if and only if: $E\Pi^- > E\Pi^+$ (condition 1 of proposition (3)).
- If $E\Pi^- < E\Pi^+$, we are in the configuration where proposing $\tau^*(N, S^+)$ is optimal for N. Since that is the highest transfer that the Southern group can obtain in negotiations in complete information, it will accept it whatever its type.
- If $E\Pi^- = E\Pi^+$, the sequential equilibrium consists for N in adopting the mixed strategy $\tau^*(N, S^+)$ with the probability α and the strategy $\tau^*(N, S^-)$ with the probability $1 - \alpha$ with $\alpha \in [0, 1]$. b^+ will accept the proposition if it is at least equal to $\tau^*(N, S^+)$ while b^- will accept, in the first period, any proposition satisfying (13). □

480

Proof of Proposition 4

Since (13) does not hold, for $\tau_N = \tau^*(N, S^-)$, the expected payoff of N is written:

$$E\Pi^* = (1 - p^+)[B - \tau^*(N, S^-)C] + p^+\delta_N[B - \tau^*(S^+, N)C]$$

By comparing the expected payoffs associated to each strategy, we obtain the searched result. □

[22]

Environment and Development Economics 18: 69–92 © Cambridge University Press 2012
doi:10.1017/S1355770X12000411

Forming coalitions to negotiate North–South climate agreements

ALEJANDRO CAPARRÓS
Consejo Superior de Investigaciones Científicas (CSIC), Institute for Public Goods and Policies (IPP), Madrid, Spain.
E-mail: alejandro.caparros@csic.es

JEAN-CHRISTOPHE PÉREAU
Université Montesquieu-Bordeaux IV, GREThA (UMR CNRS 5113), Avenue Léon Duguit, 33608 Pessac cedex, France.
E-mail: jean-christophe.pereau@u-bordeaux4.fr

Submitted November 8, 2011; revised June 21, 2012, August 29, 2012; accepted September 7, 2012; first published online October 26, 2012

ABSTRACT. This paper analyzes North-South negotiations over climate change abatement. We consider that northern countries have an incentive to negotiate over a transfer to the southern countries in exchange for their abatement efforts rather than reducing their emissions at home. We study the incentives for northern and southern countries to form negotiation-coalitions at each side of the bargaining table and the impact of these negotiation-coalitions on the final outcome. We show that the incentives can be separated into direct efficiency gains, as fixed costs savings, and indirect bargaining power gains. Depending on the relative values of these gains, we determine the equilibrium of the game. We also show that bargaining power gains encourage southern countries to negotiate separately while they encourage northern countries to unite, and that this hinders the formation of the grand coalition.

1. Introduction

In 1992 virtually all the countries in the world signed and ratified the United Nations Framework Convention on Climate Change. Five years later, the Kyoto Protocol was signed, engaging countries included in Annex I of this Protocol (OECD countries and economies in transition, hereinafter 'northern countries') to reduce their overall emission by five per cent in 2012 compared to 1990, but leaving non-Annex I countries ('southern

This work was carried out with the financial support of the CNRS-CSIC PICS program under the project SEIME.

countries') without abatement commitments.[1] Nevertheless, every year southern countries represent a more relevant part of the total emissions with the result that the current negotiation rounds are focusing on the best way to convince these countries to accept binding abatement efforts.[2] However, given their focus on development and the historical responsibility of northern countries, it is unlikely that southern countries will accept binding targets without some kind of transfer from northern countries. This 'transfer' could take the form of money (unlikely in large sums) or any other form of benefit (technology transfers or better conditions in other negotiations).

Coalitions play a significant role in these negotiations (Hampson and Hart, 1995). During all the negotiation rounds, countries have organized themselves into coalitions, such as the Umbrella Group (US and similarly minded countries during the 1990s), the European Union, the G77 and China (a large coalition of developing countries), the Least Developed Countries, the AOASIS (small island countries) and, more recently, the Rainforest Coalition (a coalition including almost all countries with rainforests, except Brazil). These coalitions have been relatively stable, although some variations have occurred over the already long history of negotiations. Nevertheless, a constant throughout the process has been that coalitions have been formed between northern or between southern countries.[3] This concept of 'coalition' differs from the one used in most of the literature on International Environmental Agreements (IEA), as in this literature a coalition is a group of countries that have decided to determine jointly their greenhouse gas (GHG) abatement effort (this is usually done by assuming that they have merged and now maximize their welfare jointly). In our framework, however, a coalition is a group of countries, either northern or southern, that have decided to join forces in international negotiations over a public good. In other words, for us, 'G77 and China' or the 'Rainforest Coalition' are coalitions, whereas they are not for most of the remaining literature on IEA. To avoid confusion, we will use hereafter the term 'negotiation-coalition' (NC) to refer to the coalitions set up for negotiation purposes that we are going to analyze (our concept is similar to the one used in Hampson and Hart (1995)).

[1] In other words, right from the beginning, the international community recognized that there were two types of fundamentally different countries: so-called Annex I countries (developed countries that accepted a 'historical responsibility' in the current levels of atmospheric CO_2) and non-Annex I countries (developing countries with a reduced 'historical responsibility' and an urgent need for development).

[2] Abatement made by one country, i.e its reduction of GHG emissions, is a public good that generates positive benefits for other countries, and this explains the difficulties encountered in reaching a global agreement for climate change (Carraro and Siniscalco, 1993; Barrett, 1994).

[3] In the official website of the United Nations Framework Convention on Climate Change, one can find the description of the different groupings of countries during the negotiations. See http://unfccc.int/parties_and_observers/parties/negotiating_groups/items/2714.php.

The question which we are looking to answer is whether or not there is a natural tendency for this type of negotiation to be carried out between NCs and whether the reason for this possible tendency is based on *efficiency gains* or *bargaining power gains* (or both). *Efficiency gains* refer to the direct gains which countries can expect when they negotiate together minus what they can expect negotiating separately. These gains can come from fixed-costs savings[4] but also from the different amounts of abatement to which the southern players rationally expect they are going to commit themselves while acting united instead of separated. If these gains are positive, countries will have an incentive to enter the negotiations as a NC. Nevertheless, there are reasons for negotiating together, or not, which are not defined by the comparison of the valuation of the two expected outcomes. In fact, even if efficiency gains did not exist, because the value of the expected outcomes under both scenarios is the same, countries would not be indifferent about negotiating together or separately. We refer to these additional reasons to form, or not, a NC as 'bargaining power' reasons. We show below that these bargaining power reasons are explained by the indirect gains obtained by countries not taking part in partial agreements (even though these partial agreements will never be the outcome of the negotiation).

Existing literature adds relatively little to the question of the formation of NCs. The bulk of the literature on IEA focuses on the formation of a single coalition, chiefly using the concept of internal and external stability from non-cooperative game theory (Finus, 2001). In the initial papers, all countries were assumed to be identical (Carraro and Siniscalco, 1993; Barrett, 1994), although this assumption has been relaxed in recent years (McGinty, 2007). Another branch of the literature initiated by Chander and Tulkens (1997) uses cooperative game theory to determine transfer schemes that ensure the non-emptiness of the core, but without modeling negotiations explicitly and without analyzing the role of bargaining power (see Breton *et al.* (2006) for a discussion of the differences between these two approaches). However, this literature does not analyze the negotiation process itself and the concept of 'coalition' used also differs, as already mentioned. In addition, contrary to our framework, these coalitions can be formed by northern and southern countries. Thus, our contribution to the literature on IEA is to focus on the bargaining *process* between northern and southern countries. More related to our approach, Rotillon *et al.* (1996) and Caparrós *et al.* (2004) model international negotiations on climate change abatement efforts using a bargaining procedure *à la* Rubinstein, in the latter case focusing on asymmetries of information. Both assume a unique North (see Carraro *et al.* (2005) or Harstad (2010) for a survey on bargaining theory and the use of this theory in the analysis of international negotiations), which reduces the relevance of the public good aspect of the problem because the northern coalition can never free ride. In addition, these papers do not distinguish, as we

[4] Acting together may also provide the opportunity to share abatement technologies (between the southern players) and this may reduce marginal abatement costs.

do below, between bargaining power and efficiency reasons to form NCs. This distinction has only been analyzed by Chipty and Snyder (1999) in the context of bilateral bargaining between one upstream firm and several downstream firms over a private good. Finally, our focus on NCs also distinguishes our paper from the now large literature that focuses on the formation of coalitions, such as Bloch (1996), Ray and Vohra (2001), Maskin (2003) or de Clippel and Serrano (2008). These papers take into account the interactions between coalitions when deciding the coalition structure to be formed, but they do not focus on the negotiations between coalitions and they are therefore not interested in NCs which are only set up for the negotiation process (bargaining power reasons are also not isolated in this literature).

This article is organized as follows. Section 2 presents a simple model for the negotiation between two northern countries and two southern countries. Both groups of countries can decide whether or not to form NCs and, once this decision has been taken, they engage in a negotiation *à la* Rubinstein to decide transfers from northern to southern players. Section 3 solves the model by backward induction, analyzing the incentives to form NCs and distinguishing between efficiency and bargaining power motivations. Section 4 discusses the implications of our model for the analysis of climate negotiations. Section 5 concludes.

2. The model

In the basic version of the model[5] we consider a world with two northern countries, denoted by 1 and 2, and two southern countries, denoted by 3 and 4. As described in the introduction, we reserve the term 'negotiation-coalition' for the strategic alliance between two northern (or two southern) countries before the bilateral negotiations take place. Since the empirical evidence briefly discussed in the introduction shows that northern and southern countries never join the same NC, we exclude the possibility of the formation of a NC between northern and southern players. When both northern countries decide to form a NC we use the notation (12), reserving the notation (34) for the case when both southern countries decide to form a NC. We use the term 'player' to describe the countries or NC that can eventually participate in the bilateral negotiations (i.e., the possible northern players are 1, 2 and (12) while the southern players are 3, 4 and (34)).

We denote by τ_N the northern NC coalition structure and by τ_S the southern NC structure, reserving the notation Ω_N for the set of all possible northern *NC structures* (with Ω_S for the southern NC). Since the northern and southern countries can only go separately or together in our simple model, Ω_N has only two possible elements, either $\tau_N = \{1, 2\}$ or $\tau_N = \{(12)\}$; and the set Ω_S has also only two possible elements, either southern countries negotiate separately, $\tau_S = \{3, 4\}$, or together, $\tau_S = \{(34)\}$.

[5] A version with n northern countries and m southern countries can be found in Caparrós and Péreau (2010).

For each pair (τ_N, τ_S) an *agreement structure* $\phi(\tau_N, \tau_S)$ is a set of bilateral agreements.[6] In our simplified framework, there are eight possible agreement structures: $\phi_1 = \{[(12), (34)]\}$, $\phi_2 = \{[(12), 3], [(12), 4]\}$, $\phi_3 = \{[1, (34)], 2\}$, $\phi_4 = \{[2, (34)], 1\}$, $\phi_5 = \{[1, 3], [1, 4], 2\}$, $\phi_6 = \{[2, 3], [2, 4], 1\}$, $\phi_7 = \{[1, 3], [2, 4]\}$, $\phi_8 = \{[1, 4], [2, 3]\}$. Square brackets are used to denote bilateral agreements and brackets are used, as before, to denote NC. For instance, $\phi_1 = \{[(12), (34)]\}$ means that the northern NC (12) reaches an agreement with the southern NC (34). In addition, there can be no agreement for each possible pair of NC structures. We call the set including all possible agreement structures $\Phi(\tau_N, \tau_S)$.

2.1. *Game tree and assumptions*

The game Γ is under perfect information and has four stages: (i) stage one, formation of the northern NC structure τ_N^*; (ii) stage two, formation of the southern NC structure τ_S^*; (iii) stage three, choice of the bilateral agreements to be negotiated between northern and southern players, i.e., the agreement structure ϕ^*; and (iv) stage four, the simultaneous negotiation over the transfer in the different bilateral agreements, following Rubinstein's alternating-offers protocol (Rubinstein, 1982).

The resolution of the game by backward induction, using the subgame perfect equilibrium (SPE) concept, is based on several assumptions:[7]

A1: The NC structure τ_N (respectively τ_S) is decided by the northern (southern) players maximizing the aggregated worth for the northern (southern) players
A2: For each pair (τ_N, τ_S), the agreement structure $\phi(\tau_N, \tau_S)$ with the highest aggregated worth is selected out of $\Phi(\tau_N, \tau_S)$.
A3: The tie-breaking rule in A1 and A2 is random selection.
A4: The distribution of the surplus of each bilateral agreement is negotiated following Rubinstein's alternating-offer procedure, with the northern

[6] Note that our 'agreements' are similar to the 'coalitions' in the standard IEA game, as they are formed by countries that have decided to determine jointly the abatement to be done. The main difference is that we allow for several agreements to co-exist and that we allow northern players to take part in more than one agreement, while there is a unique coalition in the basic IEA game (Carraro and Siniscalco, 1993) and countries can only participate in one agreement in the extensions based on coalition formation theory (Carraro, 2005). An additional difference, already mentioned, is that we model the whole process and determine the distribution of the worth within the agreement, although not within the NC.

[7] Most of the IEA literature is focused on the analysis of the stability of the coalitions. We do not analyze the stability of the agreements in our framework (the equivalent to the coalitions in the IEA literature). Nevertheless, it is important to highlight that the equilibrium agreements in our framework are SPE and, therefore, no player would have at any stage an incentive to deviate. In addition, once an agreement is reached there is no future period in our model where the agreement could be broken. As pointed out in our conclusion, analyzing this issue in more detail is an interesting future research topic.

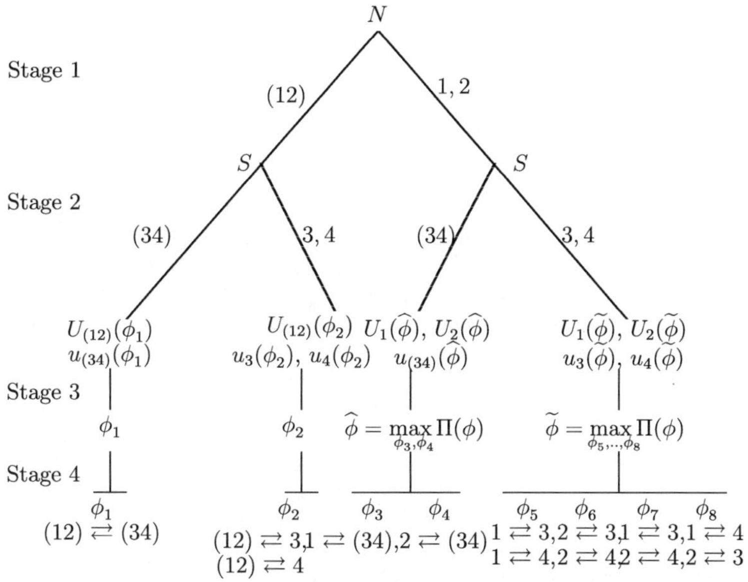

Figure 1. *Four-stage game tree with two northern and southern countries*

players proposing the first offer and under the belief that the optimal agreements are implemented in the remaining simultaneous negotiations. **A5**: The southern players can participate in only one bilateral agreement.

We will now discuss the different assumptions while describing the decision tree of the game Γ shown in figure 1. In the first stage (analyzed in section 3.4), the northern players decide whether they prefer to go to the negotiations separately, as 1 and 2, or forming the NC (12). In the second stage (analyzed in section 3.3), the southern players decide whether to go to the negotiations separately, as 3 and 4, or jointly as the NC (34). In both cases, the southern players take into account the decision taken by the northern players in the previous stage. In these two stages, assumption A1 allows us to select the particular NC structures that the northern and the southern players will choose. More complex coalition formation procedures could also be relevant in a world with many countries (see, e.g., Ray and Vohra, 1999); however, in a world with only two northern countries and two southern countries Assumption A1 simply implies that when a NC is formed it compensates internally any potential losers as long as sufficient surplus is available (nevertheless, we do not analyze explicitly the distribution of the worth within the NC).

In the third stage (analyzed in section 3.2), the agreement structure to be negotiated at stage four is decided, out of all the possible agreement structures in each branch of the game tree. In the first branch starting from the LHS in figure 1, the northern NC structure is $\tau_N = \{(12)\}$ and the southern NC structure is $\tau_S = \{(34)\}$. In this case the only possible agreement is [(12), (34)] and the negotiation in the fourth stage will necessarily take place between (12) and (34), represented by (12) $\rightleftarrows$ (34) in figure 1, i.e., the agreement structure selected at stage three is necessarily $\phi_1 = \{[(12), (34)]\}$.

In the second branch of the tree starting from the LHS there is also only one possibility, as by assumption A5 each southern player can only sign one agreement (although northern players can sign more than one agreement).[8] That is, in this case at stage three the NC structures are $\tau_N = \{(12)\}$ and $\tau_S = \{3, 4\}$, and the agreement structure selected is necessarily $\phi_2 = \{[(12), 3], [(12), 4]\}$. In the third branch from the LHS, where $\tau_N = \{1, 2\}$ and $\tau_S = \{(34)\}$, there are two possible agreement structures, either $\phi_3 = \{[1, (34)], 2\}$ or $\phi_4 = \{[2, (34)], 1\}$. We use assumption A2 to select the agreement structure generating the largest surplus to be negotiated in stage four. The reason for this assumption is that the higher the aggregated worth of the agreement structure, the higher will be the aggregated worth which the players will get from the negotiation in the next stage.[9] We call the agreement structure selected $\widehat{\phi}$. In the last branch from the LHS, where $\tau_N = \{1, 2\}$ and $\tau_S = \{3, 4\}$, we assume again that the agreement structure generating the largest surplus will be selected amongst $\phi_5, \ldots, \phi_8$ and we call this agreement structure $\widetilde{\phi}$. Assumption A3 ensures that only one structure is selected when several structures yield the maximum worth.

In the fourth stage (analyzed in section 3.1), the different bilateral agreements in the agreement structure are negotiated simultaneously. In the first branch of the game tree starting from the LHS, the only agreement negotiated is [(12), (34)]. The outcome of this negotiation is a payoff for the northern player involved, $U_{(12)}$, and a payoff for the southern player involved, $u_{(34)}$. In the second branch, the two agreements in $\phi_2 = \{[(12), 3], [(12), 4]\}$ are negotiated simultaneously. The outcome of these two negotiations determines the payoffs for the different players involved: $U_{(12)}$, u_3 and u_4. In the third branch the bilateral agreements in $\widehat{\phi}$ are negotiated simultaneously (and $\widetilde{\phi}$ in the fourth branch). In all cases the negotiations in this fourth stage take place following Rubinstein's offer

[8] Allowing southern countries to reach partial agreements with different northern countries would complicate the expressions derived below without any significant gain.

[9] To justify assumption A2 we will focus on the third branch, but the need for an assumption instead of a complete analysis of the different cases becomes clear in the fourth branch. Our assumption implies that ϕ_3 will be selected in stage three if the aggregate payoff of ϕ_3 is higher than that of ϕ_4. This relationship will hold either because the gains obtained by the members of the agreement are larger if 1 signs an agreement with (34) than if 2 signs it, in which case (34) will have an incentive to sign an agreement with 1, or because player 2 obtains a large benefit from staying outside of the agreement, which will see its interest in bargaining a deal by itself reduced, which will again favor an agreement between 1 and (34). Although under some conditions 1 may prefer that 2 sign the agreement and 2 that 1 sign it, as long as the aggregate payoff of ϕ_3 is higher than that of ϕ_4, player 2 could compensate player 1 for signing the agreement with (34). Analyzing these compensations, especially in the fourth branch, would significantly complicate the analysis in section 3.2, without any real gain for the main goal of the paper (to analyze NCs).

and counter-offer protocol (Rubinstein, 1982), with the northern players proposing the first offer (assumption A4). If there are two simultaneous negotiations (i.e., in ϕ_2, ϕ_5, ϕ_6, ϕ_7 and ϕ_8), both negotiations take place under the assumption that the optimal agreement is reached in the other negotiation. It means that in every negotiation the northern and the southern players believe that their failure to reach an agreement would not affect the other optimal agreement, which is negotiated simultaneously. It is unclear whether simultaneous or sequential bargaining better describes real-life negotiations and, as we point out at the end of the paper, comparing these two bargaining protocols is an interesting avenue for future research (see also Aghion *et al.*, 2007). Nevertheless, we believe that simultaneous bargaining better captures international negotiations on climate change, since currently several negotiations are taking place at the same time without one negotiation waiting until the previous one has finished (the US with China outside of the Kyoto Protocol, Europe with several developing countries within the Clean Development Mechanism of the Kyoto Protocol, Norway with Indonesia on Reducing Emissions from Deforestation and Degradation, etc.).

2.2. *Payoff functions*

We denote the abatement performed by the southern player j by q_j. The aggregated abatement performed by the southern players is given by the vector $Q = \{q_3, q_4\}$ if they remain independent and $Q = \{q_{(34)}\}$ if they form a NC (we also use Q to denote the sum of the components of the vector where appropriate). The abatement performed by the northern players i is denoted by $\bar{q}_i$ and the aggregated abatement by the northern players is $\overline{Q} = \{\bar{q}_1, \bar{q}_2\}$ when they act independently and $\overline{Q} = \{\overline{q}_{(12)}\}$ if they form a NC. The total abatement is then $\overline{Q} + Q$.

We further assume that the emission abatements made by the northern players are an optimal reaction to the abatement of the southern players, after the agreements discussed below have been signed. That is, we assume that northern players act as Stackelberg leaders and decide their abatement levels according to the reaction function $\overline{q}_i = f_i(Q)$ with $\frac{\partial \overline{q}_i}{\partial Q} < 0$.

The gross payoff of the northern players is given by the function $V_i(Q, \overline{Q}, \overline{q}_i)$ which depends on the total abatement done by the South, Q; the total abatement done by the North, $\overline{Q}$; and the abatement done at home by the northern player i, $\overline{q}_i$. That is, all costs and benefits, except the transfer T_{ij}, are included in V_i. As $\overline{q}_i$ and $\overline{Q}$ are functions of total southern abatement Q, we can omit them in the gross payoff expressions and use $V_i(Q)$. Slightly abusing notation whenever we are interested in detailing the components of the vector Q, we write $V_i(q_3, q_4)$ instead of $V_i(\{q_3, q_4\})$. Hence the net payoff function for the northern players $i = 1, 2, (12)$ is:

$$U_i = V_i(Q) - \sum_{j \in J_i} T_{ij}, \qquad (1)$$

where J_i is the set of the bilateral agreement signed by i with the southern players j. The set J_i can consist in one agreement, $\{3\}$, $\{4\}$ or $\{(34)\}$; in two, $\{3, 4\}$; or in none, $\{\emptyset\}$.

Although this is not necessary to derive the results below, in order to facilitate the interpretation and to connect with the IEA literature, we assume that the value function takes the form $V_i(.) = B_i(\overline{Q}, Q) - C_i(\overline{q}_i)$, where i benefits from the total emission reductions (B_i) but only bears the cost (C_i) of the abatement that it performs at home ($\frac{\partial C_i}{\partial \overline{q}_i} > 0$). Nevertheless, when a northern player signs an agreement with some (at least one) southern player j, it also bears the cost of the transfer T_{ij} for each one of the agreements it signs. The abatement q_j performed by a southern player j in exchange for the transfer granted by the northern player i, benefits i directly ($\frac{\partial B_i}{\partial q_j} \geq 0$) and reduces the abatement to be done by i at home (i.e., $\frac{\partial C_i}{\partial \overline{q}_i} \frac{\partial \overline{q}_i}{\partial q_j} \leq 0$).

For southern players we also assume that all costs and benefits except the transfer are included in v_j and we assume that $v_j(Q, \overline{Q}, q_j) = b_j(\overline{Q}, Q) - c_j(q_j)$, where b_j stands for the benefits from the total abatement obtained by j and c_j for its abatement costs (we assume $\frac{\partial b_j}{\partial q_j} \geq 0$ and $\frac{\partial c_j}{\partial q_j} \geq 0$). We simplify the notation of v_j by using $v_j(Q)$, as $\overline{Q}$ is a function of Q and the latter vector incorporates the information of the total abatement done by the southern players and the amount done at home by southern player j, i.e., q_j (as for V, when detailing the different components of Q we write $v_i(q_3, q_4)$). As by assumption A5 southern players can only be involved in one bilateral agreement, the net payoff function for a southern player j is given by:

$$u_j = v_j(Q) + T_{ij}. \tag{2}$$

The total gross payoff for all the players in each agreement structure ϕ_k is denoted by $\Pi(\phi_k) = \sum_i V_i(Q(\phi_k)) + \sum_j v_j(Q(\phi_k))$.

We also assume that northern and southern players are fundamentally different and in particular that there is at least one level of abatement $q_j > q_j^d$ for which $V_i(q_j, q_l^*) - V_i(q_j^d, q_l^*) > v_j(q_j^d, q_l^*) - v_j(q_j, q_l^*)$ for any pair of a northern player i and a southern player j, where q_j^d is the abatement that player j would perform when maximizing its payoff independently (in both cases under the assumption that the other southern player, if there is one, does its optimal abatement). This can happen because an increase in southern abatement benefits more the northern than the southern player, through B_i and b_j, or because the reduction in costs for the North, $\frac{\partial C_i}{\partial \overline{q}_i} \frac{\partial \overline{q}_i}{\partial q_j} \leq 0$, is larger than the increase in the costs in the South (this latter case is probably the most relevant one for the climate change game). Of course, a combination of the two reasons just described can also explain why northern players are 'buyers' of the abatement made by southern players. This assumption implies that the northern player is interested in granting the southern player a transfer in order to increase the abatement of the southern player, and that all southern countries will sign an agreement in equilibrium.

We also assume that northern and southern players have different discount factors, δ_N and δ_S respectively, although all northern (southern) players share the same discount factor, with $1 > \delta_N, \delta_S > 0$. That is, in addition to the differences in the valuation functions described in the previous paragraph, we are assuming that northern players (respectively southern players) are relatively similar one to another in their valuation of the time needed to reach an agreement, but that northern and southern players have different perceptions of the costs of delay.[10]

3. Results

3.1. *Bilateral bargaining (stage four)*

Since we solve the model by backward induction we analyze first the last stage of the game. As indicated in assumption A4, the transfer is determined by the Rubinstein Bargaining Solution (RBS). At this stage, the two parts of each bilateral agreement are already known and so is the abatement effort that the South must perform if the agreement is reached.

To illustrate our framework, we focus initially on a particular agreement structure $\phi_5 = \{[1,3],[1,4],2\}$. Let us consider the bargaining process between 1 and 3, both assuming that 1 and 4 will reach an optimal agreement at the same time. In case of agreement,[11] we have

$$U_1^a = V_1(q_3^*, q_4^*) - T_{13} - T_{14}^*$$
$$u_3^a = v_3(q_3^*, q_4^*) + T_{13},$$

while in case of non-agreement, we have

$$U_1^{na} = V_1(q_3^d, q_4^*) - T_{14}^*$$
$$u_3^{na} = v_3(q_3^d, q_4^*),$$

with $q_3^d = \arg\max_x v_3(x, q_4^*)$, that is, in case of disagreement country 3 determines its deviation abatement level maximizing its own payoff function (under the assumption that player 4 performs the optimal abatement level coming out from its agreement with player 1).

We define the net payoff functions (R) from an agreement at period τ as:

$$R_1(T_{13}, \tau) = \delta_N^\tau (U_1^a - U_1^{na}) = \delta_N^\tau (V_1(q_3^*, q_4^*) - V_1(q_3^d, q_4^*) - T_{13})$$
$$R_3(T_{13}, \tau) = \delta_S^\tau (u_3^a - u_3^{na}) = \delta_S^\tau (v_3(q_3^*, q_4^*) - v_3(q_3^d, q_4^*) + T_{13}).$$

Let $T_{13}^{(1)}$ be the offer (or the counter-offer) made by player 1 over the transfer T_{13} to be received by 3, and $T_{13}^{(3)}$ the counter-offer made by 3 (although the counter-offer is made by 3 it is still a transfer from player 1

[10] Let us note that discount factors in bargaining models are not necessarily related to financial discount rates but capture all the reasons why one player is more or less impatient about reaching an agreement (Muthoo, 1999).

[11] The abatement performed by the southern players depends on the agreement structure, i.e $q_3(\phi_5)$ and $q_4(\phi_5)$ but for simplicity we omit these terms.

to 3). Rubinstein's bargaining procedure mentioned above is as follows: 1 makes an offer of a transfer $T_{13}^{(1)}$ to 3. If 3 accepts then the bargaining is over. If 3 declines the offer, 3 proposes a counteroffer $T_{13}^{(3)}$. If 1 accepts this offer the agreement is struck, and if 1 does not accept, 1 makes a new counter-offer. The alternating-offers procedure continues until an agreement is reached. The RBS that we are looking for is the unique SPE given by the following two conditions (Rubinstein, 1982; Muthoo, 1999): $R_1(T_{13}^{(3)}, 0) = R_1(T_{13}^{(1)}, 1)$ and $R_3(T_{13}^{(1)}, 0) = R_3(T_{13}^{(3)}, 1)$. This yields:

$$V_1(q_3^*, q_4^*) - V_1(q_3^d, q_4^*) - T_{13}^{(3)} = \delta_N[V_1(q_3^*, q_4^*) - V_1(q_3^d, q_4^*) - T_{13}^{(1)}]$$

$$v_3(q_3^*, q_4^*) - v_3(q_3^d, q_4^*) + T_{13}^{(1)} = \delta_S[v_3(q_3^*, q_4^*) - v_3(q_3^d, q_4^*) + T_{13}^{(3)}].$$

The first equation asserts that 1 is indifferent in terms of expected payoffs as regards accepting 3's offer $T_{13}^{(3)}$ in the current period or rejecting it and making in the following period the counteroffer $T_{13}^{(1)}$ that will be accepted by 3 (as the negotiation is stationary, comparing period 1 and 2 is the same as comparing periods 0 and 1). The second equation reflects the same indifference for 3. Since we assume (assumption A4) that 1 makes the first offer, the equilibrium transfer is

$$T_{13}^* = \varphi[V_1(q_3^*, q_4^*) - V_1(q_3^d, q_4^*)] + (1 - \varphi)[v_3(q_3^d, q_4^*) - v_3(q_3^*, q_4^*)],$$

with $\varphi = \frac{\delta_S(1-\delta_N)}{1-\delta_N\delta_S}$. As shown by Rubinstein (1982), under perfect information the equilibrium offer is proposed by player 1 in the first round and it is immediately accepted by player 3, i.e., $T_{13}^{(1)} = T_{13}^*$ (delays in reaching an agreement can only take place with incomplete information (Caparrós *et al.*, 2004)).

The transfer $T_{13}^* > 0$ to be paid by 1 to 3 depends positively on the marginal contribution of 3 to the surplus obtained by 1 (first term in brackets) and on the free-riding payoff of 3 (second term in brackets).

At the same time, a similar program between 1 and 4 yields the agreement payoff functions $U_1^a = V_1(q_3^*, q_4^*) - T_{13}^* - T_{14}$ and $u_4^a = v_4(q_3^*, q_4^*) + T_{14}$, and the impasse point in case of a failure of the negotiation $U_1^{na} = V_1(q_3^*, q_4^d) - T_{13}^*$ and $u_4^{na} = v_4(q_3^*, q_4^d)$, with $q_4^d = \arg\max_x v_4(x, q_3^*)$. Hence, simultaneously 1 reaches an agreement with 4 based on a similar transfer:

$$T_{14}^* = \varphi[V_1(q_3^*, q_4^*) - V_1(q_3^*, q_4^d)] + (1 - \varphi)[v_4(q_3^*, q_4^d) - v_4(q_3^*, q_4^*)].$$

The resulting payoffs for all the players are

$$\begin{aligned} U_1(\phi_5) &= V_1(q_3^*, q_4^*) - T_{13}^* - T_{14}^* \\ U_2(\phi_5) &= V_2(q_3^*, q_4^*) \\ u_3(\phi_5) &= v_3(q_3^*, q_4^*) + T_{13}^* \\ u_4(\phi_5) &= v_4(q_3^*, q_4^*) + T_{14}^*. \end{aligned}$$

To deal with all the other agreement structures, let us assume, without loss of generality, that the northern player i is bargaining with the southern player j. An agreement between them yields:

$$U_i^a = V_i(q_j^*, Q_{-j}^*) - T_{ij} - \sum_{l \in J_i/j} T_{il}^* \tag{3}$$

$$u_j^a = v_j(q_j^*, Q_{-j}^*) + T_{ij}, \tag{4}$$

where $Q_{-j} = Q - q_j$ denotes the abatement made by the southern player not involved in the bilateral agreement, if there is any (i.e., if player j is for example 3 then Q_{-j} is q_4, but if j is (34) then Q_{-j} is empty). In case of disagreement, southern players will implement their deviation abatement $q_j^d = \arg\max_x v_j(x, Q_{-j}^*)$, under the assumption that the other southern player will implement the abatement coming out from its optimal agreement. Hence, the impasse point in case of non-agreement is:

$$U_i^{na} = V_i(q_j^d, Q_{-j}^*) - \sum_{l \in J_i/j} T_{il}^* \tag{5}$$

$$u_j^{na} = v_j(q_j^d, Q_{-j}^*). \tag{6}$$

Since we assume (assumption A4) that i makes the first offer, we obtain the following lemma.

Lemma 1. *The equilibrium transfer is defined by:*

$$T_{ij}^* = \varphi[V_i(Q^*) - V_i(q_j^d, Q_{-j}^*)] + (1 - \varphi)[v_j(q_j^d, Q_{-j}^*) - v_j(Q^*)]. \tag{7}$$

As for the particular case discussed above, the transfer T_{ij}^* to be paid by i to j depends positively on the marginal contribution of j to the surplus obtained by i (first term in brackets) and on the free-riding payoff of j (second term in brackets). That is, the more i gets out of the agreement and the more j benefits from deviating from the agreement, the more willing i is to pay. As before, this offer is proposed in the first period and immediately accepted. The resulting payoffs for i and j are, for every agreement structure ϕ_k:

$$U_i(\phi_k) = V_i(Q^*) - T_{ij}^* - \sum_{l \in J_i/j} T_{il}^* = V_i(Q^*) - \sum_{j \in J_i} T_{ij}^*$$

$$u_j(\phi_k) = v_j(Q^*) + T_{ij}^*.$$

3.2. Agreement structure and abatement (stage three)

In stage three, for each agreement in every agreement structure ϕ_k, both players involved know that they will distribute the worth on stage four according to the RBS. Thus, they are interested in maximizing the worth of the bilateral agreement by choosing the level of abatement q_j^* that maximizes the surplus. Hence for each possible bilateral agreement within every agreement structure ϕ_k, we have

$$q_j^*(\phi_k) = \arg\max_x [V_i(x, Q_{-j}^*) + v_j(x, Q_{-j}^*)]. \tag{8}$$

This gives the abatement effort that would be performed for each possible bilateral agreement (recall that each southern player j can only participate in one agreement within ϕ_k). As stated during the presentation of the model, we assume that $V_i(q_j^*, Q_{-j}^*) - V_i(q_j^d, Q_{-j}^*) > v_j(q_j^d, Q_{-j}^*) - v_j(q_j^*, Q_{-j}^*)$ with $q_j^d = \arg\max_x v_j(x, Q_{-j}^*)$.

Substituting in $\Pi(\phi_k) = \sum_i V_i(Q(\phi_k)) + \sum_j v_j(Q(\phi_k))$, we obtain the total value of each agreement structure ϕ_k. As stated above, for the first two branches on the LHS of the game tree there is only one agreement structure so that no further selection is needed. Nevertheless, for the two branches on the RHS we use assumption A2 to select the agreement structure ϕ_k with the largest aggregated value for the players involved, i.e.:

$$\widehat{\phi} = \arg\max_{\phi_3, \phi_4} \Pi(\phi)$$

$$\widetilde{\phi} = \arg\max_{\phi_5, \ldots, \phi_8} \Pi(\phi),$$

where $\widehat{\phi}$ is the agreement structure selected for $\tau_N = \{1, 2\}$ and $\tau_S = \{(34)\}$ and $\widetilde{\phi}$ for $\tau_N = \{1, 2\}$ and $\tau_S = \{3, 4\}$. This selection procedure ensures that, for each branch of the game tree in figure 1, only one agreement structure is selected and that an abatement effort in case of agreement is selected for each potential bilateral agreement within each agreement structure.

3.3. *Formation of a southern negotiation-coalition (stage two)*

We now analyze the implications of the formation of a southern NC τ_S at stage two. In particular we analyze the case in which the southern countries 3 and 4 face the question of whether or not to form a NC (34). This is the lower part in the tree shown in figure 1. Equations (1)–(8) have to be calculated for the two countries acting separately and for the NC (note that the impasse point for the NC (34) differs from that of the separated countries 3 and 4).

The formation of a southern NC at stage two is profitable under the condition that a southern NC (34) is able to get a higher payoff than two southern countries separately (this condition is equivalent to assumption A1 in a world with two countries):

$$u_{(34)} > u_3 + u_4. \tag{9}$$

Let us start by considering a particular example. Suppose that, if southern countries form a NC, (34), the outcome of stages three and four would be that (34) will sign an agreement with the northern player 1 (i.e., that agreement structure ϕ_3 will be reached) and that, if southern countries act separately, southern player 3 signs an agreement with northern player 1, while 4 signs an agreement with 2 (i.e., that agreement structure ϕ_7 will be reached if southern countries act separately). In this case, condition (9) reads:

$$v_{(34)}(q^*_{(34)}) + T^*_{1(34)} > v_3(q^*_3, q^*_4) + T^*_{13} + v_4(q^*_3, q^*_4) + T^*_{24},$$

with T^*_{ij} given by the equivalent to equation (7). We can further rewrite this condition as

$$SE + NE + (NBP - SBP) > 0$$

with

$$\begin{aligned} SE &= [v_{(34)}(q^*_{(34)}) + Dv_{(34)}(q^d_{(34)})] \\ &\quad - [v_3(q^*_3, q^*_4) + Dv_3(q^d_3, q^d_4)] - [v_4(q^*_3, q^*_4) + Dv_4(q^d_3, q^d_4)] \qquad (10) \\ NE &= [V_1(q^*_{(34)}) - V_1(q^d_{(34)})] \\ &\quad - [V_1(q^*_3, q^*_4) - V_1(q^d_3, q^d_4)] - [V_2(q^*_3, q^*_4) - V_2(q^d_3, q^d_4)] \qquad (11) \\ SBP &= D[v_3(q^d_3, q^*_4) - v_3(q^d_3, q^d_4)] + D[v_4(q^*_3, q^d_4) - v_4(q^d_3, q^d_4)] \qquad (12) \\ NBP &= [V_1(q^d_3, q^*_4) - V_1(q^d_3, q^d_4)] + [V_2(q^*_3, q^d_4) - V_2(q^d_3, q^d_4)], \qquad (13) \end{aligned}$$

and $D = \frac{1-\varphi}{\varphi} = \frac{1-\delta_S}{\delta_S(1-\delta_N)}$.

From equation (10), *SE* represents the *'southern efficiency gain'*, that is, the gain which the southern players can expect when they act together (first square bracket) minus what they can expect acting separately (second and third square brackets). *SE* focuses on direct gains[12] obtained by the different southern players. Note that to compare the two situations the southern players add what they would obtain in the event of an agreement and their net benefit in the event of a disagreement, since their expected payoff is a function of both eventualities. From equation (11), *NE* represents the *'northern efficiency gain'*, the direct gain which the northern players would obtain from an agreement with the NC minus the gain they would obtain when dealing with two separated southern players. Unlike for SE, for NE only the net benefit for the northern players from the agreement is relevant (i.e., the three square brackets show the difference before and after the different agreements). The reason is that the southern players, when deciding whether or not to form a NC, are only interested in the part of the benefit for the northern players that is relevant to them, i.e., the part that explains the transfer they can expect to obtain.

We now turn to the most interesting terms for our analysis, *NBP* and *SBP*. The first square bracket in equation (12) shows the indirect gain that an agreement between the northern player 2 and the southern player 4 induces for the southern player 3 as long as 3 signs no agreement (the

12 As mentioned in the introduction, direct net gains can come from fixed-costs savings, from the different amounts of abatement and from the opportunity to share abatement technologies (between the southern players). We call these benefits 'efficiency gains' as most of them fall into this category.

second square bracket shows the indirect gain for 4 of an agreement between 1 and 3). The first square bracket in equation (13) shows the indirect gain which the agreement between 2 and 4 induces for the northern player 1 if this player does not sign any agreement (the second square bracket shows the indirect gain for 2 of an agreement between 1 and 3). We refer to *SBP* and *NBP* as 'bargaining power' terms and, as all the gains involved are positive, we have $NBP \geq 0$ and $SBP \geq 0$. These gains are irrelevant in equilibrium (where all players sign their optimal agreements) but, as we show below, they determine whether or not forming a NC increases bargaining power.

Finally, D collects the discount terms and has the following properties: (i) for a given δ_S, the larger δ_N the larger D (ii) for a given δ_N, the larger δ_S the smaller D (iii) if $\delta_N = \delta_S$, the larger the discount factor the smaller D, and (iv) $D > 0$. In other words, a decrease in D implies more patient players or a more patient southern player compared to the northern player.[13]

To deal now with all possible agreement structures, let us assume, without loss of generality, that the outcome of stages three and four is that the northern player $i = 1, 2, (12)$ signs an agreement with the southern NC (34) if the South acts united, while the northern player $k = 1, 2, (12)$ (respectively p) signs an agreement with the southern country $j = 3, 4$ (respectively $l \neq j$) when they are separate. Thus, condition (9) yields:

$$v_{(34)}(q^*_{(34)}) + T^*_{i(34)} > v_j(q^*_3, q^*_4) + T^*_{kj} + v_l(q^*_3, q^*_4) + T^*_{pl}, \tag{14}$$

with $T^*_{i(34)}$, T^*_{kj} and T^*_{pl} given by the equivalent to equation (7). This expression can be manipulated[14] to obtain a form in which the motives for the formation of a NC can be distinguished. To this end, we define:

$$\begin{aligned}\underline{SE} &= [v_{(34)}(q^*_{(34)}) + Dv_{(34)}(q^d_{(34)})] - [v_l(q^*_3, q^*_4) + Dv_l(q^d_3, q^d_4)]\\ &\quad - [v_j(q^*_3, q^*_4) + Dv_j(q^d_3, q^d_4)] \qquad (15)\\ \underline{NE} &= [V_i(q^*_{(34)}) - V_i(q^d_{(34)})] - [V_p(q^*_3, q^*_4) - V_p(q^d_3, q^d_4)]\\ &\quad - [V_k(q^*_3, q^*_4) - V_k(q^d_3, q^d_4)] \qquad (16)\\ \underline{SBP} &= D[v_l(q^d_l, q^*_j) - v_l(q^d_3, q^d_4)] + D[v_j(q^d_j, q^*_l) - v_j(q^d_3, q^d_4)] \qquad (17)\\ \underline{NBP} &= [V_p(q^d_l, q^*_j) - V_p(q^d_3, q^d_4)] + [V_k(q^d_j, q^*_l) - V_k(q^d_3, q^d_4)]. \qquad (18)\end{aligned}$$

13 In most economic analysis, discount factors are assumed to be smaller for southern players than for northern players (or discount rates larger). Nevertheless, the terms included in D are not necessarily related to financial discount rates but incorporate all 'political' reasons that may explain why one player is more impatient than another to reach an agreement. For example, internal political pressure may be stronger in northern players, implying that they are more impatient to reach an agreement, and this would yield $\delta_N < \delta_S$.

14 We have added and subtracted in (9) $Dv_j(q^d_3, q^d_4)$, $Dv_l(q^d_3, q^d_4)$, $V_k(q^d_3, q^d_4)$ and $V_p(q^d_3, q^d_4)$.

Based on these terms, condition (19) shows that southern countries have an incentive to form a NC if

$$\underline{SE} + \underline{NE} + (\underline{NBP} - \underline{SBP}) > 0. \tag{19}$$

An increase in the efficiency terms $\underline{SE}$ and $\underline{NE}$ will always encourage southern countries to form a NC. Nevertheless, even if the sum of the efficiency terms is positive, the bargaining power terms can invert the condition above. In order to focus only on the bargaining power terms, condition (19) shows even if forming a southern NC has no efficiency gains (i.e., $\underline{SE} = \underline{NE} = 0$), southern countries have a incentive to form a NC if $\underline{NBP} > \underline{SBP}$.

That is, even in the eventuality that $\underline{SE} = \underline{NE} = 0$ (i.e., that the countries involved in the agreements obtain the same direct gain by passing from the initial situation to the final situation with all the optimal agreements in place), the southern countries may have incentives to form a NC, depending on the relative values of the bargaining power terms $\underline{NBP} \geq 0$ and $\underline{SBP} \geq 0$. In other words, southern countries have an incentive to form a NC if, while acting separately, a partial agreement induces a larger indirect gain for northern countries that do not participate than for southern countries that are not part of the agreement (taking into account in the latter case the impact of D). The intuition for this result is that, by acting together, southern countries can expect to obtain part of the benefit obtained by a northern NC not participating in the agreement while they know that the additional rent that they would potentially need to share, the benefit obtained by a non-participating southern NC, is comparatively small.

3.4. Formation of a northern negotiation-coalition (stage one)

This section analyzes the implications of the formation of a northern NC at stage one. The formation of a northern NC will be profitable if:

$$U_{(12)} > U_1 + U_2. \tag{20}$$

The main difference from the previous stage is that now we have to consider that northern players can sign multiple agreements, unlike southern players. Without loss of generality, we assume that when the northern countries act separately, 1 signs an agreement with the set J_1 of southern players and 2 signs an agreement with the set J_2 of southern players. The sets J_1 or J_2 can be empty and are disjoint such that $J_1 \cap J_2 = \emptyset$. The set J_1 can cover one agreement $\{3\}$, $\{4\}$ or $\{(34)\}$, two agreements $\{3, 4\}$ or no agreement $\{\emptyset\}$. The same holds for J_2. For the northern NC, we can have one agreement $\{(34)\}$ or two agreements $\{3, 4\}$. We use the notation $|J_i|$ to denote the number of elements in the set. Thus, we can rewrite (20) as:

$$V_{(12)}(q_w^*) - \sum_{w \in J_{(12)}} T_{(12)w}^* > V_1(q_z^*, q_y^*) - \sum_{z \in J_1} T_{1z}^* + V_2(q_z^*, q_y^*) - \sum_{y \in J_2} T_{2y}^*, \tag{21}$$

with $T^*_{(12)w}$, T^*_{1z} and T^*_{2y} given by the equivalent to equation (7). To get the motives that drive the formation of a northern NC, we define:[15]

$$\overline{SE} = \sum_{w\in J_{(12)}} [v_w(q^*_w) - v_w(q^d_w)] - \sum_{z\in J_1}[v_z(q^*_z, q^*_y) - v_z(q^d_z, q^d_y)] - \sum_{y\in J_2}[v_y(q^*_z, q^*_y) - v_y(q^d_z, q^d_y)] \quad (22)$$

$$\overline{NE} = \frac{(1-\varphi|J_{(12)}|)V_{(12)}(q^*_w) - (1-\varphi|J_1|)V_1(q^*_z, q^*_y) - (1-\varphi|J_2|)V_2(q^*_z, q^*_y)}{1-\varphi} + \frac{\varphi}{1-\varphi}\left(\sum_{w\in J_{(12)}} V_{(12)}(q^d_w) - \sum_{z\in J_1} V_1(q^d_z, q^d_y) - \sum_{y\in J_2} V_2(q^d_z, q^d_y)\right) \quad (23)$$

$$\overline{SBP} = \sum_{z\in J_1}(v_z(q^d_z, q^*_y) - v_z(q^d_z, q^d_y)) + \sum_{y\in J_2}(v_y(q^d_y, q^*_z) - v_y(q^d_z, q^d_y)) \quad (24)$$

$$\overline{NBP} = \frac{\sum_{z\in J_1}(V_1(q^d_z, q^*_y) - V_1(q^d_z, q^d_y)) + \sum_{y\in J_2}(V_2(q^d_y, q^*_z) - V_2(q^d_z, q^d_y))}{D}. \quad (25)$$

The fact that each northern player can sign multiple agreements (unlike southern players) explains the differences from the definitions of $\underline{SE}$, $\underline{NE}$, $\underline{NBP}$ and $\underline{SBP}$ in the previous section. If each northern player signs an agreement with only one player ($|J_1| = |J_2| = |J_{(12)}| = 1$), equations (22)–(25) are similar to (15)–(18): (i) $\overline{SE}$ and $\overline{NE}$ are focused on direct gains while $\overline{NBP}$ and $\overline{SBP}$ are given by indirect gains (ii) D appears in $\overline{NE}$ and $\overline{NBP}$ and not in $\overline{SE}$ and $\overline{SBP}$ (iii) $\overline{SE}$ focuses on incremental benefits as does $\underline{NE}$ and (iv) $\overline{NE}$ adds up agreement and disagreement outcomes as does $\underline{SE}$. Interpretations are therefore symmetrical.

Based on these terms, condition (26) shows that northern countries have an incentive to form a NC if

$$\overline{NE} + \overline{SE} - (\overline{NBP} - \overline{SBP}) > 0. \quad (26)$$

As in condition (19), we obtain that an increase in the efficiency terms tends to favor a united North. However, the bargaining power terms ($\overline{NBP} - \overline{SBP}$) appear in condition (26) with the opposite sign. The interpretation of $\overline{NBP} \geq 0$ and $\overline{SBP} \geq 0$ is similar to that of $\underline{NBP}$ and $\underline{SBP}$, although now the expressions have to take into account the fact that northern players can sign agreements with different southern players. Focusing again on the eventuality that $\overline{SE} = \overline{NE} = 0$, the incentive for the northern countries to form a NC depends on the relative values of the bargaining terms $\overline{NBP}$

[15] By adding and subtracting $\sum_{z\in J_1} v_z(q^d_z, q^d_y)$, $\sum_{y\in J_2} v_y(q^d_z, q^d_y)$, $D^{-1}\sum_{z\in J_1} V_1(q^d_z, q^d_y)$ and $D^{-1}\sum_{y\in J_2} V_2(q^d_z, q^d_y)$ in (21).

and $\overline{SBP}$. Northern countries have an incentive to form a NC if $\overline{SBP} > \overline{NBP}$ (i.e., if a partial agreement induces a larger indirect gain in southern players that do not participate than in northern players that are not part of the partial agreement). That is, if bargaining power favors one of the sides of the negotiations acting in a united way, the general tendency will go in the opposite direction for the other side. The intuition is the same as for the formation of a southern NC, but since we are comparing the potential gains from non-participating countries from the northern perspective the balance that favors forming a NC is the opposite. In addition, as in our framework the grand coalition can only be the outcome of a negotiation between a northern and a southern NC, bargaining power terms make the formation of the grand coalition more difficult.

3.5. Equilibrium

Collecting previous results, we can write the following proposition that describes the equilibrium of the game and shows that the equilibrium is unique:

Proposition 1. *The game* Γ *has a unique SPE in which transfers are given by Lemma 1. The equilibrium NC structure and agreement structure are*

1. $(\tau_N^*; \tau_S^*) = \{(12); (34)\}$ *with* $\phi_1^* = \{[(12), (34)]\}$ *iif*

$$u_{(34)}(\phi_1) > u_3(\phi_2) + u_4(\phi_2) \quad \textit{and}$$

$$\begin{cases} U_{(12)}(\phi_1) > U_1(\widehat{\phi}) + U_2(\widehat{\phi}) & \textit{if } u_{(34)}(\widehat{\phi}) > u_3(\widetilde{\phi}) + u_4(\widetilde{\phi}) \\ U_{(12)}(\phi_1) > U_1(\widetilde{\phi}) + U_2(\widetilde{\phi}) & \textit{if } u_{(34)}(\widehat{\phi}) < u_3(\widetilde{\phi}) + u_4(\widetilde{\phi}) \end{cases}$$

with $\widehat{\phi} = \arg\max_{\phi_3,\phi_4} \Pi(\phi)$ *and* $\widetilde{\phi} = \arg\max_{\phi_5,\ldots,\phi_8} \Pi(\phi)$.

2. $(\tau_N^*; \tau_S^*) = \{(12); 3, 4\}$ *with* $\phi_2^* = \{[(12), 3], [(12), 4]\}$ *iif*

$$u_{(34)}(\phi_1) < u_3(\phi_2) + u_4(\phi_2) \quad \textit{and}$$

$$\begin{cases} U_{(12)}(\phi_2) > U_1(\widehat{\phi}) + U_2(\widehat{\phi}) & \textit{if } u_{(34)}(\widehat{\phi}) > u_3(\widetilde{\phi}) + u_4(\widetilde{\phi}) \\ U_{(12)}(\phi_2) > U_1(\widetilde{\phi}) + U_2(\widetilde{\phi}) & \textit{if } u_{(34)}(\widehat{\phi}) < u_3(\widetilde{\phi}) + u_4(\widetilde{\phi}) \end{cases}$$

3. $(\tau_N^*; \tau_S^*) = \{1, 2; (34)\}$ *with* $\widehat{\phi}^* = \arg\max_{\phi_3,\phi_4} \Pi(\phi)$ *iif*

$$u_{(34)}(\widehat{\phi}) > u_3(\widetilde{\phi}) + u_4(\widetilde{\phi}) \textit{ and}$$

$$\begin{cases} U_{(12)}(\phi_1) < U_1(\widehat{\phi}) + U_2(\widehat{\phi}) & \textit{if } u_{(34)}(\phi_1) > u_3(\phi_2) + u_4(\phi_2) \\ U_{(12)}(\phi_2) < U_1(\widehat{\phi}) + U_2(\widehat{\phi}) & \textit{if } u_{(34)}(\phi_1) < u_3(\phi_2) + u_4(\phi_2) \end{cases}$$

4. $(\tau_N^*; \tau_S^*) = \{1, 2; 3, 4\}$ *with* $\widetilde{\phi}^* = \arg\max_{\phi_5,\ldots,\phi_8} \Pi(\phi)$ *iif*

$$u_{(34)}(\widehat{\phi}) < u_3(\widetilde{\phi}) + u_4(\widetilde{\phi}) \textit{ and}$$

$$\begin{cases} U_{(12)}(\phi_1) < U_1(\widetilde{\phi}) + U_2(\widetilde{\phi}) & \textit{if } u_{(34)}(\phi_1) > u_3(\phi_2) + u_4(\phi_2) \\ U_{(12)}(\phi_2) < U_1(\widetilde{\phi}) + U_2(\widetilde{\phi}) & \textit{if } u_{(34)}(\phi_1) < u_3(\phi_2) + u_4(\phi_2). \end{cases}$$

The uniqueness of the SPE of the game Γ comes directly from the uniqueness of the SPE of the Rubinstein bargaining model. Since the equilibrium transfers are unique, it implies that both the NC structure and the agreement structure are also unique (taking our assumptions into account).

If, for example, $u_{(34)}(\phi_1) > u_3(\phi_2) + u_4(\phi_2)$, $u_{(34)}(\widehat{\phi}) > u_3(\widetilde{\phi}) + u_4(\widetilde{\phi})$ and $U_{(12)}(\phi_1) > U_1(\widehat{\phi}) + U_2(\widehat{\phi})$, the equilibrium is the one described in Proposition (1.1). In this case, both northern and southern countries will form a NC, and the outcome of the negotiations will be a global agreement (i.e., $\phi_1^* = \{[(12), (34)]\}$). The abatement in this case is $q^*_{(34)}(\phi_1) = \arg\max_x[V_{(12)}(x) + v_{(34)}(x)]$, which is the abatement that a central planner would decide. This global agreement corresponds to the grand coalition using the standard definition of coalition used in the IEA literature.

The resulting payoffs are given by the worth for each player of the abatement, plus or minus the transfer from the North to the South:

$$U_{(12)}(\phi_1) = V_{(12)}(q^*_{(34)}(\phi_1)) - T^*_{(12),(34)}(\phi_1)$$
$$u_{(34)}(\phi_1) = v_{(34)}(q^*_{(34)}(\phi_1)) + T^*_{(12),(34)}(\phi_1)$$

with

$$\begin{aligned} T^*_{(12),(34)}(\phi_1) = {} & \varphi[V_{(12)}(q^*_{(34)}(\phi_1)) - V_{(12)}(q^d_{(34)}(\phi_1))] \\ & + (1-\varphi)[v_{(34)}(q^d_{(34)}(\phi_1)) - v_{(34)}(q^*_{(34)}(\phi_1))]. \end{aligned}$$

However, if $u_{(34)}(\widehat{\phi}) < u_3(\widetilde{\phi}) + u_4(\widetilde{\phi})$, $U_{(12)}(\phi_1) < U_1(\widetilde{\phi}) + U_2(\widetilde{\phi})$ and $u_{(34)}(\phi_1) > u_3(\phi_2) + u_4(\phi_2)$, the equilibrium will be the one described in Proposition (1.4), and the agreement structure will be the one with highest aggregated payoff out of $\phi_5, \ldots, \phi_8$. If this is, for example $\phi_5 = \{[1,3],[1,4],2\}$, the equilibrium abatement $q_3^*(\phi_5)$ and $q_4^*(\phi_5)$ will be $q_j^*(\phi_5) = \arg\max_x[V_1(x, q_k^*(\phi_5)) + v_k(x, q_k^*(\phi_5))]$ with $j \neq k = 3, 4$ and the equilibrium payoffs will be:

$$U_1(\phi_5) = V_1(q_3^*(\phi_5), q_4^*(\phi_5)) - T^*_{1,3}(\phi_5) - T^*_{1,4}(\phi_5)$$
$$U_2(\phi_5) = V_2(q_3^*(\phi_5), q_4^*(\phi_5))$$
$$u_3(\phi_5) = v_3(q_3^*(\phi_5), q_4^*(\phi_5)) + T^*_{1,3}(\phi_5)$$
$$u_4(\phi_5) = v_4(q_3^*(\phi_5), q_4^*(\phi_5)) + T^*_{1,4}(\phi_5)$$

with

$$\begin{aligned} T^*_{1,3}(\phi_5) = {} & \varphi[V_1(q_3^*(\phi_5), q_4^*(\phi_5)) - V_1(q_3^d(\phi_5), q_4^*(\phi_5))] \\ & + (1-\varphi)[v_3(q_3^d(\phi_5), q_4^*(\phi_5)) - v_3(q_3^*(\phi_5), q_4^*(\phi_5))] \\ T^*_{1,4}(\phi_5) = {} & \varphi[V_1(q_3^*(\phi_5) + q_4^*(\phi_5)) - V_1(q_3^*(\phi_5), q_4^d(\phi_5))] \\ & + (1-\varphi)[v_4(q_3^*(\phi_5), q_4^d(\phi_5)) - v_4(q_3^*(\phi_5), q_4^*(\phi_5))]. \end{aligned}$$

In this case, neither the North nor the South will be interested in forming any NC, and in equilibrium northern country 1 will sign one agreement with southern country 3 and another agreement with southern country 4, $\phi_5 = \{[1,3],[1,4],2\}$. Using the standard concept of coalitions in the IEA literature, the coalitions will now be $\{1,3\}$ and $\{1,4\}$, showing that in our framework northern countries can take part in more than one 'coalition' (in the IEA literature countries can generally participate in only one coalition, whether there is only one possible coalition (Carraro and Siniscalco, 1993) or several coalitions (Carraro, 2005)).

4. Discussion

Let the northern players be the US (player 1) and the EU[16] (player 2) and the southern players be China (player 3) and the remaining developing countries forming the G77 (player 4). Are northern (southern) players going to act united or separately during the climate change negotiations? Before answering this question, let us see to what extent our assumptions are reasonable for describing these negotiations.

The main limitation is that we have assumed that only bilateral agreements are possible, meaning that the only way to reach a global agreement would be a deal between a northern NC and a southern NC. Fortunately, this is not as far from reality as it could appear at first glance since the North (US–EU) and the South (China–G77) would probably need to reach a common position before they negotiate between themselves if a final global agreement is to be reached. If the US proposes a different deal to that of the EU, then a final global agreement is unlikely. In other words, our bilateral agreement framework does not preclude a global agreement; it only imposes a precise structure for the negotiations in order to reach this agreement. This brings us to the second strong assumption of our framework, the sequential NC structure formation: first the North, then the South. As pointed out above, this assumption is not essential for our analysis but it is probably reasonable if the North acts as a Stackelberg leader, a plausible assumption for a negotiation in which the North 'grants' a transfer to the South in exchange for an abatement effort which ultimately benefits the South more than the North (as climate change is expected to harm southern countries more). Furthermore, the EU has approved its strategy until 2020 by committing to a 30 per cent reduction in its 1990 emissions if other industrialized countries join the effort, or 20 per cent if they do not.[17] Since this was done before negotiations for the post-Kyoto agreement really started, and the EU has good chances to meet its Kyoto targets, the commitment of the EU is *a priori* and credible (almost defining a Stackelberg strategy).

Then, are the countries going to act united or separately? First they will check whether the agreement they can expect by joining forces is better or worse than the agreement they can expect by acting separately. This is captured in the direct efficiency terms given by $\underline{SE}$, $\underline{NE}$, $\overline{SE}$ and $\overline{NE}$ and is

[16] The EU and G77 are already groups of countries, but we will treat them as individual players.

[17] Brussels European Council 8/9 March 2007.7224/1/07Rev1. Brussels, 2 May 2007.

hardly surprising. As conditions (19) and (26) show, the different efficiency terms that we have defined favor southern and northern players forming NCs and, as proposition (1) has shown, the only way to reach the grand coalition is to have a negotiation between a united North and a united South. Hence, the larger the efficiency terms, the larger the options to reach a grand coalition.

Let us now move to the bargaining power terms, which are probably the most interesting ones as they may offset the efficiency terms. Condition (19) tells us that if $\underline{NBP}$ is larger than $\underline{SBP}$, then China and the G77 will have an additional reason to act together, not related to the different agreements that would actually be signed while acting together or separately. Should we expect $\underline{NBP}$ to be larger than $\underline{SBP}$? Probably not. The major beneficiaries of GHG abatement efforts are the countries of the South since they are expected to be the most damaged by climate change (IPCC, 2007). Therefore, the G77 would benefit from a partial agreement involving only China and an agreement signed only by the G77 would probably benefit China (i.e., $\underline{SBP}$ will presumably be large). On the other hand, the EU or the US would benefit relatively little from a partial agreement involving only the other northern player. Furthermore, a partial agreement might even harm one of them in political terms, as the Bush administration in the US probably suffered when the EU and the remaining countries included in Annex I of the Protocol continued with the Kyoto Protocol (i.e., $\underline{NBP}$ is probably small).

And what about the EU and the US? Do they have an incentive based on bargaining power to unite in climate negotiations? Condition (26) gives us the terms which the EU and the US have to consider while deciding whether or not to form a NC. Assume, for the sake of the argument, that each northern NC signs only one agreement and that both discount factors tend to one (i.e., $|J_1| = |J_2| = |J_{(12)}| = 1$; $\delta_S = \delta_N \to 1$ and hence $D \to 1$). In this case, $\overline{NBP} \simeq \underline{NBP}$ and $\overline{SBP} \simeq \underline{SBP}$. Owing to the fact that these terms enter with the opposite sign in condition (19) in condition (26), the same arguments used to support the lack of incentives for the merging of China and the G77 can be used to show that the EU and the US do indeed have an incentive to create a northern NC for bargaining power reasons. Thus, in our framework bargaining power gains hinder the formation of the grand coalition while they encourage southern countries to negotiate separately and northern countries to act united (equilibrium number 2 in Proposition (1)). In real-life negotiations, having a united North and a separated South probably does not preclude the formation of the grand coalition, but it certainly renders a global agreement more difficult. In any case, the take-home message is that, as shown by Hampson and Hart (1995) in their political science based discussion, which NC are formed is key to predicting the outcome of the negotiations.

Have we seen this behavior in past climate change negotiations? On the road to Kyoto in 1997 and Marrakech in 2001, not really. G77 and China negotiated together for most of the negotiation process whereas the EU and the US held opposite positions on most issues. However, the Kyoto Protocol and the Marrakech Accords include no obligations for the southern countries, meaning that we can hardly talk of a North–South agreement, in our sense. Nevertheless, current climate change negotiations are more in line with the predictions of the bargaining power part of our argument. Before the failure in Copenhagen in 2009, the G8 members agreed in July 2009 upon a common abatement target for 2050 (80 per cent reduction of GHG emissions compared to 1990 levels and a 2°C increase in temperature target) and proposed a 50 per cent global abatement target that would imply abatement efforts for southern countries[18] (G8 countries also accepted that they would need to contribute financially). The opposition of China and India prevented an agreement within the G17 on this proposal and in Copenhagen no agreement was reached either (i.e., the North held a common position and offered the South a deal, which was rejected). In Cancun a modest agreement was achieved in 2010 but it is more a deal to save the negotiation process than a deal to tackle climate change once and for all. Northern countries this time had a more or less common position (with the relative exception of Japan), while southern countries were more divided, with China bargaining hard before accepting the deal and with Bolivia refusing the deal altogether until the last minute.

5. Conclusion

This paper has studied the reasons why northern and southern countries decide to form NCs when negotiating over climate change mitigation. The main question that we have analyzed is whether or not these negotiations have a natural tendency to be carried out between NCs and whether the reason for this possible tendency is based on efficiency gains or bargaining power gains (or both). Whether or not bargaining power gains induce the formation of NCs depends on the gains obtained by those players not involved in the eventual partial agreements (which, in our framework, never take place). Furthermore, if bargaining power encourages the formation of a NC on one side of the table, on the other side the tendency will go in the opposite direction. We have also shown that, for the climate change negotiations, we should expect bargaining power terms to favor a united North and a separated South. In our framework this precludes any possibility of reaching the grand coalition. Although this result can probably not be transposed directly to real-life negotiations, what holds is that having the South acting separately will always render a global agreement more difficult. In other words, as in Hampson and Hart (1995), we have shown that the stages where the NCs are formed have a great impact on the final outcome that one should expect.

[18] 'There has been an important convergence in G8 positions, which has provided a strong impulse to the upcoming negotiations to reach an ambitious and effective global agreement in Copenhagen. [...] the G8 countries have committed to reduce their greenhouse gas emissions by 80 per cent or more by 2050 with reference to 1990 or more recent years. [...] The G8 countries confirmed their willingness to contribute their fair share of the financing [...] to ensure the necessary actions to combat climate change also in developing countries.' G8 Fact Sheet – Climate Change ⟨www.g8italia2009.it⟩.

We have left possible extensions of this paper for future research. One alternative would be to assume sequential bargaining of the different agreements instead of simultaneous bargaining. This would allow, for example, analysis of the type of bargaining protocol preferred by the North or the South, respectively. As already pointed out, the NC formation protocol which we have assumed in stages one and two could be replaced by a more complex procedure. This would probably not change the results obtained but it would allow us to study the incentives to deviate within the NCs which we have not analyzed. Finally, the model presented here could be extended to a context with asymmetric information, which would advance the model nearer to reality and would avoid one of the shortcomings of the analysis presented here, namely that agreements are reached without delay.

References

Aghion, P., P. Antràs, and E. Helpman (2007), 'Negotiating free trade', *Journal of International Economics* **73**(1): 1–30.

Barrett, S. (1994), 'Self enforcing international environmental agreements', *Oxford Economic Papers* **46**: 878–894.

Bloch, F. (1996), 'Sequential formation of coalitions in games with externalities and fixed payoff division', *Games and Economic Behavior* **14**: 90–123.

Breton, M., K. Fredj, and G. Zaccour (2006), 'International cooperation, coalitions stability and free riding in a game of pollution control', *Manchester School* **74**(1): 103–122.

Caparrós, A. and J.-C. Péreau (2010), 'Coalition formation and bargaining power: theory and application to international negotiations on public goods', Instituto de Políticas y Bienes Públicos (IPP) Working Paper No. 17, Madrid.

Caparrós, A., J.-C. Péreau, and T. Tazdaït (2004), 'North–south climate change negotiations: a sequential game with asymmetric information', *Public Choice* **121**(3–4): 455–480.

Carraro, C. (2005), 'Institution design for managing global commons: lessons from coalition theory', in G. Demange and M. Wooders (eds), *Group Formation in Economics: Networks, Clubs and Coalitions*, Cambridge, UK: Cambridge University Press, pp. 354–380.

Carraro, C. and D. Siniscalco (1993), 'Strategies for the international protection of the environment', *Journal of Public Economics* **2**: 309–328.

Carraro, C., C. Marchiori, and A. Sgobbia (2005), 'Advances in negotiation theory: bargaining, coalitions and fairness', FEEM Working Paper No. 66.05, Milan.

Chander, P. and H. Tulkens (1997), 'The core of an economy with multilateral environmental externalities', *International Journal of Game Theory* **26**: 379–401.

Chipty, T. and C.M. Snyder (1999), 'The role of firm size in bilateral bargaining: a study of the cable television industry', *Review of Economics and Statistics* **81**(2): 326–340.

de Clippel, G. and R. Serrano (2008), 'Bargaining, coalitions and externalities: a comment on Maskin', Working Paper No. 2008-16, Department of Economics, Brown University, Providence, RI.

Finus, M. (2001), *Game Theory and International Environmental Cooperation*, Cheltenham: Edward Elgar.

Hampson, O. and M. Hart (1995), *Multilateral Negotiations: Lessons from Arms Control, Trade, and the Environment*, Baltimore and London: Johns Hopkins University Press.

Harstad, B. (2010), 'How to negotiate and update climate agreements', in J.E. Aldy and R.N. Stavins (eds), *Post-Kyoto International Climate Policy: Implementing Architectures for Agreement*, Cambridge, UK: Cambridge University Press, pp. 273–299.

IPCC (2007), *Climate Change 2007: Impacts, Adaptation and Vulnerability. Contribution of Working Group II to the Fourth Assessment Report of the Intergovernmental Panel on Climate Change*, M.L. Parry, O.F. Canziani, J.P. Palutikof, P.J. van der Linden and C.E. Hanson (eds), Cambridge, UK: Cambridge University Press.

Maskin, E. (2003), 'Bargaining, coalitions and externalities', Presidential Address to the Econometric Society, Institute for Advanced Study, Princeton.

McGinty, M. (2007), 'International environmental agreements among asymmetric nations', *Oxford Economic Papers* **59**: 45–62.

Muthoo, A. (1999), *Bargaining Theory with Applications*, Cambridge, UK: Cambridge University Press.

Ray, D. and R. Vohra (1999), 'A theory of endogenous coalition structures', *Games and Economic Behavior* **26**: 286–336.

Ray, D. and R. Vohra (2001), 'Coalitional power and public goods', *Journal of Political Economy* **109**(6): 1355–1384.

Rotillon, G., T. Tazdaït, and S. Zeghni (1996), 'Bilateral or multilateral bargaining in the face of global environmental change?', *Ecological Economics* **18**: 177–187.

Rubinstein, A. (1982), 'Perfect equilibrium in a bargaining model', *Econometrica* **50**: 97–109.

[23]

Review of Economic Studies (2012) **79**, 1527–1553 doi:10.1093/restud/rds011

Advance access publication 22 March 2012

Climate Contracts: A Game of Emissions, Investments, Negotiations, and Renegotiations

BÅRD HARSTAD

Kellogg School of Management, Northwestern University; University of Oslo; and National Bureau of Economic Research

First version received June 2010; final version accepted September 2011 (*Eds.*)

The paper presents a dynamic game where players contribute to a public bad, invest in technologies, and write incomplete contracts. Despite the $n+1$ stocks in the model, the analysis is tractable and the symmetric Markov perfect equilibrium unique. If only the contribution levels are contractible, then investments are suboptimally small if the contract is short term or close to its expiration date. To encourage investments, the optimal contract is more ambitious if it is short term, and it is tougher to satisfy close to its expiration date and for players with small investment costs. If renegotiation is possible, such an incomplete contract implements the first-best. The framework helps to analyse emissions, investments, and international environmental agreements, and the results have important lessons for how to design a climate treaty.

Key words: Dynamic games, Incomplete contracts, Hold-up problems, Renegotiation design, Climate change, Environmental agreements

JEL Codes: Q54, D86, H87, F53

1. INTRODUCTION

This paper develops a dynamic model of private provision of public goods. The agents can also invest in substitute technologies, leading to $n+1$ stocks, but the analysis is nevertheless tractable. I derive and characterize a unique symmetric Markov perfect equilibrium (MPE) for the non-cooperative game as well as for situations where the agents can negotiate and contract on contribution levels.

The model is general and could fit various contexts, but the most important application might be climate change. Consistent with the model's assumptions, a country can reduce its emissions in multiple ways: a short-term solution is to simply consume less fossil fuel today, while a more long-term solution might be to invest in renewable energy sources or abatement technology. The Kyoto Protocol is a bargaining outcome limiting countries' emission levels, but it does not specify the extent to which a country should invest or simply reduce its consumption. At the same time, the Protocol is relatively short term since the commitments expire in 2012. This short duration may reflect the difficulties or costs of committing to the distant future.

With this motivation, I refer to the agents as countries, contributions as emissions, and the public bad as greenhouse gases. To explore the potential for a technological solution, I rule out technological spillovers and the associated free-riding problem. To isolate the interaction between the incentive to invest and negotiated emission quotas, I abstract from the difficulties of motivating participation and compliance.

Models with $n+1$ stocks and infinite time are often solved numerically.[1] To get analytical solutions, I make two simplifying assumptions. First, the investment cost is assumed to be linear. Second, the benefit of technology is additive and does not influence marginal emissions from consumption. With these assumptions, the continuation value turns out to be linear in *all* the $n+1$ stocks, making the analysis tractable. Furthermore, a unique symmetric MPE satisfies these conditions. This equilibrium is stationary and coincides with the unique symmetric subgame perfect equilibrium (SPE) if time were finite but approached infinity. These attractive equilibrium properties hold for every scenario studied in the paper.

In contrast to most of the literature, I do not focus on self-enforcing agreements sustained by trigger strategies. Instead, I derive the equilibrium outcome if contracts can be signed. For climate agreements, for example, countries may be able to commit at least to the near future since domestic stakeholders can hold the government accountable if it has ratified an international agreement. Rather than taking a stand on the countries' ability to commit, I derive the equilibrium contract as a function of this ability.

One extreme benchmark is the complete contracting environment. If all emissions and investments can be negotiated, the first-best is easily implemented.

At the other extreme, consider the non-cooperative outcome. Although the technology is private and investments are selfish, each country's technology stock is, in effect, a public good since its role is to substitute for the country's contribution to the public bad. If one country happens to pollute a lot, the other countries are induced to pollute less in the future since the problem is then more severe; they will also invest more in technology to be able to afford the necessary cuts in emissions. If a country happens to invest a lot in abatement technology, it can be expected to pollute less in the future. This induces the other countries to raise their emissions and reduce their own investments. Anticipating these effects, each country pollutes more and invests less than it would in an otherwise similar static model. This dynamic common-pool problem is thus particularly severe.

The real-world contracting environment probably lies between these two benchmarks: countries may be able to contract on certain variables but not on others. When analyzsing incomplete contracts, I let countries contract on emissions but not investments. There are several justifications for this choice: first, this assumption is in line with the literature on incomplete contracts (Segal and Whinston, 2010), making it possible to draw on that literature as well as to clarify my contribution to it. Second, the existing climate treaty (*i.e.* the Kyoto Protocol) does indeed specify levels of emission but not investments. Intuitively, if investment levels had been negotiated, it would be difficult for the enforcing party to distinguish actual investments from short-term abatement efforts.[2]

To begin, suppose countries commit to one period at a time. If there were no further periods, contracting on emission levels would be first-best since investments in technology are selfish. With multiple periods, however, the technology that survives to the next period is, in effect, a public good. The reason for this is that a hold-up problem arises when the countries negotiate emission levels: if one country has better technology and can cut its emissions fairly cheaply, then its opponents may ask it to bear the lion's share of the burden when collective emissions are reduced. Anticipating this, countries invest less when negotiations are coming up (as in Buchholtz and Konrad, 1994). On the one hand, with smaller investments, it is *ex post* optimal to allow for larger emission levels. On the other, since the countries are underinvesting, it is beneficial to encourage more investments and the parties can do this by negotiating a contract

1. See, *e.g.* the literature on industry dynamics, surveyed by Doraszelski and Pakes (2007).

2. Golombek and Hoel (2006, p. 2) observe that "it would hardly be feasible for a country (or some international agency) to verify all aspects of other countries' R&D policies". Thus, this assumption is standard in the literature.

that is tough and allows few emissions. Thus, the best (and equilibrium) contract is tougher and stipulates *lower* emissions compared to the *ex post* optimum, particularly if the length of the contract is relatively short and the technology long-lasting.

If a country has contracted on the emission levels for several periods, then its technology is enjoyed privately during the agreement periods but thereafter becomes, essentially, a public good, thanks again to the hold-up problem. Thus, investments decline towards the end of the contract. In anticipation of this, and to further motivate investments, the optimal and equilibrium contract becomes tougher to satisfy over time. At the same time, the suboptimally low investments make it optimal to emit more *ex post*. Surprisingly, the two effects cancel: in every period, the equilibrium emission quotas are identical to the emission levels that would have been first-best if investments had been first-best.

However, none of these contracts are renegotiation-proof. Once the investments are sunk, countries have an incentive to negotiate *ex post* optimal emission levels rather than sticking to overambitious contracts. When renegotiation is possible and cannot be prevented, an investing country understands that in the end, it does not have to comply with an overambitious contract. Nevertheless, with renegotiation, all investments and emissions are *first-best*. Intuitively, emission levels are renegotiated to *ex post* optimal levels. Countries with poor technology find it particularly costly to comply with an initially ambitious agreement and will be quite desperate to renegotiate it. This gives them a weak bargaining position and a bad outcome. To avoid this fate, countries invest more in technology, particularly if the initial contract is very ambitious. To take advantage of this effect, the optimal contract is tougher if it has a relatively short duration or if it is close to its expiration date, just as in the case without renegotiation.

If investment costs are asymmetric, the first-best requires that only low-cost countries invest. This is also the non-cooperative equilibrium outcome, but non-investing countries pollute far too much since, as a consequence, the low-cost countries will invest more. The optimal incomplete contract specifies smaller emission quotas for low-cost countries in order to motivate them (and not high-cost countries) to invest. Paradoxically, countries with renewable energy sources end up consuming less energy than countries without.

The lessons for climate policies are important. The bad news is that the possibility of developing technology leads to a lot of free riding, even when technological spillovers are absent. The good news is that investments can be encouraged by negotiating emission quotas, but the treaty must be very tough, particularly towards its end and for countries with low investment costs. Efficiency is best achieved by long-lasting agreements that are renegotiated over time. This suggests that climate negotiators have something to learn from international trade policy negotiators since trade agreements are typically long-lasting, although they can expand or be renegotiated over time.

This paper reports on the relatively technical parts of a larger project on climate agreements. The current paper lays out the general theory and focuses on renegotiation, heterogeneity, and robustness. The companion paper (Harstad, 2011) rests on quadratic utility functions and goes further in analysing short-term agreements, when they are actually worse than business-as-usual, and it characterizes the optimal agreement length. The companion paper is also allowing for firms, imperfect property rights, technological trade, investment subsidies, as well as uncertainty and stochastic shocks.

The literature on climate and environmental agreements is better reviewed elsewhere (Kolstad and Toman, 2005; Andy and Stavins, 2009; Harstad, 2011). The paper's contribution to the earlier literature on dynamic games and incomplete contracts is clarified in the next section. The model is presented in Section 3. When solving the model in Section 4, I gradually increase the possibilities for negotiations and contracts by analysing the non-cooperative game, one-period contracts, multiperiod contracts, and contracts permitting renegotiation. Section 5 permits heterogeneous and convex investment costs and imposes non-negativity constraints on the levels of technology and emissions. Section 6 concludes and the Appendix contains all proofs.

2. CONTRIBUTIONS TO THE LITERATURE

2.1. *Dynamic games and the environment*

The private provision of public goods is often studied in differential games (or a difference game, if time is discrete) where each player's action influences the future stock or state parameter.[3] The result that dynamic common-pool problems are worse than their static versions is known from Fershtman and Nitzan (1991), who present a differential game where agents privately provide public goods in continuous time. Differential games are, however, often difficult to analyse. Many authors restrict attention to linear-quadratic functional forms.[4] While some papers arbitrarily select the linear MPE (*e.g.* Fershtman and Nitzan, 1991), there are typically multiple equilibria (Wirl, 1996; Tsutsui and Mino, 1990) and many scholars manage to construct more efficient non-linear MPEs.[5] In a model of climate change, Dutta and Radner (2009) explore various equilibria and analyse agreements enforced by the threat of reverting to the bad (linear) MPE. The cost of pollution is linear in their model, however, and this eliminates the dynamic effect emphasized below. Dutta and Radner (2004, 2006) also allow countries to invest in technologies that reduce the emission factor associated with their consumption. As in this paper, the cost of investment is linear. But since the cost of pollution is also assumed to be linear, their equilibrium is "bang-bang" where countries invest either zero or maximally in the first period and never thereafter.[6]

The first contribution of this paper is the development of a tractable model that can be used to analyse investments as well as emissions. Under the assumption that technology has a linear cost and an additive impact (unlike Dutta and Radner), I find that the continuation values must be linear in all the $n+1$ stocks, which implies a single symmetric MPE, sharp predictions, and a relatively simple analysis.[7] The second and most important contribution, made possible by the first, is the incorporation of incomplete contracts in dynamic games.[8]

3. Thus, such games are subclasses of stochastic games. For overviews, see Başar and Olsder (1999) or Dockner *et al.* (2000).

4. Fershtman and Nitzan (1991) is one example. For a comprehensive overview of linear-quadratic differential games, see Engwerda (2005).

5. See, *e.g.* Dockner and Long (1993), Dockner and Sorger (1996), Sorger (1998), and Marx and Matthews (2000).

6. Investments are also permitted by Hong and Karp (2010), studying the effect on participation, and Barrett (2006) and Hoel and de Zeeuw (2010), focusing on breakthrough technologies.

7. Potentially, this trick can also be applied when studying other economic problems. In the literature on industry dynamics (see the survey by Doraszelski and Pakes, 2007), *e.g.* analytical solutions are rare and numerical simulations typical. A firm typically overinvests in capacity to get a competitive advantage. While Reynolds (1987) restricts attention to the linear MPE in a linear-quadratic model, simple two-stage games are used by d'Aspremont and Jacquemin (1988) to discuss the benefits of cooperation and by Gatsios and Karp (1992) to show that firms may invest *more* if they anticipate future merger negotiations.

8. Few papers allow for policies or negotiation in stochastic games. Exceptions include Hoel (1993), studying a differential game with an emission tax; Yanase (2006), deriving the optimal contribution subsidy; Houba, Sneek and Vardy (2000), analysing negotiations over (fish) quotas lasting forever; and Sorger (2006), studying one-period agreements. Although Ploeg and de Zeeuw (1992) even allow for R&D, contracts are either complete or absent in all these papers. In Buchholz and Konrad (1994), legislators negotiate spending on "pork" and a long-lasting public good. The equilibrium public-good level is suboptimally but strategically low (as in this paper) to discourage future coalitions from wasting money on pork. However, their mechanism relies on majority rule, and the contract incompleteness is related to future policies rather than current investments.

2.2. *Contract theory*

By permitting contracts on emissions but not on investments, this paper is in line with the literature on incomplete contracts (*e.g.* Hart and Moore, 1988). It is well known that with only one period, selfish investments, and no uncertainty regarding the optimal quantity, the first-best is implemented simply by committing to the optimal quantity before the investment stage (see Proposition 5, Segal and Whinston, 2010). This is in line with my results if technology depreciates completely after each period or the countries can commit to the end of time. If these assumptions are violated, however, I find that investments are lower if the contract length is short and that investments decrease towards the end of a contract. To encourage more investments, the optimal and equilibrium contract is tougher to comply with if the contract is short term or close to its expiration date, particularly if the technology is long-lasting compared to the length of the agreement. These results have not been detected earlier, to the best of my knowledge.

The literature on repeated moral hazard is also somewhat related to the present paper. This literature, as surveyed by Chiappori *et al.* (1994), derives a principal's optimal contract when motivating an agent to exert effort. If the optimal long-term contract happens to be "memoryless", then it is also implementable by short-term or spot contracts (see their Proposition 2). This is consistent with my results if the technology depreciates completely after each period.[9] In other dynamic settings, hold-up problems may be solved if the parties can invest while negotiating and agreements can be made only once (Che and Sakovics, 2004) or if there are multiple equilibria in the continuation game (Evans, 2008). Neither requirement is met in this paper, however.

When considering renegotiation, the paper is closer to the existing literature. Chung (1991) and Aghion *et al.* (1994) have shown how the initial contract can provide incentives by affecting the bargaining position associated with particular investments. If investments are selfish, Edlin and Reichelstein (1996) find that the first-best is achieved by initially contracting on the expected optimal quantity and thereafter allowing for renegotiation. When introducing direct investment externalities, Segal and Whinston (2002, Figure 3) show that the larger the externality, the larger the optimal default quantity compared to the expected *ex post* first-best quantity. Renegotiation is then necessary to ensure *ex post* efficiency. My model needs neither uncertainty nor externalities for renegotiation to be beneficial: with an infinite time horizon, I find that the optimal contract is very aggressive if its length is relatively short and the technology long-lasting, and renegotiation is necessary to ensure *ex post* efficiency. One country's investment affects everyone's continuation value when the agreement expires, creating an intertemporal externality. The dynamic model with selfish investments is thus different from its static counterpart but similar to one with positive direct externalities.

Guriev and Kvasov (2005) present a dynamic moral hazard problem emphasizing the termination time. Their contract is renegotiated at every point in time, to keep the remaining time horizon constant. Contribution levels are not negotiated, but contracting on time is quite similar to contracting on quantity: to increase investments, Guriev and Kvasov let the contract length increase, while Segal and Whinston (2002) let the contracted quantity increase. In this paper, agents can contract on quantity (of emissions) as well as on time, which permits the study of how the two interact. I also allow an arbitrary number of agents, in contrast to the buyer–seller situations in most papers. It is still realistic to assume that unanimity is required in the negotiations, even though a smaller majority requirement would have strengthened the incentives to invest (Harstad, 2005).

9. In general, however, Chiappori *et al.* show that the optimal long-term contract is not implementable by spot contracts because of the benefit from intertemporal consumption smoothing. This benefit does not exist in the present paper (since there is no uncertainty and since linear investment costs make agents risk neutral). On the other hand, while Chiappori *et al.* assume that current effort does not influence future outcomes, I let investments cumulate to a long-lived technology stock. This prevents spot contracts from implementing the first-best in the present paper.

3. THE MODEL

3.1. *Description of the game*

This section presents a game where a set N of $n = |N| > 1$ agents contributes over time to a public bad while they also invest in technology. The public bad is represented by the stock G. Let $1 - q_G \in [0, 1]$ measure the fraction of G that "depreciates" from one period to the next. The stock G may nevertheless increase, depending on each agent's contribution level, g_i:

$$G = q_G G_{-} + \sum_N g_i. \tag{3.1}$$

Parameter G_{-} represents the level of the public bad left from the previous period; subscripts for periods are thus skipped. To fix ideas and illustrate the importance of the results, I will henceforth refer to the agents as countries, the public bad as greenhouse gases, and contributions as emissions.

Each country $i \in N$ benefits privately from emitting g_i. In order to reduce the necessity of contributing to the public bad, i may invest in a private substitute technology. For example, $i \in N$ can consume fossil fuel (measured by g_i) or renewable energy. Let the technology stock R_i measure the capacity of i's renewable energy sources (*e.g.* the windmills). The total consumption of i is then

$$y_i = g_i + R_i. \tag{3.2}$$

As an alternative interpretation of R_i, it may measure country i's "abatement technology", *i.e.* the amount by which i can reduce (or clean) its potential emissions at no cost. If energy production, measured by y_i, is otherwise polluting, the actual emission level of country i is given by $g_i = y_i - R_i$, which again implies (3.2).

The stock R_i might also depreciate over time, at the rate $1 - q_R \in [0, 1]$. If $R_{i,-}$ represents i's technology stock in the previous period, while r_i measures i's recent investments in this stock, then i's current technology level is given by

$$R_i = q_R R_{i,-} + r_i.$$

The investment stages and the consumption stages alternate over time. A period is defined as the shortest time in which investments and thereafter consumptions take place (as indicated by Figure 1).

The per-unit investment cost is $K > 0$. Let the benefit of consumption be given by the increasing and concave function $B(y_i)$, while the individual disutility from the public bad is represented by the increasing and convex function $C(G)$. The one-period utility can then be written as

$$u_i = B(y_i) - C(G) - Kr_i.$$

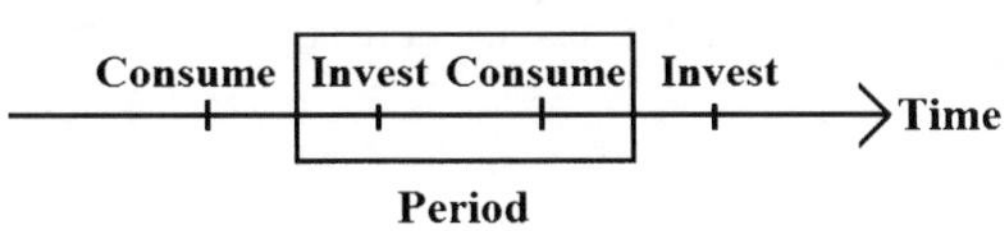

FIGURE 1
The definition of a period

Country i's objective is to maximize the present discounted value of future utilities,

$$V_{i,t} \equiv \sum_{\tau=t}^{\infty} u_{i,\tau}\delta^{\tau-t},$$

where δ is the common discount factor and $V_{i,t}$ is i's continuation value as measured at the start of period t. As mentioned, subscripts denoting period t are typically skipped when this is not confusing.

To allow for alternative applications and to simplify the analysis, note that I have *not required* g_i or r_i to be positive. Even for climate policies, a negative g_i may be technically feasible if carbon capture is an option, although this may be costly, as captured by a large $B'(\cdot)$ for such a small g_i. Similarly, r_i may be negative if the abatement technology and the associated infrastructure can be put to other uses. Nevertheless, non-negativity constraints are imposed in Section 5.2, where I discuss conditions under which the equilibria below generalize. Section 5.1 allows investment costs to be heterogeneous, Section 5.3 lets them be convex (rather than linear) functions, and Section 5.4 introduces uncertainty.

3.2. *The equilibrium concept*

As in most stochastic games, attention is restricted to Markov perfect equilibria where strategies are conditioned on the physical stocks only. As do Maskin and Tirole (2001), I look for the coarsest set of such strategies. Maskin and Tirole (2001, pp. 192–193) defend MPEs since they are "often quite successful in eliminating or reducing a large multiplicity of equilibria" and they "prescribe the simplest form of behavior that is consistent with rationality" while capturing the fact that "bygones are bygones more completely than does the concept of subgame-perfect equilibrium".

Because the investment cost is linear in the model above, multiple MPEs exist, distinguished by how the sum of investment is allocated. Since the model is otherwise symmetric, in the non-cooperative game I will restrict attention to symmetric MPEs where identical countries invest identical amounts. The asymmetric equilibria would, in any case, not survive if the investment cost functions were slightly but strictly convex, rather than linear (see Section 5).

If the countries are negotiating a contract, I assume that the bargaining outcome is efficient and symmetric *if* the pay-off-relevant variables are symmetric across countries. These assumptions are weak and satisfied in several situations. For example, we could rely on co-operative solution concepts, such as the Nash bargaining solution (with or without side transfers). Alternatively, consider a non-cooperative bargaining game where one country can make a take-it-or-leave-it offer to the others, and side transfers are feasible. If every country has the same chance of being recognized as the proposal-maker, the equilibrium contract is exactly as described below.[10]

With this equilibrium concept, there turns out to be a unique equilibrium for each of the contracting environments investigated below; furthermore, it coincides with the unique symmetric SPE if time were finite but approached infinity. This result is desirable; in fact, Fudenberg and Tirole (1991, p. 533) have suggested that "one might require infinite-horizon MPE to be limits of finite-horizon MPE".

10. For the negotiation game with side transfers, there exists multiple asymmetric MPEs since it is irrelevant if i is induced to invest more than j as long as i is compensated for this. But the pay-offs must be the same across these equilibria since the pay-offs in the bargaining outcome are pinnned down by the threat of reverting to the symmetric non-cooperative MPE.

All countries participate in the contract in equilibrium since there is no stage at which they can commit to not negotiating with the others.

4. ANALYSIS

Six contracting environments are analysed in the following subsections.

Contract	Complete	None	Incomplete	+ Renegotiations
One period	1	2	3	5
Multiperiod			4	6

Naturally, different contracting environments generate different outcomes. Nevertheless, all the contracting environments share a number of attractive features. Rather than repeating these features for each case, let me start by emphasizing them since they constitute the main methodological contribution of the paper.

Theorem. *For each contracting environment* (1–6), *it is true that*

(1) *The individual continuation value V is a function of only G_- and $R_- \equiv \sum_N R_{i,-}$.*
(2) *Continuation values are linear in the stocks and satisfy*

$$\frac{\partial V(G_-, R_-)}{\partial G_-} = -q_G(1-\delta q_R)\frac{K}{n};$$
$$\frac{\partial V(G_-, R_-)}{\partial R_-} = q_R\frac{K}{n}.$$

(3) *There is a unique symmetric MPE.*
(4) *This MPE coincides with the unique symmetric SPE if time is finite, $t \in \{1, \ldots, \bar{t}\}$, and $t < \bar{t} - 1$.*

Part (1) is easily illustrated. In principle, the continuation value V_i is a function of the $n+1$ stocks G and $\mathbf{R} \equiv [R_1, R_2, \ldots, R_n]$. However, note that choosing g_i is equivalent to choosing y_i, once the R_is are sunk. Thanks to the assumption that technology and emissions are perfect substitutes, by combining (3.1) and (3.2), we get

$$G = q_G G_- + \sum_N y_i - R.$$

This way, the R_is are eliminated from the model: they are *pay-off irrelevant* as long as R is given, and i's Markov perfect strategy for y_i is therefore not conditioned on them.[11] Hence, a country's continuation value V_i is a function of G_- and R_-, and not of $R_{i,-} - R_{j,-}$, and we can write that value as $V(G_-, R_-)$, without the subscript i.

The remaining parts of the theorem are best understood after the analysis of each scenario and by following the proofs in the Appendix. For example, in every scenario, the value of another unit of technology is that each of the n countries can invest q_R/n units less in the following period: this explains the level $\partial V/\partial R_-$ and why this is constant across scenarios. The fact that V is linear in *all* stocks hinges on the linear investment costs, revealing the power of this assumption.

11. This follows from the definition by Maskin and Tirole (2001, p. 202), where Markov strategies are measurable with respect to the coarsest partition of histories consistent with rationality.

The linearity of V makes the model tractable and permits only one equilibrium. To simplify notation, equations often rely on the conventional definitions $V_G \equiv \partial V/\partial G_-$, $V_R \equiv \partial V/\partial R_-$, $B' \equiv dB(y_i)/dy_i$, $C' \equiv dC(G)/dG$, etc.

4.1. *The first-best outcome*

For future reference, I start by describing the first-best outcome, *i.e.* the outcome if every $g_{i,t}$ and $r_{i,t}$ were contractible or chosen by a planner maximizing the sum of utilities $\sum_N V_{i,t}$.

Proposition 1.

(1) *The first-best consumption* $y_i = g_i + R_i$ *is independent of* R_i*, given* R*:*

$$B'(g_i^*(\mathbf{R}) + R_i) - n(C' - \delta V_G) = 0. \tag{4.1}$$

(2) *The first-best investment ensures that* R_i *satisfies*

$$B'(g_i^*(\mathbf{R}^*) + R_i^*) = (1 - \delta q_R)K. \tag{4.2}$$

(3) *The first-best pollution level* G *satisfies*

$$C'(G^*) = (1 - \delta q_G)(1 - \delta q_R)K/n. \tag{4.3}$$

Part (1) simply states that, at the emission stage, the (private) marginal benefit of consumption should equal the (social) cost of pollution. Intuitively, this requires identical consumption levels, $y_i^* = y_j^*, \forall i, j \in N^2$, regardless of the differences in technologies. Technological differences do imply, however, that the *ex post* first-best *emission* level should be smaller for i if R_i is large:

$$g_i^*(\mathbf{R}) \equiv y_i^* - R_i.$$

Part (2) states that the first-best investment level equalizes the marginal benefit to the marginal cost, recognizing that more investments today reduce the need to invest in the next period. By substituting (4.2) in (4.1), we find the first-best pollution level, stated in part (3). Interestingly, note that both y_i^* and G^* are independent of the past stocks: G^* is pinned down by (4.3) and, together with G_-, this pins down $\sum_N g_i$. Since y_i is pinned down by (4.2), $R_i = y_i - g_i$ follows as a residual, and so does $r_i = R_i - q_R R_{i,-}$.

4.2. *The non-cooperative equilibrium*

Suppose now that countries act non-cooperatively at every decision node. At the emission stage, a marginally larger g_i gives country i the benefit B', but i's increased cost of present pollution is C' and the cost for the future is δV_G, where V is now i's continuation value in the non-cooperative environment. Of course, each country internalizes only $1/n$ of the total harm.

Proposition 2 (emission). (1) *The equilibrium consumption* y_i^{no} *is independent of* R_i *and suboptimally large, given* R*:*

$$B'(y_i^{\text{no}}) = C' - \delta V_G. \tag{4.4}$$

(2) *Country* i *pollutes less but* $j \neq i$ *pollutes more if* R_i *is larger, fixing* R_j, $\forall j \neq i$*:*

$$\partial g_i^{\text{no}}/\partial R_i = -\frac{C''(n-1) - B''}{nC'' - B''} < 0,$$

$$\partial g_j^{\text{no}}/\partial R_i = \frac{C''}{nC'' - B''} > 0, \quad \forall j \neq i.$$

Part (1) verifies that all countries choose the same y_i^{no}, regardless of the R_is. While perhaps surprising at first, the intuition is straightforward. Since (3.2) assumes that technology has an additive impact on consumption, the returns to technology are always inframarginal and the level of technology does not change the *marginal* cost of consumption. Emissions increase one by one in consumption, no matter what the level of technology is.[12]

Since $C\left(q_G G_- + \sum_N y_i - R\right)$ is convex, (4.4) implies that a larger R must increase every y_i. This means that if R_i increases but R_j is constant, then $g_j = y_j - R_j$ must increase. Furthermore, substituting (3.2) in (4.4) implies that if R_i increases, g_i must decrease, as shown in part (2). In words, a country that has better technology pollutes less but—because of this—other countries pollute more.

These results mean that a country's technology stock is, in effect, a public good. A larger R_i raises every country's consumption and, in addition, reduces every investment in the following period. Since each country captures only $1/n$ of the benefits, it invests less than optimally.

Proposition 3 (investments).

(1) *Equilibrium investments are suboptimally low and ensure that*

$$B'(R_i^{\text{no}} + g_i^{\text{no}}) = (1 - \delta q_R/n)\left[\frac{nC'' - B''}{C'' - B''}\right] K, \tag{4.5}$$

assuming that the following second-order condition holds:

$$C'' \geq (n-1)\left[\frac{B'''(C'')^2/B'' - C'''B''}{(C'' - B'')(nC'' - B'')}\right] B'. \tag{4.6}$$

(2) *Every country invests more if R_- is small and G_- large:*

$$\partial r_i^{\text{no}}/\partial R_- = -q_R/n,$$
$$\partial r_i^{\text{no}}/\partial G_- = q_G/n.$$

(3) *The resulting pollution level is suboptimally large and given by*

$$C'(G^{\text{no}}) = (1 - \delta q_G)(1 - \delta q_R)K/n + \left[1 + \frac{(n - \delta q_R)C''}{C'' - B''}\right](1 - 1/n)K. \tag{4.7}$$

Compared to the first-best (4.2), the right-hand side of (4.5) is larger. This implies that country i's consumption, $R_i^{\text{no}} + g_i$, is *smaller* than at the first-best. Since Proposition 2 states that consumption levels are suboptimally large, given $\mathbf{R}$, it follows that investments must be suboptimally small: $g_i^{\text{no}}(\mathbf{R}^{\text{no}}) + R_i^{\text{no}} < g_i^*(\mathbf{R}^*) + R_i^*$ and $g_i^*(\mathbf{R}^{\text{no}}) > g_i^*(\mathbf{R}^{\text{no}})$ implies $\mathbf{R}^{\text{no}} < \mathbf{R}^*$. Furthermore, the effect of suboptimally small investments dominate the suboptimally large emissions, leading to lower equilibrium consumption than at the first-best.[13]

12. Obviously, this would not necessarily be true if I instead had focused on technologies that reduce the emission *factor* of *each* produced unit (as do Dutta and Radner, 2004). The additive form (3.2) is chosen not only because it simplifies the analysis tremendously but also because the resulting crowding-out effects might be reasonable in reality.

13. Note the qualifying second-order condition, which is violated if C''' is positive and large and/or if B''' is very negative. In both situations, an increase in R raises y_j at a decreasing rate. Thus, the discouragement effect when i invests (*i.e.* the fact that other countries pollute more) is reduced when i invests more, and this implies that an interior solution for r_i may not exist. However, the second-order condition holds if C''' and $-B'''$ are small relative to C'' and $-B''$, and it always holds for quadratic functions (implying $C''' = B''' = 0$).

Part (2) first states that, if the technology stock is small today, every country invests more in the next period. Naturally, this contributes to explaining the underinvestment result. Second, if the pollution level is large, all countries invest more. This contributes to explaining the suboptimally large emissions, as described by Proposition 2. Part (3) states that, when all these effects are taken into account, the resulting pollution level is larger than the first-best level.

4.3. *One-period incomplete contracts*

From now on, I let the countries negotiate and contract on their contributions to the public bad. In each period, the timing is the following. First, the countries negotiate a vector of emission levels. Thereafter, each country invests and, finally, every country complies with the contract. The bargaining outcome is assumed to be efficient and symmetric *if* the game itself is symmetric.

The bargaining game is indeed symmetric, even if technology levels differ across the countries. Just as before, the $R_{i,-}$s are eliminated from the model and the continuation value is a function of only G_- and $R_- \equiv \sum_N R_{i,-}$. Moreover, the linear investment cost implies that the continuation value must be linear in both stocks, pinning down a unique equilibrium.

When investing, country i prefers a larger stock of technology if its quota, g_i^{co}, is small since, otherwise, its consumption level would be very low. Consequently, r_i^{co} decreases in g_i^{co}.

Proposition 4 (investments).

(1) *Investments are larger if the contract is tough:* $\partial r_i/\partial g_i^{\text{co}} = -1$.
(2) *For any quota* g_i^{co}, *i's investment is suboptimally low if* $\delta q_R > 0$ *and ensures that*

$$B'(g_i^{\text{co}} + R_i^{\text{co}}) = (1 - \delta q_R/n)K. \tag{4.8}$$

In contrast to the non-cooperative game, a country's technology stock is no longer a "public good": once the emission levels are pinned down, i's investment increases y_i but not y_j, $j \neq i$. However, the technology that survives to the next period, $q_R R_i$, does become a public good since, for a fixed R, the continuation value at the start of every period is independent of R_i. Intuitively, if the agreement does not last forever, a country anticipates that having good technology will worsen its bargaining position in the future, once a new agreement is negotiated. At that stage, good technology leads to a lower quota $g_{i,+}$ since the other countries can hold up i when it is cheap for i to reduce its emissions.[14] In fact, the future consumption $y_{i,+}$ will be the same across countries, regardless of any difference in technology. This discourages a country from investing now, particularly if the current agreement is relatively short (δ large), the technology likely to survive (q_R large), and the number of countries (n) large.

Thus, if n, δ, and q_R are large, it is beneficial to encourage more investments. On the one hand, this can be achieved by a *small* g_i^{co}. On the other hand, the *ex post* optimal g_i^{co} is *larger* when equilibrium investments are low. The optimal quota must trade off these concerns. As shown in the Appendix, the equilibrium and optimal quotas are identical to the emission levels in the first-best![15] However, since the equilibrium investments are less than the optimal ones, the g_i^{co}s are suboptimally low *ex post*.

14. Alternatively, if no agreement is expected in the future, a large $R_{i,+}$ reduces $g_{i,+}$ and increases $g_{j,+}$, as stated in Proposition 2.

15. Intuitively, since the investment cost is linear, consumption is constant and independent of g_i^{co}. Increasing the quota thus has only two effects. On the one hand, investments decrease one by one in g_i^{co} (since $\partial R_i/\partial g_i^{\text{co}} = -1$) and the marginal reduction in investment cost is constant, thanks again to the linear investment cost. On the other hand, the marginal cost of pollution is $C'(G)$. Therefore, setting the optimal quota generates exactly the same trade-off as in the first-best case, and the optimal G is therefore identical.

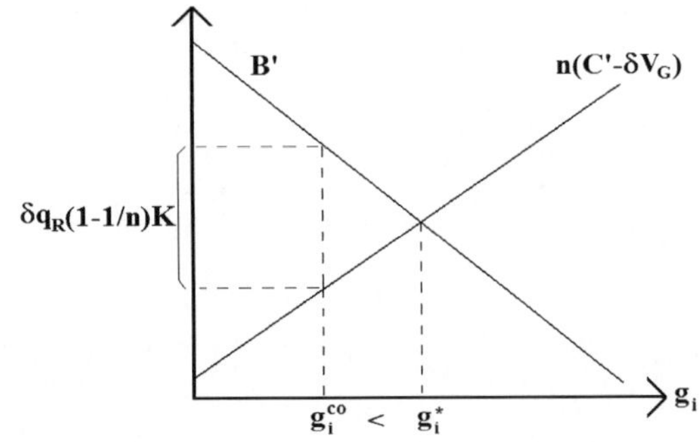

FIGURE 2
The shorter the agreement, the lower is the contracted emission level relative to the *ex post* optimum

Proposition 5 (emission).

(1) *The contracted emission levels are equal to the levels at the first-best:* $g_i^{co} = g_i^*(\mathbf{R}^*) \Leftrightarrow G^{co} = G^*$.
(2) *Since* $R_i^{co} < R_i^*$ *if* $\delta q_R > 0$, $g_i^{co} < g_i^*(\mathbf{R}^{co})$ *and the emission levels are lower than what is ex post optimal:*

$$B'(g_i^{co} + R_i^{co}) - n(C' - \delta V_G) = \delta q_R(1 - 1/n)K > 0. \quad (4.9)$$

Not only is the shadow value of polluting, B', larger than in the non-cooperative case, but it is even larger than it would be in the first-best, (4.1). For a fixed investment level, optimally g_i^{co} should have satisfied $B' - n(C' - \delta V_G) = 0$ rather than (4.9). Only then would marginal costs and benefits be equalized. Relative to this *ex post* optimal level, the quotas satisfying (4.9) must be *lower* since $B' - n(C' - \delta V_G)$ decreases in g_i^{co}. If n, q_R, and δ are large, the additional term $\delta q_R(1 - 1/n)K$ is large, and g_i^{co} must decline. This makes the contract more demanding or *tougher* to satisfy at the emission stage compared to what is *ex post* optimal. In equilibrium, the quotas play a dual role: they control current emissions and mitigate the underinvestment in technology. The optimal emission quotas are therefore particularly stringent if the contract is short term, the technology long-lasting, and the number of countries large, as illustrated in Figure 2.

On the other hand, if $\delta q_R = 0$, the right-hand side of (4.9) is zero, meaning that the commitments under the best long-term agreement also maximize the expected utility *ex post*. In this case, the countries are not concerned with how current technologies affect future bargaining power either because the existing agreement is lasting forever ($\delta = 0$) or because the technology will not survive the length of the contract ($q_R = 0$). Investments are first-best and there is no need to distort the g_i^{co}s downwards.

4.4. *Multiperiod contracts*

Assume now that at the beginning of period 1, the countries negotiate the quotas for every period $t \in \{1, 2, \ldots, T\}$. The time horizon T may be limited by the countries' ability to commit to future promises.

At the start of period 1, the pay-off-relevant stocks are G_- and R_- only. Once again, this simplifies the analysis. There is a unique MPE, the continuation value at the start of period 1 (and $T+1$) is linear in the stocks and has the same slopes as before.

When investing in period $t \in \{1, 2, \ldots, T\}$, countries take the quotas $g_{i,t}^{mc}$s as given, and the continuation value in period $T+1$ is $V(G_T, R_T)$, just as before. At the last investment stage, the problem is the same as in Section 4.3 and a country invests until (4.8) holds. Investments are then suboptimally low since the technology that survives to the next period is basically a public good.

However, by investing more in period $T-1$, a country can invest less in period T. The net investment cost for a country is then $(1-\delta q_R)K$, which leads to first-best investments. The same logic applies to every previous period since, when the quotas for the following period are fixed, technology is a private rather than a public good. Investments fall only towards the end of the commitment period, right before a new round of negotiations and hold-up problems is anticipated.[16]

Proposition 6 (investments). *Investments are first-best at $t < T$ but suboptimally low in the last period for any set of quotas:*

$$B'(g_{i,t}^{mc} + R_{i,t}^{mc}) = (1-\delta q_R)K, \quad t < T, \tag{4.10}$$
$$B'(g_{i,t}^{mc} + R_{i,t}^{mc}) = (1-\delta q_R/n)K, \quad t = T.$$

All this is anticipated when the countries negotiate the quotas. As shown in the Appendix, the optimal and equilibrium quotas must satisfy (4.3) for every $t \leq T$: the equilibrium pollution level is similar to the first-best level, for every period!

In the beginning of the agreement, when $t < T$, the $g_{i,t}^{mc}$s are *ex post* optimal as well since the investments are first-best. In the last period, however, investments decline and the contracted emission levels are lower than the *ex post* optimal levels. In other words, the optimal contract becomes tougher to satisfy towards its end.

Proposition 7 (emission).

(1) *The negotiated quotas equal the first-best emission levels in every period: $g_{i,t}^{\text{co}} = g_i^*(\mathbf{R}^*) \Leftrightarrow G_t^{mc} = G^*$.*
(2) *Thus, quotas are ex post first-best ($g_{i,t}^{mc} = g_i^*(\mathbf{R}_t^{mc})$) for $t < T$, but for $t = T$, since $R_{i,T}^{mc} < R_{i,T}^*$, the quotas are suboptimally low ex post and satisfying (4.9). Consequently, the contract becomes tougher to satisfy toward the end.*

4.5. *Renegotiation*

The contracts above are not renegotiation-proof since they specify emission levels that are less than what is optimal *ex post*, after the investments are sunk. The countries may thus be tempted to renegotiate the treaty. This section derives equilibria when renegotiation is costless.

Starting with one-period contracts, the timing in each period is the following. First, the countries negotiate the initial commitments, the g_i^{de}s, referred to as "the default". If these negotiations fail, it is natural to assume that the threat point is no agreement. Thereafter, the countries invest. Before carrying out their commitments, the countries get together to renegotiate the g_i^{de}s.

16. The sudden fall in investments from period $T-1$ to period T is due to the linear investment cost. If the investment cost were instead a convex function, then one should expect investments to gradually decline when approaching the expiration date.

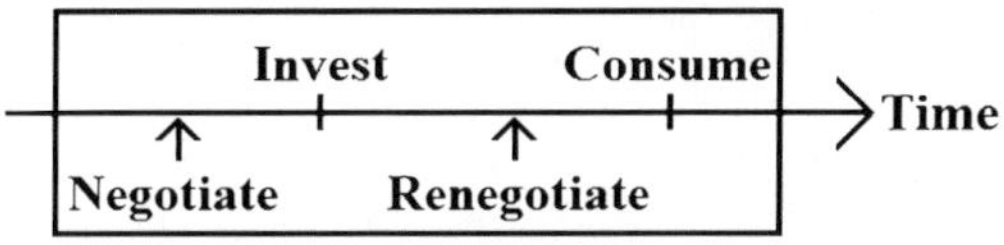

FIGURE 3
The timing when renegotiation is possible

Relative to the threat point g_i^{de}, the bargaining surplus is assumed to be split equally in expectation (this is standard in the literature, see Segal and Whinston 2010). This bargaining outcome is implemented by the Nash bargaining solution if side transfers are available or, alternatively, by randomly letting one country make a take-it-or-leave-it offer regarding quantities and transfers.

Renegotiation ensures that emission levels are *ex post* optimal, in contrast to the contracts discussed above. When investing, a country anticipates that it will not, in the end, have to comply with an overambitious contract. Will this jeopardize the incentives to invest?

The answer is no and the explanation as follows: when renegotiating a very ambitious agreement, countries that have invested little are desperate to reach a new agreement that would replace the tough initial commitments. In other words, they have a high willingness to pay when renegotiating the default outcome. Since the bargaining surplus is split evenly (in expectation) in equilibrium, the other countries take advantage of this willingness to pay and require side transfers in return. The fear of such a poor bargaining position motivates all countries to invest more, particularly if the default emission levels are small.

Proposition 8 (investments). *Country i's investment level r_i decreases in the initial quota g_i^{de}.*

This is anticipated when negotiating the initial agreement. The more ambitious this agreement is, the more the countries invest. This is desirable if the countries are otherwise tempted to underinvest. Thus, the agreement should be more ambitious if δ and q_R are large. Since the investments are influenced by the initial agreement, the g_i^{de}s can always be set such that the investments are first-best. In any case, the emission levels remain optimal, thanks to renegotiation. In sum, renegotiation provides a second instrument when the countries attempt to mitigate two common pool problems, and this permits the first-best.

Proposition 9 (emission).

(1) *The initial contract is tougher than what is ex post optimal if $\delta q_R > 0$ and is given by*

$$B'(g_i^{\text{de}} + R_i^*) = K \tag{4.11}$$
$$\Rightarrow g_i^{\text{de}} = g_i^*(\mathbf{R}) - [B'^{-1}([1-\delta q_R]K) - B'^{-1}(K)] < g_i^*(\mathbf{R}).$$

(2) *In equilibrium, all investments and emissions are first-best.*

If the technology depreciates completely after each period ($q_R = 0$) or if the players are impatient or there is no further period ($\delta = 0$), then $g_i^{\text{de}} = g_i^*(\mathbf{R})$. Otherwise, the quota g_i^{de} decreases if the length of the agreement is short while the technology is long-lasting since countries fear that more technology today will hurt their bargaining position in the near future. They thus invest less than what is optimal, unless the agreement is more ambitious. This

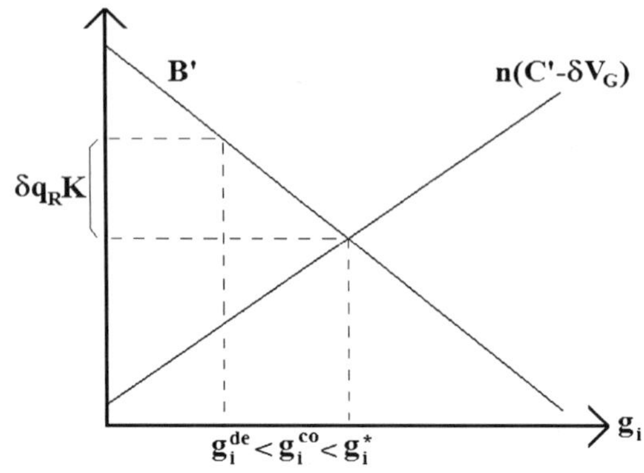

FIGURE 4

The shorter the agreement, the lower is the contracted emission level relative to the *ex post* optimum

result, illustrated in Figure 4, confirms the comparative static for the case without renegotiation (Proposition 5).[17]

Compared to the optimal contract without renegotiation, given by (4.9), the initial agreement should be tougher when renegotiation is possible: $g_i^{\text{de}} < g_i^{\text{co}}$. Intuitively, without renegotiation the contract balances the concern for investments (by reducing g_i^{co}) and for *ex post* efficiency (where g_i^{co} should be larger). The latter concern is irrelevant when renegotiation ensures *ex post* optimality, so the initial contract can be tougher—indeed, so tough that investments are first-best.

Corollary 1. *The initial contract under renegotiation* (4.11) *is tougher and specifies lower emission levels than the equilibrium contract when renegotiation is not possible* (4.9).

4.6. *Multiperiod contracts and renegotiation*

If the countries can negotiate and commit to a T-period agreement, we know from Section 4.4 that investments (and consumption) are first-best in every period—except for the last. Thus, the contracted quantities are also *ex post* optimal, and there is no need to renegotiate them. It is only in the last period that the quantities are lower than what is optimal *ex post*, and only then is there an incentive to renegotiate the contract.

Thus, when renegotiation is possible, for every period but the last the optimal and equilibrium initial contract specifies the *ex post* optimal quotas, and these are also equal to the first-best quantities since investments are optimal for $t < T$. The initial contract for the last period, $t = T$, is given by (4.11), just as in the one-period contract with renegotiation. As in Section 4.4, the initial contract becomes tougher to satisfy towards its end since the initial quotas are smaller for the last period than for the earlier periods.[18]

Proposition 10. *The equilibrium initial contract is given by* $g_{i,t}^{\text{de}} = g_i^*(\mathbf{R}^*)$ *for* $t < T$, *but for* $t = T$, *the contract is tougher and given by* (4.11). *This contract implements the first-best and is renegotiated only after the investment stage in the last period.*

17. In practice, there might be a limit to how tough the agreement can be if it must be self-enforcing or the enforcement capacity is small. Furthermore, one may require $g_i^{\text{de}} \geq 0$, as discussed in Section 5.2. With such constraints, multiperiod agreements may be necessary.

18. To see this, just compare the last-period contract (4.11) to (4.10).

In principle, there cannot be a unique equilibrium when contracts last multiple periods and renegotiation is possible. Instead of renegotiating only in the last period, the countries do equally well by, *e.g*. renegotiating in every period. However, Segal and Whinston (2010, p. 31) use the term *renegotiation* only when renegotiation changes the equilibrium pay-offs. Furthermore, with an arbitrarily small cost of renegotiating contracts, the countries would always prefer to renegotiate as little and as late as possible. This would imply a unique prediction: the outcome described by Proposition 10.

Interestingly, the equilibrium outcome is observationally equivalent to a time-inconsistency problem where the countries make ambitious plans for the future, while repeatedly backing down from promises made in the past. But rather than reflecting a time-inconsistency problem, this actually leads to the first-best.

Corollary 2. *In equilibrium, the countries repeatedly promise to pollute little in the future but when the future arrives, they renege on these promises. This procedure implements the first-best.*

5. GENERALIZATIONS AND EXTENSIONS

5.1. *Heterogeneous investment costs*

This section relaxes the assumption on identical investment costs. Instead, let country $i \in N$ face the linear investment cost K_i. Define $\underline{K} \equiv \min_i K_i$, $\underline{N} \equiv \{i \mid K_i = \underline{K}\}$, and $\underline{n} \equiv |\underline{N}|$. Thus, $\underline{n}$ is the number of countries that face the smallest cost, $\underline{K}$. I will assume that $\underline{K}$ is so small that it is optimal with $R_i > 0$ for at least some country.[19] Each country can invest as much as it wants to, subject to the constraint that its technology cannot be negative: $r_i \geq -q_R R_{i,-}$.

The first-best outcome is simple: the relevant investment cost for the planner is $\underline{K}$, as before. Thus, only the low-cost countries should invest, while countries with higher investment costs should not.[20] Further, the first-best g_i^* requires that marginal utilities are equalized. In sum, Proposition 1 holds as before.

The equilibria are a bit more complicated. When countries are heterogeneous, it matters a great deal whether they can use side transfers at the negotiation stage; it is assumed here that they can and that the bargaining surplus is split evenly, in expectation.[21] I will continue to restrict attention to "symmetric" equilibria where identical countries invest the same amount. But since K_i may vary across all countries, this condition does not have much bite. However, for each contracting environment, it turns out that exactly one symmetric MPE exists where only low-cost countries invest. Furthermore, this equilibrium is identical to the unique symmetric SPE if time $t \in \{1, 2, \ldots, \bar{t}\}$ were finite and $t < \bar{t} - 1$.

Proposition 11. *For each contracting environment, an MPE exists which coincides to the unique symmetric SPE if $t < \bar{t} - 1 < \infty$. In this equilibrium, the following is true:*

19. The requirement on $\underline{K}$ is strongest in the non-cooperative case, since investments are then smallest. An implicit condition for the requirement on $\underline{K}$ is given by Assumption A1 in the proof of Proposition 11. When contracts are possible, the requirement on $\underline{K}$ is weaker and as in Proposition 12.

20. Of course, this hinges on the linear investment cost (see Section 5.3)

21. This would be the case with the Nash bargaining solution or if a randomly drawn country makes a take-it-or-leave-it offer.

(1) *Only $i \in \underline{N}$ invest and the continuation values satisfy*

$$\frac{\partial V_i(G,R)}{\partial G} = -\frac{q_G K(1-\delta q_R)}{\underline{n}} \quad \text{and} \quad \frac{\partial V_i(G,R)}{\partial R} = \frac{q_R K}{\underline{n}} \quad \text{if } i \in \underline{N}; \tag{5.1}$$

$$\frac{\partial V_i(G,R)}{\partial G} = 0 \quad \text{and} \quad \frac{\partial V_i(G,R)}{\partial R} = 0 \quad \text{if } i \in N \backslash \underline{N}.$$

(2) *With no contract, low-cost countries consume less:*

$$B'(y_i^{\text{no}}) = C' - \delta \frac{\partial V_i(G,R)}{\partial G} \quad \forall i \in N, \tag{5.2}$$

assuming a decreasing $B'(y_i)/(-B''(y_i))$ and the following second-order condition:

$$C'' \geq B'(y_i)\left[\frac{C'''/B'' - B'''(C'')^2/(B'')^3}{1 - C''/B''(y_i)} - \frac{\sum_N [C'''/B'' - (C'')^2 B'''/(B'')^3]}{1 - \sum_N C''/B''}\right]. \tag{5.3}$$

(3) *With one-period contracts, $G = G^*$ and quotas are ex post optimal, $g_i^{\text{co}} = g_i^*(\boldsymbol{R}^{\text{co}})$, for $i \in N \backslash \underline{N}$. Low-cost countries consume less and pollute suboptimally little ex post:*

$$B'(g_i^{\text{co}} + R_i^{\text{co}}) - \sum_N (C' - \delta \partial V_j / \partial G) = \delta q_R (1 - 1/\underline{n}) K > 0, \quad i \in \underline{N}. \tag{5.4}$$

(4) *With multiperiod contracts, at $t < T$, $G = G^*$ and emissions are ex post optimal, $g_{i,t}^{mc} = g_{i,t}^*(\boldsymbol{R}^{mc}) = g_{i,t}^*(\boldsymbol{R}^*)$, for all $i \in N$. At $t = T$, the outcome is as in part* (3).

(5) *With one-period contracts and renegotiation, the first-best is implemented, the initial contract is $g_i^{\text{de}} = g_i^*$ for $i \in N \backslash \underline{N}$, while for $i \in \underline{N}$, g_i^{de} is given by*

$$B'(g_i^{\text{de}} + R_i^*) = \left(1 - \delta q_R \frac{n - \underline{n}}{\underline{n}(n-1)}\right) K \tag{5.5}$$

$$\Rightarrow g_i^{\text{de}} = g_i^*(\boldsymbol{R}^*) - \left[B'^{-1}([1 - \delta q_R]K) - B'^{-1}\left(\left[1 - \delta q_R \frac{n - \underline{n}}{\underline{n}(n-1)}\right] K\right)\right] < g_i^*(\boldsymbol{R}^*).$$

(6) *With multiperiod contracts and renegotiation, the first-best is implemented and the initial contract is $g_{i,t}^{\text{de}} = g_i^*(\boldsymbol{R}^*)$, $\forall i \in N$ for $t < T$. At $t = T$, the contract is as in part* (5).

Part 1 states that, for each contracting environment, only low-cost countries invest. Intuitively, the marginal value of adding more technology is similar for all countries, but the marginal cost is always smaller for $i \in \underline{N}$. If the stock of inherited technology is small and the pollution large, the need to invest is larger and the low-cost countries' continuation value smaller. Non-investing countries, in contrast, face a continuation value that depends on neither past technology nor past pollution. With this asymmetry, the results from Section 4 generalize in interesting ways.

Part 2 shows that, in the non-cooperative setting, the first-order condition for emission is as in Proposition 2. But only the investing countries internalize the larger future investment cost if pollution increases today. Given the difference in continuation values, the investing countries try harder to reduce emissions than do non-investing countries. Consequently, and perhaps paradoxically, they end up consuming less than countries that do not have green technology at all. The impact of the stocks on future investments is divided by the number of investing countries, $\underline{n}$. Thus, if $\underline{n}$ decreases, the free-riding problem between the investors decreases, and one may conjecture that the investors will invest more and pollute less. This conjecture is indeed true if $B(\cdot)$ and $C(\cdot)$ are quadratic functions, as shown in the Appendix.[22]

Part 3 says that, with optimal one-period contracts, only low-cost countries are encouraged to invest. This necessitates small quotas for low-cost countries, even though this is suboptimal *ex post*. The optimal contract is particularly tough for investing countries if $\underline{n}$ is large, since every $i \in \underline{N}$ understands that it benefits only $1/\underline{n}$ from the technology that survives the current contract. If $\underline{n} = 1$, however, $i = \underline{N}$ invests optimally and there is no need to reduce i's quota below the *ex post* optimal level. Countries with higher investment costs are not encouraged to invest and they consume and pollute the *ex post* optimal amount.

Multiperiod contracts share the same logic, part 4 finds. As in Section 4.4, every country invests optimally in the periods before the last and it is not necessary to have a suboptimally tough contract for these periods.

Parts 5 and 6 affirm that, with renegotiation, the emission levels are *ex post* optimal, as before. The equilibrium (and optimal) initial default quotas are set so as to motivate first-best investments. Since only low-cost countries invest in the first-best, the default quota must be tougher than the *ex post* optimal quota only for the low-cost countries. Furthermore, the initial demand on these countries should be particularly tough if their number, $\underline{n}$, is large, just as in the case without renegotiation.[23]

5.2. *Requiring non-negative emissions and technologies*

The main body of this paper does not consider any non-negativity constraints on g_i or r_i. This may be reasonable, as argued in Section 3.2, but the main motivation for ignoring non-negativity constraints is simplicity: at the start of Section 3, the $n+1$ stocks were reduced to only 2 after arguing that the R_is were pay-off-irrelevant, given R. With non-negativity constraints, however, this argument is invalid and one can no longer prove that the continuation value is necessarily a function of only G and R. Nevertheless, it is still possible to show that the equilibria described above continue to be an MPE, under specific conditions.

As in the previous subsection, technology stocks are required to be non-negative but countries are allowed to put their technology to other uses: $r_i \geq -q_R R_{i,-}$ (the alternative constraint $r_i \geq 0$ is briefly discussed below). In addition, emission levels are required to be positive: $g_i \geq 0$. To save space, attention is restricted to the four scenarios with incomplete contracts.

Proposition 12. *Consider the constraints $R_i \geq 0$ and $g_i \geq 0 \Rightarrow G \geq 0$. For each contracting environment, the equilibrium described in Section 4 continues to be an MPE under certain conditions:*

(1) *If there is no renegotiation, the condition is*

$$K \leq \frac{C'(q_G G_- + n B'^{-1}(K - \delta V_R))}{1 - \delta q_R / n - \delta q_G (1 - \delta q_R)}. \tag{5.6}$$

(2) *For contracts that can be renegotiated, the condition is $K \leq B'(R_i^*)$.*

22. For the non-cooperative case, I also need to assume a decreasing $B'/(-B'')$ to guarantee that only low-cost countries invest. The reason is that while B' measures the benefit of consuming more, the actual increase in consumption induced by a larger R decreases in $-B''$. This assumption is sufficient but not necessary and it always holds for quadratic $B(\cdot)$, *e.g.* The second-order condition is a generalization of the one in Proposition 3.

23. In principle, multiple contracts implement the first-best when renegotiation is possible, so the equilibrium is technically not unique, for the same reasons as in Section 4.6. In particular, any initial g_i^{de} so large that $i \in N \backslash \underline{N}$ does not invest can implement the first-best.

Part 1 is explained as follows. Remember that for one-period agreements, R_i decreased in g_i^{co}. Thus, the constraint $R_i \geq 0$ binds when the negotiated g_i^{co} is sufficiently generous, *i.e.* if $g_i^{\text{co}} \geq \widehat{g}_i^{\text{co}}$ for some threshold $\widehat{g}_i^{\text{co}}$. Thus, when the quotas are negotiated, it is anticipated that the benefit of reducing g_i^{co} below the *ex post* optimal level (*i.e.* that i invests more) ceases to exist when $g_i^{\text{co}} > \widehat{g}_i^{\text{co}}$. Above this threshold, the optimal g_i^{co}s are larger and set so as to be *ex post* optimal, without the need of motivating investments, and R is then zero. Hence, when the quota g_i^{co} is to be chosen, the second-order condition may not hold at the threshold $\widehat{g}_i^{\text{co}}$. Permitting $g_i^{\text{co}} > \widehat{g}_i^{\text{co}}$ is nevertheless not an equilibrium if investments are sufficiently cheap and pollution costly. This is the intuition behind condition (5.6), derived in the Appendix. For multiperiod agreements, the incentive to invest is lowest in the last period, and the condition ensuring that $R_i > 0$ in the last period, (5.6), is sufficient to ensure that $R_i > 0$ and $g_i > 0$ in the earlier periods as well.

Part 2 is even simpler. When renegotiation is possible, the trick in Section 4.5 was to negotiate a tough default agreement. The agreement had to be particularly tough (in that g_i^{de} was small) for K large and $B'(\cdot)$ small. When we require that $g_i^{\text{de}} \geq 0$, however, there is a limit to how tough the agreement can be. The condition for the first-best to be implemented by some $g_i^{\text{de}} \geq 0$ is simply $K \leq B'(R_i^*)$, as shown in the Appendix. With more than one period, the first-best was implemented in the earlier periods simply by initially committing to the *ex post* optimal emission levels. The default requirement $g_{i,t}^{\text{de}}$ is therefore at the lowest in the last period, leading to the same condition.[24]

Things are more complicated if the constraint $r_i \geq -q_R R_{i,-}$ is replaced by $r_i \geq 0$. The most interesting consequence of such a constraint might be for the multiperiod agreement. Since Proposition 6 predicts a (possibly large) fall in R_i from period $T-1$ to period T, the constraint $r_{i,T} \geq 0$ may then bind. If it does, it may be optimal to also invest zero in period $T-1$ and perhaps also in period $T-2$. Thus, as we approach the final period, investments may fall to zero for several of the last periods.[25]

5.3. *Nonlinear investment costs*

The framework above rests on the assumption that investment costs are linear. This is certainly a strong assumption: realistically, investment costs are given by a convex function (if the cheapest investments are made first) or a concave function (with returns to scale).

Let i's cost of investing r_i units be given by an increasing and convex function, $k(r_i)$. The reasoning that reduced the number of pay-off-relevant stocks from $n+1$ to 2 continues to hold. However, it is no longer possible to derive constant derivatives for the continuation value. Thus, the linear cost assumption is crucial for the simplicity above.

While a general analysis would be difficult, the steady state at the first-best is still characterized by (4.1) together with the analogous equation for (4.2):

$$B'(g_i^* + R_i^*) = (1 - \delta q_R)k'(r_s^*),$$

where $r_s^* \equiv (1 - q_R)R^*/n$ and R_i^* is the first-best level of technology. Furthermore, the most optimistic result from the above analysis continues to hold: the first-best is still implementable

24. With multiperiod contracts and renegotiation, the first-best can be implemented even if $K > B'(R_i^*)$, although such a contract is different from the one described by Proposition 10. For example, a sequence of quotas equal to the *ex post* optimal level induces first-best investments as long as the countries renegotiate (and postpone) the expiration date no later than in period $T-1$.

25. This would be more likely if n is large while δ and q_R are large, *i.e.* if the period length is short. While it is easy to derive such investment levels as functions of the quotas, it is not straightforward to characterize the equilibrium (or optimal) quotas when $r_i \geq 0$ may bind.

by one-period contracts if renegotiation is permitted, and the necessary initial contract is similar to the one described by Proposition 9.

Proposition 13. *Suppose the investment–cost function is increasing and convex, countries can negotiate one-period contracts, and these can be renegotiated.*

(1) *Country i's investment level r_i decreases in the initial quota g_i^{de}.*
(2) *In steady state, where $r = (1 - q_R)R/n$, the initial contract is given by*

$$B'(g_i^{\text{de}} + R_i^*) = k'(r_s^*)$$

$$\Rightarrow g_i^{\text{de}} = g_i^*(\boldsymbol{R}^*) - [B'^{-1}([1 - \delta q_R]k'(r_s^*)) - B'^{-1}(k'(r_s^*))] < g_i^*(\mathbf{R}^*) \quad \text{if } \delta q_R > 0.$$

(3) *In equilibrium, all investments and emissions are first-best.*

A convex investment cost can in principle help to further generalize the results above. In particular, we no longer need to restrict attention to *symmetric* MPEs in the equilibria above: the existence of asymmetric equilibria (where some countries invest more than others) hinge on the linear investment cost; for a convex investment cost function, these would cease to exist and the equilibria emphasized above are unique. Regarding the non-negativity constraints, the constraint $r_i \geq 0$ would never bind if $k'(0) = 0$. Thus, for a multiperiod contract (without renegotiation), we should expect that $r_{i,t}$ would gradually decrease over time as t approaches T. If investment costs were heterogeneous but convex functions, it would be optimal to let every country invest and not only the lowest cost countries as in Section 5.1. Deriving the optimal contract in this situation is certainly both important and interesting, suggesting a promising path for future research.

6. CONCLUSIONS

This paper presents a dynamic game in which n agents contribute to a public bad while also investing in substitute technologies. If the impact and the cost of investments are assumed to be linear, the symmetric MPE is unique and the analysis tractable. While the unique equilibrium rules out self-enforcing agreements, the framework can be employed to analyse incomplete contracts in a dynamic setting.

The model's assumptions fit well with the context of climate change and the results have important consequences for how to design a treaty. For those hoping for a technological solution, the *bad news* is that there will be as much free-riding on green investments as on emissions. Furthermore, countries have an incentive to invest little today in order to induce other countries to pollute less and invest more tomorrow. Finally, countries with large investment costs prefer to pollute a lot since, as a consequence, the technology leaders will simply invest more. In sum, green technology has the characteristic of a public good even when abstracting from technological spillovers and even if countries can sign contracts on emission levels.

The *good news* are, first, that investments are larger if the negotiated quotas are small. Second, the investments increase and can be socially optimal if the agreement is long-lasting. Third, the more long-lasting the agreement is, the less ambitious it needs to be, although it should be tougher to satisfy towards its end and for technology-leading countries. Fourth, permitting renegotiation does not reduce incentives to invest but rather enables the countries to design an agreement that motivates efficient investments as well as emissions.

A comparison to trade agreements is illuminating. Currently, the commitments made under the Kyoto Protocol expire in 2012 and the threat point for present negotiations is no agreement at all. In contrast, when the Doha-round trade negotiations broke down, the default outcome was not the non-cooperative equilibrium but the *existing* set of long-term trade agreements. Trade negotiations can thus be viewed as attempts to renegotiate existing treaties. The procedure used for negotiating trade agreements is more efficient than the one currently used for climate, according to the above analysis.

This paper has abstracted from firms, technological trade, intellectual property rights, and uncertainty (all analysed in Harstad, 2011). I have also ignored the problems of nonparticipants (discussed in Harstad, 2012). Instead, the paper has isolated the two-way interaction between investments in green technology and the design of international climate agreements. As a result, the analysis has detected and explored challenges that arise *even if* we abstract from domestic politics, technological spillovers, private information, monitoring, compliance, coalition formation, and the possibility of opting out of the agreement. While the effects discussed in this paper are likely to persist, allowing for such complications will certainly generate several new results and thereby enhance our understanding of the best agreement design. Relaxing these assumptions is thus the natural next step.

APPENDIX A

Let each i's interim continuation value (after the investment stage) be represented by $W(\cdot)$. Note that, just like $V(\cdot)$, W is a function of only G_- and R and thus is independent of i or $R_i - R_j, (i,j) \in N^2$.

Proof of Proposition 1. I start with the following lemma.

Lemma A1. $V_R^* = q_R K/n$.

Proof. The $1/n$ continuation value at the start of each period can be written as

$$V(G_-, R_-) = \max_R W(G_-, R) - (R - q_R R_-)K/n. \tag{A.1}$$

The lemma follows from the envelope theorem. ||

(1) When the objective is to maximize the sum of utilities, the first-order condition at the emission stage is

$$B'(y_i) - nC'\left(q_G G_- + \sum_N y_i - R\right) + n\delta V_G\left(q_G G_- + \sum_N y_i - R, R\right) = 0. \tag{A.2}$$

Thus, $y_i \equiv y$ is the same for every i. The second-order condition is

$$B''(y_i) - nC''\left(q_G G_- + \sum_N y_i - R\right) + n\delta V_{GG}\left(q_G G_- + \sum_N y_i - R, R\right) \le 0,$$

which holds when $V_{GG} = 0$, as implied by the following lemma.

Lemma A2. $V_G^* = -(1 - \delta q_R) q_G K/n$.

Proof. From Lemma A1, $V_{GR} = 0$ and V_G is independent of R. Therefore, (A.2) implies that $y_i = y$ is a function only of $\xi^* \equiv q_G G_- - R$, as is $G = q_G G_- - R + \sum_N y_i$, which can be written as the function $G = \chi^*(\xi^*)$. Furthermore, we can write $B(y_i) - C(G) \equiv \gamma^*(\xi^*)$, where $\gamma^*(\cdot)$ is a function of $\xi^* \equiv q_G G_- - R$. Since $R = q_G G_- - \xi^*$, the continuation value at the start of each period can be written as

$$V(G_-, R_-) = \max_{\xi^*} \gamma^*(\xi^*) + \delta V(\chi^*(\xi^*), q_G G_- - \xi^*) - (q_G G_- - \xi^* - q_R R_-)K/n$$

$$\Rightarrow \partial V/\partial G_- = \delta q_G V_R - q_G K/n,$$

by the envelope theorem. Substituting for V_R concludes the proof. ||

Part (3) is proven before part (2). The interim continuation value can be written as

$$W(G_-, R) = \max_y B(y) - C(q_G G_- + ny - R) + \delta V(q_G G_- + ny - R, R)$$
$$\Rightarrow \partial W/\partial R = C' - \delta V_G + \delta V_R = C' + \delta(1-\delta q_R) K q_G/n + \delta q_R K/n,$$

by the envelope theorem. From (A.1), the first-order condition for the investment level is

$$\partial W/\partial R - K/n = 0 \Rightarrow C' = (1-\delta q_R)(1-\delta q_G)K/n. \tag{A.3}$$

The second-order condition is $-C'' \leq 0$, which is always satisfied.

(2) Note that (A.3) can be written as $C' - \delta V_G = K/n - \delta V_R$. From (A.2), we have $C' - \delta V_G = B'(y_i)/n$. Combined, $B'(y_i) = K - n\delta V_R = (1-\delta q_R)K$.

Proof of Proposition 2. I start with the following lemma.

Lemma A3. $V_R^{\text{no}} = q_R K/n$.

Proof. In the symmetric equilibrium, i anticipates that every $j \neq i$ invests the same amount, $(R^{\text{no}} - q_R R_-)/n$, where R^{no} is the equilibrium R. At the start of each period, i maximizes its interim continuation value w.r.t. R, anticipating that $r_i = R - q_R R_- - \sum_{j\neq i}(R^{\text{no}} - q_R R_-)(n-1)/n$. Hence, i's continuation value at the start of the period can be written as

$$V(G_-, R_-) = \max_R W(G_-, R) - [R - q_R R_- - (R^{\text{no}} - q_R R_-)(1-1/n)]K.$$

Clearly, the equilibrium R (and thus R^{no}) is independent of R_-. Applying the envelope theorem concludes the proof. ||

(1) At the emission stage, i's first-order condition for y_i is

$$B'(y_i) - C'\left(q_G G_- - R + \sum_N y_i\right) + \delta V_G\left(q_G G_- - R + \sum_N y_i, R\right) = 0, \tag{A.4}$$

implying that all y_is are identical. The second-order condition is

$$B'' - C'' + \delta V_{GG}(\cdot) \leq 0,$$

which holds in equilibrium since $V_{GG} = 0$, as implied by the following lemma.

Lemma A4. $V_G^{\text{no}} = -q_G(1-\delta q_R)K/n$.

From Lemma A3, $V_{GR} = 0$ and V_G is independent of R. Therefore, (A.4) implies that y_i is a function of only $\xi^{\text{no}} \equiv q_G G_- - R$, as is $G = q_G G_- - R + \sum_N y_i$, which can be written as the function $G = \chi^{\text{no}}(\xi^{\text{no}})$. Furthermore, we can write $B(y_i) - C(G) \equiv \gamma^{\text{no}}(\xi^{\text{no}})$, where $\gamma^{\text{no}}(\cdot)$ is a function of $\xi^{\text{no}} \equiv q_G G_- - R$.

Let $\overline{\xi}^{\text{no}}$ be the equilibrium ξ^{no}. In equilibrium, j invests $r_j = (q_G G_- - \overline{\xi}^{\text{no}} - q_R R_-)/n$. Anticipating this, i's maximization problem when choosing r_i is equivalent to choosing ξ^{no}, recognizing $r_i = R - q_R R_- - \sum_{j\neq i} r_j = q_G G_- - \xi^{\text{no}} - q_R R_- - (q_G G_- - \overline{\xi}^{\text{no}} - q_R R_-)(n-1)/n$. At the start of each period, i's continuation value can be written as

$$V(G_-, R_-) = \max_{\xi^{\text{no}}} \gamma(\xi^{\text{no}}) + \delta V(\chi(\xi^{\text{no}}), q_G G_- - \xi^{\text{no}})$$
$$- [q_G G_- - \xi^{\text{no}} - q_R R_- - (q_G G_- - \overline{\xi}^{\text{no}} - q_R R_-)(1-1/n)]K.$$

Since V_R is a constant, it follows that the equilibrium ξ^{no} (which equals $\overline{\xi}^{\text{no}}$) is a constant, independent of the stocks. By the envelope theorem, it follows that

$$V_G = \delta V_R q_G - q_G K/n = -q_G(1-\delta q_R)K/n. \qquad ||$$

(2) Differentiating (A.4) w.r.t. R_i or, equivalently, w.r.t. $R = R_i + \sum_{j \neq i} R_j$ gives

$$B''\frac{dy_i}{dR} - C''\frac{ndy_i - dR}{dR} = 0 \Rightarrow \frac{dy_i^{\text{no}}}{dR} = \frac{C''}{nC'' - B''} \tag{A.5}$$

$$\Rightarrow \frac{dg_i}{dR_j} = \frac{C''}{nC'' - B''} > 0, \quad \text{if } j \neq i, \text{ while}$$

$$\frac{dg_i}{dR_i} = \frac{C''}{nC'' - B''} - 1 = -\frac{C''(n-1) - B''}{nC'' - B''} < 0.$$

Proof of Proposition 3.

(1) The first-order condition w.r.t. r_i $\left(\text{or, equivalently, w.r.t. } R = R_i + \sum_{j \neq i} R_j\right)$ is

$$B'(y_i)\frac{dy_i}{dR} - [C'(G) - \delta V_G] \sum_{j \in N} \frac{dg_j}{dR} + \delta V_R = K \Rightarrow (4.5), \tag{A.6}$$

when substituting for (A.4) and (A.5). The right-hand side of (4.5) is larger than the first-best (4.2), so $y_i^{\text{no}} < y_i^*$ when the second-order condition holds. Since $g_i^{\text{no}}(\mathbf{R}^{\text{no}}) + R_i^{\text{no}} = y_i^{\text{no}} < y_i^* = g_i^*(\mathbf{R}^*) + R_i^*$ while $g_i^{\text{no}}(\mathbf{R}^*) > g_i^*(\mathbf{R}^*)$ from Proposition 2, we have $g_i^{\text{no}}(\mathbf{R}^{\text{no}}) + R_i^{\text{no}} < g_i^{\text{no}}(\mathbf{R}^*) + R_i^*$. This implies that $R_i^{\text{no}} < R_i^*$ because $\partial(g_i^{\text{no}}(\mathbf{R}) + R_i)/\partial R = \partial y_i^{\text{no}}/\partial R > 0$ from (A.5).

Lemma A5. *The second-order condition at the investment stage holds if*

$$C'' \geq (n-1)B'(y_i)\left[\frac{-B'''(C'')^2/(B'') - C'''B''}{(C'' - B'')(nC'' - B'')}\right].$$

Proof. The second-order condition of (A.6) is

$$B''(y_i)\left(\frac{dy_i}{dR}\right)^2 + B'(y_i)\frac{d^2y_i}{(dR)^2} - C''(G)\left[n\frac{dy_i}{dR} - 1\right]^2 - [C'(G) - \delta V_G]\left[n\frac{d^2y_i}{(dR)^2}\right] \leq 0$$

$$\Rightarrow \frac{B''C''(C'' - B'')}{(nC'' - B'')^2} + (B' - n[C' - \delta V_G])\left[\frac{d^2y_i}{(dR)^2}\right] \leq 0, \tag{A.7}$$

we substitute with (A.5). By differentiating (A.5), we get

$$0 = \left(\frac{dy_i}{dR}\right)^2 B''' + B''\frac{d^2y_i}{(dR)^2} - \left(\frac{ndy_i - dR}{dR}\right)^2 C''' - C''n\frac{d^2y_i}{(dR)^2}$$

$$\Rightarrow \frac{d^2y_i}{(dR)^2} = \frac{(C'')^2B''' - (B'')^2C'''}{(nC'' - B'')^3}.$$

Substituted in (A.7),

$$\frac{B''C''(C'' - B'')}{(nC'' - B'')^2} - B'(n-1)\left[\frac{(C'')^2B''' - (B'')^2C'''}{(B'' - nC'')^3}\right] \leq 0 \Rightarrow (4.6). \quad \|$$

(2) From the proof of Lemma A4, we learned that, for the equilibrium R, $R = q_G G_- - \xi^{\text{no}}$, where ξ^{no} was a constant. Since $r = (R - q_R R_-)/n$, $\partial r/\partial G_- = q_G/n$ and $\partial r_i/\partial R_- = -q_R/n$.

(3) By substituting (4.5) in (A.4), we can write

$$C' = B' + \delta V_G = (1 - \delta q_R/n)K\frac{nC'' - B''}{C'' - B''} + \delta V_G,$$

which, with some algebra, can be rewritten as (4.7).

Proof of Proposition 4.

(1) When the level g_i^{co} is already committed to, the first-order condition for i's investment is

$$0 = B'(g_i^{co} + R_i) + \delta V_R - K \tag{A.8}$$

$$\Rightarrow y_i^{co} = B'^{-1}(K - \delta V_R), \quad R_i^{co} = B'^{-1}(K - \delta V_R) - g_i^{co}, \tag{A.9}$$

$$r_i^{co} = B'^{-1}(K - \delta V_R) - g_i^{co} - q_R R_{i,-}. \tag{A.10}$$

Substituting $V_R^{co} = q_R K/n$ (confirmed by Lemma A6) into (A.8) and (A.10) gives part (1). Since V_R^{co} is constant, the second-order condition is $B'' \leq 0$, which holds by assumption. A comparison to the first-best completes the proof of part (2).

Proof of Proposition 5. First, note that if the negotiations fail, the default outcome is the non-cooperative outcome, giving everyone the same utility. Since the r_is follow from the g_is by (A.10), negotiating the g_is is equivalent to negotiating the *equilibrium* r_is. Since all countries have identical preferences w.r.t. the r_is (and their default utility is the same), symmetry requires that r_i be the same for every i. Thus, preferences are aligned, and efficiency requires utility to be maximized w.r.t. the g_is or, equivalently, to the r_is, or to G. Anticipating (A.9)–(A.10), we can write the continuation value at the start of the period as

$$V^{co}(G_-, R_-) = \max_G B(B'^{-1}(K - \delta V_R)) - C(G) - K r_i + \delta V, \quad \text{where} \tag{A.11}$$

$$r_i = [R - q_R R_-]/n \quad \text{and}$$

$$R = \sum_{j \in N} (y_j - g_j) = n B'^{-1}(K - \delta V_R) - (G - q_G G_-).$$

Lemma A6. *(i)* $V_R^{co} = q_R K/n$. *(ii)* $V_G^{co} = -q_G K(1 - \delta q_R)/n$.

Proof. Part (i) follows from applying the envelope theorem to (A.11). Part (ii) follows from applying the envelope theorem to (A.11) and substituting for V_R^{co}. ‖

(1) The first-order condition from (A.11) is

$$0 = -C' + K/n + \delta V_G - \delta V_R \Rightarrow C' = (1 - \delta q_G)(1 - \delta q_R)K/n. \tag{A.12}$$

The second-order condition is $-C'' \leq 0$, which holds by assumption. By comparing to (4.3), $G^{co} = G^*$, which, given G_-, is equivalent to $\sum_N g_i^{co} = \sum_N g_i^*(\mathbf{R}^*)$. This, in turn, is equivalent to $g_i^{co} = g_i^*(\mathbf{R}^*)$ since both the r_is and the y_is are the same across the is for both scenarios.

(2) Combining (A.12) and (A.8) gives (4.9).

Proof of Proposition 6. The investment level in the last period is given by (A.10) for the same reasons as in Proposition 4. Anticipating the equilibrium $R_{i,T}$ (and $R_{j,T}$), i can invest q_R units fewer in period T for each unit invested in period $T-1$. Thus, in period $T-1$, the first-order condition for i's investment is

$$0 = B'(g_{i,T-1}^{mc} + R_{i,T-1}) - K + \delta q_R K$$

$$\Rightarrow R_{i,t} = B'^{-1}((1 - \delta q_R)K) - g_{i,t}^{mc}, \quad t = T-1. \tag{A.13}$$

The same argument applies to every period $t \in \{1, \ldots, T-1\}$, and the investment level is given by the analogous equation for each period $t < T$. The second-order condition, $B'' \leq 0$, holds by assumption. A comparison to the first-best completes the proof.

Proof of Proposition 7. At the start of $t = 1$, countries negotiate emission levels for every period $t \in \{1, \ldots, T\}$. In equilibrium, all countries enjoy the same default utilities. Just as before, they will therefore negotiate the quotas such that the equilibrium investment will be the same for all is. Using (A.13) for each period $t < T$, this implies

$$R_t = \sum_N (B'^{-1}(K(1-\delta q_R)) - g_{i,t}) = nB'^{-1}(K(1-\delta q_R)) - (G_t - q_G G_{t-1})$$

$$\begin{aligned} \Rightarrow r_{i,t} &= (R_t - q_R R_{t-1})/n, \quad \forall t \\ &= B'^{-1}(K(1-\delta q_R)) - (G_t - q_G G_{t-1})/n \\ &\quad - q_R B'^{-1}(K(1-\delta q_R)) + q_R(G_{t-1} - q_G G_{t-2})/n, \quad t \in \{2,\ldots,T-1\}. \end{aligned} \tag{A.14}$$

At the start of the first period, i's continuation value can be written as

$$V = \max_{\{G_t\}_t} \frac{1-\delta^T}{1-\delta} B(B'^{-1}(K(1-\delta q_R))) - \sum_{t=1}^{T} \delta^{t-1}[K r_{i,t} + C(G_t)] + \delta^T V(G_T, R_T), \tag{A.15}$$

where $r_{i,t}$ is given by (A.14) and R_T is given by (A.9).

Lemma A7. *(i)* $V_R = q_R K/n$. *(ii)* $V_G = -q_G K(1-\delta q_R)/n$.

Proof. Both parts follow from (A.15) by applying the envelope theorem. ||

(1) The first-order condition of (A.15) w.r.t. any G_t, $t \in \{1,\ldots,T\}$, gives (4.3). As in the proof of Proposition 5, $G_t^{mc} = G^* \Leftrightarrow g_{i,t}^{mc} = g_i^*(\mathbf{R}^*)$. The second-order condition, $-C'' < 0$, always holds.

(2) Since $\mathbf{R}_t^{mc} = \mathbf{R}^*$ for $t < T$, $g_{i,t}^{mc} = g_i^*(\mathbf{R}_t^{mc})$. In the last period, however, investments are as described by Proposition 4 and the equilibrium quotas are suboptimally low, just as described by Proposition 5.

Proof of Proposition 8. In the default outcome (*i.e.* if the renegotiations fail), i's interim utility is

$$W_i^{\text{de}} = B(g_i^{\text{de}} + R_i) - C\left(q_G G_- + \sum_N g_j^{\text{de}}\right) + \delta V.$$

Since i also expects $1/n$ of the renegotiation surplus, i's expected utility at the start of the period is

$$W_i^{\text{de}} + \frac{1}{n}\sum_N (W_j^{\text{re}} - W_j^{\text{de}}) - K r_i, \tag{A.16}$$

where W_j^{re} is j's utility after renegotiation. Maximizing (A.16) w.r.t. r_i gives the first-order condition:

$$K = [B'(g_i^{\text{de}} + R_i) + \delta V_R]\left(1 - \frac{1}{n}\right) + \frac{1}{n}\partial\left(\sum_N W_j^{\text{re}}\right)\Big/\partial R - \sum_{j \in N\setminus i} \frac{\delta V_R}{n}. \tag{A.17}$$

Note that V_R drops out. The term $\sum_N W_j^{\text{re}}$ can be written as

$$\sum_N W_j^{\text{re}} = \max_{\{g_1^{\text{re}},\ldots,g_n^{\text{re}}\}} \sum_N B(g_j^{\text{re}} + R_j) - C\left(q_G G_- + \sum_N g_j^{\text{re}}\right) + \delta V.$$

Since the right-hand side is concave in R_i, the second-order condition of (A.17) holds. Furthermore, $\partial(\sum_N W_j^{\text{re}})/\partial R$ is not a function of g_i^{de} (as can be seen when applying the envelope theorem). Thus, if g_i^{de} increases, the right-hand side of (A.17) decreases and, to restore equality, R_i must decrease.

Proof of Proposition 9. Part (2) is proven before part (1): Renegotiation implies that the g_i^{re}s are first-best, conditional on the stocks. Furthermore, the R_is decrease in the g_i^{de}s, as shown in the previous proof. Finally, negotiating the initial quotas ensures that these are set so as to motivate the first-best R_is, if possible.

(1) First-best investments are ensured if $\partial(\sum W_i^{\text{re}})/\partial R = K$. With this, (A.17) can be rewritten as:

$$B'(g_i^{\text{de}} + R_i^*) = K \Rightarrow g_i^{\text{de}} = B'^{-1}(K) - R_i^*.$$

Since, for the default outcome, the continuation value is the same for every country and $\sum_N V_R$ and $\sum_N V_G$ are the same as in the first-best, implemented here, the bargaining surplus at the start of the period is independent of the stocks. Thus, V_R and V_G are the same as in the non-cooperative situation.

Proof of Proposition 10. From Propositions 6–7, we know the first-best is not possible if $T < \infty$ without renegotiation. Renegotiation must take place at least in the last period. Assume first that renegotiation is not expected in period $t < T$. It is then easy to see that $R_{i,t}$ will be first-best for the same reason as in Proposition 6, conditional on, but regardless of, the quotas $g_{i,t}^{\text{de}}$s. For $t = T$, investments are first-best if $g_{i,T}^{\text{de}} = g_i^{\text{de}}$, given by Proposition 9, for the same reasons as given there.

Regarding the emission levels, renegotiation ensures that $g_{i,T} = g_i^*(\mathbf{R})$, while, for $t < T$, $g_{i,t} = g_i^*(\mathbf{R})$ is ensured simply by setting $g_{i,t}^{\text{de}} = g_i^*$.

Assume instead that $g_{i,t}^{\text{de}}$, $t < T$, are also expected to be renegotiated in period t. If $g_{i,t}^{\text{de}} = g_i^*(\mathbf{R}^*)$, a reasoning similar to that for Proposition 9 shows that investments are first-best. Thus, the renegotiated $g_{i,t}$s will be identical to the initial contract, $g_{i,t}^{\text{de}} = g_i^*$, and such renegotiation at $t < T$ changes neither emission levels nor incentives.

Proof of Theorem.

(1) The first part is proved in the text.

(2)–(3) For each contracting environment, the above proofs derived necessary conditions for a symmetric MPE. This pinned down V_G, V_R, and a unique equilibrium.

(4) Let time $t \in \{1,\ldots,\bar{t}\}$ be finite and $V^t(G_{t-1}, R_{t-1})$ denote the continuation value for period t. Consider the proofs above: for each scenario, except for the multiperiod agreement scenario, the continuation value was derived in two steps. First, it was first proven that $\partial V^t/\partial R_{t-1} = q_R K/n$, regardless of V^{t+1}. This certainly also holds for the last period, if time is finite. Second, it was proven that $\partial V^t/\partial G_{t-1} = -q_G K(1-\delta q_R)/n$ if one could substitute for $\partial V^{t+1}/\partial R_t = q_R K/n$ for the next period. Since $\partial V^t/\partial R_{t-1} = q_R K/n$ in every period $t \in \{1,\ldots,\bar{t}\}$, $\partial V^t/\partial G_{t-1} = -q_G K(1-\delta q_R)/n$ in every period $t \le \bar{t}-1$. Thus, the continuation value is as above for every period $t \le \bar{t}-1$ and, anticipating this, the equilibrium play must be as described above for all periods $t \le \bar{t}-2$. For multiperiod contracts, it was proven that, before the negotiation stage, $V_R = q_R K/n$ and $V_G = -q_G K(1-\delta q_R)/n$, independently of the continuation value following the expiration of these contracts.

Proof of Proposition 11.

(1) Follows from the proofs of (2)–(6).

(2) Although continuation values depend on i, the proofs are analogous to those above: at the emission stage, the first-order condition is (5.2). If $V_{i,G} \equiv \partial V_i/\partial G_-$ is a constant (verified below), then differentiating (5.2) gives

$$\frac{dy_i}{dR} = -\frac{C''/B''(y_i)}{1-\sum_N C''/B''(y_j)}. \tag{A.18}$$

Substituting this into the first-order condition w.r.t. $R_i \ge 0$ gives

$$\frac{B'(y_i)}{-B_i''}\left(\frac{C''}{1-\sum_N C''/B''(y_j)}\right)+\left[\frac{C'(G)-\delta\partial V_{i,t+1}/\partial G}{1-\sum_N C''/B''(y_j)}\right] \le K_i - \delta\frac{\partial V_{i,t+1}}{\partial R} \tag{A.19}$$

$$\Rightarrow B'(y_i) \le \left(K_i - \delta\frac{\partial V_{i,t+1}}{\partial R}\right)\left(\frac{1-C''/B''(y_i)}{1-\sum_N C''/B''(y_j)}\right), \tag{A.20}$$

with equality if $R_i > 0$.

Lemma A8. *The second-order condition holds if and only if* (5.3) *is satisfied.*

Proof. The second-order condition of (A.6) is

$$B''(y_i)\left(\frac{dy_i}{dR}\right)^2 + B'(y_i)\frac{d^2 y_i}{(dR)^2} - C''(G)\left(\sum_N \frac{dy_i}{dR} - 1\right)^2 - [C'(G) - \delta V_G]\left[\sum_N \frac{d^2 y_i}{(dR)^2}\right] \le 0. \tag{A.21}$$

By differentiating (A.18), we get

$$\sum \frac{d^2 y_i}{dR^2} = \frac{\sum_N [C'''/B''(y_j) - (C'')^2 B'''(y_j)/(B''(y_j))^3]}{\left(1-\sum_N C''/B''(y_j)\right)^3},$$

$$\frac{d^2 y_i}{dR^2} = \frac{C'''/B''(y_i) - (C'')^2 B'''(y_i)/(B''(y_i))^3}{\left(1-\sum_N C''/B''(y_j)\right)^2} + \frac{C''}{B''(y_i)}\left(\frac{\sum_N [C'''/B''(y_j) - (C'')^2 B'''(y_j)/(B''(y_j))^3]}{\left(1-\sum_N C''/B''(y_j)\right)^3}\right).$$

Substituting these equations and (A.18) into (A.21) gives (5.3) after some algebra. ||

As for the other scenarios, I assume that *some* K_i is small enough to invest a positive amount.

Assumption A1. (A.20) would be violated for $R = 0$ and $\min_i K_i$, *i.e.*

$$K \le B'(g_i)\left(\frac{1-\sum_N C''/B''(g_j)}{1-C''/B''(g_i)}\right) + \delta \frac{\partial V_{i,t+1}}{\partial R},$$

where the g_js are determined by (5.2) when $R = 0$.

In the last period, $\partial V_{i,t+1}/\partial G = \partial V_{i,t+1}/\partial R = 0$, (5.2) implies that y_i is the same for every i, and (A.20) can bind only for the smallest K_i. Then, as in the earlier proofs, $q_G G_- - R$ will be a constant, in equilibrium, and the equilibrium R will be independent of R_-. If every $i \in \underline{N}$ invests the same amount, then, for the same reasons as before, it holds for $i \in \underline{N}$ that

$$\partial r_i/\partial R_- = -q_R/\underline{n}, \quad \partial V_i/\partial R_- = q_R K/\underline{n} \quad \text{and} \quad \partial V_i/\partial G_- = -q_G K/\underline{n} + q_G \delta \partial V_{i,t+1}/\partial R. \tag{A.22}$$

On the other hand, $i \in N\backslash\underline{N}$ anticipates that $q_G G_- - R$, and thus y_i and G will be independent of R_- and G_-, implying that $\partial V_i/\partial R_- = \partial V_i/\partial G_- = 0$ for $i \in N\backslash\underline{N}$.

Finally, take *any* period $t < \bar{t}$ where $V_{i,t+1,R} \equiv \partial V_{i,t+1}/\partial R = q_R K/n$ and $\partial V_{i,t+1}/\partial G > \partial V_{j,t+1}/\partial G$, $i \in \underline{N}$, $j \in N\backslash\underline{N}$. Together with (5.2), this implies that, in period t, y_i is larger for $i \in N\backslash\underline{N}$. The first-order condition w.r.t. R_i can bind only for the smallest K_i if $B'(y_i)/(-B''(y_i))$ is decreasing in y_i, as can be seen from (A.19). Thus, $j \in N\backslash\underline{N}$ ensures that $R_j = 0$, while $R_i > 0$ for $i \in \underline{N}$ under Assumption A1. As before, $q_G G_- - R$ is going to be a constant and (A.22) holds, implying that $\partial V_i/\partial G_- = -q_G K(1-\delta q_R)/n$, while for $i \in N\backslash\underline{N}$, $\partial V_i/\partial R_- = \partial V_i/\partial G_- = 0$. This argument is certainly valid for $t = \bar{t} - 1$, and the argument will then also hold for period $t - 1$. The proof is completed by induction.

If B and C are quadratic functions, B'' and C'' are constant, $B'(y_i)/(-B''(y_i))$ does indeed decrease in y_i, and (A.20) binds for $K_i = K$. Combined with (5.2), we get

$$C'(G) - \delta V_{i,t+1,G} = \left(K - \delta\frac{\partial V_i}{\partial R}\right)\frac{1-\sum_N C''/B''(y_j)}{1-C''/B''(y_i)}, \quad i \in \underline{N}. \tag{A.23}$$

If $\underline{n}$ increases, $V_{i,t+1,G}$ increases and $V_{i,t+1,R}$ decreases, implying that G must also increase. Then, however, (5.2) shows that $y_i = g_i$ decreases for $i \in N\backslash\underline{N}$. Consequently, g_i must increase in $\underline{n}$ for $i \in \underline{N}$. Furthermore, substituting (A.23) in (A.19) implies that $y_i = g_i + R_i$ must decrease in $\underline{n}$ for $i \in \underline{N}$. It follows that R_i must decrease in $\underline{n}$ for $i \in \underline{N}$. ||

(3) Given g_i^{co}, the first-order condition w.r.t. $R_i \ge 0$ is

$$B'(g_i^{\mathrm{co}} + R_i) + \delta V_{i,R} - K_i \le 0 \tag{A.24}$$

$$\Rightarrow R_i = \max\{0, B'^{-1}(K_i - \delta V_{i,R}) - g_i^{\mathrm{co}}\}. \tag{A.25}$$

The second-order condition holds if $V_{i,R}$ is a constant (confirmed by the next lemma). Anticipating (A.25), the problem at the negotiation stage is

$$\max_{\{g_i^{\mathrm{co}}\}_i} \sum_N \left[B(g_i^{\mathrm{co}} + R_i) - (R_i - q_R R_{i,-})K_i - C\left(q_G G_- + \sum g_i^{\mathrm{co}}\right) + \delta V_i(G, R)\right].$$

When substituting for R_i, it is clear that $R_i > 0$ (implying that $y_i = B'^{-1}(K_i - \delta V_{i,R})$) is optimal only if $K_i = K$. Thus, the first-order condition w.r.t. g_i^{co} is

$$B'(g_i^{\text{co}}) - \sum_N (C' - \delta V_{i,G}) = 0 \quad \text{if } i \in N\backslash \underline{N}; \tag{A.26}$$

$$K - \sum_N (C' - \delta V_{i,G} + \delta V_{i,R}) = 0 \quad \text{if } i \in \underline{N}. \tag{A.27}$$

The second-order conditions hold if V_i is linear. Thus, $g_i^{\text{co}} = g_i^*$ for $i \in N\backslash \underline{N}$, while combining a binding (A.24) for $i \in \underline{N}$ with (A.27) gives (5.4). The proof is completed by substituting for $V_{i,G}$ and $V_{i,R}$ from the following lemma.

Lemma A9. *For any* $t < \bar{t}$, *(5.1) holds.*

Proof. For the same reasons as before, $\sum_N V_{i,R} = q_R K$. Then, (A.27) pins down G. Given G, (A.26) pins down g_i^{co}, $i \in N\backslash\underline{N}$. Consequently, (A.25) and $\sum_{\underline{N}} g_i^{\text{co}} = G - q_G G_- - \sum_{N\backslash\underline{N}} g_i^{\text{co}}$ together imply that $\partial R/\partial G_- = q_G$ and that $\sum_N V_{i,G} = -q_G K + \delta q_G \sum_N V_{i,R} = -q_G K(1 - \delta q_R)$ for $t < \bar{t}$. Hence, the sum of the continuation values' derivatives are the same as in the case with no contract. This implies that the surplus from cooperation is independent of the stocks! The bargaining outcome gives i its continuation value in the case of no agreement plus $1/n$ of the (constant) total surplus. Thus, $V_{i,G}$ and $V_{i,R}$ are the same as in the case of no agreement (in other words, any necessary side transfers are independent of the stocks). ||

(4) The proof is similar to the one above and is thus omitted.

(5) Following the proof of Proposition 8, the first-order condition w.r.t. $R_i \geq 0$ becomes

$$K_i \geq [B'(g_i^{\text{de}} + R_i) + \delta V_{i,R}](1 - 1/n) + \frac{1}{n}\partial\left(\sum_N W_j^{\text{re}}\right)/\partial R - \sum_{j \in N\backslash i} \frac{1}{n}\delta V_{j,R}, \tag{A.28}$$

with equality if $R_i > 0$. Since B is concave, the second-order condition holds as in the proof of Proposition 8. The first-best requires that (A.28) binds only if $K_i = K$ and $\partial\left(\sum_N W_i^{\text{re}}\right)/\partial R = K$. Furthermore, at the first-best, $V_{i,R}$ is as in the non-cooperative case for the same reason as in the proof of Lemma A9. With this fact, (A.28) can be written as (5.5). For $i \in N\backslash\underline{N}$, the first-best investment level (zero) is ensured by setting $g_i^{\text{de}} = g_i^*$. All this holds for any period $t < \bar{t}$.

(6) The proof is analogous to the one above and is thus omitted.

Proof of Proposition 12.

(1) The proof is by induction: I show that *if* the equilibrium is as described in Section 3 (with the same V_G and V_R) for period $t+1$, then this can be true for period t as well. I will start with one-period contracts.

Lemma A10. $R_i > 0$, $\forall i$, *if* $g_i^{\text{co}} < \widehat{g}_i \equiv B'^{-1}(K - \delta V_R)$.

Proof. For a given contract in period t, the first-order condition for $r_i \geq -R_i$ is

$$B'(g_i^{\text{co}} + R_i) - K + \delta V_R \leq 0,$$

with equality if $R_i > 0$, but $R_i = 0$ if $B'(g_i^{\text{co}}) - K + \delta V_R \leq 0 \Rightarrow g_i^{\text{co}} \geq B'^{-1}(K - \delta V_R)$. In either case, the second-order conditions hold since $B(\cdot)$ is concave. ||

Lemma A11. *In equilibrium,* $g_i^{\text{co}} < \widehat{g}_i$ *if (5.6) holds.*

Proof. At the negotiation stage, it is anticipated that $g_i^{\text{co}} \geq \widehat{g}_i$ implies $R_i = 0$. In a symmetric equilibrium, $g = g_i^{\text{co}}$ solves

$$\max_{g \geq 0} \left\{ \begin{array}{c} B(B'^{-1}(K - \delta V_R)) - C(q_G G_- + ng) - K[B'^{-1}(K - \delta V_R) - g - q_R R_-/n] + \delta V \\ \text{if } g \leq B'^{-1}(K - \delta V_R), \\ B(g_{i,t}) - C(q_G G_- + ng) + \delta V \text{ if } g > B'^{-1}(K - \delta V_R) \end{array} \right\}. \tag{A.29}$$

If $g < B'^{-1}(K - \delta V_R)$, then $R_i > 0$, and increasing g reduces R_i (exactly as in the proof of Proposition 4). If $g \geq B'^{-1}(K - \delta V_R)$, $R_i = 0$, and increasing g increases $B(g)$. When $g_i^{\text{co}} < \widehat{g}_i$, the second-order condition holds since $-C$

is concave; when $g_i^{\text{co}} > \widehat{g}_i$, the second-order condition holds since B is concave. However, the utility function in (A.29) has a kink at $g = B'^{-1}(K - \delta V_R)$ and may not be single-peaked. Single-peakedness is ensured if, at $g = B'^{-1}(K - \delta V_R)$, an increase in g reduces $B(g) - C(q_G G_- + ng) + \delta V$. This is guaranteed if

$$B'(B'^{-1}(K - \delta V_R)) + n\delta V_G \le nC'(q_G G_- + ng) \Rightarrow$$
$$K[1 - \delta q_R/n - \delta q_G(1 - \delta q_R)] \le C'(q_G G_- + nB'^{-1}(K - \delta V_R)). \quad \|$$

Lemma A12. *In equilibrium, the constraint $g_i \ge 0$ does not bind.*

Proof. When (5.6) holds, the first-order condition w.r.t. g is, from (A.29),

$$-C'(q_G G_- + ng) + K + \delta V_G - \delta V_R \le 0,$$

with equality if $g > 0$ (the second-order condition holds since $-C'' < 0$). Thus, $g \ge 0$ does not bind if

$$-C'(q_G G_-) + K + \delta V_G - \delta V_R \ge 0 \Rightarrow C'(q_G G_-) \le (1 - \delta q_G)(1 - \delta q_R)K/n.$$

For the equilibrium in Proposition 4, we know that $C'(G_-) = (1 - \delta q_G)(1 - \delta q_R)K/n$. Since $q_G \le 1$ and C is convex, $C'(q_G G_-) \le C'(G_-) = (1 - \delta q_G)(1 - \delta q_R)K/n$. $\|$

Together, Lemmas A1–A12 show that if (5.6) holds, the equilibrium described by Propositions 3–4 continues to be an equilibrium also when the constraints $g_i \ge 0$ and $r_i \ge -R_i$ are imposed. For the same reasons as in Section 4.3, V_G and V_R are as stated by the Theorem.

For multi-period contracts, the proof is similar. To see this, note that the last period is exactly as in part (1). Compared to the last period, for every period $t < T$ investments are larger and the condition for $R_{i,t} > 0$ weaker (and it always holds when (5.6) holds). $\|$

(2) Consider first one-period contracts with renegotiation. At the renegotiation stage, countries renegotiate every $g_i \ge 0$ and these will be equal to the first-best, as before. At the investment stage, the first-order condition is

$$[B'(g_i^{\text{de}} + R_i) + \delta V_R](1 - 1/n) + \frac{1}{n}\partial\left(\sum_N W_j^{\text{re}}\right) \Big/ \partial R - \sum_{j \in N \setminus i} \frac{1}{n}\delta V_R - K \le 0,$$

with equality if $R_i > 0$. Define $\overline{g}_i^{\text{de}} \equiv \sup(g_i^{\text{de}}: R_i > 0)$. If $g_i^{\text{de}} < \overline{g}_i^{\text{de}}$, $R_i > 0$ and the second-order condition is as before.

At the initial negotiation stage, if g_i^{de}, as described by Proposition 9, is such that $g_i^{\text{de}} \in [0, \overline{g}_i^{\text{de}}]$, then g_i^{de} implements the first-best and these quotas represent an equilibrium contract. Note that $g_i^{\text{de}} \le \overline{g}_i^{\text{de}}$ is indeed satisfied since $g_i^{\text{de}} > \overline{g}_i^{\text{de}}$ implies $R_i = 0$, while g_i^{de} induces first-best investments if R_i is unconstrained, and $R_i^* > 0$ by assumption. However, the g_i^{de} implementing the first-best is positive only if $B'(g_i^{\text{de}} + R_i^*) = K$ for *some* $g_i^{\text{de}} \ge 0$, which requires that $B'(R_i^*) \ge K$. The proof is analogous if there are multiple periods.

Proof of Proposition 13.

(1) As in the proof of Proposition 8, the first-order condition w.r.t. r_i is

$$k'(r_i) = [B'(g_i^{\text{de}} + R_i) + \delta V_R]\left(1 - \frac{1}{n}\right) + \frac{1}{n}\frac{\partial \sum_N W_j^{\text{re}}}{\partial R} - \sum_{j \in N \setminus i} \frac{\delta V_R}{n}. \tag{A.30}$$

Again, V_R drops out. The term $\sum_N W_j^{\text{re}}$ can be written as

$$\sum_N W_j^{\text{re}} = \max_{\{g_1^{\text{re}}, \ldots, g_n^{\text{re}}\}} \sum_N B(g_j^{\text{re}} + R_j) - C\left(q_G G_- + \sum_N g_j^{\text{re}}\right) + \delta V. \tag{A.31}$$

When the first-best is implemented, $V(G, R) = \max_{R_+} W(G, R_+) - k((R_+ - q_R R)/n)$ which is concave in R (as can be seen from the envelope theorem). Thus, (A.31) is concave in R_j, implying that the second-order condition of (A.30) holds. Furthermore, $\partial(\sum_N W_j^{\text{re}})/\partial R$ is not a function of g_i^{de}. Hence, if g_i^{de} increases, the right-hand side of (A.30) decreases and, to restore equality, R_i must decrease.

(2)–(3) Renegotiation implies that the g_i^{re}s are first-best, conditional on the stocks. Negotiating the initial quotas implies that these are set such as to induce the first-best R_is, if possible. First-best investments are ensured if $\partial(\sum W_i^{re})/\partial R = k'(r_i^*)$. With this equality, (A.30) can be written as $B'(g_i^{de} + R_i^*) = k'(r_i^*)$. A comparison to the steady-state first-best, $B'(g_i^* + R_i^*) = (1 - \delta q_R)k'(r_i^*)$, concludes the proof.

Acknowledgment. I am indebted to the excellent suggestions of the Editor, four expert referees, Philippe Aghion, Marco Battaglini, David Besanko, Jeff Campbell, Yeon-Koo Che, Guillaume Cheikbossian, Rolf Golombek, Faruk Gul, Michael Hoel, Johannes Hörner, Larry Karp, Charles Kolstad, Matti Liski, Jiang Ning, Wolfgang Pesendorfer, Roberto Veneziani, and the audiences at the 2009 American Economic Association meeting, the University of British Columbia, Caltech, Center for Economic Studies Ifo, the Chicago Area Political Economy conference, Columbia University, the University of Copenhagen, Emory University, the Federal Reserve Bank of Chicago, the Fondazione Eni Enrico Mattei conference on Coalitions for Climate Cooperation, Helsinki School of Economics, the Marshall School of Business, McGill University, the Mathematical Sciences Research Institute, Berkeley conference on climate change, Norwegian School of Economics, Northwestern University, University of Oslo, University of Pennsylvania, Princeton University, Paris School of Economics, Simon Fraser University, Stanford Graduate School of Business, Stockholm University, Toulouse School of Economics, the World Bank, and Yale University. Judith N. Levi assisted with the editing.

REFERENCES

AGHION, P., DEWATRIPONT, M. and REY, P. (1994), "Renegotiation Design with Unverifiable Information", *Econometrica*, **62**, 257–282.

ALDY, J. and STAVINS, R. (eds). (2009), *Post-Kyoto International Climate Policy* (Cambridge University Press).

BAŞAR, T. and OLSDER, G. J. (1999), *Dynamic Noncooperative Game Theory* (Philadelphia: SIAM).

BARRETT, S. (2006), "Climate Treaties and 'Breakthrough' Technologies", *American Economic Review*, Papers and Proceedings, **96**, 22–25.

BATTAGLINI, M. and COATE, S. (2007), "Inefficiency in Legislative Policymaking: A Dynamic Analysis", *American Economic Review*, **97**, 118–149.

BUCHHOLZ, W. and KONRAD, K. (1994), "Global Environmental Problems and the Strategic Choice of Technology", *Journal of Economics*, **60**, 299–321.

CHE, Y.-K. and SAKOVICS, J. (2004), "A Dynamic Theory of Holdup", *Econometrica*, **72**, 1063–1103.

CHIAPPORI, P.-A., MACHO, I., REY, P. and SALANIE, B. (1994), "Repeated Moral Hazard: The Role of Memory, Commitment, and the Access to Credit Markets", *European Economic Review*, **38**, 1527–1553.

CHUNG, T.-Y. (1991), "Incomplete Contracts, Specific Investment, and Risk Sharing", *Review of Economic Studies*, **58**, 1031–1042.

D'ASPREMONT, C. and JACQUEMIN, A. (1988), "Cooperative and Noncooperative R&D in Duopoly with Spillovers", *American Economic Review*, **78**, 1133–1137.

DOCKNER, E. J., JØRGENSEN, S., VAN LONG, N. and SORGER, G. (2000), *Differential Games in Economics and Management Science* (Cambridge: Cambridge University Press).

DOCKNER, E. J. and SORGER, G. (1996), "Existence and Properties of Equilibria for a Dynamic Game on Productive Assets", *Journal of Economic Theory*, **71**, 209–227.

DOCKNER, E. J. and VAN LONG, N. (1993), "International Pollution Control: Co-operative versus Noncooperative Strategies", *Journal of Environmental Economics and Management*, **24**, 13–29.

DORASZELSKI, U. and PAKES, A. (2007), "A Framework for Applied Dynamic Analysis in IO", *Handbook of Industrial Organization*, Vol. 3 (Amsterdam: North-Holland) 1887–1966.

DUTTA, P. K. and RADNER, R. (2004), "Self-enforcing Climate-Change Treaties", *Proceedings of the National Academy of Sciences of the United States of America*, **101**, 4746–4751.

DUTTA, P. K. and RADNER, R. (2006), "A Game-Theoretic Approach to Global Warming", *Advances in Mathematical Economics*, **8**, 135–153.

DUTTA, P. K. and RADNER, R. (2009), "A Strategic Analysis of Global Warming: Theory and Some Numbers", *Journal of Economic Behavior & Organization*, **71**, 187–209.

EDLIN, A. S. and REICHELSTEIN, S. (1996), "Hold-ups, Standard Breach Remedies, and Optimal Investment", *American Economic Review*, **86**, 478–501.

ENGWERDA, J. C. (2005), *LQ Dynamic Optimization and Differential Games* (West Sussex, UK: Wiley-Blackwell).

EVANS, R. (2008), "Simple Efficient Contracts in Complex Environments", *Econometrica*, **76**, 459–491.

FERSHTMAN, C. and NITZAN, S. (1991), "Dynamic Voluntary Provision of Public Goods", *European Economic Review*, **35**, 1057–1067.

FUDENBERG, D. and TIROLE, J. (1991), *Game Theory* (Cambridge, MA: MIT Press).

GATSIOS, K. and KARP, L. (1992), "How Anti-merger Laws Can Reduce Investment, Help Producers, and Harm Consumers", *Journal of Industrial Economics*, **40**, 339–348.

GOLOMBEK, R. and HOEL, M. (2006), "Second-Best Climate Agreements and Technology Policy", *Advances in Economic Analysis & Policy*, **6**, Article 1, BE Press.

GURIEV, S. and KVASOV, D. (2005), "Contracting on Time", *American Economic Review*, **95**, 1269–1385.

HARSTAD, B. (2005), "Majority Rules and Incentives", *Quarterly Journal of Economics*, **120**, 535–568.

HARSTAD, B. (2011), "The Dynamics of Climate Agreements" (Mimeo, Northwestern).

HARSTAD, B. (2012), "Buy Coal! A Case for Supply-Side Environmental Policy", *Journal of Political Economy*, forthcoming.

HART, O. D. and MOORE, J. (1988), "Incomplete Contracts and Renegotiation", *Econometrica*, **56**, 755–785.

HOEL, M. (1993), "Intertemporal Properties of an International Carbon Tax", *Resource and Energy Economics*, **15**, 51–70.

HOEL, M. and DE ZEEUW, A. (2010), "Can a Focus on Breakthrough Technologies Improve the Performance of International Environmental Agreements?", *Environmental and Resource Economics*, **47**, 395–406.

HONG, F. and KARP, L. (2010), "International Environmental Agreements with Mixed Strategies and Investment" (Mimeo).

HOUBA, H., SNEEK, K. and VARDY, F. (2000), "Can Negotiations Prevent Fish Wars?" *Journal of Economic Dynamics and Control*, **24**, 1265–1280.

KOLSTAD, C. D. and TOMAN, M. (2005), "The Economics of Climate Policy", *Handbook of Environmental Economics*, **3**, 1562–1593.

MARX, L. M. and MATTHEWS, S. A. (2000), "Dynamic Voluntary Contribution to a Public Project", *Review of Economic Studies*, **67**, 327–358.

MASKIN, E. and TIROLE, J. (2001), "Markov Perfect Equilibrium: I. Observable Actions", *Journal of Economic Theory*, **100**, 191–219.

REYNOLDS, S. S. (1987), "Capacity Investment, Preemption and Commitment in an Infinite Horizon Model", *International Economic Review*, **28**, 69–88.

SEGAL, I. and WHINSTON, M. D. (2002), "The Mirrless Approach to Mechanism Design with Renegotiation", *Econometrica*, **70**, 1–45.

SEGAL, I. and WHINSTON, M. D. (2010), "Property Rights", in Gibbons, R. and Roberts, J. (eds) *Handbook of Organizational Economics* (Elsevier).

SORGER, G. (1998), "Markov-Perfect Nash Equilibria in a Class of Resource Games", *Economic Theory*, **11**, 79–100.

SORGER, G. (2006), "Recursive Nash Bargaining Over a Productive Asset", *Journal of Economic Dynamics and Control*, **30**, 2637–2659.

TSUTSUI, S. and MINO, K. (1990), "Nonlinear Strategies in Dynamic Duopolistic Competition with Sticky Prices", *Journal of Economic Theory*, **52**, 136–161.

VAN DER PLOEG, F. and DE ZEEUW, A. (1992), "International aspects of pollution control", *Environmental and Resource Economics*, **2**, 117–139.

WIRL, F. (1996), "Dynamic Voluntary Provision of Public Goods: Extension to Nonlinear Strategies", *European Journal of Political Economy*, **12**, 555–560.

YANASE, A. (2006), "Dynamic Voluntary Provision of Public Goods and Optimal Steady-State Subsidies", *Journal of Public Economic Theory*, **8**, 171–179.

[24]

Environ Resource Econ (2009) 42:411–425
DOI 10.1007/s10640-009-9264-x

Endogenous Minimum Participation in International Environmental Treaties

Carlo Carraro · Carmen Marchiori · Sonia Oreffice

Received: 7 August 2008 / Accepted: 16 January 2009 / Published online: 12 February 2009

Abstract Many international treaties come into force only after a minimum number of countries have signed and ratified the treaty. Minimum participation constraints are particularly frequent in the case of environmental treaties dealing with global commons, where free-riding incentives are strong. Why do countries that know they have an incentive to free-ride accept to "tie their hands" through the introduction of a minimum participation constraint? This article addresses the above issues by modeling the formation of an international treaty as a three-stage non-cooperative coalition formation game. Both the equilibrium minimum participation constraint and the number of signatories—the coalition size—are determined. This article, by showing that a non-trivial partial coalition, sustained by a binding minimum participation constraint, forms at the equilibrium, explains the occurrence of minimum participation clauses in most international environmental agreements. It also analyses the endogenous equilibrium size of the minimum participation constraint.

Keywords Agreements · Climate · Negotiations · Policy · Participation rule

JEL Classification H0 · H4 · O3

C. Carraro (✉)
Department of Economics, University of Venice, CEPR, CEPS, CESifo and Fondazione ENI E. Mattei, S. Giobbe 873, 30121 Venice, Italy
e-mail: ccarraro@unive.it; carraro@feem.it

C. Marchiori
London School of Economics and Fondazione ENI E. Mattei, Venice, Italy

S. Oreffice
Departamento de Fundamentos del Análisis Económico,
Universidad de Alicante, Ctra. San Vicente del Raspeig, 03080 Alicante, Spain
e-mail: sonia@merlin.fae.ua.es

1 Introduction

Several articles have provided a thorough economic analysis of a country's incentives to sign an international treaty.[1] Particular attention has been given to international environmental treaties, because they are often aimed at providing a global public good and are therefore characterized by strong free-riding incentives. This literature, by using a non-cooperative game-theoretic approach, has also explored which policy strategies and negotiation rules can be designed to increase the number of signatories of an international environmental treaty (see, for example, Barrett 2003, 2007 and Finus 2003, 2008).

A feature of many international treaties that has been largely neglected by the theoretical literature is the clause that a treaty will only come into force if a minimum number of signatories or a minimum degree of effectiveness is achieved. For example, in the case of the Kyoto Protocol, the clause is twofold: the Protocol comes into force only if at least 55 countries (out of the Annex 1 list of the 1992 UNFCCC) sign and ratify it and if these countries represent at least 55% of reference year emissions (Article 25). More generally, almost all international environmental treaties contain a minimum participation clause. According to Rutz (2001), only two out of the 122 multilateral environmental agreements provided by the Center for International Earth Science Information Network do not contain any minimum participation rule. In 81 cases, the participation rule asks for a minimum number of signatories. In 22 cases, unanimity is required for the treaty to come into force, namely all negotiating countries must sign and ratify the agreement for it to be effective. In the remaining 17 cases, the minimum participation rule is coupled to other requirements, i.e., for these agreements it is not sufficient that a certain number of countries ratify the treaty, but these countries have to satisfy additional criteria.[2]

Despite the importance of the minimum participation rule (and of the unanimity rule as a particular case), very few theoretical studies have analyzed this issue. A previous study by Black et al. (1992) shows that the introduction of a minimum participation constraint increases the number of signatories of an international environmental agreement. However, it does not prove that countries actually have an incentive to introduce a minimum participation constraint among the rules of a given treaty.[3] Similarly, Rutz (2001) shows, in a very specific case, that a minimum participation rule, if it exists, can result in an increase in the number of signatories. However, the rule is imposed exogenously and is not shown to be an equilibrium outcome of the game.

This article endogenizes the minimum participation rule as the equilibrium outcome of a three-stage coalition formation game. Our goal is to address the following questions: Why do countries agree to have a minimum participation constraint among the clauses characterizing an international treaty? Is a minimum participation rule a way to offset free-riding incentives? Why do countries that know they have an incentive to free-ride, accept to "tie their hands" through the introduction of a minimum participation constraint? What is the endogenous equilibrium level of the minimum participation constraint?

This article answers the above questions by analyzing a non-cooperative game of coalition formation. In the theoretical literature on international agreements, the decision of countries

[1] Refer to the surveys in Barrett (2003, 2007); Carraro and Marchiori (2003); Finus (2003, 2008); Rubio and Ulph (2006).

[2] For example, the Montreal Protocol on the ozone layer asks for 11 instruments of ratification representing at least two-thirds of 1986 estimated global consumption of the controlled substances.

[3] Moreover, Black et al. (1992) design the minimum participation constraint to solve a profitability problem (cooperation becomes profitable only if a minimum number of countries sign the agreement), whereas in this article we also solve a stability problem (cooperation is undermined by the presence of free-riders).

to sign a treaty is often modeled as a two-stage game (see Barrett 1997b, 2003; Carraro and Siniscalco 1998; Tulkens 1998; Carraro and Marchiori 2003; Finus 2003, 2008; Rubio and Ulph 2006; for a few surveys). In the first stage (the *coalition stage*), countries non-cooperatively decide whether or not to sign the treaty by anticipating the consequences of their decision on the economic variables under their control. Then, in the second stage of the game (the *policy stage*), the equilibrium value of these variables is set. In this article, we introduce an additional stage preceding the coalition and policy stages, to endogenize the choice of the minimum participation level. In this first stage, called the *minimum participation stage*, countries select the number of countries that must sign the treaty for it to come into force.[4] This decision is taken non-cooperatively and unanimously by anticipating its implications on the second and third stage of the game.

Using the above theoretical framework, this article shows that, at the equilibrium, a non-trivial coalition will form, and will be sustained by a binding minimum participation constraint. It also shows that the number of signatories is increased by the presence of a minimum participation rule, even when this rule is endogenous. This explains the occurrence of minimum participation clauses in most international environmental agreements. In addition, the article explores under which conditions it is optimal for countries to adopt the unanimity rule (in many treaties, as discussed in Rutz 2001, the minimum participation constraint coincides with unanimous participation).

The idea that a non-trivial partial coalition can belong to the equilibrium structure of a coalition formation environmental game is not new. The aforementioned literature on international environmental agreements (see, for example, Carraro and Marchiori 2003 or Finus 2008 for a summary of the main results) shows that: (i) even in cases where there are positive spillovers (e.g., in the case of public good provision) and even without any commitment to cooperation, countries may form a coalition, i.e., may decide to sign a treaty in order to cooperate to achieve a common target (Carraro and Siniscalco 1992; Barrett 1994 among others); (ii) the coalition is usually formed by a subgroup of the negotiating countries (Barrett 1994), but the small initial coalition can be expanded by means of transfers or through "issue linkage," under some restrictive conditions (Carraro and Siniscalco 1993; Carraro and Siniscalco 1995; Botteon and Carraro 1997); (iii) the outcome of the game is found to strictly depend on the membership rules (open membership, exclusive membership, coalition unanimity, ...) that are adopted by the negotiating countries (Carraro and Marchiori 2003; Finus 2003); (iv) finally, when multiple coalitions form, the equilibrium is characterized by several, often small, coalitions (Bloch 1997, 1996; Ray and Vohra 1997, 1999; Yi 1997, 2003).

The novelty of this article is the endogenization of the minimum participation constraint and the analysis of a country's reasons to adopt it despite the presence of incentives to free-ride on the other countries' emission abatement.

The structure of the article is as follows. Section 2 provides the main assumptions and definitions. Section 3 analyses the coalition game for a given value of the participation constraint. Section 4 endogenizes the participation constraint, discusses the incentives to adopt a participation rule and analyses its equilibrium value. A concluding section summarizes the main results and describes the scope for further research.

[4] Results are robust to considering a given proportion of the externality globally produced (i.e., percentage of total emissions) as minimum participation constraint that countries signing the treaty must represent for the treaty to come into force.

2 Assumptions and Definitions

The three-stage game proposed in this article involves n countries, $i = 1, \ldots, n$ with $n \geq 3$. Countries are assumed to be ex-ante identical. This assumption is necessary for each country to have the same strategy space and payoff function in the third stage of the game (*policy stage*).[5] The structure of the game can be formally described as follows. In the first stage—the *minimum participation stage*—countries non-cooperatively choose the minimum participation rule α by unanimity voting. They unanimously select the α representing the share of the n countries that must sign the treaty for it to come into force, $\alpha \in [0, 1]$ being the strategy set. Let $c \in \{1, \ldots, n\}$ denote the number of cooperating countries; that is, the number of countries that decide to sign the treaty.[6] At the equilibrium and for a given α, a coalition c can form only if $c \geq \alpha n$. In the second stage—the *coalition stage*—countries decide non-cooperatively whether or not to sign the environmental treaty. The strategy set is thus binary: whether or not to join the coalition. The third stage—the *policy stage*—is as follows. Those countries that have signed the agreement play as a single player and divide the resulting payoff according to a given burden-sharing rule (any of the rules derived from cooperative game theory). The remaining countries play non-cooperatively against the coalition and against one another.[7] In this last stage, the strategies are represented by the specific policy values that each country can set. In many IEAs such as the Kyoto Protocol, these are the emission reduction targets that each country decides to achieve.

The third stage of the game can be reduced to a partition function under suitable assumptions (Bloch 1997; Ray and Vohra 1997, 1999; Yi 1997, 2003 , corresponding to assumptions A.4 and A.5 below). The gains from cooperation can thus be described by a *per-member partition function* $\mathrm{p} : \Pi \rightarrow \Re^n$, a mapping which associates each coalition structure π with a vector of individual payoffs in $\Re^n$. In particular $\mathrm{p}_j(\mathrm{C}_i; \pi)$ represents the payoff of player j belonging to the coalition C_i in the coalition structure π. Therefore, our study of coalition formation consists of the analysis of the first and second stage of the game.

As to the first stage game, if $\alpha = 0$, no restriction is introduced and a treaty comes into force whatever the number of its signatories. This is the so-called "open membership" rule, one of the most commonly assumed membership rules (Yi 1997, 2003). If $\alpha = 1$, a treaty comes into force only if all countries decide to sign the treaty. In this case, we have what is termed the "coalition unanimity" rule (Tulkens 1998; Carraro and Marchiori 2003) or full participation constraint. If $\alpha = 0.55$, and countries are symmetric, we have the 55% rule introduced in the Kyoto Protocol. Therefore, by endogenizing α, we can determine whether it is optimal for countries to opt for the open membership rule, the coalition unanimity rule or any other minimum participation rule defined by $\alpha \in [0, 1]$.

In what follows, we formally state the assumptions characterizing the three-stage game.

---

[5] The assumption of symmetry is standard and broadly used in the literature of international environmental agreements (see Barrett 2007; Finus 2008; Rubio and Ulph 2006). The advantage of this assumption is to reasonably simplify the policy stage of the game, which allows us to focus on the coalition formation process.

[6] When $c = 1$, the coalition structure is trivial. All players are singleton and play non-cooperatively against one another.

[7] These behavioral assumptions coincide with those defining the PANE equilibrium of Chander and Tulkens (1995). They can be contrasted with the traditional cooperative game approach (e.g., Chander and Tulkens 1997) and with the repeated game approach (Barrett 1994, 1997). Moreover, note that the regulatory approach often proposed in public economics is not applicable given the absence of a supranational authority.

A.1. All players decide simultaneously at all stages.[8]

A.2. All players are ex-ante identical.[9]

Developments of the theory of endogenous coalition formation (Bloch 1997; Yi 2003) have stressed the implications of allowing players to form different coalitions. However, in many environmental treaties (e.g., the case of climate change such as the Kyoto protocol, or of the Montreal protocol), the negotiating agenda focuses on a single agreement. Therefore, we assume that:

A.3. Players are invited to sign a single agreement. Hence, those that do not sign cannot propose a different agreement.

From a game-theoretic viewpoint, A.3 implies that only one coalition can form, the remaining defecting players playing as singletons.

A.4. The third stage emission game has a unique Nash equilibrium for any coalition structure.

A.4 is a necessary assumption for the existence of a well-defined partition function and allows us to focus on the second stage of the game; that is, on the process of coalition formation. In order to convert the strategic form game into a partition function, the competition among the coalition and the singletons must also be specified. The standard and perhaps most natural assumption is that:

A.5. Inside a coalition, players make their policy decisions cooperatively in order to maximize the coalitional surplus, whereas the coalition and the singletons compete with one another in a non-cooperative way.

Given A.5 and the symmetry among players, a coalition C_i can be identified just by its size, c_i. Therefore, a coalition structure with only one coalition $\pi = \{c; 1_{n-c}\}$ can be denoted by $\pi = \{c\}$, the size of the only coalition that can possibly form. The partition function is then obtained as a Nash equilibrium payoff of the game played by the coalition and the singletons.

Following Bloch (1996), we assume that the payoff inside a coalition is divided *equally* among the coalition members. In fact, in the case of identical players, equal sharing is rather a result than an assumption. As shown by Ray and Vohra (1999), in a symmetric context any of the sharing rules derived from cooperative game theory (e.g., the Nash bargaining solution or the Shapley value) approximates the equal division rule. Therefore, in the context of the present model each player inside a coalition receives the same payoff as the other members. This also implies that players' payoff depends only on the coalition size and not on the identities of the coalition members. The per-member partition function (partition function hereafter) can thus be denoted by $p(c; 1_{n-c})$, which represents the payoff of a player belonging to the size-c coalition.

We also introduce an assumption which is quite natural in the case of international environmental agreements and simplifies the analysis of the coalition formation game:

A.6. Each player's payoff function increases monotonically with respect to the coalition size (the number of signatories).[10]

[8] By contrast, Barrett (1994) assumes that the group of signatories is Stackelberg leader with respect to non-signatories. In Bloch (1997), it is assumed that players play sequentially.

[9] As previously explained, this assumption means that each country has the same strategy space and payoff function in the third stage of the game.

[10] Note that the monotonically increasing payoff function is a weaker requirement than superadditivity, since it does not impose any structure on the rate of increase. Monotonicity of the payoff function excludes the

Since the global environment is a public good, a country that finds it convenient to reduce its own emissions provides a positive contribution to the welfare of all countries, both inside and outside the coalition. Formally, A.6 implies that both members and non-members' partition functions are increasing in c for all values of $c \in \{1, \ldots, n\}$. As for those outside a coalition, the assumption that the effort undertaken by cooperators makes them better off is typically referred to as 'positive externality'. The presence of positive externalities, in turn, gives room to the well known 'free-riding problem', which is a key feature of the game analyzed in this article. Players' payoffs will be fully characterized in Sect. 3, where the coalition stage is formally described.

Finally, the first stage of our game is a sort of constitutional stage in which the rules of the game are set. Therefore:

A.7. In the minimum participation stage, decisions are taken under unanimity.

Players decide the minimum participation rule by unanimity voting, as no player can be forced to accept a minimum participation rule he does not agree upon. The decision is taken non-cooperatively and simultaneously. Each player's strategy consists of choosing a rule $\alpha \in [0, 1]$. $[0, 1]$ represents the strategy set, where $\alpha = 0$ entails the absence of participation constraint, $\alpha = 1$ implies that all players are required to join the coalition for it to come into force, and any fraction $\alpha \in (0, 1)$ represents a minimum participation constraint of αn countries. A player's utility associated to these strategies is represented by the expected value between a cooperator's and a free rider's partition functions of the coalition formed under α. The exact payoffs will be defined in Sect. 4 , which specifically analyses the minimum participation stage. Finally, the decisions made in the first stage of the game are irreversible. Given that this is the constitutional stage, it is assumed that once a minimum participation rule is set, it cannot be renegotiated in the later stages of the game. This is a standard assumption in sequential games.

In the next sections, we will analyze the equilibrium of the three-stage game with endogenous minimum participation. Given the partition function described above, we can directly focus on the coalition stage of the game and then move, by backward induction, to the minimum participation stage, in which the optimal minimum participation constraint is determined.

3 The Coalition Stage

The second stage of the game is a binary choice game (joining the coalition or behaving as a lone free-rider) whose outcome is a single coalition structure $\pi = \{c\}$, where $n-c$ players remain singletons. Given the above assumptions, a coalition member's partition function is increasing in c for all values of $c \in \{1, \ldots n\}$ and the presence of positive spillovers implies that the partition function of any player outside the coalition is also increasing in c for all values of $c \in \{1, \ldots n\}$.

Let us denote by Q(c) the relative non-member partition function $p(1; c) - p(1; 1_n)$, where $p(1; 1_n)$ is a player's partition function when no coalition is formed. Similarly, the relative per-member payoff function P(c) is equal to $p(c; 1_{n-c}) - p(1; 1_n)$. From A.6, we focus on the case in which P(c) and Q(c) are monotonic. As said, this case is relevant because it characterizes environmental negotiations and more generally the provision of public goods.

Footnote 10 continued
possibility of "exclusive membership equilibria", where the group of cooperating players can refuse entry to a player who wants to join the coalition (Cf. Yi 1997). Our main results can also be extended to non-monotonic payoff functions.

Let us start the analysis by considering the case in which no minimum participation rule is introduced. In this case, $\alpha = 0$ (open membership rule) and the equilibrium of the coalition stage is completely characterized by the following two properties, first derived in the cartel literature (d'Aspremont et al. 1983) and then often used in the environmental literature (Carraro and Siniscalco 1993; Barrett 1994; and many others).

3.1 Profitability

A coalition c is profitable if each cooperating player gets a payoff larger than the one he would obtain when no coalition forms. Formally:

$$P(c) \geq P(1) = 0, \tag{1}$$

where P(c) is the relative payoff of a cooperating player member of the coalition of size c, and P(1) is the relative payoff associated to the singleton structure.

3.2 Stability

A coalition is stable if it is both internally and externally stable. It is internally stable if no cooperating player is better off by defecting in order to form a singleton.[11] Formally[12]:

$$P(c) \geq Q(c-1). \tag{2a}$$

It is externally stable if no singleton is better off by joining the coalition c. Formally:

$$Q(c) > P(c+1) \tag{2b}$$

To determine the stability of a coalition we use the stability function $L(c) = P(c) - Q(c-1)$ proposed in Carraro and Siniscalco (1992). If L(c) is positive, there is no incentive to free-ride on the coalition of size c.

The size of the stable coalition is determined by conditions (2a)(2b),[13] i.e., the largest integer c* smaller than ĉ, with ĉ defined by $L(\hat{c}) = 0$ and $L'(\hat{c}) < 0$ (assumption A.6 guarantees the uniqueness of ĉ).[14] The size c* identifies the minimal externally stable coalition and the maximal internally stable one. Hence, the equilibrium coalition structure is $\pi^* = \{c^*\}$, with $n - c^*$ singletons. Note that c* is not necessarily greater than one. There exists an extensive literature (e.g., Barrett 1994) showing that an equilibrium stable coalition generally exists under A.1–A.6.[12] Let us denote by c_0^* the coalition formed under the open membership rule $\alpha = 0$. Then:

Proposition 1 *Let A.1–A.7 hold and $\alpha^* = c_0^*/n$. The equilibrium of the coalition game for any value of $\alpha \in [0, 1]$ is the coalition structure:*

$$\pi^*(\alpha^*) = \{c_0^*, 1_{n-c_0^*}\} \tag{4a}$$

for any minimum participation rule such that $0 \leq \alpha \leq \alpha^*$, and

[11] Yi (1997) denotes this condition by the term "stand alone stable".

[12] We assume that if a player is indifferent as to joining the coalition or defecting, then he joins the coalition.

[13] (1) is only a necessary condition implied by internal stability under A.6.

[14] This result, as well as many others on the size of stable coalitions, is shown in Carraro and Marchiori (2003).

$$\pi(\alpha) = \{c_\alpha = \alpha n,\ 1_{n-c_\alpha}\} \tag{4b}$$

for any minimum participation rule such that $\alpha^* < \alpha \leq 1$.

Proof If $\alpha \leq \alpha^*$, the equilibrium coalition is not modified by the presence of a minimum participation constraint α. Indeed, if in the first stage players set a minimum participation rule $\alpha \leq c_0^*/n$, the number of signatories necessary for the treaty to come into force is smaller than or equal to the number of countries that would sign the treaty anyway. Hence, the equilibrium coalition remains c_0^*. On the other hand, if in the first stage players agree on a minimum participation rule $\alpha > \alpha^* = c_0^*/n$, then in the second stage they have the choice either to form a coalition of size $c_\alpha = \alpha n$, or not to form any coalition at all. Indeed, all coalitions larger than c_α are unstable, because all coalitions larger than $c_0{}^*$ are unstable and $c_\alpha > c_0^*$. Moreover, no coalition smaller than $c_\alpha = \alpha n$ can form, by definition of minimum participation. Given that, by assumption A.6, $Q(c-1) > P(c) > P(1) = 0$, players prefer to form the coalition $c_\alpha = \alpha n$. □

Proposition 1 identifies the relationship between the minimum participation rule decided in the first stage of the game and the equilibrium of the coalition stage. This relationship between the value of α and the equilibrium of the coalition stage, jointly with the incentive structure that supports the equilibrium coalition c_0^*, is crucial to determine the equilibrium value of α in the minimum participation stage.

4 The Minimum Participation Stage

When countries deal with the choice of the membership rule, they consider two aspects:

(1) *If* it is convenient to introduce a minimum participation constraint rather than to negotiate under the open membership regime;
(2) *Which* minimum participation constraint is optimal.

As previously discussed, the decision in the first stage of the game is taken under the unanimity rule, as no player (country) can be forced to accept a participation constraint he does not agree upon. In order to achieve unanimity on a value of α, all players must (weakly) prefer the equilibrium supported by the agreement on a minimum participation constraint to the one that would emerge without a minimum participation rule. This is the case if, whatever the choice of a player in the second stage, a player's payoff when the minimum participation constraint is introduced is larger than (or equal to) the payoff he/she would receive in the absence of any minimum participation constraint. Specifically, we can state the following:

Proposition 2 *Let A.1–A.7 hold. In the first stage of the game, all players agree to introduce a minimum participation constraint* $\alpha = c/n$ *iff* :

$$P(c) \geq Q(c_0^*) \tag{5}$$

Proof For all players to agree on a minimum participation constraint, we must have:

- $P(c) \geq P(c_0^*)$, i.e., all players who cooperate in c and in c_0^* are better off in c;
- $Q(c) \geq P(c_0^*)$, i.e., all players who are free-riders with respect to c but cooperators in c_0^* are better off free riding with c;
- $P(c) \geq Q(c_0^*)$, i.e., all players who cooperate in c but free-ride with respect to c_0^* are better off in c;

– Q(c) ≥ Q(c_0^*), i.e., all players who free-ride both with respect to c and to c_0^* are better off free riding with c.

If these four conditions are satisfied, all players find it convenient to agree on the minimum participation constraint α that sustains the coalition of size c. Notice that c > c_0^* (otherwise the coalition is of size c_0^*) by Proposition 1. Hence, P(c) > P(c_0^*) by Assumption A.6. Moreover, Q(c) is also monotonically increasing, which implies Q(c) > Q(c_0^*) when c > c_0^*. These two results, the definition of c_0^*, and c > c_0^*, imply Q(c) > Q(c − 1) > P(c) > P(c_0^*). As a consequence, Eq. 5, i.e., P(c) ≥ Q(c_0^*), is a necessary and sufficient condition for all players to agree on the selection of a minimum participation rule such that the equilibrium coalition structure becomes {c, 1_{n-c}}. □

Can inequality (5) be met for c ≤ n? The answer is positive. Given the monotonicity of the functions P(c) and Q(c), it is possible—if n is sufficiently large—to find a coalition c large enough to satisfy P(c) ≥ Q(c_0^*). In particular, assume that n is such that there is a $c^+ < n$ that satisfies P(c^+) = Q(c_0^*). The monotonicity of P(c) implies that (5) is met for all $c \in \{c^+, \ldots, n\}$.

Let us consider a simple example, as an illustration of Proposition 2 and of the scope of condition (5). The following basic economic-environmental model with linear cost and quadratic benefit functions is also used in Barrett (1994, 1997); Carraro and Siniscalco (1992); Chander and Tulkens (1997) and many others.

Let n be the number of identical countries which emit a pollutant that damages a common environmental resource. The utility u of each country depends on the benefits arising from the use of the environment for production and consumption activities and on the damage resulting from polluting emissions:

$$u = B(x_i) - D(X) \quad i = 1, 2, \ldots, n$$

where x_i is country i's emission level and X is the total amount of polluting emissions $(X = \Sigma_{i=1}^{n} x_i)$. Let the benefit function $B(x_i)$ be quadratic and the cost function (or damage function) be linear:

$$B(x_i) = k\left(dx_i - \frac{x_i^2}{2}\right)$$

where d is the maximum level of emissions and k is a technological parameter which parameterizes total benefits from emissions (the higher k, the larger the benefit);

$$D(X) = \frac{m}{2}\left[x_i + (X - x_i)\right]$$

where m parameterizes the level of perceived damage from pollution.

A country's payoff can, therefore, be written as follows:

$$u = k\left(dx_i - \frac{x_i^2}{2}\right) - \frac{m}{2}\left[x_i + (X - x_i)\right]$$

In a coalition structure $\pi = \{c, 1_{n-c}\}$, countries inside the coalition act cooperatively and choose their emission level by maximizing the sum of the payoffs of all members. Hence, for a coalition structure $\pi = \{c, 1_{n-c}\}$, the payoff of a cooperating player can be written as:

$$u_c = k\left(dx_c - \frac{x_c^2}{2}\right) - \frac{m}{2}\left[x_c + \left((c-1)\,x_c + (n-c)\,x_f\right)\right]$$

where x_c and x_f are the emission levels of cooperators and free-riders, respectively. Differentiating u_c with respect to x_c yields $x_c^* = d - mc/2k$, which is the optimal emission level for a cooperating country.

A free-rider's payoff function is given by the following expression:

$$u_f = k\left(dx_f - \frac{x_f^2}{2}\right) - \frac{m}{2}\left[x_f + (n - c - 1)\,\hat{x}_f + cx_c\right].$$

Differentiating u_f with respect to x_f yields $x_f^* = d - m/2k$. By replacing the values of x_c^* and x_f^* into the payoff functions u_c and u_f, we obtain the payoffs for cooperators and free-riders as a function of the coalition size c:

$$\begin{aligned} P(c) &= k\left[d\left(d - \frac{mc}{2k}\right) - \frac{1}{2}\left(d - \frac{mc}{2k}\right)^2\right] + \frac{m}{2}\left[\left(d - \frac{mc}{2k}\right) + (c-1)\left(d - \frac{mc}{2k}\right)\right. \\ &\quad \left. + (n-c)\left(d - \frac{m}{2k}\right)\right] \\ Q(c) &= k\left[d\left(d - \frac{m}{2k}\right) - \frac{1}{2}\left(d - \frac{m}{2k}\right)^2\right] + \frac{m}{2}\left[\left(d - \frac{m}{2k}\right) + (n-c-1)\left(d - \frac{m}{2k}\right)\right. \\ &\quad \left. + c\left(d - \frac{mc}{2k}\right)\right] \end{aligned}$$

Let $m = 2k$ and $k = d = 1$. By computing the stability function $L(c) = P(c) - Q(c-1)$, equalizing it to zero and solving with respect to c, the following result can be derived: $c_0^* = 3$. It can also be shown that condition (5) is verified for $c \geq 5$. Therefore, Proposition 2 holds in this setting.[15]

However, (5) is only a necessary condition to identify the minimum participation constraint which is chosen by the n countries. The inequality (5) defines the set of minimum participation rules that would be voted with the unanimous consensus of all countries. However, which rule is actually going to be adopted?

To answer this question, we must write the optimization problem that each country solves when deciding the optimal minimum participation rule. Given symmetry, in the first stage of the game a country does not know whether in the coalition stage it will be a cooperator or a free rider. Therefore, it determines its optimal choice in the first stage (the value of α in the strategy set [0,1]) by solving:

$$\begin{aligned} &\max_c EX(c) = \frac{c}{n}P(c) + \frac{n-c}{n}Q(c) \\ &\text{s.t. } P(c) \geq Q[c_0^*], \quad \text{with } 1 \leq c \leq n \end{aligned} \tag{6}$$

where $EX(c)$ is a country's expected payoff in the first stage of the game, c/n is the probability of being a cooperator in the second stage (given that the rule $\alpha = c/n$ is chosen in the first stage), whereas $(n-c)/n$ is the probability of being a free-rider in the second stage of the game.

The equilibrium minimum participation rule $\alpha^\circ = c^\circ/n$ is the solution to (6), where c° is the maximizer of $EX(c)$ subject to (5). The optimization problem in (6) identifies the trade-off that countries face. On the one hand, a high value of c, namely of α, increases the payoff of both cooperators and free-riders. On the other hand, a high value of α reduces

[15] Condition (5) may hold also in a quadratic benefit-quadratic damage model. Using for example the model proposed in Diamontoudi and Sartzetakis (2006), where the cooperators' payoff function $\omega_s(.)$ corresponds to our payoff function P(.), and the free riders' payoff function $\omega_{ns}(.)$ corresponds to Q(c), condition (5) holds for $c \geq 4$.

the probability of being a free-rider in the second stage, thus reducing the probability of reaping the highest payoff. Note that the equilibrium in the first stage always exists and is unique. In particular, $\alpha = 0$ represents the equilibrium solution when countries do not unanimously agree on any $\alpha > 0$, i.e., no country finds any $\alpha > 0$ more profitable than $\alpha = 0$. If at least one country finds a $\alpha^{\circ} > \alpha^{*}$ optimal, then by symmetry all countries will: α° will be unanimously voted and will represent the unique equilibrium (α° dominating strategy for all players).

Under what conditions is $EX(c)$ increasing in c for all values of c satisfying (5)? This question is relevant because it identifies the case in which $c^{\circ} = n$, or $\alpha^{\circ} = 1$, i.e., in which the minimum participation rule requires the ratification by all countries and therefore coalition unanimity. A sufficient condition is provided by the following:

Proposition 3 *Let A1–A7 hold. The equilibrium minimum participation rule is* $\alpha^{\circ} = 1$, *i.e., the treaty comes into force only when all countries sign and ratify it, if:*

$$P(n) \geq Q[c_0^*] \tag{7a}$$

$$(1 + \gamma_P)P(c+1)/Q(c+1) \geq 1 \tag{7b}$$

$\forall c \geq c^{+}$, *where* c^{+} *is the solution of* $P(c^{+}) = Q(c_0^*)$ *and* $\gamma_P = (c/P(c+1)) * (\Delta P(c)/\Delta c)$ *is the quasi-elasticity of the function* $P(c)$.

Proof The first condition, (7a), is the necessary condition already established by Proposition 2. It is met for n sufficiently large. The second condition, (7b), can be proved as follows. A discrete change of n*EX(c) with respect to c can be written as follows:

$$\begin{aligned}
&\Delta\left[cP(c) + (n-c)Q(c)\right]/\Delta c \\
&= [\Delta c * P(c+1) + c * \Delta P(c) + \Delta(n-c) * Q(c+1) + (n-c) * \Delta Q(c)]/\Delta c \\
&= P(c+1) + c * \Delta P(c)/\Delta c - (\Delta c/\Delta c) * Q(c+1) + n * \Delta Q(c)/\Delta c - c * \Delta Q(c)/\Delta c \\
&= P(c+1) * [1 + (c/P(c+1)) * \Delta P(c)/\Delta c] - \\
&\quad -Q(c+1)\,[1 - ((c/c) * n/Q(c+1)) * (\Delta Q(c)/\Delta c) + (c/Q(c+1)) * (\Delta Q(c)/\Delta c)]
\end{aligned}$$

Then, let $\gamma_P = (c/P(c+1)) * (\Delta P(c)/\Delta c)$ and $\gamma_Q = (c/Q(c+1)) * (\Delta Q(c)/\Delta c)$ be the quasi-elasticity of P(c) and Q(c), respectively. Substituting above, we get:

$$\Delta\left[cP(c) + (n-c)Q(c)\right]/\Delta c = P(c+1) * \left[1 + \gamma_P\right] + Q(c+1) * \left[\gamma_Q * (n/c - 1) - 1\right]$$

The equilibrium minimum participation constraint is $\alpha = 1$ if $\Delta\,[cP(c) + (n-c)Q(c)]\,/\Delta c$ is positive for all $c \in \left[c^{+}, n\right]$, i.e.,

$$P(c+1) * \left[1 + \gamma_P\right] + Q(c+1) * \left[\gamma_Q * (n/c - 1) - 1\right] \geq 0 \tag{8}$$

Dividing by Q(c+1) both sides (Q(c+1) is a positive number), we get:

$$(P(c+1)/Q(c+1)) * [1 + \gamma_P] + \gamma_Q * (n-c)/c \geq 1 \tag{9}$$

Note that both terms on the left hand side are positive.
Hence, if $(1 + \gamma_P) * (P(c+1)/Q(c+1)) \geq 1$, then (9) holds. □

Note that $P(c+1)/Q(c+1) < 1$ for $c > c_0^*$ by definition of stability, and $(n-c)/c < 1$ if c is relatively high with respect to n. Both $(n-c)/c$ and $P(c+1)/Q(c+1)$ decrease with c when P and Q are monotonic and $c_0^* \geq 2$. Therefore, unless the quasi-elasticities strongly increase with c, (9) is less likely to be met for high values of c. The strong condition to be met for EX(c) to be increasing for all $c \geq c^+$, including $c = n$, is (7b), i.e.,

$$(1 + \gamma_P)P(c+1)/Q(c+1) \geq 1$$

which coincides with (9) for $c = n$. This condition is satisfied only if γ_P is sufficiently large, i.e., the profitability function increases sufficiently rapidly with c. The curvature of the coalition's payoff function is therefore crucial. The intuition is that high and increasing values of P(c) when c is close or equal to n offset the loss induced by the impossibility of free-riding in the second stage.

However, in situations in which there are increasing free-riding incentives for larger coalition sizes, condition (7b) is not likely to be satisfied. Suppose, for instance, that P(c) is concave. In such a case, the quasi-elasticity of P(c) becomes smaller and smaller as c moves toward n. At the same time, by definition of stability, the difference between Q(c) and P(c) increases with c. Therefore, $P(c+1)/Q(c+1) < 1$ also becomes smaller and smaller. As a consequence, $(1 + \gamma_P)P(c+1)/Q(c+1)$ may become smaller than one and (7b) may not be met. If benefits from cooperation do not increase with the number of cooperators, or increase too slowly, then it is *not* optimal to set a minimum participation constraint such that all countries must sign and ratify the treaty for it to come into force (that is, $\alpha = 1$). Instead, the optimal minimum participation rule will be smaller than one, which in turn implies that, at the equilibrium, some countries free-ride on the cooperative behavior of the others. By contrast, if γ_P is relatively large—e.g., because P(c) is convex—the ratio $P(c+1)/Q(c+1)$ is closer to one and (7b) is more likely to be satisfied for $c = n$. If the benefits from cooperation are rapidly increasing with the number of cooperators, no player has an incentive to take the risk of being a free-rider. Note that this conclusion crucially depends on the assumption of symmetric countries.

The evidence actually shows that several IEAs met the condition of high and increasing profitability for high levels of participation, leading to the selection of the unanimity participation rule. Rutz (2001) reports that more than one fifth of multilateral environmental agreements with a minimum participation provision indeed require full cooperation for the treaty to come into force.[16]

In the case identified by (7b), all countries not only prefer to "tie their hands" in the first stage, but also decide to "tie all hands", thus giving up the possibility of free-riding in the second stage in order to reap the benefits arising from full cooperation. As said, 22 out of 122 multilateral environmental agreements share this feature.

Using the simple economic model previously introduced, we can compute the expected utility function *EX*(*c*) and find that it is proportional to the following expression:

$$\left[c\left(d - \frac{mc}{2k}\right)\left(\frac{kd}{2} + \frac{mc}{4} - \frac{mN}{2}\right) + (N - c)\left(d - \frac{m}{2k}\right)\left(\frac{kd}{2} + \frac{m}{4} - \frac{mN}{2}\right)\right]$$

[16] In some international environmental agreements, the equilibrium number of signatories is above the minimum participation clause introduced in these treaties. Indeed, there may be circumstances related to the specific shape of players' payoff functions in which the minimum participation rule is not binding. Consider, for instance, a case discussed in Carraro and Marchiori (2003). In that case, beyond a certain participation threshold k, all players have the incentive to join the coalition. As such, a minimum participation rule equal to k/n would lead to the grand coalition. In this case, the minimum participation rule simply induces a bandwagon effect. This is an example of why the number of signatories can be larger than those identified by the minimum participation rule.

Moreover, this function increases monotonically with c within the interval $\left[c^{+}, n\right]$ if and only if the following inequality holds:

$$EX(c) - EX(c-1) > 0 \quad for c \in [c^{+}, n]$$

That is:

$$c\left(d-\frac{mc}{2k}\right)\left(\frac{kd}{2}+\frac{mc}{4}-\frac{mn}{2}\right)-(c-1)\left(d-\frac{m(c-1)}{2k}\right)\left(\frac{kd}{2}+\frac{m(c-1)}{4}-\frac{mn}{2}\right)$$
$$-\left(d-\frac{m}{2k}\right)\left(\frac{kd}{2}+\frac{m}{4}-\frac{mn}{2}\right)>0$$

It can be shown that the above inequality holds for $1 < c < \frac{4}{3}n$. Therefore, the expected utility function $EX(c)$ increases monotonically with c within the interval $\left(1, \frac{4}{3}n\right)$. As a consequence, the equilibrium minimum participation constraint is $\alpha^{\circ} = 1$ (coalition unanimity) and the equilibrium coalition size is $c^{\circ} = n$ (full cooperation).

It is worth noting that in the presence of structural asymmetries, the minimum participation constraint $\alpha = 1$ does not necessarily guarantee full cooperation because some countries may balk from signing the agreement (the profitability condition may not be met for all countries if they are different from one another). In such a situation, the requirement of full cooperation may even be counterproductive because it may lead to no cooperation, unless coupled with a transfer scheme that guarantees the profitability of the agreement for all countries (e.g., the one proposed in Chander and Tulkens 1995, 1997). Allowing for players' heterogeneity in the context of endogenous minimum participation rules would require a more complicated game structure with strategy sets also including compensation schemes across different players. We leave this extension to future research.

5 Conclusions

In this article, we endogenised the choice of a minimum participation rule in international (environmental) treaties. A minimum participation rule was modeled as the equilibrium outcome of a three-stage coalition formation game. In the first stage, countries unanimously set the minimum coalition size that is necessary for the treaty to come into force. In the second stage, they decide whether to sign the treaty. In the last stage, they set the equilibrium values of the policy variables.

When choosing a minimum participation rule, countries face the following trade-off: on the one hand, by setting a high participation constraint, they can increase the payoff of both cooperators and free-riders. On the other hand, a high minimum participation rule reduces the probability of being a free-rider in the second stage, thus reducing the probability of reaping the highest payoff.

In the first part of the analysis, we identified the condition under which all countries find it strategically convenient to limit their freedom to free-ride and introduce a minimum participation constraint. This condition is represented by (5). We also showed that the endogenously chosen minimum participation rule achieves the goal of enlarging the equilibrium number of signatories. This prediction is consistent with the results of traditional models with exogenous minimum participation rules and explains why most environmental treaties include a minimum participation rule. In the second part, we focused on the stringency of the participation constraint. We characterized the optimal minimum participation rule and analyzed under which conditions it coincides with coalition unanimity. We showed that the choice of full participation as a minimum participation rule crucially depends on the quasi-elasticity of the payoff function of cooperating countries. When this is sufficiently large for c close to n, then all countries prefer a minimum participation constraint that induces full cooperation.

These results are meant to be useful for the design of actual policy agreements. A minimum participation rule is generally helpful to enhance the environmental effectiveness of an international agreement. The optimal rule can even be coalition unanimity, in particular when the number of negotiating countries is not too large. However, the unanimity constraint is effective only if the profitability condition is met for all countries, otherwise it may be counterproductive. Indeed, a country for which the agreement is not profitable would not sign it. As a consequence, unanimity would prevent the treaty to come into force, even if all other countries would sign it. This case, however, can be a feature of games with asymmetric players, and does not arise in the symmetric player framework considered in this article. Extensions of our results to asymmetric players would also be useful to identify the characteristics of countries that belong to the equilibrium coalition.

Acknowledgments The Authors are grateful to Scott Barrett, Maria Paz Espinosa, Michael Finus, Jeff Frenkel, Jim Hammitt, Marita Laukkanen, Cynthia Lin, Rob Stavins, Martin Weitzman, two anonymous reviewers and participants at the Coalition Theory Network Workshop and at seminars in Paris and Harvard. The usual disclaimer applies.

References

Barrett S (1994) Self-enforcing international environmental agreements. Oxf Econ Pap 46:878–894

Barrett S (1997) Towards a theory of international cooperation. In: Carraro C, Siniscalco D (eds) New directions in the economic theory of the environment. Cambridge University Press, Cambridge

Barrett S (2003) Environment and statecraft. Oxford University Press, Oxford

Barrett S (2007) Why cooperate? the incentive to supply global public goods. Oxford University Press, Oxford

Black J, Levi MD, de Meza D (1992) Creating a good a atmosphere. Minimum participation for tackling the greenhouse effect. Economica 60:281–293

Bloch F (1996) Sequential formation of coalition with fixed payoff division. Games Econ Behav 14:90–123

Bloch F (1997) Non-cooperative models of coalition formation in games with spillovers. In: Carraro C, Siniscalco D (eds) New directions in the economic theory of the environment. Cambridge University Press, Cambridge

Botteon M, Carraro C (1997) Strategies for environmental negotiations: issue linkage with heterogeneous countries. In: Folmer H, Hanley N (eds) Game Theory and the Global Environment. E. Elgar, Cheltenham

Carraro C (ed) (1997) International environmental agreements: strategic policy issues. E. Elgar, Cheltenham

Carraro C, Marchiori C (2003) Stable coalitions. In: Carraro C (ed) The endogenous formation of economic coalitions. E. Elgar, Cheltenham

Carraro C, Siniscalco D (1992) The international protection of the environment: voluntary agreements among sovereign countries. In: Dasgupta P, Maler KG, Vercelli A (eds) The economics of transnational commons. Clarendon Press, Oxford

Carraro C, Siniscalco D (1993) Strategies for the international protection of the environment. J Public Econ 52:309–328

Carraro C, Siniscalco D (1995) Policy coordination for sustainability: commitments, transfers, and linked negotiations. In: Goldin I, Winters A (eds) The economics of sustainable development. Cambridge University Press, Cambridge

Carraro C, Siniscalco D (1998) International environmental agreements. Incentives and political economy. Eur Econ Rev 42:561–572

Chander P, Tulkens H (1995) A core-theoretical solution for the design of cooperative agreements on transfrontier pollution. Int Tax Public Financ 2:279–293

Chander P, Tulkens H (1997) The core of an economy with multilateral environmental externalities. Int J Game Theory 26:379–401

d'Aspremont CA, Jacquemin A, Gabszewicz JJ, Weymark J (1983) On the stability of collusive price leadership. Can J Econ 16:17–25

Diamantoudi E, Sartzetakis E (2006) Stable international environmental agreements: an analytical approach. J Public Econ Theory 8:247–263

Finus M (2003) Stability and design of international environmental agreements: the case of global and transboundary pollution. In: Folmer H, Tietenberg T (eds) International yearbook of environmental and resource economics, 2002/3. Edward Elgar, Cheltenham

Finus M (2008) Game theoretic research on the design of international environmental agreements: insights, critical remarks and future challenges. Int Rev Environ Resour Econ 2:1–39

Ray D, Vohra R (1997) Equilibrium binding agreements. J Econ Theory 73:30–78

Ray D, Vohra R (1999) A theory of endogenous coalition structure. Games Econ Behav 26:286–336

Rubio S, Ulph A (2006) Self-enforcing international environmental agreements revisited. Oxf Econ Pap 58:233–263

Rutz S (2001) Minimum participation rules and the effectiveness of multilateral environmental agreements, Working paper 01/22, Centre for Economic Research, SFIT, Zurich

Tulkens H (1998) Cooperation versus Free-riding in international environmental affairs: two approaches. In: Folmer H, Hanley N (eds) Game theory and the global environment, E. Elgar, Cheltenham. Also in Carraro C (ed) Governing the global environment, E. Elgar, Cheltenham, 2002

Yi SS (1997) Stable coalition structures with externalities. Games Econ Behav 20:201–237

Yi SS (2003) Endogenous formation of economic coalitions. A survey of the partition function approach. In: Carraro C (ed) The endogenous formation of economic coalitions. E. Elgar, Cheltenham

[25]

Economica (2014) **81**, 257–275
doi:10.1111/ecca.12073

Participation and Commitment in Voluntary Coalitions to Provide Public Goods

By ASTRID DANNENBERG†, ANDREAS LANGE‡ and BODO STURM††

†*Columbia University* ‡*University of Hamburg and Centre for European Economic Research, Germany* ††*Leipzig University of Applied Sciences and Centre for European Economic Research, Germany*

Final version received 8 August 2013.

This paper reports experimental evidence on the voluntary formation of coalitions to provide a public good. Participation and commitment in a coalition are either exogenously imposed or endogenously determined by the players themselves. We find larger voluntary participation rates when commitments in the coalition are endogenously determined using a minimum contribution rule rather than exogenously determined. However, due to a trade-off between participation and commitment, coalitions with voluntary participation are less effective in facilitating cooperation compared to when all players are forced to participate. This paper therefore confirms the rather pessimistic conclusions from coalition formation theories.

INTRODUCTION

The provision of many public goods is adversely impacted by conflicts between social and individual interests. These conflicts may lead to a severe underprovision of a public good, in particular when contributions are voluntary and cannot be enforced. This is the case, for example, for private fundraising campaigns or international efforts to provide global public goods. This paper studies how the provision of public goods can be enhanced through institutions that rely on voluntarily participation.

Our starting point is the observation that many people do voluntarily contribute to public goods (for reviews on experimental findings, see Ledyard 1995; Chaudhuri 2011), but often behave as conditional cooperators. They are willing to cooperate provided that they know or believe that others do so too (Keser and van Winden 2000; Fischbacher *et al.* 2001). As such, pre-play commitments that are *credible* and enable players to mutually commit may improve cooperation by eliciting reciprocal contributions. Chen and Komorita (1994) and Chen (1996) show that the exchange of pledges before contribution decisions are made improves cooperation when pledges are binding and restrict the whole group rather than the single player. Kurzban *et al.* (2001) find similar results in a setting with incremental commitments where participants can amend their binding pledges over time after observing one another's decisions. In these institutions, the risk of being exploited is effectively reduced when players contribute just a little more than the binding amount. In the same vein, Orzen (2008) shows that pledges substantially improve cooperation provided that the smallest pledge becomes the binding lower limit for everyone in the group. A player is thus bound by his or her pledge only if the other players also commit at the same or a higher level.[1]

These previous experiments make two implicit assumptions that are often unproven in real-world situations. First, it is assumed that the institutions under which the players act are exogenously imposed: subjects cannot choose between different institutions or between participation and non-participation. Second, it is assumed that commitments can be enforced: once the subjects commit to contributing to the public good, they are forced to comply by design. In this paper, we relax the first assumption and enquire whether an institution that allows for credible commitment can also help to improve efficiency if participation in this institution is voluntary and potentially incomplete. We do not address the second question, but rather adopt the assumption that compliance can be enforced.

Our experimental treatments are guided by a series of early theoretical papers on coalition formation in international environmental agreements (Barrett 1994; Hoel 1992; Carraro and Siniscalco 1993) and cartel formation (d'Aspremont *et al.* 1983) in which subgroups (coalitions) implement an institution to address the public goods problem. In a first stage, players decide whether or not to participate in a cooperative coalition. In the second stage, members of the coalition then provide the public good according to the terms of the coalition, while the non-members act in their own self-interest. This literature derives rather pessimistic predictions. If the coalition maximizes the joint benefits of its members and individual players have a strong incentive to free-ride, only a few players are predicted to form a coalition. Consequently, no substantial efficiency gains compared to the voluntary contribution mechanism are predicted. A coalition may, however, attract more members by lowering the required public good provision levels. Thus an institution that only partially internalizes the joint benefits of its members may be acceptable to more players and thereby generate efficiency gains. This leads to a trade-off between the number of players agreeing to the institution and the degree to which they internalize their joint benefits and provide the public good (Barrett 2002; Finus and Maus 2008).[2]

Researchers have recently begun to study the emergence of coalitions in experiments (Burger and Kolstad 2009; Kosfeld *et al.* 2009; McEvoy 2010; McEvoy *et al.* 2011). To a large part, these experiments focus on the membership decision and the effects of a minimum participation threshold. The minimum participation threshold specifies a minimum number of members for a coalition to be formed. If fewer subjects than the minimum participation threshold are willing to join, the coalition does not form and the standard voluntary contribution mechanism is played. A key result of this literature is that people are more likely to accept the coalition and to provide the public good if the coalition requires full participation rather than partial participation. This confirms that people are often willing to commit provided that others do similarly.

However, none of the aforementioned studies offers coalition members the opportunity to negotiate and to agree on a (possibly suboptimal) contribution level. Once subjects have stated their willingness to join and implement a coalition, collectively rational behaviour of the members is simply enforced by design. Kosfeld *et al.* (2009), McEvoy *et al.* (2011) and McEvoy (2012) allow for non-compliance in the contribution stage, but the decision about whether or not to comply is an uncoordinated individual decision rather than a collective decision.

In this paper, we compare the ability of different institutions to address both voluntary participation and the commitment level. The experiment involves an initial decision of players to join or abstain from a cooperative coalition. Besides experimentally testing the standard coalition game in which all members are forced to maximize their joint pay-off, we consider a minimum contribution rule that functions as follows. After the initial membership decision, all members of the coalition can each suggest a minimum public good contribution level whereby the smallest suggested level becomes the binding lower limit for all members. We compare such institutions that rely on voluntarily formed coalitions with institutions that are exogenously imposed on all players, a voluntary contribution mechanism and a mechanism in which *all* players are subject to the minimum contribution rule.

All our treatments consider a pay-off structure that is linear in the benefits of the public good, but involves convex costs of individual contributions. Such increasing marginal provision costs of the public good may arise from budget constraints and decreasing marginal utility from numeraire consumption.

Our treatments contribute to the understanding of the determinants of successful international cooperation. For example, international agreements may be ratified by only a subset of countries. While some countries form a coalition to cooperate, others may free-ride on the coalition's efforts. The minimum contribution rule particularly reflects tendencies in international negotiations. When countries disagree about the strength of commitment, such as the magnitude of emission abatement, and agreements involve (more or less) uniform obligations to countries, they may compromise by agreeing on the smallest common denominator to hold the coalition together (Endres and Finus 1999).

We find that the minimum contribution scheme frequently achieves close to efficient public good provision levels when it involves *all* players. Our findings are therefore in line with previous experiments (Chen 1996; Orzen 2008). However, much less cooperation results if participation is voluntary. As not all players participate, the coalition structure creates inequality between members and non-members. Perhaps because of this inequality, coalition members do not commit to substantial contributions to the public good. We identify a trade-off between participation and commitment in the coalition. As a result, the coalition with the minimum contribution rule performs only slightly better than the voluntary contribution mechanism and the standard coalition treatment in which members are exogenously forced to internalize their joint benefits. Overall, our results confirm the rather pessimistic conclusions from the coalition formation theory.

I. Theory and Predictions

We consider n identical agents, $i = 1, \ldots, n$, with utility

$$u_i = y_i + \gamma Q, \tag{1}$$

where y_i is a numeraire, $Q = \sum_{j=1}^{n} q_j$ represents the total provision level of the public good, and γ denotes the (constant) marginal utility from the public good. Subjects can allocate their initial income w to personal consumption or public good provision q_i with the budget constraint given by

$$y_i + q_i^2 \leq w. \tag{2}$$

The pay-off structure given by (1) and (2) is used to derive analytical predictions for this experiment. This specification corresponds to the utility function that has been used in numerous previous coalition formation studies (Barrett 1994; Carraro and Siniscalco 1993). It deviates, however, from a large part of the experimental literature on voluntary public good provision, which often considers linear provision costs and implies a dominant strategy of contributing zero in the Nash equilibrium such that any variance in the data could mistakenly be interpreted as altruism (Ledyard 1995). In comparison, our non-linear structure generates positive Nash contributions. Throughout, we assume interior solutions that require $w \geq n\gamma/2$.

Voluntary contribution mechanism (VCM) and social optimum

Individual utility maximization immediately yields the individual Nash provision level

$$q^{\text{NE}} = \gamma/2 = \arg\max_{q_i}\left[-q_i^2 + \gamma\left(q_i + \sum_{j\neq i} q_j\right)\right],$$

with the total contributions given by $Q^{\text{NE}} = n\gamma/2$. It should be noted that the Nash equilibrium involves a dominant strategy such that each individual's pay-off-maximizing action does not depend on the provision levels chosen by the remaining players. The social optimum maximizes the aggregate pay-off to players and is given by

$$q^{\text{SO}} = n\gamma/2 = \arg\max_q n[-q^2 + \gamma n q]$$

such that $Q^{\text{SO}} = n^2\gamma/2$.

Minimum contribution mechanism (VCMmin)

The implementation of the minimum contribution mechanism involves two stages. First, all players are requested to suggest a minimum public good contribution level. Second, after these minimum proposals $q_i^{\min}$ are received, all agents are required to provide at least the smallest suggested level $\min_j q_j^{\min}$. That is, agents are bound to provide $q_i \geq q^{\min} = \min_j q_j^{\min}$.

As usual, the game is solved backwards. When limited to provide $q_i \geq q^{\min} = \min_j q_j^{\min}$, the individually pay-off-maximizing contribution level at the last stage is given by $q_i = \max\,[q^{\min}, \gamma/2]$, as no player has an incentive to provide more than is required if $q^{\min}$ exceeds the individually pay-off-maximizing level q^{NE}. When making their minimum proposal $q_i^{\min}$ in the first stage, agents therefore need to recognize their potential impact on the contribution levels of the other players. Each player has a weakly dominant strategy to suggest a minimum contribution level of $q_i^{\min} = n\gamma/2$, the socially optimal level. On the one hand, $q_i^{\min} = n\gamma/2$ maximizes profits if all other players suggest $q_j^{\min} \geq n\gamma/2$. On the other hand, a player should not propose a minimum below $\min_{j\neq i} q_j^{\min}$ if another player suggests $q_j^{\min} < n\gamma/2$, as this would lower the binding minimum and thereby reduce profits. Here, suggesting $q_i^{\min} = n\gamma/2$ does not adversely affect the player. While $q_i^{\min} = n\gamma/2$ forms a weakly dominant strategy, there are other (Pareto-dominated) equilibria. In fact, any binding minimum $q_i^{\min} \leq n\gamma/2$ is established as equilibrium if at least two players suggest that level while other players suggest a larger minimum. If players play their weakly dominant strategy, however, we expect them to implement the social optimum.

Coalition formation (COAL, COALmin)

The coalition formation treatments involve a first stage in which each subject decides about membership in the coalition. We denote the set of k players ($1 \leq k \leq n$) who are members of the coalition by S. The standard coalition treatment (*COAL*) then involves a second stage in which coalition members are exogenously forced to internalize their joint benefits, so that for $i \in S$ we calculate

$$q_i = q^C(k) = k\gamma/2 = \arg\max_q k[-q^2 + \gamma kq].$$

Non-members are free to choose their contribution level. Due to the assumed linearity of the public good's benefits, their pay-off-maximizing decision does not depend on the coalition effort and is again given by $q^{NC} = q^{NE} = \gamma/2$.

In the treatment *COALmin*, we combine the participation decision with the minimum contribution mechanism that represents negotiations about the commitment level inside the coalition. Here, only members of the coalition ($i \in S$) first each suggest a minimum contribution level $q_i^{\min}$, and are then required to provide at least the smallest suggested level, $q_i \geq q^{\min} = \min_{j \in S} q_j^{\min}$. Non-members are again free to choose their contribution level such that we predict $q^{NC} = q^{NE} = \gamma/2$. Similar to the result of *VCMmin*, coalition members in the contribution stage maximize their pay-offs by choosing $q_i = \max\,[q^{\min}, \gamma/2]$. Taking this into account, they have a weakly dominant strategy to suggest a minimum contribution level of $q_i^{\min} = q^C(k) = k\gamma/2$, the level that maximizes the joint pay-off of coalition members. On the one hand, a player will reduce his pay-off by suggesting $q_i^{\min} < k\gamma/2$ if all other players in the coalition suggest $q_j^{\min} \geq k\gamma/2$, as then all members are predicted to reduce their contributions. On the other hand, suggesting $q_i^{\min} = k\gamma/2$ does not affect later contributions if $\min_{j \in S, j \neq i} q_j^{\min} < k\gamma/2$, as in this case, another player determines the binding minimum. However, besides this (Pareto-dominant) equilibrium in weakly dominant strategies, any binding minimum $q^{\min} < k\gamma/2$ can be implemented as equilibrium if at least two players suggest that level while other players suggest a larger minimum.

To measure the strength of the binding minimum for a coalition of size k, we define an internalization ratio by $\alpha(k) = q^{\min}/q^C(k) = q^{\min}/(k\gamma/2)$ (see also Finus and Maus 2008).[3] This takes the value 1 if joint benefits are fully internalized (as in the predicted equilibrium in weakly dominant strategies in *COALmin* or exogenously imposed in *COAL*). In *COALmin*, this internalization ratio is endogenous. We will explore later the extent to which this depends on the size of the coalition.

Membership decision

We now turn to the decisions of agents in the first stage, the choice of joining the coalition or not. Deciding the provision level inside the coalition leads to specific incentives of agents to join the coalition. Consequently, different coalition sizes may result.

In general, we denote the pay-off to members of the coalition given a coalition size k by $\pi^C(k)$, and the pay-offs to non-members by $\pi^{NC}(k)$.[4] Using the terminology from the coalition formation literature (Barrett 1994), a coalition of size k is internally stable if no member has an incentive to leave unilaterally, that is, if $\pi^C(k) \geq \pi^{NC}(k-1)$. It is externally stable if no player outside the coalition has an incentive to join unilaterally, that is, if $\pi^{NC}(k) > \pi^C(k+1)$. We assume that a player would join the coalition if he or she is indifferent as this increases pay-offs to all other agents.

Denoting the internalization ratio of a coalition of size k by $\alpha(k)$, we obtain the following condition for internal stability:

$$0 \leq \pi^C(k) - \pi^{NC}(k-1) = \left(-(\alpha(k)\, q^C(k))^2 + (q^{NC})^2\right) + \gamma\big(k\alpha(k)\, q^C(k) - q^{NC} - (k-1)\,\alpha(k-1)\, q^C(k-1)\big). \quad (3)$$

Here, the first bracket corresponds to the additional costs of providing the public good, while the second bracket equals the changes in benefits from the public good showing the degree to which free-riding will be punished by the remaining $k - 1$ members through a reduction of their contribution level.

Using the definitions above, condition (3) for internal stability can easily be restated as

$$(4) \qquad k^2\alpha(k)\,(2 - \alpha(k)) \geq 1 + 2(k-1)^2\alpha(k-1).$$

Since the weakly dominant strategy in the subgame following the membership decision involves full internalization of mutual benefits ($\alpha(k - 1) = \alpha(k) = 1$), the only subgame-perfect equilibrium in weakly dominant strategies in *COALmin* as well as the exogenously imposed full internalization in *COAL* implies $k = 3$, as can easily be derived from condition (4). This implies that only three agents are predicted to form a coalition if coalition members fully internalize their mutual benefits.[5]

However, if players implement other equilibria in the minimum stage that correspond to less than full internalization, coalitions of larger sizes could be stabilized. For example, assuming $\alpha(k - 1) = \alpha(k) = \alpha$, condition (4) implies internal stability if $k \leq (2 + \sqrt{3 - 2\alpha})/\alpha$ (see Finus and Maus 2008). Here, the predicted size of stable coalitions depends on α.[6] For example, coalitions of $k = 6$ players could be stabilized for $\alpha = 0.5$.

More generally, the multiplicity of equilibria in the minimum stage allows the stabilizing of any coalition size. For example, the grand coalition is stabilized when players choose the following subgame-perfect (though deviating from weakly dominant) strategies: all agents coordinate on full internalization in the grand coalition, while in all smaller coalitions, players suggest the minimum contribution at the Nash level, $q^{\min} = \gamma/2$. In this case, $\alpha(n) = 1$ and $\alpha(k) = 1/k$ for $k < n$ such that condition (4) is satisfied for $k = n$.

We can summarize these results in the following proposition.

Proposition 1. In the coalition game in which negotiating parties agree to implement provision obligations at the smallest minimum level suggested by a member of the coalition, the social optimum in a grand coalition (as well as any other coalition size) can be stabilized in a subgame-perfect equilibrium. The only equilibrium in weakly dominant strategies corresponds to a coalition in which three members fully internalize their mutual benefits.

Although precise predictions cannot be made, the multiplicity of equilibria in this minimum proposal game gives one reason to explicitly test the behaviour of players in an experimental setting.

II. Experimental Design

The pay-off function for each player was given by $\pi_i = -q_i^2 + \gamma Q = -q_i^2 + \gamma\sum_{j=1}^{n} q_j$ with $\gamma = 10$, $n = 10$ and $q_i \in [0, 100]$, and was common knowledge.[7] We begin with the traditional *VCM* as a control treatment. Subjects simultaneously and independently contributed to the public good and were subsequently informed about the average group contribution and their pay-off.

In the treatment *VCMmin*, all subjects took part in the negotiation about a minimum contribution. Players first simultaneously and independently proposed a minimum contribution of between 0 and 100. Players were informed about all minimum proposals (arranged in descending order), and the smallest proposal became the binding lower limit for the contributions of all players. In the contribution stage, all players simultaneously and independently chose their contributions to the public good, which had to be equal to or greater than the binding minimum.

The coalition formation treatments *COAL* and *COALmin* introduced an initial membership stage in which subjects decided whether or not to participate in a coalition. Decisions to join a coalition were made simultaneously and independently. Coalition members in the *COAL* treatment were forced to contribute the collectively optimal amount, while non-members could freely choose their level of contribution. After being told the number of subjects in the coalition, members in the *COALmin* treatment negotiated the minimum contribution that each member should contribute to the public good (minimum stage). This allowed them to condition their minimum proposal on the size of the coalition such that the internalization ratios $\alpha(k)$ may depend on the coalition size. All coalition members simultaneously and independently proposed a minimum contribution of between 0 and 100. The smallest proposal then became the binding lower limit for the contributions of all coalition members. Members were informed about all minimum proposals (arranged in descending order). Non-members did not take any decision in this stage and were only informed about the coalition size. In the contribution stage, members and non-members chose their contributions to the public good. While non-members could freely choose their contributions, members were bound to provide at least the binding minimum.

Table 1 summarizes the key features of our experimental design and the number of participants in each session. The experiment was run at a computer laboratory at the University of Magdeburg, Germany. In total, 400 students participated in the experiment, whereby 100 subjects took part in each treatment. No student participated in more than one treatment. Sessions lasted between 60 and 90 minutes. For each session, we recruited 20 subjects who were seated at networked computer terminals that were used to transmit all decision and pay-off information. We used z-Tree (Fischbacher 2007) for programming and ORSEE (Greiner 2004) for recruiting. Once the individuals were seated and logged into the terminals, a set of instructions with detailed information about the pay-off function and a simulator were distributed. By entering their own contribution and the expected average contribution of all other players, subjects could calculate their individual pay-off. Experimental instructions included several numerical examples and control questions in order to ensure that all subjects understood the games (see the Appendix).

Each session consisted of 12 rounds, the first two being for practice. Subjects were instructed that the practice rounds would not affect earnings. At the beginning of the

TABLE 1
SUMMARY OF EXPERIMENTAL DESIGN

Treatment	Stages	Coalition structure	n	γ	No. of subjects
VCM	Contribution	No	10	10	100
COAL	Membership contribution	Yes	10	10	100
COALmin	Membership minimum contribution	Yes	10	10	100
VCMmin	Minimum contribution	No	10	10	100

experiment, subjects were randomly assigned to groups consisting of ten participants. In line with previous experiments, subjects were not aware of the identity of the rest of their group, but they did know that they remained within the same group of players throughout the rounds (partner matching). At the end of the experiment, one of the non-practice rounds was chosen at random to determine earnings with an exchange rate between euro and token of 1:100. On average, subjects earned €11.50 in the games, which is slightly above the average hourly wage for student jobs. Additionally, all subjects received €1.00 as a show-up fee.[8]

III. EXPERIMENTAL RESULTS

We craft the results summary by both pooling the data across all periods and reporting behaviour in the first five and last five periods. Table 2 reports average contribution and pay-off levels for each of our treatments, and Figure 1 provides a graphical depiction of the data. The first panel presents the average contribution levels across treatments, and the second reports the resulting pay-off levels. As can be seen from the table and figure, implementing a minimum stage in a coalition formation framework, *COALmin*, leads to only small increases in average contributions compared with the standard *VCM* (14.8 vs. 12.3) and the standard coalition (12.1). Contributions in *VCMmin* are substantially larger (22.1). On average, the coalition treatment *COAL* gives similar contributions to *VCM* (12.1 vs. 12.3).

Figure 2 indicates that the contributions in *VCM* decrease over time as they are smaller in the last five periods than in the initial five periods (see also Table 2). This downward trend, which has been observed in many other experimental settings, primarily for linear public goods, is less pronounced for *COAL* and *COALmin*. In *VCMmin*, contributions increase over time and reach two-thirds of the efficient amount at the end of the game. A series of Mann–Whitney tests with the average contribution by one group across all periods being taken as the unit of observation shows that the differences in contributions

TABLE 2
SUMMARY STATISTICS

Treatment	q	π	k	*eff*
Total				
VCM	12.3	905.2		0.21
COAL	12.1	959.3	3.5	0.24
COALmin	14.8	1060.1	5.1	0.29
VCMmin	22.1	1418.6		0.47
First 5 periods				
VCM	15.7	1098.4		0.31
COAL	13.3	1030.1	3.7	0.27
COALmin	16.3	1160.1	5.3	0.34
VCMmin	16.8	1187.9		0.35
Last 5 periods				
VCM	8.9	711.9		0.12
COAL	10.9	888.5	3.2	0.20
COALmin	13.4	960.1	4.8	0.24
VCMmin	27.5	1649.2		0.58

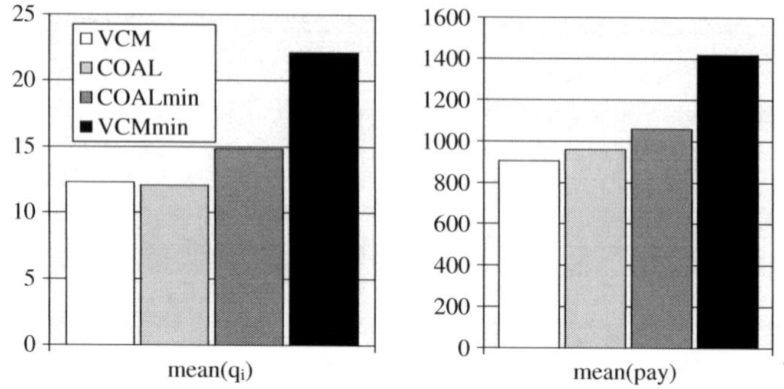

FIGURE 1. Average contribution and pay-off levels.

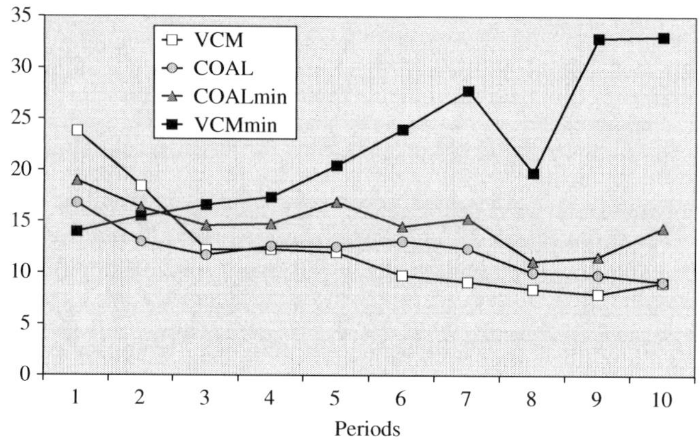

FIGURE 2. Average contribution levels over time.

between *VCM*, *VCMmin* and *COALmin* are not statistically significant ($p > 0.05$ each). Contributions in *COAL* do not differ significantly from those in *VCM* ($p = 0.94$), but are significantly smaller than those in *COALmin* ($p = 0.03$). Across the last five periods, contributions in both *VCMmin* and *COALmin* are significantly larger than in *VCM* ($p = 0.01$ each). Contributions in *COAL* are weakly significantly larger than those in *VCM* ($p = 0.10$), and the difference between *COALmin* and *COAL* is no longer significant ($p = 0.13$).

Similar comparisons follow for the efficiency level of the respective treatments (see also Table 2). Tables 3 and 4 present linear regressions of the individual contribution and pay-off levels. They show that both minimum institutions *VCMmin* and *COALmin* perform significantly better than the *VCM* ($p < 0.01$). *COAL* generates benefits relative to *VCM* only in the last periods, while performing worse in the beginning ($p < 0.10$). We can therefore formulate the following result.

Result 1

Average contribution levels in the standard coalition game (*COAL*) are similar to those in the voluntary contribution mechanism (*VCM*). The coalition with minimum contribu-

TABLE 3
LINEAR REGRESSION OF PUBLIC GOOD CONTRIBUTIONS FOR ALL TREATMENTS

Variables	All periods q_i	All periods q_i	Last 5 periods q_i
COAL	−0.202 (0.960)	−2.42* (1.343)	2.016* (1.031)
COALmin	2.551** (1.012)	0.582 (1.502)	4.520*** (1.372)
VCMmin	9.833*** (3.267)	1.046 (2.902)	18.62*** (4.492)
per6_10		−6.880*** (1.002)	
per6_10_*COAL*		4.436*** (1.430)	
per6_10_*COALmin*		3.938* (2.042)	
per6_10_*VCMmin*		17.57*** (3.806)	
Constant	12.298*** (0.616)	15.738*** (0.953)	8.858*** (0.595)
Observations	4000	4000	2000
R-squared	0.078	0.129	0.229

Notes
Random effects estimation clustered at group level; standard errors in parentheses.
***, **, * indicate significance at $p < 0.01, p < 0.05, p < 0.1$, respectively.
Definition of variables: q_i = subject's contribution; *COAL* = 1 if subject played in the *COAL* treatment, 0 otherwise; *COALmin* = 1 if subject played in the *COALmin* treatment, 0 otherwise; *VCMmin* = 1 if subject played in the *VCMmin* treatment, 0 otherwise; per6_10 = 1 for the last five periods, 0 for the first five periods; per6_10_*treatment* = interaction term of time dummy and treatment dummy.

TABLE 4
LINEAR REGRESSION OF PAY-OFF LEVELS FOR ALL TREATMENTS

Variables	All periods π_i	All periods π_i	Last 5 periods π_i
COAL	54.2 (67.65)	−68.28 (83.77)	176.6** (70.43)
COALmin	154.9*** (58.95)	61.70 (73.58)	248.2*** (82.21)
VCMmin	513.4*** (165.78)	89.52 (156.82)	937.3*** (217.72)
per6_10		−386.5*** (30.80)	
per6_10_*COAL*		244.9*** (75.06)	
per6_10_*COALmin*		186.5* (102.16)	
per6_10_*VCMmin*		847.8*** (184.31)	
Constant	905.2*** (42.69)	1098.4*** (48.29)	711.9*** (42.32)
Observations	4000	4000	2000
R-squared	0.056	0.092	0.171

Notes
Random effects estimation clustered at group level; standard errors in parentheses.
***, **, * indicate significance at $p < 0.01, p < 0.05, p < 0.1$, respectively.
Definition of variables: π_i = subject's pay-off level; *COAL* = 1 if subject played in the *COAL* treatment, 0 otherwise; *COALmin* = 1 if subject played in the *COALmin* treatment, 0 otherwise; *VCMmin* = 1 if subject played in the *VCMmin* treatment, 0 otherwise; per6_10 = 1 for the last five periods, 0 for the first five periods; per6_10_*treatment* = interaction term of time dummy and treatment dummy.

tion rule (*COALmin*) slightly increases average contributions relative to *VCM* and *COAL*. The minimum contribution rule best facilitates public good provision when involving all agents (*VCMmin*), especially in the later periods of the game.

Differences between *VCM* and the other treatments become more pronounced over time (see Figure 2 and Table 2). While contributions in *VCM* start high and decrease over time, they are relatively stable for both coalition structures. The coalitions thereby generate small but significant gains in the later periods of the game, consistent with the theory.

The beneficial impact of *VCMmin* over time, as described in Result 1, is due to an adjustment of the proposed minimum contribution levels. Increasing proposals and therefore increasing the binding minimum allows agents in the *VCMmin* treatment to gradually coordinate to larger provision levels of the public good. However, as illustrated in Figure 3, four out of ten groups are not able to achieve large provision levels, as a few players do not play the weakly dominant strategy but consistently make suboptimal minimum suggestions. The resulting low binding minimum keeps the group trapped at a low provision level, as most players contribute an amount close to the binding minimum: 65% of the contribution decisions are exactly at the minimum, and 83% exceed the minimum by 5 tokens at most. The binding minimum serves as a focal contribution, indicating that the players are willing to increase their contribution only if the requirement for all players increases.

The few players refusing to increase their minimum pledges almost certainly could have increased their pay-offs by making higher suggestions. This type of player has also been observed by Kurzban *et al.* (2001), indicating that this phenomenon is not just a feature of this particular experiment. The authors speculate that these 'strong free-riders' may be extremely competitive. These players seem to ignore the effects of their behaviour on others' decisions, and (wrongly) believe that the other players would coordinate on a high provision level without their assistance. That such coordination among a subset of players does not take place also becomes obvious in the more successful groups. In four out of the six successful groups, the binding minimum contribution increases over time, coming close to the efficient level, and drops sharply in the eighth period (see Figure 3),

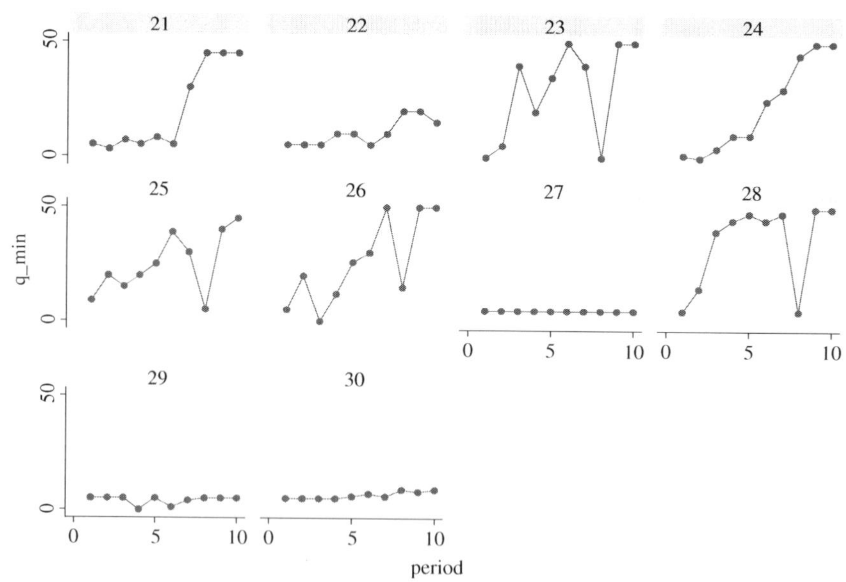

FIGURE 3. Binding minimum in *VCMmin* for each group over time.

arguably because some players believe that the group would stay at the high contribution level even with a lower binding limit. However, contributions do not stay high, but rather follow the drop in the binding minimum as these players quickly revise their decision in the following period.

Result 2
Although the minimum contribution rule generally works well when applied to all agents, it is sensitive to agents who refuse to increase their minimum suggestions. The group performance crucially depends on the existence of this type of player since most players' contributions are equal or close to the binding minimum.

While the existence of players who make small minimum proposals holds down contributions in *VCMmin*, these players could potentially abstain from the coalition in *COALmin*, thereby allowing coalition members to better internalize their mutual benefits. However, making participation endogenous creates a different kind of problem, as will be discussed below.

Figure 4 shows the binding minimum levels implemented by the coalition members in the *COALmin* treatment. The minimum levels are more erratic and less convergent than those in *VCMmin*. A sensible measure to assess the average binding minimum in the coalition is the internalization ratio—the ratio of the implemented minimum compared with the level that fully internalizes the mutual benefits of the coalition members, $\alpha(k) = q^{\min}/q^{C}(k)$. On average, the ratio is 62%. Like in the *VCMmin* treatment, the binding minimum in *COALmin* serves as a focal contribution: 58% of the members' contributions are exactly at the minimum level, and 77% are close with the difference being 5 tokens or less. The internalization ratio of the actual contributions, $\sum_{i\in S} q_i/(kq^{C}(k))$, is 83% on

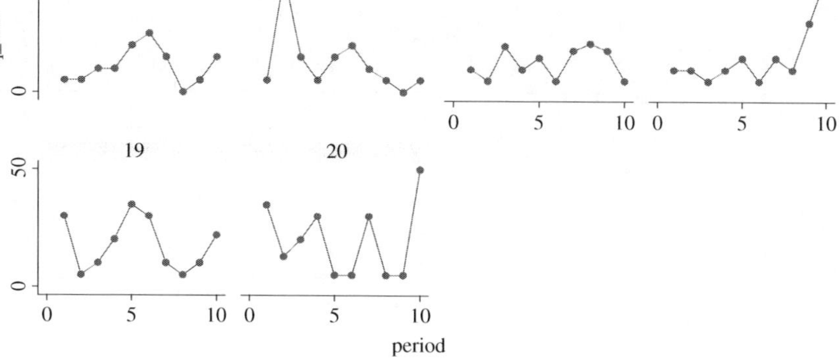

FIGURE 4. Binding minimum in *COALmin* for each group over time.

average. Interestingly, the internalization ratios depend on the size of the coalition. Figure 5 shows that the ratios based on the average suggested minimum, the binding minimum, and the contribution level ultimately chosen decrease in the coalition size k. This is confirmed by a linear regression model for the binding minimum as reported in Table 5.

Result 3
When negotiations in the coalition are institutionalized by a minimum contribution rule, the extent to which mutual benefits are internalized decreases with the size of the coalition.

Result 3 describes a clear trade-off between participation and the strength of commitment when both are endogenous. Note that although commitments are chosen *after* the entry decision, the endogenously determined internalization ratio is close to the one needed to

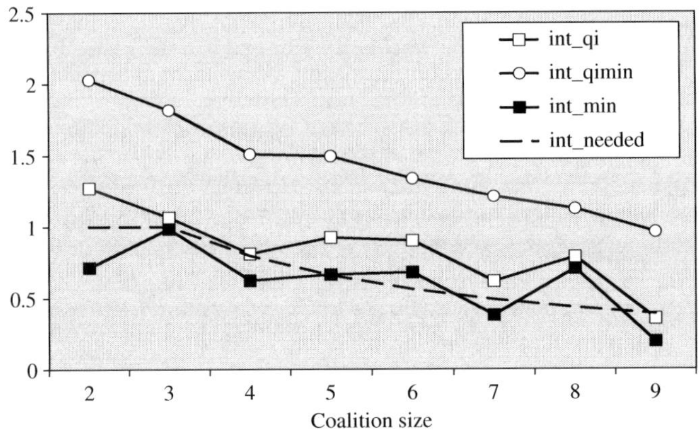

FIGURE 5. Average internalization ratios conditional on coalition size in *COALmin*: suggested minimum ratio (int_qimin), binding minimum ratio (int_min), actual contribution ratio (int_qi); the dashed line (int_needed) shows the internalization ratios theoretically required to stabilize the respective coalition size.

TABLE 5
LINEAR REGRESSION OF INTERNALIZATION RATIOS FOR COAL*MIN*

Variables	*COALmin int_min*
k	−0.0721** (0.0290)
Period	−0.0166 (0.0174)
Constant	1.112*** (0.198)
Observations	100
R-squared	0.063

Notes
Random effects estimation clustered at group level; standard errors in parentheses.
***, ** indicate significance at $p < 0.01$, $p < 0.05$, respectively.
Definition of variables: *int_min* = internalization ratio based on the binding minimum $q\min/(k\gamma/2)$; k = coalition size.

stabilize a coalition of the respective size (Figure 5). In particular, the internalization ratios for coalitions that comprise more than three members are smaller than 100% (t-test, $p < 0.01$). That is, these coalitions do not fully internalize the benefits of their members and therefore do not maximize their pay-off. A plausible explanation is that the coalition members want to avoid unfavourable pay-off differences to free-riders and therefore suggest a lower minimum. Large coalitions mean few outsiders but, at the same time, high free-rider benefits if the coalition members fully internalize their mutual benefits. Previous experiments have shown that even profitable coalitions are often rejected if they benefit outsiders (e.g. Kosfeld *et al.* 2009). Our experiment shows a different, yet related mechanism: members reduce the effort level inside the coalition in order to limit the free-rider benefits.

In turn, the possibility of choosing provision efforts *after* observing the number of coalition members provides an interesting feature of the acceptance of the coalition. The membership decision does not bind members to a specific contribution level, which apparently increases their incentives to join. The resulting average coalition size in *COALmin* is 5.07, which is significantly larger than the 3 (t-test, $p < 0.01$) that would have been predicted based on weakly dominant strategies. It is also significantly larger than 3.49, which is the average coalition size in the standard coalition game *COAL* (Mann–Whitney test, $p = 0.00$).

Result 4

The number of agents in the standard coalition (*COAL*) is close to the theoretical prediction. Making the commitment endogenous using a minimum contribution rule for the coalition (*COALmin*) triggers the entry of more agents.

Finally, we look closer at the determinants of individual decisions to enter a coalition. In Table 6 we show average marginal effects from a probit estimation model that explains the decision to join the coalition by the same individual's decision in the previous period and the lagged coalition size. We see that the individual's decision is largely driven by his or her behaviour in the previous period (the probability to join the coalition in period t is about 51% higher for a coalition member in $t - 1$ than for a non-member, $p < 0.01$), which suggests that the decision to enter is relatively stable. In fact, the likelihood of

TABLE 6
PROBIT ESTIMATION OF DECISION TO JOIN THE COALITION

Variables	*COALmin* c_i	*COAL* c_i	All c_i
c_i_lag	0.510*** (0.031)	0.371*** (0.000)	0.446*** (0.023)
k_lag	−0.007 (0.011)	0.008 (0.011)	−0.001 (0.008)
int_min_lag	−0.094* (0.041)		−0.088** (0.039)
COAL			−0.067* (0.030)
Observations	900	900	1800

Notes
Average marginal effects reported; standard errors in parentheses.
***, **, * indicate significance at $p < 0.01, p < 0.05, p < 0.1$, respectively.
Definition of variables: $c_i = 1$ if subject joined the coalition, 0 otherwise; $c_i\_lag = 1$ if subject joined the coalition in the previous period, 0 otherwise; k_lag = coalition size in the previous period; $COAL = 1$ if subject played in the *COAL* treatment, 0 otherwise; $COALmin = 1$ if subject played in the *COALmin* treatment, 0 otherwise; int_min_lag = internalization ratio $qmin/(k\gamma/2)$ in the previous period.

joining the coalition is not significantly influenced by the coalition size in the previous period. In *COALmin*, a larger internalization ratio implemented in the previous period reduces the likelihood of an agent joining the coalition ($p < 0.10$). This again suggests a trade-off between participation and commitment, which is consistent with theory. The provision level inside the coalition affects the willingness to join the coalition. While this trade-off limits the efficiency gains relative to *VCMmin* (see Result 1), the third column in Table 6 indicates a benefit of *COALmin* over *COAL*. In monitoring for the implemented internalization ratio (which is exogenously fixed at 1 in *COAL*), more agents enter the coalition in *COALmin* ($p < 0.05$).[9] This finding complements the literature on benefits from endogenous institution formation (Sutter *et al.* 2010; Tyran and Feld 2006). Allowing players to choose their commitment level appears to make joining an institution more likely.

IV. Conclusions

The provision of public goods often relies on voluntary contributions and cooperation. In the absence of strong enforcement mechanisms, it is challenging to design institutions that facilitate the provision of public goods while also being voluntarily accepted by the relevant players. International environmental agreements, for example, may be ratified by only a subset of countries. While some countries form a coalition to cooperate, others may free-ride on the coalition's efforts. Forming a coalition thereby involves (at least) two challenges: on the one hand, the institutional arrangement needs to attract members to the coalition; on the other hand, any given coalition should be able to increase the provision of the public good.

The principle of the minimum contribution scheme reflects many real-world institutional arrangements where players must find a common denominator when formulating international agreements. This paper compares the effects when participation in such an institution is either exogenously enforced or voluntary and potentially incomplete. We also compare the effects when the commitment inside a voluntarily formed coalition is either exogenously imposed or endogenously chosen by the members.

We confirm previous findings that the minimum contribution scheme effectively facilitates cooperation when it applies to *all* agents. Nevertheless, success is not guaranteed, as the scheme is sensitive to players who consistently make low minimum proposals.

Importantly, the minimum contribution rule loses its virtue when applied only to members of a voluntarily formed coalition, as participation is not complete and the coalition structure creates inequality between members and non-members. The members of the coalition do not commit to substantial contributions, and a clear trade-off between participation and commitment can be observed. The larger the coalition, the less willing the members are to implement a high provision level, arguably to limit the free-rider benefits of the non-members. Similarly, the coalition that exogenously forces members to fully internalize their mutual benefits does not attract many participants. With this, our experiment confirms the pessimistic findings from the coalition formation theory in that cooperation is either deep and narrow or shallow and broad (Barrett 2002; Finus and Maus 2008).

Our results refute previous findings from institution formation experiments (e.g. Kosfeld *et al.* 2009) in which the existence of fair-minded players led to larger coalitions and substantial efficiency gains. One potential reason is that players in our experiment could not prevent (smaller) coalitions from forming, but could only influence the coalition commitment level in order to reduce the free-rider profits. The resulting trade-off between commitment level and participation rate limits the efficiency gains through the formation of cooperative coalitions. Relative to the exogenously imposed joint pay-off maximization, the minimum contribution rule generates benefits by attracting more members. We thereby add to the recent literature on beneficial endogenous choices of rules in social dilemma situations (Sutter *et al.* 2010; Tyran and Feld 2006).

Naturally, our experiment investigates coalition formation in a rather abstract setting. Climate negotiations, for example, face important problems due to heterogeneities of countries and uncertainties about benefits and costs of emission reductions. Nonetheless, our experiment captures the basic incentive structure of international cooperation and identifies important problems for the formation of cooperative coalitions. After two decades of climate negotiations, countries still debate the path that should be taken to either establish smaller coalitions that formulate their own agreements or lower the requirements to achieve unanimity among all countries.

APPENDIX: EXPERIMENTAL INSTRUCTIONS FOR THE COALMIN TREATMENT

Welcome to the Magdeburg Experimental Laboratory MAXLAB!

Please read these instructions carefully and should you have any questions please signal us by opening the door or a show of hands. In the laboratory experiment you are taking part in, you can win money depending on your decisions and the decisions of your fellow players. Your payout from the experiment will be calculated in LabDollars (LD). The conversion rate between € and LD is **1:100**. All your decisions made in the experiment will remain **anonymous**. Only the experimenter will know your identity, but your data will be treated confidentially.

RULES OF THE GAME

Now you will learn more about the rules of the game you will be participating in. Altogether **10 players** take part in the game, so besides you there are 9 more players. Every participant faces the same decision making problem. Your task in the game, and also your fellow players' task, is to decide how many points you would like to contribute to a **joint project**. Your **payout** will be calculated as follows:

$$\text{Your payout} = -(\text{your contribution to the project})^2 + 10 \cdot (\text{sum of all contributions of all players to the project})$$

Example: If all other players have contributed an amount of 90 points to the project and you contribute an amount of 10 points, then your payout will be:

$$-(10)^2 + 10 \cdot (10 + 90) = 900\text{LD}$$

If, however, all other players contribute a total amount of 50 points and you do not contribute anything, your payout will be:

$$-(0)^2 + 10 \cdot (0 + 50) = 500\text{LD}$$

To simplify the calculation of your payout, you will find an excel-file called **'Simulator'** on your screen. You can enter your contribution and the **average** contribution of all other players and so quickly determine your payout.

There are **two stages** in this game. In **stage 1** you can decide whether you want to become a member of a coalition, i.e. if you want to join a coalition or not. Should you decide that you want to join a coalition you additionally can decide which amount should be the **minimum amount** each member of the coalition should contribute to the project. Also all other members of the coalition can state their desired minimum amount. The members will be informed about the proposals for the minimum amount of all members. If you are member of a coalition, **stage 2** will be to decide for yourself which amount you want to contribute. In this decision the **smallest** minimum amount of all members will form your **lower** limit of contribution. If you have decided not to join a coalition, **stage 2** for you will be to state your contribution to the project without any limitation.

The game consists of **10 separate rounds** in each of which you will play the same two-stage game. The nine other players you will interact with will be the same in every round. If the experiment is complete you will receive the **payout of one of the rounds** in € (according to the conversion rate stated above). The round to be paid out will be determined **randomly**. This means you should behave in **each** round as if it were the round relevant for payout. In the beginning, **two trial rounds** will be played which are **not relevant for payout**. Independent of the course of the game you will receive €1 for your participation.

CONTROL QUESTIONS

If you have read the instructions and do not have any questions, please answer the following control questions (hint: use the simulator).

1. Please assume your contribution to the project is 10 points and the average contribution of all the other players is 15 points. How much LD will be your payout of this round?

   My payout is: _______

2. Please assume the average contribution of all other players is 5 points, which of the following amounts will result in the highest payout for you?

   O 5 points O 10 points O 20 points O 30 points

3. Please assume you want to maximise your payout, does it make sense to not contribute at all (meaning zero points) to the project?

   O yes O no

4. Please assume you and three other players have joined a coalition and all members have stated the following minimum contribution: 4, 88, 22, 56. In which range does your contribution to the project have to be?

   More than or equal ____ and less than or equal _____.

5. Is it possible that a member of a coalition has to contribute more than he proposed as his minimum contribution?

   O yes O no

6. Please assume all players chose the same amount, which of the following contributions results in the highest payout for all players (please check the according box)?

   O 10 points O 30 points O 50 points O 70 points O 100 points

If you have answered all questions, please signal us. We will then check your answers. The game begins when all participants in the experiment have successfully completed the test. Good luck in the experiment! The MaXLab-Team

ACKNOWLEDGMENTS

Seminar participants at several universities and conferences provided excellent comments.

NOTES

1. In the context of environmental agreements, Endres (1997) and Endres and Finus (1999) discuss a version of this minimum contribution scheme.
2. More optimistic results on coalition formation are obtained if players are conditional cooperators (Lange and Vogt 2003; Lange 2006).
3. Note that Finus and Maus (2008) also consider a pre-stage in which all players decide on the internalization ratio, before entering the coalition game.
4. Based on the discussion in the last subsection, these pay-off levels may depend on the specific design of the last stage. For ease of notation, we do not explicitly include such a design variable in the subsequent discussion.
5. This result is well-known in the literature (e.g. Barrett 1994; Carraro and Siniscalco 1993) and is driven by the assumed linear–quadratic functional forms.
6. We depict the internalization ratio that is required to stabilize a coalition of a given size in Figure 5.
7. We chose the parameter to obtain a stark predicted difference between Nash behaviour ($q_i = 5$) and efficient commitment levels ($q_i = 50$) while also limiting the likelihood of players generating negative profits. For example, a player who chooses the socially efficient level would still make positive profits if at least half the players behaved similarly.
8. Overall, 9 out of 400 subjects earned negative pay-offs in the games. In these cases, pay-offs were restricted to zero and the subjects received only the show-up fee.
9. In order to explore this finding, further research could try to exogenously fix the internalization ratio at the endogenously determined commitment levels in *COALmin*.

REFERENCES

D'ASPREMONT, C., GABSZEWEIZ, J. J. and WEYMARK, J. A. (1983). On the stability of collusive price leadership. *Canadian Journal of Economics*, **16**, 17–25.

BARRETT, S. (1994). Self-enforcing international environmental agreements. *Oxford Economic Papers*, **46**, 878–94.

——— (2002). Consensus treaties. *Journal of Institutional and Theoretical Economics*, **158**, 529–47.

BURGER, N. and KOLSTAD, C. D. (2009). Voluntary public goods provision, coalition formation, and uncertainty. NBER Working Paper no. 15543.

CARRARO, C. and SINISCALCO, D. (1993). Strategies for the international protection of the environment. *Journal of Public Economics*, **52**, 309–28.

CHAUDHURI, A. (2011). Sustaining cooperation in laboratory public goods experiments: a selective survey of the literature. *Experimental Economics*, **14**(1), 47–83.

CHEN, X. P. (1996). The group-based binding pledge as a solution to public goods problems. *Organizational Behavior and Human Decision Processes*, **66**, 192–202.

——— and KOMORITA, S. S. (1994). The effects of communication and commitment in a public goods social dilemma. *Organizational Behavior and Human Decision Processes*, **60**, 367–86.

ENDRES, A. (1997). Negotiating a climate convention—the role of prices and quantities. *International Review of Law and Economics*, **17**, 147–56.

——— and FINUS, M. (1999). International environmental agreements: how the policy instrument affects equilibrium emissions and welfare. *Journal of Institutional and Theoretical Economics*, **155**(3), 527–50.

FINUS, M. and MAUS, S. (2008). Modesty may pay. *Journal of Public Economic Theory*, **10**, 801–26.

FISCHBACHER, U. (2007). z-Tree: Zurich toolbox for ready-made economic experiments. *Experimental Economics*, **10**, 171–8.

———, GÄCHTER, S. and FEHR, E. (2001). Are people conditionally cooperative? Evidence from a public goods experiment. *Economics Letters*, **71**, 397–404.

GREINER, B. (2004). An online recruitment system for economic experiments. In K. Kremer and V. Macho (eds), *Forschung und wissenschaftliches Rechnen 2003*, GWDG Bericht 63, Göttingen, Ges. für Wiss. Datenverarbeitung, pp. 79–93.

HOEL, M. (1992). International environmental conventions: the case of uniform reductions of emissions. *Environmental and Resource Economics*, **2**, 41–59.

KESER, C. and van WINDEN, F. (2000). Conditional cooperation and voluntary contributions to public goods. *Scandinavian Journal of Economics*, **102**(1), 23–39.

KOSFELD, M., OKADA, A. and RIEDL, A. (2009). Institution formation in public goods games. *American Economic Review*, **99**, 1335–55.

KURZBAN, R., MCCABE, K., SMITH, V. L. and WILSON, B. J. (2001). Incremental commitment and reciprocity in a real-time public goods game. *Personality and Social Psychology Bulletin*, **27**(12), 1662–73.

LANGE, A. (2006). The impact of equity-preferences on the stability of international environmental agreements. *Environmental and Resource Economics*, **34**, 247–67.
——— and VOGT, C. (2003). Cooperation in international environmental negotiations due to a preference for equity. *Journal of Public Economics*, **87**, 2049–67.
LEDYARD, J. O. (1995). Public goods. A survey of experimental research. In J. H. Kagel and A. E. Roth (eds), *The Handbook of Experimental Economics*. Princeton, NJ: Princeton University Press, pp. 111–94.
MCEVOY, D. M. (2010). Not it: opting out of voluntary coalitions that provide a public good. *Public Choice*, **142**, 9–23.
——— (2012). Paying to punish on a free ride: member and nonmember punishments in voluntary coalitions. *Applied Economics Letters*, **19**, 285–8.
———, MURPHY, J. J., SPRAGGON, J. M. and STRANLUND, J. K. (2011). The problem of maintaining compliance within stable coalitions: experimental evidence. *Oxford Economic Papers*, **63**, 475–98.
ORZEN, H. (2008). Fundraising through competition: evidence from the lab. CeDEx Discussion Paper no. 2008-11, University of Nottingham.
SUTTER, M., HAIGNER, S. and KOCHER, M. G. (2010). Choosing the stick or the carrot? Endogenous institutional choice in social dilemma situations. *Review of Economic Studies*, **77**, 1540–66.
TYRAN, J.-R. and FELD, L. P. (2006). Achieving compliance when legal sanctions are nondeterrent. *Scandinavian Journal of Economics*, **108**, 135–56.

[26]

KYOTO AND BEYOND: ALTERNATIVE APPROACHES TO GLOBAL WARMING†

Climate Treaties and "Breakthrough" Technologies

By SCOTT BARRETT*

An effective climate change treaty must promote the joint supply of *two* global public goods: climate change mitigation *and* knowledge of new technologies that can lower mitigation costs. R&D is especially needed to bring about substantial, long-term reductions in atmospheric concentrations of greenhouse gases, for this will require the development and diffusion of revolutionary, "breakthrough" technologies (Martin I. Hoffert et al., 2002). In principle, such an outcome could be realized by the Kyoto Protocol approach, if that agreement were strengthened over time. However, that response may be inadequate (Kyoto makes no provision for R&D)—and, as I shall demonstrate, unlikely to succeed.

Can a treaty system relying directly on targeted R&D and the adoption of breakthrough technologies perform better *in this same setting* of anarchic international relations? I will show that, as a general rule, the answer is no. Essentially, the same forces that undermine Kyoto also challenge the R&D and technology approach.

There is one exception to this rule: R&D leading to breakthrough technologies exhibiting increasing returns can improve dramatically on the Kyoto approach, even when these technologies are otherwise inferior to the alternatives available. This suggests that our approach to treaty design should be strategic.

I. The Kyoto Approach

Begin by considering the abatement decisions of countries in the absence of a multilateral agreement. Let q_i denote country i's abatement and let Q denote aggregate abatement; with N countries, $Q = \sum_{i=1}^{N} q_i$. Let country i's payoff be given by $\pi_i = bQ - cq_i^2/2$. If countries choose their abatement levels independently, there exists a unique Nash equilibrium in which every country, i, plays $q_i = b/c$. If countries were able to cooperate fully, each would play $q_i = bN/c$.

Kyoto strove to restructure these incentives. Under the rules of international law, however, participation in such an agreement is voluntary. A treaty can be represented as the equilibrium of a stage game with countries deciding whether to participate in stage 1; with parties choosing their abatement levels collectively in stage 2; and with nonparties choosing their abatement levels independently in stage 3. As is usual, we solve the game backwards.

Since the equilibrium of the earlier abatement game, $q_i = b/c$, is in dominant strategies, it must also be the equilibrium of the stage 3 game. Letting k denote the number of signatories to the agreement, collective maximization by signatories implies that each signatory must play bk/c in stage 2. Finally, letting π_s and π_n denote the payoff to a signatory and nonsignatory, respectively, a Nash equilibrium of the stage 1 game is a participation level k^* satisfying $\pi_s(k^*) \geq \pi_n(k^* - 1)$ and $\pi_n(k^*) \geq \pi_s(k^* + 1)$. Upon substitution, it is easy to show that the equilibrium participation level is $k^* = 3$ for $N \geq 3$. A treaty consisting of just three countries will not make much of a difference when N is large. The agreement increases aggregate abatement from bN/c to just $b(N + 6)/c$. Full cooperation requires that abatement increase to bN^2/c.

Though this result emerges from a special model, it can be shown to be qualitatively robust (Barrett, 2005). It is also consistent with the Kyoto Protocol's history thus far. Kyoto has more than three parties, but it has failed to secure the participation of the world's biggest emitter, the United States, and the participation of other countries came at a price of their not being subject to any emission limits or being given a surplus of emission entitlements ("hot air"). If parties take full advantage of the treaty's trading mechanism, their

† *Discussants:* Jeffrey Frankel, Harvard University; David Victor, Stanford University.

* School of Advanced International Studies, Johns Hopkins University, 1619 Massachusetts Avenue NW, Washington, DC 20036, and Yale Center for the Study of Globalization, Yale University (e-mail: sbarrett@jhu.edu). I am grateful to Michael Hoel and Lee Lane for comments.

collective emissions may not fall at all. If they eschew trading, however, then compliance may be in jeopardy. Because Article 18 says that any binding compliance mechanism must be agreed to by an amendment, and no such amendment has been agreed to, the Kyoto emission limits are more political than legal. Kyoto may not be self-enforcing.

II. Alternative Treaty System

An alternative to Kyoto is a system of two treaties, one promoting cooperative R&D and the other encouraging collective adoption of a breakthrough technology emerging from this R&D. The decision to adopt the new technology depends on the attractions inherent in the technology itself, not the cost of developing the technology (once incurred, this cost is sunk). The returns to R&D depend on the prospects of any new technology being adopted, however. Analysis of this treaty system therefore must begin with the technology adoption treaty.

III. Technology *Y* Treaty

Breakthrough technologies will be sector-specific, not applicable to the whole economy. The emission limits emerging from the previous analysis, however, can be decomposed by sector. Suppose that the sector of interest has the same cost function as the whole economy, only with a different cost parameter, c_0. All the results obtained previously will go through as before (for this sector), provided we substitute c_0 for c. The problem, then, is for every country, i, to choose whether to adopt the new technology or abate its emissions using the "old" technology (in which case marginal cost will equal $c_0 q_i$).

Call the new technology Y and let country i's payoff be given by

$$(1)\quad \pi_i = b_y\left(y_i + \sum_{j\neq i}^{N} y_j\right) - c_y y_i + b\left((1-y_i)q_i + \sum_{j\neq i}^{N}(1-y_j)q_j\right) - c_0(1-y_i)q_i^2/2,$$

with $y_i, y_j \in \{0,1\}$. As formulated, these technologies are mutually exclusive. Moreover, adoption of the new technology results in a discrete drop in emissions. It is perhaps easiest to think of Y as a zero-emissions technology (a new renewable energy technology, or a system for carbon capture and storage), and b_y as representing the environmental benefit of adopting Y.[1] Of course, c_y is the cost of adopting Y. Assuming $b_yN - c_y > b^2(2N-1)/2c_0$, all countries will be collectively better off when each adopts Y than when each abates an amount b/c_0 using the old technology. If Y is costly enough, however—if, in particular, $b_y - c_y < b^2/2c_0$—then each country will choose independently to abate b/c_0 rather than to adopt Y.

Could a treaty promote adoption of Y? Consider a stage game with every country deciding in stage 1 whether to be a party to a technology adoption agreement; with parties deciding collectively in stage 2 whether to adopt the new technology; and with nonparties deciding individually in stage 3 whether to adopt the technology.

By the assumptions already given, each nonsignatory will choose to reduce its emissions by b/c_0, rather than adopt Y, in stage 3. In stage 2, signatories will choose to adopt Y if and only if their collective payoff for doing so is at least as large as their collective payoff for not doing so (where the alternative is for each country to abate its sector emissions by b/c_0). Let there be k_y signatories. If these signatories adopt the technology, they each gain $b_yk_y - c_y$. If they do not adopt the technology, they each gain $b^2(2k_y - 1)/2c_0$. Adoption is collectively rational provided $b_yk_y - c_y \geq b^2(2k_y - 1))/2c_0$. Rearranging, it will be collectively rational for k_y signatories to adopt Y provided $k_y \geq k_y^0$, where k_y^0 is the smallest integer at least as large as $(2c_0c_y - b^2)/2(c_0b_y - b^2)$. By our previous assumptions, k_y^0 lies between 2 and N.

Notice that $\pi_s(k_y^0) \geq \pi_n(k_y^0 - 1) \Rightarrow b_yk_y^0 - c_y \geq b^2(2k_y^0 - 1)/2c_0$, and this relation is true by definition. Also, $\pi_n(k_y^0) \geq \pi_s(k_y^0 + 1) \Rightarrow b^2/2c_0 \geq b_y - c_y$, and this inequality is true by assumption. Hence, $k_y^* = k_y^0$ is an equilibrium of the stage 1 game, and this equilibrium can be shown to be unique.

[1] Denote each country's emissions in the absence of abatement using the old technology by $\bar{e}$. Then, if Y is a zero emissions technology, $b_y = b\bar{e}$.

In contrast to the Kyoto approach, the technology adoption agreement can, under the right circumstances, sustain a large number of parties. However, k_y^* will be large only when a large number of countries must adopt the technology in order for adoption to make these countries collectively better off. Of course, it is precisely under such circumstances that the treaty will *not* make all countries substantially better off.

IV. R&D Financing Treaty

Since the knowledge of how to develop this breakthrough technology is a public good, how can it be financed? Suppose that the R&D cost of Y is $\overline{M}_y$. Since financing must be supplied before the decision to adopt has been made, no country will know in this preliminary stage which countries will adopt the technology; each can foresee only that k_y^* countries will adopt Y. It seems reasonable to suppose that each country will decide to finance R&D to develop new technology based on its *expected* payoff. If country i contributes an amount m_i to the collective R&D effort, i's expected payoff will be

(2)

$$\pi_i = Y\left[\frac{k_y^*}{N}\left((Nb_y - c_y) - \frac{b^2}{2c_0}(2N-1)\right)\right] - m_i$$

with $Y = 1$ if $\Sigma_{i=1}^N m_i \geq \overline{M}_y$; $Y = 0$, otherwise. The term in brackets represents the expected gain to i of k_y^* countries adopting technology Y (by an earlier assumption, this value is positive).

Financing this R&D is a "threshold" public good and it is easy to see that full financing will be an equilibrium, provided the bracketed term in (2) exceeds $\overline{M}_y/N$. A treaty is required to secure the needed finance—but only to coordinate contributions (coordination is needed provided $\overline{M}_y$ exceeds the value of the bracketed term in (2), for, in this case, zero financing is a symmetric Nash equilibrium).

Financing will be forthcoming, however, only if the collective gain to having Y exceeds $\overline{M}_y$, and I have already shown that the former value will be small when adoption must be sustained by a self-enforcing agreement. If $\overline{M}_y$ is nontrivial, the combination of an R&D and technology Y treaty system would likely not be pursued.

V. Technology X Treaty

Consider now a different kind of technology—call it technology X. Unlike Y, X exhibits increasing returns to adoption. Specifically, let country i's payoff now be given by

$$(3) \quad \pi_i = b_x\left(x_i + \sum_{j\neq i}^{N} x_j\right) - \frac{c_x}{N}\left(N - \sum_{j\neq i}^{N} x_j\right)x_i$$

$$+ b\left((1-x_i)q_i + \sum_{j\neq i}^{N}(1-x_j)q_j\right)$$

$$- c_0(1-x_i)q_i^2/2$$

with $x_i, x_j \in \{0,1\}$. Assume $c_x + b^2/2c_0 > b_x > c_x/N^2 + b^2(2N-1)/2Nc_0$. This implies that X may not be adopted spontaneously (nonadoption is a Nash equilibrium) but that all countries would be better off collectively if X were adopted universally.

By the assumptions above, every country will adopt X provided enough others do. Let z denote the tipping point—the minimum number of other countries that must adopt X for it to become attractive for every country to adopt X. From (3), z is the smallest integer greater than or equal to $N(b^2/2c_0c_x + 1 - b_x/c_x)$. Under our assumptions, z must lie between 1 and $N - 1$.

Participation in a treaty mandating adoption of technology X is a coordination problem. To sustain full participation, such a treaty must specify that parties adopt X, but that the agreement enter into force only if ratified by at least z countries. No country can lose by signing such an agreement, while every country will gain by signing once the threshold z is crossed. Moreover, we know from before that full financing of the R&D costs of X can also be sustained by a self-enforcing agreement provided the total cost, $\overline{M}_x$, is sufficiently low.

This treaty system can sustain full participation even when it would be in the collective interest of a small number of countries to cooperate independently—and, thus, even when the aggregate gains to cooperation are substantial (assuming that $\overline{M}_x$ is small enough). This means that, even if technology Y were superior to X (taking into account R&D costs), *given universal adoption*, a technology X treaty system may be preferred to a technology *Y* treaty system, *given the need for treaties to be self-enforcing.*

VI. Discussion

The reasoning developed in this paper favors a strategic approach to treaty design—one that includes collective financing of R&D and the adoption, in key sectors, of breakthrough technologies exhibiting increasing returns. Such a system may not be fully cost-effective in the usual sense of the term, but this may be the price that must be paid to sustain greater cooperation. In a world characterized by technological lock-in (path dependence), it cannot be assumed that markets will pick the best technologies simply because the price signals are right (Paul A. David, 1985), whether these are transmitted directly through taxes or indirectly through permit prices. Even without lock-in, emission taxes are not guaranteed to sustain an efficient outcome when abatement costs are nonconvex (Michael Hoel, 1998).

The value of having a new technology to address global climate change depends on more than its intrinsic merits; it also depends on the extent to which the technology will be diffused, subsequent to its development. An improved solar technology (type *Y*), capable of producing energy at a lower cost than fossil fuels, would be ideal—but may not be forthcoming. The next best option is to consider technologies that, though perhaps more costly than the type *Y* alternatives, exhibit increasing returns. It is at these (type *X*) technologies, theory suggests, that current R&D should be directed.

Are there reasonable prospects for type *X* technologies in the two most important sectors, transportation and electricity generation? A shift to a breakthrough automobile technology such as hydrogen is very likely to be characterized by increasing returns, because of knowledge spillovers, economies of scale, and especially the need to combine a new automobile technology with a supporting fuel infrastructure. Increasing returns are likely to be less important for the production of hydrogen fuel, however.

Increasing returns also do not feature large in electricity generation. The light water standard may dominate nuclear reactor design (Robin Cowan, 1990), but nuclear power is not generally favored over alternative sources of generation. Similarly, carbon capture and storage, being essentially an add-on technology, seems not to have desirable properties for diffusion. There may exist other possibilities for this sector. Hoffert et al. (2002), for example, explain that renewable energy would become more attractive if the electricity transmission grid were reengineered. Overall, however, the prospects for a technology *X* in the electricity supply sector do not appear bright.

The approach developed here thus fails to offer a ready solution to the climate change mitigation problem. It does, however, commend a refocusing of current efforts.

REFERENCES

Barrett, Scott. *Environment and statecraft: The strategy of environmental treaty-making.* Oxford: Oxford University Press, 2005.

Cowan, Robin. "Nuclear Power Reactors: A Study in Technological Lock-In." *Journal of Economic History,* 1990, *50*(3), pp. 541–67.

David, Paul A. "Clio and the Economics of QWERTY." *American Economic Review,* 1985(*Papers and Proceedings*), *75*(2), pp. 332–37.

Hoel, Michael. "Emission Taxes versus Other Environmental Policies." *Scandinavian Journal of Economics,* 1998, *100*(1), pp. 79–104.

Hoffert, Martin I.; Caldeira, Ken; Benford, Gregory; Criswell, David R.; Green, Christopher; Herzog, Howard; Jain, Atul K., et al. "Advanced Technology Paths to Global Climate Stability: Energy for a Greenhouse Planet." *Science,* 2002, *298*(5595), pp. 981–87.

[27]

Environ Resource Econ (2010) 47:395–406
DOI 10.1007/s10640-010-9384-3

Can a Focus on Breakthrough Technologies Improve the Performance of International Environmental Agreements?

Michael Hoel · Aart de Zeeuw

Accepted: 13 May 2010 / Published online: 30 May 2010

Abstract In a recent paper, Barrett (Am Econ Rev 96(2):22–25, 2006) reaches the conclusion that in general the answer to the question in the title is no, except for a special case in which technology adoption involves increasing returns (network externalities). We show in this paper that a focus on the R&D phase in the development of breakthrough technologies can also increase the possibilities for cooperation.

Keywords International environmental agreements · Breakthrough technology

JEL Classification Q28 · C72 · F42

1 Introduction

In a recent paper, Barrett (2006) investigates whether an international environmental agreement, such as a climate treaty, can perform better if it relies directly on targeted R&D and the adoption of breakthrough technologies. This is important because the literature shows that treaties that just rely on abatement, such as the Kyoto Protocol, cannot be expected to

This article was initiated during our participation in the 'Environmental economics: policy instruments, technology development and international cooperation' project at the Centre for Advanced Study (CAS) at the Norwegian Academy of Science and Letters in Oslo in 2005–2006. We are grateful for many useful comments and suggestions from the participants. We are also grateful for comments on earlier versions of this paper given by Geir Asheim, Scott Barrett, Bard Harstad, Chuck Mason, participants of the NBER Summer Institute, and two anonymous referees.

M. Hoel (✉)
Department of Economics, University of Oslo, P.O. Box 1095, Blindern, 0317 Oslo, Norway
e-mail: m.o.hoel@econ.uio.no

A. de Zeeuw
Department of Economics and TSC, Tilburg University, P.O. Box 90153, 5000 LE Tilburg, The Netherlands
e-mail: a.j.dezeeuw@uvt.nl

achieve much (Barrett 2003; Finus 2003). Furthermore, the international debate circles to some extent around the question whether international treaties should focus on technology development rather than on emission reduction. Barrett (2006) reaches the conclusion that in general the answer to the question whether such a treaty can perform better is no, with the exception of breakthrough technologies that exhibit increasing returns to scale. In this paper we will show that the picture is not necessarily so grim. By introducing the option that the adoption costs of a breakthough technology vary with the level of R&D, we will show that a large stable coalition can result that leads to a substantial improvement in average welfare. The basic idea is that a coalition can improve on a non-cooperative equilibrium in which the countries invest sufficiently much in the public good R&D to trigger a breakthrough technology. Either a coalition forms that invests even more in R&D (accepting that the other countries free-ride on this investment) because it pays to further lower the costs of adoption. Or a coalition forms that invests less in R&D but is large enough to realize a substantial improvement in average welfare by itself. In the last case the non-cooperative equilibrium would lead to an overinvestment in R&D.

The literature on international environmental agreements has investigated various approaches for enlarging the size of the stable coalition. If coalition members can secure extra positive externalities between them, for example by linking the agreement to an R&D agreement with larger spillovers between coalition members, the size of the stable coalition will grow (Carraro and Siniscalco 1997). In general, if the issue of negotiation can be linked to a second issue, the bargaining space may get larger with more opportunities to reach an agreement (Cesar and de Zeeuw 1997). Another approach is extending the stability concept. For example, the concept of farsightedness takes into account that countries consider further defections as a consequence of their own defection and this stabilizes larger coalitions (Diamantoudi and Sartzetakis 2002). A similar effect is achieved by adding a first stage to the game in which countries decide on a minimum participation constraint, because this also prevents one-by-one defections down to the standard small-size stable coalition (Carraro et al. 2009). Finally, learning in case of uncertainty can enlarge the size of the stable coalition as well (Kolstad 2007). This paper takes another angle by focusing on R&D and the adoption of breakthrough technologies. Cooperation may not be needed to achieve sufficient R&D and a switch to the new technology but cooperation may be needed to prevent under- or overinvestment in R&D.

2 Adoption Costs, R&D and the Social Optimum

Barrett (2006) assumes a fixed cost of adoption, c, which requires a certain level of R&D, M. In the R&D phase countries decide how much they want to contribute to this public good. In his approach the R&D phase precedes coalition formation and the adoption of new technology. If the expected average net benefits are sufficiently high, in the R&D phase an equilibrium exists that yields the required R&D level M. The reason is simply that each country does not want to contribute less, because then the benefits are not realized, but does not want to contribute more either, because then the costs become higher without higher expected benefits. This is an example of voluntary provision of a public good (see e.g. Bagnoli and Lipman 1989).

We change the set-up somewhat and link the cost of adoption c to the level of R&D M by a function $c = c(M)$ with $c' < 0$, $c'' > 0$. In this way a trade-off is introduced between R&D costs and costs of adoption. The idea is that the more research is done the lower are the expected future costs of using this technology. Examples of the type of technologies we are thinking of are Carbon Capture and Storage (CCS), capturing CO_2 from air (see e.g. Lackner et al. 2001, and Pielke 2009), and developing small batteries for electrical cars. In case of cooperation, groups of countries may set up research joint ventures with the purpose to develop these technologies, and the more they invest the lower the costs will be of implementing these technologies. For instance, smaller batteries require more R&D but reduce the costs of implementing electrical cars, and increased energy efficiency of CCS or air capture requires more R&D but reduces the operating costs of these technologies. We assume that for values of M giving $c(M) \leq 1$, $c(M) + M$ is increasing in M, i.e. $-c' < 1$. This last assumption is reasonable: If the opposite were true, one dollar of investment in this type of R&D would reduce the adoption costs for this technology by more than one dollar *for all countries*.

In our setting it makes sense to position coalition formation before the choice of R&D, so that decisions will balance R&D costs and adoption costs. This implies that we change the order of decisions as compared to Barrett (2006): countries first decide whether they want to join the coalition or not before they decide on investment in R&D and adoption. We will show below that the change in the order of decisions does not affect the qualitative results. One of the cases below is not much different from Barrett's model and also leads to a stable coalition that does not achieve much. The only difference in that case is that in our setting R&D costs are borne by the coalition, whereas Barrett (2006) assumes that R&D is financed by all countries not knowing whether they will be in or out of the coalition. Qualitative differences occur in the other cases and are only driven by the assumption that adoption costs depend on R&D expenditures.

Barrett (2006) combines the option of adopting a breakthrough technology with a standard international abatement game which can be seen as a choice between a new clean technology and abatement options within the context of the old technology. The last situation is in fact a default option and we abstract from that by focusing only on the choice of adopting the new technology or not. It is easy to show that by doing this the qualitative results remain the same, but in this way the analysis becomes more transparent. As in Barrett (2006) we assume that R&D is a public good, but the same results can be derived in a setting with imperfect R&D spillovers. The countries are assumed identical, but it would of course be interesting to extend the analysis to a heterogeneous set of countries; this is left for further research. Finally, our model is deterministic, but it would of course also be interesting to extend the analysis to uncertainty and learning by doing.

We normalise the benefits for a single country of adopting the new technology to 1. This is equal to the damage (for the single country) caused by the pollution that will be generated continuing with the old technology (for example the damage caused by further emitting large amounts of CO_2). These benefits can be realized by the country itself but can also be free-rider benefits from adoption of the new technology in another country. Once the technology is developed, it is individually rational for each country to adopt if and only if $c \leq 1$. The problem is that usually we have that $c(M)+M > 1$ for all M, so that it is not optimal for an individual country to develop and adopt the new technology just by itself. In the setting with a constant c that Barrett (2006) uses, $c > 1$ simply means that without coalition formation the new technology will not be adopted: a group of countries is needed to generate sufficient benefits to cover the costs of adoption. Note that this is similar to the standard linear international abatement game (e.g. Kolstad 2007). In our setting, however, it is in principle possible that the countries together invest so much in R&D that it becomes individually rational to adopt the new technology, because c depends on M. We denote that level by $\bar{M}$, which means that $c(\bar{M}) = 1$.

If the countries operate together, they minimize $Nc(M) + M$, and if the resulting costs are sufficiently low to make it collectively rational to adopt the new technology, each country will have a net benefit equal to

$$V(N) = N - \min_{M}\left[c(M) + \frac{M}{N}\right] \tag{1}$$

We assume that $V(N)$ is positive so that the social optimum leads to full adoption.

3 Non-Cooperative Equilibria

Without any cooperation, the outcome is given as a subgame perfect equilibrium of a two-stage game. In the first stage each country chooses its level of R&D investments. The properties (i.e. the costs) of the new technology are determined by the sum of R&D investments as explained in Sect. 2. In the second stage each country chooses whether to adopt the new technology. It is clear from our assumptions that one subgame perfect equilibrium of this game is for all countries to have no R&D investment, and thus no adoption of any new technology. However, there may exist a second subgame perfect symmetric equilibrium. This will be the case if $\bar{M}$ exists and $\bar{M} \leq N(N-1)$. In this case an equilibrium exists in which each of the N countries invests $\bar{M}/N$ in R&D, and all countries adopt the new technology. Each country will in this case have a net benefit equal to

$$\tilde{V} = N - 1 - \frac{\bar{M}}{N} \tag{2}$$

which is non-negative for $\bar{M} \leq N(N-1)$. It will not be beneficial for a country to deviate unilaterally since the total level of R&D will end up below $\bar{M}$, so that none of the countries will adopt the new technology anymore. Moreover, since we have assumed that $c(M) + M$ is increasing in M it does not pay for an individual country to increase investments above $\bar{M}$. We can thus conclude that when $\bar{M} \leq N(N-1)$, a symmetric Nash equilibrium in R&D investments with full adoption exists. Notice that this game also has asymmetric equilibria, but we do not analyze these. Furthermore, notice that even though we have full adoption of the new technology in this equilibrium, the outcome differs from the first-best social optimum. In the latter, the choice of R&D was determined so that the sum of investment and adoption costs for each country was minimized. In the non-cooperative outcome, the level of R&D is exactly as large as it must be for all countries to want to adopt the new technology. This level will generally differ from the first-best R&D level, and generally we do not know whether it is lower or higher than the first-best level.

Since even this second non-cooperative equilibrium (if it exists) is inferior to the social optimum, we want to investigate if a stable coalition can improve upon the outcome compared to the non-cooperative equilibria. We therefore now turn to coalition formation and show that substantial improvement is possible in case the non-cooperative equilibrium with full adoption exists.

4 Coalition Formation

We use the model for coalition formation that was introduced into the literature on international environmental agreements by Hoel (1992), Carraro and Siniscalco (1993), and Barrett (1994), and which is based on a model for cartel stability developed by d'Aspremont et al. (1983). The idea is that countries first decide whether they want to join the coalition or not before they take action. In the equilibrium of this two-stage game a member of the coalition does not have an incentive to quit (internal stability) and an outsider does not have an incentive to join (external stability).

This literature usually provides a rather grim picture: Stable coalitions are large if there is not much to gain from cooperation and are otherwise typically very small (Barrett 1994). This result was challenged from different angles. Chander and Tulkens (1995), for example, use the gamma-core concept from cooperative game theory to show that the grand coalition is stable in the sense of being an element of the core. The basic idea is very similar to the idea of trigger strategies in repeated games, where the threat of loosing cooperative benefits prevents countries to cheat. The essential difference with the standard literature on international environmental agreements is the assumption on the behaviour of the other coalition members in case a country defects. In the stability concept above, the remaining coalition stays intact, whereas in the gamma-core concept the original coalition falls apart in case a country (or a group of countries) defects. Diamantoudi and Sartzetakis (2002) apply the idea of farsightedness to show that a set of stable coalitions exists, among which large ones. The basic idea is that if a country defects, other coalition members defect as well, but only up to the point where the remaining coalition is stable again. This threat proves to be sufficient to sustain large coalitions, although in a dynamic context, where detection of deviations takes time, the idea may not work (de Zeeuw 2008).

In this paper we base ourselves on the concept of internal and external stability, because the underlying assumption seems reasonable for the current practice of international environmental agreements. We will show, however, that if we replace the second stage of this game by the two-stage game of investment and adoption, decribed in the previous section, a large stable coalition may arise. We distinguish two cases.

Case 1: $\bar{M} > N(N-1)$

It is immediately clear that if the non-cooperative equilibrium with full adoption ($M = \bar{M}$) does not exist, outsiders will not invest in R&D. In that case $\tilde{V}$ is negative, so that outsiders do not have an incentive to invest in R&D if the coalition does not invest more than $k\bar{M}/N$ in total. The coalition could consider to invest more in R&D in order to induce the outsiders to invest as well, but then the coalition's net benefit becomes negative. This implies that the coalition is on its own, and adopts the new technology if and only if $c(M) \leq k$. Individual outsiders will not adopt, since total R&D expenditures M will be smaller than $\bar{M}$. Formally, we have the following three-stage game:

Stage 1: each country chooses whether or not to join the coalition
Stage 2: the coalition chooses R&D expenditures
Stage 3: the coalition chooses whether or not to adopt the new technology.

Note that there is no new stage of coalition formation between stages 2 and 3: Once a coalition is formed in stage 1, it remains a coalition after the investment decision has been made in stage 2.

Suppose that a coalition of size k forms, $0 < k < N$. Similar to the full-cooperative case in the previous section, the coalition minimizes $kc(M) + M$, and if the resulting costs are sufficiently low to make it collectively rational to adopt the new technology, each member of the coalition will have a net benefit equal to

$$\phi(k) = k - \min_{M}\left[c(M) + \frac{M}{k}\right] \tag{3}$$

The solution to the minimization in (3) is denoted by $M^*(k)$, and $c'' > 0$ implies that M^* is increasing in k. Note that the decision to join the coalition or not must indeed be binding after the second stage. Otherwise, if countries could choose again after the R&D investments are made, an individual country would prefer to be outside the coalition of size $k - 1$, because the adoption costs are larger than 1 in the case we are now considering.

Since $\phi(k)$ is increasing in k (using the envelope theorem) and since we have assumed that $\phi(1)$ is negative and $\phi(N) = V(N)$ is positive, $\phi(k)$ must be 0 for some k^0 between 1 and N. Consider the coalition of size K^0, defined as the smallest integer at least as high as k^0. The net benefit to each member of this coalition is $\phi(K^0)$, which is non-negative but small. This coalition is internally stable: if one country instead of joining the coalition chooses to be an outsider, the coalition of size $K^0 - 1$ will not develop the new technology, so that it is better to stay in the coalition. Consider now the coalition of size K satisfying $k^0 + 1 \leq K \leq N$. The net benefit to each member of such a coalition is $\phi(K)$, which is smaller than $K - 1$ since $c(M^*(K)) > 1$. Therefore this coalition is not internally stable: if one country instead of joining the coalition chooses to be an outsider, the coalition of size $K - 1$ will develop and adopt the new technology, so that the country will get a benefit of $K - 1$ without costs. Consequently, the coalition of size K^0 is also externally stable and thus the only stable coalition.

Note that the stable coalition achieves very little. This is in line with the pessimistic result that Barrett (2006) derives. In this case our model is only slightly different. We allocate total R&D costs to the coalition, whereas Barrett (2006) assumes that R&D is financed by all the countries not knowing whether they will be in or out of the coalition. Otherwise, the models are essentially the same in this case. However, in our setting a non-cooperative equilibrium with full adoption ($M = \bar{M}$) may exist and then the picture changes.

Case 2: $\bar{M} \leq N(N-1)$

If the non-cooperative equilibrium with full adoption ($M = \bar{M}$) exists, the coalition has a choice to dissolve again and rely on this equilibrium, with a non-negative net benefit $\tilde{V}$ for its members, or to act as a coalition. If the coalition acts as a coalition, then either the coalition plays the investment game in the R&D phase with the outsiders or the coalition is on its own in that phase. Although this investment game has equilibria in which a coalition member invests less than or more than an outsider, we disregard these equilibria, just as we did not consider asymmetric non-cooperative equilibria in the previous section. This implies that the coalition either relies on the non-cooperative equilibrium with full adoption ($M = \bar{M}$), or improves on that by acting as a coalition and doing all the investment on its own. Formally, we have the following four-stage game now:

Stage 1: each country chooses whether or not to join the potential coalition
Stage 2: the potential coalition decides whether or not to act as a coalition
Stage 3: either each individual country or the coalition chooses R&D expenditures
Stage 4: either individual countries or coalition and outsiders choose whether or not to adopt the new technology.

Compared with the case studied previously (with $\bar{M} > N(N-1)$), stage 2 is new. However, stage 2 was not relevant in the previous case, since in that case there could be no disadvantages from acting as a coalition.

We first focus on stages 3 and 4 of the game in case the coalition stays intact and chooses R&D expenditures. The situation is more complicated than above because the coalition of size k may choose M to be equal to or larger than $\bar{M}$. Three possibilities occur. The first

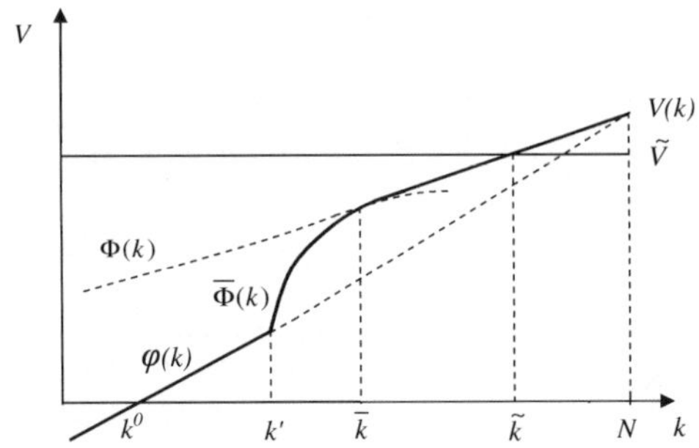

Fig. 1 Value function of a coalition member as a function of the coalition size

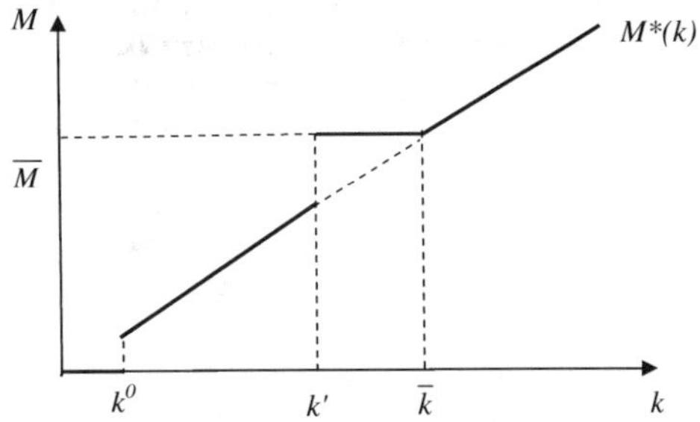

Fig. 2 Investments in R & D as a function of the coalition size

possibility is that the optimal investment in R&D is sufficiently high so that all countries adopt the new technology, which means that each member of the coalition will have a net benefit equal to

$$\Phi(k) = N - \min_{M}\left[c(M) + \frac{M}{k}\right] \tag{4}$$

The second possibility is that the coalition invests $\bar{M}$ at its own cost in order to induce the outsiders to adopt the new technology as well, which means that each member of the coalition will have a net benefit equal to

$$\bar{\Phi}(k) = N - \left[1 + \frac{\bar{M}}{k}\right] \tag{5}$$

The final possibility is that it does not pay for the coalition to invest so much, which puts us back in the situation above with $\phi(k)$, given by (3), as the net benefit. The value function of a member of the coalition of size k in this case is simply the maximum of the three possibilities $\Phi(k)$, $\bar{\Phi}(k)$ and $\phi(k)$ and is denoted by $V(k)$. Two typical outcomes are depicted in Figs. 1 and 3 with the corresponding levels of R&D in Figs. 2 and 4, respectively. In Figs. 1 and 2 $M^*(N)$ is larger than or equal to $\bar{M}$, so that all three possibilities can occur, and in Figs. 3 and 4 $M^*(N)$ is smaller that $\bar{M}$, so that only the last two possibilities can occur.

It is easy to verify that $\tilde{V} > \bar{\Phi}(k)$ for all $k < N$, with $\tilde{V}$ given by (2). In other words, the net benefit of a country in the non-cooperative equilibrium with $M = \bar{M}$ exceeds the net benefit of a member of a coalition investing $\bar{M}$ in R&D. This is illustrated in Figs. 1 and 3. The intersection point between $V(k)$ and $\tilde{V}$ is denoted by $\tilde{k}$. It follows that for $k < \tilde{k}$ the coalition

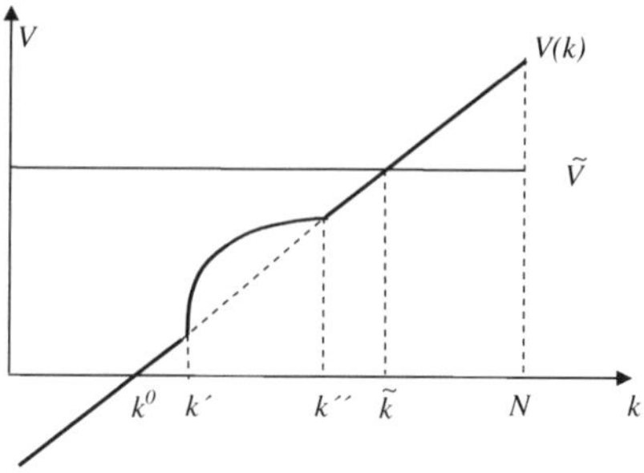

Fig. 3 Value function of a coalition member as a function of the coalition size

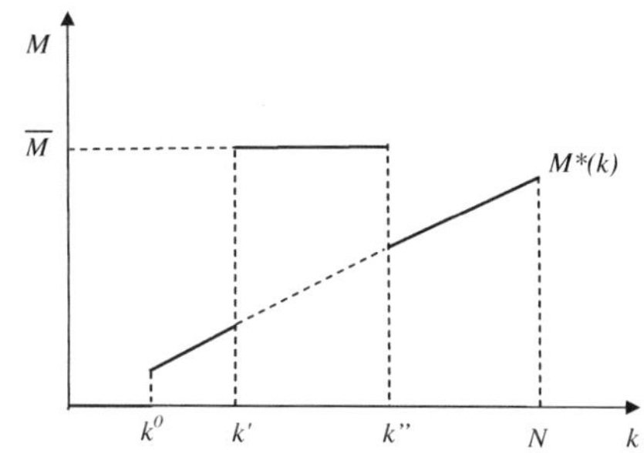

Fig. 4 Investments in R & D as a function of the coalition size

cannot do better on its own, compared to the non-cooperative equilibrium. Hence, a coalition of this size decides in stage 2 of the game not to act as a coalition, so that the symmetric non-cooperative equilibrium with $M = \bar{M}$ and with net benefits equal to $\tilde{V}$ results. For $k > \tilde{k}$, however, the coalition will invest $M^*(k)$ in R&D on its own and adopt the new technology, giving its members a net benefit $V(k) > \tilde{V}$. If $M^*(k) > \bar{M}$, as in Fig. 1, outsiders will adopt the new technology as well, and each of them achieves a net benefit equal to $N - c(M^*(k))$. If $M^*(k) < \bar{M}$, as in Fig. 3, outsiders will not adopt the new technology, and each of them receives a benefit equal to k. Outsiders are always better off than coalition members, so that both coalition members and outsiders are better off than in the non-cooperative equilibrium with full adoption.

Define $\tilde{K}$ as the lowest integer satisfying $\tilde{K} \geq \tilde{k}$. A coalition of size $\tilde{K} - 1$ will not form in stage 2 of the game. It is in the interest of those countries to stay apart and to end up in the non-cooperative equilibrium because $\tilde{V} > V(\tilde{K} - 1)$. This implies that a coalition of size $\tilde{K}$ is internally stable, because leaving this coalition will give a net benefit $\tilde{V}$ which is smaller than $V(\tilde{K})$.

Under reasonable conditions, a coalition of size $K \geq \tilde{k} + 1$ is not internally stable. For the situation of Figs. 3 and 4, it is easy to see that a coalition of size $K \geq \tilde{k} + 1$ is never internally stable: An outsider to the coalition of size $K - 1$ is better off than a member of the coalition of size K, since the costs of adoption are higher than 1 (which is the benefit of one more country adopting the technology), and the outsider is not investing in R&D. It follows that in this situation the coalition of size $\tilde{K}$ is also externally stable, and thus the largest stable coalition.

In the situation of Figs. 1 and 2, it is not quite so obvious that an outsider to the coalition of size $K - 1$ is better off than a member of the coalition of size K, provided $K \geq \tilde{k} + 1$.

In this case full adoption occurs both for the coalition of size $K-1$ and K, and the difference in payoffs is thus only in costs. The cost of an outsider to the coalition of size $K-1$ is simply $c(M^*(K-1))$, while the cost of a member of the coalition of size K is $c(M^*(K))+\frac{M^*(K)}{K}$. Consequently, a coalition of size $K \geq \tilde{k}+1$ is internally stable if and only if

$$c(M^*(K-1)) \geq c(M^*(K)) + \frac{M^*(K)}{K}$$

or

$$c(M^*(K-1)) - c(M^*(K)) \geq \frac{M^*(K)}{K} \tag{6}$$

For this inequality to hold, adoption costs must decline quite significantly if the coalition is enlarged by one member. From the definition of M^*, given by (3), it is clear that

$$c(M^*(K-1)) + \frac{M^*(K-1)}{K-1} \leq c(M^*(K)) + \frac{M^*(K)}{K-1}$$

implying that the l.h.s. of (6) cannot exceed $\frac{1}{K-1}\left[M^*(K) - M^*(K-1)\right]$. A necessary condition for (6) to hold is therefore that

$$\frac{M^*(K) - M^*(K-1)}{M^*(K)} \geq \frac{K-1}{K}$$

If for instance $K = 10$, the decline in R&D investment when the coalition size drops from 10 to 9 must be at least 90%. Although such cases cannot theoretically be ruled out, we find them quite implausible. Under reasonable conditions, a coalition of size $K \geq \tilde{k}+1$ is therefore not internally stable; the coalition of size $\tilde{K}$ is thus the largest stable coalition.

The coalition $\tilde{K}$ gives its members only a slightly higher net benefit than they would have achieved in the non-cooperative equilibrium with $M = \bar{M}$. In contrast, the outsiders get a much higher net benefit in this case; the average net benefit may therefore be substantially higher than in that non-cooperative equilibrium. The uneven distribution of benefits is typical for these models and requires further attention (just as the uneven distribution in the theoretical equilibrium for the Ultimatum game). In this paper, however, we focus on what coalition formation can achieve in terms of average welfare. Before discussing the results, we first give a numerical illustration.

5 A Numerical Illustration

Suppose that the function for adoption costs is given by

$$c(M) = \frac{\gamma}{M} \tag{7}$$

This implies that

$$M^*(k) = \sqrt{\gamma k} \tag{8}$$

and

$$\phi(k) = k - 2\sqrt{\frac{\gamma}{k}} \tag{9}$$

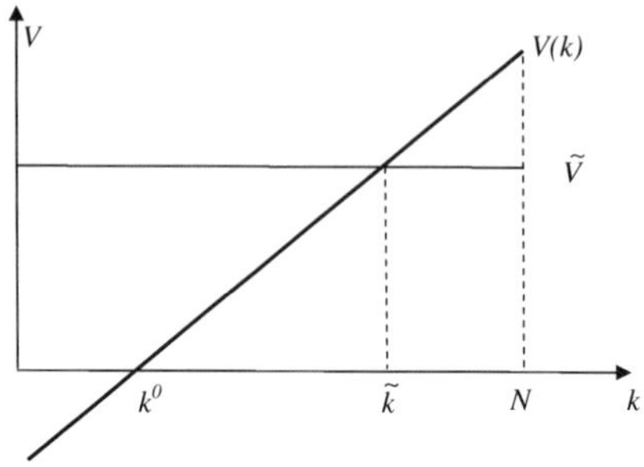

Fig. 5 Value function of a coalition member as a function of the coalition size

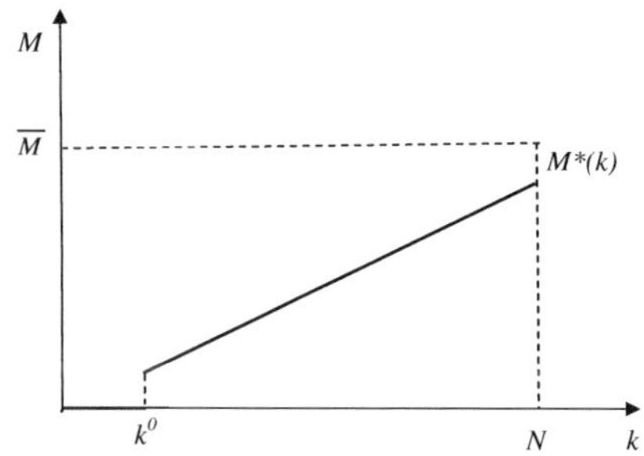

Fig. 6 Investments in R & D as a function of the coalition size

Suppose that the number of countries $N = 20$[1] and that $\gamma = 225$. It follows that the first-best R&D investment level is 67, giving each country a net benefit equal to 13.29. Furthermore, $\bar{M} = \gamma = 225$, so that a non-cooperative equilibrium with positive R&D exists, giving each country a net benefit equal to 7.75. In this case the non-cooperative equilibrium with $M = \bar{M}$ leads to an overinvestment in R&D, implying that this net benefit is considerably lower than the one in the social optimum.

Since $M^*(N) < \bar{M}$, Figs. 3 and 4 apply. It is easily verified that in this example the value function $V(k)$ does not have the kinks depicted in Fig. 3, but yields the simpler Figs. 5 and 6. The smallest coalition giving its members a non-negative net benefit is $K^0 = 10$. For this coalition size, coalition members get a net benefit equal to 0.51, while outsiders get 10. The average net benefit is 5.26, which is considerably lower than the net benefit in the social optimum and also lower than the net benefit in the non-cooperative equilibrium with $M = \bar{M}$. The largest stable coalition size in our setting, however, is $\tilde{K} = 16$, giving coalition members a net benefit equal to 8.5. The 4 outsiders get a net benefit equal to 16, so that the average net benefit is 10. Although this is smaller than the average net benefit in the social optimum, it is considerably larger than for the coalition of size K^0, and also larger than the average net benefit in the non-cooperative equilibrium with $M = \bar{M}$.

[1] In 2005 the 20 countries with the largest carbon emissions from energy use stood for 88% of total emissions (counting the EU as one country).

6 Conclusion

International environmental agreements are usually viewed upon as institutions with the purpose to overcome negative transboundary externalities. Due to free-rider incentives, however, treaties with a large number of signatories cannot be expected to be stable, so that a large part of the benefits of cooperation cannot be achieved. Barrett (2006) extended this framework to collective financing of R&D and adoption of breakthrough technologies, but effectively reaches the same conclusion, unless these technologies exhibit increasing returns to scale. In this paper we connect the level of R&D to the cost of adoption, and argue first that non-cooperative behaviour may lead to full adoption with a sufficiently high level of the public good R&D. This does not mean, however, that international treaties are not needed anymore. A treaty can do better, as it can either invest more in R&D in order to reduce the cost of adoption, or it can prevent overinvestment in R&D. We reach the important conclusion that in this setting the stability properties of international environmental agreements are much better than in the setting where treaties only focus on emission reduction.

Our result implies that it can indeed be beneficial for international environmental agreements to consider breakthrough technologies and R&D whenever this is appropriate, not only for the reason Barrett (2006) puts forward, but also for reasons put forward in this paper. For example, the Kyoto Protocol could consider setting up an R&D joint venture in order to develop carbon-free energy technologies instead of leaving this to the individual countries. Either a cheaper technology, adopted by all countries, will be developed, or a more expensive one, adopted by the signatories only. In both cases global welfare will increase. The first best will not be achieved, but the size of the coalition will be large and substantial benefits of cooperation will be realized.

References

Bagnoli M, Lipman B (1989) Provision of public goods: fully implementing the core through private contributions. Rev Econ Stud 56:583–601
Barrett S (1994) Self-enforcing international environmental agreements. Oxf Econ Pap 46:878–894
Barrett S (2003) Environment and statecraft: the stategy of environmental treaty-making. Oxford University Press, Oxford
Barrett S (2006) Climate treaties and "breakthrough" technologies. Am Econ Rev 96(2):22–25
Carraro C, Siniscalco D (1993) Strategies for the international protection of the environment. J Public Econ 52(3):309–328
Carraro C, Siniscalco D (1997) R&D cooperation and the stability of international environmental agreements. In: Carraro C (ed) International environmental negotiations: strategic policy issues. Edward Elgar, Cheltenham pp 71–96
Carraro C, Marchiori C, Oreffice S (2009) Endogenous minimum participation in international environmental treaties. Environ Resour Econ 42(3):411–425
Cesar H, de Zeeuw A (1997) Issue linkage in global environmental problems. In: Xepapadeas A (ed) Economic policy for the environment and natural resources. Edward Elgar, Cheltenham, pp 158–173
Chander P, Tulkens H (1995) A core-theoretic solution for the design of cooperative agreements on transfrontier pollution. Int Tax Public Finance 2:279–293
d'Aspremont C, Jacquemin A, Gabszewicz J, Weymark J (1983) On the stability of collusive price leadership. Can J Econ 16(1):17–25
de Zeeuw A (2008) Dynamic effects on the stability of international environmental agreements. J Environ Econ Manag 55(2):163–174

Diamantoudi E, Sartzetakis E (2002) International environmental agreements—the role of foresight. mimeo
Finus M (2003) Stability and design of international environmental agreements: the case of transboundary pollution. In: Folmer H, Tietenberg T (eds) The international yearbook of environmental and resource economics 2003/2004. Edward Elgar, Cheltenham, pp 82–158
Hoel M (1992) International environmental conventions: the case of uniform reductions of emissions. Environ Resour Econ 2(2):141–159
Kolstad C (2007) Systematic uncertainty in self-enforcing international environmental agreements. J Environ Econ Manag 53(1):68–79
Lackner K, Grimes P, Ziock H-J (2001) Capturing carbon dioxide from air. In: Proceedings of the first national conference on carbon sequestration, Washington, DC, USA
Pielke RA Jr (2009) An idealized assessment of the economics of air capture of carbon dioxide in mitigation policy. Environ Sci Policy 12:216–225

Part V
Transfers, Sharing and Fairness

[28]

Oxford Economic Papers 59 (2007), 45–62 **45**
doi:10.1093/oep/gpl028

International environmental agreements among asymmetric nations

By Matthew McGinty

Department of Economics, University of Wisconsin-Milwaukee, Milwaukee, WI 53201-0413, USA; e-mail: mmcginty@uwm.edu

This paper generalizes the benchmark model of self-enforcing international environmental agreements (IEAs) by allowing for all possible coalitions of n asymmetric nations. Asymmetries introduce gains from trade in pollution permits, reducing the incentive to deviate from a properly designed agreement. Coalitions are stable when the aggregate payoff to members is greater than the sum of individual payoffs from leaving the coalition. A benefit-cost ratio rule is proposed which distributes any remaining surplus after each coalition member receives their payoff as a non-signatory. Simulations of 20 asymmetric nations illustrate that even when the gains to cooperation are large, IEAs can achieve substantial emissions reductions. For example, when the benefit-cost ratio is one, stable coalitions can result in 47% of the difference between the full and no cooperation outcomes, compared with 5% for symmetric nations. Furthermore, 72% of the global payoff difference is obtained, relative to 9% for symmetry.

JEL classifications: H41, Q20, F02.

1. Introduction

The theoretical literature on international environmental agreements (IEAs) reaches pessimistic conclusions. IEAs suffer from the free-rider problem since reduced emissions are a global public good. Furthermore, the free-rider problem becomes more severe as the potential gains to an IEA increase (Barrett, 1994). Given this theoretical foundation, the difficulties implementing the Kyoto Treaty are hardly surprising.

This paper reverses some of that pessimism. It argues that the source of the conventional wisdom is the convenient, but highly unrealistic, assumption that nations are identical. When that assumption is relaxed by letting the marginal costs and benefits of abatement vary across nations, we find that a much higher level of abatement can be sustained by an IEA. This paper is the first to provide a solution to the benchmark model of IEAs (Hoel, 1991; Carraro and Siniscalco, 1993; Barrett, 1994) for all possible coalitions of n asymmetric signatories.

Coalitions are stable when the aggregate payoff to members is greater than the sum of individual payoffs from leaving the coalition. This means that a zero-sum system of transfers exists such that no member has an incentive to leave the coalition. A benefit-cost ratio rule is proposed which distributes any remaining surplus after each coalition member receives their payoff as a non-signatory. This rule is extremely simple and directly incorporates internal stability, two improvements over cooperative game theory rules such as the Shapley value. The transfers are implemented through a system of tradable pollution permits among members similar to the Kyoto Treaty.

Simulations of 20 asymmetric nations illustrate that even when the gains to cooperation are large IEAs can achieve substantial emissions reductions. For many parameter constellations stable coalitions result in approximately ten times the abatement obtained with symmetric nations. When the high benefit share nations have flat marginal cost curves coalitions with eight members are stable, compared with the largest coalition of three symmetric nations. The conventional wisdom that self-enforcing IEAs can not achieve substantial gains when the gains to cooperation are large does not hold when nations are asymmetric.

This paper extends the results of Barrett (1994) and Barrett (1997). Barrett (1994) performs simulations on the symmetric benchmark model of IEAs. He finds that self-enforcing IEAs can not substantially improve on the non-cooperative outcome. When abatement benefits are concave there is an inverse relationship between the number of signatories and the gains to cooperation. Concave benefits result in downward sloping reaction functions. An increase in abatement by signatories is met by a reduction in non-signatory abatement. For IEAs this situation is referred to as carbon leakage, but more generally as the free-rider problem. While nations would prefer to have others provide abatement, it is rational to provide an amount greater than the non-cooperative level when others' provision is low. Alternatively, linear benefit functions result in orthogonal reaction functions so changes in abatement by signatories to an IEA do not influence non-signatory behavior (Mäler, 1987; Hoel, 1992; Barrett, 1999). However, the standard free-rider problem still emerges in which each player chooses the dominant strategy of not cooperating.

Additional research has investigated if certain types of asymmetry can improve the success of IEAs (Barrett, 1997; Botteon and Carraro, 2001). Barrett (1997) extends his 1994 paper by allowing for two types of nations and testing 20 coalitions for stability. Still, he concludes that an IEA can achieve the least when the gains to cooperation are large, supporting the earlier symmetric model. Botteon and Carraro (2001) investigate IEA stability in a model of five asymmetric players. Abatement under the IEA is determined for two burden sharing rules: Shapley value and the Nash bargaining solution. They find higher cooperation under the Shapley value rule with transfer payments in a model calibrated using parameter estimates from Musgrave (1995). Their findings call for additional research providing a general solution to the asymmetric model, while allowing for transfers among a greater number of nations.

Recently, a great deal of attention has been given to coalition formation in IEAs (Finus, 2003; Ray and Vohra, 1999; Carraro and Marchiori, 2003; Finus and

Rundshagen, 2003). IEAs typically use the concept of internal and external stability from oligopoly literature (D'Aspremont *et al.*, 1983). This notion of stability means that no non-signatory has an incentive to join an IEA and no signatory has an incentive to leave. Coalition formation in IEAs can take various forms. In open membership games (Finus and Rundshagen, 2003) any nation may join the IEA. By contrast, in exclusive membership IEAs existing members may block the accession of a new nation. Furthermore, there may be a single coalition or multiple coalitions with different IEAs. In the single coalition framework all non-signatories behave as singletons maximizing individual payoffs. This paper analyses an open membership single coalition game similar to the Kyoto Treaty.

Transfers have been shown to increase participation when nations can commit to IEAs (Hoel, 1992; Carraro and Siniscalco, 1993). However, the problem with this approach is that commitment is not a best-response and thus violates individual rationality. Even without commitment, transfers allow nations to be compensated when they have an incentive to leave the IEA. Barrett (1992) considers pollution permit trading schemes as a system of side-payments. Nations that abate an amount greater than their requirement under the agreement receive a positive transfer, while nations that purchase permits meet their requirements at a lower cost. The required abatement under the agreement is an allocation of permit rights, defining a system of transfers. Transfers allow the signatory maximization problem to be unconstrained by the payoffs members would earn outside the IEA (Hoel, 1992). Barrett (2001) highlights the incentives for high benefit nations to provide transfers to low benefit nations to induce participation, called 'cooperation for sale'. The model presented below utilizes pollution permits as a zero-sum system of transfers among signatories.[1] A simple rule is specified which divides the remaining surplus once each coalition member receives their payoff as a single defector from the IEA. This rule specifies the required abatement, and thus transfers, given the efficient level of abatement. High benefit nations will contribute more to the total cost of abatement, while high cost nations can meet their required abatement levels through permit purchases.

The remainder of the paper is organized as follows. Section 2 derives the non-cooperative and full participation outcomes, as well as global abatement for any coalition of signatories. Section 3 provides the conditions under which any given coalition of signatories is stable, including a zero-sum system of transfers implemented through tradable pollution permits. Simulations in Section 4 illustrate the main results of the paper. Section 5 concludes and suggests additional research.

2. No, full, and partial cooperation

Let $N = \{1, \ldots, n\}$ denote the set of asymmetric nations. Global benefit is assumed to be a concave function: $B(Q) = b(aQ - (Q^2/2))$ of the worldwide quantity of

[1] The model does not consider multiple issues as in Folmer *et al.* (1993) or Cesar and de Zeeuw (1996). They show that linking cooperation across different issues can serve the same role as side-payments.

abatement $Q = \sum_{i \in N} q_i$. Greenhouse gases tend to mix uniformly in the upper atmosphere making the location of abatement irrelevant. Parameters a and b are strictly positive so the marginal benefit of the first unit of abatement is ab and the marginal benefit of the ath unit is zero. Nation $i \in N$ receives benefit share α_i, where $\alpha_i > 0 \; \forall i \in N$ and $\sum_{i \in N} \alpha_i = 1$.[2] Benefit for nation i is: $B_i(Q, \alpha_i) = b\alpha_i(aQ - (Q^2/2))$.

All nations are assumed to have convex abatement cost functions: $C_i(q_i, c_i) = (c_i(q_i)^2/2)$ generating marginal abatement cost (MAC) curves with asymmetric slopes $c_i > 0$. The atmosphere is a global public good that currently has no price to use. Since nations may freely emit effluents, MAC curves begin at the origin. However, the rates of increase vary immensely. Nations that currently use a relatively clean mix of fuels, such as nuclear or natural gas, face a much greater c_i due to fewer substitution possibilities. Nations which use a relatively high amount of coal can undertake substantial abatement at very low cost. For example, estimates by Ellerman, *et al.* (1998) suggest that the rate of marginal cost increase for Japan is more than ten times greater than the United States and more than 50 times greater than China. Differences of this magnitude suggest large gains from an IEA that efficiently allocates abatement levels. Net benefit for nation i is: $\pi_i = B_i - C_i$.

$$\pi_i(\alpha_i, c_i, q_i, Q) = b\alpha_i\left(aQ - \frac{Q^2}{2}\right) - \frac{c_i(q_i)^2}{2} \tag{1}$$

2.1 Non-cooperative outcome

In the non-cooperative outcome each nation chooses q_i to maximize π_i taking as given the sum of the others abatement, $Q_{-i} \equiv \sum_{j \neq i \in N} q_j$. The reaction function for nation i is:

$$q_i = \frac{b\alpha_i(a - Q_{-i})}{c_i + b\alpha_i} \tag{2}$$

Defining the benefit share MAC slope ratio as $\theta_i \equiv (\alpha_i/c_i)$ the slope of the reaction function is: $\partial q_i/\partial Q_{-i} = -1/(1 + 1/b\theta_i) \in (-1, 0)$. The slope of the reaction function approaches -1 as θ_i approaches ∞. Such a nation outside an agreement will reduce abatement by exactly the increase in abatement from the IEA, in other words complete carbon leakage. Carbon leakage is greatest when high benefit, low cost nations are outside an IEA. Conversely, as θ_i approaches 0 the reaction function becomes orthogonal and there is no carbon leakage.

[2] Assuming each person receives equal benefit from abatement implies α_i is global population share. GDP share is an alternative interpretation of α_i.

Abatement is a function of the coalition of signatories to an IEA. Let $K = 2^n$ be the set of all possible coalitions, and k an element of K.[3] Global abatement as a result of coalition k is $Q(k)$. The non-cooperative (or empty set) global level of abatement $Q(\emptyset) = \sum_{i \in N} q_i(\emptyset)$ is:

$$Q(\emptyset) = \frac{ab \sum_{j \in N} \theta_j}{1 + b \sum_{j \in N} \theta_j}$$
$$q_i(\emptyset) = \frac{ab\theta_i}{1 + b \sum_{j \in N} \theta_j} \tag{3}$$

The non-cooperative level of abatement is increasing in a, b, and $\sum_{j \in N} \theta_j$. With identical[4] nations $\theta = 1/nc$ and $\sum_{j \in N} \theta_j = 1/c$. However, with mean-preserving asymmetry $\sum_{j \in N} \theta_j$ may be either greater or smaller than $1/c$. Typically, asymmetry increases $Q(\emptyset)$ when the high benefit share nations are also low cost. In general, $Q(\emptyset)$ is decreasing in the covariance of α and c. Negative (positive) covariance typically indicates a non-cooperative level of abatement that is greater (lower) than symmetry, however counter-examples exist.

Equation (3) also implies any two nations abate in proportion to their $\theta' s$: $q_i/\theta_i = q_j/\theta_j$. An efficient allocation of abatement levels occurs when the MAC on the last units of abatement in each nation is equalized. Symmetry imposes allocative efficiency: $c_i q_i = c_j q_j$ in the non-cooperative outcome. Any benefit share asymmetry leads to an inefficient allocation of abatement at the non-cooperative level.

2.2 Full cooperation outcome

Full cooperation is obtained when all nations are signatories to an IEA. Abatement by one nation results in a positive externality that accrues to all other $n - 1$ nations. The grand coalition level of abatement internalizes this externality. Full cooperation maximizes the global net benefit function: $B(Q) - \sum_{i \in N} C_i(q_i)$ given the grand coalition MAC curve: $Q/\sum_{i \in N}(1/c_i)$. Abatement by the grand coalition $Q(N)$ is:

$$Q(N) = \frac{ab \sum_{i \in N} \frac{1}{ci}}{1 + b \sum_{i \in N} \frac{1}{ci}}$$
$$q_i(N) = \frac{ab}{c_i \ 1 + b \sum_{j \in N} \frac{1}{c_j} \Big)} \tag{4}$$

[3] The empty set and singletons are elements of K.

[4] Defining $\gamma \equiv c/b$ the symmetric $Q(\emptyset) = a/(1 + \gamma)$.

$Q(N)$ is increasing in a, b and $\sum_{i\in N}(1/c_i)$. For all mean-preserving distributions $\sum_{i\in N}(1/c_i)$ is minimized for symmetric c_i (proof in the Appendix). With cost asymmetry the global MAC curve is flatter, therefore $Q(N)$ is strictly greater when nations are asymmetric. The full cooperation level of abatement for each nation is unique and independent of the benefit share distribution. The full cooperation outcome is strictly greater than the non-cooperative level since $\sum_{i\in N}(1/c_i) > \sum_{i\in N}\theta_i$ for all $0<\alpha_i<1$, $\sum_{i\in N}\alpha_i = 1$.

The simulations of the symmetric model in Barrett (1994) show that the critical parameters are b and c, while the parameter a does not affect the number of signatories to an agreement. The potential gains to an IEA is the difference between the non-cooperative and full cooperation abatement levels. This difference is:

$$Q(N) - Q(\emptyset) = \frac{ab\left[\sum_{i\in N}\frac{1}{c_i} - \sum_{i\in N}\theta_i\right]}{\left[1 + b\sum_{i\in N}\frac{1}{c_i}\right]\left[1 + b\sum_{i\in N}\theta_i\right]} \qquad (5)$$

For a given distribution of MAC slopes and benefit shares, the slope of the global marginal benefit curve determines this difference. $Q(N) - Q(\emptyset)$ is strictly positive, achieves a maximum at $b = 1/\sqrt{\sum_{i\in N}(1/c_i)\sum_{i\in N}\theta_i}$, and is decreasing (increasing) in b for $b > (<) 1/\sqrt{\sum_{i\in N}(1/c_i)\sum_{i\in N}\theta_i}$. Of course, $Q(N) - Q(\emptyset)$ is strictly increasing in a.

2.3 Coalition outcomes

This sub-section determines abatement for any arbitrary partition of signatories and non-signatories to an IEA. For an arbitrary coalition, abatement by signatories is $Q_s(k) = \sum_{s\in k} q_s$, by non-signatories is $Q_t(k) = \sum_{t\notin k} q_t$, and total abatement is $Q(k) = Q_s(k) + Q_t(k)$. Coalition formation follows the Finus and Rundshagen (2003) open membership single coalition game described in the introduction. Signatories choose abatement to maximize collective net benefits and set abatement in each nation to minimize cost. Transfer payments allow the abatement choice in each signatory nation to be unconstrained by the payoff earned outside an IEA. Non-signatories behave as singletons, maximizing individual payoffs.

The reaction function for any non-signatory (2) can be decomposed into abatement by signatories $Q_s(k) = \sum_{s\in k} q_s$, and other non-signatories $Q_{-j}(k) = \sum_{t\neq j\notin k} q_t$. The aggregate reaction function of non-signatories to abatement undertaken by coalition k is:

$$Q_t = \frac{b(a - Q^s)\sum\limits_{j\notin k}\theta_j}{1 + b\sum\limits_{j\notin k}\theta_j} \qquad (6)$$

Recognizing the reaction of non-signatories the level of abatement by signatories to an IEA is:

$$
\begin{aligned}
Q_s(k) &= \frac{ab\sum\limits_{i\in k}\frac{1}{c_i}\sum\limits_{i\in k}\alpha_i}{\left[1+b\sum_{j\notin k}\theta_j\right]^2+b\sum\limits_{i\in k}\frac{1}{c_i}\sum\limits_{i\in k}\alpha_i} \\
q_s(k) &= \frac{ab\sum\limits_{i\in k}\alpha_i}{c_s\left(\left[1+b\sum\limits_{j\notin k}\theta_j\right]^2+b\sum\limits_{i\in k}\frac{1}{c_i}\sum\limits_{i\in k}\alpha_i\right)}
\end{aligned}
\tag{7}
$$

Non-signatory abatement is:

$$
\begin{aligned}
Q_t(k) &= \frac{ab\sum\limits_{j\notin k}\theta_j\left(1+b\sum\limits_{j\notin k}\theta_j\right)}{\left(1+b\sum\limits_{j\notin k}\theta_j\right)^2+b\sum\limits_{i\in k}\frac{1}{c_i}\sum\limits_{i\in k}\alpha_i} \\
q_t(k) &= \frac{ab\theta_t\left(1+b\sum\limits_{j\notin k}\theta_j\right)}{\left(1+b\sum\limits_{j\notin k}\theta_j\right)^2+b\sum\limits_{i\in k}\frac{1}{c_i}\sum\limits_{i\in k}\alpha_i}
\end{aligned}
\tag{8}
$$

Global abatement given coalition k is $Q(k) = Q_s(k) + Q_t(k)$:

$$
Q(k) = \frac{ab\left(\sum\limits_{j\notin k}\theta_j\left[1+b\sum\limits_{j\notin k}\theta_j\right]+\sum\limits_{i\in k}\frac{1}{c_i}\sum\limits_{i\in k}\alpha_i\right)}{\left(1+b\sum\limits_{j\notin k}\theta_j\right)^2+b\sum\limits_{i\in k}\frac{1}{c_i}\sum\limits_{i\in k}\alpha_i}
\tag{9}
$$

3. Coalition stability

The efficient level of signatory abatement $q_s(k)$ may not result in a higher payoff than leaving the agreement. Transfers among signatories can overcome this problem when there is an internal surplus among signatories. First, transfers are specified for a stable coalition according to a simple rule for dividing the surplus. Then coalition stability shows when this surplus exists.

3.1 Transfers

Transfers are implemented through a system of tradable pollution permits which will equate the MAC on the last unit of abatement with the permit price $p(k)$. Under an IEA, signatory s will abate until $p(k) = c_s q_s(k)$. Transfers among signatories are distributed by abatement requirements $q_r(k)$ relative to the efficient level $q_s(k)$. Signatory s supplies $q_s(k) - q_r(k)$ permits, receiving a transfer of $\tau_s(k) = p(k)[q_s(k) - q_r(k)]$. A zero-sum system of transfers equates the sum of required abatement to the aggregate signatory level for that particular coalition, $\sum_{r \in k} q_r = Q_s(k)$.

3.2 Allocation rule and its justification

For all stable coalitions there is a non-negative surplus among signatories where q_r determines the distribution of this surplus. Previous researchers (Botteon and Carraro, 2001; Barrett, 1997) have used burden sharing rules from cooperative game theory to determine q_r. The core, Shapley value, and Nash bargaining solution allocate payoffs, but are rigid and do not directly address the stability of an IEA.

The allocation should distribute the surplus in accordance with the open-membership single coalition game. Existing coalition members may not bar accession by a potential signatory. The core is the set of imputations remaining after eliminating those blocked by all possible sub-coalitions. In an open membership single coalition game members can not block accession, and therefore the core is not an appropriate solution concept.[5] The Shapley value is the weighted average of the marginal contributions of that nation across all sub-coalitions. The Shapley value is unique and always exists, but is inflexible and does not address stability. There is no reason to expect the Shapley value, or the Nash bargaining solution, to be greater than the payoff as a single deviator for all members. Furthermore, the Shapley value is difficult to compute since the value of 2^n coalitions must be determined, as well as each nations individual payoff after leaving each coalition. Similarly, the Nash bargaining solution maximizes the product of differences from threat points, rather than focusing on stability. Botteon and Carraro (2001) illustrate this point by showing that the set of stable coalitions depends on the choice of Shapley or Nash bargaining rules.

The optimal allocation rule is such that transfers are just sufficient to quell any incentive to deviate from the agreement. The rule proposed here is remarkably simple: Distribute the surplus among signatories in proportion to their benefit-cost ratio θ. Each member receives a transfer just sufficient to make coalition

[5] Chander and Tulkens (1995) propose two characteristic functions α and γ, with associated cores. The shortcoming is that they both rely on non-credible threats. The α core assumes coalition members will choose abatement to minimize the payoff of a deviator, and the γ core assumes that coalitions break up into singletons if one member deviates, even if sub-coalitions provide a higher payoff. As stated by Finus (2003), in the γ core members behave as masochists, in the α core they behave as sadists and masochists.

membership incentive compatible and the remaining surplus is distributed by θ share. Coalitions are stable when the aggregate payoff to members is greater than the sum of individual payoffs from leaving. The theta proportion rule distributes any remaining surplus after each coalition member receives their payoff as a single deviator from the agreement. The theta proportion rule is both flexible and simple.

Surplus share for signatory s is: $x_s(k) = (\theta_s / \sum_{s\in k}\theta_s)[\sum_{s\in k}\pi_s(k) - \sum_{t\in k}\pi_t(k\backslash\{t\})]$, where $\pi_t(k\backslash\{t\})$ is the payoff signatory s would earn outside coalition k choosing $q_t(k\backslash\{t\})$ from equation (8). Using this surplus sharing rule $q_r(k)$ solves $\pi_s(k) - \pi_t(k\backslash\{t\}) + \tau_s(k) = x_s(k)$.

$$q_r(k) = q_s(k) + \frac{\pi_s(k) - \pi_t(k\backslash\{t\}) - x_s(k)}{p(k)} \tag{10}$$

3.3 Stability

Let $k\backslash\{i\}$ denote the remaining coalition when signatory i leaves the IEA, and $k \cup \{j\}$ denote the coalition when non-signatory j accedes to the IEA. The subscripts s and t denote signatories choosing abatement level $q_s(k)$ and non-signatories choosing abatement level $q_t(k)$ respectively. The following conditions define individual and coalition stability.

$$\begin{aligned} \pi_s(k) - \pi_t(k\backslash\{t\}) + \tau_s(k) &\geqslant 0, \forall s = t \in k \\ \pi_t(k) - \pi_s(k \cup \{s\}) - \tau_s(k \cup \{s\}) &> 0, \forall t = s \notin k \end{aligned} \tag{11}$$

$$\Phi = \left\{ k \in K : \sum_{s\in k} \pi_s(k) - \sum_{t\in k} \pi_t(k\backslash\{t\}) \geqslant 0 \right\} \tag{12}$$

The conditions for internal and external stability are given by eq. (11). Internal stability ensures each signatory at least as high a payoff choosing $q_s(k)$ with transfer $\tau_s(k)$ than as a single defector from the IEA choosing $q_t(k\backslash\{t\})$. External stability indicates that no non-signatory earns a higher payoff joining coalition k and choosing $q_s(k \cup \{s\})$ with transfer $\tau_s(k \cup \{s\})$. Equation (12) defines the set of all stable coalitions: $\Phi \subset K$. This set consists of members such that the sum of the signatory payoffs exceeds the sum of payoffs as a single defector from the IEA. For all elements $\phi \in \Phi$ there exists a zero-sum system of transfers that simultaneously satisfies the internal stability condition for all signatories $s \in k$, defined by abatement requirements (10).

Permits have multiple purposes. First, they allow each nation to meet their required abatement at the lowest cost. Second, they act as a transfer scheme given the abatement requirements under the IEA. Therefore, $q_r(k)$ is a required contribution to total cost, not required domestic abatement. Condition (12) defines the existence of a self-enforcing IEA with a credible system of side-payments. Non-signatories are excluded from permit trading. Furthermore, (12) shows that there is no incentive to make transfers to nations outside the coalition in exchange for an increase in abatement as in Hoel and Schneider (1997). If the coalition had an incentive to do so, they would simply bring that nation into the coalition. The remaining issues are the size and membership of coalitions, and how much of the difference between the full and no cooperation outcomes can be obtained by stable coalitions.

4. Simulations

Two different simulations show the effectiveness of stable coalitions. First, seven nation simulations on all $2^7=128$ possible coalitions are presented to compare with the results of Barrett (1997). Mean-preserving asymmetry investigates the impact of allowing all nations to be asymmetric, rather than simply of two types. The largest abatement stable coalition is determined with transfer payments given by the theta proportion rule at the pollution permit price. Second, twenty nation simulations are conducted on all $2^{20}=1{,}048{,}576$ possible coalitions for different values of b and c. The twenty nation simulations provide further insight into the role of asymmetry. They allow for differences in the gains to cooperation, and the covariance between cost and benefits, for a far greater number of coalitions.

4.1 Seven nation simulations

Barrett (1997) investigates an asymmetric IEA for two types of nations.[6] Four nations are low benefit, low cost (type 1) and the remaining three are high benefit, high cost (type 2). With side payments determined by Shapley values he finds that two out of 20 potential coalitions are stable, one with three type 1 nations and zero type 2 nations and the other with one type 1 nation and two type 2 nations.

One way to measure IEA performance is the proportion of the difference between the no and full cooperation outcomes. Barrett (1997) finds: $(Q(k) - Q(\emptyset))/(Q(N) - Q(\emptyset)) = 30\%$ and 18% for the two coalitions respectively. However, the importance of this percentage depends on how different the full and no cooperation outcomes are. An IEA that closes a large share of a very small difference may be irrelevant. The effectiveness of an IEA accounts for both the percentage gain and the magnitude of the difference between full and no cooperation: $[(Q(k) - Q(\emptyset))/Q(N) - Q(\emptyset))][Q(N)/Q(\emptyset)]$. Abatement effectiveness for the two coalitions is then: 0.52 and 0.32. Alternatively, an IEA can be judged by the proportion of the global payoff difference and payoff effectiveness obtained. The global payoff gain is: $(\Pi(k) - \Pi(\emptyset))/(\Pi(N) - \Pi(\emptyset)) = 46\%$ and 30%, and payoff effectiveness is $(\Pi(k) - \Pi(\emptyset)/\Pi(N) - \Pi(\emptyset))(\Pi(N)/\Pi(\emptyset)) = 0.56$ and 0.37,

---

[6] The high benefit nations are high cost in Barrett (1997). Three type 1 nations have $c_1 = 50$ and four type 2 nations have $c_2=100$ so $\sum_{i\in N} c_i = 500$ and $\bar{c} = 71.43$. The type 1 nations have $\alpha_1 = 9/66$ and the type 2 nations have $\alpha_2 = 10/66$, so $\sum_{\alpha\in N} = 1$ and $b = 66$. The parameter a is normalized to one.

Table 1 Symmetry

	α_i	c_i	θ_i	$q(\emptyset)$	$q(N)$	$\pi(\emptyset)$	$\pi(N)$
$i = 1, \ldots, 7$	0.1429	71.43	0.0020	0.0686	0.1237	3.273	4.083
$\sum_{i \in N}$	1	500	0.0140	0.4802	0.8661	22.91	28.58

Table 2 Symmetric coalition stability

Elements $\in k$	Q	$\pi_s(k)$	$\pi_t(k \backslash \{t\})$
1	0.4640	3.304	3.273
2	0.4945	3.279	3.181
3	0.5663	3.376	3.350
4	0.6562	3.553	3.710
5	0.7419	3.752	4.084
6	0.8124	3.933	4.359
7	0.8661	4.083	4.526

respectively, where $\Pi = \sum_{i \in N} \pi_i$. Allowing for any degree of asymmetry will increase both effectiveness measures.

The seven nation simulations that follow compare symmetry and mean preserving asymmetry of the Barrett (1997) parameters where $c_i = ic_1$ and $\alpha_i = i\alpha_1$ for $i = 1, \ldots, n$. Table 1 presents abatement and payoffs at full and no cooperation for the symmetric parameterization. The difference between c_i and c_{i+1} is $d = 15$, and the difference between α_i and α_{i+1} is $\delta = .0442$. The high benefit nations are also high cost increasing the difference between non-cooperative and grand coalition abatement. For this parameterization asymmetry reduces the non-cooperative and increases the grand coalition levels of abatement.

Including the null set there are eight coalitions of symmetric nations. Let ϕ^* denote the largest abatement stable coalition: $\phi^* \in \Phi \subset K$. Table 2 shows ϕ^* consists of three members resulting in $Q(\phi^*) - Q(\emptyset)/Q(N) - Q(\emptyset) = 22\%$ of the global abatement difference and $\Pi(\phi^*) - \Pi(\emptyset)/\Pi(N) - \Pi(\emptyset) = 36\%$ of the global payoff difference. Global payoff for the three member IEA is $\Pi(\phi^*) = 24.97$. Abatement effectiveness is 0.40 and payoff effectiveness is 0.45.

Table 3 illustrates the impact of allowing for full asymmetry. Under asymmetry the no cooperation abatement and payoffs are lower and the full cooperation levels of abatement and payoffs are greater. Furthermore, $q(N)$ is less than $q(\emptyset)$ for the two highest cost and benefit nations. In the absence of side payments the two lowest cost and benefit nations are worse off at the grand coalition outcome. Both of these results are impossible under symmetry and highlight the role of transfers.

Using eq. (12) all $2^7 = 128$ potential coalitions are tested for stability. Table 4 shows that all coalitions containing one, two and three members are stable, while no coalitions containing more than four members are stable. The largest abatement stable coalition is $\{1, 2, 3, 7\}$ with global abatement 0.6894.

Table 3 Asymmetry, $d = 15$, $\delta = 0.0442$

i	α_i	c_i	θ_i	$q(\emptyset)$	$q(N)$	$\pi(\emptyset)$	$\pi(N)$
1	0.0102	26.43	0.0004	0.0141	0.2725	0.2312	-0.6483
2	0.0544	41.43	0.0013	0.0479	0.1738	1.200	1.149
3	0.0986	56.43	0.0017	0.0638	0.1276	2.146	2.757
4	0.1429	71.43	0.0020	0.0730	0.1008	3.084	4.295
5	0.1871	86.43	0.0022	0.0790	0.0833	4.018	5.800
6	0.2313	101.4	0.0023	0.0832	0.0710	4.950	7.286
7	0.2755	116.4	0.0024	0.0863	0.0619	5.880	8.761
$\sum_{i\in N}$	1	500	0.0123	0.4472	0.8909	21.51	29.40

Table 4 Asymmetric coalition stability

Elements $\in k$	Stable coalitions Φ	Largest abatement
1	all	0.4438, $\{1\}$
2	all	0.5382, $\{1, 7\}$
3	all	0.6399, $\{1, 6, 7\}$
4	$\{1, 2, 3, 7\}$	0.6894, $\phi^* = \{1, 2, 3, 7\}$
5	none	
6	none	
7	none	

The abatement and payoff increases are: $(Q(\phi^*) - Q(\emptyset))/(Q(N) - Q(\emptyset)) = 55\%$ and $(\Pi(\phi^*) - \Pi(\emptyset))/(\Pi(N) - \Pi(\emptyset)) = 80\%$. Coincidently, abatement and payoff effectiveness are both 1.09. Asymmetric effectiveness is higher due to both a higher proportion and a greater difference. The conventional wisdom from both symmetric models and those with two types of nations is that a self-enforcing IEA can not substantially improve on the non-cooperative outcome when the gains to cooperation are large. However, introducing full asymmetry has caused the effectiveness to more than double.

ϕ^* consists of the highest cost and benefit nation (7) and the three lowest cost and benefit nations (1, 2, 3). Table 5 shows the system of side payments $\tau_s(k) = p(k)\{q_s(k) - q_r(k)\}$ consistent with the theta proportion rule from Section 3. The pollution permit price that equates the marginal cost on the last unit of abatement for each coalition member is: $p(\{1, 2, 3, 7\}) = 6.311$. In Table 5 the high benefit nation purchases permits from all three low cost nations, and all four coalition members receive a higher payoff than they would as a single deviator from the agreement.

Additional simulations (not reported) show the reducing the degree of asymmetry by setting $d = 10$ and $\delta = 2/49$ reduces both the abatement gain: $(Q(\phi^*) - Q(\emptyset))/(Q(N) - Q(\emptyset)) = 47\%$, and the global payoff gain: $(\Pi(\phi^*) - \Pi(\emptyset))/(\Pi(N) - \Pi(\emptyset)) = 72\%$, but does not change the stable coalitions in Table 4. Smaller asymmetry reduces abatement and payoff effectiveness to 0.90

Table 5 Largest abatement asymmetric coalition: $\{1, 2, 3, 7\}$ and transfers

Elements $\in \phi^*$	$\pi_s(\phi^*)$	$\pi_t(\phi^* \backslash \{t\})$	$q_s(\phi^*)$	$q_r(\phi^*)$	$\tau_s(\phi^*)$	$\pi_s(\phi^*) + \tau_s(\phi^*)$
1	−0.4494	0.2785	0.2388	0.1227	0.7327	0.2833
2	1.142	1.505	0.1523	0.0923	0.3792	1.521
3	2.588	2.705	0.1118	0.0899	0.1388	2.727
7	8.043	6.763	0.0542	0.2524	−1.251	6.793
$\sum_{s \in \phi^*}$	11.32	11.25	0.5572	0.5572	0	11.32
$t \notin \phi^*$	$\pi_t(\phi^*)$		$q_t(\phi^*)$			
4	4.199		0.0410			
5	5.493		0.0444			
6	6.785		0.0468			
$\sum_{t \notin \phi^*}$	16.48		0.1321			
$\sum_{i \in N}$	27.80		0.6894			

and 0.95. In general, increasing the variance of the α and c distributions increases both ϕ^* abatement and payoff gain.

4.2 20 nation simulations

Next, twenty nation simulations for various combinations of b and c test the generality of the previous results.[7] Two types of mean-preserving asymmetry are compared with symmetry for all $2^{20} = 1,048,576$ potential coalitions. For the HiLo simulations the high benefit share nations are low cost. The HiLo simulations have: $c_i = ic_1$ and $\alpha_1/i = \alpha_i$ for $i = 1, \ldots, n$. The HiHi simulations reverse the benefit shares making the high benefit nations high cost: $c_i = ic_1$ and $\alpha_i = i\alpha_1$ for $i = 1, \ldots, n$. Furthermore, in order to mimic symmetry the mean preserving spread of benefits and costs in the HiHi simulations is such that $\theta_i = (1/nc)\ \forall i \in N$. Constant thetas increase the gains to cooperation since $Q(\emptyset)$ is the same as symmetry while $Q(N)$ is greater.[8]

Table 6 shows the largest abatement stable coalition ϕ^*. For HiLo simulations ϕ^* is typically not the stable coalition with the highest number of signatories. For example, along the main diagonal where $b/\bar{c} = 1$ many eight member coalitions are stable, but global abatement is less than the stable five member coalition. This is not the case for the HiHi ϕ^*, where the largest abatement stable coalition always contains the greatest number of signatories. In the HiLo case ϕ^* always contains the two highest benefit and lowest cost nations $(1, 2)$. Similarly, the HiHi ϕ^* always contains the highest benefit (20) and lowest cost $(1, 2, 3)$ nations.

[7] Barrett (1997) also conducts n=100 simulations without transfers for 97 type 1 nations and three type 2 nations. With full asymmetry there are $2^{100} = 1.27 \times 10^{30}$ possible coalitions for each of the nine combinations of b and c.

[8] In order to conserve on space, stability results for the more than 18 million coalitions tested in Table 6 are omitted. The full simulations were conducted in Maple and are available from the author by request. Furthermore, the programs may be augmented to test any value of n.

Table 6 N = 20 ϕ^* largest abatement stable coalitions

		$\bar{c}$		
		0.01	**1**	**100**
			Symmetry	
	0.01	{3 members}	{2 members}	{2 members}
	1	{17 members}	{3 members}	{2 members}
	100	{N}	{17 members}	{3 members}
			Asymmetry: HiLo	
	0.01	{1,2,3,10,20}†	{1,2,5}††	{1,2,3}††
b	1	{N}	{1,2,3,10,20}†	{1,2,5}††
	100	{N}	{N}	{1,2,3,10,20}†
			Asymmetry: HiHi	
	0.01	{1,2,3,4,20}	{1,2,3,5,20}	{1,2,3,5,20}
	1	{N}−{2,18,19}	{1,2,3,4,20}	{1,2,3,5,20}
	100	{N}	{N}−{2,18,19}	{1,2,3,4,20}

† (††) denotes coalitions of up to and including eight (five) members are stable but with $Q\{k\} < Q(\phi^*)$.

Table 6 shows that asymmetry increases the number of signatories to a self-enforcing IEA, but participation falls short of the grand coalition when the gains to cooperation are large.

The asymmetric abatement and effectiveness levels on or above the main diagonal of Table 7 are remarkable. When $b/\bar{c} = 1$ a coalition of symmetric nations results in only 5% of the difference between the full and no cooperation outcomes. When the high benefit nations are also high cost (HiHi) a stable coalition results in 47% of the difference, even though the full cooperation level of abatement is greater. The effectiveness measure multiplies the percentage difference by the ratio of full to no cooperation. The effectiveness measure for HiHi is 10 times the symmetric result, capturing the fact that the asymmetry results in a larger percentage of a larger difference. The noncooperative level is much greater when the high benefit nations are low cost (HiLo). For HiLo $b/\bar{c} = 1$ the difference falls to 34% and the effectiveness to 0.45, even though abatement is higher. Surprisingly, the more effective agreement is when the gains to cooperation are larger. The HiHi simulations have more effective agreements due to greater gains from permit trading.

When $b/\bar{c} = 0.01$ symmetry results in 1% of the difference between no and full cooperation. The difference is 11% for both types of asymmetry, and the effectiveness measures are dramatically higher since the grand coalition is nearly twice the level for symmetry. Below the main diagonal there is essentially no difference between $Q(\emptyset)$ and $Q(N)$ so asymmetry has little impact. Consequently, the effectiveness measures below the main diagonal are essentially the same as $(Q(\phi^*) - Q(\emptyset))/(Q(N) - Q(\emptyset))$, and have no importance.

Table 7 N = 20 ϕ^* global abatement for symmetry and asymmetry

		$\bar{c}$								
		0.01			**1**			**100**		
		Sym	HiLo	HiHi	Sym	HiLo	HiHi	Sym	HiLo	HiHi
		50.00	73.53	50.00	0.990	2.703	0.990	0.010	0.028	0.010
		52.23	81.69	72.06	1.086	5.451	3.805	0.011	0.062	0.040
	0.01	95.24	97.42	97.42	16.67	27.42	27.42	0.200	0.376	0.376
		(5%)	(34%)	(47%)	(1%)	(11%)	(11%)	(1%)	(10%)	(8%)
		(0.09)	(0.45)	(0.91)	(0.10)	(1.13)	(2.95)	(0.11)	(1.34)	(3.06)
		99.01	99.64	99.01	50.00	73.53	50.00	0.990	2.703	0.990
		99.06	99.97	99.43	52.23	81.69	72.06	1.086	5.451	3.805
b	1	99.95	99.97	99.97	95.24	97.42	97.42	16.67	27.42	27.42
		(5%)	(100%)	(44%)	(5%)	(34%)	(47%)	(1%)	(11%)	(11%)
		(0.05)	(1.00)	(0.44)	(0.09)	(0.45)	(0.91)	(0.10)	(1.13)	(2.95)
		99.99	100.0	99.99	99.01	99.64	99.01	50.00	73.53	50.00
		100.0	100.0	100.0	99.06	99.97	99.43	52.23	81.69	72.06
	100	100.0	100.0	100.0	99.95	99.97	99.97	95.24	97.42	97.42
		(100%)	na	(100%)	(5%)	(100%)	(44%)	(5%)	(34%)	(47%)
		(1.00)	na	(1.00)	(0.05)	(1.00)	(0.44)	(0.09)	(0.45)	(0.91)

The entries in each cell are $Q(\emptyset)$, $Q(\phi^*)$, $Q(N)$, $(Q(\phi^*) - Q(\emptyset))/(Q(N) - Q(\emptyset))$ and $((Q(\phi^*) - Q(\emptyset))/(Q(N) - Q(\emptyset))(Q(N)/Q(\emptyset)))$.

The same pattern emerges for the global payoff gains presented in Table 8. When $b/\bar{c} = 1$ HiHi asymmetry results in 72% of the payoff gain, compared to 9% with symmetry. For all cells on or above the main diagonal the asymmetric differences and effectiveness substantially improve the abatement and payoff gains attainable by an IEA. Contrary to conventional wisdom, when the gains to cooperation are largest, asymmetry results in substantially more abatement and higher payoffs. IEAs can substantially improve on the non-cooperative outcome when the gains to cooperation are large.

5. Conclusion

This paper generalizes the benchmark model of IEAs by incorporating different MAC slopes and benefit shares. Dramatically more encouraging results emerge from relaxing the convenient, but unrealistic, assumption of symmetry. Allowing for asymmetry does not overturn the fundamental result that there is a trade-off between the gains to an IEA and the number of signatories. However, symmetric models may vastly understate the degree of abatement achievable by a self-enforcing IEA regardless of the size of the gains to cooperation. In general, when the covariance between the benefit shares and the MAC slopes is negative the non-cooperative level and stable coalition abatement is greater than symmetry would indicate. The full cooperation level of abatement is unambiguously higher when nations differ.

Table 8 N = 20 ϕ^* global payoffs for symmetry and asymmetry

		$\bar{c}$								
		0.01			1			100		
		Sym.	HiLo	HiHi	Sym.	HiLo	HiHi	Sym.	HiLo	HiHi
		36.88	45.70	36.88	0.961	2.558	0.961	0.010	0.027	0.010
		37.88	47.23	45.35	1.049	4.732	3.495	0.011	0.055	0.037
	0.01	47.62	48.71	48.71	8.333	13.71	13.71	0.100	0.188	0.188
		(9%)	(51%)	(72%)	(1%)	(19%)	(20%)	(1%)	(17%)	(15%)
		(0.12)	(0.54)	(0.95)	(0.10)	(1.04)	(2.84)	(0.11)	(1.23)	(2.97)
		4997.06	4998.47	4997.06	3688	4570	3688	96.07	255.8	96.07
		4997.10	4998.68	4998.37	3788	4723	4535	104.9	473.3	350.0
b	1	4997.50	4998.68	4998.68	4762	4871	4871	833.3	1371	1371
		(10%)	100%	(81%)	(9%)	(51%)	(72%)	(1%)	(19%)	(20%)
		(0.10)	(1.00)	(0.81)	(0.12)	(0.54)	(0.95)	(0.10)	(1.04)	(2.84)
		499997	499999	499997	499706	499847	499706	368750	456982	368750
		499998	499999	499999	499710	499868	499837	378788	472269	453455
	100	499998	499999	499999	499750	499868	499868	476190	487106	487106
		(100%)	na	(100%)	(10%)	(100%)	(81%)	(9%)	(51%)	(72%)
		(1.00)	na	(1.00)	(0.10)	(1.00)	(0.81)	(0.12)	(0.54)	(0.95)

The entries in each cell are $\Pi(\emptyset)$, $\Pi(\phi^*)$, $\Pi(N)$, $(\Pi(\phi^*) - \Pi(\emptyset))/(\Pi(N) - \Pi(\emptyset))$ and $(\Pi(\phi^*) - \Pi(\emptyset))/(\Pi(N) - \Pi(\emptyset))(\Pi(N)/\Pi(\emptyset))$.

The size and scope of an agreement is increased by exploiting asymmetries, which introduce the possibility of IEAs that can substantially overcome the free-rider problem given the appropriate set of abatement requirements. Coalitions are stable when the aggregate payoff to members exceeds the sum of payoffs from individually leaving the coalition. Nations receive a net transfer sufficient to remain a signatory. The theta proportion rule is a simple method of distributing any remaining surplus, which determines required abatement for a given coalition. Transfers are zero-sum and coalition specific. No assumptions have been made regarding which stable coalitions will arise. However, external stability implies coalitions will continue to expand membership when it is profitable to do so, reducing the set of expected outcomes. Regardless, nothing ensures that the maximum abatement stable coalition will form from this subset of stable coalitions. This suggests an important role for international institutions.

Despite the encouraging results from including asymmetry, the difficulties in implementing the Kyoto Treaty remain evident. Even with asymmetry there exists the fundamental trade-off between the gains to an IEA and the number of signatories. Moreover, the abatement requirements under the Kyoto Treaty are not conducive to promoting internal stability. Under Kyoto, nations are required to reduce emissions of greenhouse gases by a minimum of 5.5% from their 1990 levels. A nation such as the United States, which accounts for roughly 1/5 of world emissions, would be responsible for 1/5 of total abatement. If the benefit shares are determined by population, it is not difficult to find parameter values where the Kyoto requirements violate incentive compatibility for the United States.

Alternative coalition formation rules is a potential topic for future research. With the single coalition open membership game (Finus and Rundshagen, 2003) there is no way to ensure that the largest abatement stable coalition ϕ^* is obtained. Other coalition formation sequences may occur that do not lead to ϕ^*. Exclusive membership rules may lead to higher abatement in this situation. It is also possible that the formation of multiple coalitions can achieve more than a single IEA. Alternatively, linking emissions reductions with other issues has been shown by Cesar and de Zeeuw (1996) and Folmer *et al.* (1993) to increase the degree of cooperation sustainable by self-enforcing agreements. Linkages in conjunction with side-payments in the form of pollution permits should prove to be more effective than either alone. Increasing the variance of the benefit shares and marginal cost slopes is another area for future research. The highest benefit nation has 20 times the share of the lowest nation in these simulations. This is a relatively small dispersion given empirical estimates (Ellerman *et al.*, 1998). Increasing variance should lead to even higher levels of abatement and payoffs attainable by a self-enforcing IEA, even if the number of signatories does not dramatically increase.

Acknowledgements

This work derives from the first chapter of my Ph.D. dissertation, completed June, 2002 at the University of California, Santa Cruz. This paper has greatly benefited from comments by Dan Friedman, John Heywood, seminar participants at the University of Minnesota's Applied Economics series, and anonymous referees.

References

Barrett, S. (1992) International environmental agreements as games, in C. Carraro (ed.) *International Environmental Negotiations*, Edward Elgar, Cheltenham.

Barrett, S. (1994) Self-enforcing international environmental agreements, *Oxford Economic Papers*, **46**, 878–94.

Barrett, S. (1997) Heterogeneous international environmental agreements, in C. Carraro (ed.) *International Environmental Negotiations. Strategic Policy Issues*, Edward Elgar, Cheltenham.

Barrett, S. (1999) A theory of full international cooperation, *Journal of Theoretical Politics*, **11**, 519–41.

Barrett, S. (2001) International cooperation for sale, *European Economic Review*, **45**, 1835–50.

Botteon, M. and Carraro, C. (2001) Environmental coalitions with heterogenous countries: burden sharing and carbon leakage, in A. Ulph (ed.) *Environmental Policy, International Agreements, and International Trade*, Oxford University Press, New York.

Carraro, C. and Marchiori, C. (2003) Stable coalitions, in C. Carraro (ed.) *Endogenous Formation of Economic Coalitions*, Edward Elgar, Cheltenham, 156–98.

Carraro, C. and Siniscalco, D. (1993) Strategies for the international protection of the environment, *Journal of Public Economics*, **52**, 309–28.

Cesar, H. and de Zeeuw, A. (1996) Issue linkage in global environmental problems, in A. Xepapadeas (ed.) *Economic Policy for the Environment and Natural Resources*, Edward Elgar, Cheltenham, 158–73.

Chander, P. and Tulkens, H. (1995) A core-theoretic solution for the design of cooperative agreements on transfrontier pollution, *International Tax and Public Finance*, **2**, 279–93.

D'Aspremont, C., Jacquemin, A., Gabszewicz, J.J., and Weymark, J. (1983) On the stability of collusive price leadership, *Canadian Journal of Economics*, **16**, 17–25.

Ellerman, A.D., Jacoby, H., and Decaux, A. (1998) The effect on developing countries of the Kyoto Protocol and CO_2 emissions trading, MIT Joint Program on the Science and Policy of Global Climate Change Report #41, MIT, Cambridge, MA.

Finus, M. (2003) Stability and design of international environmental agreements: the case of transboundary pollution, in H. Folmer, and T. Tietenberg (eds) *International Yearbook of Environmental and Resource Economics, 2003/4*, Edward Elgar, Cheltenham, 82–158.

Finus, M. and Rundshagen, B. (2003) Endogenous coalition formation in global pollution control: a partition function approach, in C. Carraro, (ed.) *Endogenous Formation of Economic Coalitions*, Edward Elgar, Cheltenham, 199–243.

Folmer, H., Mouche, P., and Ragland, S.(1993) Interconnected games and international environmental problems, *Environmental and Resource Economics*, **3**, 313–35.

Hoel, M. (1991) Global environmental problems: the effects of unilateral actions taken by one country, *Journal of Environmental Economics and Management*, **20**, 55–70.

Hoel, M. (1992) International environmental conventions: the case of uniform reductions of emissions, *Environmental and Resource Economics*, **2**, 141–59.

Hoel, M. and Schneider, K. (1997) Incentives to participate in an international environmental agreement, *Environmental and Resource Economics*, **9**, 153–70.

Mäler, K.G. (1987) The acid rain game, in H. Folmer, and E. Van Ierland (eds) *Valuation Methods and Policy Making in Environmental Economics*, Elsevier, Amsterdam, 80–108.

Musgrave, P. (1995) Pure global externalities: international efficiency and equity, in L. Bovenberg, and S. Cnossen (eds) *Public Economics and the Environment in an Imperfect World*, Kluwer Academic Press, Dordrecht, 237–59.

Ray, D. and Vohra, R. (1999) A theory of endogenous coalition structures, *Games and Economic Behavior*, **26**, 286–336.

Appendix

The $\sum_{i\in N}(1/c_i)$ is minimized for identical c_i.

Proof Let d_i be the deviation from the mean $\bar{c} \equiv \sum_{i\in N}(c_i/n)$. A mean preserving distribution of c_i implies the constraint: $\sum_{i\in N} d_i = 0$. The Lagrangian is: $\min_{d_1,\ldots,d_n} \{L = \sum_{i\in N}(1/(\bar{c}+d_i) + \lambda(0 - \sum_{i\in N} d_i)\}$, subject to $\sum_{i\in N} d_i=0$. The n first order conditions are of the form: $-1/(\bar{c}+d_i)^2=\lambda$. For all $c_i>0$ the second order condition for a minimum is satisfied. Any two first order conditions imply $1/(\bar{c}+d_i) = 1/(\bar{c}+d_j)$, therefore the sum is minimized for identical deviations, $d_1 = d_2 =,\ldots, = d_n$. The constraint $\sum_{i\in N} d_j = 0$, implies $d_1 = d_2 =,\ldots, = d_n = 0$, and the sum $\sum_{i\in N} 1/(\bar{c}+d_i)$ is minimized for identical c_i. □

[29]

The Manchester School Vol 77 No. 5 September 2009
1463–6786 575–593

CARTEL STABILITY UNDER AN OPTIMAL SHARING RULE*

by
HANS-PETER WEIKARD†
Wageningen University

In this paper I analyse the stability of cartels in games with heterogeneous players and externalities. I introduce a class of sharing rules for coalition pay-offs, called 'optimal sharing rules', that stabilize all cartels that are possibly stable under some arbitrary sharing rule. To illustrate the impact of optimal sharing with heterogeneous players and positive externalities I analyse a public goods game. In contrast to games with identical players that are common in the study of international environmental agreements, my results show that large coalitions may well be stable when optimal sharing is applied.

1 Introduction

Cartels are widespread. What immediately comes to mind are industry cartels. But there are many others. A cartel forms when agents form a unique coalition to coordinate behaviour *vis-à-vis* other agents. International environmental agreements (IEAs), for instance, can be considered as cartels in this sense. Industry cartels are usually considered inefficient because they hamper competition and, hence, there is little concern to reflect upon measures that help stabilize those cartels. In contrast, IEAs are formed in order to provide international public goods. Here the issue is setting incentives for participation. The main concern of this paper is to propose a family of sharing rules that distribute the pay-offs among cartel members to stabilize agreements and, thus, to improve the provision of public goods. Although the main motivation for this paper comes from the literature on IEAs, some of the formal results obtained apply to cartel games in general. Hence, I do not restrict the analysis to IEAs. However, I only consider open membership games where every player is free to join a cartel. Obviously, this setting is not always convincing for industry cartels and related situations.

* Manuscript received 26.7.07; final version received 25.6.08.
† I discussed my idea of an optimal sharing rule with Michael Finus at an early stage of this research. Harold Houba has provided helpful comments on a draft of this paper. I wish to thank the members of the STACO project, Wageningen University, for support. I am indebted to three anonymous referees whose suggestions have helped to improve the paper. I am grateful to the International Institute of Applied Systems Analysis (IIASA) for providing a stimulating research environment in the summer of 2004 when I obtained the main results.

Published by Blackwell Publishing Ltd, 9600 Garsington Road, Oxford OX4 2DQ, UK, and 350 Main Street, Malden, MA 02148, USA.

Recent work on the formation and stability of coalitions has focused on the partition function approach. A partition function is a generalization of the characteristic function of a coalition formation game. The characteristic function, introduced by von Neumann and Morgenstern (1944), assigns a pay-off to each coalition which is independent of the actions taken by non-members. In many economic environments this assumption is too restrictive. The partition function approach, first introduced by Thrall and Lucas (1963) and further developed by Ray and Vohra (1999), relaxes this assumption. The coalition pay-off may depend on the entire coalition structure. This allows us to consider externalities. Because coalition pay-offs are insufficient to identify Nash equilibria of coalition formation games, Bloch (1996) and Yi (1997) have introduced a per-member partition function. The recent literature in this domain has been surveyed by Bloch (2003) and Yi (2003). The move from coalition pay-offs to individual pay-offs requires the use of a sharing rule. In their coalition formation models Bloch (1996) and Yi (1997) assume identical players and equal sharing of coalition pay-offs. Here I relax this symmetry assumption. In a more general setting with differences in players' characteristics equal sharing is less convincing. Ray and Vohra (1999) develop a sequential model of coalition formation where coalition members bargain over the distribution of the coalition pay-off. The merit of this approach is its generality. However, the difficulty of computing equilibria in empirical settings has hampered its application.

In applied work a cartel formation game with positive spillovers introduced by d'Aspremont *et al.* (1983) has received growing attention. A cartel formation game allows for at most one non-trivial coalition. The analysis focuses on stability of cartels. A cartel is stable when no member has an incentive to leave (internal stability) and no non-member has an incentive to join (external stability). In a cartel formation game a player's strategy set consists of the options whether or not to become a coalition member. The pay-offs of all players—not only coalition members' pay-offs—depend on the coalition formed. There are spillovers to non-members. The model of d'Aspremont *et al.* (1983) has been re-examined by Donsimoni *et al.* (1986) to study market structures and used by Poyago-Theotoky (1995) to study R&D joint ventures. Recently, it has received growing attention in studies of IEAs by, for example, Hoel (1992), Carraro and Siniscalco (1993), Barrett (1994), Hoel and Schneider (1997), Finus *et al.* (2005) and Diamantoudi and Sartzetakis (2006). The usefulness of cartel formation games in these domains, however, has so far been hampered by the restrictive assumption of identical players. There are few papers relaxing the symmetry assumption. Donsimoni (1985) analyses cartel stability with heterogeneous firms. However, she obtains her results for specific cost and demand functions.[1] In

[1]Related to this work is Belleflamme's (2000) analysis of coalition stability with heterogeneous players where all players join one of two competing associations.

a study on IEAs Barrett (2001) relaxes the symmetry assumption by introducing two types of countries. In the same context Botteon and Carraro (1997) resort to numerical simulations using a five-region model of greenhouse gas abatement with asymmetric regions. In an extension of this approach Botteon and Carraro (2001) observe a need for a research programme 'to move as much as possible from numerical to general theoretical results on the non-cooperative formation of coalitions in the presence of spillovers and heterogeneous countries' and to assess the impact of institutions such as sharing rules. In this paper I take some steps in this direction.

If the identical-players assumption is relaxed, the sharing of the coalition pay-off is crucial for individual incentives to join a cartel and, hence, for cartel stability. I consider the class of sharing rules that distributes the coalition pay-off giving priority to the satisfaction of outside-option demands. I call these rules 'optimal sharing rules'. Optimal sharing will guarantee internal stability of a coalition whenever it is at all feasible, i.e. whenever the coalition pay-off equals or exceeds the sum of the outside option pay-offs. It can be shown that if a coalition is internally stable under some arbitrary sharing rule, then it is internally stable under an optimal sharing rule. This has also been noticed by Eyckmans and Finus (2004).[2]

Two main results are obtained. Both results require a superadditive pay-off function. Superadditivity means that enlargement of the cartel generates a gain compared with initial pay-offs of the cartel and the joining member. First, I consider the case of negative spillovers where non-members lose when others form a coalition. I prove that the grand coalition is the unique stable cartel under optimal sharing. Second, for the case of positive spillovers where non-members gain when others form a coalition, I show that the set of stable cartels can be generated by a simple algorithm, provided the pay-off function satisfies a 'non-essential player' property. A player is non-essential in an internally stable coalition if the remaining coalition without that player is internally stable. The intuition behind this result is that every stable coalition can be 'built up' stepwise by a sequence of internally stable enlargements starting from a small internally stable coalition. If the non-essential player property does not hold, i.e. if all players of a coalition are essential, then this coalition cannot be built up by a sequence of internally stable enlargements.

The paper is organized as follows. The next section introduces a cartel formation game. Section 3 introduces the class of optimal sharing rules and shows that these rules 'maximize' stability. Section 4 contains the main results of the paper: the characterization of stable coalitions in cartel formation

[2]Eyckmans and Finus (2004) have established a result equivalent to Theorem 1 of this paper; however, they do not attempt to analyse the equilibria that emerge under optimal sharing. The results of this paper have been obtained completely independently.

games with heterogeneous players. Section 5 provides an application. A typical situation where such game structure arises is the provision of public goods. It is well known from the literature on private provision of public goods that—due to free-riding—inefficiently low levels of public goods are provided since players do not take others' benefits into account. The inefficiency problem may be mitigated to some extent if we allow for coalition formation. Barrett (1994, p. 891) concludes from the analysis of a symmetric cartel game that self-enforcing agreements (i.e. stable coalitions) 'may not be able to improve substantially upon the non-cooperative outcome'. The finding in this paper is that Barrett's result does not generalize to cases where players differ with respect to their marginal benefits of the public good. In fact, any coalition can be stable if one player's marginal benefit is sufficiently large. In Section 5 I show for a public goods game with linear benefits, quadratic costs and heterogeneous players that it has the non-essential player property and, hence, the set of stable cartels is easy to compute even for games with a large number of players. Section 6 concludes.

2 The Cartel Game

Let N be a finite set of players and $|N| \geq 3$. Players are denoted by i, j and k. Each player j chooses a strategy $\sigma_j \in \{0, 1\}$. If player j chooses $\sigma_j = 1$, he will be a member of a (unique) coalition; if he chooses $\sigma_j = 0$, he will not be a member. Let $C \subseteq N$ be the set of coalition members. A cartel partition function assigns a pay-off to the cartel and to every non-member.

Definition 1 (cartel partition function): A cartel partition function is a mapping v that assigns pay-offs $v(C)$ to C and $v_j(C)$ to all $j \in N \backslash C$ for every $C \subseteq N$.

To arrive at individual pay-offs the partition function is supplemented by a sharing rule that distributes the coalition pay-off between coalition members. Denote by $v_i^r(C)$ the pay-off of player i under coalition C and sharing rule r.

Definition 2 (sharing rule): A sharing rule is a mapping $r: 2^N \to \mathbb{R}^{|C|}$ such that $r(C) = (v_i^r)_{i \in C}$ and $\Sigma_{i \in C} v_i^r(C) = v(C)$. The class of all sharing rules is denoted by $\mathcal{R}$.

Suppose $i \in C$, then C_{-i} is the shorthand notation for coalition $C \backslash \{i\}$, i.e. the remaining coalition when i 'leaves'. Similarly, C_{+j} is the shorthand for $C \cup \{j\}$; $j \notin C$. We assume that, if $j \notin C$, then $v_j^r(C) = v_j(C)$ for every sharing rule $r \in \mathcal{R}$. Players outside the coalition are unaffected by the sharing rule applied by coalition members. $v_i(C_{-i})$ is i's outside option pay-off which I will also call his claim. Furthermore, we assume that, for all $i, j \in N$, $v_j(\{i\}) = v_j(\varnothing)$.

If only a single player announces '1', the (trivial) coalition formed is not effective.[3]

It is obvious that the sharing rule applied to distribute the coalition pay-off will affect the stability of a coalition, the incentives to join or stay out. We now define the concepts of internal and external stability. The condition for internal stability is that each coalition member gets at least as much as he gets when leaving the coalition, while others stay on. External stability means that no player outside the coalition would gain by joining the coalition. Denote by $\hat{S}_r$ the set of internally stable coalitions under sharing rule r and by $\breve{S}_r$ the set of externally stable coalitions under r.

Definition 3 (internal stability): $C \in \hat{S}_r$ if and only if for all $i \in C$ it holds that

$$v_i^r(C) \geq v_i(C_{-i}) \tag{1}$$

Definition 4 (external stability): $C \in \breve{S}_r$ if and only if for all $j \in N \backslash C$ it holds that

$$v_j(C) > v_j^r(C_{+j}) \tag{2}$$

Coalition C is called a stable coalition if and only if it is internally and externally stable. The set of stable coalitions under r is denoted by S_r.

Definition 5 (stability): $C \in S_r$ if and only if $C \in \hat{S}_r \cap \breve{S}_r$.

Definition 4 is (slightly) non-standard. The tie-breaking rule is here that a player would join the coalition if he is indifferent between joining and staying out. I adopt this definition because incentives for cartel membership are the main concern of the paper. Clearly, if (1) and (2) hold, we have a Nash equilibrium of the cartel formation game as no player has an incentive to deviate from his chosen strategy. The set of stable coalitions S is a subset of all Nash equilibria. The full set is obtained if we weaken the stability definition and replace the strict inequality sign in (2) by a weak inequality sign. In the remainder of the paper I will use the stronger stability notion of Definitions 3–5.[4]

3 Optimal Sharing Rules

Consider the class of sharing rules where each coalition member receives his claim $v_i(C_{-i})$. Of the residual $v(C) - \Sigma_{i \in C} v_i(C_{-i})$ coalition members receive arbitrary shares. That is, if the coalition pay-off exceeds the sum of the

[3]Note that this assumption is not satisfied in the game considered by d'Aspremont *et al.* (1983).
[4]Hence, by this definition, the empty set ∅ is not externally stable, and a trivial coalition is internally stable.

claims, every member receives at least his claim. If not all claims can be met, then no one's claim will be met. This class of rules is inspired by rights-egalitarian sharing, axiomatized by Herrero *et al.* (1999), where individual claims have priority and it is a collective responsibility to meet them. Rights-egalitarian sharing emphasizes the importance of individual rights. In the context of coalition formation where membership is voluntary it is a natural assumption to grant each member a right to a position no worse than his outside option. The class of optimal sharing rules is characterized by the following condition.

Claim Rights Condition: For all $i \in C$ and all $C \subseteq N$, $v_i^r(C) \geq v_i(C_{-i})$ if and only if $v(C) \geq \Sigma_{i \in C} v_i(C_{-i})$.

Definition 6 (optimal sharing): A sharing rule r is an optimal sharing rule if and only if it meets the Claim Rights Condition. The class of optimal sharing rules is denoted by $\mathcal{R}^o$.

Of all possible sharing rules the Claim Rights Condition selects the class of surplus sharing rules where outside option pay-offs define an 'original position' for a bargain to distribute a surplus or share a loss.[5] Note that the Claim Rights Condition is weaker than rights-egalitarian sharing which proposes a unique solution to the sharing problem.

The results that follow in the remainder of this paper hold for every sharing rule $r \in \mathcal{R}^o$. The Claim Rights Condition implies that incentives to join or leave a coalition do not depend on the distribution of the residual $v_i(C) - \Sigma_{i \in C} v_i(C_{-i})$ between coalition members.

It can be shown that any rule $r \in \mathcal{R}^o$ produces the largest possible set of internally stable coalitions.

Theorem 1: In a cartel formation game, if coalition C is internally stable under some sharing rule $r' \in \mathcal{R}$, then C is internally stable under any sharing rule $r \in \mathcal{R}^o$.

Proof: By definition of internal stability $v_i^{r'}(C) \geq v_i(C_{-i})$ holds for all $i \in C$. It follows that an internally stable coalition must earn at least the sum of the outside option pay-offs. Hence, $\Sigma_{i \in C} v_i^{r'}(C) \geq \Sigma_{i \in C} v_i(C_{-i})$. For every sharing rule r it must hold that $v(C) = \Sigma_{i \in C} v_i^r(C)$. Hence, we have $v(C) \geq \Sigma_{i \in C} v_i(C_{-i})$. And it follows directly that C will be internally stable under every rule that satisfies the Claim Rights Condition. ■

Theorem 1 tells us that, given a partition function, whenever internal stability is feasible for some coalition, it will be achieved by any optimal sharing rule

[5]In applied models of coalition formation the Claim Rights Condition is usually violated (cf. e.g. Weikard *et al.* (2006) and the references cited therein).

$r \in \mathcal{R}^o$. In this sense optimal sharing 'maximizes' internal stability. Theorem 1 allows us to restrict further analysis of internal stability to an arbitrary rule $r \in \mathcal{R}^o$ and still be sure that no internally stable coalition escapes our attention. Other possible sharing rules need not be considered because no rule can generate an internally stable coalition that is not also internally stable under optimal sharing.

4 Cartel Stability

To identify the set of stable coalitions we first prove a lemma that sheds light on the relationship between a coalition C and its enlargement C_{+j}.

Lemma 1: In a cartel formation game with optimal sharing, coalition C is externally unstable if and only if there exists $j \in N \backslash C$ such that C_{+j} is internally stable.

Proof: To prove the necessity part, note that internal stability implies $v_j^r(C_{+j}) \geq v_j(C)$. Hence it does not hold that $v_j^r(C_{+j}) < v_j(C)$ and hence C is externally unstable. To prove the sufficiency part, suppose C is externally unstable under $r \in \mathcal{R}^o$; then there exists $j \in N \backslash C$ such that $v_j(C) \leq v_j^r(C_{+j})$. Note now that $v_j(C)$ is j's outside option pay-off with respect to coalition C_{+j}. Under $r \in \mathcal{R}^o$, if $v_j^r(C_{+j}) \geq v_j(C)$ holds for j, then $v_i^r(C_{+j}) \geq v_i(C_{+j-i})$ for all $i \in C \cup \{j\}$. Every member of the enlarged coalition $C \cup \{j\}$ receives at least his or her outside option pay-off and hence the enlarged coalition is internally stable. ■

Lemma 1 implies that either every coalition C is externally stable or there exists an enlarged coalition C_{+j} which is internally stable. d'Aspremont *et al.* (1983) observe a similar feature in their cartel game with identical players.

For what follows I impose restrictions on the partition function. First, pay-offs are required to be superadditive. A coalition C joined by an additional member j can achieve at least as much as C and j can achieve independently.

Definition 7 (superadditivity): A cartel partition function is superadditive if and only if, for all coalitions $C \subset N$ and all $j \in N \backslash C$, it holds that $v(C_{+j}) \geq v(C) + v_j(C)$.

Superadditivity is a natural assumption as all options open to a set of individual players should be open to a team. The following lemma holds.

Lemma 2: In a cartel formation game with optimal sharing, if pay-offs are superadditive, then every two-player coalition is internally stable.

Proof: Consider $C = \{i\}$. Suppose $i \neq j$. It follows from superadditivity that $v(\{i, j\}) \geq v(\{i\}) + v_j(\{i\})$. Since, for all i, $v(\{i\}) = v_i(\{j\}) = v_i(\emptyset)$, the superadditivity property can be rewritten as $v(\{i, j\}) \geq v_i(\{j\}) + v_j(\{i\})$. This says that the coalition pay-off is at least as much as the sum of the outside option pay-offs. Hence, under every $r \in \mathcal{R}^o$ every player receives at least his or her outside option pay-off and, hence, all two-player coalitions are internally stable under optimal sharing. ■

In order to prove the existence of non-trivial stable cartels I introduce the notion of an internally stable enlargement of C.

Definition 8 (internally stable enlargement): C_{+j} with $j \in N \backslash C$ is an internally stable enlargement of C if both C and C_{+j} are internally stable.

Theorem 2: In a cartel formation game with optimal sharing, a non-trivial ($|C| \geq 2$) stable coalition always exists if pay-offs are superadditive.

Proof: The proof is by contradiction. Suppose no non-trivial stable coalition exists. By Lemma 2 all two-player coalitions C are internally stable. By assumption they are unstable. Hence, they must be externally unstable. Then, by Lemma 1, there exists an internally stable enlargement C_{+j}. Again, by assumption, C_{+j} is not stable and, hence, not externally stable. Then, Lemma 1 applies again, and so on until all possibilities of internally stable enlargements are exhausted, i.e. when the enlarged internally stable coalition reaches size $|N|$. As N is externally stable by definition, N must be stable. This contradicts our assumption. Hence, a stable coalition C must exist such that $2 \leq |C| \leq |N|$. ■

Theorem 2 corresponds to the theorem by d'Aspremont *et al.* (1983, p. 22) which establishes the existence of stable cartels.[6]

To derive further results I impose restrictions concerning the nature of spillovers. I will examine two cases: negative spillovers and positive spillovers. With negative spillovers non-members lose as the coalition grows. In this case the grand coalition ($C = N$) is stable. In the case of positive spillovers non-members gain when others join the coalition. Here the issue is more complex. Usually an efficient outcome (the grand coalition) cannot be obtained as there are strong free-rider incentives. An example is discussed in Section 5. The simpler case of negative spillovers is discussed in turn.

Definition 9 (negative spillovers): Pay-offs exhibit negative spillovers if and only if for all coalitions $C \subset N$ and all $j, k \in N \backslash C$ it holds that $v_j(C_{+k}) < v_j(C)$.

[6]In their Proposition 2 d'Aspremont *et al.* (1983) establish that per-firm profits increase with cartel size. Superadditivity is stronger than this property. However, it is obvious from their proof that superadditivity also holds. Note that their theorem allows for stable cartels of size 1 because they allow an individual firm to take the role of a leader. Hence, in their game a cartel of size 1 can be effective. This is ruled out in the game considered in this paper.

The following theorem holds.

Theorem 3: In a cartel formation game with an optimal sharing rule $r \in \mathcal{R}^o$, if pay-offs are superadditive and exhibit negative spillovers, then the grand coalition is the unique stable coalition, i.e. $S_r = \{N\}$.

Proof: The proof is by induction. There are three steps. First, we show that if a coalition C is internally stable, then C_{+j} is also internally stable.

Step 1: Suppose C is internally stable. Then, under any sharing rule $r \in \mathcal{R}^o$, everyone receives at least his claim, as $v(C) \geq \Sigma_{i\in C} v_i(C_{-i})$. Suppose $i \in C$ and $j \in N \backslash C$. By superadditivity $\Sigma_{i\in C} v_i^r(C_{+j}) + v_j^r(C_{+j}) \geq \Sigma_{i\in C} v_i^r(C) + v_j(C)$. Then, it also holds that $\Sigma_{i\in C} v_i^r(C_{+j}) + v_j^r(C_{+j}) \geq \Sigma_{i\in C} v_i(C_{-i}) + v_j(C)$. Note now that the sum of the claims of coalition C_{+j} is $\Sigma_{i\in C} v_i(C_{+j-i}) + v_j(C)$. By negative spillovers we have $\Sigma_{i\in C} v_i(C_{+j-i}) < \Sigma_{i\in C} v_i(C_{-i})$. Hence, $\Sigma_{i\in C} v_i^r(C_{+j}) + v_j^r(C_{+j}) \geq \Sigma_{i\in C} v_i(C_{-i}) + v_j(C) > \Sigma_{i\in C} v_i(C_{+j-i}) + v_j(C)$. This establishes that the pay-off of coalition C_{+j} is larger than the sum of the claims of the coalition members. We conclude that if coalition C is internally stable, then C_{+j} is internally stable.

Step 2: By Lemma 2 we know that an internally stable two-player coalition exists. By step 1 we know that every enlargement of an n-player coalition to an $n+1$-player coalition generates an internally stable coalition. Hence, all coalitions are internally stable.

Step 3: By Lemma 1, if there exists $j \in N \backslash C$ such that $C \cup \{j\}$ is internally stable, then C is externally unstable. Such j always exists unless C is the grand coalition. Hence, the grand coalition is the unique stable coalition. ■

Negative spillovers imply that outside option pay-offs decrease when the cartel increases. This gives an incentive to join the cartel. Note that, with negative spillovers, superadditivity is not sufficient for efficiency of the grand coalition—unlike in a characteristic function game where we do not have spillovers. Efficiency of the grand coalition is guaranteed, however, if a 'full cohesiveness' condition is met. Full cohesiveness requires that the pay-off of any coalition in a partition is at least as large as the sum of the pay-offs achieved by any collection of subcoalitions (cf. Montero, 1999; Cornet, 2003).

The result of Theorem 3 comes with a cautious note. Step 3 of the proof uses the necessity part of Lemma 1. Hence, the grand coalition, although the unique stable coalition according to Definitions 3–5, may not be the unique Nash equilibrium.

Furthermore, note that Yi (1997, Proposition 4.1) also finds the grand coalition to be the unique stable coalition in a multiple coalition game with negative spillovers and identical players.

We now turn to the more interesting case of positive spillovers. Positive spillovers imply that there are free-rider incentives. Thus, we expect stable coalitions that are generally smaller than the grand coalition, but according to Theorem 2 stable coalitions have at least two members.

Definition 10 (positive spillovers): Pay-offs exhibit positive spillovers if and only if, for all coalitions $C \subset N$ and all j, $k \in N \backslash C$ with $j \neq k$, it holds that $v_j(C_{+k}) \geq v_j(C)$.

While superadditive pay-offs give an incentive to join the coalition, positive spillovers work in the opposite direction. A larger coalition does not only achieve more for its members but it also generates larger benefits to non-members. These benefits are an obstacle to forming large and successful coalitions. Generally, superadditivity and positive spillovers do not limit the set of possibly stable coalitions. If free-rider incentives are very strong, the set of stable coalitions may just be the set of two-player coalitions. If, however, only the grand coalition achieves a substantial gain while smaller coalitions are largely ineffective, then the grand coalition may be in the set of stable coalitions. One constraint on the set of stable coalitions is that, if C is stable, no subset of C of size $|C| - 1$ is stable. This is an immediate consequence of Lemma 1.

Without further restrictions on the pay-off structure, there is little more to say. However, we can identify a class of cartel formation games for which all stable coalitions can be generated in a simple straightforward manner. This class of games is characterized by the fact that *not all* players are essential for the success of the coalition. A player is essential if without this player the remaining coalition is internally unstable and would fall apart. I formally define the property of non-essentiality. A lemma follows.

Definition 11 (non-essentiality): Player $i \in C$ is non-essential for coalition $C \in \hat{S}_r$ $(r \in \mathcal{R}^o)$ if and only if $C_{-i} \in \hat{S}_r$.

Lemma 3: In a superadditive cartel formation game with an optimal sharing rule $r \in \mathcal{R}^o$, if every internally stable coalition has a non-essential member, then there exists a sequence of internally stable enlargements from $|C| = 2$ for every stable coalition.

Proof: We assume that every internally stable coalition has a non-essential member. Consider a stable coalition C of size $|C| = n > 2$. As C is stable, it is also internally stable and has a non-essential member. Hence, there exists a coalition $C_{-i} \in \hat{S}_r$ of size $n - 1$. As $C_{-i} \in \hat{S}_r$, there exists $C_{-i-j} \in \hat{S}_r$, and so forth until the size of the coalition is 2. This establishes the result. ■

The existence of a non-essential player rules out pay-off structures with specific synergies. Consider a large stone that can only be moved by four persons. Any three would not be strong enough to move the stone. In this case, a coalition of four could be stable, while there is no internally stable coalition of three. The existence of a non-essential member in every internally stable coalition rules out the possibility of stable coalitions of size $n + 1$ if there does not exist an internally stable coalition of size n.

I call a sequence of internally stable enlargements leading to C 'complete' if C can be described by a sequence of internally stable enlargements and if there is no $j \in N \backslash C$ such that C_{+j} is internally stable. A sequence of internally stable enlargements leading to C is incomplete if C is externally unstable. It can now be shown that the set of complete sequences of internally stable enlargements coincides with the set of stable coalitions and that the set of stable coalitions can be generated by the following simple algorithm.

Definition 12 (simple algorithm): (i) Start with $n = 2$. (ii) Check each internally stable n-player coalition C for internally stable enlargements for all $j \in N \backslash C$. If there is no internally stable enlargement of C, then the sequence is complete. If there is an internally stable enlargement C_{+j}, put it on a list of internally stable coalitions of size $n + 1$. (iii) If the check is completed, proceed with the listed internally stable $n + 1$-player coalitions at step (ii). (iv) Stop when the list of internally stable $n + 1$-player coalitions is empty.

Theorem 4: (a) In a superadditive cartel formation game with optimal sharing, if every internally stable coalition has a non-essential member, then the set of complete sequences of internally stable enlargements coincides with the set of stable coalitions. (b) The simple algorithm generates the set of stable coalitions.

Proof: (a) First, notice that a complete sequence of internally stable enlargements leads to a stable coalition: if C is generated by a sequence of internally stable enlargements, then it is, by definition, internally stable. As the sequence is complete there does not exist $j \in N \backslash C$ such that C_{+j} is internally stable. Hence, C is also externally stable and, therefore, stable. Second, Lemma 3 establishes that every stable coalition can be constructed by a sequence of internally stable enlargements. This sequence is in fact a complete sequence since, by Lemma 1, there cannot be an internally stable enlargement of a stable coalition. Hence, the set of complete sequences of internally stable enlargements coincides with the set of stable coalitions.

(b) By construction the simple algorithm generates all complete sequences of internally stable enlargements. Part (a) establishes that the set of complete sequences of internally stable enlargements coincides with the set of stable coalitions. ■

Note that the converse of Theorem 4(a) is not true. The set of complete sequences of internally stable enlargements may coincide with the set of stable coalitions while there exist internally stable coalitions consisting of essential members only.

The importance of Theorem 4 is that it identifies a class of coalition formation games for which the set of stable coalitions is easy to compute. The simple algorithm generates the stable coalitions for the class of cartel formation games with non-essential members in every internally stable coalition. This set of stable coalitions has a specific feature. There is exactly one stable coalition on every possible path of coalition enlargements. Suppose some coalition C is stable, then all larger coalitions containing C are (internally) unstable and all proper subsets of C are (externally) unstable. If one would draw all the paths of enlargements leading from the singletons to the grand coalition, the set of stable coalitions forms a sharp boundary between the area of internally stable but externally unstable coalitions, and the area of externally stable but internally unstable coalitions.

5 Cartel Formation and the Provision of Public Goods

In this section I show that Theorem 4 has important applications. I consider the case of the provision of public goods when benefits are linear and costs are quadratic. More specifically, I consider a two-stage game similar to the cartel game introduced by d'Aspremont *et al.* (1983) and applied to public goods problems by Carraro and Siniscalco (1993, 1997), Barrett (1994), Hoel and Schneider (1997), Carraro and Marchiori (2003), Diamantoudi and Sartzetakis (2006) and others. The important difference is that I relax the identical-players assumption usually invoked in this literature and I apply the optimal sharing rule developed in this paper. Recently, similar analyses have been conducted by Fuentes-Albero and Rubio (2009) and McGinty (2007). The former, however, restrict the analysis to two types of players while the latter resorts to numerical simulations to obtain results.

There are N players. At stage 1 each player considers whether or not to join a binding agreement. Those who join, the members of coalition C, fully cooperate in the provision of the public good at the second stage. They maximize the coalition pay-off and share the pay-off according to some sharing rule $r \in \mathcal{R}^o$. The remaining players $N \backslash C$, the free-riders, maximize their individual pay-offs.

I show that this game satisfies the condition that there exists a non-essential player for every internally stable coalition. Accordingly, all stable coalitions can be computed by the simple algorithm.

Let $b_i > 0$ denote individual marginal benefits, $a_i > 0$ denote a cost parameter and $z_i \geq 0$ denote the amount of public good provided by i. The total amount provided is $z = \Sigma_{i \in N} z_i$.

At stage 2 of the game a free-rider $k \notin C$ maximizes

$$v_k(C) = b_k z_k + b_k \sum_{j \in N \setminus \{k\}} z_j - \tfrac{1}{2} a_k z_k^2 \tag{3}$$

It follows from a free-rider's first-order condition that

$$z_k^* = \frac{b_k}{a_k} \tag{4}$$

The coalition maximizes the joint pay-off which is then available for distribution,

$$v(C) = \sum_{i \in C} b_i \sum_{k \in N} z_k - \sum_{i \in C} \tfrac{1}{2} a_i z_i^2 \tag{5}$$

From the set of first-order conditions we have

$$z_i^* = \frac{\sum_{j \in C} b_j}{a_i} \tag{6}$$

for all $i \in C$ (Samuelson's rule).

Given the first-order conditions the equilibrium coalition pay-off is

$$v(C) = \sum_{i \in C} \left[b_i \sum_{j \in C} \frac{\sum_{k \in C} b_k}{a_j} + b_i \sum_{l \notin C} \frac{b_l}{a_l} - \frac{1}{2} \frac{\left(\sum_{k \in C} b_k\right)^2}{a_i} \right] \tag{7}$$

When we do not consider redistribution, the term in brackets is the individual net benefit for coalition member i. The first addend in the bracket is the benefit from the public good provided by coalition members; the second addend gives the benefit from the provision by free-riders; the third addend is i's cost of optimal provision.

A coalition C applying some $r \in \mathcal{R}^o$ will be internally stable if and only if the coalition pay-off (7) is at least as much as the sum of the claims of the coalition members. Recall that player i's claim is the pay-off i would receive under coalition C_{-i}. Player i's claim ($i \in C$) is given by

$$v_i(C_{-i}) = b_i \frac{b_i}{a_i} + b_i \sum_{j \in C_{-i}} \frac{\sum_{k \in C_{-i}} b_k}{a_j} + b_i \sum_{l \notin C} \frac{b_l}{a_l} - \frac{1}{2} \frac{b_i^2}{a_i} \tag{8}$$

In order to check for internal stability, we have to compare (7) and the sum of individual claims $\Sigma_{i \in C} v_i(C_{-i})$. To simplify notation, let $\Phi(C) \equiv \Sigma_{i \in C} v_i(C_{-i})$.

Then, internal stability requires

$$v(C) \geq \Phi(C) \tag{9}$$

First, note that the following proposition holds; cf. d'Aspremont *et al.* (1983, p. 23) and Barrett (1994, Proposition 2).

Proposition 1: In a simple public goods game with linear benefits functions and quadratic costs functions, if players are identical ($a_i = a_j$ and $b_i = b_j$ for all $i, j \in N$), then every coalition of size 3 is stable.

Proof: To show that internal stability is satisfied, we first show that $v(C) - \Phi(C) = 0$ for $|C| = 3$. From (7) and (8) we obtain

$$v(C)-\Phi(C)=-\frac{1}{2}\left(\sum_{i\in C} b_i\right)^2\sum_{i\in C}\frac{1}{a_i}-\frac{3}{2}\sum_{i\in C}\frac{b_i^2}{a_i}+\sum_{i\in C} b_i\sum_{i\in C}\frac{b_i}{a_i}+\sum_{i\in C} b_i^2\sum_{i\in C}\frac{1}{a_i} \quad (10)$$

Setting $a_i = a$, $b_i = b$ and $|C| = 3$, we obtain from (10) that $v(C) - \Phi(C) = 0$.

External stability is satisfied for $|C| = 3$ for identical players because any four-player coalition is internally unstable: setting $a_i = a$, $b_i = b$ and $|C| = 4$, we obtain from (10) that $v(C) - \Phi(C) = -6b^2/a < 0$. ■

In order to demonstrate that a public goods game with heterogeneous players has a non-essential member in every internally stable coalition, we have to show that, if C is internally stable, then there exists i such that C_{-i} is internally stable.

The following proposition holds.

Proposition 2: In a simple public goods game with linear benefits functions ($b_i \geq 0$, for all $i \in N$) and quadratic costs functions ($a_i = 1$, for all $i \in N$), every internally stable coalition has a non-essential member.

Proof: The proof is given in the Appendix.

Theorem 4 has established how the set of stable coalitions can be determined for any numerical specification.

Another important insight follows easily. If we allow for heterogeneous players large coalitions may well be stable. It is easy to construct examples with a large number of players with different marginal benefits where the grand coalition is stable. This claim is established formally in the Appendix.

6 Conclusions

The study of coalition formation is burdened with the complexity of the strategic situation. In order to get a grip on the problem, many theoretical models have assumed that players are identical. The identical-players assumption has so far seriously limited the empirical relevance of coalition theory. This paper tries to overcome some of the difficulties in cartel formation games. Sharing rules satisfying the Claim Rights Condition guarantee that a coalition will in fact be internally stable whenever it is at all possible to stabilize the coalition. I show that under optimal sharing the grand coalition is the unique stable cartel in games with negative spillovers. For games with positive spillovers I identify a class of stable cartels that are easy to compute—even in games with many players. Optimal sharing rules perform best to stabilize coalitions for the provision of public goods. Hence, there is no need for trial and error approaches—prevalent in the literature—to identify successful sharing rules. Furthermore, I find that the often quoted claim that only small cartels are stable (e.g. Hoel and Schneider, 1997) does not hold in a generalized class of games with heterogeneous players.

It should be noticed that all results have been obtained using internal and external stability as a solution concept. The stability concept, a slightly stronger version of Nash equilibrium, considers only single deviations. Clearly, the study coalition formation suggests the possibility of multiple deviations. Thus, the use of solution concepts such as strong Nash equilibrium (Aumann, 1959) or coalition-proof Nash equilibrium (Bernheim *et al.*, 1987) is commendable. How Thoron's (1998) analysis of coalition proof cartel formation can be generalized to heterogeneous players remains an open problem.

Appendix

Proof of Proposition 2: We prove that every internally stable coalition in a simple public goods game with linear benefit functions and quadratic cost functions has a non-essential member. We have to show that

if C is internally stable, then there exists $j \in C$ such that
C_{-j} is internally stable (*)

Notice first that (*) is satisfied for every C if $|C| \leq 3$ (cf. Lemma 2). For the remainder of the proof we assume that $|N| \geq |C| \geq 4$. Suppose C is internally stable. Then, the coalition pay-off is at least as much as the sum of the claims. Internal stability requires $v(C) \geq \Phi(C)$. We have to show that a non-essential member j exists in every internally stable coalition such that

$$v(C_{-j}) - \Phi(C_{-j}) \geq 0 \tag{A1}$$

To simplify notation denote $n \equiv |C|$. Assuming $a_i = 1$ for all $i \in N$ and internal stability of C we obtain from (10)

$$v(C) - \Phi(C) = -\frac{n-2}{2}\left(\sum_{i\in C} b_i\right)^2 + \frac{2n-3}{2}\sum_{i\in C} b_i^2 \geq 0 \tag{A2}$$

This can be rewritten as

$$\frac{n-1}{2}b_j^2 - (n-2)\sum_{i\in C_{-j}} b_i b_j + \frac{2n-3}{2}\sum_{i\in C_{-j}} b_i^2 - \frac{n-2}{2}\left(\sum_{i\in C_{-j}} b_i\right)^2 \geq 0 \tag{A3}$$

Furthermore, (A1) can be written as

$$-\left(\sum_{i\in C_{-j}} b_i\right)^2+\frac{2n-5}{n-3}\sum_{i\in C_{-j}} b_i^2\geq 0 \tag{A4}$$

By appropriate manipulation of (A3) we obtain

$$-\left(\sum_{i\in C_{-j}} b_i\right)^2+\frac{2n-5}{n-3}\sum_{i\in C_{-j}} b_i^2-\frac{1}{(n-3)(n-2)}\sum_{i\in C_{-j}} b_i^2-2\sum_{i\in C_{-j}} b_ib_j+\frac{n-1}{n-2}b_j^2\geq 0 \tag{A5}$$

(A5) holds by assumption of internal stability of C. The first two addends of (A5) are equivalent to the condition for internal stability of C_{-j}. This is clear from (A4). We denote the last three addends of (A5) by

$$\Omega\equiv-\frac{1}{(n-3)(n-2)}\sum_{i\in C_{-j}} b_i^2-2\sum_{i\in C_{-j}} b_ib_j+\frac{n-1}{n-2}b_j^2 \tag{A6}$$

Then (A5) can be written as

$$0\leq v(C_{-j})-\Phi(C_{-j})+\Omega \tag{A7}$$

To show that (A1) holds, it suffices to show that $\Omega\leq 0$. It is easy to establish that $\Omega<0$ for $b_j=0$. Furthermore, $d\Omega/db_j\leq 0$ in the interval

$$\left[0,\frac{n-2}{n-1}\sum_{i\in C_{-j}} b_i\right)$$

Hence, $\Omega<0$ for

$$0\leq b_j\leq\frac{n-2}{n-1}\sum_{i\in C_{-j}} b_i$$

Now, note that we can choose j such that $b_j\leq b_i$ for all $i\in C_{-j}$. Hence, we have

$$b_j\leq\frac{1}{n-1}\sum_{i\in C_{-j}} b_i$$

Furthermore,

$$\frac{1}{n-1}\sum_{i\in C_{-j}} b_i<\frac{n-2}{n-1}\sum_{i\in C_{-j}} b_i$$

for $n\geq 4$. Hence, we can always find a player j in C such that

$$b_j\leq\frac{n-2}{n-1}\sum_{i\in C_{-j}} b_i$$

For this value of b_j we have $\Omega<0$. This establishes (A1). ■

Claim: In a simple public goods game with linear benefits functions and quadratic costs functions, where marginal benefits b_i and the cost parameter a_i are given for all $i\in N\setminus\{j\}$, the grand coalition is stable under every $r\in\mathcal{R}^o$ if player j's marginal benefit is sufficiently large.

Proof: We show that if a player's marginal benefit is sufficiently large, then the grand coalition ($C = N$) will be stable. By definition the grand coalition is externally stable. We only need to check for internal stability. Consider player j. From (10) we obtain

$$v(C) - \Phi(C) = -\frac{1}{2}\left(\sum_{i \in C_{-j}} b_i\right)^2 \sum_{i \in C_{-j}} \frac{1}{a_i} - \frac{3}{2}\sum_{i \in C_{-j}} \frac{b_i^2}{a_i} + \sum_{i \in C_{-j}} b_i \sum_{i \in C_{-j}} \frac{b_i}{a_i} + \sum_{i \in C_{-j}} b_i^2 \sum_{i \in C_{-j}} \frac{1}{a_i}$$

$$+ \frac{1}{2} b_j^2 \sum_{i \in C_{-j}} \frac{1}{a_i} - b_j \sum_{i \in C_{-j}} b_i \sum_{i \in C_{-j}} \frac{1}{a_i} + b_j \sum_{i \in C_{-j}} \frac{b_i}{a_i}$$

$$-\frac{1}{2}\left(\sum_{i \in C_{-j}} b_i\right)^2 \frac{1}{a_j} + \sum_{i \in C_{-j}} b_i^2 \frac{1}{a_j} \tag{A8}$$

Inspection of (A8) reveals that $v(C) - \Phi(C) \geq 0$ for sufficiently large b_j. ■

Remark: In the simple case where $a_i = 1$ for all $i \in N$, internal stability of the grand coalition is satisfied if and only if there exists j such that

$$b_j \geq \frac{n-2}{n-1} \sum_{i \in C_{-j}} b_i + \sqrt{\left[\frac{(n-2)^2}{(n-1)^2} + \frac{n-2}{n-1}\right]\left(\sum_{i \in C_{-j}} b_i\right)^2 - \frac{2n-3}{n-1} \sum_{i \in C_{-j}} b_i^2} \tag{A9}$$

References

d'Aspremont, C., Jaquemin, A., Gabszewicz, J. J. and Weymark, J. (1983). 'On the Stability of Collusive Price Leadership', *Canadian Journal of Economics*, Vol. 16, pp. 17–25.

Aumann, R. J. (1959). 'Acceptable Points in General Cooperative *n*-person Games', *Annals of Mathematics Studies*, Vol. 40, pp. 287–324.

Barrett, S. (1994). 'Self-enforcing International Environmental Agreements', *Oxford Economic Papers*, Vol. 46, pp. 878–894.

Barrett, S. (2001). 'International Cooperation for Sale', *European Economic Review*, Vol. 45, pp. 1835–1850.

Belleflamme, P. (2000). 'Stable Coalition Structures with Open Membership and Asymmetric Firms', *Games and Economic Behavior*, Vol. 30, pp. 1–21.

Bernheim, B. D., Peleg, B. and Whinston, M. D. (1987). 'Coalition-proof Nash Equilibria I. Concepts', *Journal of Economic Theory*, Vol. 42, pp. 1–12.

Bloch, F. (1996). 'Sequential Formation of Coalitions with Fixed Payoff Division and Externalities', *Games and Economic Behavior*, Vol. 14, pp. 90–123.

Bloch, F. (2003). 'Non-cooperative Models of Coalition Formation in Games with Spillovers', in C. Carraro (ed.), *The Endogenous Formation of Economic Coalitions*, Cheltenham, Edward Elgar, pp. 35–79.

Botteon, M. and Carraro, C. (1997). 'Burden-sharing and Coalition Stability in Environmental Negotiations with Asymmetric Countries', in C. Carraro (ed.), *International Environmental Negotiations: Strategic Policy Issues*, Cheltenham, Edward Elgar, pp. 26–55.

Botteon, M. and Carraro, C. (2001). 'Burden-sharing and Coalition Stability in Environmental Negotiations with Asymmetric Countries', in C. Carraro (ed.),

Environmental Policy, International Agreements and International Trade, Cheltenham, Edward Elgar, pp. 38–65.
Carraro, C. and Marchiori, C. (2003). 'Stable Coalitions', in C. Carraro (ed.), *The Endogenous Formation of Economic Coalitions*, Cheltenham, Edward Elgar, pp. 156–198.
Carraro, C. and Siniscalco, D. (1993). 'Strategies for International Protection of the Environment', *Journal of Public Economics*, Vol. 52, pp. 309–328.
Carraro, C. and Siniscalco, D. (1997). 'Transfers, Commitments, and Issue Linkage in International Environmental Negotiations', in A. Ulph (ed.), *International Environmental Negotiations: Strategic Policy Issues*, Cheltenham, Edward Elgar, pp. 19–37.
Cornet, M. F. (2003). 'Partition Function Bargaining with Public Demands', *Economic Theory*, Vol. 22, pp. 845–862.
Diamantoudi, E. and Sartzetakis, E. S. (2006). 'Stable International Environmental Agreements: an Analytical Approach', *Journal of Public Economic Theory*, Vol. 8, No. 2, pp. 247–263.
Donsimoni, M.-P. (1985). 'Stable Heterogeneous Cartels', *International Journal of Industrial Organization*, Vol. 3, pp. 451–467.
Donsimoni, M.-P., Economides, N. S. and Polemarchakis, H. M. (1986). 'Stable Cartels', *International Economic Review*, Vol. 27, No. 2, pp. 317–327.
Eyckmans, J. and Finus, M. (2004). 'An Almost Ideal Sharing Scheme for Coalition Games with Externalities', *Working Paper 2004-2014*, Katholieke Universiteit Leuven, Center for Economic Studies.
Finus, M., Altamirano-Cabrera, J.-C. and van Ierland, E. C. (2005). 'The Effect of Membership Rules and Voting Schemes on the Success of International Climate Agreements', *Public Choice*, Vol. 125, pp. 95–127.
Fuentes-Albero, C. and Rubio, S. J. (2009). 'Can the International Environmental Cooperation be Bought?', *European Journal of Operational Research*, forthcoming.
Herrero, C., Maschler, M. and Villar, A. (1999). 'Individual Rights and Collective Responsibility: the Rights-egalitarian Solution', *Mathematical Social Sciences*, Vol. 37, pp. 59–77.
Hoel, M. (1992). 'International Environmental Conventions: the Case of Uniform Reductions of Emissions', *Environmental and Resource Economics*, Vol. 2, pp. 141–159.
Hoel, M. and Schneider, K. (1997). 'Incentives to Participate in an International Environmental Agreements', *Environmental and Resource Economics*, Vol. 9, pp. 153–170.
McGinty, M. (2007). 'International Environmental Agreements among Asymmetric Nations', *Oxford Economic Papers*, Vol. 59, pp. 45–62.
Montero, M. (1999). 'Coalition Formation in Games with Externalities', *Center for Economic Research Discussion Paper 99-121*, Tilburg, the Netherlands.
von Neumann, J. and Morgenstern, O. (1944). *Theory of Games and Economic Behavior*, 2nd edn, Princeton, NJ, Princeton University Press.
Poyago-Theotoky, J. (1995). 'Equilibrium and Optimal Size of a Research Joint Venture in an Oligopoly with Spillovers', *Journal of Industrial Economics*, Vol. 43, No. 2, pp. 209–226.
Ray, D. and Vohra, R. (1999). 'A Theory of Endogenous Coalition Structures', *Games and Economic Behavior*, Vol. 26, No. 2, pp. 286–336.
Thoron, S. (1998). 'Formation of a Coalition-proof Stable Cartel', *Canadian Journal of Economics*, Vol. 31, No. 1, pp. 63–76.

Thrall, R. M. and Lucas, W. F. (1963). 'N-person Games in Partition Function Form', *Naval Research Logistics Quarterly*, Vol. 10, pp. 281–298.

Weikard, H.-P., Finus, M. and Altamirano-Cabrera, J.-C. (2006). 'The Impact of Surplus Sharing on the Stability of International Climate Agreements', *Oxford Economic Papers*, Vol. 58, No. 2, pp. 209–232.

Yi, S.-S. (1997). 'Stable Coalition Structures with Externalities', *Games and Economic Behavior*, Vol. 20, pp. 201–237.

Yi, S.-S. (2003). 'Endogenous Formation of Economic Coalitions: a Survey of the Partition Function Approach', in C. Carraro (ed.), *The Endogenous Formation of Economic Coalitions*, Cheltenham, Edward Elgar, pp. 80–127.

[30]

Rev Int Org (2006) 1: 379–396
DOI 10.1007/s11558-006-0162-5

Optimal transfers and participation decisions in international environmental agreements

Carlo Carraro · Johan Eyckmans · Michael Finus

Received: 13 March 2006 / Revised: 29 June 2006 / Accepted: 30 June 2006

Abstract The literature on international environmental agreements has recognized the role transfers play in encouraging participation in international environmental agreements. However, the results achieved so far are overly specific. Therefore, we develop a more general framework that enables us to study the role of transfers in a systematic way. We propose transfers using both internal and external financial resources for making "welfare optimal agreements" self-enforcing. To illustrate the relevance of our transfer scheme, we use a stylized integrated assessment simulation model of climate change to show how appropriate transfers may induce almost all countries into signing a self-enforcing climate treaty.

Keywords self-enforcing international environmental agreements · climate policy · transfers

JEL codes C72 · H23 · Q25 · Q28

1 Introduction

Transfers play a prominent role in the analysis of self-enforcing international environmental agreements (IEAs) (e.g., Carraro & Siniscalco, 1998 and Finus, 2001,

C. Carraro (✉)
University of Venice and Fondazione Eni Enrico Mattei FEEM. FEEM, Palazzo Querini Stampalia, Campo S.Maria Formosa, 30124 Venezia, Italy
e-mail: carraro@feem.it

J. Eyckmans
European University College Brussels EHSAL, Brussels, Belgium
Center for Economic Studies, Katholieke Universiteit Leuven, Leuven, Belgium

M. Finus
Department of Economics, University of Hagen, Hagen, Germany
Department of Economics, National University of Singapore, Singapore, Singapore

2003a). There are two reasons why this is not surprising. First, large asymmetries in the cost and benefit structure between countries may lead to a highly asymmetric distribution of the gains from cooperation that may hamper successful treaty-making. Second, IEAs provide a public good and therefore face strong free-rider incentives. Both of these sources of potential instability of IEAs might be mitigated through the use of appropriate transfers.

The majority of recent contributions to the literature on IEAs uses the tools of non-cooperative coalition theory (see Bloch, 2003 for an overview). The analysis is based on the valuation function that assigns an individual payoff to every coalition member taking into account the entire coalition structure, i.e., the partition of players inside and outside a coalition. The main focus is on explaining free-riding behavior in the context of externalities, identifying the main economic factors that determine the relative success of partial cooperation and suggesting instruments for discouraging free-riding.

In the context of the non-cooperative approach, transfers have been analyzed in their ex-ante and ex-post forms (see Finus, 2003a for an overview). Ex-ante means that countries commit to a certain transfer rule before they decide upon their participation in an IEA. Ex-post means that after an agreement has formed, transfers are used to broaden an existing coalition. The first putative paper on transfers goes back to Carraro and Siniscalco (1993). They analyze various forms of commitment that enables coalitions to expand via ex-post transfers and they suggest two types of transfers that may improve upon the status quo: 1) insiders (coalition members) "bribing" outsiders (non-coalition members) to join their coalition and 2) outsiders "bribing" other outsiders to join the coalition.

The idea of various forms of commitment and ex-post transfers was also pursued in later papers by Botteon and Carraro (1997), Jeppesen and Andersen (1998) and Petrakis and Xepapadeas (1996), though commitment is certainly not an assumption in line with the notion of self-interested players. Therefore, it was important to illustrate that expansions of coalitions via ex-post transfers may also be possible without commitment as this has been done by Botteon and Carraro (1997) in a simple empirical model with five heterogeneous countries.

Later papers have looked at the effect of various ex-ante and ex-post transfers rules on the success of coalition formation and on the possibility of expanding stable coalitions (e.g., Altamirano-Cabrera & Finus, 2006, Bosello, Buchner, & Carraro, 2003, 2004, Carraro & Siniscalco, 2001, Eyckmans & Finus, 2004a, Finus, Sáiz, & Hendrix, 2004 and Weikart et al., 2006). Most of these papers used a more elaborate empirical model, looked not only at stylized transfer schemes derived from cooperative game theory, but also considered schemes that are based on various moral motives for "fair sharing," considered transfers via permit trading and allowed for the possibility of multiple coalitions. Roughly speaking, all papers basically confirm earlier studies in concluding that transfers can be conducive to the success of self-enforcing agreements, but that outcomes crucially depend on the particular transfer rule, the details of the model and the data set.

The mixed evidence and the specific nature of results motivate us to look for a more general approach to study the role of transfers in the context of non-cooperative coalition theory. In particular, we want to determine the "full potential of transfers." To this end, we go back to the roots of the analysis of IEAs, removing in this paper any unnecessary complication as a first step. That is, we assume a simple cartel formation game and apply the concept of internal and external stability. We employ a notion of "optimal" transfer schemes meaning transfers designed to maximize global welfare under the constraint that the underlying IEA is self-enforcing.

The importance of this research issue for policy making is evident when considering for instance the crucial role financial or technology transfers are playing in recent discussions on possible post-Kyoto climate policies in the framework of the United Nations Framework Convention on Climate Change (UNFCCC). First, multilateral financial transfers take place under the so-called Global Environment Facility (GEF) program. Though not large in absolute terms, GEF allocates and disburses currently about 250 million dollars per year in projects for improving energy efficiency, fostering renewable energy use, and pushing for sustainable transportation. Second, the flexible mechanisms (i.e., Emissions Trading ET, Joint Implementation JI and Clean Development Mechanism CDM) foreseen in the Kyoto Protocol are expected to generate important flows of financial resources among the Protocol member states and even between signatories and non-signatories in the case of CDM projects. These financial and flexibility mechanisms can be interpreted as multilateral transfer schemes. Therefore, it is important to get new insights on how transfer schemes should be designed in order to achieve maximum success of future, voluntary international climate policy agreements.

In what follows, we present our model in "Section 2." This comprises not only a theoretical but also an empirical part. The empirical part is based on a modified version of RICE (Nordhaus & Yang, 1996)—a well-known integrated assessment model of climate change policy—which is used to illustrate our concepts. In "Section 3," we analyze the impact of different transfer designs on the formation of stable coalitions. "Section 4" concludes.

2 Model

2.1 Theoretical Background

Coalition formation is modeled as a two-stage game. There are n players and $N = \{1, ..., n\}$ denotes the set of countries or regions which we simplify refer to as countries in the sequel. In the first stage, countries choose their membership: a country can either join coalition $S \subseteq N$ and become a signatory or remain a singleton and non-signatory. These decisions lead to coalition structure $C = \{S, \{\ell\}, \ldots, \{n\}\}$, i.e., a partition of players, with s signatories (s denotes the cardinality of S) and n-s non-signatories. Given the simple structure of the first stage, a coalition structure C is fully characterized by coalition S.[1]

In the second stage, countries choose their economic strategies. Economic strategies include for instance policy variables like emission abatement and capital investment as described in "Section 2.2." At this stage, it suffices to denote the vector of economic strategies by $\omega(S) = (\omega_1(S), ..., \omega_n(S))$ with associated individual payoffs $\pi_i(\omega(S))$, given that a coalition S has formed in the first stage.

We compute the subgame-perfect equilibria of this two-stage game by backward induction.[2] To do this, it is sufficient for strategies to constitute a Nash equilibrium at every stage. For the second stage, this entails that economic strategies form a Nash equilibrium between coalition S and the n-s non-signatories. This implies that non-signatories $i \notin S$ choose their economic strategies so as to maximize their individual payoff $\pi_i(\omega)$, whereas all signatories $i \in S$ jointly maximize the aggregate payoff of their coalition, $\sum_{i \in S} \pi_i(\omega)$. Strategically, this means that the behavior of non-signatories towards all other countries is selfish and non-cooperative; signatories behave cooperatively towards their fellow members, but non-cooperatively towards outsiders. Economically, this means strategies are efficient within coalition S (but not globally). Hence, the equilibrium economic strategy vector $\omega^*(S)$ corresponds to the "social optimum" if coalition S comprises all countries $(S = N)$, i.e., the grand coalition forms, and corresponds to the "Nash equilibrium" if coalition S comprises only one member $(S = \{i\})$. Hence, any inefficiency, i.e., global welfare loss compared to the global Pareto optimum, stems from the fact that S is not the grand coalition.

Given that the second stage of the game has been solved, we define $v_i(S) = \pi_i(\omega^*(S))$ as the valuation of country i if coalition S forms. This definition summarizes all information relevant to the second stage.

For the first stage, we define a Nash equilibrium in terms of participation:[3]

$$\text{internal stability}: \quad v_i(S) \geq v_i(S \backslash \{i\}) \quad \forall i \in S. \tag{1}$$

$$\text{external stability} \quad v_i(S) \geq v_i(S \cup \{i\}) \quad \forall i \notin S. \tag{2}$$

That is, in equilibrium, no signatory belonging to coalition S has an incentive to leave its coalition in order to become a non-signatory, given the participation decisions of all other countries. By the same token, no non-signatory has an incentive to join coalition S, given the decisions of all other countries. Note that by definition coalition $S = \{i\}$ is internally stable and coalition $S = N$ externally stable by definition.

[1] The assumption of a single coalition seems to be the most obvious and realistic in the context of IEAs. Therefore, it has been frequently adopted in the literature. Exceptions are considered in Bosello et al. (2003, 2004), Carraro (2000), Eyckmans and Finus (2006) and Finus (2003b).

[2] Our assumption is in line with the mainstream of the literature on coalition theory. For an overview see Bloch (2003) and Yi (2003).

[3] This definition is due to d'Aspremont et al. (1983) and has been frequently applied in the literature on IEAs as for instance by Barrett (1994), Carraro and Siniscalco (1993), Hoel (1992) and by many scholars afterwards.

Regardless of whether we consider ex-ante or ex-post transfers, we assume that optimal economic strategies are not affected by transfers.[4] Valuations with transfers $\hat{v}_i(S)$ are related to those without transfers $v_i(S)$ simply through the relation $\hat{v}_i(S) = v_i(S) + t_i$ where $t_i > 0$ means receiving and $t_i < 0$ means paying a transfer. We make the standard assumption that transfers balance, i.e., $\sum_{i \in N} t_i = 0$ and hence $\sum_{i \in N} \hat{v}_i(S) = \sum_{i \in N} v_i(S)$ (e.g., Carraro & Siniscalco, 1993).

2.2 Simulation Model

In order to illustrate the importance of transfers for the success of coalition formation, we derive valuations from the CLIMNEG World Simulation Model (hereafter abbreviated as CWSM). CWSM is an integrated assessment, economy-climate model that captures the endogenous feedback of climate change damages on production and consumption.[5] The strategic policy variables in the CWSM are investment and carbon emission reduction.

In the CWSM, the world is divided into six regions: *USA*, *JPN* (Japan), *EU* (European Union), *CHN* (China), *FSU* (Former Soviet Union) and *ROW* (Rest of the World). In every region i, and at every time t, the following budget equation describes how "potential GDP," $Y_{i,t}$, can be allocated to consumption, $Z_{i,t}$, investment, $I_{i,t}$, emission abatement costs, $Y_{i,t}\, C_i\,(\mu_{i,t})$, and climate change damages, $Y_{i,t}\, D_i\,(\Delta T_t)$:

$$Y_{i,t} = Z_{i,t} + I_{i,t} + Y_{i,t}\, C_i(\mu_{i,t}) + Y_{i,t}\, D_i(\Delta T_t) \tag{3}$$

Output $Y_{i,t}$ is produced with capital and labor. Capital is built up through investment which is the first strategic policy variable in the model. Labour supply is assumed to be inelastic. Abatement costs $Y_{i,t}\, C_{i,t}\,(\mu_{i,t})$ are expressed as "loss of potential GDP": C_i is the share of "potential GDP" devoted to abatement, which is a function of $\mu_{i,t} \in [0,1]$, a variable that measures the relative emission reduction compared to the business-as-usual scenario without any abatement policy. Damages $Y_{i,t}\, D_i\,(\Delta T_t)$ are also expressed as "loss of potential GDP": D_i is the share of "potential GDP" destroyed by climate change damages, which is a function of temperature change ΔT_t. Temperature change depends on the stock of greenhouse gases, which in turn depends on emissions that accumulate in the atmosphere. Finally, emissions are proportional to production, but can be reduced by the abatement rate $\mu_{i,t}$ which constitutes the second strategic choice variable in the model.

Both choice variables (investment and abatement) affect output, abatement costs, damage costs and therefore also consumption, not only domestically but also abroad. This is immediately evident with regard to abatement because remaining

---

[4] That is, we follow the standard assumption of a transferable utility (TU) framework in the literature on coalitions for simplicity. This means to neglect income effects induced by transfers. For our simulation model this seems a reasonable assumption, given the relatively small size of transfers compared to total GDP.

[5] Almost all parameter values are taken from the RICE model of Nordhaus and Yang (1996). A complete overview of the equations and parameter values of the CWS model can be found in Eyckmans and Tulkens (2003). The CWS model code and and GAMS program to check coalition stability is available at http://www.climneg.be.

emissions (after abatement) increases the stock of greenhouse gases, which affects environmental damages in every country. However, it is also true for investment, since capital is an input in the production process and emissions are proportional to production.

For every of the 58 possible coalitions S, we compute the Nash equilibrium $\omega^*(S)$ between S and $N\backslash S$ in order to derive valuations $v_i(S)=\pi_i(\omega^*(S))$. That is, we determine investment and abatement strategies that maximize welfare (i.e., lifetime discounted consumption) of coalition S and of singletons in $N\backslash S$ for given investment and abatement strategies of the other players as described in "Section 2.1."[6]

2.3 Basic Incentive Structure

In this section, we briefly discuss the basic incentive structure implied by the calibration of CWSM. First, USA, JPN and EU face steep abatement cost curves, while CHN, FSU and ROW face flat ones. The regional differences in abatement costs mainly reflect differences in energy efficiency. Energy efficient regions face higher marginal costs when cutting back emissions than regions characterized by low energy efficiency because they have already exploited the cheapest energy saving techniques.

Second, damage functions are particularly steep in EU and ROW, less steep in USA and JPN and relatively flat in FSU and CHN. The high damage estimate (as a percentage of "potential GDP") for ROW is due to the fact that climate change is believed to affect developing countries strongly because their economies tend to depend on climate related production processes like agriculture, fishery and forestry (IPCC, 2001).[7] The high damage of EU is due to a high potential GDP; the low damage estimate for FSU is due to some expected benefits from moderate temperature increase, like the increased availability of arable land.

Third, in a given coalition S, the steeper the marginal damage cost curves and the flatter the marginal abatement cost curves of the members of S are, the higher the optimal abatement of coalition members will be, which follows from the first-order conditions of joint welfare maximization of coalition S. In every period, coalition members should abate up to the point where their marginal abatement costs are equal to the discounted sum of all coalition members' avoided future marginal climate change damages.

Fourth, coalition members with a flatter marginal abatement cost curve have to contribute more than those with a steeper curve, all else being equal, which also follows from the first order conditions, implying cost-effectiveness within coalition S.

[6] This implicitly assumes that some harmonized carbon tax or a tradable permit system is implemented within coalition S such that every polluting firm in S is confronted with the coalitional damage costs of its emission behaviour. Thus, the policy instruments coordinates private investment and emission reduction decisions within coalition S.

[7] Because ROW comprises many heterogenous countries, the climate change damage parameters have been revised downward in all coalition structures in which ROW acts as a singleton following Nordhaus and Yang (1996). Without this modification, non-cooperative emission control rates by ROW would be unrealistically high.

2.4 Properties of Valuations

To conclude this section, we discuss two general properties which determine the incentive structure in the coalition formation game. Both properties hold for the valuations derived from our simulation model CWSM. The first property is called superadditivity and means that the aggregate valuation of country j and coalition S increases if country j joins coalition S.

Definition 1: Superadditivity

A coalition game is superadditive if and only if for all $S \subset N$ *and* $j \notin S$*:*

$$\sum\nolimits_{i \in S \cup \{j\}} v_i(S \cup \{j\}) > \sum\nolimits_{i \in S} v_i(S) + v_j(S).$$

That is, there is "coalitional gain" from cooperation and hence cooperation is "group rational" or "*coalitionally rational*". This makes cooperation attractive for those countries participating in an IEA. This property means that starting from any coalition S and increasing the degree of cooperation by moving to $S \cup \{j\}$ or even larger coalitions, it is generally possible to allocate the coalition gain such that it constitutes a Pareto-improvement to all regions involved in cooperation.

The second property is called positive spillovers, meaning that if country j joins coalition S, all countries that do not belong to $S \cup \{j\}$ are better off.

Definition 2: Positive Spillovers

A coalition game exhibits positive spillovers if and only if for all $S \subset N, j \notin S$ *and all* $\ell \notin S \cup \{j\} : v_\ell(S \cup \{j\}) > 1 v_\ell(S)$.

Consequently, there is an "external gain" or a positive spillover from cooperation, making free-riding (i.e., not joining S) attractive. From a non-signatory's point of view, the most favorable condition is the one in which all other countries participate in the agreement.[8]

It is then clear that a region's decision to join a coalition—as well as the stability of an IEA—depends on the magnitude of both properties. On the one hand, the intensity of the superadditivity effect which, together with the sharing rule of the coalitional gain, determines the payoff of cooperation. On the other hand, the intensity of the positive externality effect which determines the free riding payoff. We will study these effects in more detail in "Section 3," but note here that superadditivity and positive externality together imply that global welfare increases through cooperation. That is, given a coalition S, whenever a single or several countries join coalition S, global welfare is raised because both the coalition members (through superadditivity) and the outsiders (through positive spillovers) benefit.

Table 1 illustrates the magnitudes at stake for the CWSM model. It displays—for a selection of coalitions—global welfare and two environmental

[8] The importance of positive (and negative) spillovers has been stressed in the recent literature on coalition theory (Bloch, 2003, Yi, 2003 and Maskin, 2003).

variables (carbon concentration and global emissions) in absolute and relative terms. The relative magnitudes can be interpreted as a "closing the gap index," measuring how close a coalition comes to the global optimum where the performance in the global optimum is 100 percent and the performance with no cooperation is 0 percent by normalization. Apart from stressing that both full and partial cooperation can make a large difference in welfare and ecological terms compared to no cooperation, Table 1 illustrates that not only the size of a coalition matters for the global success of cooperation as suggested by Barrett (1994). In the context of our model with heterogenous players, the identity of members is more important. For instance, coalition no. 32 including five members (USA, JPN, EU, CHN, FSU) ranks lower than many coalition structures comprising coalitions of only three or four members and even lower than coalition no. 31 with only two members (JPN and ROW).

From Table 1 it is also clear that, as a general tendency, the importance for *global* welfare of participation of particular countries decreases with the following sequence: ROW, CHN, EU, USA, FSU and JPN. ROW's and CHN's important role stems from the fact that they are characterized by a relatively flat emission abatement cost curve; hence they can provide cheap abatement. However, apart from emission abatement costs, also environmental damages matter. If a given coalition maximizes its joint welfare, the higher the marginal damages of coalition members are, the higher joint abatement efforts will be (and the higher are positive spillovers), all else being equal. This explains the global importance of EU for cooperation. Summarizing, cooperation between countries with low marginal abatement costs (i.e., cheap suppliers of abatement effort) and countries with high marginal willingness to avoid climate change (i.e., high demand for abatement effort), proves important to achieve substantial emission reductions in an IEA.[9]

3 Stable Coalitions

In this section, we analyze coalition formation both under the assumption of no transfers and various forms of transfers. Results and arguments are illustrated with the CWSM model. In "Section 3.1," we start with no transfers. "Section 3.2" deals with ex-ante transfers and shows the advantage of optimal transfers, as defined in Eyckmans and Finus (2004b), compared to simple transfer schemes, as assumed in most papers up to now. "Section 3.3" analyzes the expansion of coalitions via ex-post transfers.

3.1 No Transfers

In the case of no transfers, there is no coalition that is both internally and externally stable at the same time, see Table 2. In the absence of transfers, although co-

[9] Similar conclusions can also be found in Buchner, Carraro, and Cersosimo (2002).

Table 1 Welfare and Environmental Implications of Different Coalitions*

Coalition			Welfare	Concentration	Cumulative Emissions
N°	Size	Membership			
1	6	Grand Coalition (Full Cooperation)	100.0	100.0	100.0
2	5	USA, EU, CHN, FSU, ROW	99.1	92.2	93.0
3	5	USA, JPN, EU, CHN, ROW	96.6	90.0	91.1
4	5	JPN, EU, CHN, FSU, ROW	95.6	80.6	81.9
5	4	USA, EU, CHN, ROW	94.5	82.0	83.2
6	5	USA, JPN, CHN, FSU, ROW	93.2	73.2	74.8
7	4	EU, CHN, FSU, ROW	91.3	72.3	73.6
8	4	JPN, EU, CHN, ROW	89.6	69.8	71.5
9	4	USA, CHN, FSU, ROW	87.4	64.1	65.7
10	4	USA, JPN, CHN, ROW	85.9	61.8	63.9
11	3	EU, CHN, ROW	85.9	61.8	62.6
12	3	USA, CHN, ROW	84.0	60.7	62.6
13	4	JPN, CHN, FSU, ROW	78.4	50.3	52.6
14	5	USA, JPN, EU, FSU, ROW	70.3	66.0	67.1
15	4	USA, EU, FSU, ROW	69.1	61.0	62.0
	...	...	...	...	...
31	2	JPN, ROW	46.4	24.7	26.8
32	5	USA, JPN, EU, CHN, FSU	31.0	26.9	27.5
33	4	USA, EU, CHN, FSU	29.0	24.5	25.0
...	...	...	...	...	...
57	2	JPN, EU	0.6	0.2	0.3
58	1	Only Singleton Coalitions (No Cooperation)	0.0	0.0	0.0

**N°*: coalition number according to welfare ranking. *Size*: number of coalition members. *Membership*: composition of coalition. *Welfare*: global welfare expressed in relative terms: $\left(\sum_{i=1}^{N}\left(w_i(c^P) - w_i(c^N)\right)\right)/\left(\sum_{i=1}^{N}\left(w_i(c^F) - w_i(c^N)\right)\right)$ where welfare is discounted lifetime consumption integrated over 19902300, global welfare with full cooperation is $\sum_{i=1}^{N} w_i(c^F) = 339{,}831$ bill US\$$_{1990}$ (billion US dollars expressed in 1990 levels), global welfare with no cooperation is $\sum_{i=1}^{N} w_i(c^N) = 338{,}060$ bill US\$$_{1990}$ and global welfare with partial cooperation is denoted by $\sum_{i=1}^{N} w_i(c^P)$. *Concentration*: atmospheric carbon concentration M at time $t = 2{,}300$ expressed in relative terms: $\left(M(c^N) - M(c^P)\right)/\left(M(c^N) - M(c^F)\right)$ where concentration with full cooperation is $M(c^F) = 1{,}912.907$ GtC (giga tons carbon), concentration with no cooperation is $M(c^N) = 4{,}550.202$ GtC and concentration with partial cooperation is denoted by $M(c^P)$. *Cumulative Emissions*: cumulative emissions of carbon over time interval 19902300 expressed in relative terms: $\left(\mathrm{CE}(c^N) - \mathrm{CE}(c^P)\right)/\left(\mathrm{CE}(c^N) - \mathrm{CE}(c^F)\right)$ where cumulative emissions with full cooperation are $\mathrm{CE}(c^F) = 772.529$ GtC, cumulative emissions with no cooperation are $\mathrm{CE}(c^N) = 1{,}593.398$ GtC and cumulative emissions with partial cooperation are denoted by $\mathrm{CE}(c^P)$.

operation may be coalitionally and globally rational, it may not be individually rational to all coalition members. The reason is that an efficient allocation of abatement burdens within a coalition S would result in a highly asymmetric allocation of the net gains from cooperation among coalition members that face a markedly heterogeneous benefit and cost structure. For instance, regions facing a flat marginal abatement cost and damage cost curve are typically losers from cooperation without transfers. These regions are required to contribute substantial abatement efforts but they do not value strongly the ensuing improvement in environmental quality. Thus, without transfers, China faces the strongest free-rider incentives among all six regions in CWSM.

Table 2 Stable Coalition Structures Under Different Transfer Schemes*

	N°	Membership	Size	Welfare	Concentration	Cumulative Emissions
	1	Grand Coalition (Full Cooperation)	6	100.00	100.00	100.00
No Transfers	–	–	–	–	–	–
Shapley	27	CHN,ROW	2	54.57	21.36	25.19
Nash Bargaining	16	CHN,FSU,ROW	3	68.21	39.15	41.63
Chander-Tulkens	26	EU,ROW	2	57.42	40.79	42.11
	27	CHN,ROW	2	54.57	21.36	25.19
	29	USA,ROW	2	54.07	35.18	36.70
	30	FSU,ROW	2	47.18	26.75	28.41
	31	JPN,ROW	2	46.39	24.72	26.80
OPTS	5	USA,EU,CHN,ROW	4	94.50	81.96	83.18
	7	EU,CHN,FSU,ROW	4	91.17	72.26	73.61
	8	JPN,EU,CHN,ROW	4	89.41	69.75	71.53
	9	USA,CHN,FSU,ROW	4	87.31	64.08	65.75
	10	USA,JPN,CHN,ROW	4	85.99	61.80	63.91
	13	JPN,CHN,FSU,ROW	4	78.28	50.29	52.58
	15	USA,EU,FSU,ROW	4	68.96	61.02	62.01
	18	USA,JPN,EU,ROW	4	66.80	59.47	60.54
	19	JPN,EU,FSU,ROW	4	66.12	53.62	54.78
	21	USA,JPN,FSU,ROW	4	64.67	48.90	50.21
	58	Only Singleton Coalitions (No Cooperation)	1	0.00	0.00	0.00

*The same legend as with Table 1 applies.

3.2 Ex-Ante Transfer Schemes

In the light of the above negative conclusion, we now turn to ex-ante transfer schemes. That is, the membership decision in the first stage of the game is based on the assumption that coalition S will not only choose its optimal economic strategies in the second stage, but will also allocate the coalition gain from cooperation among its members according to a particular transfer scheme. We start by considering three transfer schemes that have played an important role in previous analyses of self-enforcing IEAs (see the literature cited in the Section 1). We call these schemes "simple" in order to distinguish them from our "optimal" transfer schemes that we introduce subsequently. Through the illustration of simple and optimal transfer schemes, the full potential of an optimal design of transfers will become apparent.

3.2.1 Simple Transfer Schemes

All three simple transfer schemes that we consider originate from cooperative coalition theory. It requires only a slight modification of their original definitions to account for the fact that coalition S may not only be the grand coalition but any subcoalition of N (Aumann & Drèze, 1974). The following formulas describe valuations of player i being a member of a given coalition $S \subseteq N$.

The first transfer scheme is the *S*hapley *V*alue and implies valuations of the following form:

$$\hat{v}_i^{\mathrm{SV}} = \sum_{T \subseteq S_{i \notin T}} \frac{t!(s-t-1)!}{s!} \left[\sum_{k \in T \cup \{i\}} v_k(T \cup \{i\}) - \sum_{k \in T} v_k(T) \right] \qquad \forall i \in S \qquad (4)$$

with coalition $S \subseteq N$, $T \subseteq S$ a subgroup of S and t and s the size of group S and T. Roughly speaking, the *S*hapley *V*alue gives every country a valuation according to its average marginal contribution over every possible subcoalition T of S (term between square brackets in (4)), weighted by the probability that this subpartition forms (first factor in (4)).

The second simple transfer scheme is the *N*ash *B*argaining solution (with equal weights):

$$\hat{v}_i^{\mathrm{NB}} = v_i(\{i\}) + \frac{1}{s} \left[\sum_{j \in S} v_j(S) - \sum_{j \in S} v_j(\{i\}) \right] \qquad \forall i \in S \qquad (5)$$

Every member in S receives its valuation under no cooperation (first term) plus an equal share of the coalitional surplus compared to no cooperation (second term). Thus, no cooperation serves as the threat point.

The third simple transfer scheme is the *C*hander and *T*ulkens' transfer scheme (Chander & Tulkens, 1997) in the version as applied in Eyckmans and Tulkens (2003):

$$\hat{v}_i^{\mathrm{CT}} = v_i(\{i\}) + \frac{D_j^{'}}{\sum_{j \in S} D_j^{'}} \left[\sum_{j \in S} v_j(S) - \sum_{j \in S} v_j(\{i\}) \right] \qquad \forall i \in S \qquad (6)$$

with $D_i^{'}$ discounted marginal damages of member i. It is evident that this scheme is a version of the Nash bargaining rule with unequal weights. This rule gives a higher share of the gains from cooperation to those that are most affected by climate change.

It is straightforward to show that all three simple transfer schemes are all coalitionally rational, i.e., $\forall S \subseteq N : \sum_{j \in S} \hat{v}_j(S) \geq \sum_{j \in S} v_j(S)$, and individually rational, i.e., $\forall i \in S : \hat{v}_i(S) \geq v_i(\{i\})$ since superadditivity holds (and hence the term in square brackets is positive).

The results of our simulations, as summarized in Table 2, confirm our intuition and also many previous studies on transfers: simple transfers improve upon the outcome without transfers. However, it is important to realize that this result is by no means general. First, we cannot generally conclude that simple transfer schemes would always enhance stability. Second, we cannot conclude that the Nash bargaining solution performs better than the other transfer schemes. Though in our example, stable coalition no. 16 generates higher global welfare than in any other stable coalition under the other two transfer schemes, other parameter values or other models could yield different conclusions. Finally, and most importantly, we have no information about whether we could do any better than the simple transfer schemes considered here and if so what the "best" transfer scheme would be. This question will be addressed in the following section.

3.2.2 Optimal Transfer Schemes (OPTS)

In order to address the issues raised above in a systematic way, we first focus on internal stability. For this purpose, we introduce the concept of a "Potentially Internally Stable Coalition" (PISC) which we define as follows (see Eyckmans and Finus, 2004b and Botteon and Carraro (1997) for a similar concept):

Definition 3: Potentially Internally Stable Coalition

A coalition S is said to be potentially internally stable (PIS) if and only if $\sum_{i\in S} v_i(S) \geq \sum_{i\in S} v_i(S\backslash\{i\})$.

Thus, if a coalition S is PIS, this means that it has sufficient resources to guarantee each of its members at least its free riding payoff. Hence, if S is PIS there does always exist a transfer scheme that can ensure internal stability to all members of S. All what is required to construct such a transfer scheme is to give every member of S at least his free-rider payoff or outside option $v_i(S\backslash\{i\})$. The remaining surplus can then be shared in a proportional way:

Definition 4: Optimal Transfer Schemes

A transfer scheme is called optimal if it satisfies

$$\forall S \subseteq N, \forall i \in S: \hat{v}_i^{\mathrm{OP}}(S) = v_i(S\backslash\{i\}) + \lambda_i(S)\left[\sum_{j\in S} v_j(S) - \sum_{j\in S} v_j(S\backslash\{i\})\right]$$
$$\text{with } \lambda(S) \in \Delta^{s-1} = \left\{\lambda \in {}^{s}_{+} \middle| \sum_{j\in s} \lambda_j = 1\right\}.$$

By the construction of OPTS, it is easy to see that any transfer scheme that belongs to the class of OPTS will make any PISC internally stable. The design of OPTS suggests that we have much leeway in choosing weights λ_i (S). As long as the surplus of cooperation over the free-riding valuations, as well as all weights λ_i (S), are positive, the resulting allocation will be internally stable regardless of the choice of weights. This means that the set of internally stable coalitions does not depend on the choice of weights. Interesting, though less evident, this robustness result also carries over to the set of *externally* stable coalitions. Intuitively, through OPTS a link between internal and external stability is established. If S is PIS, OPTS makes it internally stable and hence all coalitions S without a member i are externally unstable. If S is not PIS, S will not be internally stable and S without i is externally stable.

Result 1: Invariance of stability for surplus sharing weights

Any family of weights λ_i (S) of an OPTS will lead to the same set of internally and externally stable coalitions after transfers.

A second result is that the coalition with the highest global welfare among all PISC, say coalition S^*, is not only internally stable (because it is PISC) but also externally stable and hence stable.

Result 2: Welfare optimality and stability

Under OPTS, the coalition with the highest global welfare among all PISC is stable.

Intuitively, if S^* were not externally stable, then there would exist a larger PISC with higher global welfare (because of superadditivity and positive spillovers), violating the initial assumption that S^* generates the highest global welfare.

Result 1 and 2 are proved in Eyckmans and Finus (2004b). They are illustrated in Table 2 with our simulations. First, the three simple transfer schemes lead to a different set of stable coalitions. In particular, despite the Nash Bargaining solution and the Chander and Tulkens's transfer scheme only differ in weights, they produce very different results. In contrast, any set of weights for OPTS leads to the same set of stable coalitions as displayed in Table 2. Second, OPTS exploits the full potential of self-enforcing cooperation. In our example, OPTS achieves 94.5 percent of total maximum welfare for coalition no. 5 {USA, EU, CHN, ROW} whereas for the simple transfer schemes the maximum is achieved under the Nash Bargaining solution with only 68.7 percent.

3.3 Ex-Post Transfer Schemes

In this subsection, we consider ex-post transfers. This means that after a coalition S has formed, transfers are used to modify the status quo. The status quo is a stable coalition that has emerged from the coalition formation process without transfers or with some ex-ante transfer scheme. In our example, these are the coalitions listed in Table 2. Following Carraro and Siniscalco (1993), we cite two cases. In the first case, a coalition S uses transfers to expand its agreement by "bribing" outsiders to join the coalition; in the second case, an outsider $j \notin S$ uses transfers to bribe another outsider $k \notin S$ to join coalition S.

3.3.1 Expansion of Coalitions through Internal Means

The standard procedure to analyze the expansion of coalitions through internal means is to pick a stable coalition S as starting point and to check whether expansion of this coalition is possible when current coalition members pay some outsider j for joining them (Botteon & Carraro, 1997). It is natural to require that expansion of coalition S is feasible if and only if 1) the expansion constitutes a Pareto-improvement to all members of S and for the newcomer j, and 2) the enlarged coalition $S \cup \{j\}$ is internally stable.

The first requirement of a Pareto improvement follows from the presumption that current members of S will only bribe an outsider to join if they are better off after the expansion. Likewise, region j will only accept the deal if it can improve itself. The second requirement simply follows from the definition of stability. Now, it turns out that the first requirement is the condition of superadditivity, which holds in our global emission game (as in many other models), and hence it is non-binding. For the second requirement it can be shown that a necessary condition is that the enlarged coalition $S \cup \{j\}$ is potentially internally stable. Since coalition S^* with the highest global welfare among all PISC is stable under (ex ante) OPTS, this means that no expansion via internal means is possible from this coalition to form $S^* \cup \{j\}$.

Result 3: OPTS exhausts all possibilities for expansion by internal means

Coalition S^*which achieves the highest global welfare among all PISC cannot be improved upon by expansions using internal means.

Result 3 is proved in the Appendix. It implies that coalition S^* can be regarded as the best possible outcome that can be achieved by means of expansion using internal means. This can also be interpreted that if an optimal ex ante transfer scheme is employed, the analysis of ex post transfers to enlarge a coalition with internal means is redundant. In our example, this means that we cannot do better than coalition no. 5, i.e., {USA, EU, CHN, ROW}.

3.3.2 Expansion of Coalitions by External Means

We now turn to the question of whether the expansion of a coalition S is possible through external means. This means that an outsider $k \notin S$ "bribes" another outsider $j \notin S$ to join coalition S. Stable coalitions for which expansion via internal means are not possible are the obvious, although not exclusively, potential candidates for expansions via external means.[10]

Again, two conditions must be satisfied that expansions is successful. First, expansion must constitute a Pareto-improvement to all involved players, which we know is not binding if superadditivity holds. Second, the enlarged coalition $S \cup \{j\}$ must be internally stable. Since by assumption enlargement starts from S which is internally and externally stable, under OPTS, $S \cup \{j\}$ is not PIS. Consequently, expansion is only possible if and only if the enlarged coalition receives sufficient transfers to compensate for the lack of PIS and, in addition, the external player is better off despite providing these resources. That is,

$$\sum_{i \in S \cup \{j\}} v_i(S \cup \{j\}) + \sum_{i \in S \cup \{j\}} t_i \geq \sum_{i \in S \cup \{j\}} v_i(S \cup \{j\} \backslash \{i\}) \tag{7.a}$$

$$v_k(S \cup \{j\}) + t_k \geq v_k(S) \tag{7.b}$$

[10] As long as expansion via internal means is possible, it is very unlikely that expansion via external means are conducted. This is because outsiders benefit from expansion via internal means through positive spillovers without any contribution, knowing that this expansion is in the interest of all current members of coalition S.

must hold where typically $t_i \geq 0$ and $t_k \leq 0$ and due to budget neutrality $\sum_{i \in S \cup \{j\} \cup \{k\}} t_i$. That is, the positive spillover effect accruing to region k—$v_k(S \cup \{j\}) - v_k(S) \geq 0$—must be larger than the free-riding effect—$\sum_{i \in S \cup \{j\}} v_i(S \cup \{j\} \backslash \{i\}) - \sum_{i \in S \cup \{j\}} v_i(S \cup \{j\}) > 0$—when coalition S is expanded to coalition $S \cup \{j\}$. In general, we cannot derive conclusive theoretical results using general properties like superadditivity or positive spillovers for expansions using external means. Therefore, conditions (7.a) and (7.b) have to be checked for all possible expansions starting from a PIS coalition.

Result 4: Possibility of expansion using external means

Starting at a stable coalition S, expansion using external means of S is possible only if the positive spillover to the external player k∈N\S that pays another outsider j∈N\S to join S, is greater than the increase in the free-rider effect of the members of the expanded coalition S ∪ {j}.

Figure 1 shows all possibilities of expansion from coalitions with four members that are stable in our example. Figure 1 illustrates that from coalition no. 5, which is the coalition with the highest global welfare using internal means, no expansion via external means is possible. However, from some other coalitions with four members, expansion that raises global welfare is possible. It is interesting to note that only the US and the EU are potential candidates that have sufficient resources to pursue a unilateral policy of "bribing" other countries to participate in a climate agreement. Given that coalition no. 5 can already be achieved without external means, it is only JPN, FSU and the US that should pursue such a policy in the interest of the global good, though among this group only the US has the means and the incentive to do so. The maximum "degree of optimality" which can be achieved by means of a clever transfer strategy is 95.6 percent (coalition no. 4 comprising all countries except USA). Of course, full participation cannot be achieved since expansion via external means requires that at least one country remains an outsider.

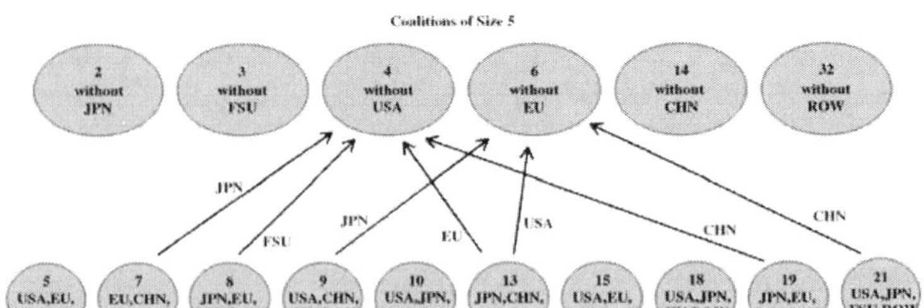

Fig. 1 Enlarging participation through external means. *Bold numbers* refer to the ranking of coalitions in terms of global welfare; description of coalition at level 5 means a coalition of 5 members, i.e., full participation without region *j*; description of coalition at level 4, the description includes the 4 coalition members that are listed; *arrow* means enlargement through bribing is possible from coalition at level 4 to coalition at level 5; region *i* attached to an arrow is the region that can be bribed to join coalition at level 4 to form coalition at level 5 through region *j* listed under coalition structure at level 5 (i.e., "without *j*").

More generally, our results stress that—in the context of incentives to free-ride—countries can play an important role, even if they do not actively participate in an IEA. By subsidizing emission abatement projects abroad, those outsiders might help other outsiders to become a formal member of an IEA and to comply with its requirements. This result may be particularly useful for the design of future IEAs like for instance post-Kyoto negotiations.

4 Summary and Conclusions

The non-cooperative game theoretic literature on international environmental agreements has largely neglected the role of transfers as a tool to enhance participation in international treaties. The few existing results on transfers and international environmental agreements (IEAs) are very specific and do not exploit the full potential of transfers for successful treaty-making.

In this paper, we have proposed a transfer mechanism that is capable of maximizing global welfare given that an IEA has to be internally and externally stable. This result can be achieved by using either ex-ante or ex-post transfers where in the latter case we distinguished between internal and external transfers. We have shown that ex-post internal transfers are redundant if ex-ante transfers are optimally designed. By contrast, ex-post external transfers can help to achieve a welfare improving, self-enforcing IEA (compared to a situation without ex-post transfers). All results depend only on very general properties of the underlying valuation functions, in particular, only on the properties superadditivity and positive spillovers, which hold for most public good models, not only those that deal with environmental issues.

To show the relevance of the theoretical results, we used a stylized integrated assessment model of climate change policy. Our simulation results highlighted the difficulty of reaching an effective agreement on climate policy because of the large asymmetries across countries and incentives to free-ride on the provision of the public good "climate change control". We showed that the situation can be largely improved by adopting an appropriate transfer scheme. In the case of transfer using internal resources, our "optimal" transfer scheme leads to far better results than the "simple" transfer schemes well-known in existing literature. In the case of transfers using external resources, we demonstrated that it is possible to reach a self-enforcing agreement involving all but one country but where this outsider contributes financially, in its self interest, to stabilize a successful agreement.

Both results have important implications for post-Kyoto climate policy negotiations. On the one hand, they stress how important is to devote sufficient time to the design of transfer schemes so that participation can be raised in the future. On the other hand, they highlight the important role that current outsiders like the USA and Australia can play. Although they did not ratify the Kyoto Protocol and may not even join this agreement at a later stage, they might nevertheless be involved in future protocols via new financial mechanisms through which they could sponsor developing countries to promote energy efficiency and, eventually perhaps to assume quantitative emission ceilings and to become a full member of the Kyoto Protocol. These transfers may well be in the interest of the outsiders if the resulting increase in global greenhouse gas emission control and ensuing reduction of climate change damages outweighs the cost of the transfers and the cost of taking domestic emission control measures.

The analysis of this paper could be extended in several directions. For instance, it would be interesting to analyze the role of transfers when multiple coalitions can form or when coalitions instead of single players could deviate from an existing coalition. Moreover, the relationship between the optimal transfer scheme proposed in this paper and the transfer scheme implicit in the implementation of an emission trading market is certainly also a research topic for future research (see Bosello, Buchner, Carraro, & Raggi, 2004 for some preliminary numerical results).

Appendix

The first requirement of Pareto-improvement means that

$$\forall i \in S: \qquad v_i(S \cup \{j\}) + t_i \geq v_i(S) \qquad \text{(A1.a)}$$

$$v_j(S \cup \{j\}) + t_j \geq v_j(S) \qquad \text{(A1.b)}$$

must hold where typically $t_i \leq 0$ and $t_j \geq 0$. By adding (A1.a) and (A1.b), summing over all regions $S \cup \{j\}$ involved in the expansion and noting that transfers should balance (i.e., $\sum_{i \in S \cup \{j\}} t_i = 0$), we obtain: $\sum_{i \in S \cup \{j\}} v_i(S \cup \{j\}) \geq \sum_{i \in S \cup \{j\}} v_i(S)$ which implies that superadditivity should hold in the move from coalition S to $S \cup \{j\}$. The second requirement of internal stability of the expanded coalition means:

$$\forall i \in S: \quad v_i(S \cup \{j\}) + t_i \geq v_i(S \cup \{j\} \backslash \{i\}) \qquad \text{(A2.a)}$$

$$v_j(S \cup \{j\}) + t_j \geq v_j(S) \qquad \text{(A2.b)}$$

A similar manipulation as above reveals that a necessary condition (A2.a) and (A2.b) to hold is $\sum_{i \in S \cup \{j\}} v_i(S \cup \{j\}) \geq \sum_{i \in S \cup \{j\}} v_i(S \cup \{j\} \backslash \{i\})$ which is the condition of potential internal stability (PIS).

Acknowledgements Constructive comments and suggestions by three anonymous referees are gratefully acknowledged.

References

Altamirano-Cabrera, J.-C., & Finus, M. (2006). Permit trading and stability of international climate agreements. *Journal of Applied Economics*, *9*(1), 19–48.

Aumann, R. J., & Drèze, J. H. (1974). Cooperative games with coalition structures. *International Journal of Game Theory*, *3*(4), 217–237.

Barrett, S. (1994). Self-enforcing international environmental agreements. Oxford Economic Papers 46 (October), 804–878.

Bloch, F. (2003). Non-cooperative models of coalition formation in games with spillovers. In C. Carraro (Ed.), *Endogenous formation of economic coalitions*, (pp. 35–79). Cheltenham, UK: Elgar.

Bosello, F., Buchner, B., & Carraro, C. (2003). Equity, development, and climate change control. *Journal of the European Economic Association*, *1*(2–3), 601–611.

Bosello, F., Buchner, B., Carraro, C., & Raggi, D. (2004). Can equity enhance efficiency? Some lessons from climate negotiations. In C. Carraro & V. Fragnelli (Eds.), *Game practice and the environment*, (pp. 37–64). Cheltenham, UK: Elgar.

Botteon, M., & Carraro, C. (1997). Burden-sharing and coalition stability in environmental negotiations with asymmetric countries. In C. Carraro (Ed.), *International environmental negotiations: strategic policy issues*, (pp. 26–55). Cheltenham, UK: Elgar.

Buchner, B., Carraro, C., & Cersosimo, I. (2002). Economic consequences of the US withdrawal from the Kyoto Protocol. *Climate Policy, 76*(2), 1–20.

Carraro, C. (2000). Roads towards international environmental agreements. In H. Siebert (Ed.), *The economics of international environmental problems*, (pp. 169–202). Tübingen: Mohr Siebeck.

Carraro, C., & Siniscalco, D. (1993). Strategies for the international protection of the environment. *Journal of Public Economics, 52*(3), 309–328.

Carraro, C., & Siniscalco, D. (1998). International environmental agreements: Incentives and political economy. *European Economic Review, 42*(3–5), 561–572.

Carraro, C., & Siniscalco, D. (2001). Transfers, commitments and issue linkage in international environmental negotiations. In A. Ulph (Ed.), *Environmental policy, international agreements and international trade*, (pp. 19–37). Oxford: Oxford University Press.

Chander, P., Tulkens, H. (1997). The core of an economy with multilateral environmental externalities. *International Journal of Game Theory, 26*(3), 379–401.

d'Aspremont, C., Jacquemin, A., Gabszewicz, J. J., & Weymark, J. A. (1983). On the stability of collusive price leadership. *Canadian Journal of Economics, 16*(1), 17–25.

Eyckmans, J., & Finus, M. (2004a). Measures to enhance the success of global climate treaties. Forthcoming in International Environmental Agreements: Politics, Law and Economics.

Eyckmans, J., & Finus, M. (2004b). *An almost ideal sharing scheme for coalition games with externalities*. CLIMNEG Working Paper Nr. 62. Leuven, Belgium: Katholieke Universiteit Leuven.

Eyckmans, J., & Tulkens, H. (2003). Simulating coalitionally stable burden sharing agreements for the climate change problem. *Resource and Energy Economics, 25*(4), 299–327.

Eyckmans, J., & Finus, M. (2006). Coalition formation in a global warming game: how the design of protocols affects the success of environmental treaty-making. *Natural Resource Modeling, 19*(3), 323–358.

Finus, M. (2001). *Game theory and international environmental cooperation*. Cheltenham, UK: Elgar.

Finus, M. (2003a). Stability and design of international environmental agreements: The case of transboundary pollution. In H. Folmer & T. Tietenberg (Eds.), *International yearbook of environmental and resource economics, 2003/4*, (pp. 82–158). Cheltenham, UK: Elgar.

Finus, M. (2003b). New developments in coalition theory: An application to the case of global pollution. In L. Marsiliani, M. Rauscher & C. Withagen (Eds.), Environmental policy in an international perspective, (pp. 19–49). Dordrecht, The Netherlands: Kluwer.

Finus, M., Sáiz, E., & Hendrix, E. M. T. (2004). *An empirical test of new developments in coalition theory for the design of international environmental agreements*. Mansholt Working Paper Series, Nr. 14. Wageningen, The Netherlands: University of Wageningen.

Hoel, M. (1992). International environment conventions: The case of uniform reductions of emissions. *Environmental and Resource Economics*, 2(2), 141–159.

IPCC (2001). *Climate change 2001: Mitigation. Contribution of working group III to the third assessment report of the intergovernmental panel on climate change*. Cambridge, UK: Cambridge University Press.

Jeppesen, T., & Andersen, P. (1998). Commitment and fairness in environmental games. In N. Hanley & H. Folmer (Eds.), *Game theory and the environment*, (pp. 65–83). Cheltenham, UK: Elgar.

Maskin, E. (2003). Bargaining, coalitions, and externalities. Mimeo. Paper presented at the EEA–ESEM Meeting in Stockholm, August.

Nordhaus, W. D., & Yang, Z. (1996). A regional dynamic general-equilibrium model of alternative climate-change strategies. *American Economic Review, 86*, 741–765 (September).

Petrakis, E., & Xepapadeas, A. (1996). Environmental consciousness and moral hazard in international agreements to protect the environment. *Journal of Public Economics, 60*(1), 95–110.

Weikart, H.-P., Finus, M., & Altamirano-Cabrera, J.-C. (2006). The impact of surplus sharing on the stability of international climate coalitions. *Oxford Economic Papers, 58*(2), 209–232.

Yi, S.-S. (2003). Endogenous formation of economic coalitions: a survey of the partition function approach. In C. Carraro (Ed.), *Endogenous formation of economic coalitions*, (pp. 80–127). Cheltenham, UK: Elgar.

[31]

Environ Resource Econ (2012) 52:327–345
DOI 10.1007/s10640-011-9530-6

Coalition Stability in Public Goods Provision: Testing an Optimal Allocation Rule

Matthew McGinty · Garrett Milam · Alejandro Gelves

Accepted: 17 November 2011 / Published online: 13 December 2011

Abstract We test the canonical model of international environmental agreements (IEAs) in a laboratory setting with asymmetric agents. IEA participation represents coalition formation and public good provision where there are gains to cooperation, but an incentive to free-ride. We test four competing methods of dividing the coalition's worth: a recently proposed optimal rule which accounts for subjects' payoffs as a single free-rider, the Shapley value, the Nash bargaining solution, and an equal split. Each treatment generates the theoretically predicted coalition size more often than not. The shares of the potential gains to cooperation achieved by each rule are: 51, 36, 40 and 13%, respectively. These results highlight the importance of using an optimal rule to improve IEAs, and more broadly for voluntary public good provision.

Keywords IEAs · Coalition formation · Experimental economics · Public goods · Shapley value · Nash bargaining solution

JEL Classification H41 · C92 · C72 · C78 · D63 · D64 · D74

1 Introduction

Coalition stability depends on the method of dividing the coalition's worth and on each member's payoff outside the coalition. Both the worth and the outside payoff vary when

M. McGinty (✉)
Department of Economics, University of Wisconsin-Milwaukee, PO Box 413, Milwaukee, WI 53201, USA
e-mail: mmcginty@uwm.edu

G. Milam
Department of Economics, University of Puget Sound, Tacoma, WA, USA

A. Gelves
Department of Economics, Midwestern State University, Wichita Falls, TX, USA

members have different characteristics, or when externalities are present. Thus, traditional sharing rules used in the international environmental agreement (IEA) literature, such as the Shapley value and the Nash bargaining solution, have different implications for stability among asymmetric players (Barrett 1997; Botteon and Carraro 2001). The Shapley value and the Nash bargaining solution do not account for the interaction between asymmetry and the outside payoff for a coalition member. Recently, optimal allocation rules have emerged that directly address the outside payoff and theoretically show that the set of stable coalitions is larger than previously thought (McGinty 2007; Weikard 2009). Yet, these rules have not been taken into the laboratory and tested. This paper tests the sharing rule proposed in McGinty (2007) to see if, indeed, IEA participation is greater than using the Shapley value, the Nash bargaining solution, and an equal split of the coalition's worth.

An important application of coalition formation and stability with asymmetry involves IEAs, such as the Kyoto treaty on greenhouse gas emissions. The theoretical literature on IEAs shows participation by individually rational nations is trivial when there are significant gains to cooperation. Barrett (1994) shows only three out of a hundred nations will participate in an IEA when the marginal benefit and marginal cost of abatement have slopes of equal magnitude. In response, there is a growing body of applied IEA literature which examines coalition stability under different rules of coalition formation and division of worth[1] (Finus 2003; Weikard and Dellink 2010; Weikard et al. 2010). This paper presents an experimental test of four methods of dividing a coalition's worth using payoffs generated by the canonical model of IEAs (Carraro and Siniscalco 1993; Barrett 1994). We show that subjects form larger coalitions under an optimal sharing rule.

The Shapley value and the Nash bargaining solution are appropriate for convex games where the grand coalition (consisting of all players) is stable and there are no externalities. However, abatement of pollutants under IEAs are public goods, and as such they typically suffer from the free-rider problem. The other $n-1$ nations are the primary beneficiaries of a nation's costly abatement. Given the positive externalities associated with abatement, traditional burden sharing rules fare poorly in the context of IEA stability. Typically, in these types of games there are gains to cooperation and the grand coalition is Pareto optimal. However, as Barrett (1997) and Botteon and Carraro (2001) show, it is not internally stable when there are significant gains to cooperation since at least one member earns a greater payoff as a free-rider. As a result there is an incentive for nations to deviate, and coalitions break down. Barrett (1997) obtains a maximum stable coalition of only three members (out of seven) using the Shapley value to determine a zero-sum system of transfers among members.

McGinty (2007) obtains more encouraging results by addressing IEA stability via an optimal rule. This identifies an IEA as internally stable when there is a positive surplus, meaning the worth of a coalition is at least as great as the sum of payoffs that each member would earn as a single defector (Weikard 2009; Weikard et al. 2010).[2] When there is a positive surplus a zero-sum system of transfers, implemented in the form of abatement requirements and tradable pollution permits, exists such that no member has an incentive to free-ride.[3] Each

[1] Methods of dividing a coalition's worth are often referred to as burden sharing rules in public goods games. These burden sharing rules implicitly define a zero-sum system of transfers among coalition members.

[2] Similar allocations with at least the outside payoff have recently emerged. These have different names in the literature. Weikard (2009) has a "claims rights condition", Carraro et al. (2006) refer to a "potentially internally stable coalition," and McGinty (2011) shows that an equal share of the surplus "risk-dominates" all other allocations.

[3] For a given coalition, actual abatement is the same for every allocation rule since the worth is a result of minimizing the coalition's aggregate abatement cost. This requires equating the marginal abatement cost of the last unit for all coalition members. However, the rules differ in how they allocate the worth, even though

member earns her payoff outside a given coalition, plus a share of the surplus determined by a positive parameter θ. This θ rule explicitly addresses the relevant alternative payoff in the presence of externalities, which the Shapley value and Nash bargaining solution (NBS) do not. McGinty (2007) shows stable coalitions defined in this fashion can obtain 72% of the aggregate payoff difference between the grand and singleton coalitions when there are twenty fully asymmetric nations. By contrast, with symmetric nations only 9% of this difference is obtained by the largest stable coalition.

This paper presents a controlled laboratory test of coalition stability comparing the θ rule to the Shapley value, the Nash bargaining solution, and an equal split of the coalition's worth. The coalition's worth is not arbitrarily chosen, but rather given by the benchmark model of IEAs (Carraro and Siniscalco 1993; Barrett 1994) in the presence of asymmetry (Barrett 1997; McGinty 2007). Previous coalition experiments focus on subjects' division of a coalition's worth and then compare actual behavior with competing theoretical predictions (Croson et al. 2004, Bolton et al. 2003, Leopold-Wildburger 1992). Instead, we focus on coalition stability by providing subjects with the division of the coalition's worth and their payoff outside a given coalition. Our approach eliminates uncertainty with respect to the bargaining process and reduces subjects' choice to either being in or out of a given coalition. The typical coalition experiment essentially assumes the grand coalition will form. By contrast, we look at coalition stability in the presence of a free-rider incentive since the grand coalition is not stable in our experiment. Rather than measure the distance between the actual outcomes and traditional solution concepts as Croson et al. (2004) we impose the solution concept and address coalition formation and stability in isolation. In this respect our paper answers Bolton (1998) call for subsequent laboratory research linking bargaining and dilemma games.

We find coalition membership is broadly consistent with the theoretical predictions across treatments. In our four player game, the θ rule results in the predicted three member coalitions 56% of the time, whereas Shapley and NBS result in the predicted two member coalitions 57 and 59%, respectively. No coalition formed 64% of the time in the equal split treatment. Our performance metric is the proportion of the payoff gap that is closed by coalition formation. Efficiency is defined as the proportion of the potential aggregate payoff difference between the grand and singleton coalitions. The θ rule results in the largest average efficiency of 51%, compared to 36% for Shapley and 40% for NBS. Efficiency for the equal split treatment was only 13%. These results highlight the importance of using the appropriate abatement requirements, implicitly defining a burden sharing rule, for the Kyoto treaty.

Our results are consistent with recent work on IEAs and coalitions. Dannenberg et al. (2010) conduct an IEA experiment with symmetric subjects. Under three different institutional arrangements subjects choose both coalition membership and a level of public good provision. As predicted by theory, they find that only small coalitions results when members are forced to choose the coalition's optimal abatement level. They find larger coalitions emerge when abatement requirements only partially internalize the positive externalities from abatement. Carraro et al. (2006), McGinty (2007) and Weikard (2009) show that coalition membership is largest when players' receive an allocation that exceeds the payoff that would be earned from individually leaving the coalition. McGinty (2011) shows that a risk-dominant

Footnote 3 continued
each rule implies a zero-sum system of transfers. One transfer method is a system of tradable pollution permits, where the permit price p is the common marginal abatement cost of the last unit. Under a permit system, coalition member s is required to abate an amount q_s^r. Actual abatement is q_s, thus the transfer τ for member s in coalition k, is: $\tau_s(k) = p(q_s - q_s^r)$. McGinty (2007) and Sect. 5.2 of McGinty (2011) provide additional details.

optimal allocation rule may result in larger stable coalitions than the Shapley value or the Nash bargaining solution. An example applies this risk-dominant rule to the fully asymmetric IEA model in McGinty (2007), and details the associated abatement requirements with transfers implemented via a system of tradable pollution permits.

The remainder of the paper is organized as follows. The theoretical model is presented in Sect. 2, detailing the underlying model, definitions of stable coalitions and the four treatments. Section 3 presents the experimental environment describing the data set, experimental procedures and theoretical predictions. Descriptive and regression results from fixed effects models are presented in Sect. 4. Section 5 concludes and suggests further research.

2 Theoretical Model and Treatments

Following Barrett (1994, 1997) and Carraro and Siniscalco (1993) we implement the standard public goods model from the IEA literature, but allow for fully asymmetric players as in McGinty (2007). The set of players is denoted $N = \{1, \ldots, n\}$. Social benefit is a concave function $B(Q) = b\left(aQ - \frac{Q^2}{2}\right)$ of the total quantity of the public good, $Q = \sum_{i \in N} q_i$. Parameters a and b are assumed to be strictly positive resulting in declining marginal benefit. Benefit shares are given by α_i, where $\alpha_i > 0\, \forall i \in N$ and $\sum_{i \in N} \alpha_i = 1$. The benefit for player i is: $B_i(Q, \alpha_i) = b\alpha_i\left(aQ - \frac{Q^2}{2}\right)$. Players have convex cost functions $C_i(q_i, c_i) = \frac{c_i q_i^2}{2}$, generating marginal cost curves that are rays from the origin with asymmetric slopes $c_i > 0$. In the absence of transfers the payoff for player i is: $\pi_i = B_i(Q, \alpha_i) - C_i(q_i, c_i)$.

Variations of this model appear in the experimental literature using the voluntary contribution mechanism (VCM) where subjects contribute a portion of their endowment to a public good (see Ledyard 1995; Laury and Holt 2005 for surveys). Several VCM experiments test either declining marginal benefit, increasing marginal cost (Isaac et al. 1985; Keser 1996) or asymmetric public and private good values (Burlando and Guala 2005; Chan et al. 1999). Two stylized facts emerge from this literature. First, subjects systematically over-contribute to public goods beyond the individually rational level. Second, this over-contribution decays as the experiment progresses. Our experimental environment examines an alternative model of public goods provision with less uncertainty regarding others' actions and a binary decision space. However, our design features an incentive structure analogous to VCM games which allows examination of the issues of over-contribution and time-decay.

We begin by identifying the singleton and grand coalition outcomes. The singleton coalition, denoted $\{i\}$, is obtained by each player acting in his own self-interest choosing q_i to maximize π_i. The singleton individual and group levels of public good provision are:

$$q_i^{\{i\}} = \frac{ab\theta_i}{1 + b\sum_{j=1}^{n} \theta_j} \quad \forall i \in N$$
$$Q^{\{i\}} = \frac{ab\sum_{j=1}^{n} \theta_j}{1 + b\sum_{j=1}^{n} \theta_j} \tag{1}$$

where $\theta_i \equiv \frac{\alpha_i}{c_i}$ is the ratio of benefit share to marginal cost slope. Each player provides the public good in proportion to their θ in the singleton outcome. The grand coalition (social optimum) maximizes the sum of payoffs, $\Pi \equiv \sum_{i=1}^{n} \pi_i$, where each player internalizes the positive externality that accrues to others. The grand coalition, denoted $\{N\}$, results in the social optimum, $q_i^{\{N\}}$ and $Q^{\{N\}}$:

$$q_i^{\{N\}} = \frac{ab}{c_i\left[1 + b\sum_{j=1}^{n}\frac{1}{c_j}\right]} \quad \forall i \in N$$

$$Q^{\{N\}} = \frac{ab\sum_{j=1}^{n}\frac{1}{c_j}}{1 + b\sum_{j=1}^{n}\frac{1}{c_j}} \tag{2}$$

All coalitions are assumed to allocate public good provision across members to maximize the coalition's worth. This occurs where the marginal cost of the last units are equalized across coalition members. The division of the coalition's worth then entails a zero-sum system of transfers among members. In the context of IEAs these transfers may take the form of tradable pollution permits, equating the marginal cost of the last unit of provision with the permit price. Abatement requirements under the IEA relative to $q_i^{\{N\}}$ then determine the sign and magnitude of the transfer for each player. The solution to all other potential coalition structures is presented in McGinty (2007).

2.1 Coalition Stability

The central issue surrounds the conditions under which a given coalition achieves stability. The literature on IEAs uses internal and external stability from the cartel literature (d'Aspremont et al. 1983) to describe individual incentives. We model an open membership single coalition game (Finus 2003). Thus, existing members may not prevent new members from joining the coalition and each non-member acts as an individual. Let k be an element of the set of all possible coalitions K. Also, let $k\setminus\{i\}$ denote the remaining coalition when member i leaves the coalition, and $k \cup \{j\}$ denote the resulting coalition when non-member j joins. The subscripts s and t denote signatories and non-signatories (free-riders) choosing provision levels to maximize coalition worth and individual payoff, respectively. The following conditions from McGinty (2007) define individual and coalition stability.

$$\pi_s(k) - \pi_t(k\setminus\{t\}) + \tau_s(k) \geq 0, \forall s = t \in k \tag{3}$$

$$\pi_t(k) - \pi_s(k \cup \{s\}) - \tau_s(k \cup \{s\}) > 0, \forall t = s \notin k \tag{4}$$

$$\Phi = \left\{k \in K : \sum_{s\in k}\pi_s(k) - \sum_{t\in k}\pi_t(k\setminus\{t\}) \geq 0\right\} \tag{5}$$

The conditions for internal and external stability are given by Eqs. (3) and (4). Internal stability ensures each member at least as high a payoff choosing $q_s(k)$ with transfer $\tau_s(k)$ than as a single defector from the coalition choosing $q_t(k\setminus\{t\})$. External stability indicates that no non-member earns a higher payoff joining coalition k and choosing $q_s(k \cup \{s\})$ with transfer $\tau_s(k \cup \{s\})$. Equation (5) defines the set of all stable coalitions: $\Phi \subset K$, given the appropriate (zero-sum) transfers. This set consists of members such that the coalition's worth, $v(k) \equiv \sum_{s\in k}\pi_s(k)$, exceeds the sum of payoffs that each member would earn individually leaving the coalition. For all elements $\phi \in \Phi$ there exists a zero-sum system of transfers that simultaneously satisfies the internal stability condition for all members $s \in k$.

To illustrate the central issues of superadditivity and coalition stability consider the following three player characteristic function game with positive externalities from Croson et al. (2004). While the payoffs are not generated by an underlying model, the structure is similar to the public goods game tested in the present paper.

$$
\begin{aligned}
v\{i\} &= 50, i = A, B, C \\
v\{AB\} &= 240, \pi_C(\{AB\}) = 140 \\
v\{AC\} &= 210, \pi_B(\{AC\}) = 150 \\
v\{BC\} &= 180, \pi_A(\{BC\}) = 180 \\
v\{ABC\} &= 400
\end{aligned} \tag{6}
$$

Each row represents one of the five possible coalition structures. The first row is the singleton payoffs, $v\{i\} = 50, i = A, B, C$. The worth of the two member coalitions are in rows two through four, with the payoff of the player outside the coalition denoted $\pi_t, t = A, B, C$. The worth of the grand coalition in row five. Note, π_t is the payoff that a player would earn if they were to leave the grand coalition. This game is superadditive since the worth of each coalition exceeds the sum of members worth in all sub-coalitions. Formally, strict superadditivity is defined as: $v\{k \cup t\} > v\{k\} + v\{t\}$. For the two member coalitions $v\{ij\} > v\{i\} + v\{j\}, i, j = A, B, C$ all distinct. The grand coalition: $v\{ABC\} > v\{ij\} + \pi_h(\{ij\})\ h, i, j = A, B, C$ all distinct. The two member coalitions show the positive externalities since the player outside the coalition benefits from the coalitions formation: $\pi_h(\{ij\}) > v\{h\}, h, i, j = A, B, C$ all distinct.

The grand coalition is socially optimal, however it is not internally stable since the game is concave. The worth of the grand coalition does not exceed the sum of the free-rider payoffs $v\{ABC\} < \pi_A + \pi_B + \pi_C$ indicating (5) is not satisfied. There exists no division of 400 such that all three players simultaneously earn a payoff greater than they could earn as a single free-rider. However, the two member coalitions are (potentially) internally stable since there exists a division of the coalition's worth where no member has an incentive to leave. The singletons are not externally stable since internally stable two member coalitions can form.

Surprisingly, Croson et al. (2004) find the grand coalition forms 91% of the time, even though it is not internally stable. Several possible explanations come to mind. First, subjects may be systematically over-optimistic. Each player may anticipate their own earnings to be greater than their outside payoffs, even if they recognize that this is not simultaneously possible for all coalition members. Second, subjects may have been conditioned to form the grand coalition since it was stable in all other treatments. Third, subjects may have exhibited some form of altruism.

2.2 Treatments

The system of transfers defines the division of the coalition's worth. We test four methods of distribution: the θ rule, the Shapley value, the NBS, and an equal split. Internal and external stability depends on the system of transfers. With an inappropriate system of transfers a potentially stable coalition may violate internal stability for at least one member. The worth of each coalition, each members payoff outside that coalition, $\pi_t(k \setminus \{t\}$ for $t \in k$, and each non-members payoff, $\pi_t(k)$ for $t \notin k$, are invariant across treatments.

Satisfying condition (5) indicates that the worth of the coalition is sufficient to make membership incentive compatible for all members simultaneously. The issue is whether the transfer scheme is sufficient to satisfy the condition for internal stability in (3). Under the θ rule each member receives the payoff they would earn outside the coalition and any remaining surplus is distributed by θ share. Each coalition member receives:

$$\pi_s^\theta(k) = \pi_t(k \setminus \{t\}) + \frac{\theta_s}{\sum_{s \in k} \theta_s}\left[v(k) - \sum_{t \in k} \pi_t(k \setminus \{t\})\right] \quad \forall s = t \in k \tag{7}$$

Botteon and Carraro (2001) and Barrett (1997) have used rules from cooperative game theory to distribute a coalition's worth in the context of IEAs. The Shapley value and Nash bargaining solution are unique payoff allocations, but they do not ensure internal stability (3), even if (5) holds. The Shapley value (SV) is:

$$\pi_s^{SV}(k) = \sum_{j \subseteq k} \frac{(|j|-1)!(|k|-|j|)!}{|k|!} [v(k) - v(k \backslash \{s\}] \quad \forall s \in k \tag{8}$$

where $|i|$, $i = j, k$ denotes the number of members (cardinality) in coalition i.

The Shapley value is the weighted sum of players' marginal contributions across all possible sub-coalitions j. The weights are binomial coefficients indicating the probability of each coalition formation sequence occurring, with the assumption that all sequences are equally likely. The Shapley value is unique and always exists, but there is no reason to expect that it, or the Nash bargaining solution, exceeds the free-rider payoff $\pi_t(k \diagdown \{t\})$ in a concave game. The Shapley value incorporates how much a player's inclusion adds to the worth of the coalition, but not how much a player earns outside of the coalition.

The Nash bargaining solution maximizes the product of differences from the singleton payoff. Assuming equal weights, the Nash bargaining solution (NBS) for each member π_s^{NBS} is the solution to the maximization problem:

$$\pi_s^{NBS}(k) = \max \left\{ \prod_{s \in k} [v(k) - \pi_s(\{s\})] \right\} \tag{9}$$

The NBS does not account for neither the worth of the sub-coalitions of k, nor the payoff a player would earn outside the coalition in games with externalities.

Finally, the equal split simply divides the worth of the coalition by the number of coalition members $|k|$.

$$\pi_s^E(k) = \frac{v(k)}{|k|} \tag{10}$$

Inclusion of the equal split treatment serves two purposes. The primary motivation stems from the need for a clear benchmark. This allows us to verify that agents are responding across treatments to the individual incentives generated by the payoff structure. The secondary rationale concerns previous literature which investigates equal division of surplus in bargaining games. Bolton (1997) shows that splitting equally can be an evolutionarily stable equilibrium in a bilateral bargaining game. The idea of a "fair share" comes up frequently in follow-up questionnaires administered in bargaining experiments (Roth 1987). If these are important issues in our environment, the equal split treatment should help identify their influence.

3 Experimental Environment

3.1 Information and Data Set

Six laboratory sessions were conducted, three each at Ryerson University in Toronto and the University of Wisconsin-Milwaukee. Subjects were recruited from first-year undergraduate Economics courses. Subjects were not allowed to communicate with one another, nor did they know the identity of the other group members. Coalitions were referred to as subgroups to avoid any demand effects (Friedman and Sunder 1994). Thus, subjects did not know that the context of the experiment was IEAs, or even more broadly, public goods provision. Five sessions had twelve subjects, while one session was run only with eight due to no-shows. Subjects were randomly assigned a type (A, B, C, or D) and a group consisting of one of each type. The Earnings Table in the "Appendix" shows the payoff for each type and allocation rule. Sessions consisted of four treatments each lasting ten periods, with three practice periods for the first treatment. Each subject participated in all four treatments, and the treatment order was randomly determined. Subjects were randomly assigned a new type and group at the end of each treatment. [4] The data set consists of 40 final observations for 68 subjects, resulting in 2, 660 observations.[5] All sessions were completed in under two hours. Subjects earned an average of $28, including the $5 show-up payment.

The experiment was programmed and conducted using the z-Tree software (Fischbacher 2007). Subjects were given the earnings table, showing payoffs for each group member for all permissible coalition structures, and the written instructions in the "Appendix". As in the earnings table, we adopt the shorthand notation $[A, B]$ or AB to denote coalition $\{A, B\}$ for the remainder of the paper. Subjects interacted through a series of three different computer screens. The first screen indicated their type (A, B, C, or D) and asked if they wished to join or remain out of the subgroup. The second screen revealed all group member's initial decisions showing both the subject's payoff with that subgroup and the payoff if they reversed their decision, ceteris paribus. Subjects then either confirmed or reversed their decision. If any of the subjects reversed their decision the entire group was returned to the first screen (with subjects who previously chose to join the subgroup starting in, rather than out). This process repeated until all group members confirmed a particular subgroup, or time ran out. Period length was sixty seconds, plus zero to sixty seconds randomly drawn from a uniform distribution.[6] The third screen showed the final subgroup and the subject's earnings for that and all previous periods.

Given our primary emphasis is the formation and stability of coalitions we allowed subjects to reverse or confirm after observing other's decisions. To test for both internal and external stability we allowed those outside to join, or those inside to leave, an already established coalition. This necessitated the time constraint and uncertain period length given the structure of the game. For example, with few exceptions, each subject's most preferred outcome is to be outside a three member coalition.[7] This means subjects have an incentive to try promote the grand coalition and then leave at the last second. The uncertain period length made this type of strategy impossible to implement. However, to avoid repeated game strategies subjects knew that each run would consist of exactly ten periods. The Folk Theorem (Friedman 1971) states that the one-shot Nash equilibrium is obtained via backwards induction in repeated games with a known final period. Thus, the repeated interaction in the experiment has the same predictions as the one-shot game in the theoretical section.

---

4 The matching procedure results in a "partial strangers" design.

5 One of the θ treatment sessions ended after five periods due to a computer crash resulting in a loss of 60 observations.

6 Monitors announced when the initial sixty seconds was up, indicating that the period could end at any time.

7 The exceptions are: (i) subject A's most preferred coalition structure is $[A, B, D]$ for Shapley, (ii) subject A's most preferred coalition structure is $[A, B, C, D]$ for Nash bargaining and equal split, and (iii) subject B's most preferred coalition structure is $[B, C, D]$ for equal split.

Table 1 Payoffs

Coalition k	$v(k)$	Efficiency (%)	$\pi_t(k\setminus\{t\})$
$[A], [B], [C], [D]$	[30.7, 46.0, 61.4, 76.7]	0	na
$[A, B]$	77.7	28.5	[30.7, 46.0]
$[A, C]$	94.3	24.3	[30.7, 61.4]
$[A, D]$	111.3	37.6	[30.7, 76.7]
$[B, C]$	108.5	21.6	[46.0, 61.4]
$[B, D]$	124.7	23.5	[46.0, 76.7]
$[C, D]$	139.3	16.2	[61.4, 76.7]
$[A, B, C]$	150.7	83.0	[32.9, 50.4, 66.1]
$[A, B, D]$	169.4	45.0	[33.2, 51.4, 82.6]
$[A, C, D]$	186.7	46.9	[32.9, 68.5, 84.1]
$[B, C, D]$	199.4	57.7	[49.3, 66.5, 82.3]
$[A, B, C, D]$	255.6	100	[39.0, 59.5, 79.1, 98.0]

3.2 Predictions: Stability and Efficiency

In order to best examine the gains from coalition formation our design assigns to the highest (lowest) cost player the highest (lowest) benefit share.[8] This type of asymmetry increases the potential gains to coalition formation, by creating larger gains from trade in pollution permits. By contrast, the Nash equilibrium is closer to the social optimum when the high benefit nations are also low cost. This type of asymmetry makes the sum $\sum_{j=1}^{n} \theta_j$ larger than if the high benefit nations are high cost. The Nash equilibrium in Eq. (1) shows that aggregate abatement is increasing in $\sum_{j=1}^{n} \theta_j$. Since the social optimum in Eq. (2) is independent of the benefit shares, the difference between Nash and optimal abatement is greatest when the high benefit share countries are also high cost.[9] Table 1 shows the worth of each coalition $v(k)$, efficiency as the proportion of the payoff gap that is obtained and each members payoff should they individually leave the coalition $\pi_t(k\setminus\{t\})$. Table 2 presents each member's payoff for all four treatments and indicates which coalitions are internally and externally stable. Subjects were only given the instructions and earnings Table 7 in the "Appendix".

The parameterization generates different theoretical predictions across the four treatments. The game is superadditive and the grand coalition is socially optimal. However, the grand coalition is unstable for all treatments since condition (5) is not satisfied: $v(N) = 255.6 < \sum_{t \in N} \pi_t(N\setminus\{t\}) = 275.5$. Condition (5) is satisfied for all two and three member coalitions. Thus, all three member coalitions are both internally and externally stable for the θ treatment. However, none of the three member coalitions are internally stable in the other three treatments. The Shapley value and Nash bargaining solutions have identical stability predictions, even though the magnitude of the decisions at the margin are different. Under Shapley and NBS only coalition AB is both internally and externally stable. For these two treatments all other two member coalitions are internally stable, but not externally stable. Thus, given two member coalitions that are not externally stable, we expect three member coalitions to form and then break down since the resulting three member coalitions are not internally stable. By contrast, under θ all three member coalitions are both internally and externally stable. No larger coalition is both internally and externally stable under equal split, thus the singleton outcome is the unique prediction. Some of the individual payoff differences between being

[8] The parameterization is: $n = 4$, $a = 50$, $b = 0.25$, $c = [\frac{2}{14}, \frac{3}{14}, \frac{4}{14}, \frac{5}{14}]$, $\alpha = [\frac{2}{14}, \frac{3}{14}, \frac{4}{14}, \frac{5}{14}]$, $\theta = [1, 1, 1, 1]$, for the ordering $[A, B, C, D]$.

[9] See McGinty (2007) for more on this point.

Table 2 Internal and external stability

Coalition k	$\pi_s^{\theta}(k)$	$\pi_s^{SV}(k)$	$\pi_s^{NBS}(k)$	$\pi_s^{E}(k)$
$[A], [B], [C], [D]$	[30.7,46.0,61.4,76.7]*	[30.7,46.0,61.4,76.7]*	[30.7,46.0,61.4,76.7]*	**[30.7,46.0,61.4,76.7]*** **
$[A, B]$	[31.6, 46.1]*	**[31.2, 46.5]*** **	**[31.2, 46.5]*** **	[38.9, 38.9]**
$[A, C]$	[32.6, 61.7]*	[31.8, 62.5]*	[31.8, 62.5]*	[47.1, 47.1]**
$[A, D]$	[33.8, 77.4]*	[32.6, 78.7]*	[32.6, 78.7]*	[55.6, 55.6]
$[B, C]$	[47.0, 61.5]*	[46.6, 61.9]*	[46.6, 61.9]*	[54.2, 54.2]
$[B, D]$	[47.8, 76.9]*	[47.0, 77.7]*	[47.0, 77.7]*	[62.4, 62.4]
$[C, D]$	[62.4, 76.9]*	[62.0, 77.3]*	[62.0, 77.3]*	[69.7, 69.7]
$[A, B, C]$	**[33.3, 50.8, 66.5]*** **	[35.1, 49.8, 65.8]**	[34.9, 50.2, 65.6]**	[50.2, 50.2, 50.2]**
$[A, B, D]$	**[33.9, 52.1, 83.4]*** **	[41.3, 56.1, 72.0]**	[36.0, 51.3, 82.0]**	[56.5, 56.5, 56.5]**
$[A, C, D]$	**[33.3, 68.9, 84.5]*** **	[37.3, 66.6, 82.8]**	[36.7, 67.4, 82.7]**	[62.2, 62.2, 62.2]
$[B, C, D]$	**[49.7, 66.9, 82.7]*** **	[51.2, 66.2, 82.0]**	[51.1, 66.5, 81.8]**	[66.5, 66.5, 66.5]
$[A, B, C, D]$	[34,54.5,74.2,93.0]**	[41.2,55.1,71.2 88.1]**	[40.9 56.2,71.6,86.9]**	[63.9,63.9 63.9,63.9]**

* Denotes internally stable

** Denotes externally stable

*** Denotes both internally and externally stable (in bold)

inside or outside of a given coalition are small, however this is the nature of the underlying theoretical model. The parameters were chosen with the expectation that on average subjects would earn approximately \$25 if actual behavior matched the theoretical predictions. Subjects earned a \$5 fee for showing up and earnings each period depended on which coalition formed. The theoretical prediction for the four runs (ten periods each) resulted in an expected value of \$20 for each subject.

As a performance metric we define efficiency (E) as the percentage of the potential aggregate payoff difference between the singleton and grand coalitions obtained by k forming: $E(k) = \frac{\sum_{i \in N} \pi_i(k) - \sum_{i \in N} \pi_i(\{i\})}{\sum_{i \in N} \pi_i(N) - \sum_{i \in N} \pi_i(\{i\})}$. The potential gains vary considerably across treatments. For the equal split treatment 0% efficiency gain is predicted since there are no stable coalitions larger than the singleton. Efficiency slightly improves under the Shapley value and NBS. Given that only coalition $[A, B]$, with the two lowest share agents, is both internally and externally stable we expect a gain of 28.5% under these two treatments. Predicted performance from stable coalitions is highest with the θ rule. The four three-member coalitions achieve between 45 and 83% of the efficiency gains. No clear equilibrium selection criteria allows us to make a prediction within the range, even if subjects form only stable coalitions. The inherent free-riding incentives generate a four-way "Battle of the Sexes" type structure where each subject would prefer to be outside a three member coalition.

The analysis below examines two related sets of hypotheses. The first concerns the prediction that agents will form coalitions which are stable under each of the four burden sharing rules. The second considers the relative efficiency of coalition formation under each rule. We are also interested in the actual efficiency performance of the θ rule given the wide range of predictions within the set of stable coalitions.

4 Results

Figure 1 presents the frequency distribution of the data in Table 3 across each of the twelve possible coalitions by treatment. For each treatment, those coalitions which are stable both

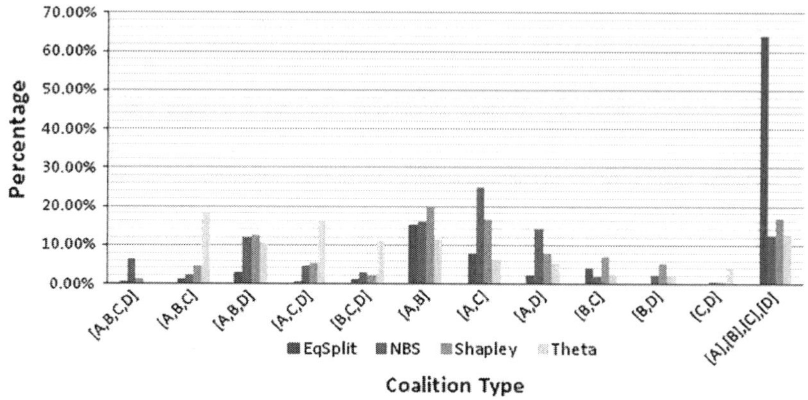

Fig. 1 Coalition frequency

Table 3 Coalition frequency (%)

	θ	Shapley	NBS	Equal
$[A],[B],[C],[D]$	12.90	17.06	12.35	64.12
$[A, B]$	11.61	20.00	15.88	15.29
$[A, C]$	6.45	16.47	24.71	7.65
$[A, D]$	5.16	7.65	14.12	2.35
$[B, C]$	2.58	7.06	1.76	4.12
$[B, D]$	1.94	5.29	2.35	0
$[C, D]$	3.87	0.59	0.59	0
$[A, B, C]$	18.06	4.71	2.35	1.18
$[A, B, D]$	10.32	12.35	11.76	2.94
$[A, C, D]$	16.13	5.29	4.71	0.59
$[B, C, D]$	10.97	2.35	2.94	1.18
$[A, B, C, D]$	0	1.18	6.47	0.59

internally and externally are the most common outcomes. The observed departures from stability varies across treatments, the most notable being a non-trivial fraction of singletons. In addition, while the grand coalition rarely forms, coalitions are clearly largest under the θ rule.

4.1 Descriptive Statistics

The theoretical predictions under Shapley value and NBS are identical. Our experimental results vary only slightly across these two treatments. Coalition formation is most evenly distributed here even though the only internally and externally stable coalition is $[A, B]$. This is the modal coalition in the Shapley value treatment, and second most frequent to $[A, C]$ for NBS. The distribution of coalitions is clearly dominated by the various two member coalitions, occurring 59 and 56% of the time for Shapley and NBS respectively. These represent the internally stable set.

The singleton outcome represents 17% of Shapley and 12% of NBS results. This raises an important issue relating the stability of a final coalition to the dynamic coalition formation

Table 4 Coalition membership frequency (%)

	θ	Shapley	NBS	Equal
Singleton	12.90	17.06	12.35	64.12
Two member coalition	31.61	57.06	59.41	29.41
Three member coalition	55.48	24.71	21.76	5.88
Grand coalition	0	1.18	6.47	0.59
A join = 1	72.26	75.29	87.06	72.35
B join = 1	58.06	55.29	45.88	34.12
C join = 1	61.94	42.94	45.29	15.29
D join = 1	50.32	35.29	43.53	7.65

process. To illustrate, consider the decision process facing agents, independent of their type. The first two agents to join the coalition increase their payoff relative to the singleton payoff under all treatments except the equal split. This stems from the internal stability of two member coalitions. When these two member coalitions are not externally stable an outside player receives a larger payoff from joining the coalition. The resulting three member coalitions are not internally stable under Shapley and NBS, thus more than one member may simultaneously leave. If two or more players leave a three-member coalition and time runs out, then the final outcome is a singleton.

The composition of coalitions we observe clearly relates to the relative strength of incentives facing different subject types. Subject A has a dominant strategy of always joining a coalition in treatments other than θ, and observed participation shown in Table 4 was high: 87.06% in NBS and 75.29% in Shapley. The stronger players (in terms of threat points) have weaker incentives which are reflected in their participation rates. For example, subject D was a member of a coalition 35.29% of the time for Shapley and 43.53% time for NBS.

Under the θ rule, coalition membership in Figure 1 is clearly the highest. The three member coalitions formed 55% of the time. Thus, a majority of the outcomes fit the theoretic stability prediction. Further, the modal coalition under this treatment $[A, B, C]$ generates the highest possible efficiency (83%) short of the grand coalition. These increases in coalition membership come in spite of the fact that subjects face the same incentives to delay their decision to join stable coalitions as under Shapley and NBS.

The θ rule allocates the coalition's worth as a function of individual θ shares and the free-rider payoff. The realigned incentives are reflected in increased willingness to join by stronger members. Table 4 shows that type A subjects chose to be members of a coalition around 72% of the time, a reduction from Shapley and NBS. However increases are seen in the percentages of B, C, and D types decisions as they rose to 58.06, 61.94, and 50.32% of the time, respectively.

From a social or policy perspective our primary interest concerns not the composition but the efficiency of coalitions under our burden sharing rules. Recall that the expected efficiency is 28.5% under Shapley and NBS. Observed efficiency was 36 and 40%, respectively. The efficiency observed under θ was 51%, within the range predicted by the theoretical model. This represents a substantial improvement over the other rules in spite of the fact that they outperform their predicted efficiency. The equal split treatment obtained only 13% of the potential gain. The θ rule generates four stable three agent coalitions; with each agent type having a unique preferred coalition. We have no prior beliefs regarding which of these might dominate but the modal coalition of $[A, B, C]$ represents the highest efficiency in the set of internally and externally stable coalitions.

The results from the equal split treatment, in which the singleton is predicted to be the only stable coalition, match the predictions. No coalition formed 64.12% of the time under equal split. We take this as an indication that individuals respond primarily to their own payoff incentives and fairness considerations do not appear to be a primary factor of concern. While the results above present a clear correspondence between the data and our theoretical expectations, we now turn to a more formal analysis.

4.2 Regression Results

To isolate the factors influencing coalition formation we estimate two regression models. The first investigates if the θ rule generates higher coalition membership than the alternative rules. The second tests if the treatments alter the decision by subjects to join a given coalition.

In order to test if the θ rule generates larger coalitions we fit an ordered logit model. The dependent variable takes one of the four ordered categories reflecting the number of coalition members. Equation (11) describes the probability of the response variable equaling a given coalition size conditional on the values of the explanatory variables, assuming a logit distribution:

$$P(y=j|x) = \exp(x\beta_j) \Big/ \left[1 + \sum_{h=2}^{4} \exp(x\beta_h)\right], \qquad j = 2, 3, 4 \qquad (11)$$

where j's are the three possible outcomes when excluding the zero members outcome, x is a vector of dummy variables, and β_j is a $k \times 1$ vector of unknown parameters. The vector x consists of three treatment dummies, three treatment order dummies, and nine period dummies. The excluded comparison groups are the θ treatment, the fourth treatment order, and the final period. Thus, we interpret the value of the coefficients on the treatment dummies (Shapley, NBS and equal split) as the effects on coalition size of these treatments relative to θ. Our theory predicts that these will be negative. The three treatment order dummies are compared to the fourth treatment order. Our theory predicts that none of these dummies will be significantly different from zero. The nine time period dummies are included to identify significant differences in coalition size for earlier interactions relative to the final period. We predict these will not be significantly different from zero, however VCM experiments typically find these to be negative, indicative of time decay. Finally, we included a geographical dummy variable taking the value one for observations from Ryerson University to capture any difference from location. We predict that the coefficient on the Ryerson dummy will not be significantly different from zero. Our estimated equation is:

$$g(p_{ij}) = u_j + x_i b, \qquad j = 2, 3, 4 \qquad (12)$$

where g is a logit cumulative probability function, u_j is the intercept that depends on category j, and i represents a certain group. Results are presented in Table 5.

The coefficients on all three treatment dummies are negative and significant at the 5% level. The magnitude of the results can be explained by calculating the odds ratios, exponentiating the estimated coefficients. These coefficients show the odds of a given treatment, relative to the odds of the θ treatment, in forming larger coalitions. For example, the equal split treatment has an exponentiated coefficient of 0.0483, meaning there is less than a 5% chance that there will be a higher number of coalition members in equal split than θ. All estimated odds ratios are less than one, indicating a smaller chance of the other three treatments generating a larger coalition size than θ.

Table 5 Estimates for coalition members as dependent variable

Parameter	Estimate	Standard error	Odds ratios
Intercept 4	−2.89 * **	0.42	
Intercept 3	0.24	0.35	
Intercept 2	2.73 * **	0.37	
Shapley	−0.91 * **	0.24	0.40
NBS	−0.61 * *	0.24	0.55
Equal split	−3.03 * **	0.26	0.05
First treatment	0.48 * *	0.24	1.62
Second treatment	0.29	0.24	1.33
Third treatment	0.10	0.23	1.11
Period 1	−0.09	0.34	0.91
Period 2	−0.10	0.34	0.91
Period 3	−0.26	0.34	0.77
Period 4	−0.31	0.34	0.73
Period 5	−0.07	0.34	0.93
Period 6	−0.37	0.34	0.69
Period 7	−0.11	0.33	0.90
Period 8	−0.12	0.33	0.90
Period 9	−0.40	0.34	0.67
Ryerson	−0.70 * **	0.15	0.50

*, **, *** indicate significant at the 10%, 5%, and 1% levels, respectively

In our second estimation we are interested in the individual decisions of subjects to join the coalition. We estimate a simple binary logistic model, where join or not join was the dependent variable. By letting $j = 1$ in Eq. (12) we obtain the probabilities for our response variable. For $j = 1$ Eq. (12) becomes:

$$g(p_{i1}) = u + x_i b \tag{13}$$

Note that in this equation the probability modeled is $join = 1$. To the vector x of treatment, treatment order and time period dummies we add the individual characteristics of the profit differential and subject type. Profit differential is defined as the incentive to free-ride: $\pi_t(k \setminus \{t\}) - \pi_s^i(k), \forall s = t \in k, i = \theta, SV, NBS, Equal$. We expect a lower (higher) probability of a subject choosing to join a coalition the greater (smaller) the differential. There are three subject type dummies with type A as the excluded comparison group. Results are given in Table 6.

The estimated coefficients for the subject type dummies are all negative, meaning that relative to subject type A, subjects B, C, and D have lower incentives to join a coalition. These estimates are all significant, and their odds ratios are all less than one. They monotonically decrease from subject B to D meaning that more powerful players have a lower probability of joining coalitions. The coefficient on the profit differential is also negative, as expected. The odds ratio shows that for every dollar of positive differential, the odds of not joining are 0.9 times the odds of joining. The treatment dummies estimates are all negative and significant. The odds ratios are all less than one, meaning that relative to the θ treatment the other treatments offer less incentive to join coalitions. The interpretation of the odds ratios is the marginal effects on the probability of a certain category occurring relative to the excluded category.

Table 6 Estimates for join (not join) as dependent variable

Parameter	Estimate	Standard error	Odds ratios
Intercept	1.43 * **	0.21	
Type B	−1.07 * **	0.13	0.34
Type C	−1.20 * **	0.13	0.30
Type D	−1.42 * **	0.14	0.24
Profit differential	−0.08 * **	0.01	0.92
Shapley	−0.31 * *	0.13	0.73
NBS	−0.12	0.13	0.89
Equal split	−0.85 * **	0.15	0.43
First treatment	0.23*	0.13	1.26
Second treatment	0.25*	0.14	1.29
Third treatment	0.12	0.13	1.13
Period 1	−0.06	0.19	0.94
Period 2	0.03	0.19	1.03
Period 3	−0.10	0.19	0.91
Period 4	−0.18	0.19	0.84
Period 5	0.09	0.19	1.10
Period 6	−0.07	0.19	0.93
Period 7	−0.04	0.19	0.97
Period 8	−0.01	0.19	0.99
Period 9	−0.12	0.19	0.89
Ryerson	−0.32 * **	0.09	0.72

*, **, *** indicate significant at the 10%, 5%, and 1% levels, respectively

There are several conclusions that are consistent across both regressions. First, period dummies are not significant, indicating there is no pattern of time decay typical of VCM public goods experiments (Ledyard 1995). Second, only the first treatment order is significant, indicating more cooperation relative to the fourth treatment. The binary decision to join or not and certainty over payoffs are potential explanations for the difference from previous VCM results.

5 Conclusion

This experiment shows that traditional cooperative allocation rules result in lower efficiency and coalition size than an optimal rule which explicitly addresses subjects' payoffs outside the coalition. The Shapley value and the Nash bargaining solution are used in the IEA literature as a zero-sum system of transfers, attempting to expand coalition membership by self-interested nations. Uniform rules fare even worse. When agents are inherently asymmetric they may not view "fair" and equal as the same thing.

Despite the strength of our results some aspects of the experimental setting might be improved. Given the sequence of actions, subjects did not have immediate information about other subjects leaving or joining a coalition. Thus, some may have been caught inside a three member coalition when they would have left if they had the choice. Our random period length attempted to mitigate this issue, but was imperfect. Since being outside a three member coalition is almost always the most preferred coalition structure, subjects may have tried to form the grand coalition and then leave. With Shapley and NBS all two member coalitions other than $[A, B]$ are internally, but not externally stable. This means that we would expect three member coalitions to form and then break down since the resulting three member coalitions are not internally stable. This "cycling" is one explanation why more three member coalitions formed for Shapley and NBS than we expected. There is no cycling effect for θ and equal split. Another possible design is to remove the time limit on periods, making subjects unanimously agree to end the period. Within period coalitions would provide additional information, but at the expense of fewer periods, particularly for the Shapley and NBS treatments where cycling should be more prevalent. Another possible design limitation is that some of the payoff differences between being in or out of a given coalition were very small. This resulted because we chose to test the standard IEA model and payoffs were scaled for appropriate subject earnings. Finally, we chose the Shapley value because of its use in the IEA literature. Croson et al. (2004) among others have noted that the Myerson-Shapley value (Myerson 1977) may be more appropriate for games with externalities, and when more than one coalition may form.

The practical implications of this experiment are clear. Efficiency in public goods games can be significantly improved by using a rule which directly addresses the free-rider payoff rather than an equal split or traditional cooperative game theory rules. The payoffs were generated using the benchmark model of IEAs, however the results apply to any pure public good. Compared to the VCM environment we have changed the subject's choice variable from a contribution, with an unknown contribution by other group members, to a binary decision to be in or out of a coalition with public information on coalition members. Our results refute two VCM stylized facts. First, we observe very little over-contribution relative to the theoretical prediction. Under θ, over-contribution (relative to the individual optimum) would imply the grand coalition. The grand coalition did not occur in any of the 170 observations. Second, we observe none of the time-decay present in VCM experiments.

One direction for future research would be to test the different optimal allocations. The optimal rules in Carraro et al. (2006), Weikard (2009) and McGinty (2007, 2011) propose different shares of the coalition surplus than the one tested in this paper. Each rule will result in the same set of stable coalitions, yet they differ in the payoff advantage to remaining a coalition member. McGinty (2011) shows theoretically that an equal share of the surplus maximizes the stability of a stable coalition, however no experimental comparison of these rules has determined which one performs better in the lab.

Acknowledgments This paper has benefited from comments by Rachel Croson, Dan Friedman, John Heywood and seminar participants at the ESA meetings. The authors wish to thank the UWM Center for Research on International Economics, the UWM Foundation and the Ryerson University Faculty of Arts for generous financial support. The usual caveat applies.

Appendix

Instructions

This experiment will consist of four different 'runs' each lasting 10 periods. In each run you will be randomly assigned a type [A, B, C, or D] and to a group of four people (you and three others). You will have the same type and the same group members for all 10 periods. At the end of each run you will randomly be assigned a new type and a new group. There is a 1/4 chance that you will be assigned a certain type each run. There are three four-person groups in this room.

In each period group members will decide to join or not join a subgroup. Your earnings for each period depend on your type (A, B, C, or D), if you are in or out of the subgroup, and the other members of the subgroup. The Earnings Table shows the earnings for each group member for all possible subgroups. You will have the information about your type and who is in the subgroup at all times on your computer screen.

In each period you will asked whether or not you wish to join the subgroup. After you and the other group members decide, you will see your earnings. You will also see who the other subgroup members are. At this point, you can click on the 'Confirm' button to confirm your decision or the 'Reverse' button to change your decision.

If anyone wishes to change their decision before confirming, everyone will be taken back to the initial decision screen. Once the person who wants to change makes their new decision, everyone returns to the confirmation screen. Everyone's earnings will be updated with the new information and you will again have the opportunity to confirm before the period ends. The period ends after all members of the group confirm their choices or when time runs out. Each period will last at least 60 s, after which it could end at any time.

At the end of the period, you will see your earnings and the subgroup members for that period and a history of subgroups and earnings for all prior periods from that run. Do not discuss your decisions with anyone in the room. Only the monitors will learn of your decisions. The other players do not know your type, only their own, and only the monitors know the identities of the group members.

Example: During run I, you are type B. The subgroup consists of $[A, D]$. Your earnings for the period are 51.4 if you stay out of the subgroup. If you decide to join, the subgroup will become $[A, B, D]$ and if that is the final subgroup your earnings will become 52.1.

Example: During run II, you are type C. The subgroup consists of $[A, C, D]$. Your earnings for the period are 66.6 if you remain in the subgroup. If you decide to leave the subgroup, it will become $[A, D]$ and your earnings will become 68.5. However, say the type A subgroup member also decides to leave. Then there is no subgroup (only D is left, so in the table it is $[A], [B], [C], [D]$) and you will earn 61.4.

Example: During run IV, you are type D. The subgroup consists of $[A, B, C,$ and $D]$. Your earnings for the period are 63.9 if this is the final subgroup. However, if both the type A and type B members leave the subgroup, your earnings will become 69.7.

See Table 7.

Table 7 Earnings table

Subgroup	Earnings for subgroup members	Earnings for non-subgroup members
Run I		
$[A], [B], [C], [D]$	NA	$[A = 30.7], [B = 46.0], [C = 61.4], [D = 76.7]$
$[A, B]$	$[A = 31.6, B = 46.1]$	$[C = 66.1], [D = 82.6]$
$[A, C]$	$[A = 32.6, C = 61.7]$	$[B = 50.4], [D = 84.1]$
$[A, D]$	$[A = 33.8, D = 77.4]$	$[B = 51.4], [C = 68.5]$
$[B, C]$	$[B = 47.0, C = 61.5]$	$[A = 32.9], [D = 82.3]$
$[B, D]$	$[B = 47.8, D = 76.9]$	$[A = 33.3], [C = 66.5]$
$[C, D]$	$[C = 62.4, D = 76.9]$	$[A = 32.8], [B = 49.3]$
$[A, B, C]$	$[A = 33.3, B = 50.8, C = 66.5]$	$[D = 98.0]$
$[A, B, D]$	$[A = 33.9, B = 52.1, D = 83.4]$	$[C = 79.1]$
$[A, C, D]$	$[A = 33.3, C = 68.9, D = 84.5]$	$[B = 59.5]$
$[B, C, D]$	$[B = 49.7, C = 66.9, D = 82.7]$	$[A = 38.9]$
$[A, B, C, D]$	$[A = 34.0, B = 54.5, C = 74.2, D = 93.0]$	NA

Table 7 continued

Subgroup	Earnings for subgroup members	Earnings for non-subgroup members
Run II		
$[A], [B], [C], [D]$	NA	[A = 30.7], [B = 46.0], [C = 61.4], [D=76.7]
$[A, B]$	$[A = 31.2, B = 46.5]$	$[C = 66.1], [D = 82.6]$
$[A, C]$	$[A = 31.8, C = 62.5]$	$[B = 50.4], [D = 84.1]$
$[A, D]$	$[A = 32.6, D = 78.7]$	$[B = 51.4], [C = 68.5]$
$[B, C]$	$[B = 46.6, C = 61.9]$	$[A = 32.9], [D = 82.3]$
$[B, D]$	$[B = 47.0, D = 77.7]$	$[A = 33, 3], [C = 66.5]$
$[C, D]$	$[C = 62.0, D = 77.3]$	$[A = 32.8], [B = 49.3]$
$[A, B, C]$	$[A = 35.1, B = 49.8, C = 65.8]$	$[D = 98.0]$
$[A, B, D]$	$[A = 41.3, B = 56.1, D = 72.0]$	$[C = 79.1]$
$[A, C, D]$	$[A = 37.3, C = 66.6, D = 82.8]$	$[B = 59.5]$
$[B, C, D]$	$[B = 51.2, C = 66.2, D = 82.0]$	$[A = 38.9]$
$[A, B, C, D]$	$[A = 41.2, B = 55.1, C = 71.2, D = 88.1]$	NA
Run III		
$[A], [B], [C], [D]$	NA	$[A = 30.7], [B = 46.0], [C = 61.4], [D = 76.7]$
$[A, B]$	$[A = 31.2, B = 46.5]$	$[C = 66.1], [D = 82.6]$
$[A, C]$	$[A = 31.8, C = 62.5]$	$[B = 50.4], [D = 84.1]$
$[A, D]$	$[A = 32.6, D = 78.7]$	$[B = 51.4], [C = 68.5]$
$[B, C]$	$[B = 46.6, C = 61.9]$	$[A = 32.9], [D = 82.3]$
$[B, D]$	$[B = 47.0, D = 77.7]$	$[A = 33.3], [C = 66.5]$
$[C, D]$	$[C = 62.0, D = 77.3]$	$[A = 32.8], [B = 49.3]$
$[A, B, C]$	$[A = 34.9, B = 50.2, C = 65.6]$	$[D = 98.0]$
$[A, B, D]$	$[A = 36.0, B = 51.3, D = 82.0]$	$[C = 79.1]$
$[A, C, D]$	$[A = 36.7, C = 67.4, D = 82.7]$	$[B = 59.5]$
$[B, C, D]$	$[B = 51.1, C = 66.5, D = 81.8]$	$[A = 38.9]$
$[A, B, C, D]$	$[A = 40.9, B = 56.2, C = 71.6, D = 86.9]$	NA
Run IV		
$[A], [B], [C], [D]$	NA	$[A = 30.7], [B = 46.0], [C = 61.4]. [D = 76.7]$
$[A, B]$	$[A = 38.9, B = 38.9]$	$[C = 66.1], [D = 82.6]$
$[A, C]$	$[A = 47.1, C = 47.1]$	$[B = 50.4], [D = 84.1]$
$[A, D]$	$[A = 55.6, D = 55.6]$	$[B = 51.4], [C = 68.5]$
$[B, C]$	$[B = 54.2, C = 54.2]$	$[A = 32.9], [D = 82.3]$
$[B, D]$	$[B = 62.4, D = 62.4]$	$[A = 33.3], [C = 66.5]$
$[C, D]$	$[C = 69.7, D = 69.7]$	$[A = 32.8], [B = 49.3]$
$[A, B, C]$	$[A = 50.2, B = 50.2, C = 50.2]$	$[D = 98.0]$
$[A, B, D]$	$[A = 56.5, B = 56.5, D = 56, 5]$	$[C = 79.1]$
$[A, C, D]$	$[A = 62.2, C = 62.2, D = 62.2]$	$[B = 59.5]$
$[B, C, D]$	$[B = 66.5, C = 66.5, D = 66.5]$	$[A = 38.9]$
$[A, B, C, D]$	$[A = 63.9, B = 63.9, C = 63.9, D = 63.9]$	NA

Runs I–IV are the theta proportion rule, Shapley value, Nash bargaining solution, and equal split, respectively. Treatment order was randomly determined

References

Barrett S (1994) Self-enforcing international environmental agreements. Oxf Econ Pap 46:878–894

Barrett S (1997) Heterogeneous international environmental agreements. In: Carraro CInternational environmental negotiations. Strategic policy issues, Edward Elgar

Bolton G (1997) The rationality of splitting equally. J Econ Beha Org 32:365–381

Bolton G (1998) Bargaining and dilemma games: from laboratory data towards theoretical synthesis. Exp Econ 1:257–281

Bolton G, Chatterjee K, McGinn K (2003) How communication links influence coalition bargaining: a laboratory investigation. Manag Sci 49(5):583–598

Botteon M, Carraro C (2001) Environmental coalitions with heterogenous countries: burden sharing and carbon leakage. In: Ulph A (ed) Environmental policy, international agreements, and international trade. Oxford University Press, New York

Burlando R, Guala F (2005) Heterogeneous agents in public goods experiments. Exp Econ 8:35–54

Carraro C, Eyckmans J, Finus M (2006) Optimal transfers and participation decisions in international environmental agreements. Rev Int Org 1(4):379–396

Carraro C, Siniscalco D (1993) Strategies for the international protection of the environment. J Public Econ 52:309–328

Chan K, Mestelman S, Moir R, Muller RA (1999) Heterogeneity and the voluntary provision of public goods. Expl Econ 2:5–30

Croson R, Gomes A, McGinn K, Nöth M (2004) Mergers and acquisitions: an experimental analysis of synergies. externalities and dynamics. Rev Financ 8:481–514

d'Aspremont C, Jacquemin A, Gabszewicz JJ, Weymark J (1983) On the stability of collusive price leadership. Can J Econ 16:17–25

Dannenberg A, Lange A, Sturm B (2010) On the formation of coalitions to provide public goods-experimental evidence from the lab. NBER working paper #15967

Finus M (2003) Stability and design of international environmental agreements: the case of transboundary pollution. In: Folmer H, Tietenberg T (eds) International yearbook of environmental and resource economics, 2003/2004. Edward Elgar, Cheltenham pp 82–158

Fischbacher U (2007) z-Tree: zurich toolbox for ready-made economic experiments. Exp Econ 10:171–178

Friedman D, Sunder S (1994) Experimental methods: a primer for economists. Cambridge University Press, Cambridge

Friedman J (1971) A non-cooperative equilibrium for supergames. Rev Econ Stud 38(113):1–12

Isaac RM, McCue K, Plott C (1985) Public goods provision in an experimental environment. J Public Econ 26:51–74

Keser C (1996) Voluntary contributions to a public good when partial contribution is a dominant strategy. Econ Lett 50:359–366

Laury S, Holt C (2005) Voluntary provision of public goods: Experimental results with interior Nash equilibria. In: Plott C, Smith V (eds) Handbook of experimental economics results, vol. 1

Ledyard J (1995) Public goods: a survey of experimental research. In: Kagel J, Roth A (eds) Handbook of experimental economics. Princeton University Press, Princeton pp 111–194

Leopold-Wildburger U (1992) Payoff divisions on coalition formation in a three-person characteristic function experiment. J Econ Behav Org 17:183–193

McGinty M (2007) International environmental agreements among asymmetric nations. Oxf Econ Pap 59(1):45–62

McGinty M (2011) A risk-dominant allocation: maximizing coalition stability. J Public Econ Theory 13(2):311–325

Myerson R (1977) Values of games in partition function form. Int J Game Theory 6(1):23–31

Roth A (1987) Bargaining phenomena and bargaining theory. In: Laboratory experimentation in economics: six points of view, Cambridge, UK, pp. 14–41

Weikard H-P (2009) Cartel stability under an optimal sharing rule. Manch Sch 77(5):575–593

Weikard H-P, Dellink R (2010) Sticks and carrots for the design of international climate agreements with renegotiations. Annal Oper Res (forthcoming)

Weikard H-P, Dellink R, Ecco van I (2010) Renegotiations in the greenhouse. Environ Resour Econ 45:573–596

[32]

Journal of Economic Theory **107**, 453–462 (2002)
doi:10.1006/jeth.2001.2949

NOTES, COMMENTS, AND LETTERS TO THE EDITOR

Sharing a River[1]

Stefan Ambec

INRA-SERD, Université de Grenoble, Grenoble, France; and CSEF, Università di Salerno, Salerno, Italy

and

Yves Sprumont[2]

CRDE and Département de Sciences Economiques, Université de Montréal, Montreal, Canada
yves.sprumont@umontreal.ca

Received July 18, 2000; final version received November 28, 2001

A group of agents located along a river have quasi-linear preferences over water and money. We ask how the water should be allocated and what money transfers should be performed. The core lower bounds require that no coalition should get less than the welfare it could achieve by using the water it controls. The aspiration upper bounds demand that no coalition enjoy a welfare higher than what it could achieve in the absence of the remaining agents. Exactly one welfare distribution satisfies the core lower bounds and the aspiration upper bounds: it is the marginal contribution vector corresponding to the ordering of the agents along the river. *Journal of Economic Literature* Classification Numbers: D62, C71.
Key Words: common property resources; fair allocation; core.

1. INTRODUCTION

The problem of river water sharing is of considerable practical importance: 148 rivers in the world flow through two countries, 30 through three,

[1] We thank E. Edmonds, M. Fleurbaey, M. Moreaux, C. Téjédo, and M. Truchon for useful comments. We are particularly indebted to an associate editor and two referees whose insightful remarks led to drastic improvements. Ambec gratefully acknowledges support from CIRANO, CRDE, and GREEN, and Sprumont thanks the Social Sciences and Humanities Research Council of Canada.

[2] Corresponding author. Département de Sciences Economiques, Université de Montréal, 3150 avenue Jean-Brillant, Montréal H3T 1N8, Canada. Fax: 514 343 5831.

0022-0531/02 $35.00

9 through four, and 13 through five or more (Barret [1]). Allocating a resource over which property rights are not well defined is notoriously problematic. In the case of river water sharing, efficiency often requires that upstream agents limit their own consumption so as to increase that of downstream agents whose marginal benefits are higher. The Coase [2] theorem suggests that agents would efficiently trade water for money if property rights over water were well defined. This is not the case, however, and conflicting views on how rights should be defined are expressed in the two main doctrines advocated in international disputes (see, e.g., Godana [4] and Kilgour and Dinar [6]). The theory of *absolute territorial sovereignty* (or ATS for short) states that a country has absolute sovereignty over the area of any river basin on its territory. On the other hand, the theory of *unlimited territorial integrity* (or UTI for short) forbids a country to alter the natural conditions on its own territory to the disadvantage of a neighboring country. In a nutshell, ATS grants a country property rights over the water originating on its territory while UTI extends these rights to the full stream of water originating upstream.

Our purpose in this note is to propose a compromise between these two doctrines. We study a model where a set of agents ordered along a river have quasi-linear preferences over water and money. Efficiency amounts to maximizing total welfare, and the property rights issue translates into a welfare distribution problem that we analyze from a cooperative game-theoretic viewpoint.[3] The location of a coalition of agents along the river determines the quantity of water they control and, thereby, the welfare they can secure to themselves. The ATS doctrine thus implies core-like constraints: each coalition should get at least its secure welfare. The interpretation of UTI is less straightforward. According to that doctrine, agents are entitled to consume the full stream of water originating upstream from their location and, therefore, have a legitimate claim to the welfare level corresponding to that consumption. But such welfare claims are typically incompatible and the UTI doctrine is silent on how to reconcile them. We propose to treat them as legitimate aspirations. Since sharing a river involves negative externalities in the precise sense that it is impossible to guarantee to every agent his aspiration welfare level, we suggest that everyone should take up a share of these externalities: no one should end up above his aspiration welfare. Pushing the argument one step further, we define the aspiration welfare level of an arbitrary coalition of agents to be the highest welfare it could achieve in the absence of the others.

[3] For a survey of the numerous applications of cooperative game theory to economic problems with a graph structure, see Sharkey [9]. A distinguishing feature of our setup is that agents are endowed with full-fledged preferences.

Our criterion requires that no coalition enjoy more than its aspiration welfare.

The very particular nature of the feasibility constraints at play (agents are ordered and water can only be transferred downstream) implies directly that at most one welfare distribution satisfies both the secure welfare lower bounds and the upper bounds defined by the coalitional aspiration levels.[4] This is the "downstream incremental" distribution, according to which an agent's welfare is just his marginal contribution to the coalition composed of his predecessors along the river. We show that this distribution does indeed meet the desired welfare bounds.

To be sure, this result is quite special to our model. Yet, other interesting allocation problems involve highly structured feasibility constraints reminiscent of the ones studied here. Multiperiod models, for instance, where agents or generations are naturally ordered along a time line, share some essential features with our setup. More generally, networks of exchange where not all participants can meet are of a similar nature. We believe that the theory of collective choices has much to gain from a systematic exploitation of the structure of the feasibility constraints in such environments.

2. A MODEL

A river flows through a number of countries, regions, cities, or irrigated corps, henceforth called agents, whose set is denoted by $N = \{1, ..., n\}$. We identify agents with their location along the river and number them from upstream to downstream: $i < j$ means that i is upstream from j. We assume that different agents are located at different points along the river. If $S, T \subset N$, we write $S < T$ if $i < j$ for every $i \in S$ and every $j \in T$. We denote by min S and max S, respectively, the smallest and largest members of S. It will be convenient to define the sets of *predecessors* and *strict predecessors* of agent i, respectively, by $Pi = \{j \in N : j \leqslant i\}$ and $P^0 i = Pi \setminus \{i\}$. We often write i instead of $\{i\}$.

The river picks up volume along its course: the flow at its source, $e_1 > 0$, is increased by the amount $e_i \geqslant 0$ between locations $i-1$ and i, say, without loss of generality, at i. A schematic representation is given in Fig. 1. Money is available in unbounded quantities to perform side-payments. Agents value money and the water from the river. Agent i's utility from consuming x_i units of water and receiving a net money transfer t_i is $u_i(x_i, t_i) = b_i(x_i) + t_i$. We call $b_i: \mathbb{R}_+ \to \mathbb{R}$ agent i's *benefit function*. It is assumed to be

[4] Thus, combining stability and fairness criteria forces a unique outcome. A similar point was made in a more general context by Dutta and Ray [3], who showed that the Lorenz partial order possesses a unique maximal element in the core of a convex game.

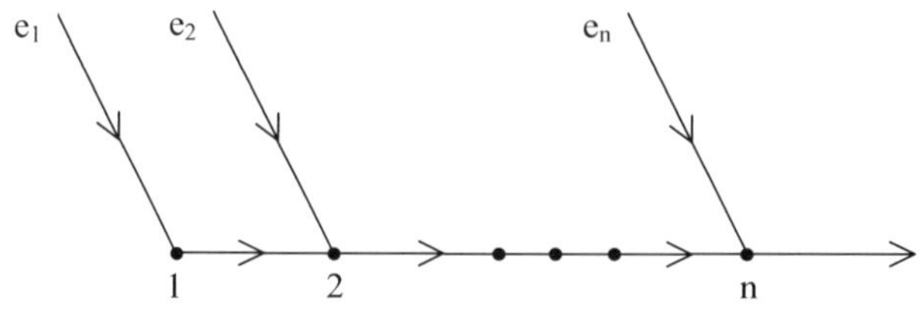

FIG. 1. A schematic representation of the model.

differentiable at every $x_i > 0$, strictly increasing, and strictly concave. We denote its derivative by b'_i and assume that $b'_i(x_i)$ goes to infinity as x_i tends to 0. The list (N, e, b), where $e = (e_1, ..., e_n)$ and $b = (b_1, ..., b_n)$, constitutes our *problem*.

A *consumption plan (for N)* is any vector $x \in \mathbb{R}^N_+$. For an arbitrary nonempty *coalition* $S \subset N$, $x_S \in \mathbb{R}^S_+$ denotes the restriction of x to S: it is a consumption plan for S. An *allocation* is a vector $(x, t) = (x_1, ..., x_n, t_1, ..., t_n) \in \mathbb{R}^N_+ \times \mathbb{R}^N$ satisfying the feasibility constraints

$$\sum_{i \in N} t_i \leqslant 0, \tag{1}$$

$$\sum_{i \in Pj} (x_i - e_i) \leqslant 0 \qquad \text{for all} \quad j \in N. \tag{2}$$

Observe that we treat water as a private good. The water stream e_i can only be consumed by the followers of i: this makes our problem different from that of allocating a stock of some standard private good with the possibility of side-payments.

A (*welfare*) *distribution* is any vector $z = (z_1, ..., z_n) \in \mathbb{R}^N$ which is the utility image of some allocation (x, t) in the sense that $z_i = b_i(x_i) + t_i$ for each agent i.

Because preferences are quasi-linear, an allocation $(x^*(N), t^*(N))$ is (Pareto) efficient if and only if it maximizes the sum of all agents' benefits and wastes no money. We call $x^*(N)$ an *optimal consumption plan*.[5] Our assumptions guarantee that this plan is unique. Its structure is simple: the marginal benefits decrease (weakly) as one moves downstream and, if two agents have different marginal benefits, some constraint must be binding between them. Formally, one easily shows that there exists a partition $\{N_k\}_{k=i, ..., K}$ of N and a list $\{\beta_k\}_{k=1, ..., K}$ of positive numbers such that

[5] This is a vector in $\mathbb{R}^N_+$. Our notation departs from the widespread convention that $x(N)$ is the sum of the components of a vector x.

$$N_k < N_{k'} \text{ and } \beta_k \geqslant \beta_{k'} \qquad \text{whenever} \quad k < k', \tag{3}$$

$$b_i'(x_i^*(N)) = \beta_k \qquad \text{for every} \quad i \in N_k \quad \text{and every} \quad k = 1, \ldots K, \qquad \text{and} \tag{4}$$

$$\sum_{i \in N_k} (x_i^*(N) - e_i) = 0 \qquad \text{for every} \quad k = 1, \ldots, K. \tag{5}$$

We now have to split the maximal welfare $\sum_{i \in N} b_i(x_i^*(N))$ among the agents.

3. LOWER AND UPPER BOUNDS ON WELFARE

Under the ATS doctrine, agents are entitled to use the water they control as they like. This generates coalitional welfare lower bounds in a straightforward way. Following Greenberg and Weber [5], call a coalition T *consecutive* if $k \in T$ whenever $i, j \in T$ and $i < k < j$. Every coalition S admits a unique coarsest partition $\mathscr{S}$ into consecutive components. Let $x^*(S)$ be the unique consumption plan for S that maximizes $\sum_{i \in S} b_i(x_i)$ subject to the constraints

$$\sum_{i \in Pj \cap T} (x_i - e_i) \leqslant 0 \qquad \text{for all} \quad j \in T \quad \text{and} \quad T \in \mathscr{S}, \tag{6}$$

and call

$$v(S) = \sum_{i \in S} b_i(x_i^*(S)) \tag{7}$$

the secure benefit of S. Observe that $v(S) = \sum_{T \in \mathscr{S}} v(T)$: coalition S cannot secure more because any water left over by one of its consecutive components cannot be guaranteed for the consumption of any other component. The game v is an example of what Greenberg and Weber [5] call a consecutive game, in which only consecutive coalitions generate surplus. We say that a welfare distribution $z = (z_1, \ldots, z_n)$ satisfies the *core lower bounds* if $\sum_{i \in S} z_i \geqslant v(S)$ for every $S \subset N$.

Next, we turn to the upper bounds implied by our interpretation of the UTI doctrine. Even in the absence of the other agents, agent i would not be able to consume more than the full stream of water running through his location, thereby enjoying his *aspiration welfare* $w(i) = b_i(\sum_{j \in Pi} e_j)$. In the Lockean tradition, agent i has a legitimate right to that welfare level, but not to more.[6] Unfortunately, the welfare distribution$(w(1), \ldots, w(n))$ is

[6] "Now, of those good things which nature hath provided in common, every one had a right, has hath been said, to as much as he could use, and property in all he could effect with his labour; all that his industry could extend to, to alter from the state nature had put it in, was his" (Locke [7, p. 24]). In our model, agent i's industry could not extend to e_j, $j > i$.

not feasible as soon as $n \geqslant 2$: in Moulin's [8] terms, the problem (N, e, b) exhibits "negative group externalities." Because no particular agent bears any distinguished responsibility for these externalities, it is natural to ask that everyone takes up a share of them: no one should end up above his aspiration welfare. This argument generalizes to coalitions in a very natural way. The *aspiration welfare* of an arbitrary coalition S is the highest welfare it could achieve in the absence of $N \setminus S$. It is obtained by choosing a consumption plan x_S maximizing $\sum_{i \in S} b_i(x_i)$ subject to the constraints

$$\sum_{i \in Pj \cap S} x_i \leqslant \sum_{i \in Pj} e_i \qquad \text{for every} \quad j \in S. \tag{8}$$

This problem has a unique solution, which we denote by $x^{**}(S)$. The aspiration welfare of S is

$$w(S) = \sum_{i \in S} b_i(x_i^{**}(S)) \tag{9}$$

and we say that a welfare distribution z *satisfies the aspiration upper bounds* if $\sum_{i \in S} z_i \leqslant w(S)$ for every $S \subset N$. The motivation for these coalitional upper bounds is the same as for the individual bounds. The aspirations of any two complementary coalitions S and $N \setminus S$ are incompatible: $w(S) + w(N \setminus S) > v(N)$. Thus, if $\sum_{i \in S} z_i > w(S)$, the opposite inequality must hold for the complement coalition: $\sum_{i \in N \setminus S} z_i < w(N \setminus S)$. This means that S benefits from the existence $N \setminus S$ while $N \setminus S$ suffers from the existence of S. We believe that there is no sound basis for such a differential treatment if none of the agents bears any responsibility for the existence of the others. Consequently, neither S nor $N \setminus S$ ought to enjoy more than their aspiration upper bounds.

An obvious but powerful relation between the core lower bounds and the aspiration upper bounds should be noted immediately: $v(Pi) = w(Pi)$ for every $i \in N$. Because of this relation, combining the two sets of bounds yields a clear-cut recommendation. Only one welfare distribution passes both tests: it is the *downstream incremental* distribution z^* defined by $z_i^* = v(Pi) - v(P^0 i)$ or, equivalently, $z_i^* = w(Pi) - w(P^0 i)$ for each $i \in N$. The theorem below establishes, in addition, that the downstream incremental distribution meets the core lower bounds and the aspiration upper bounds for *every* coalition.

Our theorem and the very definition of z^* immediately deliver two simple dual characterizations of the downstream incremental distribution: (i) among all core distributions, it is the one that lexicographically maximizes the welfare of agents $n, n-1, \ldots, 2, 1$, and (ii) among all distributions meeting the aspiration upper bounds, it lexicographically maximizes the welfare of $1, 2, \ldots, n-1, n$. In this sense, it may be regarded as a sensible compromise between the ATS and UTI doctrines.

THEOREM. *The downstream incremental distribution z^* is the unique distribution satisfying the core lower bounds and the aspiration upper bounds.*

Proof. It is easy to see that if a welfare distribution z satisfies both the core lower bounds and the aspiration upper bounds, then $z = z^*$. Since $v(1) = w(1)$, we must have $z_1 = z_1^*$. Next, fix $j < n$ and suppose $z_i = z_i^*$ for all $i \leqslant j$. Since $v(P(j+1)) = w(P(j+1))$, the core constraints and the aspiration upper bounds force $\sum_{i \in P(j+1)} z_i = v(P(j+1))$, hence $z_{j+1} = v(P(j+1)) - \sum_{i \in Pj} z_i$. By the induction hypothesis, $\sum_{i \in Pj} z_i = \sum_{i \in Pj} z_i^* = v(Pj)$. Therefore, $z_{j+1} = v(P(j+1)) - v(Pj) = z_{j+1}^*$, and we are done.

Next we turn to the converse claim, namely, that z^* does satisfy the core lower bounds and the aspiration upper bounds. Because v is a consecutive game, the core lower bounds need only be checked for the *consecutive* coalitions. But for any consecutive S, observe that $\{S, P^0 \min S\}$ is a partition of $P \max S$. Therefore

$$\sum_{i \in S} z_i^* := v(P \max S) - v(P^0 \min S) \geqslant v(S) \tag{10}$$

simply because v is superadditive.

The proof of the aspiration upper bounds is a bit more involved. It uses the following restricted form of concavity of the game w.

LEMMA. *If $S \subset T \subset N$ and $T < i$, then $w(S \cup i) - w(S) \geqslant w(T \cup i) - w(T)$.*

As a first step in the proof of this lemma, let us show that if $\varnothing \neq S \subset T \subset N$, then $x^{**}(S) \geqslant x_S^{**}(T)$. Clearly, it suffices to establish that $x^{**}(S) \geqslant x_S^{**}(S \cup t)$ whenever $\varnothing \neq S \neq N$ and $t \in N \backslash S$. Write $x^{**}(S) = x$ and $x_S^{**}(S \cup t) = y$. All agents under consideration in the argument belong to S. By definition of x and y, $\sum_{i \in S} (y_i - x_i) \leqslant 0$. Let $i_1 \leqslant \cdots \leqslant i_L$ be those i such that $x_i \neq y_i$ (if none exists, there is nothing to prove). We claim that $y_{i_L} - x_{i_L} < 0$. Suppose, by contradiction, that the opposite (necessarily strict) inequality is true. Let j be the largest predecessor of i_L such that $y_j - x_j < 0$ (which necessarily exists) and define $y_{i_L}^\varepsilon = y_{i_L} - \varepsilon$, $y_j^\varepsilon = y_j + \varepsilon$, $y_i^\varepsilon = y_i$ for $i \neq i_L, j$. Since $b_j'(y_j) > b_j'(x_j) \geqslant b_{i_L}'(x_{i_L}) > b_{i_L}'(y_{i_L})$, choosing $\varepsilon > 0$ small enough ensures that $\sum_{i \in S} (b_i(y_i^\varepsilon) - b_i(y_i)) > 0$ while y^ε meets the same constraints as y, a contradiction. Because $y_{i_L} - x_{i_L} < 0$, it now follows that $y_{i_l} - x_{i_l} < 0$ successively for $l = L-1, \ldots, 1$.

Moving to the second step, let $S \subset T \subset N$ and $T < i$. Define $x'_i = x_i^{**}(T \cup i)$ and $x'_j = x_j^{**}(T \cup i) + x_j^{**}(S) - x_j^{**}(T)$ for $j \in S$. By our first step, $x_j^{**}(T \cup i) \leqslant x_j^{**}(T) \leqslant x_j^{**}(S)$ for all $j \in S$. Therefore $0 \leqslant x_j^{**}(T \cup i) \leqslant x'_j \leqslant x_j^{**}(S)$ for all $j \in S$ and the consumption plan x' for $S \cup i$ satisfies the same constraints as $x^{**}(S \cup i)$, namely, $\sum_{k \in Pj \cap (S \cup i)} x'_k \leqslant \sum_{k \in Pj} e_k$ for all $j \in S \cup i$. Hence, $w(S \cup i) \geqslant \sum_{j \in S \cup i} b_j(x'_j)$ and

$$w(S \cup i) - w(S) \geqslant b_i(x'_i) + \sum_{j \in S} [b_j(x'_j) - b_j(x_j^{**}(S))]. \tag{11}$$

On the other hand, since $x_j^{**}(T \cup i) \leqslant x_j^{**}(T)$ for all $j \in T \setminus S$,

$$w(T \cup i) - w(T) \leqslant b_i(x'_i) + \sum_{j \in S} [b_j(x_j^{**}(T \cup i)) - b_j(x_j^{**}(T))]. \tag{12}$$

Since $x'_j - x_j^{**}(S) = x_j^{**}(T \cup i) - x_j^{**}(T)$ and $x_j^{**}(T \cup i) \leqslant x'_j$ for all $j \in S$, it follows from (11), (12), and the concavity of the benefit functions that $w(T \cup i) - w(T) \leqslant w(S \cup i) - w(S)$. This completes the proof of the lemma.

Returning to the aspiration upper bounds, we combine the above lemma with the definitions of z^* and w. For any coalition S, we obtain

$$\sum_{i \in S} z_i^* = \sum_{i \in S} [w(Pi) - w(P^0 i)] \leqslant \sum_{i \in S} [w(Pi \cap S) - w(P^0 i \cap S)] = w(S), \tag{13}$$

where the last equality holds because $P^0 i \cap S = P[\max(P^0 i \cap S)] \cap S$ for all $i \in S$, so that all terms but $w(P(\max S) \cap S) = w(S)$ and $-w(P^0(\min S) \cap S) = 0$ cancel out. ∎

4. CONCLUDING COMMENTS

We conclude by mentioning an easy extension of our work and two directions for future research.

(1) We assumed that agents are completely ordered along the river. Our theorem, however, generalizes to the case where agents are located at different points along several rivers that merge as they flow downstream. Formally, let the "being upstream from" relation $\leqslant$ be a transitive and antisymmetric relation that is "quasi-complete": for each $i \in N$, the set of *followers* $Fi := \{j \in N : i \leqslant j\}$ is completely ordered by $\leqslant$. Interpret $\min S$ and $\max S$ as the sets of minimal and maximal elements of $\leqslant$ in S. Note that $\max P^0 i$ is merely the set of "immediate" predecessors of i: in our original model, $\max P^0 i = \{i-1\}$. Define allocations as before: (1) and (2) remain the relevant constraints. Every coalition S now admits a unique

coarsest partition $\mathscr{S}$ into *connected* components, where a coalition T is connected if $\max T$ is a singleton and $k \in T$ whenever $i \in T$ and $i < k < \max T$. With this definition of $\mathscr{S}$, the core lower bounds for S remain properly defined by (7) and (6). The aspiration upper bounds are still adequately expressed by (9) and (8).

As before, $v(Pi) = w(Pi)$ for all $i \in N$. Moreover,

$$v(P^0 i) = \sum_{j \in \max P^0 i} v(Pj) = \sum_{j \in \max P^0 i} w(Pj) = w(P^0 i) \qquad \text{for all} \quad i \in N. \tag{14}$$

The two original definitions of z^* therefore remain equivalent. More important, (14) ensures that the identity $\sum_{i \in N} z_i^* = v(N)$ still holds. Our theorem extends to this model. Let us sketch the changes required in the proof.

Imposing the two sets of bounds on a distribution z forces $z_i = z_i^*$ for all $i \in N$. Simply carry the induction argument on the number of agents in the longest completely ordered subset of Pi rather than on the cardinality of Pi.

To prove the lower bounds, observe that for any connected S, $\{S, P^0 \min S\}$ is a partition of $P \max S$ (where it is understood that $P^0 \min S = \bigcup_{i \in \min S} P^0 i$), so that (10) follows again from the superadditivity of v. The extension to arbitrary coalitions is obvious.

As for the upper bounds, observe first that the last equality in (13) remains valid because $w(P^0 i \cap S) = \sum_{j \in \max(P^0 i \cap S)} w(Pj \cap S)$. The inequality in (13) rests on the lemma, whose proof is virtually unchanged. Only its first step requires a modification. Let again $\varnothing \neq S \neq N$, $t \in N \backslash S$, $x^{**}(S) = x$ and $x_S^{**}(S \cup t) = y$. Define $I_L = \max\{i \in S : x_i \neq y_i\}$. By optimality of x and y, either $y_i - x_i > 0$ for all $i \in I_L$ or all the reverse inequalities hold. In the former case, there exists a predecessor j of some $i \in I_L$ such that $y_j - x_j < 0$ and picking a maximal one yields a contradiction as before. We conclude that $y_i - x_i < 0$ for all $i \in I_L$ and it again follows easily that $y_i - x_i \leqslant 0$ for all $i \in S$.

(2) We assumed that the (net) benefit functions are increasing. It would be worthwhile exploring the case where they eventually decrease. One twist is that two meaningful definitions of the lower bounds (corresponding to soft and hard interpretations of the ATS doctrine) are possible, depending on whether a coalition is entitled to use the water that its complement does not find profitable to consume.

(3) An important aspect of river management is pollution control. Our model may help understand part of the issue by reinterpreting x_i as a measure of the level of pollution generated by agent i and e_i as a ceiling to the admissible pollution level at location i. Of course, these ceilings are not

immutable constraints and should be agreed upon. This complicates the collective choice problem substantially.

REFERENCES

1. S. Barret, "Conflict and Cooperation in Managing International Water Resources," Policy Research Working Paper 1303, World Bank, Washington, 1994.
2. R. Coase, The problem of social cost, *J. Law Econ.* **1** (1960), 1–44.
3. B. Dutta and D. Ray, A concept of egalitarianism under participation constraints, *Econometrica* **57** (1989), 615–635.
4. B. Godana, "Africa's Shared Water Resources," Frances Pinter, London, 1985.
5. J. Greenberg and S. Weber, Strong tiebout equilibrium under restricted preferences domain, *J. Econ. Theory* **38** (1986), 101–117.
6. M. Kilgour and A. Dinar, "Are Stable Agreements for Sharing International River Waters Now Possible?," Policy Research Working Paper 1474, World Bank, Washington, 1996.
7. J. Locke, "The Second Treatise of Civil Government" (J. W. Gough, Ed.), Basil Blackwell, Oxford, 1948. [Original published in 1690]
8. H. Moulin, Uniform externalities: Two axioms for fair allocation, *J. Pub. Econ.* **43** (1990), 305–326.
9. W. Sharkey, Network models in economics, *in* "Handbook in Operations Research and Management Science" (M. O. Ball *et al.*, Eds.), Vol. 8, pp. 713–765, Elsevier, Amsterdam, 1995.

[33]

ELSEVIER

Journal of Public Economics 87 (2003) 2049–2067

JOURNAL OF PUBLIC ECONOMICS

www.elsevier.com/locate/econbase

Cooperation in international environmental negotiations due to a preference for equity

Andreas Lange[a,b,*], Carsten Vogt[a]

[a]*Centre for European Economic Research (ZEW), Postfach 103443, 68034 Mannheim, Germany*
[b]*Interdisciplinary Institute for Environmental Economics, University of Heidelberg, Heidelberg, Germany*

Received 27 February 2001; received in revised form 28 August 2001; accepted 27 November 2001

Abstract

This paper demonstrates that cooperation in international environmental negotiations can be explained by preferences for equity. Within a N-country prisoner's dilemma in which agents can either cooperate or defect, in addition to the standard non-cooperative equilibrium, cooperation of a large fraction or even of all countries can establish a Nash equilibrium. In an emission game, however, where countries can choose their abatement level continuously, equity preferences cannot improve upon the standard inefficient Nash equilibrium. Finally, in a two stage game on coalition formation, the presence of equity-interested countries increases the coalition size and leads to efficiency gains. Here, even a stable agreement with full cooperation can be reached.

Keywords: International environmental negotiations; Cooperation; Equity preference; Coalition formation

JEL classification: C7; D63; H41; Q00

1. Introduction

It is well known that international cooperation is needed in order to deal effectively with environmental problems, such as global warming and the

*Corresponding author. Tel.: +49-621-1235-208; fax: +49-621-1235-226.
E-mail addresses: lange@zew.de (A. Lange), vogt@zew.de (C. Vogt).

0047-2727/02/$ – see front matter
doi:10.1016/S0047-2727(02)00044-0

depletion of the ozone layer. Since pollution crosses borders and causes negative effects on a global scale, any country that reduces emissions supplies a public good, e.g. the reduction of climate change or the protection of the ozone layer.

The impact of a single country on the global pollution level, however, is rather small. Hence, when reducing emissions, a country must bear abatement costs but benefits only little, i.e. the incentives for a country to reduce emissions are small. Rather, the country has an incentive to free-ride on (possible) emission reductions made by the rest of the world. Consequently, although it pays if all parties cooperate, each agent unilaterally has an incentive to defect from cooperation. Thus, the incentive structure can be described as a prisoner's dilemma, clearly leading to an undesired outcome. This incentive problem is rather severe, because—contrary to locally bounded pollution—there is no supranational authority that can enforce cooperative behavior.

However, the last two decades have shown an increasing number of negotiations and protocols on subjects of (global) environmental concern. Perhaps most prominent are the Montreal Protocol on Substances that Deplete the Ozone Layer (1987), the Basel Convention on the Control of Transboundary Movements of Hazardous Waste (1989), several agreements under the 1979 Convention on Long-Range Transboundary Air Pollution, and the UN Framework Convention on Climate Change (1992), which established a process leading to the Kyoto Protocol (1997) which specifies targets for the reduction of greenhouse gases. However, it is not absolutely clear whether these agreements go substantially beyond actions that countries would have done in their self-interest anyway. For example, Murdoch and Sandler (1997) argue that the Montreal Protocol does not prove cooperative behaviour, instead it might be a symbolic policy agreement. The Kyoto Protocol has not yet been ratified by most countries, although it specifies substantial emission reduction targets. However, despite the drop out of the US, the political agreement of the remaining countries that was reached at 6th Conference of the Parties (COP6) in Bonn in July 2001, seems to stabilize the protocol. Clearly, whatever the result of the ongoing negotiations will be, any adopted policy must be in the interest of each party involved.

In attempting to explain cooperation and coalition formation, most theoretical models use a two-period structure as introduced by Barrett (1992, 1994). Here, countries must first decide whether or not to join a coalition. In a second step, both the coalition and the remaining countries choose their emission levels non-cooperatively. A coalition is stable if no country wants to join the coalition and no country has an incentive to leave. Simulations by Barrett (1992, 1994, 1997), Carraro and Siniscalco (1993), and Hoel (1992) have shown that—although there is cooperation—the coalition size is rather small.

The question remains how the cooperation that is observed in prisoner's dilemma situations as well as in some international agreements can be explained. Here, besides the symbolic policy arguments mentioned above, the literature refers to repeated games: For prisoners dilemma situations cooperation can prevail due to an infinite repetition of the one stage game.[1] In this paper, however, we follow a different approach and study the effect of equity (fairness) preferences for the formation of cooperation. Fair behavior was observed in experimental economics;[2] arguments related to fairness and equity, however, can also be found in international negotiations as on the abatement of greenhouse gases.[3]

Often it is called for an equalization of per capita emission levels in the long run. On the one hand, specifically developing countries claim that developed countries with high per capita emission levels of greenhouse gases are responsible for global warming; for example the leader of the Indian delegation at the negotiations on the Rio Convention 1991 demanded that, within the negotiations, 'the principle of equity should be the touchstone for judging any proposal' (Dasgupta, 1994, p. 133). On the other hand, environmental interest groups in developed countries also demand emissions based on equal per capita levels. For example, a petition called 'The Hague Mandate', circulated by the Climate Action Network (CAN), was signed by more than 85 international organizations. In this they call for immediate action on climate change and argue that 'no citizen has a right to pollute more than any other; ...every country has a duty to ensure its emissions do not exceed its global per capita share.'[4] A government, facing voters with such preferences, clearly has to take those into account. Thus, using fairness preferences for international negotiations can be motivated by political economy arguments. The weight that a government attaches to the relativity argument will then depend on the impact these interest groups have on the national policy.

In this paper, we look at the implications of equity preferences for international cooperation in reduction of some global pollutant like the greenhouse gas CO_2. We concentrate on the case of symmetric countries, i.e. countries which are identical with respect to both their abatement costs and their benefits from abatement. For those, clearly, all the fairness arguments mentioned above coincide: equal payoffs are obtained if (per capita) emissions as well as abatement efforts coincide. We rely on a preference structure given by the ERC theory (Bolton and Ockenfels,

[1] By the different Folk theorems, basically any payoff vector can be sustained as a Nash equilibrium under certain circumstances.

[2] For a survey of bargaining experiments, see Roth (1995).

[3] An overview of alternative equity criteria for climate change policy is given by Cazorla and Toman (2001). The principle of equity is even fixed in Article 3 of the Convention on Climate Change as well as in the decision approved by the COP 6 in Bonn which states that measures shall be implemented '...with a view to reducing emissions in a manner conducive to narrowing per capita differences between developed and developing country Parties'.
Other equity arguments often used are with respect to equal burden shares or equal percentage reduction of some pollutants. For example, equal measures for the signing countries are used in the LRTAP protocol on reduction of sulphur emissions (minus 30% based on 1980 level) or in the LRTAP protocol concerning the control of volatile organic compounds (minus 30% based on 1999 level).

[4] The text can be found at www.foeeurope.org/dike/news/1001-01en.htm, the list of signers at: www.foeeurope.org/dike/news/1001-01.htm.

2000, 2001).[5] This theory explains most of the behaviour of agents observed in diverse experiments,[6] but deviates little from the traditional utility concept. The utility of an agent is not solely based on the absolute payoff but also on the relative payoff compared to the overall payoff to all agents. Given a certain relative payoff share, the utility is strictly increasing in the own absolute payoff of the agent. Given a fixed absolute payoff, the agent is best off when receiving just the equal (fair) share. To both sides of this equal share, i.e. when receiving less or more than the fair amount, utility is lower, even if the absolute payoff does not change.[7] By trading off equity and absolute payoff, this preference structure is in line with the observation by Cazorla and Toman (2001, p. 238) that 'Equity might be one motivation for countries to pursue GHG emissions policies. However, equity principles will not override other elements of national self-interest.'

We first study a symmetric N-country prisoner's dilemma game. We show that there may be Nash equilibria in which countries cooperate: If, for example, the rest of the world plays cooperatively, a country can get the equal share by choosing to cooperate as well. Hence, if it values its relative payoff being close to the equal share more than its absolute payoff, it will choose to complete the grand coalition. Clearly, partial cooperation can also occur. For such equilibria, we show that the fraction of cooperating countries is rather large ($>$50 percent).

Note, however, that in the prisoner's dilemma, the countries have only the discrete choice of cooperating or defecting, while with respect to the global warming problem, countries can choose their emission level continuously. We therefore analyse a symmetric continuous emission game. Here, ERC alone cannot improve upon the non-cooperative Nash equilibrium with standard preferences. The reason is that as long as a country receives less than the equal share and has the possibility to increase its absolute payoff, it will do so. Since this holds for all countries, the unique Nash equilibrium therefore is identical to the symmetric Nash equilibrium with standard preferences.

This result changes, however, if a stage of coalition formation is introduced. Here, in contrast to the traditional models by Barrett (1994), Carraro and Siniscalco (1993), etc., coalitions that involve a rather large fraction of countries can be stabilised. If each country puts enough weight on getting close to the equal payoff share, even the grand coalition can be obtained in equilibrium. In general,

[5] ERC hereby stands for Equity, Reciprocity, and Competition.

[6] As already noted by Bolton and Ockenfels, this theory can generate cooperation in the standard prisoner's dilemma.

[7] Note that such a preference for equity is self-centred only and is distinct from altruism (Bolton and Ockenfels, 2001). A country's utility is determined solely by its own absolute and relative payoff. For our application to the abatement of greenhouse gases, countries (voters) clearly might also care about other aspects of equity as for example inequalities within the rest of the world. However, each country can only influence its own emission levels, and thereby has an impact on its relative position to the rest of the world. Inequalities among other countries it cannot directly influence. Therefore, it seems to be appropriate to include the average of other countries only.

the presence of (some) equity-interested countries increases the incentive to join a coalition and, therefore, leads to a larger coalition size in equilibrium. The efficiency gains are maximal if countries that are interested in equity stay outside the coalition and reward a higher abatement effort of the coalition with an expansion of their own abatement level.

The paper is organised in the following way: In Section 2, the basic features of the ERC theory are introduced. We then study the discrete version of a symmetric N-country prisoner's dilemma in Section 3, whereas the emission game is analysed in Section 4. Here, we look at the Nash equilibria of the one shot game first and then study stable coalitions in the two-stage coalition formation game. The final section—as always—concludes.

2. The preference structure

The analysis in this paper relies on a preference structure in which players—along with their own absolute payoff—are motivated (non-monotonously) by the relative payoff share they receive, i.e. how their standing compares to that of others. With this, we rely on the ERC model by Bolton and Ockenfels (2000) but use a full information framework. This theory explains the behaviour of people observed in a rather large group of experiments better than the standard theory. As already mentioned, it can also be rationalised for countries (governments) whose citizens have some preference for fair sharing abatement efforts among countries.

Let the (non-negative) payoff to agent i be denoted by y_i, $i = 1, \ldots, N$, the relative share by $\sigma_i = y_i/(\Sigma_j y_j)$. The utility function is given by:

$$a_i u(y_i) + b_i r(\sigma_i)$$

where $a_i, b_i \geq 0$, $u(\cdot)$ is differentiable, strictly increasing and concave, and $r(\cdot)$ is differentiable, concave and has its maximum at $\sigma_i = 1/N$. Throughout this paper we assume that agent's disutility from disadvantageous inequality is larger than if the agent is better off than average, i.e. $r(1/N - x) \leq r(1/N + x)$ for all $x \in [0, 1/N]$. The types of countries are characterised by the relative weights a_i/b_i.

3. The prisoner's dilemma

In this section we study a simple symmetric N-country prisoner's dilemma (PD game) where each country can cooperate, 'c', or defect, 'd'. In terms of the climate change problem: The country either reduces emissions by a (pre-) specified amount or it doesn't. This situation may occur when countries have to decide whether or not to ratify an international agreement which already specifies the emission levels that must be abated.

2054 A. Lange, C. Vogt / Journal of Public Economics 87 (2003) 2049–2067

3.1. The payoff structure

The total number of cooperating countries is denoted by k. For any given k, the payoff to a country is given by $B(k)$ if the country defects (tries to free-ride). If a country plays cooperatively, it must bear some additional costs $C(k)$. Its payoff is therefore given by $B(k) - C(k)$. We assume decreasing marginal benefits for a country if the number of contributers rises, i.e. $B(\cdot)$ is increasing and concave. Furthermore, the total cost of cooperation, $kC(k)$, increases in k.

In order to generate the standard incentive structure of a PD game, we assume that $B(k+1) - B(k) < C(k+1)$, i.e. playing cooperatively reduces the absolute payoff, given an arbitrary number of 'c'-countries. To make more cooperation attractive from both the social and the individual point of view, we make the following assumption:

$$\text{(i)} \quad NB(k+1) - (k+1)C(k+1) \geq NB(k) - kC(k) \quad \text{'socially desirable'}$$

$$\text{(ii)} \quad B(k+1) - C(k+1) \geq B(k) - C(k) \quad \text{'individually desirable'}$$

Furthermore, we assume that payoffs for both cooperating and defecting countries are non-negative for all k.

3.2. The Nash equilibria

In the following, we analyse Nash equilibria in the one shot PD game where countries choose simultaneously. Assume that k countries, aside from country i, play cooperatively. Then country i chooses to play 'c' if and only if:

$$a_i u(B(k+1) - C(k+1)) + b_i r\left(\frac{B(k+1) - C(k+1)}{NB(k+1) - (k+1)\,C(k+1)}\right) \geq a_i u(B(k)) + b_i r\left(\frac{B(k)}{NB(k) - kC(k)}\right)$$

This is equivalent to country i playing 'c' if:

$$a_i / b_i \leq \delta(k) := \frac{r\left(\frac{B(k+1) - C(k+1)}{NB(k+1) - (k+1)\,C(k+1)}\right) - r\left(\frac{B(k)}{NB(k) - kC(k)}\right)}{u(B(k)) - u(B(k+1) - C(k+1))} \tag{1}$$

In other words, in order to choose 'c' the country must be overcompensated for the loss in absolute payment by moving closer to the average payment.

The general conditions for a Nash equilibrium of this ERC-PD game are then given by:

$$a_i / b_i \leq \delta(k^* - 1) \text{ for } k^* \text{ countries (playing 'c')} \tag{2}$$

$$a_i/b_i \geq \delta(k^*) \text{ for the remaining } N-k^* \text{ countries (playing 'd')} \quad (3)$$

We now have a closer look at the number of countries k^* that may possibly cooperate in a Nash equilibrium. On the one hand, as long as $\delta(k^*-1)<0$, there is no chance of having a coalition of size k^*. Here, $a_i/b_i>\delta(k^*-1)$ for all types and condition (2) cannot hold for any country. On the other hand, the conditions for a Nash equilibrium given by (2) and (3) immediately imply that if $\delta(k^*-1)>0$ then there are types $((a_i/b_i)_{i=1,\dots,N})$ of countries such that k^* countries cooperate and $N-k^*$ countries free-ride. These types—for example—could be given by $a_i/b_i=\delta(k^*-1)$ for $i=1,\dots k^*$, and $a_i/b_i=\min\{\delta(k^*-1),\delta(k^*)\}$ for $i=k^*+1,\dots,N$.[8]

Example 1. Let $B(k)=km$, $C(k)=c$, where $c>m$, $r(\sigma)=-(\sigma-1/N)^2/2$, and $u(y)=y$. Then, as easily shown: $\delta(k-1)>0$ if and only if $\frac{1}{N}-\frac{km-c}{Nkm-kc}<\frac{(k-1)\,m}{N(k-1)\,m-(k-1)\,c}-\frac{1}{N}$, or—equivalently—$[2-N/k]>0$. Therefore, if in equilibrium some countries cooperate, then at least $N/2$. Fig. 1 illustrates the function $\delta(k)$ for $m=1$, $c=1.5$, and $N=10$. An equilibrium in which k^* countries cooperate results if the ERC-type of k^* countries is below the line at $k=k^*-1$ $(a_i/b_i\leq\delta(k^*-1))$, and above the line at $k=k^*$ for the remaining $N-k^*$ countries $(a_i/b_i>\delta(k^*-1))$. In particular, if for all i: $a_i/b_i\leq\delta(N-1)=3.1\cdot10^{-4}$, the grand coalition would be stable.

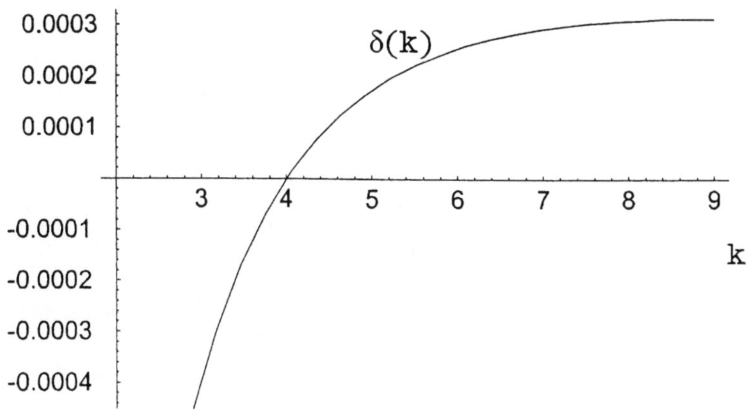

Fig. 1. Graph of $\delta(k)$ for $B(k)=k$, $C(k)=1.5k$, $N=10$.

[8] In other words, for a given distribution of ERC-types, $\delta(k^*-1)>0$ is necessary but not sufficient to get a coalition size of k^*. For a given payoff structure with $\delta(k^*-1)>0$, however, there exist ERC-types such that k^* is an equilibrium coalition size.

In order to find feasible coalition sizes, we must therefore study conditions in which $\delta(\cdot)$ is positive. Note that in (1) the denominator of $\delta(k)$ is positive, since playing 'd' always maximises the absolute payoff. The sign of the numerator, however, depends on the number k of cooperating countries. It is negative for $k=0$ and positive for $k=N-1$, since both, defection and cooperation of all countries equalise countries' payoffs and thereby maximise $r(\cdot)$. Therefore, $\delta(0)<0<\delta(N-1)$ and no country unilaterally cooperates whereas all countries playing 'c' can establish an equilibrium, provided that all countries' types (a_i/b_i) are smaller than $\delta(N-1)$.[9]

In general, however, there may also be equilibria where only a certain number k^* of countries cooperate. Remember that we assumed that countries suffer more from disadvantageous inequality than if they are better off than average, i.e. $r(1/N-x)\leq r(1/N+x)$ for all $x\in[0, 1/N]$. Therefore, in order to obtain $\delta(k)>0$, it is necessary that by choosing 'd', a country further deviates from the equal share $(1/N)$ than by playing 'c', i.e.:

$$\frac{B(k)}{NB(k)-kC(k)}-\frac{1}{N}>\frac{1}{N}-\frac{B(k+1)-C(k+1)}{NB(k+1)-(k+1)\,C(k+1)}$$

Straightforward calculus shows that this is equivalent to:

$$\begin{aligned}0&<B(k+1)\,C(k)\,Nk+B(k)\,C(k+1)\,N[k+1-N]\\&\quad+C(k)\,C(k+1)[Nk-2k(k+1)]\end{aligned}$$

or

$$\begin{aligned}0&<B(k+1)\,C(k)\frac{N}{k+1}+B(k)\,C(k+1)\,N\,\frac{k+1-N}{k(k+1)}\\&\quad+C(k)\,C(k+1)\left[\frac{N}{k+1}-2\right]\\&=\left[B(k+1)\,C(k)\frac{N}{k+1}-B(k)\,C(k+1)\frac{N}{k}\right]\\&\quad+\left[\frac{NB(k)}{k}-C(k)\right]C(k+1)\left[2-\frac{N}{k+1}\right]\end{aligned}\tag{4}$$

We can use this inequality to study the number k^* of agents that play cooperatively in equilibrium. First, note that we assumed payoffs to be non-negative and therefore $NB(k)-kC(k)>0$. Thus, the second summand is negative for $k<N/2-1$. For payoff functions that satisfy the requirement that total costs proportionally increase more than the total benefits, i.e.:

[9] Thus, if we look at the question of whether or not to ratify an already negotiated international agreement, we see that no country would unilaterally ratify if it expects the others not to do so. If, however, all except of one country have already ratified, then this country may decide to ratify as well.

$$\frac{(k+1)\,C(k+1)}{kC(k)} > \frac{NB(k+1)}{NB(k)} \tag{5}$$

the first bracket in (4) is negative as well.[10] As a consequence, inequality (4) cannot hold and $\delta(k) < 0$ for $k < N/2 - 1$. Thus, for any given vector of types, if a country plays 'c' in equilibrium, then, in total, at least half of the countries cooperate. For PD games where inequality (5) does not hold, we obtain $\delta(N/2 - 1) > 0$, suggesting that the minimal coalition size can be smaller than $N/2$. As long as the number of countries exceeds 8, however, this is not the case.[11] We can summarise our results in the following proposition:

Proposition 1. (Prisoner's Dilemma) *For any given payoff structure of the PD game with ERC preferences, there is always an equilibrium in which all countries defect. Additionally, depending on the distribution of ERC-types, there can be equilibria in which some or all countries cooperate. Then—at least—$N/2$ countries cooperate if $N > 8$ or the total costs of cooperation proportionally increase more than the benefits, i.e. inequality* (5) *holds.*

This proposition shows that if there is a coalition of cooperating countries, then it is rather large. The intuition is that in order to make cooperation attractive for a country, it has to be closer to the equal payoff share than when playing 'd'. This can only be the case if the number of countries that receive the (smaller) cooperative payoff is already large.

4. The emission game

In the last section, we assumed that countries have only the discrete choice whether or not to cooperate. Now, we look at an emission game where countries can choose their emission levels continuously. Again we concentrate on the case of countries that are homogeneous with respect to their payoff functions. It will become obvious that, here, ERC alone cannot improve upon the non-cooperative Nash equilibrium with standard preferences where only the absolute payoff matters. However, introducing more structure to the game, i.e. if countries play a coalition game as in Barrett (1994), ERC may yield a rather large coalition size or even support the grand coalition.

Let the number of countries again be denoted by N. Each country must choose

[10] Note that (5) is equivalent to $\frac{B(k+1)}{B(k)} > \frac{NB(k+1) - (k+1)\,C(k+1)}{NB(k) - kC(k)}$, i.e. a situation where private benefits of a defecting country increase faster than social benefits from cooperation.

[11] The proof is given in Appendix A. In it, we derive an upper bound for $\delta(k)$ which is negative for $k < N/2 - 1$. Therefore, $\delta(k^* - 1) > 0$ can only be the case if $k^* \geq N/2$.

its abatement level $q_i \in [0, e_i^{\max}]$ $(i = 1, \ldots, N)$. Reduction of emissions from a status quo level $e_i^{\max}$ induces some costs $C(q_i)$ that are assumed to be increasing and convex in the abatement level, $C'(\cdot) \geq 0$, $C''(\cdot) \geq 0$ Abatement creates some benefits $B(Q)$ in terms of abated damage from climate change, where $Q = \Sigma_i\, q_i$ denotes the aggregate abatement level. Benefits from abatement are increasing and (strictly) concave, $B'(\cdot) \geq 0$, $B''(\cdot) < 0$. The payoff to a country is therefore determined by $B(Q) - C(q_i)$.[12] In order to guarantee interior solutions we assume further $C'(0) = 0$ and $\lim_{q_i \to e_i^{\max}} C'(q_i) = \infty$, and $B'(0) > 0$.

4.1. Nash equilibria in the one-shot emission game

We again analyse Nash equilibria when countries act simultaneously. Taking q_j $(j \neq i)$ as given, country i chooses q_i to maximise:

$$a_i u(y_i) + b_i r(\sigma_i)$$

where $y_i = B(\Sigma_{j \neq i}\, q_j + q_i) - C(q_i)$ denotes the absolute, and $\sigma_i = y_i / \Sigma_j\, y_j$ denotes the relative payoff to country i. By choosing q_i, a country directly influences its own abatement costs and the benefits from abatement. It thereby also has an impact on the payoff to the rest of the world, which enters its own utility through the relative payoff. The first order condition is therefore given by:

$$\begin{aligned}
0 &= \left[a_i u'(\cdot) + b_i r'(\cdot) \frac{\sum\limits_{j \neq i} y_j}{\left[\sum\limits_j y_j \right]^2} \right] [B'(Q) - C'(q_i)] \\
&\quad - b_i r'(\cdot) \frac{y_i}{\left[\sum\limits_j y_j \right]^2} (N-1)\, B'(Q) \\
&= a_i u'(\cdot)[B'(Q) - C'(q_i)] \\
&\quad + b_i r'(\cdot) \left[\frac{\sum\limits_j (y_j - y_i)}{\left[\sum\limits_j y_j \right]^2} B'(Q) - \frac{\sum\limits_{j \neq i} y_j}{\left[\sum\limits_j y_j \right]^2} C'(q_i) \right] \qquad (6)
\end{aligned}$$

$$= a_i u'(\cdot)[B'(Q) - C'(q_i)] + b_i r'(\cdot) \left[\frac{1 - N\sigma_i}{\sum\limits_j y_j} B'(Q) - \frac{\sum\limits_{j \neq i} \sigma_j}{\sum\limits_j y_j} C'(q_i) \right] \qquad (7)$$

[12] Note that although using the same notation as in the previous section, the domain of the cost and benefit functions now is continuous abatement instead of discrete coalition sizes k.

The reaction of country i to a given abatement policy for the rest of the world can be calculated from this first order condition. Let us first study the two extreme cases, $a_i = 0$ and $b_i = 0$, respectively. For $b_i = 0$, i.e. an absolute payoff maximiser, the first order condition reduces to $B'(Q) - C'(q_i) = 0$. For $a_i = 0$, the country is solely interested getting the equal payoff share. Hence it would choose q_i to satisfy $NC(q_i) = \Sigma_j C(q_j)$. For a_i, $b_i \neq 0$, the chosen abatement level is between the levels for those extreme cases.

In a Nash equilibrium, the first order condition must be satisfied for all countries simultaneously. Since $r'(1/N) = 0$, it follows immediately that for all types a_i/b_i, $i = 1, \ldots, N$ there is a symmetric equilibrium where all countries choose the same abatement level, i.e. $\sigma_i = 1/N$ for all i. Here, the resulting abatement level q^* is given by the first order condition $B'(Nq^*) - C'(q^*) = 0$. It corresponds to the Nash equilibrium when agents are only interested in their absolute payoff and therefore resembles the Nash equilibrium in the PD game where all countries defect.

This equilibrium is the only one assuming that for at least one country a_i greater than 0. To see this, assume that there is an asymmetric equilibrium, i.e. some country i receives less, another country j more than the equal share. In this case, on the one hand, $\sigma_i < 1/N$ implies that $r'(\sigma_i) > 0$, and hence from Eq. (6), we obtain $B'(Q) - C'(q_i) > 0$. On the other hand, for $\sigma_j > 1/N$ we have $r'(\sigma_j) < 0$, and therefore Eq. (7) implies $B'(Q) - C'(q_j) < 0$. These two inequalities, however, imply that country j has larger (marginal) abatement costs than country i, which clearly contradicts the assumed payoff distribution. Hence, only symmetric equilibria exist. Here, if $a_i > 0$ for at least one country, we get $B'(Nq) - C'(q) = 0$ from Eq. (6). Only in the extreme (unlikely) case that all countries are solely interested in equal payoff could any arbitrary abatement level be implemented as a symmetric equilibrium.

We can summarise the result in the following proposition:

Proposition 2. (Emission game) *In the emission game for ERC preferences, the equilibrium is given by* $B'(Nq^*) = C'(q^*)$. *It is unique as long as at least one country draws utility from its absolute payoff* ($a_i > 0$).

Thus, in contrast to the prisoner's dilemma, ERC does not add any equilibria in which there is more abatement effort (cooperation). The existence of equilibria in the PD game that mimick cooperative behaviour, therefore, only arises in the presence of discrete action sets where countries have the binary choice of either getting more or less than the equal share and therefore might choose to cooperate if this brings them closer to the equal share. Having a continuous decision variable, small changes of abatement policy and—thereby—small changes of relative payoff are possible. Since $r'(1/N) = 0$, a country will always deviate from the equal share in order to increase its absolute payoff. The only situation where this is not possible, resembles the standard Nash equilibrium in which the absolute payoff is already maximised.

4.2. Coalition formation in the emission game

We now study a two-stage game of international negotiations as introduced by Barrett (1994), Carraro and Siniscalco (1993), and Hoel (1992). Let us again assume that all countries are identical with respect to their payoff function. In a first stage, countries decide whether or not to join the coalition. Here, each country takes the decisions of the other countries as given. Each country anticipates, however, that the abatement levels, which are chosen in the second stage, depend on whether it does or does not enter the coalition. In stage 2, countries inside and outside the coalition simultaneously select their abatement levels. The coalition thereby maximises its collective benefits but plays Nash against the non-signatory countries which simultaneously maximise their individual utility.[13] As shown by Carraro and Siniscalco and Hoel, the coalition size for payoff-maximising countries is 2.

We first look at the case of countries that have identical ERC-types. In a second step, we then study the case of heterogeneous ERC-types in which countries are solely interested either in their absolute payoff or in equity.

4.2.1. Cooperation of identical ERC-types

We solve the coalition formation game backwards. That is, for any coalition size k, we first study the first order conditions for the choice of the abatement level inside and outside the coalition. Then, in the second step, the equilibrium coalition size is determined by a stability condition such that that there is an incentive to neither leave nor join the coalition. In contrast to standard preferences (here corresponding to $b = 0$), for ERC preferences, the coalition size can be much larger. Instead of solving the game in general, we will show that if countries only value the relative payoff high enough, i.e. a/b is below a certain bound, then even the grand coalition can be stable.

For countries outside the coalition, the first order condition is again given by (6), whereas the coalition S maximises the utility of a representative member by choosing the abatement policy for all its members, $i \in S$. It is plain that all countries within the coalition S must have the same abatement level, q^S, since all countries are assumed to be of the same type. Hence, the absolute ($y^S = B(Q) - C(q^S)$) and the relative payoffs ($\sigma^S = y^S/(ky^S + \Sigma_{j \notin S}\, y_j)$) are equal for all members of the coalition. This leads to the following condition:

---

[13] Such an simultaneous move game for standard preferences was analyzed by Carraro and Siniscalco (1993) and Hoel (1992). This Cournot–Nash behavior at the second stage also resembles the partial-agreement Nash equilibrium concept used by Chander and Tulkens (1997). In contrast to this assumption, Barrett (1994) assumes that the coalition behaves as a Stackelberg leader.

$$0 = [au'(\cdot) + br'(\cdot)] \frac{\sum_{j \notin S} \sigma_j}{\sum_j y_j} [kB'(Q) - C'(q^S)]$$

$$- br'(\cdot) \frac{\sigma^S}{\sum_j y_j} (N-k)\, kB'(Q) \tag{8}$$

$$= au'(\cdot)[kB'(Q) - C'(q^S)] + br'(\cdot) \left[-\frac{\sum_{j \notin S} \sigma_j}{\sum_j y_j} C'(q^S) + kB'(Q) \frac{1 - N\sigma^S}{\sum_j y_j} \right] \tag{9}$$

From the last section we already know that if $\sigma_j < (>) 1/N$ for $j \notin S$, then $B'(Q) > (<) C'(q_j)$. Analogously, for the coalition, we obtain from (8) and (9) that if $\sigma^S < (>) 1/N$ then $kB'(Q) > (<) C'(q^S)$.[14] Since $B'(Q) < kB'(Q)$ $(k \geq 2)$, the first order conditions therefore imply that for countries within a coalition $\sigma^S \leq 1/N$,[15] and thus:

$$kB'(Q) \geq C'(q^S)$$

This inequality can now be used to show the following proposition:

Proposition 3. (Coalition game) *In the symmetric coalition game for identical ERC preferences (type a/b), the grand coalition is stable if a/b is sufficiently small, i.e. countries are interested enough in being close to the equal share.*

Note first, that within the grand coalition, the emission level satisfies the condition $NB'(Nq^*) = C'(q^*)$, independently of the ERC-types. Here, countries clearly receive the equal share. If country i leaves the coalition $(k = N-1)$, then from the inequalities above we obtain:

$$(N-1)\, B'((N-1)\, q^S + q_i) \geq C'(q^S) \geq C'(q_i) > B'((N-1)\, q^S + q_i) \tag{10}$$

for all ERC types a/b. Let us now look at the abatement levels that would result if the ERC-type a/b goes to zero. In this case, all countries get more and more interested in getting their equal share, and their abatement levels will converge to a level $\tilde{q}$. However, inequality (10) still must hold in the limit, i.e. $(N-1)\, B'(\tilde{Q}) \geq$

[14] Again assuming that $\sigma^S > 1/N$ then $r'(\cdot) < 0$, and (9) implies $kB'(Q) < C'(q^S)$.

[15] Assuming to the contrary that $\sigma^S > 1/N$ and thereby $\sigma_j < 1/N$ for some country j outside the coalition, the inequalities imply $C'(q_j) < B'(Q) < kB'(Q) < C'(q^S)$, which contradicts the assumption of increasing and convex abatement costs.

$C'(\tilde{q})$. This, however, implies, that $NB'(\tilde{Q}) > C'(\tilde{q})$, and therefore, in the limit, the absolute payoff of a country leaving the coalition (and of all the other countries) is smaller than within the grand coalition. Therefore, as long as a/b is small enough, the absolute payoff remains lower, and—due to the asymmetric payoff share—the utility derived from the relative payoff is also smaller than in the grand coalition. Thus, no country has an incentive to leave the grand coalition if a/b is small enough.

4.2.2. Coalition of heterogeneous ERC-types

Apparently, if one allows for heterogeneous ERC-types, starting in the grand coalition, countries that have the largest a_i/b_i will have the greatest interest to leave the coalition in order to obtain a larger absolute payoff. In the following we concentrate on the extreme case in which countries are either interested in their absolute payoff ($b_i = 0$) or in equity ($a_i = 0$). The former are referred to as A-countries, the latter as B-countries. In total, there are N_a A-countries and N_b B-countries; k_a of these A-countries and k_b B-countries form the coalition. The abatement levels are denoted by q_{as}, q_{bs} for signatory countries, q_{an} and q_{bn} for countries outside the coalition.

Let us first look at the behaviour of B-countries. Outside the coalition, any B-country can arrive at the equal share by choosing the average abatement cost level. Thus:

$$C(q_{bn}) = \frac{1}{N_a + k_b}[k_a C(q_{as}) + k_b C(q_{bs}) + (N_a - k_a)\, C(q_{an})]$$

Therefore, for a B-country inside a coalition to have no incentive to leave, it also must receive the equal share:

$$C(q_{bs}) = \frac{1}{N - k_b}[k_a C(q_{as}) + (N_b - k_b)\, C(q_{bn}) + (N_a - k_a)\, C(q_{an})] \tag{11}$$

Thus, in any equilibrium, all B-countries choose the same abatement level, $q_b := q_{bn} = q_{bs}$, and receive the equal share:

$$C(q_b) = \frac{1}{N_a}[k_a C(q_{as}) + (N_a - k_a)\, C(q_{an})]$$

A-countries outside the coalition again maximise their absolute payoff, $B(Q) - C(q_{an})$. The first order condition is given by:

$$B'(Q) = C'(q_{an}) \tag{12}$$

whereas the coalition maximises the utility of a representative A-type-member by guaranteeing its B-members the equal share, i.e. Eq. (11). The first order condition for choosing q_{as} is therefore given by:

$$0 = B'(Q)\left[k_a + k_b \frac{\partial q_{bs}}{\partial q_{as}}\right] - C'(q_{as}) \tag{13}$$

$$= B'(Q)k_a\left[1 + \frac{k_b}{N - k_b}\frac{C'(q_{as})}{C'(q_{bs})}\right] - C'(q_{as}) \tag{14}$$

By construction, for any given k_a and k_b, every B-country is indifferent to being either inside or outside the coalition. For a coalition to be stable, an A-country must not have an incentive to join or leave the coalition. In general, for any k_b there will be a certain number of A-countries, k_a, that will join the coalition. We have multiple equilibria.

The properties of the equilibria are best understood by considering an example. Consider the following quadratic payoff functions: $B(Q) = 10(Q - Q^2/2)$ and $C(q) = 5000q^2$, and $N_a = N_b = 6$.

Table 1 presents both the total abatement level and the payoffs to cooperating and non-cooperating A-countries as a function of the number of A- and B-countries inside the coalition. Here, if no B-country enters the coalition, all A-countries cooperate. If all B-countries are within the coalition, then only two A-countries enter. In general, when B-countries join the coalition, they successively drive out A-countries. This can be illustrated by the equilibria in Table 1: The first, second, fourth, and sixth entering B-country reduces the number of A-countries in the coalition and thereby lowers the payoffs to both, signatory and non-signatory countries. The payoffs to all countries increase, however, when a third or a fifth B-country enters the coalition and the number of cooperating A-countries stays at 4 and 3, respectively. Therefore, in this example, the presence of equity-interested countries leads to efficiency gains which are largest if all B-countries stay outside the coalition.

The results of the simulations—which appear to be robust to variations of the payoff functions and the total number of countries—can be summarised as follows:

Result 4. The larger the total number of equity-oriented countries (N_b) is, the higher the incentives are for A-countries to join the coalition. In other words, for a given k_b, the number of cooperating A-countries k_a increases in N_b.

Result 5. The more B-countries join the coalition, the smaller the incentives are for A-countries to do so. In other words, in equilibrium, k_b and k_a are negatively correlated.

Result 6. The total abatement level increases with the number of B-types outside the coalition. A joining B-country improves the payoffs only if it does not drive out an A-country.

2064 A. Lange, C. Vogt / Journal of Public Economics 87 (2003) 2049–2067

Table 1
Payoffs to non-cooperating, cooperating countries and aggregate abatement as a function of k_a and k_b ($N_a = 6$, $N_b = 6$)[a]

k_b		k_a 0	1	2	3	4	5	6
0	π^{an}	0.113	0.113	0.156	0.240	0.354	0.490	–
	π^{as}	–	0.113	0.142	0.202	0.284	0.382	0.492*
	Q	0.012	0.012	0.016	0.025	0.037	0.051	0.067*
1	π^{an}	0.113	0.115	0.168	0.264	0.388	0.543*	–
	π^{as}	–	0.114	0.148	0.215	0.302	0.404*	0.518
	Q	0.012	0.012	0.017	0.027	0.040	0.055*	0.073
2	π^{an}	0.113	0.118	0.185	0.293	0.430*	0.587	–
	π^{as}	–	0.115	0.157	0.230	0.321*	0.427	0.544
	Q	0.012	0.012	0.019	0.030	0.044*	0.061	0.080
3	π^{an}	0.113	0.122	0.206	0.330	0.480*	0.650	–
	π^{as}	–	0.117	0.166	0.245	0.341*	0.451	0.571
	Q	0.012	0.013	0.021	0.034	0.050*	0.068	0.088
4	π^{an}	0.113	0.130	0.233	0.375*	0.540	0.726	–
	π^{as}	–	0.120	0.176	0.260*	0.361	0.475	0.597
	Q	0.012	0.014	0.024	0.039*	0.056	0.076	0.097
5	π^{an}	0.113	0.142	0.269	0.432*	0.618	0.821	–
	π^{as}	–	0.122	0.184	0.273*	0.378	0.494	0.619
	Q	0.012	0.015	0.028	0.045*	0.064	0.086	0.110
6	π^{an}	0.113	0.161	0.318*	0.507	0.718	0.943	–
	π^{as}	–	0.124	0.188*	0.279	0.385	0.503	0.629
	Q	0.012	0.017	0.033*	0.053	0.075	0.100	0.126

[a] The equilibria are marked with *.

The rationale of Results 4 and 5 is the following: If an A-country enters the coalition and the coalition increases its abatement efforts, B-countries outside the coalition increase their abatement activities as well and thereby additionally reward the entering country. If the number of such equity-oriented B-countries outside the coalition gets larger, this external reward for joining a coalition increases and, therefore, the equilibrium coalition size increases. Analogously, if B-countries join the coalition, less countries outside the coalition reward the entering A-country by an increase of their abatement activities. Hence, the incentives for A-countries to enter the coalition decrease and the number of A-countries that are inside the coalition in equilibrium gets smaller. If all B-countries cooperate, the external reward has vanished and the incentives for A-countries to join the coalition are qualitatively the same as with standard preferences for all countries: Only two countries enter. Result 6 reflects the fact that, in general, the more countries cooperate, the higher the efficiency gains are and the closer the aggregate abatement level is to the efficient one. The impact of A- and B-countries on the decision of the coalition, however, differs in the following way: A joining A-country is interested in the absolute payoff and, consequently, the re-optimising coalition increases its abatement effort because the positive effect on one more country is now taken into account. A joining B-country, however, is not primarily interested in the absolute payoff, but in the equal share. Therefore, the coalition will not increase the total abatement level that much because the B-country refrains from deviating from the abatement level of non-cooperating countries. Consequently, the efficiency gains are larger if an A-country enters the coalition than if a B-country joins. Therefore, B-countries are welcome inside a coalition only if their entering does not drive out an A-country.

5. Conclusion

In this paper, we studied three different games that model basic features of international cooperation to provide a public good such as environmental quality. Instead of countries that are solely interested in their absolute payoff, we studied equilibria based on ERC-preferences. This theory explains the behaviour of people observed in experiments by assuming that a part of utility depends (non-monotonously) on the received relative payoff share.

We demonstrated that for a discrete prisoner's dilemma, cooperation can result. If it does, then the fraction of cooperating countries is rather large (above 50%). For the one-shot emission game, which better describes the action set with respect to environmental problems like global change, however, ERC does not change the equilibrium at all: Countries (in equilibrium) still behave as if they were exclusively maximising their own payoff.

We then studied the process of international environmental negotiations in the standard two stage coalition formation game. Here, even the grand coalition can be stable if all countries put enough weight on getting close to the equal payoff share. In general, the presence of countries that are highly motivated through obtaining the equal share leads to efficiency gains. In particular, for countries that are interested exclusively in their absolute payoff, the incentive to join a coalition increases with the number of equity-oriented countries that stay outside the coalition, but it becomes smaller when those countries enter the coalition. Thus, by entering a coalition, equity-oriented countries may drive out countries that are motivated by their own absolute payoff and thereby worsen the payoffs for all countries.

In conclusion, observing international environmental agreements that are signed by a large number of countries may indicate that (at least some) countries are not solely interested in their absolute payoff, and their behaviour could be explained by the ERC-theory. Crucial for this result, however, is the structure of the game: Countries must have a discrete choice at one stage of the game.

Note, however, that, in this paper, we exclusively analysed the case of countries which are symmetric with respect to their payoffs. In general, countries clearly are heterogeneous with respect to their benefits and costs of abatement of CO_2, and a preference for equity would probably not regard an equal payoff structure. Instead, equity might be preferred with respect to the abatement targets as percentages of the business-as-usual-emissions or in terms of CO_2 emissions per capita. The implications of such a preference for equity in the case of heterogeneous countries remain subject to further research.

Acknowledgements

We would like to thank Michael Finus, Axel Ockenfels, three anonymous referees, and Lans Bovenberg as the co-editor of this journal for helpful comments and suggestions to improve the paper. Funding of the research group 'Institutionalization of International Negotiation Systems' by the Deutsche Forschungsgemeinschaft (DFG) is gratefully acknowledged.

Appendix A. (Proof of Proposition 2)

We have to show that $\delta(k) < 0$ for $k < N/2 - 1$. Using (4), this is equivalent to:

$$\left[\frac{B(k+1)}{B(k)} - \frac{(k+1)\,C(k+1)}{kC(k)}\right] + (k+1)\,C(k+1)\left[\frac{1}{kC(k)} - \frac{1}{NB(k)}\right]\left[2 - \frac{N}{k+1}\right] < 0$$

From the monotonicity and concavity of $B(\cdot)$ it follows that $B(k+1)/(k+1) < B(k)/k$. Furthermore, total cost of cooperation $kC(k)$ increase in k. Therefore:

$$\frac{B(k+1)}{B(k)} - \frac{(k+1)\,C(k+1)}{kC(k)} \leq \frac{k+1}{k} - 1 = \frac{1}{k}$$

Since we assumed payoffs to be non-negative, $B(k) \geq C(k)$. Thus:

$$(k+1)\,C(k+1)\left[\frac{1}{kC(k)} - \frac{1}{NB(k)}\right] \geq \frac{(k+1)\,C(k+1)}{kC(k)}\,\frac{N-k}{N} \geq \frac{N-k}{N}$$

For $k < N/2 - 1$ we therefore obtain:

$$\left[\frac{B(k+1)}{B(k)} - \frac{(k+1)\,C(k+1)}{kC(k)}\right]$$
$$+ (k+1)\,C(k+1)\left[\frac{1}{kC(k)} - \frac{1}{NB(k)}\right]\left[2 - \frac{N}{k+1}\right]$$
$$\leq \frac{1}{k} + \frac{N-k}{N}\left[2 - \frac{N}{k+1}\right]$$
$$= \frac{N(k+1) + 2(N-k)\,k(k+1) - (N-k)\,Nk}{Nk(k+1)}$$

The numerator equals $-2k^3 + (3N-2)\,k^2 - N(N-3)\,k + N$ which can easily be shown to be negative for $1 \leq k < N/2 - 1$ as long as $N > 8$. Hence, for $N > 8$ we have $\delta(k-1) < 0$ for $k < N/2$.

References

Barrett, S., 1992. International environmental agreements as games. In: Pethig, R. (Ed.), Conflicts and Cooperation in Managing Environmental Resources. Springer, Berlin, pp. 11–37.

Barrett, S., 1994. Self enforcing international environmental agreements. Oxford Economic Papers 46, 878–894.

Barrett, S., 1997. Heterogeneous international environmental agreements. In: Carraro, C. (Ed.), International Environmental Agreements: Strategic Policy Issues. E. Elgar, Cheltenham, pp. 9–25.

Bolton, G.E., Ockenfels, A., 2000. ERC: a theory of equity, reciprocity, and competition. The American Economic Review 90, 166–193.

Bolton, G.E., Ockenfels, A., 2001. Self-centered fairness in games with more than two players. In: Plott, C., Smith, V. (Eds.), Handbook of Experimental Economics Results. North Holland, Amsterdam.

Carraro, C., Siniscalco, D., 1993. Strategies for the international protection of the environment. Journal of Public Economics 52, 309–328.

Cazorla, M.V., Toman, M.A., 2001. International equity and climate change policy. In: Toman, M.A. (Ed.), Climate Change Economics and Policy. Resources for the Future, Washington, pp. 235–247.

Chander, P., Tulkens, H., 1997. The core of an economy with multilateral environmental externalities. International Journal of Game Theory 26, 379–401.

Dasgupta, C., 1994. The climate change negotiations. In: Mintzer, I.M., Leonard, J.A. (Eds.), Negotiating Climate Change. The Inside Story of the Rio Convention. Cambridge University Press, Cambridge, pp. 129–148.

Hoel, M., 1992. International environmental conventions: the case of uniform reductions of emissions. Environmental and Resource Economics 2, 141–159.

Murdoch, J.C., Sandler, T., 1997. The voluntary provision of a pure public good: the case of reduced CFCs emissions and the Montreal protocol. Journal of Public Economics 63, 331–349.

Roth, A.E., 1995. Bargaining experiments. In: Kagel, J.H., Roth, A.E. (Eds.), Handbook of Experimental Economics. Princeton University Press, Princeton, NJ, pp. 253–348.

[34]

American Economic Review 2009, 99:4, 1335–1355
http://www.aeaweb.org/articles.php?doi=10.1257/aer.99.4.1335

Institution Formation in Public Goods Games

By Michael Kosfeld, Akira Okada, and Arno Riedl*

Sanctioning institutions are of utmost importance for overcoming free-riding tendencies and enforcing outcomes that maximize group welfare in social dilemma situations. We investigate, theoretically and experimentally, the endogenous formation of institutions in public goods provision. Our theoretical analysis shows that players may form sanctioning institutions in equilibrium, including those governing only a subset of players. The experiment confirms that institutions are formed and that it positively affects cooperation and group welfare. However, the data also shows that success is not guaranteed. Players are unwilling to implement equilibrium institutions in which some players have the opportunity to free ride. Our results emphasize the role of fairness in the institution formation process. (*JEL* C72, D02, H41)

Persons agree to constraints on their own liberties in exchange for comparable constraints being imposed on the liberties of others.
— James M. Buchanan and Roger D. Congleton (1998, 4)

When markets fail, the design of appropriate institutions is a key issue for economic analysis and policy. Social dilemma situations (e.g., public goods, common pool resources), in which the pursuit of individual interests conflicts with the maximization of social welfare, are a classic example. In such situations, the implementation of a sanctioning institution that castigates individual behavior if it deviates from the welfare maximizing action is a widely used solution. For example, many common pool resources regimes around the world rely on sanctions, and there is unanimous agreement in the literature that an effective sanctioning system is a major determinant of the success of such regimes (Jean-Marie Baland and Jean-Philippe Platteau 1996; Elinor Ostrom 1999).[1] Similarly, trade unions and employers' associations often have arbitration boards monitoring and enforcing the compliance of their members. Even in the international arena, famous examples exist. For instance, the EU Stability and Growth Pact was created to enforce budgetary discipline among EU member states, and the Kyoto Protocol aims to reduce global greenhouse gas emissions by implementing legally binding agreements.

* Kosfeld: Department of Economics and Business Administration, Johann Wolfgang Goethe-University, Grüneburgplatz 1, D-60323 Frankfurt am Main, Germany, and IZA (e-mail: kosfeld@econ.uni-frankfurt.de); Okada: Graduate School of Economics, Hitotsubashi University, Kunitachi, Tokyo 186-8601, Japan (e-mail: aokada@econ.hit-u.ac.jp); Riedl: Department of Economics, Faculty of Economics and Business Administration, Maastricht University, PO Box 616, NL-6200 MD Maastricht, the Netherlands, CESifo, and IZA (e-mail: a.riedl@maastrichtuniversity.nl). We thank Elinor Ostrom, Armin Falk, Ernst Fehr, Urs Fischbacher, Simon Gächter, Klaus Schmidt, Christian Zehnder, and three anonymous referees for their helpful comments. Kosfeld gratefully acknowledges financial support from the University of Zurich through the University Research Priority Program on "Foundations of Human Social Behavior: Altruism versus Egoism" and the Swiss State Secretariat for Education and Research through the EU-TMR Research Network ENABLE (MRTN CT-2003-505223). Okada gratefully acknowledges financial support from the Japan Society for the Promotion of Science under grants (A)16203011 and (S)20223001 and the Matsushita International Foundation.

[1] Examples include irrigation systems (Shui Yan Tang 1992), forests (Arun Agrawal and Gautam N. Yadama 1997), and fisheries (Edvard Hviding and Graham B. K. Baines 1994).

As diverse as these examples are, structurally, they have two important elements in common. First, the institutional arrangements are not imposed from without but are *formed from within* in the sense that, at some point in time, a set of agents voluntarily agreed to implement the particular arrangement. Second, sanctioning applies only to members of the institution; nonmembers remain free in their choices and, hence, are given a strong incentive to free-ride. Together, these two elements constitute what we term a "dilemma of endogenous institution formation": jointly, everyone profits if a sanctioning institution is formed, but each individual profits more if only the others form the institution.[2] It is exactly this dilemma of institution formation that we address in this paper.

The social dilemma situation we consider is a linear n-player public goods game. The institutional arrangement we analyze is a sanctioning institution, in which sanctions are imposed by a central authority, for example, a policeman, a court, or an arbitration board.[3] We model the process of institution formation by a three-stage noncooperative game. In the first stage of the game, each player decides whether he wants to participate in an organization that, once implemented, exerts a punishment on each member who does not contribute his full endowment to the public good. The organization is costly and only players who are members of the organization can be punished. Thus, nonmembers can free-ride on members' contributions. In the second stage, players learn how many of the other players are willing to participate. The organization is implemented if and only if all players willing to participate agree to its actual formation. In the final stage, the public goods game is played.

In the theory part of our paper, we show that two different types of subgame perfect Nash equilibria exist in this game, a so-called *organizational equilibrium*, where players successfully implement an organization, and a so-called *status quo equilibrium*, where no organization is implemented. We prove that organizations in any organizational equilibrium are of a minimum size s^*, i.e., at least s^* players participate, where s^* depends on the payoffs in the public goods game and the cost of the organization. Furthermore, using strictness (in every subgame) as an equilibrium refinement, we show that a unique strict subgame perfect equilibrium exists in terms of the organization size. In this equilibrium, exactly s^* players implement the organization and consequently contribute to the public good, whereas the remaining $n - s^*$ players do not participate and free-ride. Thus, if $s^* < n$, the organization has a proper subset of players who voluntarily commit themselves to cooperation. Although each organization member would be better off if someone else participated or if more players became members, the organization is nevertheless implemented because each member earns a higher payoff than in the status quo equilibrium, in which no organization is implemented and no players contribute to the public good. This result is related to a similar theoretical finding in the environmental economics literature. There, the notion of so-called "internal stability," a concept originally developed for the analysis of cartels and coalition formation, implies that self-enforcing environmental agreements may support only a small subgroup of signatories (Scott Barrett 1994).

While the strict subgame perfect equilibrium prediction is intuitive in terms of individual material payoff maximization, the equilibrium outcome is clearly unfavorable in terms of efficiency and equality. If players dislike payoff inequality, it seems plausible that other institutions, in particular the grand organization where all players are members, may be favored. We, therefore, also analyze the institution formation game under the assumption that some of the players

[2] The dilemma is a particular type of the so-called "second-order free-rider problem" (cf. Pamela Oliver 1980).

[3] We thereby abstract from possible enforcement problems that might arise if players themselves have to impose the sanctions. These problems represent an important research question in their own, but will not be the topic of this paper, which focuses on the *formation* of institutions. Given this, however, our analysis of institution formation is in fact rather general since most of our results can be extended to other (noncentralized) institutional arrangements (see below).

suffer from payoff inequality as is captured by the social-preference model of Ernst Fehr and Klaus M. Schmidt (1999). We show that inequality aversion may indeed select the grand organization, either as a strict or even as the unique organizational equilibrium.

The theory part shows that equilibrium selection depends on refinements and assumptions on preferences. Hence, theory alone gives only limited guidance regarding institution formation. We therefore present the results of a laboratory experiment in the second part of the paper. Our experiment goes beyond the existing literature, as it connects the classic social dilemma situation with an innovative element of political organization, i.e., the endogenous formation of institutions.[4] In each of our main treatments, subjects played 20 rounds of a four-player institution formation game as described above. We varied the marginal per capita return of the public good across treatments to yield different predictions regarding the minimum organization size s^*. We also conducted two corresponding control treatments, in which no institution could be formed and subjects played only the public goods game.[5]

Our main experimental findings are as follows. First, subjects successfully establish organizations. In both experimental treatments, from 70 to 100 percent of all groups implement an organization by the final rounds. Second and most important, the majority (on average, around 75 percent) of the organizations implemented are grand organizations, i.e., *all* players participate. This finding is consistent with the prediction based on social preferences and stands in stark contrast to the strict subgame perfect equilibrium prediction of the standard model. Further results on players' beliefs and rates of implementation show that the frequent observation of the grand organization is not driven by miscoordination among participating players. Instead, the data suggest that the institutional outcome is the result of (almost) equilibrium play. Finally, a comparison with our control treatments confirms that the opportunity to form institutions increases and stabilizes total contributions to the public good. Overall, institution formation enhances group welfare, despite the fact that it is costly.

Our theoretical and empirical results have important implications for public policy. First, since sanctioning institutions may be an effective solution in many social dilemma situations, the observation that subjects *voluntarily* implement such institutions can be taken as good news. However, subjects are very reluctant to implement (Nash equilibrium) institutions that govern only a subset of players. This is true even if participating players can earn a higher payoff compared to the nonproduction of the public good. Our result emphasizes the importance of fairness for the formation and stability of institutional arrangements, an issue that has frequently been documented, for example, in the common pool resource literature (Baland and Platteau 1996; Margaret A. McKean 2000) and has also been stressed by prominent public choice scholars (Buchanan and Congleton 1998; see the quote at the beginning of the paper). As a famous instance, it may also bring to mind the discussion of the potential impact of the withdrawal of the United States from the Kyoto Protocol on other nations' motivation to fulfill the agreement.[6] The

[4] Most of the related experimental work has focused on the effect of exogenously imposed institutions (e.g., Toshio Yamagishi 1986; Ostrom, James Walker, and Roy Gardner 1992; Yan Chen and Charles R. Plott 1996; Josef Falkinger et al. 2000; Fehr and Simon Gächter 2000a, 2002; David Masclet et al. 2003; Christopher M. Anderson and Louis Putterman 2006; Jeffrey P. Carpenter 2007a, b). Only a couple of recent papers allow, at least to some extent, for endogenous institutional choice (Walker et al. Ostrom 2000; Özgür Gürerk, Bernd Irlenbusch, and Bettina Rockenbach 2006; Matthias Sutter, Stefan Haigner, and Martin G. Kocher 2006; Jean-Robert Tyran and Lars P. Feld 2006; Stephan Kroll, Todd L. Cherry, and Jason F. Shogren 2007). Different from our study, however, the latter models do not give players the opportunity to free-ride on other players' participation in the institution, an assumption that basically eliminates the second-order free-rider problem.

[5] In addition, we conducted a third control treatment in order to check for possible experimental design effects (see details below).

[6] To give two illustrative examples—in its 2006 report on climate strategies, the Dutch Scientific Council for Government Policy (WWR) advised the Dutch government not to stick (too tightly) to the Kyoto criteria, one reason being that large countries such as the United States did not ratify the agreement (WWR 2006). Likewise, at the time of

consequences of the observed behavior are twofold. On the one hand, if the process of institution formation is successful, established institutions generally achieve a high level of efficiency because strictly more than the minimally required s^* players participate. In fact, in the most frequently observed organization in our experiment, 100 percent of the players participate and contribute to the public good. This is in stark contrast to the prediction based on s^*, yielding a participation (and cooperation) rate in the two treatments of only 50 and 67 percent, respectively. On the other hand, the risk for the process to fail is much higher than predicted as well. Institutions are rejected that from an individual as well as a social welfare perspective clearly represent a material improvement over the situation without an institution. Therefore, not taking this behavior into account not only yields misleading theoretical predictions, but may lead to the realization of highly inefficient outcomes.

While we focus on a particular institutional solution in this paper (i.e., centralized sanctioning), our analysis can easily be extended to other institutional arrangements, including alternative centralized policy instruments, such as the mechanisms proposed by Theodore Groves and John O. Ledyard (1977) and Falkinger (1996), but also noncentralized solutions, such as repeated-game trigger strategies. The only condition that must be fulfilled is that the particular institution "works," i.e., that participating players have an incentive to act in accordance with the institutional rules and contribute to the public good once the institution is formed. In game theoretic terms, the prescribed behavior must form a Nash equilibrium. This holds for the Grove-Ledyard and the Falkinger mechanism, given that parameters are chosen accordingly. It also holds for repeated-game trigger strategies, if players are sufficiently patient. The key question for any particular institutional solution, however, is whether players will actually agree to form the institution, the main problem being that each player has an incentive to free ride on others forming the institution. This second-order free-rider problem applies to any mechanism that solves the first-order free-rider problem in social dilemma situations. The contribution of our paper is to show how individuals can overcome this problem, both from a theoretical and from an experimental viewpoint, and to point out behavioral regularities that govern and limit the process of endogenous institution formation.

The paper is organized as follows. Section I theoretically analyzes the institution formation game, characterizing subgame perfect Nash equilibria both if players have standard preferences and if players have fairness preferences. Section II describes and analyzes the experiment. Section III concludes.

I. Institution Formation: Theory

A. *Model*

Consider the following n-player public goods game. There are $n \geq 2$ players, each of whom has a private endowment $w > 0$ from which he can contribute $g_i \leq w$ to a public good. Given the contribution of all players $(g_1, \ldots, g_n)$, player i's material payoff is equal to

$$\pi_i(g_1, \ldots, g_n) = w - g_i + a \sum_{j=1}^{n} g_j, \tag{1}$$

where $0 < a < 1 < na$. Parameter a models the marginal per capita return (MPCR) from contributing to the public good. Assumption $a < 1$ implies that zero contribution is the dominant

the United States withdrawal from the protocol in 2001, Australian Environment Minister Robert Hill declared that he did not think "the Kyoto Protocol will succeed without the United States" (ABC, *The World Today*, March 30, 2001; http://www.abc.net.au/worldtoday/stories/s269266.htm).

action for every player with standard preferences, i.e., each player's material welfare is maximized by contributing zero to the public good regardless of the other players' contributions. In consequence, the strategy profile $(0,\ldots,0)$ is the unique Nash equilibrium. Assumption $na > 1$ implies that all players are better off if everyone contributes his full endowment to the public good. In particular, $(w,\ldots,w)$ is the welfare maximizing strategy profile.

Generally, institution formation is a complex process. Parties are typically involved in multistage bargaining with continual updates about other parties' behavior, goals, and expectations. Often, the process is little structured ex ante and negotiations take the form of both bilateral and collective bargaining. The institution formation game we analyze in this paper is necessarily simpler. It consists of a participation stage where players announce their (un)willingness to form an institution, and an implementation stage where they can actually form the institution. The advantage of this setup is that we can provide precise game-theoretic predictions regarding institutional outcomes. At the same time, our model captures key elements of real-world negotiation processes. First, economic or political actors do not implement new institutions ad hoc, but take decisions step-by-step. Second, players receive important information about other players' willingness to form an institution before they actually decide to implement it. The precise sequence of actions in our institution formation game is as follows:[7]

Participation Stage.—Players simultaneously and independently announce whether they are willing to participate in an organization that sanctions all organization members who do not contribute their full endowment to the public good. In the following, players who declare such a willingness are called *participants*; those who do not are called *nonparticipants*.

Implementation Stage.—After players are informed about the set of participants, all participants negotiate about whether to implement an organization. Negotiations take the form that all participants simultaneously and independently either accept or reject the implementation of the organization. The organization is implemented if and only if all participants accept (unanimity rule). In case an organization is implemented, all participants become *members* of the organization. Nonparticipants cannot become members. The organization is costly. Costs arise only if an organization is implemented and are shared equally among organization members.

Contribution Stage.—All players simultaneously and independently determine their contribution to the public good. If an organization has been implemented, organization members will be sanctioned for not contributing their full endowment to the public good. Importantly, nonmembers cannot be sanctioned. If no organization has been implemented, no player is sanctioned.

A player's final payoff u_i in the institution formation game is determined as follows. Let S be the set of players who are members of the organization with $s = |S|$, and let $c \geq 0$ be the cost of the organization. There are two cases.

Case 1: If $S \neq \varnothing$, i.e., an organization is implemented, then for every player i,

$$u_i = \begin{cases} w - g_i + a\sum_{j=1}^{n} g_j - \frac{c}{s} - p(g_i) & \text{if } i \in S \\ w - g_i + a\sum_{j=1}^{n} g_j & \text{if } i \notin S, \end{cases} \tag{2}$$

[7] The institution formation game laid out in this paper is an extension of the model proposed in Akira Okada (1993).

where $p(g_i)$ is the sanction imposed on member i satisfying[8]

(3) $$p(g_i) = \begin{cases} w - g_i & \text{if } g_i < w \\ 0 & \text{if } g_i = w. \end{cases}$$

Case 2: If $S = \varnothing$, i.e., no organization is implemented, then for every player i,

(4) $$u_i = w - g_i + a \sum_{j=1}^{n} g_j.$$

Equation (2) reveals the key difference between members and nonmembers of the organization. While the former are punished for free-riding and share the costs of the organization, the latter can freely decide whatever to contribute and do not pay any part of the organization costs. Formally, the institution formation game is a n-player three-stage game with perfect information. In each stage, players choose their actions with perfect knowledge about the course of the game in previous stages.[9] In the following, we first characterize the set of subgame perfect equilibria of the institution formation game if players' preferences are captured by material payoffs u_i. We then analyze equilibria of the game if (some) players have social preferences with a taste for fairness.

B. *Standard Preferences*

In a subgame perfect equilibrium, players decide on their actions in every stage, rationally anticipating the outcome of future stages by applying backward induction. Consider first the contribution stage. If players' preferences are given by u_i, (2) and (3) imply that it is optimal for organization members to contribute w in the contribution stage once an organization has been implemented. Since nonmembers are not punished, they behave optimally by not contributing anything. Clearly, zero contribution is optimal for every player in the contribution stage if no organization has been implemented.

A key insight of this model is that, although players have an individual incentive to free-ride in the public goods game, they might increase their payoff by coordinating their contributions in the framework of an organization. Note that in equilibrium organization members earn $asw - c/s$ if an organization has been implemented, and everybody earns w if no organization is established. Players are better off joining an organization compared to the zero-contribution outcome if the number of members s is such that

(5) $$asw - \frac{c}{s} > w.$$

Let s^* denote the smallest nonnegative integer s satisfying condition (5). The threshold s^* gives the minimum size of an organization such that participants in the implementation stage have an

[8] Note that $p(g_i)$ must be larger than $(1 - a)(w - g_i)$ whenever $g_i < w$ for punishment to induce full contribution by organization members.

[9] All theoretical results are also valid if players in the implementation stage are informed only about the number of participants and if players in the contribution stage are informed only about whether the organization has been implemented or not. The subgame perfect equilibrium concept should then be replaced by a sequential equilibrium. The reason is—as will be shown below—that players' payoff functions depend on the number of participants (not their identity) and on whether an organization is established.

incentive to implement it. From $a < 1$ and $c \geq 0$ it follows that $s^* \geq 2$. Moreover, since the left-hand side of (5) is strictly increasing in s, $s^* \leq n$ exists uniquely if and only if $(an - 1)nw > c$. If the latter condition does not hold, an organization is never beneficial to the players, i.e., no group would ever have an incentive to implement it. In the following, we therefore assume that the condition holds and hence a unique threshold $s^* \leq n$ exists.

We can now characterize the set of subgame perfect equilibria of the institution formation game. For convenience, we call a subgame perfect equilibrium an *organizational equilibrium* if an organization is formed on the equilibrium path. A subgame perfect equilibrium is called a *status quo equilibrium* if no organization is formed.

PROPOSITION 1: *If players have standard preferences, there exists an organizational equilibrium with s players being members of the organization if and only if $s \geq s^*$. For any number of participants s ($1 \leq s \leq n$) there exists a status quo equilibrium.*

All proofs are given in the supplementary material (Web Appendix A, available at http://www.aeaweb.org/articles.php?doi=10.1257/aer.99.4.1335). Proposition 1 shows that players can, in principle, overcome the dilemma of endogenous institution formation. The only requirement is that at least s^* players participate. If this condition holds, there exists an organizational equilibrium for every $s \geq s^*$, in which players implement an organization of size s, rejecting all organizations of different size (cf. the proof for details). However, success is not guaranteed. For any number of participants, there always exists a status quo equilibrium, in which players refrain from implementing an organization.[10] Moreover, if s^* is strictly smaller than n, the institution formation game has multiple organizational equilibria, namely all organizations of size $s \in \{s^*, s^* + 1, \ldots, n\}$.[11] In addition to agreeing on whether an organization shall be implemented, players face two coordination problems. First, they have to solve the problem with regard to the organization size; second, if $s < n$, they also have to solve the problem with regard to who is going to become a member of the organization and who is going to stay out. Interestingly, as the following proposition shows, strictness as an equilibrium refinement might offer a possible solution to the first coordination problem. Generally, a Nash equilibrium of a strategic-form game is called *strict* if every player plays a unique best response to the other players' strategies, i.e., every player is strictly worse off by deviating from equilibrium play. A subgame perfect equilibrium of a multistage game is called strict if it induces a strict Nash equilibrium in every stage game.

PROPOSITION 2: *If players have standard preferences, the institution formation game has a unique strict subgame perfect equilibrium in terms of the organization size. In this equilibrium, exactly* s^* *players become members of the organization.*

Strictness of equilibrium yields a clear-cut prediction regarding organization size: exactly the minimum number of players s^* required for the organization to be individually profitable form the organization, while the remaining players do not participate. Unless $s^* = n$, players are thus divided into two proper subsets: those who voluntarily implement the sanctioning institution, hence contributing to the public good, and those who do not participate and do not contribute.

[10] In a certain sense, the provision of the second-order public good (i.e., the sanctioning system) in equilibrium is thus characterized by a step-level technology with critical threshold s^*. Different from common step-level public goods (see, e.g., Mark Bagnoli and Barton L. Lipman 1989; Rachel T. A. Croson and Melanie Beth Marks 2000), however, the step-level provision in our model is an equilibrium outcome. Furthermore, the threshold s^* is not given exogenously, but is determined endogenously by players' incentives to form an institution.

[11] There also exists a mixed-strategy equilibrium in this case, in which players participate with positive probability and implement the organization whenever $s \geq s^*$.

The intuition for this result is straightforward. The best that can happen in material terms to any player i is that *other* players implement an organization and contribute to the public good while i free-rides. As long as the number of participants s is strictly larger than s^* (and hence $s - 1 \geq s^*$), there is always at least one participant who can successfully choose this option. The reason is that in a strict subgame perfect equilibrium, every organization with at least s^* participants is implemented in the implementation stage. Only if the organization is at its minimum size, s^*, is free-riding no longer an option, since the remaining $s^* - 1$ players will not implement the smaller organization.[12]

Whether the organizations form and whether they have the predicted size are of course empirical questions. An alternative equilibrium outcome, which always exists and which is favored in terms of efficiency, symmetry, and equality, is the grand organization $s = n$. This organization is also consistent with the so-called *generality principle* of Buchanan and Congleton (1998) which asserts that political choices should be nondiscriminatory and must involve equal treatment of all individuals. However, the grand organizational equilibrium requires players to reject all organizations with fewer than n participants. Such rejections are weakly dominated whenever $s \geq s^*$, since each participant is better off in material terms by implementing the organization. If players maximize only their material payoff, it seems questionable whether they can credibly commit to reject all organizations smaller than n. Yet, intuition suggests that motives of fairness might induce players to reject organizations where nonmembers can free-ride. In order to see whether fairness affects equilibrium outcomes, we analyze the institution formation game if (some) players have social preferences.

C. *Social Preferences*

There exists considerable evidence that social preferences, such as a taste for fairness, equity, and efficiency, affect economic behavior in many important areas, including the provision of public goods. For an overview see, e.g., Fehr and Gächter (2000b) and Colin F. Camerer (2003). We analyze the impact of social preferences on institution formation using the inequity-aversion model suggested by Fehr and Schmidt (1999).[13] Suppose that players' material payoffs are given by $(u_1, \dots, u_n)$. Player i's utility U_i is then defined as

$$U_i = u_i - \alpha_i \frac{1}{n-1} \sum_{j \neq i} \max\{(u_j - u_i), 0\} - \beta_i \frac{1}{n-1} \sum_{j \neq i} \max\{(u_i - u_j), 0\}. \tag{6}$$

The two parameters α_i and β_i measure player i's utility loss from disadvantageous inequality and from advantageous inequality, respectively. Following Fehr and Schmidt (1999), we assume that $\beta_i \leq \alpha_i$ and $0 \leq \beta_i < 1$ for all i. We first analyze the case when players' utility loss from advantageous inequality is small ($\beta_i < 1 - a$) and behavior is driven by players' disutility from disadvantageous inequality. We then consider the case when some of the players also suffer strongly from advantageous inequality ($\beta_i > 1 - a$).

Suppose that $\beta_i < 1 - a$ for all $i = 1, \dots, n$. This implies that zero contribution to the public good is the dominant action for every player who is not a member of an organization (cf.

[12] The strict subgame perfect equilibrium prediction nicely mirrors the concept of "internal stability" developed in the cartel and coalition formation literature (cf. Claude d'Aspremont et al. 1983; Carlo Carraro and Domenico Siniscalco 1993; Barrett 1994).

[13] There exist other models of social preferences (e.g., Matthew Rabin 1993; Gary Bolton and Axel Ockenfels 2000; Gary Charness and Rabin 2002; Martin Dufwenberg and Georg Kirchsteiger 2004; Armin Falk and Urs Fischbacher 2006; James C. Cox, Daniel Friedman, and Steven Gjerstad 2007). We use the model of Fehr and Schmidt (1999) for reasons of simplicity and tractability.

Proposition 4 of Fehr and Schmidt 1999). In consequence, if an organization of size s is implemented, organization member i's utility is given by

$$(7) \qquad asw - \frac{c}{s} - \frac{\alpha_i}{n-1}(n-s)\left(w + \frac{c}{s}\right).$$

The utility of an organization member is equal to his material payoff, $asw - \frac{c}{s}$, minus some disutility that is generated by the difference between his payoff and the payoff of nonmembers. As in the case of standard preferences, we can calculate the minimum size s_i^+ of an organization to be profitable for player i, which is the smallest nonnegative integer s satisfying

$$(8) \qquad asw - \frac{c}{s} - \frac{\alpha_i}{n-1}(n-s)\left(w + \frac{c}{s}\right) > w.$$

Comparing (8) and (5) immediately reveals that the threshold s_i^+ for a player suffering from disadvantageous inequality ($\alpha_i > 0$) is larger than the one for a player with standard preferences ($\alpha_i = 0$). As players become more inequity averse, the threshold rises. When players have identical α_i's, basically all results from Section IB carry over to the social-preference case with a new threshold $s^+ \equiv s_i^+$ defined by equation (8). It is easy to see that for sufficiently strong social preferences, the grand organization is a unique organizational equilibrium ($s^+ = n$) when standard preferences predict multiple organizational equilibria ($s^* < n$).

If players differ in their concern for inequity, the situation is slightly more complex. First, Proposition 1 generalizes in the sense that an organizational equilibrium with participant set S exists if and only if equation (8) holds for every participant $i \in S$. Since there is no payoff inequality in the grand organization, the latter is always an equilibrium. The interesting question is whether motives of fairness might induce this equilibrium to be strict (i.e., players are strictly worse off when deviating from the equilibrium) or even the unique organizational equilibrium.

PROPOSITION 3: *Suppose that $\beta_i < 1 - a$ for all $i = 1,\ldots,n$. The grand organization is a strict organizational equilibrium if and only if there exist at least two players with $\alpha_i > \tilde{\alpha}$, where*

$$(9) \qquad \tilde{\alpha} = \frac{(n-1)^2\,((n-1)a-1)w - (n-1)c}{(n-1)w + c},$$

If at least $n - 1$ players satisfy $\alpha_i > \tilde{\alpha}$, implementation of the grand organization is the unique organizational equilibrium.

The intuition for Proposition 3 is as follows. In Web Appendix A we show that an organizational equilibrium with participants S is a strict equilibrium if and only if each participant is pivotal, i.e., $S\backslash\{i\}$ is no organizational equilibrium for every $i \in S$. For the grand organization, this means that no group of players is willing to implement an organization of size $n - 1$. Threshold $\tilde{\alpha}$ guarantees that a participating player suffers sufficiently from disadvantageous inequality induced by the free-riding of the nonmember to reject an organization of size $n - 1$. At least two such players are required to ensure that organizations of size $n - 1$ are never formed. Since a participating player's utility is increasing in s (cf. (7)), the grand organization is the unique organizational equilibrium if at least $n - 1$ players are sufficiently inequality averse, as this implies that no group of players is willing to implement any organization smaller than n.

Two things are worth mentioning at this point. First, the condition guaranteeing uniqueness of the grand organizational equilibrium in Proposition 3 is sufficient but not necessary. This follows because the requirement for α_i rendering an organization of size s unattractive becomes less restrictive when s falls, as participants' corresponding utility falls as well (cf. the left-hand side of (8)). For example, in one of our experimental treatments we present in the next section (IF40) the minimal degree of inequality aversion α_i to reject a three-player organization equals 0.48, whereas the minimal α_i to reject a two-player organization is even negative. There is no equilibrium in which players implement a two-player organization in this case. Thus, the grand (i.e., four-player) organization is the unique organizational equilibrium whenever condition (9) is satisfied for at least two players. Second, a ceteris paribus increase in the MPCR of the public good raises the threshold $\tilde{\alpha}$. The reason is that the material payoff of organization members rises, while the payoff inequality between members and nonmembers remains unchanged. In consequence, rejecting an organization becomes relatively more costly and players concern for inequality must be stronger to keep rejection a best response. In our experiment, for example, we implement treatments with $a = 0.4$ (IF40) and $a = 0.65$ (IF65). In the first treatment, $\tilde{\alpha} = 0.48$; in the second $\tilde{\alpha} = 2.23$.

Suppose now that there exist some players who also suffer strongly from advantageous inequality, i.e., $\beta_i > 1 - a$ for some i. Let $B = \{i \,|\, \beta_i > 1 - a\}$ denote the set of these players. Due to their concern for advantageous inequality, players in B might contribute to the public good even when they are nonmembers of the organization. As such contributions increase the payoff of organization members (both by an increase in the material payoff and by a decrease in payoff inequality), condition (8) is now relaxed. In consequence, participating players who suffer from disadvantageous inequality are less likely to reject an organization, given that nonmembers contribute to the public good, as well.

Nevertheless, organizations smaller than n may fail to be an equilibrium. The reason is that players with sufficiently large β_i prefer to be members of the grand organization. While uniqueness of the grand organization is more difficult to obtain, Proposition 4 shows that under comparable conditions as in Proposition 3 the grand organization is again a strict organizational equilibrium.

PROPOSITION 4: *Suppose* $B = \{i \,|\, \beta_i > 1 - a\} \neq \varnothing$. *The grand organization is a strict organizational equilibrium if* (*i*) *there exist at least two players with* $a_i > \tilde{\alpha}$, *and* (*ii*) *for each player* $i \in B$, $\beta_i > 1 - 1/n$. *Only condition* (*ii*) *is relevant if* $B = \{1,\ldots,n\}$.

Proposition 4 can be illustrated by the following example. Suppose there exists one player i who suffers from advantageous inequality $\beta_i > 1 - a$, while the remaining players suffer only (if at all) from disadvantageous inequality. Proposition 4 states that the grand organization is a strict organizational equilibrium if $\beta_i > 1 - 1/n$ and at least two players (possibly including player i) have $\alpha_j > \tilde{\alpha}$. The latter condition is the same as in Proposition 3 and ensures that any organization of size $n - 1$ is rejected by at least one participating player, given that the single nonparticipant does not contribute to the public good. The condition on β_i additionally guarantees that in case player i does not participate, i himself is strictly worse off compared to joining the grand organization, and hence strictly prefers to participate. Obviously, if all players are sufficiently averse to advantageous inequality, condition (i) is of no relevance since condition (ii) already guarantees that each player strictly prefers to participate in the grand organization.

Let us summarize the theoretical analysis. If players have standard preferences, i.e., maximize their material payoff, the institution formation game has multiple organizational equilibria of size $s \in \{s^*,\ldots,n\}$. In addition, there exists a status quo equilibrium for any number of participating players, in which no organization is formed. Strictness as an equilibrium refinement selects the smallest organizational equilibrium of size s^*. While this prediction is intuitive in terms of individual material payoff maximization, the resulting equilibrium outcome is unfavorable in terms of symmetry, equality, and efficiency. If players dislike payoff inequality, the prediction changes in so far as smaller organizations are no longer an equilibrium. In fact, depending on players' degree of inequity aversion, the grand organization, where all players participate, may be the unique, or at least a strict, organizational equilibrium.

The analysis shows that theory alone gives only limited guidance regarding the expected outcome of institution formation in our setup. Due to the multiplicity of equilibria and different possibilities for refinement, several outcomes are possible and equally plausible.[14] In the following, we therefore present a laboratory experiment designed to investigate the process of institution formation in public goods games.

II. Institution Formation: Experiment

A. *Procedural Details*

In order to keep the complexity of the experiment low, we slightly modified the institution formation game in the experiment. Once an organization was implemented, members of the organization did not make a decision in the contribution stage, but were bound to contribute their full endowment to the public good. Otherwise, everything else was the same as described above. The reason for this modification is that we want to focus on the problem of institution formation rather than on the separate issue of institutional enforcement.[15] The basic structure of the experimental game is as follows.[16]

At the beginning of each round of the experiment, each of four players receives an endowment of 20 points (i.e., $n = 4$ and $w = 20$). Each player then decides whether he wants to participate in an organization or not (*participation stage*).[17] After being informed about the number of players who want to participate, each participant decides whether he wants to implement the organization (*implementation stage*). The organization is implemented if and only if all participants decide to implement it. Nonparticipants do not make any decision in this stage and are informed only about the number of participants. Finally, players simultaneously determine the amount of their contributions to the public good (*contribution stage*). If an organization is implemented in the implementation stage, all members of the organization are bound to contribute their full endowment to the public good. Nonmembers, after being informed about the size of the implemented organization, freely determine the amount of their contributions. If no organization is implemented, all players freely determine the amount of their contributions.

Since the decision to participate may constitute a nontrivial coordination problem—in particular, if only a subset of players wants to form an organization—we elicited players' beliefs in the participation stage. Precisely, after players had decided whether to participate in the organization,

[14] The multiplicity of equilibrium predictions becomes even larger once we consider a repeated-game set-up. In this case, any combination of Nash equilibria of the one-shot game is a subgame perfect equilibrium of the repeated game.

[15] To examine whether the modification has an effect on the experimental results, we conducted an additional control treatment ($IF40^{+}$), where all subjects were free to decide in the contribution stage, but members of the organization were effectively punished if they did not contribute everything to the public good. In this treatment, the somewhat more complex decision environment slowed down learning; otherwise no significant differences were found in comparison to our main treatment. Details are reported in the supplementary material (Web Appendix B).

[16] Experimental instructions are provided in the supplementary material (Web Appendix C).

[17] In the experimental instructions, we did not use the terms "organization" or "institution." Instead, subjects were asked if they were willing to bind themselves to contribute their full endowment.

each player was asked to indicate his expectation about the total number of players participating in the organization. Players were rewarded for correct predictions according to the quadratic scoring rule.[18]

We implemented two experimental treatments with different minimum size s^* for an organization to be materially profitable. In both treatments, the cost of the organization was set to $c = 2$. In the first treatment (IF40), the MPCR of the public good $a = 0.4$ resulting in $s^* = 3$. In the second treatment (IF65), $a = 0.65$ yielding $s^* = 2$. In addition to these treatments, we implemented two control treatments (PG40 and PG65), in which players played the corresponding public goods game without the possibility of institution formation. Irrespective of the treatment, subjects played 20 rounds of the corresponding game with the same group of players (partner matching). All experiments were run at the CREED laboratory at the University of Amsterdam. In total, 164 subjects participated in the experiment, whereby 44 subjects participated in each of the institution formation treatments (IF40, IF65), 40 subjects participated in treatment PG40, and 36 participated in treatment PG65.[19] No subject participated in more than one treatment. Each session lasted about 120 minutes. On average, a subject earned €23.90 (about $25) in the experiment.

B. *Results*

In the results section we proceed as follows. We first analyze if subjects implement any organizations at all. Answering this question in the affirmative, we then study what kind of organizations are implemented. We also consider players' beliefs about the other players' participation decision, and investigate the probability that an organization of a particular size is implemented in the implementation stage. Finally, we analyze the overall impact of institution formation on the provision of the public good by comparing average contributions and consequential levels of efficiency in the institution formation treatments and the corresponding control treatments.

Our first result shows that players almost always initiate an organization, and also implement the initiated organization in between 43 and 61 percent of the cases.

RESULT 1: *In treatment IF40, there is always at least one player per group who wants to establish an organization, and an organization is implemented 43 percent of the time. In treatment IF65, in 98 percent of the cases, at least one player per group initiates an organization, and an organization is implemented in 61 percent of the cases.*

Support for Result 1 is presented in Table 1, which summarizes the absolute and relative number of cases in which at least one player decides to participate in the participation stage (*initiated organizations*) and in which an organization is implemented by unanimous vote in the implementation stage (*implemented organizations*). While players always initiate an organization in treatment IF40, there are four cases in treatment IF65 in which a group of players does not initiate an organization (2 percent). At the same time, slightly more organizations are implemented in treatment IF65 than in treatment IF40 (132 versus 95). None of these differences is statistically significant (Mann-Whitney test, $p > 0.14$).[20]

[18] Quadratic scoring rules are known to be incentive compatible and have successfully been used in a number of experiments, for example, by Theo Offerman (1997) and Yaw Nyarko and Andrew Schotter (2002).

[19] Another 52 subjects participated in the additional control treatment IF40$^+$.

[20] Statistical tests are based on group averages as units of observation. We report the results of two-sided tests throughout the paper.

TABLE 1—INITIATED AND IMPLEMENTED ORGANIZATIONS

	Treatment			
	IF40		IF65	
	Number	Percentage	Number	Percentage
Initiated organizations	220	100	216	98
Implemented organizations				
Total	95	43	132	61
One member	0	0	5	4
Two members	1	1	15	11
Three members	15	16	22	17
Four members	79	83	90	68

Notes: The table presents the absolute and relative number of *initiated* and *implemented organizations* over all rounds. Relative numbers are calculated as follows: initiated organizations relative to all rounds, implemented organizations relative to all initiated organizations, different size of organizations relative to all implemented organizations.

Result 1 shows that players overcome the second-order free-rider problem and successfully establish organizations. Our next result reveals which organizations are implemented.

RESULT 2: *In both IF treatments, the large majority of implemented organizations are grand organizations, i.e., all players become members. Organizations of size $s < s^*$ are very rarely observed. Moreover, the observed increase of implemented organizations over time is due solely to an increase of grand organizations.*

Table 1 (lower part) shows the distribution of implemented organizations in the two IF treatments. The data speak clearly: independent of treatment, the majority of organizations that are implemented include all four players. In treatment IF40, 79 of 95 organizations that are implemented are grand organizations (83 percent). In treatment IF65, 90 of 132 organizations are grand organizations (68 percent). In addition, players almost never implement organizations with fewer than s^* players. Recall that $s^* = 3$ in treatment IF40 and $s^* = 2$ in treatment IF65. Overall, only 1 (4) percent of the implemented organizations comprise fewer than s^* players in treatment IF40 (IF65). Thus, threshold s^* serves as a good prediction of the minimum size of an implemented organization. However, it clearly fails as a prediction of the maximum size of an organization (cf. Proposition 2). Rather than seeing only three or two players establishing an organization, we observe that most of the time an organization involves all four players. On average, the size of an implemented organization is slightly smaller in treatment IF65 than in treatment IF40 (3.49 versus 3.82), but the difference is not significant (Mann-Whitney test, $p = 0.20$).

The implementation of an organization is of course a rather complex process. It seems likely that players learn the benefits of establishing organizations in the course of the experiment, and the number of organizations implemented increases over time. Our data show that this is indeed the case. The Spearman rank order correlation between the number of implemented organizations and the round in the experiment is highly significant in both treatments (IF40: $\rho = 0.85$, $p = 0.00$; IF65: $\rho = 0.64$, $p = 0.00$). Moreover, the increase in implemented organizations is exclusively driven by the implementation of the grand organization. While the number of implemented grand organizations increases significantly over rounds (Spearman rank order correlation; IF40: $\rho = 0.87$, $p = 0.00$; IF65: $\rho = 0.72$, $p = 0.00$), the corresponding number of organizations with three or fewer members does not change significantly (Spearman rank order correlation; $p > 0.41$ in both treatments). Figure 1 illustrates the learning pattern by comparing the distribution of implemented organizations in early and late rounds. As can be seen, the share of the grand organization increases from 70 and 48 percent in rounds 1 to 5 of treatment IF40 and

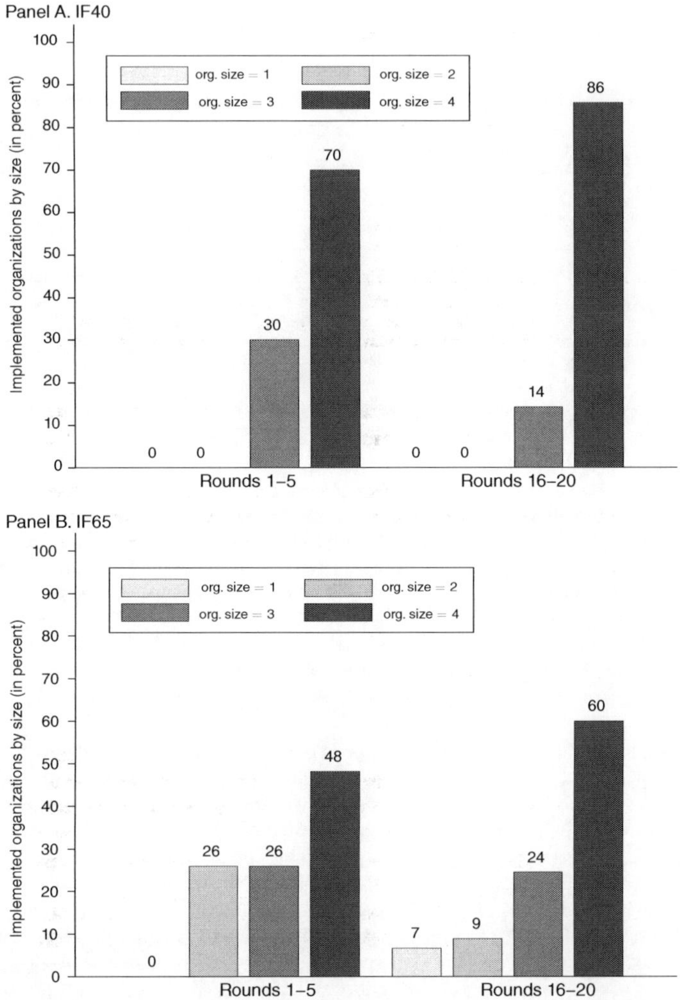

FIGURE 1. DISTRIBUTION OF IMPLEMENTED ORGANIZATIONS IN EARLY AND LATE ROUNDS

IF65, respectively, to 86 and 60 percent in rounds 16 to 20. At the same time, the relative number of smaller organizations fall.

Why do we observe so many organizations larger than s^* in the experiment? Is it that players aim to implement the s^* organization but miscoordinate in the participation stage? Or, do participating players target at the grand organization and reject organizations that are smaller in the implementation stage? The following two results shed light on the driving forces of subjects' behavior. We first show that players who participate in the participation stage mostly expect that

TABLE 2—BELIEFS AND RATE OF IMPLEMENTATION

	Treatment									
	IF40					IF65				
		Number of participants					Number of participants			
Belief	Observations	1	2	3	4	Observations	1	2	3	4
First round	26	9.42	18.19	34.50	37.88	25	19.52	14.48	23.80	42.20
Final round	35	5.29	4.83	15.86	74.03	32	2.34	5.47	21.28	70.91
All rounds	726	6.48	7.03	21.67	64.81	671	5.01	11.80	21.75	61.44
		Number of participants					Number of participants			
Implementation rate		1	2	3	4		1	2	3	4
All rounds		0.00	2.94	23.08	69.30		27.78	37.50	37.29	90.91
Observations		7	34	65	114		18	40	59	99

Notes: The upper panel of the table presents the average probability belief (in percent) of participating players in stage one of the game about the total number of participants in the organization. The lower panel presents the likelihood of implementation (in percent) of an organization depending on the number of participating players.

all other players will participate as well. Second, we show that initiated organizations comprising fewer than four participants are rejected with high probability in the implementation stage, even if no fewer than s^* players participate. These results cast doubt on the explanation that high participation rates are due to miscoordination. They suggest, rather, that players play the grand organizational equilibrium.

RESULT 3: *In both IF treatments, most players who participate in the organization in the participation stage expect that all other players will participate as well.*

Support for Result 3 is presented in the upper panel of Table 2, which shows participating players' average probability belief about the total number of players willing to participate in the organization. If players' high participation rate was mainly due to miscoordination, players should expect, with high probability, that s^* players will participate in the organization. This is not what we find. In treatment IF40 and IF65, participants hold, on average, a belief of only about 22 and 12 percent, respectively, that s^* players will participate in the organization. The belief is slightly higher in the first round of both treatments, but it decreases to 16 percent in the final round of treatment IF40, and to 5 percent in the final round of treatment IF65. As can be seen, in both treatments, participants' average belief peaks at "four participants." The average belief that all players will participate amounts to 65 and 61 percent over all rounds in treatment IF40 and IF65, respectively. In fact, the belief is already quite high in the first round and increases to over 74 and 71 percent, respectively, in the final round of the two treatments. The increase clearly mirrors the corresponding increase in the implementation of the grand organization. Overall, beliefs demonstrate that from early on players who are willing to participate rarely expect organizations of size s^* to be formed, but mostly expect that all of the players will participate in the organization. Thus, it seems unlikely that high participation rates are driven by miscoordination.

Further evidence is presented in the lower panel of Table 2, which shows the average likelihood over all rounds with which an organization is implemented, depending on the number of participating players. Note that there are many cases in both treatments in which from one to three players have to decide whether to implement an organization. As the data show, most of these organizations are not implemented. In fact, if fewer than s^* players participate, the likelihood of implementation lies below 3 percent in treatment IF40 and below 28 percent in treatment IF65. This finding corresponds to Result 2, which states that only a few organizations smaller than s^* are observed in the experiment. When the number of participants hits the threshold s^*, the likelihood of implementation rises somewhat, but still remains at a rather low level of 23 and 38 percent in treatments IF40 and IF65, respectively. Only if all players participate do organizations have a high chance to be implemented. In this case, the likelihood of implementation rises to almost 70 percent in treatment IF40 and to over 90 percent in treatment IF65.[21] One may also hypothesize that the predominant implementation of grand organizations is a consequence of the grand organization serving as a focal point in the coordination problem. Indeed, the result that most participating players expect all other players to participate as well is consistent with such a hypothesis. However, since organizations of size $s < n$ (but not smaller than s^*) are frequently rejected by players, the data suggest that the formation of grand organizations is not due solely to focal points, but that subjects' behavior is driven by additional forces such as equality and fairness.

Table 2 shows that implementation rates are, ceteris paribus, higher in treatment IF65 than in IF40. Interestingly, this is exactly what social preferences would predict (cf. the discussion following Proposition 3). While, according to standard theory, any organization larger than s^* should be implemented with probability one, fairness predicts that players reject unequal organizations but that these rejections are more costly (and hence less likely to occur) the more productive the public good. Result 4 summarizes our findings.

RESULT 4: *Organizations with fewer than four participants have a high likelihood of being rejected in the implementation stage of both IF treatments. Only the grand organization has a substantial likelihood of being implemented. Ceteris paribus, implementation rates are higher the more productive the public good.*

Once an organization is implemented, its members are bound to contribute their full endowment to the public good. Thus, if all four players participate and the organization is implemented, contribution levels reach 100 percent. Yet, as shown above, there is always the possibility that implementation fails and players end up in the status quo. It is conceivable that the failure to implement an organization might have a negative effect on voluntary contributions to the public good. However, as our final result shows, the overall impact of institution formation on average contributions and on efficiency is clearly positive.

RESULT 5: *Overall, the possibility of institution formation has a positive effect on contributions to the public good. Contributions are both higher and more stable if players are allowed to form organizations than if they are not. In consequence, achieved efficiency levels are higher in the institution formation treatments than in the control treatments.*

Support for Result 5 is presented in Figure 2, which compares the average contributions to the public good in the institution formation treatments with those in the corresponding control treatments. Consider, first, treatments IF40 and PG40 (panel A). The data clearly show that in treatment PG40 the average contribution steadily declines from 12.4 in the first round to 0.7 in the last round.

[21] The fact that 30 percent of the grand organizations are not implemented in treatment IF40 may seem surprising, but is due only to the different learning speeds in the two treatments. In IF40, the likelihood of implementation greatly increases over rounds, reaching levels similar to those in treatment IF65 in the second half of the experiment. In rounds 11 to 20 of treatment IF40, the likelihood of implementing a grand organization is 86 percent (63 observations) and even increases to 94 percent (32 observations) in the final five rounds of the experiment. As for treatment IF65, the likelihood of implementing a grand organization is 93 percent (59 observations) in the final ten rounds and 90 percent (30 observations) in the final five rounds.

In stark contrast, in treatment IF40, the average contribution increases from 9.8 in round one to a maximum of 15.3 in round 18, at which point it falls to an average of 11 in the final round. Over all rounds, players contribute an average of 10.6 in treatment IF40 and an average of 5.0 in treatment PG40. Thus, the possibility of institution formation more than doubles players' average contribution to the public good. This difference is highly significant (Mann-Whitney test, $p < 0.01$).

In the institution formation treatment IF65 (panel B), a similar pattern emerges, yet on a slightly higher level. When the MPCR equals 0.65, the average contribution is 12.9 in the first round, increases to a maximum of 17 in rounds 17 and 18, and falls to 12.7 in the final round. In line with previous studies showing a positive effect of an increase in the MPCR on contributions to the public good (see, e.g., Ledyard 1995), the average contribution is also relatively high in the control treatment PG65. From this it follows that, up to round 14, the average contributions in treatments IF65 and PG65 do not differ statistically from each other. In the final rounds, however, the average contribution falls to 7.1 in treatment PG65, but rises and stays above 12.7 in treatment IF65. Over all rounds, the average contribution in treatment PG65 is 12.4, as compared to 14.1 in treatment IF65. Thus, when players have the possibility to form organizations, their contributions are again higher than when they do not have this possibility, but the difference is not statistically significant (Mann-Whitney test, $p = 0.48$). If we consider rounds 15 to 20, the difference becomes marginally significant (Mann-Whitney test, $p = 0.10$).

Figure 2 also shows the average contribution to the public good in treatments IF40 and IF65, both when an organization has been implemented and when no organization has been implemented. Contribution levels are close to 100 percent when an organization has been implemented, since organization members are bound to contribute their full endowment and most organizations that are implemented comprise all players. What is particularly interesting is the comparison of the average contribution in the IF treatment when no organization has been established with the average contribution in the corresponding PG treatment. In both cases, all subjects are able to decide freely how much they contribute to the public good. However, in the first case, this is because subjects have rejected the implementation of an organization, while in the second case subjects do not have the possibility of institution formation. By comparing the resulting contribution levels, we can thus determine whether the failure to implement an organization has a negative effect on players' voluntary contribution to the public good. As Figure 2 shows, implementation failure has basically no effect in treatment IF40. If players can form institutions but the implementation fails, the average contribution is 4.2 compared to 5.0 in treatment PG40 (Mann-Whitney test, $p = 0.23$). In treatments IF65 and PG65, the difference is larger and marginally significant. Subjects contribute on average 8.5 in treatment IF65 if no organization exists, compared to 12.4 in treatment PG where no institution formation is possible (Mann-Whitney test, $p = 0.07$). Thus, institution formation failure has a negative effect on voluntary contributions in this treatment. Importantly, however, as we saw above, the overall effect of institution formation on contribution levels is always positive.

Finally, to evaluate efficiency we calculate the average observed level of efficiency relative to the welfare maximum in all treatments. That is, efficiency is defined as $(\Pi_{observed} - \Pi_{min})/(\Pi_{max} - \Pi_{min})$, where $\Pi_{observed}$ denotes the observed group earnings, Π_{min} the theoretical minimum group earnings, and Π_{max} the theoretical maximum group earnings. Our data show that efficiency levels are higher in the IF treatments than in the corresponding PG treatments. In treatment IF40, the average efficiency over all rounds lies at 51 percent compared to 25 percent in PG40. This difference is significant (Mann-Whitney test, $p = 0.01$). Average efficiency in treatment IF65 is 70 percent compared to 62 percent in PG65. Due to the relatively high contributions in the PG treatment this difference fails to be significant (Mann-Whitney test, $p = 0.47$). Importantly, while efficiency decreases significantly over rounds in both PG

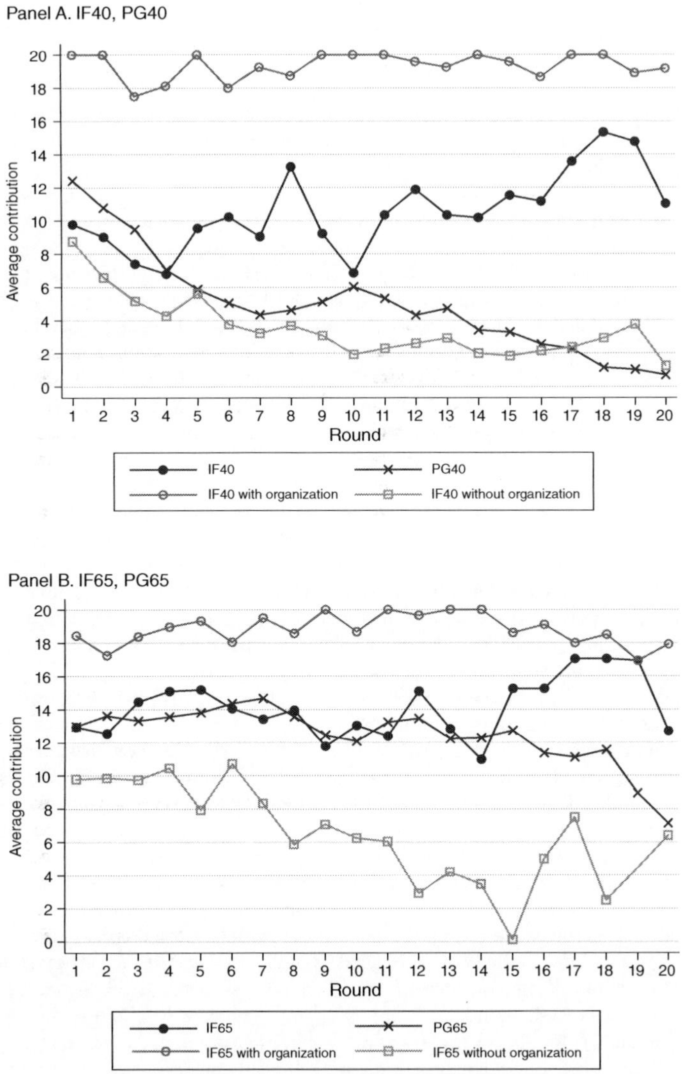

Figure 2. Average Contribution to the Public Good with and without the Possibility of Institution Formation

treatments (Spearman rank order correlation; PG40: $\rho = -0.93$, $p = 0.00$; PG65: $\rho = -0.80$, $p = 0.00$), this is not the case in the IF treatments. Here, the Spearman rank order correlation between rounds and the achieved level of efficiency is positive (IF40: $\rho = 0.73$, $p = 0.00$; IF65: $\rho = 0.30$, $p = 0.19$).

III. Conclusion

We analyze the endogenous formation of institutions in a linear public goods game. Players in our model have the possibility to establish a sanctioning organization that punishes players who do not contribute the efficient amount to the public good. While the voluntary formation of such institutions typically suffers from a second-order free-rider problem—jointly, every player profits if an institution is formed, but each player profits even more if only the other players implement an institution—we show in the theory part of this paper that players can, in principle, overcome this problem. If institution formation is efficient, a nonempty set of subgame perfect Nash equilibria exists, in which s players ($s^* \leq s \leq n$) implement a sanctioning organization. However, there also exist subgame perfect Nash equilibria in which no institution is formed. Strictness as a refinement selects the minimally profitable organization of size s^*—a result that mirrors a similar prediction based on the notion of internal stability in the environmental economics literature (Barrett 1994). Although this equilibrium prediction is intuitive in terms of individual material payoff maximization, it seems questionable if at least some players have a concern for equity and efficiency. Our analysis based on Fehr and Schmidt (1999) preferences confirms that if some players are inequity averse, the s^* organization may indeed no longer be an equilibrium, and the grand organization becomes the expected equilibrium outcome.

In the empirical part of the paper, we report the results of an experiment designed to investigate players' behavior in institution formation and its effect on public good provision. The obtained results show that organizations are formed and that the possibility of institution formation has a clear positive impact on contributions to the public good. Contributions are both higher and more stable if players are able to form institutions compared to if they are not able to do so. The results also support the theoretical prediction based on social preferences that mainly grand organizations will be formed. The likelihood of implementation of a sanctioning organization crucially depends on the number of participating players. In particular, only the grand organization has a reasonable chance of being implemented. Smaller organizations, even if more than s^* players participate, are rejected most of the time.

The main message of our paper is twofold. First, our theoretical and empirical analyses show that institution formation can be an important and effective solution in social dilemma situations. Despite the second-order free-rider problem, the institution formation process may be structured such that the implementation of a sanctioning organization is supported by a Nash equilibrium. The experimental data corroborate this result as players make frequent use of the possibility of institution formation, and thereby substantially increase efficiency. Second and most important, such success is not guaranteed. In particular, our results highlight the crucial role of fairness in the institution formation process. Individuals are very reluctant to comply with a sanctioning system that governs only a subset of individuals. This is true even if the subset of individuals can earn a higher material payoff than in the status quo with no sanctioning system, and the implementation of the system, at least from a material standpoint, would thus be optimal. The importance of fairness for the governance of public goods has been documented in the field, e.g., in studies of real-world common property regimes (Baland and Platteau 1996; McKean 2000). As one of the researchers in this literature notes, the "[d]istribution of decision-making rights and use of rights to co-owners of the commons need not be egalitarian but must be viewed as "fair." (…) If any subgroup feels cheated—denied "adequate" access of "fair" share—compared to another subgroup, the angry subgroup becomes unwilling to participate in decision making, unwilling to invest in maintaining or protecting the commons, and motivated to vandalize the commons" (McKean 2000, 47–48). Notably, fairness arguments were also put forward in the political discussion about the Kyoto Protocol, in particular after the US withdrawal from the protocol (cf. footnote 6 above). Buchanan and Congleton (1998) take such arguments—plus further efficiency considerations—to propose the so-called *generality principle* as a normative guideline for political action. The principle asserts that political choices should be general in nature, i.e., nondiscriminatory and based on equal treatment of all individuals. Our paper offers strong support for this principle both theoretically and empirically: theoretically, because our formal results show that only those institutional rules that obey the generality principle are selected as equilibrium when players have social preferences with a taste for fairness; empirically, because general rules are also the predominant outcome of institution formation in the experiment.

For economists interested in institutions, the role of fairness can be seen as good and bad news: bad news, because fairness motives act as a constraint on equilibria, which increases the risk of failure of the institution formation process (in particular, if these motives are not taken into account ex ante); good news, because once the process is successful, efficiency levels will typically be higher compared to a world in which fairness motives do not play a role.

REFERENCES

Agrawal, Arun, and Gautam N. Yadama. 1997. "How Do Local Institutions Mediate Market and Population Pressures on Resources? Forest Panchayats in Kumaon, India." *Development and Change*, 28(3): 435–65.

Anderson, Christopher M., and Louis Putterman. 2006. "Do Non-Strategic Sanctions Obey the Law of Demand? The Demand for Punishment in the Voluntary Contribution Mechanism." *Games and Economic Behavior*, 54(1): 1–24.

Bagnoli, Mark, and Barton L. Lipman. 1989. "Provision of Public Goods: Fully Implementing the Core through Private Contributions." *Review of Economic Studies*, 56(4): 583–601.

Baland, Jean-Marie, and Jean-Philippe Platteau. 1996. *Halting Degragation of Natural Resources*. Oxford: Oxford University Press.

Barrett, Scott. 1994. "Self-Enforcing International Environmental Agreements." *Oxford Economic Papers*, 46: 878–94.

Bolton, Gary E., and Axel Ockenfels. 2000. "ERC: A Theory of Equity, Reciprocity, and Competition." *American Economic Review*, 90(1): 166–93.

Buchanan, James M., and Roger D. Congleton. 1998. *Politics by Principle, Not Interest*. Cambridge: Cambridge University Press.

Camerer, Colin F. 2003. *Behavioral Game Theory: Experiments in Strategic Interaction*: Roundtable Series in Behavioral Economics. Princeton: Princeton University Press.

Carpenter, Jeffrey P. 2007a. "The Demand for Punishment." *Journal of Economic Behavior and Organization*, 62(4): 522–42.

Carpenter, Jeffrey P. 2007b. "Punishing Free-Riders: How Group Size Affects Mutual Monitoring and the Provision of Public Goods." *Games and Economic Behavior*, 60(1): 31–51.

Carraro, Carlo, and Domenico Siniscalco. 1993. "Strategies for the International Protection of the Environment." *Journal of Public Economics*, 52(3): 309–28.

Charness, Gary, and Matthew Rabin. 2002. "Understanding Social Preferences with Simple Tests." *Quarterly Journal of Economics*, 117(3): 817–69.

Chen, Yan, and Charles R. Plott. 1996. "The Groves-Ledyard Mechanism: An Experimental Study of Institutional Design." *Journal of Public Economics*, 59(3): 335–64.

Cox, James C., Daniel Friedman, and Steven Gjerstad. 2007. "A Tractable Model of Reciprocity and Fairness." *Games and Economic Behavior*, 59(1): 17–45.

Croson, Rachel T. A., and Melanie Beth Marks. 2000. "Step Returns in Threshold Public Goods: A Meta- and Experimental Analysis." *Experimental Economics*, 2(3): 239–59.

d'Aspremont, Claude, Alexis Jacquemin, Jean Jaskold Gabszewicz, and John A. Weymark. 1983. "On the Stability of Collusive Price Leadership." *Canadian Journal of Economics/Revue canadienne d'economique*, 16(1): 17–25.

Dufwenberg, Martin, and Georg Kirchsteiger. 2004. "A Theory of Sequential Reciprocity." *Games and Economic Behavior*, 47(2): 268–98.

Falk, Armin, and Urs Fischbacher. 2006. "A Theory of Reciprocity." *Games and Economic Behavior*, 54(2): 293–315.

Falkinger, Josef. 1996. "Efficient Private Provision of Public Goods by Rewarding Deviations from Average." *Journal of Public Economics*, 62(3): 413–22.

Falkinger, Josef, Ernst Fehr, Simon Gächter, and Rudolf Winter-Ebmer. 2000. "A Simple Mechanism for the Efficient Provision of Public Goods: Experimental Evidence." *American Economic Review*, 90(1): 247–64.

Fehr, Ernst, and Simon Gächter. 2000a. "Cooperation and Punishment in Public Goods Experiments." *American Economic Review*, 90(4): 980–94.

Fehr, Ernst, and Simon Gächter. 2000b. "Fairness and Retaliation: The Economics of Reciprocity." *Journal of Economic Perspectives*, 14(3): 159–81.

Fehr, Ernst, and Simon Gächter. 2002. "Altruistic Punishment in Humans." *Nature*, 415: 980–94.

Fehr, Ernst, and Klaus M. Schmidt. 1999. "A Theory of Fairness, Competition, and Cooperation." *Quarterly Journal of Economics*, 114(3): 817–68.

Groves, Theodore, and John O. Ledyard. 1977. "Optimal Allocation of Public Goods: A Solution to The "Free Rider" Problem." *Econometrica*, 45(4): 783–809.

Gürerk, Özgür, Bernd Irlenbusch, and Bettina Rockenbach. 2006. "The Competitive Advantage of Sanctioning Institutions." *Science*, 312: 108–11.

Hviding, Edvard, and Graham B. K. Baines. 1994. "Community-Based Fisheries Management, Tradition, and the Challenges of Development in Marovo, Solomon Islands." *Development and Change*, 25(1): 13–39.

Kroll, Stephan, Todd L. Cherry, and Jason F. Shogren. 2007. "Voting, Punishment, and Public Goods." *Economic Inquiry*, 45(3): 557–70.

Ledyard, John O. 1995. "Public Goods: A Survey of Experimental Research." In *The Handbook of Experimental Economics*, ed. John H. Kagel and Alvin E. Roth, 111–94. Princeton: Princeton University Press.

Masclet, David, Charles Noussair, Steve Tucker, and Marie-Claire Villeval. 2003. "Monetary and Nonmonetary Punishment in the Voluntary Contributions Mechanism." *American Economic Review*, 93(1): 366–80.

McKean, Margaret A. 2000. "Common Property: What Is It, What Is It Good For, and What Makes It Work?" In *People and Forests: Communities, Institutions, and Governance*, ed. Clark C. Gibson, Margaret A. McKean, and Elinor Ostrom, 27–55. Cambridge, MA: MIT Press.

Nyarko, Yaw, and Andrew Schotter. 2002. "An Experimental Study of Belief Learning Using Elicited Beliefs." *Econometrica*, 70(3): 971–1005.

Offerman, Theo. 1997. *Beliefs and Decision Rules in Public Good Games: Theory and Experiments*: Dordrecht; Boston and London: Kluwer Academic.

Okada, Akira. 1993. "The Possibility of Cooperation in an N-Person Prisoners' Dilemma with Institutional Arrangements." *Public Choice*, 77(3): 629–56.

Oliver, Pamela. 1980. "Rewards and Punishments as Selective Incentives for Collective Action: Theoretical Investigations." *American Journal of Sociology*, 85(6): 1356–1375.

Ostrom, Elinor. 1999. "Coping with the Tragedy of the Commons." *Annual Review of Political Science*, 2: 493–535.

Ostrom, Elinor, James Walker, and Roy Gardner. 1992. "Covenants with and without a Sword: Self-Governance Is Possible." *American Political Science Review*, 86(2): 404–17.

Rabin, Matthew. 1993. "Incorporating Fairness into Game Theory and Economics." *American Economic Review*, 83(5): 1281–1302.

Sutter, Matthias, Stefan Haigner, and Martin G. Kocher. 2006. "Choosing the Carrot or the Stick? Endogenous Institutional Choice in Social Dilemma Situations." Center for Economic Policy Research Discussion Paper 5497.

Tang, Shui Yan. 1992. *Institutions and Collective Action: Self-Governance in Irrigation*. San Francisco: ICS Press.

Tyran, Jean-Robert, and Lars P. Feld. 2006. "Achieving Compliance when Legal Sanctions Are Non-Deterrent." *Scandinavian Journal of Economics*, 108(1): 135–56.

Walker, James M., Roy Gardner, Andrew Herr, and Elinor Ostrom. 2000. "Collective Choice in the Commons: Experimental Results on Proposed Allocation Rules and Votes." *Economic Journal*, 110(460): 212–34.

WWR. 2006. *Klimaatstrategie — Tussen Ambitie en Realisme*. Amsterdam: Amsterdam University Press.

Yamagishi, Toshio. 1986. "The Provision of a Sanctioning System as a Public Good." *Journal of Personality and Social Psychology*, 51(1): 110–16.

Part VI
Multiple Coalitions

[35]

EQUITY, DEVELOPMENT, AND CLIMATE CHANGE CONTROL

Francesco Bosello
Fondazione Eni Enrico Mattei

Barbara Buchner
Fondazione Eni Enrico Mattei

Carlo Carraro
University of Venice and Fondazione Eni Enrico Mattei

Abstract

The paper analyzes the effects of different equity principles on the decision of developing countries to join a world coalition whose aim is to control greenhouse gas emissions. A game-theoretic framework is proposed to assess the incentives for different countries to sign an international treaty on climate change control. Then, the effects of different equity rules on these incentives are evaluated by using a dynamic integrated growth and climate model. Policy proposals based on transfers from developed to developing countries are also analyzed. (JEL: C7, H0, H3, Q38)

1. Introduction

The cost-effectiveness of the Kyoto Protocol—and of any treaty signed and ratified by a limited number of countries—would be enhanced by attracting as many new countries as possible among the signatories, and by achieving these additions as soon as possible. In particular, in the case of climate change control, given the expected growth rate of income and population in developing countries, it would be crucial to induce these countries to commit themselves to greenhouse gas (GHG) emission control at some stage of their economic development (IPCC 2001).

However, the state of the climate is a global public good and there are therefore strong incentives to free-ride on emission abatement. In addition, there is no supranational authority that can impose environmental regulation, which implies that emission control can only be achieved through international negotiations and environmental treaties. Finally, structural differences among countries are very large. Developing countries are likely to suffer the greatest damage from climate changes, but are those which are least responsible for the present

Acknowledgments: The authors are grateful to Davide Raggi, Igor Cersosimo, Carmen Marchiori, Marni Wood and all members of the Climate Change and Policy Unit of the Fondazione Eni Enrico Mattei for helpful comments and research assistance.

E-mail addresses: Bosello: Francesco.Bosello@feem.it; Buchner: Barbara.Buchner@feem.it; Carraro: ccarraro@unive.it.

accumulation of GHGs in the atmosphere. This obviously raises an equity issue when countries negotiate on how to share the burden of combating climate change, with developing countries rejecting any proposals to commit to emission control.

Therefore, how can developing countries be induced to undertake substantial and effective action to control their GHG emissions? Developing countries currently have no emission abatement targets under the Kyoto Protocol. However, the implementation of the Protocol (or of other climate agreements) would have significant economic impacts on them, through trade and investment links, and through the clean development mechanism, which allows flexibility in the location of emission abatement activities. There is a bidirectional relationship between climate change policy on the one hand and equity and development on the other. Climate policy affects income distribution and growth at the world level, and therefore has implications on equity and development. At the same time, equity and development affect a country's decision to sign a climate treaty and therefore have implications on the effectiveness and efficiency of climate policy.

The first side of this relationship—the effect of climate policy on equity and development—has already been studied in the recent empirical economic literature on climate change control (cf. Babiker and Jacoby 1999; Babiker, Reilly, and Jacoby 1999; Bernstein, Bromley et al. 1999; Bernstein, Montgomery, and Rutherford 1999; Ellerman, Jacoby, and Decaux 1998). The main conclusions drawn in this literature can be summarized as follows: (a) the implementation of the Kyoto Protocol negatively affects developing countries—energy exporting countries above all—particularly in the absence of emission trading; (b) losses are smaller with Annex B trading and may become gains with global trading; (c) economies in transitions belonging to Annex B, however, suffer relevant economic losses when emission trading becomes global. These results support the conclusion that climate policy generally reduces growth and welfare in developing countries. In addition, they also show that climate policy reduces equity, by inducing larger economic losses in developing countries than in developed ones.

The second side of this relationship—the effect of equity and development on climate policy—has often been advocated as an important dimension of climate negotiations. For example, Metz (2000) argues that international equity will play a major role in implementing some of the most important elements of the Kyoto Protocol, whereas Convery (2000) states that an agreement in which the burden is equitably shared is more likely to be signed by a large number of countries. However, there are only a few studies containing an empirical assessment of the link between equity and the effectiveness of climate policy (cf. McKibbin and Wilcoxen 2000; Panayotou, Sachs, and Peterson 1999). This paper provides new evidence, by explicitly adopting a game-theoretic framework, where countries' incentives to sign an international treaty on climate change control are explicitly modelled.

In particular, this paper addresses the following research question: does more equity increase the probability that more developing countries adopt the Kyoto Protocol, thus enhancing the effectiveness of climate policy? The next section describes the modelling framework adopted to answer this question and presents our optimization exercises. A concluding section summarizes our main results and their policy implications.

2. Equity and Climate Change Control

As stated, there are many papers that discuss the importance of equity in climate change negotiations and which propose equity principles designed to enhance the probability that developing countries decide to commit themselves to GHG emission control (see, in addition to Convery 2000 and Metz 2000, Cazorla and Toman 2000; Claussen and McNeilly 1998; Rose and Stevens 1993; Rose et al. 1998; Schmidt and Koschel 1998; Tol 2001). Equity proposals can be classified by distinguishing whether the applied equity criterion has been chosen according to the initial allocation of emissions ("allocation-based" equity criteria), according to the final outcome of the implementation of the policy instruments ("outcome-based" equity criteria) or according to the process by which the criterion has been chosen ("process-based" equity criteria).[1]

This paper uses a game theory framework to analyze the consequences of the implementation of different "outcome-based" equity criteria on the participation decision of developing countries. In particular, we analyze the consequences of the adoption of climate policy strategies which guarantee either one of the following three outcomes: (a) equal average abatement costs; (b) equal per capita abatement costs; (c) equal abatement costs per unit of GDP.

The analysis is based on an empirical dynamic game that captures countries' incentives to sign the Kyoto Protocol. The model used is the well-known Nordhaus and Yang's (1996) RICE optimization model.[2] The equilibrium

1. For further explanations regarding this distinction see, among others, Rose et al. (1998) and Schmidt and Koschel (1998).

2. The RICE model is a one sector optimal growth model that has been extended to incorporate the interactions between economic activities and climate. One such model has been developed for each macroregion into which the world is divided (USA, Japan, Europe, China, Former Soviet Union, and Rest of the World). Within each region a central planner chooses the optimal paths of fixed investment and emission abatement that maximize the present value of per capita consumption. Output (net of climate change) is used for investment and consumption and is produced according to a constant return Cobb–Douglas technology, which combines the inputs from capital and labor with the level of technology. Population (taken to be equal to full employment) and technology levels grow over time in an exogenous fashion, whereas capital accumulation is governed by the optimal rate of investment. There is a wedge between gross and net outputs (of climate change effects), the size of which is dependent upon the amount of abatement (rate of emission reduction) as well as the change in global temperature. The model is completed by three equations representing emissions (which are related to output and abatement), carbon cycle (which relates concentrations to emissions), and climate module (which relates the change in temperature relative to 1990 levels to carbon concentrations) respectively.

concept is the one proposed in Eyckmans and Tulkens (1999). It is called PANE (Partial Agreement Nash Equilibrium) and "can be interpreted as a special type of Nash equilibrium in which a coalition coordinates its policies taking as given the policy strategies of the free-riders who, in turn, are playing a noncooperative Nash strategy against the coalition."[3]

In our framework, the dynamic game in which countries set their policy variables (investment, emission abatement, and the demand for emission permits when trading is allowed) is preceded by a stage in which countries decide whether or not to sign the climate treaty which—if signed—will be implemented in the subsequent stages of the game. If a country does not sign the treaty, its policy variables will be set at their optimal levels given the decisions taken by the other countries (both signatories and nonsignatories). If a country does decide to sign the treaty, its policy variables maximize the joint welfare function of the signatories.

The game is solved backward. The equilibrium of the first stage of the game is computed using Yi's (1997) partition function approach. Open membership is assumed. Multiple coalitions are allowed.[4] A long term policy time horizon from 1990 to 2100 is assumed. Given that the Kyoto Protocol does not specify emission targets beyond 2012, we assume that countries commit themselves to meet the existing targets also from 2012 onward [this is the so-called "Kyoto forever" scenario adopted in most of the literature on the economic costs of the Kyoto Protocol; see, for example, Buonanno, Carraro, and Galeotti (2001) and Manne and Richels (1999); see also Chapter 8 of IPCC (2001), and the papers gathered in the collections edited by Carraro (1999, 2000)].[5]

An intermediate goal of our analysis is to identify the so-called self-enforcing agreements (Barrett 1994). In addition, we will also single out which agreements are profitable and internally stable. The definitions of profitability and stability have been derived directly from Carraro and Siniscalco (1993) (see also Eyckmans (2001) and Weyant (1999) for recent applications to climate policy). We say that an agreement is weakly profitable if the sum of the individual payoffs of the signatories is larger than the sum of their payoffs when no agreement is signed. In this case, the agreement produces a surplus (overall benefits are larger than costs), but this surplus may not be profitable for all signatories, i.e., some countries may gain, others may lose. By contrast, an agreement is strongly profitable if the payoff of all signatories is larger when the agreement is signed and implemented than when no agreement is signed. Hence,

3. Cf. Eykmans and Tulkens (1999), p. 14. In this paper, the optimization problem and the equilibrium concept are presented and the first-order conditions are derived and discussed.

4. A more detailed presentation of the theoretical framework and of the consequent empirical results is in Bosello et al. (2002).

5. The use of the "Kyoto forever" scenario may be seen a strong assumption. However, the CO2 concentration levels implicit in this assumption (if RICE is a good description of the world) coincide with those in the A1B scenario used by the International Panel on Climate Change (IPCC 2001) which can be considered the "median" scenario among those currently proposed.

TABLE 1. WEAK PROFITABILITY—SHARE OF WEAKLY PROFITABLE COALITIONS FOR EACH BURDEN-SHARING CRITERION

	Single coalitions	Multiple coalitions	Total percentage of coalitions
"Kyoto forever" implicit burden sharing	1.7%	0	0.5%
Equal average abatement costs	29.3%	0	8.4%
Equal per capita abatement costs	32.8%	0	9.4%
Equal abatement costs per unit of GDP	32.8%	0	9.4%

each single participant obtains a net benefit from the agreement. We say that an agreement is internally stable if there is no incentive to free-ride, i.e., the payoff of each signatory is larger than the payoff he/she would obtain by defecting from the group of signatories. Finally, an agreement is self-enforcing or stable if there is no incentive to free ride and no incentive to join the group of signatories, i.e., the payoff to those countries that are not signatories is larger than the one they would achieve by signing the agreement.

The ultimate goal of our analysis is to assess whether the implementation of the three aforementioned "outcome based" equity criteria increases the number of countries which participate in the climate agreement. To do this, we analyse the incentives for countries to sign a climate treaty both under the burden sharing implicit in the "Kyoto forever" scenario and under the burden sharing induced by the three "outcome based" equity criteria. To identify the burden sharing implicit in the "Kyoto forever" scenario we use an "inverse optimisation approach" (see e.g., Carraro 1988, 1989, 1997), that is, we compute the weights in the joint welfare function in an iterative way until each region's optimal investment and abatement levels are such to yield the emission targets agreed in Kyoto. In this way, the solution of the maximization process can replicate (a) the emission abatement levels for each Annex B country and (b) the share of the abatement costs born by Annex B and by Non-Annex B.

The results of our optimization experiments, where 203 different coalition structures have been examined for each burden-sharing criterion, are presented in Tables 1, 2, and 3. They can be summarized as follows:

1. All three "outcome-based" equity criteria increase the probability that a climate agreement yields a surplus. Indeed, the share of weakly profitable coalition structures is much larger with the three new equity criteria than with the burden-sharing rule implicit in the "Kyoto forever" scenario (see Table 1). Nevertheless, the possibility of multiple (regional) agreements does not improve the results: no multiple coalition structure is weakly profitable.
2. The situation is less positive when the more restrictive criterion of strong profitability is used. Indeed, even the three equity criteria proposed above fail to guarantee a large number of strongly profitable coalition structures. Nonetheless, two of the proposed equity criteria imply that the share of

TABLE 2. STRONG PROFITABILITY—SHARE OF STRONGLY PROFITABLE COALITIONS FOR EACH BURDEN-SHARING CRITERION

	Single coalitions	Multiple coalitions	Total percentage of coalitions
"Kyoto forever" implicit burden sharing	0	0	0
Equal average abatement costs	0	0	0
Equal per capita abatement costs	6.9%	0	2.0%
Equal abatement costs per unit of GDP	1.7%	0	0.5%

strongly profitable coalition structures is larger than with the burden-sharing rule implicit in the "Kyoto forever" scenario (see Table 2). When the goal is strong profitability, multiple coalitions again do not provide an incentive structure better than the one provided by single coalitions.

3. In addition, no coalition structure with multiple coalitions is internally stable (see Table 3). Hence, the only coalition structures which could be stable are the ones in which a single coalition forms. However, the share of single coalitions which are both strongly profitable and internally stable further decreases for all burden-sharing criteria (see Table 3 again). Again, only two equity criteria (equal per capita abatement costs and equal abatement costs per unit of GDP) show better results than the burden-sharing rule implicit in the "Kyoto forever" scenario.
4. As a consequence, only very few coalition structures are likely to be stable, i.e., without any incentives to leave or to enter the coalition. Only one coalition is both profitable and stable, namely self-enforcing. This equilibrium coalition structure is formed by a coalition of three countries and by three free-riders. It can be obtained only if ex ante all countries agree that abatement efforts must be such to equalize abatement costs per capita.

Summing up, the adoption of more equitable burden-sharing rules enhances the profitability of a climate agreement but not its stability, i.e., equity improves the distribution of costs and benefits but does not seem to be effective in offsetting the incentives to free-ride.

Two possible ways of addressing the problem are available. First, policy strategies could be designed to further redistribute the surplus provided by the

TABLE 3. INTERNAL STABILITY—SHARE OF INTERNALLY STABLE COALITIONS FOR EACH BURDEN-SHARING CRITERION

	Single coalitions	Multiple coalitions	Total percentage of coalitions
"Kyoto forever" implicit burden sharing	0	0	0
Equal average abatement costs	0	0	0
Equal per capita abatement costs	3.4%	0	1.0%
Equal abatement costs per unit of GDP	1.7%	0	0.5%

cooperative behavior within a coalition. This would increase the number of strongly profitable coalitions and hence the probabilities to identify a stable (i.e., self-enforcing) coalition structure. Transfer schemes designed to make a climate agreement profitable to all countries have been proposed for example in Chander and Tulkens (1997) and applied to climate models in Weyant (1999) or Eyckmans (2001).

Second, policy strategies could be designed to redistribute the surplus achieved by internally stable coalitions with the goal to induce other countries to enter the coalition. In this paper we focus on this second possibility. The first one—transfers to increase the number of strongly profitable coalitions—was previously analyzed in other papers [above all in Weyant (1999), or in Eyckmans and Tulkens (1999), and Yang (2000), where a version of the RICE model is also used].

The conditions for transfers to achieve the goal of expanding a coalition are presented in Carraro and Siniscalco (1993), Botteon and Carraro (1997). Here we would like to stress that, at the equilibrium, (a) transfers are self-financed, i.e., countries are allowed to transfer only the surplus yielded by their cooperation. Hence, we analyze how strongly profitable coalitions can be broadened through a transfer mechanism; (b) the transfer mechanism is Pareto optimal, i.e., all countries gain from using transfers to broaden the coalition. Given this latter restriction, the broadened coalition is also strongly profitable.

However, self-financing implies that there may not be enough resources to offset the incentives to free-ride for all countries which are not in the initial coalition. In addition, even countries in the initial coalition may have an incentive to free-ride when other countries join. Hence, the transfer mechanism must also be such to offset these latter incentives to free-ride. To simplify the transfer mechanism, let us start from initial coalitions that are strongly profitable and internally stable. Is there a transfer mechanisms which induces other countries to join the coalition while offsetting all countries' incentives to free-ride on the larger coalition?

Table 4 presents our results. The first column shows the initial coalitions for each of the three "outcome-based" burden sharing criteria. The second column

TABLE 4. INTERNALLY STABLE COALITIONS BEFORE AND AFTER THE USE OF TRANSFERS

Internally stable coalitions without and with transfers		
	Internally stable coalitions before transfers	Internally stable coalitions after transfers
Equalization of average abatement costs	0	0
Equalization of abatement costs/GDP	USA + CHN	USA + CHN + ROW
Equalization of per capita abatement costs	EU + CHN + ROW JPN + FSU + ROW	USA + EU + CHN + ROW USA + JPN + EU + FSU + ROW

608 Journal of the European Economic Association April–May 2003 1(2–3):601–611

shows the largest internally stable coalition that can be achieved through a transfer mechanism starting from the corresponding initial, internally stable coalition. The transfer mechanism is the one described in Botteon and Carraro (1997).

Our results can be summarized as follows:

1. No transfer mechanisms and burden sharing criteria (i.e., the three criteria that we considered) yield an incentive structure or enough resources to achieve the grand coalition, i.e., a global agreement on climate change. At least one region free-rides on the agreement.
2. The burden-sharing criteria that are most effective in guaranteeing the achievement of a large coalition without free-riding incentives are the equalization of per capita abatement costs and the equalization of abatement costs per unit of GDP. For example, with the former, a coalition formed by Japan, Former Soviet Union, and Rest of the World can offset the free-riding incentives of the EU and the United States. With the latter, a coalition formed by China and the United States can induce Rest of the World to sign the climate agreement. Notice that the United States needs a compensating transfer to enter a coalition that forms according to the criterion of equal per-capita abatement costs, whereas they belong to the initially stable coalitions if the burden-sharing criterion is equal abatement costs per unit of GDP. This is quite intuitive: the United States has high per capita emissions, but relatively small emissions per unit of GDP.

Nonetheless, a stable global agreement cannot be achieved. Hence, we wonder whether the introduction of emission trading, regardless of the ex ante burden-sharing criterion, can provide enough resources that, once transferred to free-riding countries, can induce them to sign the climate agreement. The answer to this question is positive: by solving the dynamic game, described previously, again, we find that regardless of the ex ante burden sharing criterion (equity), and regardless of the initial, internally stable coalition, the equalisation of marginal abatement costs, coupled with an appropriately designed ex post transfer mechanism, can lead to a grand coalition, i.e., a global climate agreement, signed by all countries or regions, which is stable.

In other words, through emission trading and transfers, all internally stable coalitions can be broadened to achieve a stable grand coalition. By using a twofold transfer mechanism, one designed to transform a weakly profitable coalition into a strongly profitable one, and a second one designed to make it internally stable, we could obtain an even stronger conclusion. All weakly profitable coalitions can be broadened to achieve a stable grand coalition.[6]

6. This result is implicitly shown also in Chander and Tulkens (1995) but for a different definition of stability [usually named coalition unanimity cf. Yi (1997)]. In particular, their definition of stability coincides with our definition of profitability.

3. Conclusions

The goal of this paper was to provide a theoretical framework to analyze the widespread conjecture that a more equitable sharing of the costs of controlling GHG emissions would induce developing countries to sign a global climate treaty. In addition, this paper used the proposed theoretical framework to check whether the above conjecture is supported by some empirical evidence.

The conclusion is that, at least within the limits defined by the RICE model and by the three equity criteria considered in this paper, cost-effectiveness (i.e., emission trading) seems to be more effective than equity in inducing more countries to join a climate coalition. In particular, whereas equity increases the profitability of a climate agreement, it does not offset the incentives to free-ride on emission abatement. These incentives can however be offset by appropriately designed transfers. Nevertheless, our results suggest that a better way to sustain a stable global agreement—and thus to attract as many countries as possible to combat climate change—is to couple emission trading, which provides additional financial resources to be transferred to nonsignatories, with a transfer scheme specifically designed to offset incentives to free-ride.

References

Babiker, M. and H. D. Jacoby. (1999). "Developing Country Effects of Kyoto-Type Emissions Restrictions." MIT Joint Program on the Science and Policy of Global Change, *Report No. 53,* Cambridge, Massachusetts.

Babiker, M., J. Reilly, and H. D. Jacoby (1999). "The Kyoto Protocol and Developing Countries." MIT Joint Program on the Science and Policy of Global Change, *Report No. 56,* Cambridge, Massachusetts.

Barrett, S. (1994). "Self-enforcing International Environmental Agreements." *Oxford Economic Papers,* 46, pp. 878–894.

Bernstein, M., P. Bromley, J. Hagen, S. Hassell, R. Lempert, J. Munoz, and D. Robalino (1999). *Developing Countries & Global Climate Change: Electric Power Options for Growth.* Pew Center on Global Climate Change.

Bernstein, P., W. Montgomery, and T. Rutherford (1999). "Global Impacts of the Kyoto Agreement: Results from the MS-MRT Model." *Resource and Energy Economics,* 21, 375–413.

Bosello, F., B. Buchner, C. Carraro, and D. Raggi (2002). "Can Equity Enhance Efficiency? Some Lessons from Climate Negotiations," forthcoming in C. Carraro and V. Fragnelli, eds., *Game Practice and the Environment,* E. Elgar: Cheltenham.

Botteon, M. and C. Carraro (1997). "Burden-Sharing and Coalition Stability in Environmental Negotiations with Asymmetric Countries." In *International Environmental Agreements: Strategic Policy Issues,* edited by C. Carraro, Cheltenham: E. Elgar.

Buonanno, P., C. Carraro, and M. Galeotti (2001). "Endogenous Induced Technical Change and the Costs of Kyoto." *FEEM Working Paper 80.2000,* Milan. *Resource and Energy Economics,* 25, pp. 11–34.

Carraro, C. (1988). "The Implicit Objective Function of Italian Macroeconomic Policy." *Economic Modelling,* 3, pp. 261–276.

Carraro, C. (1989). "The Tastes of European Central Bankers." In *A European Central Bank?*

Perspectives on Monetary Unification after Ten Years of the EMS, edited by M. De Cecco and A. Giovannini. Cambridge: Cambridge University Press.

Carraro, C. (1997). "Modelling International Policy Games: Lessons from European Monetary Coordination." *Empirica,* 23, pp. 163–177.

Carraro, C. (ed.) (1999). *International Environmental Agreements on Climate Change.* Dordrecht: Kluwer Academic Publishers.

Carraro, C. (ed.) (2000). *Efficiency and Equity of Climate Change Policy.* Dordrecht: Kluwer Academic Publishers.

Carraro, C. and D. Siniscalco (1993). "Strategies for the International Protection of the Environment." *Journal of Public Economics,* 52, pp. 309–328.

Cazorla, M. and M. Toman (2000). "International Equity and Climate Change Policy." *RFF Climate Issue Brief,* No. 27, Washington, D.C.

Chander, P. and H. Tulkens (1995). "A Core-Theoretic Solution for the Design of Cooperative Agreements on Transfrontier Pollution." *International Tax and Public Finance* 2(2), pp. 279–294.

Chander, P. and H. Tulkens (1997). "The Core of an Economy with Multilateral Environmental Externalities." *International Journal of Game Theory,* 26, pp. 379–401.

Claussen, E. and L. McNeilly (1998). *Equity and Global Climate Change: The Complex Elements of Global Fairness.* Pew Center on Global Climate Change.

Convery, F. J. (2000). "Blueprints for a Climate Policy: Based on Selected Scientific Contributions." In *Integrating Climate Policies in a European Environment,* edited by C. Carraro. Special issue of *Integrated Assessment,* Dordrecht: Baltzer Publishers.

Ellerman, A. D., H. D. Jacoby, and A. Decaux (1998). "The Effects on Developing Countries of the Kyoto Protocol and CO_2 Emissions Trading." MIT Joint Program on the Science and Policy of Global Change, *Report No. 41,* Cambridge, Massachusetts.

Eyckmans, J. (2001). "On the Farsighted Stability of the Kyoto Protocol, Some Simulation Results." ETE Working Paper No. 2001-03, Katholieke Universiteit Leuven, University of Leuven.

Eyckmans, J. and H. Tulkens. (1999). "Simulating Coalitionally Stable Burden Sharing Agreements for the Climate Change Problem." *Climneg Working Paper No. 18 and FEEM Working Paper* 71–99.

IPCC (2001). *Third Assessment Report and Summary for Policymakers.* Cambridge: Cambridge University Press.

Manne, A. and R. Richels (1999). "The Kyoto Protocol: A Cost-Effective Strategy for Meeting Environmental Objectives?" In *The Cost of the Kyoto Protocol: A Multi-Model Evaluation,* edited by J. Weyant. Special Issue of the *Energy Journal.*

McKibbin, W. J. and P. J. Wilcoxen. (2000). "Designing a Realistic Climate Change Policy that Includes Developing Countries." Australian National University and The Brookings Institution, University of Texas at Austin and The Brookings Institution.

Metz, B. (2000). "International Equity in Climate Change Policy." In *Integrating Climate Policies in a European Environment,* edited by C. Carraro. Special issue of Integrated Assessment, Baltzer Publishers.

Nordhaus, W. D. and Z. Yang (1996). "A Regional General-Equilibrium Model of Alternative Climate-Change Strategies." *American Economic Review,* 86(4), pp. 741–765.

Panayotou, T., J. Sachs, and A. Peterson. (1999). "Developing Countries and the Control of Climate Change: A Theoretical Perspectives and Policy Implications." CAER II Discussion Paper No. 44, Harvard Institute for International Development, Cambridge, Massachusetts.

Rose, A., B. Stevens, J. Edmonds, and M. Wise (1998). "International Equity and Differentiation in Global Warming Policy: An Application to Tradable Emission Permits," *Environment and Resource Economics,* 12(1), pp. 25–51.

Rose, A. and B. Stevens (1993). "The Efficiency and Equity of Marketable Permits for CO_2 Emissions." *Resource and Energy Economics,* 15(1), pp. 117–146.

Schmidt, T. F. N. and H. Koschel (1998). "Climate Change Policy and Burden Sharing in the European Union—Applying Alternative Equity Rules to a CGE-Framework." ZEW Discussion Paper No. 98-12.

Tol, R. S. J. (2001). "Equitable Cost-Benefit Analysis of Climate Change Policies." *Ecological Economics,* 36, pp. 71–85.

Weyant, J. (ed.) (1999). "The Cost of the Kyoto Protocol: A Multi-Model Evaluation." *Energy Journal,* Special Issue.

Yang, Z. (2000). "Forming Coalitions in International Agreements on Global Climate Change." University of Pennsylvania.

Yi, S. (1997). "Stable Coalition Structures with Externalities." *Games and Economic Behaviour,* 20, 201–223.

[36]

NATURAL RESOURCE MODELING
Volume 19, Number 3, Fall 2006

COALITION FORMATION IN A GLOBAL WARMING GAME: HOW THE DESIGN OF PROTOCOLS AFFECTS THE SUCCESS OF ENVIRONMENTAL TREATY-MAKING

JOHAN EYCKMANS
EHSAL - Europese Hogeschool Brussel
Stormstraat 2, B-1000 Brussels, Belgium &
K.U. Leuven, Centrum voor Economische Studiën
E-mail: Johan.Eyckmans@ehsal.be

MICHAEL FINUS
Department of Economics
University of Hagen
Profilstr. 8
58084 Hagen, Germany &
National University of Singapore
E-mail: Michael.Finus@fernuni-hagen.de

ABSTRACT. We combine new concepts of noncooperative coalition theory with an integrated assessment model on climate change to analyze the impact of different protocol designs on the success of coalition formation. We analyze the role of "single versus multiple coalitions," "open versus exclusive membership," "no, weak and strong consensus about membership" and "no transfers versus transfers." First, we want to find out whether and how modifications of the standard assumptions affect results that are associated with the widely applied cartel formation game in the noncooperative game theoretic analysis of international environmental agreements. Second, we discuss normative policy conclusions that emerge from the various modifications. Third, we confront our results with evidence on past international environmental treaties and derive an agenda for future research.

KEY WORDS: Design of climate treaty protocol, coalition formation, noncooperative game theory.

1. Introduction. For a long time, noncooperative game theoretic models on international environmental agreements (IEAs) have almost exclusively based their analysis on the cartel formation game of d'Aspremont et al. [1983] to study the formation and stability of such agreements.[1] A central result of this literature is that only small coalitions are stable which only marginally improve upon the noncooperative status quo. A closer inspection, see Section 2, reveals that

Received by the editors on Nov. 22, 2004, and in revised form on May 3, 2005.

this "pessimistic result" hinges on three assumptions. First, coalition formation is restricted to a single coalition. Second, players can freely join coalitions implying open membership. Third, no consensus about membership is necessary for players to form a coalition. We pursue three objectives in this paper. First, we want to find out whether and how the assumptions of the cartel formation game affect results. Additionally, we are interested in whether the transfer scheme proposed by Chander/Tulkens [1995], [1997] for enhancing stability in a cooperative global emission game performs equally well in a noncooperative game theoretic framework. In order to achieve this objective, we consider, apart from the cartel formation game, three other coalition games from the recent literature on coalition theory.[2] These coalition games constitute a continuum of assumptions where the assumptions of the cartel formation are gradually relaxed. Moreover, we distinguish between a "no transfer case" and a "transfer case" where in the latter case we adopt the Chander/Tulkens' transfer scheme for our noncooperative game theoretic setting. Together, our set of assumptions gives rise to eight agreement designs for which we determine the set of equilibrium coalition structures. Hence, we can relate the outcome of coalition formation to the issues "single versus multiple coalitions," "open versus exclusive membership" and "no, weak and strong consensus about membership" and "no transfers versus transfers." Moreover, we propose a simple procedure how the choice of the agreement design can be endogenized and analyze its implications.

Second, we want to identify lessons that we can learn for the design of future IEAs in general and for the problem of global warming in particular. This requires that our analysis captures a heterogeneous cost-benefit structure of the various actors involved in climate change and the dynamics of stock pollutants but must also allow us to evaluate different outcomes in ecological and welfare terms. Therefore, we cannot base our analysis on models typically used in the theoretical analysis of IEAs that assume symmetric players and/or a static payoff structure. Instead, our analysis employs an integrated assessment model that closely resembles the seminal RICE model by Nordhaus and Yang [1996] and conducts numerical simulations.

Third, we want to confront our results with the evidence from IEAs in order to construct an agenda of future research. Therefore, we critically relate our results to the past experience with IEAs. Subsequently, we suggest various directions how our model could be modified in order to improve upon the positive aspects of our analysis.

We believe that this paper extends the current literature in several directions. First, there is a conceptual literature that develops new noncooperative concepts for studying coalition formation.[3] To this literature, we can add a small item by showing how the endogenous choice agreement design could be modeled and operationalized in a simple way. Second, there is a theoretical literature that adopts these new concepts for the analysis of IEAs.[4] In contrast to this literature, we cannot only study the implications of different agreement designs on stability but can also evaluate outcomes in welfare and ecological terms because we depart from the assumption of symmetric players and a static payoff structure. Third, there is a literature that combines a climate model with a noncooperative game theoretic model as we do.[5] Here, we can add some of the following items that have at least partly been missing in past analyses. a) We pay much attention to the game theoretic underpinning of our model. b) Simulations are based on a fully-fledged climate model. c) We emphasize the relation between the climate and game theoretic module and discuss important properties that drive results. d) We critically confront our results with real world experience on IEAs, identify reasons for discrepancies and provide a research agenda that may help to fix things.[6]

In the following, we introduce our model (game theoretic and climate module) and discuss important properties in Section 2. In Section 3, we present and explain results of our stability analysis, and, in Section 4, we discuss interpretations and normative policy conclusions. Finally, in Section 5, we identify critical issues for a positive policy analysis and suggest a list of items that may help to improve upon these issues.

2. Model. Coalition formation is modeled as a two-stage game. In the first stage, players decide upon their membership in a coalition. In the second stage, coalition members choose their economic strategies. The game is solved by backwards induction. In the following, we describe the decision process in the first and second stage in subsection 2.1 and 2.2, respectively, the procedure of solving the game in subsection 2.3 and we discuss some general features of our model in subsection 2.4.

2.1 First stage. In our model, the set of players $P = \{1, \ldots, N\}$ comprises six elements that are countries or world regions: United States of America (USA), Japan (JPN), European Union (EU), China (CHN), Former Soviet Union (FSU) and "Rest of the World" (ROW). Players decide whether they want to remain a single player or form a coalition with other players. These decisions lead to a coalition structure $c = \{c_1, \ldots, c_M\}$ that is a partition of players where a particular coalition is denoted by c_l, $l \in \{1, \ldots, M\}$, $c_l \cap c_m = \varnothing$ for all $l \neq m$, $\cup c_l = P$ and $c \in C$ where C is the set of all possible coalition structures.

A coalition may comprise only one player (trivial coalition), several players or all players (nontrivial coalition). Similarly, a coalition structure may comprise no, one or several nontrivial coalitions. If all players form a singleton coalition, $c^N = \{\{1\}, \{2\}, \ldots, \{N\}\}$, this implies *no cooperation*; if all regions are in one coalition (grand coalition), $c^F = \{P\}$, this means *full cooperation*; any other coalition structure $c^P = \{c_1, \ldots, c_M\}$, $M < N$, is referred to as "partial cooperation." In case of six players, there are 203 different coalition structures. Apart from the trivial coalition structure, there are 57 coalition structures with only one nontrivial coalition (single coalition structures) and 145 coalition structures with at least two nontrivial coalitions (multiple coalition structures).

In the following, we consider four "coalition games" that model the decision process of participation differently. We denote these games by names that stress the rules on the formation of coalitions. Since all games assume that players make their decision simultaneously, this feature is not mentioned below anymore.

The first game is the well-known cartel formation game due to d'Aspremont et al. [1983] that we call *Open Membership Single Coalition Game* (OMSCG). This game serves as a reference point. In this game, all players can participate in an agreement; players that have no interest in cooperation remain singletons. That is, players choose their membership by announcing either "address" $\sigma_i = 0$ or $\sigma_i = 1$. Those players that have announced $\sigma_i = 0$ remain singletons and those that have announced $\sigma_i = 1$ are members of a nontrivial coalition, provided at least two players have announced $\sigma_i = 1$. Thus, in this coalition game, only single coalition structures can emerge.

Definition 1. Open membership single coalition game (OMSCG). Let the set of membership strategies of player $i \in P$ be given by $\Sigma_i = \{0,1\}$ where a particular strategy σ_i is an announcement of an address and c_l is the coalition to which i finally belongs, then

$$c_l = \begin{cases} \{i\} & \text{if } \sigma_i = 0 \\ \{j/\sigma_j = 1\} & \text{if } \sigma_i = 1. \end{cases}$$

The second game is called *Open Membership Multiple Coalition Game* (OMMCG) and is due to Yi and Shin [2000]. This game is similar to the previous game, except that coalition formation is not restricted ex-ante to only one nontrivial coalition. That is, players can freely form coalitions as long as no outsider is excluded from joining a coalition. Players choose their membership by announcing an address between 0 and N. Players that have announced the same address form a coalition, except those that have announced 0 who remain singletons.[7]

Definition 2. Open membership multiple coalition game (OMMCG). Let the set of membership strategies of player $i \in P$ be given by $\Sigma_i = \{0, 1, \ldots, N\}$ where a particular strategy σ_i is an announcement of an address and c_l is the coalition to which i finally belongs, then $c_l = \{i\} \cup \{j/\sigma_i = \sigma_j \neq 0\}$.

The third coalition game is the Δ-game of Hart and Kurz [1983] that we call *Exclusive Membership Multiple Coalition Δ-Game* (EMMCΔG). Players announce a list of players with whom they would like to form a coalition. Players with the same list form a coalition. For instance, suppose $N = 6$ and $\sigma_1 = \sigma_2 = \{1, 2, 3\}$, $\sigma_3 = \{3\}$, $\sigma_4 = \{4, 5, 6\}$ and $\sigma_5 = \sigma_6 = \{5, 6\}$, then $c = \{\{1, 2\}, \{3\}, \{4\}, \{5, 6\}\}$ forms. Players 1 and 2 have the same list and therefore form a coalition. The same is true for players 5 and 6. Players 3 and 4 remain singletons. Player 3 is a singleton by its own will and cannot be forced into a coalition with players 1 and 2. In contrast, player 4 is a singleton against his will; he cannot tie up with player 5 and 6 since he is not on their list. In both cases, these players are singletons since *membership is exclusive*.

Definition 3. Exclusive membership multiple coalition Δ-game (EMMCΔG). Let the set of membership strategies of player $i \in P$ be given by $\Sigma_i = \{c_l \subset P / i \in c_l\}$ where a particular strategy σ_i is a list of players with whom player i would like to form a coalition. Then, $c_l = \{i\} \cup \{j / \sigma_i = \sigma_j\}$.

The fourth coalition game is the Γ-game and is also due to Hart and Kurz [1983]. We call this game *Exclusive Membership Multiple Coalition* Γ-*Game* (EMMCΓG) which is similar to the previous game. Players also propose a list of players with whom they would like to tie up. However, whereas in the EMMCΓG, it suffices that a subgroup of players on a list makes the same proposal (and hence this subgroup forms a coalition), in the EMMCΓG, a coalition forms if and only if all members on a list make the same proposal. For instance, in the example above with $\sigma_1 = \sigma_2 = \{1, 2, 3\}$, $\sigma_3 = \{3\}$, $\sigma_4 = \{4, 5, 6\}$ and $\sigma_5 = \sigma_6 = \{5, 6\}$, not $c = \{\{1, 2\}, \{3\}, \{4\}, \{5, 6\}\}$ forms, but $c = \{\{1\}, \{2\}, \{3\}, \{4\}, \{5, 6\}\}$ instead. Thus, also in the fourth coalition game, membership is exclusive. However, whereas in the EMMCΔG, a *weak consensus* was required to form a coalition, now in the EMMCΓG, a *strong consensus* is necessary.

Definition 4. Exclusive membership multiple coalition Γ-game (EMMCΓG). Let the set of membership strategies of player $i \in P$ be given by $\Sigma_i = \{c_l \subset P / i \in c_l\}$ where a particular strategy σ_i is a list of players with whom player i would like to form a coalition, then, $c_l = \sigma_i$ if and only if $\sigma_i = \sigma_j$ for all $j \in \sigma_i$, otherwise $c_l = \{i\}$.

The four coalition games can be interpreted as four different "membership regimes." As the names suggest, one difference is related to the assumption whether coalition formation is restricted to only one non-trivial coalition (single coalition game) or not (multiple coalition game). A second difference is the assumption open versus exclusive membership. A third difference is related to the assumption of the degree of consensus necessary to form a coalition. In the two open membership games OMSCG and OMMCG, in the EMMCΔG weak and in the EMMCΓG, no strong consensus is needed to form a coalition. Thus, the four coalition games imply a continuum of assumptions. Along the sequence of presentation, gradually one assumption is changed at the time. For reference reason, we denote the set of coalition games by $\Phi = \{\text{OMSCG, OMMCG, EMMC}\Delta\text{G, EMMC}\Gamma\text{G}\}$.

2.2 Second stage. In the second stage, players choose their economic strategies. We follow the standard assumption of the valuation function approach, e.g., Bloch [1997], that, given a coalition structure $c = \{c_1, \ldots, c_M\}$ has formed, players belonging to the same coalition c_l jointly maximize the aggregate payoff to their coalition. In our model, this means maximizing aggregate discounted consumption of every coalition $c_l \in c$ with respect to investment $I_{i,t}$ and the level of abatement $\mu_{i,t}$ of every player i belonging to coalition c_l for every period t:

$$\max_{\{I_{i,t},\mu_{i,t}\}_{i\in c_l;t=0,\ldots,\Psi}} \sum_{i\in c_l}\sum_{t=0}^{\Psi} \frac{Z_{i,t}}{[1+\rho_i]^t} \tag{1}$$

where $Z_{i,t}$ denotes consumption, ρ_i is the discount rate of player i and Ψ denotes the time horizon. The underlying model in equation (1) is the CLIMNEG World Simulation Model, an integrated assessment economy-climate model that closely resembles the seminal RICE model by Nordhaus/Yang [1996].[8] In contrast to RICE, the CWS model allows for different regional discount rates in order to reflect huge differences between world regions in terms of economic development and openness to financial markets. As appears from the Appendix, we assume systematically higher discount rates for developing regions (CHN and ROW) than for industrialized regions (USA, JPN, EU and FSU).

The maximization of (1) is subject to:

$$Y_{i,t} = Z_{i,t} + I_{i,t} + Y_{i,t}C_i(\mu_{i,t}) + Y_{i,t}D_i(\Delta T_t). \tag{2}$$

This "budget equation" describes how "potential GDP," $Y_{i,t}$, can be allocated to consumption, $Z_{i,t}$, investment, $I_{i,t}$, emission abatement costs, $Y_{i,t}C_i(\mu_{i,t})$, and climate change damages, $Y_{i,t}D_i(\Delta T_t)$. Output $Y_{i,t}$ is produced with capital and labor. Capital is built up through investment and depreciates at some fixed rate. Labor supply is assumed to be inelastic. Therefore, investment $I_{i,t}$ is the only endogenous production input and constitutes the first choice variable in the model.

Abatement costs $Y_{i,t}C_i(\mu_{i,t})$ are expressed as "loss of potential GDP": C_i is the share of "potential GDP" devoted to abatement which is a function of $\mu_{i,t} \in [0,1]$, measuring the relative emission reduction compared to the business-as-usual scenario without any abatement policy. Damages $Y_{i,t}D_i(\Delta T_t)$ are also expressed as "loss of potential GDP": D_i is the share of "potential GDP" destroyed by climate change damages which is a function of temperature change ΔT_t. Temperature change depends on the stock of greenhouse gases which in turn depends on emissions that accumulate in the atmosphere. Finally, emissions depend on emissions released through production minus abatement. Hence, the second choice variable in this model is the abatement level $\mu_{i,t}$. Both choice variables (investment and abatement) affect output, abatement costs, damage costs and therefore also consumption not only domestically but also abroad. For abatement, this is immediately evident since remaining emissions (after abatement) increase the stock of greenhouse gases which affects environmental damages in every country. However, this is also true for investment since capital is an input in the production process and atmospheric emissions are proportional to production.

Solving (1) simultaneously for every coalition $c_l \in c$ gives the equilibrium economic strategy vector $s^* = \{I_{i,t}, \mu_{i,t}\}_{i\in P;t=0,\ldots,\Psi}$ for coalition structure c, abbreviated $s^*(c)$.[9] For six players and a period of 35 decades[10], this vector is of length $2 \times 6 \times 35$. From the assumption of joint welfare maximization, it follows that the economic strategy vector in the coalition structure with only singleton coalitions c^N ("no cooperation") corresponds to the "classical Nash equilibrium." By the same token, the strategy vector in the coalition structure with only the grand coalition c^F ("full cooperation") is identical to the "classical social optimum."

Substitution of the equilibrium strategy vector $s^*(c)$ into the payoff functions of players gives a vector of individual (discounted) payoffs for coalition structure c, called valuations, $v(c) = (v_1(c), \ldots, v_N(c))$. Performing this exercise for all 203 coalition structures in our model gives a table of valuations of dimension 203×6 which we will denote by $V(C)$ in the sequel.

The derivation of valuations above assumes *no transfers*. We denote this vector by $V^{NT}(C)$. An alternative assumption could be *transfers* leading to "corrected" valuations $V^T(C)$. Among many possibilities, we adopt the transfer scheme used in Eyckmans and Tulkens [2003] which is an extension of the scheme proposed by Chander and Tulkens

[1995], [1997]. This scheme assumes that the surplus of a coalition from cooperation is allocated according to a proportional sharing rule. The payoff of every player i after transfers, being a member of coalition c_l in a partially cooperative coalition structure c^P, is given by:

$$(3) \qquad v_i^T(c^P) = v_i^{NT}(c^N) + \lambda_i \left[\sum_{j \in c_l} (v_j^{NT}(c^P) - v_j^{NT}(c^N)) \right].$$

That is, the transfer scheme is constructed such that every player receives its noncooperative payoff (first term) and additionally a share λ_i from the total surplus of coalition c_l in coalition structure c^P compared to the singleton coalition structure c^N (second term in square brackets). Shares reflect the relation between individual and global marginal climate change damages. Hence, the second term favors countries with relatively high marginal damages since they are entitled to a larger share of the surplus of their coalition. The motivation for this transfer scheme is twofold. First, it can be argued normatively that this transfer scheme embodies a standard notion of fairness: countries that are hurt more by climate change should be compensated more. Second, Chander and Tulkens [1995], [1997] have shown that this transfer rule gives rise to an allocation in the core in a static cooperative global emission game. Hence, we are interested whether this transfer scheme performs equally well in our dynamic and non-cooperative setting. For later reference, we talk in the following about the "no transfer regime" and the "transfer regime" and denote the set of valuations by $\Omega(C) = \{V^{NT}(C), V^T(C)\}$.

From our explanations and equation (3) it appears that valuations with transfer $V^T(C)$ can be calculated ex-post from the valuations with the no transfer valuations $V^{NT}(C)$. This requires a TU-framework such that transfers do not affect the equilibrium economic strategy vectors s^*. This holds in our context because we make the following simplifying assumptions.[11] First, budget equation (2) must hold for each region separately. That is, each region can only rely on its own productive resources, at each time t, with consumption not being transferable to other regions. Second, if there are transfers, this implies a reallocation of discounted payoffs in a lump sum fashion among coalition members leaving the sum of the net present values over coalition members unaltered. To see this, note that, in the case of transfers (abbreviated trans$_i$), the maximization problem described above can alternatively be written as

$$(1') \qquad \max_{\{I_{i,t},\mu_{i,t}\}_{i\in c_l;t=0,\ldots,\Psi}} \sum_{i\in c_l} \left\{ \left[\sum_{t=0}^{\Psi} \frac{Z_{i,t}}{[1+\rho_i]^t} \right] + \text{trans}_i \right\}$$

subject to (2) and the restriction that transfers balance within coalition c_l, i.e., $\sum_{i\in c_l} \text{trans}_i = 0$. It is easy to check that problems (1) subject to (2) and (1′) subject to (2) and the transfer balance condition are equivalent.[12]

2.3 Solution of the two-stage game. From subsection 2.2, we know that the decision in the second stage leads to a vector of valuations which is either $V^{NT}(C)$ under the "no transfer regime" or $V^T(C)$ under the "transfer regime." Hence, for a given regime in the second stage of the game, players know the implications of their decision about membership in the first stage of the game. From subsection 2.1, we know that the decision process in the first stage can be modeled differently. However, all four coalition games that we consider ($\Phi = \{$OMSCG, OMMCG, EMMCΔG, EMMCΓG$\}$) share the feature that players simultaneously announce their membership strategy σ_i and the set of announcements $\sigma = (\sigma_1, \ldots, \sigma_N)$ is mapped into a coalition structure $c = \{c_1, \ldots, c_M\}$. Hence, regardless of the specific coalition game, we can define a stable coalition structure as follows:

Definition 5. Stable coalition structures. Consider a coalition game $G \in \Phi$ and a valuation $V(C) \in \Omega(C)$. Coalition structure $c^* \in C$ is called stable if there exists a membership strategy $\sigma^* \in \Sigma$ generating c^* such that for all $i \in P$ and for all $\sigma_i \in \Sigma_i : v_i(\sigma_i^*, \sigma_{-i}^*) \geq v_i(\sigma_i, \sigma_{-i}^*)$.

Definition 5 is the familiar definition of a Nash equilibrium, except that strategies are membership and not economic strategies. In the following, however, we consider a refinement of this stability concept.

Definition 6. Pareto-undominated stable coalition structures. Consider a coalition game $G \in \Phi$ and a valuation $V(C) \in \Omega(C)$. Let $c^* \in C^*$ be a stable coalition structure generated by membership strategy $\sigma^* \in \Sigma$; then c^* is called a Pareto-undominated stable coalition structure if there is no other stable coalition structure $\tilde{c}^* \in C^*$ generated by membership strategy $\tilde{\sigma}^* \in \Sigma$ where at least one player is better off and no player is worse off, i.e., there is no $\tilde{\sigma}^* \in \Sigma$ such that for all $i \in P$: $v_i(\tilde{\sigma}^*) \geq v_i(\sigma^*)$ and there exists $j \in P$: $v_j(\tilde{\sigma}^*) > v_j(\sigma^*)$.

The motivation of this refinement is that, even if players behave only in a selfish manner (as we assume), it is difficult to perceive why they should settle for a dominated equilibrium if coordination is costless, as assumed in our model.[13]

Given that we consider four membership regimes in the first stage of the coalition formation process ($\Phi = \{$OMSCG, OMMCG, EMMCΔG, EMMCΓG$\}$), and a no transfer and transfer regime in the second stage ($\Omega(C) = \{V^{NT}(C), V^T(C)\}$), there are eight possible *designs of an agreement.* For every design, we determine the set of stable coalition structures, $C^*(G, V(C))$, and the set of Pareto-undominated stable coalition structures, $C^{**}(G, V(C))$, $G \in \Phi$ and $V(C) \in \Omega(C)$. This allows us to compare stable coalition structures under various designs as is done in Section 3. However, before embarking on our stability analysis, we first clarify some general features of our model in subsection 2.4.

2.4 General features. A well-known property in the theoretical literature on coalition formation is *superadditivity.* That is, if players or coalitions merge, and hence the coalition structure changes from c to $\tilde{c}$, the aggregate payoff to those players involved in the merger increases, i.e., $\sum_{i \in c_l} v_i(c) + \sum_{i \in c_m} v_i(c) < \sum_{i \in c_l \cup c_m} v_i(\tilde{c})$. That this property may not always be satisfied in the context of externalities has been pointed out by Bloch [1997]. For instance, suppose that a merger leads to an increase of abatement of those players involved in a merger but to a decrease of abatement of outsiders due to "leakage effects." Thus, if leakage effects are sufficiently strong, superadditivity may be violated. However, we have tested numerically all valuations in our model for this property and found that it holds.

Another property that has played an important role in the more recent literature on coalition formation, see, e.g., Bloch [1997] and Yi [1997], is *positive externality.* That is, if players or coalitions merge, and hence the coalition structure changes from c to $\tilde{c}$, the payoff of outsider k increases, i.e., $v_k(c) < v_k(\tilde{c})$. Whereas a general proof of this property in a static public good model is rather straightforward, see, e.g., Finus and Rundshagen [2003], we have not been able to derive an analytical proof in our dynamic public good game. Again, we have checked numerically all valuations in our model for this property and found that it holds.

Both properties allow for the following conclusions:

1) A coalition structure $\tilde{c}$ implies a higher global welfare than a coalition structure c if $\tilde{c}$ can be derived from c by a merger or sequence of mergers of players or coalitions. In other words, an increase in the "degree of cooperation" has a positive global impact. By superadditivity, the players involved in the merger will be better off and by positive externality, also the outsiders will gain from the merger. Consequently, global welfare in any coalition structure with partial cooperation will be higher than in the coalition structure with no cooperation. By the same token, however, any form of partial cooperation will imply lower global welfare than in the case of full cooperation.

2) Every single player receives in every coalition structure at least his payoff as if there were no cooperation. This property is called *profitability* and follows from the positive externality property. This is also true for every coalition member under our transfer regime due to the superadditivity property. Recall that under our transfer regime the valuation of a player i is given by $v_i^T(c^P) = v_i^{NT}(c^N) + \lambda_i[\sum_{j \in c_l}(v_j^{NT}(c^P) - v_j^{NT}(c^N))]$ with $i \in c_l \in c$, c^N the coalition structure with only singleton coalitions (no cooperation), c^P any other coalition structure with partial cooperation and $\lambda_i > 0$. Hence, superadditivity implies that the term in square brackets is always positive. Consequently, all valuations $v^T(c) \in V^T(C)$ are profitable to all players under the transfer regime.

3) Generally, the advantage for a country participating in a coalition is that own abatement efforts are matched by other members' efforts. This reduces the externalities from emissions. This shows up in the superadditivity property which may be interpreted as follows: a coalition can achieve at least as much (and probably more) for its members than if its players remain separately. However, the positive externality property provides incentives for free-riding. Regions can

also benefit from cooperation without participating in a coalition. The relative size of both effects determines whether coalition structures are stable. Moreover, apart from transfers, the membership rule has also a crucial impact on stability. We note four relations, see Finus and Rundshagen [2003] for details:

a) In every coalition game, *profitability* is a necessary, though not sufficient, condition for stability. The reason is simple: a player can always leave his coalition to become a singleton. Regardless of how the remaining players react, this gives a player at least his payoff of the trivial coalition structure due to the positive externality property.

b) In the two open membership games (OMSCG and OMMCG) *stand alone stability* is a necessary condition for stability. That is, every coalition member must receive a higher payoff than when leaving the coalition to become a singleton while other players remain in their coalitions, i.e., for all $i \in c_l$, for all $c_l \in c$: $v_i(c_l, c) \geq v_i(\{i\}, \hat{c})$ where $\hat{c} = c \setminus \{c_l\} \cup \{c_l \setminus \{i\}, \{i\}\}$. The reason is that a player can always change his address to $\sigma_i = 0$ to become a singleton which implies that his former coalition c_l loses one member. However, stand alone stability is not a sufficient condition for stability in open membership games. Another possible deviation is that a singleton joins a coalition. Additionally, though only in the multiple coalition version, and only if a multiple coalition structure has formed, a member of one coalition can join another coalition. Hence, the number of potentially stable coalition structures is larger in the multiple than in the single coalition game by the very construction of the two games.

In the exclusive membership multiple coalition Δ-game, *stand-alone stability* is a sufficient condition for stability. The reason is that if every player announces a list of exactly only those coalition members that are in his coalition, any change of his list implies that he becomes a singleton. That is, neither a single player nor a coalition member can join another coalition by exclusive membership, as this is possible in the OMMCG and OMSCG. This is also true in the exclusive membership multiple coalition Γ-game. The only difference to the EMMCΔG is that, due to the requirement of strong consensus, any deviation leads to the breakdown of the coalition to which the deviator belonged. Hence, in the EMMCΓG, a sufficient condition for stability is *stand apart stability*: for all $i, j, \ldots, k \in c_l$, for all $c_l \in c$: $v_i(c_l, c) \geq v_i(\{i\}, \tilde{c})$ where $\tilde{c} = c \setminus \{c_l\} \cup \{\{i\}, \{j\}, \ldots, \{k\}\}$. Because of the positive externality property, stand alone stability is a stronger requirement than stand apart stability, i.e., $v_i\,(c_l,\,c) \geq v_i\,(\{i\},\,\hat{c}) \geq v_i\,(\{i\},\,\tilde{c})$. In other words, the implicit punishment in the EMMCΓG after a deviation is stronger than in the EMMCΔG and in the two open membership games.

Consequently, denoting the set of stable coalition structures in coalition game G and valuation $V(C)$ with positive externalities by $C^*(G, V(C))$, we can conclude:

$$C^*(OMSCG, V(C)) \subseteq C^*(OMMCG, V(C)) \subseteq C^*(EMMC\Delta G, V(C)) \subseteq C^*(EMMC\Gamma G, V(C)).$$

These relations suggest that multiple coalitions provide an additional option for cooperation, it is easier to sustain cooperation under exclusive membership than under open membership and for exclusive membership it is easier to sustain cooperation under the assumption of strong than under weak consensus. However, theory cannot tell us a) whether the relations above are strict and b) if they are strict, what this means for the global success of cooperation. Even in much simpler models with a static payoff structure and symmetric players, such conclusions cannot be derived which motivates and justifies our objective to analyze this problem with the help of a numerical model.[14] c) Every single coalition structure that is profitable will be stable in the EMMCΓG. This follows immediately from the arguments above that any deviation in this coalition game leads to a breakdown of the coalition to which the deviator belonged. d) The "inclusion chain" of stable coalition structures mentioned under b) is no longer valid for Pareto-undominated stable coalition structures. For instance, suppose coalition structure c is a Pareto-undominated stable coalition structure in the OMMCG. Then, c is also stable in the EMMCΔG but may be Pareto-dominated by coalition structure $\hat{c}$ that is stable in the EMMCΔG but not in the OMMCG. Also the conclusion under point c) has to be qualified under the Pareto-selection criterion. Though a profitable single coalition structure c is a member of C^*(EMMCΓG), it may not be a member of C^{**}(EMMCΓG) if it is Pareto-dominated by another coalition structure $\hat{c}$ that is a member of C^*(EMMCΓG).

3. Results. In this section, we present results of our stability analysis. The stability check comprises all 203 coalition structures for all 8 agreement designs (4 membership regimes combined with the no transfer and transfer regime, respectively). The test is conducted with an algorithm programmed in GAMS[15]. The results are displayed in Table 1 for the no transfer regime and in Table 2 for the transfer regime based on the parameter values listed in the Appendix. Only those coalition structures are listed that constitute Pareto-undominated stable coalition structures in one of the four coalition games, except for the grey rows that constitute important benchmarks. Coalition structures are listed in descending order of global welfare. From the tables, the following conclusions can be drawn.

Impact of cooperation. The difference between the first-best and the worst outcome is large in ecological terms. Carbon concentration in the trivial coalition structure (no. 203), corresponding to the classical Nash equilibrium, is more than twice as high as in the coalition structure with the grand coalition (no. 1), corresponding to the global optimum, see the legend of Table 1. Analyzing the time path of concentration would show that the "Nash equilibrium path" closely follows the business-as-usual scenario, growing steadily over time. In contrast, the "globally optimal path" grows only until 2200 though less steep and declines afterwards.[16]

In welfare terms, the difference between first-best and worst outcome is with 17.7 trillion US dollars not small in absolute terms, though it seems that 0.52 percent is rather small in relative terms, see the legend of Table 1. This is due to two main reasons. First, abatement and damage costs constitute only a small fraction of total production and consumption. Second, negative externalities of carbon emissions occur mainly in the future but receive less weight due to discounting. However, looking not only at discounted total values, reveals that differences are quite pronounced.

Figure 1 displays the evolution of the difference between full and no cooperation in terms of discounted global abatement and damage costs over time. It is evident that abatement plus damage costs are first dramatically lower under no cooperation than under full cooperation and later this is reversed. The reason is that substantial abatement efforts are required initially under full cooperation which are only

TABLE 1. Stable coalition structures under the no transfer regime*.

Rank	Coalition Structure	C^{**}(OMSCG)	C^{**}(OMMCG)	C^{**}(EMMCΔG)	C^{**}(EMMCΓG)	Welfare CGX	Concentration CGX
1	**{USA,JPN,EU,CHN,FSU,ROW}**					100.0	100.0
96	**{USA,JPN,FSU,ROW}**,{EU},{CHN}				X	64.7	48.9
101	**{EU,FSU,ROW}**,**{USA,JPN}**,{CHN}				X	63.8	48.4
109	**{USA,FSU,ROW}**,{JPN},{EU},{CHN}			X	X	60.7	43.2
113	**{USA,JPN,ROW}**,{EU},{CHN},{FSU}			X	X	59.0	41.5
117	**{USA,JPN}**,**{EU,ROW}**,{CHN},{FSU}			X	X	58.1	41.0
140	**{JPN,FSU,ROW}**,{USA},{EU},{CHN}			X	X	54.3	34.2
141	**{USA,ROW}**,{JPN},{EU},{CHN},{FSU}			X	X	54.1	35.2
148	**{USA,JPN}**,**{FSU,ROW}**,{EU},{CHN}			X	X	47.9	26.9
151	**{JPN,ROW}**,{USA},{EU},{CHN},{FSU}			X	X	46.4	24.7
203	{USA},{JPN},{EU},{CHN},{FSU},{ROW}	X	X			0.0	0.0

*Rank: rank of coalition structure in the list of all coalition structures sorted in descending order of global welfare; Coalition Structure: bold-faced regions are member of nontrivial coalitions; C**(G): Pareto-undominated stable coalition structures (see Definition 6) in a particular coalition game G (see Definitions 1 to 4); Welfare CGX: global welfare expressed in terms of closing the gap index: $(\sum_{i=1}^{N}(v_i(c^P)-v_i(c^N)))/(\sum_{i=1}^{N}(v_i(c^F)-v_i(c^N)))$ where welfare is discounted lifetime consumption integrated over 1990–2300, global welfare with full cooperation is $\sum_{i=1}^{N} v_i(c^F) = 3,398.3$ trillion US$_{1990}$ (trillion US dollars expressed in 1990 levels), global welfare with no cooperation is $\sum_{i=1}^{N} v_i(c^N) = 3,380.60$ trillion US$_{1990}$ and global welfare with partial cooperation is denoted by $\sum_{i=1}^{N} v_i(c^P)$. Concentration CGX: atmospheric carbon concentration M at time $t = 2300$ expressed in terms of closing the gap index: $(M(c^N) - M(c^P))/(M(c^N) - M(c^F))$ where concentration with full cooperation is $M(c^F) = 1912.907$ GtC (giga tons of carbon), concentration with no cooperation is $M(c^N) = 4550.202$ GtC and concentration with partial cooperation is denoted by $M(c^P)$.

TABLE 2. Stable coalition structures under the transfer regime*.

Rank	Coalition Structure	C^{**}(OMSCG)	C^{**}(OMMCG)	C^{**}(EMMCΔG)	C^{**}(EMMCΓG)	Welfare CGX	Concentration CGX
1	**{USA,JPN,EU,CHN,FSU,ROW}**				X	100.0	100.0
2	**{USA,EU,CHN,FSU,ROW}**,{JPN}				X	98.7	92.2
3	**{USA,JPN,EU,CHN,ROW}**,{FSU}				X	96.6	90.0
4	**{JPN,EU,CHN,FSU,ROW}**,{USA}				X	95.4	80.6
6	**{USA,EU,CHN,ROW}**,{JPN},{FSU}				X	94.5	82.0
7	**{USA,JPN,CHN,FSU,ROW}**,{EU}				X	93.2	73.2
9	**{EU,CHN,FSU,ROW}**,{USA},{JPN}				X	91.2	72.3
11	**{JPN,EU,CHN,ROW}**,{USA},{FSU}				X	89.4	69.8
14	**{USA,CHN,FSU,ROW}**,{JPN},{EU}				X	87.3	64.1
16	**{USA,JPN,CHN,ROW}**,{EU},{FSU}				X	86.0	61.8
18	**{USA,JPN,FSU,ROW},{EU,CHN}**				X	85.3	66.3
20	**{USA,JPN,CHN},{EU,FSU,ROW}**				X	85.0	66.0
21	**{USA,FSU,ROW},{JPN,EU,CHN}**				X	84.7	63.4
24	**{USA,JPN,ROW},{EU,CHN,FSU}**				X	84.2	62.6
25	**{EU,CHN,ROW}**,{USA},{JPN},{FSU}				X	83.9	60.7
27	**{JPN,EU,FSU,ROW},{USA,CHN}**				X	83.4	68.1
28	**{JPN,EU,CHN,FSU},{USA,ROW}**				X	82.4	58.8
32	**{USA,FSU,ROW},{EU,CHN}**,{JPN}				X	81.3	60.3
35	**{JPN,CHN,FSU,ROW},{USA,EU}**				X	79.9	50.9
36	**{USA,EU,FSU,ROW},{JPN,CHN}**				X	79.7	69.3
38	**{USA,JPN,ROW},{EU,CHN}**,{FSU}				X	79.6	58.5
41	**{USA,JPN,CHN},{EU,ROW}**,{FSU}				X	79.4	58.3
43	**{USA,CHN,ROW}**,{JPN},{EU},{FSU}				X	78.7	52.0
44	**{JPN,EU,ROW},{USA,CHN}**,{FSU}				X	78.6	60.8
46	**{JPN,EU,CHN},{USA,ROW}**,{FSU}				X	78.0	54.9
50	**{USA,EU,ROW},{JPN,CHN}**,{FSU}				X	75.8	62.4
54	**{JPN,FSU,ROW},{EU,CHN}**,{USA}				X	74.6	50.7
55	**{USA,CHN},{EU,ROW}**,{JPN},{FSU}			X		74.5	54.6
56	**{USA,ROW},{EU,CHN}**,{JPN},{FSU}			X	X	74.3	51.8
59	**{EU,FSU,ROW},{JPN,CHN}**,{USA}				X	73.9	56.1
74	**{JPN,CHN,ROW},{USA,EU}**,{FSU}				X	69.0	36.7
83	**{JPN,CHN},{EU,ROW}**,{USA},{FSU}		X	X	X	68.1	48.5
88	**{EU,CHN},{FSU,ROW}**{USA},{JPN}			X		67.0	42.8
91	**{JPN,ROW},{EU,CHN}**{USA},{FSU}			X		66.2	40.7
100	**{USA,CHN},{FSU,ROW}**,{JPN},{EU}			X		63.8	39.9
105	**{USA,CHN},{JPN,ROW}**,{EU},{FSU}			X		63.0	37.8
120	**{EU,ROW}**,{USA},{JPN},{CHN},{FSU}	X				57.4	40.8
130	**{USA,EU},{CHN,ROW}**,{JPN},{FSU}			X		56.2	22.0
139	**{CHN,ROW}**,{USA},{JPN},{EU},{FSU}	X	X			54.6	21.4
141	**{USA,ROW}**,{JPN},{EU},{CHN},{FSU}	X	X			54.1	35.2
150	**{FSU,ROW}**,{USA},{JPN},{EU},{CHN}	X	X			47.2	26.8
151	**{JPN,ROW}**,{USA},{EU},{CHN},{FSU}	X	X			46.4	24.7
203	{USA},{JPN},{EU},{CHN},{FSU},{ROW}					0.0	0.0

*See Table 1.

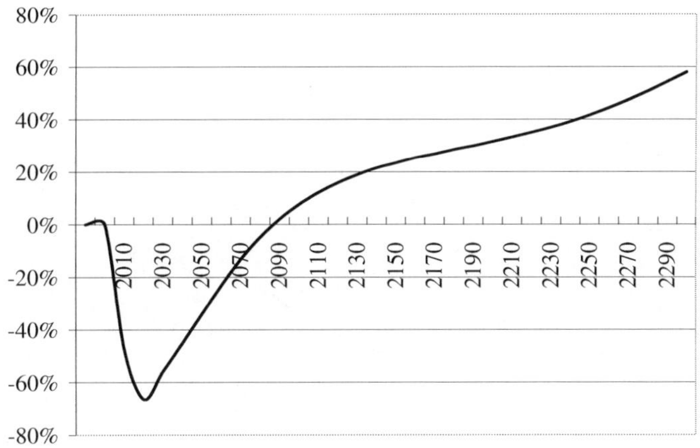

FIGURE 1. Difference of abatement and damage costs between full and no cooperation. Difference is computed as: $\Delta_t^{CD} = (CD_t^N - CD_t^F)/CD_t^N$ with $CD_t^F = \sum_{i \in N}((C_{it}(c^F) + D_{it}(c^F))/[1+\rho_i]^t)$ and $CD_t^N = \sum_{i \in N}((C_{it}(c^N) + D_{it}(c^N))/[1+\rho_i]^t)$ where the superscript N stands for no and the superscript F for full cooperation.

matched by lower damages in later periods given the stock pollutant nature of greenhouse gases. A similar, though less pronounced, picture is shown in Figure 2, which depicts the evolution of the difference in terms of global welfare. Thus, it is evident that cooperation would have a substantial impact on abatement and damage costs as well as on global welfare, although discounting and the long time horizon level off differences when looking only at net present values.

It is worthwhile to point out that our simulations are based on a discount rate of 1.5 percent for the industrialized regions and 3 percent for the developing regions. We make this assumption in order not "too overestimate the sense of governments for cooperation." This assumption seems also roughly in line with Weitzman [2001] who suggests that if a constant discount rate has to be chosen in the context of global warming, then a discount rate of 2 percent or less is appropriate to capture long-term effects. In order to test the robustness of our conclusions, we conduct at the end of this section a sensitivity analysis with discount rates that are one third of those above.[17]

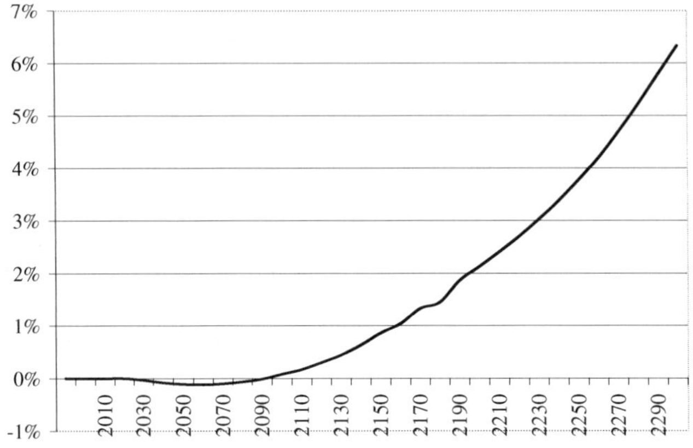

FIGURE 2. Difference of welfare between full and no cooperation. Difference is computed as: $\Delta_t = (v_t^F - v_t^N)/v_t^N$ with $v_t^F = \sum_{i \in N}(Z_{it}(c^F)/[1+\rho_i]^t)$ and $v_t^N = \sum_{i \in N}(Z_{it}(c^N)/[1+\rho_i]^t)$ where the superscript N stands for no and the superscript F for full cooperation.

Participation. From the number of participants, success cannot be inferred. Also the identity of coalition members matters. For instance, a coalition structure with only USA and ROW forming a coalition (no. 141 in Table 1) generates larger global welfare than a coalition structure with USA and JPN in one coalition and FSU and ROW in another coalition (no. 148 in Table 1). Considering in particular the top coalition structures in Table 2, together with background information on all coalition structures, it appears that the importance of regions for successful cooperation decreases along the following sequence: ROW, CHN, EU, USA, FSU, JPN. ROW's and CHN's importance derives from cheap abatement because both regions have flat marginal abatement cost curves in our model. An opposite argument holds for JPN which faces a steep marginal abatement cost curve. However, also the slope of the marginal damage cost curve is important. From the assumption of joint welfare maximization, it follows that the steeper this function is, the higher equilibrium abatement levels of coalition members will be. In our model, ROW and EU have steep and FSU and CHN flat marginal damage cost functions.

This explains the importance of EU for the success of cooperation in ecological and welfare terms.

Transfer regime. Transfers can make a difference. In every coalition game, there are many stable coalition structures listed in Table 2 that generate higher global welfare than those listed in Table 1. For instance, no nontrivial coalition structure is stable in the two open membership games under the no transfer regime whereas under the transfer regime there are five and six, respectively. In the exclusive membership multiple coalition Γ-game, the stable coalition structure with the lowest global welfare (no. 83 in Table 2) under the transfer regime generates higher global welfare than the stable coalition structure under the no transfer regime with the highest global welfare (no. 96 in Table 1).

The reason for these observations is that, in a world of heterogeneous regions, joint welfare maximization may lead to a very asymmetric distribution of the gains from cooperation without transfers. This may imply that profitability is violated for some coalition members (despite the fact that there is a surplus from cooperation), a necessary condition for stability in every coalition game. However, even when profitability holds, an asymmetric distribution also negatively affects stand alone stability and stand apart stability. Though asymmetry may imply for some participants a "slack of stability," it implies for other members a "lack of stability." That is, though some members receive a higher payoff than would be necessary to ensure stability, for other members just the opposite holds which jeopardizes stability of a coalition.

For instance, assume that China, which is characterized by a flat marginal abatement cost curve and a flat marginal damage cost curve, would form a coalition with Japan for which just the opposite holds. Then, the equilibrium abatement target will be relatively high because of high marginal damages of Japan. Additionally, China has to contribute substantially more to joint abatement than Japan because of its flatter marginal abatement cost curve, but China benefits relatively little in the form of reduced damages because of a flat marginal damage cost curve. Thus, without transfers, no single coalition structure with China as a member of a nontrivial coalition is profitable and therefore stable. Consequently, in the absence of transfers, also full cooperation is not profitable to all participants: in fact, apart from CHN also ROW would lose from cooperation. Generally, asymmetries are so strong that without transfer, no single coalition structure with a coalition of 5 members and only one single coalition structure with a coalition of 4 members is profitable. In contrast, the transfer regime always ensures profitability, see Section 2.4. Moreover, the transfer regime can balance different incentive structures, increasing overall stability.

Membership regime. The membership regime has a crucial impact on stability and the success of cooperation. Four relations are important:

a) First, if we allow for the possibility to form multiple coalitions, multiple coalition structures emerge in equilibrium.

b) Some of the stable multiple coalition structures imply higher global welfare than stable single coalition structures. For instance, comparing the outcome of the two open membership coalition games which differ only in one feature, then it is evident that in the case of transfers the set of stable coalition structures is identical, except that there is one more coalition structure in the multiple coalition game. This additional coalition structure (no. 83 in Table 2) is a multiple coalition structure that generates higher global welfare than any other stable single coalition structure under open membership. Also, in the exclusive membership multiple coalition Δ-game, all Pareto-undominated stable coalition structures are multiple coalition structures under the transfer regime of which many Pareto-dominate stable single coalition structures (e.g. no. 120, 139, 141, 150, 151 in Table 2). Only in the case of the exclusive membership multiple coalition Γ-game, the possibility of forming multiple coalitions is less important for global welfare since the implicit punishment is harsh enough to sustain large single coalition structures.

c) Exclusive membership leads to a significant number of additional stable coalition structures compared to open membership of which most imply higher global welfare. This is true without exception under the no transfer regime, Table 1. Under the transfer regime, it is only partially true, Table 2. However, among the "additional coalition structures," there are always some that lead to higher global welfare.

d) A similar conclusion can be drawn when comparing the outcome under weak, i.e., $C^{**}(EMMC\Delta G)$, versus strong, i.e., $C^{**}(EMMC\Gamma G)$, consensus: strong consensus leads to superior outcomes compared to weak consensus. This is particularly evident under the transfer regime. The requirement of strong consensus exhibits a strong enforcement power on coalitions. Basically, players face the decision "take it or leave it," and leave it means a strong implicit punishment that is conducive to stability.

Agreement design. Until now, we assumed the agreement design to be given exogenously. However, one can easily perceive that there is a stage 0 in which players decide upon the membership rule $\Phi = \{\text{OMSCG}, \text{OMMCG}, \text{EMMC}\Delta\text{G}, \text{EMMC}\Gamma\text{G}\}$ and whether there should be transfers ($\Omega(C) = \{V^{NT}(C), V^{T}(C)\}$). There are certainly many ways to model the decision process in stage 0. We choose here a simple option. First, given that there is no unique stable coalition structure under most agreement designs, we assume that all equilibria are equally likely for a particular design. Second, we compute the average expected individual welfare for every player for each of the eight agreement designs. Third, we check Pareto-dominance relations between the eight designs. Thus, our procedure means introducing a "constitutional period" in which players decide under the "veil of ignorance" upon the agreement design. It turns out in our model that the design EMMCΓG and $V^{T}(C)$ Pareto-dominates all other designs. That is, if players decide in stage 0 simultaneously upon the regime in stage 1 and 2, all players would unanimously vote for exclusive membership with strong consensus and the implementation of transfers.

Discounting. Until now, all results were based on the assumption of a discount rate of 1.5 percent for industrialized regions and 3 percent for developing regions. Assuming only one third of these values, i.e., 0.5 and 1.5 percent, respectively, has the following implications.[18] First, the impact of cooperation becomes larger. This is particularly true in terms of global welfare. Whereas previously the difference of the net present value between full and no cooperation was 0.52 percent in relative terms, it is now 2.28 percent. Thus, lower discounting makes cooperation more attractive from a global point of view. Second, stable coalition formation becomes more successful not only in absolute but also in relative terms. For each of the eight scenarios, closing the gap index of stable coalition structures is on average substantially higher with lower discount rates than displayed in Tables 1 and 2. This is in line with intuition which suggests that the incentive for cooperation increases with lower discounting because the long term benefits from cooperation receive more weight. Third, all qualitative conclusions from above are confirmed. That is, the composition of members is more important than the number of participants for the success of an agreement, transfers improve upon the prospects of successful cooperation, membership rules make a difference where multiple coalitions, exclusive membership and a strong degree of consensus have a positive impact on the outcome of stable coalition formation compared to single coalitions, open membership and a weak degree of consensus, respectively. Also endogenizing the agreement design leads to the outcome on which we reported above.

4. Interpretations and conclusions.

General remarks. The results highlight that the assumptions associated with the cartel formation game, frequently applied in the analysis of international environmental agreements, substantially affect conclusions. It is evident that all classical assumptions (single coalition structure, open membership and no consensus for coalition formation) imply rather pessimistic predictions about the success of cooperation. Relaxing each of these assumptions gives rise to additional stable coalition structures that perform better in welfare and ecological terms in our numerical simulations.

Transfer regime. Our results confirm the conjecture that cleverly designed transfers can foster the success of cooperation if actors have different interests. In particular, an efficient climate policy must be accompanied by compensation measures in order to induce participation of the key producers of cheap abatement like China and the "Rest of the World" (comprising mainly developing countries). Our simulations reveal that the Chander/Tulkens' transfer scheme adopted in this paper is capable of providing the necessary compensations. Thus, our results support the efforts in more recent IEAs, like the Convention on Biological Diversity, the Montreal Protocol and the Kyoto Protocol to set up a fund that provides financial support to developing countries in order to encourage participation and cover their additional abatement costs. The permit trading system under the Kyoto Protocol will also imply transfers that may balance asymmetries, given the fact that trading occurs voluntarily.

Membership regime. Our results suggest that the effort to get as many countries into one "climate boat" may not always be the good strategy in the presence of strong free-rider incentives. This is at least true; if not, such harsh implicit punishments are available as assumed in the exclusive membership multiple coalition Γ-game and/or if no or only limited transfers are available to balance asymmetries. If no large coalitions are stable, allowing for separate agreements among regions that have similar interests may foster the success of international agreements. In our model, regions with similar interests are solely characterized by their benefit and cost structure, though in reality, also regional and cultural similarities may play some role. In this light, the position of the USA only to ratify the Kyoto Protocol if also developing countries would take on climate responsibility has to be evaluated critically.[19] However, their announcement in the aftermaths of their withdrawal that they would pursue their own climate policy cooperating with countries of similar interests may be seen less critical. At least as an intermediate step, multiple agreements may be useful on the way to a global treaty as this is often the case with regional trade agreements that lead to more comprehensive agreements at later stages.

The results also suggest that open membership as applied in almost every international environmental treaty should not be taken for granted. In particular, considering other international agreements or organization, it turns out that exclusive membership is no exception. For instance, NATO and WTO vote by unanimity on new members. Also accession of the ten new member states in the European Union required a two-step decision procedure. First, accession had to be accepted by the European Parliament by a simple majority and subsequently the European Council had to approve the accession by unanimity. Hence, our results suggest that club good agreements, e.g., NATO, European Union and WTO, may not only enjoy a higher stability and success because the benefits from contributions are exclusive to members, but also because membership is exclusive. Though more members means more contributors, it may also mean "more instability" that may be counterproductive.

Closely related to exclusive membership is the degree of consensus needed to form a coalition. Here our results suggest that a high degree of consensus upon membership may not be an obstacle but in fact may be conducive to successful cooperation.

Participation. Ignoring the "favorable" agreement design under exclusive membership multiple coalition Γ-game and transfers (in which case large single coalition structures are stable), then our results question the commonly held view that a high participation also means success. They stress the importance to include key players in international climate policy. On the one hand, agreements should include regions that perceive the benefits from reduced environmental damages to be high and therefore push for high abatement targets. In our model, this attaches much importance to the European Union. On the other hand, regions that can provide cheap abatement should participate. In our model, this attaches a key role to China and developing countries (ROW). Obviously, this aim has to overcome free-rider incentives fundamental to climate change policy, reinforcing the need for transfer and/or differentiated abatement targets.

Discounting. It is certainly not surprising that our model confirms the well-known result that discounting is important for the absolute and relative gains from cooperation in the context of global externalities. However, it is interesting that our model shows that discounting is equally important for the success of IEAs if free-rider incentives are explicitly taken into account. This reinforces the importance and policy relevance of the discussion about "appropriate discount rates" for sustainable development as for instance discussed in Chichilnisky [1997].

5. Extensions and issues for future research. Despite our results providing some guidance for the design of future IEAs, some evidence is puzzling from a positive point of view. We name four issues.

The first issue is participation. Under the no transfer regime, predicted membership in stable coalition structure seems not in line with expectations. For instance, one may wonder why ROW and USA are frequently members of coalitions whereas EU is only part of two coalitions. Under the transfer regime, high participation, including the grand coalition, in the exclusive membership multiple coalition Γ-game makes it difficult to understand the real world problems associated for instance with the Kyoto Protocol.

The second issue is single versus multiple coalitions. Despite the fact that it is certainly true that it is easier to consider different individual interests in multiple coalitions rather than in a single coalition, this explanation may not convince all critics. For instance, one could object that a differentiation of abatement duties could also be established within one coalition. Moreover, also the assumption of joint welfare maximization leads to some form of differentiation of abatement targets, though to a very special one: all participants must comply with a cost-efficient allocation of abatement levels and with the implied aggregate abatement level.

The third issue is open versus exclusive membership. As pointed out above, almost all existing IEAs have no provision to restrict membership. Thus, one may wonder why exclusive membership has not emerged as a superior membership rule if the advantages according to our results are so obvious. The fourth issue is the degree of consensus about membership. Though consensus decisions are frequently observed in international environmental politics, they typically apply to the policy level rather than to participation.

Subsequently, we suggest some reasons that could explain these puzzles and propose some possibilities to improve upon the explanatory power of our model in terms of a positive analysis.

First, all four coalition games assume that players decide simultaneously upon their participation. However, casual empirical evidence suggests that signature and ratification of environmental treaties occur sequentially. Typically, some initiators kick off the negotiation process, a core group of countries participate in an agreement and other countries may follow suit at a later stage. However, modeling coalition formation as a sequential process introduces a couple of conceptual problems that are difficult to solve. For instance, in the case of heterogeneous players, the sequence in which players decide upon participation or make proposals for coalitions has to be endogenized. Also in models like ours with a dynamic payoff structure with substantially more than two periods, the number of strategic options for participation increases dramatically. Hence, it is not surprising that the sequentially move unanimity game of Bloch [1995] has only been applied so far to symmetric players and to games with a static payoff structure, e.g., Ray and Vohra [2001] and Yi [2000]. Moreover, this game assumes that players cannot revise their membership once they have decided upon participation. Also Rubio and Ulph [2003] who consider a sequential version of the open membership single coalition game assume symmetric players and a very simple payoff function for analytical tractability. Nevertheless, we believe that tackling these conceptual problems will add much to the understanding of the design of actual international environmental treaties.

Second, in all four coalition games, reactions after a deviation follow from the membership rule. Thus, reactions do not follow from ad hoc assumptions. Nevertheless, reactions may not be those actually taken by countries. It may be overly pessimistic to expect that a deviation does not trigger any reaction at all as implied by the two open membership games and the exclusive membership multiple coalition Δ-game. However, it may also be considered as overly optimistic that deviation triggers a breakdown of an agreement as implied by the exclusive membership multiple coalition Γ-game. Of course, it is difficult to predict what kinds of reactions are "correct." This would require an empirical analysis that obviously faces a bundle of problems associated with the gathering of data. However, also from a theoretical point of view, this issue is not straightforward. For instance, one may argue that reactions should constitute a best reply, an idea that is for instance captured by the notion of a farsighted equilibrium as proposed by Chwe [1994]. Roughly speaking, a coalition structure is stable if no player has an incentive to deviate, taking all reactions and counter reactions into account. Not surprisingly, this concept is difficult to operationalize in the context of more than three players because the chain of reactions may be very long. Thus, some simplification may be necessary as assumed for instance by the concept of an equilibrium binding agreement proposed by Ray and Vohra [1997].

Third, our empirical model divides the world into only six regions. In particular, the player "Rest of the World" (ROW) includes many countries. This implies that very heterogeneous interests within this group of countries are "homogenized" by assumption. Moreover, summing damages over many countries, even though individual countries may have low damages, leads to high damage costs estimates at the aggregate. Hence, our model probably overestimates the interest of ROW in cooperation and therefore also generally the prospects of cooperation. This qualification would call for a model with more disaggregated data. However, there is no doubt that this would require a substantial amount of additional computer resources. For instance, extending the analysis from 6 to "only" 12 players increases the number of possible coalition structures that have to be checked for stability already from 203 to over 4 billion.

Fourth, the parameter values of our empirical model are based on the RICE model by Nordhaus and Yang [1996]. As it is well known, e.g., IPCC [2001], there is a lot of uncertainty about the "correct estimates" and model assumptions vary widely. In any case, all current models are based on "objective" estimates about abatement and damage costs that may differ from the "subjective" evaluation of governments. For instance, there is some reason to believe that the subjective perception of damage costs of China and ROW is substantially lower than the objective estimates assumed in our model. This may be because these governments consider only a short planning horizon or simply value damages caused by global warming less than we assumed in our model. Consequently, China and ROW (and probably also other regions in our model) would have virtually no incentive to participate in cooperation. An obvious way to improve upon the estimates for a positive analysis would be to use data from contingent valuation studies to calibrate an empirical model. For instance, Böhringer and Vogt [2004] suggest from a survey of various studies that the willingness to pay for global greenhouse gases is even in industrialized countries rather low. However, it is also evident from their survey that the current data basis is very small, only few country studies are currently available.[20]

Fifth, following the mainstream literature on coalition formation, we assumed that players choose their economic strategies by maximizing the aggregate welfare to their coalition. However, casual empirical evidence suggests that in reality abatement allocation is neither cost-efficient nor is the aggregate abatement level optimal. That is, this assumption seems to overestimate the sense of cooperation among governments but also neglects the complicated political decision process within countries (Carraro and Siniscalco [1998], Endres [1997]). Hence, one way to improve upon the predictive power of coalition models could be to include public choice aspects in the analysis like interest groups, governmental decision processes and parliamentary voting. However, apart from increasing the complexity of models, we expect that it will be difficult to identify and quantify political variables for an empirical analysis.

Sixth, we assumed lump sum transfers for simplicity. This was consistent with our stability analysis where participation decisions are based on discounted payoffs. In reality, however, it seems reasonable to expect that transfers are paid and negotiated over time. We expect that this will imply that transfers appear to be a less straightforward instrument for improving the prospects for cooperation and hence cooperation between heterogeneous players will be more difficult to achieve. However, given the complexity of dynamic models, this conjecture may be premature and would have to be analyzed in an extended framework. A promising route of research could be along the lines of Germain et al. [2003]. However, no doubt, this would require much additional conceptual work since their cooperative game theoretic analysis is restricted to transfers within and stability of the grand coalition whereas our framework considers all forms of partial cooperation.

Seventh, we computed open-loop Nash equilibria in the second stage of the coalition formation game. A more sophisticated alternative would be closed-loop Nash equilibrium strategies as for instance considered in Dockner and Long [1993] and Rubio and Casino [2002]. The impact of our simplification is difficult to assess as there are no general results available. On the one hand, following Fudenberg and Tirole [1991], we expect the difference between both concepts to be small. They argue that open-loop strategies are good approximations of closed-loop strategies in games in which players have little influence on the other players' optimal strategies. In our simulation model, this influence is rather small because of the strong rigidity of the carbon cycle in the climate model.[21] On the other hand, if there are differences, then we expect less stable agreements for closed-loop than for open-loop strategies. This seems suggestive considering Germain et al. [2003] who note for their model that global emissions tend to be higher under closed-loop than under open-loop strategies because there is more room for strategic behavior.

Finally, though we believe that our approach of modeling the choice of the agreement design endogenously is novel and an important step, we acknowledge that it is only a first step. In particular, the decision was restricted to only eight designs of agreements, and, in the case of transfers, we considered only one compensation rule of possibly many others. Moreover, as pointed out above, the choice of abatement was fixed through the assumption of joint welfare maximization. In reality, however, one would expect that all aspects of agreement design are negotiated as a package and are therefore also strategically linked. Taking this concern seriously means to endogenize all features of the design of an agreement. This is certainly a challenging task for future research. However, first attempts to develop such a "new" framework by Maskin [2003] seem promising, though we expect that implementation in an empirical model will be anything else than trivial.

Acknowledgments. We would like to thank Thierry Bréchet (UCL-CORE) and two anonymous referees for valuable comments on an earlier version of this paper. Financial assistance from the CLIMNEG 2 project (contract CP/24), funded by the Belgian Federal Science Policy Office, is gratefully acknowledged.

Appendix

The CLIMNEG world simulation model comprises the following equations:

$$Y_{it} = Z_{i,t} + I_{i,t} + Y_{i,t}C_i(\mu_{i,t}) + Y_{i,t}D_i(\Delta T_t)$$

$$Y_{i,t} = a_{i,t}K_{i,t}^{\gamma}L_{i,t}^{1-\gamma}$$

$$C_i(\mu_{i,t}) = b_{i,1}\mu_{i,t}^{b_{i,2}}$$

$$D_i(\Delta T_t) = \theta_{i,1}\Delta T^{\theta_{i,2}}$$

$$K_{i,t+1} = [1-\delta_K]K_{i,t} + I_{i,t} \quad K_{i,0} \text{ given}$$

$$E_{i,t} = \alpha_{i,t}[1-\mu_{i,t}]Y_{i,t}$$

$$M_{t+1} = [1-\delta_M]M_t + \beta\sum_{i\in P} E_{i,t} \quad M_0 \text{ given}$$

$$F_t = \frac{4.1\ln(M_t/M_0)}{\ln(2)} + F_t^x$$

$$T_t^0 = T_{t-1}^0 + \tau_3[T_{t-1}^a - T_{t-1}^0]$$
$$T_t^a = T_{t-1}^a + \tau_1[F_t - \lambda T_{t-1}^a] - \tau_2[T_{t-1}^a - T_{t-1}^0]$$
$$\Delta T_t = \frac{T_t^a}{2.50}$$

List of variables.

$Y_{i,t}$	production (billion 1990 US$)
$Z_{i,t}$	consumption (billion 1990 US$)
$I_{i,t}$	investment (billion 1990 US$)
$K_{i,t}$	capital stock (billion 1990 US$)
$C_{i,t}$	cost of abatement (billion 1990 US$)
$D_{i,t}$	damage from climate change (billion 1990 US$)
$E_{i,t}$	carbon emissions (billion tons or gigatons of C, GtC)
$\mu_{i,t}$	emission abatement
M_t	atmospheric carbon concentration (billion tons or gigatons of C, GtC)
F_t	radiative forcing (Watt per m^2)
F_t^x	exogenous radiative forcing (Watt per m^2)
T_t^a	temperature increase in the atmosphere (^{deg} C)
T_t^0	temperature increase in the deep ocean (^{deg} C)
ΔT_t	change of temperature increase in the atmosphere (^{deg} C)

Global parameter values.

$a_{i,t}$	productivity	RICE
$L_{i,t}$	population	RICE
$\alpha_{i,t}$	emission-output rate	RICE
δ_K	capital depreciation rate	0.10
γ	capital productivity parameter	0.25
β	airborne fraction of carbon emissions	0.64

δ_M	atmospheric carbon removal rate	0.0833
τ_1	parameter temperature relationship	0.226
τ_2	parameter temperature relationship	0.44
τ_3	parameter temperature relationship	0.02
λ	parameter temperature relationship	1.41
M_0	initial carbon concentration	590
T_0^a	initial temperature atmosphere	0.50
T_0^0	initial temperature deep ocean	0.10

Regional parameter values.

	$\theta_{i,1}$	$\theta_{i,2}$	$b_{i,1}$	$b_{i,2}$	ρ_i
USA	0.01102	2.0	0.07	2.887	0.015
JPN	0.01174	2.0	0.05	2.887	0.015
EU	0.01174	2.0	0.05	2.887	0.015
CHN	0.01523	2.0	0.15	2.887	0.030
FSU	0.00857	2.0	0.15	2.887	0.015
ROW	0.02093	2.0	0.10	2.887	0.030

Variables in 1990 (*reference year*).*

	Y_i^0	(%)	K_i^0	(%)	L_i^0	(%)	E_i^0	(%)
USA	5,464.796	25.9	14,262.510	26.3	250.372	4.8	1.360	20.5
JPN	2,932.055	13.9	8,442.250	15.6	123.537	2.4	0.292	10.9
EU	6,828.042	32.4	18,435.710	34.0	366.497	7.0	0.872	28.9
CHN	370.024	1.8	1,025.790	1.9	1,133.683	21.5	0.669	3.0
FSU	855.207	4.1	2,281.900	4.2	289.324	5.5	1.066	6.8
ROW	4,628.621	22.0	9,842.220	18.1	3,102.689	58.9	1.700	29.9
World	21,078.750	100.0	54,290.380	100.0	5,266.100	100.0	5.959	100.0

* Y_i^0 and K_i^0 billion US$, L_i^0 million people and E_i^0 GtC.

ENDNOTES

1. See the extensive literature cited in Barrett [2003] and Finus [2003a].

2. For an overview, see for instance Bloch [1997] and Yi [1997].

3. See previous footnote as well as the papers compiled in Carraro [2003].

4. Examples include Carraro [2000], Carraro/Marchiori [2003], Finus [2003b] and Finus/Rundshagen [2003].

5. Examples include Bosello et al. [2003], Botteon/Carraro [1997], Buchner/Carraro [2003], Finus et al. [2005] and Tol [2001].

6. We recognize that there is also a large literature on coalition theory and fisheries. However, almost all of these papers use cooperative game theory in their analyses. A notable exception is for instance Pintassilgo [2004].

7. We consider a slightly modified version of the original game that has been proposed by Bloch [1997] and formalized by Rundshagen [2002]. In the original game, there is no possibility to announce 0. This implies that players can be forced into a coalition, an assumption not in line with the notion of voluntary participation in international environmental agreements.

8. A description of the equations of the model as well as a list of the key parameters is provided in the Appendix. An intensive discussion of the model is provided in Eyckmans and Tulkens [2003].

9. Technically, we compute open-loop Nash equilibria. For a discussion of an alternative assumption see Section 5.

10. We choose a sufficiently long period to avoid "end point bias." However, due to discounting, only a shorter period is strategically relevant for players.

11. A more formal and detailed proof of this claim can be found in Eyckmans and Tulkens [2003].

12. This means that lump sum transfers are expressed in terms of discounted consumption at time $t = 0$ and take place at $t = 0$. This assumption, though certainly a strong simplification, is in line with our assumption that players decide at time $t = 0$ in a one-shot fashion on coalition formation. For a discussion of alternative assumptions, see Section 5, in particular, point six.

13. Note that Pareto-undominance only applies with respect to the set of stable coalition structures and not with respect to all coalition structures.

14. In the environmental context, see for instance Carraro and Marchiori [2003] and Finus and Rundshagen [2003] where it is apparent that, apart from a static payoff structure and symmetric players, additional restrictive assumptions are necessary in order to derive conclusions about welfare and ecological implications of different coalition games. Similar problems are known in the context of other economic problems. See, for instance, the overview articles by Bloch [1997] and Yi [1997] and the literature cited there.

15. General Algebraic Modeling System, see http://www.gams.com.

16. See Eyckmans and Tulkens [2003] for more information concerning the reference simulations.

17. An alternative way to increase the gap between the full cooperative and non-cooperative outcome could be to change the objective function in (1) by giving more weight to the distant future as for instance suggested in Chichilnisky [1997].

18. Detailed results are available from the authors upon request.

19. In 1997, the U.S. Senate unanimously passed the Byrd-Hagel resolution which makes "meaningful" participation of developing countries a *conditio sine qua non* for ratification (The Byrd-Hagel Resolution, U.S. Senate, 12 June 1997, 105th Congress, 1st Session, Senate Resolution 98).

20. It is evident that this line of research would contrast to our exercise at the end of Section 3 where we lowered the discount rates. This highlights the controversy between normative and positive policy issues in the analysis of international environmental agreements.

21. This is an alternative way of interpreting our statement in section 2.4 that in our model leakage effects are moderate and hence superadditivity holds.

REFERENCES

S. Barrett [2003], *Environment and Statecraft. The Strategy of Environmental treaty-making*, Oxford Univ. Press, Oxford.

F. Bloch [1995], *Endogenous Structures of Associations in Oligopolies*, RAND J. Econ. **26**, 537–556.

F. Bloch [1997], *Non-Cooperative Models of Coalition Formation in Games with Spillovers*, in *New Directions in the Economic Theory of the Environment* (C. Carraro and D. Siniscalco, eds.), Cambridge Univ. Press, Cambridge, pp. 311–352.

C. Böhringer and C. Vogt [2004], *The Dismantling of a Breakthrough: The Kyoto Protocol as Symbolic Policy*, Europ. J. Polit. Econ. **20**, 597–618.

F. Bosello, B. Buchner and C. Carraro [2003], *Equity, Development, and Climate Change Control*, J. Europ. Econ. Assoc. **1**, 601–611.

M. Botteon and C. Carraro [1997], *Burden-Sharing and Coalition Stability in Environmental Negotiations with Asymmetric Countries*, in *International Environmental Negotiations: Strategic Policy Issues* (C. Carraro, ed.), Edward Elgar, Cheltenham, UK, pp. 26–55.

B. Buchner and C. Carraro [2003], *Emission Trading Regimes and Incentives to Participate in International Climate Agreements*, CATEP Policy Brief No. 3.

C. Carraro [2000], *Roads towards International Environmental Agreements*, in *The Economics of International Environmental Problems* (H. Siebert, ed.), Mohr Siebeck, Tübingen, pp. 169–202.

C. Carraro, ed. [2003], *Endogenous Formation of Economic Coalitions*, Edward Elgar, Cheltenham, UK.

C. Carraro and C. Marchiori [2003], *Stable Coalitions*, in *The Endogenous Formation of Economic Coalitions* (C. Carraro, ed.), Edward Elgar, Cheltenham, UK, pp. 156–198.

C. Carraro and D. Siniscalco [1998], *International Environmental Agreements: Incentives and Political Economy*, Europ. Econ. Rev. **42**, 561–572.

P. Chander and H. Tulkens [1995], *A Core-Theoretic Solution for the Design of Cooperative Agreements on Transfrontier Pollution*, Int. Tax Public Finance **2**, 279–293.

P. Chander and H. Tulkens [1997], *The Core of an Economy with Multilateral Environmental Externalities*, Int. J. Game Theory **26**, 379–401.

G. Chichilnisky [1997], *What is Sustainable Development?*, Land Econ. **73**, 467–491.

M.S.-Y. Chwe [1994], *Farsighted Coalitional Stability*, J. Econ. Theory **63**, 299–325.

C. d'Aspremont, A. Jacquemin, J.J. Gabszewicz and J.A. Weymark [1983], *On the Stability of Collusive Price Leadership*, Canad. J. Econ. **16**, 17–25.

E.J. Dockner and N.V. Long [1993], *International Pollution Control: Cooperative versus Non-Cooperative Strategies*, J. Environ. Econ. Manage. **25**, 13–29.

A. Endres [1997], *Negotiating a Climate Convention, The Role of Prices and Quantities*, Int. Rev. Law Econ. **17**, 147–156.

J. Eyckmans and H. Tulkens [2003], *Simulating Coalitionally Stable Burden Sharing Agreements for the Climate Change Problem*, Resource Energy Econ. **25**, 299–327.

M. Finus [2003a], *Stability and Design of International Environmental Agreements: The Case of Transboundary Pollution*, in *International Yearbook of Environmental and Resource Economics* (H. Folmer and T. Tietenberg, eds.), 2003/4, Edward Elgar, Cheltenham, UK, pp. 82–158.

M. Finus [2003b], *New Developments in Coalition Theory: An Application to the Case of Global Pollution*, in *Environmental Policy in an International Perspective* (L. Marsiliani, M. Rauscher and C. Withagen, eds.), Kluwer, Dordrecht, pp. 19–49.

M. Finus, J.-C. Altamirano-Cabrera and E. van Ierland [2005], *The Effect of Membership Rules and Voting Schemes on the Success of International Climate Agreements*, Public Choice **125**, 95–127.

M. Finus and B. Rundshagen [2003], *Endogenous Coalition Formation in Global Pollution Control. A Partition Function Approach*, in *Endogenous Formation of Economic Coalitions* (C. Carraro, ed.), Edward Elgar, Cheltenham, UK, pp. 199–241.

D. Fudenberg and J. Tirole [1991], *Game Theory*, MIT Press, Cambridge, MA.

M. Germain, P. Toint, H. Tulkens and A. de Zeeuw [2003], *Transfers to Sustain Dynamic Core-Theoretic Cooperation in International Stock Pollutant Control*, J. Econ. Dynam. Control **28**, 79–99.

S. Hart and M. Kurz [1983], *Endogenous Formation of Coalitions*, Econometrica **51**, 1047–1064.

IPCC [2001], *Climate Change* 2001: *Mitigation*, Contribution of Working Group III to the *Third Assessment Report of the Intergovernmental Panel on Climate Change*, Cambridge Univ. Press, Cambridge.

E. Maskin [2003], *Bargaining, Coalitions, and Externalities*, mimeo, presented at the EEA-ESEM Meeting in Stockholm, August 20–24.

W.D. Nordhaus and Z. Yang [1996], *A Regional Dynamic General-Equilibrium Model of Alternative Climate-Change Strategies*, Amer. Econ. Rev. **86**, 741–765.

P. Pintassilgo [2003], *A Coalition Approach to the Management of High Seas Fisheries in the Presence of Externalities*, Natur. Resource Modeling **16**, 175–197.

D. Ray and R. Vohra [1997], *Equilibrium Binding Agreements*, J. Econ. Theory **73**, 30–78.

D. Ray and R. Vohra [2001], *Coalitional Power and Public Goods*, J. Polit. Econ. **109**, 1355–1384.

S. Rubio and B. Casino [2002], *A Note on Cooperative versus Non-Cooperative Strategies in International Pollution Control*, Resource Energy Econ. **73**, 251–261.

S. Rubio and A. Ulph [2003], *An Infinite-Horizon Model of Dynamic Membership of International Environmental Agreements*, Working Paper No. 57.2003, Fondazione Eni Enrico Mattei, Milano.

B. Rundshagen [2002], *On the Formalization of Open Membership in Coalition Formation Games*, Working Paper No. 318, University of Hagen.

R. Tol [2001], *Climate Coalitions in an Integrated Assessment Model*, Comput. Econ. **18**, 159–172.

M.L. Weitzman [2001], *Gamma Discounting*, Amer. Econ. Rev. **91**, 260–271.

S.-S. Yi [1997], *Stable Coalition Structures with Externalities*, Games Econ. Behav. **20**, 201–237.

S.-S. Yi [2000], *Free-Trade Areas and Welfare: An Equilibrium Analysis*, Rev. Int. Econ. **8**, 336–347.

S.-S. Yi and H. Shin [2000], *Endogenous Formation of Research Coalitions with Spillovers*, Int. J. Indus. Organ. **18**, 229–256.

[37]

Available online at www.sciencedirect.com
SCIENCE DIRECT®

ELSEVIER Journal of Environmental Economics and Management 51 (2006) 93–109

JOURNAL OF ENVIRONMENTAL ECONOMICS AND MANAGEMENT

Regional versus global cooperation for climate control

Geir B. Asheim[a], Camilla Bretteville Froyn[b,*], Jon Hovi[b,c], Fredric C. Menz[b,d]

[a]*Department of Economics, University of Oslo, P.O. Box 1095 Blindern, 0317 Oslo, Norway*
[b]*Center for International Climate and Environmental Research – Oslo (CICERO), P.O. Box 1129 Blindern, 0318 Oslo, Norway*
[c]*Department of Political Science, University of Oslo, P.O. Box 1097 Blindern, 0317 Oslo, Norway*
[d]*Faculty of Economics and Finance, School of Business, Clarkson University, Potsdam, NY 13699, USA*

Received 22 April 2003
Available online 22 July 2005

Abstract

This paper considers whether international environmental public goods provision, such as mitigation of climate change, is better dealt with through regional cooperation than through a global treaty. Previous research suggests that, at best, a global environmental treaty will achieve very little. At worst, it will fail to enter into force. Using a simple dynamic game-theoretic model, with weakly renegotiation-proof equilibrium as solution concept, we demonstrate that two agreements can sustain a larger number of cooperating parties than a single global treaty. The model provides upper and lower bounds on the number of parties under each type of regime. It is shown that a regime with two agreements can Pareto dominate a regime based on a single global treaty. We conclude that regional cooperation might be a good alternative–or supplement–to global environmental agreements.

Keywords: Climate change; International environmental agreements; Regional cooperation; The Kyoto Protocol; Non-cooperative game theory; Public goods; Weak renegotiation proofness

*Corresponding author. CICERO, P.O. Box 1129 Blindern, 0318 Oslo, Norway, Fax: +47 22 85 87 51.
E-mail addresses: g.b.asheim@econ.uio.no (G.B. Asheim), cbf@cicero.uio.no (C.B. Froyn), jon.hovi@stv.uio.no (J. Hovi), menzf@clarkson.edu (F.C. Menz).

0095-0696/$ - see front matter
doi:10.1016/j.jeem.2005.04.004

1. Introduction

To be effective, international environmental agreements (IEAs) need to achieve broad participation. For example, no international regime on climate change will be fully effective without the involvement of countries that have not yet committed to reduce emissions levels under the Kyoto framework, such as China, India, other developing countries, the United States, and Australia.

Previous research based on non-cooperative game theory suggests that, at best, a global IEA might achieve very little and at worst, could fail to enter into force (e.g., [5,7,14]). It is therefore interesting to consider whether two or more regional treaties might be more successful than a single global treaty. In this paper we show that an arrangement with two regional agreements can sustain a larger number of cooperating parties than a single global treaty. Therefore, in a second-best world characterized by free-rider incentives, a global treaty requiring agreement among a large number of countries is not necessarily the best option.

Regional cooperation to deal with climate change is already taking place. In March 2000, the European Climate Change Programme was launched by the European Union to coordinate EU efforts to reduce greenhouse gas emissions, including an emissions trading scheme within the EU that was implemented in 2005. At the same time there is speculation about possible U.S. interest in a regional climate arrangement that would include NAFTA members (Canada, Mexico, and the United States) and could be facilitated by the North American Commission on Environmental Cooperation. Given the possibility of at least two regional climate policy arrangements and the likelihood of more in the future as developing countries commit to reduce greenhouse gas emissions, the next phase of global cooperation for climate control could evolve into a number of regional alliances.

Our formal analysis is performed within the context of a simple dynamic game-theoretic model. We show that two agreements can sustain a larger number of cooperating parties than a global treaty even though the model assumes that the cost of reducing emissions is the same with both types of regime. The model provides upper and lower bounds on the number of parties under each regime. It is shown that a system with two agreements can Pareto-dominate a regime based on a global treaty. Thus, regional cooperation might be a good alternative to global IEAs for environmental problems like climate change.

The paper is organized as follows. Section 2 provides background information and reviews the existing literature on the conditions for an effective global IEA. Section 3 compares the conditions for a single global treaty to the conditions for two regional agreements in a simple dynamic model. Finally, Section 4 offers a summary and a discussion of the main results.

2. Background and literature review

There are IEAs designed to provide pure public goods on a global scale. For example, this is the case for the Kyoto protocol. Each country's emissions of greenhouse gases add to the world's atmospheric stock and a single country's efforts to control emissions will benefit all countries in a non-exclusive and non-rival manner. All parties thus have an incentive to free ride on other countries' efforts to reduce greenhouse gas emissions.

There are two types of incentive for free riding: the incentive for a country to not sign the agreement and thus benefit from the signatories' abatement efforts; and the incentive for a signatory to not comply with commitments made in an agreement. Because of these free-rider incentives, there will generally be suboptimal equilibrium coalition structures in global pollution control [22].

There is a considerable body of literature addressing under-provision of global international pollution control.[1] An important focus of this literature is the conditions leading to the formation of multilateral agreements or coalitions. Such agreements must be self-enforcing since there is no supranational authority to enforce compliance (e.g., [3,7]).

Models of international environmental cooperation differ with respect to the specification of the utility functions of governments and the stability concept they employ. However, they can roughly be divided into two groups — reduced-stage game models and dynamic game models [22].

Reduced-stage game models depict coalition formation as a two-stage game. In the first stage, countries decide on coalition formation. In the second stage, they choose abatement levels and how the gains from cooperation will be distributed (e.g., [18,14,23]).[2] A number of reduced-stage models have addressed the possibility of giving countries the freedom to negotiate more than one climate agreement [10–12,17]. A two-stage coalition game is used to show that when more than one coalition is possible, the equilibrium coalition structure that endogenously emerges from the negotiation process is characterized by several coalitions. It has also been shown that social welfare can be higher with multiple agreements than with a single global accord [12].

Dynamic game models typically assume an infinitely repeated game where governments agree on a contract in the first stage that has to be enforced in subsequent stages through credible threats (e.g., [3,5,7]).[3] The main problem analyzed by these models is that a country can — at least temporarily — achieve net gains by free riding or by joining a smaller coalition.

The literature on dynamic game models emphasizes that, to be effective, a treaty needs to specify a strategy that can enforce compliance. This entails that it must be in the (individual) best interest of each party to act in accordance with this strategy. Furthermore, such an equilibrium agreement must be renegotiation-proof. In particular, it must be in the (collective) best interest of other countries to insist that a non-compliant country be punished before cooperation can be resumed [5,7,22]. Main results from dynamic game models include that the allocation of abatement burdens crucially affects the success of international environmental agreements, that a grand coalition is unlikely to form, and that a sub-coalition may achieve more for its members than the grand coalition.

Barrett [6] uses a repeated game framework to demonstrate that a global treaty can be broadened to incorporate all countries (a consensus treaty), but at the cost of limiting the per-country level (the 'depth') of cooperation.[4] Below we show that regional agreements can enhance participation even without limiting the depth of cooperation. Our model largely follows Barrett [5], using Farrell and Maskin's [19] weak renegotiation-proof equilibrium as solution

[1] Notable examples are [3,5,7,13,14,23,24].

[2] For overviews of the coalition literature, see [10,20].

[3] Note that it is generally easier to sustain cooperation in infinitely repeated games than in finitely repeated games.

[4] Finus and Rundshagen [21] generalize Barrett's [3] model by introducing non-linear (quadratic) cost and benefit functions. In an appendix to this paper we consider how the impact of a regime based on regional agreements is influenced by introducing non-linear benefit functions.

concept.[5] The results support the conclusions reached by Carraro and others (e.g., [11,12,17]) using a reduced-stage game framework.

3. A simple dynamic model of regional climate agreements

Consider a world containing N countries and consisting of two regions (A and B). Both the countries and the regions are identical in all relevant characteristics. Assume that there are two options for an international regime to reduce emissions of greenhouse gases. Option 1 is a global agreement on emissions reductions and option 2 is a regime consisting of two separate agreements, one for each region.[6] We assume that, regardless of which regime is chosen, each country can choose between complying at cost c leading to a fixed reduction in emissions, and not complying.

In every period of the game, each country must choose to *cooperate* (i.e., reduce emissions) or to *defect* (not reduce emissions). Following [5], in the case of a *global agreement* where there are k participating countries and $N-k$ non-participating countries, the periodic payoff of each of the k countries playing *cooperate* is

$$dk - c,$$

where d is a constant ($d>0$), while each of the $N-k$ countries playing *defect* receives

$$bk,$$

where b is another constant ($b>0$). Similarly, with two *regional agreements*, the payoffs are

$$d\cdot(k_A+k_B)-c \text{ and } b\cdot(k_A+k_B)$$

for the k_A and k_B countries playing *cooperate* in regions A and B and the $N-k_A-k_B$ countries playing *defect*, respectively. Hence, the payoff obtained by each country if it chooses *cooperate* (*defect*) is a linear function of the total number of countries that cooperates. In an appendix we show how our analysis can be extended to the case where there are decreasing returns from cooperation. Like Barrett [5, p. 527] we assume that $d\geqslant b$. Throughout, the analysis handles the general case where $d\geqslant b$, except in Section 3.3 where the restriction $d=b$ is invoked. We also assume that $b(N-1)>dN-c>0$, meaning that full participation is *not* a Nash equilibrium of the stage game, and that full participation Pareto-dominates no participation. From the assumptions that $d\geqslant b$ and $b(N-1)>dN-c$ it follows that,

$$\text{for all } k\in\{1,\ldots,N\},\ b(k-1)>dk-c, \tag{1}$$

entailing that *defect* is dominant in the stage game; in particular, $c>d$. Hence, the stage game has a unique Nash equilibrium where all countries play *defect*. Furthermore, from the assumption that $dN-c>0$, it follows that this Nash equilibrium is inefficient, since all countries would gain if they all played *cooperate*. These properties entail that the stage game is a special version of the N-person Prisoners' Dilemma. We assume throughout that all countries discount future payoffs using a common discount factor δ ($0<\delta<1$).

[5]Weakly renegotiation-proofness is defined formally in Section 3.

[6]Since the countries are modeled as symmetrical players, the real question addressed by the analysis is whether two agreements are better than one. In the case of two agreements, we label them "regional" agreements.

We assume that compliance in the global agreement is sustained by way of the "Penance" strategy. This strategy specifies that a participating country plays *cooperate* except if another participating country has been the sole deviator from Penance in the previous period, in which case *defect* is played. Non-participating countries play *defect* after any history. Penance is closely related to the Getting Even strategy, which is used in Barrett's work.[7] It is well known that a global agreement based on Getting Even admits only a very limited number of parties (e.g., [5]).

In the case of two regional agreements, compliance is likewise sustained on the basis of a close relative to Penance. This strategy — which we refer to as "*Regional* Penance" – specifies that a participating country plays *cooperate* except if another participating country in its own region has been the sole deviator from Regional Penance in the previous period, in which case *defect* is played. As in the case of a global treaty, non-participating countries play *defect* after any history. It may be noted that if all countries use this strategy, then a deviation by a participating country in region A will be punished by other parties in region A, but not by parties to the agreement in region B, and vice versa. We show below that this is part of the equilibrium behavior.

Loosely formulated, an equilibrium is renegotiation-proof if the players as a group cannot find a better alternative [8, p. 43]. More precisely, in order to be a *weakly renegotiation-proof equilibrium* (in the sense of Farrell and Maskin [19 pp. 330–331]), a strategy profile must satisfy two requirements.[8]

(1) The strategy profile must be a subgame perfect equilibrium. Hence, every weakly renegotiation-proof equilibrium is subgame perfect (but the converse does not hold). In any repeated game with discounting it is necessary and sufficient for a strategy profile to be a subgame perfect equilibrium that no player can gain by a one-period deviation after some history. Hence, even though in the subsequent analysis we consider only the case where a deviator accepts the punishment without further deviation, we are not ruling out the possibility that a player may deviate for many periods. The point is that we need only check one-period deviations, as it follows from the theory of repeated games with discounting [1, p. 390] that a player cannot gain by a multi-period deviation if he cannot gain by some one-period deviation.

(2) Since by the first requirement the strategy profile is subgame perfect, it determines after any history a subgame perfect equilibrium — a *continuation equilibrium* — in the continuation of

[7]The Getting Even strategy that Barrett [5, p. 530] uses "... requires that country i play *cooperate* unless i has played *defect* less often than any of the other players in the past ...", and entails that a participating country is offered a future reward if it does not take part in the punishment of a deviating country. This feature does not seem essential, but leads to non-trivial calculations. The Penance strategy simplifies the Getting Even strategy by *not* letting such deviating behavior be rewarded. Likewise, specifying punishment after multilateral deviations, while certainly a real-life possibility, would in the model only complicate the analysis without changing the behavior along the equilibrium path. Like Barrett we do not seek to characterize the set of weakly renegotiation-proof equilibria. However, by analyzing the Regional Penance strategy defined below we show that the set of weakly renegotiation-proof equilibria extends beyond the kind of strategy profiles considered by Barrett [5]. A complete characterization of the set of weakly renegotiation-proof equilibria in the model is an interesting problem that we leave for future research.

[8]Various concepts of renegotiation-proof equilibrium exist in the literature (e.g., [9,19,2]; see [8] for a discussion). When we use that term in this article we always refer to the concept of weakly renegotiation-proof equilibrium, formalized for 2-person games by Farrell and Maskin [19], and extended to N-person games by Bergin and MacLeod [8, Definition 3].

the game. The second requirement is that there must not exist two continuation equilibria such that all players are better off in one continuation equilibrium than in the other. The reason is that if two such continuation equilibria existed, the players would renegotiate from one to the other. In the case of Penance and Regional Penance where all punishments last only one period, this requirement is satisfied if not all players strictly gain by choosing collectively to restart cooperation at once instead of implementing the threatened punishment when a deviation has taken place in the previous period.

In the following, we compare a global agreement to a regime based on two agreements with respect to each of these two requirements.

3.1. The subgame perfection requirement

Consider first the case of a global agreement. In this case there are two kinds of histories to check. In the first kind there is no single deviator in the previous round. Penance then prescribes that all participating countries play *cooperate*. As long as all participating countries follow this prescription, a participating country j collects a payoff of $dk - c$ in each period, where k is the number of parties in the global agreement. By contrast, suppose country j deviates from Penance in period t by playing *defect* and reverts to Penance at time $t+1$. It then gets a payoff of $b(k-1)$ in period t. In period $t+1$ it is punished by the other players and receives $d-c$. From time $t+2$ onwards cooperation is restored, meaning that country j obtains $dk - c$ in that and any future period. It is individually rational to stick to Penance unless country j strictly increases its sum of discounted payoffs in periods t and $t+1$ by deviating. Formally, this condition is satisfied if

$$(1+\delta)(dk-c) \geqslant b(k-1) + \delta(d-c).$$

Solving for k gives:

$$k \geqslant \frac{\delta d - b + c}{\delta d + d - b} = 1 + \frac{c-d}{\delta d + d - b}. \tag{2}$$

If condition (2) is fulfilled, then the strategy profile where all countries play Penance is a Nash equilibrium.

The second kind of histories to check is the one where a single deviation has taken place in period $t-1$. It follows from the definition of Penance that one of the $k-1$ participating countries that did not deviate in period $t-1$, say country i, would deviate from Penance in period t if it plays *cooperate*. However, such a deviation leads to a loss for country i in period t since it follows from (1) that $b > d2 - c$. It also leads to a loss in period $t+1$ since punishment will be triggered. Hence, it suffices to check that the country that deviated in period $t-1$ has no incentive to deviate from Penance again in period t by playing *defect*. If this country accepts the punishment and reverts to Penance at time t, then it receives $d-c$ in period t and $dk-c$ in any future period. By contrast, if it deviates for one more period before reverting to Penance, then it receives 0 at time t and $d-c$ at time $t+1$. Thus, it is individually rational for the deviating country to accept the punishment at time t, provided that

$$d - c + \delta(dk-c) \geqslant 0 + \delta(d-c).$$

Solving for k gives

$$k \geqslant \frac{\delta d - d + c}{\delta d} = 1 + \frac{c-d}{\delta d}. \tag{3}$$

Note that $(c-d)/\delta d \geqslant (c-d)/(\delta d + d - b)$ since we assume that $d \geqslant b$ and, by (1), $c > d$, entailing that condition (3) implies condition (2). Hence, (3) is a sufficient and necessary condition for the strategy profile where all countries play Penance to be a subgame-perfect equilibrium. This condition places a lower bound on the number of parties needed to make an agreement sustainable.[9]

Next, consider a regime with two regional agreements. Once again, there are two kinds of histories to check in order to ensure subgame perfection. In the first kind there is no single deviation in (only) one of the regions in the previous period. If a participating country j in region A deviates from Regional Penance in period t by playing *defect* and reverts to Regional Penance at time $t+1$, then in period t it collects a payoff of $b(k_A - 1 + k_B)$, where k_S is the number of parties in the agreement for region S ($S = A, B$). In period $t+1$, the deviating party pays penance by being the only country in region A to play *cooperate*, whereas cooperation continues as before in region B. Country j thus obtains a payoff of $d(1+k_B) - c$. From time $t+2$ onwards cooperation is restored, so j receives $d(k_A + k_B) - c$ in that and in any future period. Had country j not deviated, it would have received $d(k_A + k_B) - c$ in every period. Therefore, it is individually rational to stick to Regional Penance, provided that

$$(1+\delta)(d(k_A + k_B) - c) \geqslant b(k_A - 1 + k_B) + \delta(d(1+k_B) - c). \tag{4}$$

We concentrate on equilibria with the same number of parties in each of the two agreements. For the analysis of weakly renegotiation-proof equilibria, this is without loss of generality, as reported in footnote 10. Setting $k_S = k_A = k_B$ and solving (4) for k_S gives:

$$k_S \geqslant \frac{\delta d - b + c}{\delta d + 2(d-b)} = 1 + \frac{c - d - (d-b)}{\delta d + 2(d-b)}. \tag{5}$$

Condition (5) ensures that the strategy profile where all countries use Regional Penance is a Nash equilibrium.

The second kind of histories to check is the one where a single deviation has occurred in (only) one of the two regions in the previous period. Suppose the deviation has taken place in region A in period $t-1$. It follows from the definition of Regional Penance that one of the $k_A - 1$ participating countries that did not deviate in period t, say country i, would deviate from Regional Penance in period t if it plays *cooperate*. However, such a deviation leads to a loss for country i in both period t since, by (1), $b(1+k_B) > d(2+k_B) - c$, and in period $t+1$ since punishment will be triggered. Hence, it suffices to check that the country in region A that deviated in period $t-1$ as well as all the k_B participating countries in region B have no incentive to deviate from Regional Penance in period t by playing *defect*.

Consider first the country in region A that deviated from Regional Penance in period $t-1$. If this country accepts the punishment and reverts to Regional Penance at time t, then it receives $d(1+k_B) - c$ in period t, and $d(k_A + k_B) - c$ in any future period. By contrast, if it deviates also

[9] This is one possible explanation why most multilateral treaties (including the Kyoto protocol) provide a clause stating that their entry into force requires ratification by a certain number of countries.

at time t and reverts to Regional Penance at time $t+1$, then it receives bk_B at time t and $d(1+k_B)-c$ at time $t+1$, before cooperation is restored. Hence, it is individually rational for the deviating country to accept the punishment at time t, provided that

$$d(1+k_B)-c+\delta(d(k_A+k_B)-c)\geqslant bk_B+\delta(d(1+k_B)-c). \tag{6}$$

Setting $k_S=k_A=k_B$ and solving for k_S gives

$$k_S\geqslant\frac{\delta d-d+c}{\delta d+d-b}=1+\frac{c-d-(d-b)}{\delta d+d-b}. \tag{7}$$

Turn now to the k_B participating countries in region B, given that a deviation in region A has occurred in period $t-1$. If all countries use Regional Penance from period t, then a participating country j in region B receives $d(1+k_B)-c$ in period t, and $d(k_A+k_B)-c$ in period $t+1$. By contrast, if country j deviates from Regional Penance in period t by playing *defect* and reverts to Regional Penance at time $t+1$, then it receives $b(1+k_B-1)$ at time t and $d(k_A+1)-c$ at time $t+1$. In either case, cooperation is reestablished in both regions in period $t+2$. Thus, country j has no incentive to deviate from Regional Penance if

$$d(1+k_B)-c+\delta(d(k_A+k_B)-c)\geqslant bk_B+\delta(d(k_A+1)-c).$$

Setting $k_S=k_A=k_B$ and solving for k_S yields (7).

Note that $(c-d-(d-b))/(\delta d+d-b)\geqslant(c-d-(d-b))/(\delta d+2(d-b))$ since we assume that $d\geqslant b$ and, by (1), $b>d2-c$, entailing that condition (7) implies condition (5). Hence, (7) is a sufficient and necessary condition for the strategy profile where all countries play Regional Penance to be a subgame-perfect equilibrium. This condition places a lower bound on the number of parties in a regime based on two agreements.

3.2. *The renegotiation-proofness requirement*

The requirement of renegotiation-proofness is satisfied in the present setting if not all players strictly gain by choosing collectively to restart cooperation at once instead of implementing the threatened punishment when a deviation has taken place in the previous period. Hence, at least one country must be at least as well off with punishment as with renegotiation. It is *not* necessary that all countries but the one being punished are at least as well off with punishment as with renegotiation.

Consider the global agreement, and assume that a participating country j deviates in period $t-1$. Punishment implies that all participating countries but the deviator switch from playing *cooperate* to playing *defect*. This implies that not only the deviator, but all non-participating countries are worse off with punishment than with renegotiation. Weak renegotiation-proofness thus requires that the punishing participating countries are at least as well off with punishment as with renegotiation, because otherwise all countries would gain by restarting cooperation at once. If the punishment is carried out in accordance with Penance, then each of the punishing participating countries receives a payoff of b in period t. By contrast, if cooperation is restarted immediately, then each of them receives $dk-c$ in period t. In either case, each participating country receives $dk-c$ in every future period, so only period t needs to be considered. Therefore, weak renegotiation-proofness requires that

$$b\geqslant dk-c,$$

which can be restated as

$$k \leqslant \frac{b+c}{d}. \tag{8}$$

Thus, weak renegotiation proofness requires that the number of participating countries not be too large. The intuition behind this result is that, other things being equal, a larger number of participating countries increases the payoff associated with cooperation. A large number of participating countries thus makes renegotiation attractive, thereby undermining the (collective) incentive to punish a deviation.

Next, consider regional agreements. Assume that country j is a party to the agreement in region A, and that this country has defected in period $t-1$. If the punishment is carried out in accordance with Regional Penance, then each of the punishing participating countries in region A receives $b(1+k_B)$ in period t. By contrast, if cooperation is restarted at once, each of them receives $d(k_A+k_B)-c$ in period t. Thus, weak renegotiation-proofness requires that

$$b(1+k_B) \geqslant d(k_A+k_B)-c.$$

With the same participation in both regions, this condition reduces to

$$k_S \leqslant \frac{b+c}{2d-b}. \tag{9}$$

3.3. *Summary of main results with numerical illustrations*

While if follows from conditions (3) and (7) that the subgame perfection requirement places a lower bound on the number of participating countries, by conditions (8) and (9) the renegotiation-proofness requirement places an upper bound on this number. Combining the subgame perfection and renegotiation proofness requirements, we find that a global treaty modeled by the Penance strategy profile must satisfy

$$\frac{\delta d - d + c}{\delta d} \leqslant k \leqslant \frac{b+c}{d}. \tag{10}$$

It follows from condition (10) that the strategy profile where all countries play Penance is a weakly renegotiation-proof equilibrium for some δ close (but not equal) to 1 if and only if

$$\frac{c}{d} < k \leqslant \frac{b}{d} + \frac{c}{d}. \tag{11}$$

Because $d \geqslant b$, there is at most one integer k^* satisfying (11). Hence, whenever a global agreement exists, the number of parties is always fully determined. Moreover, as the stage game is a Prisoners' Dilemma, so that $dN - c > 0$, there exists a unique integer $k^* \leqslant N$ satisfying (11) if $d = b$ (since then the right hand side of (11) becomes $1 + c/d$). This establishes that a global treaty can always be implemented as a weakly renegotiation-proof equilibrium provided that $d = b$ and the countries are sufficiently patient.

Similarly, two symmetrical regional agreements modeled by Regional Penance must satisfy

$$\frac{\delta d - d + c}{\delta d + d - b} \leqslant k_S \leqslant \frac{b+c}{2d-b}. \tag{12}$$

Condition (12) implies that the strategy profile where all countries play Regional Penance is a weakly renegotiation-proof equilibrium for some δ close (but not equal) to 1 if and only if

$$\frac{c}{2d-b} < k_S \leqslant \frac{b}{2d-b} + \frac{c}{2d-b}. \tag{13}$$

Because $d \geqslant b$, there is at most one integer k_S^* satisfying (13). Hence, whenever a regime based on two symmetrical regional agreements exists, the number of parties is always fully determined. Moreover, if $d=b$ and $N \geqslant 2(1+c/d)$, then there exists a unique integer k_S^* satisfying (13) and $k_S^*+k_S^* \leqslant N$ (since with $d=b$ (13) becomes $c/d < k_S \leqslant 1 + c/d$). Hence, with enough countries, two symmetrical agreements can always be implemented as a weakly renegotiation-proof equilibrium provided that $d=b$ and the countries are sufficiently patient.[10]

We thus see in both regimes that the renegotiation-proofness requirement in combination with the subgame perfection requirement determines the number of parties, thus leaving a very narrow set of possibilities. If $k^* < N$ in the case of a global treaty, or if $k_S^* + k_S^* < N$ in the case of two symmetrical agreements, then there exist multiple equilibria: even though condition (11) determines the *number* of participating countries in a global agreement, and condition (13) the *number* of participating countries in the two regional agreements, neither condition determines *which* countries participate.

It follows from conditions (10) and (12) that, for $d=b$, the lower and upper bounds on each regional agreement are identical to the corresponding bounds on the global agreement. In this case the number of participating countries required and admitted by the regional agreements regime is twice that of a global agreement. To illustrate, assume that $\delta = 0.95$, $c=8$, and $d=b=1$. We then find that a global regime requires at least nine participating countries. At the same time, it admits a maximum of nine countries as well. With two agreements, by contrast, the minimum and maximum number of parties is nine in each region, meaning that the total number of participating countries is 18.

If $d>b$, then the difference between the two types of regime can be smaller. For example, if $\delta = 0.95$, $c=8$, $d=1.25$, and $b=1$, then the minimum and maximum number of parties with regional agreements becomes 12 (six in each region). By contrast, with a global agreement the minimum and maximum number of parties is seven.

[10]If we had done the analysis *without* setting $k_S = k_A = k_B$, then, instead of (13), the following inequalities would have been necessary conditions:

$$\frac{c-(d-b)k_B}{d} < k_A \leqslant \frac{b}{d} + \frac{c-(d-b)k_B}{d}, \tag{14}$$

$$\frac{c-(d-b)k_A}{d} < k_B \leqslant \frac{b}{d} + \frac{c-(d-b)k_A}{d}. \tag{15}$$

However, the right inequality of (14) contradicts the left inequality of (15) if $k_A > k_B$ (and vice versa if $k_A < k_B$). It follows that two *asymmetric* agreements cannot be implemented as a weakly renegotiation-proof equilibrium. This means that we can set $k_S = k_A = k_B$ without loss of generality.

3.4. Regional agreements and Pareto dominance

The model analyzed in this paper suggests that a regime based on regional agreements can be beneficial to all countries. A global agreement with k^* participating countries is (strongly) Pareto dominated by a situation where participation is extended to n countries (where $k^* < n \leqslant N$) if this makes every country better off. It follows immediately that the k^* countries that play *cooperate* both before and after such an enlargement gain by increased participation. The same is true for the $N - n$ countries that play *defect* both before and after the enlargement. The reason is simply that, from the point of view of these countries, the enlargement increases the number of *other* countries that play *cooperate*. The question is therefore whether the $n - k^*$ countries that switch from *defect* (and receiving a periodic payoff of bk^*) to *cooperate* (and obtaining $dn - c$) gain by doing so. Hence, a global treaty with k^* participating countries is Pareto dominated by an enlargement to n members if

$$dn - c > bk^* .$$

Below we compare the following two situations,

- a global regime where k^* countries cooperate,
- a regional regime where $k_A^* > 1$ countries form a new agreement in region A, and the $k_B^* = k^*$ members of the global agreement constitute another agreement in region B,

and show that the former is always Pareto dominated by the latter.

As argued above, the k_B^* countries in region B included as parties under both regimes, as well as the $N - k_A^* - k_B^*$ countries that are non-parties in both cases, prefer the regional regime. Hence, the real test is whether the regional regime is acceptable to the k_A^* members of the additional agreement, i.e., to countries that are allowed to free ride in the first situation, but required to cooperate in the second. With a global agreement, each of these countries receives a periodic payoff of bk^*. With two regional agreements, each of them obtains $d(k_A^* + k_B^*) - c$. Hence, if

$$d(k_A^* + k_B^*) - c > bk^* , \tag{16}$$

then each of these countries is better off being a party to the new regional agreement than as a free rider on a global agreement. It follows that, if condition (16) is fulfilled, a regime based on regional agreements Pareto dominates a global agreement. We show below, through a verbal argument and an analytical demonstration, that this condition is always fulfilled.

For a verbal argument, recall that it is feasible for any participating country j in the additional agreement in region A to choose a perpetual deviation by always playing *defect*. If country j starts playing *defect* in period t, then all other countries in region A will also play *defect* from period $t+1$, thereby effectively terminating the additional agreement. Such a perpetual deviation by country j from the agreement in region A would give country j a higher payoff than it would obtain with the global agreement, since it benefits from the cooperation of the $k_A^* - 1$ other countries in region A in period t, and from the cooperation of the $k_B^* = k^*$ participating countries in region B in all periods. However, it follows from the subgame perfection requirement that such a deviation is not profitable, implying that a perpetual play of *cooperate* in the new agreement in region A is even more advantageous. Hence, whenever the subgame perfection requirement is fulfilled, a single global agreement is Pareto dominated by a regional regime where a new regional agreement is introduced.

For an analytical demonstration, multiplying (4) by $1-\delta$ and rearranging terms gives

$$d(k_A^*+k_B^*)-c \geqslant (1-\delta)b(k_A^*-1+k_B^*)+(1-\delta)\delta\big(d(1+k_B^*)-c\big) + \delta^2\big(d(k_A^*+k_B^*)-c\big). \tag{17}$$

By rearranging (6), we get

$$(1-\delta)\big(d(1+k_B^*)-c\big)+\delta\big(d(k_A^*+k_B^*)-c\big)\geqslant bk_B^*. \tag{18}$$

Combining inequalities (17) and (18) yields

$$d(k_A^*+k_B^*)-c\geqslant(1-\delta)b(k_A^*-1+k_B^*)+\delta bk_B^*,$$

which can be rewritten as

$$d(k_A^*+k_B^*)-c\geqslant bk_B^*+(1-\delta)b(k_A^*-1).$$

Condition (16) follows with *strict* inequality since $k_B^*=k^*$, $k_A^*>1$, and $\delta<1$. Hence, even the participating countries in region A *strictly* gain by forming the additional agreement.

4. Summary and discussion

This paper has used a simple dynamic game-theoretic model to analyze how a regime based on two agreements would affect the prospects for international environmental cooperation. A main finding is that two treaties are able to encompass a larger number of parties than a single global treaty, assuming that the cost of cooperation is the same with both types of regime, and that all countries are identical in all relevant characteristics. We have also shown that a system with two agreements can Pareto dominate a regime based on a global treaty. These results support the findings of previous research by Carraro and others (e.g., [11,12,17]) using a reduced-stage game framework.

We end this paper by commenting briefly on three issues that relate to the analysis in Section 3. First, while the model analyzed in this paper includes only two regions, one might reasonably ask why one should stop there. Why not design a regime with an even larger number of agreements? This idea certainly cannot be dismissed on the basis of the above model. We have seen, however, that each regional agreement requires a certain minimum number of parties. Hence, with a fixed number of countries to select from there is an upper limit to the number of agreements that can be sustained in equilibrium.

Second, an important driving force in Barrett's [5] model of a global treaty, as well as in our model of regional cooperation, is the way in which the agreement(s) are enforced. The maximum number of signatories is very limited in Barrett's model because a single violation triggers punishment by *all* other parties. This drives the number of participating countries down via the renegotiation-proofness requirement. The reason is that for each participating country the payoff obtained when all participating countries play *cooperate* must not exceed the payoff of playing *defect* when only the country being punished plays *cooperate*. Because $d>0$, this requirement can only be satisfied if (very) few countries participate. If *not all* participating countries punish a deviator, then more countries will cooperate in the punishment phase. This means that, in contrast to Barrett's model, the renegotiation proofness requirement becomes less strict and admits a larger number of participating countries. Regional cooperation means that only some of the participating countries punish a deviator: even if a violation in region A is punished by all parties to the agreement for region A, cooperation in region B continues undisturbed. It is this feature that allows the number of participating countries to become larger with a regime based on regional agreements.

Finally, in addition to the regional agreements approach highlighted in this paper, there are also other avenues for broadening the number of parties that can be sustained in weakly renegotiation-proof equilibrium. For example, the maximum number of parties in a global agreement might be increased by designing an enforcement scheme whereby a single defection triggers punishment only by a subset of the other parties.[11] Indeed, even a global agreement could — in principle — be enforced regionally, so the effects analyzed here could also be achieved within the setting of a global agreement. However, it seems rather implausible that a system of regional enforcement would be adopted except as part of a regime based on regional agreements. Moreover, informal arguments suggest that a regime based on regional agreements is likely to have other advantages as well. Countries that are in close geographic proximity also tend to be culturally close, have similar economic and political systems, and therefore have similar preferences.[12] These factors might reduce uncertainty, enhance trust, and reduce negotiation costs. Countries may thus not only be more likely to comply with a regional agreement, but also more inclined to join a regional agreement in the first place. This further strengthens the conclusions of the formal analysis in Section 3, where there are no cultural, economic, or political ties between any subsets of countries. Thus, regional cooperation might be a good alternative — or supplement — to global IEAs like the Kyoto Protocol.

Acknowledgments

We are indebted to two anonymous referees of this journal for detailed comments on previous drafts. We are also grateful to Scott Barrett, Leif Helland, Aanund Hylland, Ottar Mæstad, Kevin Siqueira, Jon Vislie, and colleagues at CICERO, in particular Odd Godal and Asbjørn Torvanger, for helpful discussions and suggestions. Finally, we thank the Research Council of Norway for financial support, Lynn Nygaard for editorial assistance, and Frode Rørvik for computer support.

[11]In fact, in the appendix to [5] this type of regime is examined, but rejected on the grounds that it fails to be strongly collectively rational.

[12]Of course, this is not always the case. In the Middle East, for instance, some neighboring countries obviously have very different preferences on certain issues.

Appendix A. Decreasing returns from cooperation

Throughout this paper, it has been assumed that the utility obtained by each country if it chooses to *cooperate* (*defect*) is a linear function of the total number of countries that cooperate. What happens if there are decreasing returns from cooperation?

Let $f(k)$ denote the gain accruing to each country if k is the total number of countries cooperating, when compared to the situation where no country cooperates. Hence, in a global agreement, the payoff of each of the k countries playing *cooperate* is

$$f(k) - c,$$

while each of the $N - k$ countries playing *defect* receives

$$f(k).$$

Likewise, with two regional agreements, the payoffs are

$$f(k_A + k_B) - c \text{ and } f(k_A + k_B)$$

for the k_A and k_B countries playing *cooperate* in regions A and B and the $N - k_A - k_B$ countries playing *defect*, respectively. This means that the setting of the main text is generalized by not imposing a linear relationship between the gains from cooperation and the number of cooperating countries. On the other hand, the gain accruing to a country depends now only on the total number of cooperating countries, not also on whether the country itself plays *cooperate* (as in the main text when $d > b$).

The function $f(\cdot)$ is assumed to satisfy

$$f(0) = 0 \text{ and for all } k \geqslant 1,\ 0 < f(k+1) - f(k) \leqslant f(k) - f(k-1), \tag{A.1}$$

meaning that the gain from an additional country cooperating is a non increasing function of the number of countries cooperating. We refer to this as non-increasing returns from cooperation. If $d = b$, then the analysis of the main text is the special case where $f(k) = dk = bk$. Moreover, consider the following assumptions:

$$f(1) < c, \tag{A.2}$$

$$f(N) - f(1) > c, \tag{A.3}$$

$$\text{There exists an even integer } N^* \leqslant N \text{ s.t. } f(N^*) - f(1 + N^*/2) > c. \tag{A.4}$$

It is straightforward to check that (A.1)–(A.3) ensure that the stage game is a Prisoners' Dilemma, while (A.1) and (A.4) imply (A.3).

Assume (A.1)–(A.3). Consider a global treaty based on Penance with k participating countries. It follows from the corresponding analysis in the main text that subgame perfection requires that

$$f(k-1) - f(k) + c \leqslant \delta(f(k) - f(1)) \text{ and } -f(1) + c \leqslant \delta(f(k) - f(1)),$$

where, by (A.1), the first inequality implies the second one. Hence, if

$$2f(k) - f(k-1) > f(1) + c, \tag{A.5}$$

there exists $\delta < 1$ such that the strategy profile where all countries play Penance is a subgame perfect equilibrium. On the other hand, for this strategy profile to be a weakly renegotiation-proof equilibrium, it must also hold that $f(1) \geqslant f(k) - c$, or equivalently,

$$f(k) \leqslant f(1) + c. \tag{A.6}$$

To see that there exists some k^* satisfying both (A.5) and (A.6), note that (A.3) implies that $f(k+1) \leqslant f(1) + c$ for $k = 0$ and $f(k+1) > f(1) + c$ for $k = N - 1$. Let k^* be the smallest integer satisfying $f(k^* + 1) > f(1) + c$, entailing that (A.6) is satisfied for $k = k^*$. It follows that (A.5) is also satisfied for $k = k^*$ since, by (A.1),

$$2f(k^*) - f(k^* - 1) \geqslant f(k^* + 1).$$

Furthermore, since $f(k)$ is strictly increasing, we have that k^* determined in this way is the largest k for which (A.5) and (A.6) are satisfied.

Assume now the stronger set of assumptions (A.1), (A.2), and (A.4). In Regional Penance with k_S participating countries in each agreement, it follows from the corresponding analysis in the main text that subgame perfection requires that

$$f(2k_S - 1) - f(2k_S) + c \leqslant \delta(f(2k_S) - f(1 + k_S)) \text{ and}$$
$$f(k_S) - f(k_S + 1) + c \leqslant \delta(f(2k_S) - f(1 + k_S)),$$

where, by (A.1), the first inequality implies the second one. Hence, if

$$2f(2k_S) - f(2k_S - 1) > f(1 + k_S) + c, \tag{A.7}$$

there exists $\delta < 1$ such that the strategy profile where all countries play Regional Penance is a subgame perfect equilibrium. On the other hand, for this strategy profile to be a weakly renegotiation-proof equilibrium, it must also hold that $f(1 + k_S) \geqslant f(2k_S) - c$, or equivalently,

$$f(2k_S) \leqslant f(1 + k_S) + c. \tag{A.8}$$

To see that there exists some k_S^* satisfying both (A.7) and (A.8), note that (A.1) and (A.4) imply that, if $0 < k_B \leqslant N^*/2$, then $f(k_A + 1 + k_B) \leqslant f(1 + k_B) + c$ for $k_A = 0$ and $f(k_A + 1 + k_B) > f(1 + k_B) + c$ for $k_A = N^*/2 - 1$. Let $\kappa(k_B)$ be the smallest integer satisfying $f(\kappa(k_B) + 1 + k_B) > f(1 + k_B) + c$, entailing that

$$f(\kappa(k_B) + k_B) \leqslant f(1 + k_B) + c. \tag{A.9}$$

It follows that

$$2f(\kappa(k_B) + k_B) - f(\kappa(k_B) - 1 + k_B) > f(1 + k_B) + c \tag{A.10}$$

also holds since $2f(\kappa(k_B) + k_B) - f(\kappa(k_B) - 1 + k_B) \geqslant f(\kappa(k_B) + 1 + k_B)$ by (A.1). By (A.1) and the definition of $\kappa(\cdot)$ we have that

$$0 < \kappa(k_B) \leqslant \kappa(k_B + 1) < N^*/2 \quad \text{if } 0 \leqslant k_B \leqslant N^*/2 - 1. \tag{A.11}$$

Consider the sequence $\{k^0, k^1, k^2, \ldots\}$ defined by $k^0 = 0$ and, for all $n \geqslant 1$, $k^n = \kappa(k^{n-1})$. By (A.10), $\{k^0, k^1, k^2, \ldots\}$ is a non-decreasing sequence that converges to k_S^* satisfying

$$0 < k_S^* = \kappa(k_S^*) < N^*/2. \tag{A.12}$$

Now, (A.9), (A.10), and (A.12) imply that (A.7) and (A.8) are satisfied for $k_S = k_S^*$.

To compare the maximum number of participating countries in a single global agreement vs. two regional agreements, assume (A.1), (A.2), and (A.4) (which, as noted above, imply (A.3)). It follows from the definition of $\kappa(\cdot)$ that the maximum number of participating countries under a global agreement is given by $k^* = \kappa(0)$, while the regional agreements regime admits $k_S^* = \kappa(k_S^*)$ participating countries in each region. Since, by (A.11), $k^* = \kappa(0) \leqslant k_S^* = \kappa(k_S^*)$, we reach the conclusion that *two regional agreements allow for at least twice as many participating countries as compared to a single global agreement*, when there are non-increasing gains from cooperation.

The intuition for this finding is as follows. With decreasing gains from cooperation, the temptation for a given number of punishing countries to restart cooperation at once after a deviation has occurred, is smaller if cooperation continues unaffected in a separate regional agreement. Therefore, participation in *each* of two regional agreements can be *strictly larger* than the maximal participation in a single global agreement.

A numerical example can be used to illustrate this possibility. It can be checked that (A.1), (A.2), and (A.4) are satisfied with $c = 1.01$, $f(k) = \sqrt{k}$, and $N = 18$. Then $k = 4$ is the largest number of participating countries satisfying (A.5) and (A.6), while the sequence $\{k^0, k^1, k^2, \ldots\}$, defined by $k^0 = 0$ and, for all $n \geqslant 1$, $k^n = \kappa(k^n - 1)$, equals $\{0, 4, 6, 7, 7, \ldots\}$. Hence, (A.7) and (A.8) are satisfied with $k_S = 7$ countries in each of the two regional agreements. Actually, $k_S = 8$ is the largest number of participating countries satisfying (A.7) and (A.8); hence, our algorithm in the case of Regional Penance does not necessarily converge to the maximum number of participating countries, but suffices to show that maximum total participation at least doubles when moving from a single global treaty to two regional agreements.

References

[1] D. Abreu, On the theory of infinitely repeated games with discounting, Econometrica 56 (1988) 383–396.
[2] G.B. Asheim, Extending renegotiation-proofness to infinite horizon games, Games Econ. Behav. 3 (1991) 278–294.
[3] S. Barrett, Self-enforcing international environmental agreements, Oxford Econ. Pap. 46 (1994) 878–894.
[5] S. Barrett, A theory of full international cooperation, J. Theoret. Polit. 11 (1999) 519–541.
[6] S. Barrett, Consensus treaties, J. Institut. Theoret. Econ. 158 (2002) 529–547.
[7] S. Barrett, Environment & Statecraft: The Strategy of Environmental Treaty-Making, Oxford University Press, New York, 2003.
[8] J. Bergin, B. MacLeod, Efficiency and renegotiation in repeated games, J. Econ. Theory 61 (1993) 42–73.
[9] B.D. Bernheim, D. Ray, Collective dynamic consistency in repeated games, Games Econ. Behav. 1 (1989) 295–327.
[10] F. Bloch, Non-cooperative models of coalition formation in games with spillovers, in: C. Carraro, D. Siniscalco, (Eds.), New Directions in the Economic Theory of the Environment, Cambridge University Press, Cambridge, UK, 1997.
[11] C. Carraro, The structure of international agreements on climate change, in: C. Carraro, (Ed.), International Environmental Agreements on Climate Change, Kluwer Academic Publishers, Dordrecht 1999.
[12] C. Carraro, The economics of coalition formation, in: J. Gupta, M. Grubb, (Eds.), Climate Change and European Leadership, Kluwer Academic Publishers 2000.
[13] C. Carraro, D. Siniscalco, The international dimension of the environmental policy, European Econ. Rev. 26 (1992) 379–387.
[14] C. Carraro, D. Siniscalco, Strategies for the international protection of the environment, J. Public Econ. 52 (1993) 309–328.
[17] C. Carraro, D. Siniscalco, International environmental agreements: Incentives and political economy, Europ. Econ. Rev. 42 (1998) 561–572.
[18] P. Chander, H. Tulkens, Theoretical foundations of negotiations and cost sharing in transfrontier pollution problems, Europ. Econ. Rev. 36 (1992) 388–398.
[19] J. Farrell, E. Maskin, Renegotiation in repeated games, Games Econ. Behav. 1 (1989) 327–60.
[20] M. Finus, Game Theory and International Environmental Cooperation, E. Elgar, Cheltenham 2001.

[21] M. Finus, B. Rundshagen, Renegotiation-proof equilibria in a global emission game when players are impatient, Environ. Resource Econ. 12 (1998) 275–306.
[22] M. Finus, B. Rundshagen, Endogenous coalition formation in global pollution control: A partition function approach, in: C. Carraro, (Ed.), Endogenous Formation of Economic Coalitions, E. Elgar, Cheltenham, 2003.
[23] M. Hoel, International environmental conventions: the case of uniform reductions of emissions, Environ. Resource Econ. 2 (1992) 141–159.
[24] H. Tulkens, An economic model of international negotiations relating to transfrontier pollution, in: K. Krippendorff, (Ed.), Communication and Control in Society, Gordon & Breach, New York, 1979.

Part VII
Uncertainty, Risk and Catastrophic Events

 Oxford Economic Papers 50 (1998), 173–185 173

International environmental agreements under uncertainty

By Seong-lin Na* and Hyun Song Shin†

* Department of Economics, Hanyang University
† Nuffield College, Oxford, OX1 1NF

Negotiations toward international environmental agreements take place under uncertainty. We address the process of coalition formation in this context. Coalitions are more likely to form among countries which are similar. Since countries are more likely to be facing similar conditions *ex ante* rather than *ex post*, (i.e. before the resolution of uncertainty rather than after it), the possibility of coalition formation is enhanced the sooner the negotiations take place. The social value of better scientific information may well be negative in such circumstances.

1. Introduction

Environmental issues such as climate change, the depletion of the ozone layer, acid rain, deforestation, and biological diversity have come to the forefront of public concern in recent years, and have focused the attention of policy makers. Environmental issues have an intrinsically trans-national, or global dimension, in which the actions of one party will affect the welfare of other interested parties. However, in the absence of a supra-national body which can issue and enforce directives on environmental policy, the potential gains from policy coordination are limited by the willing participation of diverse, and self-interested sovereign countries.

The emergence of international environmental agreements as a means towards achieving this coordination can be seen in this light. Although such agreements have historical precedents, their role has become much more prominent in the last decade or so, in particular with the Montreal Protocol on the ozone layer (and its subsequent amendments strengthening its provisions), and the Rio Convention on climate change. The incentives underlying international environmental agreements have been analysed in terms of coalition formation among self-interested players using game-theoretic tools. Examples of recent papers include Barrett (1994), Carraro and Siniscalco (1993, 1994), Chander and Tulkens (1992), Heal (1994), and Sandler (1996). The starting point of these papers is to postulate a set of payoff functions for individual countries motivated by the externalities inherent in global environmental policy, and then to analyse the payoffs which accrue to groups of countries which act to maximize the coalition's payoff. The incentives to join a coalition are then analysed in terms of the difference in the payoff to an individual country inside and outside the coalition.

Although the existing literature has generated numerous insights into the problem of international environmental agreements, it has been limited in a number of respects. First, the analysis has largely been conducted under the assumption of symmetry of the interested parties. This may not be the most promising approach to examining the strategic incentives generated by differential gains and losses across jurisdictions. A related feature is perfect information. This runs counter to the acknowledged uncertainties underlying global environmental issues and the disparitities between countries.

In some instances, uncertainty may not be an overriding concern. In this respect, it is illuminating to distinguish the different types of environmental externality.[1] For acid rain and deforestation, the problem is largely one of coordinating the actions of an interlocking set of local parties. The curbing of deforestation, for instance, yields private goods, country-specific public goods, and global public goods (Sandler, 1993), and the bargaining takes account of this mix. For such problems, any uncertainty there is plays a relatively small role, and the focus of the bargaining is on the sharing of the potential surplus to coordination. In the case of ozone depletion, it could be argued that much of the uncertainty concerning its likely effects had been resolved before the signing of the Montreal Convention. Indeed, Murdoch and Sandler (1997) argue that the support for the Montreal Convention was due to the fact that it codified reductions in CFC emissions that polluters were voluntarily prepared to accomplish as the scientific evidence came in. In this case, also, uncertainty seems to have played only a marginal role.[2]

However, the same is not true with regard to the greenhouse effect and the related question of global climate change.[3] There are two dimensions to the uncertainties associated with global warming. First, at an aggregate level, there is uncertainty concerning the broad impact of increased carbon dioxide emissions on the increase of average temperature over the globe and the likely effects thereof. Second, and more importantly, there is uncertainty concerning the distribution of the damage caused by carbon dioxide emissions as a result of localized climate change, and hence uncertainty concerning the distribution of the benefit to curtailing carbon dioxide emissions.

At an aggregate level, several estimates have been reported of the global economic costs of the greenhouse effect, for example, Nordhaus (1991), Cline (1992), and Fankhauser (1993). However, the differential impact of global warming across geographical regions of the world is less well understood. Climatologists raise the spectre of shifting currents which may turn Europe into Alaska, of the drying up of grain belts into deserts, of great rivers drying up as a result of the disappearance of

[1] We are grateful to the referee for pointing out the important differences in these cases.

[2] One qualification here has to do with mixed strategy equilibria. Sandler and Sargent (1995) show that mixed strategies may play a role in negotiations and, as such, will generate uncertainties for the players.

[3] Morrisette *et al.* (1990) consider some of the lessons from previous agreements for negotiations on global warming.

ice packs, and of the northward migration of old and new tropical pests and diseases. In short, the likelihood of catastrophic surprises due to global warming is much harder to predict than the mere increase in average temperatures.[4] For the British Isles, and northern Europe in general, global warming may not translate into a general increase in temperature, since the current climate is dependent on the prevailing currents, and such currents may be disrupted by global warming. Formal models of climate give some indication of the effects, but they are subject to the usual limitations. Thus, even if a consensus emerged as to the average impact of global warming, the uncertainty concerning regional variations is unlikely to dissipate, and this will affect the incentives of countries to join an environmental coalition in subtle and complex ways.

The main contribution of our paper is to re-examine the issue of international environmental agreements in a context (such as global warming) in which there is uncertainty concerning the distribution of the benefits of pollution abatement activity, and when countries are not identical. We analyse a game of coalition formation among three countries. The payoffs of all three countries are increasing in the aggregate pollution abatement, but the marginal benefit of aggregate abatement to an individual country is uncertain.

Within this framework, it is possible to address two important issues. First, when disparate countries are negotiating to form coalitions, what determines the membership of such coalitions? Second, what is the value (if any) of delaying environmental negotiations until better scientific information is available? The answer to the first question turns out to be that coalitions are more likely to form between countries which are similar. The second question is examined by comparing two scenarios of negotiations. In the first, negotiations take place before the uncertainty is resolved, and abatement levels are chosen in anticipation of learning the true values of abatement. In the second, the countries negotiate after the uncertainty is resolved, and only then decide on their strategies concerning coalition formation and abatement levels.

These two scenarios are formalized in terms of two distinct extensive form games based on the same fundamentals, where the differing rules of the two games capture the distinction between negotiating under uncertainty and negotiating after uncertainty has been resolved. In the first, nature's move in determining the benefit of abatement activity comes after the moves of the individual countries, and the *ex ante* payoff of an individual country is calculated by taking expectations over the possible choices of nature at the final stage. In the second game, nature moves before the individual countries. Thus, when the countries begin their negotiations for coalition formation, they are in full possession of the information concerning the value of abatement activity. For each outcome of nature's chance move, there corresponds a profile of equilibrium actions of the three countries, including the

[4] Nordhaus (1993) presents the results of a survey of scientistst' judgements concerning such catastrophic changes.

decisions on whether to join a coalition or not. Under this second scenario, the actions of individual countries are functions of the underlying state of nature, and therefore are random variables themselves. The *ex ante* payoff of an individual country is then obtained by taking the expected value of these final payoffs. By comparing the *ex ante* payoffs under the two scenarios, we may address the question of whether the countries are better off (in *ex ante* terms) to have *ex ante* negotiations or to delay negotiations until the uncertainty has been resolved.

The value of information in environmental policy has been discussed in the context of a two country model by Ulph and Ulph (1994a) and by Ulph and Maddison (1995). Although the issue of coalition formation cannot be addressed adequately in a two country model, these papers are able to demonstrate that the value of information in strategic situations can be negative, depending on the uncertainty structure underlying the game. These papers assume the strategic substitutability of pollution abatement. In our model, we do not need this assumption, and we regard this as a virtue of our model, as we explain below. If, indeed, there is strategic substitutability of abatement activity, our conclusions are strengthened.

To gain an intuition for our main results, it is instructive to contrast the role of uncertainty in a single-person decision problem from that in a strategic situation. In a one-person decision problem, the value of information can never be negative, since it is open to the decision maker to ignore any information. In general, a decision maker can do strictly better by utilizing the information concerning the state of nature. However, in strategic situations, this need not hold. In our case, there is a trade-off. On the one hand, information concerning the state of nature helps the decision makers by allowing them to take more informed decisions. However, it is also the case that cooperation is more difficult to achieve when the likely winners and losers are known when negotiation takes place. In our case, the outcome when more information is available is less efficient in terms of total payoffs than the corresponding welfare benchmark when negotiations take place in ignorance.

This argument relies on a very different mechanism from those which invoke the option value of abatement activity, arising from the fact that changes to the stocks of greenhouse gases take long periods of time to achieve (Chichilnisky and Heal, 1993; Ulph and Ulph 1994b). The validity of such arguments on the advantages to flexibility will serve to reinforce the mechanism involved in our paper.

2. A three country model

Our model describes the interactions between three countries, indexed by the set $\{1, 2, 3\}$. We denote by y_i the level of pollution abatement undertaken by country i. The payoff of country i consists of the total benefit derived from total pollution abatement by all countries, minus the cost of abatement activity of its own abatement activity. In particular, we define the final payoff of country i as

$$\pi_i = \theta_i\left(\sum_k y_k\right) - \frac{y_i^2}{2} \quad (1)$$

We note that we do not impose the strategic substitutability of abatement levels. We believe this to be a virtue of our model, since for the case of greenhouse gas emissions, the degree of such substitutability, if any, is likely to be small.[5] The marginal benefit of aggregate abatement, θ_i, is a random variable which takes values in the set

$$\{z_1, z_2, z_3\} \quad (2)$$

where each z_k is a positive number, and $z_1 < z_2 < z_3$. The realizations of the set of random variables $\{\theta_i\}$ is determined by nature in the following sequence of experiments. First, nature selects one of the three countries with uniform probability of 1/3. If country i is drawn, then nature sets $\theta_i = z_3$. In other words, country i is given the highest marginal benefit of total abatement. Next, of the two remaining countries (excluding i), nature chooses a country with equal probability of 1/2. If country j is chosen at this stage, then nature sets $\theta_j = z_2$. In other words, country j is given the second largest marginal benefit of abatement activity. For the last remaining country (called k), nature sets the realization of θ_k to z_1. This country has the lowest marginal benefit of abatement activity.

By construction, each θ_i is identically distributed over the set $\{z_1, z_2, z_3\}$, and each outcome has probability 1/3. The expected payoff of country i in its general form is given by

$$E(\pi_i) = E\left(\theta_i \sum_k y_k - \frac{y_i^2}{2}\right) \quad (3)$$

where the expectation is taken over the joint distribution over the set of random variables $\{\theta_i\}$. Each country aims to maximize this expected payoff within the particular rules of the game. We consider two sets of rules. In the first, the countries decide on their abatement levels before the uncertainty concerning $\{\theta_i\}$ is resolved. In the second, countries choose abatement after they learn the outcome of nature's experiment. An important element of the countries' decisions is the possibility of forming coalitions before choosing their abatement levels.

If a country is a member of a coalition C, it is constrained to maximize the sum of payoffs of all countries in C. A country which is not a member of a coalition chooses its abatement level to maximize its own individual payoff, without regard to the payoffs of the other countries. The payoffs of all countries (whether in a coalition or not) is then derived from the underlying payoffs of the non-cooperative game—that is, from (1). If the coalition consists of all three countries, we call

[5] Carrro and Siniscalco (1994, pp. 14–15) conjecture that 'in most cases of trans-national or global pollution, the best-reply functions should be near-orthogonal, and that the emission reduction by some countries should not be offset by greater emissions elsewhere'.

this the grand coalition. If the coalition consists of two countries, we shall refer to this as a partial coalition.

The stability of a coalitional structure is determined by the self-interested membership decisions of all countries. Since there are just three countries, it is without loss of generality to consider at most one coalition in existence. Informally, we can say that a coalition C is stable if, first, no country in C can do better by leaving C and acting individualistically, and second, no country outside C can do better by joining the members of C to form an enlarged coalition. Three coalitional structures may emerge—namely, the grand coalition, a partial coalition, or the individualistic outcome, in which there is non-trivial coalition. More formally, if we denote by $\Pi_i(C)$ the payoff of country i when it is a member of the coalition consisting of the set C of countries, and denote by $\Pi_i(\{i\})$ the corresponding payoff of country i when it acts individualistically (in the trivial coalition $\{i\}$), we say that a coalition C is stable if

$$\Pi_i(C) \geq \Pi_i(\{i\}) \quad \text{for all countries } i \in C \tag{4}$$

$$\Pi_j(C \cup \{j\}) < \Pi_j(\{j\}) \quad \text{for all countries } j \notin C \tag{5}$$

If such a coalition exists, then no existing member prefers to leave it, and no outsider wants to join it. Notice that, according to our definition of stability, if a country is indifferent between joining a coalition and staying outside it, it will decide to join.

If a coalition forms before the uncertainty concerning $\{\theta_i\}$ is resolved, then the payoff Π_i refers to the expected payoff of country i, while if a coalition forms after the random variables $\{\theta_i\}$ are realized, it refers to the *ex post* payoff π_i given by (1).

Having described the framework and the questions to be posed, we proceed to analyse the outcome under our two scenarios. We first start with the case of coalition formation before the uncertainty is resolved. We shall call this the case of *ex ante* negotiation.

3. Equilibrium with *ex ante* negotiation

Under this scenario, the countries negotiate before learning the true value of the benefit of aggregate pollution abatement. Thus, in (3) the pollution abatement levels must be chosen before the realizations of $\{\theta_i\}$ are known, so that the $\{y_i\}$ are non-random. Thus, (3) takes the form

$$E(\pi_i) = E(\theta_i)\left(\sum_{k \in C} y_k + \sum_{k \notin C} y_k\right) - \frac{y_i^2}{2} \tag{6}$$

Suppose that C is a coalition which does not include i. Let us consider the optimal abatement level of country i for two cases: when it is outside the coalition C, and when it joins the countries in C to form the enlarged coalition $C \cup \{i\}$. If country i is outside the coalition, its optimal abatement level is

$$y_i = E(\theta_i) \tag{7}$$

On the other hand, if i is a member of the coalition $C \cup \{i\}$, its abatement level is chosen to maximize $\sum_{k \in C \cup \{i\}} E(\pi_k)$. From (6), this maximand can be written as

$$\left(\sum_{k \in C \cup \{i\}} E(\theta_k) \right) \left(\sum_{k \in C \cup \{i\}} y_k + \sum_{k \notin C \cup \{i\}} y_k \right) + \tfrac{1}{2} \sum_{k \in C \cup \{i\}} y_k^2 \tag{8}$$

so that the optimal abatement level of country i which joins the countries in coalition C is given by

$$y_i = \sum_{k \in C \cup \{i\}} E(\theta_k) \tag{9}$$

By substituting the optimal abatement levels into (6), we can compare the expected payoff for country i for when it stays outside the coalition C, and for when it joins the countries in C to form the enlarged coalition $C \cup \{i\}$.

Suppose country i stays outside the coalition, and denote by $\bar{C}$ the number of countries in C. Since each country $k \in C$ chooses abatement level $\bar{C}E(\theta_k)$, country i's expected payoff is

$$E(\theta_i)\bar{C} \sum_{k \in C} E(\theta_k) + E(\theta_i) \sum_{k \notin C} E(\theta_k) - \tfrac{1}{2} E(\theta_i)^2 \tag{10}$$

On the other hand, if country i joins the countries in C to form the enlarged coalition $C \cup \{i\}$, then its expected payoff is

$$E(\theta_i)(1 + \bar{C}) \sum_{k \in C \cup \{i\}} E(\theta_k) + E(\theta_i) \sum_{k \notin C \cup \{i\}} E(\theta_k) - \tfrac{1}{2} \left(\sum_{k \in C \cup \{i\}} E(\theta_k) \right)^2 \tag{11}$$

The payoff difference between (11) and (10) then represents the incentive for country i to join the countries in C to form the enlarged coalition $C \cup \{i\}$. Let us denote this payoff difference by $\Delta_A(i, C)$. The notation indicates that it is the incentive for country i to join coalition C when there is *ex ante* negotiation. By cancelling out common terms

$$\Delta_E(i, C) = \bar{C}(E(\theta_i))^2 - \tfrac{1}{2} \left(\sum_{k \in C} E(\theta_k) \right)^2$$

$$= \bar{C}(E(\theta_i))^2 - \tfrac{1}{2} \bar{C}^2 \left(\frac{\sum_{k \in C} E(\theta_k)}{\bar{C}} \right)^2$$

Thus, ruling out the trivial case of an empty C, we have $\Delta_A(i, C) \geq 0$ if and only if

$$2(E(\theta_i))^2 \geq \bar{C} \left(\frac{\sum_{k \in C} E(\theta_k)}{\bar{C}} \right)^2 \tag{12}$$

This condition has a simple interpretation. The expression inside the brackets on the right-hand side is the average of the *ex ante* expected values of θ_k for the members of C. Since the θ_k are identically distributed, we have $E(\theta_i) = E(\theta_j) = \bar{z}$, for all i and j, where $\bar{z} = (z_1 + z_2 + z_3)/3$. Thus, provided that $\bar{z} \neq 0$, which is guaranteed by our assumptions, we can divide by $\bar{z}^2$ so that (12) is reduced to the condition that $2 \geq \bar{C}$. In other words, any country has an incentive to join an existing coalition of size 2. From eqs (4) and (5), a country will join a coalition if it is indifferent between joining and not joining, so that the grand coalition consisting of all three countries is stable under the scenario of *ex ante* negotiations. Moreover, the grand coalition is the only stable coalitional structure, since if the coalition consists of fewer than three countries, a country outside the coalition prefers to join, rather than to stay out. Let us summarize this conclusion in terms of the following lemma.

Lemma 1 With *ex ante* negotiation, the grand coalition is the uniquely stable coalition structure.

From this conclusion, we can now specify the equilibrium payoffs of the three countries. In the grand coalition, the abatement level of each country is given by $y_i = 3E(\theta_i)$. Thus, substituting into eq. (6), the *ex ante* expected payoff of country i is $E(\pi_i) = 3E(\theta_i)\sum_{k=1}^{3} E(\theta_k) - [(3E(\theta_i))^2/2]$, so that

$$E(\pi_i) = \tfrac{9}{2}E(\theta_i)^2 = \tfrac{9}{2}\bar{z}^2$$

4. Equilibrium with *ex post* negotiation

Under our second scenario, coalition formation and pollution abatement takes place after the realizations of the random variables $\{\theta_i\}$ are known. Thus, the abatement levels $\{y_i\}$ are functions of $\{\theta_i\}$, and hence are, themselves, random variable. The *ex post* payoff of country i given a coalition C of countries is given by

$$\pi_i = \theta_i\left(\sum_{k\in C} y_k + \sum_{k\notin C} y_k\right) - \frac{y_i^2}{2} \tag{14}$$

Let us compare the optimal abatement level of country i under two cases: when it is outside the coalition C, and when it joins the countries in C to form the enlarged coalition $C\cup\{i\}$. If country i is outside the coalition C, its optimal abatement level is

$$y_i = \theta_i \tag{15}$$

If i is a member of the coalition $C\cup\{i\}$, its abatement level is chosen to maximize $\sum_{k\in C\cup\{i\}} \pi_k$. Then, the optimal abatement level of country i who joins the countries in coalition C is given by

$$y_i = \sum_{k\in C\cup\{i\}} \theta_k \tag{16}$$

By substituting the optimal abatement levels into (14), we can compare the payoff for country i for when it stays outside the coalition C, and for when it joins the countries in C to form the enlarged coalition $C\cup\{i\}$. When country i stays outside the coalition, its payoff is given by

$$\theta_i\bar{C}\sum_{k\in C}\theta_k + \theta_i\sum_{k\notin C}\theta_k - \tfrac{1}{2}\theta_i^2 \tag{17}$$

where, as before, $\bar{C}$ denotes the number of countries in the coalition C. If country i joins the countries in C to form the enlarged coalition $C \cup \{i\}$, then its payoff is

$$\theta_i(1+\bar{C}) \sum_{k \in C \cup \{i\}} \theta_k + \theta_i \sum_{k \notin C \cup \{i\}} \theta_k - \tfrac{1}{2}\left(\sum_{k \in C \cup \{i\}} \theta_k \right)^2 \tag{18}$$

The difference in i's payoff between (18) and (17) then represents the incentive for country i to join the countries to C to form the enlarged coalition $C \cup \{i\}$. Let us denote this payoff difference by $\Delta_P(i, C)$. This notation is analogous to that in the previous section, and indicates that it is the incentive for country i to join coalition C when there is *ex post* negotiation. By cancelling out common terms, this payoff difference can be shown to be

$$\Delta_P(i, C) = \bar{C}\theta_i^2 - \tfrac{1}{2}\left(\sum_{k \in C} \theta_k \right)^2$$
$$= \bar{C}\theta_i^2 - \tfrac{1}{2}\bar{C}^2 \left(\frac{\sum_{k \in C} \theta_k}{\bar{C}} \right)^2$$

Thus, $\Delta_P(i, C) \geq 0$ if and only if

$$2\theta_i^2 \geq \bar{C}\left(\frac{\sum_{k \in C} \theta_k}{\bar{C}} \right)^2 \tag{19}$$

In other words, country i has a positive incentive to join the coalition C if and only if, two times the square of θ_i is no smaller than $\bar{C}$ times the square of the average θ_k of the countries in the existing coalition. Among other things, this implies that any coalition which forms at the *ex post* stage consists of the countries with the highest realizations of θ_i.

In particular, we can be certain that the grand coalition is never a stable coalitional structure given our assumption that $z_1 < z_2 < z_3$. To see this, consider the country which draws the lowest realization of θ (i.e. the country i for which $\theta_i = z_3$). In order for the grand coalition to be stable, the inequality (19) must hold when the coalition C consists of the countries which have drawn the two larger values of θ. However, this is impossible since $z_3 < (z_1 + z_2)/2$.

Having ruled out the grand coalition, we ask whether a partial coalition consisting of two of the three countries is stable. Since each θ_i takes values in the set $\{z_1, z_2, z_3\}$, the coalitional structure which is implied by (19) depends on the relative magnitude of the two largest values, namely z_2 and z_3. If $2z_2 \geq z_3$, then (19) implies that the partial coalition involving the countries with the two highest realizations of θ is stable, and is the only stable coalitional structure. However, if $2z_2 < z_3$ even the partial coalition of two countries is unstable, and the only stable outcome is the individualistic outcome in which each country chooses its Nash equilibrium level. Let us summarize these conclusions in terms of the following lemma.

Lemma 2 With *ex post* negotiation, the grand coalition is never a stable outcome. If $2z_2 \geq z_3$, a partial coalition of two countries is the uniquely stable coalition structure, but if $2z_2 < z_3$, the only stable coalitional structure is the individualistic outcome.

By virtue of this lemma, we are able to characterize completely the *ex post* coalition formation game, and the corresponding *ex post* payoffs of the three countries. Then, by taking expectations over all possible realizations of θ_i, we can calculate the *ex ante* expected payoff of country *i*. The lemma suggests that we must do this for two separate cases—the first when the partial coalition is stable, and the second when only the individualisic outcome is stable.

4.1 Case of partial coalition ($2z_2 \geq z_3$)

In this case, the countries which have drawn the two highest realizations of θ (namely, z_2 and z_3) form a partial coalition, and choose identical abatement levels of $z_2 + z_3$. The country which has drawn the value z_1 chooses abatement level z_1. Thus, if we denote by π_i the *ex post* payoff of the country which has drawn the value z_i, these *ex post* payoffs are given by

$$\pi_1 = z_1(z_1 + 2(z_2 + z_3)) - \tfrac{1}{2}z_1^2 \tag{20}$$

$$\pi_2 = z_2(z_1 + 2(z_2 + z_3)) - \tfrac{1}{2}(z_2 + z_3)^2 \tag{21}$$

$$\pi_3 = z_3(z_1 + 2(z_1 + z_3)) - \tfrac{1}{2}(z_2 + z_3)^2 \tag{22}$$

In order to calculate the *ex ante* payoff of a firm *i*, we take expectations over all possible realizations of $\{\theta_i\}$. Since a country has equal probability of drawing a value of θ from the set $\{z_1, z_2, z_3\}$, the *ex ante* expected payoff of all countries is identical, and is given by $(\pi_1 + \pi_2 + \pi_3)/3$. Thus, for any country *i*, $E(\pi_i)$ is given by $\frac{1}{3}((z_1 + z_2 + z_3)(z_1 + 2(z_2 + z_3))) \;\; -\frac{1}{2}(z_1^2 + 2(z_2 + z_3)^2)$. Simplifying this expression, we have

$$E(\pi_i) = 3\bar{z}^2 + \frac{z_1(z_2 + z_3)}{3} \tag{23}$$

We can compare this *ex ante* payoff with that which results from *ex ante* negotiation given by eq. (13), to give us the payoff advantage of *ex ante* negotiations over *ex post* negotiation. Since the *ex ante* payoff to *ex ante* negotiation is $9\bar{z}^2/2$, the payoff advantage to *ex ante* negotiation is

$$\tfrac{1}{6}(z_1^2 + (z_2 + z_3)^2) > 0 \tag{24}$$

demonstrating that *ex ante* negotiation is superior to *ex post* negotiation, even when a partial coalition is feasible. The precise magnitude of this superiority can be seen to be large if the numerical values of the two highest marginal benefits of abatement are large. In this case, the welfare loss resulting from *ex post* negotiation is very high.

4.2 Case of individualistic outcome ($2z_2 < z_3$)

In this case, even a partial coalition is infeasible, and the only stable coalition structure is that of individualistic choice in the Nash equilibrium of the abatement game. Then, each country's *ex post* choice of abatement is the realized value of its marginal benefit of abatement, θ. The country which draws the value z_i chooses abatement level z_i. Thus, using the same notation as above, the *ex post* payoffs of the three countries are given by

$$\pi_1 = z_1(z_1 + z_2 + z_3) - \tfrac{1}{2}z_1^2 \quad (25)$$

$$\pi_2 = z_2(z_1 + z_2 + z_3) - \tfrac{1}{2}z_2^2 \quad (26)$$

$$\pi_3 = z_3(z_1 + z_2 + z_3) - \tfrac{1}{2}z_3^2 \quad (27)$$

Hence, the *ex ante* expected payoff of any country i is given by

$$E(\pi_i) = 3\bar{z}^2 - \tfrac{1}{6}(z_1^2 + z_2^2 + z_3^2) \quad (28)$$

As we would expect, the *ex ante* payoff when no coalitions are possible is smaller than when a partial coalition is feasible. Moreover, the payoff advantage of *ex ante* negotiation is even greater when no coalitions form. The payoff advantage of *ex ante* negotiation then becomes

$$3\bar{z}^2 + \tfrac{1}{6}(z_1^2 + z_2^2 + z_3^2) > 0 \quad (29)$$

This term is closely related to the variance of θ_i, and suggests that the potential welfare loss resulting from *ex post* negotiations is greater when the disparities in the marginal benefits of abatement are larger.

Summarizing our discussion, we can state the main result of this paper.

Theorem For any country i, its *ex ante* payoff is strictly higher with *ex ante* negotiations than with *ex post* negotiations.

The simplicity of the uncertainty structure governing our game plays a large role in allowing us to obtain explicit solutions to the coalition formation games, and the resulting payoffs from these equilibria. There are, however, some general lessons from the analysis which would be applicable to more complex situations. One of the key planks of our argument has been the feature that coalitions are easier to form between countries which share similar characteristics. Hence, the size of a coalition depends on the degree of disparity among the negotiating countries. If negotiations are conducted among disparate countries, the resulting coalition is likely to be small. This intuition explains our main result. When negotiations take place before the uncertainty is resolved, the differences between the three countries is small. However, after the uncertainty has been resolved, the three countries are now disparate players, facing rather different incentives. More generally, the differences between the negotiating countries will be smaller in *ex ante* terms than in *ex post* terms, so that the general principle which favours coalition formation among countries which are similar will have the implication that coalitions are more likely to form before the resolution of uncertainty rather than after it. Since the overall loss of economic efficiency is directly mirrored in the extent of the failure to form the grand coalition, this leads to our general conclusion that *ex ante* negotiation is welfare enhancing.

5. Concluding remarks

In this paper, we have outlined a reason for why *ex ante* negotiations are superior to *ex post* negotiations, based on the strength of the incentives to form coalitions. We believe that this is a novel observation, and one which has implications for the conduct of the international negotiations currently in progress. There is an alternative argument for the superiority of *ex ante* negotiations which draws on the fact that pollution has many of the features of an irreversible action, so that the option value of *ex ante* abatement is positive (Chichilnisky and Heal, 1993; Ulph and Ulph, 1994b). Such an option value will serve to reinforce the superiority of *ex ante* negotiations. A fully fledged dynamic model which combines both features is a worthy subject of further study. It is hoped that future research will tackle these issues, and to inform empirical debate concerning the actual magnitudes involved.

Acknowledgements

We have benefited from the incisive comments of a refereee. We also thank Alistair Ulph for comments on an earlier version of this paper.

References

Barrett, S. (1994). 'Self-enforcing International Environmental Agreements', *Oxford Economic Papers*, **46**, 878–94.

Carraro, C. and Siniscalco, D. (1993). 'Strategies for the International Protection of the Environment', *Journal of Public Economics*, **52**, 309–28.

Carraro, C. and Siniscalco, D. (1994). 'International Coordination of Environmental Policies and Implementation of International Environmental Agreements', Paper for the 50th congress of the International Institute of Public Finance, Harvard University, August.

Chander, P. and Tulkens, H. (1992). 'Theoretical Foundations of Negotiations and Cost-Sharing in Transfrontier Pollution Problems', *European Economic Review*, **36**, 388–98.

Chichilnisky, G. and Heal, G. (1993). 'Global Environmental Risks', *Journal of Economic Perspectives*, **7**(4), 65–86.

Cline, W. (1992). *The Economics of Global Warming*, Institute of International Economics, Washington, DC.

Fankhauser, S. (1993). 'The Economic Costs of Global Warming: Some Monetary Estimates', in Y. Kaya, N. Nakicenovic, W. Nordhaus, and F. Toth (eds), *Costs, Impacts and Benefits of CO_2 Mitigation*, International Institute for Applied Systems Analysis, CP93-2, 85-105, Laxenburg, Austria.

Heal, G. (1994). 'The Formation of Environmental Coalitions', in C. Carraro (ed.), *Trade, Innovation, Environment*, Kluwer Academic Publishers, Dordrecht.

Morrisette, P.M., Darmstadter, J., Plantinga, A.J., and Toman, M.A. (1990). 'Lessons from other International Agreements for a Global CO_2 Accord', Discussion Paper ENR 91-02, Resources for the Future, Washington, DC.

Murdoch, J.C. and Sandler, T. (1997). 'The Voluntary Provision of a Pure Public Good: the Case of Reduced CFC Emissions and the Montreal Protocol', *Journal of Public Economics*, **63**, 331–49.

Nordhaus, W. (1991). 'To Slow or Not to Slow: The Economics of the Greenhouse Effect', *Economic Journal*, **101**, 920–37.

Nordhaus, W. (1993). 'Survey on Uncertainties Associated with Future Climate Change', unpublished paper, Yale University.

Sandler, Y. (1993). 'Tropical Deforestation: Markets and Market Failures', *Land Economics*, **69**, 225–33.

Sandler, T. (1996). 'A Game-theoretic Analysis of Carbon Emissions', in R. Congleton (ed.), *The Political Economy of Environmental Protection*, University of Michigan Press, Ann Arbor, MI.

Sandler, T. and Sargent, K. (1995). 'Management of Transnational Commons: Coordination, Publicness and Treaty Formation', *Land Economics*, **71**, 145–62.

Ulph, A. and Maddison, D. (1995). 'Uncertainty, Learning and International Environmental Policy Co-ordination', Discussion Paper No. 9507, Southampton University.

Ulph, A. and Ulph, D. (1994a). 'Who Gains from Learning About Global Warming?', Discussion Paper No. 9407, Southampton University.

Ulph, A. and Ulph, D. (1994b). 'The Irreversibility Effect Revisited', Discussion Paper No. 9408, Southampton University.

Journal of Public Economics 103 (2013) 29–43

Contents lists available at SciVerse ScienceDirect

Journal of Public Economics

journal homepage: www.elsevier.com/locate/jpube

The role of uncertainty and learning for the success of international climate agreements

Michael Finus [a,*], Pedro Pintassilgo [b]

[a] Department of Economics, University of Bath, Bath, BA2 7AY, UK
[b] Faculty of Economics and Research Centre for Spatial and Organizational Dynamics, University of Algarve, Faro, Portugal

ARTICLE INFO

Article history:
Received 30 October 2009
Received in revised form 8 March 2013
Accepted 6 April 2013
Available online 13 April 2013

JEL classification:
C72
D62
D81
H41
Q20

Keywords:
Voluntary provision of public goods
Transnational cooperation
Self-enforcing agreements
Uncertainty
Learning

ABSTRACT

Transnational externalities (e.g. transboundary pollution, trade, contagious diseases and terrorism) warrant coordination and cooperation between governments, but this proves often difficult. One reason for the meager success is the public good character of many of these economic problems, encouraging free-riding. Another reason one might suspect is uncertainty, surrounding most environmental problems, and in particular climate change. This provides often an excuse for remaining inactive. Paradoxically, some recent papers have concluded just the opposite: the "veil of uncertainty" can be conducive to the success of international environmental cooperation. In this paper, we explain why and under which conditions this can be true. However, we argue that those conditions are the exception rather than the rule. Most important, we suggest a mechanism for those conditions where learning has a negative effect on the success of cooperation which removes this effect or even turns it into a positive effect. Our results apply beyond the specifics of climate change to similar problems where cooperation generates positive externalities.

1. Introduction

Transnational externalities (e.g. transboundary pollution, trade, contagious diseases and terrorism; see e.g. Sandler, 2004; Yi, 1996) warrant coordination and cooperation between governments, but this proves often difficult. One of the greatest challenges to international cooperation and public good provision the world is currently facing is climate change, as emphasized by the two prominent studies, the Stern and the IPCC Reports (Stern, 2006; IPCC, 2007). International response to global warming can be traced back to 1988 when the IPCC was founded – an international body that gathers and synthesizes current worldwide scientific evidence on climate change. However, it was not until 1997 that 38 countries agreed to specific emission ceilings under the Kyoto Protocol, which was not ratified before 2002, after several concessions had been granted to various participants and after the US had declared that it would withdraw from the treaty altogether. Currently, a "Post-Kyoto" agreement for the period after 2012 is being negotiated, with the aim to tighten emission ceilings, encourage the participation of the US and the "emerging" polluters China and India.

One important problem for effective cooperation is free-riding. For a common property resource this is well known since Hardin (1968) and has been reiterated in the specific context of self-enforcing international environmental agreements (SEIEAs) by Barrett (1994), Carraro and Siniscalco (1993) and Hoel (1992). Later papers, using richer models, either with an empirical (e.g. Bosello et al., 2003; Finus and Tjøtta, 2003; Weikard et al., 2006) or theoretical (e.g. Asheim et al., 2006; Barrett, 2001, 2006; Rubio and Ulph, 2007; Ulph, 2004) focus, have suggested some possibilities to make SEIEAs more effective, but have confirmed the general negative conclusion more or less.[1]

Another important problem is the large uncertainty surrounding environmental damages caused by greenhouse gases. Predictions about mitigation costs are also difficult (IPCC, 2007; Stern, 2006). For instance, the former US President George Bush used uncertainty as one argument for his decision to withdraw from the Kyoto Protocol. In a letter to Senators, dated March 13, 2001, as quoted by Kolstad (2007), he wrote: "I oppose the Kyoto Protocol ... we must be very careful not to take actions that could harm consumers. This is especially true given the incomplete state of scientific knowledge".

* Corresponding author. Tel.: +44 1225386228; fax: +44 1225383423.
E-mail address: m.finus@bath.ac.uk (M. Finus).

[1] Surveys are provided in Barrett (2003) and Finus (2003).

0047-2727/$ – see front matter © 2013 Elsevier B.V. All rights reserved.
http://dx.doi.org/10.1016/j.jpubeco.2013.04.003

Recently, a literature has emerged, which combines free-riding with uncertainty and learning, using simple SEIEA-models with a static payoff function[2] (Kolstad, 2007; Kolstad and Ulph, 2008, 2011; Na and Shin, 1998). Their main conclusion is that in the strategic context of the formation of climate agreements, learning leads to worse outcomes (measured in terms of the aggregate payoff and public good provision level) than no learning. This "negative" result about the role of learning for the success of SEIEAs, though interesting, is intriguing as it runs counter to increased research efforts on climate change worldwide. This motivates the three research questions posed in this paper. What are the driving forces to generate this result? How general is this conclusion? Is there a way to fix the problem? The following paragraphs provide a brief introduction to how we address these questions.

1.1. What are the driving forces to generate this result?

In order to understand the driving forces it is important to note that the "standard" SEIEA-model assumes a two-stage coalition formation game. In the first stage, countries choose their membership and in the second stage their contributions to the public good. The game is solved backwards. In the second stage, non-members derive their contributions from maximizing their own payoffs but coalition members from maximizing the aggregate payoff over all their members. Hence, in this model, the social optimum is reproduced if all countries join the agreement (grand coalition) and the Nash equilibrium if all countries remain singletons (singletons coalition structure). Intermediate cases of cooperation are also captured, with some, though not all, countries joining the agreement. In the first stage, stable coalitions are determined based on the concept of internal and external stability.

In this paper, we suggest a three step procedure in order to understand how uncertainty and learning affects the success of SEIEAs. In the first step, we look at stable coalitions (their sizes and membership) with and without learning in the first stage. In the second step, we consider a generic coalition and compare total provision levels and payoffs with and without learning in the second stage. In the third step, we combine steps 1 and 2 to derive overall conclusions. In many cases, this allows already for immediate conclusions.

We say that the *second stage effect from learning* is positive (negative) if the total payoff with learning is larger (smaller) than without learning for any possible coalition that has formed in the first stage. For the understanding of this second stage effect two benchmarks are useful. The first benchmark is the grand coalition. When it forms, there is no strategic interaction, between the coalition and non-members, and the second stage effect from learning is positive, and in most cases strictly positive. This is in line with the general wisdom that "learning is good". The second benchmark is the singletons coalition structure corresponding to the Nash equilibrium. Now there is strategic interaction between players and the value of information can become negative as shown for instance in Gollier and Treich (2003) for three economic examples. For instance, suppose as in Ulph (1998) that all players have ex-ante symmetric expectations but turn out to be asymmetric ex-post. Assume that asymmetry means only different marginal benefits from the total public good provision but symmetric marginal costs from individual public good contributions. Then equilibrium contributions with learning are asymmetric but are symmetric without learning. Both equilibria are inefficient (each player sets marginal cost only equal to own but not the sum of marginal benefits) but the equilibrium under no learning is at least cost-effective (i.e. marginal contribution costs equalize). Thus, with strategic interaction, the second stage effect from learning can be negative. In our coalition model, which captures not only the grand coalition and the singletons coalition structure but also intermediate cases of cooperation, such a negative second stage effect from learning occurs if there is pure uncertainty about the distribution of the benefits from public good provision.

By the same token, we say that the *first stage effect from learning* is positive (negative) if the size of stable coalitions is larger (smaller) with than without learning. Again, the example from above is useful to illustrate this point. With learning those players with high marginal benefits gain more than proportionally from cooperation and those with low marginal benefits less than proportionally. If differences are sufficiently pronounced, no or only small coalitions will be stable.[3] Without learning, and ex-ante symmetric expectations, the distribution is symmetric and larger coalitions are stable. The intuition of this phenomenon is along the lines of Young (1994), borrowing the concept of the veil of uncertainty from Brennan and Buchanan (1985), who argues that agreements are easier if potential participants do not know the distributional consequences.

1.2. How general is this conclusion?

A natural approach to answer this question is to set up a model that takes the best ingredients of existing models and generalizes them in several ways.

First, we consider the *three learning scenarios* full, partial and no learning as proposed in Kolstad (2007) and Kolstad and Ulph (2008, 2011) where learning means to learn the values of the uncertain parameters of the payoff functions. The additional scenario of partial learning compared to Na and Shin (1998) turns out to be useful when breaking down the effects of learning between both stages.

Second, we consider a *strictly concave payoff function* which we take from Na and Shin (1998). Compared to the papers by Kolstad and Ulph, which assume a linear payoff function, this avoids boundary solutions in the second stage and allows to derive clear-cut second stage effects from uncertainty and learning.

Third, we consider *three types of uncertainty*. Taking a broader view, we argue that the models by Kolstad (2007) and Kolstad and Ulph (2008) essentially capture *pure uncertainty about the level of the benefits* from the public good provision and the model by Na and Shin (1998) *pure uncertainty about the distribution of these benefits*. We find these benchmarks useful for analytical clarity but consider also the more realistic case with *simultaneous uncertainty about the level and the distribution of these benefits*. Moreover, we neither restrict attention to only three players as in Na and Shin (1998), nor do we assume only two values of the random benefit variables as in Kolstad (2007) and Kolstad and Ulph (2008, 2011), but allow for any possible values of the random variables.

In this more general setting, we derive sufficient conditions when learning has a positive impact on the success of SEIEAs, when it has a negative impact and also point out the circumstances under which conclusions are not clear-cut. It will become apparent that those cases where learning has a negative impact on the success of SEIEAs are the exception rather than the rule.

1.3. Is there a way to fix the problem?

If the first stage effect from learning is negative because of an unequal distribution of the gains from cooperation among coalition members, then we show that an appropriate transfer mechanism can fix this problem. In fact, even larger coalitions than for symmetry can be stabilized because the relative benefit from participating in a coalition compared to free-riding increases.

In the following, we outline the general setting of our coalition model in Section 2. Section 3 derives the model solutions for stages 1 and 2 and Section 4 compares results. Section 5 summarizes the main conclusions and proposes some issues for future research.

[2] That is, it captures the public bad nature of greenhouse gases but not their dynamics as stock pollutants.

[3] This idea is also illustrated in a simple two-player model in Helm (1998) and in Kolstad (2005).

2. Model outline

2.1. Coalition formation game

2.1.1. General remarks

International environmental agreements are typical agreements where countries have to decide whether they are in or out, participation is voluntary and membership is open, i.e. a country can neither be forced into nor excluded from participation. Therefore, we model coalition formation as a two-stage open membership single coalition game. In the first stage, players (i.e. countries in our context) decide whether to join an agreement (i.e. a climate treaty in our context) or to remain outsiders as singletons. In the second stage, players choose their policy levels (i.e. contributions to the public good which we call abatement in our context). The game is solved backward, assuming that strategies in each stage must form a Nash equilibrium.

This game has also been called cartel formation game with non-members called fringe players. It originates from the literature in industrial organization (d'Aspremont et al., 1983) and has been applied widely in this literature (e.g. Deneckere and Davidson, 1985; Donsimoni et al., 1986; Poyago-Theotoky, 1995; see Bloch, 2003; Yi, 1997 for surveys) but also in the literature on self-enforcing international environmental agreements (e.g. Barrett, 1994; Carraro and Siniscalco, 1993; Rubio and Ulph, 2007; see Barrett, 2003; Finus, 2003 for surveys).

In the first stage, players' membership decisions lead to a coalition structure $K = \{S, 1_{n-m}\}$, which is a partition of players, with n being the total number of players, S the coalition of size m, $m \leq n$, 1_{n-m} the $n - m$ singletons, and N the set of players, $S \subseteq N$. Because coalition formation is only interesting if there are at least three players, we assume subsequently $n \geq 3$. Due to the simple structure of this coalition formation game, i.e. there can be at most one non-trivial coalition (i.e. all players that do not belong to S are singletons with a non-trivial coalition being a coalition of at least two players), coalition structure K is entirely determined by coalition S. We will sometimes call the members of S signatories and the non-members of S non-signatories.

In the second stage, given that some coalition S has formed, players choose their abatement levels q_i. The decision is based on the following payoff function[4]:

$$\Pi_i = B_i\left(\sum_{k=1}^{n} q_k\right) - C_i(q_i), i \in N \tag{1}$$

where B_i (•) is country i's concave benefit function from global abatement (in the form of reduced damages, e.g. measured against some business-as-usual-scenario) and C_i (•) its convex abatement cost function from individual abatement. The global public good nature of abatement is captured by the benefit function which depends on the sum of all abatement contributions. For a start, we assume that all functions and their parameters are common knowledge and introduce uncertainty later.

Working backward, we assume that the equilibrium strategy vector in the second stage constitutes a (coalitional) Nash equilibrium between coalition S with its m members and the $n - m$ singletons. Specifically, the coalition acts de facto as a single or meta player (Haeringer, 2004), maximizing the aggregate payoff of its members

$$\max_{q^S} \sum_{i \in S} \Pi_i(S) \Rightarrow \sum_{i \in S} B_i'\left(\sum_{k=1}^{n} q_k\right) = C_i'(q_i) \ \forall i \in S \tag{2}$$

[4] An alternative specification of payoff functions, comprising damage cost functions from global emissions and benefit functions from individual emissions, produces equivalent results. This equivalence holds as long as non-negativity constraints are observed, as discussed for instance in Diamantoudi and Sartzetakis (2006) and Rubio and Ulph (2006).

whereas all singletons maximize their own payoffs

$$\max_{q_i} \Pi_i(S) \Rightarrow B_i'\left(\sum_{k=1}^{n} q_k\right) = C_i'(q_i) \ \forall i \notin S \tag{3}$$

where q^S in Eq. (2) is the abatement vector of those players that belong to coalition S, B_k' and C_k' are the derivatives of B_k and C_k with respect to q_k, respectively. The simultaneous solution of the first-order conditions (F.O.C.s) in Eqs. (2) and (3) delivers equilibrium abatement levels $q^*_{i \in S}(S)$ of signatories and $q^*_{i \notin S}(S)$ of non-signatories.[5] The F.O.C.s in Eq. (2) are the Samuelson conditions for the optimal provision of a public good, though they hold only for coalition members; the F.O.C.s in Eq. (3) are those in a non-cooperative equilibrium.[6]Substituting the equilibrium abatement levels for a given coalition S into payoff functions (1) delivers the payoffs of signatories, $\Pi^*_{i \in S}(S)$, and non-signatories, $\Pi^*_{i \notin S}(S)$, in the second stage of the coalition formation game.

In the first stage, stable coalitions are determined by invoking the concept of internal and external stability, which is de facto a Nash equilibrium in membership strategies.

$$\textit{internal stability}: \quad \Pi_i^*(S) \geq \Pi_i^*(S \backslash \{i\}) \ \forall i \in S \tag{4}$$

$$\textit{external stability}: \quad \Pi_i^*(S) > \Pi_i^*(S \cup \{i\}) \ \forall i \notin S. \tag{5}$$

That is, no signatory should have an incentive to leave coalition S to become a non-signatory and no non-signatory should have an incentive to join coalition S. In order to avoid knife-edge cases, we assume that if players are indifferent between joining coalition S and remaining outside, they will join the agreement. Coalitions which are internally and externally stable are called stable.

Note that the singletons coalition structure is stable by definition and hence existence of a stable coalition is guaranteed: if all players announce not to join the agreement, then a single deviation by one player will make no difference. If there is more than one stable coalition, we apply the Pareto-dominance criterion and delete all those coalitions which are Pareto-dominated by other stable coalitions. In our model it is easy to show that the singletons coalition structure is Pareto-dominated by any other coalition structure.

2.1.2. Symmetry, asymmetry and transfers

In the case of ex-ante symmetric players, i.e., all players have the same payoff function, things simplify: payoffs among the group of signatories are identical and the same applies to the group of non-signatories. Hence, we can talk about a coalition of size m being stable, with the understanding that all combinations of m players form a stable coalition. Moreover, internal stability can be written as $\Pi_i^{*M}(m) \geq \Pi_i^{*NM}(m-1)$ and external stability as $\Pi_i^{*NM}(m) > \Pi_i^{*M}(m+1)$ with the superscript "M" standing for "Member" and the superscript "NM" standing for "Non-Member". This simplification emphasizes that if a coalition of size m is externally stable, then we can conclude that a coalition of size $m + 1$ will not be internally stable. Similarly, if a coalition of size m is internally stable, then a coalition of size $m - 1$ will not be externally stable.

If players are not ex-ante symmetric, then this simplification is not appropriate. In fact, given the assumption of joint welfare maximization of coalition members (implying an efficient allocation of abatement

[5] The simultaneous choice in the second stage has been called Nash–Cournot assumption in the literature. An alternative assumption could be a sequential choice which has been called Stackelberg assumption. In the context of coalition formation the latter assumption is not innocuous for several reasons. For the specific type of payoff functions which we introduce in Section 2.3, both assumptions are equivalent. For a discussion see Finus (2003).

[6] Note that if $S = N$ (i.e. all players form the grand coalition) the equilibrium abatement vector corresponds to the social optimum and if either $S = \{i\}$ or $S = \varnothing$ (i.e. all players act as singletons) this corresponds to the Nash equilibrium.

duties among coalition members), if players have different payoff functions, payoffs among coalition members may be quite different (implying an asymmetric distribution of the total payoff of the coalition among its members). This may upset the stability of coalitions. A way to circumvent this problem is transfers. In principle, many transfer schemes are perceivable which typically will lead to different sets of stable coalitions. In order to avoid this sensitivity, we consider transfers in its most general form, based on the concept of an almost ideal transfer scheme (AITS) proposed by Eyckmans and Finus (2009) with similar notions in Fuentes-Albero and Rubio (2009), McGinty (2007) and Weikard (2009).

Suppose we ask the question whether there exists a transfer scheme that can stabilize a coalition S. Then a necessary condition for internal stability is that the sum of payoffs in the coalition (summing over the left-hand side of inequality (4)) exceeds the sum of free-rider payoffs when leaving the coalition (summing over the right-hand side of inequality (4)). We call this potential internal stability which is reflected in inequality (6) below. By the same token, a necessary condition for external stability is that the sum of non-members' payoffs (summing over the left-hand side of inequality (5)) exceeds the sum of payoffs when joining the coalition (summing over the right-hand side of inequality (5)). We refer to this condition as potential external stability in inequality (7).

$$\textit{potential internal stability}: \quad \sum_{i\in S} \Pi_i^*(S) \geq \sum_{i\in S} \Pi_i^*(S\backslash\{i\}) \tag{6}$$

$$\textit{potential external stability}: \quad \sum_{i\in S} \Pi_i^*(S) > \sum_{i\notin S} \Pi_i^*(S\cup\{i\}) \ \ \forall i \notin S \tag{7}$$

The interesting aspect is that with the AITS these two necessary conditions become also sufficient conditions. The AITS works as follows: every coalition member $i \in S$ receives his free-rider payoff when leaving the coalition, $\Pi_i^*(S \setminus \{i\})$, plus a positive share λ_i, $\sum_{i\in S}\lambda_i = 1$, of the surplus obtained by the coalition as a whole over the sum of free-rider payoffs, $\omega(S) = \sum_{i\in S}\Pi_i^*(S) - \sum_{i\in S}\Pi_i^*(S \setminus \{i\})$. Hence, the payoff of a coalition member with transfers is given by $\Pi_i^{*T}(S) = \Pi_i^*(S \setminus \{i\}) + \lambda_i\omega(S)$ where the superscript "T" stands for transfers. Replacing $\Pi_i^*(S)$ by $\Pi_i^{*T}(S)$ in the condition for internal stability in Eq. (4) shows that if the surplus of coalition S is (weakly) positive, $\omega(S) \geq 0$, i.e., coalition S is potentially internally stable, then coalition S will be internally stable with the AITS. By the same token, replacing $\Pi_i^*(S \cup \{i\})$ by $\Pi_i^{*T}(S \cup \{i\})$ in the condition for external stability in Eq. (5) shows that if the surplus of all enlarged coalitions $S \cup \{i\}$ is negative, $\omega(S \cup \{i\}) < 0$ for all $i \notin S$, i.e. coalition S is potentially externally stable, then coalition S will be externally stable with the AITS. Hence, a similar link between internal and external stability observed above for symmetric players can be established with the AITS. That is, if coalition S is externally stable, then all coalitions $S \cup \{i\}$, for all $i \notin S$ are not internally stable; if coalition S is internally stable, then all coalitions $S \setminus \{i\}$, for all $i \in S$, are not externally stable.

There are three more interesting features of the AITS worthwhile mentioning of which we make use later. First, the design of the AITS suggests that it aims at stabilizing a coalition internally. This seems sensible as in the context of public good provision one is more concerned about players leaving a coalition than players joining a coalition. Every coalition that can be potentially internally stabilized will be internally stable with the AITS. Without transfers or with other transfer scheme (e.g. Nash-Bargaining solution or Shapley Value; see Eyckmans and Finus, 2009 for examples), this may not necessarily be the case. Hence, we can conclude that if a coalition S is internally stable without transfers, it will also be internally stable with the AITS. Conversely, if a coalition is not internally stable with the AITS, it will also not be stable without transfers (or any other transfer scheme).

Second and immediately related to this, if a coalition S is stable (internally and externally stable) without transfers, then with the AITS coalition S will either also be stable or a larger coalition with additional players R, $S \cup R$, will be stable. The reason is that if S was not externally stable with the AITS, then a coalition $S \cup \{i\}$ would be internally stable. If $S \cup \{i\}$ was also externally stable, then it would be stable. If it was not externally stable, then the argument of adding players is repeated, noting that eventually one coalition must be externally stable, simply because the grand coalition is externally stable by definition.

Third, as we will argue in Section 3, in our public good game the total payoff (summing payoffs over all players) when coalition $S \cup \{i\}$ forms is strictly higher than when coalition S forms. That is, the formation of larger coalitions translates into higher global payoffs. This explains the word "ideal" in the name of the transfer scheme. Among those coalitions which can be potentially internally stabilized, the AITS stabilizes the coalition with the highest global payoff. Because this may not be the grand coalition if free-rider incentives are too strong, the word "almost" is part of the name of the transfer scheme.

From the three features we can conclude that with AITS, stable coalitions will be at least as large and successful (in terms of aggregate payoffs) than without transfers.

2.2. *Three learning scenarios*

Suppose that some parameter values of the payoff functions are uncertain. Following Kolstad and Ulph (2008, 2011), we assume risk-neutral agents and distinguish *three learning scenarios*: 1) full learning, 2) partial learning and 3) no learning. Full Learning (abbreviated to FL) can be considered as a benchmark case in which players learn about the true parameter values before taking the membership decision in the first stage. Hence, uncertainty is fully resolved at the beginning of the game. For Partial Learning (abbreviated to PL) it is assumed that players decide about membership under uncertainty but know that they will learn about the true parameter values before deciding upon abatement levels in the second stage. Hence, the membership decision is based on expected payoffs, under the assumption that players will take the correct decision in the second stage. Finally, under No Learning (abbreviated to NL) also the abatement decision has to be taken under uncertainty. That is, players derive their abatement strategies by maximizing expected payoffs. The membership decisions are also based on expected payoffs, though these payoffs differ from those under partial learning, given that less information is available.

It is worthwhile pointing out that our assumption implies that learning takes the form of perfect learning (Kolstad and Ulph, 2008, 2011). That is, if players learn about parameter values, no uncertainty remains. Hence, partial learning is de facto delayed learning, though we stick to the terminology introduced by Kolstad and Ulph. Furthermore, partial learning requires assuming that players stick to their membership decision taken in the first stage even though additional information becomes available in the second stage.[7]Full learning is certainly an optimistic and no learning a pessimistic benchmark about the role of learning in the context of climate change. Partial learning approximates (because beliefs are not updated in a Bayesian sense) the fact that information becomes available over time.

2.3. *Three types of uncertainty*

2.3.1. *Introduction*

The assumption about the uncertain parameter values in the payoff functions of players are summarized in *three types of uncertainty*. Due to the complexity of coalition formation, the consideration of a

[7] This assumption has been called fixed membership in the literature (Rubio and Ulph, 2007; Ulph, 2004). The record of international environmental agreements shows that countries usually do not leave agreements once they have ratified them, though new members may join at a later stage (Finus, 2003). This is also suggested by Iida (1993) in other areas of international policy coordination.

particular payoff function is required. In order to avoid binary equilibrium choices "abate" or "not abate", as for instance in Kolstad (2007) and Kolstad and Ulph (2008, 2011), which renders the model not well-behaved in the first and second stage (see Section 4.2), we consider a strictly concave payoff function which is still simple enough to derive analytical results:[8]

$$\Pi_i = b_i \sum_{k=1}^{n} q_k - c_i \frac{q_i^2}{2}, i \in N, b_i > 0, c_i > 0 \tag{8}$$

where b_i is a benefit parameter, $b_i \sum_{k=1}^{n} q_k$ is the benefit from global abatement, c_i is a cost parameter, and $c_i \frac{q_i^2}{2}$ is the abatement cost from individual abatement.

Generally, the benefit as well as the cost parameters could be uncertain. However, following Kolstad (2007), Kolstad and Ulph (2008, 2011) and Na and Shin (1998), in the climate context, uncertainty about the benefits from reduced damages appears to be more important than uncertainty about abatement costs. Hence, we simplify the model, by dividing payoffs by the cost parameter c_i, define the benefit–cost ratio by $\gamma_i = b_i/c_i$, and hence payoff function (8) reads:

$$\Pi_i = \gamma_i \sum_{k=1}^{n} q_k - \frac{q_i^2}{2} \quad , i \in N, \ \gamma_i > 0. \tag{9}$$

Henceforth, we call γ_i the benefit parameter. If this parameter is uncertain, then it is represented by the random variable Γ_i, with associated distribution f_{Γ_i}. The assumptions regarding our three types of uncertainty are displayed in Table 1.

All three types of uncertainty capture an important aspect surrounding climate change (IPCC, 2007). There is much uncertainty about the absolute level of the benefits from abatement in the form of reduced damages (type 1). It is still not clear how serious the damages of climate change will be overall. But there is also much debate about the regional distribution of damages, i.e. which countries will benefit most from greenhouse gas emission reductions (type 2)? Hence, uncertainty of type 3, considering simultaneously uncertainty about the level and the distribution of the benefits from abatement, reflects the most comprehensive type of uncertainty. Nonetheless, types 1 and 2 turn out to be useful benchmarks in the analysis. As the random variable Γ_i is the benefit–cost ratio and the only variable in our simple model, it exclusively determines the gains from cooperation. Hence, we de facto model uncertainty about the level and/or distribution of the gains from cooperation – a problem which certainly applies to many economic problems with externalities.

Note that for all three types, *uncertainty is symmetric* as all players know as much or little about their own as about their fellow players' payoff functions. Moreover, *expectations are symmetric* as all players share the same beliefs about the distribution of the uncertain parameters. This requires that some coordination has taken place ex-ante. Even under learning, disagreement about optimal policy levels is not an issue (e.g. because of asymmetric benefit parameters) as coalition members maximize their joint welfare. We comment on this simplification in Section 5. Nevertheless, as we will see in Section 3, disagreement will figure indirectly into our model when it comes to decide on the participation in the agreement. Then under full learning, asymmetry may cause little participation if not balanced by transfers.

[8] A similar payoff function has been used for instance by Barrett (2006) and Na and Shin (1998) but also by many others.

2.3.2. Assumptions

2.3.2.1. Uncertainty of type 1: Pure uncertainty about the level of benefits. Uncertainty of type 1 is inspired by Kolstad (2007) and Kolstad and Ulph (2008, 2011), which the authors call systematic uncertainty as it relates to a *common parameter.* All players have the same expectations ex-ante, and once uncertainty is resolved, all players have the same benefit parameter ex-post. Regarding the distribution of the benefit parameter, Kolstad (2007) and Kolstad and Ulph (2008) assume a simple Bernoulli distribution, in which the benefit parameter can take on only two values: low and high.

For uncertainty of type 1, all benefits are equal, i.e., $\Gamma_i = \Gamma_j$ for all i, $j \in N$. This may be viewed as a draw from an urn where a random player draws a ticket with a value γ_i which applies to all players. The main idea captured by these assumptions is that there is *pure uncertainty about the level of the benefits* from global abatement. Distributional issues play no role as realizations are completely symmetric. This may also be viewed as uncertainty about the global benefits from abatement. For the later analysis it is helpful to point out that because γ_i represents the marginal benefits from abatement, uncertainty about the level of benefits translates into a positive variance of the sum of marginal benefits but a zero variance across the realized marginal benefits. That is, the sum of marginal benefits is random but, for each realization, the benefit parameters are equal across players.

Compared to the studies mentioned above, we can be more general in two respects. First, we do not have to assume any particular distribution of the uncertain benefit parameter. Second, our payoff function does not imply binary equilibrium abatement strategies and hence optimal abatement strategies are a function of the benefit parameter. This leads to more interesting effects of uncertainty and learning on second stage outcomes and also helps to derive clear-cut conclusions.

2.3.2.2. Uncertainty of type 2: Pure uncertainty about the distribution of benefits. Uncertainty of type 2 draws on the idea of Na and Shin (1998). They consider three players with ex-ante symmetric expectations about the random variable Γ_i. The ex-post realizations are three different values $\gamma_1 < \gamma_2 < \gamma_3$. This can be viewed as players sequentially drawing one ticket from an urn with three tickets, each ticket representing a different value γ_i, $i = 1,2,3$. Tickets are not returned to the urn.

The main idea captured by this experiment is that uncertainty relates to *individual parameters.* Moreover, though expectations about the benefit parameters are symmetric, their realizations are asymmetric. Each player is allocated a different value from the set $\{\gamma_1, \gamma_2, \gamma_3\}$. Thus the sum of all γ_i's over all players is known, i.e. the sum of marginal benefits is constant. As a result, the variance of the sum of marginal benefits is zero but the variance of the realized marginal benefits across players is positive.

In contrast to uncertainty of type 1, type 2 can be interpreted as *pure uncertainty about the distribution of the benefits* from global abatement. That is, the vector of random variables $\Gamma = (\Gamma_1,...,\Gamma_n)$ can be interpreted as the shares of the global benefits from abatement, as for instance modeled in Dellink et al. (2008), simply by dividing each random variable by the sum of marginal benefits, $\sum_{i=1}^{n} \Gamma_i$.

We generalize the idea of Na and Shin in four respects. First, we do not restrict attention to only three players but assume any arbitrary number of players n. Second, not all realizations γ_i have to be different, though at least two, otherwise no uncertainty would remain, i.e. we assume $\gamma_1 \leq \gamma_2 \leq ... \leq \gamma_n$ with at least one inequality being strict. Third, we also consider the scenario of partial learning, whereas Na and Shin compare only full and no learning. Fourth, we consider the possibility to mitigate asymmetries through transfers.

2.3.2.3. Uncertainty of type 3: Simultaneous uncertainty about the level and distribution of benefits. Uncertainty of type 3 combines the

Table 1
Three types of uncertainty about the benefit parameters.

Type of uncertainty	Definition of parameters	Interpretation of parameters	Ex-ante expectations of parameters	Ex-post realizations of parameters
1) Pure uncertainty about the level of benefits	One random player draws a benefit parameter for all players from some distribution	Common	Symmetric	Symmetric
2) Pure uncertainty about the distribution of benefits	Each player draws a benefit parameter from a set $\{\gamma_1,..., \gamma_n\}$ without replacement	Individual	Symmetric	Asymmetric
3) Simultaneous uncertainty about the level and the distribution of benefits	Each player draws a benefit parameter from a set $\{\gamma_1,..., \gamma_n\}$ with replacement	Common and individual	Symmetric	Asymmetric

features of uncertainty of types 1 and 2: there is *simultaneous uncertainty about the level and the distribution of the benefits* from global abatement. That is, there is uncertainty about a *common parameter and individual parameters*. This idea is captured by assuming that all random variables, Γ_i, $i = 1,..., n$, are independently distributed, though identically (because we assume symmetric expectations). This may be viewed as players sequentially drawing tickets from an urn with tickets being returned after each draw. As for uncertainty of type 2, we assume that at least two tickets are different. The assumptions of uncertainty of type 3 imply that the sum of marginal benefits is random and its variance is larger (smaller) than under uncertainty of type 2 (type 1). That is, the level effect is maximal for uncertainty of type 1, minimal for uncertainty of type 2 and intermediate for type 3. Moreover, the variance across the realized parameter values for uncertainty of type 3 is, on average, lower (larger) than for type 2 (type 1). That is, the distribution effect is maximal for uncertainty of type 2, minimal for uncertainty of type 1 and intermediate for type 3.[9]Different from Kolstad and Ulph (2011), we do not restrict the uncertain parameter values to two values (high and low), equilibrium abatement strategies are not binary and we study the role of transfers.

3. Model solution

In this section, we solve the model for the three learning scenarios and the three uncertainty cases. As a reference point, we start by considering certainty.

3.1. Certainty

According to the procedure of backwards induction, we first solve the coalition formation game for second stage outcomes. Using payoff function (9), and following the instructions of how signatories choose their equilibrium abatement levels as stated in Eq. (2) and non-signatories as stated in Eq. (3), we obtain the following equilibrium abatement levels:

$$q_i^*(S) = \sum_{\ell \in S} \gamma_\ell \; \forall i \in S. \; q_j^*(S) = \gamma_j \; \forall j \notin S. \; \sum_{k \in N} q_k^*(S) = m \sum_{i \in S} \gamma_i + \sum_{j \notin S} \gamma_j. \quad (10)$$

Casual inspection of Eq. (10) reveals that individual and total abatement increase in the benefit parameters. It is also evident that when a coalition S of size m has formed, and a non-signatory $j \notin S$ joins the coalition, total abatement will increase. We call this last property global efficiency from cooperation (GEF), implying that a sequence of accessions to a coalition S increases global abatement, with the highest global abatement being obtained in the grand coalition.

If abatement levels in Eq. (10) are substituted into payoff functions (9), we obtain payoffs without transfers of signatories $\Pi^*_{i \in S}(S)$, non-signatories $\Pi^*_{j \notin S}(S)$ and total payoffs, $\sum_{k \in N} \Pi^*_k(S) = \sum_{i \in S} \Pi_i^*(S) + \sum_{j \notin S} \Pi_j^*(S)$. The same observations made above for abatement can be observed for payoffs. That is, payoffs increase in the benefit parameter, and total payoffs increase with an enlargement of the coalition (GEF), with the highest total payoff being obtained by the grand coalition, corresponding to the social optimum.[10]

For the understanding of the incentives to form coalitions, two more properties are interesting. If a non-signatory j joins coalition S such that $S \cup \{j\}$ forms, then total payoffs of this group increases, $\sum_{i \in S} \Pi_i^*(S \cup \{j\}) > \sum_{i \in S} \Pi_i^*(S) + \Pi^*_{j \notin S}(S)$. This property, called *superadditivity* (SAD), underlines that there is generally an incentive for cooperation, at least if the gains from cooperation are shared "fairly" among coalition members. However, also all remaining non-signatories $k \notin S \cup \{j\}$ benefit from this enlargement of coalition S, $\Pi^*_{k \notin S \cup \{j\}}(S \cup \{j\}) > \Pi^*_{k \notin S \cup \{j\}}(S)$, a property called *positive externality* (PE). Hence, even if we consider transfers and find that the grand coalition is not stable, then this can be interpreted as the positive externality effect being stronger than the superadditivity effect. In the absence of transfers, an additional reason for small coalitions may be that the gains from cooperation are shared unequally as Lemma 1 illustrates.

Lemma 1. Equilibrium coalition size under certainty and no transfers

Consider a coalition S which is composed of m members with parameters $\gamma_{i_1} \leq \gamma_{i_2} \leq ... \leq \gamma_{i_m}$, and $n - m$ singletons with parameters $\gamma_{j_1} \leq \gamma_{j_2} \leq ... \leq \gamma_{j_{n-m}}$. In the absence of transfers, no stable coalition comprises more than three players, and a coalition is stable if and only if:

a) $\gamma_{i_1} = \gamma_{i_2} = \gamma_{i_3}$ *if* $m = 3 \wedge n = 3$
b) $\gamma_{i_1} = \gamma_{i_2} = \gamma_{i_3} \wedge 2\gamma_{j_{n-m}} < 3\gamma_{i_1}$ *if* $m = 3 \wedge n \geq 4$
c) $\frac{\gamma_{i_2}}{\gamma_{i_1}} \leq \sqrt{2} \wedge \gamma_{j_{n-m}} < \frac{\gamma_{i_1} + \gamma_{i_2}}{2}$ *if* $m = 2 \wedge n \geq 3$
d) $\forall \gamma_i, i \in N$ *if* $m = 1$

Proof. Follows using payoffs in Appendix 1 and applying the conditions of internal and external stability in Eqs. (4) and (5), respectively. **Q.E.D.**

The main conclusions from Lemma 1 are straightforward. First, no coalition larger than three players is stable. Second, a three-player coalition is only stable if all coalition members have the same benefit parameter (conditions a and b). If the total number of players exceeds three, then the symmetric benefit parameters of the coalition members must be sufficiently large compared to those benefit parameters of outsiders, such that this coalition is also externally stable (condition b). Third, if a coalition among different players is formed, the maximum stable coalition size is two and players cannot be too different (condition c). Moreover, the mean of the two benefit parameters of the coalition members must be sufficiently large compared to the benefit parameters of outsiders, otherwise external stability would be violated. Finally, the singleton coalition structure is stable by definition (condition d).

The upshot of Lemma 1 is that in the absence of transfers, heterogeneity is an obstacle to cooperation. This is quite different with transfers.

[9] We view assumptions about uncertainty of type 3 as the most "natural" way to combine type 1 and 2, though it is clear that an infinite number of possible combinations could be constructed which are either closer to type 1 or 2.

[10] Property GEF, as well as SAD and PE mentioned below, are easy to prove using payoffs provided in Appendix 1.

Lemma 2. Equilibrium coalition size under certainty and transfers

Consider a coalition S which is composed of m members and let the vector of benefit parameters be denoted by γ^S, the standard deviation of γ^S by σ_S, the arithmetic mean of γ^S by μ_S and define the coefficient of variation by $CV_S = \frac{\sigma_S}{\mu_S}$. Assume that coalition members apply the almost ideal transfer scheme.

a) *There is a stable coalition of at least three players; no stable coalition comprises less than three players.*

b) *A coalition of size $m \geq 4$ is internally stable if and only if $CV_S \geq A(m)$ where $A(m) = \sqrt{\frac{m^2-4m+3}{2m-3}}$ with $A(m)$ increasing in m.*

c) *A coalition of size $m \geq 4$ is externally stable if and only if for all coalitions $S \cup \{j\}, j \notin S$, $CV_{S \cup \{j\}} < A(m+1)$.*

Proof. See Appendix 2.

Comparing Lemmas 1 and 2 shows that the largest coalition without transfers comprises three players, which is the smallest coalition size in the presence of transfers. Without transfers, heterogeneity is an obstacle for forming large coalitions, with transfers it is an asset and if the coefficient of variation, CV_S, is large enough, even the grand coalition can be stable. Without transfers, payoffs among coalition members are asymmetrically distributed, making free-riding attractive. With transfers, an asymmetric distribution of payoffs is no longer an issue. Moreover, heterogeneity generates additional surplus from cooperation. More precisely, a low mean μ_S and a high standard deviation σ_S are conducive to cooperation as the coefficient of variation increases. The intuition is the following.

First, the larger the average benefit parameter, the larger will be equilibrium abatement levels of signatories and the more ambitious are coalitional abatement efforts compared to no cooperation. Ambition translates into a high free-rider incentive.

Second, the higher the standard deviation of the coalition members' benefit parameters, the larger are the gains from cooperation for coalition members, increasing the attractiveness of cooperation (or decreasing the attractiveness of free-riding). Recall, if countries behave non-cooperatively, they set their own marginal benefits equal to their own marginal costs from abatement, whereas if they form a coalition they set the sum of marginal benefits of coalition members equal to their marginal cost. Hence, the move from no to coalitional cooperation is associated with two effects. First, the externality is internalized among coalition members. This internalization effect has two components, the benefit to the coalition is the superadditivity component, but because this benefit is *non-exclusive* also non-signatories benefit from higher abatement efforts of the coalition, which is the positive externality component. Second, abatement burdens are allocated cost-effectively among coalition members. This cost-effectiveness effect is an *exclusive* benefit accruing to the coalition and hence contributes only to the superadditivity component.

Now suppose all coalition members are symmetric, i.e. have the same benefit parameter. Because all players have the same abatement cost function in our model, no cooperation already implies a cost-effective allocation of abatement burdens. Hence, coalition members benefit from the internalization effect but not from the cost-effectiveness effect. In other words, the superadditivity component is not very strong compared to the positive externality component, resulting in only small coalitions being stable.

In contrast, if players have different marginal benefit parameters γ_i, no cooperation is no longer cost-effective and cooperation is also associated with a cost-effectiveness effect. In fact, the larger the difference in benefit parameters, the larger will be the cost-effectiveness effect. In other words, the larger the asymmetry, the stronger is the superadditivity compared to the positive externality component, allowing for larger stable coalitions.

3.2. Uncertainty

3.2.1. Full learning

The scenario of full learning (FL) is (almost) a direct application of what has been derived for certainty. For the purpose of later comparison with the scenarios of partial and no learning, we will evaluate outcomes from an ex-ante perspective. This has no impact on the properties superadditivity, positive externality and global efficiency from cooperation.

In terms of equilibrium coalitions, with reference to our three types of uncertainty (see Subsection 2.3.2), the following predictions follow directly from Lemmas 1 and 2.

Proposition 1. Equilibrium coalitions under full learning

Under the full learning scenario, the expected equilibrium coalition size $E[m^{*FL}]$ is given by:

Uncertainty of type 1 (Pure uncertainty about the level of benefits)
$E[m^{*FL}] = 3$ where all possible 3-player coalitions are stable, with and without transfers.

Uncertainty of type 2 (Pure uncertainty about the distribution of benefits)
Without transfers: $1 \leq E[m^{*FL}] \leq 3$. In particular:

a1) *$E[m^{*FL}] = 1$, only the singletons coalition structure is stable if neither of conditions a, b and c in* Lemma 1 *hold;*

a2) *$E[m^{*FL}] > 1$, with at least one non-trivial coalition being stable, though no stable coalition comprises more than three players if at least one of the conditions a, b and c in* Lemma 1 *hold.*

With transfers (AITS): $E[m^{*FL}] \geq 3$, all stable coalitions comprise at least three players. In particular, recalling that the coefficient of variation of the benefit vector of coalition S,γ^S, is denoted by CV_S:

b1) *$E[m^{*FL}] = 3$, all three-player coalitions are stable if $CV_S < \sqrt{3/5}, \forall S : \#S = 4$;*

b2) *$E[m^{*FL}] > 3$, stable coalitions comprise three or more players if $\exists S, \#S = 4 : CV_S \geq \sqrt{3/5}, n \geq 4$;*

b3) *$E[m^{*FL}] \geq 4$, stable coalitions comprise four or more players if $CV_S \geq \sqrt{3/5}\ \forall S : \#S = 4, n \geq 4$.*

Uncertainty of type 3 (Simultaneous uncertainty about the level and the distribution of benefits)
Without transfers: $1 < E[m^{*FL}] \leq 3$.
With transfers (AITS): $E[m^{*FL}] \geq 3$, all stable coalitions comprise at least three players. In particular, recalling that the coefficient of variation of the benefit vector γ^S of coalition S is denoted by CV_S:

c1) *$E[m^{*FL}] = 3$, all three-player coalitions are stable if for all possible realizations of the vector of benefit parameters, $\gamma = (\gamma_1,...,\gamma_n)$, $CV_S < \sqrt{3/5}$, $\forall S : \#S = 4$.*

c2) *$E[m^{*FL}] > 3$: all stable coalitions comprise three or more players and at least one comprises four or more players if there are realizations of the vector of benefit parameters, $\gamma = (\gamma_1,...,\gamma_n)$, such that $\exists S : CV_S \geq \sqrt{3/5}$, $n \geq 4$.*

If there is pure uncertainty about the level of benefits (uncertainty of type 1), then all players will have the same benefit parameter γ_i. For symmetric players, transfers make no difference and all coalitions of three players are stable. If there is pure uncertainty about the distribution of benefits or simultaneous uncertainty about the level and distribution of benefits (uncertainty of types 2 and 3), then we assumed that at least two γ_i-values are different, but we did not rule out the possibility that some players may have the same γ_i-value. Hence, without transfers, coalitions comprising two and/or three players can be stable but also the singletons coalition structure. A subtle detail of uncertainty of type 3 is that even if the distribution of the random benefit parameters is spread over a wide range of different values, there is a positive probability that all players receive exactly the same value. That is, there is the possibility that all players draw the same ticket from the urn, as tickets are returned. This explains the statement $E[m^{*FL}] > 1$ for uncertainty of type 3. With transfers, we know from Lemma 2 that all stable coalitions will comprise at least three players. The specific conditions listed in Proposition 1 will prove useful for the comparison of results in Section 4.

3.2.2. Partial learning

Partial Learning (PL) assumes that players know the parameter values once they enter the second stage of coalition formation. Hence, equilibrium abatement levels in the second stage are exactly those stated under certainty in Eq. (10) and equilibrium payoffs are those provided in Appendix 1. Thus, PL and FL are identical in the second stage and, as a consequence, properties GEF, PE and SAD also hold for PL, which can be easily verified using payoffs in Appendix 3.

However, PL and FL are different regarding the first stage, in which countries choose their membership. Under PL, countries decide on their membership not based on actual payoffs but based on expected payoffs which are provided in Appendix 3. Since all players share the same beliefs, expected payoffs of all signatories are identical and the same applies to all non-signatories. Consequently, membership is based on symmetric ex-ante payoffs. This allows for three simplifications. First, we can discard transfers, i.e. they would not make any difference. Second, the actual identity of coalition members does not matter for the outcome. In other words, if a coalition of size m is stable, all coalitions with m members are stable. Third, there is a direct link between internal and external stability as explained in Section 2.1 for symmetric players.

Proposition 2. Equilibrium coalitions under partial learning

Let the vector of random benefit parameters, including all players, be denoted by $\Gamma = (\Gamma_1, \Gamma_2, \ldots, \Gamma_n)$, with $\Gamma_1, \Gamma_2, \ldots, \Gamma_n$ being identical distributed because of symmetric beliefs of all players. Denote the standard deviation of Γ_i by σ, the expected value by μ and define the coefficient of variation by $CV = \frac{\sigma}{\mu}$.

Under the partial learning scenario, with and without transfers, the expected equilibrium coalition size $E[m^{*PL}]$ is given by:

Uncertainty of type 1 (Pure uncertainty about the level of benefits)
$E[m^{*PL}] = 3$ where all possible 3-player coalitions are stable.

Uncertainty of type 2 (Pure uncertainty about the distribution of benefits)

- a1) *If $CV < B(m,n)$ where $B(m,n) = \sqrt{\frac{(m-3)(n-1)}{n+m-3}}$, with $B(m,n)$ increasing in m and n, then $E[m^{PL}] = 3$ where all possible 3-player coalitions are stable.*
- a2) *If $CV \geq B(m,n)$, $n \geq 4$, then $E[m^{*PL}] \geq 4$ and all possible m-player coalitions are stable for which $B(m,n) \leq CV < B(m+1, n)$ if $m < n$ and $CV \geq B(m,n)$ if $m = n$ hold.*

Uncertainty of type 3 (Simultaneous uncertainty about the level and distribution of benefits)

- b1) *If $CV < C(m)$ where $C(m) = \sqrt{m-3}$ with $C(m)$ increasing in m, then $E[m^{PL}] = 3$ where all possible 3-player coalitions are stable.*
- b2) *If $CV \geq C(m)$, $n \geq 4$, then $E[m^{PL}] \geq 4$ and all possible m-player coalitions are stable for which $C(m) \leq CV < C(m+1)$ if $m < n$ and $CV \geq C(m)$ if $m = n$ hold.*

Proof. See Appendix 4.

The qualitative conclusions under partial learning have much resemblance to what we found for certainty and full learning if transfers among coalition members are applied. Asymmetric distributions of the random benefit parameters with a large standard deviation as well as a low expected value (implying together a large coefficient of variation) are conducive to the stability of large coalitions. The mechanism is similar to the one laid out for certainty and transfers (Section 3.1). The only difference is that not the standard deviation and the mean of the random benefit parameters of coalition members but of all players that matter. This is because under PL, all players are ex-ante symmetric when deciding upon membership.

Note that the threshold of the coefficient of variation for uncertainty of type 2, $B(m, n)$, depends not only on the coalition size m but also on the total number of players as the draw of "benefit parameter tickets" from the urn is not independent (draw without replacement). Moreover, the threshold of the coefficient of variation for uncertainty of type 3, $C(m)$ is higher than for uncertainty of type 2, $B(m, n)$, i.e. $C(m) > B(m,n)$. The reason is that for any given distribution of the random benefit parameters, the variance across realized benefit parameter values is larger for uncertainty of type 2 than type 3, as explained in Section 2.3.

3.2.3. No learning

No learning (NL) assumes that players do not know the realizations of the random variables Γ_i in the second stage of coalition formation. Hence, they derive their equilibrium abatement levels from maximizing expected payoffs, i.e. payoffs in Eqs. (2) and (3) have to be replaced by expected payoffs. For our specific payoff functions in Eq. (9), replacing γ_i by the random variable Γ_i, this implies that payoffs are linear in the random variables and hence certainty equivalence holds. That is, the maximization of expected payoffs is equivalent to the maximization of payoffs under certainty for $\gamma_i = E[\Gamma_i]$. Moreover, we can make use of the fact that for all three types of uncertainty players are assumed to have identical beliefs, i.e. $E[\Gamma_i] = E[\Gamma_j]$ for all $i, j \in N$. This delivers equilibrium abatement levels:

$$q_i^{**}(S) = mE[\Gamma_k] \ \forall i \in S, q_j^{**}(S) = E[\Gamma_k] \ \forall j \notin S, \quad \sum_{k \in N} q_k^{**} = \left(m^2 - m + n\right)E[\Gamma_k] \tag{11}$$

where we use two asterisks in order to stress the difference to the abatement levels under certainty (see Eq. (10), which are also those under full and partial learning). It is evident that individual and total abatement levels increase in the expected value of the benefit parameter. Moreover, total abatement increases with the size of the coalition. Computing expected payoffs (see Appendix 5) confirms this also for payoffs. Hence, property GEF holds under NL and as it is easy to confirm, using the payoffs in Appendix 5, properties SAD and PE as well.

When it comes to determining stable coalitions, two pieces of information are useful. First, due to symmetric beliefs of all players, not only the abatement levels (as is evident from Eq. (11)) but also the payoffs (as is evident from Appendix 5) of all signatories are identical and the same is true for all non-signatories. Hence, transfers have no influence on stable coalitions. Second, because abatement levels and payoffs only depend on the (symmetric) expected value of the benefit parameters (and not on its variance), the type of uncertainty (types 1, 2 and 3) does not matter for the size of equilibrium coalitions.

Proposition 3. Equilibrium coalitions under no learning

Under the no learning scenario, with and without transfers, for all three types of uncertainty (i.e. pure uncertainty about the level of benefits, pure uncertainty about the distribution of benefits, and simultaneous uncertainty about the level and distribution of benefits), the expected equilibrium coalition size is $E[m^{*NL}] = 3$ where all possible 3-player coalitions are stable.

Proof. Inserting payoffs in Appendix 5 into the condition for internal and external stability as defined in Eqs. (4) and (5), respectively, delivers the result. **Q.E.D.**

Compared to full and partial learning, stable coalitions are robust to the variance and expected value of the benefit parameters.

4. Comparison of results

4.1. Preliminaries

In this section, we compare results for the three learning scenarios for each of the three types of uncertainty. For a sensible comparison across the three learning scenarios, it is important to evaluate outcomes from an ex-ante perspective. That is, we compute and compare expected coalitions sizes and expected total payoffs across the three learning scenarios.

Analytically, we proceed in three steps. First, we compare coalitions relating to the first stage of coalition formation. We say that the first stage effect from learning is positive (negative) if the expected coalition size with learning is larger (smaller) than without learning. For this comparison, we draw heavily on the results established in Section 3. Second, we compare expected aggregate payoffs in the second stage of coalition formation, considering a generic coalition S of size m. We say that the second stage effect from learning is positive (negative) if the expected aggregate payoff with learning is larger (smaller) than without learning. Third, we pull the information from steps 1 and 2 together in order to derive conclusions regarding the overall outcome. As we consider three scenarios of learning, the effect of learning is viewed along the sequence of no, partial and full learning (NL, PL and FL).

In many cases, conclusions are derived along the following simple reasoning. If the effect of learning is positive (negative) in both stages, the effect of learning will be positive (negative) overall. The essential link between both stages is related to the property of global efficiency from cooperation (GEF). If for any generic coalition S of size m learning scenario A performs better than scenario B in stage 2, then if scenario A leads to equal or larger coalitions than scenario B in the first stage, we can conclude that A performs better than B overall. Moreover, it is helpful to recall that FL and PL are identical in the second stage and hence any difference regarding the overall performance of these two learning scenarios must stem from the first stage.

In some cases this simple reasoning does not work, even though the break-down in steps 1 and 2 is always useful when it comes to understand the driving forces. One reason is that sometimes effects in the two stages work in the opposite direction and hence overall conclusions are either not clear-cut or need further analysis. Another reason is that in some cases under full learning and no transfers the link between stages 1 and 2 is not direct. That is, not only the expected coalition size in the first stage matters, but also the identity of players in stable coalitions. Hence, the concept of a generic coalition of size m, which means that all combinations of m players forming a coalition of size m are considered, is not useful. The expected total abatement and payoff of a generic coalition of size m may be lower than if a particular coalition with m or less players forms, provided this particular coalition comprises players with high benefit values γ_i.

Finally, regarding the overall evaluation, we exclusively focus on aggregate payoffs, neglecting aggregate abatement. Nevertheless, it is useful to be aware of the following relations regarding total expected abatement in the second stage. Taking expectations over total abatement level under PL and FL in (10) for a generic coalition S of size m and comparing it with the expected total abatement level under NL in (11) reveals that both are equal. Hence, in the second stage, expected total abatement levels of the three learning scenarios are equal for each type of uncertainty. The reason is that equilibrium abatement levels are linear in the benefit parameters and, for a given number of players and coalition size, only depend on expected values. This implies that the second stage effect from learning regarding total payoffs can be exclusively attributed to different allocations of abatement levels across players under different learning scenarios. This simplification turns out to be useful for the interpretation of different second stage outcomes.

4.2. Pure uncertainty about the level of benefits

Pure uncertainty about the level of benefits (uncertainty of type 1) allows for the derivation of straightforward results. From Section 3 we know that all three scenarios of learning lead to coalitions of size three as players are ex-ante and ex-post symmetric. However, as Lemma 3 shows, there is no equivalence in terms of second stage outcomes.

Lemma 3. First and second stage outcomes (pure uncertainty about the level of benefits)

First stage of coalition formation: The expected stable coalition size under full learning (FL), partial learning (PL) and no learning (NL) are ranked as follows:

FL = PL = NL.

Second stage of coalition formation: Consider a generic coalition S of size m, $1 \leq m \leq n$. For every generic coalition S, the total expected payoffs under the three learning scenarios are ranked as follows:

FL = PL > NL

where the difference between the two learning scenarios (FL and PL) and no learning (NL) increases in the variance of the random benefit parameter Γ_i.

Proof. First stage outcomes follow directly from Propositions 1, 2 and 3. Second stage outcomes follow from FL = PL by definition and by comparing expected payoffs under PL and NL, as provided in Appendices 3 and 5, respectively, noting that $\Gamma_i = \Gamma_j \ \forall i, j \in N$ for uncertainty of type 1. **Q.E.D.**

The intuition regarding second stage outcomes is the following. Because of symmetric realizations of all benefit parameters, all signatories choose the same abatement level under FL and PL and the same holds for non-signatories. If the realized global benefit parameter is high (low), then all players will chose high (low) abatement levels. This means that with learning, though abatement levels across signatories and non-signatories are different, collectively the abatement levels across all players will be optimally adjusted according to the realization of the global benefit parameter.

Under NL, abatement levels within the group of signatories and non-signatories are also symmetric because abatement levels are chosen based on the expected value of the benefit parameters, which are identical due to symmetric beliefs. However, due to less information compared to FL and PL, abatement levels are based on the expected value of the global benefit parameter, and hence not adjusted according to the realization of the global benefit parameter. This means that if the realized global benefit parameter is high (compared to the expected value), abatement levels across all players fall short of optimal abatement levels (undershooting) and if the realized global benefit parameter is low (compared to the expected value), abatement levels are too high (overshooting). The collective under- and overshooting under NL compared to FL and PL leads to a loss of net benefits (i.e. payoffs) which is more pronounced the larger the variance of the global benefit parameter. The reason is that the total payoff in the second stage is a strictly convex function in the benefit parameter γ.[11] Therefore, the total payoff from adopting an abatement level based on the expected

[11] This is easily proved by using total payoffs as stated in Appendix 1, assuming symmetry of all benefit parameters and deriving the first and second derivatives with respect to the common benefit parameter.

value of the global benefit parameter is lower than the expected value of total payoffs.

Proposition 4. Overall outcome (pure uncertainty about the level of benefits)

For total expected payoffs, the outcome of the two-stage coalition formation game is ranked as follows:

$FL = PL > NL.$

Proof. Follows directly from Lemma 3.

Thus, under pure uncertainty about the level of benefits, the overall outcome is that learning leads to higher expected payoffs. This contrasts with results by Kolstad (2007), and Kolstad and Ulph (2008). They find that though full learning leads to larger stable coalitions than no learning, expected total payoffs are smaller. For partial learning they find multiple equilibria for some parameter values, and conclude that the most likely equilibrium leads to lower membership and a lower expected aggregate payoff than full and no learning. Thus, in terms of payoffs, they suggest: NL > FL > PL. The question that arises is why their results are so different from ours? The following paragraphs provide a brief explanation, which is supported by a short formal exposition and examples in Appendix 7.

Kolstad and Ulph assume a linear payoff function and hence equilibrium abatement levels are corner solutions, i.e. binary strategies "abate" or "not abate". The model is calibrated such that, on the one hand, abatement pays in the social optimum and in a coalition above a threshold number of players. On the other hand, abatement does not pay for non-signatories and particularly in the Nash equilibrium. The stable coalition is a knife-edge equilibrium: once a coalition member leaves, the coalition breaks apart as for the remaining coalition members it does no longer pay to abate. This threshold depends on the benefit parameter γ; the larger γ, the higher the benefits from cooperation but less coalition members are needed to form a profitable coalition. Hence, the size of stable coalitions decreases in the benefit parameter γ.

Though this model looks simple at the outset, it appears to us that this simplicity comes at a price. Regarding the first stage effect from learning, the claim that expected membership is higher under FL than NL rests on the assumption that the equilibrium size of coalitions can be approximated as a continuous variable. However, as Karp (2011) has pointed out, accounting for the fact that the number of signatories can only be an integer value, Kolstad and Ulph's conclusion may not be valid. More important, the second stage effect from learning changes in their model depending on the coalition size and can be positive, negative and nil. Therefore, one would expect that also overall effects may not be clear-cut. This expectation is not confirmed using the approximation in Kolstad and Ulph (2008) because total expected payoffs of the overall game are strictly concave in γ. However, again, Karp (2011) has shown that the approximation may be misleading and that the ranking FL > NL is possible. Clearly, in our model, effects in the first and second stage are well-behaved and always clear-cut, and payoffs are strictly convex in γ and therefore "learning is good".

4.3. Pure uncertainty about the distribution of benefits

From Section 3 we know that pure uncertainty about the distribution of benefits (uncertainty of type 2) can lead to smaller coalitions under FL than under PL and NL if transfers are not used to balance a possible asymmetric distribution of the gains from cooperation. With transfers, this shortcoming can be fixed. In fact, if the asymmetry of the benefit parameters across players is large enough, coalitions under FL will be larger than under NL. The same is true for PL, though no transfers are needed. All of this is compactly summarized in Lemma 4 regarding the first stage of coalition formation.

Lemma 4. First and second stage outcomes (pure uncertainty about the distribution of benefits)

First stage of coalition formation: The expected stable coalition size under full learning (FL), partial learning (PL) and no learning (NL) are ranked as follows:

No transfers : $PL \geq NL > FL;$

Transfers : $PL, FL \geq NL.$

Second stage of coalition formation: Consider a generic coalition S of size m, $1 \leq m \leq n$. For every generic coalition S, the total expected payoffs under the three learning scenarios are ranked as follows:

$NL \geq FL = PL$ *with strict inequality if* $m<n$

where the difference between the two learning scenarios and no learning increases in the variance of the random benefit parameters Γ_i.

Proof. First stage outcomes follow directly from Propositions 1, 2 and 3. Second stage outcomes follow from FL = PL by definition and by comparing expected payoffs under PL and NL, as provided in Appendix 3 and 5, respectively, noting that $\sum_{k=1}^{n} \Gamma_k$ is a constant under uncertainty of type 2 and using the terms in (III) in Appendix 4. **Q.E.D.**

Interestingly, regarding second stage outcomes, results are completely reversed compared to pure uncertainty about the level of benefits (compare Lemmas 3 and 4). The intuition is along the lines of the example mentioned in the introduction. Recall from Section 4.1 that expected total abatement in the second stage is the same under all three scenarios of learning and hence total benefits, which are linear in total abatement, will be identical. Thus, differences must be related to abatement costs.

Total abatement costs will also be the same for all scenarios of learning if and only if the grand coalition forms because all players will choose abatement levels such that marginal abatement costs are equal to the sum over all marginal benefits. Since not only under PL and FL, but also under NL, the sum of marginal benefits is known (i.e. there is no uncertainty about the level of benefits), the second stage effect from learning is nil in the grand coalition, which is also the social optimum.

However, for any coalition smaller than the grand coalition, expected total costs under NL are strictly lower than under PL and FL and hence the second stage effect from learning is strictly negative. In the singletons coalition structure this is evident: under FL and PL, players will choose different abatement levels due to different realizations of the benefit parameter whereas under NL all players will chose the same abatement level due to identical expected values of the benefit parameter. Only a symmetric allocation of abatement duties is cost-effective as all players have the same abatement cost functions.

Finally, for larger coalitions than the singleton coalition structure, a similar argument applies. Signatories choose higher abatement levels than non-signatories, the difference between both groups is smaller on average under NL than under FL and PL. In other words, the allocation of abatement between signatories and non-signatories is, on average, "more symmetric" under NL than under PL and FL. The advantage of NL over PL and FL increases with the variance of the random benefit parameters.

When deriving conclusions regarding overall outcomes, it is clear that if the negative second stage effect from learning is not upset through a positive first stage effect from learning, then NL leads to better outcomes than FL and PL. PL also performs better than FL if the first stage effect from learning is negative, which can happen without transfers. All of this is summarized under point i) in Proposition 5 below. As the identity of members of stable coalition matters under FL and no transfers, the sufficient condition for NL > FL and PL > FL is rather restrictive (i.e. only the singleton coalition structure is stable under FL and no transfers). For instance, point ii) suggests a class of distributions for which also NL > PL > FL (if $n \geq 4$) holds, despite under FL a two-player coalition among the two players with the largest γ_i-values is stable. For this class of distributions, all three-player coalitions are stable under PL and NL.

In order to reverse the result that "learning is bad" (see point iii) in Proposition 5), it is clear that the negative second stage effect from learning has to be overcompensated by a positive first stage effect. Not surprisingly, because of opposing effects of asymmetry on both stages, the sufficient conditions for FL > NL appear to be rather strong. Nevertheless, there is clear evidence that asymmetry improves upon overall outcomes under both learning scenarios, though under FL transfers are required. Finally, point iv) confirms that even without transfers, "learning can be good". The example we use in Appendix 6 to show this is quite informative. It assumes 100 countries with a distribution of the benefit parameter such that 98 countries have a low benefit parameter $\gamma_i = 1$ and two countries a very large benefit parameter $\gamma_j = 10$. This large asymmetry, even without transfers, leads to a two-player coalition among the two countries with the highest benefit parameter under FL, which performs better than a generic coalition of size 3 under NL and even of size 4 under PL.

Proposition 5. Overall outcome (pure uncertainty about the distribution of benefits)

Let the outcome of the two-stage coalition formation game be evaluated in terms of expected total payoffs.

i) *Sufficient conditions for "learning being bad" are*
   - *NL > FL if $n = 3$, or if only the singletons coalition structure is stable under FL and no transfers (condition a1 in Prop. 1), or if no coalition larger than 3 players is stable under FL with transfers (condition b1 in Prop. 1).*
   - *NL ≥ FL if no coalition larger than 3 players is stable under PL (condition a1 in Prop. 2), with strict inequality if $n \geq 4$.*
   - *PL > FL if only the singletons coalition structure is stable under FL and no transfers (condition a1 in Prop. 1) or if no coalition larger than 3 player is stable under FL with transfers (condition b1 in Prop. 1) and stable coalitions are strictly larger than 3 players under PL (condition a2 in Prop. 2).*

ii) *For any discrete uniform distribution of the benefit parameter over equidistant values, i.e $\{\gamma_1, \gamma_2, \ldots, \gamma_n\}$, with $\gamma_{i+1} = \gamma_i + b$, $b > 0$, $i = 1, \ldots, n-1$ the following ranking holds under no transfers:*
   NL ≥ PL > FL with strict inequality if $n \geq 4$.

iii) *Sufficient conditions for "learning being good" are*
   - *FL > NL if stable coalitions are strictly larger than 3 players under FL and transfers (condition b3, Prop. 1).*
   - *PL > NL if stable coalitions are strictly larger than 3 players under PL (condition a2 in Prop. 2).*

iv) *There are distributions of the random benefit variable for which even without transfers "learning is good" with ranking FL > PL > NL for which a necessary condition is that at least one non-trivial coalition is stable under FL (condition a2 in Prop. 1).*

Proof. See Appendix 6.

From Proposition 5 it appears that Na and Shin's result that "learning is bad" is a rather special result. As mentioned in point i), it only holds for any distribution if there are only three players. Then, the stable coalition under NL is the grand coalition, corresponding to the social optimum, whereas FL will lead only to a two- or one-player coalition and the second stage disadvantage of FL cannot be overturned. However, considering also the learning scenario PL in Na and Shin's setting, with $n = 3$, would already lead to less clear-cut results as then NL = PL. Moreover, as Proposition 5 suggests, there are many situations when "learning can be good", even if there is pure uncertainty about the distribution of the benefits from cooperation. In particular, Na and Shin's negative result about the role of learning for the success of SEIEAs is driven by asymmetry, but as we show, in a more general setting, asymmetry can have the opposite effect.

4.4. Simultaneous uncertainty about the level and distribution of benefits

Simultaneous uncertainty about the level and the distribution of benefits constitutes a mix of the two previous types of uncertainty (see Section 2.3.2, in particular footnote 9). The uncertainty part about the distribution shows up in the same first stage outcome as for uncertainty of type 2 (compare Lemmas 4 and 5) and the uncertainty part about the level shows up in the same second stage outcome as for uncertainty of type 1 (compare Lemmas 3 and 5). In particular, the last result stresses that the negative second stage effect from learning disappears once there is not pure uncertainty about the distribution of the benefits but also some uncertainty about the level of the benefits, which is certainly true for climate change but can also be expected for many other economic problems associated with positive externalities.

Lemma 5. First and second stage outcomes (simultaneous uncertainty about the distribution of benefits)

First stage of coalition formation: The expected stable coalition size under full learning (FL), partial learning (PL) and no learning (NL) are ranked as follows:

No transfers : PL ≥ NL ≥ FL;

Transfers : PL, FL ≥ NL.

Second stage of coalition formation: Consider a generic coalition S of size m, $1 \leq m \leq n$. For every generic coalition S, the total expected payoffs under the three learning scenarios are ranked as follows:

FL = PL > NL

where the difference between the two learning scenarios and no learning increases in the variance of the random benefit parameters Γ_i.

Proof. First-stage outcomes follow directly from Propositions 1, 2 and 3. Second-stage outcomes follow from FL = PL by definition and by comparing expected payoffs under PL and NL, as provided in Appendix 3 and 5, respectively, noting that benefit parameters are independently distributed. **Q.E.D.**

Combining the effects from both stages as stated in Lemma 5, allows for straightforward conclusions.

Proposition 6. Overall outcome (simultaneous uncertainty about the level and distribution of benefits)

i) *PL > NL.*
ii) *FL > NL with transfers.*
iii) *There are distributions of the random benefit variable for which without transfers learning is bad: NL > FL.*

Proof. i) and ii) follow directly from Lemma 5. iii) assume for instance that the random benefit parameters are equidistantly distributed, e.g. $\gamma_i = \{1, 2, 3\}$, $\forall i \in N$, $n = 3$, then there are 27 possible combinations of realizations of the random variables for which stable coalitions can be determined using Lemma 1 under FL and no transfers and aggregate payoffs can be computed using Appendix 1. Computing the average aggregate payoff over all possible realizations gives an expected aggregate payoff of 39 whereas under NL, using Appendix 5, and noting that $m^{*NL} = 3$, the expected aggregate payoff is 54. **Q.E.D.**

Though without transfers FL may lead to worse outcomes than NL, this never happens with transfers and PL is anyway always better than NL. Overall, if there is some uncertainty about the level of the benefits from cooperation, then even in a strategic context "learning is good", though a compensation scheme may be needed under FL. This clearly qualifies the negative conclusions about the role of learning for the success of SEIEAs derived in previous papers.

5. Summary and conclusions

This paper addressed the role of uncertainty and learning for the formation of self-enforcing international environmental agreements (SEIEAs). The central question was whether the veil of uncertainty impacts positively or negatively on the success of such agreements. The previous literature suggested a positive impact with the conclusion that "learning is bad". From our model the answer appears to be less straightforward. In fact, we showed that the negative impact of learning on the success of SEIEAs requires a couple of special assumptions and can also be mitigated in some cases by a transfer mechanism that is designed to minimize free-rider incentives.

The essential ingredients of our SEIEA-model is a (pure) public good provision game in which there is an incentive to cooperate but also an incentive to free-ride, as the benefits of public good provision are non-exclusive. Hence, full participation may not be a stable outcome. We assumed two-stages: in the first stage countries choose their membership in an agreement (becoming a signatory or remaining a non-signatory) and in the second stage choose their contributions to the public good. Uncertainty is related to the parameters of the benefit functions of individual players, with benefits depending on the sum of contributions. We considered uncertainty about the level and the distribution of benefits. Each type of uncertainty is considered first in isolation and then combined, giving rise to three types of uncertainty altogether. Uncertainty about the level of benefits is related to a common parameter and uncertainty about the distribution of the benefits is related to individual parameters. We also considered three scenarios of learning, with the two benchmark scenarios, full and no learning, and the intermediate scenario partial learning.

It became apparent that uncertainty and learning has an impact on both stages. There are two main reasons why "learning can be bad". First, the veil of uncertainty can be helpful when it comes to decide about participation in an agreement if the gains from cooperation could be unequally distributed. The higher the uncertainty about the distribution of the benefits from cooperation compared to the level of the benefits and the more dispersed individual benefit shares are among players, the smaller will be stable coalitions under full learning. In contrast, under no and partial learning an asymmetric distribution of the gains from cooperation is not an issue. Due to identical beliefs about the distribution of the random variables, all players are symmetric ex-ante when choosing membership.

Second, when signatories and non-signatories choose their optimal contributions in the second stage, they interact strategically as contributions are strategic substitutes. More information would be beneficial if there was a social planner, which corresponds to the grand coalition in our model or if individual players acted in isolation. However, in our setting, players act strategically as long as not all players join an agreement and hence more information could lead to worse outcomes. We showed that in our model if there is pure uncertainty about the distribution of the benefits from cooperation, the aggregate payoff under partial and full learning falls short of that under no learning. The allocation of contributions among players is "more symmetric" under no learning (as it is based on symmetric expected values of the random variable) than under partial and full learning (as it is based on realized values of the random variable) which turns out to be a cost advantage for strictly convex and symmetric abatement cost functions.

Though these negative effects from learning on the success of cooperation are interesting, we argued in this paper that there are good reasons why we should expect the effect of learning to be much less negative. Regarding the negative impact of asymmetry on membership, we argued that a well-designed transfer scheme can hedge against an asymmetric distribution of the gains from cooperation under full learning. In fact, we showed that with transfers, membership under full learning never falls short of that under no learning and exceeds it for a sufficient degree of asymmetry. That is, the larger the degree of asymmetry, the larger the relative advantage of membership in an agreement over free-riding. Hence, the larger will be stable agreements under full learning, also compared to no learning. Under partial learning, a similar asymmetry-effect works, though no transfers are needed. A sufficient degree of asymmetry leads to larger agreements under partial than under no learning.

Regarding the possible negative impact of learning on total payoffs in the second stage, we concluded that it only holds in our model if there is pure uncertainty about the distribution of benefits. Once there is also uncertainty about the level of the benefits, which certainly applies to climate change, the effect from learning in the second stage becomes positive.

Taking a wider perspective, it is clear that our results apply beyond the specifics of climate change. The setting is that of a public good game but any economic problem which is characterized by positive externalities from cooperation would lead to similar qualitative conclusions. Knowing ex-ante the total size of the cooperative pie from coordinated action is advantageous as it allows to better target at the total level of efforts. In contrast, receiving confirmation that the individual slices of the pie of various participants might be quite unequally distributed may cause problems and therefore should be hedged against with an appropriate compensation mechanism. Heterogeneity among cooperating agents can be an asset if comparative advantages are exploited, but only if the gains from cooperation are fairly shared.

Despite that we generalized some aspects of previous models, our model also shares some of their limitations. We pick only three which we believe are the most policy-relevant. First, we assumed that all players share the same beliefs, which requires some agreement or coordination about the current scientific evidence. Regarding many international policy issues, as in climate change, not all parties use and believe in the same scientific evidence. Therefore, a natural extension of our model could be to capture this phenomenon through a bargaining process in the second stage of coalition formation. It would be interesting to analyze whether an evolvement to more similar views, e.g. through intensified international research collaboration, political dialog, and institutional coordination, through international bodies like for instance IPCC, would be conducive to the success of cooperation. Second, we assumed risk-neutrality but one could also look at the effects of risk-aversion on treaty outcomes, like in Boucher and Bramoullé (2010). They show that uncertainty about the level of benefits leads to lower abatement efforts than certainty, but higher participation, which may lead overall to higher aggregate payoffs. Whether this would also hold in our setting is difficult to predict as their model assumes a linear payoff function as Kolstad (2007) and Kolstad and Ulph (2008, 2011) and, as we have argued above, already in the context of risk-neutrality conclusions can be fundamentally different. Third, it would be interesting to model partial learning in a Bayesian sense. This would require to model coalition formation as a truly dynamic game in which players can revise their membership over time (Rubio and Ulph, 2007). In order to obtain new interesting economic insight compared to our analysis and that of Rubio and Ulph (2007), it would probably also be necessary to allow for the possibility that players can reduce uncertainty through investment in R&D (learning-by-research) and implementation (learning-by-doing). For instance, the more players invest in mitigating climate change, the more they will learn about their abatement costs over time. However, most likely, already even for simple extensions it will be difficult to obtain analytical solutions as first results by Breton and Sbragia (2011) suggest.

Acknowledgment

The research has been conducted while Pedro Pintassilgo was a visiting scholar at the Department of Economics, University of Stirling. He would like to acknowledge the hospitality of the department as well as the financial support by the Portuguese Foundation for Science and Technology (FCT), grant no. BSAB/735/2007. Both authors have benefited from the comments by Andrew Oswald, University of Warwick, and Ian Lange, University of Stirling, three anonymous referees and Roberton C. Williams III, the Co-Editor.

Appendix 1. Payoffs under certainty in the second stage[12]

Using equilibrium abatement levels in Eq. (10) and inserting them into payoff functions in Eq. (9), gives the following payoffs:

$$\begin{aligned}
&\Pi^*_{i\in S}(S) = \gamma_i\Big(m\sum_{\ell\in S}\gamma_\ell + \sum_{j\notin S}\gamma_j\Big) - \frac{1}{2}\Big(\sum_{\ell\in S}\gamma_\ell\Big)^2\\
&\Pi^*_{j\notin S}(S) = \gamma_j\Big(m\sum_{i\in S}\gamma_i + \sum_{k\notin S}\gamma_k\Big) - \frac{1}{2}\big(\gamma_j\big)^2\\
&\Pi^*(S) = \sum_{i\in S}\Pi^*_i(S) + \sum_{j\in S}\Pi^*_j(S) =\\
&\frac{1}{2}m\Big(\sum_{i\in S}\gamma_i\Big)^2 + \Big(\sum_{i\in S}\gamma_i\Big)\Big(\sum_{j\notin S}\gamma_j\Big)(1+m) + \Big(\sum_{j\notin S}\gamma_j\Big)^2 - \frac{1}{2}\sum_{j\notin S}\big(\gamma_j\big)^2
\end{aligned} \tag{I}$$

These payoffs are used to compute stable coalitions under certainty and full learning.

Appendix 2. Proof of Lemma 2

A coalition of size $m = 1$ is internally stable by definition. For all other coalitions S of size $m \geq 2$, we use payoffs provided in Appendix 1 and insert them in the condition for potential internal stability (6), recalling that all potentially internally stable coalitions are internally stable for the almost ideal transfer scheme, which gives after some manipulation the condition $(2m - 3)\sigma_S^2 + (-m^2 + 4m - 3)\mu_S^2 \geq 0$. This condition always holds for $m = 2$ and $m = 3$. For $m \geq 4$, we solve this condition and obtain $CV_S \geq A\,(m)$. Noting the relation between (potential) internal and (potential) external stability, external stability of a coalition S of size m follows from the failure of potential internal stability for all coalitions $S \cup \{i\}$ of size $m + 1$ where one player outside coalition S is added to S. By the same token, a coalition of size $m = 1$ and $m = 2$ cannot be externally stable because all coalitions of size $m = 2$ and $m = 3$ are internally stable. Hence, the minimum size of stable coalitions with transfers is $m = 3$. **(Q.E.D.)**

Appendix 3. Expected payoffs under partial learning in the second stage

Replacing all γ_i's by their random counterparts Γ_i's in the payoffs in Appendix 1, taking expectations over the payoffs and noting that the random variables $\Gamma_1, \Gamma_2, \ldots, \Gamma_n$ are identically distributed due to symmetric beliefs, we obtain after some manipulation:

$$\begin{aligned}
&E[\Pi^*_{i\in S}(S)] = \frac{m}{2}E\big[(\Gamma_i)^2\big] + \Big(n + \frac{m^2}{2} - \frac{3}{2}m\Big)E[\Gamma_i\Gamma_k], \quad k\in N\setminus\{i\}\\
&E\big[\Pi^*_{j\in S}(S)\big] = \frac{1}{2}E\Big[\big(\Gamma_j\big)^2\Big] + \big(m^2 + n - m - 1\big)E\big[\Gamma_j\Gamma_k\big], \quad k\in N\setminus\{j\}\\
&E\Big[\sum_{\ell\in N}\Pi^*_\ell(S)\Big] = \Big(\frac{n + m^2 - m}{2}\Big)E\big[(\Gamma_i)^2\big]\\
&+\Big(-\frac{m^3}{2} + \Big(n - \frac{1}{2}\Big)m^2 + (1-n)m + n^2 - n\Big)E[\Gamma_i\Gamma_k], i\neq k\in N
\end{aligned} \tag{II}$$

These payoffs are used to compute stable coalitions under partial learning. They also represent second stage expected payoffs under partial and full learning for a generic coalition S of size m.

Appendix 4. Proof of Proposition 2

A coalition of size $m = 1$ is internally stable by definition. For all other cases, $m \geq 2$, we use payoffs in Appendix 3 and apply the definition of internal stability in Eq. (4). After some manipulation, this condition reads $E[(\Gamma_i)^2] + (2 - m)E[\Gamma_i\Gamma_k] \geq 0$ with player i and k being any two different players. This condition can be further manipulated, depending on the type of uncertainty. For uncertainty of type 1, $\Gamma_i = \Gamma_k\ \forall i, k \in N$ and hence the condition for internal stability reads $(3 - m)E[(\Gamma_i)^2] \geq 0$ and due to the link between internal and external stability, the condition for external stability reads $(2 - m)E[(\Gamma_i)^2] < 0$ (by setting $m = m + 1$ in the condition for internal stability and assuming that it fails). Hence, $m^* = 3$. For uncertainty type 2, we make use of the fact that $\sum_{k=1}^{n}\Gamma_k$ is a constant and hence it can be shown that

$$E[\Gamma_i\Gamma_k] = \frac{n(E[\Gamma_i])^2 - E\big[(\Gamma_i)^2\big]}{n-1}. \tag{III}$$

Denoting $\mu = E[\Gamma_i]$ and $\sigma^2 = Var[\Gamma_i]$ and noting that $E[(\Gamma_i)^2] = \sigma^2 + \mu^2$, the condition for internal stability reads $(n + m - 3)\sigma^2 + (3 - m)(n - 1)\mu^2 \geq 0$. This condition always holds for $m = 2$ and $m = 3$. Because of the link between internal and external stability, $m = 2$ is not externally stable. For $m \geq 4$, the internal stability condition can be rewritten as $CV \geq B(m,n)$ and hence for external stability we have $CV < B\,(m + 1, n)$. For uncertainty of type 3, $\Gamma_1, \Gamma_2, \ldots, \Gamma_n$ are independently distributed and hence the condition for internal stability reads $\sigma^2 + (3 - m)\mu^2 \geq 0$ from which the conclusions in Proposition 2 follow immediately. **(Q.E.D.)**

Appendix 5. Expected payoffs under no learning in the second stage

Substituting γ_i by $E\,[\Gamma_i]$ in payoff function (9), inserting abatement levels from Eq. (11) into these payoff functions for signatories and non-signatories, and taking expectations over payoffs, delivers:

$$\begin{aligned}
&E[\Pi^*_{i\in S}(S)] = \Big(\frac{m^2}{2} - m + n\Big)(E[\Gamma_k])^2\\
&E\big[\Pi^*_{j\notin S}(S)\big] = \Big(m^2 - m + n - \frac{1}{2}\Big)(E[\Gamma_k])^2\\
&E\Big[\sum_{k\in N}\Pi^*_k(S)\Big] = \Big(-\frac{1}{2}m^3 + m^2 n + m\Big(-n + \frac{1}{2}\Big) + n^2 - \frac{1}{2}\Big)(E[\Gamma_k])^2
\end{aligned} \tag{IV}$$

These payoffs are used to compute stable coalitions under no learning. They also represent expected payoffs under no learning in the second stage for a generic coalition S of size m.

[12] For some proofs we only provide the intuition due to space limitations. Details are available upon request.

42 M. Finus, P. Pintassilgo / Journal of Public Economics 103 (2013) 29–43

Appendix 6. Proof of Proposition 5

Statements i) follow directly from combining first and second stage effects from learning and recalling property GEF. Statements ii) requires comprehensive calculations which are available upon request. Statement iii): If condition a2 in Proposition 2 holds, then $E[m^{*PL}] \geq 4 > 3 = E[m^{*NL}]$. Using the lower bound $m^{*PL} = 4$ (because of property GEF), inserting this into expected aggregate payoffs under PL, as provided in Appendix 3, and inserting $m^{*NL} = 3$ into expected payoffs under NL, as provided in Appendix 5, delivers the result PL > NL. FL > NL follows from $E[m^{*FL}] \geq 4$ according to condition b3 in Proposition 1 and noting that FL and PL are equal in the second stage. Statement iv): Consider $n = 100$, $\gamma_1 = \gamma_2 = \ldots = \gamma_{98} = 1$, and $\gamma_{99} = \gamma_{100} = 10$. Computing the expected payoff under NL using Appendix 5 and setting $m^{*NL} = 3$ because of Proposition 3, gives 14,673.1. Under FL and no transfers, only the two players with $\gamma_{99} = \gamma_{100} = 10$ form a stable coalition according to Lemma 1, and hence inserting this information into aggregate payoffs, as provided in Appendix 1, delivers a payoff of 15,835. Under PL, $m^{*PL} = 4$ according to Proposition 2, condition a2, and inserting this information into aggregate payoffs provided in Appendix 3, gives 15,395 and hence overall the ranking FL > PL > NL follows. **(Q.E.D)**

Appendix 7. First and second stage effects in Kolstad and Ulph's model

Consider the payoff function $\Pi_i = \gamma \sum_{k \in N} q_k - q_i$, assume $\frac{1}{n} \leq \gamma < 1$, and a normalized binary abatement strategy such that $q_i \in \{0, 1\}$. Then, in the social optimum $q_i^{*S} = 1$ and in the Nash equilibrium $q_i^{*N} = 0$. A coalition will abate and make a difference to no cooperation, provided $\gamma \geq \frac{1}{m}$ or $m \geq \frac{1}{\gamma}$. The internally and externally stable coalition of size m^* is the largest integer equal or larger than $\frac{1}{\gamma}$, i.e. $I\left(\frac{1}{\gamma}\right)$. Under uncertainty, let the probability of a low value γ_ℓ be p and of a high value γ_h be $1 - p$ and denote the expected value by $\overline{\gamma} = p\gamma_\ell + (1-p)\gamma_h$. Then, $m^{*NL} = I\left(\frac{1}{\overline{\gamma}}\right)$ and $m^{*FL} = pI\left(\frac{1}{\gamma_\ell}\right) + (1-p)I\left(\frac{1}{\gamma_h}\right)$. Moreover, non-signatories abatement levels are always zero whereas signatories abatement levels are given by:

$$q_{i \in S}^{*NL} = \begin{cases} 1 & \text{if} \quad m \geq m^{*NL} = \overline{m}^* = I\left(\dfrac{1}{\overline{\gamma}}\right) \\ 0 & \text{if} \quad m < m^{*NL} = \overline{m}^* = I\left(\dfrac{1}{\overline{\gamma}}\right) \end{cases}$$

$$q_{i \in S}^{*FL} = \begin{cases} 1 & \text{if} \quad m \geq m_s^* = I\left(\dfrac{1}{\gamma_s}\right), \quad s \in \{\ell, h\} \\ 0 & \text{if} \quad m < m_s^* = I\left(\dfrac{1}{\gamma_s}\right), \quad s \in \{\ell, h\}. \end{cases}$$

Consider possible coalition sizes, m_1, m_2, m_3 and m_4 such that $m_1 < m_h < m_2 < \overline{m} < m_3 < m_\ell < m_4$, and denote total abatement by Q and the total payoff by Π, then for m_1: $Q^{*NL} = 0 = Q^{*FL}$ and $\Pi^{*NL} = \Pi^{*FL} = 0$; m_2: $Q^{*NL} = 0 < (1-p)m_2 = Q^{*FL}$ and $\Pi^{*NL} = 0 < m_2\,(1-p)(n\gamma_h - 1) = \Pi^{*FL}$; m_3: $Q^{*NL} = m_3 > (1-p)m_3 = Q^{*FL}$ and $\Pi^{*NL} = nm_3\overline{\gamma} - m_3 > m_3(1-p)(n\gamma_h - 1) = \Pi^{*FL}$ and m_4: $Q^{*NL} = m_4 = Q^{*FL}$ and $\Pi^{*NL} = nm_4\overline{\gamma} - m_4 = (1-p)(m_4 n\gamma_h - m_4) + p(m_4 n\gamma_\ell - m_4) = \Pi^{*FL}$ which shows that the second stage effect from learning depends on the size of the coalition. Consider $\gamma_\ell = 0.3$, $\gamma_h = 0.5$, then if $p = 0.95$, $\overline{\gamma} = 0.31$ and $\overline{m}^* = m^{*NL} = 4$ whereas the expected size under FL is $m^{*FL} = 3.9$. If $p = 0.5$ expected membership under both learning scenarios is 3 whereas if $p = 0.75$, expected membership under FL is 3.5 whereas under NL only 3. This shows that first stage effects are not always clear-cut. For a more general analysis, see Karp (2011).

References

Asheim, G.B., Froyn, C.B., Hovi, J., Menz, F.C., 2006. Regional versus global cooperation for climate control. Journal of Environmental Economics and Management 51 (1), 93–109.

Barrett, S., 1994. Self-enforcing international environmental agreements. Oxford Economic Papers 46, 878–894.

Barrett, S., 2001. International cooperation for sale. European Economic Review 45 (10), 1835–1850.

Barrett, S., 2003. Environment and Statecraft: The Strategy of Environmental Treaty-making. Oxford University Press, New York.

Barrett, S., 2006. Climate treaties and "breakthrough" technologies. American Economic Review 96 (2), 22–25.

Bloch, F., 2003. Non-cooperative models of coalition formation in games with spillovers. In: Carraro, C. (Ed.), The Endogenous Formation of Economic Coalitions. Edward Elgar, Cheltenham, UK, pp. 35–79 (et al., ch. 2).

Bosello, F., Buchner, B., Carraro, C., 2003. Equity, development, and climate change control. Journal of the European Economic Association 1 (2-3), 601–611.

Boucher, V., Bramoullé, Y., 2010. Providing global public goods under uncertainty. Journal of Public Economics 94, 591–603.

Brennan, G., Buchanan, J.M., 1985. The Reason for Rules: Constitutional Political Economy. Cambridge University Press, Cambridge, UK.

Breton, M., Sbragia, L., 2011. Learning under Partial Cooperation and Uncertainty. Papers presented at the Environment and Sustainability Forum, Exeter, UK.

Carraro, C., Siniscalco, D., 1993. Strategies for the international protection of the environment. Journal of Public Economics 52 (3), 309–328.

d'Aspremont, C., Jacquemin, A., Gabszewicz, J.J., Weymark, J.A., 1983. On the stability of collusive price leadership. Canadian Journal of Economics 16 (1), 17–25.

Dellink, R., Finus, M., Olieman, N., 2008. The stability likelihood of an international climate agreement. Environmental and Resource Economics 39 (4), 357–377.

Deneckere, R., Davidson, C., 1985. Incentives to form coalitions with Bertrand competition. The Rand Journal of Economics 16 (4), 473–486.

Diamantoudi, E., Sartzetakis, E.S., 2006. Stable international environmental agreements: an analytical approach. Journal of Public Economic Theory 8 (2), 247–263.

Donsimoni, M.-P., Economides, N.S., Polemarchakis, H.M., 1986. Stable cartels. International Economic Review 27 (2), 317–327.

Eyckmans, J., Finus, M., 2009. An Almost Ideal Sharing Scheme for Coalition Games with Externalities. Stirling Discussion Paper Series, 2009-10. University of Stirling.

Finus, M., 2003. Stability and Design of International Environmental Agreements: The Case of Transboundary Pollution. In: Folmer, H., Tietenberg, T. (Eds.), International Yearbook of Environmental and Resource Economics, 2003/4. Edward Elgar, Cheltenham, UK, pp. 82–158 (ch. 3).

Finus, M., Tjøtta, S., 2003. The Oslo protocol on sulfur reduction: the great leap forward? Journal of Public Economics 87 (9–10), 2031–2048.

Fuentes-Albero, C., Rubio, S.J., 2009. Can the international environmental cooperation be bought? European Journal of Operation Research 2002, 255–264.

Gollier, C., Treich, N., 2003. Decision-making under scientific uncertainty: the economics of the precautionary principle. Journal of Risk and Uncertainty 27 (1), 77–103.

Haeringer, G., 2004. Equilibrium binding agreements: a comment. Journal of Economic Theory 117 (1), 140–143.

Hardin, G., 1968. The tragedy of the commons. Science 162 (3859), 1243–1248.

Helm, C., 1998. International cooperation behind the veil of uncertainty. Environmental and Resource Economics 12 (2), 185–201.

Hoel, M., 1992. International environment conventions: the case of uniform reductions of emissions. Environmental and Resource Economics 2 (2), 141–159.

Iida, K., 1993. Analytical uncertainty and international cooperation: theory and application to international economic policy coordination. International Studies Quarterly 37, 431–457.

IPCC, 2007. Climate Change 2007. Synthesis Report.

Karp, L., 2011. The effect of learning on membership and welfare in an international environmental agreement. Climatic Change. http://dx.doi.org/10.1007/s10584-011-0134-5.

Kolstad, C.D., 2005. Piercing the veil of uncertainty in transboundary pollution agreements. Environmental and Resource Economics 31 (1), 21–34.

Kolstad, C.D., 2007. Systematic uncertainty in self-enforcing international environmental agreements. Journal of Environmental Economics and Managament 53 (1), 68–79.

Kolstad, C.D., Ulph, A., 2008. Learning and international environmental agreements. Climatic Change 89 (1-2), 125–141.

Kolstad, C.D., Ulph, A., 2011. Uncertainty, learning and heterogeneity in international environmental agreements. Environmental and Resource Economics 50, 389–403.

McGinty, M., 2007. International environmental agreements among asymmetric nations. Oxford Economic Papers 59, 45–62.

Na, S.-L., Shin, H.S., 1998. International environmental agreements under uncertainty. Oxford Economic Papers 50 (2), 173–185.

Poyago-Theotoky, J., 1995. Equilibrium and optimal size of a research joint venture in an oligopoly with spillovers. The Journal of Industrial Economics 43 (2), 209–226.

Rubio, S.J., Ulph, A., 2006. Self-enforcing international environmental agreements revisited. Oxford Economic Papers 58 (2), 233–263.

Rubio, S.J., Ulph, A., 2007. An infinite-horizon model of dynamic membership of international environmental agreements. Journal of Environmental Economics and Management 54 (3), 296–310.

Sandler, T., 2004. Global Collective Action. Cambridge University Press, Cambridge.

Stern, N., 2006. Stern Review: The Economics of Climate Change. Report prepared for the HM Treasury in the UK. Cambridge University Press, Cambridge, UK (Published 2007).

Ulph, A., 1998. Learning about Global Warming? In: Hanley, N., Folmer, H. (Eds.), Game Theory and the Environment. Edward Elgar, Cheltenham, UK, pp. 255–286 (et al., ch. 13).

Ulph, A., 2004. Stable international environmental agreements with a stock pollutant, uncertainty and learning. Journal of Risk and Uncertainty 29 (1), 53–73.
Weikard, H.-P., 2009. Cartel stability under optimal sharing rule. The Manchester School 77, 575–593.
Weikard, H.-P., Finus, M., Altamirano-Cabrera, J.C., 2006. The impact of surplus sharing on the stability of international climate agreements. Oxford Economic Papers 58 (2), 209–232.
Yi, S.-S., 1996. Endogenous formation of customs unions under imperfect competition: open regionalism is good. Journal of International Economics 41, 153–177.
Yi, S.-S., 1997. Stable coalition structures with externalities. Games and Economic Behavior 20 (2), 201–237.
Young, O., 1994. International Governance: Protecting the Environment in a Stateless Society. Cornell University Press, Ithaca, New York.

[40]

Schmollers Jahrbuch 121 (2001), 1–26
Duncker & Humblot, Berlin

International Environmental Cooperation in the One Shot Prisoners' Dilemma*

By Alfred Endres and Cornelia Ohl

Abstract

Our analysis simultaneously deals with two types of uncertainty: first, the uncertainty of the behaviour of nature (stochastic or parametric uncertainty) and second, the uncertainty of the behaviour of nations (strategic uncertainty). This *risk-strategic* analysis points out that chances of international coalition formation to protect the global commons depend on the characteristics of the national welfare distributions and the country specific risk attitudes. Focusing on a static two-country-model and a dichotomous choice setting we point out that risk aversion is a prerequisite for transforming a static prisoners' dilemma (according to the order of expected national welfare) into a game of higher cooperation possibility. For different intensities of risk aversion we develop a typology of coopertive behaviour showing that enforcing environmental agreement is not necessarily harder than initiating it. Moreover we investigate how the design of strategies of international risk management (here: emission trading with and without trade restrictions) feeds back to the incentive structure of an international treaty, like the Kyoto protocol. We argue that the traditional judgement criteria of policy assessment in an international setting should be put into a wider context by the criterion of *"cooperative push"*. This criterion reflects the ability of instruments and technologies to initiate and self-enforce international environmental agreements. Thereby it provides the necessary link between local and global concern.

Zusammenfassung

Wir integrieren zwei Typen von Unsicherheit, die in der Literatur zu den internationalen Umweltproblemen bisher getrennt behandelt worden sind, parametrische (stochastische) und strategische Unsicherheit. Die erstgenannte Form bezieht sich auf die den Entscheidungsträgern unbekannten künftigen Zustände der Natur, die zweite darauf, dass für jeden einzelnen Staat das Verhalten der anderen Staaten ungewiss ist. Unsere *risiko-strategische* Analyse ergibt, dass die Chancen, multinationale Koalitionen zum Schutz der globalen Umwelt zu schließen, stark von den län-

* The authors gratefully acknowledge the receipt of grant no. EN 192/4 -1 by the German Research Foundation (DFG). They are also indebted to Bianca Rundshagen, University of Hagen, and two anonymous referees for their helpful comments on an earlier draft of this paper.

derspezifischen Risiko-Präferenzen abhängen. Risikoscheu erweist sich als notwendige Bedingung für die Überführung eines statischen Gefangenendilemmas in ein Spiel mit höherer Kooperationswahrscheinlichkeit. Wir zeigen, wie dieser Zusammenhang für ein internationales Risikomanagement genutzt werden kann und wenden die Analyse auf das Kyoto Protokoll an. Bisher hat sich die ökonomische Analyse umweltpolitischer Instrumente auf traditionelle Kriterien wie Effizienz, ökologische Treffsicherheit und dynamische Anreizwirkung konzentriert. Aus unserer Analyse folgt, dass der Katalog im internationalen Kontext um das Kriterium der "Kooperationsfreundlichkeit" (Cooperative Push) ergänzt werden muß. Mit diesem Kriterium gelingt es, die Fähigkeiten von verschiedenen umweltpolitischen Instrumenten, die individuelle Rationalität und die Stabilität von Vereinbarungen zu fördern, zu erfassen.

JEL-Classification: Q20, Q28, C78.

1. Introduction

To limit global environmental risks internationally coordinated action is a must. Global pollutants, defined to show uniform diffusion, are pure public bads (e.g. Siebert 1998). Therefore countries are unable to reach the nationally most preferred pollution (emission) load unilaterally. Consequently nations have to reach an agreement on a globally accepted emission target. However, agreements are difficult to attain and to protect. The provision of global public goods (here: environmental quality) is a voluntary task of sovereign nations which is usually not enforceable by global institutions. Since each country maximises national welfare as a free-rider (problem of self-enforcement; e.g. Barrett 1994, 1997; Heister 1997) globally optimal solutions are generally missing.

Because of national welfare depending on the success of international risk management i.e., on the possibility of cooperation among sovereign nations, economic literature has given close attention to the strategic behaviour of the polluting countries. Up to now the design of mechanisms which initiate and self-enforce international environmental agreements have been in the centre of interest of game theoretic analyses (e.g. Barrett 1992, 1997; Carraro/Siniscalco 1997; Endres 1997; Finus/Rundshagen 1998; Stähler 1998a, b). A contract is said to be self-enforcing when each country maximises national welfare by holding on to the agreement. The property of self-enforcement is of great importance since supranational institutions to monitor international agreements (and if necessary to punish free-riders) are missing.

To reduce complexity, game theoretic literature on this issue usually neglects an important aspect of the problem: National welfare is undetermined because of scientific lacks of knowledge regarding the relationship between environmental pollution and damage. Additional uncertainties arise from the ambiguities of economic damage evaluation. Thus game theoretic analyses only concentrate on strategic uncertainty caused by the interaction of rational players (here: the nations). In contrast to this the interaction of scientific uncertainty and the optimal level of pollution control is the heart of risk theoretic analyses (e.g. Chichilnisky/Heal 1998; Siebert 1998; Welsch 1995; Xepapadeas 1997). However these analyses generally neglect strategic aspects which are as important as aspects of stochastic (parametric) uncertainties, in the international context.

The *risk-strategic* analysis dealt with in this paper tries to bridge the coexisting literature by the integration of both: *strategic uncertainty* caused by the behaviour of nations and *stochastic uncertainty* caused by the behaviour of nature. In this overriding context we discuss the effects of the interaction of stochastic and strategic uncertainties on the possibility of international coalition formation. To keep things simple, the analysis is confined to a two-country-model and a dichotomous policy choice (cooperation versus defection).

We proceed as follows:

The following *section 2* defines the prisoners' dilemma which in the past has been the starting point of many game theoretic analyses. *Section 3* expands this basic model by the integration of stochastic uncertainty. With the help of a "classical" decision criterion, the μ-σ-principle, we link stochastic and strategic uncertainty. *Section 4* introduces the "global alliance of risk". *Section 5* discusses the influence of alternative preference patterns (risk-neutrality, risk aversion and risk-seeking) on the possibility of international cooperation. Here the focus is on how national risk attitudes determine the strategy choices of the involved countries and the stability of an international environmental agreement. Based on the results derived in part 5 of the paper, *section 6* develops a typology of cooperative behaviour in terms of game theory. *Section* 7 points out that national choice of environmental technology and policy instruments (as emission taxation or quota restriction) can be designed as catalysts of global environmental cooperation given the risk preferences of the involved countries. Here, the focus is on the consequences of the design of alternative emission trading regimes, as discussed within the Kyoto protocol. *Section 8* summarises the main findings. *Section 9* (*Epilogue*) investigates the consequences of alternative assumptions regarding the global alliance of risk. Particularly, the variability of environmental damages is modelled different from the main body of the paper.

2. The Prisoners' Dilemma

The dependence of national welfare on the behaviour of foreign nations can be described with game theoretic methods. According to the quality of real world interdependencies different types of games are illustrative, like the chicken and assurance game.[1] To characterise the interdependencies of nations suffering from a global pollutant the *prisoners' dilemma* is a relevant game. The structure of the game covers the following incentive scheme: Two nations decide on whether to coordinate environmental measures or not (choice of "cooperation" or "defection"). As argued above, binding agreements are not feasible in a supranational setting i.e., the game is non-cooperative. Therefore nations are not sure which action (cooperation or defection) each other nation takes.[2]

Given the strategy choice of the foreign country we label the maximum welfare level which is feasible for the home country by W_{DC}, W_{CC}, W_{DD} and W_{CD} respectively.

We further presuppose that the incentive structure of the prisoners' dilemma adequately describes the decision problem of the two countries under consideration. Then the following order of a home country's national welfare displays the incentive problems of international environmental agreements:

$$W_{DC} > W_{CC} > W_{DD} > W_{CD} \quad (1)$$

For the foreign country the same order holds.

Ranking (1) states that:

Referring to the status quo (*DD*), welfare of both nations can be (Pareto-) improved if bilateral environmental cooperation (*CC*) takes place ($W_{CC} > W_{DD}$). However if the foreign nation behaves as a free-rider (and the home country cooperates unilaterally) national welfare in the home country is lower than national welfare feasible in the status quo ($W_{CD} < W_{DD}$). This is so because national costs of emission control exceed the damage avoided in the home country. If on the other hand the home country behaves as a free-

1 For an introduction into game theory see e.g. Luce / Raiffa 1957; for a modern version e.g. Eichberger 1993.

2 With *C* for cooperation and *D* for defection four activity combinations have to be judged: *DC, CC, DD* and *CD*. The first capital letters point to the action of the country under consideration (the "home / first country"), the second to the action of the other country (the "foreign / second country"). If the foreign country defects, the home country is able to choose cooperation or defection leading to the combinations *CD* and *DD,* respectively. Analogously *CC* and *DC* are obtainable if the foreign country's choice is cooperation.

rider (and the foreign country cooperates unilaterally) the home country's welfare improves compared to the choice of bilateral cooperation ($W_{DC} > W_{CC}$). These coherencies also hold vice versa. Consequently (because of $W_{DD} > W_{CD}$ and $W_{DC} > W_{CC}$), each country does best when choosing to defect. Thus disregarding stochastic uncertainties the choice of defection is a dominant option (the equilibrium choice) and in the end the reason why international cooperation fails (at least in this static setting).

So given that a prisoners' dilemma is at work both nations are expected to hold on to the status quo which is unfavourable for each of them compared to bilateral cooperation ($W_{DD} < W_{CC}$). Following these game theoretic considerations international coalition formation is not feasible. However, some international environmental agreements among sovereign nations have been achieved and enforced, as the Montreal Protocol on CFCs, 1987. Moreover an agreement on the limitation of greenhouse gases has been reached in Kyoto, 1997. These phenomena (and other examples of international coalition formation) can not be explained within the simple model of the static prisoners' dilemma. Extensions of this model are needed to cover complex real world stories. As is well known, switching to dynamic games is a possibility to follow that road (e.g. Pearce 1992; Stähler 1996; Taylor 1987). Another possibility is the introduction of stochastic uncertainties, as shown in the paper at hand.

Analysing the effects of stochastic uncertainties on cooperation in a static setting instead of a dynamic one is theoretically appealing. This is so because the static context is the *worst case scenario* regarding the chances for cooperation. If the introduction of stochastic uncertainties generates incentives to cooperate in this adverse setting, the point regarding the importance of these uncertainties is particularly strong.[3] Moreover, in certain real world applications modelling a static game seems to be particularly appropriate. A dynamic game in general presupposes that the same stage game (e.g., a prisoners' dilemma) is played more than once. However, environmental issues like global warming can not always be characterised in terms of a dynamic setting.

[3] It is well known that the problems of the prisoners' dilemma can be solved in a dynamic game with infinite time horizon and a sufficiently low rate of discount. Apart from the argument in favour of using a static model, given above, it may also be doubted that the dynamic variant is more realistic in this paper's context: E.g., the targets of the Kyoto protocol have to be attained within the years 2008 and 2012 (see Oberthür / Ott 1999). After this period further measures have to be agreed upon in a subsequent protocol. To date, it is not clear which targets and measures for which countries a further protocol will contain nor can we take it for granted that there will be a subsequent agreement at all.

3. On the Integration of Probabilistic Phenomena into the Prisoners' Dilemma Game

Scientific uncertainty as well as uncertainty arising from damage evaluation (Hanley/Spash 1993; Houghton et al. 1996) complicate the forecast of the quality and the degree of manmade and natural environmental change.[4] Due to these lacks of knowledge the advantages of slowing global warming by the reduction of anthropogenic caused greenhouse gases are unknown to the nations. Consequently the assessment of national welfare when limiting environmental risks has to be expressed in terms of a national welfare distribution. In the paper at hand this is specified by the mean μ and the spread of the outcomes σ.[5]

The possibility of *manmade* environmental change was already pointed out by Svante Arrhenius at the beginning of the 20th century for the case of climate variation (Arrhenius 1903). Today this possibility enables international risk management to reduce the level as well as the probability of environmental damage by the limitation of global pollutants (endogenous risk). However anthropogenic effects are overlapped by natural phenomena; in the context of climate change, e.g. the position of earth to sun or the activity of sun spots and their relation to climate variations (exogenous risk). The impact of exogenous and endogenous risk factors on environmental change is therefore hard to distinguish but there is no doubt that human activity is unable to reduce the realisation of exogenous risks. It is also out of dispute that ecological systems, like the climate, are subject to sudden fluctuations. Thus, the possibility remains that even a small effect of an anthropogenic risk may have a significant environmental impact.[6] That is why incentives to reduce manmade risks exist, even if anthropogenically caused risks are not the dominant risk factor. We therefore expand the concept of national welfare maximisation under strategic uncertainty by the integration of national attitudes towards risk.

Risk preferences may imply the desire to minimise fluctuations of national welfare (risk aversion). Alternatively, national preference may be insensitive towards the possible sway of welfare (risk-neutrality). A further possibility is that the variation of welfare is explicitly welcome (risk-seeking). These spread preferences may be recorded by the well established

---

4 E.g., research on the climate system still lacks knowledge of the interaction between oceans and climate or between climate and earth's albedo. (Albedo indicates the intensity relation of incoming radiation from sun to earth and outgoing reflections.)

5 These measures are the basis of many risk economic analyses e.g., in investment- and portfolio theory.

6 Imagine a barrel that is brimful with exogenous risks. Then it is just one drop of endogenous risk which runs the barrel over.

μ-σ-criterion (Huang / Litzenberger 1988; Sinn 1990). In the context of this paper the decision for or against environmental protection depends on the level of expected national welfare σ and the spread of national welfare μ.[7] Both decision parameters implicitly take into account the level of damage as well as of damage probability. The paper at hand determines the preference value (ϕ) of a risky option as:

$$\phi = \mu - \alpha\sigma \,. \tag{2}$$

With the preference value ϕ we are able to transform national welfare W (now expressed in terms of μ and σ) into national risk utility. The level of ϕ is determined by the level of the mean μ which records nationally expected welfare and the standard deviation σ measuring the spread of the national welfare distribution, with σ being weighed by a national risk discount factor (α). Risk-neutrality corresponds to a value of α equal to zero, risk aversion to positive and risk-seeking to negative values. The bigger ϕ the higher is the national welfare measured in terms of risk-utility.

If we replace the national welfare levels under different activity combinations (W_{DC}, W_{CC}, W_{DD} and W_{CD}) considered to be certain in equation (1) above, by their corresponding means the following order results:

$$\mu_{DC} > \mu_{CC} > \mu_{DD} > \mu_{CD} \,. \tag{3}$$

The subscripts of μ show the activity combinations to which the means refer. Again, the first letter points to the action of the home country, the second to that of the foreign country. The order of the means corresponds to the incentive structure of a prisoners' dilemma game: Independent from the foreign action the home country's expected welfare is highest in the case of its defection ($\mu_{DC} > \mu_{CC}$ and $\mu_{DD} > \mu_{CD}$). In this sense we speak of a prisoners' dilemma under stochastic uncertainty.

Choosing D or C, however, it is not only expected welfare that varies but also the spread parameter σ. This is so because global pollutants cross over national boundaries and affect the level of damage as well as damage probability in *each* of the nations. This interaction generates the "global alliance of risk" that is measured by the following order of the standard deviation σ (which aggregates alternative levels of damage as well as damage probability):

---

[7] It should be noted that the method applied here is also relevant in connection with different decision criteria or measures of risk, like the variance or semi-variance of a welfare distribution. For alternative measures of risk e.g. see Brachinger / Weber (1997). Also see e.g. Sinn (1983) for the possibility of transforming a μ-σ-criterion into a risk utility function of the von Neumann-Morgenstern-type. On the approach presented above, also see Endres / Ohl (1998).

(4) $$\sigma_{DD} > \sigma_{CD} = \sigma_{DC} > \sigma_{CC} > 0\,.$$

Ranking (4) states that:

Starting from the status quo environmental risk can be reduced by each cooperative contribution of the countries ($\sigma_{DD} > \sigma_{CD}$; $\sigma_{DC} > \sigma_{CC}$). Thus we presuppose that the simple principle holds: Each emission unit abated or not emitted in the first place reduces manmade environmental risks. The limitation of anthropogenically caused risk is highest if both countries cooperate simultaneously. Assuming symmetric emission reduction for complexity reduction,[8] each nation unilaterally has the same impact on risk reduction (e.g. climate stabilisation) so that $\sigma_{CD} = \sigma_{DC}$ holds.

4. The Global Alliance of Risk

In the model presented here and (most likely) also in reality, the nations form a *global alliance of risk:* The environmental damages they suffer and the probabilities with which these damages occur depend upon the aggregate effect of all countries' emissions reductions.

There are (at least!) two important issues to be dealt with when modelling a global alliance of risk:

- In traditional environmental economic analysis it is assumed that environmental damage continuously varies with the level of emissions. The expected damage is taken to be a well behaved function of emissions, "preferably" with positive first and second order derivatives.[9] If we take the global environmental problem dealt with in this paper to be of this conventional kind, it is completely appropriate to model it with the von Neumann-Morgenstern approach. This is what we do in this paper, using the μ-σ-criterion. An alternative view would be to describe global environmental problems to be "catastrophic": With a very low probability they might lead to the extreme form of damage, extinguishing life on our planet. It is sometimes argued that for global warming, a certain threshold of emission concentrations might exist, the transgression of which would lead to global catastrophe. Just like in the case of nuclear power plants, this type of risk is not adequately modelled using the expected utility framework.[10]

8 In case of cooperation each cooperating country is expected to avoid the same amount of global pollutants. Of course, the model is open for a generalisation.

9 See, e.g., Endres (2000), Kolstad (2000), or any other environmental economics textbook.

10 See, e.g., Chichilnisky (2000).

Of course, whether global environmental problems are of the conventional or of the catastrophic type is an empirical matter. It is possible that the answer will depend upon the type of global environmental problem we are dealing with. In the context of global warming there is considerable scientific dispute upon whether a critical threshold exists (Roughgarden / Schneider (1999), Nordhaus (1994)). If it does exist, it is more likely to be relevant in the long run (Manne / Richels (1999)), than in the short or intermediate run (Pizer (1999)).[11] In the negotiations for the Kyoto protocol, a critical threshold of CO_2- concentrations in the atmosphere did not play any role for the definition of the emission reduction targets specified in the agreement.

All in all, even though it may not be appropriate for each specific global problem in any context, expected utility theory, as used in this paper, is probably helpful to explain and to evaluate the behaviour of countries in the international environmental arena.

- Of course, using the μ-σ-criterion does not say how the means and standard deviations of environmental damages vary with emissions.

  There will be little dispute with our assumption (3), supposing that the structure of our problem is an expected prisoners' dilemma.[12] Our assumption on the standard deviations is likely to be more controversial.[13]

  We have already given a brief rationale for our presumption that the degree of uncertainty decreases with the degree of cooperation when we introduced (4) in chapter 3, above. Let us elaborate: Emissions are closely correlated with economic variables, like the GDP. Future development of GDP is uncertain. In case the countries do not commit to stabilize their emissions (i.e., they defect, DD in (4)), the uncertainty of future emission levels will be a consequence of the GDP-uncertainty. This is different in the case of both countries agreeing to certain target levels of emissions (CC in (4)). In this case the effect of the sway in the GDP on the sway of emissions will be counterbalanced by environmental policy.[14] Thus, in the

11 Consequently, WBGU (2000) characterizes global warming by two different types of risk: One conventional (intermediate) and one catastrophic (long run).

12 The prisoners' dilemma is the economist's traditional interpretation of global (and other) environmental problems (see Brockmann / Stronzik / Bergmann (1999) for a recent application to global warming). Still, there is no natural law that environmental problems are of the prisoners' dilemma type, indeed. After all, this is an empirical matter like, alas, so many issues dealt with in this paper. E.g., Lipnowski / Maital (1983) argue that many environmental problems are described better by the Chicken Game.

13 Indeed, both referees were concerned about this assumption.

14 Stabilization of national emissions to meet the standards of an international agreement is complete if the countries use a transferable discharge permit system to regulate emissions. If fully enforced, this system operates with complete "ecological accuracy".

case of bilateral cooperation the variance of emissions is plausible to be smaller than in the case of bilateral defection. If one country cooperates and the other one defects (*CD* in (4)), the variance is plausible to be in-between.

Of course, we would be happy to present empirical evidence instead of speculations in discussing (4). Unfortunately, this kind of evidence on the relationship between the variability of damages and the level of emissions is fragmentary at best.[15] The data base for an empirically supported formulation of (4) is much too weak and spotty. Moreover, in areas of application where we have some evidence, it seems to point into different directions: A case where the standard deviation seems to have increased with the level of emissions (global temperature) is rain. In the process of global warming not only did the average amount of rain per incidence increase, but also the periods of drought on the one hand and of "flash floods" on the other hand. However, the variability of daily temperature seems to constitute a counter example: Even though average temperature increased, its daily amplitude decreased.

What should we conclude for economic analysis from the fact that sciences do not give a clear cut answer about how to specify (4), and perhaps never will?

To our opinion, economic analysis should produce *conditional* answers: If the inequality signs in (4) point one way, what is the consequence for the nations' propensity to cooperate? If they point the other way, how does the propensity change?

This is the route we have taken in the present paper. In its main body we carry our analysis *given* the standard deviation decreases with increasing levels of cooperation (i.e. with decreasing levels of emissions), as assumed in the formulation of (4) given in chapter 3, above. In the epilogue (Ch. 9) we investigate what it does to our results if the unequality signs in (4) are reversed.

5. Risk Preferences and Strategy Choice

If nations apply the μ-σ-principle introduced above and if they seek to maximise national risk utility, cooperation is preferred to defection if the former choice yields a higher ϕ than the latter. Dependent on the action chosen by the foreign country we have to distinguish two cases:

---

15 These remarks draw upon our personal correspondence with Hartmut Graßl (MPI and University of Hamburg). We would like to thank Professor Graßl for his patience in explaining scientific complications to some economists.

a) The *foreign* country defects

An incentive to cooperate exists in the *home* country if:

(5) $$\phi(CD) = \mu_{CD} - \alpha\sigma_{CD} > \phi(DD) = \mu_{DD} - \alpha\sigma_{DD} \Rightarrow$$

(I) $$\alpha > (\mu_{DD} - \mu_{CD})/(\sigma_{DD} - \sigma_{CD}) \equiv \alpha_{\mathrm{I}}^{\min} > 0$$

b) The *foreign* country cooperates

The *home* country is expected to cooperate, too, if:

(6) $$\phi(CC) = \mu_{CC} - \alpha\sigma_{CC} > \phi(DC) = \mu_{DC} - \alpha\sigma_{DC} \Rightarrow$$

(II) $$\alpha > (\mu_{DC} - \mu_{CC})/(\sigma_{DC} - \sigma_{CC}) \equiv \alpha_{\mathrm{II}}^{\min} > 0$$

Given the strategy choice of the foreign country the minimum degree of risk aversion which supports cooperation in the home country is labelled $\alpha_{\mathrm{I}}^{\min}$ and $\alpha_{\mathrm{II}}^{\min}$, respectively. Both thresholds are determined by the stochastic terms of the model. Inequality (I) refers to an incentive to cooperate unilaterally (condition of incentive compatibility: $\alpha > \alpha_{\mathrm{I}}^{\min}$). Inequality (II) refers to an incentive for bilateral cooperation (condition of self-enforcement: $\alpha > \alpha_{\mathrm{II}}^{\min}$). The conditions (I) and (II) show that internationally coordinated measures on the one hand depend on the level of the parameters μ and σ (the 'objective' risk assessment) and on the other hand on the intensity of national risk preferences (the 'subjective' risk assessment) expressed by the value of α.

Given the order of μ and σ as of (3) and (4), cooperative behaviour is only to be expected if α takes a positive value ($\mu_{DD} > \mu_{CD}$ and $\sigma_{DD} > \sigma_{CD}$ as well as $\mu_{DC} > \mu_{CC}$ and $\sigma_{DC} > \sigma_{CC}$). Thus, in equilibrium the choice of cooperation is only to be expected if the countries under consideration are risk averse ($\alpha > 0$). Neither risk-neutrality ($\alpha = 0$) nor risk-seeking ($\alpha < 0$) strengthens the possibility of international cooperation, here. Consequently country-specific risk preferences are able to explain why even nations showing identical profiles of national welfare distributions do not necessarily act in the same way.

Although risk aversion is a prerequisite for international cooperation, given (3) and (4) hold, the supposition of risk aversion is not sufficient to improve environmental quality. Only if the intensity of risk aversion exceeds one of the thresholds, incentives for cooperative environmental protection arise. In contrast to this, traditional risk economic analyses, generally excluding strategic interdependencies, lead to the result that environmental quality goes up if risk aversion is at work (e.g. Kreuzberg 1994; Siebert

1998). So integrating strategic and parametric uncertainties, as done in the paper at hand, really makes a difference.

Moreover conditions (I) and (II) reveal that the claims on the minimum degree of risk aversion necessary to foster cooperation adapt to country-specific differences in the levels of the stochastic parameters, μ and σ. This will e.g. be the case if differences in national objective risk assessment arise (the levels of μ and σ then might nationally vary because of different expertise or in case of asymmetric nations because of differences in national damage expectation). In consequence the thresholds in the home country may differ from that of the foreign country. This suggests (also see part 7 of the analysis) that an agreement among countries differing in culture and socio-economic characteristics is not necessarily harder to attain and to protect than an agreement among homogenous countries.[16]

Consequently acquiring knowledge of national risk attitudes (their nature *and* strength) is as important as gathering information on scientific uncertainties (here: on μ and σ) and their relation to the strategic (emission) behaviour of nations. Knowledge of each is a prerequisite first, to build up expectations regarding the cooperative behaviour of foreign countries and second, to predict the success of international risk management.

Having this knowledge, the condition of incentive compatibility (I) and the condition of self-enforcement (II), that determine the thresholds $\alpha_{\mathrm{I}}^{\min}$ and $\alpha_{\mathrm{II}}^{\min}$, provide a claim profile with which national risk preferences can be distinguished according to their cooperative push.

6. A Typology of Cooperative Behaviour

The *risk-strategic* analysis presented above demonstrates that even in a static setting certain risk preferences are able to push international cooperation within the unfavourable expectation structure of a prisoners' dilemma game.[17] The cooperative behaviour of nations (here: for the case of risk aversion) can be illustrated by different types of games depending on the re-

16 Consider for example two nations (1, 2) differing in national reduction costs (e.g. assume: $\mu_{CD}^1 > \mu_{CD}^2$ and $\mu_{CC}^1 > \mu_{CC}^2$ with the superscript indicating the country to which the means refer). Then the thresholds in country 1 ($\alpha_{\mathrm{I}}^{1\,\min}$; $\alpha_{\mathrm{II}}^{1\,\min}$) are lower than the thresholds in country 2 ($\alpha_{\mathrm{I}}^{2\,\min}$; $\alpha_{\mathrm{II}}^{2\,\min}$) i.e., $\alpha_{\mathrm{I}}^{1\,\min} < \alpha_{\mathrm{I}}^{2\,\min}$ and $\alpha_{\mathrm{II}}^{1\,\min} < \alpha_{\mathrm{II}}^{2\,\min}$. In this setting bilateral cooperation may still prevail. First, if $\alpha^1 = \alpha^2 \geq \alpha_{\mathrm{II}}^{2\,\min}$ and second, if $\alpha^1 \neq \alpha^2$ but $\alpha^1 \geq \alpha_{\mathrm{II}}^{1\,\min}$ as well as $\alpha^2 \geq \alpha_{\mathrm{II}}^{2\,\min}$ holds.

17 Hence problems of the prisoners' dilemma type, like the case of global warming, do not necessarily have to be redefined (according to their expected cost-utility-structure) as games of a higher cooperation possibility (as e.g. done by Edenhofer 1996 or supposed by Ostrom 1999) to explain that international measures of risk management are chosen by independent countries. Nor is it necessary to put them into a dynamic setting for this purpose.

lation between the thresholds $\alpha_{\text{I}}^{\min}$ and $\alpha_{\text{II}}^{\min}$, and the national risk attitude α. Since it can not be generally stated whether the threshold of incentive compatibility ($\alpha_{\text{I}}^{\min}$) is lower (equal or higher) than the border of self-enforcement ($\alpha_{\text{II}}^{\min}$),[18] there are three cases to distinguish: $\alpha_{\text{I}}^{\min} = \alpha_{\text{II}}^{\min}$ (case 1), $\alpha_{\text{I}}^{\min} < \alpha_{\text{II}}^{\min}$ (case 2) and $\alpha_{\text{I}}^{\min} > \alpha_{\text{II}}^{\min}$ (case 3).

Case 1: $\alpha_{\text{I}}^{\min} = \alpha_{\text{II}}^{\min}$

In the special case where $\alpha_{\text{I}}^{\min}$ equals $\alpha_{\text{II}}^{\min}$, each intensity of risk aversion of a home country lower than the thresholds ($\alpha < \alpha_{\text{I}}^{\min}; \alpha_{\text{I}}^{\min}$) leads to a higher degree of national risk utility (ϕ) if this country defects irrespective of the foreign strategy choice ($\phi_{DD} > \phi_{CD}$ and $\phi_{DC} > \phi_{CC}$). For each intensity of risk aversion higher than the thresholds ($\alpha > \alpha_{\text{I}}^{\min}; \alpha_{\text{II}}^{\min}$) the opposite holds: The preference value of a cooperating home country is always higher than its preference value in case of defecting ($\phi_{DD} < \phi_{CD}$ and $\phi_{DC} < \phi_{CC}$).

With this, case 1 delivers two general results holding in each of the three cases:

If actual risk aversion overleaps both, the threshold of incentive compatibility ($\alpha_{\text{I}}^{\min}$) and the threshold of self-enforcement ($\alpha_{\text{II}}^{\min}$), cooperation is in the self-interest of each nation irrespective of what the other country does. In terms of game theory, the thresholds $\alpha_{\text{I}}^{\min}$ and $\alpha_{\text{II}}^{\min}$ separate between a game of the prisoners' dilemma type (which is played when $\alpha < \alpha_{\text{I}}^{\min}, \alpha_{\text{II}}^{\min}$) and a no conflict game (which is played when $\alpha > \alpha_{\text{I}}^{\min}, \alpha_{\text{II}}^{\min}$). Thus if (I) and (II) are simultaneously met in each of the nations the global alliance of risk is able to completely absorb international incentives to free ride. Below the thresholds environmental cooperation does not constitute an equilibrium choice and therefore should not be expected.

We now drop the restrictive assumption of the special case where $\alpha_{\text{I}}^{\min} = \alpha_{\text{II}}^{\min}$ holds and allow the two thresholds to deviate from each other. Consider the case $\alpha_{\text{I}}^{\min} < \alpha_{\text{II}}^{\min}$.

Case 2: $\alpha_{\text{I}}^{\min} < \alpha_{\text{II}}^{\min}$

In addition to what has been said for case 1 we now have to consider a further game in the case where the country-specific risk attitude α lies in-between the thresholds ($\alpha_{\text{I}}^{\min} < \alpha < \alpha_{\text{II}}^{\min}$).

If we consider case 2 and it is only condition (I) that is met ($\alpha_{\text{I}}^{\min} < \alpha < \alpha_{\text{II}}^{\min}$) an incentive to cooperate unilaterally exists. However, since the condition of self-enforcement is not fulfilled ($\alpha < \alpha_{\text{II}}^{\min}$) this incentive is unstable as soon as the other country chooses to cooperate. In equili-

18 This crucially depends on the relation between the differences of the means and the standard deviations as given in (I) and (II). For more details on this issue see Endres / Ohl (2000a) in the context of the quadratic μ-σ-principle.

brium therefore "only" unilateral cooperation is induced. In addition to the corner solutions: prisoners' dilemma- and no conflict game, we thus have a game of chicken if $\alpha_{I}^{min} < \alpha < \alpha_{II}^{min}$ holds.

In the chicken dilemma the question is who the cooperating party is. Since each country is better off if the other country acts cooperatively no country might wish to cooperate first. This may lead to a wait-and-see-strategy where each country speculates on maximising national risk welfare as a free-rider (by the choice of defection). However sooner or later unilateral cooperation takes place (an unilateral incentive to cooperate exists; $\alpha > \alpha_{I}^{min}$). To forecast which of the countries turns out to be the "chicken", further assumptions have to be made. E.g. in the case of asymmetric nations (regarding their intensity of risk aversion), it is very suggestive that the country with the higher degree of risk aversion is pushed into cooperation first. The reason is that in case of holding on to bilateral defection a highly risk averse nation incurs a higher loss of national risk welfare than its less risk averse counterpart.[19]

Case 3: $\alpha_{II}^{min} < \alpha_{I}^{min}$

If we have $\alpha_{II}^{min} < \alpha_{I}^{min}$ and it is condition (II) that is exclusively fulfilled ($\alpha_{II}^{min} < \alpha < \alpha_{I}^{min}$) incentives to cooperate unilaterally are missing. Nevertheless, bilateral cooperation could be supported by the promise of reciprocal (conditional) cooperation. Signalling cooperation trustworthily is, here, an important factor to actually transform the prisoners' dilemma into a game of stag hunt (a variant of the assurance game) which is played in-between the thresholds.[20]

The stag hunt points to a situation where nations prefer either the status quo or a bilateral cooperation. However, since bilateral cooperation is preferred to bilateral defection each country prefers to contribute to a cooperative risk management (which is here a focal point in the sense of Schelling 1960) as long as it expects the other country to cooperate, too.

Moreover, since the stability border α_{II}^{min} is lower than α_{I}^{min}, actual values of α that are characterised as incentive compatible also fulfil the criterion of self-enforcement. Thus, given case 3 we are able to conclude that if α exceeds the border of incentive compatibility (I), nations do not play the chicken game as in case 2, but the no conflict game instead ($\alpha_{II}^{min} < \alpha_{I}^{min} < \alpha$).

---

19 We will show in section 7., below, that considering different types of policy instruments to perform environmental cooperation may also be able to solve the chicken dilemma by transforming this game into one with a higher cooperation possibility, like the no conflict game or the game of stag hunt.

20 The importance of reciprocal trust to induce bi- or multilateral cooperation under anarchy (in a non-cooperative game) has also been pointed out e.g. by Axelrod 1984 in a dynamic setting.

It has been shown that depending on the actual intensity of risk aversion (α) and its relation to the threshold values different types of games evolve. Table 1 summarises the different types of cooperative behaviour:

Table 1

A typology of cooperative behaviour

Type of game	Incentive structure	Demands on the risk attitude (α)
prisoners' dilemma	$\phi(DC) > \phi(CC) > \phi(DD) > \phi(CD)$	$\alpha < \alpha_{I}^{min}; \alpha_{II}^{min}$
chicken	$\phi(DC) > \phi(CC) > \phi(CD) > \phi(DD)$	$\alpha_{I}^{min} < \alpha < \alpha_{II}^{min}$
stag hunt	$\phi(CC) > \phi(DC) > \phi(DD) > \phi(CD)$	$\alpha_{I}^{min} > \alpha > \alpha_{II}^{min}$
no conflict	$\phi(CC) > \phi(DC) > \phi(CD) > \phi(DD)$	$\alpha_{I}^{min}; \alpha_{II}^{min} < \alpha$

The analysis presented above points out that country-specific risk preferences α as well as the stochastic parameters, μ and σ, defining the thresholds α_{I}^{min} and α_{II}^{min}, codetermine the incentive for and the stability of international environmental agreements among sovereign nations. The relation between the thresholds and the actual intensity of risk aversion determines whether an incentive to cooperate exists and if so whether this incentive is stable. Our analysis suggests that enforcing an agreement is not necessarily harder than initiating it. E.g. the possibility of $\alpha_{II}^{min} < \alpha_{I}^{min}$ rises ceteris paribus the higher reduction costs in case of unilateral cooperation are.[21] Thus the relation between the thresholds can be shifted by strategies of *risk management*. This is especially useful in cases where $\alpha < \alpha_{I}^{min}; \alpha_{II}^{min}$ holds, so that the prisoners' dilemma remains unsolved. Moreover, in the context of a global pollutant also partial cooperation (as in the game of chicken) is unsatisfactory. The reason is that defection of one country causes the threat of undermining unilateral emission reductions by emission increases in the defecting nation (e.g. Hoel 1991). Consequently enlarging cooperation by policy measures (decreasing the value of α_{II}^{min}) in the context of global environmental risk is as important as initiating it. Some possibilities to do that are discussed in the subsequent chapter.

21 In this case μ_{CD} is small compared to μ_{DD}. Consequently the border of incentive compatibility, α_{I}^{min}, is relatively high. For the effects on the threshold values in case of different policy instruments also see Endres / Ohl (2000b).

7. Strategies of Risk Management to Shift the Borders of Incentive-Compatibility and Self-Enforcement

The borders of incentive compatibility (α_{I}^{min}) and self-enforcement (α_{II}^{min}) are determined by the stochastic parameters μ and σ. Since both values are not only determined by national damage assessment but also by the choice of environmental technology or policy instrument (as measures of actually performing international cooperation) global risk management is able to move both boundaries. To show this, we clarify the relation between the thresholds (α_{I}^{min}; α_{II}^{min}), μ and σ, as defined in equations (I) and (II), in the following μ-σ-diagram:

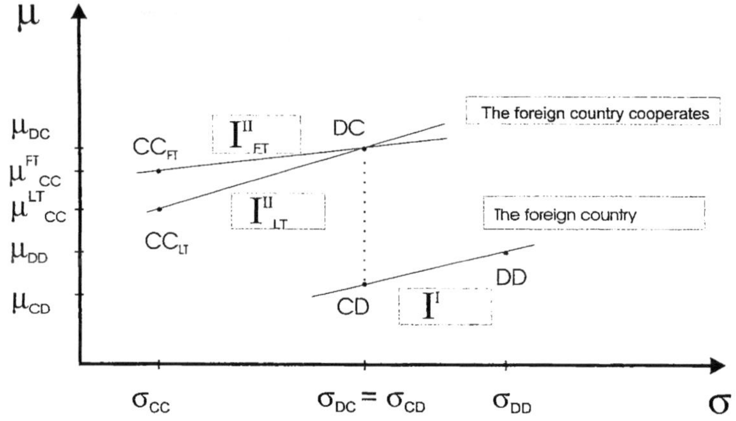

Figure 1: Designing an international environmental treaty (case 1)

Equations (I) and (II) define the slope of a straight line. Thus, in figure 1 above, the slopes of the straight lines represent the values of α_{I}^{min} and α_{II}^{min}, respectively, as given by (I) and (II). Each curvature is the location of different μ-σ-combinations leading to the same demands on the national risk attitude α given the strategy choice of the foreign country. Consequently, we can speak of I^{I} and I^{II} as of Iso-requirement-curves of cooperation.[22] Each μ-σ-tuple specifies the impact of national damage assessment, technology choice and / or policy instrument applied on the national welfare distribution. The status quo for example is displayed by *DD* which refers to a business-as-usual-scenario where no measures of international risk management are used.

22 The subscripts *LT* and *FT*, used in figure 1, will be introduced later in this section.

In case of one country or both countries choosing to cooperate each cooperating nation has to decide which environmental instrument or technology to use to perform international cooperation. Generally different types of instruments and technologies are available. Thus the values of the means as well as the values of the standard deviation which specify the μ-σ-tuples *CC* and *CD* depend on politically determined factors. Therefore usually a range of μ-σ-combinations has to be judged. For example comparing emission taxation and quota restriction, the former is (at least in theory) more efficient than the latter. Moreover quotas usually are more accurate in reaching a target (in terms of the level of emission abatement) than taxes. These instrument-specific pros and cons will influence the levels of μ and σ in different ways.[23] Consequently the slopes of the above curves (the relation of the tuples *CD* and *DD* as well as *DC* and *CC*) and hence the possibility of international coalition formation generally depend on measures of global risk management. That is why decision makers are able to shift the demands on the actual α (the slope of the curves) by the selection and the design of environmental policy instruments and technologies.

Consider for example an international quota regime as discussed within the Kyoto protocol to limit greenhouse gases (e.g. Heister et al. 1991). To date it is unclear which part of the assigned amounts[24] will be licensed for trade. Especially the US state that they will not accept any restrictions on trade. On the other hand the European countries claim for trade restrictions performed by a buyers or a sellers cap on the assigned amounts.[25] With the help of figure 1 above and figure 2 below we are able to highlight that the design of an international trade regime (that defines the endogenous risk pattern, here: of global environmental change) may be crucial for the enforcement of international agreements like the Kyoto protocol.

We assume that independent of the applied trade regime the order of the means as well as the order of the standard deviations as given by (3) and (4) hold. Thus the standard deviation is lowest in case of bilateral cooperation

23 For example $\mu_{\text{Tax}} > \mu_{\text{Quota}}$ and $\sigma_{\text{Tax}} > \sigma_{\text{Quota}}$ may hold.

24 The assigned amounts are determined by the country-specific emission targets of the protocol. The US, Japan and the European countries for example agreed to limit their individual amount of six major greenhouse gases (CO_2, CH_4, N_2O, HFCs, PFCs, SF_6) to 7%, 6% and 8%, respectively, below the level of 1990 (base year) emissions during the five-year budget period 2008 – 2012.

25 For further details on this issue see e.g. Michaelowa / Koch 1999. Also notice that in climate negotiations two group of countries can be distinguished: First, the "umbrella group" with the US as the leading country prefering modest emission controls and second, the "EU-bubble" that includes countries supporting tight emission and trade controls (e.g. the European countries and the small islands group called the AOSIS). This enables us to focus on a two-country-model (with one country reflecting a group of countries) as the most simple way to approximate the interest schemes of the different countries envolved in the process of climate negotiations.

and it increases in case of one country or both countries defecting, as is displayed at the abscissas of figures 1 and 2 ($\sigma_{CC} < \sigma_{CD} = \sigma_{DC} < \sigma_{DD}$). The order of the means is that of a prisoners' dilemma as it is shown at the ordinates of the figures ($\mu_{DC} > \mu_{CC} > \mu_{DD} > \mu_{CD}$).[26]

We first focus on the simple case (case 1, figure 1, above) where the design of the trade regime exclusively causes a difference in expected national welfare. Because of differences in national reduction costs each country's welfare improves when implementing an unlimited trade regime compared to a regime with constraints on trade. (The transformation of differences in national reduction costs into national welfare gains is not bounded by trade restrictions.) We thus have:

$$\mu_{CC}^{FT} > \mu_{CC}^{LT} \tag{7}$$

with μ_{CC}^{FT} the level of expected welfare in case of free trade (*FT*) and μ_{CC}^{LT} the level of expected welfare in case of limited trade (*LT*). The policy-specific differences in the levels of the means as given by (7) affect the slope of the *CC-DC*-curve and hence the level of the threshold $\alpha_{\mathrm{II}}^{\min}$. In case of free trade the threshold is lower than in case of limited trade i.e., the *LT*-curve is steeper than the *FT*-curve ceteris paribus.[27] Thus an unlimited trade regime may lead to the case where $\alpha_{\mathrm{I}}^{\min} > \alpha_{\mathrm{II}}^{\min}$ holds whereas trade restrictions may induce the case $\alpha_{\mathrm{I}}^{\min} < \alpha_{\mathrm{II}}^{\min}$ so that $\alpha_{\mathrm{II}FT}^{\min} < \alpha_{\mathrm{I}}^{\min} < \alpha_{\mathrm{II}LT}^{\min}$ results (see figure 1, above). Consequently, if the actual intensity of risk aversion is close to $\alpha_{\mathrm{I}}^{\min}$ ($\alpha_{\mathrm{I}}^{\min} \cong \alpha$), the selection of a quota regime supporting free trade is able to induce stable bilateral cooperation whereas a regime with barriers of trade fails to do this job ($\alpha_{\mathrm{II}FT}^{\min} < \alpha_{\mathrm{I}}^{\min} \cong \alpha < \alpha_{\mathrm{II}LT}^{\min}$). Thus, ceteris paribus, free trade has a higher cooperative push than trade restrictions: In the *FT*-case a no conflict game evolves whereas in the *LT*-case a chicken game is played. Allowing free trade therefore might enhance the chances of the protocol being ratified by the parties. This confirms the statement (e.g. made in Brockmann / Stronzik / Bergmann 1999) that only the flexibility within the mechanisms of the Kyoto protocol is able to induce countries like the US to take over legally binding emission targets. Moreover, if the US is less risk averse than the EU ($\alpha^{\mathrm{US}} < \alpha^{\mathrm{EU}}$) figure 4 suggests that a reason for ratifying a protocol supporting trade restrictions by the EU but not by the US could be that $\alpha_{\mathrm{II}FT}^{\min} < \alpha^{\mathrm{US}} < \alpha_{\mathrm{II}LT}^{\min} < \alpha^{\mathrm{EU}}$ holds. So varying degrees of risk aversion

26 For a classification of the global warming issue as a prisoners' dilemma game also see Brockmann / Stronzik / Bergmann 1999.

27 Notice that the *CD-DD*-curve is not altered by the design of the trade regime. In a two-country-modell trade restrictions only matter in the case of bilateral cooperation. In the case of unilateral emission reduction trading of the amounts avoided among countries does not happen anyway (we neglect the possibility of trading among each group of countries for complexity reduction).

are one explanation for differences in the acceptance of the mechanism design of treaties.

We now focus on the case (case 2, figure 2, below) where not only the means but also the standard deviations depend on the choice of the trade regime. Regarding the implementation of the Kyoto protocol trade restrictions are expected to have some impact on the global level of greenhouse gases. The reason is that the assigned amounts mainly of the Russian Federation and the Ukraine (as given by the zero reduction target) cause the problem of "hot air". The term "hot air" stands for emission rights that are allowed by the Kyoto protocol but will not be called on because of declining economic (and hence emission) growth in some of the eastern European countries. In the case of trade restrictions it should become impossible to sell or buy the whole amount of "hot air". Therefore the level of global emissions can be affected by the design of the trade regime.[28] Regarding these interdependencies it is not only the means that vary but also the standard deviations. In figure 2 we assume that free trade still induces the higher level of expected welfare (i.e., efficiency gains from free trade outweigh savings from damage reduction that origin from trade restrictions freezing the amount of tradable "hot air"). That is $\mu_{CC}^{FT} > \mu_{CC}^{LT}$ as given by (7) holds. However, additional to case 1 (see figure 1, above), we further assume that the expected difference in the global emission level causes a difference in the standard deviations (σ_{CC}).[29] In accordance with the global alliance of risk (as introduced in chapter 3) we assume that

$$\sigma_{CC}^{FT} > \sigma_{CC}^{LT} \tag{8}$$

holds. Figure 2 illustrates this case.[30]

If the actual α of each nation equals $\alpha_{\mathrm{I}}^{\min}(\alpha \cong \alpha_{\mathrm{I}}^{\min})$ and $\alpha_{\mathrm{II}FT}^{\min} > \alpha_{\mathrm{I}}^{\min} > \alpha_{\mathrm{II}LT}^{\min}$ holds, now, a no conflict game evolves in the *LT*-case and a chicken game is played in the *FT*-case.[31] The reason is that in the case of free trade the dis-

28 There may be other elements of the design of the trade regime affecting the level of global emissions. However, to highlight the role of hot air-trading we focus on this issue, here.

29 Herold 1998 e.g. calculates that "hot air" could affect the global emission level of greenhouse gases by an amount of 516–650 mio. tons (only CO_2) during the first budget period (2008–2012). For the problem of "hot air" also see Michaelowa/Koch 1999.

30 If the *FT*-curve is actually steeper than the *LT*-curve, as it is assumed in figure 2 below, can not be answered within this theoretical paper. Here, focusing on a trade regime like it is discussed in connection with the Kyoto protocol is only to illustrate the method applied and does not claim for any empirical validity even if the calculations of Herold 1998 (see footnote 28, above) seem to point in this direction.

31 Thus, as argued in section 6, above, the choice of instruments is also able to improve the situation by transforming the chicken into a game of a higher cooperation possibility.

advantages of the higher sway of the outcomes may overcompensate the advantages of the higher mean. If this effect enforces the agreement critically depends on the intensity of the countries' actual risk aversion. If the actual α falls short of the thresholds ($\alpha < \alpha^{\min}_{IIFT}$; $\alpha^{\min}_{IILT}$) a ratification of the protocol is not to be expected, irrespective of trade being restricted or not. However, if α exceeds the threshold $\alpha^{\min}_{IILT}$ but not the threshold $\alpha^{\min}_{IIFT}$ ($\alpha^{\min}_{IIFT} > \alpha > \alpha^{\min}_{IILT}$ the spread effect dominates and trade restrictions will be characterised by a higher cooperative push than free trade.

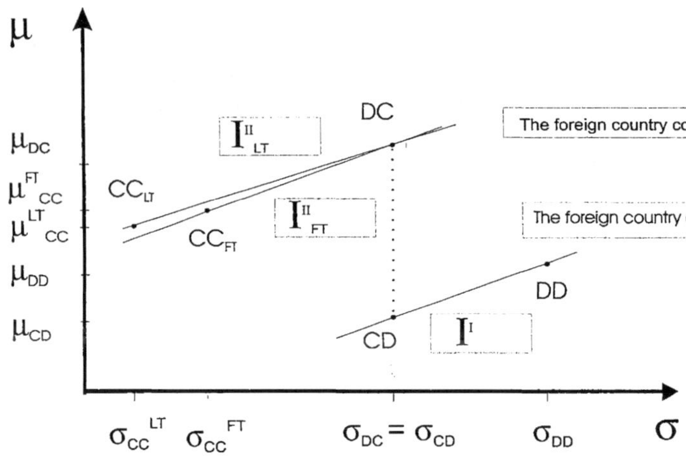

Figure 2: Designing an international environmental treaty (case 2)

In case of the Kyoto protocol it is suggestive that the US are not only less risk averse than the EU ($\alpha^{EU} > \alpha^{US}$) but in addition that their assessment of the risks and costs of global warming differs from that of the EU as well (the stochastic parameters of the model differ, too). Consequently the threshold of self-enforcement could be lower in case of free trade for the US as displayed in figure 1. The risk of "hot air" trading is not expected to alter the spread of the welfare distribution by the US. If, on the other hand, the European countries expect "hot air" trading to affect the spread of the welfare distribution, the threshold of self-enforcement could be lower when trade is restricted as is shown in figure 2. In this case we have: $\alpha^{\min}_{IIFT} < \alpha^{US} < \alpha^{\min}_{IILT}$ and $\alpha^{\min}_{IILT} < \alpha^{\min}_{IIFT} < \alpha^{EU}$. Improving the possibility of cooperation then requires to set the rules of the agreement in accordance to the incentive structure (the prerequisites of cooperation) of the country showing the lowest intensity of risk aversion. The reason is that to induce a country with low risk aversion to cooperate it is necessary to push the threshold in this country down as far as possible. Here: Only allowing free trade leads to bilateral cooperation ($\alpha^{EU} > \alpha^{min}_{IIFT}$ and $\alpha^{US} > \alpha^{min}_{IIFT}$). The price of enlarging the agreement consequently is that less risk averse countries are able to "dictate" the rules of the game (the mechanism design of the protocol).

Thus, to sum up, the cooperative power of an environmental policy instrument critically depends upon the interaction of the expected stochastic and strategic uncertainties. To make these interdependencies (i.e., the interaction between the choice of instruments / technologies and the possibility of international cooperation) work as a catalyst for stable international coalition formation we recommend to introduce a further judgement criteria for international policy assessment: *"the cooperative push"*. This criterion reflects the ability of instruments (and technologies) to initiate and self-enforce international cooperation to protect the global commons. *"The cooperative push"* provides the necessary (and still missing) link between local and global concerns.[32]

8. Conclusion

Our analysis simultaneously deals with two types of uncertainty: first, with the uncertainty of the behaviour of nature (stochastic uncertainty) and second, with the uncertainty of the behaviour of nations (strategic uncertainty). This *risk-strategic* analysis points out that both, research on national risk preferences as well as scientific uncertainties is needed to predict whether international environmental cooperation takes (should take) place or not. In a static setting we focused on two countries suffering from global environmental risks. We studied how the chances for international cooperation depend on the characteristics of national welfares.[33] To do that, we first developed a profile (in terms of the criteria-specific thresholds α^{min}_{I} and α^{min}_{II}) to be able to distinguish national risk attitudes according to their cooperative power.

The thresholds are determined by the objective risk pattern (μ and σ). Dependent on the activity chosen by the nations (cooperation versus defection) we assumed the order of expected welfare μ in each home country to follow the incentive structure of a prisoners' dilemma game. Regarding the global

32 Using the analysis of alternative designs of trade mechanisms within the Kyoto protocol as an example, limited trade has the higher cooperative push compared to free trade given the conditions of case 2. The opposite holds if the assessment of risk has to be reflected by case 1 of the analysis above.

33 To model international cooperation we chose a dichotomous choice setting with cooperation versus defection. National welfares have been characterised by the means μ, the standard deviations σ as well as the national attitudes towards risk α.

alliance of risk we assumed σ to decrease with each cooperative contribution. Given these assumptions both thresholds are positive. Thus, only risk aversion is able to foster international environmental cooperation.

If national risk preferences fall short of both thresholds ($\alpha < \alpha_{\mathrm{I}}^{\min}; \alpha_{\mathrm{II}}^{\min}$) internationally coordinated measures are not to be expected. Only, if nations are risk averse and if their actual intensity of national risk aversion exceeds one of the thresholds, a possibility of international cooperation is induced. The step over $\alpha_{\mathrm{I}}^{\min}$ ensures that incentives to cooperate unilaterally exist (condition of incentive compatibility: $\alpha > \alpha_{\mathrm{I}}^{\min}$). Stepping over $\alpha_{\mathrm{II}}^{\min}$ confirms that this incentive is stable (condition of self-enforcement: $\alpha > \alpha_{\mathrm{II}}^{\min}$). In cases where unilateral cooperation is not initiated ($\alpha < \alpha_{\mathrm{I}}^{\min}$) but the condition of self-enforcement is fulfilled ($\alpha > \alpha_{\mathrm{II}}^{\min}$), taking over unilateral actions by one country is able to initiate further cooperative measures in a second country. Conditional cooperation may happen.

Thus the integration of both, stochastic as well as strategic uncertainty also helps to identify the advantages and disadvantages of a "first mover"-policy in international risk management.

Moreover, the *risk-strategic* analysis points out that the traditional criteria of policy assessment (like efficiency and ecological accuracy that initially have been developed for local concerns) should be put into a wider context if used in an international setting:

The thresholds ($\alpha_{\mathrm{I}}^{\min}$ and $\alpha_{\mathrm{II}}^{\min}$) are determined by the stochastic terms of the model (μ and σ) that reflect the shape of the national welfare distributions. In the case of *manmade* environmental risks, μ and σ depend on national/international policy measures that alter the level of damage as well as of damage probability (endogenous risks). Consequently, if cooperative behaviour is not supported by actual risk preferences environmental policy is able to stimulate international cooperation by a shift of the endogenous parameters, here: μ and σ. Varying these terms (the national welfare distribution) by the choice of environmental policy (i.e., by a skilful selection of instruments and technologies) the thresholds, $\alpha_{\mathrm{I}}^{\min}$ and $\alpha_{\mathrm{II}}^{\min}$, can be pushed down. Hence, if environmental instruments and technologies can be ordered according to their impact on the stochastic parameters (here: μ and σ), the instrument/technology leading to the lowest demands on the actual intensity of risk attitude able to foster international cooperation (here: the minimum level of $\alpha_{\mathrm{I}}^{\min}$ and $\alpha_{\mathrm{II}}^{\min}$) could be selected. With this choice the chances for effective global risk management are at their best.

It was further shown that in the case of asymmetric nations (according to their intensity of risk aversion) that assess risk differently (the thresholds differ among the nations) the enlargement of environmental cooperation may require to set the rules of the game so as to push the threshold of the country with the lowest risk aversion as far down as possible. Hence, the desire of maximum coalition formation could strengthen the position of countries, here, with low risk aversion in the negotiation process. So they may be able to dictate the rules of the game. These results suggest to supplement the traditional criteria of policy assessment by a further criterion: *"the cooperative push"*. This criterion highlights the fact that national policy measures can be designed to stimulate self-enforcing international coalition formation to reduce global environmental risks. Implementing *"the cooperative push"* may lead to new impulses for the concept of sustainable development claimed by the UN Framework Convention agreed on in Rio de Janeiro 1992 and in many subsequent international declarations.

9. Epilogue

According to chapter 4, it is impossible to establish a general rule of how the standard deviation of environmental damages (σ) varies with the emission behaviour of nations. Therefore, even though we gave some rationale for the ordering assumed in (4), above, it seems to be wise to be prepared for things being different. So let us assume that contrary to what has been said above, the global alliance of risk is characterized by

$$(4') \qquad 0 < \sigma_{DD} < \sigma_{CD} = \sigma_{DC} < \sigma_{CC} \, .$$

Ranking (4') states that the variability of damages increases with each cooperative contribution. Consequently (see section 5, above), (I) and (II) turn into

$$(I') \qquad \alpha < (\mu_{DD} - \mu_{CD})/(\sigma_{DD} - \sigma_{CD}) \equiv \alpha_{I}^{min} < 0$$

$$(II') \qquad \alpha < (\mu_{DC} - \mu_{CD})/(\sigma_{DC} - \sigma_{CC}) \equiv \alpha_{II}^{min} < 0 \, .$$

Thus, given the reversed order of σ as assumed by (4') it is only the risk seeking nations which can be expected to cooperate.

Analogously to what has been said in the main body of the paper, it may not be possible to shift the countries' risk preference parameter (α) to meet (I'), (II'). However, also in analogy to what has been said for the case of risk averse countries, above, the critical thresholds (α_{I}^{min}, α_{II}^{min}) may be influenced by international risk management in order to make cooperation more likely.

Taking these results for a modified global alliance of risk (4') together with the results of the paper's main body assuming the initial global alliance of risk (4), the following conclusions prevail:

- No matter what form the global alliance of risk takes, (4) or (4′), it is the involved counrties' risk preferences that decide whether the road to cooperation will be taken in a static expected prisoners' dilemma game.
- No matter what form the global alliance of risk takes, it will not be enough to trigger cooperation that countries possess the favourable type of risk preference. It is an additional prerequisite that this preference is strong enough to overcome a certain threshold. There is one threshold for unilateral and one for bilateral cooperation.
- The form of the global alliance of risk decides which kind of risk preference is favourable for cooperation: If variability decreases with emission reduction, risk aversion turns out to be a necessary condition. If, on the other hand, variability increases with emission reduction, risk aversion leads to defection.

Literature

Arrhenius, S. A. (1903): Lehrbuch der kosmischen Physik, Leipzig.

Axelrod, R. (1984): The Evolution of Cooperation, New York.

Barrett, S. (1992): International Environmental Agreements as Games in: Pethig, R. (Ed.): Conflicts and Cooperation in Managing Environmental Resources, Berlin a.o., 11–37.

– (1994): Self-Enforcing International Environmental Agreements, Oxford Economic Papers, 46, 878–894.

– (1997): Towards a Theory of International Environmental Cooperation, in: Carraro, C. / Siniscalco, D. (Ed.), New Directions in the Economic Theory of the Environment, 239–280.

Brachinger, H.W. / *Weber*, M. (1997): Risk as a Primitive: A Survey of Measures of Perceived Risk, OR Spectrum, 4, 235–250.

Brockmann, K.L. / *Stronzik*, M. / *Bergmann*, H. (1999): Emissionsrechtehandel – eine neue Perspektive für die deutsche Klimapolitik nach Kioto, Heidelberg.

Carraro, C. / *Siniscalco*, D. (1997): The International Protection of the Environment: Voluntary Agreements among Sovereign Countries, in: Dasgupta, P. / Mäler, K.-G. / Vercelli, A. (Ed.), The Economics of Transnational Commons, Oxford, 192–209.

Chichilnisky, G. (2000): An Axiomatic Approach to Choice under Uncertainty with Catastrophic Risks, Resource and Energy Economics, 22, 221–231.

Chichilnisky, G. / *Heal*, G. (1998): Global Environmental Risks, in: Chichilnisky, G. / *Heal*, G. / Vercelli, A. (Ed.), Sustainability: Dynamics and Uncertainty, Dordrecht, 23–46.

Edenhofer, O. (1996): Das Management globaler Allmenden, in: Diekmann, A. / Jaeger, C.C. (Ed.), Umweltsoziologie, Sonderheft 36, 390–419.

Eichberger, J. (1993): Game Theory for Economists, San Diego.

Endres, A. (1997): Negotiating a Climate Convention: The Role of Prices and Quantities, International Review of Law and Economics, 17, 147–156.

– (2000): Umweltökonomie, Stuttgart.

Endres, A. / *Ohl*, C. (1998): Die Begrenzung globaler Umweltrisiken – eine risiko-strategische Betrachtung, Zeitschrift für Umweltpolitik und Umweltrecht, 4, 511–517.

Endres, A. / *Ohl*, C. (2000a): Das Kooperationsverhalten von Staaten bei der Begrenzung globaler Umweltrisiken: Zur Integration stochastischer und strategischer Unsicherheitsaspekte, Schweizerische Zeitschrift für Volkswirtschaft und Statistik, 136(2), 181–206.

– (2000b): Taxes versus Quotas: New Insights into an old Affair to Limit Global Environmental Risks, Environmental Economics and Policy Studies, forthcoming.

Finus, M. / *Rundshagen*, B. (1998): Toward a Positive Theory of Coalition Formation and Endogenous Instrumental Choice in Global Pollution Control, Public Choice, 96, 145–186.

Hanley, N. / *Spash* C.L. (1993): Cost Benefit Analysis and the Environment, Aldershot.

Heister, J. (1997): Der internationale CO_2-Vertrag: Strategien zur Stabilisierung multilateraler Kooperation zwischen souveränen Staaten, Tübingen.

Heister, J. et al. (1991): Umweltpolitik mit handelbaren Emissionsrechten: Möglichkeiten zur Verringerung der Kohlendioxid- und Stickoxidemissionen, Tübingen.

Herold, A. (1998): Hot air and the Kyoto Protocol, in: WWF Forum on Environment and Development (Ed.), Emission Trading in International Climate Protection, Proceedings, Workshop, June 25–26, 1998, Bonn Bad Godesberg, 50–52.

Hoel, M. (1991): Global Environmental Problems: The Effects of Unilateral Actions Taken by One Country, Journal of Environmental Economics and Management, 20, 55–70.

Houghton, J.T. et al. (1996): Climate Change 1995: The Science of Climate Change, Cambridge, UK.

Huang, C. / *Litzenberger*, R. H. (1988): Foundations for Financial Economics, Amsterdam, New York.

Kolstad, Ch. D. (2000): Environmental Economics, New York, Oxford.

Kreuzberg, P. (1994): CO_2-Emissionen unter Risikoaversion: Vermeidungskosten als Prämie für "Greenhouse Insurance", Wissenschaft und Studium, 9, 463–466.

Lipnowski, I. / Maital, S. (1983): Voluntary Provision of a Pure Public Good as a Game of Chicken, The Journal of Public Economics, 20, 381–386.

Luce, R. D. / *Raiffa*, H. (1957): Games and Decisions, New York.

Manne, A. / *Richels*, R. (1999): On Stabilizing CO_2 Concentrations – Cost-Effective Emission Reduction Strategies, in: Carraro, C. (Ed.), International Environmental Agreements on Climate Change, Dordrecht.

Michaelowa, A. / *Koch*, T. (1999): Critical Issues in Current Climate Policy: "Hot Air", Multi-Level Emission Trading Registries and Changes in Emission Commitments Due to International Conflicts, HWWA-Report 194.

Nordhaus, W.D. (1994): Expert Opinion on Climate Change, American Scientist, 82, 45–51.

Oberthür, S. / *Ott*, H.E. (1999): The Kyoto Protocol, Berlin a. o.

Ostrom, E. (1999): Die Verfassung der Allmende: jenseits von Staat und Markt, Tübingen.

Pearce, D. G. (1992): Repeated Games: Cooperation and Rationality, in: Laffont, J.-J. (Ed.), Advances in Economic Theory, Cambridge, 132–174.

Pizer, W.A. (1999): Choosing Price or Quantity Controls for Greenhouse Gases, Climate Issue Brief No. 17, www.rff.org / issue_briefs / pdf_files / ccbrf17.pdf.

Roughgarden, T. / *Schneider*, S. (1999): Climate Change Policy: Quantifying Uncertainties for Damages and Optimal Carbon Taxes – Towards More Comprehensive Calculations, Energy Policy, 27(7), 415–429.

Schelling, T. C. (1960): The Strategy of Conflict, Cambridge.

Siebert, H. (1998): Economics of the Environment: Theory and Policy, Berlin a. o.

Sinn, H.-W. (1983): Economic Decisions under Uncertainty, Amsterdam a. o.

– (1990): Expected Utility, μ-σ Preferences, and Linear Distribution Classes: A Further Result, Journal of Risk and Uncertainty, 3, 277–281.

Stähler, F. (1996): Reflections on Multilateral Environmental Agreements, in: Xepapadeas, A. (Ed.), Economic Policy for the Environment and Natural Resources: Techniques for the Management and Control of Pollution Cheltenham, 174–196.

– (1998a): Economic Games and Strategic Behaviour: Theory and Application, Cheltenham.

Stähler, F. (1998b): On the Economics of International Environmental Agreements, in: Michaelis, P. / Stähler, F. (Ed.), Recent Policy Issues in Environmental and Ressource Economics, Heidelberg, 155–173.

Taylor, M. (1987): The Possibility of Cooperation, Cambridge.

WBGU (2000): World in Transition: Strategies for Managing Global Environmental Risks, Annual Report 1998, Berlin a. o.

Welsch, H. (1995): Greenhouse Gas Abatement under Ambiguity, Energy Economics, 17, 91–100.

Xepapadeas, A. (1997): Advanced Principles in Environmental Policy, Cheltenham.

[41]

Journal of Public Economics 94 (2010) 591–603

Contents lists available at ScienceDirect

Journal of Public Economics

journal homepage: www.elsevier.com/locate/jpube

Providing global public goods under uncertainty

Vincent Boucher [a], Yann Bramoullé [b,*]

[a] CIREQ, Department of Economics, Université de Montréal, Canada
[b] CIRPÉE and GREEN, Department of Economics, Universit é Laval

ARTICLE INFO

Article history:
Received 2 February 2009
Received in revised form 22 March 2010
Accepted 7 June 2010
Available online 15 June 2010

JEL classification:
D62
D80
Q54

Keywords:
Global public goods
Climate change
International agreements
Uncertainty
Risk aversion

ABSTRACT

We study how uncertainty and risk aversion affect international agreements to supply global public goods. We consider a benchmark model with homogeneous countries and linear payoffs. When countries directly contribute to a public good, uncertainty tends to lower signatories' efforts but may increase participation. Despite risk aversion, uncertainty may improve welfare. In contrast, when countries try to reduce a global public bad, uncertainty tends to increase signatories' efforts and decrease participation. In that case, an ex-ante reduction of uncertainty may have a large positive multiplier effect on welfare.

1. Introduction

In this paper, we study how uncertainty and risk aversion affect international agreements to supply global public goods. When a public good is truly global, as with climate, disease eradication, or research on energy, the effort of one country benefits to all.[1] The problem of free-riding tends to be exacerbated by the international dimension. The principle of national sovereignty means that countries cannot be coerced into joining an agreement to supply the public good. Countries can stay out of the agreement on free-ride on those within. Given the importance of this issue, it is critical to understand the determinants of international cooperation and how it could be facilitated.

Uncertainty is a key feature of most global public goods. Consider, for instance, research on nuclear fusion. Many countries have agreed to fund coordinated work on an experimental fusion reactor.[2] Yet, the chances of success and ultimate social benefits of the project are very uncertain. Uncertainty, of course, also plays a central role in the science, economics and politics of climate change. The link between actions (CO_2 emissions) and consequences (damages due to climate change) is subject to multiple layers of uncertainty. This has greatly hindered the establishment of effective international cooperation to address "the biggest market failure the world has seen", see Stern (2008).

Under risk aversion, uncertainty itself adds significant costs. Indeed, this partly explains the magnitude of the welfare loss estimates obtained in the Stern review, see Stern (2007). More generally, economists working on integrated assessment models usually assume risk aversion and this has first order effects on the outcomes, see e.g. Nordhaus (1994), Heal and Kriström (2002). In these models, an ex-ante reduction of uncertainty has straightforward positive effects. While this property is well-grounded in situations with a unique decision-maker, its validity in a strategic setting is less clear. How does uncertainty affect the emergence of international cooperation aimed at supplying a global public good? Does an ex-ante reduction of uncertainty always help cooperation, or could it have negative effects? Through what channels does uncertainty affect the incentives to join or quit an international coalition? How does uncertainty affect welfare in a second-best context characterized by partial, endogenous cooperation? We seek to provide formal answers to these questions.

To do so, we extend the model of treaty formation with linear payoffs discussed in Barrett (2003). This model is simple and well-understood. It provides a natural benchmark to incorporate and

* Corresponding author.
E-mail address: ybramoulle@ecn.ulaval.ca (Y. Bramoullé).

[1] Barrett (2008) presents a unifying analysis of the various types and properties of global public goods.

[2] This project, ITER, is currently supported by the European Union, China, India, Japan, South Korea, Russia and the United States, see the project's website (www.iter.org) for more information.

0047-2727/$ – see front matter © 2010 Elsevier B.V. All rights reserved.
doi:10.1016/j.jpubeco.2010.06.008

analyze new features.[3] Countries face an n-player prisoner's dilemma and can join a coalition trying to take collective action. We consider two formulations of the model: one where countries contribute to a global public good, such as R&D, and one where countries seek to reduce the production of a global public bad like CO_2 emissions. In both formulations, we introduce two new assumptions. First, the benefits from collective action are subject to uncertainty.[4] This notably captures uncertainty on pollution damages. Second, countries have risk averse preferences.[5] We study how uncertainty and risk aversion affect the outcomes of the treaty formation game and especially effort, participation and welfare in equilibrium.

We find strong effects. Surprisingly, the two formulations yield radically different outcomes. Uncertainty tends to increase participation and improve welfare in the public good formulation. In contrast, it tends to reduce participation and welfare in the public bad one. This difference in outcomes arises from the different consequences of free-riding in the two models. In either case, countries who do not sign the treaty exert no effort. In the first formulation, it means that they do not contribute to the public good. Signatories' payoffs thus only depend on their own effort, and the *variability* of these payoffs increases with effort. Welfare is more uncertain for higher levels of signatories' contribution. Because of risk aversion, uncertainty then tends to decrease the contribution of signatory countries at any level of participation. Cooperation is not fixed, however. Stability of the treaty requires signatories to reach a critical mass. A treaty is stable at the precise level of participation such that an additional country signing has little effect while a country leaving has a strong impact. Because of the decreased contribution schedule, this threshold level is shifted to the right. Thus, while effort tends to decrease at given participation levels, participation itself tends to increase in equilibrium.

The mechanism is similar but leads to opposite outcomes in the public bad formulation. Countries that free-ride pollute as usual. This now generates an important background risk faced by signatory countries. As a consequence, the variability of signatories' payoffs becomes *negatively* related to signatories' effort. Signatories tend to increase their effort under uncertainty at any level of participation to shield themselves against this risk. Again, participation is not fixed and the critical mass of countries necessary to obtain a stable treaty is now shifted to the left. In the public bad formulation, participation tends to decrease under uncertainty.

In addition, we find that multiple equilibria may emerge in some circumstances.[6] A stable treaty with low action and low participation may coexist with one with high action and high participation. Multiplicity, however, appears more likely in the public good formulation.

Finally, we study the robustness of our analysis to uncertainty on the cost of collective action. We find that our main results extend to a setting with uncertain costs and that some new, and somewhat subtle, differences arise between outcomes in the two formulations.

Thus, our analysis uncovers important strategic effects associated to an ex-ante reduction of uncertainty on international agreements. When trying to provide a public good, reducing uncertainty may hurt the establishment of international cooperation. When trying to reduce a public bad, reducing uncertainty may in contrast have a positive multiplier effect due to an increase in participation.

Our analysis contributes to two literatures. First, there is an active research agenda studying international environmental agreements, see Barrett (2003). Uncertainty is mostly ignored in this literature, however, with the exception of a few papers studying the effect of learning under risk neutrality.[7] These papers usually contrast different timings for the resolution of uncertainty. For instance, Kolstad (2007) shows in a static framework that it may be better to negotiate after uncertainty is resolved rather than before. In contrast, Ulph (2004) finds that, in a model with two periods, the positive effects of learning may not be robust to the introduction of renegotiation between periods.[8] We do not look at learning here. Rather, we relax the assumption of risk neutrality. We provide the first analysis of the effect of risk aversion on international environmental agreements. To our knowledge, the finding that the two formulations yield radically different outcomes is the first of its kind in this literature.[9]

Second, a growing literature examines the effect of uncertainty and risk aversion in strategic settings.[10] Especially, Bramoullé and Treich (2009) look at global pollution under damage uncertainty. They assume that countries do not cooperate and find that uncertainty may decrease emissions and increase welfare in equilibrium. In contrast, we endogenize cooperation in this paper. In circumstances where participation is not affected, we indeed find similar results.[11] However, cooperation levels are usually affected and in ways leading to a reversal of the effect of uncertainty. Thus, reducing uncertainty before international negotiations take place could have a large, positive multiplier effect.

Our analysis is also related to the literature on discrete public goods and uncertainty. In this literature, the public good is binary and provided only if private contributions exceed a required threshold level. When this level is certain the first best outcome can be attained in equilibrium, as shown early on by Palfrey and Rosenthal (1984). Several authors have studied how uncertainty on the threshold contribution level affects the analysis. Nitzan and Romano (1990) find that uncertainty often leads to inefficiency and under-provision of the public good, see also Suleiman (1997). In contrast, McBride (2006) finds that, in some circumstances, an increase in uncertainty can increase equilibrium contributions and welfare.[12] As here, these studies emphasize the strategic effects of uncertainty in a context of public good provision. However, our analysis differs from these studies in key respects. In our setup, the public good is continuous and not subject to an exogenous contribution threshold.[13] Rather, a threshold level of *participation* emerges endogenously from stability conditions in the treaty game. Moreover, while this threshold level depends on uncertainty, it is itself not uncertain. It depends in a non-stochastic way on the level of risk affecting the economic parameters. Finally, the distinction between the public good and the public bad formulations plays a central role here, but is absent from these studies.[14]

The remainder of the paper is organized as follows. Section 2 introduces the model. Section 3 analyzes the case where countries

[3] And, indeed, many authors have built on this model, see e.g. Barrett (2001), Kolstad (2007), Rubio and Ulph (2007), Ulph (2004).

[4] We also study an extension of the baseline model where the costs of action are subject to uncertainty.

[5] Economists often view countries as risk neutral due to their size and to the possibility to pool independent risks across individuals, see Arrow and Lind (1970). This view, however, is not appropriate to think about climate change and most other global issues. Climate risks are highly correlated within communities and potential damages are large, involving a sizable portion of the economies even in conservative estimates.

[6] This complements earlier findings by Ulph (2004) and Kolstad (2007) who show the emergence of multiple stable treaties in a context with learning and risk neutrality.

[7] See Kolstad (2007, 2005), Ulph (2004), Ulph and Maddison (1997), Na and Shin (1998).

[8] Also, it may be easier to reach an agreement before the "veil of uncertainty" has been pierced. Kolstad (2005) and Na and Shin (1998) provide some supporting arguments, under risk neutrality.

[9] In all existing models, the two formulations are equivalent.

[10] See e.g. Gradstein et al. (1992), Sandler and Sterbenz (1990) on the exploitation of a renewable resource, Eso and White (2004) on auctions, and White (2008) on bargaining.

[11] This may happen, for instance, following a small enough increase in uncertainty.

[12] The intuition is that when uncertainty is greater, there is more variation in possible values of the threshold contribution level, which may increase the equilibrium probability that an agent is pivotal and his incentives to contribute.

[13] This notably means that the continuous public good is severely under-provided under certainty, in contrast to what happens in the discrete case.

[14] Also, we look at risk averse countries while the literature on discrete public goods has, so far, focused on risk neutral agents.

directly contribute to a global public good. Section 4 looks at the reduction of a global public bad. Section 5 extends the analysis to uncertain costs and Section 6 concludes.

2. The model

In this section, we introduce uncertainty and risk aversion to the model of Barrett (2003). We study two different formulations of the model. While under risk neutrality these two formulations are generally equivalent, we will show that they lead to remarkably different conclusions under risk aversion. Consider n identical countries that face a problem of collective action. In the public good formulation, each country i exerts some costly effort q_i which benefits to all other countries. Individual payoff is equal to $b\left(\sum_{j=1}^{n} q_j\right)-cq_i$ where b represents the marginal benefit from overall effort and c the marginal cost from individual effort. Because payoff is linear, we assume that there is a maximal level of effort q_{max}. This public good formulation captures situations where countries may directly contribute to a public good, such as R&D.

In contrast in the public bad formulation, countries seek to reduce the emissions of a global pollutant. Country i emits a level $e_i \in [0, q_{\max}]$ where $q_{\max}$ represents business as usual emissions. Individual payoff is equal to $ce_i - b\left(\sum_{j=1}^{n} e_j\right)$. Effort q_i represents the level of reduction of emissions: $q_i = q_{\max} - e_i$.[15] The parameter c now represents the private marginal benefit from pollution while b captures the marginal social damage. In fact, they have the same interpretation as in the public good formulation since abating one unit in one country costs c and increases the payoff of all countries by b.

Our main assumption is that the marginal social benefit b is subject to uncertainty. For simplicity, we suppose that b can take two values.[16] Marginal benefit b is equal to b_L with probability p and to $b_H > b_L$ with probability $1-p$. It is useful to introduce the expected marginal benefit $\bar{b} = pb_L + (1-p)b_H$. In contrast, uncertainty does not affect the marginal cost of effort c in our baseline model.[17] We study how our analysis is affected by cost uncertainty in Section 5.

Preferences of countries towards risk are identical, and represented by a strictly increasing and concave Von-Neumann Morgenstern utility function U. Thus, country i seeks to maximize his expected utility

$$EU(q_i, q_{-i}) = pU\left[b_L\left(\sum_{j=1}^{n} q_j\right) - cq_i\right] + (1-p)U\left[b_H\left(\sum_{j=1}^{n} q_j\right) - cq_i\right]$$

in the public good formulation, and

$$EU(e_i, e_{-i}) = pU\left[ce_i - b_L\left(\sum_{j=1}^{n} e_j\right)\right] + (1-p)U\left[ce_i - b_H\left(\sum_{j=1}^{n} e_j\right)\right]$$

in the public bad one. Since countries are ex-ante identical, we define social welfare W as the sum of the expected utilities of all countries. As in the model with certainty, we consider two restrictions on the parameters. First, $b_H < c$, so that playing $q_i = 0$ or $e_i = q_{max}$ is a strictly dominant strategy for all countries in the game without treaty. Second, $nb_L > c$, so that the first best outcome requires $q_i = q_{\max}$ or $e_i = 0$ for all i. Without treaty, there is severe under-provision of the public good.

To study the formation of an international environmental agreement, we adopt the approach pioneered by Barrett (1994). Countries have to decide whether to join an international agreement before contributing to the global public good. More precisely, the game with treaty unfolds in three stages:

Stage 1 Simultaneously and independently, all countries decide to sign or not to sign the agreement. In what follows, we denote by k, the number of signatories.

Stage 2 Signatories choose their effort in order to maximize their *collective* payoff.

Stage 3 Non-signatories independently choose the effort maximizing their individual payoff.

A key feature of our approach is that uncertainty is resolved *after* stage 3. Therefore, countries are uncertain about the state of the world when deciding whether to sign the agreement. Our main objective is to study how uncertainty and risk aversion affects this decision, and hence the existence and properties of a stable treaty. We make use of the usual notion of stability.

Definition 1. *A treaty is stable when signatories have no incentive to quit and non-signatories have no incentive to join.*

Conditions for stability can be formally summarized with a unique function Δ. Given symmetry and concavity, non-signatories will play identical actions, and this holds for signatories as well. Therefore, we denote by $q^s(k)$ and $q^{ns}(k)$ the optimal levels of effort exerted respectively by a signatory and a non-signatory country when the number of signatories is equal to k, and by $U^s(k)$ and $U^{ns}(k)$ their expected utilities. We introduce $\Delta(k) = U^s(k) - U^{ns}(k-1)$ which represents the net benefit to a signatory of staying in the agreement. The fact that a non-signatory country does not want to join at k^* is equivalent to the fact that a signatory country wants to quit at k^*+1. Thus, a treaty is stable at $k^* \in [2, n-1]$ if and only if $\Delta(k^*) \geq 0$ and $\Delta(k^*+1) \leq 0$.[18] While a treaty with no action and $q^s = 0$ is always stable, we restrict attention in what follows to treaties with strictly positive effort.

When there is no uncertainty, $b_L = b_H = \bar{b}$ and the analysis of Barrett (2003) applies. Both formulations yield the same outcomes. Non-signatories always play $q^{ns} = 0$ or $e^{ns} = q_{\max}$. The collective payoff of signatories is $k(k\bar{b} - c)q^s$ in the public good formulation and $k(k\bar{b} - c)q^s - k(nb - c)q_{max}$ in the public bad one. In either case, signatories play $q^s = 0$ if $k < c/\bar{b}$ and $q^s = q_{\max}$ if $k > c/\bar{b}$. Signatories must reach a critical mass before action becomes worthwhile. Linearity of the payoff function induces a bang-bang solution. A treaty is stable if and only if $c/\bar{b} \leq k^* \leq c/\bar{b} + 1$. This usually pins down a unique number. Social welfare in the stable treaty is strictly greater than without treaty. Interestingly, the analysis under certainty also covers the case where countries are risk neutral. This justifies the introduction of risk aversion, which we assume in the remainder of the paper.

We next analyze the treaty game under uncertainty. Since $b_L < b_H < c$, a non-signatory country always plays $q^{ns} = 0$ or $e^{ns} = q_{\max}$ no matter the number or the actions of signatories. This solves Stage 3. We study the two formulations, in turn, in what follows. In each case, we first study Stage 2 and the actions of signatories. This study is critical to understand when a treaty is stable. The analysis under certainty illustrates how the shape of q^s determines stability. Under certainty, q^s is a step function and k is the first integer situated just above the step. On the one hand, $q^s(k^*) = q_{\max}$ and $q^s(k^*-1) = 0$. If a signatory quits, the drop in collective action is very large, hence a signatory does not want to quit. On the other hand, $q^s(k^*+1) = q^s(k^*)$. If a non-signatory joins, the other signatories do not change what they do, which gives no incentives for non-signatories to join. The general intuition carries over more generally. A signatory does not want to quit if he expects that his departure would induce a high drop in the actions of the remaining

[15] Expressed in terms of efforts, individual payoff is now equal to: $b\left(\sum_{j=1}^{n} q_j\right) - cq_i - (nb - c)q_{\max}$.

[16] Some of our results are valid for any distribution of the parameter, see the discussion in conclusion.

[17] In a climate change context, this assumption reflects the fact that scientific uncertainty has little effect on private abatement costs, but is a main source of uncertainty regarding the future damages from greenhouse gas emissions.

[18] A treaty with n signatories is stable if and only if $\Delta(n) \geq 0$. In that case, all countries are signatories, and the condition on non-signatory countries disappears.

594 V. Boucher, Y. Bramoullé / Journal of Public Economics 94 (2010) 591–603

signatories. In contrast, a non-signatory does not want to join if he anticipates his decision to yield little increase, or even a decrease, in overall effort. We then make use of our study of q^s to analyze Stage 1, and the existence and properties of stable treaties in both formulations.

3. Providing a global public good

3.1. Signatories' actions

In this section, we study the actions of signatories $q^s(k)$ when countries contribute to a global public good. We have three results. First, we find that the bang-bang property which characterizes q^s under certainty partially disappears under uncertainty. Effort may take intermediate values in situations where action is desirable in one state of the world but not in the other. In that case, signatories trade-off the benefit of action in one state against its cost in the other. Second, we find that uncertainty unambiguously lowers q^s. Holding k constant, an increase in uncertainty always leads to a (weak) decrease in signatories' efforts. And third, we show that an increase in risk aversion has a similar effect and also lowers signatories' actions.

Recall, signatories jointly decide how to maximize their collective welfare, defined as the sum of their expected utilities. It means that q^s is the solution of the following program:

$$\max_{0 \le q \le q_{\max}} EU[(kb-c)q] \tag{1}$$

We first describe when the solution is at a corner. Observe that the objective function is strictly concave. Marginal expected utility is equal to $E(kb-c)U'[(kb-c)q]$. At $q=0$, this reduces to $(k\bar{b}-c)U'(0)$. Since $U'>0$, $q^s=0$ when $k \le c/\bar{b}$. This notably covers the case where $k \le c/b_H$, and signatories prefer to exert no effort in both states of the world. In contrast, when $k \ge c/b_L$, the marginal expected utility is always strictly positive. Signatories prefer to play q_{max} in both states of the world. Thus, $q^s = q_{max}$ when $k \ge c/b_L$. When k lies between $c/\bar{b}$ and c/b_L, $kb_L - c<0$ while $kb_H - c>0$, and the solution may be interior. The first order condition of program (1) can be written as follows:

$$p(c-kb_L)U'\left[(kb_L-c)q^s\right] = (1-p)(kb_H-c)U'\left[(kb_H-c)q^s\right] \tag{2}$$

This condition says that marginal utilities are equal in both states of the world. It expresses a trade-off between the positive marginal utility from action when $b=b_H$ and the negative marginal utility when $b=b_L$. This equation always has a solution $q>0$ if $U'(-\infty)=+\infty$ or $U'(+\infty)=0$. Then, q^s is equal to the solution of this first order condition if this solution is lower than q_{max}, and to q_{max} otherwise. In any case, $q^s(k)$ is continuous over $[0,n]$ when k is allowed to take real values. In summary.

Proposition 1. *Effort q^s exerted by a signatory country when k countries have signed the treaty is such that:*

(1) $q^s(k)=0$ if $k \le c/\bar{b}$ and $q^s(k)=q_{max}$ if $k \ge c/b_L$.
(2) If $c/\bar{b}<k<c/b_L$, q^s is equal to the smaller of q_{max} and of the solution to Eq. (2).

Signatories' effort under uncertainty is always lower than or equal to effort under certainty.

Observe that when $bL=\bar{b}$, there is no uncertainty and Proposition (1) reduces to the result described in the previous section.

Given that q^s varies from 0 to q^{max}, a natural question is whether q^s is necessarily increasing over $[c/\bar{b}, c/b_L]$. Does more participation always lead to a higher effort per signatory? It turns out that the answer is negative. Even with standard preferences, q^s may be decreasing over some range of participation levels. We explore this issue in more detail in the Appendix A. We show that if the level of absolute risk aversion is decreasing and decreases sufficiently rapidly, then q^s is indeed non decreasing. However, monotonicity of q^s in general is ambiguous and we provide counterexamples below.

We next study the effect of an *increase in uncertainty* on $q^s(k)$. Proposition (1) shows that $q^s(k)$ is lower under uncertainty than under certainty. Our next result shows that this property holds more generally for any increase in uncertainty in the sense of Rothschild and Stiglitz (1970). Let b and b' be two binary distributions of marginal benefit values with the same expected value $\bar{b}=\bar{b}'$. Here, b' is more risky than b if and only if $b'_L \le b_L$ and $b'_H \ge b_H$.

Proposition 2. *Suppose that b' is more risky than b. Then $\forall k \in [0,n]$, $q^s(k,b') \le q^s(k,b)$. An increase in uncertainty always reduces signatories' actions.*

The intuition behind this result can be seen by looking at the expected value and the variance of ex-post payoffs π for signatories. Here, $E(\pi)=(k\bar{b}-c)q^s$ and $Var(\pi)=p(1-p)(b_H-b_L)^2k^2(q^s)^2$. When $k \ge c/\bar{b}$, an increase in q increases the expected payoff and also increases its variance. One is desirable, but the other is not under risk aversion. The optimal choice of q results from a trade-off between these two motives.[19] Then, holding q and k constant, an increase in uncertainty leads to an increase in the payoff's variance while leaving its expected value unchanged. The marginal effect of a decrease in q on the variance is greater when uncertainty is greater, hence signatories' action is lower.

This effect is confirmed by looking at an increase in risk aversion. Recall that utility function V represents more risk averse preferences than U if there is an increasing and concave function Φ such that $V=\Phi(U)$. In our context, the effect of risk aversion is similar to the effect of uncertainty.

Proposition 3. *Suppose that countries with utility V are more risk averse than countries with utility U. Then $\forall k \in [0,n]$, $q^s(k,V) \le q^s(k,U)$. An increase in risk aversion always reduces signatories' efforts.*

For a given participation level, uncertainty and risk aversion tend to reduce signatories' actions. Signatories exert less effort in order to diminish the variability of their payoffs. While this effect may seem to be purely negative (lower level of collective action), we will see in the next section that it can have a positive indirect consequence. Lower action at all participation levels may increase the participation level that is sustainable in equilibrium.

To conclude on signatories' actions, we look at specific utility functions. We are able to obtain closed-form expressions in three important cases. These expressions are useful for applications and numerical simulations, and also allow us to obtain more precise information on the shape and monotonicity of q^s. Corresponding expressions can easily be derived for the public bad formulation. Denote by $\tilde{q}^s(k)$ the solution to Eq. (2).[20]

3.1.1. Quadratic utility

Suppose that $U(\pi)=\pi-\lambda\pi^2$. We impose that $(nb_H-c)q_{max} \le 1/(2\lambda)$ to ensure that utility is strictly increasing over the strategy space. We have:

$$\tilde{q}^s(k) = \frac{1}{2\lambda}\frac{k\bar{b}-c}{p(kb_L-c)^2+(1-p)(kb_H-c)^2}$$

We show in the Appendix A that q^s is non decreasing in k when U is quadratic.

3.1.2. CARA utility

[19] In contrast, when $k<c/\bar{b}$, an increase in q decreases the expected value *and* increases the variance, and there is no trade-off. This provides another explanation for the fact that $q^s=0$.

[20] As shown in Proposition 1, $q^s(k)=\min(\tilde{q}^s(k), q_{\max})$.

Suppose that $U(\pi) = -e^{-A\pi}$ where A denotes the level of absolute risk aversion. We obtain:

$$\tilde{q}^s(k) = \frac{1}{Ak(b_H - b_L)} ln\left[\frac{(1-p)(kb_H - c)}{p(c - kb_L)}\right]$$

We ask whether q^s is increasing in k and find that two cases appear. Either q^s is increasing, or it is first increasing, then decreasing, and again increasing over $[c/\bar{b}, c/b_L]$. Also, q^s is more likely to be increasing when p is higher. Fig. 1 illustrates.

3.1.3. CRRA utility

Suppose that $U(\pi) = \frac{1}{1-\gamma}(\pi_0 + \pi)^{1-\gamma}$ if $\gamma \neq 1$, and $ln(\pi_0 + \pi)$ if $\gamma = 1$. Introducing a baseline payoff π_0 is necessary to ensure that ex-post payoffs are always positive. We obtain:

$$\tilde{q}^s(k) = \pi_0 \frac{(1-p)^{1/\gamma}(kb_H - c)^{1/\gamma} - p^{1/\gamma}(c - kb_L)^{1/\gamma}}{p^{1/\gamma}(c - kb_L)^{1/\gamma}(kb_H - c) + (1-p)^{1/\gamma}(kb_H - c)^{1/\gamma}(c - kb_L)}$$

Especially, when $\gamma = 1$ this reduces to $\tilde{q}^s = \pi_0\left[\frac{1-p}{p(c-kb_L)} - \frac{p}{(1-p)(kb_H - c)}\right]$, which is clearly increasing in k. This may not necessarily be the case, however, when $\gamma \neq 1$.

3.2. Stable treaties

In this section, we study the properties of stable treaties. We first show that a stable treaty always exists and that participation in any stable treaty is always higher than under certainty. We then characterize participation in the best stable treaty when q_{max} is large enough. We finally discuss the impact of uncertainty on welfare and illustrate these effects through numerical simulations.

We first show that a stable treaty with positive action always exists.

Proposition 4. *In the public good formulation, a stable treaty always exists. Participation in any stable treaty is such that* $k^* \in [c/\bar{b}, c/b_L + 1]$. *Participation under uncertainty is always greater than or equal to participation under certainty.*

Our proof relies on the study of the function Δ. Recall, $\Delta(k) = U^s(k) - U^{ns}(k-1)$. On the one hand, $U^s = U^{ns} = 0$ on $[0, c/\bar{b}]$. This means that $\Delta(k) = 0$ if $k \in [1, c/\bar{b}]$ and $\Delta(k) > 0$ if $k \in]c/\bar{b}, c/\bar{b} + 1]$. On the other hand, $q^s = q_{max}$ on $[c/b_L, n]$. This implies that $\Delta(k) < 0$ on $[c/b_L + 1, n]$. If we define k^* as the greatest integer such that $\Delta(k^*) > 0$, we have: $\Delta(k^*) > 0$, $\Delta(k^* + 1) \leq 0$, and $k^* > c/\bar{b}$, which means that $q^s(k^*) > 0$.

If the number of signatory countries lies just below $c/\bar{b}$, an additional country has a strictly positive incentive to join. If he does not join, signatories do no exert any effort. If he joins, all signatories start exerting positive effort. The number of signatories then increases until an additional country does not want to join, at which point the treaty is stable.

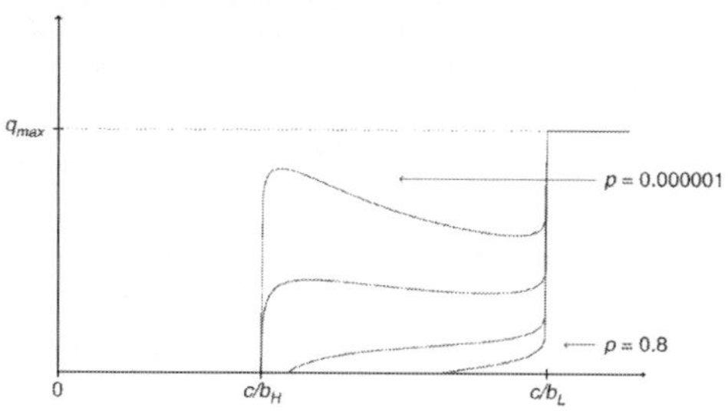

Fig. 1. Effort with uncertainty, CARA function.

Clearly, in any stable treaty with positive action, $c/\bar{b} \leq k^* < c/b_L + 1$. We next derive two further results on this participation level. First, if for some integer k, $q^s(k) < q^s(k-1)$, we have $U^s(k) < U^{ns}(k-1)$, hence $\Delta(k) < 0$. Applying the previous argument shows that there exists a stable treaty with $k^* \leq k$. Second, we show that an equilibrium around c/b_L is generally guaranteed if q_{max} is large enough. More precisely, let $[\pi]$ denote the smallest integer greater than or equal to π.

Proposition 5. *Suppose that either* $lim_{\pi \to -\infty} U'(\pi) = +\infty$ *or* $lim_{\pi \to +\infty} U'(\pi) = 0$, *and that either* c/b_L *is not an integer or* $lim_{\pi \to +\infty} U(\pi) = +\infty$. *Then, there exists* $\bar{q} > 0$ *such that if* $q_{max} \geq \bar{q}$, *a treaty with* $k^* = [c/b_L]$ *and* $q^s = q_{max}$ *is stable.*

The intuition behind this result lies in the fact that $q^s(k)$ becomes very steep when k gets close to c/b_L. Recall that at c/b_L, action becomes desirable in both states of the world, hence, given linearity of the payoffs, should be as high as possible. If q_{max} is large, the drop in collective action if one country quits is large, which gives an incentive for signatories to stay in the treaty.[21]

How does uncertainty affect welfare in a stable treaty? In this model, uncertainty affects welfare through three different channels. First, holding q^s and k fixed, it has a direct negative effect resulting from risk aversion. Second, Proposition 2 shows that, holding k fixed, it has an indirect negative effect through its reduction of $q^s(k)$. Finally, there is a third countervailing effect. Proposition 4 shows that uncertainty has a positive impact on participation k^*. This third effect may overcome the first two and lead to an overall increase in welfare. Note that when $q^s = q_{max}$, the second effect disappears but the effect of uncertainty on welfare is still ambiguous. In this case, eliminating uncertainty lowers the incentives to participate. In the end, *an ex-ante reduction of uncertainty may hurt the prospects for international cooperation on the provision of a global public good.*

We illustrate these effects by reporting results from numerical simulations in Figs. 2–4. We consider a utility function with constant relative risk aversion $U(\pi) = \frac{1}{1-\gamma}(\pi_0 + \pi)^{1-\gamma}$. We look at the effect of an increase in uncertainty on stable treaties through a decrease in b_L, holding b_H and $\bar{b}$ constant. Parameters are set at the following values: $n = 100, \gamma = 0.5, \pi_0 = 10^3, c = 1000, \bar{b} = 991, b_H = 992$. Fig. 2 represents signatories' effort in stable treaties, Fig. 3 the level of participation and Fig. 4 welfare.

Observe first that multiple equilibria emerge for $b_L \leq 100$. For any value of b_L, here, the equilibrium identified in Proposition 5 is always stable. When uncertainty is not too high, this is the unique stable treaty. It is characterized by high participation *and* high effort. When uncertainty is very high, however, a second treaty becomes stable.[22] This second treaty yield much worse outcomes: lower participation, effort and welfare. Thus, if countries are stuck in this bad equilibrium, a decrease in uncertainty could initially lead to strong increase in participation and effort if it allowed countries to leave the multiplicity domain.

Next, focus on the properties of the good treaty. Fig. 3 illustrates how participation increases when uncertainty increases. This shows that Proposition 4 extends here to increases in uncertainty. Fig. 4 illustrates both the direct negative effect of uncertainty and the indirect positive effect due to an increase in participation. When the decrease in b_L is relatively low, k^* does not change and welfare decreases as a consequence of the direct effect. In contrast, when b_L decreases below some threshold values, an additional country signs

[21] Either technical condition stated in the Proposition ensures that an arbitrarily large drop in collective action indeed translates into an incentive not to quit. The result may not hold, however, if c/b_L is an integer and U is bounded.

[22] Extensive simulations, not reported here, show that equilibrium multiplicity is indeed a robust phenomenon in the public good formulation. Multiplicity tends to be more prevalent for higher levels of risk and risk aversion.

596 V. Boucher, Y. Bramoullé / Journal of Public Economics 94 (2010) 591–603

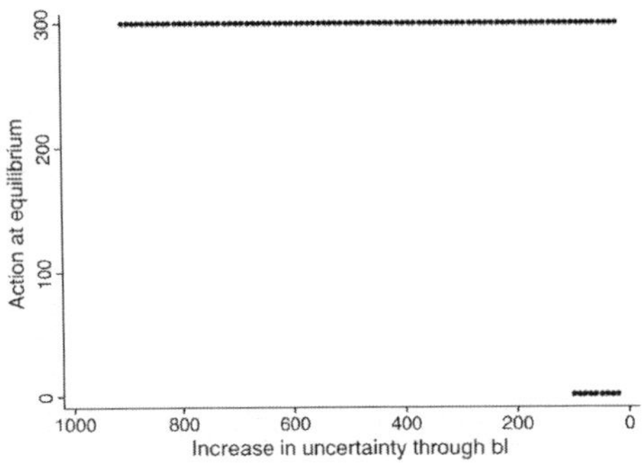

Fig. 2. Effects of uncertainty on signatories' effort (public good).

the treaty and welfare increases discontinuously. Overall, the positive effect clearly dominates and welfare may be much higher at high levels of uncertainty than under certainty.

4. Reducing a global public bad

4.1. Signatories' actions

We next turn to the analysis of the public bad formulation. To better differentiate the two models, we focus on emissions in what follows. As in the previous section, we first analyze how signatories' emissions depend on participation. We then use this understanding to study the properties of stable treaties.

When countries try to reduce a global public bad, optimal signatories' emissions solve the following program:

$$\max_{0 \le e \le q_{max}} EU[(c-kb)e-(n-k)bq_{max}] \quad (3)$$

The study of this program, in Appendix A, yields the following result.

Proposition 6. *In the public bad formulation, the emissions e^s of a signatory country when k countries have signed the treaty is such that:*

(1) $e^s(k) = q_{max}$ *if* $k \le c/b_H$ *and* $e^s(k) = 0$ *if* $k \ge c/\bar{b}$.
(2) *If* $c/b_H < k < c/\bar{b}$, e^s *is equal to the solution of the first order condition of program (3) if this solutions lies within* $[0, q_{max}]$.

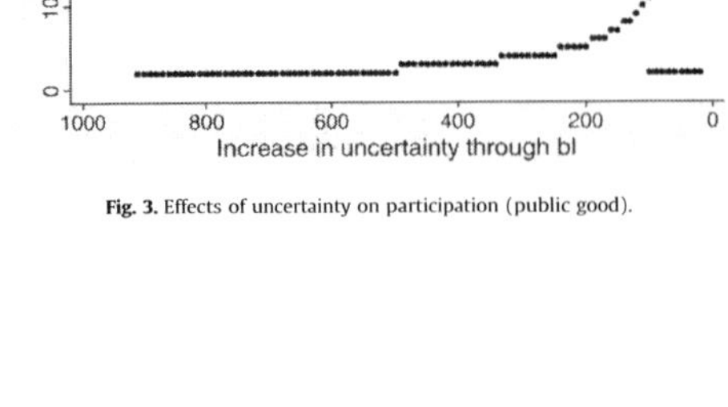

Fig. 3. Effects of uncertainty on participation (public good).

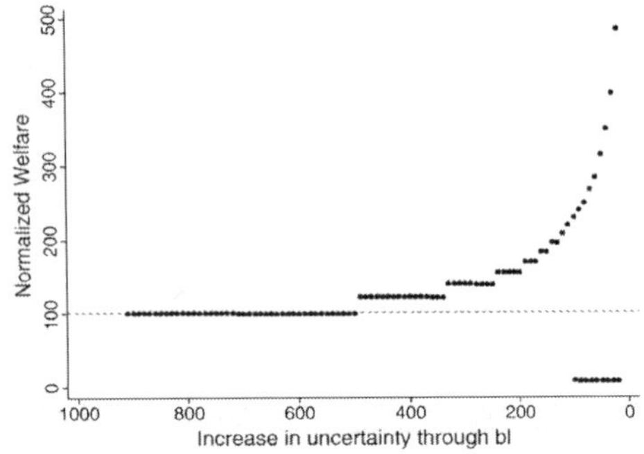

Fig. 4. Effects of uncertainty on welfare (public good).

Signatories' effort under uncertainty is always higher than or equal to effort under certainty.

Comparing Propositions 6 and 1, we see similarities as well as key differences between the two formulations. The solution still lies at a corner except maybe when action is desirable in one state of the world but not in the other. However, the precise ranges where effort may take intermediate values differ. Effort becomes maximal at different threshold participation levels: $c/\bar{b}$ in the public bad formulation, c/b_L in the public good one. A striking consequence is that uncertainty has opposite effects on signatories' efforts in both formulations. Uncertainty increases effort in the public bad formulation but lowers it in the public good one. We will see below that this also induces opposite effects on participation.

The main reason behind this difference lies in the impact of non-signatories' actions. In both cases, non-signatories do not exert any effort. When countries contribute to a public good, the baseline payoff for signatories is simply equal to zero. In contrast, in a context of global pollution, no effort yields pollution damages. Signatories thus face a baseline uncertain loss caused by non-signatories' emissions. As is well-known, the presence of a background risk may deeply affect decisions of risk averse agents, see Gollier (2001).

Our next results confirm this reversal in qualitative outcomes. We look at the impacts of an increase in uncertainty and an increase in risk aversion. We provide the counterparts to Propositions 2 and 3 for the public bad formulation.

Proposition 7. *Consider the public bad formulation and suppose that b' is more risky than b. Then $\forall k \in [0,n]$, $e^s(k,b') \le e^s(k,b)$. An increase in uncertainty always increases signatories' effort.*

Proposition 8. *Consider the public bad formulation and suppose that agents with utility V are more risk averse than agents with utility U. Then $\forall k \in [0,n]$, $e^s(k,V) \le e^s(k,U)$. An increase in risk aversion always increases signatories' efforts.*

Holding participation constant, signatories reduce their emissions and increase their efforts if uncertainty or risk aversion increases. These results are similar to results obtained in Bramoullé and Treich (2009). Indeed, they have a common explanation. The variance of the payoff is now equal to $p(1-p)(b_H-b_L)^2k^2e^2$ and is *increasing* in emissions. If uncertainty is higher, the payoff's variance is higher. To compensate, risk averse agents tend to decrease their emissions. Thus, exerting more effort reduces the payoff's variance in the public bad formulation but increases it in the public good one. This leads to opposite effects on signatories' actions hence, as we see next, on participation.

V. Boucher, Y. Bramoullé / Journal of Public Economics 94 (2010) 591–603 597

4.2. Stable treaties

We study stable treaties in a context of global pollution. We first show that the argument behind Proposition 4 extends. However, its conclusion is reversed.

Proposition 9. *In the public bad formulation, a stable treaty always exists. Participation in any stable treaty is such that* $k^* \in c/b_H, c/\bar{b}+1$. *Participation under uncertainty is always lower than or equal to participation under certainty.*

The reason for existence is essentially the same as under the public good formulation. Participation in a stable treaty must belong to the range where effort may take intermediate values. The different locations of these ranges lead to the different effects of uncertainty. The following result extends Proposition 5 to the public bad formulation.

Proposition 10. *Suppose that* c/b_H *is not an integer and that the level of absolute risk aversion of U is constant or decreasing. Then, there exists* $\bar{q}>0$ *such that if* $q_{\max} \geq \bar{q}$, *there is a unique stable treaty with* $k^* = \lfloor c/b_H \rfloor$ *and* $q^s = q_{max}$.

We emphasize two important differences between Propositions 10 and 5. One one hand, this result is more restrictive since we now need to make an assumption on the shape of preference towards risk. The reason relies, again, on the effect of non-signatories' actions. As q_{max} increases, signatories face increasingly dire prospects. The background risk generated by non-signatories' emissions becomes worse. If risk aversion is constant or decreasing, this *increases* the incentives for signatories to exert effort.[23] On the other hand, this result is also sharper since we now get *uniqueness* of the stable treaty when business as usual emissions are high enough. This indicates that equilibrium multiplicity may generally be less frequent in the public bad formulation.[24]

Finally, how does uncertainty affect welfare in a stable treaty? As in the public good formulation, uncertainty has three distinct effects on welfare. The first is similarly negative. Holding k and q^s fixed, uncertainty decreases welfare because of risk aversion. In contrast, the second effect has an opposite sign. Holding k fixed, uncertainty tends to increase signatories' efforts which improves welfare. However, the third effect also has an opposite sign. As shown in Proposition 9, uncertainty tends to decrease participation and this has a negative impact on welfare. In cases where $e=0$ and effort is maximal, the second effect disappears and uncertainty unambiguously lowers welfare. This arises, for instance, in situations identified by Proposition 10. When this happens, an ex-ante reduction of uncertainty has a positive "multiplier" effect in a context of global pollution. This multiplier effect is purely strategic in nature. It captures the increased incentives to participate caused by the reduction of uncertainty.

As in the previous section, we run numerical simulations to illustrate, see Figs. 5–7.[25] We still consider a utility function with constant relative risk aversion. Due to the different roles played by b_L and b_H in the two formulations, we now study the effect of an increase in uncertainty through an increase in b_H holding b_L and $\bar{b}$ constant. Parameters are set at the following values: $n=100, \gamma=0.5, \pi_0=3.10^7$, $c=1000, \bar{b}=111, b_L=90$.

Observe first that here the stable treaty is unique. As uncertainty increases, participation decreases. Fig. 6 shows the existence of a threshold value for b_H such participation is equal to 10 below this

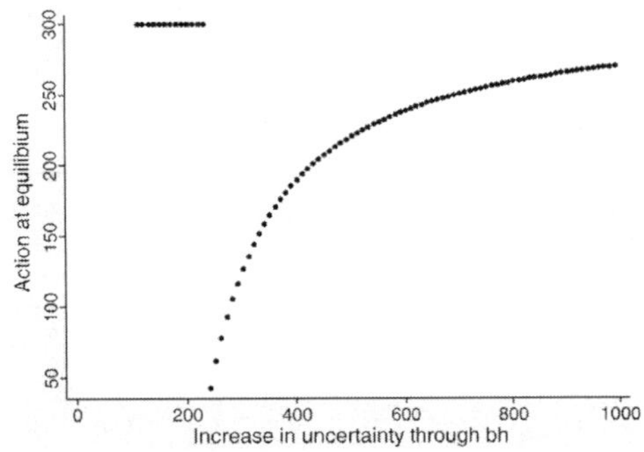

Fig. 5. Effects of uncertainty on signatories' effort (public bad).

value and drops to 9 above it. When $k^*=10$, Fig. 5 shows that effort is maximal. When it drops to $k^*=9$, signatories' effort first drops discontinuously and then increases as uncertainty increases. The initial drop is consistent with the fact that, for given parameter values, effort tends to decrease if participation decreases. The following increase in effort illustrates Proposition 7: holding participation constant, effort increases as uncertainty increases.

Next, look at welfare. As long as $k^*=10$, q and k^* are constant and welfare decreases due to the direct negative effect of uncertainty. When k^* drops from 10 to 9, welfare also drops under the combined effect of the reduction in participation and the reduction in effort. When $k^*=9$, we see a trade-off emerging between the direct negative effect and the indirect positive effect due to higher effort. Interestingly, the resulting outcome is non-monotonic. The positive effect first dominates. But the increase in effort occurs at a decreasing rate, and at some point, the indirect positive effect is overwhelmed by the direct negative one. Again, changes in participation have a crucial impact on welfare.

5. Uncertainty on the cost of effort

A restrictive assumption of our baseline model is that uncertainty only affects the benefits from action. In reality, the marginal cost of effort is often uncertain as well. For instance, when negotiating an international agreement on climate change, countries may not know

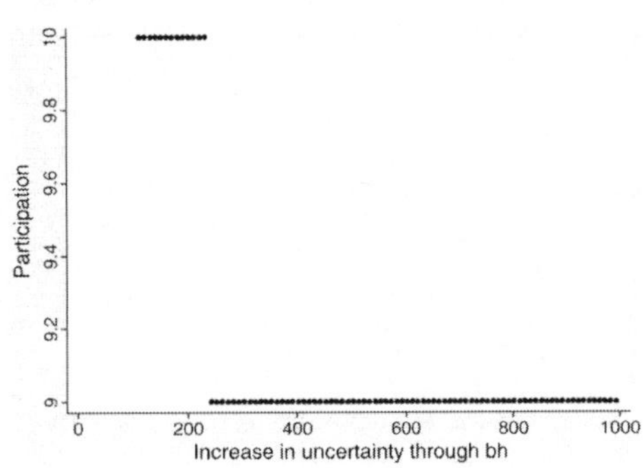

Fig. 6. Effects of uncertainty on participation (public bad).

[23] We suspect, but have not shown, that the result may not hold under increasing absolute risk aversion. Signatories would then tend to become *less* risk averse when faced with a worse baseline situation.

[24] And indeed, multiplicity did not appear in the simulations we ran under the public bad formulation.

[25] To facilitate comparison with the public good formulation, we look here at effort rather than emissions.

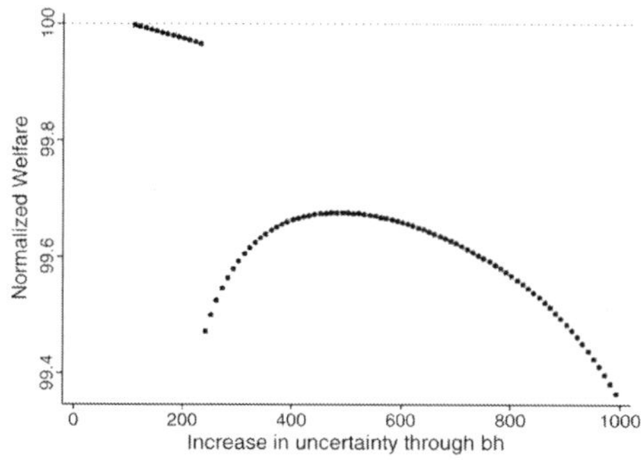

Fig. 7. Effects of uncertainty on welfare (public bad).

exactly the benefits they will derive from their future CO_2 emissions. In international R&D projects, realized costs often differ substantially from their initial estimates. In this section, we study the robustness of our results to the assumption of certain costs. We relax this assumption, and ask whether uncertainty on costs tend to counteract or to amplify uncertainty on benefits. We find that our main results extend and that some new, and somewhat subtle, differences arise between outcomes in the two formulations. Throughout this section, we assume that marginal cost of effort c is equal to c_L with probability p' and to c_H with probability $1-p'$, that cost and benefit are independent, and that $c_L > b_H$ and $nb_L > c_H$.

5.1. Providing a global public good

Suppose first that countries contribute to a global public good. Suppose that k signatories have signed the agreement, and look at how uncertainty on costs affects signatories' efforts. Signatories choose their action $q^s(k)$ to maximize $EU[(kb-c)q]$ where the expectation is now taken over all realizations of b and c. Marginal expected utility at $q=0$ is equal to $(k\bar{b}-\bar{c})U'(0)$, hence $q^s(k)=0$ if $k \le \bar{c}/\bar{b}$. Among the four possible states of the world, the one which is least favorable to action is the one where benefits are lowest $b=b_L$ while costs are highest $c=c_H$. So action is desirable in all states of the world, and $q^s(k)=q_{max}$, if $k \ge c_H/b_L$. Thus, if we compare a situation with certain cost $\bar{c}$ to one with uncertain cost, the range of participation levels for which action may be interior expands from $]\bar{c}/\bar{b}, \bar{c}/b_L[$ to $]\bar{c}/\bar{b}, c_H/b_L[$. The first order condition of signatories' decision problem still trades-off the positive marginal utilities of action in some states of the world versus its negative marginal utilities in others. This condition is of course more complicated than with two states, and cannot be solved analytically for CARA or CRRA utility functions. However, a positive solution always exists if $U'(-\infty)=+\infty$ or $U'(+\infty)=0$, as in Section 3.1.

In any case, signatories' action is necessarily (weakly) lower under uncertain costs when $k \in [\bar{c}/b_L, c_H/b_L]$. Uncertainty on costs tends to lower action at high participation levels. Does this property hold for all k? More generally, does an increase in cost uncertainty always lowers effort? And does Proposition 2 still hold in a setting with uncertain costs? To gain some insight on these questions, we can look at the expected value and variance of the payoff. Here the expected payoff is equal to $E(\pi)=(k\bar{b}-\bar{c})q$ while its variance is $Var(\pi) = \left[p(1-p)(b_H-b_L)^2k^2 + p'(1-p')(c_H-c_L)^2\right]q^2$. An increase in risk on cost increases the variance of the payoff without affecting its expected value, which gives an incentive to signatories to reduce their effort. As it turns out, this intuition does *not* carry over to all risk averse utility functions. Effects related to moments of order 3 and higher of the distribution of payoffs come into play. To obtain unambiguous results, we must impose some restrictions on the preferences, familiar from the theoretical literature on risk (Gollier (2001)). Introduce relative prudence, denoted by P_r, as:

$$P_r(\pi) = -\frac{u'''(\pi)}{u''(\pi)}\pi$$

Proposition 11. *Under uncertainty on benefits and costs, if $P_r \le 2$, then any increase in uncertainty on b or c reduces signatories' actions at all participation levels.*

Especially, relative prudence is equal to zero, and Proposition 11 holds, if utility is quadratic. In contrast, Proposition 11 may not hold when relative prudence is not everywhere lower than 2. Thus, even though cost uncertainty often amplifies the effect of benefit uncertainty by further lowering signatories' effort, it could in some cases go in the opposite direction. In contrast, the effect of an increase in risk aversion is unchanged and still clear-cut. We show in Appendix A that Proposition 3 directly extends to uncertain costs.

In turn, cost uncertainty affects stable treaties. We show in Appendix A that our proof of existence extends. Participation k in any stable treaty now lies between $\bar{c}/\bar{b}$ and c_H/b_L. More importantly, the argument behind Proposition 5 also extends. At $k=\lfloor c_H/b_L \rfloor - 1$, action is not desirable in at least one state of the world. When the interior solution exists, it is independent of q_{max}. Then, as q_{max} gets large, the drop in collective action when one signatory quits at $k=\lfloor c_H/b_L \rfloor$ becomes arbitrarily large. This guarantees that:

Proposition 12. *Consider uncertainty on benefits and costs. Suppose that either $\lim_{\pi \to -\infty} U'(\pi)=+\infty$ or $\lim_{\pi \to +\infty} U'(\pi)=0$, and that either c_H/b_L is not an integer or $\lim_{\pi \to +\infty} U(\pi)=0$. Then, there exists $\bar{q}$ such that if $q_{max} \ge \bar{q}$, a treaty with $k=\lfloor c_H/b_L \rfloor$ and $q^s=q_{max}$ is stable.*

For the treaty identified in Proposition 12, uncertainty on costs plays a role similar to uncertainty on benefits. It further increase participation, hence may further increase welfare despite its direct negative effect due to risk aversion. Interestingly, participation in this best possible treaty depends on the lowest value of benefit, but of the *highest* value of cost. The reason is that, taken together, these two values determine the worst possible state of the world and that in turn, this state determines the critical threshold level of participation.

5.2. Reducing a global public bad

Suppose next that countries try to reduce a global public bad. For signatories' actions, we show in Appendix A that $q^s=0$ if $k \le c_L/b_H$, $q^s=q_{max}$ if $k \ge \bar{c}/\bar{b}$ and q^s may be interior when $k \in]c_L/b_H, \bar{c}/\bar{b}[$. As in Section 4.1, the critical state of the world is now the one which is *most* favorable to action, that is, when $b=b_H$ and $c=c_L$. As soon as participation raises above the threshold level for which action in this state is desirable, signatories' efforts may become positive. Since $c_L/b_H < \bar{c}/b_H$, uncertainty on cost always (weakly) *increases* signatories' actions at low participation levels. As in the public good formulation, comparative statics with respect to risk are generally ambiguous. We obtain the following counterpart to Proposition 11. Introduce the level of absolute prudence, denoted by P, as:

$$P(\pi) = -\frac{u'''(\pi)}{u''(\pi)}$$

Proposition 13. *Under uncertainty on benefits and costs, if $P \le 2/[nq_{max}(c_H-b_L)]$, then any increase in uncertainty on b or c increases signatories' actions at all participation levels.*

The condition of Proposition 13 is actually more demanding than the condition of Proposition 11. The presence of a background risk implies that more restrictions have to be imposed on preferences to

V. Boucher, Y. Bramoullé / Journal of Public Economics 94 (2010) 591–603 599

obtain clear-cut comparative statics with respect to risk.[26] Interestingly, this background risk also affects the comparative statics with respect to risk aversion. If signatories become more risk averse, they do not necessarily increase their actions under uncertain costs.[27] This is related to standard results on the portfolio problem with a background risk, see Ch. 8.2 in Gollier (2001).[28] Especially, say that V is more risk averse than U in the sense of Ross (1981) if $V=\lambda U+f$ with $\lambda>0$ and f decreasing and concave. Then, we show in Appendix A that an increase in risk aversion in the sense of Ross always leads to an increase in the action of signatories.

Finally, consider stable treaties. Existence still holds, but Proposition 10 does not extend as directly as Proposition 5. We show the following result.

Proposition 14. Consider uncertainty on benefits and costs. Suppose that the level of absolute risk aversion of U is constant or decreasing. There exists $\bar{q}$ *such that if* $q_{\max}\geq\bar{q}$, *then* $\forall k\geq\bar{c}/b_H$, $q^s(k)=q_{max}$ *and participation* k^* *in any stable treaty satisfies* $[c_L/b_H]\leq k^*\leq[\bar{c}/b_H]$. *We may have* $q^s(k^*)\neq q_{max}$.

So if q_{max} is large enough, adding uncertainty on cost to uncertainty on benefits decreases participation further. However, and in contrast to what happens with in the public good formulation, it may not decrease all the way towards the threshold level $[c_L/b_H]$. To illustrate why, consider CARA utility functions. We show in Appendix A that if $k^*\in]c_L/b_H, \bar{c}/b_H, \bar{c}/b_H[$ (and only if), signatories' emissions tend to the following limit as q_{max} tends to infinity:

$$e^s(k)=\frac{1}{A(c_H-c_L)}ln\left[\frac{1-p'}{p'}\right]\frac{c_H-kb_H}{kb_H-c_L}$$

This follows from two observations. First, as q_{max} increases, the background payoff in the bad situation $-(n-k)b_Hq_{max}$ becomes increasingly worse than the background payoff in the better situation $-(n-k)b_Lq_{max}$. What happens in the bad state dominates, and signatories act as if the bad state were certain. Second, income effects are absent under CARA, which implies that optimal emissions when $b=b_H$ do not depend on q_{max}. So in that range of participation levels, the emission's limit is independent of q_{max}. This means that if $k=[c_L/b_H]-1$ the effect of one signatory joining becomes larger and larger. Yet, at $k=[c_L/b_H]$, an additional signatory may still have an incentive to join, since this may further decrease emissions, and the treaty may not be stable. However, when $k>\bar{c}/b_H$, signatories' emissions tend to zero, which means that $k^*\leq[\bar{c}/b_H]$ hence that participation is lower than under certain costs if q_{max} is high enough.

6. Conclusion

We conclude with a discussion of some limitations of our analysis and of promising directions for future research. In this paper, we introduce uncertainty and risk aversion to a simple model of international agreement to supply a global public good. We find that uncertainty and risk aversion significantly affect the analysis of treaty formation. They yield qualitative changes as well as significant quantitative changes on the outcomes of the game. This complexity provides some justification, ex-post, for the study of a simple benchmark model. It also raises the question of the robustness of our results. Five features of the model especially deserve attention: the fact that the risk is binary, linearity of the payoffs, homogeneity of the agents, the static dimension, and the simplicity of the policy framework.

Many of our results would directly extend to an arbitrary risk. For instance, the argument behind Propositions 1 and 6 hold in general. Similarly, our existence results Propositions 4 and 9 are valid for any risk. In contrast, signing the effects of risk and risk aversion in general may be more complicated. As is well-known from risk theory, comparative statics may be ambiguous and may involve conditions based on the third and higher derivatives of the utility function, see Gollier (2001). Section 5 already illustrates some of these complexities in a setup with uncertain costs of action.

Linearity of the payoffs is a critical assumption. Under certainty, non-linear payoffs may lead to very different equilibria, see Barrett (2003). An interest of the model with linear payoffs, however, is that it neatly captures the idea of critical mass. Collective action becomes worthwhile only when enough countries have joined in and once this threshold is reached, there is little benefit from an additional signatory. As such, studies of models with linear payoffs may be useful to understand what happens more generally. With any payoff function, stability captures a form of local critical mass. A treaty is stable when the drop in collective action if one country is high enough and when the addition of one signatory has little effect. Thus, we conjecture that our results may extend, under conditions to be determined, to models with non-linear payoffs. As soon as uncertainty lowers signatories' actions, corresponding threshold levels may increase which may improve participation in equilibrium. In contrast, if uncertainty increases signatories' actions the threshold levels may decrease and participation may drop.

Under heterogeneity, anonymity is lost. The idea of local critical mass is still relevant, but the identity of the signatories and non-signatories now matters. To understand the effects of heterogeneity under uncertainty, however, we would first need to understand its effects under certainty.[29] Despite some recent advances in that direction (see, e.g., Barrett (2001) and McGinty (2007)), a general analysis of international agreements under heterogeneity is still lacking. If anything, we expect the effects of uncertainty and risk aversion to be exacerbated under heterogeneity, given that countries may differ in the risk they face and in their degree of risk aversion.

Our model is static. However, the intertemporal dimension clearly matters for many global public goods, climate change being a prime example. Several non trivial issues arise in a dynamic framework. First, the agreement may have to be renegotiated over time and the set of signatories may evolve. Second, agreements have to address the intertemporal dimension. They may, for instance, specify temporal *sequences* of action. Third, the timing (and conditions) of future renegotiations may itself be an important part of the negotiations. We note that the study of these issues *under certainty* has barely started.[30] Thus, as with heterogeneity, a general understanding of the dynamics of international agreements under certainty is still lacking. Uncertainty would have interesting and complicated dynamic effects. Especially, uncertainty generally evolves over time thanks to learning, and the learning process itself may be affected by signatories' and non-signatories' actions.

Finally, our current policy framework is extremely simplified, both within and between countries. Within countries, national considerations such as elections and political constraints could very well affect countries' ability to contribute to the global public good, and hence international negotiations. And countries often interact with each other through a variety of channels such as trade, health, and foreign aid, and these other dimensions can have deep impacts on negotia-

[26] In both Propositions 11 and 13, the sufficient conditions depend on the magnitude of the increase in risk. While in the first, it is captured by the index of relative prudence, in the second, it appears directly in the bound (see the Appendix A for details).

[27] So while Proposition 3 directly extends to uncertain costs, this is not the case for Proposition 8.

[28] Most results assume independence between risks, hence cannot be applied here.

[29] Note that by continuity, we expect our results to hold under heterogeneity, as long as the extent of heterogeneity is not too high.

[30] See Rubio and Ulph (2007) for one of the first truly dynamic analysis of international environmental agreements in the presence of a stock pollutant.

tions on a specific global public good.[31] It would be interesting to introduce uncertainty to a model of international agreement incorporating some of these considerations.

Overall, there is much research to be done to better understand the determinants of international agreements in more realistic settings. In this paper, we have shown that uncertainty and risk aversion have strong strategic effects in a simple benchmark model. Given the importance of uncertainty in reality, we think that such strategic effects will also play a major role in the analysis of more realistic models incorporating arbitrary risks, general payoffs, heterogeneity, dynamics, and richer policy considerations.

Acknowledgments

A previous version of the paper was circulated under the title "Risk Aversion and International Environmental Agreements". We would like to thank Christian Gollier, Nicolas Treich, and Alistair Ulph as well as the participants in various conferences and seminars for helpful the comments and discussions. We gratefully acknowledge the financial support from IEDS, FQRSC and SSHRC.

Appendix A

Proof of claim p.11. Denote by $\tilde{q}^s(k)$ the solution to condition (2). Rewriting the condition yields: $pU_L'(kb_L-c)+(1-p)U_H'(kb_H-c)=0$ where the indices L and H stand for the function arguments $(kb_L-c)q$ and $(kb_H-c)q$. Derivating with respect to k yields:

$$pU_L'b_L + p(kb_L-c)U_L''\left[(kb_L-c)\frac{\partial\tilde{q}^s}{\partial k}(k) + b_L^s\tilde{q}^s(k)\right]$$
$$+ (1-p)U_H'b_H + (1-p)(kb_H-c)U_H''\left[(kb_H-c)\frac{\partial\tilde{q}^s}{\partial k}(k) + b_H\tilde{q}^s(k)\right] = 0$$

Rewriting using $A=-U''/U'$ shows that $\tilde{q}^s(k)$ is weakly increasing if:

$$pU_L'[-b_L + (kb_L-c)A_Lb_L\tilde{q}^s(k)] + (1-p)U_H'[-b_H + (kb_H-c)A_Hb_H\tilde{q}^s(k)]\leq 0$$

From the f.o.c., we have $\frac{U_L'}{U_H'} = \frac{(1-p)(kb_H-c)}{p(c-kb_L)}$. Substituting, we obtain that $\tilde{q}^s(k)$ is weakly increasing if:

$$\tilde{q}^s(k)(A_Lb_L-A_Hb_H)\geq\frac{-c(b_H-b_L)}{(kb_H-c)(c-kb_L)}$$

If $A(\pi_H)\leq(b_L/b_H)A(\pi_L)$, then this condition is clearly satisfied since the LHS is non-negative and the RHS is negative. This means that the level of absolute risk aversion at π_H is sufficiently lower than the level of absolute risk aversion at $\pi_L\leq\pi_H$.

Proof of Proposition 2. Consider q^s as a function of b_L and b_H, holding $\bar{b}$ constant. We have $p=(b_H-\bar{b})/(b_H-b_L)$. Substituting, condition (2) becomes: $U_L'(b_H-\bar{b})(kb_L-c)+U_H'(\bar{b}-b_L)(kb_H-c)=0$. Derivating this last expression with respect to b_H, we see that:

$$U_L'(c-kb_L)\geq k(\bar{b}-b_L)U_H'\Rightarrow\frac{\partial q^s}{\partial b_H}(k,b_H)\leq 0$$

The left-hand side holds for $k\in[c/\bar{b},\ c/b_L]$. Similarly, derivating with respect to b_L, we have:

$$U_H'(kb_H-c)\leq U_L'(b_H-\ \bar{b})k\Rightarrow\frac{\partial q^s}{\partial b_L}(k,b_H)\geq 0$$

where the left-hand side also holds when $k\in[c/\bar{b},\ c/b_L]$. Thus, holding b_H constant, q^s decreases if b_L decreases and holding b_L constant, q^s decreases if b_H increases. This means that $q^s(b_L',b_H')\leq q^s(b_L,b_H')\leq q^s(b_L,b_H)$.

Proof of Proposition 3. Denote by $F(q) = EV[(kb-c)q]$ the objective function of the signatories when agents have utility V. Since $V=\Phi(U)$, we have:

$$F'(q) = p\Phi'(U[(kb_L-c)q])U'[(kb_L-c)q] + (1-p)\Phi'(U[(kb_H-c)q])U'[(kb_H-c)q].$$

At $q=q^s(U)$, $pU'[(kb_L-c)q]+(1-p)U'[(kb_H-c)q] = 0$, and $F'(q^s(U))=(1-p)U'[(kb_H-c)q](\Phi'(U[(kb_H-c)q])-\Phi'(U[(kb_L-c)q])$. Since $(kb_H-c)q>(kb_L-c)q$, U is increasing, and Φ' is decreasing, we have $F'(q^s(U))\leq 0$. Since F is concave, F' is decreasing, and $F'(q^s(V))=0$, we have:$q^s(V)\leq q^s(U)$.

A.1. Computations for specific utilities

A.1.1. Quadratic utility

Consider $q^s(k) = \frac{1}{2\lambda}\frac{k\bar{b}-c}{p(kb_L-c)^2 + (1-p)(kb_H-c)^2}$. Introduce $b^2=pb_L^2+(1-p)b_H^2$. Derivating q^s with respect to k, we obtain:

$$\frac{\partial q^s(k)}{\partial k} = \frac{-k^2\bar{b}\bar{b}^2 + 2k\bar{b}^2c-c^2\ \bar{b}}{2\lambda\left(k^2b^2-2ck\bar{b} + c^2\right)^2}.$$

Then, q^s is increasing if $-k^2\bar{b}b^2+2kb^2c-c^2\bar{b}\geq 0$. In $k=c/\bar{b}$, this expression becomes $c^2(b^2-\bar{b}^2)/\bar{b}$, which is always positive. In addition, we know that $q^s(c/b_L)>q_{max}$ from Proposition 1. Since the previous condition is quadratic in k, there are two cases. Either q^s is increasing over $[c/\bar{b}, c/b_L]$, or q^s is first increasing and then decreasing, and $max_{[c/\bar{b},c/b_L]}q^s>q_{max}$. In either case, $min(q^s,q_{max})$ is increasing over $[c/\bar{b},\ c/b_L]$.

A.1.2. CARA utility

Let $q^s(k) = \frac{1}{Ak(b_H-b_L)}ln\left[\frac{(1-p)(kb_H-c)}{p(c-kb_L)}\right]$. Introduce the three following auxiliary functions: $f(k) = \frac{(1-p)(kb_H-c)}{p(c-kb_L)}$, $g(k) = \frac{(kb_H-c)(c-kb_L)}{k}$, and $h(k) = ln[f(k)]g(k)$. Derivating q^s with respect to k, we obtain:

$$\frac{\partial q^s}{\partial k} = \frac{1}{A(b_H-b_L)k}\left(\frac{-ln(f)}{k} + \frac{c(\bar{b}-b_L)}{(1-p)(kb_H-c)(c-kb_L)}\right)$$

This means that q^s is increasing if $\frac{c(\bar{b}-b_L)}{(1-p)(kb_H-c)(c-kb_L)}\geq\frac{ln[f(k)]}{k}$, which is equivalent to $h(k)\leq c(b_H-b_L)$. We next study the properties of h. We have: $h' = ln[f]g' + \frac{f'}{f}g$. Since $k\geq c/\bar{b}$, $f(k)\geq 1$ and $ln[f(k)]\geq 0$. In addition, $f'(k)\geq 0$ and $g'(k)=-b_Lb_H+c^2/k^2$. This means that, $h'(k)>0$ if $k\in\left[0,c/\sqrt{b_Lb_H}\right]$. >Looking at the second derivative of h gives:

$$\frac{\partial^2h}{\partial k^2}(k) = ln[f(k)]g''(k) + \frac{c(b_H-b_L)}{k^2(c-kb_L)(kb_H-c)}[c(-k(b_H + b_L) + 2c)]$$

which is strictly negative for $k\geq c/\sqrt{b_Lb_H}$. Therefore, h is increasing over $\left[0,c/\sqrt{b_Lb_H}\right]$ and strictly concave over $\left[c/\sqrt{b_Lb_H},c/b_L\right]$. In addition, we know that q^s is increasing at $c/\bar{b}$ and becomes arbitrarily large when k gets close to c/b_L. This implies that q^s is either increasing over $[c/\bar{b},\ c/b_L]$, or increasing, decreasing, and increasing again. Besides, $\partial h/\partial p\leq 0$, hence the whole h function shifts downwards when p increases. Suppose that $p_1>p_2$. If q^s is increasing for p_1, then q^s increasing for p_2. If q^s is increasing, decreasing, increasing for p_1, then

[31] See e.g.Barrett (2003, Ch. 12), and the references therein, on linkage between different issues in international negotiations.

V. Boucher, Y. Bramoullé / Journal of Public Economics 94 (2010) 591–603 601

either q^s is increasing, or the interval on which it is decreasing is smaller.

Proof of Proposition 5. Let $[x]$ denote the smallest integer greater than or equal to x. Let $k^* = [c/b_L]$. Recall, $\Delta(k^*+1)<0$ since $q^s(k^*) = q^s(k^*+1) = q_{max}$. In addition, if a signatory quits at k^*, he obtains $(k^*-1)bq^s(k-1)$. Thus,

$$\Delta\left(k^*\right) = p\left[U\left(\left(k^*b_L-c\right)q_{\max}\right)-U\left(\left(k^*-1\right)b_Lq^s\left(k^*-1\right)\right]\right.$$
$$+ (1-p)\left[U\left(\left(k^*b_H-c\right)q_{\max}\right)-U\left(\left(k^*-1\right)b_Hq^s\left(k^*-1\right)\right]\right.$$

Since $q^s(k^*-1)<c/b_L$, if q_{max} is large enough, $q^s(k^*-1)<q_{max}$. Since $k^*b_H-c>0$, if q_{max} is large enough, $(k^*b_H-c)q_{\max}>(k^*-1)b_Hq^s(k^*-1)$. If $k^* \neq c/b_L$, we also have $k^*b_L-c>0$, hence $(k^*b_L-c)q_{\max}>(k^*-1)b_Lq^s(k^*-1)$ if q_{max} is large enough. In this case, since U increasing, $\Delta(k^*)>0$. This result also holds if $k^* = c/b_L$, and $lim_{\pi\to\infty} U(\pi) = +\infty$.

Proof of Proposition 6. If $k \leq c/b_H$, $c-kb$ is less than or equal to zero in both states of the world, and strictly less than zero in one, hence signatories' emissions are clearly maximal. Next, suppose that $k \geq c/\bar{b}$. The marginal expected utility of a signatory at $e=0$ is equal to $E(c-kb)U'[-(n-k)bq_{max}]$. Observe that $c-kb$ is decreasing in b while $U'[-(n-k)bq_{max}]$ is weakly increasing in b by concavity of U. The covariance rule (Gollier, 2001) implies that $cov[(c-kb)U'[-(n-k)bq_{max}]] \leq 0$, hence that $E(c-kb)U'[-(n-k)bq_{max}] \leq E(c-kb)EU'[-(n-k)bq_{max}]$. Finally, $E(c-kb) = c-k\bar{b} \leq 0$ and $EU'[-(n-k)bq_{max}] \geq 0$, which means that the marginal expected utility at 0 is less than or equal to zero. By concavity, $e^s=0$. QED.

Proof of Proposition 7. Consider e^s as a function of b_L and b_H. We have $p = \frac{b_H-\bar{b}}{b_H-b_L}$. The f.o.c for the coalition's problem is $p(c-kb_L)U_L'+(1-p)(c-kb_H)U_H'=0$. Substituting with $p = \frac{b_H-\bar{b}}{b_H-b_L}$, it becomes:

$$(b_H-\bar{b})(c-kb_L)U_L' + (\bar{b}-b_L)(c-kb_H)U_H' = 0 \qquad (4)$$

We now derive with respect to b_L, with $\frac{\partial \bar{b}}{\partial b_L} = 0$, we have:

$$\frac{\partial e}{\partial b_L}(k)\left[\left(b_H-\bar{b}\right)(c-kb_L)^2U_L'' + \left(\bar{b}-b_L\right)(c-kb_H)^2U_H''\right] = \left(b_H-\bar{b}\right)kU_L'$$
$$+ \left(b_H-\bar{b}\right)(c-kb_L)U_L''[ne_{\max}-k[e_{\max}-e(k)]] + (c-kb_H)U_H'.$$

Replacing $(c-kb_H)U_H' = -\frac{p}{1-p}(c-kb_L)U_L'$, from the f.o.c., in the last equation leads to: $\frac{\partial e}{\partial b_L}(k) \geq 0$, for all $k \leq c/\bar{b}$. We now derivate (4) with respect to b_H, which gives us:

$$\frac{\partial e}{\partial b_L}(k)\left[\left(b_H-\bar{b}\right)(c-kb_L)^2U_L'' + \left(\bar{b}-b_L\right)(c-kb_H)^2U_H''\right] = -(c-kb_L)U_L'$$
$$+ \left(\bar{b}-b_L\right)(c-kb_H)U_H''[ne_{\max}-k(e_{\max}-e(k))] + k\left(\bar{b}-b_L\right)U_H'.$$

Again, using the f.o.c., we have: $\frac{\partial e}{\partial b_H}(k) \leq 0$, for all $k \leq c/\bar{b}$. Since $\frac{\partial e}{\partial b_L}(k) \geq 0$ and $\frac{\partial e}{\partial b_H}(k) \leq 0$, for all $k \leq c/\bar{b}$, and both derivatives equal to zero for $k > c/\bar{b}$ (from Proposition 6), this proves Proposition 7.

Proof of Proposition 8. Denote by $F(q) = EV[(c-kb)e-(n-k)be_{\max}]$ the objective function of the signatories when agents have utility V. Since $V = \Phi(U)$,we have:

$$F'(q) = p(c-kb_L)\phi_L'U_L' + (1-p)(c-kb_H)\phi_H'U_H'.$$

At $e = e^s(U)$, we have $p(c-kb_L)U_L' = -(1-p)(c-kb_H)U_H'$, and $F'(q) = p(c-kb_L)U'_L[\phi_L'-\phi_H']$ since $(c-kb_H)e<(c-kb_L)e$, U is increasing, and Φ' is decreasing, we have. $F'(e^s(U)) \geq 0$. Since F concave, F' decreasing, and $F'(e^s(V)) = 0$, we have:$e^s(V) \geq e^s(U)$.

Proof of Proposition 10. We will show that there is a $\bar{q}>0$ such that if $q_{\max} \geq \bar{q}$, then for any $k \geq c/b_H$, $e^s(k)=0$. The first order conditions of program 3 can be written

$$p(c-kb_L)U'[(c-kb_L)e-(n-k)b_Lq_{max}] = (1-p)(kb_H-c)U'[(c-kb_H)e-(n-k)b_Hq_{max}]$$

Denote by $\tilde{e}^s(k)$ its solution. When u is CARA, direct computations lead to:

$$\tilde{e}^s(k) = -\frac{1}{Ak(b_H-b_L)}ln\left[\frac{(1-p)(kb_H-c)}{p(c-kb_L)}\right] - \frac{n-k}{k}q_{max}$$

If q_{max} is large enough, this is clearly negative for any k which implies that $e^s(k)=0$ as soon as $k \geq c/b_H$.

Next, take the derivative of the f.o.c. with respect to q_{max}. Algebraic computations lead to:

$$\frac{\partial \tilde{e}^s}{\partial q_{max}} = -\frac{n-k}{k}\frac{b_HA(\pi_H)-b_LA(\pi_L)}{(kb_H-c)A(\pi_H) + (c-kb_L)A(\pi_L)}$$

where $A(\pi) = -u''(\pi)/u'(\pi)$, $\pi_H = (c-kb_H)e-(n-k)b_Hq_{max}$ and similarly for π_L. Observe that $\pi_H-\pi_L \leq 0$ if $e \geq 0$. When u satisfies DARA, $\pi_H \leq \pi_L \Rightarrow A(\pi_H) \geq A(\pi_L)$, hence $\frac{\partial \tilde{e}^s}{\partial q_{max}} \leq -\frac{n-k}{k^2}<0$. This guarantees that if $\exists q_1$ such that $\tilde{e}^s(q_1)<0$, then $\forall q_{max} \geq q_1$, $\tilde{e}^s(q_{max})<0$. Also, if $\exists q_2$ such that $\tilde{e}^s(q_2)>0$, then $\tilde{e}^s$ decreases in q_{max} til it becomes negative for $q_{max} \geq q_2$. In the end, $\exists \bar{q}>0$ such that if $q_{max} \geq \bar{q}$, then for any $k \geq c/b_H$, $e^s(k)=0$. QED.

Proof of Proposition 11. We show that any increase in uncertainty reduces the level of action. Denote q^* before the increase in uncertainty. If q^* is not interior, the proof is obvious (see section 5.1). Then, assume that that k leads to an interior solution for q^*. We use the fact that a distribution x' is more risky than$x$ iff there exist a white noise ε such that $x' = x+\varepsilon$. The first derivative of the problem of the coalition after the increase in risk, evaluated at q^*, is: $E_\varepsilon E_{b,c}U'(kb-c)q^*+\varepsilon q^*)\times(kb-c+\varepsilon)$. We can rewrite this last expression as :$E_\varepsilon E_{b,c}f(\varepsilon)$. If $f(\varepsilon)$ is concave, we have : $E_\varepsilon E_{b,c}f(\varepsilon) \leq E_{b,c}f(0) = 0$ (Jensen's inequality), so the optimal level of action has to be lower for higher risks (since U is strictly concave). We can compute $f''(\varepsilon) = U'''q^{*2}(kb-c+\varepsilon)+2U''q^*$, so $f''(\varepsilon) \leq 0$ iff $-\frac{U'''}{U''}q\,(kb-c+\varepsilon) \leq 2$.

A.2. Robustness results for the public good model

A.2.1. The effect of risk aversion

We show that an increase in risk aversion reduces the level of action. Consider the first derivative of the coalition when preferences are represented by $V(x) = \phi(U(x))$, with $\phi'>0$ and $\phi''<0$ (i.e. V more risk averse than U): $kE\phi'[U((kb-c)q)]U'[(kb-c)q](kb-c)$. Notice that, $(kb-c)\phi'(U[(kb-c)q]) \leq (kb-c)\phi'(U[0])$ for any values of k,b,c,q. Then, the first derivative is lower than $k\phi'(U[0])EU'[(kb-c)q^*(U)] = 0$ so the optimal level of action is lower for more risk averse preferences.

A.2.2. Existence

We show that there exist a stable treaty with positive action. Let k_1 be the first integer such that $q^*(k)>0$. We know that $k_1 \geq \bar{c}/\bar{b}$. The expected utility for a non-signatory at k_1-1 is $EU^n(k_1-1) = EU[0]$, while the expected utility for a signatory at k_1 is $EU^s(k_1) = EU[(k_1b-c)q^s]$. Then, at k_1-1 a non-signatory has a strict incentive to join the coalition. Now, let $k_2 = [c_H/b_L]$. The expected utility of a non-signatory at k_2 is $EU^n(k_2) = U[k_2bq_{\max}]$ while the expected utility of a signatory at k_2+1 is $EU^s(k_2+1) = U[k_2bq_{\max} + (b-c)q_{\max}]$. Then, at k_2, a non-signatory has no incentive to join the coalition. This shows that there exists a stable treaty $k^* \in [k_1, k_2]$. Since $EU's(0) > 0$ for $k \geq \bar{c}/\bar{b}$, we know that $q(k^*) > 0$.

Proof of Proposition 12. Define $k^* = [c_H/b_L]$. At k^*, we know that a non-signatory has no incentive to join the coalition. We show that if q_{max} is large enough, we have $q^*(k^* - 1) < q_{max}$. If $k^* - 1 < \bar{c}/\bar{b}$, the result is obvious, otherwise we show that the f.o.c holds. Notice that (at least) $(kb_L - c_H)$ is negative, and $(kb_H - c_L)$ is positive. Since the utility function is concave, the negative part of the first derivative is increasing in q in absolute value, while the positive part is decreasing in q. If $lim_{x \to -\infty} U'(x) = \infty$ (or if $lim_{x \to \infty} U'(x) = 0$) and q_{max} is large enough, there exists q^* such that the first derivative equals zero.

We now show that k^* is stable. The expected utility of a non-signatory at $k^* - 1$ is given by $EU^n(k^* - 1) = EU[(k^* - 1)bq^s(k^* - 1)]$ while the expected utility for a signatory at k^* is given by $EU^s(k^*) = EU[(k^*b - c)q_{\max}]$. If c_H/b_L is not an integer, the payoff of a signatory at k^* is strictly increasing in q_max. If $lim_{\pi \to \infty} U(\pi) = \infty$, there exist q_{max} such that $EU^s(k^*) > EU^n(k^* - 1)$. Then, if q_{max} is large enough, a signatory at k^* has no incentive to quit the coalition, since $q^s(k^* - 1)$ is strictly smaller than q_{max} and independent of q_{max}. Then k^* is stable.

A.3. Robustness results for the public bad model

A.3.1. Level of emissions

We show that $q(k) = 0$ if $k \leq c_L/b_H$, and $q(k) = q_{\max}$ if $k \geq \bar{c}/\bar{b}$.

The first derivative for the coalition is $kEU'[(c - kb)e - (n-k)bq_{\max}](c - kb)$, which is lower than $kEE\{U'[(c-kb)e - (n-k)bq_{\max}](\bar{c} - kb) \mid b\}$, and lower than $kEU'[(c-kb)e - (n-k)bq_{\max}]\left(\bar{c} - k\bar{b}\right)$ by the law of iterated expectations and the successive application of the covariance rule. Then, the level of emissions is null for $k > \bar{c}/\bar{b}$. Also, as in the public good model, we can show that the first derivative is strictly positive for $k \leq c_L/b_H$. Then, the level of emissions is maximal for $k \leq c_L/b_H$.

A.3.2. Risk aversion in the sense of Ross (1981)

Consider the maximization problem of the coalition when countries' preferences are represented by $v(x)$, more risk averse than $u(x)$ in the sense of Ross (1981). The first derivative can be written as a function of $u(x)$, i.e. $\lambda E(c-kb)u'[(c - kb)e - (n-k)bq_{\max}] + Ef'[(c-kb)e - (n-k)bq_{\max}](c - kb)$. By the covariance rule, we have that $E(c - kb) f'[(c-kb)e - (n-k)bq_{\max}] \leq \left(\bar{c} - k\bar{b}\right) Ef'[(c-kb)e - (n-k) bq_{\max}] \leq 0$ since $k \leq \bar{c}/\bar{b}$. Then, we have that $\lambda E(c - kb)u'[x] + Ef'[x](c - kb) \leq \lambda E(c - kb)u'[x]$. Since the RHS is equal to zero at e (u), an increase in risk aversion in the sense of Ross (1981) leads to a lower level of emissions.

A.3.3. Existence

We show that there exists a stable level of participation k^* with positive action. Let k_1 be the first integer such that $e^s(k_1) < q_{\max}$. Notice that $EU[(c - nb)q_{\max}] < EU[(c - k_1 b)e^s(k_1) - (n - k_1)bq_{\max}]$ since $e^s(k_1) < q_{\max}$ solves the problem of the coalition for $k = k_1$, and U is strictly increasing.

Now, let $k_2 = [\bar{c}/\mathrm{b}]$, we have that $EU[cq_{\max} - (n - k_2)bq_{\max}] < EU[-(n - k_2 - 1)bq_{\max}]$. Since a country has a strict incentive to join the treaty at k_1, and no incentive to join at k_2, we know that there exist a stable treaty $k \in [k_1, k_2]$. Suppose that $e(k^*) = q_{\max}$, then it means that there exists k_3 such that $e(k_3) < e(k_3 + 1)$. Then, at k_3, no country has an incentive to join the coalition. There exists a stable treaty $\tilde{k} \in [k_1, k_3]$. Since $e(k_1) < q_{\max}$, we know that there exists some stable treaty with strictly positive action.

Proof of Proposition 13. We show that any increase in uncertainty increases the level of action. Denote e the optimal level of emissions for a given level of uncertainty. If e is not interior, the solution is obvious. Then, assume that that k leads to an interior solution for e^*. The proof goes as the one of Proposition 11.

First, we look at an increase in risk through c. The first derivative of the problem of the coalition after the increase in risk, evaluated at e^*, is: $E_\varepsilon E_{b,c} U'(c + \varepsilon - kb)q^*)(c - kb + \varepsilon)$. We can rewrite this last expression as: $E_\varepsilon E_{b,c} f(\varepsilon)$. If $f(\varepsilon)$ is concave, we know that the optimal level of emissions has to be lower for higher risks (since U is strictly concave). We can compute $f''(\varepsilon) = U''' e^{*2}(c - kb + \varepsilon) + 2U'' e^*$, so $f''(\varepsilon) \leq 0$ if $-\frac{U'''}{U''} e^*(c - kb + \varepsilon) \leq 2$.

Next, we look at an increase in risk through b. Proceeding the same way, we find that $f''(\varepsilon) \leq 0$ if $-\frac{U'''}{U''}\left(ke^* + (n-k)q_{\max}\right)(c - kb - k\varepsilon)/k \leq 2$

Denote c_H and b_L, the highest possible cost, and the lowest possible benefit *after* the increase in risk, then a sufficient for both conditions is: $-\frac{U'''}{U''} \leq \frac{2}{nq_{\max}(c_H - b_L)}$.

Proof of Proposition 14. Suppose that U respects CARA or DARA. We show that if $q_{\max}$ is high enough, $q^s(k) = q_{\max}$ for all $k \geq \bar{c}/b_H$, and that any stable treaty k respects $c_L/b_H \leq k \leq \bar{c}/b_H$.

First, define $\pi(e) = E_{b,c} u((c - kb)e - (n-k)bq_{max})$. Since $\pi'' < 0$, we have $e = 0 \Leftrightarrow \pi'(0) < 0$, where $\pi'(0) = E_b(c - kb)u'(-(n-k)bq_{max})$, the expected marginal utility at $e = 0$.

Suppose first that u is CARA. Then $u'(x) = Ae^{-Ax}$ which yields $\pi'(0) = e^{A(n-k)b_L q_{max}}\left[p(\bar{c} - kb_L) - (1-p)(kb_H - \bar{c})e^{A(n-k)(b_H - b_L)q_{max}}\right]$. When q_{max} is large enough, $\pi'(0)$ has the same sign as $-(kb_H - \bar{c})$, hence negative if $k > \bar{c}/b_H$ and positive if $k < \bar{c}/b_H$. This shows that $e = 0$ if $k > \bar{c}/b_H$ and q_{max} is high enough.

Suppose next that u is DARA. We have $\pi'(0) = u'(-(n-k)b_L q_{max})\left[p(\bar{c} - kb_L) - (1-p)(kb_H - \bar{c})\frac{u'(-(n-k)b_H q_{max})}{u'(-(n-k)b_L q_{max})}\right]$.

Next, we know that $f(x) - f(y) = f'(z)(x - y)$ with $z \in [x, y]$ which means that $u'(-(n-k)b_H q_{max}) - u'(-(n-k)b_L q_{max}) = (n-k)(b_H - b_L) q_{max}(-u'')\left(-(n-k)\hat{b}q_{max}\right)$ with $\hat{b} \in [b_L, b_H]$. Thus, $\frac{u'(-(n-k)b_H q_{max})}{u'(-(n-k)b_L q_{max})} = 1 + (n-k)(b_H - b_L)q_{max}\frac{(-u'')\left(-(n-k)\hat{b}q_{max}\right)}{u'(-(n-k)b_L q_{max})}$.

Since u is DARA, prudence holds, $u''' > 0$, hence $(-u)''$ is decreasing.

Since $-(n-k)\hat{b}q_{max} \leq -(n-k)b_L q_{max}$, we have $(-u'')\left(-(n-k)\hat{b}q_{max}\right) \geq (-u'') \times (-(n-k)b_L q_{max})$. Hence $\frac{(-u'')\left(-(n-k)\hat{b}q_{max}\right)}{u'(-(n-k)b_L q_{max})} \geq A(-(n-k)b_L q_{max}) \geq A(0)$.

Thus, $\frac{u'(-(n-k)b_H q_{max})}{u'(-(n-k)b_L q_{max})} \geq 1 + (n-k)(b_H - b_L)A_0 q_{max}$, which tends to ∞ when q_{max} tends to ∞, which implies that $\pi'(0)$ also has the same sign as $-(kb_H - \bar{c})$, hence is negative if $k > \bar{c}/b_H$.

A.3.4. Remark on CARA utility functions

For CARA utility functions, we show that, as $q_{\max}$ tends to infinity, the solution with uncertainty on the cost and the benefit of action tends to the one with uncertainty on the cost, when $b = b_H$ with certainty if $k \in]c_L/b_H, \bar{c}/b[$.

Let $\pi'(0) = e^{A(n-k)b_L q_{max}}\left[p(\bar{c} - kb_L) - (1-p)(kb_H - \bar{c})e^{A(n-k)(b_H - b_L)q_{max}}\right]$. Since $kb_H - \bar{c} < 0$, $\pi'(0) > 0$ if q_{max} is large enough. Then, either $\pi'(q_{max}) > 0$, and $e = q_{max}$ or there is an interior solution $e > 0$ solving $\pi'(e) = 0$.

In that case, $pe^{Ab_L(ke + (n-k)q_{max})}E_c(c - kb_L)e^{-Ace} + (1-p)e^{Ab_H(ke + (n-k)q_{max})}E_c(c - kb_H)e^{-Ace} = 0$, so $e^{A(b_H - b_L)(ke + (n-k)q_{max})} = -\frac{p}{1-p}\frac{E_c(c - kb_L)e^{-Ace}}{E_c(c - kb_H)e^{-Ace}} > 0$ implying that $E_c(c - kb_L)e^{-Ace} > 0$ and $E_c \times (c - kb_H)e^{-Ace} < 0$. Then, $E_c(c - kb_H)e^{-Ace} = -p'(kb_H - c_L)e^{-Ac_L e} + (1 - p')(c_H - kb_H)e^{-Ac_H e}$, where $kb_H - c_L > 0$ since $k > c_L/b_H$ and $c_H - kb_H > 0$ since $k < c/b_H < c_H/b_H$.

Thus, $-p'(kb_H - c_L)e^{-Ac_L e} + (1 - p')(c_H - kb_H)e^{-Ac_H e} = M(e) \times e^{-A(b_H - b_L)(ke + (n-k)q_{\max})}$, where $M(e) = -\frac{p}{1-p}E_c(c - kb_L)e^{-Ace}$ is bounded. Therefore, the LHS tends to zero, which means that e tends to e^* which solves: $-p'(kb_H - c_L)e^{-Ac_L e} + (1 - p')(c_H - kb_H) \times$

$e^{-Ac_He}=0$, which corresponds to the f.o.c. of the problem: $maxE_cu((c-kb_H)e-kb_Hq_{max})$. Finally, observe that $e^*>0\Leftrightarrow k<\bar{c}/b_H$.

References

Arrow, K., Lind, R., 1970. Uncertainty and the evaluation of public investment decisions. The American Economic Review 60, 364–378.

Barrett, S., 1994. Self-enforcing international environmental agreements. Oxford Economic Papers 46, 878–894.

Barrett, S., 2001. International cooperation for sale. European Economic Review 45, 1835–1850.

Barrett, S., 2003. Environment & Statecraft. Oxford University Press, Oxford.

Barrett, S., 2008. Why Cooperate? The Incentive to Supply Global Public Goods. Oxford University Press, Oxford.

Bramoullé, Y., Treich, N., 2009. Can uncertainty alleviate the commons problem? Journal of the European Economic Association 7 (5), 1042–1067.

Eso, P., White, L., 2004. Precautionary bidding in auctions. Econometrica 72, 77–92.

Gollier, C., 2001. The Economics of Risk and Time. The MIT Press, Cambridge.

Gradstein, M., Nitzan, S., Slutsky, S., 1992. The effect of uncertainty on interactive behaviour. The Economic Journal 102, 554–561.

Heal, G., Kriström, B., 2002. Uncertainty and climate change. Environmental & Resource Economics 22, 3–39.

Kolstad, C.D., 2005. Piercing the veil of uncertainty in transboundary pollution agreements. Environmental & Resource Economics 31, 21–34.

Kolstad, C.D., 2007. Systematic uncertainty in self-enforcing international agreements. Journal of Environmental Economics and Management 53, 68–79.

McBride, M., 2006. Discrete public good under threshold uncertainty. Journal of Public Economics 90, 1181–1199.

McGinty, M., 2007. International environmental agreements among asymmetric nations. Oxford Economic Papers 59, 45–62.

Na, S.-L., Shin, H.S., 1998. International environmental agreements under uncertainty. Oxford Economic Papers 50, 173–185.

Nitzan, S., Romano, R.E., 1990. Private provision of a discrete public good with uncertain cost. Journal of Public Economics 42, 357–370.

Nordhaus, W., 1994. Managing the global commons. The Economics of Climate Change. MIT Press, Cambridge.

Palfrey, T., Rosenthal, H., 1984. Participation and the provision of a discrete public goods: a strategic analysis. Journal of Public Economics 24, 171–193.

Ross, S.A., 1981. Some stronger measures of risk aversion in the small and in the large with applications. Econometrica 3, 621–638.

Rothschild, M., Stiglitz, J., 1970. Increasing risk: I. A definition. Journal of Economic Theory 2, 225–243.

Rubio, S., Ulph, A., 2007. An infinite horizon model of dynamic membership of international environmental agreements. Journal of Environmental Economics and Management 54, 296–310.

Sandler, T., Sterbenz, F.P., 1990. Harvest uncertainty and the tragedy of the commons. Journal of Environmental Economics and Management 18, 155–167.

Stern, N., 2007. The Economics of Climate Change. Cambridge University Press.

Stern, N., 2008. The Economics of climate change. American Economic Review Papers and Proceedings 98 (2), 1–37.

Suleiman, R., 1997. Provision of step-level public goods under uncertainty: a theoretical analysis. Rationality and Society 9, 163–187.

Ulph, A., 2004. Stable international environmental agreements with a stock pollutant, uncertainty and learning. The Journal of Risk and Uncertainty 29, 53–73.

Ulph, A., Maddison, D., 1997. Uncertainty, learning and international environmental policy coordination. Environmental & Resource Economics 9, 451–466.

White, L., 2008. Prudence in bargaining: the effect of uncertainty on bargaining outcomes. Games and Economic Behavior 62 (1), 211–231.

Journal of Environmental Economics and Management 66 (2013) 235–250

Contents lists available at ScienceDirect

Journal of Environmental Economics and Management

journal homepage: www.elsevier.com/locate/jeem

Climate treaties and approaching catastrophes ☆

Scott Barrett [a,b,*]

[a] School of International and Public Affairs & Earth Institute, Columbia University, New York, NY 10027, United States
[b] Princeton Institute for International and Regional Studies, Princeton University, Princeton, NJ 08544, United States

ARTICLE INFO

Article history:
Received 7 November 2011
Available online 4 January 2013

Keywords:
Climate change
International environmental agreements
Catastrophe
Cooperation
Coordination
Uncertainty
Enforcement

ABSTRACT

If the threshold that triggers climate catastrophe is known with certainty, and the benefits of avoiding catastrophe are high relative to the costs, treaties can easily coordinate countries' behavior so as to avoid the threshold. Where the net benefits of avoiding catastrophe are lower, treaties typically fail to help countries cooperate to avoid catastrophe, sustaining only modest cuts in emissions. These results are unaffected by uncertainty about the impact of catastrophe. By contrast, uncertainty about the catastrophic threshold normally causes coordination to collapse. Whether the probability density function has "thin" or "fat" tails makes little difference.

1. Introduction

There is universal agreement, codified in the Framework Convention on Climate Change, that atmospheric concentrations of greenhouse gases should be stabilized "at a level that would prevent *dangerous* [my emphasis] anthropogenic interference with the climate system." The Kyoto Protocol tried to get countries to reduce their emissions without identifying a threshold for dangerous interference, but Kyoto failed, mainly for lack of an enforcement mechanism.[1] The recently negotiated Copenhagen and Cancun agreements also lack enforcement, and are legally non-binding besides, but unlike Kyoto these agreements identify a dangerous threshold—a 2° Celsius increase in average global temperature. In this paper I ask, Can the fear of crossing a catastrophic threshold overcome the enforcement challenge? Can it make climate treaties more effective?

The theory of international environmental agreements has generally offered a gloomy prognosis for cooperation, particularly on this issue.[2] To this point, however, the literature has considered only continuous abatement benefit functions. In this paper I take the benefit function to be *discontinuous* at a threshold. This simple change gives rise to a new result. If the threshold is known with *certainty*, and the loss from catastrophe vastly exceeds the costs of avoiding it, then the collective action problem changes fundamentally. Rather than *cooperate* to limit emissions, countries need only

☆ The idea for this paper came to me as I was thinking of how to respond to a question posed by William Nordhaus at a seminar I was giving at Yale a few years ago. I am grateful to Apurva Sanghi, Arthur Campbell, Astrid Dannenberg, Claude Henry, Erin Mansur, Geir Asheim, Kenneth Arrow, Mark Cane, Matthew Kotchen, Martin Weitzman, Robert Keohane, Robert Mendelsohn, Robert Socolow and Thomas Schelling for comments on a previous draft. I am also grateful to the editor, two anonymous referees, and participants at seminars given at Columbia, Harvard, Princeton, Stanford, the World Bank, and Yale (a later seminar) for their comments. An earlier version of this paper was given at the invitation of Michael Finus as a keynote lecture at the Environmental Protection and Sustainability Forum, University of Exeter, April 2011.

* Correspondence address: School of International and Public Affairs & Earth Institute, Columbia University, New York, NY 10027, United States. Fax: +1 212 854 4782.
E-mail address: sb3116@columbia.edu

[1] See Barrett [7].
[2] For surveys of the literature, see Michael Finus [13], Ulrich J. Wagner [26], and Barrett [6,7].

0095-0696/$ - see front matter © 2012 Elsevier Inc. All rights reserved.
http://dx.doi.org/10.1016/j.jeem.2012.12.004

coordinate to avoid catastrophe. Under these circumstances, climate treaties can sustain the efficient outcome. Essentially, nature herself enforces an agreement to avoid catastrophe.

Here is why: when the net benefit of steering clear of the threshold is very high, avoiding catastrophe is a Nash equilibrium. (I mainly restrict attention in this paper to symmetric equilibria in pure strategies.) Since each country is too small to avert catastrophe on its own, however, there also exists a Pareto inferior Nash equilibrium in which the threshold is crossed. Under these circumstances, the challenge is for countries to coordinate on the mutually preferred equilibrium—a task for which treaties are exquisitely suited. When these circumstances do not apply, there exists a unique Nash equilibrium that sits on the wrong side of the threshold. Since there also exist other feasible, mutually preferred outcomes, countries will want to cooperate so as to improve on this inefficient equilibrium (as in a prisoners' dilemma game). A treaty can help here, too, but credible enforcement mechanisms for supplying this kind of global public good are generally weak.[3] Relative to the inefficient Nash equilibrium, which is common to both situations, coordination, when possible, almost always succeeds better than cooperation.

Weitzman [27] has drawn our attention to "uncertain catastrophes with tiny but highly unknown probabilities," showing that, if the probability of ever larger catastrophes does not fall faster than welfare losses increase as we venture deeper into the tails of the probability density function, then the expected gain from a policy to reduce emissions will be infinite. Under these circumstances, policy should do everything possible to reduce emissions as quickly as possible, even though doing so cannot guarantee that catastrophe will be avoided.[4] He calls this result, appropriately enough, "the dismal theorem."

Nordhaus [18]: 21 argues, to the contrary, that any conceivable catastrophic outcome can be avoided by policy, but that doing so requires "solving the global public goods problem by gathering most nations together to take collective action."[5] Weitzman does not address the collective action aspects of this challenge, but his theorem should carry through in a decentralized setting. If it pays the world as a whole always to devote more resources to reducing a threat that cannot be eliminated, no matter the cost, then it should pay individual countries (having the same preferences, and facing the same uncertainty as assumed by Weitzman) always to devote more resources to reducing the same existential threat, even when the benefits of doing so are widely dispersed. That is, under Weitzman's assumptions, coordination should be unnecessary. Every country should want to put its economy on a "climate war" footing, irrespective of whether other countries join them in this effort.

Underpinning the dismal theorem is the strong assumption that utility is unbounded.[6] If this assumption is relaxed, so that preferences are not very risk averse as consumption approaches zero, a different outcome emerges, one that is consistent with Weitzman's policy recommendation of "*a relatively more cautious approach* [emphasis added]" to reducing greenhouse gas emissions ([27]: 13). In this paper, for simplicity, I take utility to be linear. While this assumption rules out Weitzman's dismal theorem, I shall show that it nonetheless implies that countries may want to adopt a much more cautious approach as compared to one in which the threat of catastrophe can be ignored. Unfortunately, under these same circumstances, I also find that the uncertain prospect of approaching catastrophe has little if any effect on the non-cooperative outcome, or the ability of an international agreement to sustain collective action. The reason is that uncertainty (about the threshold for catastrophe) makes the expected damage function continuous, rendering coordination ineffective, and restoring the main results reported previously in the literature—that cooperation is needed, but difficult to sustain. In this model, whether the probability density function (pdf) has thin or fat tails makes little difference. Uncertainty about the magnitude of catastrophic damages is also relatively unimportant. It is uncertainty in the threshold that matters.

A certain threshold can be interpreted as a *discrete* uniform distribution having a single value. An interesting extension of this model, which ties my earlier analysis of certainty together with my later analysis of uncertainty, is the *continuous* uniform distribution. This distribution exhibits a critical discontinuity (in the distribution's right support), being fat-tailed throughout its range and zero-tailed elsewhere. I show that, for this special pdf, the result I obtained previously for the model of certainty extends to uncertainty—countries may be able to coordinate on the right-side discontinuity so as to guarantee avoiding catastrophe. However, as compared with the certainty case, I find that opportunities for coordination are extremely limited. Even under the favorable conditions of this special pdf, cooperation will almost certainly be needed and, as usual, difficult to sustain.

My overall conclusion is that fear of approaching catastrophes with uncertain thresholds strengthens the imperative to cooperate without improving the prospects for collective action.

[3] This is true even in a repeated game so long as treaties must be renegotiation-proof; see Barrett [5,6,7]. The one-shot game developed in this paper obeys an analogous property.

[4] If the conditions that give rise to this result are relaxed even slightly, then the threat of catastrophe raises numerous policy issues; see Kousky et al. [15].

[5] To Nordhaus [18], three conditions must hold for catastrophe to be cause for concern: policy failure, as just noted, high temperature sensitivity to concentrations, and extremely convex damages. In this paper, because temperature is subsumed in the analysis (damages are related to aggregate emissions), the latter two conditions will be satisfied when the abatement benefit function is discontinuous (or, in the uncertainty case, when the variance is very small) and the loss due to crossing the threshold (parameter X in this paper) is very large.

[6] Arrow [2] notes that, for individuals, the assumption implies that "the value of statistical life is infinite, a conclusion clearly contrary to all empirical evidence and to everyday observation."

Although the focus of this paper is on climate change, the model can be applied, with suitable modifications, to other situations. Consider the problem of space debris. If the volume of debris in low earth orbit exceeds a critical level, one more collision dramatically increases the probability of more collisions, creating a cascading effect that could render orbital space unusable (Kessler and Cour-Palais [14]). Here, catastrophe can be avoided by limiting additions to the stock of debris, or by designing satellites and rocket boosters so that they can be maneuvered to a "graveyard" or decaying orbit when they become obsolete. Another example is antibiotic resistance. Resistant strains of a pathogen sometimes have high evolutionary fitness in the presence of a drug treatment, but suffer a fitness cost relative to drug-sensitive strains when the drug is removed (Smith [23]). If use of the drug is low, the sensitive strains will win out. If use is high, the resistant strains will thrive. In between there exists a critical threshold for drug use, with the priority for public health being to stay on the good side of the threshold. Consider, finally, exploitation of a biological resource such as a species of fish for which there exists a strictly positive minimum viable population level. Economic overexploitation of the resource may be a problem even at higher stock levels, but exploitation below the critical level would be disastrous; crossing it would cause the species to go extinct. This may be a particularly important issue for fish populations such as tuna that aggregate, if school size is independent of stock abundance, and harvest economics depend on schooling (Clark and Mangel [11]).

Notwithstanding the importance of these and other applications, my main concern in this paper is with climate change. I shall introduce my model in Section 3, but it will help to begin by putting this model in the context of a previously published experiment, which was also concerned with climate change.

2. An experiment in catastrophe avoidance

The experiment by Milinski et al. [17] simulates the "collective-risk social dilemma" in preventing "dangerous climate change." In their experiment there are six players. Each is given €40. The game is played over ten periods. In each period, every player must choose to contribute €0, €2, or €4 without communicating. If, at the end of the game, at least €120 has been contributed, dangerous climate change is averted with certainty, and each player gets a payoff equal to the amount of money he has left (there are no refunds in this game). If less than €120 has been contributed, each player loses all the money she has left with probability 0.9. In their experiment, Milinski et al. [17] played the game with ten groups of students, only half of which succeeded in avoiding the threshold.

There are two symmetric pure strategy equilibria. In one, every player contributes €0 every period, giving each player an expected payoff of €4. In the other, every player contributes €2 every period, giving each a certain payoff of €20. (Of course, there also exist many asymmetric pure strategy equilibria in which different players contribute different amounts, possibly in different periods). The latter equilibrium is efficient; the former is not.

In contrast to the conventional representation of the climate change game, the game with catastrophic damages is pleading for coordination. Why, then, did half the groups in the Milinski et al. [17] experiment fail to coordinate? The main reason is almost surely that, by construction, the players were not allowed to communicate, let alone formulate a treaty to coordinate their contributions.[7]

The usual way of modeling an international environmental agreement is in three stages.[8] In stage one, countries choose independently whether to be a party or non-party to the agreement. In stage two, parties choose their actions (in this case, contributions) so as to maximize their collective payoff. Finally, in stage three, non-parties choose their actions with the aim of maximizing their individual payoffs. A treaty is self-enforcing if, given the treaty and participation level, non-participants do not want to change their behavior; if, given the participation level, parties to the treaty do not want to change the obligations expressed in the treaty; and if, given the participation decisions of other countries, each country does not want to change its decision of whether to be a party or non-party to the treaty.

Applying this notion of a self-enforcing treaty to the Milinski et al. [17] game, it will help to assume that contributions are made in a single period and that each player can contribute any amount up to his or her endowment. If participation were full, the treaty would then tell each country to contribute €20, netting each country a payoff of €20. Were a country to drop out of this agreement, the remaining parties, choosing collectively, would change their contributions. They would reason that, if they contributed an amount Y in total, then, taking this contribution as given, the non-party would contribute an amount $Z = €120 - Y$ for $€120 \geq Y \geq €84$ and $Z = €0$ for $Y < €84$. Knowing this, the five remaining signatories could do no better than to contribute $Y = €84$ collectively (€16.80 each). This would net each of the five parties €23.20, whereas the sole non-signatory would get just €4. Recall that, were this country not to withdraw, it would get a payoff of €20. Obviously, with the treaty written in this way, no country has an incentive to withdraw, starting from a situation in

[7] A similar experiment by Tavoni et al. [25] confirms the importance of the assumption about communication.

[8] See Barrett [7]. This three-stage formulation imbues parties to the agreement with the special ability to commit to playing a collective abatement level—the so-called "Stackelberg leadership" assumption. If non-signatories have dominant strategies, this assumption is indistinguishable from the Nash assumption. In the Milinski et al. model, however, the payoff to avoiding catastrophe jumps at the threshold, implying that countries may not have dominant strategies. The main justification for the leadership assumption is the customary law principle, *pacta sunt servanda*: treaties are binding. Parties to an agreement are expected to fulfill their obligations, with custom being enforced outside the treaty (that is, outside the model). Note that this is a minimal commitment, since international law also says that countries are free to participate in treaties or not as they please—an assumption captured in Stage 1 of the above model. In this paper, the leadership assumption is also important for equilibrium selection—it ensures that, in a coordination game, the more efficient equilibrium will be supported by a self-enforcing agreement.

which participation is full. Hence, a treaty comprising six signatories, each of which contributes €20, is self-enforcing. Moreover, while there exist two Nash equilibria in pure strategies in the underlying game, the self-enforcing agreement just noted is unique and supports the Pareto efficient equilibrium.

To sum up, while Milinski et al. [17] claim that countries may fail to avert catastrophe when doing so is feasible and efficient, simple theory suggest that, so long as countries are permitted to negotiate a treaty, catastrophe should be rather easy to avoid. In the remainder of this paper I inquire into whether this result could be expected to hold in richer (analytical) environments.

3. A climate change catastrophe game

Assume that countries have symmetric payoff functions. Denoting country i's abatement by q_i and aggregate abatement by Q, where $Q=\sum_{i=1}^{N} q_i$ and N is the number of countries, country i's payoff is assumed to be given by

$$\pi_i = \begin{cases} bQ - \frac{cq_i^2}{2} \text{if} Q \geq \overline{Q} \\ bQ - X - \frac{cq_i^2}{2} \text{if} Q < \overline{Q} \end{cases}. \quad (1)$$

In (1), parameter b stands for each country's marginal benefit of abating "gradual" climate change, cq_i represents country i's marginal abatement costs, and X denotes the damages each country suffers from "catastrophic" climate change. These damages are experienced if and only if global abatement falls short of $\overline{Q}$, the threshold.[9] If we let $X=0$, (1) collapses to the standard model in the literature on international environmental agreements.[10]

It is worth noting the important differences between (1) and Milinski et al. [17] model. First, in (1) contributions (abatement levels) are beneficial even if the threshold is exceeded; abatement reduces "gradual" as well as "catastrophic" climate change. Second, Milinski et al. vary X (more specifically, the expected value of X), but not the threshold, and (1) allows us to vary both.[11] Third, in (1), marginal abatement costs are increasing, whereas Milinksi et al. implicitly assume that they are constant. Fourth, Milinski and coauthors take abatement (contributions) to be discrete, and (1) allows abatement to be varied continuously. Finally, while (1) assumes that X is certain, and Milinksi et al. let X be uncertain, later in this paper I shall let both X *and* $\overline{Q}$ be uncertain.

In common with Milinski et al. [17] and Weitzman [27], model (1) abstracts away from the dynamics of climate change. It is, however, able to capture the key difference between "gradual" and "abrupt and catastrophic" climate change. It assumes that gradual climate change is so slow that marginal benefits are constant, and that abrupt climate change is so fast that transition is instantaneous.[12] The model can be interpreted as compressing perhaps a century of decision-making into a single period. Since the impacts of abrupt change will unfold more slowly than this (some of the impacts could take a millennium or more to play out), X should be interpreted as capturing the full consequences, including into the distant future, of decisions taken this century to avoid or exceed a threshold. The parameter b should be interpreted similarly for gradual climate change.

In the full cooperative outcome, the aggregate payoff will be

$$\Pi^{FC} = \begin{cases} bQN - \sum_i \frac{cq_i^2}{2} \text{if} Q \geq \overline{Q} \\ bQN - XN - \sum_i \frac{cq_i^2}{2} \text{if} Q < \overline{Q}. \end{cases} \quad (2)$$

If $X=0$, maximization of (2) yields $Q^{FC} = bN^2/c$. Assume $\overline{Q} > bN^2/c$ (avoiding the threshold requires abating more than is optimal for addressing only "gradual" climate change).[13] Then there are two possibilities. Either it will pay all countries collectively to meet the threshold ($Q^{FC} = \overline{Q}$), just, or it will not pay to meet the threshold ($Q^{FC} = bN^2/c$). Upon substituting

[9] Damages are normally related to temperature, temperature to concentrations, and concentrations to an emissions profile. However, there is strong evidence that temperature can be related directly to cumulative emissions (Allen et al. [1]; Zickfeld et al. [28]). In this paper, I take business as usual emissions as given. My focus is on reductions from this level: the level of abatement. The greater is the level of abatement, the smaller will be cumulative (total) emissions, and the lower will be temperature and, therefore, damages. Emissions will need to be reduced to a critical level, denoted here by $\overline{Q}$, if a dangerous threshold, such as the 2 °C change in temperature, is to be avoided.

[10] See especially Barrett [7].

[11] Milinski et al. [17] vary the expected value of X by varying the probability that damages will be "catastrophic" given that contributions fall short of the threshold.

[12] In a model with increasing marginal benefits, international cooperation still achieves very little [3], so it is not the constancy of marginal benefits that matters so much as continuity in the benefit function.

[13] This assumption implies that a focus only on "gradual" climate change could cause the world to cross a dangerous threshold. In the words of Stern [24], who warns of possible tipping points, "It is important to be clear that the "climate policy ramp" advocated by some economists involves a real possibility of devastating climatic changes." Of course, it is conceivable that the full cooperative abatement level could be determined solely by "gradual" climate change ($\overline{Q} < bN^2/c$). Suppose, then, that a treaty addressing only "gradual" climate change sustained the abatement level, Q^*. Then, if $\overline{Q} < Q^* \leq bN^2/c$, consideration of "catastrophic" climate change would have no effect on the usual analysis; abatement to avoid "gradual" climate change would avoid catastrophe in the bargain. If, to the contrary, $Q^* < \overline{Q} < bN^2/c$, consideration given to "catastrophic" climate change might improve the prospects for collective action, while still falling short of sustaining the full cooperative abatement level. In this case, the analysis would proceed in the same way as in this paper.

these values into (2), it is easy to show that it will pay to meet the threshold if and only if

$$X \geq \frac{b^2N^2}{2c} - \left(b\overline{Q} - \frac{c\overline{Q}^2}{2N^2}\right). \tag{3}$$

Fig. 1 illustrates the relationship (which is non-linear due to marginal costs increasing). If the impact of catastrophic climate change, X, is "small," catastrophe is worth avoiding only if $\overline{Q}$, the threshold, is "small" (so that the costs of avoiding catastrophe are low). If $\overline{Q}$ is "large," catastrophe is worth avoiding only if X is "large."

In the non-cooperative outcome, each country i will maximize (1), taking as given the abatement choices of other countries. There are at most two symmetric Nash equilibria in pure strategies. In one, every country i plays $q_i = b/c$, and the threshold is exceeded. In the other, every country plays $q_i = \overline{Q}/N$, and the threshold is avoided, just. Suppose every country $j \neq i$ plays $q_j = \overline{Q}/N$. Then i will play either $q_i = \overline{Q}/N$ or $q_i = b/c$. Upon substituting these values in (1), it is easy to show that country i will prefer to play the former abatement level (in which case, avoiding catastrophe is a Nash equilibrium) rather than the latter if and only if

$$X \geq \frac{b^2}{2c} - \left(\frac{b\overline{Q}}{N} - \frac{c\overline{Q}^2}{2N^2}\right). \tag{4}$$

The top part of the Fig. 2 shows the space in which (3) and (4) both hold—that is, the space in which catastrophe avoidance is both (i) collectively optimal and (ii) a symmetric Nash equilibrium. The bottom part, unchanged from Fig. 1,

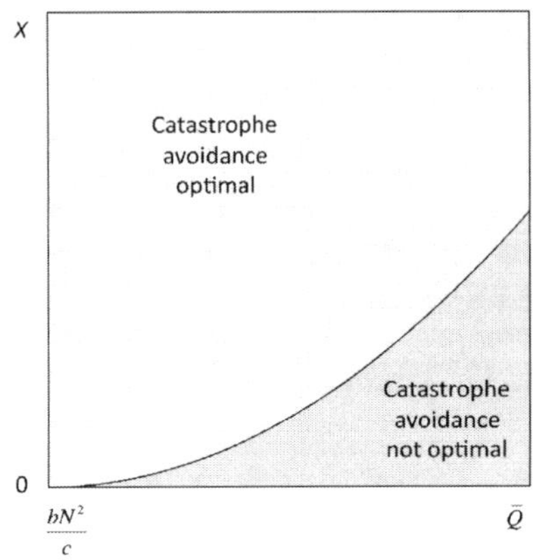

Fig. 1. Optimal catastrophe avoidance.

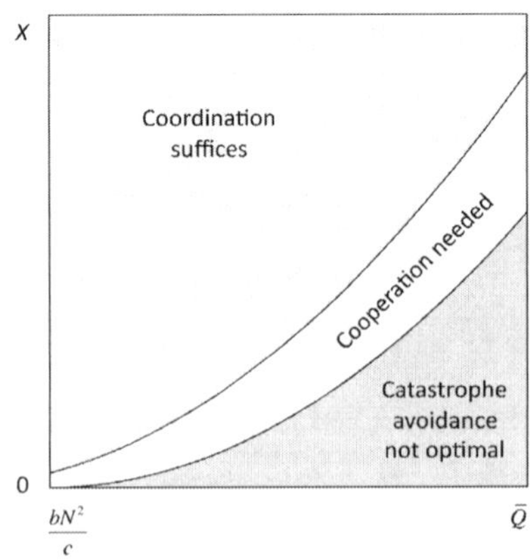

Fig. 2. When coordination suffices and cooperation is needed.

shows the combinations of X and $\overline{Q}$ for which avoiding catastrophe is inefficient. Finally, sandwiched between these two spaces is a middle area in which (3) holds but (4) does not hold—that is, catastrophe avoidance is collectively optimal but cannot be supported as a symmetric Nash equilibrium. (Note that this middle area only exists because abatement reduces "gradual" as well as "catastrophic" climate change.) The figure shows that catastrophe avoidance cannot be sustained by coordination if X is extremely small, even if $\overline{Q}$ is close to the full cooperative abatement level for gradual climate change. Only if X is "big enough" will it be possible for countries to coordinate to avoid a "catastrophe."

In comparing (3) and (4), we see that coordination can always be relied upon to avoid catastrophe when doing so is optimal provided either $N=1$ (no externality) or $b=0$ (no benefit to abatement apart from avoiding catastrophe). The former result is to be expected. The latter result is surprising.[14] Intuitively, the incentives to avoid "catastrophic" climate change should be helped when the actions needed to do this also reduce "gradual" climate change. Imagine, however, that every country i plays $q_i=\overline{Q}/N$. Then, taking the abatement levels of other countries as given, a country that deviates unilaterally will suffer a smaller loss when the abatement by the other countries reduces gradual climate change. Ironically, the effect of abatement in reducing gradual climate change increases the incentive for a country to deviate from an agreement seeking to avert catastrophic climate change.

To summarize:

Proposition 1. *When inequality (3) does not hold, so that the impact of "catastrophic" climate change is low relative to the cost of avoiding catastrophe, the collective action problem is a prisoners' dilemma game for reducing "gradual" climate change. When (4) holds, so that the impact of catastrophic climate change is high relative to the cost of preventing catastrophe, the collective action problem is a coordination game for avoiding catastrophe. In between these situations, (3) holds but (4) does not hold, and the collective action problem is a prisoners' dilemma game for avoiding catastrophe.*[15]

4. Climate treaties for avoiding catastrophes

What role is there for climate treaties in averting an approaching catastrophe? Start from a situation in which every country participates in a treaty requiring that each play $q_i=\overline{Q}/N$. What conditions must hold for such a treaty to be self-enforcing?

Consider first the incentives for a country to withdraw from the treaty. Upon withdrawing, the deviating country will take as given the behavior of the remaining $N-1$ cooperating countries *as this is specified in the self-enforcing agreement*. Suppose that, once i withdraws, the agreement instructs these countries to play Q_{-i}. Then, upon withdrawing, i will either want to play $q_i=\overline{Q}-Q_{-i}$, to ensure that the threshold is met, or it will let the threshold slip and play $q_i=b/c$. Given that the other countries abate Q_{-i}, i will prefer to play the former level rather than the latter if

$$b\overline{Q}-\frac{c}{2}\left(\overline{Q}-Q_{-i}\right)^2 \geq b\left(Q_{-i}+\frac{b}{c}\right)-X-\frac{c}{2}\left(\frac{b}{c}\right)^2 \tag{5}$$

Solving the quadratic, the minimum value of Q_{-i} which ensures that (5) is met is given by $\hat{Q}_{-i}=\overline{Q}-b/c-\sqrt{2X/c}$. Hence, off the equilibrium of full participation, if the $N-1$ cooperating countries play $\hat{Q}_{-i}$, then i will play $q_i=b/c+\sqrt{2X/c}$ and catastrophe will be avoided. If, to the contrary, they play $Q_{-i}<\hat{Q}_{-i}$, then i will play $q_i=b/c$, and the catastrophic threshold will be crossed.

How, then, will the $N-1$ remaining signatories want to play? If they play $Q_{-i}<\hat{Q}_{-i}$, then they cannot do better collectively than to play $Q_{-i}=b(N-1)^2/c$. Knowing how the deviant country will behave, they will prefer to abate $Q_{-i}=\hat{Q}_{-i}$ rather than $Q_{-i}=b(N-1)^2/c$ if and only if

$$b\overline{Q}-\frac{c}{2}\left(\frac{\overline{Q}-b/c-\sqrt{2X/c}}{N-1}\right)^2 \geq b\left(\frac{b(N-1)^2}{c}+\frac{b}{c}\right)-X-\frac{c}{2}\left(\frac{b(N-1)}{c}\right)^2. \tag{6}$$

If (6) holds, and if i should withdraw from the agreement with full participation, the treaty will instruct the other $N-1$ countries to play (not $Q_{-i}=\overline{Q}(N-1)/N$, the Nash assumption, but) $Q_{-i}=\hat{Q}_{-i}$. Under these conditions, country i will not want to withdraw if

$$b\overline{Q}-\frac{c}{2}\left(\frac{\overline{Q}}{N}\right)^2 \geq b\overline{Q}-\frac{c}{2}\left(\frac{b}{c}+\sqrt{\frac{2X}{c}}\right)^2. \tag{7}$$

Rewriting, we see that inequality (7) requires $b/c+\sqrt{2X/c}\geq\overline{Q}/N$; a country that withdraws is punished by being moved to abate *more* as a non-signatory than as a signatory (the opposite of free riding!).

---

[14] For Milinski et al. 's [17] model, there is no middle wedge; efficient catastrophe avoidance is always a Nash equilibrium.

[15] Obviously, these results apply specifically to model (1), which is linear-quadratic (linear benefits and quadratic costs). However, it is easy to see that these results also apply to the model discussed in Section 2, which is linear–linear with $b=0$, except that, in this model, there is no collective action problem for limiting gradual climate change. The paper by Barrett and Dannenberg [10], written after this paper was submitted, shows that Proposition 1 also holds for a model with linear benefits (with $b>0$) and stepwise-linear marginal costs (their model was developed for the purpose of conducting an experimental test of the theory presented in this paper); compare Fig. 1 in their paper with Fig. 2 in this one.

If (6) and (7) hold, no country can gain by withdrawing unilaterally from the agreement that avoids catastrophe. It is easy to show (see the Appendix) that, if (7) holds, then so will (6) hold. Moreover, inequality (7) reduces to (4), which is the condition for when catastrophe avoidance can be sustained by coordination. Hence, we have:

Proposition 2. *When (4) holds, so that the impact of catastrophic climate change is high relative to the cost of preventing catastrophe, a self-enforcing international environmental agreement can sustain the full cooperative outcome, ensuring that catastrophe is avoided. Under these conditions, the treaty serves as a coordinating device.*

This analysis assumes full participation. It is well known that, in most cases, treaties can only support a lower level of participation. So, might collective action to avoid catastrophe be sustained even if (7) fails to hold?

Even if (7) were violated, inequality (6) could still hold. That is, the $N-1$ signatories could have an incentive to play $\overline{Q}-b/c-\sqrt{2X/c}$, knowing that the sole non-signatory would then abate $b/c+\sqrt{2X/c}<\overline{Q}/N$ (in which case the non-signatory would be a true free rider). Moreover, this agreement comprising $N-1$ countries would be self-enforcing, *provided it could be sustained by coordination.* However, it is easy to show that the conditions that enable $N-1$ countries to coordinate to avoid catastrophe also enable N countries to coordinate to sustain the full cooperative outcome (see the Appendix). Moreover, by extending this logic to other participation levels, it is clear that, if (7) fails to hold, it will not be possible to sustain coordination to avoid catastrophe by *any* number of countries ($\leq N$).

Of course, when (7) holds, it could be possible for fewer than N countries to coordinate so as to avoid catastrophe (this is just another way of saying that there may exist a large number of asymmetric Nash equilibria for avoiding catastrophe). However, coordination by fewer than N countries (with the others acting as free riders) offers no advantage over coordination by the grand coalition—only a disadvantage, since it is more costly for $N-1$ countries to sustain the threshold than for N countries to do so (with marginal costs increasing, cost-effective abatement requires that every country abate the same quantity). Since every country's expected payoff (with the expectation taken before an agreement is adopted, when no country knows if it will be a signatory or a non-signatory) is greater for an agreement comprising N countries than one comprising $N-1$ (or fewer) countries, provided both agreements are self-enforcing, in the negotiation stage all countries will favor a universal agreement. Moreover, once such an agreement has entered into force, if (7) holds, no country will have an incentive to withdraw.

Together, these results imply:

Proposition 3. *If there exists a self-enforcing international environmental agreement that avoids catastrophe by coordinating countries' abatement, then this agreement will be unique and sustain full participation.*

What happens if avoiding catastrophe is efficient and yet coordination fails? This would put us in the middle territory of Fig. 2. Countries could still try to *cooperate* to avoid catastrophe in this region, by using strategies to overcome the prisoners' dilemma, as in the usual model of an international environmental agreement. In this case, signatories would reduce their abatement in the belief that, by doing so, the threshold would be blown. However, the cooperation problem is now different from the usual one, which assumes only "gradual" climate change. For model (1), a deviation triggers a harsher punishment—the "catastrophic" loss due to crossing the threshold. It may, therefore, sustain greater cooperation.[16]

For it to be collectively rational for the $N-1$ signatories to play $Q_{-i}=b(N-1)^2/c$ after i has withdrawn, inequality (6) must not hold. Knowing this, country i will not want to withdraw if

$$X\geq\frac{b^2[2(N-1)^2+1]}{2c}-\left(b\overline{Q}-\frac{c\overline{Q}^2}{2N^2}\right). \tag{8}$$

Condition (8) is the same as (3) except for the intercept. It is straightforward to show that the intercept term in (8) is no greater than the intercept term in (3) for $N\leq 3$. That is, when coordination fails, a treaty can still be relied upon to sustain the full cooperative outcome—meaning, avoid catastrophe—if $N\leq 3$. It is well known that, with the functional forms assumed in (1), a self-enforcing international environmental agreement can sustain the full cooperative outcome even in the absence of catastrophe for $N\leq 3$. Consideration of catastrophe thus does nothing to change this result.

For $N>3$, however, matters can be different. Cooperation may fail entirely to avoid catastrophe, or it may succeed in avoiding catastrophe but only within a portion of the middle wedge shown in Fig. 2.[17] In this region, inequality (8) holds but (6) does not hold.

Inequality (6) is awkward to work with, but an example can illustrate where and why cooperation can be effective within this middle wedge. Suppose $N=20$, $b=c=1$, and $\overline{Q}=600$.[18] Then the vertical distance of the cooperation wedge shown in Fig. 2, as defined by inequalities (3) and (4), requires $420.5>X\geq 50$ (coordination succeeds for $X\geq 420.5$;

[16] "Punishment" appears in the one-shot model of an international environmental agreement when the treaty specifies that cooperating countries will lower their per-country abatement as the number of participating countries falls. We normally think of "punishment" in the context of a repeated game. But the stage game model of an international environmental agreement is particularly suited to exploring the enforcement of participation, whereas the repeated game model is better suited to exploring the enforcement of compliance. I have shown elsewhere that there is a natural correspondence between these two approaches [5–7].

[17] Using other functional forms, self-enforcing agreements can sustain full cooperation for $N>3$ without considering the possibility of catastrophe. These include the quadratic-quadratic [3] and linear-linear [6] models. However, in both of these cases, a self-enforcing agreement can sustain the full cooperative outcome only when this outcome is nearly identical to the non-cooperative outcome.

[18] You can think of this assumption as suggesting that collective action by the G-20 group of countries could prevent catastrophic climate change.

catastrophe avoidance is inefficient for $X < 50$). Condition (8) requires $X \geq 211.5$; the impact of catastrophe must be large enough to deter withdrawal. Of course, X must also be small enough that the $N-1$ countries remaining in the agreement are willing to cross the threshold should i decline to cooperate. Upon substituting we find that (6) will not hold if $\sqrt{722(X+418.5)} < 599-\sqrt{2X}$, and it can be shown that there is *no* value for X within the relevant range ($420.5 > X \geq 211.5$) for which reversal of inequality (6) is satisfied. For these parameter values, cooperation to avoid catastrophe fails completely. Cooperation will have a better chance of avoiding catastrophe if $\overline{Q}$ is bigger. Raising the threshold helps because it makes credible the threat by the $N-1$ remaining signatories to lower their abatement dramatically should country i withdraw. For example, if the threshold is increased to 4000, the vertical range of the cooperation wedge becomes $19{,}800.5 > X \geq 16{,}200$ (coordination succeeds for $X \geq 19{,}800.5$; avoiding catastrophe is inefficient for $X < 16{,}200$), and cooperation can succeed in avoiding catastrophe—but only over a small portion of this range ($16{,}371.5 \geq X \geq 16{,}361.5$).

It can also be shown that if N countries cannot sustain cooperation to avoid catastrophe, then neither can fewer than N countries do so.[19] Of course, in the event that cooperation to avoid catastrophe fails, a self-enforcing international environmental agreement comprising three countries (for $N \geq 3$) could still succeed in reducing "gradual" climate change. However, such an agreement would reduce emissions just a little relative to the non-cooperative outcome; as is well known, each of the three signatories would abate $q_s = 3b/c$ and each non-signatory just b/c. Cooperation to limit "gradual" climate change would have only a modest impact on emissions.

To sum up:

Proposition 4. *When (3) holds but (4) does not hold, cooperation may succeed in averting catastrophe, but only under special circumstances. Moreover, even when these special circumstances apply, cooperation improves welfare relatively little.*[20] *Coordination, by contrast, succeeds spectacularly when the net benefits of avoiding catastrophe are particularly large.*

It should not surprise us that coordination makes all the difference. The Montreal Protocol succeeded partly because the strategic application of trade restrictions transformed protection of the ozone layer, ordinarily a prisoners' dilemma, into a coordination game [4,6]. Network externalities can play the same role in a treaty that strategically targets technology standards rather than emission limits [6,8]. In both of these cases, a self-enforcing treaty sustains coordination with full participation.[21] The novelty in the present paper is that here the opportunity for coordination is not strategic but rather a gift from Mother Nature.

Having established the preeminent role of coordination in averting catastrophe under conditions of certainty, in the remainder of this paper I shall limit my attention to the conditions under which coordination can reduce, if not eliminate, the probability of crossing a catastrophic threshold under conditions of uncertainty.

5. Impact and threshold uncertainty

[27] emphasizes fat-tailed uncertainty in climate sensitivity, compounded by uncertainties in translating temperature changes into welfare changes. Here I distinguish between uncertainty in the *impact* of catastrophe (X) and uncertainty about the *threshold* that triggers catastrophe $(\overline{Q})$.

The two uncertainties are both very substantial, but also very different. One way to understand the nature of threshold uncertainty is to consider the recent attention-getting paper by Rockström et al. [19]. They identify a "planetary boundary" in terms of atmospheric CO_2 concentrations of 350 ppm by volume (ppmv) so as "to ensure the continued existence of the large polar ice sheets," while noting that the paleoclimatic record implies "that there is a critical threshold between 350 and 550 ppmv." Uncertainty about this threshold is purely scientific (and much of it is substantially irreducible). Uncertainties about the impacts associated with the loss of polar ice, by contrast, are both scientific and economic. Loss of the Greenland Ice Sheet, for example, would likely cause sea level to rise between 2 and 7 m over a period of 300 to over 1000 years (Lenton [16]). The value of X will obviously depend on the extent of sea level rise and its rate of change, but it will also depend on the *values* attached to these changes, including discounting. Much has been made in the economics literature of the importance of these values and their uncertainties, but I shall now show that, as regards the prospects for collective action, uncertainty about the impact of crossing a threshold is relatively unimportant. It is scientific uncertainty about the threshold that really matters.

Consider first uncertainty about the impact of catastrophe, X. It is straightforward to demonstrate that all the results shown thus far carry through if we substitute the expected value of X, $E(X)$, for X (this, again, is assuming that countries are

[19] For example, it would only pay $N-1$ countries to cooperate so as to ensure that the threshold was avoided when the other country, the free rider, reduced emissions by b/c, if $b\overline{Q}-(c/2)((\overline{Q}-b/c)/(N-1))^2 \geq b(b(N-1)^2/c+b/c)-X-(c/2)(b(N-1)/c)^2$. But if this condition applies, then (6) surely holds, which means we are in a coordination situation.

[20] Recall that the wedge for cooperation in Fig. 2 borders the region in which catastrophe avoidance is inefficient.

[21] In Barrett [4,6], trade restrictions sustain a first best. In Barrett [6,8], technology standards may sustain only a second best. In the present paper, the prospect of catastrophe may sustain a first best; it would sustain a second best only under the conditions described towards the end of footnote 13. With threshold uncertainty, as explained in the next section, a coordinating treaty can only sustain a second best, except for a special case discussed in Section 6.

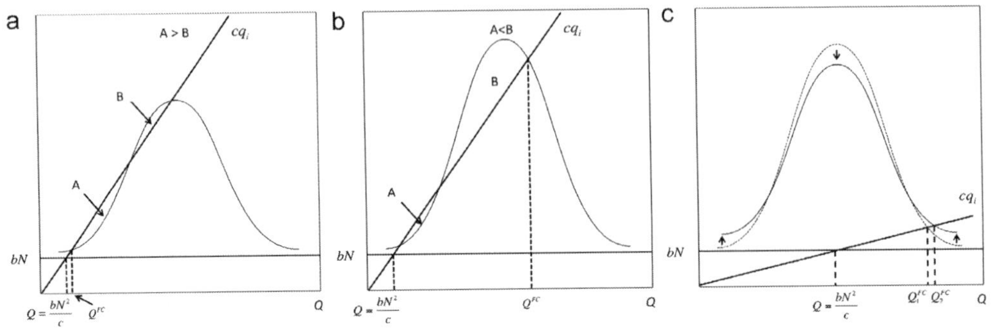

Fig. 3. (a) Prospect of catastrophe has little effect on the full cooperative outcome. (b) Prospect of catastrophe has profound effect on the full cooperative outcome. (c) Prospect of catastrophe with "thin" and "fat" tails.

risk neutral).[22] Uncertainty about the magnitude of catastrophic damages does not fundamentally alter the nature of the cooperation challenge. So long as the catastrophic threshold is certain, and $E(X)$ is large, countries will be able to coordinate so as to avert catastrophe.[23]

Consider now uncertainty about the threshold, $\overline{Q}$. In particular, let $\overline{Q}$ be a random variable with a continuous cumulative probability distribution $F(Q) = \Pr\left(\overline{Q} \leq Q\right)$. Were countries to cooperate fully, they would now maximize

$$E(\Pi^c) = bQN - \sum_i \frac{cq_i^2}{2} - XN[1 - F(Q)], \tag{9}$$

which requires

$$bN - cq_i + XNf(Q) = 0, \tag{10}$$

where $f(Q)$ is the probability density function (pdf). Note that, in general, Eq. (10) will not be sufficient for a maximum.

The effect of uncertain catastrophe depends on the function f. If the pdf has infinite supports, f will always be positive, and the prospect of catastrophe will commend greater abatement, compared to a situation in which catastrophe is ignored. Recall that when catastrophe is certain, abatement in the full cooperative outcome may or may not be affected.

Fig. 3a illustrates Eq. (10) for a plausible pdf ($f(Q)$).[24] In the figure, the bell curve placed on top of the aggregate marginal benefit of avoiding "gradual" climate change represents $XNf(Q)$, the expected aggregate marginal benefit of catastrophe avoidance. For the pdf shown in this figure, concern about catastrophic climate change increases the full cooperative abatement level very slightly above bN^2/c, the optimal abatement level for gradual climate change ignoring catastrophic climate change. If abatement is increased beyond Q^{FC}, net benefits fall. They later increase before falling again, but so long as area A exceeds area B, as it does in Fig. 3a, the net benefits of increasing abatement beyond Q^{FC} will be negative, and it will be optimal to limit aggregate abatement to Q^{FC}.

Fig. 3b looks similar to 3a, but in this case area B exceeds area A, and the optimal aggregate abatement level increases dramatically, relative to bN^2/c, due to the uncertain prospect of catastrophe. Obviously, what matters in these figures is the size of area A relative to B. These relative values depend in turn on the values of b, X, and c as well as the pdf for the threshold.

Does the possibility of "fat tails" matter? As shown in Fig. 3c, a thickening of the tails of the pdf changes the full cooperative outcome very little, with the full cooperative abatement level increasing only slightly, from Q_1^{FC} to Q_2^{FC}.[25] Weitzman, of course, gets a very different result, but as noted in the introduction, he assumes that social preferences are strongly risk averse.

This brief discussion summarizes just a few of many possible constructions. My aim here is not to be comprehensive, but to illustrate how sensitive or insensitive policy recommendations can be to small changes in the constituent parts of Eq. (10).

---

22 I am also assuming that $E(X)$ exists. For some fat-tailed distributions, expected value does not exist.

23 This result should have been anticipated from my previous discussion of the Milinski et al. [17] experiment. Recall that, in this experiment, the value of X (and not $\overline{Q}$) was uncertain. An example of a situation in which the threshold was certain but the impacts were uncertain is the Millennium Bug or Y2K problem. Of course, this turned out not to be the catastrophe expected, though we shall never know whether this is because the action taken to avert a crisis succeeded or earlier fears of calamity were overblown. The theory presented here is consistent with both interpretations.

24 All the pdfs drawn in this paper are symmetric, but my discussion and analysis does not depend on this assumption. There are reasons to believe the pdf for climate sensitivity exhibits positive skewness (Roe and Baker [20]).

25 The diagram can be thought of as comparing the normal and t-distributions, the former being thin- and the latter thick-tailed.

To sum up, we have:

Proposition 5. *Uncertainty about the impact of "catastrophic" climate change has no effect on the full cooperative abatement level. By contrast, uncertainty about the threshold may cause abatement in the full cooperative outcome to increase a lot relative to the full cooperative outcome for limiting "gradual" climate change.*

If countries fail to cooperate, each country i will choose q_i to maximize

$$E(\pi_i) = bQ - \frac{cq_i^2}{2} - X[1-F(Q)]. \tag{11}$$

The solution requires

$$b - cq_i + Xf(Q) = 0 \forall i. \tag{12}$$

What determines the gap between the full cooperative and non-cooperative outcomes? Look first at Fig. 4a. The top portion of the figure shows the full cooperative outcome. Though positioned somewhat differently, this portion of the figure is analogous to the situations depicted in Fig. 3a–c. The bottom portion of the figure shows the non-cooperative outcome. As drawn, Fig. 4a shows that the prospect of uncertain catastrophe has a substantial effect on the full cooperative outcome (raising it from bN^2/c to Q^{FC}), but virtually no effect on the non-cooperative abatement level (which, ignoring catastrophe, is bN/c, and which increases just a little to Q^{NC}, assuming that the pdf has an infinite support on the left side). Uncertainty about the threshold robs us of any chance for coordination on the full cooperative level of abatement. Indeed, compared to the certainty case (assuming that the certain threshold is equal to its expected value in Fig. 4a), uncertainty about the threshold makes countries want to abate *more*, and yet causes them to abate *less*.

Fig. 4b illustrates a situation in which the prospect of uncertain catastrophe increases abatement in both the full cooperative and non-cooperative outcomes (again, relative to a situation in which there was only gradual climate change), while at the same time narrowing the gap between the non-cooperative and full cooperative outcomes (ignoring catastrophe, this gap is $bN^2/c - bN/c$; taking catastrophe into account, it is $Q^{FC} - Q^{NC}$). In this case, even though the threshold is uncertain, the prospect of catastrophe creates an opportunity for coordination on the mutually preferred (but still inefficient) Nash equilibrium. It is obvious from the figure, however, that the circumstances that support this more cheerful situation are not necessarily to be expected. They include not only a high impact, X, but also a very low variance in the pdf and a small N—circumstances that are not favored by climate change.

Again, my aim here has not been to be comprehensive, but to indicate when collective action is likely to succeed or fail. The main results may be stated simply:

Proposition 6. *Compared to the certainty case, uncertainty about the impact of catastrophe has no effect on collective action. By contrast, uncertainty about the threshold is critical, being very likely to transform the collective action problem from a coordination game under threshold certainty to a prisoners' dilemma under threshold uncertainty.*

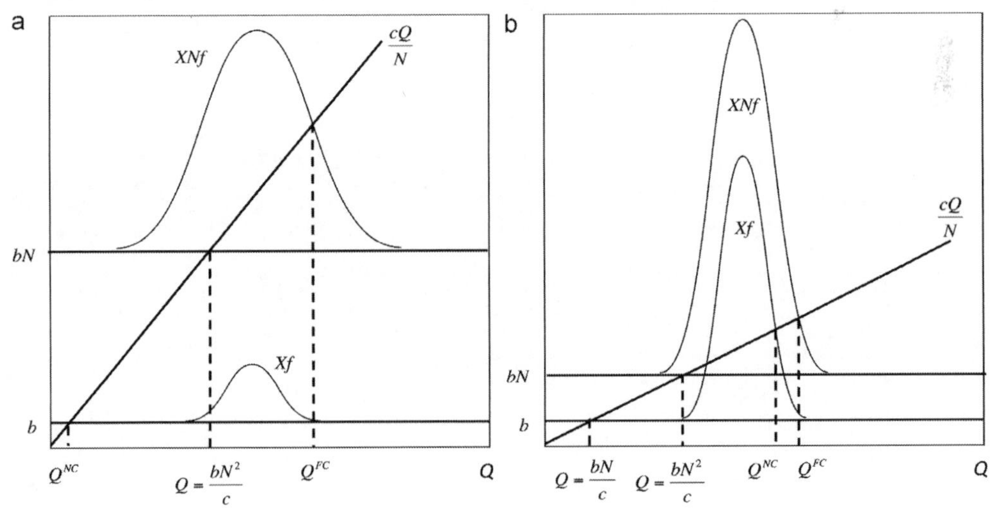

Fig. 4. (a) Uncertain prospect of catastrophe widens the gap between the non-cooperative and full cooperative outcomes. (b) Uncertain prospect of catastrophe narrows the gap between the non-cooperative and full cooperative outcomes.

Proposition 7. *Compared to the certainty case, uncertainty about the threshold can make countries want to abate more in the full cooperative outcome, and yet cause them to abate less. Indeed, threshold uncertainty can cause countries to act as if the risk of catastrophe could be ignored, even when this risk is very great.*

To obtain more concrete results, we will need to choose a particular pdf. I do so in the next section.

6. Illustration for the uniform distribution

In this section I illustrate the points just made by assuming that the threshold obeys a continuous uniform distribution. An advantage of this distribution is that it offers a kind of blend of the two approaches emphasized in this paper. As in the previous section, there is threshold uncertainty. As in the earlier part of the paper, there is a discontinuity—in this case, at both of the distribution's (finite) supports. The discontinuity on the right side means that there may be a role for coordination in supporting a first best. As explained in the previous sub-section, this would not be the case for most distributions.

Assume, then, that the threshold concentration level is distributed uniformly with pdf

$$f(Q)=\begin{cases} 0 & \text{for } Q<\overline{Q}_{\min} \\ \frac{1}{\overline{Q}_{\max}-\overline{Q}_{\min}} & \text{for } Q\in[\overline{Q}_{\min},\overline{Q}_{\max}] \\ 0 & \text{for } Q>\overline{Q}_{\max}. \end{cases} \tag{13}$$

This implies that we are certain about the *range* of values for the threshold, but uncertain about the *particular value*—that is, we have no reason to believe that any value within this range is more or less likely that any other value. In practical terms, the uniform distribution is consistent with Rockström et al.'s [19] representation of the climate problem, in which the paleoclimatic record suggests that the polar ice caps disappeared for concentrations between 350 and 550 ppmv, and preservation of the ice caps is taken to be the "planetary boundary," due presumably to the large impact of breaching it.[26]

The corresponding cumulative distribution function is

$$F(Q)=\begin{cases} 0 & \text{for } Q<\overline{Q}_{\min} \\ \frac{Q-\overline{Q}_{\min}}{\overline{Q}_{\max}-\overline{Q}_{\min}} & \text{for } Q\in[\overline{Q}_{\min},\overline{Q}_{\max}] \\ 1 & \text{for } Q>\overline{Q}_{\max}. \end{cases} \tag{14}$$

If countries cooperate fully, they will maximize

$$E(\Pi^c)=\begin{cases} bQN-\sum_i \frac{cq_i^2}{2}-XN & \text{for } Q<\overline{Q}_{\min} \\ bQN-\sum_i \frac{cq_i^2}{2}-XN\left[1-\left(\frac{Q-\overline{Q}_{\min}}{\overline{Q}_{\max}-\overline{Q}_{\min}}\right)\right] & \text{for } Q\in\left[\overline{Q}_{\min},\overline{Q}_{\max}\right] \\ bQN-\sum_i \frac{cq_i^2}{2} & \text{for } Q>\overline{Q}_{\max}. \end{cases} \tag{15}$$

Assuming $Q_{\min}>bN^2/c$, countries will either play $Q=bN^2/c$ or they will play $Q\in\left[\overline{Q}_{\min},\overline{Q}_{\max}\right]$ (with this assumption, it will never pay to abate more than $\overline{Q}_{\max}$). In the latter case, full cooperation requires that countries abate the smaller of $\overline{Q}_{\max}$ or

$$Q^{FC}=\frac{bN^2}{c}+\frac{XN^2}{c\left(\overline{Q}_{\max}-\overline{Q}_{\min}\right)}. \tag{16}$$

Fig. 5a and b illustrate the two possibilities. In Fig. 5a, (16) holds; the solution is "interior." In Fig. 5b, a corner solution holds. (Note that, with $\overline{Q}_{\min}>bN^2/c$, it will never pay to abate $\overline{Q}_{\min}$.) Both figures are analogous to Fig. 3b.

Upon substituting $Q^{FC}=bN^2/c$ and (16) into (15), it is easy to show that countries should undertake some additional abatement (over and above the amount justified for addressing "gradual" climate change) to reduce the chance of catastrophe provided

$$X>\frac{2c(\overline{Q}_{\max}-\overline{Q}_{\min})}{N^2}\;\overline{Q}_{\min}-\frac{bN^2}{c}\Bigg). \tag{17a}$$

[26] I am simplifying Rockström et al.'s [19] arguments for 350 ppmv. In addition to preservation of the ice caps, they suggest that, after taking slow feedbacks into account, a doubling in CO_2 concentrations (presumably, to around 550 ppmv), would cause equilibrium temperature to increase about 6 °C, which would "severely challenge the viability of contemporary human societies."

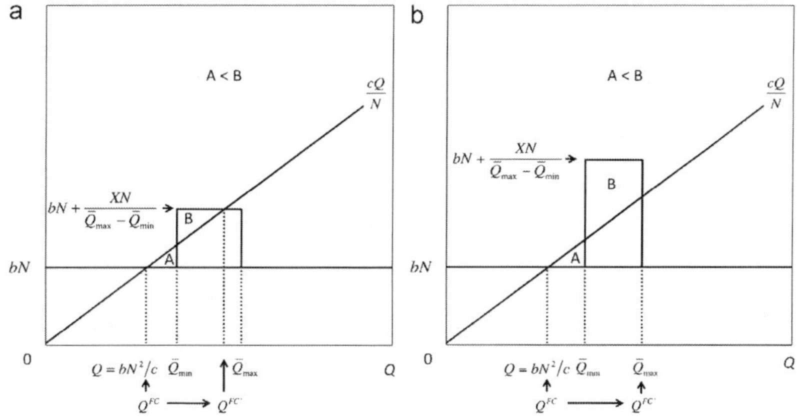

Fig. 5. (a) "Interior" solution for the full cooperative outcome with threshold uncertainty. (b) "Corner" solution for the full cooperative outcome with threshold uncertainty.

This is for an interior solution. For a corner solution the equivalent condition is

$$X > \frac{b^2N^2}{2c} - \left(b\overline{Q}_{max} - \frac{c\overline{Q}^2_{max}}{2N^2} \right). \tag{17b}$$

Fig. 5a and b again illustrate these conditions. In Fig. 5a, condition (17a) implies that triangle B is larger in area than triangle A. In Fig. 5b, condition (17b) implies that trapezoid B is larger in area than triangle A. As before, the values of areas A and B depend on b, c, X, and the pdf (in particular, $\overline{Q}_{min}$ and $\overline{Q}_{max}$).

If countries do not cooperate, each country i will maximize

$$E(\pi_i) = \begin{cases} bQ' - \frac{cq_i^2}{2} - X & \text{for } Q < \overline{Q}_{min} \\ bQ - \frac{cq_i^2}{2} - X\left[1 - \left(\frac{Q - \overline{Q}_{min}}{\overline{Q}_{max} - \overline{Q}_{min}}\right)\right] & \text{for } Q \in [\overline{Q}_{min}, \overline{Q}_{max}]. \end{cases} \tag{18}$$

In a Nash equilibrium, country will either play $q_i = b/c$ or, solving (10),

$$q_i = \frac{b}{c} + \frac{X}{c\left(\overline{Q}_{max} - \overline{Q}_{min}\right)} \tag{19}$$

or $q_i = \overline{Q}_{max}/N$, whichever of the latter two values is smaller.[27] Once again, the existence of two Nash equilibria (in pure strategies) makes abatement a coordination game. Upon comparing the payoffs, countries will abate so as to *reduce* the chance of catastrophe if

$$X > \frac{2c(\overline{Q}_{max} - \overline{Q}_{min})}{2N-1} \left(\overline{Q}_{min} - \frac{b(2N-1)}{c} \right). \tag{20a}$$

They will abate so as to *eliminate* the possibility of catastrophe if

$$X > \left[\frac{b^2}{2c} - \left(\frac{b\overline{Q}_{max}}{N} - \frac{c\overline{Q}_{max}}{2N^2} \right) \right]. \tag{20b}$$

Conditions ((17a and b)) for the full cooperative outcome and (20a and b) for the non-cooperative outcome are identical only when $N=1$. As N increases, the gap between these outcomes widens. Fig. 6a (which resembles Fig. 4a) shows that the prospect of uncertain catastrophe can have a substantial effect on the full cooperative outcome, raising it from bN^2/c to $\overline{Q}_{max}$, even as it has no effect at all on the non-cooperative outcome, which remains unchanged at $Q^{NC} = bN/c$. Fig. 6b (which is analogous to Fig. 4b) illustrates a situation in which complete avoidance of catastrophe can be sustained by coordination (only the bottom portion of the pdf for the aggregate benefit of avoiding catastrophe, drawn in broken lines, is shown in this figure). Unfortunately, this more pleasing result only holds under special circumstances. For example, X must be truly huge, especially if N is large.

[27] In this latter case I am limiting attention to the symmetric equilibrium.

S. Barrett / Journal of Environmental Economics and Management 66 (2013) 235–250 247

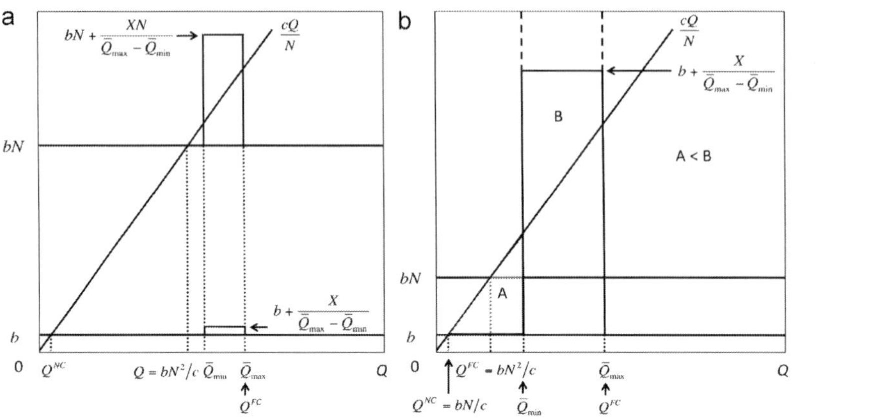

Fig. 6. (a) Uncertain prospect of catastrophe widens the gap between the non-cooperative and full cooperative outcomes. (b) Uncertain prospect of catastrophe closes the gap between the non-cooperative and full cooperative outcomes.

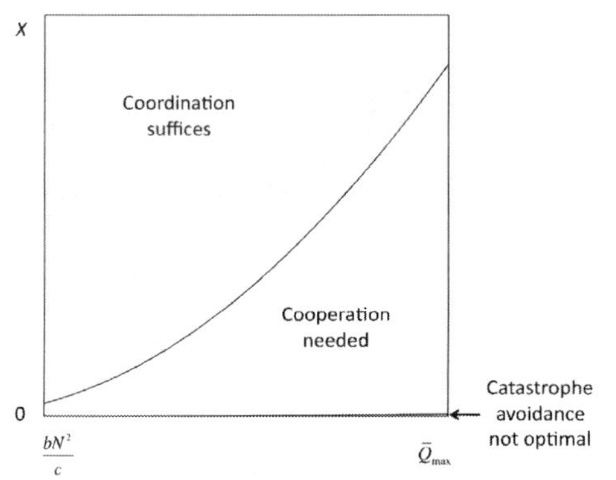

Fig. 7. When coordination suffices and cooperation is needed under threshold uncertainty.

Fig. 6a and b are both consistent with Rockström et al.'s [19] recommendation that the risk of catastrophe be eliminated, by holding concentrations down to 350 ppmv. Even without risk aversion, full cooperation can recommend a "precautionary" approach to catastrophe avoidance. However, the same conditions that support this conclusion may not enable countries to rally so as to achieve the collective action needed to sustain this outcome. If Fig. 6a described the world, countries would agree with Rockström et al.'s [19] recommendation, and yet fail to act so as to achieve it. Only if Fig. 6b portrayed the world could we expect actions to match aspirations.

Fig. 7 shows the range of values for X and $\overline{Q}_{\max}$ for which catastrophe avoidance can be sustained by coordination (Fig. 7, constructed from (17b)and (20b) is thus akin to Fig. 2, constructed from (3) and (4)).[28] It is obvious from comparing Figs. 2 and 7 that uncertainty widens the area requiring cooperation (thereby shrinking the area for which coordination suffices). However, the extent of this widening is much greater than suggested by visual inspection of the figures. As can be seen by comparing inequalities (4) and (20b), the value of X that enables catastrophe to be avoided by coordination is N times greater in Fig. 7 than in Fig. 2. Uncertainty about the catastrophic threshold makes it very likely that avoiding "catastrophic" climate change is a prisoners' dilemma rather than a coordination game.

It will help again to work through an example. Recall from before that, for $b=c=1$, $N=20$, and $\overline{Q}=600$ (threshold certainty), full cooperation requires that catastrophe be avoided for $X \geq 50$, whereas coordination can avoid catastrophe only for $X \geq 420.5$. Table 1 shows these values, and others corresponding to them, for various scenarios of threshold

[28] The simulations producing these figures assume $b=c=1$ and $N=100$.

Table 1
Minimum values of X for various values of $\overline{Q}$and$\overline{Q}_{min}-\overline{Q}_{max}$ (assuming $b=c=1$ and $N=20$).

Behavior	Consequence	$\overline{Q}$	$\overline{Q}_{min}-\overline{Q}_{max}$			
		600	590–610	550–650	500–700	450–750
Full cooperation	Reduce chance of catastrophe	–	38	75	100	75
	Eliminate chance of catastrophe	50	55	78	113	153
Coordination	Reduce chance of catastrophe	–	565	2364	4728	6323
	Eliminate chance of catastrophe	421	8703	9923	11,560	13,323

uncertainty. Reading across the top portion of the table we see a smooth transition for what is required for full cooperation as we move from threshold certainty to increasing levels of uncertainty. Notice that as the range of uncertainty increases, the size of the impact needed to justify elimination of the probability of catastrophe under full cooperation increases. This is because, holding the expected value of $\overline{Q}$ constant, a wider range for the pdf means a larger value for $\overline{Q}_{max}$; the benefit of completely avoiding catastrophe must increase to justify the higher cost of completely avoiding catastrophe. (See inequality (17b); in terms of Fig. 5b, as the range of the pdf widens, area A shrinks, while area B becomes both wider and shorter; for elimination of the risk to be justified, the value of X must increase so as to keep the northeast corner of trapezoid B above cQ/N.) By contrast, the minimum impact needed to justify reducing the chance of catastrophe below 100 percent increases and then decreases as the range for the distribution increases. Here there are two opposing forces. As the range of the distribution widens, the value of $f(Q)$ (represented by $1/(\overline{Q}_{max}-\overline{Q}_{min})$) falls, making an incremental reduction in the probability of catastrophe less valuable. (In (17a), the first multiplicand on the right increases; in Fig. 5a, the height of area B falls.) To compensate, the minimum X must increase. At the same time, a larger range causes the value of $\overline{Q}_{min}$ to fall, lowering the cost of reducing the probability of catastrophe. (In inequality (17a), the multiplicand on the far right shrinks; in Fig. 5a, area A shrinks and area B increases.) Now, coordination can be sustained for a lower minimum X.

Reading down the first column of numbers we can confirm the observation noted in Section 4 that the impact of catastrophe must be large for it to pay countries to coordinate so as to avoid the threshold (in the table, it must be nearly nine times as large, compared to the full cooperative outcome—421/50=8.4). More striking is how these requirements change as we move to the right in the table. The magnitude of the impact of catastrophe needed to give coordination a foothold increases dramatically, even for a very small range of uncertainty (in the table, even for the relatively tight range of 590–610, this value must be nearly 158 times as large compared to the full cooperative outcome—8703/55=158.2). Threshold uncertainty makes little difference to the full cooperative outcome (which increases from 50 to only 55 for these scenarios), but a huge difference to the ability of a self-enforcing agreement to sustain catastrophe avoidance by coordination. Recalling that the uniform distribution is particularly favorable to coordination, this is a disturbing finding.

7. Conclusions and implications

The standard model of a self-enforcing international environmental agreement predicts that collective action in reducing greenhouse gas emissions will be grossly inadequate. When this model is modified to incorporate a certain threshold with catastrophic damages, treaties can become highly effective. If the benefits of avoiding the threshold are high relative to the costs, the prospect of catastrophe transforms treaties into coordination devices.

Uncertainty about catastrophic damages is relatively unimportant to this calculus, whereas uncertainty about the threshold is critical. When the catastrophic threshold is certain, a slight relaxation in abatement from the threshold triggers a discontinuous change in damages—a huge deterrent to free riding. When the threshold is uncertain, a similar relaxation anywhere within the range of uncertainty for the threshold causes expected damages to increase just a little—a tiny disincentive to free ride. While the uncertain prospect of approaching catastrophes may commend substantially greater abatement in the full cooperative outcome, it may make little difference to non-cooperative behavior or to the ability of a climate treaty to sustain substantial cuts in global emissions.[29]

Multilateral efforts to reduce greenhouse gas emissions will be helped if threshold-uncertainty can be reduced very substantially (recall Figs. 4–6 especially). Though climate change is inherently uncertain, recent research suggests that signals of a generic nature may warn of an approaching threshold. A prime example is the phenomenon known as "critical slowing down," in which a system responds very slowly to perturbations as it nears a bifurcation [21]. Numerous abrupt climate shifts in the Earth's ancient history are consistent with this mathematical property [12], but this detection method is fallible; it may fail to identify some approaching thresholds, and predict others that are not actually there (Scheffer et al. [19]). As well, it only detects thresholds as we approach them, and by then the costs of stabilizing concentrations quickly so as to avoid catastrophe will be high. As shown here, under these circumstances, coordination will materialize only if

[29] Experiments relying on a modified version of the model developed here strongly support these conclusions. See Barrett and Dannenberg [10].

the impacts of breaching the threshold are truly huge. Of course, detection may also come too late for it to be possible to avoid catastrophe.

What, then, are the implications for policy? One implication, which I note with tongue in cheek, is that the collective action problem would be "solved" if we programmed a Doomsday Machine to set off a catastrophe with certainty once concentrations of greenhouse gases reached a specified level.[30] More seriously, negotiators should focus their attention on emission-reducing strategies that can promote collective action *without being conditional on climate thresholds.*[31] Finally, and most regrettably, we also need to prepare for the likelihood that we will someday cross a catastrophic threshold.[32]

Appendix A.1. ***Proof that (7) is a necessary and sufficient condition for a self-enforcing agreement to sustain the full cooperative outcome (avoiding catastrophe) when avoiding catastrophe is a Nash equilibrium.***

It will help to write out in long hand the condition for when avoiding catastrophe is a Nash equilibrium:

$$b\overline{Q}-\frac{c}{2}\left(\frac{\overline{Q}}{N}\right)^2 \geq b\left(\frac{\overline{Q}(N-1)}{N}+\frac{b}{c}\right)-X-\frac{c}{2}\left(\frac{b}{c}\right)^2. \tag{A.1}$$

Noting that inequality (7) requires $b/c+\sqrt{2X/c}\geq \overline{Q}/N$, the left hand side of (6) must be at least as large as the left hand side of (A.1). Now compare the right hand sides of these two inequalities. By assumption, $\overline{Q}/N > bN/c$, so we certainly have $\overline{Q}(N-1)/N > b(N-1)^2/c$, meaning that the first term on the right hand side of (6) must be smaller than the first term on the right hand side of (A.1). Finally, $b(N-1)/c > b/c$, so that the last term on the right hand side of (6) must be a larger negative value than its counterpart in (A.1). Taken together, this means that if (A.1) holds then (6) certainly holds; and, since (A.1) is the same as (7), this means that (7) is a sufficient condition for when catastrophe avoidance by N countries can be sustained by a self-enforcing agreement.

Appendix A.2. ***Proof that if N−1 countries can coordinate to avoid catastrophe, then so can N countries coordinate to avoid catastrophe.***

Suppose country i plays $q_i = b/c+\sqrt{2X/c}\geq \overline{Q}/N$ and each of the $j=1,\ldots,N-1$ other countries plays $q_j=(\overline{Q}-b/c-\sqrt{2X/c})/(N-1)$. This will be a Nash equilibrium provided that, if every other country plays these abatement levels, no country can gain by deviating. For i we require

$$b\overline{Q}-\frac{c}{2}\left(\frac{b}{c}+\sqrt{\frac{2X}{c}}\right)^2 \geq b\left(\overline{Q}-\frac{b}{c}-\sqrt{\frac{2X}{c}}+\frac{b}{c}\right)-X-\frac{c}{2}\left(\frac{b}{c}\right)^2 \tag{A.2}$$

and it is easy to confirm that the two sides of this expression are equal, making the weak inequality hold exactly.

For each country j we require

$$b\overline{Q}-\frac{c}{2}\left(\frac{\overline{Q}-b/c-\sqrt{2X/c}}{N-1}\right)^2 \geq b\left(\overline{Q}-\frac{(\overline{Q}-b/c-\sqrt{2X/c})}{(N-1)}+\frac{b}{c}\right)-X-\frac{c}{2}\left(\frac{b}{c}\right)^2,$$

which may be rewritten as

$$X\geq \frac{N}{(N-2)}\left\{\left[\frac{b^2}{2c}-\left(\frac{b\overline{Q}}{N}-\frac{c\overline{Q}^2}{2N^2}\right)\right]-\frac{\sqrt{2Xc}}{N^2}\left(\overline{Q}-\frac{bN}{c}\right)\right\}. \tag{A.3}$$

Constraint (A.3) lies above (4), the condition for coordination by N countries, provided

$$\left(\frac{\overline{Q}}{N}-\frac{b}{c}-\sqrt{\frac{2X}{c}}\right)\left(\overline{Q}-\frac{bN}{c}\right)\geq 0. \tag{A.4}$$

The second term on the left of (A.4) is positive by assumption and the first term is positive when (7) is violated. Hence, if it pays $N-1$ countries to coordinate to avoid catastrophe, it will also pay N countries to coordinate to avoid catastrophe.

[30] Tom Schelling told me in an email that my paper reminded him of something from fifty years ago: "There was some speculation, real or pretended, that the earth's atmosphere might have limited tolerance for nuclear explosions, and that beyond some threshold a nuclear blast might engulf the entire globe in a great flash of light that would exterminate all life on the surface. The thought didn't last long, but just long enough to stimulate the thought that if the threshold could be established with any certainty, it would be a great idea to detonate enough nuclear explosives to approach the threshold. Then nobody would dare to use a nuclear weapon. A real 'doomsday' proposal." A recent example of "strategic catastrophe-making" in the domestic context is the so-called "fiscal cliff," the combined effect of automatic budget cuts and tax hikes set to occur on midnight, December 31st, 2012 (the threshold), if the United States Congress does not pass new legislation to reduce US government debt. A problem with this strategy is that the impact of "catastrophe" is limited: the economy would not plunge immediately into recession. This is why the "fiscal cliff" has also been called the "fiscal slope."

[31] As noted previously, there may exist other ways in which treaties can coordinate behavior, including the use of trade restrictions [4,6] and the adoption of standards for technologies exhibiting network externalities [6,8].

[32] This is one reason why "geoengineering," especially the injection of reflective particles into the stratosphere, may need to be considered as a future option; see Schelling [22] and Barrett [9].

References

[1] Myles R. Allen, David J. Frame, Chris Huntingford, Chris D. Jones, Jason A. Lowe, Malte Meinshausen, Nicolai Meinshausen, Warming caused by cumulative carbon emissions towards the trillionth tonne, Nature 458 (2009) 1163–1166.
[2] Kenneth J. Arrow, A note on uncertainty and discounting in models of economic growth, Journal of Risk and Uncertainty 38 (2009) 87–94.
[3] Scott Barrett, Self-enforcing international environmental agreements, Oxford Economic Papers 46 (1994) 878–894.
[4] Scott Barrett, The strategy of trade sanctions in international environmental agreements, Resource and Energy Economics 19 (1997) 345–361.
[5] Scott Barrett, A theory of full international cooperation, Journal of Theoretical Politics 11 (1999) 519–541.
[6] Scott Barrett, Environment and Statecraft: The Strategy of Environmental Treaty-Making, Oxford University Press, Oxford, 2003.
[7] Scott Barrett, The theory of international environmental agreements, in: Karl-Göran Mäler, Jeffrey Vincent (Eds.), Handbook of Environmental Economics, vol. 3, Elsevier, Amsterdam, 2005, pp. 1457–1516.
[8] Scott Barrett, Climate treaties and 'breakthrough' technologies, American Economic Review (Papers & Proceedings) 96 (2) (2006) 22–25.
[9] Scott Barrett, The incredible economics of geoengineering, Environmental and Resource Economics 39 (2008) 45–54.
[10] Scott Barrett, Astrid Dannenberg, Climate negotiations under scientific uncertainty, Proceedings of the National Academy of Sciences 109 (43) (2012) 17372–17376.
[11] Colin W. Clark, Marc Mangel, Aggregation and fishery dynamics: a theoretical study of schooling and the purse seine tuna fisheries, Fishery Bulletin 77 (2) (1979) 317–337.
[12] Vasilis Dakos, Marten Scheffer, Egbert H. van Nes, Victor Brovkin, Vladimir Petoukhov, Hermann Held, Slowing down as an early warning signal for abrupt climate change, Proceedings of the National Academy of Sciences 105 (38) (2008) 14308–14312.
[13] Michael Finus, Game Theory and International Environmental Cooperation, Edward Elgar, Cheltenham, 2001.
[14] Donald J. Kessler, Burton G. Cour-Palais, Frequency of artificial satellites: the creation of a debris belt, Journal of Geophysical Research 83 (A6) (1978) 2637–2646.
[15] Carolyn Kousky, Olga Rostapshova, Michael Toman, Richard Zeckhauser, Responding to Threats of Climate Change Mega-Catastrophes. Resources for the Future Discussion Paper 09-45, 2009.
[16] Timothy M. Lenton, Hermann Held, Elmar Kriegler, Jim W. Hal, Wolfgang Lucht, Stefan Rahmstorf, Hans Joachim Schellnhuber, Tipping elements in the earth's climate system, Proceedings of the National Academy of Sciences 105 (6) (2008) 1786–1793.
[17] Manfred Milinksi, Ralf D. Sommerfeld, Hans-Jürgen Krambeck, Floyd A. Reed, Jochem Marotzke, The collective-risk social dilemma and the prevention of simulated dangerous climate change, Proceedings of the National Academy of Sciences 105 (7) (2008) 2291–2294.
[18] William D. Nordhaus, An Analysis of the Dismal Theorem, Cowles Foundation Discussion Paper No. 1686, Yale University, 2009 January 16.
[19] Johan Rockström, Will Steffen, Kevin Noone, Åsa Persson, Stuart Chapin III, Eric F. Lambin, Timothy M. Lenton, Marten Scheffer, Carl Folke, Hans Joachim Schellnhuber, Björn Nykvist, Cynthia A. de Wit, Terry Hughes, Sander van der Leeuw, Henning Rodhe, Sverker Sörlin, Peter K. Snyder, Robert Costanza, Uno Svedin, Malin Falkenmark, Louise Karlberg, Robert W. Corell, Victoria J. Fabry, James Hansen, Brian Walker, Diana Liverman, Katherine Richardson, Paul Crutzen, Jonathan A. Foley, A safe operating space for humanity, Nature 461 (2009) 472–475.
[20] Gerald H. Roe, Marcia B. Baker, Why is climate sensitivity so unpredictable? Science 318 (2007) 629–632.
[21] Marten Scheffer, Jordi Bascompte, William A. Brock, Victor Brovkin, Stephen R. Carpenter, Vasilis Dakos, Hermann Held, Egbert H. van Nes, Max Rietkerk, George Sugihara, Early-warning signals for critical transitions, Nature 461 (2009) 53–59.
[22] Thomas C. Schelling, The economic diplomacy of geoengineering, in: Thomas C. Schelling (Ed.), Strategies of Commitment, Harvard University Press, Cambridge, MA, 2006. 45–50.
[23] David L. Smith, The epidemiology of antibiotic resistance: policy levers, in: Ramanan Laxminarayan and Anup Malani; with David Howard and David L. Smith, Extending the Cure: Policy Responses to the Growing Threat of Antibiotic Resistance, Washington DC: Resources for the Future, 2007, pp. 39–68.
[24] Nicholas Stern, The economics of climate change, American Economic Review (Papers & Proceedings) 98 (2008) 2–37.
[25] Alessandro Tavoni, Astrid Dannenberg, Giorgos Kallis, Andreas Loeschel, Inequality, communication, and the avoidance of disastrous climate change in a public goods game, Proceedings of the National Academy of Sciences 108 (29) (2011) 11825–11829.
[26] Ulrich J. Wagner, The design of stable international environmental agreements: economic theory and political economy, Journal of Economic Surveys 15 (3) (2001) 377–411.
[27] Martin L. Weitzman, On modeling and interpreting the economics of catastrophic climate change, Review of Economics and Statistics 91 (1) (2009) 1–19.
[28] Zickfeld Kirsten, Michael Eby, H. Damon Matthews, Andrew J. Weaver, Setting cumulative emissions targets to reduce the risk of dangerous climate change, Proceedings of the National Academy of Sciences 106 (38) (2009) 16129–16134.

[43]

Inequality, communication, and the avoidance of disastrous climate change in a public goods game

Alessandro Tavoni[a,1], Astrid Dannenberg[b], Giorgos Kallis[c], and Andreas Löschel[b,d,e]

[a]Grantham Research Institute and Centre for Climate Change Economics and Policy, London School of Economics, London WC2A 2AZ, United Kingdom; [b]Department of Environmental and Resource Economics, Centre for European Economic Research, 68161 Mannheim, Germany; [c]Institució Catalana de Recerca i Estudis Avançats, and Institut de Ciencia i Tecnologia Ambientals, Universidad Autonoma de Barcelona, 08193 Bellatera, Spain; [d]Alfred Weber Institute for Economics, University of Heidelberg, 69115 Heidelberg, Germany; and [e]Centre for Climate Economics and Policy, Australian National University, Canberra, ACT 0200, Australia

Edited* by Simon A. Levin, Princeton University, Princeton, NJ, and approved May 24, 2011 (received for review February 17, 2011)

International efforts to provide global public goods often face the challenges of coordinating national contributions and distributing costs equitably in the face of uncertainty, inequality, and free-riding incentives. In an experimental setting, we distribute endowments unequally among a group of people who can reach a fixed target sum through successive money contributions, knowing that if they fail, they will lose all their remaining money with 50% probability. In some treatments, we give players the option to communicate intended contributions. We find that inequality reduces the prospects of reaching the target but that communication increases success dramatically. Successful groups tend to eliminate inequality over the course of the game, with rich players signaling willingness to redistribute early on. Our results suggest that coordination-promoting institutions and early redistribution from richer to poorer nations are both decisive for the avoidance of global calamities, such as disruptive climate change.

threshold public good | climate burden | experimental economics | cooperation | self-serving bias

Preserving the global climate commons is one of the biggest collective action problems humanity has ever faced (1); evidence suggests that we have already exceeded the planet's "safe operating space" in the climate system (2). Containing the rise in global mean temperature is a global public good, wherein the benefits of efforts to reduce emissions are shared by all, irrespective of individual contributions. Such disconnect between individual and collective interest is a prime cause of public goods underprovision (3–7). Although public goods experiments under controlled conditions oversimplify the complexity of international climate action (8), they nonetheless shed light on the relative importance of factors that affect its success (9). Standard public good games are concerned with the creation of a collective gain (10–15). Climate change, however, is about avoiding an uncertain public bad. This has been framed as a "collective-risk social dilemma," a threshold public good game of loss avoidance played with sequential contributions to a fund aimed at avoiding a probabilistic loss arising if the target is missed (16, 17). By capturing catastrophic climate change, this game has shed new light on the issue. It lacks, however, important features that may be determinant for the findings: inequality between participants and the opportunity to communicate with one another.

International progress in reducing CO_2 emissions has been remarkably slow, not least because of free-riding incentives, as partly captured by the threshold public goods game of loss avoidance (16). The challenge of this game, however, is coordination. Players are best off when synchronizing contributions in the face of multiple equilibria (3, 18). The game therefore calls for communication. The latest climate agreements negotiated in Copenhagen and Cancun introduced a pledge and review system of voluntary emission reduction commitments for 2020 (19). Can such a simple mechanism of communicating intentions be effective to enhance coordination?

Optimism from reaching a global agreement following Cancun is shadowed by concerns over implementation and particularly whether richer nations will go far enough in financing abatement and adaptation for poorer nations (20). Equity concerns over the distribution of emission cuts and associated costs are at the heart of the sustainability of international climate change action (21, 22). Inequality has been studied extensively in the context of collective action problems. The presence of inequality is often found to complicate cooperation (23–25), although communication between users tends to improve the likelihood of cooperation (26, 27). Different patterns of interaction are observed depending on the type and cause of inequality and on the type of resource at stake (28). Given these findings, we examine how inequality and potential differences in equity concerns between rich and poor affect their ability to coordinate efforts, and how this is mediated through communication of contribution intentions.

An essential feature of the global climate change game is that inequality in endowments is mirrored by inequality in past appropriation of the climate commons; roughly speaking, the richer a nation is, the more "carbon space" it has used in the atmosphere attributable to past greenhouse gas emissions (29, 30). We test its implications in the laboratory, with 240 students randomly assigned to groups of 6. Specifically, we assess the effects of inherited inequality in wealth and appropriation on coordination success in reaching a safety target, and how this is mediated through communication of contribution intentions. As in the study by Milinski et al. (16), each player was endowed with €40 that could be invested in climate protection. Players could choose between an investment of €0, €2, or €4 per round. The target was to invest €120 collectively by the 10th and final round so as to avoid simulated dangerous climate change and to secure what was left in the private account. Groups that failed to invest at least €120 lost all their savings with a 50% probability. Players did not know the identity of their team's members; after each round, they were informed about the others' contributions, the aggregate group contribution in that round, and the cumulative past contribution of each player and of the group as a whole (*Materials and Methods*).

To capture the idea of inheritance of past wealth and debt, we started the game with three inactive rounds in which players had no freedom to choose because contributions were determined by the computer. In the control treatment ("*Base*"), the computer allocated symmetrically to all players €2 per round. In the "*Base-Unequal*" treatment, the computer allocated asymmetrically to half of the group €4 per round and to the other half €0 per round. "Rich" players hence entered round 4 with €40, and

Author contributions: A.T., A.D., G.K., and A.L. designed research; A.T. and A.D. performed research; A.T. and A.D. analyzed data; and A.T., A.D., G.K., and A.L. wrote the paper.

The authors declare no conflict of interest.

*This Direct Submission article had a prearranged editor.

See Commentary on page 11733.

[1]To whom correspondence should be addressed. E-mail: a.tavoni@lse.ac.uk.

This article contains supporting information online at www.pnas.org/lookup/suppl/doi:10.1073/pnas.1102493108/-/DCSupplemental.

Table 1. End payoffs (and corresponding climate account values for the group in parenthesis) arising if the three pure strategies were adopted by all players for the seven active rounds

Treatment	€0 per round	€2 per round	€4 per round
Symmetrical	(36)	(120)	(204)
w_{all} = €34	17*	20	6
Asymmetrical	(36)	(120)	(204)
w_{rich} = €40	20*	26	12
w_{poor} = €28	14*	14	0

In the symmetrical treatments (*Base* and *Pledge*), all group members begin active play having contributed €6 in the previous three rounds, leaving them with a disposable endowment, w_{all} = €34; in the asymmetrical treatments (*Base-Unequal* and *Pledge-Unequal*), three rich players have no prior contributions and the three poor players have prior contributions of €12, leaving them with w_{rich} = €40 and w_{poor} = €28, respectively.
*Expected values based on the 50% probability of account loss when the target sum of €120 is not reached.

"poor" players entered round 4 with €28. Irrespective of the treatment, for the sake of comparability, all groups started the active phase with €36 in the climate account. In two treatments, players were given the option to announce what they planned to invest during the game: one time at the end of the three inactive rounds and again at the end of round 7. Subjects knew that pledges were nonbinding. The *"Pledge"* treatment introduced the pledges to the symmetrical case, whereas the *"Pledge-Unequal"* treatment implemented the pledges in the asymmetrical case.

The multiplicity of equilibria in the game makes classification virtually impossible. The game is a modified n-person stochastic threshold public goods game, with a total of 10 rounds, of which only 7 allow freedom of choice over the three possible actions. Both contributing nothing and contributing €2 in each round are (symmetrical) pure strategy Nash equilibria, because unilateral deviations are nonprofitable (31) (Table 1). Depending on the round and the path that led to it, a contribution of €4, bringing the individual investment above €20, may still be optimal if successful in guaranteeing that past investments are not wasted. Conversely, if at a certain stage, the target becomes out of reach because of insufficient members' contributions, one's best response is to stop contributing and play the odds.

In the symmetrical treatments, each group trajectory leading to a cumulative contribution of €120, irrespective of individual contributions provided that each subject invests at most €22 overall, is a Nash equilibrium. This is the case because the latter investment translates into a payoff of €18, which is above the €17 that is expected when all players choose not to contribute to the public good (Table 1, second column). Therefore, individuals can maximize the payoff of the game by choosing the intermediate level of contribution, invest a further €14 over rounds 4–10, and secure the €20.

In the asymmetrical treatments, as a result of the different disposable endowments of rich and poor players, the former gain the most when the climate is protected with equal burden sharing in the active rounds (€26, resulting from an investment of €14). Relative to the no-contribution equilibrium, it is more appealing because the rich will be at least as well off when investing €20 at most. The poor, on the other hand, do not stand to gain from the equal burden-sharing equilibrium in the active rounds, assuming risk neutrality. Given the early round contributions of €12, only

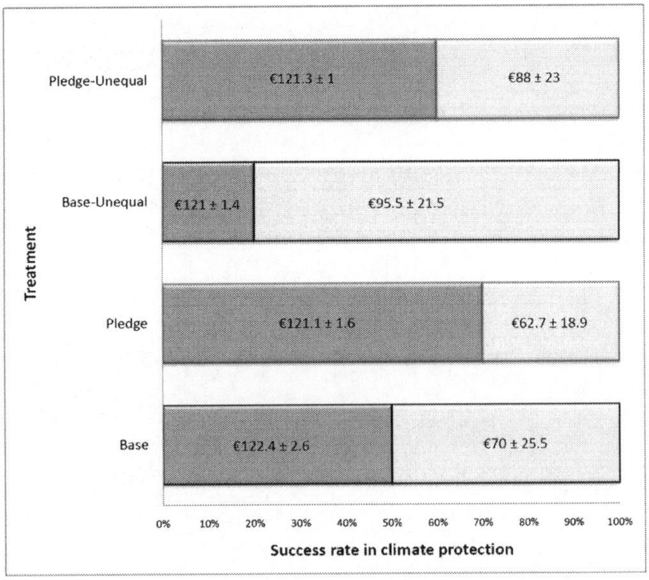

Fig. 1. Success rate in avoiding dangerous climate change. The lower two treatments are symmetrical, and the upper two are asymmetrical. The blue sections of the bars indicate the % of successful groups, whereas the gray sections indicate the corresponding failures (with red contours for the treatments with communication). (*Insets*) For both group classes, the average investments (inclusive of the €36 collected in the first three rounds) and SDs are shown. Only in *Pledge* was the outcome somewhat close to the rational prediction of all groups reaching the target (n = 10; P = 0.082, binomial test). Further analyses are available in *SI Text*.

by investing less than €14 in the active rounds (with the group still reaching the threshold) can these players have a higher expectation than by not investing in the public good.

The game design allows for such redistribution. The rich have a surplus of €12 in the €2 per active round equilibrium relative to the poor and can, in principle, forego part or all of it by investing more and allowing the poor to decrease their investment correspondingly. An average of €3 per round for the rich and €1 per round for the poor almost equalizes contributions (and expected payoffs) among the players. With full redistribution, rich and poor have a final payoff of €20, which, for the rich, is still rational in the sense of not being welfare diminishing relative to not contributing anything.

Lastly, note that a threshold public goods game like this one [and that by Milinski et al. (16)] differs from the majority of games used to investigate the climate change cooperation problem (32). In particular, the problem of enforcement is facilitated by the "disastrous" consequences of contributing less than €120, which therefore becomes a focal contribution level. The traditional formulation, in contrast, does not include catastrophic climate change but only gradual effects (mostly assuming linear benefits and nonlinear costs). The introduction of the catastrophe makes these curves discontinuous.

Results

Results show that inequality makes success harder but that the *Pledge* treatment option increases success dramatically (Fig. 1). Both *Pledge* treatments were well above the corresponding ones without pledges. Income inequality reduced the prospects of success: 5 of 10 groups succeeded in the *Base* treatment vs. 2 of 10 in the *Base-Unequal* treatment. In the latter, investment by the failing groups was €15 higher [n = 13; P = 0.039, two-sided Mann–Whitney–Wilcoxon (MWW) test], indicating that inequality also led to poorer coordination on the nonprovision outcome. The *Pledge* treatment option had more effect under conditions of inequality: Success rates tripled from 2 of 10 in the *Base-Unequal* treatment to 6 of 10 in the *Pledge-Unequal* treatment (n = 20; P = 0.085, one-sided Fisher's exact test). The latter success rate (6 of 10) is not significantly different from the 7 of 10 achieved by participants of the symmetrical *Pledge* treatment (P = 0.500), indicating that inequality is a less serious threat once a better coordination mechanism is introduced. This positive effect of communication is remarkable, given the fact that the incentives to coordinate toward a high contribution level are relatively weak. Going for the €2 per round strategy provides only moderate benefits compared with zero contribution, and, unlike the latter, it requires the cooperation of the remaining group members.

Although nonbinding, players respected pledges. Following the second pledge, average cumulative contributions in rounds 8–10 were €31.8 and €30 in *Pledge-Unequal* and *Pledge*, respectively, and the stated amounts were €32.6 and €29.6. The closer the pledges were to actual contributions, the higher was the probability of group success. As the difference between cumulative contributions and pledged amounts increases, the probability of a player being in a successful group decreases significantly (Fig. 2).

Successful groups were strikingly effective in eliminating the inherited inequality. In the two unequal treatments, the difference in contributions between rich and poor players belonging to successful groups is not significant (Fig. 3*A*; n = 16; P = 0.820, two-sided MWW test). Even in the absence of communication, participants of successful groups tacitly coordinated on an equalizing redistribution that offset the original endowment asymmetry. Conversely, the difference in contributions between rich and poor is significant in failing groups (€12.83 by the rich and €18.17 by the poor, n = 24; P = 0.014), indicating that such redistribution did not take place (Fig. 3*B*).

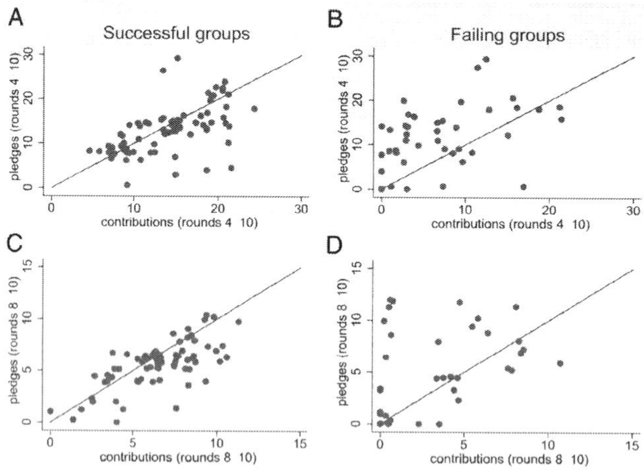

Fig. 2. Departure from announced contributions and its link to success. The probability that a player belongs to a successful group decreases with the contribution-pledge gap, i.e., with the differences between cumulative contributions and the corresponding amounts pledged both early (rounds 4–10, Probit; P = 0.002) and later (rounds 8–10, Probit; P = 0.032) in the game. All regressions have group-cluster robust SEs to take into account outcome interdependence among participants; regression tables and marginal effects interpretation are provided in *SI Text*. (*A* and *B*) Link between success and adherence to the initial pledge is visually confirmed. For the groups that provided the public good (*A*), the contribution-pledge gap is tighter than for the unsuccessful ones (*B*), as indicated by the dispersion around the bisector. Similarly for the second pledge, greater clustering around the bisector takes place in *C* than in *D*. A small random noise (5%) has been inserted to make all data points visible.

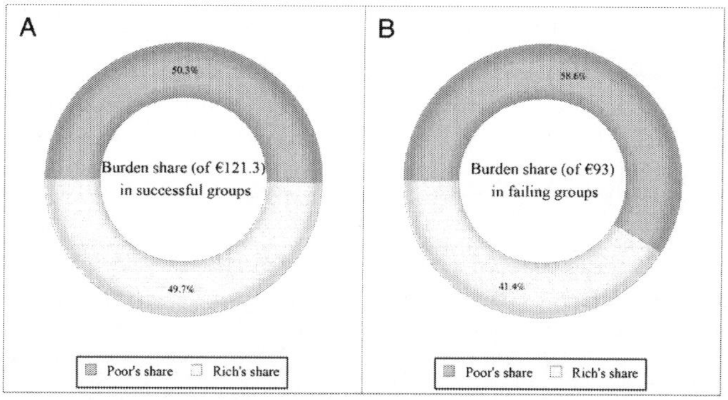

Fig. 3. Relative share of the total contributions taken up by players with different endowments, depending on which group they belong to. (*A*) In successful groups partaking of the treatments with unequal endowments (*Base-Unequal* and *Pledge-Unequal*), the rich compensated the poor by investing more in the active rounds and equalized cumulative contributions over the entire game at approximately €20. (*B*) In failing groups, such wealth redistribution did not take place to the same extent. The initial gap of €12 between the rich and poor was not fully offset because the former invested, on average, €12.83 each over the entire game (collectively contributing ~40% of the €93 provided), whereas the poor invested €18.17 (collectively contributing ~60% of €93).

Signaling willingness to invest in the public good rather than to gamble early on was critical for the fate of the game. Successful group members provided, on average, €1.92 in round 4, whereas failing group members provided €1.25 (n = 40; P = 0.000, two-sided MWW test). Early signals by the rich of willingness to redistribute were decisive in the asymmetrical games. On average, rich players in successful groups contributed €3.17 in round 4, whereas they contributed €2.06 in failing groups (n = 20; P = 0.005, two-sided MWW test). Cumulative contributions by the rich over rounds 4–6 were €9.83 in successful groups, whereas the rich in failing groups appeared to be unwilling to commit to early redistribution and invested only €6.67 (n = 20; P = 0.004).

The questionnaire that followed the game (*SI Text*) confirms that the rich's fairness perceptions and willingness to redistribute were decisive for success. Being confronted with the statement that "the rich players should contribute more during the active rounds than the poor players," 75% of the rich in successful groups but only 53% of the rich in failing groups agreed with that claim (n = 60; P = 0.071, one-sided Fisher's exact test). Therefore, the rich's opinion in that question and the group's success are significantly correlated (n = 60; P = 0.086, Spearman's correlation test). Furthermore, subjects' responses show a clear self-serving bias of fairness perceptions. The acceptance of the above claim is highly dependent on the player's wealth (n = 120; P = 0.000, Spearman's correlation test). In numbers, 90% of the poor but only 62% of the rich support the claim for redistribution (n = 120; P = 0.000, one-sided Fisher's exact test). This bias, which has been found also in climate negotiations (22), appears to be an important determinant for effective coordination.

Discussion

We find unambiguous evidence that the poor are not willing to compensate for the rich's inaction. These findings suggest that early leadership by the richer nations, in addition to appropriate coordination mechanisms, is instrumental to the avoidance of disastrous climate change. Extrapolation from our results, however, should be cautious (33). Controlling CO_2 emissions is a much more complex task than the one of coordination on a known threshold faced by the subjects of this experiment (3, 18). Climate change involves considerable uncertainty, especially the prospect of ensuing catastrophes. Many effects of global warming will be felt gradually; they may provide valuable early warning signals but, at the same time, worsen the free-rider problem. Climate change also entails not only asymmetry in wealth and carbon debt but asymmetry in adaptation capacity and risk exposure (34). Losses, even under catastrophic climate change, will be unequally distributed depending on countries' income or location (35). Different types of inequality stand to have different effects, the more so when in concert (36). The shape of the wealth distribution may also affect outcomes. Whereas an increase in inequality may well enhance the incentives for the rich to contribute more, such an increase may simultaneously reduce the incentives for the poor (29). Future research is needed to bring more realism and complexity in the collective-risk social dilemma, introducing uncertain thresholds and gradual climate change for instance. Experimental games may further help to explore the barriers to cooperation and to identify promising institutions.

Nevertheless, the finding that inequality hampers coordination and makes a coordination-promoting institution indispensable is important. Countries can be expected to coordinate their national efforts to reach a common goal, and communication and agreement on a common fairness notion are preconditions. Success in providing the global climate protection good is inextricably related to the willingness of the rich to take up a sizeable share of the burden early on. Signaling commitment to contribute early on appears decisive for coordination. Unfortunately, communication and consent are not sufficient to tackle climate change. Societies will have to do much more to solve this global collective action problem.

Materials and Methods

The experimental sessions were held in a computer laboratory at the University of Magdeburg, Magdeburg, Germany, using undergraduate and graduate students recruited from the general student population. In total, 240 students participated in the experiment, 60 of whom took part in each treatment. Subjects were seated randomly at linked computers with which they communicated their investment decisions during the game (further details and software demonstration are provided in *SI Text*). At the beginning of a session, a set of written instructions was handed out. The instructions included several numerical examples and control questions that tested subjects' understanding of the game. After reading the instructions and answering the control questions correctly, subjects were randomly assigned to a six-person group and began the game. The subjects did not know their fellows' identities, but they knew that they remained within the same group of players throughout the game. All decisions were made under completely anonymous conditions. The game comprised 10 rounds, with the first 3 rounds involving contributions predetermined by the software. After the 3 fixed rounds, 7 active rounds followed, where subjects could invest €0, €2, or €4 per round. After each round, subjects were informed about individual contributions (via nicknames) and the group contribution as well as cumulative individual contributions and cumulative group contribution up to the current round. Subjects knew that they would either lose their savings (what was left over after they invested a certain amount in climate protection) with 50% probability if the total sum of investment over all rounds fell short of the €120 target or would get their savings for sure if the total sum equaled or exceeded the target. The experimental sessions lasted about 60 min, and subjects earned, on average, €17.23 in the games. Earnings were anonymously paid in cash at the end of the session.

ACKNOWLEDGMENTS. We thank the MaXLab team of the University of Magdeburg, Magdeburg, Germany, for their support in conducting the experiment, as well as M. Milinski for providing the instructions and other information concerning a previously published experiment. This work was supported by the Gottfried Wilhelm Leibniz Scientific Community. A.T. is supported by the Centre for Climate Change Economics and Policy, which is funded by the UK Economic and Social Research Council and Munich Re.

1. Dietz T, Ostrom E, Stern PC (2003) The struggle to govern the commons. *Science* 302: 1907–1912.
2. Rockström J, et al. (2009) A safe operating space for humanity. *Nature* 461:472–475.
3. Barrett S (2007) *Why Cooperate? The Incentives to Supply Global Public Goods* (Oxford Univ Press, Oxford).
4. Carraro C, Siniscalco D, eds (1997) *New Directions in the Economic Theory of the Environment* (Cambridge Univ Press, New York).
5. Hardin G (1968) The tragedy of the commons. *Science* 162:1243–1248.
6. Levin SA (1999) *Fragile Dominion: Complexity and the Commons* (Perseus Books, Reading, MA).
7. Ostrom E, Burger J, Field CB, Norgaard RB, Policansky D (1999) Revisiting the commons: Local lessons, global challenges. *Science* 284:278–282.
8. Brekke KA, Johansson-Stenman O (2008) The behavioral economics of climate change. *Oxf Rev Econ Policy* 24:280–297.
9. Inman M (2009) The climate change game. *Nature Rep Clim Change* 3:130–133.
10. Ledyard J (1995) *Handbook of Experimental Economics*, eds Kagel JH, Roth AE (Princeton Univ Press, Princeton), pp 111–194.
11. Sturm B, Weimann J (2006) Experiments in environmental economics and some close relatives. *J Econ Surv* 20:419–457.
12. Ostrom E (2000) *Polycentric Games and Institutions: Readings from the Workshop in Political Theory and Policy Analysis*, ed McGinnis MD (Univ of Michigan Press, Ann Arbor, MI), pp 472–522.
13. Fehr E, Gächter S (2000) Cooperation and punishment in public goods experiments. *Am Econ Rev* 90:980–994.
14. Gürerk Ö, Irlenbusch B, Rockenbach B (2006) The competitive advantage of sanctioning institutions. *Science* 312:108–111.
15. Fischbacher U, Gächter S (2010) Social preferences, beliefs, and the dynamics of free riding in public goods experiments. *Am Econ Rev* 100:541–556.
16. Milinski M, Sommerfeld RD, Krambeck H-J, Reed FA, Marotzke J (2008) The collective-risk social dilemma and the prevention of simulated dangerous climate change. *Proc Natl Acad Sci USA* 105:2291–2294.
17. Dreber A, Nowak MA (2008) Gambling for global goods. *Proc Natl Acad Sci USA* 105: 2261–2262.
18. Barrett S (2003) *Environment and Statecraft: The Strategy of Environmental Treaty-Making* (Oxford Univ Press, Oxford).
19. Tollefson J (2010) Last-minute deal saves climate talks. *Nature* 468:875–876.
20. Kintisch E, Regalado A (2010) Climate negotiations. Cancún delegates see the trees through a forest of hot air. *Science* 330:1597.
21. Baer P, et al. (2000) Equity and greenhouse gas responsibility. *Science* 289:2287.
22. Lange A, Löschel A, Vogt C, Ziegler A (2010) On the self-interested use of equity in international climate negotiations. *Eur Econ Rev* 54:359–375.
23. Alesina A, La Ferrara E (2000) Participation in heterogeneous communities. *Q J Econ* 115:847–904.
24. Costa DL, Kahn ME (2003) Civic engagement and community heterogeneity: An economist's perspective. *Perspect Polit* 1:103–112.
25. Anderson LR, Mellor JM, Milyo J (2008) Inequality and public good provision: An experimental analysis. *J Socio-Econ* 37:1010–1028.
26. Ostrom E, Gardner R, Walker J (1994) *Rules, Games, and Common-Pool Resources* (Univ of Michigan Press, Ann Arbor, MI).
27. Hackett S, Schlager E, Walker J (1994) The role of communication in resolving commons dilemmas: Experimental evidence with heterogeneous appropriators. *J Environ Econ Manage* 27:99–126.
28. Baland J-M, Platteau J-P (2007) *Inequality, Cooperation and Environmental Sustainability*, eds Baland J-M, Bardhan P, Bowles S (Princeton Univ Press, Princeton), pp 10–35.
29. Roberts JT, Parks BC (2006) *A Climate of Injustice: Global Inequality, North-South Politics and Climate Policy* (MIT Press, Cambridge, MA).
30. Srinivasan UT, et al. (2008) The debt of nations and the distribution of ecological impacts from human activities. *Proc Natl Acad Sci USA* 105:1768–1773.
31. Nash JF (1950) Equilibrium points in N-person games. *Proc Natl Acad Sci USA* 36:48–49.
32. Finus M (2001) *Game Theory and International Environmental Cooperation* (Edward Elgar, Cheltenham, UK).
33. Levitt SD, List JA (2007) What do laboratory experiments measuring social preferences reveal about the real world? *J Econ Perspect* 21:153–174.
34. Kasperson RE, et al. (2005) *Ecosystems and Human Well-Being. Current State and Trends*, eds Hassan RM, Scholes R, Ash N (Island Press, Washington, DC), Vol 1, pp 143–164.
35. Adger WN, et al. (2007) *Climate Change 2007: Impacts, Adaptation and Vulnerability. Contribution of Working Group II to the Fourth Assessment Report of the Intergovernmental Panel on Climate Change*, eds Parry M, Canziani OF, Palutikof J, van der Linden P, Hanson CE (Cambridge Univ Press, Cambridge, UK), pp 717–744.
36. Baland J-M, Platteau JP (1997) Wealth inequality and efficiency in the commons. *Oxf Econ Pap* 49:451–482.

[44]

EXPERIMENTAL ANALYSIS OF RISK AND UNCERTAINTY IN PROVISIONING PRIVATE AND PUBLIC GOODS

LATA GANGADHARAN and VERONIKA NEMES*

This paper uses laboratory evidence from public goods games to examine how in payoff-equivalent situations, decision makers contribute toward private or public goods when they are exposed to different kinds of risks and uncertainties in the provision of these goods. We find that location matters with subjects moving away from the environmental uncertainty in favor of strategic uncertainty when that uncertainty is located on the private good. When the uncertainty relates to the public good, subjects face both kinds of uncertainties on the same good, leading to a significant drop in contributions. An opportunity to reduce uncertainty increases cooperation. (*JEL* C90, D81, H41)

I. INTRODUCTION

This paper investigates how individuals perceive risks and uncertainties when making decisions about private and public goods. Using the framework of public goods games and laboratory experiments, we study whether—in payoff-equivalent situations—decision makers contribute more or less toward private and public goods when outcomes involve different forms of risks and uncertainties.

The economics literature usually applies the term "uncertainty" to describe situations when the probabilities of possible future events are unknown. The term "risk" on the other hand is used to describe situations when the probabilities of future events are known.[1] The presence of uncertainty and risk can be very important in the provisioning of public goods and in policy making. For example, classic public good problems like climate change and depletion of the ozone layer can be thought to differ in the degree of knowledge about the probabilities associated with the events. While climate change is undoubtedly a case of uncertainty since less precise probability estimates exist in either science or economics, the ozone layer depletion can be characterized more as risk since detailed scientific and economic predictions exist about the depletion of the ozone layer and its impact on human health (e.g., melanoma, cataracts). Risk and uncertainty exist in the private domain as well. For example, in the area of individual health, some estimates exist about the correlation between smoking and lung cancer (could be characterized as risk); however, research is still in its infancy on the connection between working with animals and being susceptible to diseases like bird flu (a case of uncertainty).

*The authors are grateful to Ananish Chaudhuri, Pushkar Maitra, Andrew Muller, Charlie Plott, seminar participants at Deakin University, the Australian Conference of Economists, the Experimental Economics Workshop in Sydney, the Indian Statistical Institute in Delhi and Calcutta, Econometric Society Australasian Meetings, Australian National University, and Monash University for their valuable comments on earlier versions of the paper and to Jonathan Payne for his help in running the experiments. Funding from the Australian Research Council is greatly appreciated.

Gangadharan: Associate Professor, Department of Economics, University of Melbourne, VIC 3010, Australia. Phone 61-3-8344-5408, Fax 61-3-8344-6899, E-mail latag@unimelb.edu.au

Nemes: Research Associate, Department of Economics, University of Melbourne, VIC 3010, Australia. Phone 61-3-8344-5410, Fax 61-3-8344-6899, E-mail nemesv@unimelb.edu.au and Senior Economist, Department of Sustainability and the Environment, VIC 3002, Australia. Phone 61-3-9637-8804, E-mail veronika.nemes@dse.vic.gov.au

1. This follows Knight (1921) who introduced the dichotomy of known and unknown probabilities.

ABBREVIATIONS

CPR: Common Pool Resource
GLS: Generalized Least Squares
MPCR: Marginal Per Capita Return

Economic Inquiry
(ISSN 0095-2583)
Vol. 47, No. 1, January 2009, 146–164

doi:10.1111/j.1465-7295.2007.00118.x
Online Early publication January 22, 2008

In addition to the risks and uncertainties mentioned above, the process of policy making involves further uncertainties due to the unknown information about the decisions of others. In the standard noncooperative experimental setting of public goods and common pool resource games, there is always uncertainty about the ultimate size of the public good or resource request, as subjects do not know how other group members will behave. Messick, Allison, and Samuelson (1988) refer to this uncertainty as "strategic uncertainty."[2]

An implicit assumption underlying these public goods and common pool resource dilemma games is that the optimal size of the public good or the carrying capacity of the commons is known and that there is no uncertainty associated with the benefits of the public or private good.[3] However, in many problems, decision makers do not know with certainty the optimal level of public goods or the carrying capacity of the resource. For example, continuing with the example of climate change, the optimal level of carbon dioxide emissions abatement and the costs and benefits of proposed mitigation strategies are not known with certainty. Messick, Allison, and Samuelson (1988, 678–79) introduced the terminology "environmental uncertainty" to distinguish this external factor from strategic uncertainty: "Environmental uncertainty refers to environmental variables that determine which group action is best, while [strategic] uncertainty centres on how other group members will respond... The problem that is raised by the environmental uncertainty is the problem of optimality or efficiency, while the problem raised by [strategic] uncertainty is (...) coordination."

Several researchers have found that the distinction between strategic and environmental uncertainty is important in decision making.[4] Existing literature focuses mainly on uncertainty in common pool resource dilemmas where risk and uncertainty are defined in terms of threshold uncertainty. Messick, Allison, and Samuelson (1988), for example, incorporated a probabilistic destruction of the resource when the safe yield was surpassed. In their experiment, groups of four subjects participated in a single-trial task. If the group total request were below a certain threshold, each group member received his request. If the group total were above the threshold, a coin was flipped to determine whether the members would receive their requests. Messick, Allison, and Samuelson (1988) show that even when strategic uncertainty was absent (i.e., there was only one participant), the random size of the resource led to a suboptimal outcome. Rapoport et al. (1992) used mean-preserving spreads to model increasing environmental uncertainty regarding the size of the common pool resource. The authors found in a five-person common pool resource experiment that subjects dealt with the strategic uncertainty by requesting roughly one-fifth of the mean amount available. As the environmental uncertainty increased, however, subjects requested more than an equal share from the resource. Associated with the observed overexploitation of the common resources, research (Rapoport et al. 1992; Suleiman, Budescu, and Rapoport 1966) suggests that subjects' estimates of the resource size increase as the resource uncertainty increases. One explanation provided by Rapoport et al. (1992) is that people tend to perceive the variability and central tendency of probability distributions

2. The role of strategic uncertainty has been extensively modeled in the laboratory. Several factors including group size (Au and Ngai 2003; Buchanan 1968; Isaac and Walker 1988; Isaac, Walker, and Williams 1990; Loehman, Quesnel, and Babb 1996; Marwell and Ames 1979), the information structure inherent in the experimental design (Au and Ngai 2003; Budescu, Rapoport, and Suleiman 1995a, 1995b; Suleiman, Budescu, and Rapoport 1960), the incentives or contribution mechanism employed (Green and Laffont 1979; Groves and Ledyard 1977, 1980; Loehman, Quesnel, and Babb 1996), implementation of rewards and sanctions (Walker and Halloran 2004), existence and the level of provision point (Isaac, Schmidtz, and Walker 1989; Marks and Croson 1998; Schelling, 1960; Schmidtz, 1987; Suleiman, Budescu, and Rapoport 2001), and incentives for the public good relative to the private good (Isaac and Walker, 1988; Isaac, Walker, and Thomas 1984) have been shown to have a significant impact on the contribution levels. For a survey on treatment variables that affect contribution levels in public goods games, see Ledyard (1995), and for a meta-analysis of several of these factors, see Zelmer (2003).

3. For example, both the theoretical work of Bagnoli and Lipman (1989) and the experiments by Bagnoli and McKee (1991) show that if we offer a group of people the opportunity to contribute to the provision of a public good when the cost of the good, the payoffs to those in the group, and the initial wealth positions of those in the group are all common knowledge, then the Pareto-efficient outcome will emerge if their collective valuations exceed the cost of the public good.

4. Loewenstein, Thompson, and Bazerman (1989) found evidence from experiments that people's attitudes in decision processes involving risk and uncertainty may be very different in social (similar to strategic) and nonsocial (similar to environmental) domains owing to the differences between social and nonsocial preferences.

to be positively correlated. A second explanation is that when people are asked to provide their best estimate of an unknown resource, their estimate will be biased by what they hope for. Such a tendency has been labeled in the literature as "optimism," "outcome desirability bias," or simply "wishful thinking" (Hogarth 1987).[5] While there is extensive research in both the psychological and the economics literature on the effects of resource uncertainty in common pool resource situations under simultaneous and sequential protocols (Budescu, Rapoport, and Suleiman 1995a, 1995b; Gustafsson, Biel, and Gärling 1999, 2000; Rapoport and Suleiman 1992; Rapoport et al. 1992; Suleiman and Rapoport 1988), there is considerably less research on the effects of risk and uncertainty in public goods situations.[6]

Wit and Wilke (1998), for example, investigated the effects of environmental uncertainty (regarding the provision threshold) and strategic uncertainty on contribution to step-level public goods. In the "low–environmental uncertainty" treatment, subjects were informed that the provision threshold would be randomly sampled from a uniform distribution defined over the range [800, 1,000], while in the "high–environmental uncertainty" treatment, the range was increased to [400, 1,400]. The strategic uncertainty was manipulated by giving subjects false information regarding the variance of contributions of others. The authors found lower contributions under high environmental uncertainty with the most dramatic drop in the contribution levels when high environmental uncertainty was coupled with high strategic uncertainty. Au (2004) compared voluntary contributions in a public goods experiment in the case where the provision point was fixed at $m = 3$ to the case where the provision point could take the values $m = 2$, 3, and 4 with equal probability. He found that the provision rate for the public good was significantly higher when the provision point was known precisely. Gustafsson, Biel, and Gärling (2000) compared the voluntary contributions to public goods with the same expected provision threshold but different variances. In general, the average contribution was smaller in the high-variance group. While Wit and Wilke (1998) and Gustafsson, Biel, and Gärling (2000) applied a simultaneous design, Au (2004) replicated these results using a sequential decision-making protocol (i.e., subjects made decisions sequentially). These findings are consistent with the results from common pool resource studies showing that higher environmental uncertainty reduces the rate of intensity of cooperative behavior.

Suleiman, Budescu, and Rapoport (2001), however, report that the effects of environmental uncertainty in the common pool resource and public goods games exhibit qualitative differences. While all studies in the common pool resource dilemma show that individual requests increase with environmental uncertainty, Suleiman, Budescu, and Rapoport (2001) find that the effect of threshold uncertainty is moderated by the threshold mean: contributions to the public good increased as a function of uncertainty for the lower threshold mean and decreased for the higher threshold mean. McBride (2006) and Gustafsson, Biel, and Gärling (2000) also show that outcomes in public good games may be more sensitive to uncertainty and parameter changes compared to common pool resource games. The differential effect of environmental uncertainty in the two types of game was also documented by van Dijk et al. (1999, 111) who concluded that "environmental uncertainty may affect choice behaviour differently

5. An emerging literature on neuroeconomics examines the interactions between experimental economics, neuroscience, and psychology (Camerer, Loewenstein, and Prelec 2005). In this area, researchers attempt to understand decision making in different scenarios, including situations with risk and uncertainty, using medical brain imaging techniques.

6. There is some overlap between research in economics and psychology (see Rabin 1998, for a discussion of some psychological findings that are relevant to economics.) Decision making under risk and uncertainty is an area where the two disciplines have influenced each other. The study of social dilemmas has a rich tradition in both disciplines, with both sides agreeing on main empirical findings. However, they diverge in many other ways; in particular, economists and psychologists would disagree on what they feel needs further explaining (Lopes 1994). For example, experimental economic research and psychological research into the nature of the common pool resource seem to differ. While modeling *renewable* resources in the psychological literature has led to the construction of multistage decision tasks in common pool resource (CPR) games (e.g., Messick and Brewer 1983), modeling exhaustible resources in the experimental economic literature has led to the construction of a sequence of independent single-stage decision tasks (e.g., Hackett, Schlager, and Walker 1994). Exhaustible resources are a special case of renewable resources: where the renewal rate is zero. Budescu, Rapoport, and Suleiman (1995a, 172) conclude that the reason they are treated differently "may reflect the fact that, *psychologically*, the two cases seem to include different perspectives (frames), and invoke distinct considerations in the appropriators' minds."

in public good dilemmas, than in resource dilemmas."[7]

Very little research has been conducted to examine the impact of endogenous probabilities in these social dilemma games. In the common pool resource experiments of Walker and Gardner (1992), the probability of destruction of the resource was an increasing function of total group exploitation of the commons, and such destruction ended the experiment. The authors found that in a decision environment with well-defined probability and significant opportunity costs of destruction, efficiency was very low (21%) and the resource was quickly destroyed (longest experiment consisted of six periods). Using members' individual effort in a team sport as a motivation, Dickinson (1998) conducted a public good experiment where he introduced environmental uncertainty as an endogenous probability of provision into the voluntary contributions mechanism—the provision of the public good (winning in the sports event) was conditional on the aggregate contribution level. The author found that the increase in probability as a function of contribution levels, *per se*, did not significantly affect contribution levels (only through the increase in marginal incentives to contribute). In another experiment, Dickinson (2001) measured the efficiency-improving properties of giving incentives in a public good game where monetary prizes were given to high contributors defined both in terms of absolute contributions and then in terms of relative ability (relative to initial endowment). Results show that such private incentives increased the efficiency levels by 10%–25%.

In this paper, we explore, using laboratory experiments, how environmental and strategic uncertainties when coupled together influence decisions in the context of the provision of public and private goods. This paper makes several contributions. While previous research has examined different aspects of risk and uncertainty in isolation and often in a common pool resource setting, this paper distinguishes between risk and uncertainty and between variability in public versus private goods in a standard public good setting. Complimenting the literature on common pool resource dilemmas where uncertainty is defined in terms of threshold uncertainty, this paper defines risk and uncertainty in light of different probability distributions. The treatments considered allow us to isolate strategic uncertainty from several forms of environmental risk and uncertainty using known (environmental risk treatments), unknown (environmental uncertainty treatments), and endogenous (incentives treatments) probabilities. In the baseline treatment, no environmental uncertainty exists and only strategic uncertainty influences the outcome. Throughout the rest of the treatments, strategic uncertainty (e.g., group size, communication, marginal incentives to contribute) is kept constant and environmental risk and uncertainty are introduced in both private and public goods settings. A within-subject design and a multiple-period game allow us to examine these areas of research in a systematic manner to gain further insights into different aspects of decision making.

The current design was chosen to allow for comparison with existing research but also to ensure that the effects of different kinds of environmental uncertainty, on the contribution levels, are isolated. While Wit and Wilke (1998) manipulated the strategic and environmental uncertainty simultaneously, we keep the strategic uncertainty constant across treatments in order to study the effects of environmental uncertainty. Also, while previous research defined the environmental uncertainty by increasing the range from where the threshold of the provisioning of the public good was drawn (Wit and Wilke, 1998) or the range of the possible size of the common pool resource while keeping the expected values constant (Rapoport et al. 1992), we model the environmental uncertainty by defining different forms of probability distributions. Furthermore, we implemented a simultaneous decision protocol and a continuous public good framework, as opposed to a step-level public good. Also, similar to the incentives treatment in Dickinson (1998), we introduce endogenous probability and extend it to the provisioning of both the private and the public goods in two separate treatments.

II. EXPERIMENTAL DESIGN

A. Experimental Procedure

The laboratory experiment was implemented in ten sessions each involving 5 subjects

7. For further discussion of the different kinds of environmental and strategic risk and uncertainty studied in the literature, in various experimental settings, refer to Nemes (2005).

and 15 decision-making periods for each of the seven treatments.[8] We used a within-subject design; all 50 subjects participated in the seven treatments, which allowed for powerful comparisons across treatments. This design also allowed us to keep all elements of strategic uncertainty constant across treatments as the same five subjects interacted in different treatments. At the beginning of each experimental session, participants were told that they would be participating in an economic experiment in which they would make investment decisions. A group of five participants faced the same investment dilemma and made simultaneous decisions. Each subject participated only once in the experiment, that is, they participated in one session only.

In order to control for risk aversion at the individual level, prior to the computerized experiment, we asked the subjects to participate in a hand-run lottery game. Participants had to indicate their preferences between two options: a safe option that yielded \$7 with certainty or a risky option. The risky option had a probability of winning \$12 changing from 10% in the first lottery to 100% in the tenth lottery or correspondingly, the probability of winning \$2 changing from 90% in the first lottery to 0% in the tenth lottery (similar to lottery games described in Brown and Stewart 1999; Holt and Laury 2002). One of the games on the sheet was played at the end of each session. Using the information on when subjects switch from the risky option to the safe option, we can define whether they are risk lovers, risk neutral, or risk averse. Subjects who switch before or at the fourth lottery are classified as risk lovers. Subjects who switch at the fifth lottery are risk neutral, and subjects who switch after the fifth lottery are risk averse. Conducting the lottery game, therefore, helped in measuring the risk preference of the subjects, which we can then compare with their behavior in the public goods experiments where they face risk and uncertainty in the group context.[9] Furthermore, qualitative and quantitative information were also collected in the form of questionnaires from the participants. The participants completed these questionnaires after Treatments 4 and 5 and also at the end of the session. (The instructions and questionnaires are available from the authors.)

The experiments were conducted using subjects drawn from a population of undergraduate and graduate students at the University of Melbourne. Subjects made an average of 33 cents per decision period and a total of \$37 on average. Furthermore, subjects received an additional \$2, \$7, or \$12 depending on their choice and the outcome of the lottery game. In order to control for wealth effects, their income from the lottery game was only revealed to the subjects at the end of the session.

B. Decision Process

At the start of each period, each subject was given the same number of tokens or experimental dollars, ω_i. Subjects had to divide these tokens between contributing to a "private account" (x_i) and a "group account" (g_i). Investment into the private account earned α experimental dollars with probability p_x to the individual only. Contributions to the group account earned $\beta G/n$ experimental dollars to each participant with probability p_g, where $G = \sum_{i=1}^{n} g$, β is the "efficiency factor" of the group account, and n is the number of participants in the group. Contributions to the group account yielded the same return to all participants irrespective of their contributions. Therefore, the group account exhibited the public goods' nonexcludible and nonrival characteristics. At the end of each period, subjects learned the aggregate level of contribution to the public good and their return from the private as well as the group accounts.

C. Formulation of the Social Dilemma

Individual *i*'s expected payoff is given by $E(u_i) = p_x\alpha x_i + p_g\beta G/n$. The subjects must maximize this payoff function, subject to a budget constraint ($\omega_i = x_i + g_i$), a public goods identity ($G = \sum_{i=1}^{n} g_i$), and a nonnegativity constraint ($g_i \geq 0$).

Let us define M as the marginal per capita return (MPCR)—the amount that is generated for each member of the group when one individual contributes a token to the public good.[10] EM is the expected value of the MPCR:

8. The experiment was programmed and conducted with the software *z*-Tree (Fischbacher 2007).

9. While the risk attitude survey was an important part of the current research and we use the risk estimates in the multivariate estimations presented in the Results section, the results are discussed in further detail in Nemes (2005).

10. It can also be seen as the marginal rate of substitution of the private good for the public good or as a measure of the incentives to contribute to the public good.

TABLE 1
Treatment Parameterization

Treatment		α	β	p_x		p_g	
Baseline	1	1	1.5	1	Known	1	Known
Risk	2	2	1.5	0.5	Known	1	Known
	3	1	3	1	Known	0.5	Known
Uncertainty	4	2	1.5	$p_x = r_x$, uniform distribution, $r_x \in (0,1)$	Unknown, unless requested	1	Known
	5	1	3	1	Known	$p_g = r$, uniform distribution, $r_g \in (0,1)$	Unknown, unless requested
Incentives	6	2	1.5	$p_{xi} = r_x + f(x_i)$	Unknown	1	Known
	7	1	3	1	Known	$p_g = r_g + f(G)$	Unknown

$$\text{EM} = E(M) = -E(\partial u_i/\partial G)/(\partial u_i/\partial g_i) = \beta p_g/n\alpha p_x.$$

In the baseline treatment, $p_x = p_g = 1$ and EM $= M$. The values of α and β may change from one treatment to the other but remain constant throughout each treatment.

The social dilemma arises if the parameters satisfy the inequality:

$$0 < \beta p_g/n < \alpha p_x < \beta p_g.$$

In this case, a unit investment into the private account provides a subject with more payoff than a unit investment into the group account. On the other hand, taking a unit of the private investment from every participant and contributing it to the group account make each participant better off. Thus, the unique dominant strategy is to free ride (i.e., zero contribution, $g_i = 0$, $\forall i$), while the socially optimal Pareto-efficient solution requires every participant to contribute all his or her tokens to the group account (i.e., $g_i = \omega_i$, $\forall i$).

D. Treatments

We examine seven treatments in this paper. We controlled for order effects by switching the order in which subjects experienced the treatments.[11]

11. In keeping with other papers in this area (Dickinson 1998; Wit and Wilke, 1998), we used a within-subject design. The baseline treatment was always conducted first, and the risk, uncertainty, and incentives treatments followed in the same order. We used this ordering pattern specifically to ensure that subjects were familiar with the simple treatments before they made decisions in the more complex treatments. The order of the treatments varied in the sense whether it was the private or the public good that had risks or uncertainties associated with it. On analysis of the data, we do not find any evidence of ordering effects.

Baseline (Treatment 1). The baseline treatment (Treatment 1) is a standard public goods game; there is no risk regarding return from the private or the group account. All parameters (α, β, p_x, and p_g) are known with certainty to the participants and $p_x = p_g = 1$, $\alpha = 1$, $\beta = 1.5$. There is no environmental uncertainty, and the only uncertainty that is present in the game is strategic uncertainty, which is kept constant throughout the following treatments. The expected marginal per capita ratio is EM $= M = 0.3$. The group payoff in case of the Pareto-efficient outcome would be 1.5 experimental dollars for each group member for each period, while the free-riding Nash equilibrium would yield each subject 1 experimental dollar per period. The parameters for $n = 5$ and $\omega_i = 100$ are summarized in Table 1.

Environmental Risk (Treatments 2 and 3). In the risk treatments, the probability of obtaining the return from the private account (Treatment 2) or the group account (Treatment 3) is drawn from a Bernoulli distribution while keeping everything else constant (including the expected value of the marginal incentives to contribute). Participants are no longer guaranteed the return from their investments. In Treatment 2, the probability of return from the private account is reduced to $p_x = 0.5$, but it is held constant throughout the treatment and all participants know its value. For each period, the computer generated a random number, $r \in [0,1]$. If $r \leq p_x$, then the investment multiplied by the efficiency factor on the private account (αx_i) was returned. Otherwise the investment into the private account was lost to the participants. The investment into the group account was returned with certainty $p_g = 1$, and the efficiency factor on the group account remained unchanged at $\beta = 1.5$.

Similarly, in Treatment 3, the return from the group account is no longer guaranteed, but its probability $p_g = 0.5$ is held constant and is known to all participants. In order to preserve the value of EM = 0.3 and maintain the single-period dominant strategy of zero contributions and the Pareto-efficient outcome at 100% contribution, the efficiency factor on the private and group accounts was modified to $\alpha = 2$ and $\beta = 3$ in Treatments 2 and 3, respectively. The parameters of the group account were held at the values of the baseline treatment, $p_x = 1$ and $\alpha = 1$. Similar to Treatment 2, the return from the group account was determined by random number generation.[12]

Environmental Uncertainty (Treatments 4 and 5). In Treatments 4 and 5, the probability of receiving the return from the investment is drawn from a uniform distribution in each period and is not displayed to participants. Therefore, there is uncertainty regarding the return from the investment into the private (Treatment 4) and group (Treatment 5) accounts. To learn the exact probability of return, participants must pay a fee of $f = 20$ from their endowments. When participants choose to do so, the information regarding the probability of return from the private (p_x) and group (p_g) accounts is displayed on the computer screen in Treatments 4 and 5, respectively. This probability is the private information of those participants who paid the fee for the information, and they remain with $\omega_i - f = 80$ tokens to invest in the period.[13]

12. The environmental risk treatments can be thought to model decisions where, based on past events and statistical analysis, decision makers can estimate the possible outcomes and the probability of occurrence of these outcomes. For example, several studies have been undertaken to estimate the costs, health-related benefits, and associated probabilities of certain automobile emissions standards (like clean fuels, banning lead additives, reduction of carbon monoxide and volatile organic compounds). Decisions in such cases are usually based on known probabilities, that is, in the face of risk.

13. Ideally it would have been interesting to have uncertainty treatments in which subjects could not choose to convert uncertainty into risk and then add treatments in which that option was available to them. This would have increased the number of treatments and would have been very difficult to accommodate within the 2-h experimental slot. Hence, we chose the potentially more interesting (and policy-relevant) treatment where subjects were given the choice to pay to convert uncertainty into risk.

Treatments 4 and 5 allow us to divide the participants into two distinct groups: those who know the exact value of the probability (facing risk) and those who do not (facing uncertainty). Treatments 4 and 5 capture scenarios when the probabilities of the benefits or costs of an outcome are initially uncertain. An example for Treatment 4 could be a building inspection, where architects can inspect a building for a prospective buyer, for a fee, and convert uncertain situations into risky situations. We can interpret Treatment 5 as a setup where decision makers may invest in research in order to get clearer estimates about the probabilities. Research, however, is costly and will leave decision makers with less money to invest into the public good itself.

At the end of Treatments 4 and 5, a questionnaire was filled out by the participants to assess their "willingness to pay" for the information regarding the probability of return from the private and group accounts, respectively.

Incentives (Treatments 6 and 7). In Treatments 6 and 7, participants have the opportunity to increase the probability of return from the private and group accounts, respectively. In Treatment 6, participants face an uncertainty regarding the probability of return from the private account. Similar to the previous treatments, the probability initially is drawn from a uniform distribution. Participants, however, due to their individual investments into the private account, could increase the probability of return from the private account. Similarly, in Treatment 7, a higher level of aggregate investment into the group account may lead to the increase of the initial probability of return from the group account. The probability of return in each round is determined by the sum of a random number drawn from a uniform probability distribution and an increment due to the investment into the private (Treatment 6) or group account (Treatment 7) in that round. That is, $p_{xi} = r(1+x_i/2\omega)$ and $p_g = r(1 + G/2n\omega)$, respectively.

In Treatment 6, therefore, depending on the individual investments, the participants may face differing levels of probabilities. Both the initial random number and the probability increment remain unknown to the participants until the end of the period. They do know, however, that they can increase the probability by increasing their contributions. At the end of each period, the computer displays the initial

TABLE 2
Total and Mean Investment into the Private and the Group Accounts

	Investment into the				
	Private Account		Group Account		
Treatment	Total	Mean	Total	Mean	Standard Deviation (Private and Group Accounts)
1	56,539	75.39	18,461	24.61	32.66
2	36,449	48.60	38,551	51.40	38.26
3	65,664	87.55	9,336	12.45	21.74
4	41,230	54.97	33,770	45.03	41.90
5	67,269	89.69	7,731	10.31	22.89
6	63,928	85.24	11,072	14.76	26.07
7	62,976	83.97	12,024	16.03	28.31
Total	39,4055	75.06	130,945	24.94	

probability and the increase in probability of return from the private and group accounts in Treatments 6 and 7, respectively.[14]

In summary, in Treatments 2–7, we vary environmental risk and uncertainty on the private and group accounts while keeping the strategic uncertainty and the expected payoff (together with the expected value of the marginal incentives to contribute) equivalent to those in the Baseline treatment.

E. Hypotheses

We use the design described above to test the following hypotheses to verify and extend the behavioral evidence in decision making under risk and uncertainty.

Hypothesis 1: In payoff-equivalent situations, subjects contribute less to the public good when risk or uncertainty is associated with the public good than when risk or uncertainty is associated with the private good (comparing Treatment 2 with 3 and Treatment 4 with 5).

Hypothesis 2: In payoff-equivalent situations, subjects contribute more to the public good when they face endogenous probability of provision of the public good than when they face endogenous probability of provision of the private good (comparing Treatment 6 with 7).

Hypothesis 3: In payoff-equivalent situations, subjects' contribution level to the public good is higher when facing uncertainty than when facing risk, both associated with the private good (comparing Treatments 2 and 4).

Hypothesis 4: In payoff-equivalent situations, subjects' contribution level to the public good is lower when facing uncertainty than when facing risk, both associated with the public good (comparing Treatments 3 and 5).

Hypothesis 5: In payoff-equivalent situations, when probabilities are endogenous, that is, subjects can increase the probability of provision of the private good, subjects' contribution to the public good is lower than that when facing uncertainty associated with the private good (comparing Treatment 6 with 4).

Hypothesis 6: In payoff-equivalent situations, when probabilities are endogenous, that is, subjects can increase the probability of provision of the public good, subjects' contribution to the public good is higher than when facing uncertainty associated with the public good (comparing Treatment 7 with 5).

III. RESULTS

We start with an overview of the data collected from the experiments and then report results from econometric tests, which examine the impact of risk, uncertainty, and incentives on individual behavior.

A. Overview

Table 2 and Figure 1 present the average proportional contributions to the private and the public goods as a percentage of their endowments for all treatments.[15] The dark

14. The incentives treatments mainly reflect cases when technological advancement increases the probability of obtaining the benefits from the project. Treatment 7 models cases where innovations have an impact on a public scale, and Treatment 6 captures cases where the reward for innovation has a private characteristic.

15. In the uncertainty treatments (Treatments 4 and 5), subjects may choose to pay a fee of 20 experimental dollars from their endowment to reveal the probability associated with the private and the public good, respectively. The proportional contribution takes account of the fact that in these two treatments, some subjects have only 80 experimental dollars to place in the private and group accounts.

FIGURE 1
Proportional Contributions into the Private and the Group Accounts

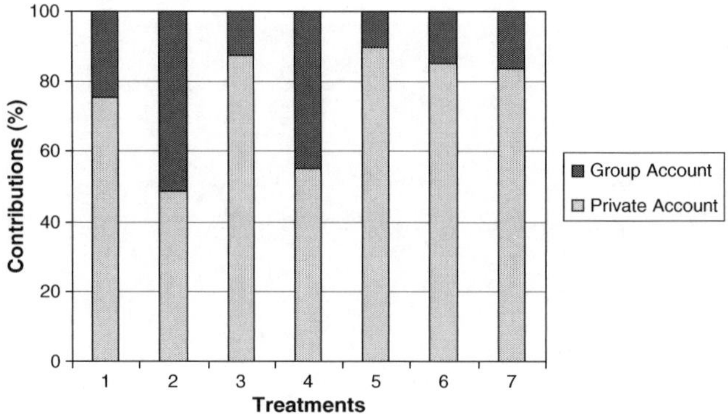

parts of the columns in Figure 1 illustrate the proportion of the endowment contributed to the public good. In all treatments except 2 and 4, the contribution to the public good is approximately 20%. In Treatments 2 and 4, where there is risk or uncertainty associated with the private good, the average contribution toward the public good reaches approximately 50%. Subjects seem to move away from the risky or uncertain private good and contribute more toward the public good.

Figure 2 tracks the average contributions to the public good over the 15 periods in each treatment and confirms the much higher contributions in Treatments 2 and 4. Contributions start at a higher point and gradually taper down by the end of the session. Figure 3 depicts the deviation from the baseline treatment. Since the baseline treatment (Treatment 1) is a pure strategic uncertainty treatment (there is no environmental uncertainty) and the strategic uncertainty is kept constant throughout Treatments 2–7, the deviation from the baseline treatment could be considered to be solely the effect of the environmental risk and uncertainty. Except in Treatments 2 and 4, when the risk and uncertainty are related to the return from the private good, the contribution levels are below that in the baseline treatment.

FIGURE 2
Average Contribution Levels over the 15 Periods

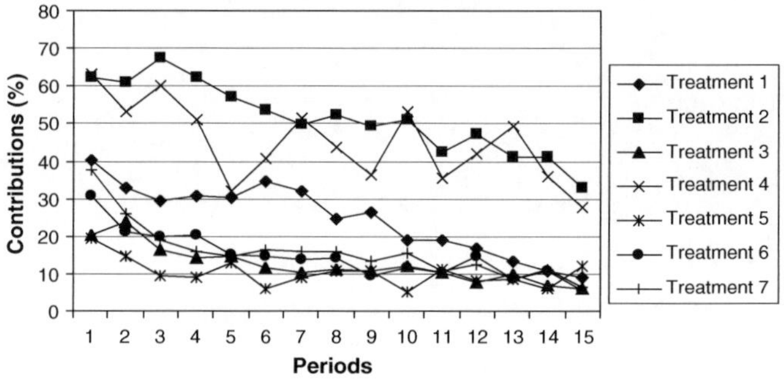

FIGURE 3
Average Contribution Levels Relative to the Baseline Treatment

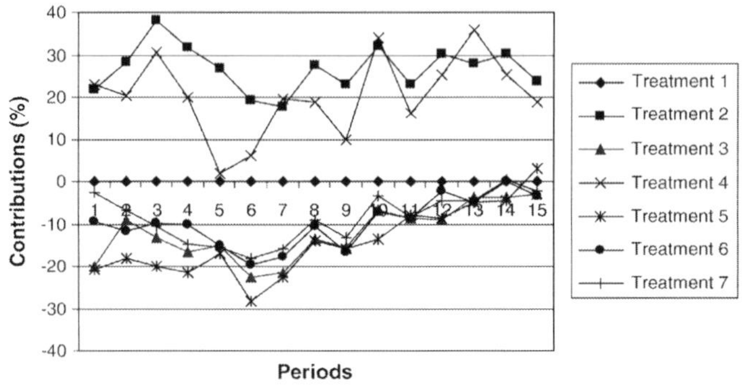

Figure 4 shows the decisions made by subjects under risk (i.e., known probability) and uncertainty (i.e., unknown probability) in Treatments 4 and 5. Note the difference in the pattern of contributions when risk is introduced to the provisioning of the private good (Treatment 4, marked by diamond shape) versus public good (Treatment 5, marked by triangles). As the probability of providing the private good increases, subjects contribute less to the public good and at $p_x = 0.8$, subjects contribute almost 0% to the public good. When risk is introduced in the provisioning of the public good, subjects divert away from contributing to the public good. Only when the probability of providing the public good is very high do subjects start to contribute, and 100% contribution levels are achieved when the probability of providing the public good is 1. Similarly, there is a difference in the patterns of contributions when subjects face uncertainty (unknown probability) related to the private good (Treatment 4, rectangular shapes) and when they face uncertainty related to the public good (Treatment 5, round shapes). Regardless of what the actual probabilities are, subjects facing uncertainty contribute around 50% to the public good when uncertainty is associated with the private good. When uncertainty is associated with the public good, however, contribution levels drop to around 5%.

FIGURE 4
Contributions to the Group Account with Environmental Risk and Uncertainty (Treatments 4 and 5)

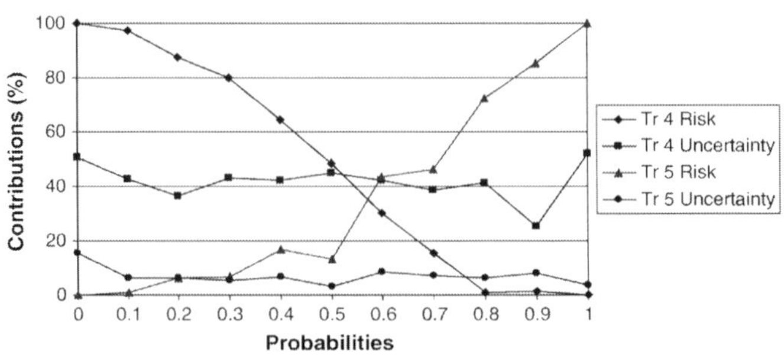

In Treatment 4, in 43% of the cases, subjects are willing to pay the fee of 20 experimental dollars to reveal the probability of return from the private good. In Treatment 5, subjects on average pay for information 12% of the time.[16] Though not recorded in Figure 4, it is interesting to note that a few of these subjects did not seem to use the revealed information in their decision making. For example, even when the probability of return from the private good was very low ($0 \leq p_x \leq 0.3$) or the probability of return from the public good was very high ($0.7 \leq p_g \leq 1$), these subjects invested either all or a substantial part of their remaining tokens into the private good. The interesting aspect is not their investment decision, *per se*, but the fact that they were willing to pay for the information about the probabilities but then not use this information in their decision making. The above cases could be understood as examples of aversion from strategic uncertainty. So even when environmental probabilities are unfavorable, they are still more attractive than strategic uncertainty (i.e., contribution to the public good).

B. Econometric Tests

To examine whether our data support the six hypotheses outlined above, we present results from very conservative pairwise *t*-tests and nonparametric Wilcoxon rank-sum tests in Table 3. These statistical tests use one observation per treatment per session. The nonparametric tests are particularly valuable, as they require a minimum of statistical assumptions. In addition, we also report results from multivariate regression models. The regression models evaluate the contribution of different factors on the decisions made by subjects. These panel regressions employ a random effects error structure, with the subject representing the random effect. We also cluster the errors at the group level to capture any unobserved heterogeneity in the group.

We present results from random effects generalized least squares regressions with clustering, where the dependent variable is the

16. This difference (43% vs. 12%) is statistically significant. Also subjects' willingness to pay, obtained from the questionnaires after Treatments 4 and 5, reveal that they were willing to pay significantly more to convert uncertainty into risk when the uncertainty was associated with the private good (17% vs. 9% of their endowment).

TABLE 3

Pairwise *t*-Tests and Nonparametric Wilcoxon Rank-Sum Tests for the Difference in Means

Treatment	1	2	3	4	5	6
2	3.379*** (2.721***)					
3	−1.873* (−1.739*)	−5.790*** (−3.628***)				
4[a]	2.899*** (2.343***)	−1.792* (−1.890*)	5.758*** (3.780***)			
5[a]	−2.237** (−2.192**)	−6.197*** (−3.704***)	−1.108 (−1.058)	−6.265*** (−3.780***)		
6	−1.540 (−1.285)	−5.521*** (−3.704***)	0.478 (0.680)	−2.755** (−2.343**)	0.947 (0.983)	
7	−1.140 (−1.361)	−4.562*** (−3.099***)	0.570 (0.151)	−4.236*** (−2.948***)	1.447 (1.512)	0.205 (−0.151)

Notes: The pairwise estimates use the horizontal as the comparison base. The first number reported in each cell is the *t*-statistic and the second number reported is the *z*-statistic from the Wilcoxon rank-sum test.

[a]In Treatments 4 and 5, we only include subjects who do not want to know the probability they face.

*Significant at the 10% level; **significant at the 5% level; ***significant at the 1% level.

TABLE 4
Summary of Variables Used in the Regressions

Variable	Description	Mean	Minimum	Maximum
Prop_pubaccount	The proportional contribution to the public account as a percentage of the endowment	0.25	0	1
Loutcome	Lag of the outcome variable (outcome = 1 if good was provided)	0.60	0	1
Ldeviation	Difference between the amount contributed to the public good by an individual in the previous period and the average group contributions in the previous period	99.79	0	400
Sex	=1 if female	0.42	0	1
Age	Age of the subjects	22.5	18	45
Risk_Averse	Subjects who switch after the fifth lottery	0.74	0	1
Risk_Neutral	Subjects who switch at the fifth lottery	0.10	0	1
Risk Lover	Subjects who switch before or at the fourth lottery	0.16	0	1
EM	The expected marginal per capita ratio	0.64	0	226.59
lprobincr	The increase in probability achieved in the previous period by the contributions in Treatments 6 and 7	8.68	0	33

individual contribution made to the public good as a proportion of the endowment.[17] Apart from the treatment dummy, the other regressors included are inverse of time period (invperiod) to capture the dynamic elements of the data (1/*t*); whether the random outcome was such that it resulted in the provision of the private or public good in the previous period, captured by a binary variable (loutcome); the difference between the amount contributed to the public good by an individual in the previous period and the average group contribution in the previous period (ldeviation); the increase in probability achieved in the previous period by the investments in Treatments 6 and 7 (lprobincr); the expected marginal per capita ratio (EM); and individual-specific characteristics.[18] The individual characteristics are age, gender, and their attitudes toward risk as measured by the lottery game conducted before the public goods experiment. As explained in the Experimental Design section, we classified subjects into three categories: risk averse, risk neutral, and risk lovers, using the results obtained from the lottery game. This is an individual-specific risk measure and allows us to control for their risk attitudes. We define dummies for risk-averse and risk-neutral subjects and include these in the regression. The variables used in the regressions are summarized in Table 4.

Since subjects participated in the seven treatments in different order, we conducted statistical tests to determine if the order in which they participated had an impact on contribution levels in the different treatments. Whether subjects participated in Treatment 2 or 3 first did not lead to significant differences in contributions in the two treatments (probability value = .98). Similarly, the order for Treatments 4 and 5 (probability value = .85) and Treatments 6 and 7 (probability value = .95) did not have any significant behavioral implications.

Tables 5, 6, and 7 report results from pairwise regressions, and Table 8 presents results from pooled regressions where we include the data for all the treatments and introduce treatment dummies (T2–T7), with Treatment 1 as the baseline dummy, to examine how subjects behave in the different treatments. In Table 8, we show results from different model

17. We also conducted random effects tobit regressions in which the dependent variable is censored and takes values between 0 and 1. The results are similar to the Generalized Least Squares (GLS) estimates.

18. The explanatory variables used to understand behavior vary depending on the treatments being considered. For example, lprobincr is used only when comparing Treatments 6 and 7 and EM is used only when comparing Treatments 2 and 4 and Treatments 3 and 5 in regressions where we separate the data for subjects who want to know the probability they face.

TABLE 5

Random Effects GLS Estimates for the Individual Contribution Levels: Pairwise Regressions, Testing Hypotheses 1 and 2

	Treatments 2 and 3	Treatments 4 and 5	Treatments 6 and 7
Invperiod	0.4607*** (0.0518)	0.2654*** (0.0899)	0.3204*** (0.0796)
Treatment	−0.3883*** (0.0518)	−0.3412*** (0.0469)	0.0307 (0.0507)
Loutcome	0.0425*** (0.0213)	0.0053 (0.0293)	−0.0118 (0.0201)
Ldeviation	0.0047*** (0.0007)	0.0038*** (0.0009)	0.0044*** (0.0010)
Sex	−0.0002 (0.0439)	0.0277 (0.0368)	0.0246 (0.0297)
Age	−0.0001 (0.0031)	0.0029 (0.0025)	0.0102*** (0.0030)
Risk_averse	0.0166 (0.0462)	0.0166 (0.0470)	−0.0347 (0.0352)
Risk_neutral	0.0432 (0.0659)	0.1506*** (0.0552)	0.0724 (0.0661)
Lprobincr			0.0024 (0.0016)
Constant	0.3944*** (0.1105)	0.2868*** (0.0950)	−0.1629** (0.0621)
Observations	1,400	1,400	1,400
Sigma_u	0.1079	0.0883	0.0982
Sigma_e	0.2308	0.2825	0.1828
Rho	0.1792	0.0890	0.2240

Notes: Numbers in parentheses are robust standard errors.
**Significant at the 5% level; ***significant at the 1% level.

specifications, which are reported to indicate the robustness of the results.[19] The pooled results are consistent with the pairwise regressions reported. Table 9 summarizes the results for all the hypotheses.

Results relating to the first two hypotheses are reported in Tables 3 and 5. These hypotheses examine the impact of the location of the risk and uncertainty, that is, whether it is associated with the private or the public good. Both the statistical tests (*t*-tests and the Wilcoxon tests) and the regression results provide strong support for Hypothesis 1, that is, in payoff-equivalent situations, subjects contribute less to the public good when risk or uncertainty is associated with the public good than when risk or uncertainty is associated with the private good. The dummies for Treatment 3 (in the regressions comparing Treatments 2 and 3) and Treatment 5 (in the regressions comparing Treatments 4 and 5) are negative and statistically significant. Applying a threshold uncertainty design, Wit and Wilke (1998), Au (2004), and Gustafsson, Biel, and Gärling (2000) find that as uncertainty thresholds increase, contribution to the public good declines.[20] Our results from Hypothesis 1 are consistent with these previous studies and show that in a continuous public good setup, increased environmental variability reduces cooperation.

Hypothesis 2, that is, in payoff-equivalent situations, subjects contribute more to the public good when they face endogenous probability of provision of the public good (Treatment 7) than when they face endogenous probability of provision of the private good (Treatment 6), is not supported by the data. The statistical tests and the regressions show that the average contributions to the public good are higher in Treatment 7 than in Treatment 6, although the difference is not statistically significant. This is the only treatment in which even though subjects face uncertainty in the public good, they contribute about the same as they do when they face uncertainty in the private good (Treatment 6). This is

19. Model 1 in Table 8 includes only time and the treatment dummies, Model 2 adds lagged variables (loutcome, ldeviation), and Model 3 adds the individual-specific characteristics. The results are robust to these additions. We also conducted these robustness tests for the pairwise regressions reported in Tables 5, 6, and 7 but do not present them in the paper to conserve space. Again, the results are robust to different model specifications.

20. Suleiman (1997) shows in a theoretical framework that threshold uncertainty in public good games can lead to lower contributions. McBride (2006) predicts that the effect of uncertainty in these games depends on the value of the public good, with a higher value leading to an increase in contributions. However, these predictions receive weak support in the experiments (McBride 2007).

TABLE 6

Random Effects GLS Estimates for the Individual Contribution Levels: Pairwise Regressions, Testing Hypotheses 3 and 4

	Treatments 2 and 4		Treatments 3 and 5	
	GLS[a]	GLS[b]	GLS[a]	GLS[b]
Invperiod	0.5129*** (0.0731)	0.4568*** (0.1038)	0.2212*** (0.0784)	0.2890*** (0.0840)
Treatment	−0.1134* (0.0581)	−0.1633*** (0.0456)	−0.0411 (0.0279)	0.1062*** (0.0333)
Loutcome	0.0786** (0.0371)	0.0618** (0.0292)	−0.0096 (0.0107)	−0.0209 (0.0143)
Ldeviation	0.0032*** (0.0006)	0.0026*** (0.0007)	0.0035*** (0.0006)	0.0035*** (0.0005)
Sex	0.0135 (0.0783)	0.0153 (0.0889)	0.0121 (0.0363)	−0.0064 (0.0407)
Age	0.0030 (0.0080)	0.0002 (0.0073)	0.0003 (0.0016)	0.0010 (0.0019)
Risk_averse	0.0662 (0.0699)	0.0683 (0.0809)	−0.0269 (0.0408)	−0.0287 (0.0448)
Risk_neutral	0.0748 (0.1166)	0.1004 (0.1059)	0.0403 (0.0287)	0.0499 (0.0388)
EM		0.7325*** (0.0818)		1.7056*** (0.2745)
Constant	0.2512 (0.1926)	0.1080 (0.1797)	0.0915 (0.0727)	−0.4336*** (0.0766)
Observations	1,097	986	1,329	771
Sigma_u	0.1522	0.1459	0.0605	0.0693
Sigma_e	0.2530	0.2678	0.1377	0.1488
Rho	0.2658	0.2290	0.1615	0.1781

Notes: Numbers in parentheses are robust standard errors.

[a]In Treatments 4 and 5, we only include subjects who do not pay to know the probability.

[b]In Treatments 4 and 5, we only include subjects who pay to know the probability.

*Significant at the 10% level; **Significant at the 5% level; ***Significant at the 1% level.

driven by the impact of the endogenous probability of provision on both the private and the public goods. This impact is stronger on the private good, that is, increasing the emphasis on the private good (allowing the probability of provision to increase with contributions), crowds out the incentive to contribute to the public good. The behavior of subjects in Treatment 7 suggests that even though they contribute a little more, they do not fully realize the incentives to contribute when these incentives are public in nature and targeted toward total group contributions.[21] We test the difference in contribution levels more specifically later using Hypothesis 5 (comparing across Treatments 4 and 6) and Hypothesis 6 (comparing across Treatments 5 and 7).

Results associated with Hypotheses 3–6, which examine whether individuals behave differently when they face risk, uncertainty, or can change the probability they face, are presented in Tables 3, 6, and 7. Hypothesis 3, that is, in payoff-equivalent situations, subjects' contribution level to the public good is higher when facing uncertainty than when facing risk, both associated with the private good, is not supported by the tests and the regression results (dummy for Treatment 4 is negative and significant).[22] The negative sign indicates positive optimism or wishful thinking on the part of the subjects. In Treatment 2, subjects know that they have only 50% chance of obtaining the private good. In Treatment 4, subjects face uncertainty on the private good, and in this case, they could put more weight on the high end of the distribution. Subjects seem to interpret uncertainty relating to the private good to justify their lower contribution to the public good. Wishful thinking is discussed in the context of uncertainty in common pool resource experiments (Budescu, Rapoport, and Suleiman 1995a, 1995b; Gustafsson, Biel, and Gärling

21. In the literature, there is little direct comparison of public versus private contributions with endogenous treatments, so it is difficult to compare our results with previous findings. Some work in this area suggests that incentives do not significantly alter contribution levels when benefits are accrued by all group members and are public in nature (Dickinson 1998) but they do when the incentives of contributing are private in nature (Dickinson 2001).

22. For this comparison, we exclude subjects who choose to pay to know the probability (converting uncertainty into risk). Hence, we compare the risk treatment with the data on pure uncertainty from Treatment 4. The dummy for Treatment 4 would be positive and significant if the data supported Hypothesis 3.

TABLE 7
Random Effects GLS Estimates for the Individual Contribution Levels: Pairwise Regressions, Testing Hypotheses 5 and 6

	Treatments 4[a] and 6	Treatments 5[a] and 7
Invperiod	0.3502*** (0.0915)	0.2063*** (0.0661)
Treatment	−0.2833*** (0.0592)	0.0549* (0.0285)
Loutcome	0.0146 (0.0269)	−0.0038 (0.0053)
Ldeviation	0.0040*** (0.0007)	0.0026*** (0.0010)
Sex	0.0500 (0.0510)	0.0024 (0.0476)
Age	0.0106** (0.0041)	0.0035** (0.0017)
Risk_averse	0.0374 (0.0434)	−0.0508 (0.0550)
Risk_neutral	0.0762 (0.0685)	0.1370 (0.0911)
Constant	0.0575 (0.1050)	0.0024 (0.0971)
Observations	1,097	1,329
Sigma_u	0.1018	0.1056
Sigma_e	0.2063	0.1522
Rho	0.1957	0.3251

Notes: Numbers in parentheses are robust standard errors.
[a]In Treatments 4 and 5, we only include subjects who do not pay to know the probabilities.
*Significant at the 10% level; **significant at the 5% level; ***significant at the 1% level.

1999; Rapoport et al. 1992; Suleiman, Budescu, and Rapoport 1960) with findings that with an increased level of uncertainty regarding the size of the resource, the resource request from the common pool resource increased.[23] It seems that when subjects estimate the probabilities, they have the tendency to over- or underestimate it depending on what they hope for.[24]

Hypothesis 4, that is, in payoff-equivalent situations, subjects' contribution level to the public good is lower when facing uncertainty than when facing risk, both associated with the public good, is not supported. The dummy for Treatment 5 is negative but not significant, indicating that the contributions to the public good are not significantly different when subjects face uncertainty as compared to risk. The negative sign suggests that subjects are fearful of wasting ones' contribution due to pessimistic beliefs about the environmental uncertainty relating to the public good.[25] This argument is complementary to the wishful thinking relating to the private good. Hence, whether subjects have optimistic or pessimistic beliefs about provision could depend on the location of uncertainty: on the private or the public good. Separating the data for subjects who want to know the probability they face in Treatment 5, we find that for subjects who choose to convert uncertainty into risk, the contributions are higher in Treatment 5.[26]

23. It is difficult to directly compare this literature here, as in our paper, we find indications of optimism when the uncertainty relates to the private good, so it is in the subject's private interest to put more weight on the high end of the distribution. In the literature cited, optimism is discussed in situations when there is uncertainty on a common pool resource and not a private good. The aim of providing these references is to connect the concept of optimism to the literature; however, we should keep in mind that the location of the uncertainty matters.

24. In Table 6, we also present results from regressions where we include only individuals who pay to know the probability in Treatment 4 and compare to behavior in Treatment 2. The dummy for Treatment 4 remains negative and significant for this subsample. It seems as if subjects who pay to know the probability facing the private good have already made a commitment toward the private good and so continue to contribute in that direction.

25. Pessimism has been discussed in the context of public good threshold experiments in Au (2004) and Wit and Wilke (1998), and they find that group members justify their noncooperation by pessimism. Bac (1996) labels a player as pessimistic if the player thinks that he or she is the only player who desires the public good and hence expects that others will not contribute.

26. Note the difference in results for private and public goods here. In Treatment 4 when we only include subjects who have paid to know the probability, we find that subjects contribute significantly less to the public good in Treatment 4. However, subjects contribute significantly more to the public good in Treatment 5 (Column 5 of Table 6). The result in the case of the private good could potentially suggest that there is an order effect in the data (as contributions decline). However, in the case of the public good, the contributions actually increase, not decline. So order effects are perhaps not a convincing explanation for our results. Instead, we suggest that there are other reasons (such as commitment toward the good in question) for this result.

TABLE 8

Random Effects GLS Estimates for the Individual Contribution Levels: Pooled Regressions

Variables	Model 1	Model 2	Model 3
Invperiod	0.226*** (0.0374)	0.369*** (0.0540)	0.369*** (0.030)
T2	0.267*** (0.0550)	0.279*** (0.0573)	0.279*** (0.014)
T3	−0.122*** (0.0264)	−0.108*** (0.0275)	−0.108*** (0.014)
T4	0.204*** (0.0599)	0.211*** (0.0639)	0.211*** (0.014)
T5	−0.143*** (0.0428)	−0.131*** (0.0459)	−0.131*** (0.014)
T6	−0.099* (0.0530)	−0.093* (0.0551)	−0.093*** (0.013)
T7	−0.086*** (0.0396)	−0.081*** (0.0394)	−0.081*** (0.014)
Loutcome		0.017 (0.0147)	0.016** (0.008)
Ldeviation		0.004*** (0.0006)	0.004*** (0.000)
Sex			0.006 (0.029)
Age			0.004 (0.003)
Risk_averse			−0.008 (0.040)
Risk_neutral			0.092 (0.059)
Constant	0.196*** (0.0570)	0.156*** (0.0586)	0.057*** (0.072)
Sigma_u	0.1486	0.0973	0.0980
Sigma_e	0.2685	0.2467	0.2467
Rho	0.2344	0.1347	0.1365

Notes: For the most inclusive model (Model 3): pairwise tests within pooled regressions: Test T2 = T3, $\chi^2(1)$ = 491.76***; Test T4 = T5, $\chi^2(1)$ = 444.49***; Test T6 = T7, $\chi^2(1)$ = 1.84; Test T2 = T4, $\chi^2(1)$ = 14.61***; Test T3 = T5, $\chi^2(1)$ = 10.08***; Test T2 = T6, $\chi^2(1)$ = 543.66***; Test T4 = T6, $\chi^2(1)$ = 392.08***; Test T3 = T7, $\chi^2(1)$ = 0.39; Test T5 = T7, $\chi^2(1)$ = 6.33***. Numbers in parentheses are robust standard errors.

*Significant at the 10% level; **significant at the 5% level; ***significant at the 1% level.

These subjects paid a fee to know the probability, so they are perhaps the subjects who would be more aware about contributions to the public good and are already committed toward the public good. This may explain the statistically significant increase in contribution levels relative to Treatment 3.

We obtain strong support for Hypothesis 5. Both the statistical tests (Table 3) and the regressions (Table 7) show that in payoff-equivalent situations, when probabilities are endogenous, that is, subjects can increase the probability of provision of the private good (Treatment 6), subjects' contribution to the public good will be lower than that when facing uncertainty on the private good (Treatment 4). In Treatment 6, subjects can increase the probability of return they face if they invest more in the private good. This incentive crowds out the contribution to the public good.

Hypothesis 6, that is, in payoff-equivalent situations, when probabilities are endogenous (Treatment 7), subjects' contribution to the public good will be higher than when facing uncertainty associated with the public good (Treatment 5), gets some support from the data (Tables 3 and 7). The point estimates show that the contributions are higher in Treatment 7 and statistically significant in the regressions.[27]

C. Other Regressors

The results from the multivariate regressions show that the time variable is positive and highly significant, implying that contributions to the public good decay over time.[28] The variable which measures whether the good was provided in the previous period (loutcome) is positive and significant in some regressions, indicating that a provision of the good in the previous period increases the contribution to the public good in the current period.[29] Deviation (ldeviation) which captures the

27. Dickinson (1998) shows that incentives do not significantly affect contribution levels in public goods games.

28. The specification of this variable captures the nonlinearity in the data. Figure 2 indicates that contributions have a nonlinear pattern with a sharper drop in the early periods and a more gradual decline in the later period for most treatments; hence, this particular specification (1/time) was used to highlight the time element of the contributions.

29. This variable, however, does not have a consistent sign across regressions.

TABLE 9
Summary of Results for the Hypotheses

Hypotheses	Individual Contributions to the Public Good	Comparison	Sign and Significance	t-Test, Wilcoxon Test and Regression Results
H1	$g_2 > g_3$ $g_4 > g_5$	Private vs. public risk Private vs. public uncertainty	Correct sign; statistically significant	Supported
H2	$g_7 > g_6$	Incentives on private vs. public	Correct sign; not significant	Not supported
H3	$g_4 > g_2$	Uncertainty vs. risk on private	Incorrect sign; statistically significant	Not supported
H4	$g_3 > g_5$	Uncertainty vs. risk on public	Correct sign; not significant	Not supported
H5	$g_4 > g_6$	Private uncertainty vs. incentives	Correct sign; statistically significant	Supported
H6	$g_7 > g_5$	Uncertainty vs. incentives on public	Correct sign; statistically significant	Supported[a]

[a]The regressions support this hypothesis. The point estimates for the statistical tests provide support; however, the difference is not statistically significant.

difference between the individual and group contribution in the previous period is very substantial in explaining contributions, with a higher level of deviation leading to significantly higher contributions in this period. This result is intuitively appealing since it suggests that people use feedback from other members of the group when deciding on future contribution levels. The feedback is used by the subjects as information about the strategic uncertainty, and such information seems to guide subjects in their decision. These results are consistent with Dickinson (1998) who also finds that the deviation of a person's contributions in the previous round from the group affects individual contributions. Furthermore, they confirm Festinger's (1954) conjecture that the more uncertain people are in their knowledge about the task, the more likely it is that they will try to reduce their uncertainty via information on how others respond. The variable lprobincr, which captures the impact of the lag of the increase in probability in the provision of the good, shows that as the probability in the previous period increases, the contributions increase in this period. The variable EM is positive and highly significant in explaining contributions.[30] This result corroborates previous findings, for example, Dickinson (1998), and suggests that even when the dominant strategy of Nash equilibrium of zero contribution is maintained, higher marginal incentives, *per se*, significantly increased contribution levels. Individual-level characteristics like age, gender, and risk behavior of the subject do not provide a consistent pattern of behavior.[31]

IV. CONCLUSIONS

We use laboratory experiments and a standard public goods game to examine the decisions of subjects when facing both strategic and environmental risks and uncertainties. We keep the strategic uncertainty constant throughout the seven treatments examined in this paper and introduce different forms of risk and uncertainty regarding the provision of private and public goods. We find that even in a payoff-equivalent situation, environmental risk and uncertainty associated with both private and public problems act as a significant factor when making decisions, with subjects moving away from the environmental uncertainty or risk. The location of the

30. Several studies (e.g., Isaac, Walker, and Thomas 1987; Kim and Walker 1984) have investigated the importance of marginal incentives (EM) on contribution levels. In the results reported in the paper, we include EM as an explanatory variable in regressions where we only have data on individuals who want to know the probability they face.

31. These individual-specific variables are sometimes significant in the unreported tobit regressions. Risk-averse subjects (74% of our sample) contribute less, and risk-neutral subjects (10% of our sample) contribute significantly more than risk-loving subjects (16% of our sample). It seems that risk-neutral subjects see the underlying payoff equivalence between the treatments, and they also see that regardless of the risk associated with the returns, the underlying problem is a public goods game with Pareto-efficient contribution levels of 100%. Risk-averse subjects, however, divert away from not only the external risk and uncertainty but also the strategic uncertainty associated with contributing to the public goods.

environmental risk or uncertainty matters: subjects move away from the environmental uncertainty in favor of strategic uncertainty when that uncertainty relates to the private good. When environmental uncertainty relates to the public good, subjects face strategic and environmental uncertainty on the same good, leading to a collapse of the contributions.

Our results could be used to understand some aspects of environmental policy making. For example, the climate change negotiations that have been taking place since 1992 involve a large degree of scientific and economic uncertainty. Our results suggest that it was not the strategic uncertainty, per se, but the presence of the environmental uncertainty that has led to the long period of negotiations on climate change.

In the incentives (endogenous probability of provision) treatments, we find that subjects contribute more than they do in the other treatments, that is, an opportunity to reduce the uncertainty relating to the public good increases cooperation (Hypothesis 6). This relates to the role of innovation in determining outcomes. The reward for research and development and innovation is captured here by the increased probability of obtaining the desired outcome in the endogenous probability treatment.

Subjects are willing to pay a fee to reveal the uncertainty in treatments where the uncertainty relates to a private good rather than a public good. Interestingly, we find that some subjects are willing to pay to avoid uncertainty (unknown probabilities) in favor of risk (known probabilities) even if such probabilities play little role in their decision making. We also find that information on fellow group members' cooperation seems to serve as a norm of how to deal with uncertainty. This indicates that transparency of international negotiations could serve as an important factor in the success of the negotiations, since participants use the contributions of other group members as a yardstick when facing strategic and environmental uncertainty.

REFERENCES

Au, W. "Criticality and Environmental Uncertainty in Step-Level Public Goods Dilemmas." *Group Dynamics: Theory, Research, and Practice*, 8, 2004, 40–61.

Au, W., and M. Ngai. "Effects of Group Size Uncertainty and Protocol of Play in a Common Pool Resource Dilemma." *Group Processes and Intergroup Relations*, 6, 2003, 265–83.

Bac, M. "Incomplete Information and Incentives to Free Ride." *Social Choice Welfare*, 13, 1996, 419–32.

Bagnoli, M., and B. Lipman. "Provision of Public Goods: Fully Implementing the Core through Private Provision." *Review of Economic Studies*, 4, 1989, 583–602.

Bagnoli, M., and M. McKee. "Voluntary Contribution Games: Efficient Private Provision of Public Goods." *Economic Inquiry*, 29, 1991, 351–67.

Brown, P. M., and S. Stewart. "Avoiding Severe Environmental Consequences: Evidence on the Role of Loss Avoidance and Risk Attitudes." *Journal of Economic Behavior and Organization*, 38, 1999, 179–98.

Buchanan, J. M. *The Demand and Supply of Public Goods*. Chicago: Rand McNally, 1968.

Budescu, D. V., A. Rapoport, and R. Suleiman. "Common Pool Resource Dilemmas under Uncertainty: Qualitative Tests of Equilibrium Solutions." *Games and Economic Behavior*, 10, 1995a, 171–201.

———. "Positional Order Effects in Social Dilemma Games with Uncertain Resources." *Organizational Behavior and Human Decision Processes*, 61, 1995b, 225–38.

Camerer, C., G. Loewenstein, and D. Prelec. "Neuroeconomics: How Neuroscience Can Inform Economics." *Journal of Economic Literature*, 43, 2005, 9–64.

Dickinson, D. L. "The Voluntary Contribution Mechanism with Uncertain Group Payoffs." *Journal of Economic Behavior and Organization*, 35, 1998, 517–33.

———. "The Carrot vs the Stick in Work Team Motivation." *Experimental Economics*, 4, 2001, 107–24.

Festinger, L. "A Theory of Social Comparison Processes." *Human Relations*, 7, 1954, 117–40.

Fischbacher, U. "z-Tree: Zurich Toolbox for Ready-Made Economic Experiments." *Experimental Economics*, 10, 2007, 171–78.

Green, J. R., and J. J. Laffont. *Incentives in Public Decision Making*. New York: North Holland, 1979.

Groves, T., and J. Ledyard. "Optimal Allocation of Public Goods: A Solution to the Free Rider Problem." *Econometrica*, 45, 1977, 783–809.

———. "The Existence of Efficient and Incentive Compatible Equilibria with Public Goods." *Econometrica*, 48, 1980, 1487–506.

Gustafsson, M., A. Biel, and T. Gärling. "Over-Harvesting of Resources of Unknown Size." *Acta Psychologica*, 103, 1999, 47–64.

———. "Egoism Bias in Social Dilemmas with Resource Uncertainty." *Group Processes and Intergroup Relations*, 3, 2000, 351–65.

Hackett, S., E. Schlager, and J. Walker. "The Role of Communication in Resolving Common Dilemmas: Experimental Evidence with Heterogeneous Appropriators." *Journal of Environmental Economics and Management*, 27, 1994, 99–126.

Hogarth, R. M. *Judgement and Choice* (2nd ed.). New York: Wiley, 1987.

Holt, C., and S. Laury. "Risk Aversion and Incentive Effects." *American Economic Review*, 92, 2002, 1644–55.

Isaac, M. R., D. Schmidtz, and J. M. Walker. "The Assurance Problem in a Laboratory Market." *Public Choice*, 62, 1989, 217–36.

Isaac, M. R., and J. M. Walker. "Group Size Effects in Public Goods Provision: The Voluntary Contributions

Mechanism." *Quarterly Journal of Economics*, February 1988, 179–200.

Isaac, M. R., J. M. Walker, and S. H. Thomas. "Divergent Evidence on Free Riding: An Experimental Examination of Possible Explanations." *Public Choice*, 43, 1984, 113–49

Isaac, M. R., J. M. Walker, and A. W. Williams. *"Group Size and Voluntary Provision of Pubic Goods: Experimental Evidence Utilizing Large Groups."* University of Arizona Working Paper No. 25, 1990.

Kim, O., and M. Walker. "The Free Rider Problem: Experimental Evidence." *Public Choice*, 43, 1984, 3–24.

Knight, F. H. *Risk, Uncertainty and Profit*. Boston: Houghton Mifflin, 1921.

Ledyard, J. O. "Public Goods: A Survey of Experimental Research," in *The Handbook of Experimental Economics*, edited by J. H. Kagel and A. E. Roth. Princeton, NJ: Princeton University Press, 1995, 111–81.

Loehman, E., F. N. Quesnel, and E. M. Babb. "Free-Rider Effects in Rent-Seeking Groups Competing for Public Goods." *Public Choice*, 86, 1996, 35–61.

Loewenstein, G. F., L. Thompson, and M. H. Bazerman. "Social Utility and Decision Making in Interpersonal Contexts." *Journal of Personality & Social Psychology*, 57, 1989, 426–41.

Lopes, L. L. "Psychology and Economics: Perspectives on Risk, Cooperation and the Marketplace." *Annual Review of Psychology*, 45, 1994, 197–227.

Marks, M., and R. Croson. "The Effect of Alternative Rebate Rules in the Provision Point Mechanism of Voluntary Contributions: An Experimental Investigation." *Journal of Public Economics*, 67, 1998, 195–220.

Marwell, G., and R. E. Ames. "Experiments on the Provision of Public Goods I: Resources, Interest, Group Size, and the Free Rider Problem." *American Journal of Sociology*, 84, 1979, 1335–60.

McBride, M. "Discrete Public Good Games under Threshold Uncertainty." *Journal of Public Economics*, 90, 2006, 1181–99.

———. "Threshold Uncertainty in Discrete Public Good Games: An Experimental Study." Working Paper, University of California, Irvine, 2007.

Messick, D. M., S. T. Allison, and C. D. Samuelson. "Framing and Communication Effects on Group Members' Responses to Environmental and Social Uncertainty," in *Applied Behavioral Economics*, Vol. II, edited by S. Maital. Brighton: Wheatsheaf, 1988, 677–700.

Messick, D. M., and M. B. Brewer. "Solving Social Dilemmas: A Review," in *Review of Personality and Social Psychology*, edited by L. Wheeler and P. Shaver. Beverly Hills, CA: Sage, 1983, 11–44.

Nemes, V. "Climate Change: From Decision Making under Uncertainty to Policy Implementation." PhD Dissertation, Department of Economics, The University of Melbourne, 2005.

Rabin, M. "Psychology and Economics." *Journal of Economic Literature*, 36, 1998, 11–46.

Rapoport, A., D. V. Budescu, R. Suleiman, and E. Weg. "Social Dilemmas with Uniformly Distributed Resources," in *Social Dilemmas: Theoretical Issues and Research Findings*, edited by W. Liebrand, D. Messick, and H. Wilke. Oxford: Pergamon Press, 1992, 43–57.

Rapoport, A., and R. Suleiman. "Equilibrium Solutions for Resource Dilemma." *Group Decision and Negotiation*, 1, 1992, 169–294.

Schelling, T. C. *The Strategy of Conflict*. Cambridge, MA: Harvard University Press, 1960.

Schmidtz, D. "Contracts and Public Goods." *Harvard Journal of Law and Public Policy*, Spring, 1987, 475–503.

Suleiman, R. "Provision of Step-Level Public Goods under Uncertainty: A Theoretical Analysis." *Rationality and Society*, 9, 1997, 163–87.

Suleiman, R., D. V. Budescu, and A. Rapoport. "Fixed Position and Property Rights in Sequential Resource Dilemmas under Uncertainty." *Acta Psychologica*, 93, 1966, 229–45.

———. "Provision of Step-Level Public Goods with Uncertain Provision Threshold and Continuous Contribution." *Group Decision and Negotiation*, 10, 2001, 253.

Suleiman, R., and A. Rapoport. "Environmental and Social Uncertainty in Single Trial Resource Dilemmas." *Acta Psychologica*, 68, 1988, 99–112.

van Dijk, E., H. Wilke, M. Wilke, and L. Metman. "What Information Do We Use in Social Dilemmas? Environmental Uncertainty and the Employment of Coordination Rules." *Journal of Experimental Social Psychology*, 35, 1999, 109–35.

Walker, J. M., and R. Gardner. "Probabilistic Destruction of Common-Pool Resources: Experimental Evidence." *Economic Journal*, 102, 1992, 1149–61.

Walker, J. M., and M. A. Halloran. "Rewards and Sanctions and the Provision of Public Goods in One-Shot Settings." *Experimental Economics*, 7, 2004, 235–47.

Wit, A., and H. Wilke. "Public Good Provision under Environmental and Social Uncertainty." *European Journal of Social Psychology*, 28, 1998, 249–56.

Zelmer, J. "Linear Public Goods Experiments: A Meta-Analysis." *Experimental Economics*, 6, 2003, 299–310.

Part VIII
Dynamic Coalition Formation

[45]

Available online at www.sciencedirect.com

ScienceDirect

ELSEVIER

Journal of Environmental Economics and Management 54 (2007) 296–310

JOURNAL OF ENVIRONMENTAL ECONOMICS AND MANAGEMENT

www.elsevier.com/locate/jeem

An infinite-horizon model of dynamic membership of international environmental agreements ☆

Santiago J. Rubio[a,*], Alistair Ulph[b]

[a]*Department of Economic Analysis and ERI-CES, University of Valencia, Spain*
[b]*Faculty of Humanities, University of Manchester, UK*

Received 11 August 2003
Available online 11 August 2007

Abstract

Much of the literature on international environmental agreements (IEAs) uses static models, although most important transboundary pollution problems involve stock pollutants. The few papers that study IEAs using models of stock pollutants do not allow for the possibility that membership of the IEA may change endogenously over time. In this paper we analyse a simple infinite-horizon version of the static model of self-enforcing IEAs, in which damage costs increase with the stock of pollution, and countries decide each period whether to join an IEA. Using a quadratic approximation of the value function of the representative country we show that there exists a steady-state stock of pollution with corresponding steady-state IEA membership, and that as the stock rises towards steady state the IEA membership falls. Moreover, we find that the greater is the cost of damage, and hence the greater are the potential gains from cooperation, the smaller is the membership of a self-enforcing IEA.

JEL classification: F02; F18; Q20

Keywords: Self-enforcing international environmental agreements; Internal and external stability conditions; Stock pollutant

1. Introduction

There is now an extensive literature on international environmental agreement (IEAs) (see [2,3,8] for excellent surveys of this literature). Yet, with a few exceptions discussed below, this literature works with simple static models of pollution despite the fact that many of the important problems (climate change, ozone depletion, acid rain) for which IEAs are relevant involve stock pollutants. Those papers that do study IEAs for stock pollutants assume that the membership of the IEA is determined once and for all at the outset of the agreement, so signatories commit themselves to an infinite horizon path of emissions. But agreements such as

☆This paper is a substantially revised version of the paper presented to VIII Coalition Theory Workshop, January 24–25, 2003, Aix-en-Provence. This version was presented at the Third World Congress of Environmental and Resource Economists, July 3–7, 2006, Kyoto.
*Corresponding author. Fax: +34 96 38 28249.
E-mail addresses: santiago.rubio@uv.es (S.J. Rubio), Alistair.Ulph@manchester.ac.uk (A. Ulph).

0095-0696/$ - see front matter
doi:10.1016/j.jeem.2007.02.004

the Kyoto protocol on climate change restrict emissions for signatories only for a relatively limited period of time, and it is not clear whether the signatories to any subsequent agreement will be the same as for the initial agreement. In this paper we introduce a simple model of a stock pollutant with an infinite number of periods, in each of which countries have to decide whether or not to join an IEA to control emissions for that period. Of course in deciding whether or not to join in any one period countries need to calculate the impact of their decisions on both the future stock of pollution and membership of future IEAs.

We now briefly summarize those papers which study the formation of IEAs with a stock pollutant. Rubio and Casino [12] use the concept of self-enforcing IEAs introduced by Carraro and Siniscalco [5] and Barrett [1] for a static pollution problem to study a stock pollutant. As in the static model, the dynamic model is analysed as a two-stage game. Countries first decide whether or not to join an IEA, and then choose their infinite-horizon paths of emissions. Emission paths are calculated by solving a differential game in either open-loop or feedback strategies assuming that the IEA signatories act to maximize their joint welfare while each non-signatory maximizes its individual welfare. The first-stage game (whether to join an IEA) uses the same stability analysis as the static model in which country assesses unilaterally whether to join or leave an IEA assuming that the membership decisions of all other countries are unaffected. So in this model the dynamics of the stock pollutant only affect emissions strategies, not IEA membership. Many of the attempts to model empirically how many countries might join an IEA to deal with climate change have the same feature that countries are assumed to make a once-for-all decision whether to join an IEA, with the dynamics of the stock pollutant affecting only emissions paths and hence present-value payoffs (see, e.g. [7]).

Petrosjan and Zaccour [11] solve a dynamic game of international environmental cooperation but assuming that all countries decide to cooperate at the initial time. Given this assumption they focus on how to allocate the burden of reducing emissions over time. Germain et al. [9] extend the cooperative framework of Chander and Tulkens [6] by showing that it is possible to devise dynamic transfers to ensure stability of the grand coalition when there is a stock pollutant. The membership of the grand coalition is thus maintained over time, although, importantly, in this case this is achieved by appropriate design of the transfers, rather than just by assuming that membership decisions are taken once-and-for-all.

Both Rubio and Casino [12] and Germain et al. [9] analyse models in which an IEA operates over the whole life of a stock pollutant. Karp and Sacheti [10] consider a two-period model of a stock pollutant but assume that an IEA will only form in one of the periods—either just in the first, because the IEA will fall apart at the end of one period, or in the second, because there may be substantial delays in forming an IEA. They then assess how the incentives to join an IEA are affected by the dynamics of the stock pollutant, as well as by differences in the extent to which the pollutant is local or global and the extent to which planners discount the future. In Karp and Sacheti membership of the IEA varies sharply over time, but in a way that is exogenously imposed.

In an earlier paper [13] we extended the model of self-enforcing IEAs found in Carraro and Siniscalco [5] and Barrett [1] to a two-period model of a stock pollutant. We studied two ways in which IEA membership might be decided. In the fixed membership model we followed Rubio and Casino and assumed that countries made a once-for-all decision at the outset whether or not to join an IEA. In the variable membership model we assumed that countries decided each period whether or not to join an IEA. We showed that in the variable membership case the number of countries would rise over time, and that for a wide range of parameter values, variable membership gave higher global welfare than fixed membership. So allowing membership to vary over time matters. In one version of the variable membership model there was an extreme case of the rising membership model in which no country would join in the first period, but some countries would join in the second period which provided an explanation for one of the patterns of membership imposed exogenously by Karp and Sacheti.

However, the two-period model has the undesirable characteristic that, even if unit damage costs are an increasing function of the stock of pollution, emissions generated in the second period are less damaging than emissions generated in the first period simply because they affect only the second period, while emissions in period one affect both periods. It was this feature of the model which drove the result that membership would rise over time.

In this paper we extend the two-period model of Rubio and Ulph [13] to an infinite-horizon model. Section 2 sets out the model and provides as benchmarks the non-cooperative and cooperative equilibria. In Sections 3 and 4 we analyse IEAs where membership varies over time. In Section 3 we set up the model of IEA formation in a dynamic setting. Using a random assignment rule, we show how an implicit equation for the value function of the representative country can be obtained, and how this function can be approximated using a quadratic function. In Section 4 a numerical analysis is developed. Our numerical results show that there exists a steady-state stock of pollution, which lies between the cooperative and non-cooperative steady-states, with a corresponding steady-state IEA membership, and that as the stock rises towards steady state the membership of the IEA falls. Moreover, they show that the gains from cooperation that can be achieved through a dynamic self-enforcing IEA and the corresponding membership decrease as the environmental damages rise for a given stock. Both results can be summarized by saying that the membership and the gains from cooperation decrease as the environmental damages rise.

2. The model

There are an infinite number of time periods, indexed by $t = 0, 1, \ldots,$ and N identical countries, $i = 1, \ldots, N$.[1] Ignoring subscripts for the moment, in any period, each country has to choose a level of emissions q. We assume that there is an upper bound on emissions, which we normalize to 1, so $q \in [0, 1]$. Total emissions by all other countries is denoted by Q, so total emissions are $q + Q$. Suppose that at the start of the period the cumulative stock of emissions is z. Then the cumulative stock of emissions at the start of the next period is $\rho z + q + Q$ where $\rho \in (0, 1)$ depends on the *natural decay factor* per period. We denote the initial stock of emission by z_0 and we assume without loss of generality that $z_0 = 0$.[2]

If a country pollutes it derives a unit of benefit per unit of emissions, which we shall normalize to 1 and assume constant over time. The stock of pollution at the start of the period generates costs in that period for each country which we assume to be a simple quadratic: γz^2. Thus the flow of net benefits to a country in the period is given by $\pi(q, z) = q - \gamma z^2$. Finally, the *discount factor* per period is δ also between zero and one.

2.1. The fully non-cooperative equilibrium

We assume that all countries decide their emissions simultaneously and in a non-cooperative way. Thus the representative country takes Q_t as given and chooses q_t in each period in order to

$$\max_{\{q_t\}_0^\infty} \sum_{t=0}^{\infty} \delta^t (q_t - \gamma z_t^2) \quad \text{s.t. } z_{t+1} = \rho z_t + Q_t + q_t, \ \ z_0 = 0 \text{ and } q_t \in [0, 1]. \tag{1}$$

We solve this problem by dynamic programming. Let $U(z_t)$ stands for the present value of current and all future net benefits to a country when the stock of emissions at the start of the period t is z_t and each country selects its optimal emissions in each period. For this problem the optimal emissions must satisfy the Bellman equation:

$$U(z_t) = \max_{q_t \in (0,1)} \{\pi(q_t, z_t) + \delta U(z_{t+1})\} = \max_{q_t \in (0,1)} \{q_t - \gamma z_t^2 + \delta U(\rho z_t + Q_t + q_t)\}.$$

It is easy to show that there exists a critical value for γ given by

$$\bar{\gamma} = \frac{(1 - \delta\rho)(1 - \rho)}{2\delta N}, \tag{2}$$

such that if $\gamma \leqslant \bar{\gamma}$ the countries select the maximum level of emissions in each period. The following result summarizes for this condition the solution to the non-cooperative difference game defined by (1).[3]

---

[1]A referee criticized the assumption of identical countries, noting that the IEA literature has moved on considerably from the early models made where a similar assumption were made. We accept this point, but emphasize the point made in the introduction: there are no papers which model the effect of the stock of pollution on membership dynamics and we believe it is a legitimate research strategy to start the analysis of this extension within the context of some of the simplified earlier models.

[2]For $z_0 > 0$ the same qualitative results are obtained but the presentation becomes much more cumbersome. In the Conclusions we comment on what would occur in this case.

[3]The details of the proofs of all results in this section are provided in a separate Appendix A that is available through JEEM's on-line archive of supplementary material, which can be accessed at http://www.are.org/journal/index.html.

Result 1. *If γ is less than or equal to $\bar{\gamma}$ the optimal emissions for each country and each period are equal to 1 and total emissions equal to N. The stock converges asymptotically to steady state $\hat{z} = N/(1-\rho)$ and the present value of net benefits for each country is given by*

$$U(0) = \frac{1}{1-\delta} - \frac{\delta\gamma N^2(1+\delta\rho)}{(1-\delta\rho)(1-\delta\rho^2)(1-\delta)}.$$

Sketch of the proof. First we write the Bellman equation as if the solution for optimization problem (1) was given by $q_t = 1$ for all t and all countries. Then we guess that U is a quadratic function and calculate its parameters by equating coefficients. We then use the Kuhn–Tucker conditions to obtain the constraint that the parameters must satisfy in order to get as the solution for problem (1) the corner solution $q_t = 1$ for all t. This constraint depends inversely on the stock but as the stock is increasing and converges to steady state $\hat{z} = N/(1-\rho)$, when total emissions are equal to N, it is possible to find an upper bound on γ defined by (2) that applies for all periods. Finally, for $z_0 = 0$ the quadratic value function yields the expression reported in Result 1 for the present value of net benefits. □

2.2. The fully cooperative equilibrium

Now we suppose that all countries cooperate and coordinate their emissions in order to maximize the present value of the aggregate net benefits:

$$\max_{\{q_{1t},\ldots,q_{Nt}\}_0^\infty} \sum_{t=0}^{\infty} \delta^t \left(\sum_{i=1}^{N} q_{it} - N\gamma z_t^2 \right) \tag{3}$$

$$\text{s.t.} \quad z_{t+1} = \rho z_t + \sum_{i=1}^{N} q_{it}, \quad z_0 = 0 \text{ and } q_{it} \in [0,1] \text{ for all } i. \tag{4}$$

We also solve this problem by dynamic programming. Let $V(z_t)$ stands for the present value of current and all future *aggregate* net benefits when the stock of emissions at the start of the period t is z_t and in each period the emissions of each country is optimally selected. Then the optimal emissions must satisfy the Bellman equation:

$$\begin{aligned} V(z_t) &= \max_{\{q_{1t},\ldots,q_{Nt}\}} \left\{ \sum_{i=1}^{N} q_{it} - N\gamma z_t^2 + \delta V(z_{t+1}) \right\} \\ &= \max_{\{q_{1t},\ldots,q_{Nt}\}} \left\{ \sum_{i=1}^{N} q_{it} - N\gamma z_t^2 + \delta V\left(\rho z_t + \sum_{i=1}^{N} q_{it} \right) \right\} \end{aligned}$$

subject to $q_{it} \in [0,1]$ for all i.

It is easy to show that there exists a critical value for γ given by

$$\underline{\gamma} = \frac{1-\rho\delta}{2\delta N^2}, \tag{5}$$

such that if $\gamma > \underline{\gamma}$ the countries select a level of emissions between zero and one in the initial period that leads to the steady state in one period. The following result summarizes for this condition the solution to the optimization dynamic problem defined by (3) and (4).

Result 2. *If γ is greater than $\underline{\gamma}$ the optimal emissions for each country in the initial period are equal to $\underline{\gamma}/\gamma < 1$ and the steady state, $\hat{z} = N\underline{\gamma}/\gamma$ and $\hat{q} = (1-\rho)\underline{\gamma}/\gamma$, is reached in one period. The present value of net benefits for a representative country is given by*

$$\frac{V(0)}{N} = \frac{(1-\delta\rho)^2}{4\delta\gamma N^2(1-\delta)}.$$

Sketch of the proof. First we obtain from the first order condition for an interior solution the optimal policy assuming that the value function is quadratic. Then, after substituting the optimal policy in the Bellman equation, we calculate the parameters of the value function by equating coefficients so that finally the optimal policy can be rewritten in terms of the parameters of the model. For this linear optimal policy we find that the steady-state is reached in one period and that emissions are strictly positive independently of the value of γ, with the initial emissions greater than the steady-state emissions. Finally we impose the condition that the initial emissions must be lower than unity and we obtain the lower bound on γ defined by (5). To obtain the expression reported in Result 2 for the present value of net benefits we use the quadratic value function for $z_0 = 0$. □

It is obvious that Results 1 and 2 are compatible only if $\bar{\gamma}$ is bigger than $\underline{\gamma}$ so that we have to check if this is the case. From (2) and (5) it is easy to conclude that both results are compatible for $N > 1/(1-\rho)$.[4] We focus on this case from now on.

The comparison between the fully non-cooperative equilibrium and the fully cooperative equilibrium is given in the following corollary.

Corollary 1. *In each period emissions are higher in the non-cooperative equilibrium, steady-state stock is lower in the cooperative equilibrium than in the non-cooperative equilibrium and the discounted present value of net benefits per country is higher in the cooperative equilibrium than in the non-cooperative equilibrium.*

Sketch of the proof. The comparison of emissions is straightforward. The comparison between the steady-state stocks depends on γ, but it is easy to check that if γ is greater than $\underline{\gamma}$ (as required by Result 2), the steady-state stock in the cooperative equilibrium is less than the steady-state stock in the non-cooperative equilibrium. Finally, according to Results 1 and 2 the sign of the difference between the present value of net benefits for cooperative and non-cooperative solution depends on a strictly convex quadratic function of γ that is positive for all γ. □

Finally, we would like to highlight that for this model the dynamics of the two regimes is drastically different. In the non-cooperative regime, the steady state is approached asymptotically, whereas in the cooperative regime the steady state is reached in one period. Thus, cooperation in a dynamic setting leads to a significantly different time path of variables in comparison with the time paths obtained from the non-cooperative regime.

3. IEA with dynamic membership

We now consider the formation of an infinite sequence of IEAs, in which, in each period, countries are free to join or leave the agreement. This model is designed to capture the fact that countries can only make commitments for periods which are short relative to the infinite life of the stock of the pollutant. The model of IEA formation in each period is a dynamic version of the model of self-enforcing IEAs introduced in the work of Carraro and Siniscalco [5] and Barrett [1]. In each period, given the initial stock, there is a two-stage game. In the first stage countries decide whether or not to join an IEA. In the second-stage emission game, non-signatory countries choose emissions in a non-cooperative fashion similar to Section 2.1, while signatory countries act in a cooperative fashion similar to 2.2, choosing emissions to maximize the total net benefits of all signatory countries, taking as given the behaviour of non-signatories.[5] In the first-stage membership game we look for a stable IEA in which no signatory country would wish to switch to be a non-signatory (internal stability), and no non-signatory would wish to switch to being a signatory (external stability).

However, since we have assumed that all countries are identical, all we can determine is the number of signatories of the self-enforcing IEA in each period as a function of the stock of the pollutant. We cannot say which countries become signatories in any period. To make progress, we assume that in each period there is a random process for determining which countries become signatories, such that the probability of any country being a signatory in that period is simply the membership of the stable IEA in that period divided by the total number of countries. This probability of being a signatory is clearly the same for all countries, and is independent of whether a country was a signatory or non-signatory in previous periods. Since the membership

[4]Notice that this is not a very restrictive condition if the natural decay rate is not very small. For example, if the natural decay rate is 0.05, ρ is equal to 0.95 and the number of countries has to be greater than 20.

[5]This constitutes what Chander and Tulkens [6] call a partial agreement Nash equilibrium (PANE) of the emission game.

can vary over time, obviously the probability of being a signatory in any period varies over time. So each country has the same *expected* present value of current and future net benefits, which will depend on the stock of the pollutant at the start of next period.

We therefore define $W(z_t)$ as the *expected* present value of current and future net benefits of the representative country when: (a) the stock of emissions at the start of the current period is z_t; (b) in each period the number of signatories of a self-enforcing IEA is given by the stability conditions defined above; (c) in each period emissions are given by the PANE of the emissions game; and (d) in each period there is a random process for determining which countries act as signatories, with each country having an equal probability of being a signatory.

In each period we solve first the two-stage game by backward induction. Then given the number of signatories and the emissions for signatories and non-signatories, all of which depend on z_t, we obtain the value functions for signatories and non-signatories. Finally, assuming that at the start of each period each country has the same probability of being proposed as a signatory it is possible to derive an equation which implicitly defines the expected value function $W(z_t)$ as a weighted sum of the value functions for signatories and non-signatories.

Unfortunately, there does not exist a quadratic value function for $W(z_t)$ that satisfies the implicit equation we have obtained. However, we have devised an algorithm that allows us to approximate the value function by a quadratic function of the stock and derive some numerical results. We begin the presentation of this procedure by obtaining the solution for the second-stage emissions game.

3.1. Second-stage emissions game

Suppose that in a period with initial stock of the pollutant z_t, there are n signatories to an IEA. Then total emissions can be written as $\sum_{j=1}^{N-n} q_{jt}^f + \sum_{k=1}^{n} q_{kt}^s$, where f stands for free-rider or non-signatory countries and s stands for signatory countries. In each period, given n, signatories and non-signatories choose their emissions optimally so that the following Bellman equation must be satisfied for the non-signatories:

$$W_j^f(z_t) = \max_{q_{jt}^f \in [0,1]} \left\{ q_{jt}^f - \gamma z_t^2 + \delta W\left(\rho z_t + \sum_{l=1}^{N-n} q_{lt}^f + \sum_{k=1}^{n} q_{kt}^s \right) \right\}, \quad j = 1, \ldots, N-n. \tag{6}$$

At this point we assume that the dominant strategy for non-signatories is to select the maximum emissions for all z_t and all $n<N$.[6] Then, according to Kuhn–Tucker conditions, the following condition must be satisfied for all z_t and all $n<N$:

$$1 + \delta W'\left(\rho z_t + N - n + \sum_{k=1}^{n} q_{kt}^s \right) \geqslant 0. \tag{7}$$

Given that this condition holds, the Bellman equation for signatories is written as follows:

$$V(z_t) = \max_{\{q_{1t}^s, \ldots, q_{nt}^s\}} \left\{ \sum_{k=1}^{n} q_{kt}^s - n\gamma z_t^2 + \delta n W\left(\rho z_t + N - n + \sum_{k=1}^{n} q_{kt}^s \right) \right\} \tag{8}$$

subject to $q_{kt}^s \in [0, 1]$ for $k = 1, \ldots, n$. We write nW in the Bellman equation because W stands for the expected value function of an individual country whereas V stands for the value function of all signatory countries. Thus to define this equation properly it is necessary to multiply W by n. The Kuhn–Tucker conditions for the maximization problem in (8) are

$$1 + n\delta W'(\rho z_t + N - n) \leqslant 0, \quad q_t^s = 0, \tag{9}$$

[6] In Section 3.3 we derive the conditions on the parameters of the quadratic approximation that guarantee that this assumption holds in equilibrium, see Result 3. In Section 4 we run our numerical analysis taking these conditions into account. Our results show that for all the combinations of the model parameter values studied in this paper, there always exists a quadratic approximation that satisfies these conditions so that the dominant strategy for non-signatories in equilibrium is to select the maximum of emissions for all z_t and all $n<N$.

$$1 + n\delta W'(\rho z_t + N - n + nq_t^s) = 0, \quad 0 \leqslant q_t^s \leqslant 1, \tag{10}$$

$$1 + n\delta W'(\rho z_t + N) \geqslant 0, \quad q_t^s = 1. \tag{11}$$

In the second-stage equilibrium where signatories behave symmetrically, the Bellman equation can be written as

$$W^f(z_t, n) = 1 - \gamma z_t^2 + \delta W[\rho z_t + nq^s(z_t, n) + N - n], \tag{12}$$

$$W^s(z_t, n) = V(z_t, n)/n = q^s(z_t, n) - \gamma z_t^2 + \delta W[\rho z_t + nq^s(z_t, n) + N - n], \tag{13}$$

where $q^s(z_t, n)$ is implicitly defined by (9)–(11).

3.2. First-stage membership game

We now introduce our definition of the equilibrium of the first-stage membership game. In a period with initial stock of the pollutant z_t we define an IEA with n^* signatories as *stable or self-enforcing* if it satisfies the two following properties:

Internal stability: $W^s(z_t, n^*) \geqslant W^f(z_t, n^* - 1)$, so no signatory has any incentive to switch to being a non-signatory.

External stability: $W^f(z_t, n^*) \geqslant W^s(z_t, n^* + 1)$, so no non-signatory has any incentive to switch to being a signatory.

They can be summarized in the following way. We write the internal stability condition as $W^s(z_t, n^*) - W^f(z_t, n^* - 1) \geqslant 0$ and the external stability $W^s(z_t, n^* + 1) - W^f(z_t, n^*) \leqslant 0$. Defining the function $\Omega(z_t, n) = W^s(z_t, n) - W^f(z_t, n - 1)$, internal stability corresponds to $\Omega(z_t, n^*) \geqslant 0$, while external stability corresponds to $\Omega(z_t, n^* + 1) \leqslant 0$. We say that an IEA consisting of n^* signatories is *self-enforcing* if it is both internally and externally stable.

Using (12) and (13), we have

$$\Omega(z_t, n) = q^s(z_t, n) - 1 + \delta W[\rho z_t + nq^s(z_t, n) + N - n] - \delta W[\rho z_t + (n - 1)q^s(z_t, n - 1) + N - n + 1]. \tag{14}$$

We assume for the moment that for all admissible values of the stock z_t there exists a self-enforcing IEA consisting of n^* signatories, so that we can define the function $n^*(z_t)$.[7] If this is the case, we can rewrite (12) and (13) as functions only of the stock

$$W^f(z_t) = 1 - \gamma z_t^2 + \delta W[\rho z_t + N - n^*(z_t)(1 - q^s(z_t, n^*(z_t)))],$$

$$W^s(z_t) = q^s(z_t, n^*(z_t)) - \gamma z_t^2 + \delta W[\rho z_t + N - n^*(z_t)(1 - q^s(z_t, n^*(z_t)))].$$

Then assuming that at the start of each period each country has the same probability $n^*(z_t)/N$ of being a signatory, the expected present-value of current and future net benefits when the initial stock of emissions is z_t is given by

$$\begin{aligned} W(z_t) &= \frac{n^*(z_t)}{N} W^s(z_t) + \left(1 - \frac{n^*(z_t)}{N}\right) W^f(z_t) \\ &= 1 - \frac{n^*(z_t)}{N}(1 - q^s(z_t, n^*(z_t))) - \gamma z_t^2 + \delta W[\rho z_t + N - n^*(z_t)(1 - q^s(z_t, n^*(z_t)))]. \end{aligned} \tag{15}$$

Thus (15) provides an equation which implicitly defines the value function $W(z_t)$. Then, if there exists a value function that satisfies this equation along with conditions (7) and (9)–(11) we know that at least for certain values for the parameters of the model, the solution of the model implies that the emissions for non-signatories are equal to the unity for all z_t and all $n < N$.

[7]There could be more than one self-enforcing IEA. In this case the function $n^*(z_t)$ is defined by associating to each z_t the highest value for n^* for which the IEA is self-enforcing.

3.3. Approximating the value function $W(z_t)$

It is clear from (15) that if the term $n^*(z_t)(1-q^s(z_t,n^*(z_t)))$ was linear in z_t, then $W(z_t)$ would be a quadratic function of z_t. However, $n^*(z_t)(1-q^s(z_t,n^*(z_t)))$ is not linear. Therefore, we have developed an algorithm that allows us to obtain a quadratic approximation, $A-Bz_t-Cz_t^2$, of the value function $W(z_t)$. Before presenting the algorithm, we investigate the conditions that the parameters of the quadratic approximation must satisfy in order to ensure that the dominant strategy for the non-signatories is always to emit the maximum as we have assumed in Section 3.1.

Suppose that the parameters B and C of this approximation satisfy

$$\delta B+\frac{2\delta CN}{1-\rho}\leqslant 1<N\delta B. \tag{16}$$

For a given stock z_t, define threshold values for n, denoted by $\underline{n}(z_t)$ and $\bar{n}(z_t)$ and given by

$$\underline{n}(z_t)=\frac{1}{\sigma(z_t)},\quad \bar{n}(z_t)=\frac{\sigma(z_t)-\sqrt{\sigma(z_t)^2-8\delta C}}{4\delta C}, \tag{17}$$

where $\sigma(z_t)=\delta(B+2C(N+\rho z_t))$. Notice that $1<\underline{n}(z_t)<\bar{n}(z_t)<N$. We then have

Result 3. *If conditions* (16) *hold, the optimal emissions by a non-signatory in each period are always* 1, *independent of the number of signatories and the level of stock. The optimal emissions by the representative signatory are*

$$q^s(z_t,n)=\begin{cases}0 & \text{if } \bar{n}(z_t)\leqslant n\leqslant N,\\ \dfrac{1}{n}\left(\dfrac{1}{2\delta nC}-\dfrac{B}{2C}-N+n-\rho z_t\right) & \text{if } \underline{n}(z_t)\leqslant n\leqslant \bar{n}(z_t),\\ 1 & 1\leqslant n\leqslant \underline{n}(z_t),\end{cases} \tag{18}$$

where $\underline{n}(z_t)$ *and* $\bar{n}(z_t)$ *are given by* (17).

Proof. See Appendix A.

If the parameters B and C satisfy (16), the dominant strategy for non-signatory countries is always to pollute at the maximum level, as in the non-cooperative equilibrium. If there are a large enough number of signatories, $n\geqslant\bar{n}(z_t)$, the optimal strategy for signatory countries is to set their emissions to zero. As the number of signatories falls below $\bar{n}(z_t)$ it will pay signatories to set positive emissions, and as the number of signatories falls below $\underline{n}(z_t)$ signatories will also set their emissions at the maximum level. When this occurs both signatories and non-signatories select the maximum level of emissions and the PANE of the emission game is identical to the fully non-cooperative equilibrium.

We now describe the algorithm, which consists of the following steps:

(i) For each combination of parameters (δ,ρ,N) such that $N>1/(1-\rho)$, we calculate 10 values for γ according to the following rule:

$$\gamma_i=\underline{\gamma}+\frac{i}{10}(\bar{\gamma}-\underline{\gamma}),\quad i=1,2,\dots,10, \tag{19}$$

where $\bar{\gamma}$ and $\underline{\gamma}$ are given by (2) and (5).[8] In this way, we restrict our analysis to the set of parameter values for which Results 1 and 2 apply.

(ii) For each combination of parameters (δ,ρ,N,γ) obtained according to the previous rule we propose as initial values for A, B and C, the values of the parameters of the quadratic value function corresponding to the fully non-cooperative equilibrium.[9]

[8] As discussed in Section 2.2 $\underline{\gamma}<\bar{\gamma}$ when $N>1/(1-\rho)$.

[9] We describe in Appendix A how we modify the algorithm if these values of B and C do not satisfy (16). Appendix A is available through JEEM's on-line archive of supplementary material, which can be accessed at http://www.are.org/journal/index.html.

(iii) We select 201 values for z in the interval $(0, N/(1-\rho))$ according to the following rule:

$$z_j = \frac{j}{200}\frac{N}{1-\rho}, \quad j = 0, 1, \ldots, 200,$$

where $N/(1-\rho)$ is the steady-state stock of the fully non-cooperative equilibrium.

(iv) For each value for z we calculate $\underline{n}(z)$ and $\bar{n}(z)$ according to (17) and $q^s(z, n)$ and $q^s(z, n-1)$ according to (18) and rewrite (14) for $W(z_t) = A - Bz_t - Cz_t^2$ which yields

$$\begin{aligned}\Omega(z,n) = {} & q^s(z,n) - 1 - \delta(B + 2C(\rho z + N - n))(nq^s(z,n) - (n-1)q^s(z,n-1)) \\ & - \delta C(n^2 q^s(z,n)^2 - (n-1)^2 q^s(z,n-1)^2) + 2\delta C(n-1)q^s(z,n-1) \\ & + 2\delta C(\rho z + N - n) + \delta(B + C).\end{aligned}$$

(v) Then for each value for z, beginning from N, we look for the greatest integer that satisfies $\Omega(z,n) \geqslant 0$, i.e. we look for the self-enforcing IEA with the greatest membership, $n^*(z)$, and once we have found it we use it to calculate $q^s(z, n^*(z))$ according to (18).[10] In this way for the 201 values of z we have the corresponding values for $n^*(z)$ and $q^s(z, n^*(z))$.

(vi) Using these values we estimate a linear regression $\alpha + \beta z \approx n^*(z)(1 - q^s(n^*(z), z))$ for the 201 values of z.

(vii) We substitute $n^*(z)(1 - q^s(n^*(z), z))$ by $\alpha + \beta z$ in the value function $W(z_t)$ given by (15) and equating coefficients of z we find a new estimate of A, B and C given by the following expressions:

$$C = \frac{\gamma}{1 - \delta(\rho - \beta)^2},$$

$$B = \frac{1}{1 - \delta(\rho - \beta)}\left(\frac{\beta}{N} + 2\delta C(\rho - \beta)(N - \alpha)\right),$$

$$A = \frac{1}{1 - \delta}\left(\frac{N - \alpha}{N} - \delta(N - \alpha)B - \delta(N - \alpha)^2 C\right).$$

(viii) We iterate this procedure until we find an "approximate" fixed point, i.e., until the variation of A, B and C is less than 5%.[11]

3.4. Dynamics of the stock and membership

To complete the analysis, we describe in this section how using the approximation of the value function given by the previous algorithm we can obtain numerically the dynamics of the stock of the pollutant and hence the number of signatories and emissions in each period.

We know that as A, B and C satisfy conditions (16), the emissions are given by (18) so that the stock evolves according to difference equation

$$z_{t+1} = \rho z_t + N - n^*(z_t) + n^*(z_t)q^s(n^*(z_t), z_t), \tag{20}$$

with $z_0 = 0$. Then we can obtain the dynamics of the stock using steps (iv) and (v) of the algorithm for $z_0 = 0$ and all the subsequent values for the stock given by difference equation (20). In more detail, for $z_0 = 0$ we calculate firstly $\underline{n}(0)$ and $\bar{n}(0)$ according to (17) and derive $\Omega(0, n)$, step (iv) but now for the initial value of the stock. After that, we look for the highest integer that satisfies $\Omega(0, n) \geqslant 0$, i.e. we look for n^* and use it to calculate $q^s(n^*(0), 0)$ according to (18), step (v). Finally, we calculate z_1 using the difference equation

$$z_1 = N - n^*(0) + n^*(0)q^s(n^*(0), 0),$$

[10] Notice that if n^* is the greatest integer that satisfies $\Omega(z, n^*) \geqslant 0$ then for $n > n^*$, $\Omega(z, n)$ must be negative which implies that $\Omega(z, n^* + 1) = W^s(z, n^* + 1) - W^f(z, n^*)$ must be negative, i.e. that the external stability condition is also satisfied for n^*.

[11] If we impose a convergence in a range of 1%, the algorithm works for all the combinations of the parameters values evaluated in the numerical analysis of the next section except for three cases for which B and C do not satisfy conditions (16). However, for a convergence criterion of 5%, B and C always satisfy conditions (16).

and we repeat the calculations for z_1. This procedure can be repeated as often as may be necessary to determine the dynamics of the model. If a steady state exists, the expression for the steady-state stock will depend on the steady-state emissions for signatories. Recall that the emissions of non-signatories are always unity. If the emissions of signatories are zero, (20) implies that the steady-state stock is

$$\hat{z}^{\text{iea}} = \frac{N - n^*(\hat{z}^{\text{iea}})}{1-\rho}, \tag{21}$$

where the superscript iea stands for the self-enforcing IEA. If signatories emissions are not zero, (18) implies that their steady-sate emissions are

$$q^s(\hat{z}^{\text{iea}}, n^*(\hat{z}^{\text{iea}})) = \frac{1}{n^*(\hat{z}^{\text{iea}})}\left(\frac{1}{2\delta C n^*(\hat{z}^{\text{iea}})} - \frac{B}{2C} - N + n^*(\hat{z}^{\text{iea}}) - \rho \hat{z}^{\text{iea}}\right). \tag{22}$$

Substitution into (20) yields the steady-state stock

$$\hat{z}^{\text{iea}} = \frac{1}{2\delta C n^*(\hat{z}^{\text{iea}})} - \frac{B}{2C}. \tag{23}$$

Notice that in the first case we obtain an expression very similar to the one that defines the steady-state stock of the fully non-cooperative equilibrium whereas in the second case the expression is closer to the one that defines the steady-state stock of the fully cooperative equilibrium.

4. Numerical results

We now report on some of the numerical modelling we have undertaken to solve for the value function for the dynamic self-enforcing IEA model.[12]

As we explain in the presentation of the algorithm, we first selected 216 combinations of values for the parameters ρ, δ and $N(\rho \geqslant 0.7, \delta \geqslant 0.9, N \geqslant 75)$ and for each of these combinations we calculated 10 values of γ using (19). For each of the resulting 2160 combinations of parameter values we first applied our algorithm to compute the parameters of the quadratic approximation of the value function, and then calculated the dynamics of the stock for the first 151 time periods using the difference equation (20). As a result of these calculations, we obtained for each period t the value of the stock, z_t, the highest stable level of cooperation, $n^*(z_t)$, the emissions for signatories, $q^s(z_t, n^*(z_t))$, and the threshold values for the number of signatories, $\underline{n}(z_t)$ and $\bar{n}(z_t)$.

Next, we present the main findings of our numerical simulations.

4.1. The size of self-enforcing IEAs

An important finding from our numerical simulations is that for all 2160 combinations of parameter values and for all values of the stock in the first 151 periods, there always exists a unique self-enforcing IEA given by the smallest integer larger than $\underline{n}(z_t)$.[13] The intuition for why this is the equilibrium number of signatories is that if the number of signatories fell below $\underline{n}(z_t)$ the optimal emissions for signatories is to select the maximum level of emissions and this implies a serious threat that deters the signatories of a self-enforcing IEA from leaving the agreement. This threat does not operate when the number of signatories does not fall bellow $\underline{n}(z_t)$ after a country leaves the agreement. If after the exit of one country from the IEA n is greater than $\bar{n}(z_t)$, signatories keep their emissions at the minimum level according to (18), while if n lies in the interval $(\underline{n}(z_t), \bar{n}(z_t))$ signatories will reduce their emissions, but not enough to deter such an exit.

[12]The results are generated using Ox version 4.02 (Linux) (C) J.A. Doornik, 1994–2006. The algorithm is available through Santiago Rubio's web page http://www.uv.es/~srubio. The complete numerical analysis is available through JEEM's on-line archive of supplementary material, which can be accessed at http://www.are.org/journal/index.html.

[13]It is easy to show that for $n < \underline{n}(z_t)$ there exist other agreements that are self-enforcing. However, these kind of agreements are irrelevant because they do not imply any difference with respect to the fully non-cooperative solution since for these agreements the signatories select the maximum level of emissions as the non-signatories.

For all the stock values analysed in this paper, the interval $[\underline{n}(z_t), \bar{n}(z_t)]$ either contains only one integer or it does not contain any. The latter case was more important in our simulation results, so that the optimal policy for signatories is most often to emit the minimum when $n > \bar{n}(z_t)$ or the maximum when $n < \underline{n}(z_t)$.

4.2. Dynamics of IEA

Our numerical simulations have identified three possible patterns of dynamic behaviour. The first, the more common case, is illustrated in Fig. 1.[14] In this figure the first bar stands for the stock as a percentage of the steady-state value and the second one for the number of signatories of a self-enforcing IEA. The figure shows that the stock is monotonically increasing whereas the number of signatories is non-increasing. This behaviour is explained by the fact that our numerical simulations yield a monotonically decreasing value for $\underline{n}(z_t)$. Since n^* is the smallest integer larger than $\underline{n}(z_t)$, whenever this critical value falls below an integer number, the membership of a self-enforcing IEA also falls. Thus we have derived a non-positive relationship between the stock and the number of signatories selected by the stability conditions.

In the first five periods n^* goes from 15 to 10 and for $t = 20$ the steady-state number of signatories is reached and the stock value is equal to 85% of its steady-state value. The emissions of signatories are always zero except for the first and twentieth periods. Thus, from period 20 the steady-state stock is approached asymptotically according to the difference equation $z_{t+1} = 0.9z_t + 94$ which yields a steady-state stock value equal to 940. Observe that these dynamics are similar to the one that characterized the fully non-cooperative equilibrium.

The second pattern of dynamics we have observed is characterized by a constant number of signatories.[15] When this occurs, the self-enforcing IEA consists only of two countries. Finally, the third pattern of dynamics we find are characterized by convergence to the steady state in a finite number of periods and positive steady-state emissions for signatories.[16] These dynamics are similar to the dynamics of the fully cooperative equilibrium in the sense that when the steady-state number of signatories is reached, the steady-state stock is reached in one period.

Summarizing, we can say that our numerical analysis allows us to conclude that the stock is always increasing, that it converges to a steady state and that the profile of the number of signatories of a self-enforcing IEA is a non-increasing function of the stock.

4.3. Impact of variations in damage cost parameter

In this section we explore the impact of variations in γ. The results are shown in Fig. 2.

On the horizontal axis we present the 10 values of γ_i using (19). The first bar shows the *potential gains from cooperation* (the difference in the present value of net benefits between the cooperative and non-cooperative solutions) normalized so that this gain equals 100 when $\gamma = \gamma_{10}$. The second bar shows the *partial gains in cooperation* (the difference in the present value of net benefits between the IEA solution and the non-cooperative solution) as a percentage of the potential gains. We note that the potential gains from full cooperation rise sharply as the damage cost parameter rises whereas the partial gains achieved by dynamic IEAs falls equally sharply, so that for $\gamma \geqslant \gamma_3$ less than 10% of the potential gains are realized.[17]

The initial membership and its steady-state value are given above the third and fourth bars. Increasing the damage cost parameter causes both the initial and steady-state membership of the IEA to fall. In absolute terms the reduction in membership caused by the increase of γ means that the dynamic aspects of the model of membership become less significant, with membership varying between 3 and 2 when γ takes its maximum value.

[14] We have found this kind of dynamics for the 93.75% of the combinations studied.

[15] We found this pattern of dynamics in 5.55% of the cases studied. A sufficient condition for this result is that γ must be big enough, i.e. γ_9 or γ_{10}, and ρ cannot be very big.

[16] We have found only fifteen cases that behave according to this type of dynamics. Necessary conditions are a value for ρ not very big, in fact no bigger than 0.80, and γ equal to γ_1 for the corresponding combination of δ, ρ and N.

[17] This explains why the average gain coming from a self-enforcing IEA for all the combinations of the parameter values analysed in this Section is 7.5% of potential gains whereas the maximum gain is 22.5%.

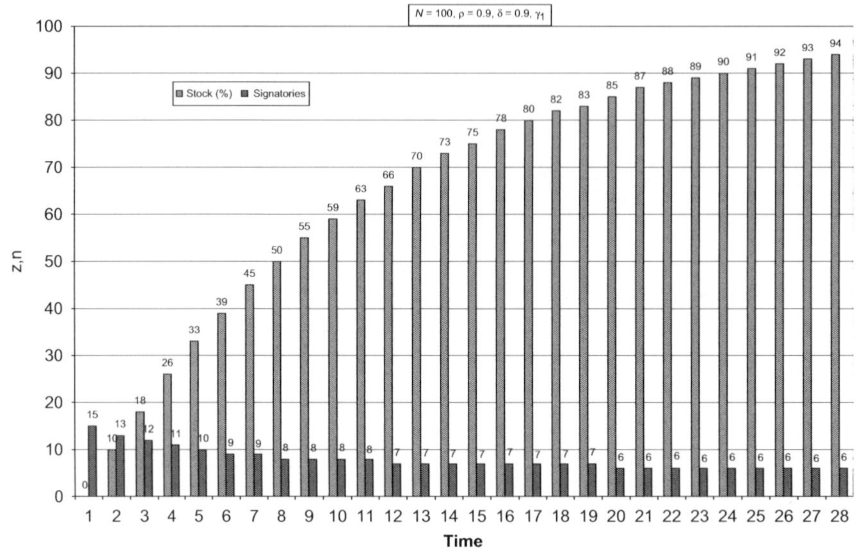

Fig. 1. Dynamics of stock and number of signatories.

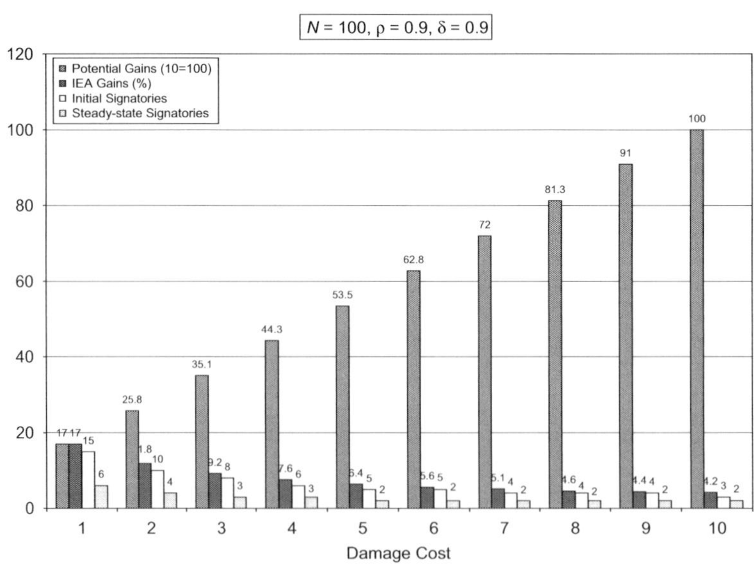

Fig. 2. Gains from co-operation and number of signatories.

These relationships are a dynamic generalization of the pessimistic results of Barrett [1], but apply both to a variation across models due to changes in the damage cost parameter, and within a model, so that for a given damage cost parameter IEA membership falls as the stock of the pollutant increases.

5. Conclusions

In this paper we have extended the familiar model of self-enforcing IEAs from the usual setting of a static pollution problem to a dynamic setting of a stock pollutant. Unlike previous models, which assumed that countries make a once-for-all decision at the outset whether or not to join an IEA, we have explored the implications of allowing IEA membership to vary over time as the stock of pollution changes. Our numerical analysis shows that there exists a steady-state stock of pollution and corresponding steady-state IEA membership, and that as the stock rises towards its steady-state the IEA membership falls towards its steady-state. Moreover, it establishes that the gains from cooperation arising from a dynamic self-enforcing IEA and the corresponding membership decrease as the cost of damage rises.[18] The intuition behind this result is that it is simply the dynamic generalization of the pessimistic static results that the greater is the cost of damage, and hence the greater are the potential gains from cooperation, the smaller is the membership of a self-enforcing IEA.

We acknowledge that our model is an extremely simple first attempt to model how stock dynamics influence membership dynamics. There are many possible lines for further research. Within the confines of the existing model, one obvious question is how global welfare in this variable membership model would compare with global welfare in a fixed membership model.[19]

The more important extensions are to move outside the framework of this model. Our assumption of identical countries is a limitation of our model, and the use of a random process for determining which countries become signatories gives a relatively limited role for stock dynamics to influence dynamic membership decisions. Even if we continued to assume identical countries it would be desirable to explore the implications of using what we have called in Rubio and Ulph [13] the Status Quo Assignment Rule, in which current membership has a more direct influence on future membership.

Finally, it will be important for future research to move away from the model of self-enforcing agreements used here so as to exploit the dynamics of the stock pollutant model by introducing concepts of threats from dynamic game theory. There is now a range of competing models of coalition stability (see [4,9] for excellent surveys), and some of these models involve the concept of sequential formation of coalitions (one consequence being the possibility of more than one coalition of countries). However, in these models the underlying pollution problem is static, and the notion of sequential formation is a conceptual rather temporal one. It would be interesting to consider how far such concepts could be integrated with the underlying dynamics of the pollution problem to assess how coalition structures might evolve in real time as the stock of the pollutant varies.

Acknowledgments

We are grateful to Henry Tulkens and two referees for helpful comments. Comments from the Associate Editor, Larry Karp, have also been extremely useful. Suggestions from the Editor, Charles Mason, have made the paper read much more smoothly. We are grateful as well to José C. Pernías from the Laboratory for Research in Experimental Economics (LINEEX) at the University of Valencia for his assistance with the program used to do the numerical analysis. The usual caveat applies. Research of Santiago J. Rubio was supported by the *Ministerio de Educación y Ciencia* under Grant SEJ2004-05704 and by the *Fundación BBVA*.

[18] In the paper we have focused on the case of an initial stock of pollutant equal to zero in order to simplify the exposition. Nevertheless, for $z_0 > 0$ the same qualitative results apply except that if the initial stock of pollution is above the steady state then the stock will fall and the IEA membership will rise.

[19] In our earlier two-period model we found that the variable membership model generally gave higher welfare than the fixed membership model. But in the two-period setting variable IEA membership rose over time and it would be interesting to know whether the welfare ranking of the two types of models would carry over to the infinite-horizon case.

Appendix A. Proof of Result 3

If we approximate the value function $W(z_t)$ using a quadratic function $A - Bz_t - Cz_t^2$, condition (7) can be rewritten as

$$1 \geqslant \delta\left(B + 2C\left(\rho z_t + \sum_{l=1}^{N-n} q_{lt}^f + \sum_{k=1}^{n} q_{kt}^s\right)\right). \tag{24}$$

Then as the maximum level of emissions is equal to the unity, it must be true that

$$\rho z_t + \sum_{l=1}^{N-n} q_{lt}^f + \sum_{k=1}^{n} q_{kt}^s \leqslant \frac{N}{1-\rho},$$

where the r.h.s is the steady-state stock of the fully non-cooperative solution. Notice that the fully non-cooperative solution, as characterized in Result 1, must be an upper bound of any policy that implies that the emissions for signatories are lower than or equal to unity. Then if (24) is satisfied for $N/(1-\rho)$, the dominant strategy for the non-signatories is to select for each value of the stock between zero and $N/(1-\rho)$, the maximum level of emissions independently of the level of cooperation n. By substitution of $N/(1-\rho)$ in (24) we obtain

$$1 \geqslant \delta B + \frac{2\delta CN}{1-\rho}, \tag{25}$$

which yields the l.h.s. of condition (16) in the main text.

On the other hand, for a linear-quadratic approximation of the value function Kuhn–Tucker conditions (9), (10) and (11) yield

$$1 \leqslant n\delta(B + 2C(\rho z_t + N - n)), \quad q_t^s = 0, \tag{26}$$

$$1 = n\delta(B + 2C(\rho z_t + N - n + nq_t^s)), \quad 0 \leqslant q_t^s \leqslant 1, \tag{27}$$

$$1 \geqslant n\delta(B + 2C(\rho z_t + N)), \quad q_t^s = 1. \tag{28}$$

These expressions allow us to calculate threshold values $\underline{n}$ and $\bar{n}$ for the number of signatories. Eq. (26) defines the frontier

$$1 = n\delta(B + 2C(\rho z_t + N - n)),$$

which gives us a quadratic equation for n

$$2\delta Cn^2 - \sigma(z_t)n + 1 = 0, \tag{29}$$

where $\sigma(z_t) = \delta(B + 2C(N + \rho z_t))$. This equation has two positive real roots given by

$$0 < \bar{n}_1(z_t) = \frac{\sigma(z_t) - \sqrt{\sigma(z_t)^2 - 8\delta C}}{4\delta C} < \bar{n}_2(z_t) = \frac{\sigma(z_t) + \sqrt{\sigma(z_t)^2 - 8\delta C}}{4\delta C}$$

that satisfy $\bar{n}_1(z_t) < N < \bar{n}_2(z_t)$ provided that $1 < \delta BN$. According to Fig. 3, Eq. (29) has two positive real roots if the maximum of function $\sigma(z_t)n - 2\delta Cn^2$ is greater than unity, then to obtain that $\bar{n}_1(z_t) < N < \bar{n}_2(z_t)$ we need that $\sigma(z_t)N - 2\delta CN^2 > 1$ is also satisfied. Developing this expression we obtain that this inequality holds if $\delta BN + 2\delta\rho CNz_t > 1$ which is guaranteed for $z_t \geqslant 0$ if $\delta BN > 1$. This inequality is the one that appears in the r.h.s. of condition (16) in the main text. Thus this condition, $\delta BN > 1$, guarantees two things at the same time, that Eq. (29) has two positive real roots and that N belongs to the interior of the interval defined by these two roots. From now we identify $\bar{n}_1(z_t)$ with $\bar{n}(z_t)$ in the main text.

On the other hand, (28) defines the frontier

$$1 = n\delta(B + 2C(\rho z_t + N)),$$

which directly yields

$$\underline{n}(z_t) = \frac{1}{\delta(B + 2C(\rho z_t + N))} = \frac{1}{\sigma(z_t)}.$$

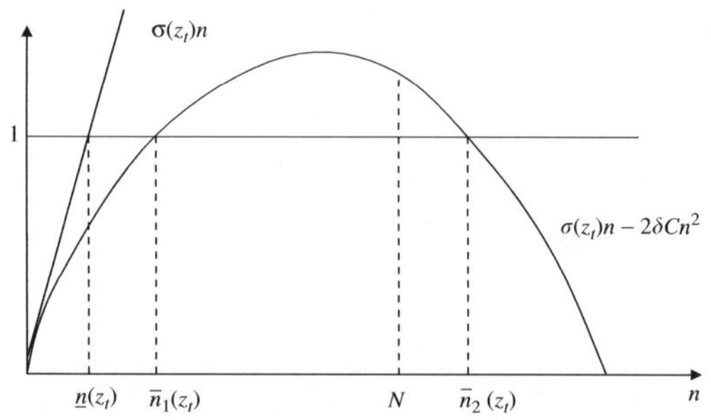

Fig. 3. Threshold values for n, to determine signatories' emissions.

It is obvious from Fig. 3 that $\bar{n}(z_t)$ is greater than $\underline{n}(z_t)$. Moreover, it is also straightforward to check that condition (25) guarantees that $\underline{n}(z_t)$ is greater than unity.

Thus for given z_t if n belongs to interval $[\bar{n}(z_t), N]$ condition (26) is satisfied and the optimal policy for signatories is to choose the minimum level of emissions, $q_t^s = 0$. However, if n belongs to interval $[1, \underline{n}(z_t)]$ condition (28) is satisfied and the optimal policy is to select the maximum level of emissions, $q_t^s = 1$. Finally, for any n between $\underline{n}(z_t)$ and $\bar{n}(z_t)$ the optimal policy is defined by (27) which yields

$$q^s(n, z_t) = \frac{1}{n}\left(\frac{1 - n\delta B}{2Cn\delta} - N + n - \rho z_t\right).$$

Thus, given condition (16), the optimal policy for the non-signatories is always 1 whereas the optimal policy for signatories depends on n as summarized by (18) in Result 3.

References

[1] S. Barrett, Self-enforcing international environmental agreements, Oxford Econ. Pap. 46 (1994) 878–894.

[2] S. Barrett, Towards a theory of international environmental cooperation, in: C. Carraro, D. Siniscalco (Eds.), New Directions in the Economic Theory of the Environment, Cambridge University Press, Cambridge, 1997, pp. 239–280.

[3] S. Barrett, Environment & Statecraft. The Strategy of Environmental Treaty-Making, Oxford University Press, Oxford, 2003.

[4] F. Bloch, Non-cooperative models of coalition formation in games with spillovers, in: C. Carraro, D. Siniscalco (Eds.), New Directions in the Economic Theory of the Environment, Cambridge University Press, Cambridge, 1997, pp. 311–352.

[5] C. Carraro, D. Siniscalco, Strategies for the international protection of the environment, J. Public Econ. 52 (1993) 309–328.

[6] P. Chander, H. Tulkens, A core-theoretic solution for the design of cooperative agreements on transfrontier pollution, Int. Tax Public Finance 2 (1995) 279–293.

[7] J. Eyckmans, On the farsighted stability of the Kyoto protocol, CLIMNEG Working Paper 40, CORE, Universite Catholique de Louvain, 2001.

[8] M. Finus, Game Theory and International Environmental Cooperation, Edward Elgar, Cheltenham, UK and Northampton, USA, 2001.

[9] M. Germain, P. Toint, H. Tulkens, A. de Zeeuw, Transfers to sustain core-theoretic cooperation in international stock-pollutant control, J. Econ. Dynam. Control 28 (2003) 79–99.

[10] L. Karp, S. Sachetti, Dynamics and limited cooperation in international environmental agreements, mimeo, University of California, Berkeley, 1997.

[11] L. Petrosjan, G. Zaccour, Time-consistent Shapley value allocation of pollution cost reduction, J. Econ. Dynam. Control 27 (2003) 381–398.

[12] S. Rubio, B. Casino, Self-enforcing international environmental agreements with a stock pollutant, Spanish Econ. Rev. 7 (2005) 89–109.

[13] S. Rubio, A. Ulph, A simple dynamic model of international environmental agreements with a stock pollutant, Discussion Paper in Economics and Econometrics 0209, Department of Economics, University of Southampton, 2002.

[46]

Available online at www.sciencedirect.com

ScienceDirect

ELSEVIER

Journal of Environmental Economics and Management 55 (2008) 163–174

JOURNAL OF ENVIRONMENTAL ECONOMICS AND MANAGEMENT

www.elsevier.com/locate/jeem

Dynamic effects on the stability of international environmental agreements

Aart de Zeeuw[a,b,*]

[a]*Department of Economics and CentER, Tilburg University, P.O. Box 90153, 5000 LE Tilburg, The Netherlands*
[b]*The Beijer Institute of Ecological Economics, Stockholm, Sweden*

Received 6 June 2005
Available online 4 November 2007

Abstract

In terms of the number of signatories, one observes both large and small international environmental agreements. The theoretical literature, based on game theory, discusses different concepts and mechanisms regarding stability of coalitions. The conclusion has been reached that in all circumstances, under farsightedness, both large and small stable coalitions can occur. This theory is based on behavioural reaction patterns but does not take account of the interaction with the dynamics of emission adjustments. This paper shows that when these two dynamical processes are integrated, large and small stable coalitions can still occur but only if the costs of emissions are relatively unimportant as compared with the costs of abatement.

JEL classification: Q2; C70; F42

Keywords: IEAs; Coalitional stability; Dynamics

1. Introduction

Global environmental problems such as ozone depletion and climate change require voluntary cooperation by sovereign states to internalize the negative externalities of cross-border emissions. In the last decades, so-called International Environmental Agreements (IEAs) were signed, but the number of signatories varies considerably (e.g., [11]). The Montreal Protocol (1987) with the purpose to phase out CFCs (that cause depletion of the ozone layer) has been signed and ratified by 181 countries. However, the Kyoto Protocol (1997) with the purpose to reduce greenhouse gas emissions (that cause climate change) is a different story. The number of countries that will reduce emissions in the first phase is much smaller, and especially after the withdrawal of the USA in 2001, the basis of this agreement is very weak. It came into force only recently by the ratification of Russia.

*Fax: +31 13 4663042.
E-mail address: A.J.deZeeuw@uvt.nl

0095-0696/$ - see front matter © 2007 Elsevier Inc. All rights reserved.
doi:10.1016/j.jeem.2007.06.003

The theoretical literature on this issue has mainly focused on stability concepts for coalitions (see [11] for an overview). The dominant strand in this literature is based on the idea of internal and external stability, meaning that an individual country neither has an incentive to leave nor to join the coalition [2,5,13]. For most models in this tradition, the outcome is rather grim: the size of the stable coalition is very small. A more optimistic view on the size of stable coalitions results if some trigger mechanism is introduced that can change the position or the behaviour of the other countries. If deviation of one country would induce the coalition to fall apart, the mere threat of loosing all cooperative benefits may deter deviations in the first place and support cooperation. This idea is at the heart of the trigger mechanism in repeated games [12]. It is also the main idea for the γ-core concept that is rooted in cooperative game theory and that is used to investigate stability of the grand coalition [6].

These two strands of literature take opposite assumptions on what happens in case a country leaves the coalition. Either it is assumed that the rest of the coalition remains intact or it is assumed that the whole coalition breaks down. Farsightedness (e.g., [16]) is a concept that advocates behavioural assumptions that are in a sense midway between these two models. The idea is that if a country leaves, it may trigger other countries to leave as well until a new stable situation is reached. In this case, a country compares its initial position with its position at the end of this process. It can be shown that farsightedness leads to a set of large and small stable coalitions [8,10]. In a way the two approaches, described earlier, are reconciled by this concept. Moreover, an important conclusion is that countries can coordinate on a large coalition that is stable.

Dynamic aspects are mostly ignored in this literature. Behavioural reaction patterns are dynamic, of course, but abatement processes are usually dynamic as well. In most of the models, it is assumed that countries reduce emissions in one step, but this is not realistic and also not rational, as can be seen from a standard optimal control model with convex costs where the target is reached in steps. In practice, countries take time to adjust their emission levels. For example, in the Kyoto Protocol the 15 countries of the European Union plan to reduce greenhouse gas emissions with 8% by 2008–2012, as compared with 1990 levels. This paper investigates how the interaction between the behavioural reaction pattern and this adjustment process affects the results on the stability of coalitions.

The concept of farsightedness is sufficiently rich to allow for a set of large and small stable coalitions. Suppose that the countries coordinate on the largest stable coalition. Deviation is deterred because under farsightedness it is not in the interest of a country to become an outsider to a smaller stable coalition. If welfare is discounted over time, this statement is still true, provided the discount factor is high enough. However, if the other members of the coalition make adjustments before a deviation is detected, the game changes over time and then it is not so clear that deviations will be deterred.

This paper uses an optimal control model with the objective to phase out emissions in the long run at the lowest costs. At each time step, the countries choose some level of abatement, like investment in green technology, in order to lower the emission levels. In the absence of cooperation, this adjustment process is slower and has higher costs than in case (some) countries cooperate. It will be shown that a one-shot game of this type always has large stable coalitions under farsightedness. However, in the dynamic game it depends on the relative costs of abatement and emission levels whether large coalitions can be sustained or not. This result may give more insight into why certain IEAs are successful, in terms of membership, and others are not.

2. The model

The basic model is a simple abatement model. In the initial state of the world, global emissions are at a level e_0. Under business-as-usual, without environmental concern, global emissions will grow over time with a factor b.[1] There are n countries ($n > 2$). At each point in time, each country can choose an abatement level a_i, $i = 1,2,\ldots,n$, like an investment in green technology or a structural change. This yields the following dynamics

[1]The idea is that output or income grows and related to that, emissions grow; in order to lower emissions, decoupling is needed which means that abatement has to realize a larger percentage decrease in emissions than the percentage increase in emissions that results in the absence of environmental concern.

for the level of emissions:

$$e(t+1) = (1+b)e(t) - \sum_{i=1}^{n} a_i(t), \quad t = 0, 1, \ldots, e(0) = e_0. \tag{1}$$

The emission level is damaging but it is also costly to abate. A simple cost indicator is given by

$$C_i = \sum_{t=0}^{\infty} \delta^t \left[\frac{1}{2} a_i^2(t) + \frac{1}{2} p e^2(t) \right], \quad i = 1, 2, \ldots, n, \quad p > 0, \tag{2}$$

where p denotes the relative weight attached to the damage costs as compared to the abatement costs and δ denotes the discount factor.

This is an optimal control model where the level of emissions is driven down to zero, regardless of whether the countries cooperate or not. In that sense, it differs from the typical international pollution control model with a stock pollutant (e.g., [15]) where the cooperative steady state is not the same as the non-cooperative steady state. In the case of climate change, it would be more appropriate to formulate a stock pollutant model, because damage is related to the stock of greenhouse gases. The message of the paper would not change, however, but the analysis is much more complicated. Moreover, the Kyoto Protocol is an agreement on the level of emissions, and not on the level of the stock of greenhouse gases. Note that the model still has a "state". The state of the system is the information that is needed to determine future optimal or equilibrium strategies. In this model, the level of emissions is the state. In a stock pollutant model, the state would be the stock of greenhouse gases. The model in this paper describes a situation where the target in the long run is to reduce emissions to zero but under cooperation this target is reached faster and with lower costs than in the absence of cooperation. This will become clear in Section 4 where the model is solved. Furthermore, note that in order to lower the level of emissions, first of all abatement has to compensate the business-as-usual growth in emissions. This implies that the value of the growth factor b will affect the levels of abatement, but it will not affect the qualitative results of the model. Finally, the n countries are assumed to be identical. This is not very realistic, of course, especially in the situation of climate change, but in this way we can abstract from issues like transfers between countries within a coalition or a possible relationship between size and membership. We want to focus on the number of countries that join the coalition. An extensive literature exists (e.g., [4,16,17]) on general coalitional structures but we restrict the analysis to one coalition with all other countries as individual outsiders. This corresponds to the current practice of IEAs, although recently competing agreements with a different focus are developing such as the technology agreement, led by the USA, as an alternative for the Kyoto Protocol.

3. Stability concepts

In order to be able to discuss the different stability concepts in a simple framework, we start with a one-shot static version of the model with emission level e, $b = 0$, and cost indicators

$$C_i = \frac{1}{2} a_i^2 + \frac{1}{2} p \left(e - \sum_{j=1}^{n} a_j \right)^2, \quad i = 1, 2, \ldots, n. \tag{3}$$

Suppose that m ($<n$) countries form the coalition. This implies that these m countries (called members) jointly minimize the sum of their costs, whereas the other countries (called outsiders) just minimize their own costs. It follows that first-order conditions for the equilibrium levels of abatement for members i and outsiders j are given by

$$a_i - mp \left(e - \sum_{k=1}^{n} a_k \right) = 0, \quad i = 1, \ldots, m, \tag{4}$$

$$a_j - p \left(e - \sum_{k=1}^{n} a_k \right) = 0, \quad j = m+1, \ldots, n. \tag{5}$$

Adding up and some simple manipulations yield the resulting level of emissions and the equilibrium abatement levels:

$$e - \sum_{k=1}^{n} a_k = \frac{1}{1 + (m^2 + n - m)p} e, \tag{6}$$

$$a_i = \frac{mp}{1 + (m^2 + n - m)p} e, \quad i = 1, \ldots, m, \tag{7}$$

$$a_j = \frac{p}{1 + (m^2 + n - m)p} e, \quad j = m + 1, \ldots, n. \tag{8}$$

It follows that the costs in equilibrium for a coalition member (C^c) and for an outsider (C^o) are given by

$$C^c = \frac{1}{2} \frac{p + m^2 p^2}{(1 + (m^2 + n - m)p)^2} e^2, \tag{9}$$

$$C^o = \frac{1}{2} \frac{p + p^2}{(1 + (m^2 + n - m)p)^2} e^2. \tag{10}$$

A few things are immediately clear: for $m = 0$ or 1 one gets the standard Nash equilibrium between individual countries and for $m = n$ one gets the full-cooperative outcome, with higher abatement levels and lower total costs. Note that for any size of the coalition $m > 1$ the costs of the outsiders are lower than the costs of the coalition members and also lower than the costs in the Nash equilibrium (free-rider benefits).

The usual approach to the theory on stable IEAs (e.g., [2,5]) is based on the ideas developed for cartel stability [1] and requires what is called *internal* and *external* stability. Internal stability means that a country does not have an incentive to leave the coalition, that is the costs of an outsider to the coalition of size $m-1$ must be larger than (or equal to) the costs of a member of the coalition of size m. External stability means that a country does not have an incentive to join the coalition, that is the costs of a member of the coalition of size $m+1$ must be larger than the costs of an outsider to the coalition of size m. This corresponds to the Nash equilibrium of the so-called open-membership game in which the countries first decide on whether they want to be a member of the coalition or not and then choose their abatement levels [4,11,17]. It is straightforward to derive from (9) and (10) that

$$C^c(m) > C^o(m - 1), \tag{11}$$

for all p, n, and $m > 2$, so that all coalitions larger than 2 are not internally stable. The same inequality (for $m = 3$) also implies that a coalition of size 2 is externally stable. It follows that the size of the coalition that is both internally and externally stable is either 1 or 2. Straightforward calculations show that if $p \leqslant 1/(n - 4 + 2\sqrt{(n^2 - 3n + 3)})$ the size is 2 and otherwise the size is 1.

The general conclusion of this analysis is that the size of the stable coalition is small. In this model the size is not larger than 2. Moreover, the size is only 1 if the relative weight p of the damage costs as compared with the abatement costs is high. As can be seen from (4) and (5), the parameter p is also an indicator of the steepness of the best-reply function. Hoel [13] and Carraro and Siniscalco [5] note that the slope of the best-reply function indicates by how much the outsiders offset the extra abatement of the coalition members (leakage effect). Based on this it is argued that the steeper the slope, the larger the incentive to deviate from the coalition. Finus [11] extends this to a broader set of models and concludes the steeper the best-reply function, the lower the number of participants. Barrett [2] finds the opposite result but with a Stackelberg model in which the coalition moves first and can take the reaction of the outsiders into account. Moreover, the Stackelberg model allows for a wider range of outcomes, not only 1 or 2. Barrett [2] uses this result to give an explanation for the difference in participation rate between the Montreal Protocol and the Kyoto Protocol.

The conclusion that the size of the stable coalition is very small was challenged from different angles. Chander and Tulkens [6] use the γ-core concept from cooperative game theory to argue that the grand coalition can be stable. It is assumed that when a sub-coalition of size m deviates the initial coalition falls apart and a game is played between the sub-coalition of size m and the other countries individually. In the symmetric case, this is precisely the game described above with costs $C^c(m)$ and $C^o(m)$, respectively. Because $C^c(m) > C^c(n)$ for all $m < n$, sub-coalitions do not have incentives to deviate and the grand coalition is stable in that sense. This mechanism is similar to trigger strategies in repeated games [12]. In that model, it is assumed that individual deviations are punished by a switch to non-cooperative behaviour. The threat of loosing the cooperative benefits in the future is sufficient to deter deviations and to support the grand coalition, provided the discount factor is high enough. These models are criticized because the assumptions are strong, but weaker assumptions also prove to support large stable coalitions.

Recently the concept of *farsightedness* was introduced (e.g., [7,16]) and applied to IEAs [8,10]. The basic idea is that countries realize that a deviation may trigger further deviations but they do not necessarily assume a full breakdown. In that perspective, the mechanism of internal stability is too weak, since this assumes that no further deviations take place, but the mechanism of the γ-core is too strong, since this assumes that the initial coalition falls apart completely. The idea is furthermore that such a sequence of deviations comes to an end when a (new) stable situation is reached. If an outsider in that new situation is worse off than a coalition member in the initial situation, deviations are deterred.

It is useful for the sequel to label these different stability concepts.[2] The requirement of internal and external stability will be called *myopic stability* and the requirement of farsightedness will be called *static farsighted stability*. For the game in this section, it was found that for p sufficiently small, only the size 2 coalition is myopically stable. The coalition of size 2 therefore satisfies static farsighted stability, if it is assumed that the coalition of size 1 belongs to the set of static farsighted stable coalitions.[3] However, coalitions larger than 2 may be elements of this set as well. The coalition of size 3 is not myopically stable and therefore also not stable in the sense of static farsightedness, but the coalition of size 4 may be stable in that sense. It may be better to be a member of that coalition than an outsider to the coalition of size 2. It is not difficult to show that this is indeed the case. If the total number of countries n is large enough, this process continues and usually generates a bigger set of stable coalitions. The largest one is usually close in number to the grand coalition. Further details will not be investigated here but will be left for the analysis of the dynamic model in Section 4.

The general conclusion from this section is that reasonable behavioural assumptions lead to large stable coalitions. The usually grim picture that (in a Nash equilibrium context) the size of the stable coalition is only 1 or 2 has disappeared. However, in Section 4, it will be shown that this conclusion does not generally hold in a dynamic model in which detection of deviations takes time. This is unfortunate but also has an advantage: the size of the stable coalition depends again on the parameter p, as in the model with myopic stability, but now in a much richer framework that takes account of farsightedness and dynamic emission adjustments. The results can be connected to the characteristics of the environmental problem and thus used to understand the size of IEAs in practice.

4. Stability in a dynamic context

Let us return to the model (1)–(2) presented in Section 2. This is called a difference game because the state transition is given by the difference equation (1). It is assumed that the countries can observe the emission levels and can therefore condition their abatement activities a_i at time t on the level of emissions e at time t. Furthermore, it is assumed that the countries cannot credibly commit to a strategy beforehand and may adjust their strategy at each point in time. This implies that we look for the feedback Nash or Markov perfect equilibrium. The parameters in the state transition (1) and in the cost indicators (2) are time independent. The solution of this difference game will therefore be stationary, meaning that the equilibrium abatement activities will only be a function of the level of emissions and not a function of time. The levels will change over time, of course, but the functional relationship between abatement and emissions remains the same. The feedback Nash or Markov perfect equilibrium can be found by solving the Hamilton–Jacobi–Bellman or dynamic programming equations in the value functions of the countries (e.g., [9]).

Suppose again that m countries form a coalition and that a Nash equilibrium results between the coalition and the other countries individually. For the time being, m is fixed in order to derive the value functions as a

[2]Thanks are due to one of the referees for suggesting these names.

[3]This assumption is not self-evident as can be seen in [14].

function of m. These value functions will subsequently be used to analyse possible changes in the coalition size m, as a consequence of deviations. The dynamic programming equations are

$$\frac{1}{2}mk^{c}e^{2}=\min\left[\sum_{i=1}^{m}\frac{1}{2}a_i^2+\frac{1}{2}mpe^2+\frac{1}{2}\delta mk^{c}\left((1+b)e-\sum_{k=1}^{n}a_k\right)^2\right], \tag{12}$$

$$\frac{1}{2}k^{o}e^{2}=\min\left[\frac{1}{2}a_j^2+\frac{1}{2}pe^2+\frac{1}{2}\delta k^{o}\left((1+b)e-\sum_{k=1}^{n}a_k\right)^2\right],\quad j=m+1,\ldots,n, \tag{13}$$

where $V^{c}(e)=1/2k^{c}e^{2}$ and $V^{o}(e)=1/2k^{o}e^{2}$ denote the value functions for a member of the coalition and for an outsider, respectively. It follows that the first-order conditions for the equilibrium levels of abatement are

$$a_i-\delta mk^{c}\left((1+b)e-\sum_{k=1}^{n}a_k\right)=0,\quad i=1,\ldots,m, \tag{14}$$

$$a_j-\delta k^{o}\left((1+b)e-\sum_{k=1}^{n}a_k\right)=0,\quad j=m+1,\ldots,n. \tag{15}$$

Adding up and simple manipulations lead to the next-period level of emissions and to the feedback Nash equilibrium abatement strategies:

$$(1+b)e-\sum_{k=1}^{n}a_k=\frac{1}{1+\delta(m^2k^{c}+(n-m)k^{o})}(1+b)e, \tag{16}$$

$$a_i=\frac{\delta mk^{c}}{1+\delta(m^2k^{c}+(n-m)k^{o})}(1+b)e,\quad i=1,\ldots,m, \tag{17}$$

$$a_j=\frac{\delta k^{o}}{1+\delta(m^2k^{c}+(n-m)k^{o})}(1+b)e,\quad j=m+1,\ldots,n. \tag{18}$$

Substitution in the dynamic programming equations (12) and (13) yields the following set of two equations for the (positive) parameters k^{c} and k^{o} of the two value functions:

$$k^{c}=p+\frac{\delta k^{c}(1+\delta m^2k^{c})}{(1+\delta(m^2k^{c}+(n-m)k^{o}))^2}(1+b)^2, \tag{19}$$

$$k^{o}=p+\frac{\delta k^{o}(1+\delta k^{o})}{(1+\delta(m^2k^{c}+(n-m)k^{o}))^2}(1+b)^2. \tag{20}$$

As in the static case in Section 3, for $m=0$ in Eq. (20) one gets the equation for the parameter k^{na} of the value function for the feedback Nash equilibrium between individual countries, namely,

$$k^{\mathrm{na}}=p+\frac{\delta k^{\mathrm{na}}(1+\delta k^{\mathrm{na}})}{(1+\delta nk^{\mathrm{na}})^2}(1+b)^2 \tag{21}$$

with abatement strategies

$$a_i^{\mathrm{na}}=\frac{\delta k^{\mathrm{na}}}{1+\delta nk^{\mathrm{na}}}(1+b)e,\quad i=1,\ldots,n \tag{22}$$

and for $m=n$ in Eq. (19) one gets the equation for the parameter k^{fc} of the value function for the full-cooperative solution, namely,

$$k^{\mathrm{fc}}=p+\frac{\delta k^{\mathrm{fc}}}{1+\delta n^2k^{\mathrm{fc}}}(1+b)^2 \tag{23}$$

with abatement strategies

$$a_i^{\mathrm{fc}} = \frac{\delta n k^{\mathrm{fc}}}{1 + \delta n^2 k^{\mathrm{fc}}}(1+b)e, \quad i = 1, \ldots, n. \tag{24}$$

It is not difficult to show (without explicitly solving Eqs. (21) and (23)) that

$$k^{\mathrm{fc}} < k^{\mathrm{na}} < nk^{\mathrm{fc}}. \tag{25}$$

The first inequality in (25) implies that for any initial global emission level, total costs are lower under full cooperation than in the feedback Nash equilibrium between individual countries. The second inequality in (25), with Eqs. (22) and (24), implies that abatement levels are always higher under full cooperation. This is to be expected and confirms the claims made when introducing the model in Section 2. The situation for a coalition of size m with individual outsiders, however, is much more complicated.

The structure of the model is relatively simple. The value functions have a quadratic form without linear or constant terms, so that only one parameter per value function has to be determined. Nevertheless, the set of Eqs. (19) and (20) is still complicated and cannot be solved analytically. We have to resort to a numerical solution, but first a precise definition of what we mean by stability in a dynamic context is needed.

In principle, the value functions $V^{\mathrm{c}}(e_0) = 1/2k^{\mathrm{c}}e_0^2$ and $V^{\mathrm{o}}(e_0) = 1/2k^{\mathrm{o}}e_0^2$ can be analysed in the same way as the cost functions C^{c} and C^{o}, given by Eqs. (9) and (10). Again, it can be shown that the requirement of myopic stability is only satisfied for very small coalitions but static farsighted stability yields a set of large and small stable coalitions. This approach, however, does not take account of the dynamics of the problem. More specifically, it takes time to detect deviations. Therefore, a deviator enjoys extra free-rider benefits before the deviation is detected and this may change the situation. In a repeated game, the stability properties would not change, if the discount factor is high enough. For example, suppose that the static game of Section 3 is repeated and that the coalitions of size m and m^+ ($m > m^+$) are subsequent elements in the set of coalitions that satisfy static farsighted stability. It follows that $C^{\mathrm{c}}(m) < C^{\mathrm{o}}(m^+)$. Furthermore, suppose that a deviator has costs $C^{\mathrm{d}}(m) < C^{\mathrm{c}}(m)$ before detection and becomes an outsider to the coalition of size m^+ afterwards. Deviation is deterred if

$$\frac{1}{1-\delta}C^{\mathrm{c}}(m) < C^{\mathrm{d}}(m) + \frac{\delta}{1-\delta}C^{\mathrm{o}}(m^+) \tag{26}$$

and this holds if the discount factor is high enough, i.e. if

$$\delta > \frac{C^{\mathrm{c}}(m) - C^{\mathrm{d}}(m)}{C^{\mathrm{o}}(m^+) - C^{\mathrm{d}}(m)}. \tag{27}$$

However, the game introduced in Section 2, and analysed in this section, changes over time. Owing to the abatement activities of the countries in the first period, the level of emissions and therefore the costs decrease. The future values of coalition members and outsiders do not only depend on the parameters k^{c} and k^{o} of the value functions but also on the future levels of emissions. Therefore, it may happen that the threat of becoming an outsider to the smaller stable coalition is not sufficiently strong to deter deviations, even if the discount factor is close to one. Based on this, we introduce the concept of *dynamic farsighted stability*. A coalition satisfies this type of stability if deviations are deterred, assuming that the deviator becomes an outsider to the smaller stable coalition that results *in the next period*, for a *different* level of emissions. The idea is similar to static farsighted stability. The set of stable coalitions is constructed from below. The size 1 coalition or the feedback Nash equilibrium between individual countries belongs to this set. By subsequently checking $m = 2$, 3 (if necessary), etcetera, the next element in the set can be found, in case it exists. If a next element is found, the procedure is continued until the full set of stable coalitions is characterized. This will become clear in the example that follows, but first we have to characterize the value function of the country that deviates from a size m coalition (which is similar to the right-hand side of inequality (26)).

Suppose therefore that a member of the coalition of size m considers deviating but the other countries in the coalition cannot detect this deviation and react to it before the next point in time. This implies that the deviating country can enjoy free-rider benefits for one period of time but also has to face the consequences afterwards. The deviating country becomes an outsider to a smaller coalition of size m^+ at the next point in time. The numbers m and m^+ still have to be determined. Denote the parameter of the value function of an outsider to such a size m^+ coalition by k^{o+}. It follows that the optimal behaviour of a deviating country is described by the dynamic programming equation

$$\frac{1}{2}k^{d}e^2 = \min\left[\frac{1}{2}a_d^2 + \frac{1}{2}pe^2 + \frac{1}{2}\delta k^{o+}\left((1+b)e - \sum_{k=1}^{n} a_k\right)^2\right], \tag{28}$$

where a_d denotes the optimal abatement level of the deviating country and where $V^d(e) = 1/2k^de^2$ denotes the value function of this country. Note that the last term in the dynamic programming equation (28) shows the different aspects that were discussed earlier. It indicates the discounted costs of becoming an outsider to a smaller coalition but for a lower level of emissions. The first-order condition becomes

$$a_d - \delta k^{o+}\left((1+b)e - a_d - \sum_{k\neq d} a_k\right) = 0. \tag{29}$$

Because (in this time period) the other countries are assumed to stick to their strategy as a member of the coalition of size m or as an outsider to that coalition (Eqs. (17) and (18), respectively), this leads to

$$a_d = \frac{\delta k^{o+}}{1+\delta k^{o+}}\frac{1+\delta m k^c}{1+\delta(m^2k^c + (n-m)k^o)}(1+b)e, \tag{30}$$

$$(1+b)e - \sum_{k=1}^{n} a_k = \frac{1}{1+\delta k^{o+}}\frac{1+\delta m k^c}{1+\delta(m^2k^c + (n-m)k^o)}(1+b)e. \tag{31}$$

It follows that the parameter of the value function of the deviating country is given by

$$k^d = p + \frac{\delta k^{o+}}{1+\delta k^{o+}}\frac{(1+\delta m k^c)^2}{(1+\delta(m^2k^c + (n-m)k^o))^2}(1+b)^2. \tag{32}$$

Deviations are deterred if $k^c < k^d$, where k^d is given by Eq. (32) and k^c (together with k^o) is given by the set of Eqs. (19) and (20).

In Section 3, it was shown that for a static version of the game, a set of coalitions exists that satisfy static farsighted stability, with the largest coalition close to the grand coalition; in this section it was shown that this set of coalitions is also stable if the static game is repeated and if the discount factor is high enough. The question is whether a set of coalitions (larger than 1) exists that satisfy dynamic farsighted stability. In the sequel, the discount factor δ is first taken equal to 1. Solutions to the dynamic programming equations still exist because of the special structure of the problem. At the end, it is shown what the implications are of lowering the discount factor δ.

The set of Eqs. (19) and (20) cannot be solved analytically and therefore we have to resort to a numerical analysis. Values have to be given to three parameters: the factor b, indicating the business-as-usual growth in emissions, the number of countries n, and finally p, denoting the relative weight between damage and abatement costs. The value of b affects the abatement levels but it does not affect the qualitative results of the analysis. Therefore, we only present the case $b = 0$ here. The number of countries n is not essential for the qualitative results either. We have chosen $n = 8$, because then the full spectrum of results can be presented. This number is also not unrealistic, since international negotiations usually take place between blocks of countries. The value of p, however, is essential for the results. It will be shown that only for very low values of p a set of coalitions (larger than 1) exists that satisfy dynamic farsighted stability.

Tables 1–4 show deviation costs associated with coalitions of size 1–8, with 8 representing the grand coalition, for four values of p. The first two rows in each table represent the solutions to the set of Eqs. (19) and (20) for each value of m. For $m = 1$, k^c and k^o are equal because this represents the feedback Nash equilibrium between individual countries. For $m = 8$, there are no outsiders. Note that a static analysis with respect to the value functions V^c and V^o gives the usual picture. Myopic stability gives a stable coalition of size 1 in all these cases, since $k^c(m) > k^o(m-1)$ for all $m > 1$ (compare with Eq. (11) and the text thereafter). Static farsighted stability, however, always gives a set of stable coalitions larger than 1. In Table 1, this set consists of

Table 1
Insider/outsider/deviator costs, coalition size m, $p = 1$ ($b = 0$, $n = 8$)

	$m = 1$[a,b]	$m = 2$	$m = 3$	$m = 4$	$m = 5$	$m = 6$[a]	$m = 7$	$m = 8$
k^c	1.0245	1.0425	1.0456	1.0391	1.0311	*1.0244*	1.0192	1.0154
k^o	1.0245	1.0161	1.0085	1.0043	1.0023	1.0013	1.0007	
k^d		1.0379	1.0363	1.0287	1.0216	1.0162	1.0124	1.0097

[a]Static farsighted stable.
[b]Dynamically farsighted stable.

Table 2
Insider/outsider/deviator costs (10^{-2}), coalition size m, $p = 10^{-2}$ ($b = 0$, $n = 8$)

	$m = 1$[a,b]	$m = 2$	$m = 3$[a]	$m = 4$	$m = 5$	$m = 6$	$m = 7$[a]	$m = 8$
k^c	3.0169	3.1472	*3.0143*	2.7150	2.4231	2.1841	*1.9953*	1.8463
k^o	3.0169	2.6455	*2.1598*	1.8128	1.5913	1.4449	1.3434	
k^d		3.0050	2.8304	2.5850	2.3464	2.1370	1.9594	1.8105

[a]Static farsighted stable.
[b]Dynamically farsighted stable.

Table 3
Insider/outsider/deviator costs (10^{-3}), coalition size m, $p = 10^{-4}$ ($b = 0$, $n = 8$)

	$m = 1$[a,b]	$m = 2$	$m = 3$[a,b]	$m = 4$	$m = 5$	$m = 6$[a]	$m = 7$[b]	$m = 8$
k^c	2.6215	2.7386	*2.5829*	2.2505	1.9307	*1.6702*	1.4644	1.3010
k^o	2.6215	2.2307	*1.6960*	1.3082	1.0582	0.8896	0.7692	
k^d		2.6192	*2.5945*	1.6899	1.6612	1.6317	*1.6022*	0.7687

[a]Static farsighted stable.
[b]Dynamically farsighted stable.

Table 4
Insider/outsider/deviator costs (10^{-4}), coalition size m, $p = 10^{-5}$ ($b = 0$, $n = 8$)

	$m = 1$[a,b]	$m = 2$	$m = 3$[a,b]	$m = 4$	$m = 5$	$m = 6$[a,b]	$m = 7$	$m = 8$
k^c	8.2042	8.5716	*8.0749*	7.0175	6.0010	*5.1741*	4.5212	4.0032
k^o	8.2042	6.9650	*5.2650*	4.0319	3.2379	2.7027	2.3207	
k^d		8.2019	*8.1765*	5.2585	5.2287	*5.1977*	2.7013	2.6851

[a]Static farsighted stable.
[b]Dynamically farsighted stable.

size 1 and size 6 (since $k^c(6) < k^o(1)$). In Table 2, this set consists of size 1, size 3 (since $k^c(3) < k^o(1)$) and size 7 (since $k^c(7) < k^o(3)$). In Tables 3 and 4 this set consists of size 1, size 3 (since $k^c(3) < k^o(1)$) and size 6 (since $k^c(6) < k^o(3)$). Note that the stability properties get better when the parameter p gets smaller. The stable set is larger in Tables 2–4 than in Table 1. Although the countries can coordinate on a larger coalition (size 7) in Table 2 than in Tables 3 and 4 (size 6), this is due to the fixed number of countries n: for a sufficiently large n, the stable set will prove to be larger in Tables 3 and 4 than in Table 2. The general conclusion is that in all these cases static farsighted stability allows the countries to coordinate on a large stable coalition.

In order to be able to investigate dynamic farsighted stability, we have to calculate k^d for the different coalition sizes (with Eq. (32)) and at the same time construct the stable set. The size 1 coalition or the feedback

Nash equilibrium between individual countries belongs to this set. Therefore, in the first step of the algorithm $m^+ = 1$ and $k^{o+} = k^o(1)$. Then $k^d(2)$ can be calculated and compared with $k^c(2)$. If $k^d(2) > k^c(2)$, it follows that deviations are deterred and the coalition of size 2 is stable. If not, $k^d(3)$ can be calculated and compared to $k^c(3)$, etcetera. Two things can happen: either an element m^* of the stable set is found or $k^d(m) < k^c(m)$ for all $m > 1$. In the first case m^+ is changed to m^* and k^{o+} is changed to $k^o(m^*)$ and the procedure is continued for larger m. In the second case, the stable set consists of only the size 1 coalition. This last situation occurs in both Tables 1 and 2. However, in Table 3 first the coalition of size 3 is found to satisfy dynamic farsighted stability, since $k^d(3) > k^c(3)$, and then also the coalition of size 7, since $k^d(7) > k^c(7)$. In Table 4 the stable set consists of size 3 and size 6, since $k^d(3) > k^c(3)$ and $k^d(6) > k^c(6)$. In contrast to the conclusion above under static farsighted stability, now it is found that countries can only coordinate on a large stable coalition, if the parameter p is very small (Tables 3 and 4).

It is time to take stock of what we have found. Initially the usual picture reappears. The concept of myopic stability gives very small stable coalitions, but the concept of static farsighted stability gives both small and large stable coalitions and allows the countries to coordinate on the largest stable coalition. This conclusion partly changes, however, if it is assumed that detection of deviations takes some time. The reason is that in that time abatement occurs and therefore the level of emissions decreases prior to detection. It follows that costs will have decreased, and therefore the threat of triggering a smaller coalition may not be sufficient to deter defections anymore. It can be seen in Tables 1–4 that deviations are only deterred when the parameter p is very small (as in Tables 3 and 4). It is interesting to see that the value of the parameter p again determines whether a coalition is stable or not, as in the case of myopic stability for the simple static model in Section 3. While the intuition in the simple model was based on the steepness of the best-reply function, here there is an additional aspect. The parameter p denotes the relative weight of the costs attached to the emission levels. If p is small, the level of emissions is slowly adjusted and therefore the game changes slowly over time. If p is large, however, the level of emissions and therefore costs decrease substantially in each time period. Apparently, in the last situation the trigger mechanism does not sustain large coalitions anymore. Note that in the case of myopic stability for the simple static model in Section 3, the distinction that p made was only between a size 2 and a size 1 coalition; with dynamic farsighted stability in a full dynamic model, the distinction is between a large coalition and a size 1 coalition.

It is interesting to see what the implication is of lowering the discount factor δ. It has an effect on all the costs, of course, but it can also have an effect on the set of stable coalitions. If the parameter p is relatively large, as in the context of Tables 1 and 2, a reduction of the discount factor δ from 1.0 to 0.9 does not change the set of coalitions that satisfy dynamic farsighted stability: it still only consists of the size 1 coalition. If, however, p is sufficiently small, as in the context of Tables 3 and 4, the set of stable coalitions consists of size 1, size 2, size 4 and size 7, following a reduction of δ to 0.9. Thus, if a set of stable coalitions larger than 1 exists, this set contains more elements for $\delta = 0.9$ than for $\delta = 1$.

In some sense, the circle is closed. The static model in Section 3 led to the following conclusion, formulated by Finus [11]: the steeper the best-reply function, the lower the number of participants. Since the parameter p is an indicator for the steepness of the best-reply function, the conclusion can be reformulated as follows: the higher p, the lower the size of the coalition. Under static farsightedness, the dependence on p had disappeared. This paper shows, however, that the conclusion is re-established by integrating the behavioural reaction patterns and the emissions adjustment processes in one dynamic context: only if p is very small, large coalitions can be sustained.

It is encouraging that this conclusion gets wide support with different types of models. On the other hand, the conclusion does not seem to reflect the practice of IEAs. It is generally argued [3,11] that the Montreal Protocol can be characterized by a high value of p, since abatement costs are relatively small as compared to the costs of holes in the ozone layer. The Kyoto Protocol, on the other hand, can be characterized by a medium value of p, since both abatement costs and costs of climate change are high. Therefore, it should be easier to sustain the Kyoto Protocol, but the opposite has occurred. It can be argued that p for the Kyoto Protocol is not small enough to sustain a large coalition, but this does not explain why the Montreal Protocol has been a success. The reason is probably that due to technological development, the issue disappeared [2]: it proved not to be difficult to do without CFCs. The Agreement was interesting for many reasons but not so much because it solved a problem. Theory may also give an explanation. Barrett [2] shows for the static model that the result turns around if for the game between the coalition and the outsiders, the Stackelberg concept is used instead of the Nash equilibrium. Finus [11] also concludes the Stackelberg assumption yields a larger number of participants, the steeper is the best-reply function; this result is in line with the Montreal Protocol. Finus [11] is critical of the Stackelberg concept in this context, but it may anyhow be interesting to investigate whether the result for the dynamic model also turns around, if the Stackelberg concept is used. This question is left for further research. At this point, it is concluded that theory predicts that it will be hard to sustain a large coalition, unless damage costs are very small as compared with abatement costs. This is bad news for IEAs, but it only stresses the importance to design agreements in a way that may overcome these mechanisms. This is also left for further research.

5. Conclusion

Stability of IEAs is a heavily debated issue. Some are optimistic and expect that the grand coalition can be sustained. Their arguments are based on the core concept in cooperative game theory or on trigger mechanisms in repeated games. Others are pessimistic and expect that only very small coalitions can be sustained. They base themselves on the concept of internal and external stability, which is called myopic stability in this paper. The idea of farsightedness can be used to reconcile these approaches: both large and small stable coalitions can occur, and countries may be able to coordinate on a large stable coalition. This is good news for the optimists.

This paper first clarifies these issues and then takes the analysis one step further. The main issue is compliance or deterrence of deviations. This paper shows that if it takes time to detect deviations, trigger mechanisms may not work. The reason is simply that the game changes over time. This leads to the result that large coalitions can only be sustained, if damage costs are very small as compared with abatement costs. In static models under myopic stability a similar result appeared, but now the result is stronger and it occurs in a much richer dynamic framework. The result implies that in the most important situations, in which environmental damage costs are relatively substantial, large coalitions will not easily arise. Unfortunately, the pessimistic view still prevails. The challenge remains to find ways in which stability properties of large coalitions can be improved. This is left for further research.

Acknowledgments

The author is grateful to Gérard Gaudet, Reyer Gerlagh, Michael Hoel, Eric Maskin, Alistair Ulph, three anonymous referees, the editor, and seminar participants in Victoria, Paris, New Delhi, Antwerp and Rethymnon for comments on earlier versions of this paper; also the hospitality of CAS in Oslo is gratefully acknowledged.

References

[1] C. d'Aspremont, J. Jacquemin, J. Gabszewicz, J. Weymark, On the stability of collusive price leadership, Can. J. Econ. 16 (1983) 17–25.
[2] S. Barrett, Self-enforcing international environmental agreements, Oxford Econ. Papers 46 (1994) 878–894.
[3] S. Barrett, Environment and Statecraft, The Strategy of Environmental Treaty-making, Oxford University Press, Oxford, 2003.
[4] F. Bloch, Non-cooperative models of coalition formation in games with spillovers, in: C. Carraro, D. Siniscalco (Eds.), New Directions in the Economic Theory of the Environment, Cambridge University Press, Cambridge, 1997, pp. 311–352.
[5] C. Carraro, D. Siniscalco, Strategies for the international protection of the environment, J. Publ. Econ. 52 (1993) 309–328.
[6] P. Chander, H. Tulkens, A core-theoretic solution for the design of cooperative agreements on transfrontier pollution, Int. Tax Publ. Finance 2 (1995) 279–293.
[7] M. Chwe, Farsighted coalitional stability, J. Econ. Theory 63 (1994) 299–325.
[8] E. Diamantoudi, E. Sartzetakis, International environmental agreements—the role of foresight, Working Paper, 2002.
[9] E. Dockner, S. Jørgensen, N. Long, G. Sorger, Differential Games in Economics and Management Science, Cambridge University Press, Cambridge, UK, 2000.
[10] J. Eyckmans, On the farsighted stability of international climate agreements, Working Paper, 2003.

[11] M. Finus, Stability and design of international environmental agreements: the case of transboundary pollution, in: H. Folmer, T. Tietenberg (Eds.), The International Yearbook of Environmental and Resource Economics 2003/2004, Edward Elgar, Cheltenham, 2003, pp. 82–158.
[12] J. Friedman, Game Theory with Applications to Economics, Oxford Economic Press, New York, 1986.
[13] M. Hoel, International environmental conventions: the case of uniform reductions of emissions, Environ. Resource Econ. 2 (1992) 141–159.
[14] L. Karp, J. Zhao, A proposal to reform the Kyoto Protocol: the role of escape clauses and foresight, Working Paper, 2007.
[15] F. van der Ploeg, A. de Zeeuw, International aspects of pollution control, Environ. Resource Econ. 2 (1992) 117–139.
[16] D. Ray, R. Vohra, Coalitional power and public goods, J. Political Econ. 109 (2001) 1355–1384.
[17] S.-S. Yi, Stable coalition structures with externalities, Games Econ. Behav. 20 (1997) 201–237.

Available online at www.sciencedirect.com
SCIENCE DIRECT®

ELSEVIER

Journal of Economic Dynamics & Control 28 (2003) 79–99

JOURNAL OF Economic Dynamics & Control

www.elsevier.com/locate/econbase

Transfers to sustain dynamic core-theoretic cooperation in international stock pollutant control

Marc Germain[a,*], Philippe Toint[b], Henry Tulkens[a,c], Aart de Zeeuw[d]

[a] *CORE, Université Catholique de Louvain, voie du roman pays 34, 1348 Louvain-la-Neuve, Belgium*
[b] *Département de Mathématique, Facultés Universitaires Notre-Dame de la Paix, 61, rue de Bruxelles, 5000 Namur, Belgium*
[c] *Facultés Universitaires Saint-Louis, Boulevard du Jardin botanique 43, 1000 Bruxelles, Belgium*
[d] *Department of Economics and CentER, Tilburg University, 5000 LE Tilburg, The Netherlands*

Abstract

For international environmental agreements aiming at world efficiency in the presence of transboundary flow pollution, it is known that, in a static context, efficiency and stability in the sense of the core of a cooperative game can be achieved using appropriately defined transfers between the countries involved. However, for accumulating pollutants, such as CO_2 in the atmosphere, a dynamic analysis is required.

This paper provides a transfer scheme for which a core property is proved analytically in a dynamic (closed-loop) game theoretic context. The characteristic function of the cooperative dynamic game yielding this result is discussed and an algorithm to compute the transfers numerically is presented and tested on an example. The transfers are also compared with an open-loop formulation of the model.

JEL classification: C73; D62; F42; H4; Q3

Keywords: Transfrontier pollution; Stock pollutant; Dynamic cooperative games; Coalitions; Core solution

* Corresponding author. CORE, voie du roman pays, 1348 Louvain-la-Neuve, Belgium. Tel.: +32-10-474321; fax: +32-10-474301.

E-mail address: germain@core.ucl.ac.be (M. Germain).

0165-1889/02/$ - see front matter
PII: S0165-1889(02)00107-0

1. Introduction

This paper presents a cooperative game theoretic analysis of the economics of international agreements on transnational pollution control, when the environmental damage arises from stock pollutants that accumulate (and possibly decay). The issues raised by the necessity of cooperation amongst the countries involved, if a social optimum is to be achieved, have already been addressed in the literature in terms of game theory concepts; see e.g. Mäler (1989). However, most of these contributions have only dealt with static (one shot) games, which are only suitable for flow pollution models. With stock pollutants the problem acquires an intertemporal dimension. In this case, dynamic game theory is a more appropriate tool of analysis, as is done in, e.g. van der Ploeg and de Zeeuw (1992), Kaitala et al. (1992), Hoel (1992), Tahvonen (1994), Petrosjan and Zaccour (1995).

With exception of the last paper, these contributions leave aside the issue of the *voluntary* implementation of the international optimum. This is an important drawback since no supranational authority can be called upon to impose the optimum in a context where the countries' interest in cooperation diverges strongly between one another, and especially if some countries loose when the social optimum is implemented. In view of ensuring such implementation, it has often been suggested that financial transfers between the countries involved would provide incentives towards cooperation. [1]

This property, understood in the sense of the core of a cooperative game, has in effect been demonstrated by Chander and Tulkens (1995, 1997), who propose a particular transfer scheme based on parameters reflecting the relative intensities of the countries' environmental preferences.

This result, established for flow pollutants only (that is, in a static game model), has been extended by Germain et al. (1998a) to the larger context of open-loop dynamic games. But this extension suffers from the fact that it implies the restrictive assumption that negotiations take place once and for all and bear on emissions trajectories that extend until the end of the planning horizon. Agreements are thus binding all along. The aim of the present paper is precisely to relax this assumption: *here, cooperation is negotiated at each period*, i.e. players reevaluate at each time the interest of cooperation, taking account of the current stock of pollutant. This approach is one of closed-loop (or feedback) dynamic games.

In this context, financial transfers are considered again. They are now formulated in such a way that they induce cooperation in the core-theoretic sense at each period. The model thus yields a sequence of cooperative international agreements. This result is obtained under more general convexity conditions than in the earlier attempt by Germain et al. (1998b) that was restricted to linear damage cost functions.

Other approaches to the cooperation issue concentrate on the stable number of signatories by considering the trade-off for an individual country between joining some

[1] In a two-countries framework, Petrosjan and Zaccour (1995) use financial transfers to obtain a cooperative solution as a Nash equilibrium. However, being limited to two countries, they cannot deal with the issue of coalitions, which is one of our aims here. Kverndokk (1994) deals both with coalitions and financial transfers for the greenhouse gas problem, but he resorts neither to dynamic games, nor to cooperative ones.

coalition and staying out of that coalition (see Carraro and Siniscalco, 1993; Barrett, 1994). A critical comparison between these approaches and the one of this paper is given in Tulkens (1998). On the other hand, one must emphasize that this paper does not intend to describe the way actual international negotiations are conducted (e.g. the Kyoto protocol). It chooses instead a normative approach, identifying transfers that make agreements less likely to be broken. For an alternative approach in this sense, see Bahn et al. (1998).

The structure of the paper is as follows. Section 2 presents the international stock pollutant model and characterizes the international optimum as well as the closed-loop Nash non-cooperative equilibrium. In Section 3, financial transfers are formulated so as to ensure that each country is not worse off when it participates, either in the finite or in the infinite horizon case. In Section 4, transfers are further specified so as to achieve stability in the core-theoretic sense, followed by a discussion of the characteristic function of dynamic cooperative games yielding that result.

Section 5 describes an algorithm that solves the model developed in Sections 3 and 4, and presents numerical results obtained by application of the algorithm to a simple climatic change model with three regions, including a comparison with the open-loop case. Section 6 is a conclusion.

2. International optimality and non-cooperative equilibrium

Our economic model is written in discrete time. Consider n countries indexed by $i \in \mathcal{N} = \{1,2,\dots,n\}$ and some planning period $\mathcal{T} = \{1,2,\dots,T\}$ (T, the planning horizon, is a positive integer, possibly infinite). In each country, pollution is entailed by economic activity: let $\mathbf{E}_t = (E_{1t},\dots,E_{nt})'$ denote the vector of the different countries' emissions of a certain pollutant at time t. These emissions spread uniformly in the atmosphere and contribute to a stock of pollutant S according to the equation

$$S_t = [1-\delta]S_{t-1} + \sum_{i=1}^{n} E_{it}, \tag{1}$$

where the initial stock of pollutant S_0 is given and where δ is the pollutant's natural rate of degradation ($0 < \delta < 1$). This model describes, for example, the basics of the climate change problem where the flows of emissions of greenhouse gases accumulate into a stock, which only gradually assimilates and which is the cause of the climate change.

This stock of pollutant causes damages to each country's environment. For country i ($i \in \mathcal{N}$), these damages during period t are measured in monetary terms by the function $D_i(S_t)$, where D_i is supposed to be a differentiable, increasing and convex function of the current stock S_t ($D_i' > 0$, $D_i'' \geqslant 0$). As described in (1), this current stock is a function of the inherited stock S_{t-1} and of the current emissions $\mathbf{E}_t$. The only way to control the stock of pollutant is through the control of emissions.[2] More precisely, each country i can reduce its own emissions, and the cost of doing this is described

[2] That is reducing pollution is done through the reduction of emissions, and not through the cleaning of the environment.

by a differentiable, decreasing and strictly convex function $C_i(E_i)$ ($C_i' < 0, C_i'' > 0$). $C_i(E_i)$ measures the total costs incurred by country i from limiting its emissions to E_i. The convexity of the function reflects the intuitive idea that the marginal cost of reducing emissions is higher for lower levels of emissions.

From Eq. (1), it is clear that the damages to the environment of country i will depend on the emissions of all countries. In what follows, we will consider two different modes of behaviour of the countries. A first one assumes that they behave in an internationally optimal way, i.e. that each of them takes account of the impact of its pollution not only on itself but on all other countries as well. In this case, countries *jointly* choose at each period their emission levels in order to minimize total discounted costs, i.e. for each $t \in \mathscr{T}$, they solve the following problem:

$$\min_{\{\mathbf{E}_s\}_{s\in\{t,\dots,T\}}} \left\{ \sum_{s=t}^{T} \sum_{i=1}^{n} \beta^s [C_i(E_{is}) + D_i(S_s)] \right\}$$

subject to the constraints (1) and $E_{it} \geqslant 0$ ($\forall t \in \mathscr{T}, \forall i \in \mathscr{N}$), β is the discount factor ($0 < \beta \leqslant 1$). According to Bellman's principle of optimality, the solution can be found by solving the dynamic programming equations

$$W(T, S_{T-1}) = \min_{\mathbf{E}_T} \left\{ \sum_{i=1}^{n} [C_i(E_{iT}) + D_i(S_T)] \right\}, \tag{2}$$

$$W(t, S_{t-1}) = \min_{\mathbf{E}_t} \left\{ \sum_{i=1}^{n} [C_i(E_{it}) + D_i(S_t)] + \beta W(t+1, S_t) \right\},$$

$$t = 1, 2, \dots, T-1 \tag{3}$$

subject to the constraints (1) and $E_{it} \geqslant 0$ ($\forall t \in \mathscr{T}, \forall i \in \mathscr{N}$). In (2) and (3), W is called the *value function*. As the total costs of all countries are minimized, the resulting trajectories of emissions and stock constitute the *international optimum*. The convexity of the functions C_i and D_i ($\forall i \in \mathscr{N}$) suffices to guarantee that the minimum exists and is unique.

In an alternative mode of behaviour, one may assume that countries behave non-cooperatively in the sense of a Nash equilibrium, where each of them minimizes at each period only its own discounted costs, taking as given the emissions of the other countries. Formally, at each $t \in \mathscr{T}$, each country $i \in \mathscr{N}$ solves the following problem:

$$\min_{\{E_{is}\}_{s\in\{t,\dots,T\}}} \left\{ \sum_{s=t}^{T} \beta^s [C_i(E_{is}) + D_i(S_s)] \right\},$$

subject to the constraints (1), $E_{it} \geqslant 0$ and $E_{jt}\, j \neq i$ given. Within the framework of dynamic programming this leads to the value functions

$$N_i(T, S_{T-1}) = \min_{E_{iT}} \{[C_i(E_{iT}) + D_i(S_T)]\}, \tag{4}$$

$$N_i(t, S_{t-1}) = \min_{E_{it}} \{[C_i(E_{it}) + D_i(S_t)] + \beta N_i(t+1, S_t)\}, \quad t = 1, 2, \dots, T-1 \tag{5}$$

under the constraints (1) and $E_{it} \geqslant 0$ $\forall t \in \mathscr{T}$. The convexity of the functions C_i and D_i suffices to guarantee that the Nash equilibrium exists and is unique.

The family of trajectories of emissions thus determined for each i in N, together with the resulting stock, constitute a *non-cooperative feedback Nash equilibrium* (Başar and Olsder, 1995).[3] The value functions N_i characterize the costs-to-go at this equilibrium for each country i. The emissions are functions of the current stock of pollutant and the equilibrium has the property of *Markov perfectness* (Maskin and Tirole, 1988), in the sense that it remains an equilibrium if the game is restarted at any intermediate year t from any value of the stock of pollutant.

At the international optimum, and contrary to what happens at the Nash equilibrium, each country takes account of the impact of its pollution on the environment of all other countries. Therefore, from a collective point of view, the international optimum is better than the feedback Nash equilibrium. Nothing ensures, however, that this is also true at the individual level. Indeed, countries being different, it is possible that some country at sometime t is better off at the non-cooperative equilibrium than at the optimum, so that cooperation is not profitable for this country (at least at time t). The same can occur for subsets of countries—i.e. coalitions—in the sense that, by limiting cooperation to such coalitions, the members of the latter could be better off than at the international optimum. The aim of this paper is to show that there exist financial transfers between countries that can make each one of them interested in achieving the international optimum at all periods t (individual rationality), and are such that, in addition, no sub-group of countries has ever an incentive to form a coalition and enact an optimum for itself only (coalitional rationality). We are thus aiming at a *cooperative* international optimum.

To make the above argument precise, it is important to state explicitly what the fallback position is for each country when no transfers occurs. At each time t one could take the non-cooperative feedback Nash equilibrium from t onwards as such point of reference, and determine the transfers accordingly. However, one should not neglect the fact that countries know that later on, thanks to the cooperative transfers to which they will have access, they will be better off than at the non-cooperative Nash equilibrium. Hence, a better point of reference at time t is: *non-cooperation at time t, followed by cooperation afterwards.*[4]

This is a rational expectations argument, that we introduce formally in Section 3 to deal with individual rationality, and that we use again in Section 4 to deal with coalitional rationality in the framework of a dynamic cooperative game.

3. Transfers to sustain individual rationality at the international optimum

3.1. Transfers at final time T

Let us start by determining which transfers yield gains for all countries when they cooperate in the last period, for any level of the stock of pollutant S inherited from the

[3] See Mäler and de Zeeuw (1998) and van der Ploeg and de Zeeuw (1992) for applications of such an equilibrium to acid rains and climate change, respectively.

[4] This idea was already put forward in Houba and de Zeeuw (1995) in the framework of a dynamic bargaining problem.

past. In the non-cooperative equilibrium the countries are supposed to solve problem (4). Country i's total cost is then

$$N_i(T,S) = C_i(E_{iT}^N) + D_i(S_T^N), \quad i \in \mathcal{N}, \tag{6}$$

where E_{iT}^N denotes the emission equilibrium level and S_T^N denotes the resulting stock of pollutant given by

$$S_T^N = [1-\delta]S + \sum_{i=1}^{n} E_{iT}^N. \tag{7}$$

If countries cooperate, they jointly solve problem (2). Country i's total cost is

$$W_i(T,S) = C_i(E_{iT}^*) + D_i(S_T^*), \tag{8}$$

where E_{iT}^* is the optimal emission level and S_T^* is the optimal stock of pollutant, given by

$$S_T^* = [1-\delta]S + \sum_{i=1}^{n} E_{iT}^*. \tag{9}$$

By definition of the optimum, one verifies that

$$W(T,S) \triangleq \sum_{i=1}^{n} W_i(T,S) \leqslant \sum_{i=1}^{n} N_i(T,S) \triangleq N(T,S). \tag{10}$$

The difference between the two sides of this inequality measures the *ecological surplus* resulting from international cooperation.

However, (10) is not sufficient to ensure cooperation. Indeed, if $\exists i \in \mathcal{N}$ such that $W_i(T,S) > N_i(T,S)$, then country i will not cooperate without financial compensation for the higher cost it incurs. Since dynamic programming reduces the choice of emissions to one period at the time, one can use the transfers formula proposed by Chander and Tulkens (1997) in a static framework. Let

$$\theta_i(T,S) = -[W_i(T,S) - N_i(T,S)] + \mu_{i,T}[W(T,S) - N(T,S)] \tag{11}$$

be the transfer (<0 if received, >0 if paid) to country i at time T, where $\mu_{i,T} \in]0,1[$, $\forall i \in \mathcal{N}$ and $\sum_{i=1}^{n} \mu_{i,T} = 1$. Then country i's total cost *including* transfers becomes

$$\tilde{W}_i(T,S) = W_i(T,S) + \theta_i(T,S). \tag{12}$$

By construction, the budget of the transfers defined by (11) is balanced (i.e. $\sum_{i=1}^{n} \theta_i(T,S) = 0$). Since

$$\tilde{W}_i(T,S) - N_i(T,S) = \mu_{i,T}[W(T,S) - N(T,S)] \leqslant 0, \quad \forall i \in \mathcal{N} \tag{13}$$

cooperation with transfers is *individually rational* at time T, in the sense that each country has interest to participate whatever the inherited stock of pollutant S.[5]

[5] The fact that $\mu_{i,T}$ cannot be equal to 0 ensures that country i will benefit from cooperation if $W(T,S) < N(T,S)$. The fact that $\mu_{i,T}$ cannot be equal to 1 excludes that country i monopolizes all the gains of cooperation.

3.2. Transfers at earlier periods

Countries know that, whatever they do at $T-1$, financial transfers exist (defined by (11)) that make the international optimum at T preferable for each of them with respect to the non-cooperative equilibrium. Let us assume that these transfers induce cooperation,[6] *and that countries therefore expect, at $T-1$, that they will cooperate in period T*. This is the rational expectations assumption announced at the end of Section 2. The problem we wish to consider now is whether under this assumption transfers can be designed that make the countries interested to cooperate at $T-1$ as well.

In the absence of cooperation at $T-1$, each country i minimizes its own discounted total costs over the two periods $T-1$ and T, expecting cooperation and transfers in T. Thus, given the emissions of the other countries, country i solves problem (5) for $t=T-1$ with N_i under the *min* operator replaced by $\tilde{W}_i$. This leads to

$$V_i(T-1,S)=\min_{E_{i,T-1}}\{[C_i(E_{i,T-1})+D_i(S_{T-1})]+\beta\tilde{W}_i(T,S_{T-1})\},\quad i\in\mathcal{N} \tag{14}$$

under the constraints (1) and $E_{i,T-1}\geqslant 0$, $\forall i\in\mathcal{N}$.[7] This yields an equilibrium characterized at time $T-1$ by emission levels $\mathbf{E}^V_{T-1}$ as functions of the initial stock S at time $T-1$. The value function

$$V_i(T-1,S)=C_i(E^V_{i,T-1})+D_i(S^V_{T-1})+\beta\tilde{W}_i(T,S^V_{T-1}),\quad i\in\mathcal{N}, \tag{15}$$

where

$$S^V_{T-1}=[1-\delta]S+\sum_{i=1}^{n}E^V_{i,T-1} \tag{16}$$

denotes country i's discounted equilibrium costs. We will call this equilibrium the *fallback non-cooperative equilibrium* at time $T-1$.

In the case where all countries cooperate, they solve problem (3) for $t=T-1$. Optimal levels of emissions and of the resulting stock of pollutant are denoted by $\mathbf{E}^*_{T-1}$ and S^*_{T-1}, respectively. Both are functions of S. This yields

$$W_i(T-1,S)=C_i(E^*_{i,T-1})+D_i(S^*_{T-1})+\beta\tilde{W}_i(T,S^*_{T-1}) \tag{17}$$

which is country i's part in the optimal total discounted costs, *taking into account the transfers and the resulting cooperation expected in T*.

As in period T (see (10)), one verifies that

$$W(T-1,S)\triangleq\sum_{i=1}^{n}W_i(T-1,S)\leqslant\sum_{i=1}^{n}V_i(T-1,S)\triangleq V(T-1,S). \tag{18}$$

[6] Note that, following Chander and Tulkens (1997, Section 5), one could indeed obtain the cooperative optimum with transfers as an equilibrium, called *ratio-equilibrium*.

[7] Since the value functions $\tilde{W}_i(T,S_{T-1})$ contain transfers that sum up to zero, convexity of the cost functions C_i and D_i does not ensure that the objectives in (14) are convex, unlike what happens for T. In Section 5, we present a numerical algorithm that allows to check the second-order conditions at each period.

$V(T-1,S) - W(T-1,S)$ measures the ecological surplus induced by extending cooperation to period $T-1$, with respect to the alternative scenario where cooperation is limited to T. However, (18) is again not sufficient to induce cooperation at time $T-1$. If $\exists i \in \mathcal{N}$ such that $W_i(T-1,S) > V_i(T-1,S)$, then country i will not want to extend cooperation to period $T-1$ without financial compensation.

To induce country i to participate in $T-1$, we proceed as in period T. Let

$$\theta_i(T-1,S) = -[W_i(T-1,S) - V_i(T-1,S)] \\ + \mu_{i,T-1}[W(T-1,S) - V(T-1,S)] \qquad (19)$$

be the transfer paid or received by country i at time $T-1$, where $\mu_{i,T-1} \in]0,1[, \forall i \in \mathcal{N}$ and $\sum_{i=1}^{n} \mu_{i,T-1} = 1$. Then country i's total cost *including* transfers becomes

$$\tilde{W}_i(T-1,S) = W_i(T-1,S) + \theta_i(T-1,S). \qquad (20)$$

By construction, the budget of the transfers defined by (19) is balanced (i.e. $\sum_{i=1}^{n} \theta_i(T-1,S) = 0$). Furthermore, it is easy to see that these transfers make cooperation individually rational at time $T-1$, whatever the inherited stock of pollutant S.

The preceding analysis can be repeated for all earlier periods. The final result will be that the countries cooperate in each period. This determines the emission levels in each period and also the trajectory of the stock of pollutant, given its initial value S_0. In turn this trajectory determines the values of the functions V_i, W_i and $\tilde{W}_i$, and therefore also the values of the transfers θ_i. In Section 4 it will be shown that these transfers yield an interesting property from the perspective of cooperative game theory for specific values of μ_i.

In the infinite horizon case, the backward reasoning considered above applies no more. However, we can consider the stationary solution by taking advantage of the fact that the cost functions C_i and D_i as well as the sharing parameters μ_i do not depend directly on time. The functional forms of the solutions thus only vary in time through the varying stock of pollutant S. The structure of the problem is then the same as in the finite horizon case.

4. Transfers to sustain coalitional rationality at the international optimum

The only condition so far on the μ_i's appearing in (11) or (19) is to take values between 0 and 1 and to sum up to 1. The aim of this section is to utilize the degrees of freedom left to obtain the property of "coalitional rationality" suggested by the core concept in cooperative game theory. We achieve this goal by extending to dynamic games the results obtained by Chander and Tulkens (1995, 1997) in a static framework.

4.1. Associating a dynamic cooperative game with the economic model

In general, a cooperative[8] game (with transferable utility) is defined as a pair $[\mathcal{N}, w(\cdot)]$, where $\mathcal{N} = \{1,\dots,n\}$ is the set of players (i.e. the n countries), and $w(\cdot)$ is the characteristic function of the game, that is, a function that associates to every subset (i.e. coalition) of $\mathcal{N}$ a real number called the *worth of the coalition*. In the present dynamic context, we consider such a game *at each point in time* of our economic model and we therefore specify a characteristic function $w_t(\cdot, S_t)$, at each t, where S_t is the inherited stock of pollutant at that time. We then consider the sequence over time of such cooperative games $[\mathcal{N}, w_t(\cdot, S_t)]$, $t = 1,\dots,T$; this sequence defines the dynamic cooperative game that we[9] associate with our economic model.

4.2. Specifying the components of the dynamic game at each t

For each of the games $[\mathcal{N}, w_t(\cdot, S_t)]$ to be fully specified, the characteristic functions $w_t(\cdot, S_t)$ should further state: (i) what the strategies are that the players (countries) have access to, individually and as coalitions; (ii) which strategy they will actually choose; and (iii) the worth each coalition may expect from using its chosen strategy. As all this will be defined for a given time t, we drop from now on the time index until further notice, letting all emissions and stock variables be those of time t, except for $\bar{S}$ which denotes the stock inherited from the past, S_{t-1}.

As to (i), we assume that the strategy space for each country i is the interval of possible emission levels E_i at time t, i.e. $[0, \infty[$, and for each coalition $U \subseteq \mathcal{N}$ the product of $|U|$ of these intervals, where $|U|$ denotes the number of countries in the coalition U. As to (ii), when some coalition $U \subseteq \mathcal{N}$ forms (including singletons), we consider that the vector $\mathbf{E}^U$ of the strategies adopted by all players is as follows:

(a) The emissions of the coalition members are the vector $\{E_i^U : i \in U\}$ which is the solution of the optimization problem

$$\min_{\{E_i\}_{i\in U}} \sum_{i\in U} [C_i(E_i) + D_i(S) + \beta \tilde{W}_i(S)] \quad \text{s.t. } S = [1-\delta](\bar{S}) + \sum_{i=1}^{n} E_i, \tag{21}$$

where $\forall j \in \mathcal{N} \setminus U$, $E_j = E_j^U$ as defined by (b), and $\tilde{W}_i(S)$ is defined as in (20).

(b) The emissions E_j^U, $j \in \mathcal{N} \setminus U$, of the countries outside of the coalition, are the solutions that simultaneously solve the optimization problems

$$\min_{E_j} C_j(E_j) + D_j(S) + \beta \tilde{W}_j(S), \quad j \in \mathcal{N} \setminus U, \quad \text{s.t. } S = [1-\delta]\bar{S} + \sum_{i=1}^{n} E_i, \tag{22}$$

where $\forall i \in U$, $E_i = E_i^U$ as defined by (a) and $\tilde{W}_j(S)$ is also as in (20).

[8] See, e.g. Osborne and Rubinstein (1994, p. 857). These authors use the more recent—and actually more appropriate—alternative terminology of "coalitional game".

[9] Other concepts of dynamic cooperative games are discussed in Section 4.5.

There are two assumptions present in this specification. The first one is the "γ-assumption", [10] according to which if a coalition forms, its members together minimize the sum of their discounted total costs (item (a)) while each country outside of it reacts by minimizing its own individual discounted total cost, that is, by turning to a natural fallback position (item (b)). The vector $\mathbf{E}^U$ is thus of the nature of a Nash equilibrium of a non-cooperative game where the players are the coalition U and the individual countries outside the coalition taken as singletons. [11] Our second assumption is the rational expectations assumption, introduced in the preceding section; it is expressed here by the inclusion of $\tilde{W}_i(S)$ and $\tilde{W}_j(S)$ in the objective functions of (21) and (22), respectively, to be minimized.

Finally, and on the basis of the above, the worth of any coalition U reads as

$$w(U;S) = \sum_{i \in U} [C_i(E_i^U) + D_i(S^U) + \beta \tilde{W}_i(S^U)], \tag{23}$$

where $S^U = [1-\delta]\bar{S} + \sum_{i=1}^n E_i^U$. Note that $w(\mathcal{N};S)$ is equal to $W(S)$ as defined in (17), with $t = T-1$, i.e. to the optimal total cost for all countries together. The explicit form of the characteristic function of the games $[\mathcal{N}, w_t(\cdot, S_t)]$ is thus (23).

4.3. *Imputations at each t*

For any game $[\mathcal{N}, w(\cdot;S)]$ so specified, any n-dimensional vector of which the components sum up to $w(\mathcal{N};S)$ is called an *imputation* of the game. An imputation can thus be seen as a way of sharing the optimal total cost between the players.

The vector $(W_1(S), \dots, W_n(S))$, where $W_i(S)$ is defined as in (17), is an example of such an imputation, in which each country bears its own abatement and damage costs as induced by the optimal strategy $\{\mathbf{E}^*\}$. However, the possibility of financial transfers between the countries implies that (an infinite number of) other imputations exist, associated with the same strategy. Indeed, all vectors $(\tilde{W}_1(S), \dots, \tilde{W}_n(S))$ defined like (20), with $\sum_i \theta_i(S) = 0$, are imputations as well.

A *solution* of the game is an imputation that satisfies certain properties. Among the imputations defined as (20), the ones that satisfy

$$\sum_{i \in U} \tilde{W}_i(S) \leqslant w(U;S) \quad \forall U \subseteq \mathcal{N} \tag{24}$$

are said *to belong to the γ-core* of the game. In words, the core is the set of imputations with the property that each possible coalition bears a fraction of the total costs $W(S)$ less than or equal to $w(U;S)$, i.e. the least cost that this coalition can guarantee for itself. Imputations belonging to the core of the game defined above are therefore called "rational in the sense of coalitions" since the members of any coalition would suffer a total cost higher than the one at the optimum with transfers.

[10] Borrowed from the concept of "γ-characteristic function" introduced by Chander and Tulkens (1995, 1997).

[11] This equilibrium is called by Chander and Tulkens a *partial agreement Nash equilibrium with respect to the coalition U*. The authors show that such an equilibrium exists and is unique under assumptions similar to those of our economic model.

4.4. Transfers inducing γ-core imputations at each t

Recall from Section 3 that $\mathbf{E}^*$ was defined as the vector of optimal emission levels and S^* as the optimal stock level, that $\mathbf{E}^V$ was the vector of emission levels and S^V the stock characterizing the fallback non-cooperative equilibrium, and that $(W_1(S),\ldots,W_n(S))$ and $(V_1(S),\ldots,V_n(S))$ were, respectively, the vectors of discounted total costs at the optimum and at the fallback non-cooperative equilibrium (cf. (17) and (15)), with $W(S)$ and $V(S)$ the sum of these vectors' components. Recall also that all variables mentioned here are defined at time t.

For the game $[\mathcal{N}, w(\cdot;S)]$, consider now the imputation $(\tilde{W}_1(S),\ldots,\tilde{W}_n(S))$ defined by

$$\tilde{W}_i(S) = W_i(S) + \tilde{\theta}_i(S) \quad \forall i \in \mathcal{N}, \tag{25}$$

where $\tilde{\theta}_i(S)$ is a transfer of the form

$$\tilde{\theta}_i(S) = -[C_i(E_i^*) - C_i(E_i^V)] + \tilde{\mu}_i(S^*)\sum_{j=1}^{n}[C_j(E_j^*) - C_j(E_j^V)] \tag{26}$$

with $S^* = (1-\delta)S + \sum_i E_i^*$ and where

$$\tilde{\mu}_i(S^*) = \frac{F_i'(S^*)}{\sum_{j=1}^{n} F_j'(S^*)} \tag{27}$$

with

$$F_i(S^*) = D_i(S^*) + \beta\tilde{W}_i(S^*). \tag{28}$$

We can now state the following.

Theorem. *The imputation* (25) *belongs to the* γ*-core of the game* $[\mathcal{N}, w(\cdot;S)]$, *if one of the following two conditions is satisfied*:

(i) $\forall i \in \mathcal{N}$, D_i *is linear, or*
(ii) $\forall i \in \mathcal{N}$, $\tilde{W}_i$ *is monotonically increasing and convex and,* $\forall U \subseteq \mathcal{N}$,

$$\sum_{i\in U}[C_i(E_i^U) - C_i(E_i^V)] \geqslant 0, \tag{29}$$

where $\mathbf{E}^U$ *is the vector of emission levels at the partial agreement Nash equilibrium with respect to coalition* U.

Proof. Case (i) linear damage costs—is proven in Germain et al. (1998b). The proof of case (ii) is given in Germain et al. (1998c) available from the authors upon request.[12] □

[12] Notice that the transfers (26) are formulated differently than in (11) or (19). The first part of the proof of case (ii) shows that the former with (27) and (28) are equivalent to the latter.

4.5. Cooperative international optimum and alternative cooperative dynamic games

Turning back now to the cooperative dynamic game defined in Section 4.1, i.e. the sequence of games $[\mathcal{N}, w_t(\cdot, S_t)]$, $t = 1, \ldots, T$ associated with the economic model of Section 2, consider the internationally optimal trajectories of emissions and stock defined there *together* with, at each time t, the transfers defined by (26)–(28).

These optimal trajectories, combined with γ-core-theoretic transfers are the solution we propose for the cooperative dynamic game. We call this combination the *cooperative international optimum*. As we have argued above, this optimum is cooperative in a rational expectations sense.

Recently, Kranich et al. (2001) have proposed three other core concepts for cooperative dynamic games, called, respectively, the "classical", the "weak sequential" and the "strong sequential" cores. Each one of these core concepts derives from a different form of the characteristic function, the differences residing in the assumptions made on how "deviations" by coalitions may (or may not) occur over time. Our characteristic function (and hence our core concept) is actually a fourth one, in its dynamic aspects (independently of its γ nature). It is based on the assumption that if deviations occur at time t, they do not continue afterwards because from time $t+1$ on, cooperation with our transfers is more beneficial for all than non-cooperation. This rational expectations attitude assumed to be adopted by the players at time t, was expressed summarily above as "noncooperation (i.e. deviation) at time t, followed by cooperation afterwards". Our characteristic function is thus a "rational expectations characteristic function", and hence our core concept a "rational expectations core".

Referring to the formulation proposed in another paper by Filar and Petrosjan (2000), it seems to us that the economic model to which our dynamic games are associated implies (through the role played by the stock externality) that, in the notation of these authors (their Eq. (2.1)), the function v at time $t+1$ is determined by the solution retained at time t, that is, along the cooperative international optimum we have defined, by the Pareto optimal stock prevailing at t. We thus present an application of the model of dynamic games of these authors.

4.6. Two further comments

The transfers (26) used in the theorem are the sum of two components. The first term in the right-hand side of (26) is either (if > 0) an amount received by country i equal to the increase of its abatement costs due to its cooperating behaviour during this period, or (if < 0) a payment made by country i equal to its savings in abatement costs, in the case where cooperation allows for an increase in this country's emissions. The second term (always < 0) is country i's contribution to cover the total cost of the aggregate abatement effort of the cooperating countries.

Other remarks concern condition (ii). Notice that it is a sufficient, not a necessary one. It requires, with (29), that if a coalition forms, the total of the abatement costs of its members are larger than what this total would be at the fallback non-cooperative equilibrium. How likely are such coalitions? In order for a coalition U

to form—irrespective to its power to oppose (25)—it should at least satisfy

$$\sum_{i\in U}[C_i(E_i^U)-C_i(E_i^V)]+\sum_{i\in U}[F_i(S^U)-F_i(S^V)]\leqslant 0 \tag{30}$$

since otherwise its members are all better off at the fallback non-cooperative equilibrium. In (30) the magnitudes in the bracketed differences under the second summation sign are always negative, because S^U is smaller than S^V. Condition (ii) thus states that a coalition U that forms can only oppose (25) if in addition the first sum in (30) is negative, i.e. if the aggregate abatement costs of its members are lower than what they are at the fallback non-cooperative equilibrium.[13] However, one may think that the existence of such coalitions is unlikely because, as it is shown in the proof of the theorem, the sum of the emissions of the members of any coalition is always lower than what it is at the fallback non-cooperative equilibrium.

5. Computability issues

The aim of this section is to describe an original numerical algorithm that calculates the value functions and the transfers when the cost functions C_i and D_i are convex. An original algorithm is indeed needed because in the backward induction with the complex transfer parameters, the problem is not analytically tractable, contrary to number of linear-quadratic problems in the literature (such as in Mäler and de Zeeuw, 1998, for example). It is written for finite horizon problems, which is appropriate for most practical applications. The emissions and stock trajectories associated with the international optimum can be calculated by traditional nonlinear programming techniques. The transfers are more difficult to calculate, because they make use of values calculated at the fallback non-cooperative equilibrium, so that one must proceed by backward induction. The basic idea is to construct explicit approximations for the surfaces $\tilde{W}_i(t,S_{t-1})$, $i\in\mathcal{N}$, as polynomial functions of S_{t-1}, by using classical regression.

5.1. Statement of the algorithm

The algorithm consists of several steps:

Step 1: The algorithm starts by solving the optimization problem (2) associated with the international optimum. The first-order conditions are

$$C_i'(E_{it}^*)+\sum_{\tau=t}^{T}\beta^{\tau-t}[1-\delta]^{\tau-t}\sum_{j=1}^{n}D_j'(S_\tau^*)=0 \quad \forall i\in\mathcal{N},\forall t\in\mathcal{T} \tag{31}$$

and

$$S_t^*=[1-\delta]S_{t-1}^*+\sum_{i=1}^{n}E_{it}^* \quad \forall t\in\mathcal{T}, \tag{32}$$

[13] In a static framework, Chander and Tulkens (1997) propose a condition on the marginal damages that ensures that all members of a coalition decrease their emissions w.r.t. the non-cooperative equilibrium. Adapted to our framework, this condition is that $\forall U\subset\mathcal{N}$, $U\neq\mathcal{N}$, $|U|\geqslant 2$, $\sum_{i\in U}F_i'(S^*)\geqslant F_j'(S^V)$, $j\in U$. This condition guarantees (29).

where C_i' and D_j' are the derivatives of functions C_i and D_j, respectively. Solving this $T(n+1)$ equations yields the optimal values of the emissions and of the stock of pollutant for all periods.

Step 2(a): To calculate the transfers, one must first solve the dynamic programming problems associated with the fallback non-cooperative equilibrium. The algorithm proceeds backwards, starting from the last period. At the final time T, given the inherited stock of pollutant S, the fallback non-cooperative equilibrium (which coincides with the Nash equilibrium at the last period) has to satisfy the conditions

$$C_i'(E_{iT}^N) + D_i'\left([1-\delta]S + \sum_{j=1}^{n} E_{jT}^N\right) = 0, \quad i \in \mathcal{N}. \tag{33}$$

Once this system of equations has been solved for the variables E_{jT}^N, and knowing by step 1 the optimal emissions at time T, it is possible to calculate the transfers $\theta_i(S,T)$ using (26) (with $\tilde{\mu}_i(S_T^*) = D_i'(S_T^*)/(\sum_{j=1}^{n} D_j'(S_T^*))$) as well as the value functions $\tilde{W}_i(S,T)$ using (25).

Step 2(a) is done for a set Ω_T of given values of the inherited stock of pollutant S. This set is chosen to be representative of the interval of possible values of S. This interval is bounded below by the value of S that would be obtained with zero emissions during periods $1,2,\ldots,T-1$, given S_0, and bounded above by the value of S that would be obtained with maximum emissions during the same period.

Step 2(b): We now assume that total costs with transfers at time T have the following polynomial form:

$$\tilde{W}_i(T,S) = k_{iT,m}S^m + k_{iT,m-1}S^{m-1} + \cdots + k_{iT,0}, \quad i \in \mathcal{N}, \tag{34}$$

where m is the order of the polynomial chosen so that the fit is good enough. To identify the parameters $k_{iT,m}, k_{iT,m-1}, \ldots, k_{iT,0}$, we use an ordinary least square regression method implemented in the software Matlab, so that the functions $\tilde{W}_i(T,S)$ are now approximated with *explicit analytical functions* of S.

Steps 2(a) and (b) are then repeated for period $T-1$. Given (14) and the inherited stock of pollutant S, first-order conditions associated to the fallback non-cooperative equilibrium are

$$C_i'(E_{it}^V) + D_i'\left([1-\delta]S + \sum_{j=1}^{n} E_{j,T-1}^V\right) + \beta \tilde{W}_i'\left(\left[[1-\delta]S + \sum_{j=1}^{n} E_{j,T-1}^V\right], T\right) = 0, \quad i \in \mathcal{N}. \tag{35}$$

Once this system of equations has been solved for the $E_{j,T-1}^V$, knowing the optimal emissions at time $T-1$ (by step 1) and the value functions $\tilde{W}_i(S,T)$ (by application of steps 2(a) and (b) for time T), it is possible to calculate the transfers $\theta_i(S,T-1)$ using (26)–(28) and the value functions $\tilde{W}_i(S,T-1)$ using (25). This is done for a set Ω_{T-1} of given values of the inherited stock of pollutant S, so that by regression of

Table 1
Optimal abatement rates (%) and atmospheric temperature changes (°C)

t	η_1	η_2	η_3	ΔT
1995	14.29	30.45	22.49	0.87
2005	14.47	28.74	23.08	1.00
2015	14.67	27.46	23.61	1.15
2025	14.94	26.50	24.09	1.29
2035	15.19	25.81	24.54	1.44
2045	15.37	25.31	24.96	1.59
2055	15.51	24.95	25.34	1.74
2065	15.63	24.70	25.68	1.89
2075	15.74	24.53	25.99	2.03
2085	15.83	24.42	26.28	2.18
2095	15.91	24.34	26.53	2.32
2105	15.98	24.29	26.76	2.47

the calculated values of the value functions $\tilde{W}_i(S, T-1)$ on Ω_{T-1}, the value functions $\tilde{W}_i(S, T-1)$ can be approximated with explicit analytical functions of S.

The algorithm continues backwards by repeating steps 2(a) and (b) until period 1.

Step 3: Once the value functions $\tilde{W}_i$ are known as functions of S for all times of the planning period $\mathscr{T} = \{1, \ldots, T\}$, the algorithm calculates the actual values of these value functions and of the transfers for all countries all along the optimal trajectory calculated at step 1.

5.2. A numerical example

In the following, we show some numerical results obtained by application of the algorithm to a simple climatic change model with three regions and 30 periods (originally considered by Germain and van Ypersele (1999) in a six regions framework). The three regions considered are Industrialized Countries (IC),[14] China and Rest of the World (RW). Periods are decades, starting with decade 1990–2000. Details of the equations and values of the parameters are given in Appendix A. The polynomials used are of degree three.

First, by regressing the value functions $\tilde{W}_i$ on a large set of possible values of the inherited stock of pollutant S, we were able to verify that the objective function of each country is well behaved and yields existence and uniqueness of the fallback non-cooperative equilibrium at each period.

We show the results for a planning horizon extending over 11 periods (i.e. until the end of the 21st century), but to avoid boundary problems, computations were made until $T = 30$.[15] Table 1 shows the evolution of the optimal abatement rates of emissions η_i (i.e. the reduction of emissions induced by the optimal solution w.r.t. the BAU scenario, see Appendix A) and atmospheric temperature (which is itself a

[14] The region IC groups Europe, US, Japan and Former Soviet Union.

[15] As t approaches T, results become meaningless because one ignores environmental damages after T.

Table 2
Individual pay-off differences between fallback equilibrium (V_i) and optimal trajectories without (W_i) and with ($\tilde{W}_i$) transfers (θ_i) between the regions $i = 1$–3 (in billion 1990 US$)

t	$V_1 - W_1$	$V_2 - W_2$	$V_3 - W_3$	θ_1	θ_2	θ_3	$V_1 - \tilde{W}_1$	$V_2 - \tilde{W}_2$	$V_3 - \tilde{W}_3$
1995	41.63	−12.76	51.65	10.34	−17.25	6.91	31.30	4.48	44.73
2005	41.26	−11.44	51.17	11.09	−16.13	5.04	30.17	4.69	46.13
2015	40.42	−10.30	50.58	11.56	−15.16	3.60	28.86	4.86	46.98
2025	39.48	−9.34	49.67	11.97	−14.34	2.37	27.51	5.00	47.30
2035	38.44	−8.54	48.50	12.28	−13.64	1.36	26.17	5.09	47.14
2045	37.15	−7.86	47.20	12.35	−13.01	0.66	24.81	5.15	46.54
2055	35.62	−7.27	45.72	12.22	−12.42	0.20	23.41	5.15	45.52
2065	33.90	−6.75	44.04	11.93	−11.86	−0.07	21.97	5.11	44.11
2075	32.03	−6.28	42.16	11.51	−11.31	−0.20	20.52	5.03	42.36
2085	30.04	−5.86	40.09	10.98	−10.75	−0.23	19.05	4.90	40.32
2095	27.97	−5.45	37.83	10.38	−10.18	−0.20	17.59	4.72	38.04
2105	25.85	−5.06	35.43	9.72	−9.57	−0.14	16.14	4.51	35.57

function of the stock of CO_2 in the atmosphere, see Appendix A). The evolutions of the abatement rates are contrasting: increasing slowly for IC (indexed as 1 in the table) and RW (indexed as 3), decreasing slowly for China (indexed as 2). Despite the control of emissions, atmospheric temperature continues to increase monotonically during the whole century. The increase of temperature amounts to around 2.5°C at the end of the 21st century compared to its preindustrial level.

The first three columns of Table 2 give the differences between the pay-offs of the three regions along the international optimum (without current transfers) and at the fallback non-cooperative equilibrium, ($V_i - W_i$). Positive (negative) values mean that the region is better (worse) off at the optimum. While IC and RW are better off in each period, China is worse off in every period. Thus, the international optimum does not satisfy individual rationality.

The following three columns of Table 2 give the transfers θ_i received or paid by each region to make the international optimum trajectory a cooperative one. A negative (positive) value means that the transfer is received (paid). China receives at each period a transfer that induces this country to cooperate. The principal donor is IC, but RW is also a donor until the middle of the century. After 2060 RW also becomes a receiver, but to a far less extent than China.

That these transfers are sufficient to induce cooperation from all regions at all periods is shown by the last three columns of Table 2, which give for each region the difference between its pay-off at the cooperative optimum (i.e. with current transfers) and at the fallback non-cooperative equilibrium ($V_i - \tilde{W}_i$). All values are positive which means that all regions (in particular China) win from cooperation at all periods. In other words, taking into account that cooperation prevails after t, all regions are induced to extend cooperation to current period t thanks to current transfers at period t.

Table 3 reports on differences between the pay-offs obtained by coalitions along the international optimum (with and without transfers) and at the partial agreement Nash equilibria w.r.t. these coalitions (as defined in Section 4.1). The purpose is to check

Table 3
Coalition pay-off differences between partial agreement Nash equilibria w.r.t. coalitions ($V_{(\cdot)}$) and optimal trajectories without ($W_{(\cdot)}$) and with ($\tilde{W}_{(\cdot)}$) transfers (In billion 1990 US$. Coalitions: I, R: IC and RW; C, R: China and RW; I, C: IC and China.)

t	$V_{C,R} - W_{C,R}$	$V_{C,R} - \tilde{W}_{C,R}$	$V_{I,R} - W_{I,R}$	$V_{I,R} - \tilde{W}_{I,R}$	$V_{I,C} - W_{I,C}$	$V_{I,C} - \tilde{W}_{I,C}$
1995	27.75	38.08	68.69	51.45	21.32	28.23
2005	28.27	39.36	66.88	50.75	23.18	28.22
2015	28.56	40.12	65.26	50.10	24.24	27.84
2025	28.39	40.36	63.80	49.47	24.85	27.23
2035	27.89	40.17	62.42	48.78	25.11	26.47
2045	27.24	39.58	60.93	47.93	24.91	25.57
2055	26.407	38.62	59.26	46.84	24.34	24.55
2065	25.39	37.31	57.35	45.49	23.47	23.40
2075	24.21	35.72	55.17	43.85	22.36	22.16
2085	22.90	33.88	52.70	41.95	21.08	20.85
2095	21.47	31.85	49.97	39.79	19.67	19.47
2105	19.97	29.69	46.99	37.42	18.20	18.06

coalitional rationality. It turns out that, according to Columns 1, 3 and 5, no coalition can do better by itself than what it obtains at the international optimum. Coalitional rationality is thus satisfied without transfers. But this property is particular to this example and parameter values: indeed, it is not satisfied in other similar cases, as for example in Eyckmans and Tulkens (1999), where coalitional rationality only obtains with financial transfers.

5.3. Comparing open-loop and close-loop models

It is interesting to compare the negotiation process described in this paper with the open-loop case, where the countries negotiate once and for all at the beginning of the planning period. Germain et al. (1998a) have defined the following transfers that ensure that all countries are interested to cooperate in the open-loop framework:

$$\Theta_i = -\sum_{t=1}^{T} \alpha^t [C_{it}(E_{it}^*) - C_{it}(E_{it}^{\mathrm{o}})]$$

$$+\sum_{t=1}^{T} \alpha^t \frac{D'_{i,t+1}(S_{t+1}^*)}{\sum_{j=1}^{n} D'_{j,t+1}(S_{t+1}^*)} \sum_{j=1}^{n} [C_{jt}(E_{jt}^*) - C_{jt}(E_{jt}^{\mathrm{o}})], \tag{36}$$

where E_{it}^{o} ($i \in \mathcal{N}, t \in \mathcal{T}$) is the emission of country i at time t characterizing the *non-cooperative open-loop Nash equilibrium*.[16] Table 4 summarizes the results.

Columns 1 and 2 give the total discounted costs for each region at the optimum ($W_i(S_0)$) and at the non-cooperative open-loop Nash equilibrium ($O_i(S_0)$). It is clear that China has no interest to cooperate without financial compensation. Column 3 gives

[16] Not to be confused with the *feedback* Nash equilibrium defined in Section 3.

Table 4
Comparison with the open-loop case

i	$W_i(S_0)$	$O_i(S_0)$	$\Theta_i(S_0)$	$W_i(S_0)+\Theta_i(S_0)$	$\sum_t \theta_i(t,S_{t-1})$
IC	10352.0	10839.4	284.4	10636.4	202.3
CHINA	2321.5	2159.0	−240.1	2081.5	−222.1
RW	20078.9	20627.6	−44.3	20034.6	19.9

the transfers $\Theta_i(S_0)$ calculated by formula (36). With these transfers, it appears from the comparison of columns 2 and 4 that all regions gain from cooperation. Finally, column 5 gives the sum of transfers given or received by the three regions all along the planning period in the closed-loop case. Comparison of columns 3 and 5 shows that although of the same magnitude, the transfers given by IC and RW to compensate China are different in the closed-loop framework than what they are in the open-loop case.

The closed-loop case in general leads to different results than the open-loop case. The reason is that the closed-loop case allows for more strategic behaviour because the countries observe the stock of pollution at each point in time and condition their emissions on these observations. When a country considers to increase emissions, this country will argue that it leads to a higher stock and thus to less emissions by the other countries, which then partly offsets the increase. Therefore, in equilibrium, emissions will be higher in the closed-loop case than in the open-loop case.

In the model of this paper two effects influence the transfers when compared to the open-loop case. On the one hand, the closed-loop case leads to more emissions so that more transfers are needed. On the other hand, since the construct with a fallback equilibrium assumes cooperation in the future, it brings the outcome closer to the full-cooperative one, so that less transfers are needed.

One cannot say a priori whether the closed-loop case leads to higher or lower transfers than the open-loop case. It is clear that when the solutions coincide, the transfers are the same. This occurs when the damage function is linear, as shown in Germain et al. (1998b), because in that case emissions do not depend on the stock.

6. Conclusion

This paper develops transfer schemes that yield both individual and coalitional rationality for the design of international agreements that seek to achieve global optimality in stock pollutant control problems, in case these agreements can in principle be negotiated at each point in time. The paper also provides an algorithm to calculate the transfers, whose functioning is illustrated by a numerical example applied to the climate change problem in the context of a partition in three regions of the world.

Our approach presents nevertheless some limits, that invite to further research. First, the transfer scheme is supposed to guarantee the cooperative outcome simply because it makes cooperation beneficial for all countries. It would be interesting to drop this hypothesis and try to sustain the cooperative outcome as a non-cooperative equilibrium. Another track of research would be to study what happens when, for one reason or another, a country deviates from the international agreement at a certain period. Technical work could also be done on the algorithm. Its main strengths are its relative simplicity and its ability to solve problems that are not linear-quadratic, but it is limited to one state variable problems and several mathematical questions could be deepened (for example the question of the propagation of numerical errors due to the successive regressions).

Acknowledgements

This paper was first circulated as CLIMNEG Working Paper No. 6 and CORE Discussion Paper No. 9832. It was initiated as part of a research project (project 7-29039) financed by the Fonds de Développement Scientifique of the Université catholique de Louvain. It was completed as part of the CLIMNEG research project (project CG/DD/241) financed at UCL by the Belgian Government. The fourth author's contribution was made possible thanks to a Human Capital Mobility programme (CHRX CT93) of the European Union. The authors gratefully acknowledge comments and suggestions from C. d'Aspremont, J.-P. van Ypersele, P.-A. Jouvet, Ph. Michel, A. Ulph, Y. Nesterov and especially R. Amir, as well as from two referees of this journal.

Appendix A. equations and parameters used in the numerical example

The model:

$$\eta_{it} = 1 - \frac{E_{it}}{\sigma_{it} Y_{it}}, \tag{A.1}$$

$$S_{t+1} - S_0 = [1-\delta][S_t - S_0] + \beta \sum_{i=1}^{n} E_{it}, \tag{A.2}$$

$$\Delta T_t = \gamma \ln\left(\frac{S_t}{S_0}\right), \tag{A.3}$$

$$C_{it}(E_{it}) = a_{i1}\eta_{it}^{a_{i2}} Y_{it} = a_{i1}\left[1 - \frac{E_{it}}{\sigma_{it} Y_{it}}\right]^{a_{i2}} Y_{it}, \tag{A.4}$$

$$D_{it}(S_t) = b_{i1}\Delta T_t^{b_{i2}} Y_{it} = b_{i1}\left[\gamma \ln\left(\frac{M_t}{M_0}\right)\right]^{b_{i2}} Y_{it}, \tag{A.5}$$

where E_{it} is the CO_2 emissions of region i at time t, σ_{it} the CO_2 emissions/output ratio (exogenous), Y_{it} the output (exogenous), η_{it} the control rate of the emissions, S_t the stock of CO_2 in the atmosphere, ΔT_t the variation of the atmospheric temperature (w.r.t. to its preindustrial level), C_{it} the abatement costs, D_{it} the damage costs.

The value of the parameters $S_0 = 590$ billion tons of carbon (preindustrial level), $\delta = 0.0833$ (rate of decay of CO_2 in the atmosphere between two periods), $\beta = 0.64$ (marginal atmospheric retention ratio of CO_2), γ=2.5/ln(2), annual discount rate=0.02.

Parameters of the cost and damage functions are as follows:

i	IC	CHI	RW
a_{i1}	0.07	0.15	0.1
a_{i2}	2.887	2.887	2.887
b_{i1}	0.01102	0.015523	0.02093
b_{i2}	1.5	1.5	1.5

Output (Y_{it}) and CO_2 emission/output ratio time series are borrowed from Germain and van Ypersele (1999).[17]

References

Bahn, O., Haurie, A., Kypreos, S., Vial, J.-P., 1998. Advanced mathematical programming modeling to assess the benefits from international CO_2 abatement cooperation. Environmental Modeling and Assessment 3, 107–115.

Barrett, S., 1994. Self-enforcing international environmental agreements. Oxford Economic Papers 46, 878–894.

Başar, T., Olsder, G.-J., 1995. Dynamic Non-cooperative Game Theory. Academic Press, London.

Carraro, C., Siniscalco, D., 1993. Strategies for the international protection of the environment. Journal of Public Economics 52, 309–328.

Chander, P., Tulkens, H., 1995. A core-theoretic solution for the design of cooperative agreements on transfrontier pollution. International Tax and Public Finance 2, 279–293.

Chander, P., Tulkens, H., 1997. The core of an economy with multilateral environmental externalities. International Journal of Game Theory 26, 379–401.

Eyckmans, J., Tulkens, H., 1999. Simulating with RICE coalitionally stable burden sharing agreements for the climate change problem. CLIMNEG Working Paper No. 18, Université Catholique de Louvain, Louvain-la-Neuve.

Filar, J.A., Petrosjan, L.A., 2000. Dynamic cooperative games. International Game Theory Review 2 (1), 47–65.

Germain, M., van Ypersele, J.-P., 1999. Financial transfers to sustain international cooperation in the climate change framework. CLIMNEG Working Paper No. 19, Université Catholique de Louvain, Louvain-la-Neuve.

Germain, M., Toint, Ph., Tulkens, H., 1998a. Financial transfers to ensure cooperative international optimality in stock pollutant abatement. In: Fauchaux, S., Gowdy, J., Nicolaï, I. (Eds.), Sustainability and Firms: Technological Change and the Changing Regulatory Environment. Edward Elgar, Cheltenham.

Germain, M., Tulkens, H., de Zeeuw, A., 1998b. Stabilité stratégique en matière de pollution internationale avec effet de stock: le cas linéaire. Revue Economique 49 (6), 1435–1454.

Germain, M., Toint, P., Tulkens, H., de Zeeuw, A., 1998c. Transfers to sustain core-theoretic cooperation in international stock pollutant control. CORE Discussion Paper No. 9832, Université Catholique de Louvain, Louvain-la-Neuve.

Hoel, M., 1992. Emission taxes in a dynamic international game of CO_2 emissions. In: Pethig, R. (Ed.), Conflicts and Cooperation in Managing Environmental Resources, Microeconomic Studies. Springer, Berlin.

Houba, H., de Zeeuw, A., 1995. Strategic bargaining for the control of a dynamic system in state-space form. Group Decision and Negotiation 4, 71–97.

[17] Data of these authors are mainly issued from different versions of the wellknown RICE model, developed by Nordhaus and Yang (1996) and Nordhaus and Boyer (2000).

Kaitala, V., Pohjola, M., Tahvonen, O., 1992. Transboundary air pollution and soil acidification: a dynamic analysis of an acid rain game between Finland and the USSR. Environmental and Resource Economics 2, 161–181.

Kranich, L., Perea, A., Peters, H., 2001. Core concepts for dynamic TU games. METEOR Research Memorandum RM/01/024, University of Maastricht.

Kverndokk, S., 1994. Coalitions and side payments in international CO_2 treaties. In: Van Ierland, E. (Ed.), International Environmental Economics, Theories, Models and Application to Climate Change, International Trade and Acidification, Developments in Environmental Economics, Vol. 4. Elsevier, Amsterdam.

Mäler, K.-G., 1989. The acid rain game. In: Folmer, H., van Ierland, E. (Eds.), Valuation Methods and Policy Making in Environmental Economics. Elsevier, Amsterdam.

Mäler, K.-G., de Zeeuw, A., 1998. The acid rain differential game. Environmental and Resource Economics 12 (2), 167–184.

Maskin, E., Tirole, J., 1988. A theory of dynamic oligopoly. I: overview and quantity competition with fixed wage costs. Econometrica 56 (3), 549–569.

Nordhaus, W., Boyer, J., 2000. Warming the World: Economic Models of Global Warming. MIT Press, Cambridge, MA.

Nordhaus, W., Yang, Z., 1996. A regional dynamic general-equilibrium model of alternative climate-change strategies. American Economic Review 86, 741–765.

Osborne, M., Rubinstein, A., 1994. A Course in Game Theory. MIT Press, Cambridge, MA.

Petrosjan, L., Zaccour, G., 1995. A multistage supergame of downstream pollution. G-95-14, GERAD, Ecole des Hautes Etudes Commerciales. Université de Montréal.

van der Ploeg, F., de Zeeuw, A., 1992. International aspects of pollution control. Environmental and Resource Economics 2, 117–139.

Tahvonen, O., 1994. Carbon dioxide abatement as a differential game. European Journal of Political Economy 19, 685–705.

Tulkens, H., 1998. Cooperation vs. free riding in international environmental affairs: two approaches. In: Hanley, N., Folmer, H. (Eds.), Game Theory and the Environment. Edward Elgar, Cheltenham.

[48]

Environ Resource Econ (2010) 45:573–596
DOI 10.1007/s10640-009-9329-x

Renegotiations in the Greenhouse

Hans-Peter Weikard · Rob Dellink · Ekko van Ierland

Accepted: 7 October 2009 / Published online: 5 November 2009

Abstract International climate policies are being shaped in a process of ongoing negotiations. This paper develops a sequential game framework to explore the stability of international climate agreements allowing for multiple renegotiations. We analyse how the incentives to reach an international climate agreement in the first period will be impacted by the prospect of further negotiations in later periods and by the punishment options related to renegotiations. For this purpose we introduce a dynamic model of coalition formation with twelve world regions that captures the key features of the climate-economy impacts of greenhouse gas emissions. For a model with one round of renegotiations we find that a coalition of China and the United States is the unique renegotiation proof equilibrium. In a game with more frequent renegotiations we find that the possibility to punish defecting players helps to stabilise larger coalitions in early stages of the game. Consequently, several renegotiation proof equilibria emerge that outperform the coalition of China and USA in terms of abatement levels and global payoff. The Grand Coalition, however, is unstable.

Keywords International climate agreements · Self-enforcing international environmental agreements · STACO model · Coalition formation with renegotiation

JEL Classification C73 · D74 · Q54

H.-P. Weikard (✉) · R. Dellink · E. van Ierland
Environmental Economics and Natural Resources Group, Wageningen University, Hollandseweg 1, 6706 KN, Wageningen, The Netherlands
e-mail: hans-peter.weikard@wur.nl

R. Dellink
Institute for Environmental Studies, VU University Amsterdam, Amsterdam, The Netherlands

1 Introduction

The control of greenhouse gases (GHGs) is a global public benefit. In the absence of global institutions underprovision of GHG abatement is expected and well understood. Less understood are conditions under which effective international climate agreements (ICAs) might emerge and which mechanisms might help to stabilise agreements. Recent literature on the stability of ICAs has focused on 'once and forever' models of negotiations, e.g. Hoel (1992), Carraro and Siniscalco (1993), Botteon and Carraro (1997), and Finus et al. (2006). However, the Conference of Parties meetings of the UNFCCC clearly illustrate that negotiating GHG emission controls is a process rather than a matter of striking an agreement. Furthermore, as the commitment period 2008–2012 agreed upon in the Kyoto protocol has started, there is a need to look ahead. Hence, the formation of ICAs is probably best understood in a sequential game framework.

The starting point of our analysis is that the prospect of renegotiations at a later stage might change the incentives to join a coalition and, therefore, it might change participation at an earlier stage. In this paper we examine whether the possibility of renegotiations hamper or help the formation of stable coalitions, and whether renegotiations will lead to agreements with higher or lower levels of abatement. The stability concept frequently used in coalition formation games is that of internal and external stability where no coalition member has an incentive to leave and no player outside the coalition has an incentive to join (d'Aspremont et al. 1983). This concept of stability is, however, no longer adequate in a game with renegotiations. As we study a sequential game we employ the equilibrium concept of Renegotiation Proof Equilibrium (RPE). The initial idea was developed by Bernheim (1987), Bernheim and Ray (1989), and Farrell and Maskin (1989). It has been applied to study international environmental agreements by Finus and Rundshagen (1998) in a two-player game. In our context a RPE is a sequence of coalitions[1] (or ICAs) such that (i) the path of play is subgame perfect, i.e. a Nash equilibrium is played at every subgame, and (ii) there is no subgame where a Pareto dominated Nash equilibrium is played. Subgame perfectness ensures that a player who defects from an equilibrium can be punished in the subsequent play of the game. In order to be credible, the punishment path of play must be such that no player has an incentive to deviate, i.e. the punishment will be carried out if a player deviates. Pareto dominated equilibria are ruled out because they would not survive renegotiation.

The topic of renegotiations has attracted a lot of attention in the recent literature on international environmental agreements. Barrett (1994, 1999) considers an infinitely repeated game. For such games, according to the folk theorem, the full cooperative outcome is generally supported by an equilibrium if discount rates are sufficiently close to zero (Fudenberg and Maskin 1986). The folk theorem applies when players can commit to some future path of play. If a commitment mechanism is lacking, renegotiations limit the possibility of credible punishment of free-rider behaviour. Hence, Barrett (1994, 1999) finds that infinitely repeated play does not guarantee full participation in international environmental agreements. Barrett's (1999) model has recently been extended by Asheim et al. (2006). They show that two regional agreements may outperform a single global agreement when large global coalitions cannot be stabilised. Na and Shin (1998) consider coalition formation with uncertain environmental benefits. Analysing a three-country, two-period model where new information becomes available in the second period they argue that new information may be detrimental to coalition stability. While a coalition of three countries, which are assumed to be identical ex ante, is stable, the coalition might break apart when information becomes

[1] We refer to the situation where no agreement is formed and all players are singletons as a 'trivial' coalition.

available revealing differences between countries. Rubio and Ulph (2002) examine coalition stability in a two-period model with a stock pollutant. Ulph (2004) and Kolstad (2007) further extend this model type to examine stability of ICAs under uncertainty and learning. The later models, unlike Na and Shin's model, are general with respect to the number of countries. Recently Rubio and Ulph (2007) and De Zeeuw (2008) have studied difference games of coalition membership in an infinite time horizon model. De Zeeuw uses farsighted stability as the solution concept. Rubio and Ulph's (2007) work is closest to our study. Their model and ours are both cartel games, i.e. it is assumed that the coalition formed is unique. Rubio and Ulph employ the solution concept of internal and external stability; we use a refinement of the same solution concept. There are two major differences, however. Rubio and Ulph work with a model with identical players and an infinite time horizon. Our model allows for heterogeneous players but the time horizon is finite. The latter allows us to employ backward induction. Rubio and Ulph find that coalition membership declines over time. Although this finding concurs with our results, the reasons for this finding are very different. This will be explained in more detail below.

Common to all these modelling approaches is, however, the assumption that all countries are identical, at least ex ante. This is a serious limitation for most practical applications of models of coalition stability and in particular for models of ICAs.[2] There are two important motives to relax the assumption of identical countries. First, differences in abatement cost and benefits can induce cooperation as cheap abatement options can be better exploited. Second, asymmetry makes transfers effective that further help to stabilise agreements. Applied modelling approaches have employed integrated assessment models to obtain simulation results for heterogeneous players. Ciscar and Soria (2002), for example, have considered a sequential game of the formation of climate policies, but they do not consider coalition formation. In an empirically rich model with 26 regions Babiker (2001) finds that a coalition of OECD countries is not stable, but he cannot provide general insights in the existence of stable coalitions. A systematic screening for stable climate agreements has been conducted by Eyckmans and Tulkens (2003) who check stability for all 56 coalitions in a 6-regions model. Finus et al. (2006) use a larger 12-regions model to examine stability of ICAs. However, these latter approaches use a one-shot game model. Germain et al. (2003) use a dynamic model to explore core-stability. However, their analysis is confined to three regions.

What is lacking in the literature, however, are dynamic models of coalition formation that allow for heterogeneous players. Hence, this paper aims at filling a gap between the theoretical models (with general results and identical players) and empirical models (which allow for different costs and benefits of abatement of GHGs across regions, but which give little general insight in the stability of ICAs). Note that we leave the study of the impact of uncertainty and learning to future work.[3] The merit of our model is that we present a dynamic GHG abatement game with an empirically relevant specification, asymmetric players and renegotiations.

We develop a sequential game where at each stage (commitment period) the players (regions) announce whether they join a unique international climate coalition or not.[4] Then, for given membership of the ICA, the ICA and the singletons choose abatement efforts to maximise their respective net benefits (cf. Chander and Tulkens 1995). We identify the

[2] How different incentives to join an ICA do in fact influence coalition membership, is demonstrated by Weikard et al. (2006) and McGinty (2007).

[3] Dellink et al. (2008) is a starting point in the domain of applied modelling.

[4] Due to computational complexity we do not consider multiple coalitions in our model with renegotiations. See Sáiz et al. (2006) for an analysis of multiple coalitions in a one-shot coalition formation game.

renegotiation proof equilibria for a finite sequential game, which reflects the fact that fossil fuels—the major source of GHGs—are depletable. We extend the Stability of Coalitions model (STACO) introduced by Finus et al. (2006) and refined by Nagashima et al. (2009). STACO is a 12-regions integrated assessment model that consists of abatement and damage costs modules. STACO calculates the pay-offs of all possible coalitions and performs single-deviation stability checks. Details of the calibration of STACO are described in Sect. 3 below. We investigate the impact of the frequency of renegotiations, ranging from two to five over our time horizon of 100 years and identify renegotiation proof sequences of ICAs.

Our findings can be summarised as follows. If there is no or just one round of renegotiations, i.e. not more than two commitment periods, we have a unique RPE consisting of USA and China. The uniqueness of the equilibrium does not allow for any punishment. With more than two commitment periods we find multiple equilibria so that threats of punishment can be installed. A region that is supposed to cooperate in an equilibrium, but free-rides, can be punished by playing the RPE that gives the worst payoff to that region in the subgame that follows.

The paper is organised as follows. Section 2 presents our model of negotiations and renegotiations on greenhouse gas emissions reduction. Section 3 provides the calibration of the STACO model. Section 4 presents results for the one-shot game as a benchmark and examines games with multiple negotiation rounds. We consider the effects of timing and frequency of renegotiations. Section 5 concludes.

2 The Game

We consider a game with a finite number $R (1 \leq R < \infty)$ of renegotiation stages where a cartel formation game with open membership is the stage game. At each stage $r = 1, \ldots, R$ players decide whether or not to join a unique coalition (the cartel). Then the coalition and the remaining players simultaneously set their GHG abatement levels. Our game is not a repeated game as the stage games are not identical. This is due to the fact that we allow for growth in the economy and technical progress leading to a reduction of abatement costs over time. More formally, let $N = \{1, 2, \ldots, n\}$ be a set of players (regions). We adopt a discrete time model with a finite planning horizon T. Periods are denoted by $t = 1, 2, \ldots, T$. Every player i chooses whether or not to sign an ICA at an initial stage ($r = 1$ and $t(r) = 1$) and at each renegotiation stage $r = 2, \ldots, R$. In what follows we label the initial negotiation also 'renegotiation'. Hence $r \in \{1, \ldots, R\}$ and we can denote a moment of renegotiation as $t(r) \in \{1, \ldots, T\}$. There are $R \leq T$ renegotiations. The model allows for multiple renegotiations at arbitrary renegotiation times.

The stage game consists of a membership game followed by an abatement game. In the membership game, at each stage $r = 1, \ldots, R$ each region $i \in N$ adopts a strategy $\sigma_{i,r} \in$ {no,yes}, where $\sigma_{i,r}$ =no means that i is not joining the coalition at stage r and $\sigma_{i,r}$ =yes means that i is joining the coalition at stage r. This membership game determines a coalition $K^r \subseteq N$ that forms at time $t(r)$.

The membership game is followed by an abatement game. We assume that each region i has a path of baseline emissions $(\bar{e}_{i,1}, \ldots, \bar{e}_{i,T})$ specifying uncontrolled emissions $\bar{e}_{i,t}$ at each time t up to the planning horizon T. For ease of reference we denote such paths by $\langle \bar{e}_i \rangle_1^T \equiv (\bar{e}_{i,1}, \ldots, \bar{e}_{i,T})$. Each player adopts an abatement path (pollution control strategy) $\langle q_i \rangle_1^T \equiv (q_{i,1}, \ldots, q_{i,T})$ from the set of feasible abatement paths. A path is feasible if for all regions $i \in N$, region i's abatement never exceeds its baseline emissions, i.e. $q_{i,t} \in [0, \bar{e}_{i,t}]$ for all t. The abatement path is determined in a sequence of decisions at times $t(r)$, $r = 1, \ldots, R$. For example, if we have three renegotiation stages, then $R = 3$. Suppose renegotiation periods are $t(1) = 1$, $t(2) = s$ and $t(R) = s'$, then the relevant partition of the abatement path is written as

$$\langle q_i \rangle_1^T = \left(\langle q_i \rangle_1^{s-1}, \langle q_i \rangle_s^{s'-1}, \langle q_i \rangle_{s'}^T \right).$$

For the costs and benefits of abatement we consider that greenhouse gas abatement is a pure public good. Each region receives benefits B_i from global abatement $\langle q \rangle_1^T \equiv \langle \sum_{i \in N} q_i \rangle_1^T$ and incurs costs $C_i(\langle q_i \rangle_1^T)$ for own abatement. These benefits and costs determine a gross payoff before transfers. Denoting the vector of regional abatement levels by $\langle \boldsymbol{q} \rangle_1^T \equiv (\langle q_1 \rangle_1^T, \ldots, \langle q_n \rangle_1^T)$ the gross payoff can be written as

$$\pi_i \left(\langle \boldsymbol{q} \rangle_1^T \right) = B_i \left(\langle q \rangle_1^T \right) - C_i \left(\langle q_i \rangle_1^T \right). \quad (1)$$

We assume that members of a given coalition K^r formed at time $t(r)$ will adopt abatement paths to maximise the joint payoff of the coalition. Non-members seek to maximise their own payoff. This defines a difference game played by the coalition and the singleton players. In Appendix 1 we show that the abatement game has a unique interior solution for the STACO specification of benefit and cost functions (cf. Sect. 3).[5] This solution of the stage r abatement game when coalition K^r is formed is denoted by $\left\langle q_1^{K^r} \right\rangle_{t(r)}^{t(r+1)-1}, \ldots, \left\langle q_n^{K^r} \right\rangle_{t(r)}^{t(r+1)-1}$. The uniqueness of the solution allows us, for convenience, to transform gross payoffs defined on the domain of stage r abatement paths $\pi_i^r \langle \boldsymbol{q} \rangle_{t(r)}^{t(r+1)-1}$ to payoffs defined on the domain of coalitions. Furthermore, we assume that a coalition $K^r \subseteq N$ can arrange financial transfers $F_i(K^r)$ to redistribute gross payoffs between members. We require that $\sum_{i \in K} F_i(K^r) = 0$ and $F_j(K^r) = 0$ for $j \notin K^r$. We obtain, then, a stage r valuation function that gives the payoff after transfers for every player and for every stage r coalition.

$$v_i^r(K^r) \equiv B_i \left(\left\langle q^{K^r} \right\rangle_{t(r)}^{t(r+1)-1} \right) - C_i \left(\left\langle q_i^{K^r} \right\rangle_{t(r)}^{t(r+1)-1} \right) + F_i(K^r). \quad (2)$$

Hence, the sequential game we specify has R stages where at every stage r a n-player membership game is played, followed by $n - |K^r| + 1$-player abatement game where the coalition payoff is distributed among its members.

The game is solved by backward induction. Consider the renegotiations at times $t(1), \ldots, t(r), \ldots, t(R)$. At the time of the final renegotiation $t(R)$ there is an inherited stock of GHGs resulting from past emissions. Benefits from abatement depend on the stock of GHGs. In general this precludes a simple application of backward induction as the game in the final stage depends on the path of play in earlier stages due to the stock pollutant. In the STACO specification, however, although benefits in later stages depend on stock of GHGs and, therefore, the earlier path of play, *marginal* benefits are independent of the stock of GHGs. In this setting it is possible to determine the Nash equilibrium abatement path adopted at the final renegotiation stage, $\left\langle q_i^{K^R} \right\rangle_{t(R)}^T$.

Next, given the payoffs for all coalitions $K^R \subseteq N$ we determine the equilibria of the membership game. At the final stage we have a one-shot game where the standard definitions of stability apply (cf. d'Aspremont et al. 1983):

[5] This type of solution has been called 'Partial Agreement Nash Equilibrium' by Chander and Tulkens (1995). Coalition members have signed a binding agreement. The coalition and the remaining singleton players play a Cournot-Nash game.

Internal stability: A coalition K^R is internally stable if and only if $v_i^R\left(K^R\right) \geq v_i^R\left(K^R\backslash\{i\}\right)$, for all $i \in K^R$.
External stability: A coalition K^R is externally stable if and only if $v_i^R\left(K^R\right) \geq v_i^R(K^R \cup \{i\})$ for all $i \in N\backslash K^R$.
Stability: A coalition K^R is stable if and only if it is internally and externally stable.
Obviously, the set of stable coalitions coincides with the set of (subgame perfect) Nash equilibria of the stage-R game. Renegotiation proofness requires the following:
Renegotiation Proof Equilibrium: A sequence of coalitions is a RPE if and only if (i) the corresponding strategy profiles are a subgame perfect equilibrium and (ii) in every subgame the equilibrium outcome is Pareto undominated by any other equilibrium outcome of the subgame.

In general, we denote the set of RPE of the r subgame by Ψ^r. Typical elements of Ψ^r are sequences of coalitions denoted by $\psi^r = (K^r, \ldots, K^R)$. Consider a sequence $\psi^r \in \Psi^r$. We will write the valuation of the r subgame as $V_i^r(\psi^r) \equiv v_i^r(K^r) + \cdots + v_i^R(K^R)$.

Now consider the renegotiation at an earlier stage $r < R$. We have to distinguish two situations. If the RPE of the $r+1$ subgame is unique, i.e. Ψ^{r+1} contains a single element, backward induction is straightforward to apply. In this case, at time $t(r)$ each region knows that it will receive the unique RPE payoff of the $r+1$ subgame at stage $r+1$. Then, for a given coalition K^r formed at stage r the relevant payoffs are

$$v_i^r(K^r) + V_i^{r+1}\left(\psi^{r+1}\right) = \pi_i^r\left(\left\langle \boldsymbol{q}^{K^r}\right\rangle_{t(r)}^{t(r+1)-1}\right) + F_i(K^r) + V_i^{r+1}\left(\psi^{r+1}\right). \quad (3)$$

The first order conditions to determine the Nash equilibrium stage r abatement path $\left\langle \boldsymbol{q}^{K^r}\right\rangle_{t(r)}^{t(r+1)-1}$ are independent of earlier play as discussed before. Since there is a unique equilibrium of the $r+1$ subgame, later play cannot affect the decision at stage r either.

If the $r+1$ subgame equilibrium is not unique, we have to consider the set of RPE, Ψ^{r+1}. In this case, the subgame perfect equilibria of the r subgame may include equilibria where membership of player i is induced by the threat that an equilibrium which is bad for i will be played in the continuation game. Denote by w_i^{r+1} the worst equilibrium path of play for player i, such that $V_i^{r+1}(\psi^{r+1}) \geq V_i^{r+1}(w_i^{r+1})$ for all $\psi^{r+1} \in \Psi^{r+1}$. Then a coalition K^r followed by ψ^{r+1} is a subgame perfect equilibrium of the stage r subgame if and only if for all $i \in K^r$

$$v_i^r(K^r\backslash\{i\}) - v_i^r(K^r) \leq V_i^{r+1}\left(\psi^{r+1}\right) - V_i^{r+1}\left(w_i^{r+1}\right) \quad (4)$$

and for all $i \in N\backslash K^r$

$$v_i^r(K^r \cup \{i\}) - v_i^r(K^r) \leq V_i^{r+1}\left(\psi^{r+1}\right) - V_i^{r+1}\left(w_i^{r+1}\right). \quad (5)$$

Equation (4) requires that the gain a coalition member receives from defecting from coalition K^r at stage r must be no greater than the loss from a punishment in the form w_i^{r+1} being played instead of ψ^{r+1}. Equation (5) requires that the gain of a singleton player from entering coalition K^r must be no greater than the loss from a punishment in the form of w_i^{r+1} being played instead of ψ^{r+1} from stage $r+1$ onwards. Notice that the punishment play w_i^{r+1} is in fact a credible punishment, because, by definition, w_i^{r+1} is a RPE. As an RPE, w_i^{r+1} is Pareto undominated and, hence, there must be at least one player who does not have a better option than to play w_i^{r+1} and, therefore, the punishment will be carried out. Equations (4) and (5) are generalisations of the notions of internal and external stability, respectively.

The set of RPE of the stage-r subgame are the subgame perfect equilibria that are Pareto undominated by any other equilibrium outcome of the stage-r subgame.

3 Calibration of the STACO Model

In our setting with heterogeneous players general results on the composition and size of stable coalitions cannot be obtained. Hence, for further analysis we employ a specification of benefits and costs of abatement and a transfer scheme with an empirically meaningful parameter calibration and obtain further insights through numerical simulations. This section explains the calibration of our model which we label STACO-2.2. The model refines STACO 2.1. (Nagashima et al. 2009) in order to examine the impact of renegotiations. Here, we focus on the main features of the model. We consider twelve world regions: USA, Japan (JPN), European Union-15 (EU15), other OECD countries (OOE), Eastern European countries (EET), former Soviet Union (FSU), energy exporting countries (EEX), China (CHN), India (IND), dynamic Asian economies (DAE), Brazil (BRA) and rest of the world (ROW). We account for the benefits of abatement to infinity, but adopt a shorter planning horizon of 100 years, ranging from 2011 to 2110, for determining abatement paths. In this setting the intertemporal aspects of climate change are well reflected.[6] Within the planning horizon there are R rounds of renegotiations. This limits the duration of the initial and any further international climate agreement. Below we present empirical results for $R = 1, 2, 3, 4, 5$ with equal length of stages. Hence for $R = 5$, for example, we have renegotiations at $t = 1, 21, 41, 61, 81$ where coalitions are formed and abatement decisions are taken.

The basic model equations are presented in Box 1. Appendix 2 gives an overview of the model parameters. First we calibrate the benefit function for each region and each period. Benefits from abatement are avoided damages which, in turn, depend on stock of CO_2. Damages are derived from the damage cost module of the DICE model (Nordhaus 1994) and the climate module by Germain and Van Steenberghe (2003). The stock of CO_2 for each period M_t is given in Eq. 6. Stock depends on the stock in the previous period M_{t-1}, a natural equilibrium stock $\bar{M}$, a decay rate δ, baseline (or uncontrolled) emissions $\bar{e}_{i,t}$, abatement $q_{i,t}$ and the airborne fraction of emissions that remains in the atmosphere ω.[7] We use data for CO_2 emissions derived from the EPPA model (Reilly 2005). The damage function is a function of the stock of CO_2 and can be approximated by a linear function. In Eq. 7 y_t denotes global GDP in year t. We assume a GDP growth rate of about 2% annually, using data from the DICE model. Parameters γ_1 and γ_2 are estimated by OLS-regression (Dellink et al. 2004); for the global damage parameter γ_D, we use the estimate by Tol (1997) that damages amount to 2.7% of GDP for a doubling of CO_2 stock over pre-industrial levels. Equation (8) shows the global benefits of abatement depending on stock.

Each region receives a share θ_i of the global benefits, as displayed in Appendix 2, Table 8; see Eq. 9. Benefits are the sum of per-period benefits discounted at rate ρ; see Eq. 10. Equation 11 gives the marginal benefits from current abatement (discounted back to period t).

We specify the abatement cost functions following the estimates of the EPPA model by Ellerman and Decaux (1998). In Eq. 12 α and β are regional cost parameters. We assume exogenous technological progress which is modelled as a reduction of current abatement

6 Admittedly, we assume that no thresholds or irreversibilities in the climate system are relevant for the time period under consideration.

7 This representation of the carbon cycle is extremely simplified but, as our focus is on the analysis of stability of ICAs, it suffices our needs (cf. Nagashima et al. 2009).

costs at an annual rate ς. Equations (13) and (14) specify discounted and marginal abatement costs, respectively.

Based on this specification of benefits and costs, Eq. 15 gives the gross payoff in net present value terms of the stage r abatement game.

For our investigation of the impact of transfers among the regions within a coalition we assume surplus sharing between coalition members; see Weikard et al. (2006) and Nagashima et al. (2009).[8] As explained in Sect. 2, the coalition maximises its joint payoff in the abatement game. Equilibrium abatement levels are unique (see Appendix 1) and labelled $\boldsymbol{q}_{i,t}^{K^r}$. The coalition surplus $S_t(K^r)$ Eq. 16 is defined as the joint gain of the coalition members over their joint payoff if no coalition is formed (labelled 'All-Singletons'). The sharing rule assigns a share $\lambda_{i,t} \in [0, 1]$ of the coalition surplus $S_t(K^r)$ to every coalition member $i \in K$ such that $\sum_{i \in K} \lambda_{i,t} = 1$ for all t. The shares are proportional to baseline emissions as specified in Eq. 17.[9] Implementing surplus sharing involves a financial transfer between coalition members. In Eq. 18 stage-r transfers are specified as the benchmark payoff each player receives under All-Singletons plus a share of the coalition surplus accumulated over all stage r periods minus its gross payoff. One of the main advantages of surplus sharing is that individual rationality is always satisfied as long as a coalition is profitable at all, i.e. coalition members will always gain compared to All-Singletons. Note that all transfers are arranged within a given commitment period r. Moreover, Eq. 19 states that singletons do not participate in the transfer scheme. Hence, the stage r payoff is given by the gross payoffs from equilibrium abatement plus the transfer; see Eq. 20. Equation (21) gives total payoff for a region where ψ^1 is the sequence of coalitions $K^1, \ldots, K^R$.

4 Results

This section presents the results from STACO 2.2. We first present, as a benchmark, the one-shot game without renegotiations (Sect. 4.1). We investigate incentives for cooperation and identify which coalitions are stable. Then, we study the effects of renegotiations in a model with a first round of initial negotiations and one round of renegotiations (Sect. 4.2.), and a first round of initial negotiations and several rounds of renegotiations (Sect. 4.3). Section 4.4 examines the impacts of different transfer schemes in the two-stage game.

We identify the equilibria by checking players' incentives to leave or join the coalition for every possible coalition. We are considering only a unique ICA and single deviations.

4.1 The One-shot Game

We start by presenting the results for the benchmark case with an initial negotiation and no further renegotiations ($R = 1$). Table 1 gives results for two reference cases, the All-Singletons and the Grand Coalition, and for the only stable and Pareto undominated (renegotiation proof) equilibrium, consisting of USA and China. The All-Singletons case reflects the non-cooperative solution where no ICA is signed. It is a Nash equilibrium, but as it is Pareto

[8] Nagashima et al. (2009) discuss and compare several transfer schemes. Their analysis shows that "optimal transfers", as suggested by e.g. Weikard (2009), would imply larger global abatements than other transfer schemes. It is, however, not trivial how this extends to a setting with multiple commitment periods. How optimal transfers can be designed in a setting with renegotiations is the topic of a companion paper, see Weikard and Dellink (2008).

[9] Nagashima et al. (2009) show that a transfer scheme based on the baseline path of emissions outperforms transfers based on a historical base year in terms of stability, global payoffs and abatement.

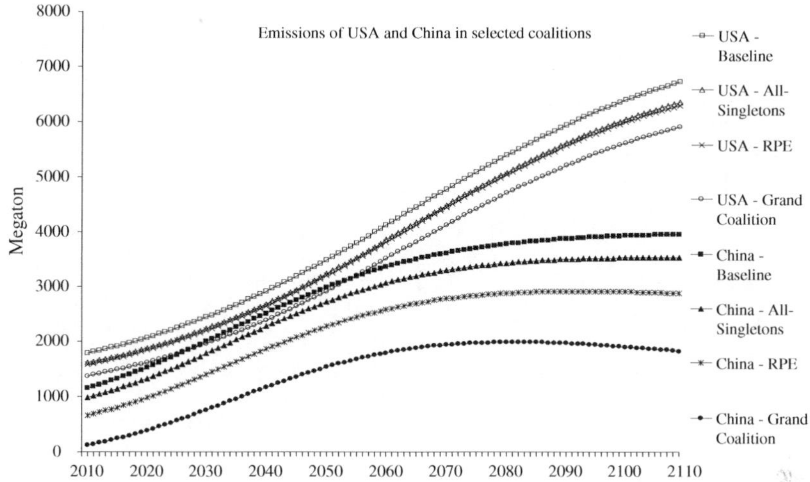

Fig. 1 Emission paths for USA and China—key results for selected coalitions. *Note*: emissions for USA-All Singletons and USA-RPE almost overlap

dominated by the equilibrium coalition (USA, China), it is not renegotiation proof. The Grand Coalition represents full cooperation. It is not an equilibrium since all regions (except EET, EEX and China) have positive incentives to free-ride (indicated in column 6 of Table 1).

The first observation from Table 1 is that abatement and payoffs differ substantially across regions. Under All-Singletons China abates the most, both in absolute and relative terms, i.e. as percentage of its own baseline emissions. The main beneficiary of the global abatement efforts is the EU-15, as they have the highest marginal benefits from abatement (see Appendix 2). This result does not include transfers as under All-Singletons each region bears the full cost of their abatement efforts.

Secondly, the three regions with the lowest marginal abatement costs, China, USA and India, adopt high relative levels of abatement in all coalitions. This reflects the maximisation of regional net benefits in the non-cooperative case. Under cooperation marginal abatement costs are equalised across participating regions. This requires a larger effort by regions with low abatement cost. In all scenarios, the USA remains the largest emitter, even though their abatement levels are high. This reflects the large share of the USA in global emissions of GHGs in the benchmark projection. Global emission levels are increasing over time and a stabilisation of emissions will not occur, not even under the Grand Coalition. Figure 1 shows emissions paths for USA and China for the Baseline, under All-Singletons, the coalition (USA, China; indicated as RPE) and the Grand Coalition.

Thirdly, under All-Singletons the differences in marginal abatement costs (not reported in the table) imply that benefits from cooperation can be reaped. The gains from cooperation are large: the net present value of payoff from abatement of the Grand Coalition increases roughly threefold compared to the non-cooperative case; abatement levels of the Grand Coalition are almost five times larger than All-Singletons abatement. These increased abatement efforts are not in the interest of all regions, however. Though abatement percentages are decreasing over time, especially for quickly growing regions such as China, the huge abatement efforts put on the regions with low marginal abatement costs may be optimal from a global perspective,

Box 1 Equations in STACO-2.2

Regions $i,\ j \in N \equiv \{1,\ldots i, j, \ldots, n\}$; Periods $s,t \in \{1,\ldots,T\}$ Stages $r \in \{1,\ldots,R\}$		
Stock of CO_2	$M_t(\langle q\rangle_1^t) = \bar{M} + (1-\delta)\cdot(M_{t-1}-\bar{M}) + \omega\cdot\sum_{i=1}^{n}\left(\bar{e}_{i,t}-q_{i,t}\right)$	(6)
Current damages	$d_t(M_t(\langle q\rangle_1^t)) = \left[\gamma_1+\gamma_2\cdot\left(\frac{M_t}{\bar{M}}\right)\right]\cdot\gamma_D\cdot y_t$	(7)
Current benefits	$b_t(\langle q\rangle_1^t) = d_t(M_t(\langle 0\rangle_1^t)) - d_t(M_t(\langle q\rangle_1^t))$	(8)
	$b_{i,t}(\langle q\rangle_1^t) = \theta_i b_t$	(9)
Discounted benefits	$B_i(\langle q\rangle_1^T) \equiv \sum_{t=1}^{\infty}\left\{(1+\rho)^{-t}\cdot b_{i,t}(\langle q\rangle_1^T)\right\}$	(10)
Marginal benefits from current abatement	$B'_{i,t}(q_t) \equiv \sum_{s=t}^{\infty}\left\{(1+\rho)^{t-s}\cdot\frac{\partial b_{i,s}}{\partial q_t}\right\}$	(11)
Abatement cost	$c_{i,t}(q_{i,t}) = \frac{1}{3}\cdot\alpha_i\cdot(1-\varsigma)^t\cdot q_{i,t}^3 + \frac{1}{2}\cdot\beta_i\cdot(1-\varsigma)^t\cdot q_{i,t}^2$	(12)
Discounted abatement costs	$C_i(\langle q_i\rangle_1^T) \equiv \sum_{t=1}^{T}(1+\rho)^{-t}\cdot c_{i,t}(q_{i,t})$	(13)
Marginal abatement costs	$c'_{i,t}(q_{i,t}) \equiv \frac{\partial c_{i,t}}{\partial q_{i,t}}$	(14)
Stage r gross payoff (no transfers)	$\pi_i^r\left(\langle \boldsymbol{q}\rangle_{t(r)}^{t(r+1)-1}\right) = B_i\left(\langle q\rangle_{t(r)}^{t(r+1)-1}\right) - C_i\left(\langle q_i\rangle_{t(r)}^{t(r+1)-1}\right)$	(15)
Coalitional surplus	$S_t(K^r) = \sum_{i\in K}\pi_i^r\left(\left\langle \boldsymbol{q}^{K^r}\right\rangle_{t(r)}^{t(r+1)-1}\right) - \sum_{i\in K}\pi_i^r\left(\left\langle \boldsymbol{q}^{N}\right\rangle_{t(r)}^{t(r+1)-1}\right)$	(16)
Coalition member shares	$\lambda_{i,t} \equiv \frac{\bar{e}_{i,t}}{\sum_{i\in K}\bar{e}_{i,t}}, \quad \forall i\in K$	(17)
Financial transfer	$F_i(K^r) = \pi_i^r\left(\left\langle \boldsymbol{q}^{N}\right\rangle_{t(r)}^{t(r+1)-1}\right) + \sum_{\tau=t(r)}^{t(r+1)-1} S_\tau(K^r)\cdot\lambda_{i,\tau}$	
	$-\pi_i^r\left(\left\langle \boldsymbol{q}^{K^r}\right\rangle_{t(r)}^{t(r+1)-1}\right), \quad \forall i\in K$	(18)
	$F_i(K^r) = 0 \quad \forall i\notin K$	(19)
Stage r payoff	$v_i^r(K^r) \equiv \pi_i^r\left(\left\langle \boldsymbol{q}^{K^r}\right\rangle_{t(r)}^{t(r+1)-1}\right) + F_i(K^r)$	(20)
Total payoff	$V_i(\psi^1) = \sum_{r=1}^{R} v_i^r(K^r)$	(21)

Table 1 One-shot game (no renegotiations)—key results for selected coalitions

	All-singletons		Grand coalition			Coalition of USA and China[a]		
	Abatement in 2011	Payoff (NPV)	Abatement in 2011	Payoff (NPV)	Incentive to change announcement (NPV)	Abatement in 2011	Payoff (NPV)	Incentive to change announcement (NPV)
	% Baseline emissions	Billion US$	% Baseline emissions	Billion US$	Billion US$	% Baseline emissions	Billion US$	Billion US$
USA	9.9	1,117	22.8	3, 460	750	11.4	1,319	−201
JPN	2.5	943	11.7	1,325	2,323	2.5	1,454	−421
EU15	7.6	1,240	18.3	2,387	2,243	7.6	1,940	−388
OOE	5.6	188	29.9	728	50	5.6	290	−42
EET	4.4	71	47.6	375	−78	4.4	110	−7
FSU	6.7	362	26.2	1,264	190	6.7	562	−78
EEX	1.9	164	28.4	750	−81	1.9	253	−23
CHN	14.8	298	89.3	2,053	−1,104	42.5	449	−151
IND	10.5	268	65.8	888	182	10.5	417	−64
DAE	1.9	136	34.7	533	28	1.9	210	−30
BRA	0.1	84	6.2	251	109	0.1	130	−30
ROW	6.3	365	30.7	1,196	265	6.3	566	−87
Global	8.0	5,238	36.6	15,211		12.6	7,700	

[a] Stable and Pareto undominated (renegotiation proof) equilibrium

but will not be in the interest of China unless financial compensation by regions that benefit from abatement efforts can be arranged. The compensation payments defined by the surplus sharing rule, however, are so large that Japan and the European Union have strong incentives to leave the Grand Coalition.[10] The only regions that have no incentive to leave the Grand Coalition are Eastern European Countries (EET), Energy Exporters (EEX) and China; all other 9 regions would be better off when free-riding. This sheds some light on the possibility of issue linking. In order to stabilise the Grand Coalition benefits from cooperation in other fields such as removal of trade barriers, for example, must offset the incentive to change membership for those 9 regions that prefer to leave.

Fourthly, the stable coalition of USA and China improves over the All-Singletons case, both in terms of abatement and payoff. Abatement efforts are larger for the coalition members. Other regions do not change their abatement efforts as they have dominant strategies, given the constant marginal benefits in our specification. All regions do, however, benefit from the additional abatement by the coalition members and thus payoffs are higher than in the All-Singletons case for all regions. In fact, the region that benefits most from this coalition is the European Union, as its benefits increase the most and its abatement costs remain the same as in the All-Singletons case. A large gap remains between the stable coalition (USA, China) and the Grand Coalition and only about 25% of the potential gains from cooperation can be reaped.

Fifthly, all other possible coalitions violate either internal or external stability. This indicates strong free-rider incentives. All regions benefit from other players' provision of the public good, but have limited incentives to contribute. Only a few coalitions are internally stable and these consist of no more than three regions.

4.2 Results for the Two-stage Renegotiation Game with Transfers

When we introduce a renegotiation round, the possibility to change membership between both periods emerges. When more than one RPE arises in the second stage of the game, there are credible possibilities to force regions to cooperate in the first stage (cf. Sect. 2). It turns out that in the two-stage game the equilibria that emerge in the second stage are all dominated by the equilibrium (USA, China). Thus, we have a unique RPE at stage two. Consequently, there are no credible punishment strategies and the same equilibrium arises in the first stage. The RPE of the two-stage game is a sequence of coalitions consisting of USA and China in both stages. As reported in Table 2, this result is robust with respect to the renegotiation time (i.e. whether renegotiations occur after 20, 40, 50, 60 or 80 years), but it is not robust with respect to the frequency of renegotiations as we will show below.

Table 3 offers a closer look at the coalition (USA, China) in the game with two periods of 50 years each. Due to low marginal abatement costs, China will take on a large share of the abatement effort. In percentages, these are especially large in the early years when China's emissions are still relatively low (compare Tables 1 and 3). In contrast, the abatement percentage of the USA in 2011 is only a little above the level of the All-Singletons case; cf. Fig. 1. Thus, the main mechanism that governs this result is that China takes the additional benefits in the USA into account and abates more, financed by the USA, similar to the adoption of the Clean Development Mechanism in the Kyoto Protocol.

[10] The gross payoff without transfers for these two regions is positive and large in the Grand Coalition.

Table 2 Two-stage game ($R = 2$): renegotiation proof equilibria and associated global payoff for different moments of renegotiation, $t(R) = 21, 41, 51, 61, 81$

Renegotiation after	20 Years	40 Years	50 Years	60 Years	80 Years
Coalition 1st Stage (payoff)	USA/China (1,710 bln)$	USA/China (3,413 bln)$	USA/China (4,229 bln)$	USA/China (5,010 bln)$	USA/China (6,447 bln)$
Coalition 2nd Stage (payoff)	USA/China (5,990 bln)$	USA/China (4,287 bln)$	USA/China (3,471 bln)$	USA/China (2,690 bln)$	USA/China (1,253 bln)$
Total payoff	7,700 bln $	7,700 bln $	7,700 bln $	7,700 bln $	7,700 bln $

Note: the reported renegotiation proof equilibria are unique

Table 3 Two-stage game ($R = 2$): key results for the sustained coalition of USA and China in the game with renegotiations after 50 years

	Abatement		Payoff (NPV)		Incentive to change announcement (NPV)		Marginal abatement cost	
	% Baseline emissions		Billion US$		Billion US$		US$/ton	
	2011	2061	1st Stage	2nd Stage	1st Stage	2nd Stage	2011	2061
USA	11.4	7.5	716	603	−109	−92	28.5	46.6
JPN	2.5	2.6	800	654	−234	−187	17.1	27.9
EU15	7.6	5.7	1,067	873	−213	−175	23.4	38.1
OOE	5.6	2.9	160	131	−25	−17	3.4	5.6
EET	4.4	2.8	60	49	−3	−3	1.3	2.1
FSU	6.7	5.0	309	253	−38	−40	6.7	10.9
EEX	1.9	1.7	139	114	−12	−11	3.0	4.8
CHN	42.5	23.0	250	199	−87	−64	28.5	46.6
IND	10.5	5.1	229	187	−38	−27	4.9	8.1
DAE	1.9	1.6	116	95	−16	−14	2.5	4.0
BRA	0.1	0.1	71	58	−16	−13	1.5	2.5
ROW	6.3	4.4	312	255	−47	−40	6.7	11.0

As the same coalition arises in both stages of the game, these results are directly comparable to the RPE in the one-shot game: for instance, the net present value of the payoff of both stages adds up to the payoff in the one-shot game. The differences in regional emissions over time makes the incentive structures different for both stages, but these differences are not decisive and the same renegotiation proof coalition exists in both stages.

Finally, Table 3 shows the marginal abatement costs for each region in 2011 and 2061. Optimality requires that the marginal abatement costs of the coalition members are equal. They are higher than the marginal abatement costs under All-Singletons because coalition members account for the positive externalities accruing to other members. The development of the marginal abatement costs over time is driven by the increase in marginal benefits of abatement in a growing economy. Recall that damages from GHG emissions are a share of global GDP; see Eq. 7. With decreasing marginal abatement costs (for any given abatement level) due to technological progress abatement levels will rise over time.

4.3 Extensions to Three and More Stages—Increasing the Frequency of Negotiations

Increasing the frequency of renegotiations may have an impact on the equilibria that emerge in the different stages. With more frequent renegotiations, the stages are shorter. The calculation of payoffs using all future benefits from current abatement (cf. Sect. 3) guarantees that the shorter stages do not lead to myopic behaviour by the players. We expect that increasing the number of renegotiations may enhance stability of larger coalitions in the earlier stages when multiple subgame RPE arise in later stages. In this case, consider the RPE of the stage r subgame that gives the lowest payoff of all RPEs of that subgame to a certain region. This RPE can be used as a threat to enforce the cooperation of that region at the preceding stage $r - 1$.

In fact, once we incorporate multiple rounds of renegotiations ($R > 2$), each with equal length, several renegotiation proof equilibria emerge. The RPEs that perform best in terms of net present value of global payoff aggregated over all stages are reported in Table 4.

The game with three rounds of negotiations has a unique RPE in the third stage, again consisting of USA and China. In the second stage this RPE is, however, no longer unique. The coalition of EU15 with the Eastern European countries and China emerges as a second equilibrium. Even though it is inferior from a global perspective, it is preferred by all regions except EU15. Hence, it is Pareto undominated. This implies that all regions other than EU15 can be threatened by forming the coalition (USA, China) in the second stage. Thus, a number of RPEs with EU15, EET and China in the second and USA and China in the third stage emerge. In contrast, we find only two equilibria with USA and China in the second stage, as only if EU15 can be punished in this setting. In the 3-stage game an initial announcement of the USA to sign in the second stage if and only if EU15 signs in the first stage is credible. If EU15 would not sign, then the best RPE of the stage-2 subgame for the USA is the coalition (EU15, EET, China) followed by (USA, China) in stage 3. Hence, the USA do not have an incentive to deviate from this punishment strategy.

The 4-stage game leads to similar results as the 3-stage game. In the later part of the century, i.e. in the last two stages, a unique RPE of USA and China exists while there is room for multiple RPEs in the earlier stages and thus a range of equilibria in the first stage. Although the numerical results differ, the incentive mechanism is the same as for the 3-stage game. The uniqueness of the RPE at the stage-3 subgame (stretching over stages 3 and 4) can be seen as an extended stage 3. In our setting this implies that the relatively ambitious ICAs that emerge in first commitment period are maintained only for a shorter time (25 years instead of 33 years) and total payoffs over the entire model horizon are slightly lower than in the three-stage game. Hence, more frequent renegotiations do not necessarily lead to better outcomes.

The situation changes when there are five stages in the game, i.e. renegotiations every 20 years. In this case, the coalition of USA and China is again unique in the last two stages of the game, while two RPE arise for the third stage: USA and China or EU15, EET and China. The main difference with the games with fewer stages is that there are now three consecutive stages with multiple RPE. Essentially, the two equilibria in the third stage induce a range of equilibria in the second stage, and thereby the possibilities to set incentives for cooperation in the first stage become much larger.

The equilibria reported in Table 4 are the best performing ones in terms of net present value of global payoff (and given the characteristics of our model also in terms of GHG concentrations). In most cases, these involve coalitions with China, as the marginal abatement costs are lowest in this country. There are, however, in total 1,542 RPEs, with many different coalition members in the first stage. Even some stable 10-player coalitions emerge at the initial stage.

It is worth noting that the Grand Coalition is not stable at any stage, which is hardly surprising as free-rider incentives increase with the number of coalition members. Punishment strategies are insufficient to overcome these in the 5-stage game. At the first stage the best performing RPE achieves 64% of the gains the Grand Coalition would achieve at that stage but over a century it achieves only 36% of the gains of the Grand Coalition.

Finally, in Table 5 the incentives to change announcement are given for the best performing RPE in the 5-stage game. These incentives are expressed in net present value, calculated back to 2010, and can, hence, be directly compared. The development of these incentives is a mixture of several mechanisms, including technological progress and increasing emissions and abatement levels over time.

Table 4 Several rounds of negotiations: selected renegotiation proof equilibria, ranked for each game according to global payoff

Coalition members 1st stage	Coalition members 2nd stage	Coalition members 3rd stage	Coalition members 4th stage	Coalition members 5th stage	Global payoff NPV billion US $
2-Rounds of negotiations, $t(r) = 1, 51$					
USA/CHN	USA/CHN	–	–	–	7,700
3-Rounds of negotiations $t(r) = 1, 34, 67$					
EU15/EET/EEX/CHN	USA/CHN	USA/CHN	–	–	8,100
USA/EET/CHN	EU15/EET/CHN	USA/CHN	–	–	8,020
JPN/EEX/CHN	EU15/EET/CHN	USA/CHN	–	–	7,862
USA/CHN	EU15/EET/CHN	USA/CHN	–	–	7,842
JPN/EET/CHN	EU15/EET/CHN	USA/CHN	–	–	7,804
USA/CHN	USA/CHN	USA/CHN	–	–	7,700
{40 RPE sequences in total}					
4-Rounds of negotiations $t(r) = 1, 26, 51, 76$					
EU15/EET/EEX/CHN	USA/CHN	USA/CHN	USA/CHN	–	8,004
USA/EET/CHN	EU15/EET/CHN	USA/CHN	USA/CHN	–	7,950
JPN/EEX/CHN	EU15/EET/CHN	USA/CHN	USA/CHN	–	7,828
USA/CHN	EU15/EET/CHN	USA/CHN	USA/CHN	–	7,814
JPN/EET/CHN	EU15/EET/CHN	USA/CHN	USA/CHN	–	7,785
USA/CHN	USA/CHN	USA/CHN	USA/CHN	–	7,700
{42 RPE sequences in total}					
5-Rounds of negotiations $t(r) = 1, 21, 41, 61, 81$					
USA/EET/FSU/EEX/ CHN/DAE/ROW	EU15/EEX/CHN	EU15/EET/CHN	USA/CHN	USA/CHN	8,824
USA/OOE/EET/FSU/ CHN/DAE/ROW	EU15/EEX/CHN	EU15/EET/CHN	USA/CHN	USA/CHN	8,801
USA/EET/FSU/EEX/ CHN/DAE/ROW	EU15/EET/CHN	EU15/EET/CHN	USA/CHN	USA/CHN	8,785
EU15/OOE/EET/EEX/ CHN/IND/ROW	USA/EET/CHN	EU15/EET/CHN	USA/CHN	USA/CHN	8,758

Table 4 Continued

Coalition members 1st stage	Coalition members 2nd stage	Coalition members 3rd stage	Coalition members 4th stage	Coalition members 5th stage	Global payoff NPV billion US $
EU15/EET/EEX/CHN/ IND/DAE/ROW	USA/EET/CHN	EU15/EET/CHN	USA/CHN	USA/CHN	8,755
EU15/OOE/EET/FSU/ EEX/CHN/IND	USA/EET/CHN	EU15/EET/CHN	USA/CHN	USA/CHN	8,754
EU15/EET/FSU/EEX/ CHN/IND/DAE	USA/EET/CHN	EU15/EET/CHN	USA/CHN	USA/CHN	8,751
EU15/OOE/EET/CHN/ IND/DAE/ROW	USA/EET/CHN	EU15/EET/CHN	USA/CHN	USA/CHN	8,731
EU15/OOE/EET/FSU/ CHN/IND/DAE	USA/EET/CHN	EU15/EET/CHN	USA/CHN	USA/CHN	8,727
USA/OOE/EET/EEX/ CHN/DAE/ROW	EU15/EEX/CHN	EU15/EET/CHN	USA/CHN	USA/CHN	8,718
{ 1,542 RPE sequences in total }					

Table 5 Five-stage game: incentives to change announcement (NPV; billion US$) in the best performing RPE (coalition members in bold)

	1st Stage	2nd Stage	3rd Stage	4th Stage	5th Stage
USA	**86**	−33	−19	**−38**	**−34**
JPN	−290	−115	−97	−78	−67
EU15	−309	**−5**	**−16**	−73	−62
OOE	−31	−8	−4	−7	−6
EET	**3**	1	**−2**	−1	−1
FSU	**37**	−12	−12	−17	−15
EEX	**11**	**−1**	0	−5	−4
CHN	**−58**	**−58**	**−52**	**−27**	**−21**
IND	−54	−12	−7	−11	−10
DAE	**17**	−5	−3	−6	−5
BRA	−19	−7	−5	−5	−5
ROW	**50**	−16	−13	−17	−14

From the table the impact of the punishment strategies can be clearly seen. In the first stage, 6 out of the 7 coalition members are threatened into collaboration: their incentives to leave the coalition are positive, but smaller than the threat that an inferior coalition is played in later stages. The table also shows that the European Union and Japan are much better off outside the coalition in the first stage than inside.

4.4 Changing the Transfer Scheme

Previous studies, including Nagashima et al. (2009), have shown that the transfer scheme adopted can have major impacts on the stability of coalitions. The outcomes of our 2-stage game with a surplus sharing rule are compared with outcomes of transfers resulting from tradable emission permits and with the no-transfers situation. The tradable permit system is interesting as it is part of the Kyoto Protocol. The no-transfers situation serves as a benchmark. In order to keep the analysis comparable across scenarios we assume that the permits are distributed to coalition members according to baseline emissions.

Table 6 shows that the transfer scheme influences the coalition members in the RPE, but in all cases one unique RPE emerges with two coalition members. Introducing a tradable emission permit system as the transfer scheme substantially alters the incentive structures of the regions. Regions with low marginal abatement costs will be able to sell emission permits and thus increase their net benefits, while regions with high marginal abatement costs may want to buy permits. Therefore, EU15, with high marginal benefits and high marginal abatement costs, prefers to trade with China (low marginal abatement costs). EU15 buys emission permits to combine high global abatement efforts with low regional abatement levels, whereas China can gain from selling permits. The USA is a much less attractive partner for China, as the USA has ample opportunities to reduce emissions at low costs domestically. Therefore, the equilibrium permit price in a market consisting of USA and China will be lower.

When no transfer scheme is adopted China can no longer benefit from the transfers, and thus it has no incentives to join a coalition. Consequently, the two regions with the largest incentives to stimulate abatement, Japan and EU15, are forced to cooperate without partners, and as their marginal abatement costs are relatively high, their ambition level is low. The coa-

Table 6 Influence of the transfer scheme on the Renegotiation Proof Equilibria

	RPE 1st stage	RPE 2nd stage	NPV of global payoff (bln$)	Concentrations in 2110 (Gton)
Surplus sharing (base case)	USA, China	USA, China	7,700	1,425
Tradable permits	EU-15, China	EU-15, China	7,667	1,425
No transfers	Japan, EU-15	Japan, EU-15	5,486	1,445

Note: Each simulation generates one unique Renegotiation Proof Equilibrium

lition of Japan and EU15 hardly improves the global payoff compared to the non-cooperative outcome.

5 Discussion and Conclusions

In this final section we would like to turn to the policy implications of our findings. Before we do so we summarise the main findings and we discuss some limitations of our modelling approach.

This paper develops a framework to determine the RPE coalitions in a dynamic model of coalition formation. In order to gain a better understanding of international climate agreements we consider heterogeneous players, reflecting the differences of marginal benefits and costs of abatement between regions. Coalition formation helps to exploit low cost abatement options in some regions. We focus on the impacts of renegotiations on the stability of international climate agreements in a setting of asymmetric countries explicitly including the option of transfers among coalition members. Because a full analytical analysis of stability with asymmetric regions and renegotiations is so far impeded by the complexity of the situation, we develop a dynamic version of the STACO model (STACO 2.2) that comprises costs and benefits of greenhouse gas abatement for 12 world regions and allows for renegotiations and transfers. We apply a numerical analysis to test whether coalitions are renegotiation proof in a number of settings.

We find that in the absence of transfers renegotiations do not lead to large or ambitious coalitions. In a setting with transfers and with a limited number of renegotiations we find a unique RPE, the coalition of the USA and China. More interesting are the results for scenarios with a larger number of rounds of renegotiations. We find multiple renegotiation proof equilibria with more ambitious ICAs in early stages of the game. In the later stages only the coalition of the USA and China survives as an equilibrium of the subgame. The general pattern—large coalitions initially, but small coalitions later—concurs with the findings of Rubio and Ulph (2007). However, what drives this result is quite different in our setting. Rubio and Ulph study a difference game with identical players, an infinite time horizon, a stock pollutant and a convex damage function. Because players are assumed to be identical, only the number, not the identity, of coalition members matters for the development of the stock of the pollutant. Rubio and Ulph use a "random assignment rule" to determine which player is a member at any given stage. This rules out the use of punishment schemes. Their result, declining membership over time, is entirely driven by an increase of the stock of the pollutant over time which implies increasing marginal damages and, hence, increasing (potential) gains from cooperation and increasing free-rider incentives.[11] In our case the decline of participation and abatement over time is driven by declining threats. With a finite time horizon—reflecting finite fossil fuel reserves—there is no credible threat of punishment at the last stage. The stability of larger coalitions at an early stage is driven by the existence of multiple equilibria of the relevant subgame that can serve as credible threats.

Turning now to the limitations of our analysis we would like to mention four major issues. The first two issues concern the rules of the game, the third concerns the solution concept and the fourth is more policy oriented. First, notice that we obtain results for a given transfer rule—surplus sharing proportional to baseline emissions. Transfers are limited to coalition members and are arranged within a given commitment period. We do not explore in this paper how a transfer scheme may emerge from the negotiation process and we not explore how transfers can be used to create incentives for coalition membership. The former can be addressed in a bargaining framework, cf. Caparrós et al. (2004). The latter is explored in a companion paper to this paper, cf. Weikard and Dellink (2008).

Second, the rules of our game do not allow for the co-existence of multiple coalitions. A full numerical analysis of all possible multi-coalition structures with several rounds of renegotiations is practically impossible due to combinatorial complexity. The case where two agreements coexist has been explored by Asheim et al. (2006). Their finding is that global abatement levels would be higher with two regional agreements in place of one global agreement. The challenge for future research is to push forward model analyses that are rich enough to capture the main options available in climate policy-making, such as regional agreements and well-designed transfer schemes, while maintaining analytical tractability. Clearly, "small" analytical models and "large" integrated assessment models must complement each other.

Third, and related to the previous issue, we employ a refined Nash equilibrium as a solution concept. The refinements, subgame perfectness and renegotiation proofness, maintain the presumption that players consider deviations always individually, never collectively. Of course, in climate politics there is nothing that rules out collective action of various groups of countries such as e.g. the collaboration of G77 and China.

Fourth, in our model there is no cost of leaving the ICA. In practice a country that joins makes a political commitment. Typically, leaving the ICA would involve a cost and this would support internal stability of any ICA.

Given the limitations of our scenario analysis, conclusions for climate policies must be drawn with care. We can however point to a number of insights from our study. The first is, of course, that the prospect of renegotiation matters. In our setting, future play offers options to impose threats that set incentives to join an ICA at an earlier stage. Many studies have argued that effective climate coalitions will not be stable (e.g. Barrett 1994; Finus et al. 2006; Altamirano-Cabrera and Finus 2006). Our analysis shows that with heterogeneous players and a transfer scheme that shares the gains from cooperation it is possible to exploit some of the low cost abatement options. This fact, in combination with additional incentives that emerge from credible threats under renegotiations leads to ICAs that can capture 36% of the potential gains from cooperation over a century, with 64% of the potential gains of the first stage in our model with five renegotiation stages. These results are far less bleak than what is suggested in the literature on self-enforcing environmental agreements. Moreover, even better results seem to be possible if the transfer scheme is specifically designed to stimulate membership and if multiple agreements are considered.

[11] It is well-known in the literature on international environmental agreements that gains from cooperation are coupled with free-rider incentives, cf. Barrett (1994).

The inspection of our results suggests that more frequent negotiations generate a larger number of equilibria that can be used to set incentives. Although this is generally true, there is a drawback. More frequent negotiations do not necessarily lead to more effective agreements. Although threats are larger with more frequent negotiations, the corresponding commitment periods are shorter. The gains from stabilising a larger coalition (if there is any) can be captured only for a shorter period. This can be seen when comparing 3 rounds with 4 rounds of negotiation (see Table 4).

Furthermore, in our applied setting we find powerful incentives for the EU15 and USA to exploit the prospect of renegotiations strategically and coerce cooperative behaviour of the other (while free-riding themselves). Such strategic possibilities will only be temporary, however, as the incentives to cooperate change with every round of renegotiation.

The bottom line of our analysis is that a clear prospect for future renegotiations may help to bring about an early effective agreement.

Acknowledgement An earlier version of this paper was presented at the Third World Congress of Environmental and Resource Economists, July 3-6, 2006, Kyoto, Japan. We gratefully acknowledge the support of our collaborators in the STACO project.

Appendix 1

Each stage game of our greenhouse renegotiation game explained in Sect. 2 consists of a coalition formation game and a GHG abatement game. Here we show that the latter has a unique interior solution under STACO specifications (see Box 1). The GHG abatement game is a difference game where each player fixes a sequence of abatement levels for the relevant renegotiation period. Hence, we consider the abatement levels $q_{i,t}$ for each player i at each time t, with $t(r) \leq t < t(r+1)$. First notice that the abatement game is played by the coalition (acting as one player) and $|N| - |K|$ other players. Each player chooses an abatement path $\langle q_i \rangle_{t(r)}^{t(r+1)-1}$ which is a best response to other players' abatement paths. In the STACO specification, according to Eqs. 6 and 7, the stock of GHGs is linear in abatement and damages are linear in stock of GHGs. Therefore, marginal benefits Eq. 11 do not depend on others' abatement, i.e. each player has a dominant strategy. Along an optimal abatement path it must hold that discounted marginal benefits from current abatement equal discounted marginal abatement cost from current abatement at any time t, i.e.

$$(1+\rho)^{-t} \cdot B'_{i,t}(q_t) = (1+\rho)^{-t} \cdot C'_{i,t}(q_{i,t}). \tag{A1}$$

Furthermore, we have $C'_{i,t}(0) = 0$ while $B'_{i,t} > 0$ for all $q_{i,t} \in [0, \bar{e}_{i,t}]$ and $C'_{i,t}(\bar{e}_{i,t}) > B'_{i,t}(\bar{e}_{i,t})$. Together with the strict convexity of the abatement cost function this guarantees a unique interior solution.

Appendix 2

See Tables 7 and 8.

Table 7 Global parameters

Symbols	Description	Value	Unit	Source
$\bar{M}$	Pre-industrial level of CO_2 stock	590	Gton	Nordhaus (1994)
δ	Natural annual removal rate of CO_2 stock	0.00866	–	Nordhaus (1994)
ω	Airborne fraction of emissions remaining in the atmosphere	0.64	–	Nordhaus (1994)
ρ	Discount rate	0.02	–	Assumption
θ_i	Share of region i in global benefits	See Appendix 2, columns 3 and 4		Own calculation based on Fankhauser (1995)
α_i	Abatement cost parameter of region i	See Appendix 2, column 5		Own calculation based on Ellerman and Decaux (1998)
β_i	Abatement cost parameter of region i	See Appendix 2, column 6		Own calculation based on Ellerman and Decaux (1998)
ς	Technological progress parameter	0.005	–	Assumption
γ_D	Scale parameter of damage and benefit function	0.027	–	Tol (1997)

Table 8 Regional parameters in the benefit and abatement cost function

Regions	Emission in 2010 Gton (share)	Share of global benefits calibration I θ_i	Parameter of abatement cost α_i	Parameter of abatement cost β_i
USA	1.763 (0.238)	0.226	0.0005	0.0398
JPN	0.344 (0.046)	0.173	0.0155	1.8160
EU15	0.943 (0.127)	0.236	0.0024	0.1503
OOE	0.360 (0.049)	0.035	0.0083	0
EET	0.226 (0.030)	0.013	0.0079	0.0486
FSU	0.774 (0.104)	0.068	0.0023	0.0042
EEX	0.469 (0.063)	0.030	0.0032	0.3029
CHN	1.127 (0.152)	0.062	0.00007	0.0239
IND	0.344 (0.046)	0.050	0.0015	0.0787
DAE	0.316 (0.043)	0.025	0.0047	0.3774
BRA	0.122 (0.016)	0.015	0.5612	8.4974
ROW	0.637 (0.086)	0.068	0.0021	0.0805
World	7.425 ($\Sigma = 1$)	($\Sigma = 1$)		

References

Altamirano-Cabrera J-C, Finus M (2006) Permit trading and stability of international climate agreements. J Appl Econ 9:19–47

Asheim GB, Bretteville Froyn C, Hovi J, Menz FC (2006) Regional versus global cooperation for climate control. J Environ Econ Manage 51:93–109

Babiker MH (2001) The CO2 abatement game: costs, incentives, and the enforceability of a sub-global coalition. J Econ Dyn Control 25:1–34

Barrett S (1994) Self-enforcing international environmental agreements. Oxf Econ Pap 46:878–894

Barrett S (1999) A theory of full international cooperation. J Theor Polit 11(4):519–541

Bernheim BD, Ray D (1989) Collective dynamic consistency in repeated games. Games Econ Behav 1: 295–326

Bernheim BD, Peleg B, Whinston MD (1987) Coalition-proof Nash equilibria I. Concepts. J Econ Theor 42:1–12

Botteon M, Carraro C (1997) Burden-sharing and coalition stability in environmental negotiations with asymmetric countries. In: Carraro C (ed) International environmental negotiations: strategic policy issues. Edward Elgar, Cheltenham pp 26–55

Caparrós A, Pereau J-C, Tazdaït T (2004) North–south climate change negotiations: a sequential game with asymmetric information. Public Choice 121(3–4):455–480

Carraro C, Siniscalco D (1993) Strategies for international protection of the environment. J Public Econ 52:309–328

Chander P, Tulkens H (1995) A core-theoretic solution for the design of cooperative agreements on transfrontier pollution. Int Tax Public Finance 2:279–293

Ciscar JC, Soria A (2002) Prospective analysis of beyond Kyoto climate policy: a sequential game framework. Energy Policy 30:1327–1335

d'Aspremont C, Jaquemin A, Gabszewicz JJ, Weymark JA (1983) On the stability of collusive price leadership. Can J Econ 16(1):17–25

Dellink RB, Altamirano-Cabrera J-C, Finus M, van Ierland EC, Ruijs A, Weikard H-P (2004) Empirical background paper of the STACO model, mimeo, Wageningen University

Dellink RB, Finus M, Olieman N (2008) The stability likelihood of an international environmental agreement. Environ Resour Econ 39:357–377

De Zeeuw A (2008) Dynamic effects on the stability of international environmental agreements. J Environ Econ Manage 55:163–174

Ellerman AD, Decaux A (1998) Analysis of post-Kyoto CO2 emissions trading using marginal abatement curves. Joint program on the science and policy of global change Report 40, MIT, Cambridge
Eyckmans J, Tulkens H (2003) Simulating coalitionally stable burden sharing agreements for the climate change problem. Resour Energy Econ 25:299–327
Fankhauser S (1995) Valuing climate change. Earthscan, London
Farrell J, Maskin E (1989) Renegotiation in repeated games. Games Econ Behav 1:327–360
Finus M, Rundshagen B (1998) Renegotiation-proof equilibria in a global emission game when players are impatient. Environ Resour Econ 12:275–306
Finus M, van Ierland EC, Dellink RB (2006) Stability of climate coalitions in a cartel formation game. Econ Gov 7:271–291
Fudenberg D, Maskin E (1986) The folk theorem in repeated games with discounting or with incomplete information. Econometrica 54:533–554
Germain M, Van Steenberghe V (2003) Constraining equitable allocations of tradable CO2 emission quotas by acceptability. Environ Resour Econ 26:469–492
Germain M, Toint P, Tulkens H, de Zeeuw A (2003) Transfers to sustain dynamic core-theoretic cooperation in international stock pollutant control. J Econ Dyn Control 28:79–99
Hoel M (1992) International environmental conventions: the case of uniform reductions of emissions. Environ Resour Econ 2:141–159
Kolstad CD (2007) Systematic uncertainty in self-enforcing international environmental agreements. J Environ Econ Manage 53: 68–79
McGinty M (2007) International environmental agreements among asymmetric nations. Oxf Econ Pap 59: 45–62
Na S, Shin HS (1998) International environmental agreements under uncertainty. Oxf Econ Pap 50:173–185
Nagashima M, Dellink RB, van Ierland EC, Weikard H-P (2009) Stability of international climate coalitions—a comparison of transfer schemes. Ecol Econ 68:1476–1487
Nordhaus WD (1994) Managing the global commons: the economics of climate change. MIT Press, Cambridge, MA
Reilly J-M (2005) Emission paths in the MIT–EPPA model, personal communication
Rubio SJ, Ulph A (2002) A simple dynamic model of international environmental agreements with a stock pollutant. Discussion paper in economics and econometrics 0209, Department of Economics, University of Southampton
Rubio SJ, Ulph A (2007) An infinite horizon model of dynamic membership of international environmental agreements. J Environ Econ Manage 54:296–310
Sáiz ME, Hendrix EMT, Olieman NJ (2006) On the computation of stability in multiple coalition formation games. Comput Econ 28:251–275
Tol RSJ (1997) A decision-analytic treatise of the enhanced greenhouse effect. Ph.D. thesis, Vrije Universiteit, Amsterdam
Ulph A (2004) Stable international environmental agreements with a stock pollutant, uncertainty and learning. J Risk Uncertain 29:53–73
Weikard H-P (2009) Cartel stability under an optimal sharing rule. Manchester Sch 77(5):599–616
Weikard H-P, Dellink RB (2008) Sticks and carrots for the design of climate agreements with renegotiations. FEEM working paper 26.2008. Venice, Itlay
Weikard H-P, Finus M, Altamirano-Cabrera J-C (2006) The impact of surplus sharing on the stability of international climate agreements. Oxf Econ Pap 58:209–232

[49]

Environ Resource Econ (2010) 45:25–48
DOI 10.1007/s10640-009-9304-6

A Dynamic Model for International Environmental Agreements

Michèle Breton · Lucia Sbragia · Georges Zaccour

Accepted: 17 June 2009 / Published online: 9 July 2009

Abstract In this paper we develop a model that uses a dynamic framework to analyze the process through which countries join international environmental agreements (IEAs). In the model, while all countries suffer from the same environmental damage as a result of total global emissions, non-signatory countries decide on their emission levels by maximizing their own welfare, whereas signatory countries decide on their emission levels by maximizing the aggregate welfare of all signatory countries. It is assumed that signatory countries will be able to punish non-signatories, at some cost to themselves. When countries decide on their pollution emissions, they account for the evolution of the stock of pollution over time. Moreover, we propose a mechanism to describe how countries reach a stable IEA. The model is able to capture situations characterized by partial cooperation within an IEA that is stable over time. It also captures situations where all countries participate in a stable agreement, and situations where no stable agreement is feasible. Where more than one possibility coexist, the long-term outcome of the game depends on the initial conditions (i.e., the initial number of signatory countries and pollution level).

Keywords Environment · International agreements · Dynamic game · Replicator dynamics · Non-cooperative game

M. Breton (✉)
GERAD and HEC Montréal, 3000 Chemin de la Côte Sainte-Catherine, Montréal, QC H2T 2A7, Canada
e-mail: michele.breton@hec.ca

L. Sbragia
GERAD and Durham University, Durham, UK
e-mail: lucia.sbragia@hec.ca

G. Zaccour
GERAD and Chair in Game Theory & Management, HEC Montréal, Montréal, QC, Canada
e-mail: georges.zaccour@gerad.ca

1 Introduction

Environmental problems are often international in scope. The welfare of one country depends not only on its own policies but also on those of other countries. Examples of such problems include acid rain, pollution emissions, and the harvesting of fishing grounds or rain forests. In the case of pollution emissions, airborne pollutants mingle in the atmosphere, and therefore, any increase in the concentration of these gases depends on the total emissions from various countries. Countries have realized that these kinds of problems have to be solved on a global basis, through the use of international environmental agreements (IEAs).

The issue of whether or not countries will participate in IEAs is complex for two main reasons. First, countries are sovereign and their participation in international agreements is voluntary. There is no supra-national authority that forces countries to participate in such an agreement, nor is there an international environmental court system powerful enough to guarantee compliance with an IEA (Endres 2004; Wagner 2001; Finus 2000). Second, each country may have an incentive to opt for a free ride. While the costs for reducing emissions are borne by each country that is taking action, the benefits of any reduction in emissions are shared by all countries. Thus, each country stands to gain from waiting for the others to reduce their emissions. This problem is commonly known as "The Tragedy of the Commons" (Hardin 1968). Because of the interactions they involve, global environmental problems can be modeled using a game-theoretic framework. The literature on this approach has been developed along two main streams: the cooperative and non-cooperative approaches. In this paper, we adopt the non-cooperative point of view, which assumes that players cannot make binding agreements and that they act as rivals and in their own best interest.[1] Therefore, successful agreements must be *self-enforcing* (d'Aspremont et al. 1983). For an agreement to be self-enforcing, it must first be profitable, to ensure that adhering to the agreement is rational for each country; and second, it must be stable, to ensure that the group of signatory countries is in equilibrium.

The main body of literature on non-cooperative games and self-enforcing agreements uses a static framework to describe the problem of pollution emissions, that is, it assumes that environmental damage depends on the flow of emissions. In this literature, following d'Aspremont et al. (1983), equilibrium is defined by two conditions: internal stability, which implies that members have no incentive to leave the agreement, and external stability, which implies that non-members have no incentive to join the agreement. Underlying these concept is the idea that governments can re-optimize their decisions. However, in static models, this is only hypothetical, because, since the game will not be played again, there is no adjustment towards a stable solution and no change of state. This framework makes it possible to consider several stages in which specific decisions must be made. A typical example is a two-stage game: in the first stage countries decide whether or not to become members of an IEA (membership game); and, in the second stage they decide on their emission levels (emissions game).

Many static models using this stability concept have reached the conclusions that successful cooperation among a large number of countries is difficult to achieve, and that the number of members in a stable IEA is inversely related to the extent of the environmental damage. In order to explain the participation levels observed in actual IEAs, for

[1] Cooperative game theory assumes that players realize that, if they act as a group and coordinate their actions, they can obtain mutual benefits. Any given agreement is therefore the result of questioning the circumstances under which the agreement can be established, the results that can be achieved by the group, and the ways in which the benefits of cooperation will be redistributed among participants (Dockner et al. 2000; Wagner 2001).

example, the Montreal Protocol, [2] static models have incorporated several features that would increase the likelihood of cooperation, including the following: Stackelberg leadership of signatories (Barrett 1994; Diamantoudi and Sartzetakis 2006; Rubio and Ulph 2006), transfers (Carraro and Siniscalco 1993; Hoel and Schneider 1997), reputation effects (Hoel and Schneider 1997; Jeppensen and Andersen 1998; Cabon-Dhersin and Ramani 2006), issue linkages (Botteon and Carraro 1998; Le Breton and Soubeyran 1997; Barrett 1997; Katsoulacos 1997; Carraro and Siniscalco 1997, 1998; Mohr and Thomas 1998), including a minimum participation clause (Carraro et al. 2003), and considering modest emission-reduction targets (Finus 2004).

While all models gloss over reality to some extent, static games applied to pollution emissions and IEAs may be criticized on two important aspects. The first is the stock externality: transboundary environmental damage is usually related to the accumulation (stock) of pollution, rather than to emissions levels (flow). The second aspect is that countries can reconsider their membership decision at various points in time, especially if environmental damage changes over time.

In the literature considering pollution-stock dynamics, internal and external stability conditions are conceptually linked to renegotiation-proof agreements (in repeated games) or dynamically consistent agreements (in dynamic games). Part of this literature develops differential games that compare different solution concepts without addressing the membership problem (see for instance Long 1992; van der Ploeg and de Zeeuw 1992, Dockner and Long (1993); Dockner and Nishimura (1999), and Rubio and Casino 2002). Papers that do take membership decisions into account include those of Rubio and Casino (2005), Germain et al. (2003), de Zeeuw (2008), and Rubio and Ulph (2007). In Rubio and Casino (2005), the membership game is played only once, for good. Then, on the basis of the results, signatory and non-signatory countries decide on their emission levels by solving an infinite-horizon differential game in both open-loop and feedback strategies. Thus, the dynamics of the pollution stock affects the emissions strategies but not the membership decision. In Germain et al. (2003), a transfer scheme is proposed in a dynamic game, such that full cooperation is maintained over time. In this case, pollutant-stock dynamics is included in the membership decision, but the transfer scheme is able to ensure full cooperation so that the number of IEA members does not change over time. In de Zeeuw (2008), the farsightedness concept is applied to ensure the dynamic stability of coalitions aimed at reducing emissions over time. Finally, in Rubio and Ulph (2007) the emergence and stability of IEAs over time is studied using a difference game with a dynamic pollution stock. In each period of time, countries solve both an emissions game and a membership game so that, over time, the number of signatories changes with the pollution stock. The authors assume an upper bound on non-signatory emissions. They also presume that signatories are randomly selected in each period. Therefore, the stability concept implies that signatories and non-signatories have the same welfare expectations. This results in a unique steady-state pollution stock, with a corresponding steady-state for the size of the stable IEA. As in the static case, the size of the self-enforcing IEA is inversely related to the damage costs.

In this paper, we recognize that countries can join or quit an agreement over time and that it takes time for an IEA to achieve stability. We propose a possible mechanism to describe how this could happen. We choose a discrete time setting because, even though pollution evolves continuously over time, countries tend to make decisions about whether to join or leave an IEA at discrete moments.

[2] The Montreal Protocol on Substances that Deplete the Ozone Layer, signed in 1987, currently includes 191 nations.

The agreement modeled in this paper includes two clauses. The first, which is commonly used in the literature (see, e.g., Barrett 1994; Rubio and Casino 2005), is that a signatory country agrees to decide on its emission levels by maximizing the aggregate welfare of the coalition. A non-signatory country, however, decides on its emission levels by maximizing its own welfare. In both cases, the countries solve a dynamic emissions game, that is, they use the total discounted welfare over an infinite horizon, taking into account the evolution of the pollution stock. The second clause requires that signatory countries punish non-signatory countries. This means that, in becoming a member of the agreement, each country agrees to inflict a cost onto non-members. In particular, we assume that the punishment is a function of the level of accumulated pollution. This is motivated by the further assumption that, as pollution levels increase, so does the level of environmental concern of signatory countries. Therefore, the punishment they agree to inflict becomes stronger.

Including some form of punishment in an IEA as a tool to increase and sustain cooperation is not uncommon in static games (Barrett 2003, 1997). This non-environmental cost suffered by non-signatories can be interpreted as some form of sanction related to international trade, or as a social norm. Indeed, the measures used by the Montreal Protocol to boost participation in the agreement include trade sanctions against non-members (banning trade between signatories and non-signatories). We first consider an exogenous punishment, as in Hoel and Schneider (1997), and then we compute the "optimal" punishment as being the one that makes the long-run, full-cooperation solution impervious to unilateral deviations.

Moreover, given that any punishment (and more specifically a trade restriction) is not without adverse effects, we also include a cost borne by the signatory countries for punishing the non-signatories. We assume this cost to be proportional to the punishment. This was not addressed Hoel and Schneider (1997), nor has it been addressed elsewhere, in the context of IEAs to reduce pollution emissions,[3] to the best of our knowledge. Our motivation for including this additional cost is that restricting relationships with other countries will generate some welfare loss. To illustrate the impact of this modelling choice, we also analyze the limiting case where punishment can be inflicted at no cost.

Under our assumptions about the clauses included in the agreement, we first solve the dynamic emissions game for a given membership size. We then use a dynamic version of the internal and external stability concept of d'Aspremont et al. (1983) to define a stable agreement. Using this stability concept, we then solve the (dynamic) membership game and obtain the stable agreement size, as a function of the pollution-stock level. Finally, we propose an evolutionary process to describe how countries reach a stable situation, in which the pollution-stock level and the IEA size no longer change.

It is important to highlight the differences between our assumptions about how countries reach a stable IEA and those of other dynamic game models. Rubio and Ulph (2007), for instance, assume that, at each decision stage, countries solve both an emissions game and a membership game. Thus, at each stage, the size of the stable coalition is computed based on the pollution-stock level, and a group of signatory countries is randomly selected. This process evolves towards a steady state of the pollution stock, with a corresponding steady state for the size of a stable IEA.

In this paper, we do not assume that countries renegotiate their membership in the agreement at each stage. Negotiations only take place at an "initial" moment, but membership in the agreement is not closed. Countries may decide at any stage to join or quit the coalition, following an evolutionary process in which individual welfare is considered. This process

[3] It has however been considered in a fisheries context (see, e.g., Sethi and Somanathan 1996 and Bischi et al. 2004).

evolves towards a steady state of the pollution stock, with a corresponding steady state for the size of the stable IEA.

The main strength of an evolutionary process is that it is able to depict the stickiness and delays that may characterize this convergence. The motivation for a gradual convergence to stability is justified by practice. Often, an IEA is promoted and signed by a first group of founding countries (which are, for a variety of reasons, more sensitive to the environmental problem) and then, over time, other countries come to sign the agreement. [4] For example, the Montreal Protocol was ratified by an initial group of countries, and was then joined by new members every year.

In particular, we adopt replicator dynamics in discrete time, which creates evolutionary pressures in favor of the group obtaining the highest payoff. Following the spirit of evolutionary games, the group that performs better is joined by a fraction of new agents. The speed at which countries switch to the superior strategy is related to the difference in welfare, and it reflects the intangible or political cost of changing behaviors. It is worthwhile noticing that, in this process, the evolution of the players' welfare over time depends not only on the dynamics of emissions and pollution, but also on how the composition of the different groups evolves.

Our model is able to capture different long-run situations: where no stable agreement is reached; where all countries join the agreement; and (more realistically) where some countries take part in a stable IEA, and others do not.

The paper is organized as follows. Section 2 presents the model and the general dynamics governing the evolution of the pollution stock. In Section 3, we propose and solve a dynamic game in which countries optimize their welfare over an infinite horizon by taking into account the evolution of the pollution stock. Section 4 introduces the stability concept, solves the corresponding membership game, and proposes a replicator dynamics for the number of signatory countries. Section 5 presents numerical illustrations and various analyses. Finally, Section 6 concludes the paper. Computational details are provided in the Appendix.

2 The Model

Let us consider N identical countries. A fraction s of them, identified as "signatory countries," decides to join an international environmental agreement, according to which their production activity is decided by maximizing the aggregate welfare of the coalition. We denote by S the set of signatory countries. The countries in the remaining fraction $(1-s)$, identified as "non-signatory countries" or "defectors," act individually, that is, each one decides its production activity by maximizing its individual welfare. We denote by D the set of defectors. Note that this assumption about the behavior of signatories and non-signatories is common to both the cooperative and noncooperative approaches to dealing with IEAs.

Each country has a production activity that generates benefits as well as pollution. Let us indicate by e_{jt} the emissions generated by the production of country j in time period t. We suppose that the net revenue (i.e., gross revenue minus costs) derived from country j's production activity in a given period is an increasing concave function of its emissions, and is given by the quadratic function

$$R(e_{jt}) = e_{jt}\left(b - \frac{1}{2}e_{jt}\right).$$

[4] Of course, it may also happen that a member country decides to abandon the agreement.

Countries suffer a level of environmental damage resulting from (global) pollution, that is assumed linear and given by

$$D(P_t) = dP_t,$$

where $d > 0$ is the constant marginal damage and P_t is the stock of pollution at time t.

As part of the agreement, we assume that each signatory country has to punish a non-signatory for its irresponsible behavior (Hoel and Schneider 1997). Moreover, we suppose that this punishment is directly proportional to the level of pollution, to reflect that the signatories' environmental concern increases with the pollution stock. The non-environmental cost incurred by a defector punished by Ns signatories is given by

$$\phi(s, P_t) = Ns\alpha P_t.$$

This sanction can be interpreted as trade restrictions imposed on a non-signatory country or a carbon tax levied on that country's exports.

Given that punishing a country also has a negative impact on signatory countries, we presume a cost for punishing. That cost is proportional to the punishment αP_t imposed on the $N(1-s)$ non-signatory countries,[5] so that each signatory incurs a non-environmental cost given by

$$\omega(s, P_t) = N(1-s)\tau\alpha P_t,$$

where $\tau \geq 0$. As a consequence, the welfare of signatory country $j \in S$ in time period t is given by

$$W_t^S(e_{jt}, P_t, s) = e_{jt}\left(b - \frac{e_{jt}}{2}\right) - dP_t - \tau\alpha N(1-s)P_t,$$

whereas the welfare of non-signatory $j \in D$ is given by

$$W_j^D(e_{jt}, P_t, s) = e_{jt}\left(b - \frac{e_{jt}}{2}\right) - dP_t - \alpha NsP_t.$$

In the sequel, we will use the following convenient abbreviated notation

$$c_S \equiv d + \tau\alpha N(1-s) \quad (1)$$

$$c_D \equiv d + \alpha Ns \quad (2)$$

to represent the marginal (environmental and non-environmental) impacts of pollution on the welfare of signatory and non-signatory countries, respectively (which are however not constants, but linear functions of s). Notice that the difference in marginal costs $c_S - c_D$ is a linear function of s which is positive for $s < \frac{\tau}{1+\tau}$ and negative for $s > \frac{\tau}{1+\tau}$.

Finally, the evolution over time of the pollution stock is assumed to be governed by the discrete time equation

$$P_t = P_{t-1}(1-\delta) + \sum_{i\in S} e_{it} + \sum_{k\in D} e_{kt} \quad (3)$$

where $\delta \in (0, 1)$ is the natural decay, $\sum_{i\in S} e_{it}$ is the total emissions of signatory countries, and $\sum_{k\in D} e_{kt}$ is the total emissions of non-signatories at time period t.

This model deliberately simplifies reality, to allow us to derive closed-form solutions and theoretical properties. It nevertheless accounts for two important features of the stability

[5] The marginal cost of punishing is assumed to be independent from the number of signatory countries. This is the case, for example, with a trade sanction: the cost to a member's welfare depends only on the number of countries to which the sanctions are applied.

problem facing environmental agreements: first, the dynamics of the pollution stock and the related damage cost; and, second, the negative externalities arising from global pollution, that is, from the emissions of all players.

The assumption that all players are identical is not a crucial one. Indeed, the model can easily be extended to include the case of asymmetric countries, and the equilibrium strategies are readily obtained—but we then have to provide specific values for the parameters (see for instance Bahn et al. 2008 for an illustration of this model, calibrated to nine different regions using the MERGE policy-assessment model).

Assuming a quadratic revenue function is rather popular in the environmental economics literature (see, e.g., Rubio and Casino 2005; Diamantoudi and Sartzetakis 2006; Rubio and Casino 2002; Dockner and Long 1993), and captures the diminishing returns of production activity. The most debatable assumption is clearly the linear environmental damage. This simplification is not uncommon (see for instance Hoel and Schneider 1997; Finus 2004) and is supported by some empirical estimations (see Labriet and Loulou 2003). It can be justified with the assumption that players approximate the damage function using local marginal information. However, it does not take into account the potential catastrophic consequences of significant increases in the pollution stock. A convex damage function is a more realistic choice, albeit at the expense of tractability; using a quadratic damage function in our model yields qualitatively similar results that are no longer in closed-form. In the Conclusion, we give further consideration to the impact, on both the qualitative results and on the problem's tractability, of adopting a quadratic damage function rather than a linear one.

3 A Dynamic Game of Emissions

In this section, we solve the dynamic emissions game. We assume that, for a given fixed number of signatories, countries optimize their welfare by taking into account the evolution of the pollution stock. The total discounted welfare of players is maximized over an infinite horizon, where $\beta \in (0, 1)$ is the one-period discount factor assumed to be common to all players. The welfare-maximization problem for a signatory country $j \in S$ is thus given by

$$\max_{(e_j), j \in S} W^S = \sum_{j \in S} \sum_{t=0}^{\infty} \beta^t \left(e_{jt} \left(b - \frac{e_{jt}}{2} \right) - P_t c_S \right)$$

s.t.

$$P_t = P_{t-1} (1 - \delta) + \sum_{i \in S} e_{it} + \sum_{k \in D} e_{kt}, \ P_0 \text{ given,} \tag{4}$$

where e_{jt} is the emissions of country j during period t and e_j denotes the sequence of emissions $\{e_{jt}\}_{t=0,\ldots,\infty}$. In the same way, the welfare-maximization problem for a defector country $j \in D$ is

$$\max_{e_j} W^D = \sum_{t=0}^{\infty} \beta^t \left(e_{jt} \left(b - \frac{e_{jt}}{2} \right) - P_t c_D \right),$$

subject to (4).

In order to characterize the players' optimal reaction strategies and the dynamic equilibrium, we use a dynamic programming formulation where the state variable is P, that is, the pollution-stock level in the preceding time period. We solve the game for a Nash equilibrium in stationary feedback strategies between the group of signatories, acting as a single player, and the non-signatories, acting as $N(1-s)$ individual players, where $Ns \in [1, N-1]$. We call κ the constant representing the combined effect of the discount factor and the natural pollution decay,[6] that is,

$$\kappa \equiv \frac{1}{1-\beta(1-\delta)} > 1.$$

We first compute the optimal reaction of the set of signatory countries to a given stationary strategy vector for the defectors and we denote by $E^D(P)$ the resulting total emissions of the non-signatory countries as a function of P. For each signatory country, the value function $V^S(P; E^D)$ represents a signatory country's optimal total welfare, given $E^D(P)$. It satisfies

$$V^S(P; E^D) = \max_e \Big\{ e\left(b - \frac{e}{2}\right) - \left(P(1-\delta) + Nse + E^D(P)\right) c_S + \beta V^S\left(P(1-\delta) + Nse + E^D(P); E^D\right)\Big\}. \quad (5)$$

Proposition 1 *The value function of a signatory country is linear in P. The optimal reaction of signatory countries is independent of the level of pollution and of the defectors' strategy, and is given by*

$$e^S = b - Ns\kappa c_S, \quad (6)$$

assuming $b > Ns\kappa c_S$.

Proof Assume that $V^S(P; E^D) = k^S - m^S P$. Then, first-order sufficient conditions yield

$$e^S = b - Ns\left(m^S\beta + c_S\right),$$

which achieves the maximum in (5) and which does not depend on P or $E^D(P)$. Replacing e^S in (5) we obtain an expression that is linear in P and that verifies our assumption. It is now straightforward to obtain, by identification (see Appendix 7.1):

$$m^S = c_S\kappa(1-\delta)$$

$$k^S = Ns\kappa c_S \frac{Ns\kappa c_S - 2b}{2(1-\beta)} + \frac{b^2}{2(1-\beta)} - \frac{\kappa E^D(P) c_S}{(1-\beta)}.$$

The optimal emissions of a signatory country are therefore

$$e^S = b - Ns\kappa c_S.$$

□

Notice that the emissions of the signatory countries endogenize the damage of the entire coalition and are convex in s. It is then straightforward to compute the total emissions of signatory countries as $E^S = Ns(b - Ns\kappa c_S)$.

In the same way, we now express the optimal reaction of a defector to a given stationary-strategy vector for the other countries. We denote by $E^d(P)$ the total emissions of all other non-signatory countries as a function of P, and by E^S the total emissions of the signatory

[6] A value of $\kappa = 1$ corresponds to myopic limiting cases, with either $\beta = 0$ (no value for the future) or $\delta = 1$ (no stock accumulation).

countries. The value function V^D of a defector country represents a defector's optimal total welfare, given P, $E^d(P)$ and E^S. It satisfies

$$V^D(P; E^S, E^d) = \max_e \left\{ e\left(b - \frac{e}{2}\right) - \left(P(1-\delta) + E^S + e + E^d(P)\right)c_D \right.$$
$$\left. + \beta V^D\left(P(1-\delta) + E^S + e + E^d(P); E^S, E^d\right)\right\}. \quad (7)$$

Proposition 2 *The value function for a defector country is linear in P. The optimal reaction of non-member countries is independent of the level of pollution and of the other players' strategy, and is given by*

$$e^D = b - \kappa c_D, \quad (8)$$

assuming $b > \kappa c_D$.

Proof Assume that $V^D(P; E^S, E^d) = k^D - m^D P$. Then, first-order sufficient conditions yield

$$e^D = b - c_D - m^D\beta.$$

Substituting e^D in (7) we obtain a linear function of P and (see Appendix 7.2)

$$m^D = c_D\kappa(1-\delta)$$
$$k^D = \frac{b^2 - \kappa c_D\left(2b - \kappa c_D + 2\left(E^S + E^d(P)\right)\right)}{2(1-\beta)},$$

so that the optimal emissions for a defector country are

$$e^D = b - \kappa c_D.$$

□

Notice that the optimal emissions[7] of a defector are linear decreasing in s, independent of P and of the other players' strategies. It is then straightforward to compute the total emissions of the other defector countries as $E^d = (N - Ns - 1)(b - \kappa c_D)$.

Combining these results, the equilibrium strategy pair for both kinds of players is given by $\left(e^S, e^D\right)$ and the total emissions at equilibrium are

$$E(s) \equiv N\left(b - \kappa\left(c_D(1-s) + Ns^2 c_S\right)\right).$$

The consequence of assuming that marginal environmental damage is constant, is that each country's optimal emissions are independent of the others' (orthogonal free-riding). However, they are still linked because they are both functions of s.

It is straightforward to verify that the full defection equilibrium, where all players maximize their individual welfare, that is,

$$\max_{e_j} W^D = \sum_{t=0}^{\infty} \beta^t \left(e_{jt}\left(b - \frac{e_{jt}}{2}\right) - P_t d\right),$$

[7] The emissions of a defector are higher than those of a signatory for $Ns \in [1, N-1]$ iif $d > \alpha$ and $\tau > \frac{1}{N-1}$ (see Appendix 7.3). Condition $d > \alpha$ is satisfied in our numerical illustrations. This means that the punishment inflicted on a non-member country by an individual signatory is not significant compared to the damage it suffers from pollution. If countries were able to inflict more significant punishments at little cost, then they could make the non-signatories pollute even less than the cooperative optimum.

subject to (4), can be obtained by setting s as equal to 0 in (7). Similarly, the cooperative solution, where all players agree to maximize their aggregate welfare, that is,

$$\max_{(e_j)} W^S = \sum_{j=1}^{N}\sum_{t=0}^{\infty}\beta^t \left(e_{jt}\left(b-\frac{e_{jt}}{2}\right)-P_t d\right)$$

subject to (4), can be obtained by setting s as equal to 1 in (5). We extend the domain of the value functions in the obvious way: when $s = 0$, V^D represents the total discounted welfare and e^D the players' equilibrium strategy when there is no coalition; and when $s = 1$, V^S represents the total discounted welfare and e^S the players' optimal strategy under full cooperation.

Replacing (6) and (8) in (3), the dynamics of the pollution stock becomes

$$P_t = P_{t-1}(1-\delta) + N\left(b - \kappa\left(c_D(1-s) + Ns^2 c_S\right)\right). \tag{9}$$

For a given s, the steady state of the pollution stock is

$$P^*(s) = \frac{E(s)}{\delta}, \tag{10}$$

and the individual total discounted welfare of a signatory and a non-signatory country when the pollution stock is P are, respectively (see Appendix 7.4),

$$V^S(s,P) = \frac{b^2 - N^2\kappa^2 c_S^2 s^2}{2(1-\beta)} - \kappa c_S\left(P(1-\delta) + \frac{E(s)}{1-\beta}\right) \tag{11}$$

$$V^D(s,P) = \frac{b^2 - \kappa^2 c_D^2}{2(1-\beta)} - \kappa c_D\left(P(1-\delta) + \frac{E(s)}{1-\beta}\right). \tag{12}$$

4 Evolution and Stability of IEAs

We now consider a dynamic version of the stability concept introduced by d'Aspremont et al. (1983), which is widely used in static games. We first solve the membership game under this stability concept, that is, we state the necessary conditions characterizing a stable coalition for a given level of the pollution stock. We then propose a corresponding discrete-time replicator dynamics for the proportion of signatories that will allow players to achieve a stable IEA in the long run.

4.1 Unilateral Deviation and Foresight

In a static context, the internal- and external-stability concept assumes that each player is able to compare his current welfare with what he would achieve if he unilaterally switched groups. In a dynamic context, where each players' criterion is his total discounted welfare, the conditions for internal and external stability become, respectively,

$$V^S(s,P) \geq V^D\left(\tfrac{Ns-1}{N}, P\right) \tag{13}$$

$$V^S\left(\tfrac{Ns+1}{N}, P\right) < V^D(s,P). \tag{14}$$

Inequality (13) means that, at P, the total discounted welfare over an infinite horizon of a country in a coalition of Ns signatories is greater than what it would attain as a defector standing in opposition to a coalition of $Ns - 1$ countries; and therefore there is no incentive for the signatory to defect. Inequality (14) means that, at P, the total discounted welfare over an infinite horizon of a defector country standing in opposition to a coalition of Ns signatories is greater than what it would attain as a signatory in a coalition of $Ns+1$ countries; and therefore there is no incentive for the defector to join the coalition.

Define the stability function

$$\Psi(s,P)=V^S(s,P)-V^D\left(\frac{Ns-1}{N},P\right).$$

It is easy to check (see Appendix 7.5) that the extension of $\Psi(s,\cdot)$ to $\left[\frac{1}{N},1\right]$ is continuous in s, and that a necessary condition for IEA stability with $\lfloor Ns\rfloor$ signatories when the stock of pollution is $\bar{P}$ is obtained when $\Psi(s,\bar{P})=0$. This is given by

$$\begin{aligned}\bar{P}\,(1-\beta)\,(1-\delta)&=-Nb+\kappa\frac{(Ns-1)^2\,(c_S+\alpha\tau)\,(c_D-\alpha)}{(c_D-\alpha-c_S)}\\&\quad+\kappa\frac{(2N\,(1-s)+1)\,(c_D-\alpha)^2-Nc_S\left(2c_D\,(1-s)+Ns^2c_S\right)}{2\,(c_D-\alpha-c_S)}.\end{aligned}\tag{15}$$

Note that (15) characterizes the solution of the membership game. For any level of pollution, it provides a condition that must be satisfied by the size of a stable coalition.

4.2 Dynamics

As the pollution-stock level changes over time, the number of signatories required for a stable IEA also changes. In order to describe this dynamics, it would be possible, for instance, to follow the idea proposed in Rubio and Ulph (2007) where, in each period, countries solve a membership game. This amounts to finding the number of signatories that satisfies the stability condition (15) at P_t; in practice, this would require, for each time period, either an exogenous intervention or some time-consuming negotiation process.

Instead, we introduce a mechanism that describes how countries gradually reach a stable IEA.[8] We assume that the proportion of signatory countries evolves over time following the discrete-time replicator dynamics[9]

$$s_{t+1}=s_t\frac{V^S(s_t,P_t)}{s_tV^S(s_t,P_t)+(1-s_t)V^D(s_t-\frac{1}{N},P_t)}\quad\text{if}\quad s_t\in[1/N,1],\tag{16}$$

where V^S and V^D are given by (11)–(12) and are assumed positive.[10]

The denominator of (16) represents a weighted average of the (total discounted) welfare of a signatory country and that of a country that unilaterally defects from the agreement. The weights are determined by current proportions of signatories and defectors.

8 This means that, so long as the level of the pollution-stock levels are unstable, the coalition size is not determined by a membership game.

9 Replicator dynamics in continuous time is commonly used for common-pool resource games, particularly in the area of fisheries, to describe the evolution of a population of agents, where two behaviors can be adopted (see for instance Sethi and Somanathan 1996; Noailly et al. 2007; Osés-Eraso and Viladrish-Grau 2007). In Xepapadeas and Passa (2004), the replicator dynamics is used to model the participation and compliance of firms to voluntary environmental agreements.

10 It is straightforward to adapt the adjustment mechanism if the welfare of one or both groups of players is negative, or if the proportion of signatories vanishes (see Appendix 7.6).

Whenever the welfare obtained from one behavior is greater than that obtained from the opposite behavior, some of the countries will switch behaviors to join the group that is performing better. Equation (16) captures the notion that a strategy yielding profits that are above *(below)* average increases *(decreases)* the relative share of the population using that strategy[11] and that the "speed" of change depends on the relative welfare inequality. This update mechanism ensures that shifts in population shares will be gradual, and demonstrates that there may be delays or inertia involved in countries' behavior, for instance due to the political or intangible costs of switching.

In this case, (9) and (16) give rise to a two-dimensional dynamic system that describes the evolution over time of the pollution stock and of the proportion of signatories. It is easy to check that a steady state (s^*, P^*) of this dynamic system satisfies the stability condition for an IEA at $\bar{P} = P^*(s^*)$ (see Appendix 7.6). Players reach a stable IEA by comparing total discounted welfares achieved by remaining in, or unilaterally leaving, a coalition, without taking into account the dynamics of s.

5 Steady-State and Sensitivity Analysis

We now study the dynamics of the pollution and of countries' shares when the players behavior is given by (16). We are specifically interested in finding out whether or not this model can have the following outcomes: full cooperation; the coexistence of cooperators and defectors; and, no cooperation at all. We also want to determine the conditions under which these outcomes would occur. In our numerical simulations, we use parameter values that give rise to positive individual emissions.

When they exist, the equilibrium steady-state values of the pollution stock and of the proportion of signatory countries are indexed by υ. We denote $\xi_\upsilon = (s_\upsilon, P_\upsilon)$ an equilibrium steady state of the dynamic system (9)–(16).

We first notice that the boundary equilibria ξ_n, corresponding to $s_n = 0$, and ξ_c, corresponding to $s_c = 1$, may be reached by the dynamic system. These are given by

$$\xi_n = \left(0, \frac{N(b - \kappa d)}{\delta}\right)$$

$$\xi_c = \left(1, \frac{N(b - \kappa d N)}{\delta}\right),$$

where $P_n > P_c$. The dynamic system may also reach inner steady states corresponding to IEAs with partial cooperation.

5.1 Partial Cooperation

Depending on the parameter values, there may be 0, 1 or 2 coexisting inner steady states, corresponding to situations where the necessary stability condition is satisfied at the steady-state pollution level. These steady states are defined by the intersection of $P^*(s)$ with $\bar{P}(s)$. When two inner steady states exist, the one with the lower percentage of signatory countries, denoted by ξ_l, is a saddle point[12] and the one with the higher percentage of signatory coun-

[11] This idea is not uncommon in economics. For example, when new strategies or technologies are introduced on the market, firms will tend to imitate the most successful ones, or the ones that yield a 'satisfactory' profit level. Here, we assume that governments follow the same kind of behavior.

[12] A saddle point corresponds to a root where the stability function is increasing. Entry and exit conditions are not satisfied at a saddle point.

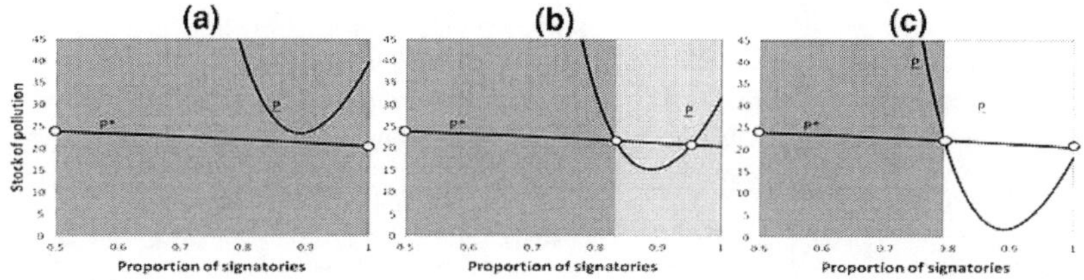

Fig. 1 Impact of increasing punishment. The parameter values are: $N = 100$, $b = 200$, $d = 0.3$, $\tau = 0.92$, $\delta = 0.8$, $\beta = 0.9$. In (**a**), $\alpha = 0.00031$; in (**b**), $\alpha = 0.000313$; in (**c**), $\alpha = 0.000318$. The *dark area* represents the set of initial conditions of s and P generating trajectories converging to the fully non-cooperative outcome. The *lightly shaded area* represents the set of initial conditions of s and P generating trajectories converging to the partial-cooperation outcome. The *white area* represents the set of initial conditions of s and P generating trajectories converging to the full-cooperative outcome

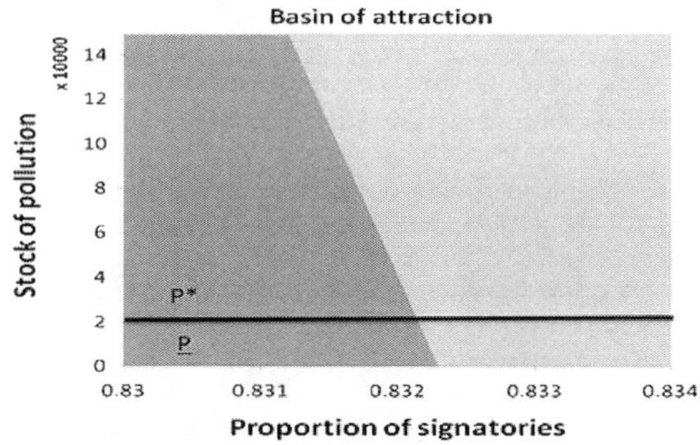

Fig. 2 Zoom-in on the basin of attraction

tries, denoted by ξ_h, is a stable node, corresponding to a stable IEA with partial cooperation. Figures 1, 2, and 3 illustrate three representative examples in the (s, P) plane, and how these situations can appear when the values of parameters α and τ are changed.

Figure 1 illustrates the impact of an increase in punishment α. In Fig. 1a (with $\alpha = 0.00031$), functions P^* and $\bar{P}$ do not intersect: there is no value of s that allows the stability criteria to be satisfied at the steady-state pollution stock. From any initial conditions (s_0, P_0), the dynamic system converges to ξ_n, that is, no stable agreement is possible. In Fig. 1b (with $\alpha = 0.000313$), if the initial group of signatory countries is large enough (that is, if the initial conditions are in the lightly shaded area), then the steady-state ξ_h is reached, corresponding to a stable IEA with partial cooperation. An interesting observation is that the minimum number of initial signatories that allows an inner steady state is decreasing with the initial pollution stock (see Fig. 2 for a zoom-in). This means that, when the level of accumulated pollution is high, we only need a small number of founding countries to reach a stable IEA.

In general, an increase of α decreases both $P^*(s)$ and $\bar{P}(s)$; it determines a higher level of cooperation and a lower level of pollution at the steady state, and it enlarges the basin of initial states generating trajectories converging to ξ_h. In Fig. 1c (with $\alpha = 0.000318$),

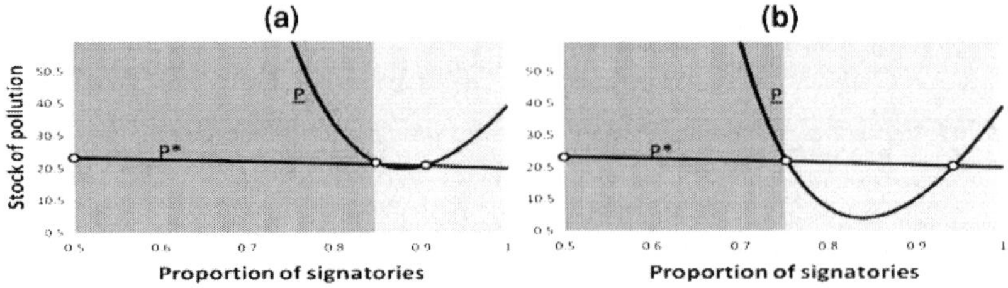

Fig. 3 Impact of decreasing cost for punishment. The parameter values are: $N = 100$, $b = 200$, $d = 0.3$, $\alpha = 0.00031$, $\delta = 0.8$, $\beta = 0.9$. In (**a**), $\tau = 0.90$; in (**b**), $\tau = 0.82$. The *dark area* represents the set of initial conditions of s and P generating trajectories converging to the fully non-cooperative outcome. The *lightly shaded area* represents the set of initial conditions of s and P generating trajectories converging to the partial-cooperation outcome

starting from any initial condition in the white area, the system converges to ξ_c. Further increases in punishment do not have any effect on the long-run values of the dynamic variables, because full cooperation has already been achieved. However, increased punishment makes full cooperation more robust (i.e., supported by a greater number of initial states).

Figure 3 illustrates the impact of decreasing τ, starting from the case depicted in Fig. 1a (with $\tau = 0.92$). A decrease in τ decreases both P^* and $\bar{P}$. Figure 3a (with $\tau = 0.90$) shows an inner steady state with partial cooperation. Further decreases in τ have a positive effect on the long-run values of the dynamic variables (higher cooperation and lower pollution stock), and especially, on the basin of the stable steady state (Fig. 3b). This means that, if the mandatory punishment set out by the agreement is not too expensive, the initial group of founding countries does not need to be very large to arrive at a stable agreement. Reducing the cost of punishing may or may not generate full cooperation, depending on the value of the other parameters (this is discussed further in Section 5.2).

A sensitivity analysis of the other parameters shows that increasing the profitability of emissions (parameter b) has a positive impact on the number of signatories at the steady state, as well as on the set of initial states that generate trajectories converging to partial or full cooperation; however, this is at the expense of a higher steady-state pollution stock. (This is because, even if cooperators become more numerous, they also have a higher incentive to emit more pollution). It is even more interesting that an increase in the marginal environmental cost (parameter d) negatively impacts on the size of the steady-state coalition as well as on the size of the basin of attraction leading to it. This result is consistent with what has been observed in static games: when the potential gain from cooperation is high, membership in an IEA is likely to be small. This is because there is an incentive to opt for a free ride. This in turn implies that a stable IEA is the smallest group such that emitting less is welfare-enhancing.

Figure 4 represents various trajectories for the pollution stock and the number of signatory countries over time for the case depicted in Fig. 1b, where an inner steady state exists. Depending on the initial conditions, trajectories converge either to a situation with no cooperation or to a stable IEA. The way this stable solution is reached also depends on the initial condition. Possible evolutions in the space (s, P) are illustrated in Fig. 5. Notice how the evolutionary pressure increases s above $\bar{P}$ (where signatories fare better) and decreases s otherwise.

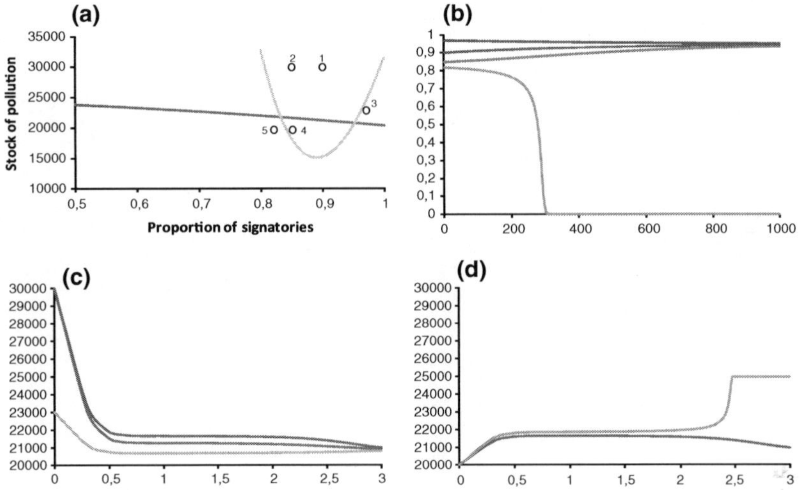

Fig. 4 Trajectories corresponding to various initial conditions. Parameters are the same as in Fig. 1b. *Panel a* shows the five initial conditions. In *Panel b*, trajectories for the proportion of signatories are represented over time: from top to bottom, they correspond to initial conditions 3, 1, 2 and 4 (merged), and 5, which converges to 0. In *Panels c* and *d*, trajectories for the pollution stock are represented as a function of log-time. From top to bottom, they correspond in *Panel c* to initial conditions 2, 1 and 3, and in *Panel d* to initial conditions 5 and 4

5.2 No Cost for Punishing

We now consider the special case that exists when enforcing a sanction does not entail any cost (as in Hoel and Schneider 1997). In this case, the marginal cost of pollution for a non-signatory country is always higher than for a signatory country, and a non-signatory emits more than a signatory for $Ns\,(d-\alpha) > d$.

In addition, studying the stability function shows that the entry and exit conditions can be satisfied at a maximum of one point (see Appendix 7.7). If punishing has no adverse effect on the welfare of coalition members, then there is, at most, one inner equilibrium in which defectors and signatories coexist. Moreover, the system never admits complete defection as a solution: the dynamic system converges either to an inner steady state with partial cooperation, or to a situation with full cooperation.

5.3 No Punishment

An interesting question is how the dynamic model performs when there is no punishment clause in the agreement ($\alpha = \tau = 0$). In this case, the exit condition

$$V^S(s, P) \geq V^D\left(\frac{Ns-1}{N}, P\right)$$

reduces to $(Ns-3)\,(Ns-1) \leq 0$, while the entry condition

$$V^S\left(\frac{Ns+1}{N}, P\right) < V^D(s, P)$$

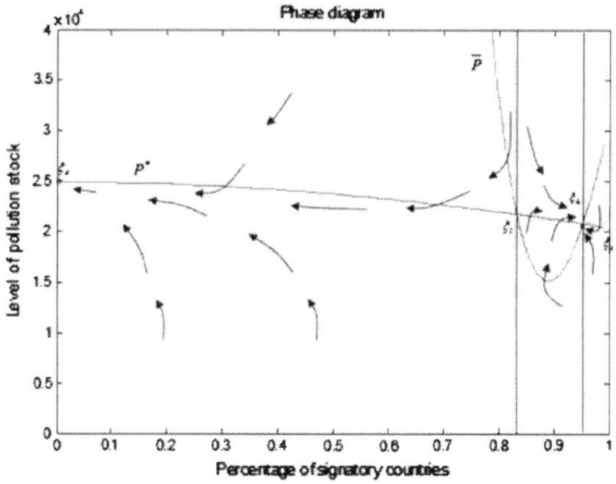

Fig. 5 Evolution of s and P

reduces to $Ns > 2$. Consequently, for any pollution level, the only stable IEA involves 3 members, which corresponds to the result obtained for static games (see Barrett 1994).

5.4 Endogenous Punishment

In the above analysis, we assumed that the punishment agreed upon by the signatories was exogenous. Recognizing that the best outcome is full cooperation, it is possible for the initial group of signatories to compute a value for α that would allow full cooperation to be achieved in the long run. The smallest level of α for which full cooperation is achieved at the steady state of pollution satisfies (15) at $\bar{P} = N\left(\frac{b-Nd\kappa}{\delta}\right)$ and $s = 1$, or equivalently solves (see Appendix 7.8)

$$\begin{aligned}&\alpha^2\kappa^2\delta(N-1)(2\tau(N-1)+1)\\&\quad+2\alpha\left(-Nb+d\kappa\left(\kappa\delta(\tau-2)(N-1)+N^2\right)\right)\\&\quad+d^2\kappa^2\delta(N-3)=0.\end{aligned}$$

For instance, with the parameter values used in Fig. 1, one obtains $\alpha^* = 3.1713 \times 10^{-4}$.

6 Conclusions

In this paper, we dealt with the problem of stability in international environmental agreements on pollution emissions. Stock externalities were included, as was the possibility for countries to abide by the agreement or to defect at any time. We considered an agreement that includes a provision allowing signatory countries to punish non-signatory countries, even if sanctioning the defectors entails a cost. We developed a model in which countries optimize their welfare over an infinite horizon, taking into account the evolution of the pollution stock. In defining a stable coalition, we applied well-known internal- and external-stability conditions (d'Aspremont et al. 1983). And lastly, we proposed an evolutionary mechanism that might be used to reach a stable IEA over time. This idea is validated by actual practice, where IEAs show growth over time. A inspiring example is that of the Montreal Protocol, to which many countries have adhered in the 20 years since it was first ratified. This shows that IEAs with high participation rates do occur, and that membership in the agreement may change over time, even without formal renegotiation.

The main results of our numerical simulations can be summarized as follows. The outcome in which no country joins the IEA is always possible, but provided that sanctions are strong enough and/or that the cost of punishing is not too high, this outcome coexists with either a partial-cooperation solution or with full cooperation. When two outcomes coexist, initial conditions are decisive: if the initial coalition is not large enough for a given initial level of pollution stock, then the equilibrium solution is full defection. The basin boundary can be interpreted as a minimum participation clause, where the minimum number of members is affected by the level of punishment included in the agreement and/or by its cost. However, the greater the initial level of pollution, the fewer initial signatory countries are needed for convergence to a stable agreement. This indicates that it is easier to endogenously bring other countries into the IEA when the global environmental damage is high. Lastly, as in the static models, the number of signatories in a stable IEA is inversely related to the environmental cost or equivalently, to the benefits of cooperation. This is to be expected from self-enforcing agreements, where stability is attained when signatories and defectors have similar welfare levels. When the benefits from cooperation are greater, the welfare of signatories increases more rapidly with the size of the coalition, and it takes fewer signatories to attain the same level of welfare as that of defectors.

To complete the analysis, we considered the special case that occurs when enforcing a sanction doesn't entail any cost. This case never results in full defection. The only possible outcome, independent of the initial conditions but depending on the value of the parameters, is either partial or full cooperation. We also considered the special case in which punishment is not part of the agreement, and we found that stable coalitions with large numbers of players cannot be implemented in such a case, regardless of the pollution stock. Lastly, we observed that it is easy to obtain a punishment level that ensures that full cooperation will be reached in the long run.

Our assumption of linear environmental damage makes the problem tractable, since the players' strategies become independent of the state and of the other players' actions. Linear damage may be interpreted as a local approximation of the damage function by players. When a quadratic damage function is used, we can still show that the equilibrium value function is quadratic, and that the players' equilibrium strategies are linear (both in the pollution stock and in the other players' emissions). However, the equilibrium solution can no longer be obtained in closed form, because the coefficients in the equilibrium strategies have to be obtained numerically. Our experiments using a quadratic damage function yielded qualitatively similar results.

7 Appendix

7.1 Equilibrium Emissions of Signatory Countries for Fixed s

Assume that $V^S(P; E^D) = k^S - m^S P$. We then have

$$V^S(P; E^D) = \max_e \left\{ e\left(b - \frac{e}{2}\right) - \left(P(1-\delta) + Nse + E^D(P)\right) c_S + \beta V^S\left(P(1-\delta) + Nse + E^D(P); E^D\right) \right\}$$

$$= \max_e \left\{ e\left(b - \frac{e}{2}\right) - \left(P(1-\delta) + Nse + E^D(P)\right) c_S \right.$$
$$\left. + \beta\left(k^S - m^S\left(P(1-\delta) + Nse + E^D(P)\right)\right)\right\}.$$

Differentiating with respect to emissions yields:

$$\frac{d}{de}\left(e(b - e/2) - \left(P(1-\delta) + Nse + E^D(P)\right) c_S \right.$$
$$\left. + \beta\left(k^S - m^S\left(P(1-\delta) + Nse + E^D(P)\right)\right)\right) = b - e - Ns\left(c_S + m^S\beta\right),$$
$$\frac{d}{de}\left(b - e - Ns\left(c_S + m^S\beta\right)\right) = -1,$$

so that the first order conditions are necessary and sufficient, provided that the solution is interior. The FOC are satisfied at

$$e^S = b - Ns\left(m^S\beta + c_S\right).$$

Replacing e^S in (5), we obtain

$$V^S(P; E^D) = \left(b - Ns\left(m^S\beta + c_S\right)\right)\left(b - \frac{(b - Ns(m^S\beta + c_S))}{2}\right)$$
$$- \left(P(1-\delta) + Ns\left(b - Ns\left(m^S\beta + c_S\right)\right) + E^D(P)\right) c_S$$
$$+ \beta\left(k^S - m^S\left(P(1-\delta) + Ns\left(b - Ns\left(m^S\beta + c_S\right)\right) + E^D(P)\right)\right)$$
$$= k^S\beta - E^D(P)\left(c_S + m^S\beta\right) + \frac{1}{2}Ns\left(c_S + m^S\beta\right)\left(-2b + Nsc_S + Nm^Ss\beta\right)$$
$$+ \frac{1}{2}b^2 - P(1-\delta)\left(c_S + m^S\beta\right) = k^S - m^S P$$

yielding

$$m^S = c_S\kappa(1-\delta),$$
$$k^S(1-\beta) = -E^D(P)c_S\kappa + \frac{1}{2}Nsc_S\kappa(Nsc_S\kappa - 2b) + \frac{1}{2}b^2,$$
$$e^S = b - Ns\kappa c_S,$$
$$E^S = Ns(b - Ns\kappa c_S),$$

where, using (1) $e^S = N^2\alpha\kappa\tau s^2 - N\kappa(d + N\alpha\tau)s + b$ is a convex function of s.

7.2 Equilibrium Emissions of Non-Signatory Countries for a Fixed s

Assume that $V^D(P; E^S, E^d) = k^D - m^D P$. We then have

$$V^D(P; E^S, E^d) = \max_e \left\{ e\left(b - \frac{e}{2}\right) - \left(P(1-\delta) + E^S + e + E^d(P)\right) c_D \right.$$
$$\left. + \beta\left(k^D - m^D\left(P(1-\delta) + E^S + e + E^d(P)\right)\right)\right\}.$$

Differentiating w.r.t. emissions yields:

$$\frac{d}{de}\Big(e\,(b-e/2)-\Big(P\,(1-\delta)+E^S+e+E^d\,(P)\Big)\,c_D$$
$$+\beta\Big(k^D-m^D\Big(P\,(1-\delta)+E^S+e+E^d\,(P)\Big)\Big)\Big)=b-e-c_D-m^D\beta,$$
$$\frac{d}{de}\Big(b-e-c_D-m^D\beta\Big)=-1,$$

so that the optimal emissions are given by

$$e^D=b-c_D-m^D\beta,$$

provided that the solution is interior. Substituting e^D in (7), we obtain

$$\begin{aligned}V^D(P;E^S,E^d)&=\Big(b-c_D-m^D\beta\Big)\left(b-\frac{(b-c_D-m^D\beta)}{2}\right)\\&-\Big(P\,(1-\delta)+E^S+\Big(b-c_D-m^D\beta\Big)+E^d\,(P)\Big)\,c_D\\&+\beta\Big(k^D-m^D\Big(P\,(1-\delta)+E^S+\Big(b-c_D-m^D\beta\Big)+E^d\,(P)\Big)\Big)\\&=-\frac{1}{2}\Big(c_D+m^D\beta\Big)\Big(2b-c_D+2E^S+2E^d-m^D\beta\Big)+k^D\beta+\frac{1}{2}b^2\\&-P\,(1-\delta)\Big(c_D+m^D\beta\Big)\\&=k^D-m^DP,\end{aligned}$$

so that

$$\begin{aligned}m^D&=c_D\kappa(1-\delta),\\k^D(1-\beta)&=-\frac{1}{2}c_D\kappa\Big(2b+2E^S+2E^d(P)-c_D\kappa\Big)+\frac{1}{2}b^2,\\e^D&=b-\kappa c_D,\\E^D&=(N\,(1-s))\,(b-\kappa c_D)\,,\end{aligned}$$

where, using (1), $e^D=b-\kappa\,(d+\alpha Ns)$ is a linear decreasing function of s.

7.3 Comparison of Emissions of Signatories and Defectors

Compute

$$\begin{aligned}e^D-e^S&=b-\kappa\,(d+\alpha Ns)-(b-Ns\kappa\,(d+\tau\alpha N\,(1-s)))\\&=-\kappa\left(d-Ns\,(d+\alpha\,(N\tau-1))+N^2s^2\alpha\tau\right),\end{aligned}\tag{17}$$

which is a concave function of s.

At $s=\frac{1}{N}$, (17) is equal to $\kappa\alpha\,(\tau\,(N-1)-1)\,,$ which is positive iff $\tau>\frac{1}{N-1}$; assume this condition holds.

At $s=\frac{N-1}{N}$, (17) is equal to

$$\begin{aligned}\kappa\,(d\,(N-2)+\alpha\,(\tau-1)\,(N-1))&>\kappa\left(d\,(N-2)+\alpha\left(\frac{1}{N-1}-1\right)(N-1)\right)\\&=\kappa\,(N-2)\,(d-\alpha)\,,\end{aligned}$$

which is positive if $d>\alpha$. Since the function is concave, it cannot be negative in $\left[\frac{1}{N},\frac{N-1}{N}\right]$.

7.4 Equilibrium Solution at a Fixed s

Total emissions:

$$E(s) = E^S + E^D = N\left(b - \kappa\left(c_D(1-s) + Ns^2 c_S\right)\right)$$

Pollution stock

$$P_t = P_{t-1}(1-\delta) + N\left(b - \kappa\left(c_D(1-s) + Ns^2 c_S\right)\right)$$

Steady-state

$$P^*(s) = \frac{E(s)}{\delta} = \frac{N\left(b - \kappa\left(c_D(1-s) + Ns^2 c_S\right)\right)}{\delta}$$

Total discounted welfares

$$\begin{aligned}
V^S(s, P) &= -c_S\kappa(1-\delta)P + \frac{2N\kappa c_S\left(-b + \kappa\left(c_D(1-s)\right)\right) + N^2s^2\kappa c_S^2 + b^2}{2(1-\beta)} \\
&= -c_S\kappa(1-\delta)P - \frac{\kappa c_S}{(1-\beta)}E(s) + \frac{b^2 - N^2s^2\kappa^2c_S^2}{2(1-\beta)}, \\
V^D(s, P) &= -c_D\kappa(1-\delta)P + \frac{\kappa c_D\left(-2Nb + \kappa\left(2N^2s^2c_S + c_D\left(2N(1-s)-1\right)\right)\right) + b^2}{2(1-\beta)} \\
&= -c_D\kappa(1-\delta)P - \frac{\kappa c_D}{(1-\beta)}E(s) + \frac{b^2 - \kappa^2c_D^2}{2(1-\beta)}.
\end{aligned}$$

7.5 Stability Condition

At a given P, consider the continuous function

$$w^S(s) = -c_S\kappa(1-\delta)P + \frac{N\kappa c_S\left(-2b + \kappa\left(2c_D(1-s) + Ns^2c_S\right)\right) + b^2}{2(1-\beta)}$$

defined on $\mathbb{R}$. Recalling that c_S and c_D are linear functions of s, it follows that w^S is a polynomial of degree 4 in s. Notice that $w^S(s)$ coincides with the total discounted equilibrium welfare of a signatory country when the pollution stock is P and the proportion of signatories is s for integer values of $Ns \in [1, N]$.

In the same way, the function

$$w^D(s) = -c_D\kappa(1-\delta)P + \frac{\kappa c_D\left(-2Nb + \kappa\left(2N^2s^2c_S + c_D\left(2N(1-s)-1\right)\right)\right) + b^2}{2(1-\beta)}$$

is also a polynomial of degree 4 in s, which coincides with the total discounted equilibrium welfare of a non-signatory country when the pollution stock is P and the proportion of signatories is s for integer values of $Ns \in [0, N-1]$. It follows that the stability function $\Psi(s) = w^S(s) - w^D(\frac{Ns-1}{N})$ is a polynomial function of degree 4 in s which admits at most four real roots. It is straightforward to check that a root $s^0 \in \left[\frac{1}{N}, 1\right]$ of the stability function satisfies the entry and exit conditions of a coalition $s^* = \lfloor Ns^0 \rfloor$ if the stability function is decreasing at s^0.

At a given $\bar{P}$, compute

$$w^D\left(\frac{Ns-1}{N}\right) = \\ -(c_D-\alpha)\kappa(1-\delta)\bar{P} \\ +\frac{\kappa(c_D-\alpha)\left(-2Nb+\kappa\left(2(Ns-1)^2(c_S+\alpha\tau)+(c_D-\alpha)\left(2N\left(1-\left(\frac{Ns-1}{N}\right)\right)-1\right)\right)\right)}{2(1-\beta)} \\ +\frac{b^2}{2(1-\beta)}.$$

The stability function is equal to 0 when

$$-c_S\kappa(1-\delta)\bar{P}+\frac{N\kappa c_S\left(-2b+\kappa\left(2c_D(1-s)+Ns^2c_S\right)\right)+b^2}{2(1-\beta)} \\ = -(c_D-\alpha)\kappa(1-\delta)\bar{P} \\ +\frac{\kappa(c_D-\alpha)\left(-2Nb+\kappa\left(2(Ns-1)^2(c_S+\alpha\tau)+(c_D-\alpha)(2N(1-s)+1)\right)\right)+b^2}{2(1-\beta)}.$$

Rearranging yields:

$$\begin{aligned} 2(1-\beta)(1-\delta)\bar{P}(c_D-\alpha-c_S) = & -2bN(c_D-\alpha-c_S) \\ & +2\kappa(Ns-1)^2(c_S+\alpha\tau)(c_D-\alpha) \\ & +\kappa(2N(1-s)+1)(c_D-\alpha)^2 \\ & -N\kappa c_S\left(2c_D(1-s)+Ns^2c_S\right). \end{aligned}$$

A necessary condition for stability at $\bar{P}$ is therefore:

$$\begin{aligned} (1-\delta)(1-\beta)\bar{P} = -bN + & \frac{\kappa(Ns-1)^2(c_S+\alpha\tau)(c_D-\alpha)}{(c_D-\alpha-c_S)} \\ & +\kappa\frac{(2N(1-s)+1)(c_D-\alpha)^2-Nc_S\left(2c_D(1-s)+Ns^2c_S\right)}{2(c_D-\alpha-c_S)}. \end{aligned}$$

7.6 Steady-State of the Dynamic System

At the steady-state, if $s^* \in \left[\frac{1}{N}, 1\right]$ and $P^* = P^*(s^*)$, then

$$V^S(s^*, P^*) = s^*V^S(s^*, P^*) + (1-s^*)V^D\left(s^*-\frac{1}{N}, P^*\right)$$

or equivalently

$$V^S(s^*, P^*) = V^D\left(s^*-\frac{1}{N}, P^*\right),$$

which corresponds to a root of the stability function at (s^*, P^*). Now, notice that in the neighborhood of a root in the stability function, the update mechanism moves the proportion of signatories away from the root when the stability function is increasing, and towards the root when the stability function is decreasing. A coalition size $\lfloor Ns^*\rfloor \in \left[\frac{1}{N}, 1\right]$ thus satisfies the entry and exit conditions.

An equivalent adjustment mechanism, which is well defined on $\left[0, \frac{N-1}{N}\right]$ is the following

$$s_{t+1} = s_t \frac{V^S(s_t + \frac{1}{N}, P_t)}{s_t V^S(s_t + \frac{1}{N}, P_t) + (1 - s_t) V^D(s_t, P_t)}. \tag{18}$$

One can use (18) whenever the proportion of signatories falls below $1/N$, and revert to (16) if it increases past $\frac{N-1}{N}$, for instance. A steady state of (18) corresponds to a root of the stability function, where the coalition size satisfying the entry and exit conditions is $\lceil Ns^* \rceil \in \left[0, \frac{N-1}{N}\right]$.

7.7 No Cost for Punishing

In this case, the difference between emissions of defectors and signatories is given by

$$e^D - e^S = \kappa\left(-d + Ns\left(d - \alpha\right)\right).$$

Notice that the function

$$w^S(s) = -d\kappa\left(1 - \delta\right)P + \frac{N\kappa d\left(-2b + \kappa\left(2c_D\left(1 - s\right) + Ns^2 d\right)\right) + b^2}{2\left(1 - \beta\right)}$$

is quadratic, while

$$w^D(s) = -c_D\kappa(1 - \delta)P + \frac{\kappa c_D\left(-2Nb + \kappa\left(2N^2 s^2 d + c_D\left(2N\left(1 - s\right) - 1\right)\right)\right) + b^2}{2(1 - \beta)}$$

is cubic. It follows that the stability function $\Psi(\frac{Ns+1}{N}) = w^S(\frac{Ns+1}{N}) - w^D(s)$ is a polynomial function of degree 3 in s.

The coefficient of s^3 in the stability function is $N^3\alpha\kappa^2\frac{\alpha-d}{1-\beta}$, and it has the same sign as $\alpha - d$. At $s = 0$, compute

$$\Psi\left(\frac{1}{N}\right) = d\alpha\frac{\kappa^2}{1 - \beta}\left(N - 1\right) > 0$$

and

$$\Psi'\left(\frac{1}{N}\right) = N\frac{\kappa}{1 - \beta}\left(P\alpha\left(1 - \delta\right)\left(1 - \beta\right) + d\kappa\left(d - \alpha\right) + N\alpha\left(b - d\kappa\right)\right).$$

1 Under the assumption that $b > \kappa\left(d + \alpha Ns\right)$ (interior solution), than if $d > \alpha$, the stability function is increasing and positive at $s = 0$. Since the coefficient of s^3 is negative, there is at most one root of the stability function where it is decreasing which is greater than 0; since a stable coalition is an integer value not smaller than that root, $s = 0$ is not a stable coalition.
2 If on the other hand $d < \alpha$, the coefficient of s^3 is positive, and the stability function is positive at $s = 0$. There is at most one root of the stability function where it is decreasing which is greater than 0, and $s = 0$ is not a stable coalition.
3 If $d = \alpha$, the stability function is quadratic, concave, and positive at $s = 0$. There is at most one root of the stability function where it is decreasing which is greater than 0, and $s = 0$ is not a stable coalition.

7.8 Endogenous Punishment

The smallest level of α for which full cooperation is achieved at the steady state of pollution satisfies (15) at $\bar{P} = N\left(\frac{b-Nd\kappa}{\delta}\right)$ and $s = 1$, or equivalently solves

$$\begin{aligned}2(1-\beta)(1-\delta)N\left(\frac{b-Nd\kappa}{\delta}\right)(\alpha(N-1)) = &-2bN(\alpha(N-1)) \\ &+2\kappa(d+\alpha\tau)(N-1)^2(d+\alpha(N-1)) \\ &+\kappa(d+\alpha(N-1))^2 \\ &-N^2d^2\kappa,\end{aligned}$$

which reduces to

$$\begin{aligned}&\alpha^2\kappa^2\delta(N-1)(2\tau(N-1)+1) \\ &\quad+2\alpha\left(-bN+d\kappa\left(\kappa\delta(\tau-2)(N-1)+N^2\right)\right) \\ &\quad+d^2\kappa^2\delta(N-3)=0.\end{aligned}$$

Acknowledgements This research was supported by NSERC (Canada) and FQRSC (Québec). We would like to thank the referees and editor for their helpful comments.

References

d'Aspremont C, Jacquemin A, Gabszewicz JJ, Weymark JA (1983) On the stability of collusive price leadership. Can J Econ 16(1):17–25

Bahn O, Breton M, Sbragia L, Zaccour G (2008) Stability of international environmental agreements: an illustration with asymmetrical countries. Int Trans Operational Res 16(3):307–324. doi:10.1111/j.1475-3995.2008.00678

Barrett S (1994) Self-enforcing international environmental agreements. Oxf Econ Pap 46:878–894

Barrett S (1997) The strategy of trade sanctions in international environmental agreements. Resour Energy Econ 19:345–361

Barrett S (2003) Increasing participation and compliance in international climate change agreements. Int Environ Agreem Politics Law Econ 3:349–376

Bischi GI, Lamantia F, Sbragia L (2004) Competition and cooperation in natural resources exploitation: an evolutionary game approach. In: Carraro C, Fragnelli V (eds) Game practice and the environment. E. Elgar Publishing, Cheltenham pp 187–221

Botteon M, Carraro C (1998) Strategies for environmental negotiations: issue linkage with heterogeneous countries. In: Hanley N, Folmer H (eds) Game theory and the environment. E. Elgar Publishing, Cheltenham pp 181–203

Cabon-Dhersin ML, Ramani S (2006) Can social externalities solve the small coalitions puzzle in international environmental agreements ?Econ Bull 17:1–8

Carraro C, Marchiori C, Orefice S (2003) Endogenous minimum participation in international environmental treaties. Working paper No. 113.2003 FEEM

Carraro C, Siniscalco D (1993) Strategies for the international protection of the environment. J Public Econ 52:309–328

Carraro C, Siniscalco D (1997) R&D cooperation and the stability of international environmental agreements. In: Carraro C (ed) International environmental negotiations strategic policy issues. E. Elgar Publishing, Cheltenham pp 71–96

Carraro C, Siniscalco D (1998) International environment agreements: incentives and political economy. Eur Econ Rev 42:561–572

Diamantoudi E, Sartzetakis ES (2006) Stable international environmental agreements: an analytical approach. J Public Econ Theory 8(2):247–263

Dockner E, Jorgensen S, Long NV, Sorger G (2000) Differential games in Economics and Management science. Cambridge university press, Cambridge

Dockner E, Long NV (1993) International pollution control: cooperative versus noncooperative strategies. J Environ Econ Manag 24:13–29
Dockner E, Nishimura K (1999) Transboundary pollution in a dynamic game model. Jpn Econ Rev 50(4):443–456
Endres A (2004) Game theory and global environmental policy. Poiesis & Praxis 3:123–139
Finus M (2000) Game theory and international environmental co-operation: a survey with an application to the Kyoto Protocol. Nota di lavoro 86.2000, Fondazione Eni Enrico Mattei
Finus M (2004) Modesty pays: sometimes. Nota di lavoro 68.2004, Fondazione Eni Enrico Mattei
Germain M, Toint P, Tulkens H, Zeeuw A (2003) Transfers to sustain dynamic core-theoretic cooperation in international stock pollutant control. J Econ Dyn Control 28:79–99
de Hardin G (1968) The tragedy of the commons. Science 162(3859):1243–1248
Hoel M, Schneider K (1997) Incentives to participate in an international environmental agreement. Environ Resour Econ 9:153–170
Jeppensen T, Andersen P (1998) Commitment and fairness in environmental games. In: Hanley N, Folmer H (eds) Game theory and the environment. E. Elgar Publishing, Cheltenham, pp 65–83
Katsoulacos Y (1997) R&D spillover, cooperation, subsidies and international agreements. In: Carraro C (ed) International environmental negotiations: strategic policy issues. E. Elgar Publishing, Cheltenham, pp 97–109
Labriet M, Loulou R (2003) Coupling climate damages and ghg abatement costs in a linear programming framework. Environ Model Assess 8(3):261–274
Le Breton M, Soubeyran A (1997) The interaction between international environmental and trade policies. In: Carraro C (ed) International environmental negotiations—strategic policy issues. E. Elgar Publishing, Cheltenham, pp 126–149
Long NV (1992) Pollution control: a differential game approach. Ann Operational Res 37:283–296
Mohr E, Thomas J (1998) Pooling sovereign risks: the case of environmental treaties and international debt. J Dev Econ 55(1):153–169
Noailly J, Withagen CA, Van Den Bergh JCJM (2007) Spatial evolution of social norms in a common-pool resource game. Environ Resour Econ 36:113–141
Osé s-Eraso N, Viladrich-Grau M (2007) On the sustainability of common property resources. J Environ Econ Manag 53:393–410
Rubio S, Casino B (2002) A note on cooperative versus non-cooperative strategies in international pollution control. Resour Energy Econ 24:251–261
Rubio S, Casino B (2005) Self-enforcing international environmental agreements with a stock pollutant. Span Econ Rev 7:89–109
Rubio S, Ulph A (2006) Self-enforcing environmental agreements revisited. Oxf Econ Pap 58:233–263
Rubio S, Ulph A (2007) An infinite-horizon model of dynamic membership of international environmental agreements. J Environ Econ Manag 54(3):296–310
Sethi R, Somanathan E (1996) The evolution of social norms in common property resource use. Am Econ Rev 86(4):766–788
van der Ploeg F, de Zeeuw AJ (1992) International aspects of pollution control. Environ Resour Econ 2:117–139
Wagner UJ (2001) The design of stable international environmental agreements: economic theory and political economy. J Econ Surv 15(3):377–411
Xepapadeas A, Passa C (2004) Participation in and compliance with public voluntary environmental programs: an evolutionary approach. Nota di Lavoro 67.2004, Fondazione Eni Enrico Mattei
Zeeuw de A (2008) Dynamic effect on the stability of international environmental agreements. J Environ Econ Manag 55:163–174